Springer Series in Statistics

Springer Series in Statistics

Alho/Spencer: Statistical Demography and Forecasting
Andersen/Borgan/Gill/Keiding: Statistical Models Based on Counting Processes
Atkinson/Riani: Robust Diagnostic Regression Analysis
Atkinson/Riani/Ceriloi: Exploring Multivariate Data with the Forward Search
Berger: Statistical Decision Theory and Bayesian Analysis, 2nd edition
Borg/Groenen: Modern Multidimensional Scaling: Theory and Applications, 2nd edition
Brockwell/Davis: Time Series: Theory and Methods, 2nd edition
Bucklew: Introduction to Rare Event Simulation
Cappé/Moulines/Rydén: Inference in Hidden Markov Models
Chan/Tong: Chaos: A Statistical Perspective
Chen/Shao/Ibrahim: Monte Carlo Methods in Bayesian Computation
Coles: An Introduction to Statistical Modeling of Extreme Values
Devroye/Lugosi: Combinatorial Methods in Density Estimation
Diggle/Ribeiro: Model-based Geostatistics
Dudoit/Van der Laan: Multiple Testing Procedures with Applications to Genomics
Efromovich: Nonparametric Curve Estimation: Methods, Theory, and Applications
Eggermont/LaRiccia: Maximum Penalized Likelihood Estimation, Volume I: Density Estimation
Fahrmeir/Tutz: Multivariate Statistical Modeling Based on Generalized Linear Models, 2nd edition
Fan/Yao: Nonlinear Time Series: Nonparametric and Parametric Methods
Ferraty/Vieu: Nonparametric Functional Data Analysis: Theory and Practice
Ferreira/Lee: Multiscale Modeling: A Bayesian Perspective
Fienberg/Hoaglin: Selected Papers of Frederick Mosteller
Frühwirth-Schnatter: Finite Mixture and Markov Switching Models
Ghosh/Ramamoorthi: Bayesian Nonparametrics
Glaz/Naus/Wallenstein: Scan Statistics
Good: Permutation Tests: Parametric and Bootstrap Tests of Hypotheses, 3rd edition
Gouriéroux: ARCH Models and Financial Applications
Gu: Smoothing Spline ANOVA Models
Gyöfi/Kohler/Krzyźak/Walk: A Distribution-Free Theory of Nonparametric Regression
Haberman: Advanced Statistics, Volume I: Description of Populations
Hall: The Bootstrap and Edgeworth Expansion
Härdle: Smoothing Techniques: With Implementation in S
Harrell: Regression Modeling Strategies: With Applications to Linear Models, Logistic Regression, and Survival Analysis
Hart: Nonparametric Smoothing and Lack-of-Fit Tests
Hastie/Tibshirani/Friedman: The Elements of Statistical Learning: Data Mining, Inference, and Prediction
Hedayat/Sloane/Stufken: Orthogonal Arrays: Theory and Applications
Heyde: Quasi-Likelihood and Its Application: A General Approach to Optimal Parameter Estimation
Huet/Bouvier/Poursat/Jolivet: Statistical Tools for Nonlinear Regression: A Practical Guide with S-PLUS and R Examples, 2nd edition
Iacus: Simulation and Inference for Stochastic Differential Equations

(continued after index)

Friedrich Liese · Klaus-J. Miescke

Statistical Decision Theory

Estimation, Testing, and Selection

Springer

Friedrich Liese
Universität Rostock
Institut für Mathematik
Universitätsplatz 1
18051 Rostock
Germany
friedrich.liese@mathematik.uni-rostock.de

Klaus-J. Miescke
Department of Mathematics, Statistics
& Computer Science
University of Illinois at Chicago
851 South Morgan Street
Chicago IL 60607-7045
USA
klaus@uic.edu

ISBN: 978-0-387-73193-3 e-ISBN: 978-0-387-73194-0
DOI: 10.1007/978-0-387-73194-0

Library of Congress Control Number: 2008924221

Printed on acid-free paper

9 8 7 6 5 4 3 2 1

springer.com

To Gabi and Madelyn

Preface

This monograph is written for advanced Master's students, Ph.D. students, and researchers in mathematical statistics and decision theory. It should be useful not only as a basis for graduate courses, seminars, Ph.D. programs, and self-studies, but also as a reference tool.

At the very least, readers should be familiar with basic concepts covered in both advanced undergraduate courses on probability and statistics and introductory graduate-level courses on probability theory, mathematical statistics, and analysis. Most statements and proofs appear in a form where standard arguments from measure theory and analysis are sufficient. When additional information is necessary, technical tools, additional measure-theoretic facts, and advanced probabilistic results are presented in condensed form in an appendix. In particular, topics from measure theory and from the theory of weak convergence of distributions are treated in detail with reference to modern books on probability theory, such as Billingsley (1968), Kallenberg (1997, 2002), and Dudley (2002).

Building on foundational knowledge, this book acquaints readers with the concepts of classical finite sample size decision theory and modern asymptotic decision theory in the sense of LeCam. To this end, systematic applications to the fields of parameter estimation, testing hypotheses, and selection of populations are included. Some of the problems contain additional information in order to round off the results, whereas other problems, equipped with solutions, have a more technical character. The latter play the role of auxiliary results and as such they allow readers to become familiar with the advanced techniques of mathematical statistics.

The central theme of this book is what optimal decisions are in general and in specific decision problems, and how to derive them. Optimality is understood in terms of the expected loss, i.e. the risk, or some functional of it. In this regard estimators, tests, and selection rules are initially considered in the book side by side, and then individually in the last three chapters.

Originally we were motivated to write this book by the lack of any noticeable coverage of selection rules in books on decision theory. In over more than 50 years' worth of scholarship, the majority of the over 1000 published articles on selection rules do not utilize a rigorous decision-theoretic approach. Instead, many articles on selection rules restrict themselves to a specific parametric family, propose an ad hoc rule, study its performance characteristics, and (at the very best) compare its performance with another competing selection rule. By contrast, this book offers a fuller point of view, and the last chapter provides a thorough presentation of optimal selection rule theory.

Two other justifications for including selection theory are as follows. First, in modern medium-level books on mathematical statistics, the decision-theoretic approach is usually presented in a rather restricted and concise manner. This practice, combined with an emphasis on estimation under the squared error loss and on testing under the zero–one loss, fails to explain why extra efforts should be made to become familiar with decision theory and to use it. Of course, dealing with selection rules requires new types of loss structures, and learning more about them leads to a better understanding of the wide range of powerful tools that decision theory has to offer. Second, permutation invariance plays an important role in selection theory. The structure of the problem of optimal permutation invariant selection rules, along with its multisample statistical model, is quite unique. Indeed, it provides a rather different setting in decision theory when compared to estimation and testing problems based on a single sample, and we wished to make those differences more readily available to our readers. In addition, as we wrote the first parts of the book, which we began in the spring of 1999, it became clear that two additional aspects of decision theory were important to us: asymptotic decision theory and the coexistence of the frequentist and Bayes approaches in decision theory. With this final realization, we settled on our main topics for the book, and they have carried us along ever since.

This book combines innovation and tradition in ways that we hope can usefully extend the line of scholarship that starts with classical monographs on decision theory by Wald (1950), Blackwell and Girshick (1954), Ferguson (1967), and DeGroot (1970) and continues with modern works by Pfanzagl and Wefelmeyer (1982, 1985), Strasser (1985), Janssen, Milbrodt, and Strasser (1985), LeCam (1986), LeCam and Yang (1990), Torgersen (1991), Bickel, Klaasen, Ritov, and Wellner (1993), Rieder (1994), and Shiryaev and Spokoiny (2000). Most of these recent publications focus primarily on fundamental structural relationships in finite and asymptotic decision theory. By contrast, we have chosen to include parts of mathematical statistics as they have been represented by Witting (1985), Lehmann (1986), Pfanzagl (1994), Witting and Müller-Funk (1995), Lehmann and Casella (1998), and Lehmann and Romano (2005). As a result, this monograph is uniquely able to synthesize otherwise disparate materials, while establishing connections between classical and modern decision theory and inviting readers to explore their interrelationships.

The importance of creating a bridge between the classical results of mathematical statistics and the modern asymptotic decision theory founded by

LeCam should not be underestimated. So far, LeCam's theory has been applied primarily to estimation and testing problems, which we now also present in the last part of Chapter 9 with treatments of selection problems. We also include new applications of this theory, which we hope demonstrate its broad and powerful applicability. The prominent monographs in that area are by Strasser (1985), LeCam (1986), Torgersen (1991), Bickel, Klaasen, Ritov, and Wellner (1993), and LeCam and Yang (2000). These are written for mathematical researchers in decision theory, and they are only partially accessible to graduate students. Representations of parts of modern decision theory, mainly applications of LAN theory to estimation and testing problems, that are accessible to graduate students can be found in the books by Behnen and Neuhaus (1989), Witting and Müller-Funk (1995), Hájek, Šidák, and Sen (1999), and Lehmann and Romano (2005). In these works, however, considerations are restricted to the asymptotic behavior of the log-likelihood under, say, a null hypothesis and local alternatives. As a consequence, the general theory of statistical models and their convergence are deliberately excluded, as are central statements of modern asymptotic decision theory. These statements provide the fundamental link between the convergence of the distributions of the likelihood ratio, the decision-theoretically motivated concept of convergence of models, and the closely related randomization criterion. They make it possible to establish the asymptotic lower Hájek–LeCam bound on the risk. The combination of this asymptotic lower bound, linearization techniques for the log-likelihood, projection techniques for the statistics, and the lemmas of LeCam constitute the backbone of modern asymptotic statistical theory. In this book we wish to present the fundamental facts and their relations to each other on an intermediate level in a form that is mathematically self-contained. This style of presentation will, we hope, enable the reader to gain deep insight into and appreciation for the structure of modern decision theory.

Another goal of this book is to provide a broad coverage of both the frequentist and the Bayes approaches in decision theory. Most existing books seem to prefer one or the other. We consider the Bayes approach to be a useful decision-theoretic framework among others, and we use it heavily throughout the book; however, we do so without extra nonmathematical philosophical justification. In this spirit we distinguish between the average risk, where the randomness of parameters is not an issue, and the Bayes risk. This distinction allows us also to treat settings with improper priors just mathematically with the average risk. Readers who are interested in contemporary presentations of Bayesian analysis, including its philosophical foundation, reasoning, and justification, are referred to the fundamental books on Bayesian analysis by Berger (1985), Bernardo and Smith (1994), Robert (2001), and Ghosh and Ramamoorthi (2003).

Chapter 1. The fundamental probabilistic concepts and technical tools are provided here. These are in the first section properties of exponential families, where we have used Brown (1986) as a guideline for our presentation.

At first glance, the importance of this class of distributions seems to be more or less due to its favorable analytical form. However, several deeper reaching characterization theorems show that, roughly speaking, finite optimal decisions are only possible for this class of distributions. The class of conjugate priors that are important for Bayes decisions, which arises in a natural way from an exponential family, is studied systematically after the Bayes framework has been introduced in Section 1.2. Tools that are used later on for Bayes estimation, testing, and selection are also prepared here.

Distances between distributions play a central role. They reflect, for example, the degree of information content in a binary model, and they explain why a decision between distributions that are farther apart is easier than a decision between distributions that are closer together. Moreover, some of the distances or transforms, (e.g., the variational and the Hellinger transforms) and their mutual relations are utilized to introduce and establish the concepts of the strong and weak convergence of statistical models. The variational and Hellinger distance, as well as the Kullback–Leibler distance, the χ^2-distance, and the Bayes error for testing hypotheses in binary models are special members of the class of v-divergences that were independently introduced by Csiszár (1963) and Ali and Silvey (1966) and constructed with the help of a convex function v. The behavior under randomization and interrelations of these functionals for different convex functions studied in Section 1.3 provides a deeper understanding of these functionals and prepares for applications in subsequent chapters. Information in Bayes models is considered next, and the chapter concludes with an introduction to $\mathbb{L}_2$-differentiability, where we have used Witting (1985) as a guideline for our presentation.

Chapter 2. The central topic is the Neyman–Pearson lemma and its extensions. Links between Neyman–Pearson, minimax, and Bayes tests are discussed and studied in detail. After a consideration of statistical models with stochastic ordering, especially with a monotone likelihood ratio, which include exponential families, Neyman–Pearson's lemma is extended to tests for composite one-sided hypotheses.

Chapter 3. An introduction to the general framework of decision theory is given, followed by a discussion of its components. The concept of convergence of decisions and the sequentially weak compactness of the set of all decisions for a given model are central topics. Here and at several other places of the book, we restrict ourselves to compact decision spaces and dominated models. This practice helps keep technical tools at the graduate level, and it usefully restricts references to results in other literature.

Special properties of the risk as a function of the parameter as well as of the decision are studied to prepare for theorems of the existence of Bayes and minimax decisions. Furthermore, the interrelations between Bayes and minimax decisions are studied in preparation of proofs of minimaxity of estimators and tests later on that are based on Bayes properties and a constant risk. Γ-minimax decisions, which are analogues to minimax decisions in the

Bayes approach, are also briefly considered in Section 3.6, and the chapter concludes with special versions of the minimax theorem and the complete class theorem. For readers interested in further results, references are made to the fundamental monographs by Strasser (1985) and LeCam (1986).

Chapter 4. The chapter begins with examples in which randomizations of models appear in a natural way. The concept of ε-deficiency due to LeCam (1964), which is a comparison of the risk function "up to ε", is essential for the approximation and convergence of models and takes the center stage in this chapter. Another fundamental result is the randomization theorem of decision theory. It shows that the decision-theoretic concept of ε-deficiency is identical with the variational distance between one model and a suitable randomization of the other model. A transition to standard models gives the statement that finite models are uniquely determined by their standard distributions and the Hellinger tranforms. The characterization of the ε-deficiency via Bayes risks leads to the concept of standard decision problems for which the associated risk is just a special v-divergence. This is the concave function criterion of decision theory, and it connects concepts from information and decision theory.

In the second part of the chapter, sufficient statistics are characterized by the fact that the induced model is equivalent to the original model. The v-divergences are used to give for the sufficiency an information-theoretic characterization due to Csiszár (1963), the test-theoretic characterization due to Pfanzagl (1974), and the well-known factorization criterion by Neyman. A discussion of the different concepts of sufficiency such as pairwise sufficiency, Blackwell sufficiency, and Bayes sufficiency is included. A brief discussion of ancillarity, which includes Basu's theorem, concludes the chapter.

Chapter 5. The treatment of the reduction by invariance is kept concise by mainly considering the groups of permutations, location-scale transforms, and rotations. Whereas permutation invariance is especially relevant for selection rules, the other groups are utilized to prove the Hunt–Stein theorem on the minimaxity of best invariant tests. Hereby the existence of the Haar measure can be established directly in a simple manner without having recourse to further literature. The connection between best equivariant estimators and minimax estimators is provided by the Girshick–Savage theorem. With the conclusion of this chapter all tools from finite decision theory that are necessary for our purposes have been collected.

Chapter 6. The previous results on ε-deficiency and the randomization theorem are used to develop a theory of convergence of models within our fixed framework. Asymptotically normal models play a central role. Whereas the term *model* is standard in mathematical statistics, the term *experiment* is more common in modern decision theory. As both concepts are essentially the same (see Lehmann and Romano (2005) p. 550), we use the term *model* throughout the book. The transition to standard models makes it possible to get for finite models the well-known bounds on the ε-deficiency in terms of the Dudley metric of standard distributions, which leads to the characterization of the convergence of finite models in terms of the distributions of the likelihood ratios. For binary models the concepts of contiguity and entire separation are introduced through the accumulation points of a sequence of models. As in Jacod and Shiryaev (1987, 2002) and Liese (1986), we use Hellinger integrals to get the results on the contiguity and the entire separation of sequences of binary models, especially the results of Oosterhoff and van Zwet (1979) for triangular arrays of independent models. In the study of the asymptotic normality of double sequences of binary models, we follow the ideas of LeCam and Yang (1990).

After the introduction and brief discussion of Gaussian models the LAN- and ULAN-properties are introduced and established for localized sequences of differentiable models. From the start, after Witting and Müller-Funk (1995) and Rieder (1994), regression coefficients that satisfy the Noether condition are used. Special cases then are the row-wise i.i.d. case, the two-sample problem, and regression models with deterministic covariables. Suitable versions of the third lemma of LeCam are given. These results allow us to study the risks of sequences of decisions in a shrinking sequence of the localization point of the models, providing a comparison of the efficiency of different sequences of decisions.

In the remainder of the chapter, the lower Hájek–LeCam bound is derived. To avoid advanced techniques from topology, the bound is established here only for compact decision spaces and dominated limit models. This proves sufficient for our purposes, as the models considered here are nearly always parametric models. The lower Hájek–LeCam bound makes it possible to break up the proof of asymptotic optimality of estimators, tests, and selection rules into separate steps. The first step consists of finding in the asymptotic Gaussian model the optimal solution, which depends only on the sufficient central variable. By replacing the central variable with the central sequence a sequence of decisions is obtained. Under additional regularity assumptions the convergence of the risks to the lower Hájek–LeCam follows, and this in turn guarantees the optimality of the sequence of decisions.

Chapter 7. The chapter on parameter estimation begins with the Cramér–Rao inequality and the result, which has been proved by various authors under different regularity assumptions: namely that equality only holds for exponential families. This result corroborates the importance of exponential families for statistical analyses under finite sample sizes and it distinguishes a need for asymptotic considerations. Classical results on UMVU estimators, selected topics on Bayes estimators, and considerations regarding the admissibility of estimators conclude the first part of this chapter.

The second part is devoted to the study of the asymptotic properties of estimators of parameters. For all asymptotic considerations it is mandatory to deal first with the question of the consistency of estimators. Only for estimators with this property can classical and modern linearization techniques be utilized. From a variety of possible approaches to consistency we have chosen the concept of M-estimators, and we follow here to some extent the presentation in Pfanzagl (1994). Besides a treatment of the consistency of M-estimators and the MLEs, and a discussion of the existence of MLEs in exponential families, we study location and regression models. Techniques from convex analysis, due to Hjort and Pollard (1993), allow us to verify consistency without assumptions regarding compactness for convex criterion functions. The part on consistency is completed with the consistency in Bayes models. In giving the fundamental results of Doob (1949) and Schwartz (1965) we follow Ghosh and Ramamoorthi (2003). One way of proving the asymptotic normality of M-estimators is based on the classical Taylor expansion. However, for the treatment of regression models with not necessarily differentiable criterion functions it is preferable to follow linearization techniques for convex criterion functions based on Hjort and Pollard (1993). Doing so avoids conditions regarding differentiability. The necessity of taking the second way arises in $\mathbb{L}_1$-regression and more generally in quantile regressions, as they are represented in Jurečková and Sen (1996). The asymptotic normality of the posterior distribution (i.e., the Bernstein–von Mises theorem) is established and used to prove the asymptotic normality of the Bayes estimator.

The last section of this chapter deals with the asymptotic optimality of the MLE. The result by Bahadur (1964) on the majorization of the covariance matrix of the limit distribution of an asymptotically normal estimator over the inverse of Fisher's information matrix is presented. Then the estimation problem is treated systematically as a decision problem, and the lower bound on the risks is derived under different conditions by utilizing the general results from Chapter 6. This is done in the finite-dimensional case for the asymptotically median unbiased estimators. In the multivariate case, an asymptotic minimax bound is derived. It is shown that in each case, under weak assumptions the MLE achieves the respective lower bound. With these main theorems in asymptotic estimation theory this chapter is completed.

Chapter 8. At the beginning uniformly best unbiased level α tests for two-sided hypotheses in one-parameter exponential families are characterized. Then there follows a section on testing linear hypotheses in multivariate normal distributions with a common known covariance matrix. These results constitute, from an asymptotic point of view, the solution of the decision problem in the limit model. Uniformly best unbiased level α tests in d-parameter exponential families, which are conditional tests, are derived next. Selected topics on uniformly best invariant level α tests and Bayes tests conclude the first part of this chapter.

The second part is devoted to the study of the asymptotic properties of tests. It begins with the study of exponential rates of error probabilities in binary models, which leads to the theorems of Stein and Chernoff. The major treatment of asymptotic tests starts with a problem that is of importance of its own: the central question about the linearizations of statistics. Whereas in the area of parameter estimation such linearizations are the result of the linearization of equations, supporting tools of this type are not available for tests. For the latter, the projection techniques due to Hájek are fundamental. This has been used already for U-statistics in the special case of Hoeffding. The usefulness of these projection techniques is demonstrated on U-statistics and rank statistics, which serve as preparation for the results on the local asymptotic optimality of linear rank tests. Projection techniques are also used to study statistics that include estimated nuisance parameters.

The results on the linearization of statistics are used to establish the asymptotic normality of the test statistics under the null hypothesis. A combination of the asymptotic upper Hájek–LeCam bound for the power with the third lemma of LeCam allows the characterization of locally asymptotically most powerful tests and the calculation of the relative efficiency of given tests. For one-dimensional and multivariate parameters of interests, in models with or without nuisance parameters, we characterize the locally asymptotically optimal tests. In particular, we study Neyman's score test, the likelihood ratio tests, and tests that are based on the MLE known as Wald tests. The asymptotic relative efficiency of selected rank tests, especially for the two-sample problem, is determined by investigating the local asymptotic power along parametric curves in the space of the distributions. For given rank tests, the parametric models are determined for which these rank tests are locally asymptotically best.

Chapter 9. Selection rules are presented here within the decision theoretic framework of the book. The goal is to select a best, or several of the best, of k independent populations. The foundation of finite sample size selection rules goes back to Paulson (1949, 1952), Bahadur and Goodman (1952), Bechhofer (1954), Bechhofer, Dunnett, and Sobel (1954), Gupta (1956, 1965), Lehmann (1957a,b, 1961, 1963, 1966), and Eaton (1967a,b). The first research monographs were written by Bechhofer, Kiefer, and Sobel (1968), Gibbons, Olkin, and Sobel (1977), and Gupta and Panchapakesan (1979).

After an introduction of the selection models, optimal point selection rules are derived for parametric and especially for exponential families. For equal sample sizes the fundamental Bahadur–Goodman–Lehmann–Eaton theorem states that the natural selection rule is the uniformly best permutation invariant decision. For unequal sample sizes the situation changes dramatically and the natural selection rule loses many of its qualities (see Gupta and Miescke (1988)). Bayes selection rules in explicit form are not always readily available. For exponential families, conjugate priors can be chosen such that the posterior distributions are balanced and provide Bayes solutions of a simple form. Combining selection with the estimation of the parameter of the selected population is also considered. The next section deals with subset selections and especially with Gupta's subset selection rule. Γ-minimax selections are also considered here. Section 9.3 deals with multistage selection rules that improve the efficiency by combining the approaches of the previous two sections (see, e.g., Miescke (1984a, 1999)). Selected results, including Bayes designs for stagewise sampling allocations, are presented in detail.

The second part of the chapter is on asymptotic properties of selection rules, and it starts with the exponential rates of the error probabilities of selection rules from Liese and Miescke (1999a). These results are related to results of Chernoff (1952, 1956) and Krafft and Puri (1974). Then localized parametric models are considered. It is shown that under equal sample sizes the natural selection rule based on the central sequence is both locally asymptotically uniformly best in the class of all permutation invariant selection rules and locally asymptotically minimax in terms of the pointwise comparison of the asymptotic risks. Because the statistics used by the selection rules have a specific difference structure, which is similar to the situation of two-sample problems, the localization point that appears in the central sequence can be replaced by an estimator without changing the asymptotic efficiency. The same holds true for additional nuisance parameters. In the nonparametric selection model we study selection rules that are based on rank statistics. Here we use results that have been prepared previously for nonparametric tests.

There are a number of people and institutions that we would like to thank for supporting this book project. Several rounds of reviews over the past three years have given us immeasurable help getting the book into shape, and we are deeply indebted to all of the experts who were willing to review our material and provide critical comments and suggestions. We are very grateful to the Mathematical Research Institute at Oberwolfach for letting us stay and work within its RIP program for two weeks in both 2004 and 2005. The support we have received from Springer Verlag and the guidance we have received from John Kimmel and his technical staff have greatly facilitated our work, and we are especially appreciative. We also thank the colleagues in our departments who have contributed to countless discussions throughout the progress of the book. Their input as well as their understanding of our long preoccupation with this project are very much appreciated. Additionally, thanks are due to our departments and universities for the time and working space that they have provided us. We thank Peter Dencker and Jin Tan for proofreading parts of the book and Jenn Fishman for helping us with the revision of the preface.

Our special thanks go to Ingo Steinke, who proofread several versions of the book, pointed out many inaccurate details, and provided valuable suggestions for improving the book's overall layout. His continuous interest and help in this project is highly appreciated.

Finally, we would like to say some words in memory of Shanti S. Gupta, who passed away in 2002. His inspiration, support, and encouragement have deeply affected our lives and, in particular, our work on this book.

Rostock and Chicago,
October 2007

Friedrich Liese
Klaus-J. Miescke

Contents

1

Statistical Models

The starting point of all statistical inference is the observation of data that are subject to unavoidable random errors. The intention is to draw conclusions from the data in such a way that the information that is contained in the data is exploited as much as possible. For this purpose we need a mathematical model that explains the fluctuation of the observations from measurement to measurement, then a mathematical frame for possible conclusions, and finally a tool for the assessment of the quality of concrete conclusions. Although usually error-free conclusions from disturbed data cannot be drawn, we can improve, or even optimize, the inference by utilizing our knowledge of the probabilities of the random events that are relevant for the statistical problem at hand.

The basic object is a suitably chosen space $\mathcal{X}$ in which all concrete measurements can be observed. Following standard practice in probability theory let there be given a σ-algebra $\mathfrak{A}$ of subsets of $\mathcal{X}$ so that $\mathfrak{A}$ contains all subsets of $\mathcal{X}$ that are relevant for the problem. The pair $(\mathcal{X}, \mathfrak{A})$ is called the *sample space.* If $\mathcal{X}$ is a metric space, then we use the Borel sets as the σ-algebra $\mathfrak{A}$. On the other hand, if $\mathcal{X}$ is finite or countably infinite, then we use the power set $\mathfrak{P}(\mathcal{X})$ for $\mathfrak{A}$.

To explain the fluctuation of the observations we assume that each observation $x \in \mathcal{X}$ is the realization of a random variable X with values in $\mathcal{X}$ that is defined on some underlying abstract *probability space* $(\Omega, \mathfrak{F}, \mathbb{P})$, where $\mathbb{P}$ is a probability measure on $(\Omega, \mathfrak{F})$. By definition, such a random variable is a mapping $X : \Omega \to \mathcal{X}$ that is $\mathfrak{F}$-$\mathfrak{A}$ measurable, i.e., $X^{-1}(A) \in \mathfrak{F}$, $A \in \mathfrak{A}$, where $X^{-1}(A) = \{\omega : X(\omega) \in A, \, \omega \in \Omega\}$. To indicate that X is measurable we use the notation $X : \Omega \to_m \mathcal{X}$.

To be able to work on concrete problems a link to a family of concrete probability spaces, say $(\mathcal{X}, \mathfrak{A}, P_\theta)$, $\theta \in \Delta$, has to be established by means of possible distributions P_θ of X at $\theta \in \Delta$ that include the true but unknown distribution of X. This leads to the concept of a *statistical model.* The first step toward a statistical model is to choose a suitable family $(P_\theta)_{\theta \in \Delta}$ of distributions of X on $(\mathcal{X}, \mathfrak{A})$. This can be a difficult and challenging task, depending

F. Liese, K.-J. Miescke, *Statistical Decision Theory*,
DOI: 10.1007/978-0-387-73194-0_1,

on the experimental situation. The choice has to be made based on the initial information that is available about the random behavior of X in the experiment. To be mathematically consistent we assume that there is a family of probability measures $(\mathbb{P}_\theta)_{\theta\in\Delta}$ on $(\Omega, \mathfrak{F})$ such that for every $\theta \in \Delta$ the distribution of X under $\mathbb{P}_\theta$ is given by $P_\theta = \mathbb{P}_\theta \circ X^{-1}$; i.e., $P_\theta(A) = \mathbb{P}_\theta(X \in A)$, $A \in \mathfrak{A}$. By combining the sample space with the set of possible distributions of X we arrive at the statistical model

$$\mathcal{M} = (\mathcal{X}, \mathfrak{A}, (P_\theta)_{\theta\in\Delta}). \tag{1.1}$$

If the *parameter set* Δ is finite, then we call $\mathcal{M}$ a *finite model.* The simplest models are *binary models* where Δ consists only of two elements.

1.1 Exponential Families

Many of the frequently used parametric families of distributions $(P_\theta)_{\theta\in\Delta}$ in a statistical model $\mathcal{M} = (\mathcal{X}, \mathfrak{A}, (P_\theta)_{\theta\in\Delta})$ are special cases of exponential families. Examples are the normal, binomial, Poisson, beta, and gamma families. Because all of these families share properties that are typical for an exponential family, it is natural and proves useful to study first this important general statistical model, and to collect analytical properties that are used throughout this book.

We are following here the tradition set by Lehmann (1959, 1983) in his classical books on testing and estimation, and continued in their respective second editions: Lehmann (1986), Lehmann and Casella (1998), and Lehmann and Romano (2005). More general treatments of exponential families are provided in Barndorff-Nielsen (1978) and Brown (1986). We also refer to Hoffmann-Jørgensen (1994), Johansen (1979), and Küchler and Sørensen (1997).

Let $(\mathcal{X}, \mathfrak{A})$ be a given measurable space and $T : \mathcal{X} \to_m \mathbb{R}^d$ be a *statistic.* For any $\boldsymbol{\mu} \in \mathcal{M}^\sigma(\mathfrak{A})$, we take

$$\Delta = \{\theta : \int \exp\{\langle\theta, T\rangle\} d\boldsymbol{\mu} < \infty\} \subseteq \mathbb{R}^d, \quad \text{and} \tag{1.2}$$

$$K(\theta) = \ln(\int \exp\{\langle\theta, T\rangle\} d\boldsymbol{\mu}), \quad \theta \in \Delta, \tag{1.3}$$

where $\langle\theta, T\rangle = \theta^T T = \sum_{i=1}^d \theta_i T_i$ is the Euclidean scalar product of the vectors $\theta = (\theta_1, ..., \theta_d)^T$ and $T = (T_1, ..., T_d)^T$. Given $0 < \alpha < 1$, we set $p = 1/\alpha$ and $q = 1/(1-\alpha)$. Then $1/p + 1/q = 1$ and by Hölder's inequality (see Lemma A.13) it holds for $\theta_1, \theta_2 \in \Delta$,

$$\begin{aligned}\exp\{K(\alpha\theta_1+(1-\alpha)\theta_2)\} &= \int \exp\{\langle\alpha\theta_1+(1-\alpha)\theta_2, T\rangle\}d\boldsymbol{\mu} \\ &= \int \exp\{\langle\alpha\theta_1, T\rangle\}\exp\{\langle(1-\alpha)\theta_2, T\rangle\}d\boldsymbol{\mu} \\ &\le \Big(\int \exp\{\langle\theta_1, T\rangle\}d\boldsymbol{\mu}\Big)^{\alpha}\Big(\int \exp\{\langle\theta_2, T\rangle\}d\boldsymbol{\mu}\Big)^{1-\alpha} \\ &= \exp\{\alpha K(\theta_1)+(1-\alpha)K(\theta_2)\}.\end{aligned} \tag{1.4}$$

This means that the set Δ in (1.2) is a convex set, and that the function K in (1.3) is convex. For every $\theta \in \Delta$,

$$P_\theta(A) = \int_A \exp\{\langle\theta, T\rangle - K(\theta)\}d\boldsymbol{\mu}, \; A \in \mathfrak{A}, \tag{1.5}$$

is a probability measure on $(\mathcal{X}, \mathfrak{A})$, and the family of distributions $(P_\theta)_{\theta\in\Delta}$ is called an *exponential family*. We denote by

$$f_\theta(x) := \frac{dP_\theta}{d\boldsymbol{\mu}}(x) = \exp\{\langle\theta, T(x)\rangle - K(\theta)\}, \quad x \in \mathcal{X}, \tag{1.6}$$

the density of P_θ with respect to $\boldsymbol{\mu}$, $\theta \in \Delta$.

It should be noted that, in general, the parameter set Δ is neither closed nor open. An exponential family $(P_\theta)_{\theta\in\Delta}$ is called *regular* if $\Delta = \Delta^0$, where here and in the sequel Δ^0 denotes the interior of Δ.

Throughout the book, whenever an exponential family is considered, the following two assumptions are made to make sure that the dimensions of $\mathbb{R}^d$ and Δ can not be reduced.

(A1) The statistics $T_1, ..., T_d$ are linearly independent in the sense that for $a_0, a_1, ..., a_d \in \mathbb{R}$, the relation $a_1T_1 + \cdots + a_dT_d = a_0$, $\boldsymbol{\mu}$-a.e., implies that $a_i = 0$, $i = 0, 1, ..., d$.

(A2) The interior Δ^0 of Δ is nonempty.

If the condition (A1) is fulfilled, then the parameter θ is *identifiable*; that is, $P_{\theta_1} = P_{\theta_2}$ implies $\theta_1 = \theta_2$. If not already achieved from the very beginning, the technical tools of reparametrization and a suitable choice of the measure $\boldsymbol{\mu}$ are available for this purpose.

Definition 1.1. *Under the assumptions (A1) and (A2) made on T and Δ, respectively, the family of distributions $(P_\theta)_{\theta\in\Delta}$ given by* (1.5) *is called a d-parameter exponential family in natural form, with natural parameter θ and generating statistic T. The statistical model*

$$\mathcal{M}_{ne} = (\mathcal{X}, \mathfrak{A}, (P_\theta)_{\theta\in\Delta}) \tag{1.7}$$

with $(P_\theta)_{\theta\in\Delta}$ from (1.5) *is called a natural exponential model. It is called regular if Δ is open.*

An exponential family in natural form is also called an *exponential family in canonical form* in the literature.

Problem 1.2.* The representation of an exponential family in natural form by (1.5) is not unique in the triplet $(T, \theta, \boldsymbol{\mu})$.

The ambiguity pointed out in the above problem is often utilized to find representations that are better adapted to the problem under consideration.

As the density in (1.5) is positive we see that the distributions from the exponential family are measure-theoretically equivalent to $\boldsymbol{\mu}$; that is, $P_\theta(B) = 0$ if and only if $\boldsymbol{\mu}(B) = 0$, or in short $\boldsymbol{\mu} \ll\gg P_\theta$. This implies

$$P_\theta \ll\gg P_{\theta_0}, \quad \theta_0, \theta \in \Delta, \tag{1.8}$$

and that the density of P_θ with respect to P_{θ_0} is given by

$$\frac{dP_\theta}{dP_{\theta_0}} = \exp\{\langle \theta - \theta_0, T \rangle - K(\theta) + K(\theta_0)\}, \quad \theta_0, \theta \in \Delta.$$

For $d = 1$ the condition (A1) only means that the statistic T is not $\boldsymbol{\mu}$-a.e. constant and therefore in view of (1.8) T is not P_θ-a.s. constant. For $d > 1$ the condition (A1) excludes the cases where P_θ-a.s. the statistic T takes on values in a lower-dimensional subspace. We show later that $\mathsf{E}_\theta \|T\|^2 < \infty$, $\theta \in \Delta$. For such a random vector the fact that only values from a subspace are attained can be characterized with the help of the covariance matrix.

Problem 1.3.* Let $Y_1, ..., Y_d$ be random variables with finite second moments. There exist $a_0, a_1, ..., a_d \in \mathbb{R}$ with $\sum_{i=1}^d a_i^2 > 0$ and $a_1 Y_1 + \cdots + a_d Y_d = a_0$, $\mathbb{P}$-a.s., if and only if the covariance matrix of $(Y_1, ..., Y_d)$ is singular.

For some purposes it proves convenient to study the family of induced distributions $Q_\theta = P_\theta \circ T^{-1}$. The statistical model

$$\mathcal{M}_{re} = (\mathbb{R}^d, \mathfrak{B}_d, (Q_\theta)_{\theta \in \Delta}), \tag{1.9}$$

is called the *reduced model* or the *model in minimal form*. For every $B \in \mathfrak{B}_d$ and $\boldsymbol{\nu} = \boldsymbol{\mu} \circ T^{-1}$,

$$Q_\theta(B) = \int I_B(T) \exp\{\langle \theta, T \rangle - K(\theta)\} d\boldsymbol{\mu} = \int_B \exp\{\langle \theta, t \rangle - K(\theta)\} \boldsymbol{\nu}(dt),$$

so that

$$g_\theta(t) := \frac{dQ_\theta}{d\boldsymbol{\nu}}(t) = \exp\{\langle \theta, t \rangle - K(\theta)\}, \quad t \in \mathbb{R}^d. \tag{1.10}$$

When passing from the natural form to the reduced form we changed the sample space with the consequence that the new generating statistic (i.e., the identical mapping) is very simple. Later we show that the two models, the natural model and the reduced model, are identical from the decision-theoretic point of view.

It is an important property of exponential families that the distributions of a sample of size n form again an exponential family where the new generating statistic is the sum. The following proposition presents the precise statement which is a consequence of the fact that the density of a product measure with respect to another product measure is simply the product of the individual densities; see Proposition A.29.

Proposition 1.4. *Let $(P_\theta)_{\theta\in\Delta}$ be a natural exponential family with respect to $\boldsymbol{\mu}$. Then $(P_\theta^{\otimes n})_{\theta\in\Delta} \subseteq \mathcal{P}(\mathfrak{A}^{\otimes n})$ is a natural exponential family with respect to $\boldsymbol{\mu}^{\otimes n}$ with generating statistic $T_{\oplus n}(x_1,...,x_n) := \sum_{i=1}^n T(x_i)$ and it holds that*

$$\frac{dP_\theta^{\otimes n}}{d\boldsymbol{\mu}^{\otimes n}} = \exp\{\langle\theta, T_{\oplus n}\rangle - nK(\theta)\}.$$

If X and Y are independent random vectors with distributions P and Q, respectively, then the distribution of $X+Y$ is given by the convolution of the two distributions P and Q, defined by

$$(P * Q)(B) := \int P(B-x)Q(dx), \quad B \in \mathfrak{B}_d.$$

According to (1.9) the reduced version of $P_\theta^{\otimes n}$ is given by $Q_{n,\theta} := P_\theta^{\otimes n} \circ T_{\oplus n}^{-1}$. As $T_{\oplus n}$ is the sum of n independent identically distributed (i.i.d.) random vectors we see that the reduced model is given by

$$Q_{n,\theta} = \mathcal{L}(T_{\oplus n}|P_\theta^{\otimes n}) = (P_\theta \circ T^{-1})^{*n},$$

where $*n$ denotes the n-fold convolution.

For practical purposes we may also change the parameter set. Such a reparametrization can often be made to get new parameters that allow for a better statistical interpretation. Let $\Lambda \subseteq \mathbb{R}^d$ and $\kappa : \Lambda \to \Delta$ be a mapping. Then (1.5) can be reparametrized to

$$\begin{aligned} P_{pe,\eta}(A) &:= P_{\kappa(\eta)}(A) = \int_A \exp\{\langle\kappa(\eta), T\rangle - K(\kappa(\eta))\}d\boldsymbol{\mu}, \quad A \in \mathfrak{A}, \quad (1.11)\\ h_\eta(x) &:= \frac{dP_{pe,\eta}}{d\boldsymbol{\mu}}(x) = \exp\{\langle\kappa(\eta), T(x)\rangle - K(\kappa(\eta))\}, \quad x \in \mathcal{X}, \end{aligned}$$

where $\eta \in \Lambda$. The statistical model

$$\mathcal{M}_{pe} = (\mathcal{X}, \mathfrak{A}, (P_{pe,\eta})_{\eta\in\Lambda}), \tag{1.12}$$

with $(P_{pe,\eta})_{\eta\in\Lambda}$ from (1.11), is called a *reparametrized exponential model*. Whenever the representation (1.12) is used, we assume without loss of generality that the mapping $\kappa : \Lambda \to \Delta$ is a one-to-one mapping of Λ into Δ. This guarantees that for any two parameter points $\eta_1, \eta_2 \in \Lambda$, $P_{pe,\eta_1} = P_{pe,\eta_2}$ implies $\eta_1 = \eta_2$. In this case, the parameter η in the family $(P_{pe,\eta})_{\eta\in\Lambda}$ is identifiable. Moreover, we use $\gamma = \kappa^{-1}$ in the sequel. A concrete statistical model usually is introduced by specifying $(P_{pe,\eta})_{\eta\in\Lambda}$, where the parameter η admits a direct statistical interpretation.

In the following examples, we look at some common parametric families of distributions and represent them as exponential families. As the natural parameter is not necessarily the parameter that admits a statistical interpretation we often introduce another more meaningful parameter.

Here and in the sequel, whenever an at most countable sample space $\mathcal{X}$ appears we use the power set $\mathfrak{P}(\mathcal{X})$, i.e., the system of all subsets of $\mathcal{X}$, as σ-algebra $\mathfrak{A}$ in our statistical model. Unless explicitly mentioned otherwise, we use the counting measure as the dominating measure so that we have only to deal with the probability mass function (p.m.f.), $f(x) := P(\{x\})$, $x \in \mathcal{X}$, which is the density of P with respect to the counting measure. We set

$$\begin{aligned}
\mathbb{S}_{d-1} &= \{(p_1, ..., p_{d-1}) : p_i > 0,\ i = 1, ..., d-1,\ \textstyle\sum_{j=1}^{d-1} p_j < 1\},\\
\mathbf{S}_d^o &= \{(p_1, ..., p_d) : p_i > 0,\ i = 1, ..., d,\ \textstyle\sum_{j=1}^{d} p_j = 1\},\\
\mathbf{S}_d^c &= \{(p_1, ..., p_d) : p_i \geq 0,\ i = 1, ..., d,\ \textstyle\sum_{j=1}^{d} p_j = 1\}.
\end{aligned} \tag{1.13}$$

Example 1.5. Let $X_1, ..., X_n$ be a sample of i.i.d. observations from an experiment with d possible outcomes that have probabilities p_i, $i = 1, ..., d$. The sample space is $\mathcal{X} = \{(\varepsilon_1, ..., \varepsilon_n) : \varepsilon_i \in \{1, ..., d\},\ i = 1, ..., n\}$ and it holds

$$\begin{aligned}
\mathbb{P}(X_1 = \varepsilon_1, ..., X_n = \varepsilon_n) &= \prod\nolimits_{i=1}^{n} p_{\varepsilon_i}\\
&= \exp\{\sum\nolimits_{j=1}^{d} T_j(x) \ln p_j\}, \quad x = (\varepsilon_1, ..., \varepsilon_n) \in \mathcal{X}, \quad \text{where}\\
T_j(x) &= |\{i : \varepsilon_i = j,\quad i = 1, ..., n\}|
\end{aligned}$$

is the number of observations with outcome j, $j = 1, ..., d$. As $\sum_{j=1}^{d} p_j = 1$ the assumption (A2) is not met. However, by a reduction to $d-1$ parameters we can get an exponential family (1.5) in natural form. Put for $i = 1, ..., d-1$,

$$\begin{aligned}
&\theta_i = \kappa_i(p) := \ln(p_i/p_d), \quad p_i = \gamma_i(\theta) = \exp\{\theta_i\}(1 + \textstyle\sum_{j=1}^{d-1} \exp\{\theta_j\})^{-1},\\
&p_d = 1 - \textstyle\sum_{i=1}^{d-1} p_i, \quad p_d = \gamma_d(\theta) = 1 - \textstyle\sum_{i=1}^{d-1} \gamma_i(\theta),\\
&T = (T_1, ..., T_{d-1}), \quad K(\theta) = n \ln(1 + \textstyle\sum_{j=1}^{d-1} \exp\{\theta_j\}),\\
&\theta = (\theta_1, ..., \theta_{d-1}) \in \Delta = \mathbb{R}^{d-1}, \quad p = (p_1, ..., p_d) \in \mathbf{S}_d^o.
\end{aligned}$$

As $\sum_{i=1}^{d} k_i \ln p_i = \sum_{i=1}^{d} k_i \theta_i$ we see that $P_\theta = \mathcal{L}(X_1, ..., X_n)$ has the p.m.f.

$$\begin{aligned}
f_\theta(x) &= \exp\{\sum\nolimits_{i=1}^{d} T_i(x) \ln p_i\}\\
&= \exp\{\sum\nolimits_{i=1}^{d-1} T_i(x) \ln p_i + (n - \sum\nolimits_{i=1}^{d-1} T_i(x)) \ln(1 - \sum\nolimits_{i=1}^{d-1} p_i)\}\\
&= \exp\{\sum\nolimits_{i=1}^{d-1} \theta_i T_i(x) - K(\theta)\}.
\end{aligned}$$

With $\boldsymbol{\mu}$ being the counting measure we see that $(P_\theta)_{\theta\in\Delta}$is a regular $(d-1)$ parameter exponential family with natural parameter $\theta \in \mathbb{R}^{d-1}$ that satisfies (A1) and (A2). The distributions in the reduced model are then

$$P_\theta \circ T^{-1} = \mathsf{M}(n, \gamma(\theta)), \quad \theta \in \mathbb{R}^{d-1},$$

where $\mathsf{M}(n,p)$ denotes the multinomial distribution with parameters n and $p = (p_1, ..., p_d) \in \mathbf{S}_d^o$.

Problem 1.6. Verify the statements in the previous example regarding (A1) and (A2).

Problem 1.7. Let $X_1, ..., X_n$ be i.i.d. Bernoulli variables with success probability $p \in (0,1)$. Then the joint distribution on $\mathcal{X} = \{0,1\}^n$ is given by $((1-p)\boldsymbol{\delta}_0 + p\boldsymbol{\delta}_1)^{\otimes n}$, where $\boldsymbol{\delta}_a$ is the δ-distribution that is concentrated at point a. Set

$$\begin{aligned}
\theta = \kappa(p) &:= \ln(p/(1-p)), \quad p = \gamma(\theta) := \frac{\exp\{\theta\}}{1+\exp\{\theta\}}, \\
K(\theta) &= n\ln(1+e^\theta), \quad \Delta = \mathbb{R}, \\
T(x) &= \sum\nolimits_{i=1}^n x_i, \quad x = (x_1, ..., x_n) \in \{0,1\}^n.
\end{aligned}$$

Then the family of distributions $(P_\theta)_{\theta\in\Delta} = ((1-\gamma(\theta))\boldsymbol{\delta}_0 + \gamma(\theta)\boldsymbol{\delta}_1)^{\otimes n}$ has the p.m.f. $f_\theta = \exp\{\theta T - K(\theta)\}$ and is thus a one-parameter exponential family with natural parameter θ and generating statistic T. The distributions in the reduced model are $P_\theta \circ T^{-1} = \mathsf{B}(n, \gamma(\theta))$, $\theta \in \mathbb{R}$.

Problem 1.8.* Sometimes, the parameter set $(0,1)$ of the binomial distribution $\mathsf{B}(n,p)$ is extended by putting $\mathsf{B}(n,0) = \boldsymbol{\delta}_0$ and $\mathsf{B}(n,1) = \boldsymbol{\delta}_n$. Show that the extended family $\mathsf{B}(n,p)$, $p \in [0,1]$, cannot be represented as an exponential family.

Problem 1.9.* The family of Poisson distributions $(\mathsf{Po}(\lambda))_{\lambda>0}$ with p.m.f.

$$\mathsf{po}_\lambda(k) = \frac{\lambda^k}{k!}\exp\{-\lambda\}, \quad k \in \mathbb{N}, \quad \lambda > 0,$$

can be represented as a one-parameter exponential family in natural form.

Example 1.10. The exponential families in Example 1.5 and in the Problems 1.7 and 1.9 are regular, i.e., their natural parameter sets are open. This property is often met, but there is an important exponential family that does not share this property. Let $W(t)$, $t > 0$, be a standard Wiener process and ν and σ be fixed positive constants . For $a > 0$ we denote by $T_a = \inf\{t : \nu t + \sigma W(t) \geq a\}$ the first passage time at which the process $\nu t + \sigma W(t)$ crosses the level a. It can be shown (see Seshadri (1993)) that T_a is finite with probability one and that it has a distribution, called the inverse Gaussian distribution $\mathsf{Gi}(\lambda, m)$, that has the Lebesgue density

$$\mathsf{gi}_{\lambda,m}(x) = \sqrt{\frac{\lambda}{2\pi x^3}}\exp\{-\frac{\lambda}{2m^2}\frac{(x-m)^2}{x}\}I_{(0,\infty)}(x), \tag{1.14}$$

where $\lambda = (a/\sigma)^2$ and $m = a/\nu$. Letting $m \to \infty$ we get as the density of the first passage time of the standard Wiener process

$$\mathsf{gi}_{\lambda,\infty}(x) = \sqrt{\frac{\lambda}{2\pi x^3}}\exp\{-\frac{\lambda}{2x}\}I_{(0,\infty)}(x). \tag{1.15}$$

To present the densities $\mathsf{gi}_{\lambda,m}$ in standard exponential form we set $\mathcal{X} = (0,\infty)$ and introduce the measure $\boldsymbol{\mu}$ by $\boldsymbol{\mu}(dx) = (2\pi x^3)^{-1/2}\boldsymbol{\lambda}(dx)$ on $\mathfrak{B}_{(0,\infty)}$. If $T_1(x) = x$, $T_2(x) = 1/x$,

$$(\theta_1,\theta_2) = (-\frac{\lambda}{2m^2}, -\frac{\lambda}{2}), \quad K(\theta_1,\theta_2) = 2\sqrt{\theta_1\theta_2} - \frac{1}{2}\ln(-2\theta_2),$$

then

$$\frac{d\mathsf{Gi}(\lambda,m)}{d\boldsymbol{\mu}}(x) = \exp\{\theta_1 x + \theta_2\frac{1}{x} - K(\theta_1,\theta_2)\}.$$

The natural parameter set is $\Delta = (-\infty,0] \times (-\infty,0)$ which is not open. This set corresponds to the set $(0,\infty) \times (0,\infty]$ in the original parametrization.

Normal distributions are exponential families. The two-parameter case is studied in the next example. The one-parameter cases, where either the variance or the mean is known, are considered in Lemma 1.37 and Example 1.38, respectively.

Example 1.11. Let X be an observation from a normal distribution $\mathsf{N}(\mu,\sigma^2)$, where $(\mu,\sigma^2) \in \Lambda = \mathbb{R} \times (0,\infty)$ is unknown. The density φ_{μ,σ^2} of the distribution $\mathsf{N}(\mu,\sigma^2)$ with respect to the Lebesgue measure $\boldsymbol{\lambda}$ on $\mathbb{R}$ is

$$\begin{aligned}\varphi_{\mu,\sigma^2}(x) &= (2\pi\sigma^2)^{-1/2}\exp\left\{-(2\sigma^2)^{-1}(x-\mu)^2\right\} \\ &= \exp\left\{\kappa_1(\mu,\sigma^2)T_1(x) + \kappa_2(\mu,\sigma^2)T_2(x) - (1/2)[\mu^2/\sigma^2 + \ln(2\pi\sigma^2)]\right\},\end{aligned}$$

where

$$\begin{aligned}(T_1(x),T_2(x)) &= \left(x,x^2\right), \\ (\theta_1,\theta_2) &= \left(\kappa_1(\mu,\sigma^2),\kappa_2(\mu,\sigma^2)\right) := \left(\mu/\sigma^2, -1/(2\sigma^2)\right), \\ \left(\mu,\sigma^2\right) &= (\gamma_1(\theta),\gamma_2(\theta)) := (-\theta_1/(2\theta_2), -1/(2\theta_2)).\end{aligned} \tag{1.16}$$

Hence $\mathsf{N}(\mu,\sigma^2)$, $(\mu,\sigma^2) \in \mathbb{R} \times (0,\infty)$, is a reparametrized exponential family with generating statistic $T = (T_1,T_2)$ and φ_{μ,σ^2} turns into

$$\begin{aligned}f_\theta(x) &= \exp\left\{\theta_1 T_1(x) + \theta_2 T_2(x) - K(\theta)\right\}, \quad \text{where} \\ K(\theta) &= -(1/2)[-\theta_1^2/(2\theta_2) + \ln(-\theta_2/\pi)], \quad \theta \in \mathbb{R} \times (-\infty,0).\end{aligned} \tag{1.17}$$

The set $\Delta = \mathbb{R} \times (-\infty,0)$ is the natural parameter set as

$$\int \exp\left\{\theta_1 T_1(x) + \theta_2 T_2(x)\right\}\boldsymbol{\lambda}(dx) < \infty$$

if and only if $(\theta_1,\theta_2) \in \mathbb{R} \times (-\infty,0)$. Thus we have a regular two-parameter exponential family, represented in natural form by f_θ, $\theta \in \mathbb{R} \times (-\infty,0)$, and represented in reparametrized form by φ_{μ,σ^2}, $(\mu,\sigma^2) \in \mathbb{R} \times (0,\infty)$. The latter is based on the statistically relevant parameters μ and σ^2.

Suppose now that we have a sample of size n; i.e., let $X_1,...,X_n$ be i.i.d. with distribution $\mathsf{N}(\mu,\sigma^2)$. Then by Proposition 1.4 $\mathsf{N}^{\otimes n}(\mu,\sigma^2)$, $(\mu,\sigma^2) \in \mathbb{R} \times (0,\infty)$, is again an exponential family, but now with the generating statistic

$$T_{\oplus n}(x_1,...,x_n) = (\sum\nolimits_{i=1}^n x_i, \sum\nolimits_{i=1}^n x_i^2).$$

Problem 1.12.* The family of distributions in the reduced model $(\mathsf{N}^{\otimes n}(\mu,\sigma^2))\circ T_{\oplus n}^{-1}$ has the Lebesgue density $\sigma^{-2}\varphi_{n\mu,n\sigma^2}(s_1)\mathsf{h}_{n-1}(s_2/\sigma^2 - s_1^2/(n\sigma^2))$, where h_{n-1} is the Lebesgue density of a χ^2-distribution with $n-1$ degrees of freedom.

Next we consider some exponential families that appear as distributions of nonnegative random variables.

Example 1.13. Let $(\mathsf{Ga}(\lambda,\beta))_{\lambda,\beta>0}$ be the family of gamma distributions which have the Lebesgue densities

$$\mathsf{ga}_{\lambda,\beta}(x) = \frac{\beta^\lambda}{\Gamma(\lambda)}x^{\lambda-1}\exp\{-\beta x\}I_{(0,\infty)}(x), \quad x\in\mathbb{R}, \quad \lambda,\beta>0.$$

We introduce the measure $\boldsymbol{\mu}$ by $\boldsymbol{\mu}(dx) = I_{(0,\infty)}(x)\,x^{-1}\boldsymbol{\lambda}(dx)$, and set $T_1(x)=\ln x$, $T_2(x)=-x,\quad x>0$. The $\boldsymbol{\mu}$-density is then given by (1.6), with $K(\lambda,\beta)=\ln\Gamma(\lambda)-\lambda\ln\beta$, and $(\mathsf{Ga}(\lambda,\beta))_{\lambda,\beta>0}$ becomes a two-parameter exponential family in natural form with natural parameter $\theta=(\lambda,\beta)\in\Delta=(0,\infty)\times(0,\infty)$ and generating statistic $T(x)=(\ln x,-x)$.

Problem 1.14. Represent the family $(\mathsf{Ga}(\lambda,\beta))_{\lambda,\beta>0}$ for a fixed known λ, as well as for a fixed known β, as a one-parameter exponential family in natural form. Extend this representation to the case of an i.i.d. sample $X_1,...,X_n$ where the distribution of X_1 belongs to the gamma family.

Problem 1.15.* Let $(\mathsf{Be}(\alpha,\beta))_{\alpha,\beta>0}$ be the family of beta distributions, which have the Lebesgue densities

$$\mathsf{be}_{\alpha,\beta}(x) = \frac{\Gamma(\alpha+\beta)}{\Gamma(\alpha)\Gamma(\beta)}x^{\alpha-1}(1-x)^{\beta-1}I_{(0,1)}(x), \quad x\in\mathbb{R}, \quad \alpha,\beta>0.$$

It can be represented as a two-parameter exponential family in natural form.

Perhaps the most important and useful analytic property of an exponential family in natural form is that in its expectations, differentiations with respect to the coordinates of $\theta=(\theta_1,...,\theta_d)\in\Delta^0$ and integration with respect to $x\in\mathcal{X}$ can be exchanged. Denote by $\mathbb{C}=\{z: z=u+iv,\ u,v\in\mathbb{R}\}$ the set of complex numbers, where u and v are the real and imaginary parts of z, respectively. Similarly, we set $\mathbb{C}^d=\{z: z=u+iv,\ u,v\in\mathbb{R}^d\}$ and again denote by u and v the real and imaginary parts of the vector $z\in\mathbb{C}^d$. A function $\psi=\psi_1+i\psi_2$ is called measurable if ψ_1 and ψ_2 are real-valued measurable functions and denote this again by $\psi:\mathcal{X}\to_m\mathbb{C}$. We set for any $\boldsymbol{\mu}\in\mathcal{M}(\mathfrak{A})$

$$\mathbb{U}=\{u: u\in\mathbb{R}^d, \quad \int|\psi(x)|\exp\{\langle u,T(x)\rangle\}\boldsymbol{\mu}(dx)<\infty\}.$$

Then with $z=u+iv$ the relation $|\exp\{i\alpha\}|=1$ yields $|\exp\{\langle z,T(x)\rangle\}| = |\exp\{\langle u,T(x)\rangle\}|$ so that the function

$$M_\psi(z)=\int\psi(x)\exp\{\langle z,T(x)\rangle\}\boldsymbol{\mu}(dx)$$

is well defined on $\mathbb{F} = \mathbb{U} + i\mathbb{R}^d = \{z : z = u + iv,\ u \in \mathbb{U},\ v \in \mathbb{R}^d\}$. For brevity, we introduce the notation

$$D^\alpha := \frac{\partial^{m_1+\cdots+m_d}}{\partial z_1^{m_1}\cdots\partial z_d^{m_d}}, \qquad \alpha = (m_1, ..., m_d) \in \mathbb{N}^d,$$
$$|\alpha| = \sum\nolimits_{l=1}^d m_l, \quad z^\alpha = z_1^{m_1}\cdots z_d^{m_d}.$$

We recall that for an open set $A \subseteq \mathbb{C}^d$ a function $f : A \to \mathbb{C}$, is called *analytic* if, for every $z_0 \in A$, f can be expanded in a power series

$$f(z) = \sum\nolimits_{k=0}^{\infty} \sum_{\alpha:|\alpha|=k} \frac{1}{m_1!\cdots m_d!} a_\alpha (z - z_0)^\alpha$$

which is absolutely convergent in some neighborhood of z_0. In this case f is infinitely often differentiable and it holds

$$D^\alpha f(z_0) = a_\alpha. \tag{1.18}$$

The following result has been established in the literature in several different versions. Presumably, the first proof was presented in Lehmann (1959).

Lemma 1.16. *For every $\theta_0 \in \mathbb{U}^0$ there exists some $\varepsilon > 0$ such that*

$$\int \exp\{\varepsilon \|T(x)\|\} |\psi(x)| \exp\{\langle\theta_0, T(x)\rangle\}\boldsymbol{\mu}(dx) < \infty. \tag{1.19}$$

The function $M_\psi(z) = \int \psi(x)\exp\{\langle z, T(x)\rangle\}\boldsymbol{\mu}(dx)$ is analytic in the interior $\mathbb{F}^0 = \mathbb{U}^0 + i\mathbb{R}^d$ of $\mathbb{F}$, and it holds for $\alpha = (m_1, ..., m_d)$,

$$\begin{aligned} D^\alpha M_\psi(z) &= \int \psi(x) T_1^{m_1}(x)\cdots T_d^{m_d}(x)\exp\{\langle z, T(x)\rangle\}\boldsymbol{\mu}(dx) \\ &= \int \psi(x) D^\alpha \exp\{\langle z, T(x)\rangle\}\boldsymbol{\mu}(dx), \ z \in \mathbb{F}^0. \end{aligned}$$

Proof. Fix $z_0 = (z_{1,0}, ..., z_{d,0}) \in \mathbb{U}^0 + i\mathbb{R}^d$ and $z = (z_1, ..., z_d)$ and denote by u_i and $u_{i,0}$ the real parts of z and z_0, respectively. The inequalities $\|T\| \le \sum_{i=1}^d |T_i|$ and $\exp\{|x|\} \le \exp\{x\} + \exp\{-x\}$ imply (1.19). The latter inequality and $\sum_{k=0}^n |w|^k/k! \le \exp\{|w|\}$ yields for $\|z - z_0\| \le \delta$

$$\begin{aligned} &|\psi(x)\exp\{\langle z_0, T(x)\rangle\}| \sum\nolimits_{l=0}^n \frac{|\langle z - z_0, T(x)\rangle|^l}{l!} \\ &\le |\ \psi(x)\exp\{\sum\nolimits_{j=1}^d u_{j,0}T_j(x)\}\ | \exp\{\delta \sum\nolimits_{j=1}^d |T_j(x)|\} \\ &\le \sum_{\varepsilon_1,\dots,\varepsilon_d \in \{-1,0,1\}} |\psi(x)| \exp\{\sum\nolimits_{j=1}^d (u_{j,0} + \varepsilon_j\delta)T_j(x)\}. \end{aligned}$$

For sufficiently small δ and the vectors $(u_{1,0}+\varepsilon_1\delta, ..., u_{d,0}+\varepsilon_d\delta)$ belong to $\mathbb{U}^0$ so that the function on the right-hand side of the above inequality is integrable with respect to $\boldsymbol{\mu}$. Hence by Lebesgue's theorem (see Theorem A.18),

$$M_\psi(z) = \int \psi(x) \exp\{\langle z_0, T(x)\rangle\} \sum\nolimits_{k=0}^{\infty} \frac{\langle z - z_0, T(x)\rangle^k}{k!} \boldsymbol{\mu}(dx)$$
$$= \sum\nolimits_{k=0}^{\infty} \sum_{|\alpha|=k} \frac{1}{m_1! \cdots m_d!} a_\alpha (z - z_0)^\alpha,$$

where

$$a_\alpha = \int \psi(x) T_1^{m_1}(x) \cdots T_1^{m_d}(x) \exp\{\langle z_0, T(x)\rangle\} \boldsymbol{\mu}(dx).$$

The relation $a_\alpha = D^\alpha f(z_0)$ in (1.18) with $f = M_\psi$ completes the proof. ■

Theorem 1.17. *Let $(P_\theta)_{\theta\in\Delta}$ be an exponential family in natural form as given by (1.5). Then for every $\theta \in \Delta^0$ there exists an $\varepsilon > 0$ with*

$$\mathsf{E}_\theta \exp\{\varepsilon \|T\|\} < \infty, \tag{1.20}$$

so that

$$\mathsf{E}_\theta \|T\|^a < \infty \quad \textit{for every } a > 0. \tag{1.21}$$

The function K is infinitely often differentiable in Δ^0 and it holds for every $\alpha = (m_1, ..., m_d) \in \mathbb{N}^d$,

$$\mathsf{E}_\theta T_1^{m_1} \cdots T_d^{m_d} = \exp\{-K(\theta)\} D^\alpha \exp\{K(\theta)\}. \tag{1.22}$$

Proof. The statement (1.20) follows from (1.19), and (1.21) is implied by (1.20). From Lemma 1.16 we can see for $\psi(x) \equiv 1$ that the real-valued function $\exp\{K(\theta)\} = \int \exp\{\langle\theta, T\rangle\} d\boldsymbol{\mu}$ is infinitely often differentiable in $\mathbb{U}^0 = \Delta^0$. Because $\int \exp\{\langle\theta, T\rangle\} d\boldsymbol{\mu} \neq 0$, the function $K(\theta)$ is also infinitely often differentiable. To prove (1.22) we note that by Lemma 1.16

$$\exp\{K(\theta)\}\mathsf{E}_\theta T_1^{m_1} \cdots T_d^{m_d} = \int T_1^{m_1}(x) \cdots T_d^{m_d}(x) \exp\{\langle\theta, T(x)\rangle\} \boldsymbol{\mu}(dx)$$
$$= \int D^\alpha \exp\{\langle\theta, T(x)\rangle\} \boldsymbol{\mu}(dx) = D^\alpha \int \exp\{\langle\theta, T(x)\rangle\} \boldsymbol{\mu}(dx)$$
$$= D^\alpha \exp\{K(\theta)\}.$$

■

Remark 1.18. In the previous lemma we have proved the existence of all moments of T provided the parameter belongs to the interior of Δ. For the boundary points, in general, this statements is no longer true as the following example shows. Consider the inverse Gaussian distribution $\mathsf{Gi}(\lambda, m)$ with natural parameters $(\theta_1, \theta_2) = (-\lambda/(2m^2), -\lambda/2) \in (-\infty, 0] \times (-\infty, 0)$ and Lebesgue density $\mathsf{gi}_{\lambda,m}$ from (1.14) for $\theta_1 = 0$, i.e., $\mathsf{gi}_{\lambda,\infty}$ in (1.15). Obviously $\mathsf{E}_{0,\theta_2} T_1 = \int_0^\infty x\, \mathsf{gi}_{\lambda,\infty}(x) dx = \infty$.

There is a simple explanation for this effect. We have pointed out in Example 1.10 that $\mathsf{gi}_{\lambda,m}$ is the density of the first passage time at which the process $\nu t + \sigma W(t)$ crosses the level a, where $\lambda = (a/\sigma)^2$ and $m = a/\nu$. The case $m = \infty$ corresponds to $\nu = 0$, i.e., there is no positive drift. In this case the Wiener process hits the level a very late so that the hitting time is finite with probability one, but the expected value is infinite.

For brevity, we introduce the notation

$$\nabla = \left(\frac{\partial}{\partial\theta_1}, ..., \frac{\partial}{\partial\theta_d}\right)^T \quad \text{and} \quad \nabla\nabla^T = \left(\frac{\partial^2}{\partial\theta_i\partial\theta_j}\right)_{1\le i,j\le d}.$$

The following formulas for calculating the means and covariances of $T_1, ..., T_d$ are direct consequences of (1.22).

Corollary 1.19. *Under the assumptions of Theorem* 1.17, *and conditions (A1) and (A2), for every $\theta \in \Delta^0$ the mean vector and the covariance matrix of T are given by*

$$\mathsf{E}_\theta T = \nabla K(\theta), \quad \mathsf{C}_\theta(T) = \nabla\nabla^T K(\theta). \tag{1.23}$$

The matrix $\nabla\nabla^T K(\theta)$ is nonsingular for every $\theta \in \Delta^0$ and the infinitely often differentiable function K is strictly convex in Δ^0.

Proof. Let $\theta \in \Delta^0$. From (1.22) we get for any $\theta \in \Delta^0$ and $D^\alpha = \frac{\partial}{\partial\theta_i}$,

$$\mathsf{E}_\theta T_i = \exp\{-K(\theta)\}\frac{\partial}{\partial\theta_i}\exp\{K(\theta)\} = \frac{\partial K(\theta)}{\partial\theta_i}.$$

This proves the first statement. Similarly with $D^\alpha = \frac{\partial^2}{\partial\theta_i\partial\theta_j}$,

$$\begin{aligned} \mathsf{E}_\theta T_i T_j &= \exp\{-K(\theta)\}\frac{\partial^2}{\partial\theta_i\partial\theta_j}\exp\{K(\theta)\} \\ &= \frac{\partial K(\theta)}{\partial\theta_i}\frac{\partial K(\theta)}{\partial\theta_j} + \frac{\partial^2 K(\theta)}{\partial\theta_i\partial\theta_j} = \frac{\partial K(\theta)}{\partial\theta_i}\frac{\partial K(\theta)}{\partial\theta_j} + \frac{\partial^2 K(\theta)}{\partial\theta_i\partial\theta_j}. \end{aligned}$$

The nonsingularity of $\nabla\nabla^T K(\theta)$ follows from $\mathsf{C}_\theta(T) = \nabla\nabla^T K(\theta)$ and the fact that by assumption (A1) the components of T are not a.s. linearly dependent, and Problem 1.3. We already know from (1.4) that K is convex. The nonsingularity of $\nabla\nabla^T K(\theta)$ implies that K is strictly convex. ■

We illustrate the above results by examples.

Example 1.20. It has been shown in Example 1.13 that $(\mathsf{Ga}(\alpha,\beta))_{\alpha,\beta>0}$ is a two parameter exponential family in natural form with natural parameter (λ,β) and generating statistic $T(x) = (T_1(x), T_2(x))$, where by $K(\lambda,\beta) = \ln\Gamma(\lambda) - \lambda\ln\beta$, $\lambda,\ \beta > 0$. From (1.23) we get, with $\Psi = \Gamma'/\Gamma$,

$$\mathsf{E}_\theta T = (\Psi(\lambda) - \ln\beta, -\frac{\lambda}{\beta}) \quad \text{and} \quad \mathsf{C}_\theta(T) = \begin{pmatrix} \Psi'(\lambda) & -\frac{1}{\beta} \\ -\frac{1}{\beta} & \frac{\lambda}{\beta^2} \end{pmatrix}.$$

Example 1.21. Let $X_1, X_2, ...$ be a Bernoulli sequence with success probability p, where $p \in (0,1)$. For a fixed given $k \in \{1, 2, ...\}$ let $X = \min\{n : X_1 + \cdots + X_n = k\} - k$. Thus, $X + k$ is the number of times one has to play a game with winning probability p in independent repetitions until k games have been won. X follows a negative binomial distribution $\mathsf{Nb}(k, \eta)$ with p.m.f.

$$\mathsf{nb}_{k,p}(x) = \frac{(x+k-1)!}{x!(k-1)!} p^k (1-p)^x, \quad x = 0, 1, 2, ...$$

Put $\theta = \kappa(p) = \ln(1-p)$, and $\boldsymbol{\mu}(\{x\}) = (x+k-1)!/(x!(k-1)!)$. Then the distribution of X has the density $f_\theta(x) = \exp\{\theta x + k \ln(1 - \exp\{\theta\})\}$ with respect to $\boldsymbol{\mu}$. This shows that $\mathsf{Nb}(k, 1 - e^\theta)$ is a one-parameter exponential family with $T(x) = x$ and $K(\theta) = -k \ln(1 - \exp\{\theta\})$. From (1.23) we get

$$\mathsf{E}_{\kappa(p)} T = k \frac{1-p}{p} \quad \text{and} \quad \mathsf{V}_{\kappa(p)}(T) = k \frac{1-p}{p^2}.$$

In the previous examples we have already studied different ways of parametrizing an exponential family. However, among all parametrizations there is one in particular, not mentioned so far, that has a special meaning. This is the so-called *mean value parametrization* which is considered at the conclusion of this section. To prepare for this parametrization we need the following well-known result (see, e.g., Brown (1986) and Witting (1985)).

Theorem 1.22. *Under the assumptions of (A1) and (A2) the mapping*

$$\gamma_{\mathsf{m}} : \theta \mapsto \nabla K(\theta) = \mathsf{E}_\theta T \tag{1.24}$$

is a diffeomorphism of Δ^0 onto the open set $\gamma_{\mathsf{m}}(\Delta^0)$.

Proof. We already know from Corollary 1.19 that K is strictly convex. This yields for every $\theta_1, \theta_2 \in \Delta^0$ with $\theta_1 \neq \theta_2$,

$$K(\theta_1) > K(\theta_2) + \langle(\theta_1 - \theta_2), \nabla K(\theta_2)\rangle,$$
$$K(\theta_2) > K(\theta_1) + \langle(\theta_2 - \theta_1), \nabla K(\theta_1)\rangle.$$

Hence, $\langle \theta_2 - \theta_1, \nabla K(\theta_1) - \nabla K(\theta_2)\rangle < 0$, so that $\theta_1 \neq \theta_2$ implies $\nabla K(\theta_1) \neq \nabla K(\theta_2)$ and γ_{m} is a bijection. As by Proposition 1.16 K is infinitely often differentiable we see that the mapping γ_{m} is continuously differentiable. An application of the global inverse function theorem (see, e.g., Theorem 3.2.8 in Duistermaat and Kolk (2004)) completes the proof. ■

Brown (1986) proved under the so-called steepness condition a stronger result which at the same time characterizes the range $\gamma_{\mathsf{m}}(\Delta^0)$. We come back to this result later when we study maximum likelihood estimators in exponential families in Section 7.5.

By denoting the inverse mapping of γ_{m} by κ_{m}, we can represent the exponential family in the mean value parametrization, at least for $\theta \in \Delta^0$, by

$$P_{\kappa_{\mathsf{m}}(\mu)}, \quad \mu \in \gamma_{\mathsf{m}}(\Delta^0).$$

The name of this particular parametrization reflects the obvious fact that $\mathsf{E}_{\kappa_{\mathsf{m}}(\mu)}T = \mu$. If we have a reparametrized exponential family $(P_{\kappa(\eta)})_{\eta\in\Lambda}$, then we get the mean value parametrization if we use $\mu = \gamma_{\mathsf{m}}(\kappa(\eta))$. Instead of calculating the functions κ and γ_{m} it is often easier to express μ directly by η via the relation $\mathsf{E}_{\kappa(\eta)}T = \mu$.

Example 1.23. We have seen in Example 1.7 that the binomial distribution $\mathsf{B}(n,p)$, $0 < p < 1$, is the reduced form of a one-parameter exponential family and therefore again an exponential family where the generating statistic T is the identical mapping. Hence with $\mu = \mathsf{E}_p T = np$ the p.m.f. in the mean value parametrization is given by

$$\binom{n}{k}\left(\frac{\mu}{n}\right)^k\left(1-\frac{\mu}{n}\right)^{n-k}, \quad \mu \in (0,n).$$

Problem 1.24. Find the Lebesgue density of $\mathsf{N}(\mu,\sigma^2)$, where both μ and σ^2 are unknown, in the mean value parametrization.

When inspecting the structure of the density f_θ of an exponential family one might get the impression that the factor $\exp\{-K(\theta)\}$ plays only a subordinate role as a normalizing factor. However, the reduced form of the exponential family in (1.10) reveals the meaning of $\exp\{-K(\theta)\}$. Only this function and the underlying measure $\boldsymbol{\nu}$ vary from one family to another. Therefore $\exp\{-K(\theta)\}$ carries the full structure of the exponential family. Furthermore, we show that $K(\theta)$ determines the measure $\boldsymbol{\nu} = \boldsymbol{\mu}\circ T^{-1}$ uniquely.

Proposition 1.25. *Suppose $\boldsymbol{\mu}_1$ and $\boldsymbol{\mu}_2$ are σ-finite measures on $(\mathcal{X},\mathfrak{A})$ and $T:\mathcal{X}\to_m \mathbb{R}^d$. If there are $a_h < b_h$ for $h = 1,...,d$ such that*

$$\mathsf{X}_{h=1}^d(a_h,b_h) \subseteq \{\theta : \int \exp\{\langle\theta,T\rangle\}\,d\boldsymbol{\mu}_1 = \int \exp\{\langle\theta,T\rangle\}\,d\boldsymbol{\mu}_2\},$$

then $\boldsymbol{\mu}_1\circ T^{-1} = \boldsymbol{\mu}_2\circ T^{-1}$.

Proof. Use any $\theta_0 = (\theta_{0,1},...,\theta_{0,d}) \in \mathsf{X}_{h=1}^d(a_h,b_h)$ and turn to the measures

$$\widetilde{\boldsymbol{\mu}}_j(dx) := \exp\{\langle\theta_0,T\rangle\}\,\boldsymbol{\mu}_j(dx), \quad j = 1,2.$$

After a normalization by the same constant c we may assume that $Q_j = c\widetilde{\boldsymbol{\mu}}_j\circ T^{-1}$ are probability measures so that for $u \in \mathsf{X}_{h=1}^d(a_h-\theta_{0,h},b_h-\theta_{0,h})$,

$$\int \exp\{\langle u,t\rangle\}\,Q_1(dt) = \int \exp\{\langle u,t\rangle\}\,Q_2(dt).$$

The functions $M_j(z) := \int \exp\{\langle z,t\rangle\}\,Q_j(dt)$ with $z = u+iv$ are analytic in $(\mathsf{X}_{h=1}^n(a_h-\theta_{0,h},b_h-\theta_{0,h}))+i\mathbb{R}^d$ in view of Lemma 1.16. By the uniqueness theorem for analytic functions of several variables (see Theorem 9.4.2 in Dieudonné (1960)) we get $M_1(z) = M_2(z)$ for every $z \in (\mathsf{X}_{h=1}^n(a_h-\theta_{0,h},b_h-\theta_{0,h}))+i\mathbb{R}^d$. Hence especially for real parts equal to zero

$$\int \exp\{i\langle v,t\rangle\} Q_1(dt) = \int \exp\{i\langle v,t\rangle\} Q_2(dt)$$

for every $v \in \mathbb{R}^d$. From the uniqueness of characteristic functions (see Theorem A.51) we get $Q_1 = Q_2$ and by $(\boldsymbol{\mu}_j \circ T^{-1})(dt) = (1/c)\exp\{\langle\theta_0,t\rangle\} Q_j(dt)$, $j = 1,2$, the statement. ■

One of many other reasons for the wide applicability of exponential families in several applications and especially in physics (see von der Linden et al. (1999)) is that these distributions maximize the Shannon entropy under linear constraints. To be more precise we introduce the Shannon entropy as a measure that describes how much the probability mass of a distribution is scattered on the sample space. Assume $\nu \in \mathcal{M}^\sigma(\mathfrak{A})$ and consider the set of distributions defined by

$$\mathcal{P}_{\mathsf{S}} = \{P : P \in \mathcal{P}(\mathfrak{A}) : P \ll \nu, \quad \int \left|\frac{dP}{d\nu}\ln\frac{dP}{d\nu}\right| d\nu < \infty\}.$$

For every $P \in \mathcal{P}_{\mathsf{S}}$ we call

$$\mathsf{S}_\nu(P) := -\int \frac{dP}{d\nu}\ln\frac{dP}{d\nu}d\nu \tag{1.25}$$

the *Shannon entropy* of P.

Example 1.26. Consider the natural exponential family $(P_\theta)_{\theta\in\Delta}$ with generating statistic T and $\boldsymbol{\mu} = \gamma$. Then with $\mathsf{E}_\theta T = \nabla K(\theta)$ from (1.23),

$$\begin{aligned}\mathsf{S}_\nu(P_\theta) &= -\mathsf{E}_\theta \ln\frac{dP_\theta}{d\nu} = -\langle\theta, \mathsf{E}_\theta T\rangle + K(\theta) \\ &= -\langle\theta, \nabla K(\theta)\rangle + K(\theta).\end{aligned} \tag{1.26}$$

Problem 1.27. If ν is the Lebesgue measure and $P_\theta = \mathsf{N}(\mu,\sigma^2)$, then

$$\begin{aligned}\mathsf{S}_\nu(\mathsf{N}(\mu,\sigma^2)) &= -\int \varphi_{\mu,\sigma^2}(t)\ln\varphi_{\mu,\sigma^2}(t)dt \\ &= \frac{1}{2}\ln(2\pi\sigma^2) + \int \varphi_{\mu,\sigma^2}(t)\frac{(t-\mu)^2}{2\sigma^2}dt = \frac{1}{2} + \frac{1}{2}\ln(2\pi\sigma^2).\end{aligned} \tag{1.27}$$

We see from the last example that the Shannon entropy may assume any real value and is not bounded. This means that we need constraints when searching for distributions that maximize the Shannon entropy. Let $T = (T_1,...,T_d) : \mathcal{X} \to_m \mathbb{R}^d$ and consider for fixed $\theta_0 \in \Delta^0$ the following subset of $\mathcal{P}_{\mathsf{S}}$ defined by linear constraints for the expectation.

$$\begin{aligned}\mathcal{P}_{\theta_0} &= \{P : P \in \mathcal{P}_{\mathsf{S}}, \ \int \|T\|\, dP < \infty, \ \int \langle\theta_0, T\rangle\, dP \geq \int \langle\theta_0, T\rangle\, dP_{\theta_0}\} \\ &= \{P : P \in \mathcal{P}_{\mathsf{S}}, \ \mathsf{E}_P\|T\| < \infty, \ \langle\theta_0, \mathsf{E}_P T\rangle \geq \langle\theta_0, \mathsf{E}_{\theta_0} T\rangle\}.\end{aligned}$$

The distribution from the natural exponential family is singled out by the fact that it maximizes the Shannon entropy in $\mathcal{P}_{\theta_0}$.

Theorem 1.28. *Suppose $(P_\theta)_{\theta\in\Delta}$ is a natural exponential family with generating statistic T and $\boldsymbol{\mu} = \gamma$. If $\theta_0 \in \Delta^0$, then the distribution P_{θ_0} maximizes the Shannon S_ν entropy in the class $\mathcal{P}_{\theta_0}$ and is uniquely determined by this property.*

Proof. Suppose that the distribution P belongs to $\mathcal{P}_{\theta_0}$ and has the ν-density $f_{\theta_0} = dP_{\theta_0}/d\nu$. Then by (1.26) and

$$\begin{aligned}
&\mathsf{S}_\nu(P_{\theta_0}) - \mathsf{S}_\nu(P) = -\langle\theta_0, \mathsf{E}_{\theta_0}T\rangle + K(\theta_0) - \mathsf{S}_\nu(P)\\
&\geq -\langle\theta_0, \mathsf{E}_P T\rangle + K(\theta_0) - \mathsf{S}_\nu(P) = -\int(\langle\theta_0, T\rangle - K(\theta_0))f d\nu - \mathsf{S}_\nu(P)\\
&= \int f\ln(f/f_{\theta_0})d\nu = \int[(f/f_{\theta_0})\ln(f/f_{\theta_0}) + 1 - f/f_{\theta_0}]f_{\theta_0}d\nu.
\end{aligned}$$

The function $g(x) = x\ln x + 1 - x$, $x > 0$, satisfies $g(1) = g'(1) = 0$ and $g''(x) = x^{-1} > 0$. Hence $g(x) \geq 0$ and $g(x) = 0$ if and only if $x = 1$. This means $\mathsf{S}_\nu(P_{\theta_0}) - \mathsf{S}_\nu(P) \geq 0$ where the equality holds if and only if $f = f_{\theta_0}$ ν-a.e. ■

Example 1.29. Let $\mathcal{X} = \mathbb{R}, \mathfrak{A} = \mathfrak{B}, \nu = \boldsymbol{\lambda}$. We characterize the set of distributions in which the normal distribution attains maximum Shannon entropy. We have according to (1.16), $(\theta_1, \theta_2) = (\mu/\sigma^2, -(2\sigma^2)^{-1})$. Furthermore, $T(x) = (x, x^2)$ and $\mathsf{E}_\theta T = \nabla K(\theta) = (\mu, \mu^2 + \sigma^2)$. Hence $\mathcal{P}_{\theta_0}$ is the set of all distributions with finite second moment and

$$\int\langle\theta_0, T\rangle\, dP = \frac{\mu_0}{\sigma_0^2}\int tP(dt) + (-\frac{1}{2\sigma_0^2})\int t^2P(dt) \geq \frac{\mu_0^2}{2\sigma_0^2} - \frac{1}{2},$$

or

$$\mathcal{P}_{\theta_0} = \{P : \int t^2P(dt) < \infty,\ \frac{1}{2}\int t^2P(dt) - \mu_0\int tP(dt) \leq \frac{\sigma_0^2}{2} - \frac{\mu_0^2}{2}\}.$$

Especially if we consider the subclass of all distributions from $\mathcal{P}_\theta$ that satisfy $\int tP(dt) = \mu_0$ we see that the normal distribution $\mathsf{N}(\mu_0, \sigma_0^2)$ maximizes the Shannon entropy in the class of all distributions with expectation μ_0 and a variance that does not exceed σ_0^2.

Problem 1.30. The exponential distribution with expectation θ_0 maximizes the Shannon entropy in the class of all distributions on $\mathbb{R}_+$ that are absolutely continuous with respect to the Lebesgue measure and have an expectation that does not exceed θ_0.

1.2 Priors and Conjugate Priors for Exponential Families

In a Bayes model the parameter is treated as a random variable and $(P_\theta)_{\theta\in\Delta}$ is interpreted as a family of conditional distributions. To deal with such situations we present some basic facts on conditional distributions and the construction of distributions on product spaces with given marginal and transition probabilities. Suppose $(\mathcal{X}, \mathfrak{A})$ and $(\Delta, \mathfrak{B}_\Delta)$ are measurable spaces. Later on we use $(\mathcal{X}, \mathfrak{A})$ as the sample space and Δ as the parameter set. At this stage, however, the two spaces are still kept arbitrary.

If (X, Θ) is a random vector with values in $\mathcal{X} \times \Delta$, then a stochastic kernel $\mathsf{P} : \mathfrak{A} \times \Delta \to_k [0,1]$ (see Definition A.35) is called a *regular conditional distribution* of X, given $\Theta = \theta$, if

$$\mathbb{P}((X, \Theta) \in C) = \int [\int I_C(x, \theta) \mathsf{P}(dx|\theta)] \Pi(d\theta), \quad C \in \mathfrak{A} \otimes \mathfrak{B}_\Delta,$$

where $\Pi = \mathbb{P} \circ \Theta^{-1}$ is the distribution of Θ. The standard extension technique of measure theory, see Lemma A.6, shows that this condition is equivalent to

$$\mathbb{E}h(X, \Theta) = \int [\int h(x, \theta) \mathsf{P}(dx|\theta)] \Pi(d\theta)$$

for every $h : \mathcal{X} \times \Delta \to_m \mathbb{R}_+$. Furthermore by the disintegration lemma (see Lemma A.41) it holds

$$\mathbb{E}(h(X, \Theta)|\Theta = \theta) = \int h(x, \theta) \mathsf{P}(dx|\theta), \quad \Pi\text{-a.s.} \tag{1.28}$$

So far we have analyzed the distribution of (X, Θ) by decomposing it into a conditional distribution and a marginal distribution. On the other hand, we often start with a given family of distributions $(P_\theta)_{\theta \in \Delta}$ that satisfies the following condition; see also Definition A.35.

(A3) $\mathsf{P} : \mathfrak{A} \times \Delta \to [0,1]$, defined by $\mathsf{P}(A|\theta) = P_\theta(A)$, $A \in \mathfrak{A}$, $\theta \in \Delta$, is a stochastic kernel.

Then for any distribution $\Pi \in \mathcal{P}(\mathfrak{B}_\Delta)$

$$(\mathsf{P} \otimes \Pi)(C) := \int [\int I_C(x, \theta) \mathsf{P}(dx|\theta)] \Pi(d\theta), \quad C \in \mathfrak{A} \otimes \mathfrak{B}_\Delta, \tag{1.29}$$

is a distribution on $\mathfrak{A} \otimes \mathfrak{B}_\Delta$, and by Fubini's theorem for stochastic kernels (see Proposition A.40) it holds

$$\int h(x, \theta)(\mathsf{P} \otimes \Pi)(dx, d\theta) = \int [\int h(x, \theta) \mathsf{P}(dx|\theta)] \Pi(d\theta) \tag{1.30}$$

for every $h : \mathcal{X} \times \Delta \to_m \mathbb{R}_+$. If the condition (A3) is satisfied and $\mathcal{L}(X, \Theta) = \mathsf{P} \otimes \Pi$, then (X, Θ) is called a *Bayes model* and Θ is called a *model for the parameter* θ. The marginal distributions are then given by Π and

$$\mathsf{M}(A) := (\mathsf{P}\Pi)(A) = (\mathsf{P} \otimes \Pi)(A \times \Delta).$$

When we turn to a sample of size n, then the assumption of independence of the observations that is made in the classical (frequentist) model has to be replaced by the conditional independence in the Bayes model. The Bayes model for n observations $X_1,...,X_n$ consists of the random variables $X_1,...,X_n,\Theta$, where $X_1,...,X_n$ are conditionally i.i.d., given $\Theta=\theta$, for $\theta\in\Delta$. This means that for any $A\in\mathfrak{A}^{\otimes n}\otimes\mathfrak{B}_\Delta$,

$$\begin{aligned}&\mathbb{P}((X_1,...,X_n,\Theta)\in A)\\&\quad=\int(\int(\cdots\int I_A(x_1,...,x_n,\theta)P_\theta(dx_1)\cdots)P_\theta(dx_n))\Pi(d\theta),\end{aligned}$$

which is equivalent to

$$\mathbb{E}h(X_1,...,X_n,\Theta)=\int(\int(\cdots\int h(x_1,...,x_n,\theta)P_\theta(dx_1)\cdots)P_\theta(dx_n))\Pi(d\theta),$$

for every $h:\mathcal{X}^n\times\Delta\to_m\mathbb{R}_+$. Using the notation with the symbol $\otimes$ from Definition A.36 we may write in short

$$\mathcal{L}(X_1,...,X_n,\Theta)=\mathsf{P}^{\otimes n}\otimes\Pi, \tag{1.31}$$

which is in accordance with (1.29) if we replace there P with $\mathsf{P}^{\otimes n}$. It should be pointed out that in the Bayes model for a sample of size n the marginal distribution of $X=(X_1,...,X_n)$ is not a product distribution. More specifically, $\mathcal{L}(X_1,...,X_n)$ is a mixture of product measures,

$$(\mathsf{P}^{\otimes n}\Pi)(B)=\mathbb{P}((X_1,...,X_n)\in B)=\int P_\theta^{\otimes n}(B)\Pi(d\theta),\quad B\in\mathfrak{A}^{\otimes n}.$$

Because (1.29), when P is replaced with $\mathsf{P}^{\otimes n}$, is equivalent to (1.31) we subsequently deal mainly with the case of $n=1$.

An important step is the disintegration of $\mathsf{P}\otimes\Pi$ with respect to M; see Lemma A.41. To guarantee the existence of regular conditional distributions we establish the following condition.

(A4) The space $(\Delta,\mathfrak{B}_\Delta)$ is a Borel space.

For the definition of a Borel space see Definition A.7. Under condition (A4), by Theorem A.37, there exists a stochastic kernel $\boldsymbol{\Pi}:\mathfrak{B}_\Delta\times\mathcal{X}\to_k[0,1]$ such that

$$\int[\int h(x,\theta)\mathsf{P}(dx|\theta)]\Pi(d\theta)=\int[\int h(x,\theta)\boldsymbol{\Pi}(d\theta|x)]\mathsf{M}(dx) \tag{1.32}$$

for every $h:\mathcal{X}\times\Delta\to_m\mathbb{R}_+$. Instead of (1.30) and (1.32) we also make use of the intuitive and short notation

$$(\mathsf{P}\otimes\Pi)(dx,d\theta)=\mathsf{P}(dx|\theta)\Pi(d\theta)=\boldsymbol{\Pi}(d\theta|x)\mathsf{M}(dx). \tag{1.33}$$

Often, the joint distribution of (X,Θ) is constructed by assuming that the distribution of Θ has a density π with respect to some $\boldsymbol{\tau}\in\mathcal{M}^\sigma(\mathfrak{B}_\Delta)$, and that the family $(P_\theta)_{\theta\in\Delta}$ satisfies the following condition.

(A5) $P_\theta \ll \boldsymbol{\mu}$ for some $\boldsymbol{\mu} \in \mathcal{M}^\sigma(\mathfrak{A})$ and $(x,\theta) \mapsto f_\theta(x) := \frac{dP_\theta}{d\boldsymbol{\mu}}(x)$ is $\mathfrak{A}\otimes\mathfrak{B}_\Delta$-$\mathfrak{B}$ measurable.

Under condition (A5) $f_\theta(x)\pi(\theta)$ is the $\boldsymbol{\mu}\otimes\boldsymbol{\tau}$ density of (X,Θ), i.e., it holds

$$\begin{aligned}\mathbb{E}h(X,\Theta) &= \int\Big[\int h(x,\theta)f_\theta(x)\boldsymbol{\mu}(dx)\Big]\pi(\theta)\boldsymbol{\tau}(d\theta)\\ &= \int\Big[\int h(x,\theta)\mathsf{P}(dx|\theta)\Big]\pi(\theta)\boldsymbol{\tau}(d\theta)\end{aligned}$$

for every $h : \mathcal{X}\times\Delta \to_m \mathbb{R}_+$, where P defined by $\mathsf{P}(dx|\theta) = f_\theta(x)\boldsymbol{\mu}(dx)$ satisfies (A3) and is a regular version of the conditional distribution of X, given $\Theta = \theta$. Moreover, the joint distribution of (X,Θ) is given by $\mathsf{P}\otimes\Pi$. By construction $\mathsf{P}\otimes\Pi \ll \boldsymbol{\mu}\otimes\boldsymbol{\tau}$, and with $\mathsf{M} = \mathsf{P}\Pi$ as the marginal distribution of X,

$$\frac{d(\mathsf{P}\otimes\Pi)}{d(\boldsymbol{\mu}\otimes\boldsymbol{\tau})}(x,\theta) = f_\theta(x)\pi(\theta), \quad \boldsymbol{\mu}\otimes\boldsymbol{\tau}\text{-a.e.,} \quad \text{and}$$

$$\mathsf{m}(x) := \frac{d\mathsf{M}}{d\boldsymbol{\mu}}(x) = \int f_\theta(x)\pi(\theta)\boldsymbol{\tau}(d\theta), \quad \boldsymbol{\mu}\text{-a.e.} \tag{1.34}$$

We set

$$\pi(\theta|x) = \begin{cases} f_\theta(x)\pi(\theta)/\mathsf{m}(x) & \text{if } \mathsf{m}(x) > 0 \\ \pi(\theta) & \text{if } \mathsf{m}(x) = 0\end{cases}, \quad \text{and} \tag{1.35}$$

$$\boldsymbol{\Pi}(B|x) = \int_B \pi(\theta|x)\boldsymbol{\tau}(d\theta), \quad B \in \mathfrak{B}_\Delta.$$

Then, without any additional assumption on $(\Delta, \mathfrak{B}_\Delta)$, $\boldsymbol{\Pi}$ is a stochastic kernel and (1.32) holds. $\pi(\theta|x)$ is called the conditional density of Θ, given $X = x$. Now (1.33) can also be written in terms of densities,

$$f_\theta(x)\pi(\theta) = \pi(\theta|x)\mathsf{m}(x), \quad \boldsymbol{\mu}\otimes\boldsymbol{\tau}\text{-a.e.,}$$

and

$$\mathbb{E}(h(X,\Theta)|X) = \int h(X,\theta)\pi(\theta|X)\boldsymbol{\tau}(d\theta), \quad \mathbb{P}\text{-a.s.}$$

Finally we remark that whenever the joint distribution $\mathsf{P}\otimes\Pi$ is absolutely continuous with respect to a σ-finite product measure, then it can be dominated by the product of the marginal distributions of $\mathsf{P}\otimes\Pi$.

Problem 1.31.* If $\mathsf{P}\otimes\Pi \ll \boldsymbol{\mu}\otimes\boldsymbol{\tau}$ holds for some $\boldsymbol{\mu} \in \mathcal{M}^\sigma(\mathfrak{A})$ and $\boldsymbol{\tau} \in \mathcal{M}^\sigma(\mathfrak{B}_\Delta)$, then $\mathsf{P}\otimes\Pi \ll \mathsf{M}\otimes\Pi$ with $\mathsf{M} = \mathsf{P}\Pi$.

In Bayes analysis the observation X and the parameter Θ are both considered to be random variables. The basic idea is that after the unobservable random variable Θ has been realized and provided a value θ, a second random variable X is realized and provides a value x, where the conditional distribution of X, given $\Theta = \theta$, is P_θ for every $\theta \in \Delta$. To construct the joint distribution of X and Θ, we start with two measurable spaces $(\mathcal{X}, \mathfrak{A})$ and $(\Delta, \mathfrak{B}_\Delta)$, with the understanding that the values of X and Θ belong to $\mathcal{X}$ and Δ, respectively. It is clear that the dependence of X on Θ, as specified by the linking conditional distributions P_θ, $\theta \in \Delta$, and the marginal distribution Π of Θ, are the crucial ingredients. Π is called the *prior distribution*, in short *prior*, as it controls the outcome of $\Theta = \theta$ prior to the experiment. For the inference on Θ based on the observation $X = x$, however, the conditional distribution $\boldsymbol{\Pi}(\cdot|x)$ of Θ, given $X = x$, usually turns out to be the crucial tool. It is called the *posterior distribution*, or in short, *posterior*.

For the remainder of this section let us assume that $(P_\theta)_{\theta\in\Delta}$ is an exponential family in natural form (1.5), where P_θ has the density $f_\theta(x)$ in (1.6) with respect to $\boldsymbol{\mu}$. For special priors both the marginal and the posterior distribution can be explicitly evaluated. We use the symbol $f \propto g$ to express that the functions f and g are identical up to a constant factor.

Problem 1.32.* If $\mathsf{Ga}(\lambda, \beta)$ is the conditional distribution of X given $\Theta = \beta$, and $\pi(\beta) = \mathsf{ga}_{a,b}(\beta)$, $\beta \in \mathbb{R}$, $a, b > 0$, then $\pi(\beta|x) = \mathsf{ga}_{a+\lambda,b+x}(\beta)$, $\beta \in \mathbb{R}$, $x > 0$, and $\mathsf{m}(x) \propto x^{\lambda-1}(b+x)^{-(a+\lambda)}I_{(0,\infty)}(x)$.

As a special class of priors we introduce now the so-called *conjugate priors* for exponential families that are fundamental to Bayes analysis. This concept makes many Bayes decision problems feasible and explicitly tractable. Moreover, as the posterior distribution turns out here to be of the same type as the prior distribution it can be assessed what has been learned about the unknown parameter θ after $X = x$ has been observed.

Because $(x, \theta) \mapsto \exp\{\langle \theta, T(x)\rangle\}$ is $\mathfrak{A} \otimes \mathfrak{B}_d$ measurable the natural parameter set $\Delta \subseteq \mathbb{R}^d$ in (1.2) is a Borel set. Let $\mathfrak{B}_\Delta$ stand for the σ-algebra of Borel subsets of Δ. Then $f_\theta(x)$ from (1.6) is measurable as a function of (x, θ), i.e., assumption (A5) is satisfied, and $\mathsf{P}(\cdot|\theta) := P_\theta(\cdot)$ is a stochastic kernel for P_θ as defined by (1.5).

Fix $\boldsymbol{\tau} \in \mathcal{M}^\sigma(\mathfrak{B}_\Delta)$. As in (1.4) one can see easily that

$$\Upsilon = \{(a, b) : a \in \mathbb{R},\, b \in \mathbb{R}^d,\ \int \exp\{\langle b, \theta\rangle - aK(\theta)\}\boldsymbol{\tau}(d\theta) < \infty\} \tag{1.36}$$

is a convex subset of $\mathbb{R}^{d+1}$. The set Υ is later used as the parameter set of a family of prior distributions. Therefore the structure of Υ is of special interest. A complete characterization of Υ was given by Brown (1986) when $\boldsymbol{\tau}$ is the Lebesgue measure. To formulate that result we need some properties of the measure $\boldsymbol{\nu} = \boldsymbol{\mu} \circ T^{-1}$. Let $\mathsf{S}(\boldsymbol{\nu})$ denote the support of $\boldsymbol{\nu}$, i.e.,

$$\mathsf{S}(\boldsymbol{\nu}) = \{t : \boldsymbol{\nu}(\{y : \|y - t\| > \varepsilon\}) > 0 \text{ for every } \varepsilon > 0\}.$$

It is the smallest closed set B with $\boldsymbol{\nu}(\mathbb{R}^d \backslash B) = 0$. Furthermore,

$$\mathsf{CS}(\boldsymbol{\nu}) := \text{ closure of the convex closure of } \mathsf{S}(\boldsymbol{\nu})$$

is called the *convex support* of $\boldsymbol{\nu}$. For a proof of the following result we refer to Brown (1986), p. 113.

Theorem 1.33. *If $(P_\theta)_{\theta\in\Delta}$ is a d-parameter exponential family in natural form (1.10) that satisfies conditions (A1) and (A2), and if $\boldsymbol{\tau}$ is the Lebesgue measure, then*

$$\Upsilon = \{(a,b) : a \in \mathbb{R},\, b \in \mathbb{R}^d,\ a > 0,\ \frac{1}{a}b \in (\mathsf{CS}(\boldsymbol{\nu}))^0\}, \tag{1.37}$$

where $(\mathsf{CS}(\boldsymbol{\nu}))^0$ is the interior of $\mathsf{CS}(\boldsymbol{\nu})$.

Analogously to (1.3) we set

$$L(a,b) = \ln(\int \exp\{\langle b,\theta\rangle - aK(\theta)\}\boldsymbol{\tau}(d\theta)), \quad (a,b) \in \Upsilon, \tag{1.38}$$

and introduce the family of $\boldsymbol{\tau}$-densities $\pi_{a,b}(\theta)$ and the family of distributions $\Pi_{a,b}$, $(a,b) \in \Upsilon$, respectively, on $(\Delta, \mathfrak{B}_\Delta)$ by

$$\pi_{a,b}(\theta) = \exp\{\langle b,\theta\rangle - aK(\theta) - L(a,b)\}, \tag{1.39}$$
$$\Pi_{a,b}(B) = \int_B \pi_{a,b}(\theta)\boldsymbol{\tau}(d\theta), \quad B \in \mathfrak{B}_\Delta.$$

Definition 1.34. *Suppose that Υ in (1.37) is nonempty. For an exponential family $(P_\theta)_{\theta\in\Delta}$ in natural form (1.5), and $\boldsymbol{\tau} \in \mathcal{M}^\sigma(\mathfrak{B}_\Delta)$, the family of distributions $(\Pi_{a,b})_{(a,b)\in\Upsilon}$ given by (1.39) is called the family of conjugate priors for the natural parameter $\theta \in \Delta$.*

The notation of conjugate prior is not unique in the literature. Some authors call $(\Pi_{a,b})_{(a,b)\in\Upsilon}$ the family of natural conjugate priors and any family of priors a conjugate family if the posterior distributions belong to the same family. For further details and references we refer to Diaconis and Ylvisaker (1979), Robert (2001), and Gutierréz-Peña and Smith (1997).

With $\Pi_{a,b}$ as the prior, and $\mathsf{P}(\cdot|\theta) := P_\theta(\cdot)$ with $\boldsymbol{\mu}$-density f_θ from (1.6), we get

$$\begin{aligned}(\mathsf{P}\otimes\Pi_{a,b})(C) &= \int [\int I_C(x,\theta)\mathsf{P}(dx|\theta)]\Pi_{a,b}(d\theta)\\ &= \int I_C(x,\theta) f_\theta(x)\pi_{a,b}(\theta)(\boldsymbol{\mu}\otimes\boldsymbol{\tau})(dx,d\theta), \quad C \in \mathfrak{A}\otimes\mathfrak{B}_\Delta.\end{aligned}$$

By (1.34) the density $\mathsf{m}_{a,b}$ of the marginal distribution $\mathsf{M}_{a,b} = \mathsf{P}\Pi_{a,b}$ of X is

$$\begin{aligned}\mathsf{m}_{a,b}(x) &= \int f_\theta(x)\pi_{a,b}(\theta)\boldsymbol{\tau}(d\theta)\\ &= \exp\{L(a+1, b+T(x)) - L(a,b)\}.\end{aligned} \tag{1.40}$$

As

$$\int [\int f_\theta(x)\pi_{a,b}(\theta)\boldsymbol{\tau}(d\theta)]\boldsymbol{\mu}(dx) = \int \mathsf{m}_{a,b}(x)\boldsymbol{\mu}(dx) = 1,$$

we see that $(a+1, b+T(x)) \in \Upsilon$, $\boldsymbol{\mu}$-a.e. The conditional density of Θ, given $X = x$, denoted by $\pi_{a,b}(\theta|x)$, with respect to $\boldsymbol{\tau}$ is

$$\begin{aligned}\pi_{a,b}(\theta|x) &= \frac{f_\theta(x)\pi_{a,b}(\theta)}{\mathsf{m}_{a,b}(x)} \qquad (1.41)\\ &= \exp\{\langle b+T(x), \theta\rangle - (a+1)K(\theta) - L(a+1, b+T(x))\}\\ &= \pi_{a+1,b+T(x)}(\theta), \quad \theta \in \Delta,\ x \in \mathcal{X},\ (a,b) \in \Upsilon.\end{aligned}$$

The next statement is a consequence of (1.41) and shows that the posterior distribution can be obtained in a simple manner. Presumably this was one of the reasons for introducing conjugate priors in the literature.

Lemma 1.35. *If $\Pi_{a,b}$, $(a,b) \in \Upsilon$, is a conjugate prior for θ in the Bayes model with $\mathcal{L}(X,\Theta) = \mathsf{P} \otimes \Pi_{a,b}$, then the stochastic kernel*

$$\boldsymbol{\Pi}_{a,b}(\cdot|x) = \Pi_{a+1,b+T(x)}(\cdot) \qquad (1.42)$$

is a version of the conditional distribution $\mathcal{L}(\Theta|X = x)$, $x \in \mathcal{X}$.

For every $n = 1, 2, ...$, the family of conjugate priors for $(P_\theta^{\otimes n})_{\theta\in\Delta}$ is again $(\Pi_{a,b})_{(a,b)\in\Upsilon}$. If $\Pi_{a,b}$, $(a,b) \in \Upsilon$, is a conjugate prior for Θ in the Bayes model with $\mathcal{L}(X_1, ..., X_n, \Theta) = \mathsf{P}^{\otimes n} \otimes \Pi_{a,b}$, then the stochastic kernel

$$\boldsymbol{\Pi}_{n,a,b}(\cdot|x_1, ..., x_n) := \Pi_{a+n,b+T_{\oplus n}(x_1,...,x_n)} \qquad (1.43)$$

is a version of the conditional distribution $\mathcal{L}(\Theta|X_1 = x_1, ..., X_n = x_n)$, $(x_1, ..., x_n) \in \mathcal{X}^n$.

Proof. The first statement follows from (1.41). To prove the second statement we note that by Proposition 1.4 $(P_\theta^{\otimes n})_{\theta\in\Delta}$ is an exponential family with $\boldsymbol{\mu}^{\otimes n}$-density $f_{n,\theta} = \exp\{\langle\theta, T_{\oplus n}\rangle - K_n(\theta)\}$, where $K_n(\theta) = nK(\theta)$, $\theta \in \Delta$. Set

$$\begin{aligned}\Upsilon_n &= \{(\alpha,\beta) : \alpha \in \mathbb{R},\ \beta \in \mathbb{R}^d, \qquad (1.44)\\ &\qquad L_n(\alpha,\beta) = \ln\left(\int \exp\{\langle\beta,\theta\rangle - \alpha K_n(\theta)\}\boldsymbol{\tau}(d\theta)\right) < \infty\},\\ \pi_{n,\alpha,\beta}(\theta) &= \exp\{\langle\beta,\theta\rangle - \alpha K_n(\theta) - L_n(\alpha,\beta)\},\\ \Pi_{n,\alpha,\beta}(B) &= \int_B \pi_{n,\alpha,\beta}(\theta)\boldsymbol{\tau}(d\theta), \quad B \in \mathfrak{B}_\Delta,\ (\alpha,\beta) \in \Upsilon_n.\end{aligned}$$

Then with $L_n(\alpha,\beta) = L(n\alpha,\beta)$ we get at $x = (x_1, ..., x_n) \in \mathcal{X}^n$,

$$\begin{aligned}\mathsf{m}_{n,\alpha,\beta}(x) &= \int f_{n,\theta}(x)\pi_{n,\alpha,\beta}(\theta)\boldsymbol{\tau}(d\theta)\\ &= \exp\{L_n(\alpha+1, \beta+T_{\oplus n}(x)) - L_n(\alpha,\beta)\}\\ &= \exp\{L(n\alpha+n, \beta+T_{\oplus n}(x)) - L(n\alpha,\beta)\},\\ \frac{f_{n,\theta}(x)\pi_{n,\alpha,\beta}(\theta)}{\mathsf{m}_{n,\alpha,\beta}(x)} &= \exp\{\langle\beta, \theta+T_{\oplus n}(x)\rangle - n\alpha K(\theta) - L(n\alpha+n, \beta+T_{\oplus n}(x))\}.\end{aligned}$$

Apparently, $\pi_{n,\alpha,\beta} = \pi_{a,b}$ for $(a,b) = (n\alpha, \beta)$, and it holds $(\alpha,\beta) \in \Upsilon_n$ if and only if $(a,b) \in \Upsilon$. The proof is thus completed. ■

Suppose now that there is another parameter set Λ that is equipped with a σ-algebra, say $\mathfrak{B}_\Lambda$, and there is a mapping $\kappa : \Lambda \to \Delta$ from Λ onto Δ that is one-to-one. Let the inverse mapping be denoted by $\gamma : \Delta \to_m \Lambda$ and suppose that both mappings, κ and γ, are measurable with respect to the corresponding σ-algebras. This is called bimeasurability of κ and denoted by $\kappa : \Lambda \leftrightarrow_m \Delta$. At any time, depending on the concrete situation, one may choose to work with the exponential model in natural or reparametrized form. Any prior of Θ in the first model determines the prior of $\Xi = \gamma(\Theta)$ in the reparametrized model and vice versa. Sometimes it is easier to work with a prior of Ξ, and sometimes it is easier to work with a prior of Θ. However, in applications the final results are usually reported for Ξ if $\eta \in \Lambda$ is the parameter that admits a better statistical interpretation. We call

$$(\Gamma_{a,b})_{(a,b)\in\Upsilon} \quad \text{with } \Gamma_{a,b} := \Pi_{a,b} \circ \gamma^{-1},\ (a,b) \in \Upsilon,$$

the family of *conjugate priors for* $\eta \in \Lambda$. To evaluate the prior $\Gamma_{a,b}$ we establish a transformation for the prior densities. To this end we introduce the measure $\boldsymbol{\varkappa}$ on $(\Lambda, \mathfrak{B}_\Lambda)$ by $\boldsymbol{\varkappa} = \boldsymbol{\tau} \circ \gamma^{-1}$. The situation simplifies further if $\boldsymbol{\tau} = \boldsymbol{\lambda}_d$ is the Lebesgue measure. Then $\boldsymbol{\lambda}_d(\partial\Delta) = 0$; see Lang (1986). Consequently for all priors that are absolutely continuous with respect to $\boldsymbol{\lambda}_d$ the boundary points of Δ are irrelevant. Therefore we may restrict the family $(P_\theta)_{\theta\in\Delta}$ and consider only parameter values from the interior Δ^{C} which is not empty by assumption (A2). We assume that Λ is open and

$$\kappa : \Lambda \leftrightarrow \Delta^0 \text{ a diffeomorphism with Jacobian } J(\eta) = \left(\frac{\partial \kappa_i}{\partial \eta_j}\right)_{1\le i,j\le d}. \tag{1.45}$$

Proposition 1.36. *If* $\kappa : \Lambda \leftrightarrow_m \Delta$, *then the conjugate prior* $\Gamma_{a,b}$ *for* η *has the* $\boldsymbol{\varkappa}$*-density*

$$\frac{d\Gamma_{a,b}}{d\boldsymbol{\varkappa}}(\eta) = \exp\{\langle\kappa(\eta), b\rangle - aK(\kappa(\eta)) - L(a,b)\} = \pi_{a,b}(\kappa(\eta)), \quad \eta \in \Lambda. \tag{1.46}$$

If $\boldsymbol{\tau} \ll \boldsymbol{\lambda}_d$ *and* (1.45) *is satisfied, then*

$$\frac{d\Gamma_{a,b}}{d\boldsymbol{\lambda}_d}(\eta) = \exp\{\langle\kappa(\eta), b\rangle - aK(\kappa(\eta)) - L(a,b)\}\frac{d\boldsymbol{\tau}}{d\boldsymbol{\lambda}_d}(\kappa(\eta))\,|J(\eta)|, \tag{1.47}$$

where $|J(\eta)|$ *is the absolute value of the determinant of* $J(\eta)$. *If* $\Gamma_{a,b} = \Pi_{a,b} \circ \gamma^{-1}$, $(a,b) \in \Upsilon$, *is a conjugate prior for* η *in the Bayes model with* $\mathcal{L}(X_1,...,X_n,\Xi) = \mathsf{P}^{\otimes n} \otimes \Gamma_{a,b}$, *then the stochastic kernel*

$$\boldsymbol{\Gamma}_{n,a,b}(\cdot|x_1,...,x_n) := \Gamma_{a+n,b+T_{\oplus n}(x_1,...,x_n)} \tag{1.48}$$

is a version of the posterior distribution $\boldsymbol{\Gamma}_{n,a,b}$ *of* Ξ, *given* $(X_1,...,X_n) = (x_1,...,x_n) \in \mathcal{X}^n$.

Proof. As $\kappa : \Lambda \to \Delta$ is one-to-one, and both mappings κ and γ are measurable, we have $\boldsymbol{\tau} = \boldsymbol{\varkappa} \circ \kappa^{-1}$, and for every $h : \Lambda \to_m \mathbb{R}_+$ it holds

$$\int h(\eta)\Gamma_{a,b}(d\eta) = \int h(\gamma(\theta))\pi_{a,b}(\theta)\boldsymbol{\tau}(d\theta) = \int h(\eta)\pi_{a,b}(\kappa(\eta))\boldsymbol{\varkappa}(d\eta).$$

This proves (1.46). If (1.45) is satisfied, then by the transformation theorem for the Lebesgue measure (see Theorem A.23),

$$\frac{d\boldsymbol{\varkappa}}{d\boldsymbol{\lambda}_d}(\eta) = \frac{d\boldsymbol{\tau}}{d\boldsymbol{\lambda}_d}(\kappa(\eta))\,|J(\eta)|\,, \quad \boldsymbol{\lambda}_d\text{-a.e.}$$

Let Q be the kernel $P_{\kappa(\eta)}$ and $h : \mathcal{X}^n \times \Lambda \to_m \mathbb{R}_+$. Then it holds in view of (1.43) and $\mathsf{P}^{\otimes n}\Pi_{a,b} = \mathsf{Q}^{\otimes n}\Gamma_{a,b}$,

$$\begin{aligned}
&\int [\int h(x,\eta)P^{\otimes n}_{\kappa(\eta)}(dx)]\Gamma_{a,b}(d\eta) = \int [\int h(x,\kappa(\theta))P^{\otimes n}_{\theta}(dx)]\Pi_{a,b}(d\theta) \\
&= \int [\int h(x,\kappa(\theta))\Pi_{a+n,b+T_{\oplus n}(x)}(d\theta)](\mathsf{P}^{\otimes n}\Pi_{a,b})(dx) \\
&= \int [\int h(x,\eta)\Gamma_{a+n,b+T_{\oplus n}(x)}(d\eta)](\mathsf{Q}^{\otimes n}\Gamma_{a,b})(dx),
\end{aligned}$$

which completes the proof. ■

Subsequently we find families of conjugate priors for important classes of distributions. We start with normal distributions where one of the two parameters is assumed to be known.

Lemma 1.37. *The family* $(\mathsf{N}(\sigma^2\theta,\sigma^2))_{\theta\in\mathbb{R}}$, *where* $\sigma^2 > 0$ *is known, is a one-parameter exponential family with the density* $f_\theta(x) = \exp\{\theta T(x) - K(\theta)\}$, *where* $T(x) = x$ *and* $K(\theta) = \sigma^2\theta^2/2$, *with respect to the dominating measure*

$$\boldsymbol{\mu}(dx) = (2\pi\sigma^2)^{-1/2}\exp\{-(2\sigma^2)^{-1}x\}\boldsymbol{\lambda}(dx).$$

If $\boldsymbol{\tau} = \boldsymbol{\lambda}$, *then the family of conjugate priors for* θ *in* $(\mathsf{N}^{\otimes n}(\sigma^2\theta,\sigma^2))_{\theta\in\mathbb{R}}$ *is the family of all normal distributions* $\mathsf{N}(\nu,\delta^2)$, $\nu\in\mathbb{R}$, $\delta^2>0$. *If* $\Xi \sim \mathsf{N}(\nu,\delta^2)$ *is a prior for* $\mu = \sigma^2\theta$ *in the Bayes model* $\mathcal{L}(X_1,...,X_n,\Xi) = \mathsf{N}^{\otimes n}(\mu,\sigma^2)\otimes \mathsf{N}(\nu,\delta^2)$, *then*

$$\begin{aligned}
&\mathcal{L}(\Xi|X_1 = x_1,...,X_n = x_n) = \mathsf{N}(\mu(x_1,...,x_n),\tau^2), \quad \textit{where} \\
&\mu(x_1,...,x_n) = \frac{1}{n\delta^2+\sigma^2}(\delta^2\sum\nolimits_{i=1}^n x_i + \nu\sigma^2) \quad \textit{and } \tau^2 = \frac{\sigma^2\delta^2}{n\delta^2+\sigma^2}.
\end{aligned}$$

Proof. The first statement follows from the structure of $\varphi_{\mu,\sigma^2}(x)$. To find the family of conjugate priors we note that by (1.36) and (1.39),

$$\Upsilon = \{(a,b) : a,b\in\mathbb{R}\,,\ \int \exp\{\theta b - a\frac{\sigma^2}{2}\theta^2\}\boldsymbol{\lambda}(d\theta) < \infty\} = (0,\infty)\times\mathbb{R},$$

$$\pi_{a,b}(\theta) \propto \exp\{b\theta - a\frac{\sigma^2}{2}\theta^2)\}, \quad (a,b)\in(0,\infty)\times\mathbb{R}.$$

But this is the family of densities φ_{γ,ρ^2} if we set $\gamma = b/(a\sigma^2)$ and $\rho^2 = 1/(a\sigma^2)$. Now we set $\mu = \kappa(\theta) = \sigma^2\theta$. Then $\Gamma_{a,b} = \mathsf{N}(\gamma, \rho^2)\circ\kappa^{-1} = \mathsf{N}(b/a, \sigma^2/a)$. Hence for $a = \sigma^2/\delta^2$ and $b = \nu\sigma^2/\delta^2$ we obtain $\Gamma_{a,b} = \mathsf{N}(\nu, \delta^2)$. The statement on $\mathcal{L}(\Xi|X_1 = x_1, ..., X_n = x_n)$ follows from (1.48). Indeed,

$$\Gamma_{a+n,b+T_{\oplus n}} = \mathsf{N}((b + T_{\oplus n})/(a+n), \sigma^2/(a+n))$$

is a normal distribution with the parameters $(\nu\sigma^2 + T_{\oplus n}\delta^2)/(\sigma^2 + n\delta^2)$ and $\sigma^2\delta^2/(\sigma^2 + n\delta^2)$. ■

Next we consider conjugate priors in other models. As we have seen in Lemma 1.37 it is sufficient to consider the case $n = 1$. Then the conjugate prior for arbitrary n can be obtained via (1.48).

Example 1.38. Consider the family $(\mathsf{N}(\mu, \sigma^2))_{\sigma^2>0}$, where $\mu \in \mathbb{R}$ is known. Put $T(x) = (x-\mu)^2$, $\theta = -(2\sigma^2)^{-1} \in \Delta = (-\infty, 0)$ and $K(\theta) = (1/2)\ln(-\pi/\theta)$. Then the Lebesgue densities are

$$f_\theta(x) = \frac{d\mathsf{N}(\mu,\sigma^2)}{d\boldsymbol{\mu}}(x) = \exp\{\theta T(x) - K(\theta)\},$$

so that $\mathsf{N}(\mu, -1/(2\theta))$,where $\mu \in \mathbb{R}$ is known, is an exponential family. Hence, by (1.36) and (1.39), with $\boldsymbol{\tau} = \boldsymbol{\lambda}$,

$$\Upsilon = \{(a,b) : a, b \in \mathbb{R}, \int_{-\infty}^{0} (-\theta)^{a/2} \exp\{\theta b\}\boldsymbol{\lambda}(d\theta) < \infty\} = (-2, \infty) \times (0, \infty),$$

$$\pi_{a,b}(\theta) \propto \exp\{b\theta - aK(\theta)\} \propto I_{(-\infty,0)}(\theta)(-\theta)^{a/2} \exp\{\theta b\}.$$

As the distribution of $\Xi = -(2\Theta)^{-1}$ is the conjugate prior for σ^2 we see from the transformation rule (1.47) that the family of inverse gamma distributions $\mathsf{Ig}(\frac{a}{2}+1, \frac{b}{2})$, $(a,b) \in (-2,\infty) \times (0,\infty)$, with Lebesgue densities

$$\mathsf{ig}_{\alpha,\beta}(t) = \frac{\beta^\alpha}{\Gamma(\alpha)} t^{-\alpha-1} \exp\{-\frac{\beta}{t}\} I_{(0,\infty)}(t), \quad (\alpha, \beta) = (\frac{a}{2}+1, \frac{b}{2}) \in (0,\infty)^2,$$

is the family of conjugate priors for $\mathsf{N}(\mu, \sigma^2)$, $\sigma^2 > 0$, when $\mu \in \mathbb{R}$ is known.

Remark 1.39. In Bayes analysis quite often normal distributions are parametrized as $\mathsf{N}(\mu, \rho^{-1})$, $\mu \in \mathbb{R}$, $\rho > 0$, where ρ is called the *precision* of the normal distribution. Besides the intuitive interpretation, this has the advantage of dealing with gamma distributions, rather than with inverse gamma distributions, in the conjugate prior.

Problem 1.40. Revise the results of Example 1.38 to fit the parametrization $\mathsf{N}(\mu, \sigma^2(\rho))$ with $\sigma^2(\rho) = 1/\rho$, where $\rho > 0$ is unknown, but $\mu \in \mathbb{R}$ is known. Determine also the posteriors of $\sigma^2 > 0$ and the posteriors of $\rho > 0$. Extend the results to an i.i.d sample.

Finally we consider normal distributions where both the mean and variance are unknown. In contrast to the previous two examples we have to deal now with bivariate priors.

Example 1.41. We know from Example 1.11 that $\mathsf{N}(\mu, \sigma^2)$, $\mu \in \mathbb{R}$, $\sigma^2 > 0$, is a two parameter exponential family. Using (1.17) we get from the definition of the conjugate prior in (1.39), that $\Pi_{a,b}$ has the density

$$\begin{aligned}\pi_{a,b}(\theta) &\propto \exp\{\langle b, \theta\rangle - aK(\theta)\} \\ &\propto (-\theta_2)^{a/2} \exp\{b_1\theta_1 + b_2\theta_2 + a\theta_1^2/(4\theta_2)\}, \quad \theta \in \Delta,\end{aligned} \tag{1.49}$$

with respect to the Lebesgue measure on $\mathbb{R} \times (-\infty, 0)$. From here we see that

$$\begin{aligned}\Upsilon &= (0, \infty) \times \{(b_1, b_2) : \int_0^\infty s^{(a+1)/2} \exp\{-s(b_2 - b_1^2/a)\}ds < \infty\} \\ &= \left\{(a, b_1, b_2) : ab_2 > b_1^2, a > -3\right\}.\end{aligned}$$

For practical purposes, the conjugate prior density (1.49) offers less useful interpretations than the associated prior of Ξ which is the random version of (μ, σ^2). The absolute value of the determinant of the Jacobian $J(\mu, \sigma^2)$ of the mapping $\kappa(\mu, \sigma^2) = (\mu/\sigma^2, 1/(2\sigma^2))^T$ is given by $J(\eta) = 1/(2\sigma^6)$. If now $\pi_{a,b}$ from (1.49) is the prior density of Θ, then the prior density $\gamma_{a,b}$ of Ξ is determined by

$$\begin{aligned}\gamma_{a,b}(\mu, \sigma^2) &= \pi_{a,b}(\kappa(\mu, \sigma^2)) \left|J(\mu, \sigma^2)\right| \propto \exp\{\langle b, \kappa(\mu, \sigma^2)\rangle - aK(\kappa(\mu, \sigma^2))\}\frac{1}{2\sigma^6} \\ &\propto \exp\{\frac{b_1\mu}{\sigma^2} - \frac{b_2}{2\sigma^2} - \frac{a\mu^2}{2\sigma^2}\}(\sigma^2)^{-(a/2)-3} \\ &\propto \frac{1}{\sigma} \exp\{-\frac{a}{2\sigma^2}(\mu - \frac{b_1}{a})^2\}(\sigma^2)^{-(a+5)/2} \exp\{-\frac{1}{2}(b_2 - \frac{b_1^2}{a})(\sigma^2)^{-1}\} \\ &\propto \varphi_{b_1/a, \sigma^2/a}(\mu)\mathsf{ig}_{(a+3)/2, (ab_2 - b_1^2)/(2a)}(\sigma^2).\end{aligned}$$

This prior of $\Xi = (\Xi_1, \Xi_2)$ can be interpreted as follows. Given $\Xi_2 = \eta_2$, Ξ_1 has a normal distribution $\mathsf{N}\left(b_1/a, \sigma^2/a\right)$, and marginally, Ξ_2 has an inverse gamma distribution $\mathsf{Ig}((a+3)/2, (ab_2 - b_1^2)/(2a))$, $(a, b_1, b_2) \in \Upsilon$.

To prepare for the next result, which deals with multinomial distributions, we determine the Jacobian of a special transformation.

Problem 1.42.* The determinant of the $d \times d$ matrix

$$\begin{pmatrix} a_1 & 1 & \cdots & 1 \\ 1 & a_2 & & \cdot \\ \cdot & & \cdot & \cdot \\ \cdot & & & \cdot \\ 1 & & \cdot & 1 \\ 1 & \cdot & \cdots 1 & a_d \end{pmatrix}$$

with $a_1, ..., a_d$ in the main diagonal and 1 in all other positions is given by

$$D_d(a_1, ..., a_d) = \prod\nolimits_{j=1}^d (a_j - 1) + \sum\nolimits_{i=1}^d \prod\nolimits_{j=1, j\neq i}^d (a_j - 1).$$

Problem 1.43.* Consider the differentiable mapping $\kappa : \mathbb{S}_{d-1} \to \mathbb{R}^{d-1}$ with

$$\kappa_i(p) = \ln{(p_i)} - \ln(1 - \sum\nolimits_{i=1}^{d-1} p_i), \quad i = 1, ..., d-1.$$

The Jacobian, i.e., $J(\eta) = \left(\frac{\partial \kappa_i}{\partial p_j}\right)_{1\le i,j\le d-1}$, has the determinant

$$\det(J(p)) = (1 - \sum_{i=1}^{d-1} p_i)^{-1} \prod_{j=1}^{d-1} \frac{1}{p_j}.$$

The next result is a continuation of Example 1.5. As we have seen there the multinomial distribution

$$\mathsf{M}(n, (p_1, p_2, ..., p_{d-1}, 1 - \sum_{i=1}^{d-1} p_i)), (p_1, ..., p_{d-1}) \in \mathbb{S}_{d-1},$$

can be represented as a $(d-1)$-parameter exponential family. Let $\mathbf{M}_n$ be the stochastic kernel given by $\mathbf{M}_n(B|p_1, p_2, ..., p_{d-1}) = \mathsf{M}(n, (p_1, p_2, ..., p_{d-1}, 1 - \sum_{i=1}^{d-1} p_i))(B)$, $B \in \mathfrak{B}_+^{\otimes d}$, $(p_1, ..., p_{d-1}) \in \mathbb{S}_{d-1}$.

Lemma 1.44. *With $\boldsymbol{\tau} = \boldsymbol{\lambda}_{d-1}$ as the dominating measure on the parameter space $\Delta = \mathbb{R}^{d-1}$ the family of all Dirichlet distributions on the Borel sets of $\mathbb{S}_{d-1}$ with the Lebesgue densities*

$$\mathsf{di}_{(\alpha_1,...,\alpha_d)}(t_1, ..., t_{d-1}) = \frac{\Gamma(\sum_{i=1}^d \alpha_i)}{\prod_{i=1}^d \Gamma(\alpha_i)} \prod_{i=1}^d t_i^{\alpha_i - 1} I_{\mathbb{S}_{d-1}}((t_1, ..., t_{d-1})),$$
$$\alpha_i > 0,\ i = 1, ..., d,\quad t_d = 1 - \sum_{i=1}^{d-1} t_i,$$

is the family of conjugate priors for the family of multinomial distributions $\mathsf{M}(n, (p_1, p_2, ..., p_{d-1}, 1 - \sum_{i=1}^{d-1} p_i))$, $(p_1, ..., p_{d-1}) \in \mathbb{S}_{d-1}$. If $\mathcal{L}(X_1, ..., X_d, \Xi) = \mathbf{M}_n \otimes \mathsf{Di}_{(\alpha_1,...,\alpha_d)}$, then for every $x_1, ..., x_d \in \mathbb{N}$ with $\sum_{i=1}^d x_i = n$,

$$\mathcal{L}(\Xi | X_1 = x_1, ..., X_d = x_d) = \mathsf{Di}_{(\alpha_1 + x_1, ..., \alpha_d + x_d)}. \tag{1.50}$$

Proof. From Proposition 1.36 and Problem 1.43 we see that

$$\frac{d\Gamma_{a,b}}{d\boldsymbol{\lambda}_d}(p) \propto \exp\{\langle \kappa(p), b\rangle - aK(\kappa(p))\} |J(p)|$$
$$\propto \exp\{\langle \kappa(p), b\rangle - an \ln(p_d)\} \prod_{j=1}^d \frac{1}{p_j} \propto \prod_{i=1}^d p_i^{b_i - 1},$$

where $p_d = 1 - \sum_{i=1}^{d-1} p_i$ and $b_d = an - \sum_{i=1}^{d-1} b_i$. As the integral of the right-hand term with respect to the Lebesgue measure over $\mathbb{S}_{d-1}$ is finite if and only if $b_i > 0$, $i = 1, ..., d$, we get

$$\Upsilon = \{(b_1, ..., b_{d-1}, a) : b_i > 0,\ i = 1, ..., d-1,\ an - \sum_{i=1}^{d-1} b_i > 0\},$$
$$\Gamma_{a,b} = \mathsf{Di}(\alpha_1, ..., \alpha_d), \quad \alpha_i > 0,\ i = 1, ..., d.$$

The statement regarding the posterior distribution of Ξ follows from (1.48) and $an - \sum_{i=1}^{d-1}(b_i + x_i) = an - \sum_{i=1}^{d-1} b_i - (n - x_d) = (a+1) - (b_d + x_d)$. ■

An interesting special case is conjugate priors of binomial distributions.

Example 1.45. We have seen in Example 1.7 that the binomial distribution is an exponential family. This family is, of course, a special case of a multinomial distribution. We set $d = 1$, $p_1 = p$, $p_2 = 1-p$, $\alpha_1 = \alpha$, and $\alpha_2 = \beta$. Then by Lemma 1.44 the family of beta distributions $\mathsf{Be}(\alpha, \beta)$, $\alpha, \beta > 0$, is the family of conjugate priors with Lebesgue density

$$\mathsf{be}_{\alpha,\beta}(p) = \frac{\Gamma(\alpha+\beta)}{\Gamma(\alpha)\Gamma(\beta)} p^{\alpha-1}(1-p)^{\beta-1} I_{(0,1)}(p), \quad \alpha, \beta > 0.$$

If Ξ has a beta distributions $\mathsf{Be}(\alpha, \beta)$, then by (1.50)

$$\mathcal{L}(\Xi|X = x) = \mathsf{Be}(\alpha + x, \beta + n - x).$$

The expectation of the distribution $\mathsf{Be}(\alpha, \beta)$ is $\alpha/(\alpha+\beta)$, and thus we get

$$\mathbb{E}(\Xi|X = x) = \frac{\alpha+\beta}{\alpha+\beta+n}\frac{\alpha}{\alpha+\beta} + \frac{n}{\alpha+\beta+n}\frac{x}{n},$$

i.e., $\mathbb{E}(\Xi|X = x)$ is on $\{0, 1, ..., n\}$ a convex linear combination of the prior expectation $\alpha/(\alpha+\beta)$ and x/n.

In the last example it turned out that the conditional expectation $\mathbb{E}(\Xi|X = x)$ is a convex linear combination of the prior expectation and x/n. A similar fact has been established previously for the normal means model in Lemma 1.37. This is no coincidence. The following more general result regarding $\mathsf{E}_\theta T \equiv \nabla K(\theta)$ (see Corollary 1.19) is due to Diaconis and Ylvisaker (1979); see also Brown (1986).

$$\mathbb{E}(\nabla K(\Theta)|X = x) = \frac{b + T_{\oplus n}(x)}{a+n}, \quad \mathsf{M}_{a,b}\text{-a.s.}, \quad x \in \mathcal{X}^n, \tag{1.51}$$

which is a convex linear combination of $\mathbb{E}(\mathsf{E}_\Theta T) = \mathbb{E}(\nabla K(\Theta)) = b/a$ and $n^{-1}\sum_{i=1}^n T(x_i)$. Conversely, every prior with a Lebesgue density that has this property (1.51) must have one of the densities from (1.39) with respect to $\boldsymbol{\tau} = \boldsymbol{\lambda}_d$.

Problem 1.46. Show that the priors for the gamma distributions $\mathsf{Ga}(\lambda, \theta)$, $\theta > 0$, where $\lambda > 0$ is fixed known, that have been derived in Problem 1.32 are the family of conjugate priors.

For many concrete exponential families the concept of conjugate priors may lead to families of priors that are exotic and unheard of. However, for several exponential families that are commonly in use the families of conjugate priors are known and well-established families of distributions. A comprehensive list of commonly used conjugate priors can be found in Bernardo and Smith (1994). We also refer to Diaconis and Ylvisaker (1979) for further details on conjugate priors.

For reference purposes later on some of the commonly used conjugate priors are listed below under (1.52).

$\mathcal{L}(X\|\Xi)$	Prior $\mathcal{L}(\Xi)$	Posterior $\mathcal{L}(\Xi\|X=x)$
$\mathsf{B}(n,p)$	$\mathsf{Be}(\alpha,\beta)$	$\mathsf{Be}(\alpha+x,\beta+n-x)$
$\mathsf{M}(n,p)$	$\mathsf{Di}(\alpha_1,...,\alpha_d)$	$\mathsf{Di}(\alpha_1+x_1,...,\alpha_d+x_d)$
$\mathsf{Po}(\lambda)$	$\mathsf{Ga}(\alpha,\beta)$	$\mathsf{Ga}(\alpha+x,\beta+1)$
$\mathsf{Ex}(\lambda)$	$\mathsf{Ga}(\alpha,\beta)$	$\mathsf{Ga}(\alpha+1,\beta+x)$
$\mathsf{Ga}(\alpha_0,\lambda)$, α_0 known	$\mathsf{Ga}(\alpha,\beta)$	$\mathsf{Ga}(\alpha+\alpha_0,\beta+x)$
$\mathsf{N}(\mu,\sigma_0^2)$, σ_0^2 known	$\mathsf{N}(\nu,\tau^2)$	$\mathsf{N}\left(\frac{\sigma_0^2\nu+\tau^2x}{\tau^2+\sigma_0^2},\frac{\tau^2\sigma_0^2}{\tau^2+\sigma_0^2}\right)$
$\mathsf{N}(\mu_0,\sigma^2)$, μ_0 known	$\mathsf{Ig}(\alpha,\beta)$	$\mathsf{Ig}(\alpha+\frac{1}{2},\beta+\frac{1}{2}(\mu_0-x)^2)$

(1.52)

The problem of choosing the right prior in a model is a field of intensive research and discussion in Bayes statistics. As the model is only complete after the prior has been chosen this problem is in fact a model choice problem. Often it is desired to find a rich and flexible parametric class of priors for which the posterior densities of Θ or Ξ and the marginal densities of X are explicitly available. This goal is in fact met with the conjugate family of priors, as long as the function $L(a,b)$ can be calculated explicitly. This is usually the case when $\boldsymbol{\tau}$ is the Lebesgue measure. However, since the measure $\boldsymbol{\tau}$ is completely arbitrary, we actually may choose $\boldsymbol{\tau}=\Pi$ for any given prior Π, say. In this case the set Υ in (1.36) contains the point $(0,0)$ and thus is nonempty. If $(\Pi_{a,b})_{(a,b)\in\Upsilon}$ is the conjugate family with respect to $\boldsymbol{\tau}=\Pi$, then, of course, $\Pi=\Pi_{0,0}\in(\Pi_{a,b})_{(a,b)\in\Upsilon}$. To summarize, we have found a method to embed any given prior Π in a family of conjugate priors. Two examples for that situation are given below.

Example 1.47. We have seen in Lemma 1.37 that the conjugate priors for the normal distributions $\mathsf{N}(\mu,\sigma^2)$ with a known $\sigma^2>0$ are the normal distributions $\mathsf{N}(\nu,\delta^2)$, $\nu\in\mathbb{R}$, $\delta^2>0$, if $\boldsymbol{\tau}$ is the Lebesgue measure $\boldsymbol{\lambda}$. Suppose now instead that for some $\alpha<\beta$,

$$\boldsymbol{\tau}(dt)=\frac{1}{\beta-\alpha}I_{[\alpha,\beta]}(t)\boldsymbol{\lambda}(dt). \tag{1.53}$$

Although the conjugate prior and posterior densities are still the same from (1.39) and (1.41), respectively, they are now with respect to the measure $\boldsymbol{\tau}$ given by (1.53). We conclude that the family of conjugate priors for $\mu\in\mathbb{R}$ consists of all normal distributions $\mathsf{N}^{(\alpha,\beta)}(\nu,\delta^2)$, $\nu\in\mathbb{R}$, $\delta^2>0$, that are truncated at α and β. Their Lebesgue densities are

$$\varphi_{\nu,\delta^2}^{(\alpha,\beta)}(\mu) = [\Phi((\beta-\nu)/\delta) - \Phi((\alpha-\nu)/\delta)]^{-1}\varphi_{\nu,\delta^2}(\mu)I_{[\alpha,\beta]}(\mu), \quad \mu \in \mathbb{R}.$$

As in Lemma 1.37 we get the posterior distributions

$$\mathsf{N}^{(\alpha,\beta)}\left(\frac{\sigma^2\nu + \delta^2\sum_{i=1}^n x_i}{\sigma^2 + n\delta^2}, \frac{\sigma^2\delta^2}{\sigma^2 + n\delta^2}\right).$$

Problem 1.48. In the setting of Example 1.45, find the conjugate priors of the binomial distributions $\mathsf{B}(n,p)$, $p \in (0,1)$, when $\boldsymbol{\tau}$ is given by (1.53) with $0 < \alpha < \beta < 1$.

Another feature of conjugate priors is that this class is large enough to be used to approximate any prior by a mixture of conjugate priors. General results in this direction are due to LeCam (1986). We refer to Robert (2001) for a detailed discussion of this topic and illustrate the situation only by an example.

Example 1.49. In Example 1.37 the family of conjugate priors of $\mathsf{N}(\mu,\sigma^2)$, $\mu \in \mathbb{R}$, where $\sigma^2 > 0$ is known, has been seen to be the family of all normal distributions on $\mathbb{R}$ for $\boldsymbol{\tau} = \boldsymbol{\lambda}$. Let now Π be any prior. The question is whether Π can be approximated by mixtures of normal distributions. Consider the mixture of normal densities

$$\pi_{\tau^2}(t) = \int \varphi_{s,\tau^2}(t)\Pi(ds) = \int \varphi_{0,\tau^2}(t-s)\Pi(ds),$$

and denote by Π_{τ^2} the associated distributions. Then for any bounded and continuous function ψ,

$$\begin{aligned}\int \psi(t)\Pi_{\tau^2}(dt) &= \int[\int \psi(t)\frac{1}{\sqrt{2\pi}\tau}\exp\{-\frac{1}{2}(\frac{t-s}{\tau})^2\}\Pi(ds)]dt \\ &= \int[\int \psi(s+\tau x)\frac{1}{\sqrt{2\pi}}\exp\{-\frac{x^2}{2}\}dx]\Pi(ds).\end{aligned}$$

By the continuity of ψ and Lebesgue's theorem (see Theorem A.18) we obtain

$$\lim_{\tau^2\to 0}\int \psi(t)\Pi_{\tau^2}(dt) = \int \psi(t)\Pi(dt),$$

which means that Π_{τ^2} converges weakly to Π. Moreover, if Π has a Lebesgue density π, say, then we have even convergence in variational distance. Indeed,

$$\begin{aligned}\|\Pi_{\tau^2} - \Pi\| &= \int |\pi_{\tau^2}(t) - \pi(t)|dt = \int |\int \varphi_{0,\tau^2}(s-t)\pi(s)ds - \pi(t)|dt \\ &\le \int[\int |\pi(t+\tau x) - \pi(t)|dt]\varphi_{0,1}(x)dx \to 0,\end{aligned}$$

which follows from the fact that every function that is integrable with respect to the Lebesgue measure is $\mathbb{L}_1$-continuous, i.e., $\int |\pi(s+\varepsilon) - \pi(s)|ds \to 0$, as $\varepsilon \to 0$ (see, e.g., Dudley (2002)).

1.3 Divergences in Binary Models

In this section we introduce and study classes of distances in the space of probability distributions which originated from different roots. Some of them were introduced in information theory to describe the amount of information, or the amount of uncertainty, delivered by a random sample. When the sample size tends to infinity the investigation of the rate of convergence of error probabilities leads to special information functionals. Other functionals are linked to the Cramér–Rao inequality and its generalizations. Hellinger integrals are Laplace transforms of the log-likelihood and thus describe completely the structure of a binary model. Information functionals are also used later on to characterize the sufficiency and the approximate sufficiency of a statistic. More results on information functionals can be found in Vajda (1989), Liese and Vajda (1987), and in the references that are given there.

First we collect well-known properties of convex functions that are be used in the sequel. A function $\mathsf{v} : (a,b) \to \mathbb{R}$ is called a *convex function* if for every $x, z \in (a,b)$ and $0 \le \alpha \le 1$ it holds

$$\mathsf{v}(\alpha x + (1-\alpha)z) \le \alpha \mathsf{v}(x) + (1-\alpha)\mathsf{v}(z).$$

For $\alpha = (z-y)/(z-x)$ it is easy to see that this condition is equivalent to

$$\frac{\mathsf{v}(y)-\mathsf{v}(x)}{y-x} \le \frac{\mathsf{v}(z)-\mathsf{v}(x)}{z-x} \le \frac{\mathsf{v}(z)-\mathsf{v}(y)}{z-y}, \quad a < x < y < z < b. \tag{1.54}$$

The next problem presents further well-known properties of convex functions.

Problem 1.50.* Every convex function $\mathsf{v} : (a,b) \to \mathbb{R}$ is continuous in (a,b), and it has at each $x \in (a,b)$ a derivative from the left $D^-\mathsf{v}(x)$ which is left continuous and a derivative from the right $D^+\mathsf{v}(x)$ which is right continuous. Both $D^-\mathsf{v}$ and $D^+\mathsf{v}$ are nondecreasing and it holds

$$\mathsf{v}(y) - \mathsf{v}(x) \ge (y-x)D^+\mathsf{v}(x), \quad a < x < y < b, \tag{1.55}$$

$$D^-\mathsf{v}(x) \le D^+\mathsf{v}(x) \le D^-\mathsf{v}(y) \le D^+\mathsf{v}(y) \quad a < x < y < b, \tag{1.56}$$

$$D^-\mathsf{v}(z) = D^+\mathsf{v}(z-0), \quad \text{and} \quad D^-\mathsf{v}(z+0) = D^+\mathsf{v}(z), \quad z \in (a,b). \tag{1.57}$$

The inequalities in (1.54) show that $(\mathsf{v}(y)-\mathsf{v}(x))/(y-x)$ is nondecreasing in both x and y. Hence

$$\frac{\mathsf{v}(c)-\mathsf{v}(c-\varepsilon)}{\varepsilon} \le \frac{\mathsf{v}(t)-\mathsf{v}(s)}{t-s} \le \frac{\mathsf{v}(d+\varepsilon)-\mathsf{v}(d)}{\varepsilon},$$

for $a < c-\varepsilon < s < t < d+\varepsilon < b$. Putting $L = \varepsilon^{-1}\max(|\mathsf{v}(d+\varepsilon)-\mathsf{v}(d)|, |\mathsf{v}(c)-\mathsf{v}(c-\varepsilon)|)$ it follows that

$$|\mathsf{v}(t)-\mathsf{v}(s)| \le L|t-s|, \quad c \le s \le t \le d.$$

This implies especially that v is absolutely continuous. It is well-known (see Theorem A.24) that the derivative of every absolutely continuous function exists $\boldsymbol{\lambda}$-almost everywhere. But it follows from Problem 1.50 that for a convex function the derivative exists up to an at most countable set which is just the set of points of discontinuities of $D^+\mathsf{v}$ or, equivalently, of $D^-\mathsf{v}$. Theorem A.24 implies

$$\mathsf{v}(y) - \mathsf{v}(x) = \int_x^y D^+\mathsf{v}(s)ds = \int_x^y D^-\mathsf{v}(s)ds, \quad a < x < y < b. \tag{1.58}$$

The next problem gives a direct proof of (1.58).

Problem 1.51.* If $\mathsf{v} : (a,b) \to \mathbb{R}$ is convex, then (1.58) holds. Conversely, if $\mathsf{v}(y) - \mathsf{v}(x) = \int_x^y g(s)ds$, $a < x < y < b$, for some nondecreasing function g, then v is convex.

The second statement in Problem 1.51 implies in particular that every twice continuously differentiable function v with $\mathsf{v}''(x) \geq 0$ is convex. Furthermore, for a convex v the limits $\lim_{x\downarrow a} \mathsf{v}(x)$ and $\lim_{x\uparrow b} \mathsf{v}(x)$ exist and have values from $[-\infty, \infty]$. We thus can extend v by setting $\mathsf{v}(a) = \lim_{x\downarrow a} \mathsf{v}(x)$ and $\mathsf{v}(b) = \lim_{x\uparrow b} \mathsf{v}(x)$.

As $D^+\mathsf{v}$ is continuous from the right there is a uniquely determined σ-finite measure γ_v on the Borel sets of (a,b) that satisfies

$$\gamma_\mathsf{v}((x,y]) = D^+\mathsf{v}(y) - D^+\mathsf{v}(x), \quad a < x < y < b. \tag{1.59}$$

For a twice continuously differentiable v the function $D^+\mathsf{v}$ is continuously differentiable, so that for $0 < x < y$,

$$D^+\mathsf{v}(y) - D^+\mathsf{v}(x) = \int_x^y \mathsf{v}''(t)dt \quad \text{and} \quad \gamma_\mathsf{v}(B) = \int_B \mathsf{v}''(t)dt. \tag{1.60}$$

Therefore the measure γ_v can be viewed as a measure for the curvature of v. We use this measure for the curvature to establish a generalized second-order Taylor expansion.

Lemma 1.52. *If* $\mathsf{v} : (a,b) \to \mathbb{R}$ *is convex, then for* $a < x, y < b$

$$\mathsf{v}(y) - \mathsf{v}(x) - D^+\mathsf{v}(x)(y-x) = \begin{cases} \int (y-t)I_{(x,y]}(t)\gamma_\mathsf{v}(dt) & \text{if } x < y, \\ \int (t-y)I_{(y,x]}(t)\gamma_\mathsf{v}(dt) & \text{if } y < x. \end{cases} \tag{1.61}$$

If $\mathsf{v} : (0,\infty) \to \mathbb{R}$, *then the function*

$$\mathsf{v}_0(x) = \mathsf{v}(x) - \mathsf{v}(1) - (x-1)D^+\mathsf{v}(1) \tag{1.62}$$

has the representation

$$\mathsf{v}_0(x) = \begin{cases} \int (x - t\wedge x)I_{(1,\infty)}(t)\gamma_\mathsf{v}(dt) & \text{if } x > 1, \\ \int (t - t\wedge x)I_{(0,1]}(t)\gamma_\mathsf{v}(dt) & \text{if } 0 < x \leq 1. \end{cases} \tag{1.63}$$

Proof. We have for $x < y$ from (1.58) and Fubini's theorem (see Theorem A.26)

$$\begin{aligned}\mathsf{v}(y) - \mathsf{v}(x) - D^+\mathsf{v}(x)(y-x) &= \int_x^y (D^+\mathsf{v}(s) - D^+\mathsf{v}(x))ds \\ = \int [\int I_{(x,y]}(s) I_{(x,s]}(t)\gamma_{\mathsf{v}}(dt)]ds &= \int (y-t) I_{(x,y]}(t)\gamma_{\mathsf{v}}(dt).\end{aligned}$$

By interchanging the roles of x and y we get for $x > y$,

$$\begin{aligned}&\mathsf{v}(y) - \mathsf{v}(x) - D^+\mathsf{v}(x)(y-x) \\ &= -(\mathsf{v}(x) - \mathsf{v}(y) - D^+\mathsf{v}(y)(x-y)) + (D^+\mathsf{v}(x) - D^+\mathsf{v}(y))(x-y) \\ &= -\int (x-t) I_{(y,x]}(t)\gamma_{\mathsf{v}}(dt) + \int (x-y) I_{(y,x]}(t)\gamma_{\mathsf{v}}(dt) \\ &= \int (t-y) I_{(y,x]}(t)\gamma_{\mathsf{v}}(dt).\end{aligned}$$

The statement (1.63) follows from (1.61) as $\mathsf{v}_0(1) = D^+\mathsf{v}_0(1) = 0$. ■

A convex function v is called *strictly convex at* x_0 if the function v is not linear in $(x_0 - \varepsilon, x_0 + \varepsilon)$ for any $\varepsilon > 0$. v is called *strictly convex in* (a, b) if it is strictly convex at every $x_0 \in (a, b)$.

Problem 1.53.* A convex function $\mathsf{v} : (a, b) \to \mathbb{R}$ is strictly convex at $x_0 \in (a, b)$ if and only if $\gamma_{\mathsf{v}}((x_0 - \varepsilon, x_0 + \varepsilon)) > 0$ for every $\varepsilon > 0$. If v is twice continuously differentiable and $\mathsf{v}''(x) > 0$, $0 < x < \infty$, then the function v is strictly convex in (a, b). A convex function v is strictly convex in (a, b) if and only if

$$\mathsf{v}(\alpha x + (1-\alpha)y) < \alpha\mathsf{v}(x) + (1-\alpha)\mathsf{v}(y), \quad a < x, y < b, \quad x \neq y, \quad 0 < \alpha < 1.$$

The next problem collects properties of the function v_0 in (1.62).

Problem 1.54.* Let $\mathsf{v} : (0, \infty) \to \mathbb{R}$ be convex. Then v_0 from (1.62) is nonnegative and $\mathsf{v}_0(1) = 0$. It holds $D^+\mathsf{v}_0(x) = D^+\mathsf{v}(x) - D^+\mathsf{v}(1)$. The function v_0 is nonincreasing in $(0, 1]$ and nondecreasing in $(1, \infty)$. The function v is strictly convex at 1 if and only if at least one of the following two cases holds: $\mathsf{v}_0(x) > 0$, $x \in (0, 1)$, or $\mathsf{v}_0(x) > 0$, $x \in (1, \infty)$.

For later purposes we define the *-conjugate function by

$$\mathsf{v}^*(x) = x\mathsf{v}(\frac{1}{x}), \quad x > 0. \tag{1.64}$$

Problem 1.55.* If $\mathsf{v} : (0, \infty) \to \mathbb{R}$ is convex, then $\mathsf{v}^* : (0, \infty) \to \mathbb{R}$ is convex, too, and

$$(\mathsf{v}^*)^* = \mathsf{v}, \quad \mathsf{v}^*(0) = \lim_{x\to\infty} \frac{1}{x}\mathsf{v}(x),$$

where $\mathsf{v}^*(0) := \lim_{x\downarrow 0} \mathsf{v}^*(x) \in (-\infty, \infty]$.

The concept of the likelihood ratio, introduced below, plays a major role in many areas of statistics. Let $(\mathcal{X},\mathfrak{A})$ be a given measurable space and $P_0, P_1 \in \mathcal{P}(\mathfrak{A})$. Suppose that P_0 and P_1 are dominated by $\mu \in \mathcal{M}^\sigma(\mathfrak{A})$ with f_0 and f_1 as the respective densities. We set

$$P_{1,a}(A) = P_1(A \cap \{f_0 > 0\}) \quad \text{and} \quad P_{1,s}(A) = P_1(A \cap \{f_0 = 0\}).$$

Then we get the *Lebesgue decomposition* of P_1 with respect to P_0,

$$\begin{array}{ll} P_1 = P_{1,a} + P_{1,s}, & P_{1,a} \perp P_{1,s} \\ P_{1,a} \ll P_0, & P_{1,s} \perp P_0 \end{array}, \tag{1.65}$$

where $P_{1,a} \perp P_{1,s}$ means that $P_{1,a}(A) = P_{1,s}(\mathcal{X}\backslash A) = 0$ for some $A \in \mathfrak{A}$, and $P_{1,s} \perp P_0$ is meant analogously. $P_{1,a}$ is called the absolute continuous part and $P_{1,s}$ the singular part of P_1 with respect to P_0.

Problem 1.56.* The measures $P_{1,a}$ and $P_{1,s}$ are uniquely determined by the decomposition (1.65).

Now we are ready to introduce the concept of the likelihood ratio of P_1 with respect to P_0, regardless of whether P_1 is absolutely continuous with respect to P_0.

Definition 1.57. *Every function $L_{0,1} : \mathcal{X} \to_m [0,\infty]$ that satisfies*

$$P_1(B) = \int_B L_{0,1} dP_0 + P_1(B \cap \{L_{0,1} = \infty\}), \quad B \in \mathfrak{A}, \tag{1.66}$$

is called the likelihood ratio of P_1 with respect to P_0

For a proof of the next lemma we refer, for example, to Strasser (1985).

Lemma 1.58. *$L_{0,1}$ is by the condition* (1.66) *$\{P_0, P_1\}$-a.s. uniquely determined, and it holds*

$$L_{0,1} = \frac{dP_{1,a}}{dP_0}, \quad P_0\text{-}a.s., \quad P_{1,s}(B) = P_1(B \cap \{L_{0,1} = \infty\}), \quad B \in \mathfrak{A}. \tag{1.67}$$

Using the densities f_0 and f_1 it is easy to see that

$$L_{0,1}(x) := \frac{f_1(x)}{f_0(x)} I_{\{f_0>0\}}(x) + \infty I_{\{f_0=0, f_1>0\}}(x), \quad x \in \mathcal{X}, \tag{1.68}$$

is a likelihood ratio.

Problem 1.59. It holds $P_0(L_{0,1} < \infty) = 1$.

Now we introduce a general class of information functionals. Let $P_0, P_1 \in \mathcal{P}(\mathfrak{A})$ be dominated by $\mu \in \mathcal{M}^\sigma(\mathfrak{A})$ and f_0, f_1 be their respective μ-densities.

Definition 1.60. *For every convex function* $\mathsf{v} : (0, \infty) \to \mathbb{R}$ *the functional*

$$\mathsf{I}_\mathsf{v}(P_0, P_1) := \int \mathsf{v}(f_0/f_1) f_1 I_{\{f_0>0, f_1>0\}} d\mu \qquad (1.69)$$
$$+ \mathsf{v}(0) P_1(f_0 = 0) + \mathsf{v}^*(0) P_0(f_1 = 0)$$

is called the v*-divergence of* P_0 *with respect to* P_1.

To see that the right-hand term is well defined we remark that by $\mathsf{v}_0 \geq 0$ it holds $\mathsf{v}(0), \mathsf{v}^*(0) > -\infty$ and refer to inequality (1.55) that gives

$$\mathsf{v}(f_0/f_1) I_{\{f_0>0\}} f_1 \geq \mathsf{v}(1) f_1 + (D^+\mathsf{v}(1))(f_0 - f_1).$$

As the right-hand function is integrable we see that the integral in (1.69) is well defined but may take on the value $+\infty$. Note that $P_1(f_0 = 0)$ and $P_0(f_1 = 0)$ are the weights of the singular parts of P_1 and P_0 with respect to P_0 and P_1, respectively. They are independent of the special choice of the dominating measure μ. This follows also for the integral in (1.69) by the chain rule; see Proposition A.28. Therefore the definition of $\mathsf{I}_\mathsf{v}(P_0, P_1)$ is independent of the special choice of μ.

Problem 1.61.* The values $\mathsf{v}(0)$ and $\mathsf{v}^*(0)$ appearing in (1.69) can be expressed in terms of γ_v. It holds

$$\lim_{x \downarrow 0} \mathsf{v}^*(x) = \gamma_\mathsf{v}((1, \infty)) + D^+\mathsf{v}(1), \qquad (1.70)$$

$$\lim_{x \downarrow 0} \mathsf{v}(x) = \int t I_{(0,1]}(t) \gamma_\mathsf{v}(dt) + \mathsf{v}(1) - D^+\mathsf{v}(1). \qquad (1.71)$$

The concept of v-divergence was independently introduced by Csiszár (1963) and Ali and Silvey (1966). This general class of functionals includes special cases which appeared in Bhattacharyya (1946), Kakutani (1948), Kullback and Leibler (1951), Chernoff (1952), Matusita (1955), Rényi (1960), and others.

The functional $\mathsf{I}_\mathsf{v}(P_0, P_1) - \mathsf{v}(1)$ depends only on the nonlinear part of v.

Problem 1.62.* If $\mathsf{w}(x) = \mathsf{v}(x) + ax + b$, then

$$\mathsf{I}_\mathsf{v}(P_0, P_1) - \mathsf{v}(1) = \mathsf{I}_\mathsf{w}(P_0, P_1) - \mathsf{w}(1), \qquad (1.72)$$

and especially v_0 in (1.62) satisfies $\mathsf{I}_\mathsf{v}(P_0, P_1) - \mathsf{v}(1) = \mathsf{I}_{\mathsf{v}_0}(P_0, P_1)$.

Although the functional $\mathsf{I}_\mathsf{v}(P_0, P_1)$ does not satisfy the axioms of a metric in general, it has several properties that allow this functional to be interpreted as a measure of distance.

Proposition 1.63. *If* $\mathsf{v} : (0, \infty) \to \mathbb{R}$ *is convex, then* $\mathsf{I}_\mathsf{v}(P_0, P_1) - \mathsf{v}(1) \geq 0$, *with equality holding for* $P_0 = P_1$. *If* v *is strictly convex at* $x_0 = 1$, *then* $\mathsf{I}_\mathsf{v}(P_0, P_1) - \mathsf{v}(1) = 0$ *implies* $P_0 = P_1$. *The functional* $\mathsf{I}_{\mathsf{v}^*}$ *is dual to* I_v *in the sense that*

$$\mathsf{I}_\mathsf{v}(P_0, P_1) = \mathsf{I}_{\mathsf{v}^*}(P_1, P_0). \qquad (1.73)$$

Proof. The function v_0 is nonnegative so that $\mathsf{I}_{\mathsf{v}_0}(P_0, P_1) = \mathsf{I}_{\mathsf{v}}(P_0, P_1) - \mathsf{v}(1)$ is nonnegative as well. If $P_0 = P_1$, then $f_0 = f_1$, μ-a.e., and $P_1(f_0 = 0) = P_0(f_1 = 0) = 0$ so that the integral on the right-hand side of (1.69) has the value $\mathsf{v}(1)$.

Assume now that v is strictly convex at $x_0 = 1$. In view of Problem 1.62 it is sufficient to consider v_0. Then by Problem 1.54 either $\mathsf{v}_0(x) > 0$ for every $x > 1$ or $\mathsf{v}_0(x) > 0$ for every $0 < x < 1$. Suppose that the first condition holds. Then $\mathsf{v}_0(x) \geq 0$ and $\mathsf{I}_{\mathsf{v}}(P_0, P_1) - \mathsf{v}(1) = 0$, together with (1.69) for $\mathsf{v} = \mathsf{v}_0$, show that $\mu(f_0 > f_1) = 0$. This implies

$$0 = \int (f_1 - f_0)d\mu = \int_{\{f_1 > f_0\}} (f_1 - f_0)d\mu,$$

and therefore $\mu(f_1 > f_0) = 0$. Hence $\mu(f_1 \neq f_0) = 0$ and $P_1 = P_0$. The case where $\mathsf{v}_0(x) > 0$ holds for every $0 < x < 1$ can be treated similarly. The statement (1.73) is an immediate consequence of (1.69) and (1.64). ∎

$\mathsf{I}_{\mathsf{v}}(P_0, P_1) - \mathsf{v}(1)$ does not satisfy the triangular inequality and in general it is not symmetric in (P_0, P_1). From Definition 1.60 it follows that symmetry in (P_0, P_1) holds if $\mathsf{v}(x) = \mathsf{v}^*(x) := x\mathsf{v}(1/x)$. To keep the notation simple we use the symbol $\mathsf{I}_{\mathsf{v}}(P_0, P_1)$ also if v is concave.

Now we present some special parametrized classes of functions, which are either convex or concave, and provide well-known information functionals.

v	$\mathsf{I}_{\mathsf{v}}(P_0, P_1)$
$\mathsf{m}_s = \begin{cases} \lvert x-1\rvert^s \text{ if } 1 \leq s < \infty \\ \lvert x^s - 1\rvert^{\frac{1}{s}} \text{ if } 0 < s < 1 \end{cases}$	$\chi^s(P_0, P_1)$
$\mathsf{v}_s = x^s$, if $s > 0$, $s \neq 1$	$\mathsf{H}_s(P_0, P_1)$
$\mathsf{w}_s = \begin{cases} \frac{x^s - sx - (1-s)}{s(s-1)}, \text{ if } s > 0, \neq 1 \\ x \ln x - x + 1, \text{ if } s = 1 \end{cases}$	$\begin{cases} \mathsf{K}_s(P_0, P_1) \\ \mathsf{K}(P_0, P_1) \end{cases}$
$\mathsf{k}_\pi = \pi \wedge (1-\pi) - (\pi x) \wedge (1-\pi)$, $0 < \pi < 1$	$\mathsf{B}_\pi(P_0, P_1)$

(1.74)

The functionals $\chi^s(P_0, P_1)$ are called χ^s-*divergences*. Especially,

$$\chi^2(P_0, P_1) = \int \frac{(f_1 - f_0)^2}{f_1} I_{\{f_1 > 0\}} d\mu + \infty P_0(f_1 = 0)$$

is the well-known χ^2-*distance*. Furthermore, $\chi^{\frac{1}{2}}(P_0, P_1) = \int (\sqrt{f_0} - \sqrt{f_1})^2 d\mu$ is the square of the *Hellinger distance* of P_0 and P_1 which is defined by

$$\mathsf{D}(P_0, P_1) = [\int (\sqrt{f_0} - \sqrt{f_1})^2 d\mu]^{1/2}. \tag{1.75}$$

It is clear that $\mathsf{D}(P_0, P_1)$ is a metric on $\mathcal{P}(\mathfrak{A})$ because $\mathsf{D}(P_0, P_1)$ is the $\mathbb{L}_2(\mu)$-distance of the square roots of the densities. For $s = 1$ we get

$$\chi^1(P_0, P_1) = \int |f_1 - f_0| d\mu =: \|P_0 - P_1\|, \tag{1.76}$$

the *variational distance* of P_0 and P_1. Clearly, $\|P_0 - P_1\|$ is also a metric on $\mathcal{P}(\mathfrak{A})$. We note that the functionals $\chi^s(P_0, P_1)$, $\mathsf{H}_s(P_0, P_1)$, and $\mathsf{K}_s(P_0, P_1)$ are symmetric in (P_0, P_1) for $s = 1/2$ in each family. The functionals

$$\mathsf{H}_s(P_0, P_1) = \begin{cases} \int f_0^s f_1^{1-s} d\mu & \text{if } \ 0 < s < 1, \\ \int f_0^s f_1^{1-s} I_{\{f_1 > 0\}} d\mu + \infty P_0(f_1 = 0) & \text{if } \ s > 1, \end{cases} \tag{1.77}$$

from (1.74) are called *Hellinger integrals.* In the literature they are mainly used for $0 < s < 1$, but for some purposes an extension to $s > 1$ proves useful. The Hellinger integral of order 1/2 and the Hellinger distance are related by

$$\mathsf{D}^2(P_0, P_1) = 2[1 - \mathsf{H}_{1/2}(P_0, P_1)]. \tag{1.78}$$

Some reasons for the wide applicability of Hellinger integrals are that they appear in many problems in statistics as lower bounds for the risks and that they can be evaluated explicitly for important classes of distributions.

Problem 1.64. For normal distributions it holds for $s \neq 1$,

$$\begin{aligned} &\mathsf{H}_s(\mathsf{N}(\mu_1, \sigma_1^2), \mathsf{N}(\mu_2, \sigma_2^2)) \\ &= [\frac{\sigma_1^{2(1-s)} \sigma_2^{2s}}{s\sigma_2^2 + (1-s)\sigma_1^2}]^{1/2} \exp\{-\frac{1}{2} s(1-s) \frac{(\mu_1 - \mu_2)^2}{s\sigma_2^2 + (1-s)\sigma_1^2}\}. \end{aligned} \tag{1.79}$$

Likewise, for Poisson distributions it holds for $s \neq 1$,

$$\mathsf{H}_s(\mathsf{Po}(\lambda_1), \mathsf{Po}(\lambda_2)) = \exp\{\lambda_1^s \lambda_2^{1-s} - s\lambda_1 - (1-s)\lambda_2\}.$$

The functionals $\mathsf{K}_s(P_0, P_1)$ from (1.74) are simple transformations of the Hellinger integrals which are studied in more detail in the next section. It holds

$$\mathsf{K}_s(P_0, P_1) = \frac{1}{s(1-s)} [1 - \mathsf{H}_s(P_0, P_1)], \quad s \neq 1, \ s > 0. \tag{1.80}$$

The functional $\mathsf{K}(P_0, P_1)$ from (1.74) is called the *Kullback–Leibler distance* of P_0 and P_1.

$$\begin{aligned} \mathsf{K}(P_0, P_1) &= \int [(f_0/f_1) \ln(f_0/f_1) - (f_0/f_1) + 1] f_1 I_{\{f_1 > 0\}} d\mu + \infty P_0(f_1 = 0). \\ &= \begin{cases} \int (\ln(f_0/f_1)) dP_0 & \text{if } \ P_0 \ll P_1 \\ \infty & \text{otherwise.} \end{cases} \end{aligned} \tag{1.81}$$

Problem 1.65. It holds

$$\mathsf{K}(\mathsf{N}(\mu_1,\sigma_1^2),\mathsf{N}(\mu_2,\sigma_2^2)) = \frac{1}{2}[\ln(\sigma_2^2/\sigma_1^2) + (\sigma_1^2/\sigma_2^2) - 1 + (\mu_1-\mu_2)^2/\sigma_2^2].$$

The Kullback–Leibler distance appears in many problems in probability and statistics theory. One reason for this is that it is, up to the sign, the expectation of the log-likelihood. This expectation comes into consideration when we study in Chapter 8 the exponential rate at which the second-kind error probabilities of level α tests tend to zero. By construction the functions w_s form a continuous family of convex functions, and especially $\mathsf{w}_s \to \mathsf{w}_1$ as $s \to 1$. The Kullback–Leibler distance is closely related to the Shannon entropy that was introduced in (1.25). Indeed we see from (1.81), by interchanging the roles of P_0 and P_1, that $\mathsf{K}(P_1,P_0) = -\mathsf{S}_{P_0}(P_1)$. This means that for fixed P_0 the search for distributions P_1 that maximize the Shannon entropy can be done by minimizing $\mathsf{K}(P_1,P_0)$ with respect to P_1. The question of whether such a distribution P_1 exists for a given convex set $\mathcal{P}_0$ is closely related to the lower semicontinuity of $\mathsf{K}(P_1,P_0)$ with respect to the convergence in variational distance. Distributions that minimize $\mathsf{K}(\cdot,P_0)$ are called projections and were first studied by Csiszár (1963).

To give a statistical interpretation of $\mathsf{B}_\pi(P_0,P_1)$ in (1.74) we consider the problem of testing the simple null hypothesis $\mathsf{H}_0 : P_0$ versus the alternative $\mathsf{H}_A : P_1$. A statistical test φ is a measurable mapping $\varphi : \mathcal{X} \to_m [0,1]$ and the value $\varphi(x)$ is the probability of rejecting H_0. If H_0 is true, then $\int \varphi dP_0$ is the probability of falsely rejecting H_0, which is called the *error probability of the first kind*. Similarly, if H_A is true, then $\int(1-\varphi)dP_1$ is the probability of falsely rejecting H_A, which is called the *error probability of the second kind.* For any $0 \le \pi \le 1$ the average

$$\pi \int \varphi dP_0 + (1-\pi)\int (1-\varphi)dP_1$$

is called the *Bayes risk.* In the next chapter we systematically study tests for the binary model and especially tests that minimize the Bayes risk. Here we calculate only the minimal Bayes risk.

Lemma 1.66. *In the binary model $(\mathcal{X},\mathfrak{A},\{P_0,P_1\})$ the minimal Bayes risk*

$$\mathsf{b}_\pi(P_0,P_1) = \inf_\varphi(\pi \int \varphi dP_0 + (1-\pi)\int (1-\varphi)dP_1)$$

is given by

$$\mathsf{b}_\pi(P_0,P_1) = \int (\pi f_0) \wedge ((1-\pi)f_1)d\mu, \tag{1.82}$$

where μ is a σ-finite dominating measure and $f_i = dP_i/d\mu$.

Proof. It holds

$$\pi \int \varphi dP_0 + (1-\pi) \int (1-\varphi) dP_1 = \int [\pi \varphi f_0 + (1-\pi)(1-\varphi) f_1] d\mu$$
$$= (1-\pi) + \int \varphi [\pi f_0 - (1-\pi) f_1] d\mu.$$

The right-hand side becomes minimal if we set $\varphi_0 = 1$ if $\pi f_0 < (1-\pi) f_1$ and 0 else. The Bayes risk of this test is given by

$$\int [\pi \varphi_0 f_0 + (1-\pi)(1-\varphi_0) f_1] d\mu = \int (\pi f_0) \wedge ((1-\pi) f_1) d\mu.$$

■

We see from (1.82) that the minimal Bayes risk is related to the divergence $\mathsf{I}_{\mathsf{k}_\pi}(P_0, P_1)$ defined by k_π in (1.74) by

$$\mathsf{B}_\pi(P_0, P_1) = \pi \wedge (1-\pi) - \mathsf{b}_\pi(P_0, P_1). \tag{1.83}$$

The functional $\mathsf{B}_\pi(P_0, P_1)$ admits the following interpretation. $\pi \wedge (1-\pi)$ can be considered as the minimal Bayes error probability that is made if no data are available, and $\mathsf{b}_\pi(P_0, P_1)$ as the minimal Bayes error probability after an observation is made. Therefore $\mathsf{B}_\pi(P_0, P_1)$ may be considered as an information gain.

Problem 1.67.* The function $\pi \mapsto \mathsf{b}_\pi(P_0, P_1)$ is continuous in $(0, 1)$.

Using (1.63) we show in the next theorem that $\mathsf{I}_\mathsf{v}(P_0, P_1) - \mathsf{v}(1)$ is a superposition of the functionals of the information gains $\mathsf{B}_\pi(P_0, P_1)$ with respect to a *curvature measure* ρ_v on $(0, 1)$, defined by

$$\rho_\mathsf{v}(B) = \int (1+t) I_B(\frac{1}{1+t}) \gamma_\mathsf{v}(dt), \tag{1.84}$$

which is equivalent to

$$\int h(\pi) \rho_\mathsf{v}(d\pi) = \int (1+t) h(\frac{1}{1+t}) \gamma_\mathsf{v}(dt), \quad h : (0,1) \to_m \mathbb{R}_+. \tag{1.85}$$

The representations of v-divergences in the next theorem have been established by Österreicher and Feldman (1981) for twice-differentiable functions v, and by Torgersen (1991) for the special case of Hellinger integrals. The general case was treated in Liese and Vajda (2006). Such representations connect the concept of the distance of distributions measured by the v-divergence with decision-theoretic concepts based on the minimal Bayes risk.

Theorem 1.68. *For every convex function* $\mathsf{v} : (0, \infty) \to \mathbb{R}$ *and every distribution* P_0, P_1 *it holds*

$$\mathsf{I}_\mathsf{v}(P_0, P_1) - \mathsf{v}(1) = \int_{(0,1)} \mathsf{B}_\pi(P_0, P_1) \rho_\mathsf{v}(d\pi). \tag{1.86}$$

Corollary 1.69. *It holds*

$$\mathsf{K}_s(P_0,P_1)=\int_{(0,1)}\frac{\mathsf{B}_\pi(P_0,P_1)}{(1-\pi)^{1+s}\pi^{2-s}}d\pi,\qquad -\infty<s<\infty,$$
$$\mathsf{H}_s(P_0,P_1)=s(1-s)\int_{(0,1)}\frac{\mathsf{b}_\pi(P_0,P_1)}{(1-\pi)^{1+s}\pi^{2-s}}d\pi,\quad 0<s<1.$$

Proof. Due to the invariance property (1.72) the left-hand term in (1.86) remains unchanged if we turn from v to v_0 in (1.62). As $\gamma_{\mathsf{v}}=\gamma_{\mathsf{v}_0}$ the right-hand term also remains unchanged. Hence we may assume $\mathsf{v}(1)=D^+\mathsf{v}(1)=0$ without loss of generality. We see from (1.63) that

$$\begin{aligned}
&\int I_{(0,\infty)}(f_0)\mathsf{v}(\frac{f_0}{f_1})dP_1\\
&=\int(\int\{I_{(1,\infty)}(t)[f_0-(tf_1)\wedge f_0)]+I_{(0,1]}(t)[tf_1-(tf_1)\wedge f_0)]\}\gamma_{\mathsf{v}}(dt))\\
&\qquad\qquad\times I_{(0,\infty)}(f_0\wedge f_1)d\mu\\
&=\int[P_0(f_1>0)-(1+t)\mathsf{b}_{1/(1+t)}(P_0,P_1)]I_{(1,\infty)}(t)\gamma_{\mathsf{v}}(dt)\\
&\quad+\int[tP_1(f_0>0)-(1+t)\mathsf{b}_{1/(1+t)}(P_0,P_1)]I_{(0,1]}(t)\gamma_{\mathsf{v}}(dt).
\end{aligned}$$

Now we use (1.70) and (1.71) and obtain

$$\begin{aligned}
\mathsf{I}_{\mathsf{v}}(P_0,P_1)&=\int I_{(1,\infty)}(t)[1-(1+t)\mathsf{b}_{1/(1+t)}(P_0,P_1)]\gamma_{\mathsf{v}}(dt)\\
&\quad+\int I_{(0,1]}(t)[t-(1+t)\mathsf{b}_{1/(1+t)}(P_0,P_1)]\gamma_{\mathsf{v}}(dt)\\
&=\int I_{(0,\infty)}(t)(1+t)[\frac{1}{1+t}\wedge\frac{t}{1+t}-\mathsf{b}_{1/(1+t)}(P_0,P_1)]\gamma_{\mathsf{v}}(dt).
\end{aligned}$$

To complete the proof of the theorem we only have to utilize (1.85).

In order to prove the corollary we use $\mathsf{w}_s(x)$ from (1.74). Then $\gamma_{\mathsf{w}_s}(dx)=x^{s-2}dx$. Hence for every Borel set $B\subseteq(0,1)$,

$$\rho_{\mathsf{w}_s}(B)=\int I_B(\frac{1}{1+t})(1+t)t^{s-2}dt=\int I_B(\pi)(1-\pi)^{s-2}\pi^{-1-s}d\pi,$$

which proves the first statement of the corollary. For the second statement we use in addition

$$s(1-s)\int_{(0,1)}[\pi\wedge(1-\pi)](1-\pi)^{s-2}\pi^{-1-s}d\pi=1,\quad 0<s<1,$$

and (1.77). ■

In addition to the binary model $\mathcal{M} = (\mathcal{X}, \mathfrak{A}, \{P_0, P_1\})$ let now $(\mathcal{Y}, \mathfrak{B})$ be another measurable space and $\mathsf{K} : \mathfrak{B} \times \mathcal{X} \to_k [0,1]$ be a stochastic kernel. By setting for $i = 0, 1$,

$$(\mathsf{K}P_i)(B) = \int \mathsf{K}(B|x)P_i(dx), \quad B \in \mathfrak{B},$$

$\mathcal{N} = (\mathcal{Y}, \mathfrak{B}, \{Q_0, Q_1\})$ with $Q_i = \mathsf{K}P_i$, $i = 0, 1$, is again a binary model. Intuitively it is clear that the model $\mathcal{N}$ is less informative than $\mathcal{M}$ as it is harder to distinguish between $\mathsf{K}P_0$ and $\mathsf{K}P_1$ than to distinguish between P_0 and P_1. Thus we can anticipate that the inequality $\mathsf{I}_\mathsf{v}(\mathsf{K}P_0, \mathsf{K}P_1) \le \mathsf{I}_\mathsf{v}(P_0, P_1)$ holds true. This inequality is the content of the following monotonicity theorem which goes back to Csiszár (1963). In preparation for this theorem we study first the information gain $\mathsf{B}_\pi(P_0, P_1)$ in (1.83). Let us consider the hypothesis testing problem $\mathsf{H}_0 : \mathsf{K}P_0$ versus $\mathsf{H}_A : \mathsf{K}P_1$. For any test $\varphi : \mathcal{Y} \to_m [0,1]$ for the model $\mathcal{N}$ we get from Fubini's theorem for stochastic kernels (see Proposition A.40)

$$\int \varphi d(\mathsf{K}P_0) = \int [\int \varphi(y)\mathsf{K}(dy|x)]P_0(dx),$$

where $x \mapsto \int \varphi(y)\mathsf{K}(dy|x)$ is a test for the model $\mathcal{M}$. As $\mathsf{b}_\pi(\mathsf{K}P_0, \mathsf{K}P_1)$ is the minimal Bayes risk we arrive at

$$\begin{aligned} \mathsf{B}_\pi(\mathsf{K}P_0,\mathsf{K}P_1) &= \sup_\varphi[\pi \wedge (1-\pi) - \pi \int \varphi d(\mathsf{K}P_0) - (1-\pi)\int (1-\varphi)d(\mathsf{K}P_1)] \\ &\le \sup_\psi[\pi \wedge (1-\pi) - \pi \int \psi dP_0 - (1-\pi)\int (1-\psi)dP_1], \end{aligned}$$

where the first and second supremum are taken over all tests for $\mathcal{N}$ and $\mathcal{M}$, respectively. Hence

$$\mathsf{B}_\pi(\mathsf{K}P_0, \mathsf{K}P_1) \le \mathsf{B}_\pi(P_0, P_1). \tag{1.87}$$

This inequality says that the information gain by taking an observation is smaller in the model $\mathcal{N}$ than in the model $\mathcal{M}$.

We now establish the monotonicity property of v-divergences. This helps us to discuss the concept of sufficiency and to specify the approximate sufficiency of a statistic. The basic idea is as follows. Suppose we are faced with two distributions P_0 and P_1 and employ a statistic T for data compression. By doing so we are aware of the fact that the distance between P_0 and P_1 has been reduced and that it is now harder to distinguish between $P_0 \circ T^{-1}$ and $P_1 \circ T^{-1}$ than it has been before to distinguish between P_0 and P_1. The question arises as to how much information has been lost, and how to quantify that. An answer is provided by the following monotonicity theorem.

Theorem 1.70. *If $(\mathcal{X}, \mathfrak{A})$ and $(\mathcal{Y}, \mathfrak{B})$ are measurable spaces and $\mathsf{K} : \mathfrak{B} \times \mathcal{X} \to_k [0,1]$ is a stochastic kernel, then for $P_0, P_1 \in \mathcal{P}(\mathfrak{A})$ and every convex function $\mathsf{v} : (0, \infty) \to \mathbb{R}$,*

$$\mathsf{I}_{\mathsf{v}}(\mathsf{K}P_0, \mathsf{K}P_1) \le \mathsf{I}_{\mathsf{v}}(P_0, P_1), \tag{1.88}$$

with equality holding for

$$\mathsf{b}_{\pi}(\mathsf{K}P_0, \mathsf{K}P_1) = \mathsf{b}_{\pi}(P_0, P_1), \ 0 < \pi < 1. \tag{1.89}$$

Conversely, if v *is strictly convex in* $(0, \infty)$, *then* $\mathsf{I}_{\mathsf{v}}(\mathsf{K}P_0, \mathsf{K}P_1) = \mathsf{I}_{\mathsf{v}}(P_0, P_1) < \infty$ *implies (1.89).*

Corollary 1.71. *If* $T : \mathcal{X} \to_m \mathcal{Y}$, $\mathfrak{G}$ *is a sub-*σ*-algebra of* $\mathfrak{A}$, *and* $P_i^{\mathfrak{G}}$ *is the restriction of* P_i *to* $\mathfrak{G}$, *then*

$$\mathsf{I}_{\mathsf{v}}(P_0^{\mathfrak{G}}, P_1^{\mathfrak{G}}) \le \mathsf{I}_{\mathsf{v}}(P_0, P_1) \tag{1.90}$$
$$\mathsf{I}_{\mathsf{v}}(P_0 \circ T^{-1}, P_1 \circ T^{-1}) \le \mathsf{I}_{\mathsf{v}}(P_0, P_1), \tag{1.91}$$

with equality holding in both inequalities if (1.89) holds with $\mathsf{K}P_i$ *replaced by* $P_i^{\mathfrak{G}}$ *and* $P_1 \circ T^{-1}$, *respectively. For strictly convex* v *and* $\mathsf{I}_{\mathsf{v}}(P_0, P_1) < \infty$ *the equality in (1.90) or (1.91) implies (1.89).*

Proof. The inequality (1.88) follows directly from (1.87) and Theorem 1.68. If (1.89) is satisfied, then the equality in (1.88) follows from Theorem 1.68. Suppose now that v is strictly convex on $(0, \infty)$. In view of Problem 1.53 this requirement is equivalent to the condition that $\gamma_{\mathsf{v}}((a, b)) > 0$ for every $0 < a < b < \infty$, which by the definition of ρ_{v} is equivalent to

$$\rho_{\mathsf{v}}((a, b)) > 0, \quad 0 < a < b < 1. \tag{1.92}$$

Suppose now that $\mathsf{I}_{\mathsf{v}}(\mathsf{K}P_0, \mathsf{K}P_1) = \mathsf{I}_{\mathsf{v}}(P_0, P_1) < \infty$. Then by Theorem 1.68,

$$0 = \mathsf{I}_{\mathsf{v}}(P_0, P_1) - \mathsf{I}_{\mathsf{v}}(\mathsf{K}P_0, \mathsf{K}P_1) = \int [\mathsf{b}_{\pi}(\mathsf{K}P_0, \mathsf{K}P_1) - \mathsf{b}_{\pi}(P_0, P_1)] \, \rho_{\mathsf{v}}(d\pi).$$

The integrand is nonnegative in view of (1.87). Consequently,

$$\rho_{\mathsf{v}}(\{\pi : \mathsf{b}_{\pi}(\mathsf{K}P_0, \mathsf{K}P_1) \neq \mathsf{b}_{\pi}(P_0, P_1)\}) = 0. \tag{1.93}$$

As by Problem 1.67 the function $\pi \mapsto \mathsf{b}_{\pi}(\mathsf{K}P_0, \mathsf{K}P_1) - \mathsf{b}_{\pi}(P_0, P_1)$ is continuous the relations (1.92) and (1.93) provide $\mathsf{b}_{\pi}(\mathsf{K}P_0, \mathsf{K}P_1) = \mathsf{b}_{\pi}(P_0, P_1)$ for every $0 < \pi < 1$, which completes the proof. The corollary follows from the fact that measurable mappings are special kernels. ■

The next problem deals with a simple application of (1.88).

Problem 1.72.* The variational distance and the Hellinger distance satisfy

$$\|\mathsf{K}P_0 - \mathsf{K}P_1\| \le \|P_0 - P_1\| \quad \text{and} \quad \mathsf{D}(\mathsf{K}P_0, \mathsf{K}P_1) \le \mathsf{D}(P_0, P_1). \tag{1.94}$$

Csiszár (1963) not only proved inequality (1.88). He showed also that for a strictly convex function v equality implies that the kernel K is sufficient, which means especially for a kernel that is induced by a statistic that this statistic has to be sufficient. To study this problem in more detail we need auxiliary results.

Lemma 1.73. *Let* X_0, X_1 *be nonnegative random variables on* $(\Omega, \mathfrak{F}, \mathbb{P})$ *with* $\mathbb{E}X_i < \infty$, $i = 0, 1$. *If* $\mathfrak{F}_0 \subseteq \mathfrak{F}$ *is a sub-*σ*-algebra of* $\mathfrak{F}$, *then*

$$\mathbb{E}((X_0 X_1)^{1/2}|\mathfrak{F}_0) \le (\mathbb{E}(X_0|\mathfrak{F}_0)\mathbb{E}(X_1|\mathfrak{F}_0))^{1/2}, \quad \mathbb{P}\text{-}a.s. \tag{1.95}$$

The $\mathbb{P}$*-a.s. equality holds if and only if*

$$X_0\mathbb{E}(X_1|\mathfrak{F}_0) = X_1\mathbb{E}(X_0|\mathfrak{F}_0), \quad \mathbb{P}\text{-}a.s. \tag{1.96}$$

Proof. We put $Y_i = \mathbb{E}(X_i|\mathfrak{F}_0)$ and $A_i = \{Y_i = 0\}$. Then it holds $\mathbb{E}I_{A_i}X_i = \mathbb{E}(I_{A_i}X_i|\mathfrak{F}_0) = 0$ and

$$\mathbb{E}I_{A_i}\mathbb{E}((X_0X_1)^{1/2}|\mathfrak{F}_0) = \mathbb{E}I_{A_i}(X_0X_1)^{1/2} = 0,$$

so that both sides of (1.95) and (1.96) are $\mathbb{P}$-$a.s.$ zero on $A_1 \cup A_2$. Thus for the rest of the proof we may assume that Y_1 and Y_2 are positive. Put $Z_i = X_i/Y_i$. Then $\mathbb{E}(Z_i|\mathfrak{F}_0) = 1$ and

$$\begin{aligned} 0 &\le \mathbb{E}((Z_0^{1/2} - Z_1^{1/2})^2|\mathfrak{F}_0) = 2 - 2\mathbb{E}((Z_0Z_1)^{1/2}|\mathfrak{F}_0) \\ &= 2 - 2(Y_0Y_1)^{-1/2}\mathbb{E}((X_0X_1)^{1/2}|\mathfrak{F}_0), \end{aligned}$$

which proves (1.95), where equality holds if and only if $Z_0 = Z_1$, $\mathbb{P}$-a.s., which is equivalent to (1.96). ■

Suppose $(\mathcal{X}, \mathfrak{A})$ and $(\mathcal{Y}, \mathfrak{B})$ are two measurable spaces. Let $\mathsf{K} : \mathfrak{B} \times \mathcal{X} \to_k [0, 1]$ be a stochastic kernel and P a distribution on $(\mathcal{X}, \mathfrak{A})$. Denote by $S : \mathcal{X} \times \mathcal{Y} \to_m \mathcal{Y}$ the projection onto $\mathcal{Y}$. Recall that $\mathsf{K}\otimes P$ and $\mathsf{K}P$ are distributions on $(\mathcal{X} \times \mathcal{Y}, \mathfrak{A} \otimes \mathfrak{B})$ and $(\mathcal{Y}, \mathfrak{B})$, respectively, defined by

$$\begin{aligned} (\mathsf{K} \otimes P)(C) &= \int [\int I_C(x, y)\mathsf{K}(dy|x)]P(dx), \quad C \in \mathfrak{A} \otimes \mathfrak{B}, \\ (\mathsf{K}P)(B) &= (\mathsf{K} \otimes P)(\mathcal{X} \times B), \quad B \in \mathfrak{B}. \end{aligned}$$

If $T : \mathcal{X} \to_m \mathcal{Y}$ and $\mathsf{K}(\cdot|x) = \boldsymbol{\delta}_{T(x)}(\cdot)$, then obviously $\mathsf{K}P = P \circ T^{-1}$. The next problem gives the density of $\mathsf{K} \otimes P$ with respect to $\mathsf{K} \otimes Q$, and that of $\mathsf{K}P$ with respect to $\mathsf{K}Q$.

Problem 1.74.* If P and Q are any distributions on $(\mathcal{X}, \mathfrak{A})$ with $P \ll Q$, then with $L = dP/dQ$ it holds

$$\frac{d(\mathsf{K} \otimes P)}{d(\mathsf{K} \otimes Q)}(x, y) = L(x), \quad \mathsf{K} \otimes Q\text{-a.s.}, \tag{1.97}$$

$$\frac{d(\mathsf{K}P)}{d(\mathsf{K}Q)}(y) = \mathsf{E}_{\mathsf{K}\otimes Q}(L|S = y), \quad \mathsf{K}Q\text{-a.s.}, \tag{1.98}$$

$$\frac{d(P \circ T^{-1})}{d(Q \circ T^{-1})}(y) = \mathsf{E}_Q(L|T = y), \quad Q\text{-a.s.} \tag{1.99}$$

Set

$$\overline{P} = \frac{1}{2}(P_0 + P_1), \quad L_i = \frac{dP_i}{d\overline{P}}, \ i = 0, 1. \tag{1.100}$$

Then $L_1 = 2 - L_0$, $\overline{P}$-a.s., and by (1.75)

$$\mathsf{D}^2(P_0, P_1) = \mathsf{E}_{\overline{P}}(L_0^{1/2} - (2 - L_0)^{1/2})^2. \tag{1.101}$$

Moreover (1.97) implies

$$\mathsf{D}^2(P_0, P_1) = \mathsf{D}^2(\mathsf{K} \otimes P_0, \mathsf{K} \otimes P_1). \tag{1.102}$$

Now we formulate conditions on the density L_0 so that equality holds in $\mathsf{D}(\mathsf{K}P_0, \mathsf{K}P_1) \leq \mathsf{D}(P_0, P_1)$. Later in Chapter 4 we show that this condition is closely related to the concept of sufficiency. In view of (1.102) and $\mathsf{K}P_i = (\mathsf{K} \otimes P_i) \circ S$ it is sufficient to study the problem under which conditions the Hellinger distance is preserved under a measurable mapping.

Proposition 1.75. *It holds* $\mathsf{D}(P_0 \circ T^{-1}, P_1 \circ T^{-1}) = \mathsf{D}(P_0, P_1)$ *if and only if*

$$\mathsf{E}_{\overline{P}}(L_0|T) = L_0, \quad \overline{P}\text{-}a.s. \tag{1.103}$$

Proof. It holds

$$\mathsf{D}^2(P_0, P_1) = \mathsf{E}_{\overline{P}}(L_0^{1/2} - (2 - L_0)^{1/2})^2 = \mathsf{E}_{\overline{P}}(2 - 2(L_0(2 - L_0))^{1/2}),$$

and similarly by (1.99)

$$\mathsf{D}^2(P_0 \circ T^{-1}, P_1 \circ T^{-1}) = \mathsf{E}_{\overline{P}}(2 - 2(\mathsf{E}_{\overline{P}}(L_0|T)(2 - \mathsf{E}_{\overline{P}}(L_0|T)))^{1/2}).$$

Hence $\mathsf{D}(P_0 \circ T^{-1}, P_1 \circ T^{-1}) = \mathsf{D}(P_0, P_1)$ if and only if

$$\mathsf{E}(\mathsf{E}_{\overline{P}}(L_0|T)(2 - \mathsf{E}_{\overline{P}}(L_0|T)))^{1/2} = \mathsf{E}_{\overline{P}}(L_0(2 - L_0))^{1/2}.$$

An application of Lemma 1.73 with $X_0 = L_0$ and $X_1 = 2 - L_0$ completes the proof. ■

The condition $\mathsf{E}_{\overline{P}}(L_0|T) = L_0$ is equivalent to L_0 being $\sigma(T)$-measurable. But then $2 - L_0$, which is the density of P_1, is also $\sigma(T)$-measurable. Later in Chapter 4 we study how this measurability condition is related to the concept of sufficiency.

There are many concepts to reduce or condense a large sample $X_1, ... X_n$. One of them is to choose a partition $\mathfrak{p} = \{A_1, ..., A_n\}$ of the sample space $\mathcal{X}$ and then to use only the relative frequencies at which the observations have appeared in the cells of the partition. Hereby it is understood that a partition $\mathfrak{p}$ of $\mathcal{X}$ is a collection $\{A_1, ..., A_n\}$ of subsets of $\mathcal{X}$ with $A_i \in \mathfrak{A}$, $A_i \cap A_j = \emptyset$ for $i \neq j$, and $A_1 \cup \cdots \cup A_n = \mathcal{X}$. Instead of the original sample space $(\mathcal{X}, \mathfrak{A})$ we use now the sample space $(\mathcal{X}, \sigma(\mathfrak{p}))$, where $\sigma(\mathfrak{p})$ is the σ-algebra generated by the partition $\mathfrak{p}$. Suppose that we have a nondecreasing sequence of partitions $\mathfrak{p}_n$ so that the sequence of σ-algebras $\mathfrak{A}_n := \sigma(\mathfrak{p}_n)$ generates $\mathfrak{A}$. Then we can approximate $\mathfrak{A}$-measurable tests by $\mathfrak{A}_n$-measurable tests. Consequently, we can get to the minimal Bayes risk approximately if we only know the cells to which the observations belongs, provided that n is large enough so that the partition is sufficiently fine.

Lemma 1.76. *Let $\mathfrak{A}_1 \subseteq \mathfrak{A}_2 \subseteq \cdots$ be a nondecreasing sequence of sub-σ-algebras of $\mathfrak{A}$ which generates $\mathfrak{A}$, and let $P_0^{\mathfrak{A}_n}$ and $P_1^{\mathfrak{A}_n}$ be the restrictions of P_0 and P_1, respectively, to $\mathfrak{A}_n$, $n = 1, 2, ...$ Then $\mathsf{B}_\pi(P_0^{\mathfrak{A}_n}, P_1^{\mathfrak{A}_n}) \uparrow \mathsf{B}_\pi(P_0, P_1)$ as n tends to infinity.*

Proof. The monotonicity follows from (1.87). Set $\overline{P} = \frac{1}{2}(P_0 + P_1)$ and consider the densities $L_i = dP_i/d\overline{P}$, $i = 0, 1$, as random variables on $(\mathcal{X}, \mathfrak{A}, \overline{P})$. The conditional expectation $\mathsf{E}_{\overline{P}}(L_i|\mathfrak{A}_n) =: L_{i,n}$ with respect to $\overline{P}$ satisfies for every $A \in \mathfrak{A}_n$

$$\int_A \mathsf{E}_{\overline{P}}(L_i|\mathfrak{A}_n) d\overline{P} = \int_A L_i d\overline{P} = P_i^{\mathfrak{A}_n}(A),$$

which implies $L_{i,n} = dP_i^{\mathfrak{A}_n}/d\overline{P}^{\mathfrak{A}_n}$. Hence by Levy's martingale convergence theorem (see Theorem A.34), $\mathsf{E}_{\overline{P}}|L_i - L_{i,n}| \to 0$. Using the elementary inequality $|a \wedge b - c \wedge d| \le |a - b| + |c - d|$ we arrive at

$$\int |(\pi L_0) \wedge ((1-\pi)L_1) d\overline{P} - (\pi L_{0,n}) \wedge ((1-\pi)L_{1,n}| d\overline{P} \to 0, \quad \text{as } n \to \infty.$$

By combining this fact with (1.82) we get $\mathsf{B}_\pi(P_0^{\mathfrak{A}_n}, P_1^{\mathfrak{A}_n}) \uparrow \mathsf{B}_\pi(P_0, P_1)$. ■

Theorem 1.77. *If $\mathfrak{A}_1 \subseteq \mathfrak{A}_2 \subseteq \cdots$ is a nondecreasing sequence of sub-σ-algebras of $\mathfrak{A}$ which generates $\mathfrak{A}$, then*

$$\lim_{n\to\infty} \mathsf{I}_\mathsf{v}(P_0^{\mathfrak{A}_n}, P_1^{\mathfrak{A}_n}) = \mathsf{I}_\mathsf{v}(P_0, P_1). \tag{1.104}$$

Corollary 1.78. *It holds*

$$\mathsf{I}_\mathsf{v}(P_0, P_1) = \sup_{\mathfrak{p}} \sum\nolimits_{A\in\mathfrak{p}} \mathsf{v}(\frac{P_0(A)}{P_1(A)}) P_1(A),$$

where the supremum is taken over all partitions $\mathfrak{p}$ with $\mathfrak{p} \subseteq \mathfrak{A}$, and where the conventions $\mathsf{v}(0/0)0 = 0$ and $\mathsf{v}(a/0)0 = a\mathsf{v}^*(0)$ for $a > 0$ are used.

Proof. The statement (1.104) follows from Lemma 1.76, the representation (1.86), and the monotone convergence theorem; see Theorem A.16.

To prove the corollary we first note that by (1.90)

$$\sup_{\mathfrak{p}} \mathsf{I}_\mathsf{v}(P_0^{\sigma(\mathfrak{p})}, P_1^{\sigma(\mathfrak{p})}) = \sup_{\mathfrak{p}} \sum\nolimits_{A\in\mathfrak{p}} \mathsf{v}(\frac{P_0(A)}{P_1(A)}) P_1(A) \le \mathsf{I}_\mathsf{v}(P_0, P_1).$$

To show that in fact equality holds we set $\mathfrak{F} = \sigma(f_0, f_1)$. As $f_i = dP_i^{\mathfrak{F}}/d\mu^{\mathfrak{F}}$ we get from (1.82),

$$\mathsf{b}_\pi(P_0^{\mathfrak{F}}, P_1^{\mathfrak{F}}) = \mathsf{b}_\pi(P_0, P_1), \quad 0 < \pi < 1,$$

and $\mathsf{I}_\mathsf{v}(P_0^{\mathfrak{F}}, P_1^{\mathfrak{F}}) = \mathsf{I}_\mathsf{v}(P_0, P_1)$ by Corollary 1.71. As open intervals (a, b) with rational endpoints generate $\mathfrak{B}$, the σ-algebra of Borel sets on $\mathbb{R}$, and the complete images of (a, b) under f_0 and f_1 generate $\mathfrak{F}$ we see that $\mathfrak{F}$ is countably generated. This means that we find a nondecreasing sequence of algebras $\mathfrak{A}_n$ that generate $\mathfrak{F}$. If $\mathfrak{p}_n$ is the system of atoms of $\mathfrak{A}_n$, then $\mathfrak{p}_n$, $n = 1, 2, ...,$ is a nondecreasing sequence of partitions with $\mathfrak{F} = \sigma(\cup_{n=1}^\infty \mathfrak{p}_n)$. Hence,

$$\mathsf{I}_\mathsf{v}(P_0, P_1) = \mathsf{I}_\mathsf{v}(P_0^{\mathfrak{F}}, P_1^{\mathfrak{F}}) = \lim_{n\to\infty} \mathsf{I}_\mathsf{v}(P_0^{\mathfrak{A}_n}, P_1^{\mathfrak{A}_n})$$

by (1.104) if we replace $\mathfrak{A}$ by $\mathfrak{F}$ there. ■

Problem 1.79.* It holds,

$$\|P_0 - P_1\| = \sup_{\mathfrak{p}} \sum\nolimits_{A\in\mathfrak{p}} |P_0(A) - P_1(A)| = 2 \sup_{A\in\mathfrak{A}} |P_0(A) - P_1(A)|$$

$$\mathsf{H}_s(P_0, P_1) = \inf_{\mathfrak{p}} \sum\nolimits_{A\in\mathfrak{p}} P_0^s(A)P_1^{1-s}(A) \le \sum\nolimits_{A\in\mathfrak{p}} P_0^s(A)P_1^{1-s}(A), \quad 0 < s < 1.$$

Problem 1.80.* If P_0 and P_1 are distributions on $(\mathcal{X}, \mathfrak{A})$ and $h : \mathcal{X} \to_m \mathbb{R}$ is bounded, say $\|h\|_u \le c$, then

$$\left| \int h dP_0 - \int h dP_1 \right| \le c \|P_0 - P_1\| \text{ and } \sup_{\|h\|_u \le 1} \left| \int h dP_0 - \int h dP_1 \right| = \|P_0 - P_1\|,$$

where $\|h\|_u = \sup_{x\in\mathcal{X}} |h(x)|$. If $\mathcal{X}$ is a metric space and $\mathfrak{A}$ the σ-algebra of Borel sets, then the supremum may be taken only over all continuous functions h with $\|h\|_u \le 1$.

In the remainder of this section we collect properties of Hellinger integrals that are needed in the following chapters.

Recall that by (1.69), (1.74), and (1.80),

$$\mathsf{H}_s(P_0, P_1) = \int f_0^s f_1^{1-s} d\mu, \quad 0 < s < 1, \tag{1.105}$$

$$\mathsf{H}_s(P_0, P_1) = \int f_0^s f_1^{1-s} I_{\{f_1>0\}} d\mu + \infty P_0(f_1 = 0), \ 1 < s < \infty, \tag{1.106}$$

$$\mathsf{K}_s(P_0, P_1) = \frac{1}{s(1-s)}(1 - \mathsf{H}_s(P_0, P_1)), \quad s \ne 1, \ s > 0.$$

In the next problem we collect some simple properties of Hellinger integrals that are used frequently.

Problem 1.81.* Suppose $L_i, i = 0, 1$ and $\overline{P}$ are defined in (1.100) and denote by $L_{0,1}$ the likelihood ratio of P_1 with respect to P_0. Then

$$\mathsf{H}_s(P_0, P_1) = \mathsf{E}_{P_0} L_{0,1}^{1-s} \tag{1.107}$$

$$= \mathsf{E}_{\overline{P}} L_0^s (2 - L_0)^{1-s}, \quad s \ne 1, \ s > 0, \tag{1.108}$$

where we used the convention $0^{-a} = \infty$ for $a > 0$. It holds

$$(\mathsf{H}_{s_1}(P_0, P_1))^{1/(1-s_1)} \geq (\mathsf{H}_{s_2}(P_0, P_1))^{1/(1-s_2)}, \quad 0 < s_1 < s_2 < 1. \tag{1.109}$$

The Hellinger distance and the Hellinger integral are related by

$$2(1 - \mathsf{H}_{1/2}(P_0, P_1)) = \mathsf{D}^2(P_0, P_1) = \mathsf{E}_0(L_{0,1}^{1/2} - 1)^2 + (1 - \mathsf{E}_0 L_{0,1}). \tag{1.110}$$

For $0 < s < 1$ it holds

$$0 \leq \mathsf{H}_s(P_0, P_1) \leq 1, \tag{1.111}$$

$$\mathsf{H}_s(P_0, P_1) = 0 \Leftrightarrow P_0 \perp P_1 \quad \text{and} \quad \mathsf{H}_s(P_0, P_1) = 1 \Leftrightarrow P_0 = P_1. \tag{1.112}$$

For $1 < s < \infty$ it holds

$$1 \leq \mathsf{H}_s(P_0, P_1) \leq \infty, \tag{1.113}$$

$$\mathsf{H}_s(P_0, P_1) = 1 \Leftrightarrow P_0 = P_1 \quad \text{and} \quad \mathsf{H}_s(P_0, P_1) < \infty \Rightarrow P_0 \ll P_1.$$

We recall the Lebesgue decomposition (1.65), where $dP_{1,a} = f_1 I_{(0,\infty)}(f_0) d\mu$ is the part of P_1 that is absolutely continuous with respect to P_0. That its total mass $P_1(f_0 > 0)$ can be obtained by Hellinger integrals was first noticed by Nemetz (1967, 1974).

Problem 1.82.* It holds

$$\lim_{s \downarrow 0} \mathsf{H}_s(P_0, P_1) = P_1(f_0 > 0) = P_1(L_{0,1} < \infty),$$

$$P_1 \ll P_0 \quad \Leftrightarrow \quad \lim_{s \downarrow 0} \mathsf{H}_s(P_0, P_1) = 1. \tag{1.114}$$

Statement (1.114) has been used by several authors to find conditions that guarantee the absolute continuity of stochastic processes; see, e.g., Jacod and Shiryaev (1987) and Liese and Vajda (1987) for details and references.

Sometimes it is necessary to consider the Hellinger integral as a function of a complex variable.

Problem 1.83.* If P_0 and P_1 are not mutually singular, then the function $z \mapsto \mathsf{H}_{s+it}(P_0, P_1) := \int (f_0/f_1)^{s+it} f_1 d\mu$, $z = s + it$, is analytic for $s \in (0, 1)$, $t \in \mathbb{R}$.

Often the information functionals do not appear directly in a problem under consideration. Then inequalities between the different expressions are useful. This especially concerns the situation where one statistical model has to be approximated by another model in the strong sense of variational distance. If that is possible, then optimal decisions in the approximating model are at least approximately optimal in the original model.

Inequalities that provide bounds on the variational distance $\|P_0 - P_1\|$ by means of the tractable Hellinger distance and Kullback–Leibler distance originated from different areas of probability theory, information theory, and statistics, and thus were independently established by several authors; see Nemetz (1967), Kailath (1967), Vajda (1971), LeCam (1974), Strasser (1985), Reiss (1989), and Jongbloed (2000).

Proposition 1.84. *It holds,*

$$\begin{aligned}\mathsf{D}^2(P_0,P_1) &\le \|P_0-P_1\| \le \left[4-\mathsf{D}^2(P_0,P_1)\right]^{1/2}\mathsf{D}(P_0,P_1)\\ &\le 2\mathsf{D}(P_0,P_1),\\ \mathsf{D}^2(P_0,P_1) &\le 2(1-\exp\{-\frac{1}{2}\mathsf{K}(P_0,P_1)\}) \qquad (1.115)\\ \|P_0-P_1\| &\le 2\sqrt{\mathsf{K}(P_0,P_1)}. \qquad (1.116)\end{aligned}$$

Proof. The first inequality follows from $(\sqrt{a_1}-\sqrt{a_2})^2 \le |a_1-a_2|$. To get the second and third inequality we apply the Schwarz inequality to $|f_0-f_1| = |\sqrt{f_0}-\sqrt{f_1}||\sqrt{f_0}+\sqrt{f_1}|$ and use

$$\int |\sqrt{f_0}+\sqrt{f_1}|^2 d\mu = 4-\mathsf{D}^2(P_0,P_1) \le 4.$$

To prove (1.115) we may assume $P_0 \ll P_1$ as otherwise $\mathsf{K}_1(P_0,P_1)=\infty$ and the inequality becomes trivial. If now $P_0 \ll P_1$, then by (1.81), $\mathsf{K}(P_0,P_1) = \mathsf{E}_{P_0}\ln(dP_0/dP_1)$. Then with $Y=-(1/2)\ln(dP_0/dP_1)$,

$$\begin{aligned}\mathsf{D}^2(P_0,P_1) &= 2(1-\mathsf{H}_{1/2}(P_0,P_1)) = 2(1-\mathsf{E}_{P_0}(dP_0/dP_1)^{-1/2})\\ &= 2(1-\mathsf{E}_{P_0}\exp\{Y\}) \le 2(1-\exp\{\mathsf{E}_{P_0}Y\}),\end{aligned}$$

by the convexity of the exponential function and Jensen's inequality. ■

The inequality $\|P_0-P_1\| \le c\sqrt{\mathsf{K}(P_0,P_1)}$ with some constant c has a long history and was independently established by many authors; see Liese and Vajda (1987) for details. There one can also find improved bounds. An important application of inequality (1.116) can be found in Reiss (1993), where the following result has been established.

Proposition 1.85. *The Hellinger distance and the variational distance of the binomial distribution and the Poisson distribution satisfy*

$$\begin{aligned}\mathsf{D}(\mathsf{B}(n,\lambda/n),\mathsf{Po}(\lambda)) &\le \sqrt{3}\lambda/n \quad \textit{and}\\ \|\mathsf{B}(n,\lambda/n)-\mathsf{Po}(\lambda)\| &\le 2\lambda/n.\end{aligned}$$

These inequalities imply as a special case the well-known convergence of $\mathsf{B}(n,\lambda/n)$ to $\mathsf{Po}(\lambda)$ for fixed λ. But they also allow an approximation of binomial distribution $\mathsf{B}(n,p_n)$ by Poisson distributions if p_n tends to 0 at a lower rate. For applications and extensions of Proposition 1.85 to binomial processes and curve estimations we refer to Reiss (1993). Other applications of the inequalities of Proposition 1.84 can be found in Jacod and Shiryaev (2002) and in Liese (1986), where Hellinger integrals of distributions of stochastic processes have been evaluated and used to examine the variational distance between two such distributions. Hellinger integrals for independent observations behave as characteristic functions of independent random variables.

Problem 1.86.* Suppose $(\mathcal{X}_i, \mathfrak{A}_i)$, $i = 1, 2, \ldots$, are measurable spaces and $P_i, Q_i \in \mathcal{P}(\mathfrak{A}_i)$, $i = 1, 2, \ldots$ Then for every $n = 1, 2, \ldots$,

$$\mathsf{H}_s(\bigotimes_{i=1}^n P_i, \bigotimes_{i=1}^n Q_i) = \prod\nolimits_{i=1}^n \mathsf{H}_s(P_i, Q_i), \quad 0 < s < 1, \tag{1.117}$$

$$\mathsf{H}_s(\bigotimes_{i=1}^\infty P_i, \bigotimes_{i=1}^\infty Q_i) = \prod\nolimits_{i=1}^\infty \mathsf{H}_s(P_i, Q_i), \quad 0 < s < 1,$$

$$\mathsf{D}^2(\bigotimes_{i=1}^n P_i, \bigotimes_{i=1}^n Q_i) \le \sum\nolimits_{i=1}^n \mathsf{D}^2(P_i, Q_i). \tag{1.118}$$

Definition 1.87. *Given a finite model $(\mathcal{X}, \mathfrak{A}, \{P_1, \ldots, P_m\})$, where P_k is dominated by $\mu \in \mathcal{M}^\sigma(\mathfrak{A})$ and has the density $f_k = dP_k/d\mu$, $1 \le k \le m$, we call the function*

$$s \mapsto \mathsf{H}_s(P_1, \ldots, P_m) = \int f_1^{s_1} \cdots f_m^{s_m} d\mu, \quad s = (s_1, \ldots, s_m) \in \mathbf{S}_m^0, \tag{1.119}$$

the Hellinger transform of the family $\{P_1, \ldots, P_m\}$.

If $m = 2$, then we have

$$\mathsf{H}_{(s,1-s)}(P_1, P_2) = \mathsf{H}_s(P_1, P_2), \tag{1.120}$$

so that for $0 < s < 1$ the functional $\mathsf{H}_{(s,1-s)}(P_1, P_2)$ is the Hellinger integral $\mathsf{H}_s(P_0, P_1)$ of order s in (1.105). Set

$$\overline{P} = m^{-1} \sum\nolimits_{i=1}^m P_i, \quad L_i = \frac{dP_i}{d\overline{P}}, \quad i = 1, \ldots, m. \tag{1.121}$$

Then by the chain rule (see Proposition A.28)

$$\mathsf{H}_s(P_1, \ldots, P_m) = \mathsf{E}_{\overline{P}} \prod\nolimits_{i=1}^m L_i^{s_i},$$

which shows the definition of $\mathsf{H}_s(P_1, \ldots, P_m)$ is independent of the choice of the dominating measure.

Example 1.88. Let $(P_\theta)_{\theta \in \Delta}$ be an exponential family with $\boldsymbol{\mu}$-densities $f_\theta(x) = \exp\{\langle \theta, T(x) \rangle - K(\theta)\}$, $\theta \in \Delta$. For fixed $\theta_1, \ldots, \theta_m \in \Delta$ we get for $s \in \mathbf{S}_m^0$,

$$\begin{aligned} \mathsf{H}_s(P_{\theta_1}, \ldots, P_{\theta_m}) &= \int \exp\{\sum\nolimits_{i=1}^m s_i \langle \theta_i, T \rangle - \sum\nolimits_{i=1}^m s_i K(\theta_i)\} d\boldsymbol{\mu} \\ &= \exp\{K(\sum\nolimits_{i=1}^m s_i \theta_i) - \sum\nolimits_{i=1}^m s_i K(\theta_i)\}. \end{aligned} \tag{1.122}$$

Especially for normal distributions $(\mathsf{N}(\mu, \sigma^2))_{\mu \in \mathbb{R}}$ with a known variance σ^2 we get from Lemma 1.37, with $\theta = \mu/\sigma^2$ and $K(\theta) = \sigma^2\theta^2/2$,

$$\mathsf{H}_s(\mathsf{N}(\mu_1, \sigma^2), \ldots, \mathsf{N}(\mu_m, \sigma^2)) = \exp\{\frac{1}{2\sigma^2}(\sum\nolimits_{i=1}^m s_i \mu_i)^2 - \frac{1}{2\sigma^2} \sum\nolimits_{i=1}^m s_i \mu_i^2\}.$$

Problem 1.89. If $\mathsf{N}(\mu, \Sigma)$ is a d-dimensional normal distribution with known nonsingular covariance matrix Σ, then for $s \in \mathbf{S}_m^0$

$$\mathsf{H}_s(\mathsf{N}(\mu_1, \Sigma), \ldots, \mathsf{N}(\mu_m, \Sigma)) = \exp\{\frac{1}{2} \parallel \sum\nolimits_{i=1}^m s_i \Sigma^{-1/2} \mu_i \parallel^2 - \frac{1}{2} \sum\nolimits_{i=1}^m s_i \mu_i^T \Sigma^{-1} \mu_i\}.$$

Hellinger integrals and Hellinger transforms behave as characteristic functions for sums of independent random variables when turning to product models. The reason for this fact is that

$$\mathsf{H}_s(P_1,...,P_m) = \mathsf{E}_{\overline{P}} \exp\{\sum\nolimits_{i=1}^m s_i \ln L_i\}$$

is, up to the sign, the Laplace transform of the vector of the log-likelihoods.

Proposition 1.90. *If $P_1,...,P_m \in \mathcal{P}(\mathfrak{A})$ and $Q_1,...,Q_m \in \mathcal{P}(\mathfrak{B})$, then*

$$\mathsf{H}_s(P_1 \otimes Q_1,...,P_m \otimes Q_m) = \mathsf{H}_s(P_1,...,P_m)\mathsf{H}_s(Q_1,...,Q_m), \quad s \in \mathbf{S}_m^0.$$

Proof. Introduce $\overline{Q}$ and $M_1,...,M_m$ analogously to $\overline{P}$ and $L_1,...,L_m$; see (1.121). Then by Proposition A.29 $d(P_i \otimes Q_i) = L_i(x)M_i(y)d(\overline{P} \otimes \overline{Q})$, so that $(L_1,...,L_m)$ and $(M_1,...,M_m)$ are independent with respect to $\overline{P} \otimes \overline{Q}$ and

$$\begin{aligned}\mathsf{H}_s(P_1 \otimes Q_1,...,P_m \otimes Q_m) &= \mathsf{E}_{\overline{P}\otimes\overline{Q}} \prod\nolimits_{i=1}^m L_i^{s_i} M_i^{s_i} \\ &= (\mathsf{E}_{\overline{P}} \prod\nolimits_{i=1}^m L_i^{s_i})(\mathsf{E}_{\overline{Q}} \prod\nolimits_{i=1}^m M_i^{s_i}), \quad s \in \mathbf{S}_m^0.\end{aligned}$$

■

The next result corresponds to Corollary 1.77, where we have considered v-divergences for the restrictions of two distributions to a nondecreasing sequence of sub-σ-algebras.

Proposition 1.91. *If $\mathfrak{G}_1 \subseteq \mathfrak{G}_2 \subseteq \cdots$ is a nondecreasing sequence of sub-σ-algebras and $\mathfrak{G} = \sigma(\cup_{i=1}^\infty \mathfrak{G}_i)$, then for $s \in \mathbf{S}_m^0$ it holds*

$$\mathsf{H}_s(P_1^{\mathfrak{G}_n},...,P_m^{\mathfrak{G}_n}) \to \mathsf{H}_s(P_1^{\mathfrak{G}},...,P_m^{\mathfrak{G}}) \quad as \quad n \to \infty.$$

Proof. For $i = 1,...,m$ we get $L_{i,n} := \mathsf{E}_{\overline{P}}(L_i|\mathfrak{G}_n) = dP_i^{\mathfrak{G}_n}/d\overline{P}^{\mathfrak{G}_n}$ from (1.99). Levy's martingale convergence theorem (see Theorem A.34) gives $\lim_{n\to\infty} \mathsf{E}_{\overline{P}}|L_{i,n} - L_i| = 0$. Using the inequality $|a^s - b^s| \le |a-b|^s$, $a,b \ge 0$, $0 < s < 1$, we get

$$\begin{aligned}&|\mathsf{H}_s(P_1^{\mathfrak{G}_n},...,P_m^{\mathfrak{G}_n}) - \mathsf{H}_s(P_1^{\mathfrak{G}},...,P_m^{\mathfrak{G}})| \\ &\le \mathsf{E}_{\overline{P}}|L_{1,n} - L_1|^{s_1} L_{2,n}^{s_2} \cdots L_{m,n}^{s_m} + \cdots + \mathsf{E}_{\overline{P}}(L_1^{s_1} \cdots L_{m-1}^{s_{m-1}})|L_{m,n} - L_m|^{s_m}.\end{aligned}$$

Hence, by the generalized Hölder inequality (see Lemma A.13)

$$|\mathsf{H}_s(P_1^{\mathfrak{G}_n},...,P_m^{\mathfrak{G}_n}) - \mathsf{H}_s(P_1^{\mathfrak{G}},...,P_m^{\mathfrak{G}})| \le \sum\nolimits_{i=1}^m (\mathsf{E}_{\overline{P}}|L_{i,n} - L_i|)^{s_i},$$

which completes the proof. ■

Proposition 1.92. *It holds for $s \in \mathbf{S}_m^0$,*

$$\mathsf{H}_s(P_1,...,P_m) = \inf_{\mathfrak{p}} \sum\nolimits_{A\in\mathfrak{p}} P_1^{s_1}(A) \cdots P_m^{s_m}(A),$$

where the infimum is taken over all finite partitions of the sample space into sets from $\mathfrak{A}$.

Proof. Let $\mathfrak{p} = \{A_1, ..., A_k\}$, $A_i \in \mathfrak{A}$, be any partition. Then by the general Hölder inequality it holds

$$\begin{aligned}\mathsf{H}_s(P_1, ..., P_m) &= \sum\nolimits_{i=1}^{k} \int I_{A_i} L_1^{s_1} \cdots L_m^{s_m} d\overline{P} \\ &\le \sum\nolimits_{i=1}^{k} (\int I_{A_i} L_1 d\overline{P})^{s_1} \cdots (\int I_{A_i} L_m d\overline{P})^{s_m} = \sum\nolimits_{i=1}^{k} P_1^{s_1}(A_i) \cdots P_m^{s_m}(A_i).\end{aligned}$$

Denote by $\sigma(\mathfrak{p})$ the σ-algebra generated by $\mathfrak{p}$ and set $L_{l,\mathfrak{p}} = dP_l^{\sigma(\mathfrak{p})}/d\overline{P}^{\sigma(\mathfrak{p})}$. Then by $\sum_{i=1}^{k} s_i = 1$ and

$$L_{l,\mathfrak{p}}(x) = \sum\nolimits_{i=1}^{k} I_{A_i}(x) \frac{P_l(A_i)}{\overline{P}(A_i)},$$

$$\sum\nolimits_{i=1}^{k} P_1^{s_1}(A_i) \cdots P_m^{s_m}(A_i) = \int (\prod\nolimits_{l=1}^{m} L_{l,\mathfrak{p}}(x)) \overline{P}(dx) = \mathsf{H}_s(P_1^{\sigma(\mathfrak{p})}, ..., P_m^{\sigma(\mathfrak{p})}).$$

Denote by $\mathfrak{G}$ the smallest σ-algebra with respect to which $L_1, ..., L_m$ are measurable. Then $dP_l^{\mathfrak{G}}/d\overline{P}^{\mathfrak{G}} = dP_l/d\overline{P}$ and $\mathsf{H}_s(P_1^{\mathfrak{G}}, ..., P_m^{\mathfrak{G}}) = \mathsf{H}_s(P_1, ..., P_m)$. As $\mathfrak{G}$ is countably generated there is a nondecreasing sequence of partitions $\mathfrak{p}_n$ such that $\sigma(\mathfrak{p}_n)$ generates $\mathfrak{G}$. It remains to apply Proposition 1.91. ■

Now we study Hellinger transforms of finite models that are models reduced by a statistic or by randomization via a stochastic kernel. Let $(\mathcal{X}, \mathfrak{A})$ and $(\mathcal{Y}, \mathfrak{B})$ be measurable spaces, $\mathsf{K} : \mathfrak{B} \times \mathcal{X} \to_k [0, 1]$ a stochastic kernel, and $\{P_1, ..., P_m\} \subseteq \mathcal{P}(\mathfrak{A})$. The next proposition is the monotonicity property of Hellinger transforms (see, e.g., Strasser (1985)).

Proposition 1.93. *Suppose that $(\mathcal{X}, \mathfrak{A})$ and $(\mathcal{Y}, \mathfrak{B})$ are measurable spaces, $\mathsf{K} : \mathfrak{B} \times \mathcal{X} \to_k [0, 1]$ is a stochastic kernel, and $\mathcal{P} = \{P_1, ..., P_m\} \subseteq \mathcal{P}(\mathfrak{A})$. Then*

$$\mathsf{H}_s(\mathsf{K}P_1, ..., \mathsf{K}P_m) \ge \mathsf{H}_s(P_1, ..., P_m), \quad s \in \mathbf{S}_m^0. \tag{1.123}$$

Corollary 1.94. *If $T : \mathcal{X} \to_m \mathcal{Y}$, then*

$$\mathsf{H}_s(P_1 \circ T^{-1}, ..., P_m \circ T^{-1}) \ge \mathsf{H}_s(P_1, ..., P_m), \quad s \in \mathbf{S}_m^0.$$

Proof. The statement in the corollary follows from Proposition 1.92. The relation (1.97) yields

$$\mathsf{H}_s(\mathsf{K} \otimes P_1, ..., \mathsf{K} \otimes P_m) = \mathsf{H}_s(P_1, ..., P_m).$$

If $S : \mathcal{X} \times \mathcal{Y} \to_m \mathcal{Y}$ is the projection onto $\mathcal{Y}$, then $\mathsf{K}P_i = (\mathsf{K} \otimes P_i) \circ S^{-1}$, so that (1.123) follows from the corollary. ■

1.4 Information in Bayes Models

In this section we consider divergences of distributions on a product space constructed with different conditional distributions but with the same marginal distribution. More precisely, let $(\mathcal{X}, \mathfrak{A}), (\mathcal{Y}, \mathfrak{B})$ be measurable spaces, $\mathsf{K}, \mathsf{L} : \mathfrak{B} \times \mathcal{X} \to_k [0,1]$ stochastic kernels, and P a distribution on $(\mathcal{X}, \mathfrak{A})$. Subsequently we give a representation of $\mathsf{I}_\mathsf{v}(\mathsf{K} \otimes P, \mathsf{L} \otimes P)$.

Proposition 1.95. *If* $\mathfrak{B}$ *is countably generated, then*

$$\mathsf{I}_\mathsf{v}(\mathsf{K} \otimes P, \mathsf{L} \otimes P) = \int \mathsf{I}_\mathsf{v}(\mathsf{K}(\cdot|x), \mathsf{L}(\cdot|x))P(dx).$$

Proof. As $\mathfrak{B}$ is countably generated we find an increasing sequence of partitions $\mathfrak{p}_n = \{B_{n,1}, ..., B_{n,m_n}\}$, $B_{n,i} \in \mathfrak{B}$, where $\mathfrak{B} = \sigma(\cup_{n=1}^\infty \mathfrak{p}_n)$. Let $\mathsf{M}(B|x) = \frac{1}{2}(\mathsf{K}(B|x) + \mathsf{L}(B|x))$, and denote by $(\mathsf{M} \otimes P)^{\sigma(\mathfrak{p}_n) \otimes \mathfrak{B}}$ the restriction of $\mathsf{M} \otimes P$ to $\sigma(\mathfrak{p}_n) \otimes \mathfrak{B}$. Then

$$\frac{d(\mathsf{K} \otimes P)^{\mathfrak{A} \otimes \sigma(\mathfrak{p}_n)}}{d(\mathsf{M} \otimes P)^{\mathfrak{A} \otimes \sigma(\mathfrak{p}_n)}}(x,y) = \sum\nolimits_{i=1}^{m_n} \frac{\mathsf{K}(B_{i,n}|x)}{\mathsf{M}(B_{i,n}|x)} I_{B_{i,n}}(y) = \frac{d\mathsf{K}^{\sigma(\mathfrak{p}_n)}(\cdot|x)}{d\mathsf{M}^{\sigma(\mathfrak{p}_n)}(\cdot|x)}(y)$$

$$\frac{d(\mathsf{L} \otimes P)^{\mathfrak{A} \otimes \sigma(\mathfrak{p}_n)}}{d(\mathsf{M} \otimes P)^{\mathfrak{A} \otimes \sigma(\mathfrak{p}_n)}}(x,y) = \sum\nolimits_{i=1}^{m_n} \frac{\mathsf{L}(B_{i,n}|x)}{\mathsf{M}(B_{i,n}|x)} I_{B_{i,n}}(y) = \frac{d\mathsf{L}^{\sigma(\mathfrak{p}_n)}(\cdot|x)}{d\mathsf{M}^{\sigma(\mathfrak{p}_n)}(\cdot|x)}(y).$$

It follows from Definition 1.60 that

$$\mathsf{I}_\mathsf{v}((\mathsf{K} \otimes P)^{\mathfrak{A} \otimes \sigma(\mathfrak{p}_n)}, (\mathsf{L} \otimes P)^{\mathfrak{A} \otimes \sigma(\mathfrak{p}_n)}) = \int \mathsf{I}_\mathsf{v}(\mathsf{K}^{\sigma(\mathfrak{p}_n)}(\cdot|x), \mathsf{L}^{\sigma(\mathfrak{p}_n)}(\cdot|x))P(dx).$$

An application of Theorem 1.77 to the left-hand side gives

$$\mathsf{I}_\mathsf{v}(\mathsf{K} \otimes P, \mathsf{L} \otimes P) = \lim_{n \to \infty} \mathsf{I}_\mathsf{v}((\mathsf{K} \otimes P)^{\mathfrak{A} \otimes \sigma(\mathfrak{p}_n)}, (\mathsf{L} \otimes P)^{\mathfrak{A} \otimes \sigma(\mathfrak{p}_n)}).$$

We know that $\mathsf{I}_\mathsf{v}(\mathsf{K}^{\sigma(\mathfrak{p}_n)}(\cdot|x), \mathsf{L}^{\sigma(\mathfrak{p}_n)}(\cdot|x))$ is bounded below by $\mathsf{v}(1)$; see Proposition 1.63. By Theorem 1.77 $\mathsf{I}_\mathsf{v}(\mathsf{K}^{\sigma(\mathfrak{p}_n)}(\cdot|x), \mathsf{L}^{\sigma(\mathfrak{p}_n)}(\cdot|x))$ tends increasingly to $\mathsf{I}_\mathsf{v}(\mathsf{K}(\cdot|x), \mathsf{L}(\cdot|x))$. An application of the monotone convergence theorem (see Theorem A.16) completes the proof. ■

To study the dependence between the random variables X and Y with values in $(\mathcal{X}, \mathfrak{A})$ and $(\mathcal{Y}, \mathfrak{B})$, respectively, we compare the joint distribution $\mathcal{L}(X,Y)$ with the product of the marginal distributions $\mathcal{L}(X) \otimes \mathcal{L}(Y)$. It is clear that the smaller this distance is the weaker is the dependence between X and Y. To specify the distance between the distributions we use the divergences introduced in Definition 1.60,

$$\mathsf{I}_\mathsf{v}(X||Y) := \mathsf{I}_\mathsf{v}(\mathcal{L}(X,Y), \mathcal{L}(X) \otimes \mathcal{L}(Y)),$$

and call $\mathsf{I}_\mathsf{v}(X||Y)$ the *mutual information of* X *and* Y. It is obvious that $\mathsf{I}_\mathsf{v}(X||Y)$ is symmetric in X and Y.

Suppose $\mathcal{L}(X,Y)$ is absolutely continuous with respect to $\mu \otimes \nu$, where μ and ν are σ-finite measures on $(\mathcal{X},\mathfrak{A})$ and $(\mathcal{Y},\mathfrak{B})$, respectively. Denote by $f_{Y,X}$, f_X, f_Y the corresponding densities. By Definition 1.60

$$\mathsf{I}_{\mathsf{v}}(X||Y) = \int [\int \mathsf{v}(\frac{f_{X,Y}(x,y)}{f_X(x)f_Y(y)})f_X(x)\mu(dx)]f_Y(y)\nu(dy). \tag{1.124}$$

If $\mathsf{v}(x) = x\ln x$, then I_{v} is the Kullback–Leibler distance; see (1.74) and (1.81). In this case $\mathsf{I}(X||Y) := \mathsf{I}_{x\ln x}(X||Y)$ satisfies

$$\begin{aligned}\mathsf{I}(X||Y) &= \int [\int \ln(\frac{f_{Y,X}(x,y)}{f_X(x)f_X(y)})f_{X,Y}(x,y)\mu(dx)]\nu(dy)\\ &= \int f_{X,Y}\ln f_{X,Y}d(\mu\otimes\nu) - \int f_X \ln f_X d\mu - \int f_Y \ln f_Y d\nu\\ &= \mathsf{S}_\mu(\mathcal{L}(X)) + \mathsf{S}_\nu(\mathcal{L}(Y)) - \mathsf{S}_{\mu\otimes\nu}(\mathcal{L}(X,Y)),\end{aligned}$$

where S_ν is the Shannon entropy introduced in (1.25). As

$$\mathsf{S}_\nu(\mathcal{L}(Y)) + \mathsf{S}_\mu(\mathcal{L}(X)) = \mathsf{S}_{\nu\otimes\mu}(\mathcal{L}(Y)\otimes\mathcal{L}(X)), \tag{1.125}$$

we can say that $\mathsf{I}_{\mathsf{v}}(Y||X)$ is the reduction of entropy due to the dependence of X and Y.

Example 1.96. Suppose that (X,Y) has a normal distribution $\mathsf{N}(a,\Sigma)$ with expectation $a = (a_1,a_2)$ and covariance matrix

$$\Sigma = \begin{pmatrix} \sigma_1^2 & \rho\sigma_1\sigma_2 \\ \rho\sigma_1\sigma_2 & \sigma_2^2 \end{pmatrix},$$

where we assume $-1 < \rho < 1$ and $\sigma_i^2 > 0$, $i = 1,2$. Then

$$f_{X,Y}(x,y) = (2\pi\sigma_1\sigma_2)^{-1}(1-\rho^2)^{-1/2}\exp\{-\frac{1}{2(1-\rho^2)}\left[s^2 - 2\rho st + t^2\right]\},$$

where $s = (x-a_1)/\sigma_1$ and $t = (y-a_2)/\sigma_2$. Let $\mu = \nu = \boldsymbol{\lambda}$. Then

$$\begin{aligned}\mathsf{S}_{\boldsymbol{\lambda}_2}(\mathsf{N}(a,\Sigma)) &= \ln(2\pi\sigma_1\sigma_2) + \frac{1}{2}\ln(1-\rho^2)\\ &+\frac{1}{2(1-\rho^2)}[\mathbb{E}(\frac{X-a_1}{\sigma_1})^2 - 2\rho\mathbb{E}(\frac{X-a_1}{\sigma_1})(\frac{Y-a_2}{\sigma_2}) + \mathbb{E}(\frac{Y-a_2}{\sigma_2})^2]\\ &= \ln(2\pi\sigma_1\sigma_2) + \frac{1}{2}\ln(1-\rho^2) + 1.\end{aligned}$$

On the other hand, $\mathsf{S}_{\boldsymbol{\lambda}}(\mathsf{N}(a_i,\sigma_i^2)) = \frac{1}{2} + \frac{1}{2}\ln(2\pi\sigma_i^2)$ by (1.27), so that $\mathsf{I}(X||Y) = -\frac{1}{2}\ln(1-\rho^2)$.

The mutual information can also be expressed in terms of the conditional densities. Indeed, (1.124) yields

$$\mathsf{I}_{\mathsf{v}}(X||Y) = \mathsf{I}_{\mathsf{v}}(Y||X) = \int [\int \mathsf{v}(\frac{f_{Y|X}(y|x)}{f_Y(y)}) f_Y(y)\nu(dy)] f_X(x)\mu(dx)$$
$$= \int \mathsf{I}_{\mathsf{v}}(\mathsf{K}(\cdot|x), P_Y) P_X(dx),$$

where $f_{Y|X}(y|x) = f_{Y,X}(x,y)/f_X(x)$ is the conditional density of Y, given $X = x$, and

$$\mathsf{K}(B|x) = \int_B f_{Y|X}(y|x)\nu(dy)$$

is the conditional distribution of Y, given $X = u$. The next proposition shows that a similar statement holds without the existence of densities, provided that regular conditional distributions exist.

Proposition 1.97. *Suppose that $(\mathcal{Y}, \mathfrak{B})$ is a Borel space, and that $\mathsf{K}(\cdot|x) = \mathcal{L}(Y|X = x)$ is a regular conditional distribution. Then*

$$\mathsf{I}_{\mathsf{v}}(Y||X) = \int \mathsf{I}_{\mathsf{v}}(\mathsf{K}(\cdot|x), P_Y) P_X(dx),$$

where P_X and P_Y are the distributions of X and Y, respectively.

Proof. Introduce the kernel L by $\mathsf{L}(B|x) = P_Y(B)$ and apply Proposition 1.95. ■

We consider the model $(\mathcal{X}, \mathfrak{A}, (P_\theta)_{\theta\in\Delta})$ that satisfies the condition (A3) and suppose that Π is a prior on $(\Delta, \mathfrak{B}_\Delta)$ with $\mathcal{L}(X, \Theta) = \mathsf{P} \otimes \Pi$. It is clear that we can make better inferences on Θ based on X if there is a stronger dependence between the random variables. In this sense we can characterize the informativeness of the model $(P_\theta)_{\theta\in\Delta}$ by making the dependence between X and Θ as large as possible. More precisely, for a family of priors $\mathcal{P}$ we set

$$C_{\mathsf{v}}(\mathsf{P}, \mathcal{P}) = \sup_{\Pi\in\mathcal{P}} \mathsf{I}_{\mathsf{v}}(\Theta||X), \quad \mathcal{L}(\Theta) = \Pi, \ \Pi \in \mathcal{P}.$$

For $\mathsf{v}(x) = x \ln x$ the value $C_{\mathsf{v}}(\mathsf{P}, \mathcal{P})$ is called the *channel capacity* in information theory; see Cover and Thomas (1991). It can be shown that $C_{\mathsf{v}}(\mathsf{P}, \mathcal{P})$ is the maximum amount of information that can be transmitted through the channel $\mathsf{P} = (P_\theta)_{\theta\in\Delta}$. In our context we use $C_{\mathsf{v}}(\mathsf{P}, \mathcal{P})$ as a numerical value that characterizes the maximum dependence between X and Θ. We call $\Pi_0 \in \mathcal{P}$ a *most informative prior* if

$$\mathsf{I}_{\mathsf{v}}(\Theta_0||X) = C_{\mathsf{v}}(\mathsf{P}, \mathcal{P}), \quad \mathcal{L}(\Theta_0) = \Pi_0.$$

We recall that by Proposition 1.97

$$\mathsf{I}_{\mathsf{v}}(\Theta||X) = \mathsf{I}_{\mathsf{v}}(\mathsf{P} \otimes \Pi, (\mathsf{P}\Pi) \otimes \Pi) = \int \mathsf{I}_{\mathsf{v}}(P_\theta, \mathsf{P}\Pi)\Pi(d\theta),$$

so that a prior that maximizes $\mathsf{I}_{\mathsf{v}}(X||\Theta)$ guarantees a maximum dependence between X and Θ, i.e., a largest distance between the joint distribution and the product of the marginal distributions. The next theorem shows that the conjugate prior is most informative in Gaussian models.

Theorem 1.98. *For the model* $(\mathcal{X}, \mathfrak{A}, (P_\theta)_{\theta\in\Delta}) = (\mathbb{R}, \mathfrak{B}, (\mathsf{N}(\theta,\sigma^2))_{\theta\in\mathbb{R}})$, *with a known* $\sigma^2 > 0$, *and* $\mathsf{v}(x) = x\ln x$, *the prior* $\Pi_0 = \mathsf{N}(0,\tau^2)$ *is most informative in the class*

$$\mathcal{P}_0 = \{\Pi : \Pi \in \mathcal{P}(\mathfrak{B}_\Delta),\ \Pi \ll \boldsymbol{\lambda},\ \int \theta\Pi(d\theta) = 0,\ \int \theta^2\Pi(d\theta) \le \tau^2\}.$$

The channel capacity is given by $C_{\mathsf{v}}(\mathsf{P},\mathcal{P}) = \frac{1}{2}\ln(1+\tau^2/\sigma^2)$.

Proof. Put $\pi(\theta) = (d\Pi/d\boldsymbol{\lambda})(\theta)$. Then $\mathcal{L}(X,\Theta)$ has a Lebesgue density that is given by $\varphi_{\theta,\sigma^2}(x)\pi(\theta)$. We evaluate the Shannon entropy of $\mathcal{L}(X,\Theta)$.

$$\begin{aligned}
&\mathsf{S}_{\boldsymbol{\lambda}_2}(\mathcal{L}(X,\Theta))\\
&= -\int [\int \varphi_{\theta,\sigma^2}(x)\pi(\theta)\ln(\varphi_{\theta,\sigma^2}(x)\pi(\theta))dx]d\theta\\
&= -\int [\int \varphi_{0,\sigma^2}(x-\theta)\ln(\varphi_{0,\sigma^2}(x-\theta))dx]\pi(\theta)d\theta - \int \pi(\theta)\ln\pi(\theta)d\theta\\
&= -\int \varphi_{0,\sigma^2}(x)\ln(\varphi_{0,\sigma^2}(x))dx - \int \pi(\theta)\ln\pi(\theta)d\theta\\
&= \mathsf{S}_{\boldsymbol{\lambda}}(\mathsf{N}(0,\sigma^2)) + \mathsf{S}_{\boldsymbol{\lambda}}(\Pi).
\end{aligned}$$

Hence by (1.125),

$$\begin{aligned}
\mathsf{I}(X\|\Theta) &= \mathsf{S}_{\boldsymbol{\lambda}}(\mathcal{L}(X)) + \mathsf{S}_{\boldsymbol{\lambda}}(\mathcal{L}(\Theta)) - \mathsf{S}_{\boldsymbol{\lambda}}(\mathsf{N}(0,\sigma^2)) - \mathsf{S}_{\boldsymbol{\lambda}}(\Pi)\\
&= \mathsf{S}_{\boldsymbol{\lambda}}(\mathcal{L}(X)) - \mathsf{S}_{\boldsymbol{\lambda}}(\mathsf{N}(0,\sigma^2)).
\end{aligned}$$

If Z and Θ are independent and $\mathcal{L}(Z) = \mathsf{N}(0,\sigma^2)$, then $\mathcal{L}(X) = \mathcal{L}(Z+\Theta)$. The independence of Z and Θ yields

$$\mathbb{E}(Z+\Theta)^2 = \mathbb{E}Z^2 + \mathbb{E}\Theta^2 = \sigma^2 + \tau^2.$$

We know from Example 1.29 that $\mathsf{N}(0,\sigma_0^2)$ maximizes the Shannon entropy in the class of all distributions with expectation 0 and variance σ_0^2. Hence

$$\mathsf{S}_{\boldsymbol{\lambda}}(\mathcal{L}(Z+\Theta)) \le \mathsf{S}_{\boldsymbol{\lambda}}(\mathsf{N}(0,\sigma^2+\tau^2)),$$

where for $\mathcal{L}(\Theta) = \mathsf{N}(0,\tau^2)$ the equality is attained. Finally,

$$\begin{aligned}
C_{\mathsf{v}}(\mathsf{P},\mathcal{P}_0) &= \max_{\mathcal{L}(\Theta)\in\mathcal{P}_0} \mathsf{I}(X\|\Theta)\\
&= \mathsf{S}_{\boldsymbol{\lambda}}(\mathsf{N}(0,\sigma^2+\tau^2)) - \mathsf{S}_{\boldsymbol{\lambda}}(\mathsf{N}(0,\sigma^2)) = \frac{1}{2}\ln(1+\tau^2/\sigma^2).
\end{aligned}$$

■

We recall the concept of Markov dependence and a Markov chain. Suppose U, V, W are random variables on $(\Omega,\mathfrak{F},\mathbb{P})$ which take values in $(\mathcal{U},\mathfrak{U})$, $(\mathcal{V},\mathfrak{V})$, and $(\mathcal{W},\mathfrak{W})$, respectively, which we assume to be Borel spaces. Let $\mathsf{K}(\cdot|u)$ be a regular conditional distribution of V given $U = u$, and $\mathsf{L}(\cdot|v)$ a regular conditional distribution of W given $V = v$. Denote by $P_{(U,V,W)} = \mathbb{P} \circ (U, V, W)^{-1}$ and $P_{(U,V)} = \mathbb{P} \circ (U, V)^{-1}$ the joint distributions of (U, V, W) and (U, V), respectively, and let $P_U = \mathbb{P} \circ U^{-1}$ be the marginal distribution of U. The sequence U, V, W is called a *Markov chain* if

$$P_{(W,V,U)} = \mathsf{L} \otimes (\mathsf{K} \otimes P_U). \tag{1.126}$$

Equivalently, we can say that U, V, W is a Markov chain if

$$\mathcal{L}(W|U = u, V = v) = \mathcal{L}(W|V = v), \quad P_{(U,V)}\text{-a.s.}$$

This means that the conditional distribution of W depends on the "past" U, V only through V. In general, any sequence of random variables $X_0, X_1, ...$ is called a Markov chain if for every $n \geq 1$ the random variables $U = (X_0, ...X_{n-1})$, $V = X_n$, and $W = (X_{n+1}, X_{n+2}, ...)$ form a Markov chain. The interpretation is that the future behavior depends only on the presence V, and for a fixed presence the strict past does not have any influence.

Problem 1.99.* The following statements are equivalent.

- (A) U, V, W is a Markov chain.
- (B) W, V, U is a Markov chain.
- (C) U and W are conditionally independent in the sense $\mathcal{L}((U, W)|V = v) = \mathcal{L}(U|V = v) \otimes \mathcal{L}(W|V = v)$, $\mathcal{L}(V)$-a.s.

Problem 1.100. If U, V, W is a Markov chain, $\mathsf{K}(\cdot|u)$ a regular conditional distribution of V given $U = u$, and $\mathsf{L}(\cdot|v)$ a regular conditional distribution of W given $V = v$, then

$$\mathsf{M}(B|u) := (\mathsf{LK})(B|u) = \int \mathsf{L}(B|v)\mathsf{K}(dv|u)$$

is a regular conditional distribution W, given $U = u$. The marginal distributions P_U, P_V, and P_W are related by

$$P_V(A) = (\mathsf{K}P_U)(A) := \int \mathsf{K}(A|u)P_U(du), \quad A \in \mathfrak{V}$$

$$P_W(B) = (\mathsf{L}P_V)(B) := \int \mathsf{L}(B|v)P_V(dv), \quad B \in \mathfrak{U}.$$

In a Markov chain the influence of the initial distributions P_U becomes weaker in the future. This result is known as data processing inequality in information theory; see Cover and Thomas (1991).

Proposition 1.101. *If U, V, W is a Markov chain, then*

$$\mathsf{I_v}(W||U) \leq \mathsf{I_v}(V||U).$$

Proof. It follows from Proposition 1.97 and Problem 1.100 that

$$\mathsf{I_v}(W||U) = \int \mathsf{I_v}(\mathsf{M}(\cdot|u), P_W)P_U(du) = \int \mathsf{I_v}(\mathsf{LK}(\cdot|u), \mathsf{L}P_V)P_U(du).$$

Now we apply Theorem 1.70 to get

$$\mathsf{I}_{\mathsf{v}}(\mathsf{L}\mathsf{K}(\cdot|u), \mathsf{L}P_V) \leq \mathsf{I}_{\mathsf{v}}(\mathsf{K}(\cdot|u), P_V)$$

for every u. Hence by Proposition 1.97 again

$$\mathsf{I}_{\mathsf{v}}(W||U) \leq \int \mathsf{I}_{\mathsf{v}}(\mathsf{K}(\cdot|u), P_V) P_U(du) = \mathsf{I}_{\mathsf{v}}(V||U).$$

■

In the previous part of this section we have used the information-theoretic approach to find priors which maximize the dependence between X and Θ. Now we study the structure of hierarchical Bayes models. Suppose that $(\Upsilon, \mathfrak{B}_\Upsilon)$ is another measurable space, and that Π_ξ, $\xi \in \Upsilon$, is a family of prior distributions on $\mathfrak{B}_\Delta$ so that the mapping $\xi \mapsto \Pi_\xi(B)$ is $\mathfrak{B}_\Upsilon$-$\mathfrak{B}$ measurable for every $B \in \mathfrak{B}_\Delta$. Hence Π_ξ defines a stochastic kernel $\mathsf{\Pi} : \mathfrak{B}_\Delta \times \Upsilon \to_k [0,1]$. Let now Γ be a distribution on $(\Upsilon, \mathfrak{B}_\Upsilon)$. By a *hierarchical Bayes model* we mean a random vector (Ξ, Θ, X) for which $\mathcal{L}(X, \Theta, \Xi) = \mathsf{P} \otimes \mathsf{\Pi} \otimes \Gamma$, or equivalently, for every $h : \mathcal{X} \times \Delta \times \Upsilon \to_m \mathbb{R}_+$,

$$\mathbb{E}h(X, \Theta, \Xi) = \int [\int [\int h(x, \theta, \xi) P_\theta(dx)] \Pi_\xi(d\theta)] \Gamma(d\xi). \tag{1.127}$$

The random variable Θ is a random version of the parameter θ, and the random variable Ξ is a random version of the *hyperparameter* ξ. Suppose now that $(\mathcal{X}, \mathfrak{A})$, $(\Upsilon, \mathfrak{B}_\Upsilon)$, and $(\Delta, \mathfrak{B}_\Delta)$ are Borel spaces. By construction the conditional distribution of X, given $\Theta = \theta$ and $\Xi = \xi$, depends only on θ so that (Ξ, Θ, X) is a Markov chain in the sense of (1.126). Let $\mathsf{K} = \mathcal{L}(\Theta|X = x)$ and $\mathsf{L} = \mathcal{L}(\Xi|\Theta = \theta)$ be regular conditional distributions. Then by (1.127),

$$\mathcal{L}((X, \Xi)|\Theta = \theta) = P_\theta \otimes \mathsf{L}(\cdot|\theta),$$

so that X and Ξ are conditionally independent, given Θ.

The idea of a hyperparameter arises in view of the fact that the choice of the prior affects the inference in Bayes analysis. Often it is more appropriate to make the choice flexible and to fix the prior only in a second step of the Bayes hierarchy. The fact that this second step (i.e., the choice of the hyperparameter) has less influence on the inference can be made mathematically rigorous by comparing the mutual information of Θ and X with that of Ξ and X. The next result is a direct consequence of Proposition 1.101.

Proposition 1.102. *If* $(\mathcal{X}, \mathfrak{A})$, $(\Upsilon, \mathfrak{B}_\Upsilon)$, *and* $(\Delta, \mathfrak{B}_\Delta)$ *are Borel spaces, then*

$$\mathsf{I}_{\mathsf{v}}(\Xi||X) \leq \mathsf{I}_{\mathsf{v}}(\Theta||X). \tag{1.128}$$

The inequality (1.128) shows that the random hyperparameter Ξ has less influence on the inference based on X than Θ has. One special case, for the convex function $\mathsf{v}(x) = x \ln x$, appears in Lehmann and Casella (1998), and another one, for Hellinger integrals, in Goel and DeGroot (1981).

1.5 $\mathbb{L}_2$-Differentiability, Fisher Information

In this section we introduce a differentiability concept for parametrized statistical models. Our starting point is the statistical model $(\mathcal{X}, \mathfrak{A}, (P_\theta)_{\theta\in\Delta})$, $\Delta \subseteq \mathbb{R}^d$. We recall from Definition 1.57 that for $\theta_0, \theta \in \Delta$ any measurable function $L_{\theta_0,\theta}$ with values in $[0,\infty]$ is called the likelihood ratio of P_θ with respect to P_{θ_0} if

$$P_\theta(A) = \int_A L_{\theta_0,\theta} dP_{\theta_0} + P_\theta(A \cap \{L_{\theta_0,\theta} = \infty\}).$$

It holds $P_\theta(L_{\theta_0,\theta} = \infty) = 1 - \mathsf{E}_{\theta_0} L_{\theta_0,\theta}$, and $L_{\theta_0,\theta}$ is a probability density of P_θ with respect to P_{θ_0} if and only if $P_\theta \ll P_{\theta_0}$. If $\mu \in \mathcal{M}^\sigma(\mathfrak{A})$ dominates the subfamily $\{P_{\theta_0}, P_\theta\}$ with corresponding densities f_{θ_0}, f_θ, then

$$L_{\theta_0,\theta} = (f_\theta/f_{\theta_0}) I_{\{f_{\theta_0}>0\}} + \infty I_{\{f_{\theta_0}=0, f_\theta>0\}}$$

is the likelihood ratio which is $\{P_{\theta_0}, P_\theta\}$-a.s. uniquely determined; see Lemma 1.58. In the following $\theta_0 \in \Delta^0$ remains fixed, and we introduce the notation $L_{\theta_0}(u) := L_{\theta_0,\theta_0+u}$, for $\theta_0+u \in \Delta$. To introduce a differentiability concept, the first idea would be to require that there be a vector-valued measurable function $\dot{L}_{\theta_0}$ such that for any $u \to 0$ the remainder $R(u) = L_{\theta_0}(u) - 1 - \langle u, \dot{L}_{\theta_0}\rangle$ satisfies $R(u) = o_{P_{\theta_0}}(\|u\|)$. In this case $\dot{L}_{\theta_0}$ would be considered as a gradient where the degree of approximation is specified by stochastic convergence. As it turns out, however, this type of approximation is too weak for many purposes. This is mainly due to the fact that stochastic convergence does not imply convergence in the squared mean, or more generally in the rth mean for $r \geq 1$. A stronger concept could require that $R(u)$ tends to zero in the sense of $\mathbb{L}_r(P_{\theta_0})$, $r \geq 1$. However, this approach has the shortcoming that for $r > 1$, and especially for the important case of $r = 2$, additional moment assumptions become necessary as not every probability density is squared integrable. The following simple fact shows how to get out of this dilemma. If $t \to 1$, then $r(t^{1/r} - 1)/(t-1) \to 1$. Thus $r(L_{\theta_0}^{1/r}(u) - 1)$ has the same local behavior as $L_{\theta_0}(u) - 1$, but in addition $r(L_{\theta_0}^{1/r}(u) - 1)$ has a finite moment of order r. This is the basic idea of the concept of $\mathbb{L}_r$-differentiability. The special case of $r = 2$ is of utmost importance as $\mathbb{L}_2(P_{\theta_0})$ is a Hilbert space, and thus we utilize primarily this fact in the sequel. Set

$$\mathbb{L}_{2,d}(P_{\theta_0}) = \{T : \quad T : \mathcal{X} \to_m \mathbb{R}^d, \;\; \mathsf{E}_{\theta_0} \|T\|^2 < \infty\}. \tag{1.129}$$

Definition 1.103. *The family $(P_\theta)_{\theta\in\Delta}$ is called $\mathbb{L}_2$-differentiable at $\theta_0 \in \Delta^0$ if there exists a neighborhood $U(\theta_0)$ of θ_0 such that*

$$P_\theta \ll P_{\theta_0}, \quad \theta \in U(\theta_0), \tag{1.130}$$

and there exists some $\dot{L}_{\theta_0} = (\dot{L}_{\theta_0,1}, ..., \dot{L}_{\theta_0,d})^T \in \mathbb{L}_{2,d}(P_{\theta_0})$, called the $\mathbb{L}_2$-derivative (of the model) at θ_0, such that as $u \to 0$,

$$\mathsf{E}_{\theta_0}([L_{\theta_0}^{1/2}(u) - 1] - \frac{1}{2}\langle \dot{L}_{\theta_0}, u\rangle)^2 = o(\|u\|^2) \tag{1.131}$$

The matrix $\mathsf{I}(\theta_0) = \mathsf{E}_{\theta_0}\dot{L}_{\theta_0}\dot{L}_{\theta_0}^T$ *is called the Fisher information matrix.*

For detailed discussions and results related to $\mathbb{L}_2$-differentiability and other differentiability concepts and approaches we refer to Strasser (1985), LeCam (1986), Witting (1985), Pfanzagl and Wefelmeyer (1985), Bickel, Klaassen, Ritov, and Wellner (1993), and Janssen (1998).

Remark 1.104. Often, instead of (1.130), it is only required that the total mass of the part of P_{θ_0+u} that is singular to P_{θ_0} satisfies $P_{\theta_0+u}(L_{\theta_0}(u) = \infty) = o(\|u\|^2)$. It is easy to see that in this case there are distributions $\widetilde{P}_\theta \ll P_{\theta_0}$, $\theta \in U(\theta_0)$, with $\mathsf{D}^2(P_{\theta_0+u}, \widetilde{P}_{\theta_0+u}) = o(\|u\|^2)$. Then $(\widetilde{P}_\theta)_{\theta\in U(\theta_0)}$ is again $\mathbb{L}_2$-differentiable and has the same $\mathbb{L}_2$-derivative. This is the reason why we directly require (1.130).

Typically, the distributions P_θ, $\theta \in \Delta$, are defined by a family of densities f_θ, $\theta \in \Delta$, with respect to some σ-finite measure μ. Then it proves useful to deal directly with the densities.

Problem 1.105.* If (1.130) is satisfied, then $(P_\theta)_{\theta\in\Delta}$ is $\mathbb{L}_2$-differentiable at θ_0 if and only if there is some $\dot{f}_{\theta_0} \in \mathbb{L}_{2,d}(\mu)$, defined analogously to (1.129), with

$$\int (f_{\theta_0+u}^{1/2} - f_{\theta_0}^{1/2} - \frac{1}{2}\langle u, \dot{f}_{\theta_0}\rangle)^2 d\mu = o(\|u\|^2).$$

In this case it holds $\dot{L}_{\theta_0} = \dot{f}_{\theta_0}/f_{\theta_0}^{1/2}$, P_{θ_0}-a.s.

For any sequence of random variables a convergence in $\mathbb{L}_2(P_{\theta_0})$ implies the P_{θ_0}-stochastic convergence. For the reverse direction an additional condition is needed that makes the sequence uniformly squared integrable. By Vitali's theorem (see Theorem A.21) this condition is equivalent to the norm convergence. This splitting of the $\mathbb{L}_2(P_{\theta_0})$-convergence is sometimes useful and is more or less the content of the next lemma.

Lemma 1.106. *If* $P_\theta \ll P_{\theta_0}$, $\theta \in U(\theta_0)$, *is satisfied, then the family* $(P_\theta)_{\theta\in\Delta}$ *is* $\mathbb{L}_2$*-differentiable at* $\theta_0 \in \Delta^0$ *with* $\mathbb{L}_2$*-derivative* $\dot{L}_{\theta_0}$ *if and only if the following two conditions are met.*

$$L_{\theta_0}(u) - 1 = \langle u, \dot{L}_{\theta_0}\rangle + o_{P_{\theta_0}}(\|u\|) \tag{1.132}$$

$$\mathsf{E}_{\theta_0}(L_{\theta_0}^{1/2}(u) - 1)^2 = \frac{1}{4}u^T\mathsf{I}(\theta_0)u + o(\|u\|^2). \tag{1.133}$$

Under the assumption (1.132) *the condition* (1.133) *is equivalent to the uniform integrability of* $\|u\|^{-2}\,(L_{\theta_0}^{1/2}(u) - 1)^2$ *as* $u \to 0$.

Proof. The expansion $\sqrt{1+x} = 1 + \frac{1}{2}x + o(x)$, as $x \to 0$, shows that (1.132) is equivalent to

$$L_{\theta_0}^{1/2}(u) - 1 = \frac{1}{2}\langle u, \dot{L}_{\theta_0}\rangle + o_{P_{\theta_0}}(\|u\|).$$

The rest follows from Vitali's theorem; see Theorem A.21. ■

The norm given by $\mathsf{E}_{\theta_0}(L_{\theta_0}^{1/2}(u) - 1)^2$ is closely related to the Hellinger distance. Indeed, (1.110) and $P_\theta \ll P_{\theta_0}$ yield

$$\mathsf{E}_{\theta_0}(L_{\theta_0}^{1/2}(u) - 1)^2 = \mathsf{D}^2(P_{\theta_0+u}, P_{\theta_0}) = \mathsf{D}^2(P_{\theta_0}, P_{\theta_0+u}).$$

From (1.133) we obtain

$$\begin{aligned} \mathsf{D}^2(P_{\theta_0+u}, P_{\theta_0}) &= \frac{1}{4}u^T\mathsf{I}(\theta_0)u + o(\|u\|^2) \\ \mathsf{H}_{1/2}(P_{\theta_0}, P_{\theta_0+u}) &= 1 - \frac{1}{8}u^T\mathsf{I}(\theta_0)u + o(\|u\|^2), \quad \text{as } u \to 0. \end{aligned} \tag{1.134}$$

Similar expansions may be obtained for $\mathsf{H}_s(P_{\theta_0}, P_{\theta_0+u})$, $0 < s < 1$.

Problem 1.107.* If the family $(P_\theta)_{\theta\in\Delta}$ is $\mathbb{L}_2$-differentiable at $\theta_0 \in \Delta^0$, then

$$\mathsf{H}_s(P_{\theta_0}, P_{\theta_0+u}) = 1 - \frac{1}{2}s(1-s)u^T\mathsf{I}(\theta_0)u + o(\|u\|^2), \quad \text{as } u \to 0.$$

A more general result can be established for the functionals $\mathsf{I}_\mathsf{v}(P_0, P_1)$ if the function v tends to infinity only moderately.

Problem 1.108.* If the convex function v is twice continuously differentiable and satisfies $\mathsf{v}(0) + \mathsf{v}^*(\infty) < \infty$, then

$$\mathsf{I}_\mathsf{v}(P_{\theta_0+u}, P_{\theta_0}) - \mathsf{v}(1) = \frac{1}{2}\mathsf{v}''(1)u^T\mathsf{I}(\theta_0)u + o(\|u\|^2).$$

The results of the last two problems show that the local distance of distributions is determined by the Fisher information. The question arises as to whether one can use this fact to consider differentiable models as a differentiable manifold for which all geometric properties are expressed in terms of the Fisher information. This was the idea of Amari (1985) and other authors. For details we refer to Amari (1985).

The following technical result is useful for the next propositions.

Problem 1.109.* If X, X_n, Y_n, $n = 1, 2, ...$, are random variables with $\mathbb{E}X^2 < \infty$, $\mathbb{E}(X_n - X)^2 \to 0$, and $\sup_n \mathbb{E}Y_n^2 < \infty$, then the sequence X_nY_n, $n = 1, 2, ...$, is uniformly integrable.

Next we show that the $\mathbb{L}_2$-differentiability implies the $\mathbb{L}_1$-differentiability specified in the next proposition. Furthermore it is shown that the expectation of $\dot{L}_{\theta_0}$ at θ_0 is zero.

Proposition 1.110. *If the family $(P_\theta)_{\theta\in\Delta}$ is $\mathbb{L}_2$-differentiable at $\theta_0 \in \Delta^0$ with $\mathbb{L}_2$-derivative $\dot{L}_{\theta_0}$, then*

$$\mathsf{E}_{\theta_0}|(L_{\theta_0}(u) - 1) - \langle u, \dot{L}_{\theta_0}\rangle| = o(\|u\|) \quad \textit{and} \quad \mathsf{E}_{\theta_0}\dot{L}_{\theta_0} = 0.$$

Proof. The first statement is equivalent to

$$\lim_{n\to\infty} \|u_n\|^{-1} \mathsf{E}_{\theta_0}|(L_{\theta_0}(u_n) - 1) - \langle u_n, \dot{L}_{\theta_0}\rangle| = 0$$

for every sequence $u_n \to 0$ with $\|u_n\|^{-1} u_n \to a$. The relation (1.132) implies

$$Z_n := \|u_n\|^{-1} ((L_{\theta_0}(u_n) - 1) - \langle u_n, \dot{L}_{\theta_0}\rangle) = o_{P_{\theta_0}}(1).$$

To prove the first statement it remains to show that the sequence Z_n is uniformly integrable. The uniform integrability of $\|u_n\|^{-1} \langle u_n, \dot{L}_{\theta_0}\rangle$ follows from $|\,\|u_n\|^{-1} \langle u_n, \dot{L}_{\theta_0}\rangle| \leq\| \dot{L}_{\theta_0} \|$ and $\mathsf{E}_{\theta_0} \| \dot{L}_{\theta_0} \|^2 < \infty$. Set $X = \frac{1}{2}\langle a, \dot{L}_{\theta_0}\rangle$, $X_n = \|u_n\|^{-1} (L_{\theta_0}^{1/2}(u_n) - 1)$, and $Y_n = (L_{\theta_0}^{1/2}(u_n) + 1)$. Then the uniform integrability of Z_n follows from Problem 1.109. By condition (1.130) $\mathsf{E}_{\theta_0}(L_{\theta_0}(u_n) - 1) = 0$ for all sufficiently large n. Hence $\mathsf{E}_{\theta_0}\langle a, \dot{L}_{\theta_0}\rangle = 0$. By using the sequence $u_n = \varepsilon_n \mathsf{E}_{\theta_0}\dot{L}_{\theta_0}$, $\varepsilon_n \to 0$, we get $\mathsf{E}_{\theta_0}\langle \mathsf{E}_{\theta_0}\dot{L}_{\theta_0}, \dot{L}_{\theta_0}\rangle = \| \mathsf{E}_{\theta_0}\dot{L}_{\theta_0} \|^2 = 0$. ■

For exponential families we have shown in Theorem 1.17 that the function $\theta \mapsto \mathsf{E}_\theta T$ is differentiable and that the derivation can be carried out under the integral sign. A similar statement holds for $\mathbb{L}_2$-differentiable families.

Proposition 1.111. *If the family $(P_\theta)_{\theta\in\Delta}$ is $\mathbb{L}_2$-differentiable at $\theta_0 \in \Delta^0$ with $\mathbb{L}_2$-derivative $\dot{L}_{\theta_0}$, $T : \mathcal{X} \to_m \mathbb{R}$, and $\sup_{\theta\in U(\theta_0)} \mathsf{E}_\theta T^2 < \infty$, then $g(\theta) = \mathsf{E}_\theta T$ is differentiable at θ_0 and it holds*

$$\nabla g(\theta_0) = \mathsf{E}_{\theta_0} T\dot{L}_{\theta_0}.$$

Proof. Fix a sequence $u_n \to 0$ with $\|u_n\|^{-1} u_n \to a$. Then for all sufficiently large n

$$\begin{aligned}&\|u_n\|^{-1} [g(\theta_0 + u_n) - g(\theta_0)] - \mathsf{E}_{\theta_0} T\langle a, \dot{L}_{\theta_0}\rangle \\ &= \mathsf{E}_{\theta_0}(\|u_n\|^{-1} [L_{\theta_0}(u_n) - 1]T - T\langle a, \dot{L}_{\theta_0}\rangle).\end{aligned}$$

As the sequence under the expectation is $o_{P_{\theta_0}}(1)$, in view of (1.132), we have only to show the uniform integrability of this sequence. To apply the result of Problem 1.109 we write

$$\|u_n\|^{-1} [L_{\theta_0}(u_n) - 1]T = [\|u_n\|^{-1} (L_{\theta_0}^{1/2}(u_n) - 1)][(L_{\theta_0}^{1/2}(u_n) + 1)T] = X_n Y_n$$

and set $X = \langle a, \dot{L}_{\theta_0}\rangle$. The $\mathbb{L}_2$-differentiability of $(P_\theta)_{\theta\in\Delta}$ at $\theta_0 \in \Delta^0$ yields the $\mathbb{L}_2$-convergence of X_n to X. The moment condition on Y_n is obtained from

$$\mathsf{E}_{\theta_0} Y_n^2 \leq 2\mathsf{E}_{\theta_0}(L_{\theta_0}(u_n) + 1)T^2 \leq 2 \sup_{\theta\in U(\theta_0)} \mathsf{E}_\theta T^2$$

for all sufficiently large n. ■

Sometimes a reparametrization of the family $(P_\theta)_{\theta\in\Delta}$ is useful. Suppose $\theta_0 \in \Delta^0$ and $U(\theta_0) \subseteq \Delta^0$ is an open neighborhood of θ_0. Let $\Lambda \subseteq \mathbb{R}^k$ be an open set and let $\kappa = (\kappa_1, ..., \kappa_d)^T : \Lambda \to U(\theta_0)$ be differentiable. Then with the Jacobian

$$J_\kappa(\eta) = \left(\frac{\partial \kappa_i}{\partial \eta_j}(\eta)\right)_{i=1,...,d,\ j=1,...,k} \tag{1.135}$$

the first-order Taylor expansions of $\kappa(\eta)$ can be written as

$$\kappa(\eta + w) - \kappa(\eta) = J_\kappa(\eta)w + o(\|w\|), \quad \eta \in \Lambda,\ w \in \mathbb{R}^k. \tag{1.136}$$

Proposition 1.112. *Let κ be differentiable at $\eta_0 \in \Lambda$. If $(P_\theta)_{\theta\in\Delta}$ is $\mathbb{L}_2$-differentiable at $\theta_0 = \kappa(\eta_0) \in \Delta^0$ with $\mathbb{L}_2$-derivative $\dot{L}_{\theta_0}$ and Fisher information matrix $\mathsf{I}(\theta_0)$, then $(P_{\kappa(\eta)})_{\eta\in\Lambda}$ is $\mathbb{L}_2$-differentiable at η_0 with $\mathbb{L}_2$-derivative $J_\kappa^T(\eta_0)\dot{L}_{\kappa(\eta_0)}$ and Fisher information matrix $J_\kappa^T(\eta_0)\mathsf{I}(\kappa(\eta_0))J_\kappa(\eta_0)$.*

Proof. For $\eta \to \eta_0$ and $u(\eta) = \kappa(\eta) - \kappa(\eta_0)$ it follows from (1.131) that

$$\mathsf{E}_{\theta_0}[(L^{1/2}_{\kappa(\eta_0)}(\kappa(\eta)) - 1) - \frac{1}{2}\langle\kappa(\eta), \dot{L}_{\kappa(\eta_0)}\rangle]^2 = o(\|\kappa(\eta)\|^2).$$

Hence with (1.136) and $\mathsf{E}_{\theta_0} \parallel \dot{L}_{\kappa(\eta_0)} \parallel^2 < \infty$,

$$\mathsf{E}_{\theta_0}[(L^{1/2}_{\kappa(\eta_0)}(\kappa(\eta)) - 1) - \frac{1}{2}(J_\kappa^T(\eta_0)\dot{L}_{\kappa(\eta_0)})^T(\eta - \eta_0)]^2 = o(\|\eta - \eta_0\|)^2.$$

■

The $\mathbb{L}_2$-differentiability is preserved if we turn to a finite number of independent observations.

Problem 1.113.* If $(\mathcal{X}_i, \mathfrak{A}_i, (P_{i,\theta})_{\theta\in\Delta})$ is $\mathbb{L}_2$-differentiable at $\theta_0 \in \Delta^0$ with derivatives $\dot{L}_{i,\theta_0}$ and Fisher information matrices $\mathsf{I}_i(\theta_0)$, $i = 1, ..., n$, then

$$(\times_{i=1}^n \mathcal{X}_i, \bigotimes_{i=1}^n \mathfrak{A}_i, (\bigotimes_{i=1}^n P_{i,\theta})_{\theta\in\Delta})$$

is $\mathbb{L}_2$-differentiable with derivative $\dot{L}_{\otimes n,\theta_0}(x_1, ..., x_n) = \sum_{i=1}^n \dot{L}_{i,\theta_0}(x_i)$ and Fisher information matrix $\mathsf{I}_{\otimes n}(\theta_0) = \sum_{i=1}^n \mathsf{I}_i(\theta_0)$.

In many situations one is confronted with the following problem. Suppose the model $(P_\theta)_{\theta\in\Delta}$ is $\mathbb{L}_2$-differentiable and T is some statistic which takes values in $\mathcal{Y}$, where $(\mathcal{Y}, \mathfrak{B})$ is another measurable space. Then the question arises if the reduced model $(Q_\theta)_{\theta\in\Delta} = (P_\theta \circ T^{-1})_{\theta\in\Delta}$ is again $\mathbb{L}_2$-differentiable, and if so, how the new $\mathbb{L}_2$-derivative is related to the previous one.

Theorem 1.114. *Assume that the family $(P_\theta)_{\theta\in\Delta}$ is $\mathbb{L}_2$-differentiable at $\theta_0 \in \Delta^0$ with $\mathbb{L}_2$-derivative $\dot{L}_{\theta_0}$ and Fisher information matrix $\mathsf{I}(\theta_0)$. Then for any statistic T the family $(Q_\theta)_{\theta\in\Delta} = (P_\theta \circ T^{-1})_{\theta\in\Delta}$ is again $\mathbb{L}_2$-differentiable at $\theta_0 \in \Delta^0$ with $\mathbb{L}_2$-derivative and Fisher information matrix, respectively,*

$$\dot{L}_T, \theta_0(t) = \mathsf{E}_{\theta_0}(\dot{L}_{\theta_0}|T = t), \quad Q_{\theta_0}\text{-a.s.},$$
$$\mathsf{I}_T(\theta_0) = \mathsf{E}_{\theta_0}([\mathsf{E}_{\theta_0}(\dot{L}_{\theta_0}|T)][\mathsf{E}_{\theta_0}(\dot{L}_{\theta_0}|T)]^T).$$

Corollary 1.115. *The Fisher information matrix of the reduced model is not larger, in the Löwner semiorder, than the Fisher information matrix of the original model, i.e.,* $\mathsf{I}_T(\theta_0)$ *satisfies* $u^T[\mathsf{I}(\theta_0) - \mathsf{I}_T(\theta_0)]u \geq 0$ *for every* $u \in \mathbb{R}^d$.

Proof. It is clear that condition (1.130) implies $Q_\theta \ll Q_{\theta_0}$, $\theta \in U(\theta_0)$. The relation (1.99) gives $dQ_\theta/dQ_{\theta_0} = \mathsf{E}_{\theta_0}(L_{\theta_0}(u)|T)$. Proposition 1.110 gives for $u \to 0$,

$$\mathsf{E}_{\theta_0}|[\mathsf{E}_{\theta_0}(L_{\theta_0}(u)|T)-1]-\langle\mathsf{E}_{\theta_0}(\dot{L}_{\theta_0}|T), u\rangle| \leq \mathsf{E}_{\theta_0}|[L_{\theta_0}(u)-1]-\langle\dot{L}_{\theta_0}, u\rangle| = o(\|u\|),$$

so that

$$\mathsf{E}_{\theta_0}(L_{\theta_0}(u)|T) - 1 - \langle\mathsf{E}_{\theta_0}(\dot{L}_{\theta_0}|T), u\rangle = o_{P_{\theta_0}}(\|u\|).$$

For uniformly integrable sequences of random variables we may carry out the limit under the expectation. Therefore it remains to show that

$$\|u\|^{-2}\left[(\mathsf{E}_{\theta_0}(L_{\theta_0}(u)|T))^{1/2} - 1\right]^2$$

is uniformly integrable for $u \to 0$. From $\mathsf{E}_{\theta_0}(L_{\theta_0}^{1/2}(u)|T) \leq (\mathsf{E}_{\theta_0}(L_{\theta_0}(u)|T))^{1/2}$ we get

$$\begin{aligned}
&\|u\|^{-2}\left[(\mathsf{E}_{\theta_0}(L_{\theta_0}(u)|T))^{1/2} - 1\right]^2 \\
&= \|u\|^{-2}\left[\mathsf{E}_{\theta_0}(L_{\theta_0}(u)|T) - 2(\mathsf{E}_{\theta_0}(L_{\theta_0}(u)|T))^{1/2} + 1\right] \\
&\leq \|u\|^{-2}\left[\mathsf{E}_{\theta_0}(L_{\theta_0}(u) - 2L_{\theta_0}^{1/2}(u) + 1)|T)\right] \\
&= \mathsf{E}_{\theta_0}(\|u\|^{-2}\left[L_{\theta_0}^{1/2}(u) - 1\right]^2|T).
\end{aligned}$$

Therefore we get from Lemma A.32, that the terms on the right-hand side are uniformly integrable as $u \to 0$. To conclude the proof we apply Lemma 1.106. ■

The following result provides a technical tool for proving the next theorem. The subsequent general formulation is taken from Srivastava (1998).

Problem 1.116.* Let $(\mathcal{X}, \mathfrak{A})$ be a measurable space and $\mathcal{S}, \mathcal{T}$ be metric spaces with the corresponding σ-algebras of Borel sets $\mathfrak{S}$ and $\mathfrak{T}$. Suppose that $\mathcal{S}$ is separable. Suppose that $\psi : \mathcal{S} \times \mathcal{X} \to \mathcal{T}$ has the following properties. $\psi(s, \cdot)$ is $\mathfrak{A}$-$\mathfrak{T}$ measurable for every s from a dense subset of $\mathcal{S}$ and $\psi(\cdot, x)$ is continuous for every $x \in \mathcal{X}$. Then ψ is $(\mathfrak{S} \otimes \mathfrak{A})$ -$\mathfrak{T}$ measurable.

The differentiability concept discussed above refers to the $\mathbb{L}_2$-convergence. Quite often the densities $f_\theta(x)$ are differentiable with respect to the parameter θ in the common sense for almost all x. Therefore criteria that link the usual differentiability with the $\mathbb{L}_2$-differentiability turn out to be useful whenever local approximations for a given model are desired. We use the notation $\nabla = (\partial/\partial\theta_1, ..., \partial/\partial\theta_d)^T$ and formulate suitable conditions that imply the $\mathbb{L}_2$-differentiability.

(A6) Given the model $\mathcal{M} = (\mathcal{X}, \mathfrak{A}, (P_\theta)_{\theta\in\Delta})$ we require that

(A) $(P_\theta)_{\theta\in\Delta}$, $\Delta \subseteq \mathbb{R}^d$, is dominated by some $\mu \in \mathfrak{M}^\sigma(\mathfrak{A})$.

(B) An $A \in \mathfrak{A}$ exists with $\mu(\mathcal{X}\backslash A) = 0$, $f_\theta(x) = \dfrac{dP_\theta}{d\mu}(x) > 0$, $x \in A$, $\theta \in \Delta$.

(C) The function $\theta \mapsto f_\theta(x)$ is differentiable
and $\nabla f_\theta(x)$ is continuous in Δ^0, $x \in A$.

The following criterion goes back to Hájek (1972); see also Strasser (1985) and Witting (1985).

Theorem 1.117. *Suppose the condition (A6) is fulfilled. Assume that there is an open neighborhood $U(\theta_0)$ of θ_0 such that*

$$\int \|\nabla \ln f_\theta\|^2 f_\theta d\mu < \infty, \quad \theta \in U(\theta_0), \quad \textit{and} \tag{1.137}$$

$$\theta \mapsto \int (\nabla \ln f_\theta)(\nabla \ln f_\theta)^T f_\theta d\mu \quad \textit{is continuous in } U(\theta_0). \tag{1.138}$$

Then $(P_\theta)_{\theta\in\Delta}$ is $\mathbb{L}_2$-differentiable at θ_0 with $\mathbb{L}_2$-derivative $\dot{L}_{\theta_0} = \nabla \ln f_{\theta_0}$ and Fisher information

$$\mathsf{I}(\theta_0) = \int (\nabla \ln f_{\theta_0})(\nabla \ln f_{\theta_0})^T f_{\theta_0} d\mu.$$

Proof. Put $\dot{L}(u) := \nabla \ln f_{\theta_0+u}$. Then for $u \to 0$ by assumption (1.138),

$$\mathsf{E}_{\theta_0} \parallel \dot{L}(u) L_{\theta_0}^{1/2}(u) \parallel^2 = \mathsf{E}_{\theta_0+u} \parallel \dot{L}(u) \parallel^2 \to \mathsf{E}_{\theta_0} \parallel \dot{L}(0) \parallel^2 .$$

A componentwise application of Vitali's theorem (see Theorem A.21) and the continuity of ∇f_θ yields $\lim_{u\to 0} \mathsf{E}_{\theta_0} \parallel \dot{L}(u) L_{\theta_0}^{1/2}(u) - \dot{L}(0) \parallel^2 = 0$ and therefore

$$\lim_{\delta\to 0} \sup_{\|u\|\le\delta} \mathsf{E}_{\theta_0} \parallel \dot{L}(u) L_{\theta_0}^{1/2}(u) - \dot{L}(0) \parallel^2 = 0. \tag{1.139}$$

As

$$\frac{d}{ds} L_{\theta_0}^{1/2}(su) = f_{\theta_0}^{-1/2} \frac{d}{ds} f_{\theta_0+su}^{1/2} = \frac{1}{2} L_{\theta_0}^{1/2}(su)\langle u, \dot{L}(su)\rangle$$

is continuous in s for every $x \in A$ we get from Problem 1.116 that the right-hand term is a measurable function of (s, x). Hence

$$\begin{aligned}
\mathsf{E}_{\theta_0}[L_{\theta_0}^{1/2}(u) - 1 - \frac{1}{2}\langle u, \dot{L}(0)\rangle]^2 &= \frac{1}{4}\mathsf{E}_{\theta_0}[\int_0^1 L_{\theta_0}^{1/2}(su)\langle u, \dot{L}(su)\rangle ds - \langle u, \dot{L}(0)\rangle]^2 \\
&\le \frac{1}{4}\int_0^1 \mathsf{E}_{\theta_0}[L_{\theta_0}^{1/2}(su)\langle u, \dot{L}(su)\rangle ds - \langle u, \dot{L}(0)\rangle]^2 ds \\
&\le \frac{1}{4}\|u\|^2 \int_0^1 \mathsf{E}_{\theta_0} \parallel L_{\theta_0}^{1/2}(su)\dot{L}(su) - \dot{L}(0) \parallel^2 ds \to 0,
\end{aligned}$$

by (1.139). ■

Remark 1.118. If the regularity conditions (A6) are satisfied and (1.138) holds we know from Theorem 1.117 that $(P_\theta)_{\theta\in\Delta}$ is $\mathbb{L}_2$-differentiable with Fisher information matrix $I(\theta_0)$ as specified in 1.103. Independent of this situation, if the regularity conditions are fulfilled, then regardless of whether (1.138) holds, as long as (1.137) holds we call

$$\mathsf{I}(\theta) = \int (\nabla \ln f_\theta)(\nabla \ln f_\theta)^T f_\theta d\mu$$

the Fisher information matrix.

Example 1.119. Suppose $f : \mathbb{R} \to (0, \infty)$ is a continuously differentiable Lebesgue density with

$$\int [f'(x)]^2 \frac{1}{f(x)} \boldsymbol{\lambda}(dx) < \infty \quad \text{and} \quad \int x^2 [f'(x)]^2 \frac{1}{f(x)} \boldsymbol{\lambda}(dx) < \infty.$$

Let P_θ be the distribution with the Lebesgue density

$$f_\theta(x) = \frac{1}{\sigma} f(\frac{x-\mu}{\sigma}), \quad \theta = (\mu, \sigma) \in \Delta = \mathbb{R} \times (0, \infty).$$

Then $f_\theta(x)$ is continuously differentiable with respect to θ, and it holds

$$(\nabla \ln f_\theta(x))^T = \frac{1}{f_\theta(x)} \left(-\frac{1}{\sigma^2} f'(\frac{x-\mu}{\sigma}), -\frac{1}{\sigma^2} f(\frac{x-\mu}{\sigma}) - \frac{x-\mu}{\sigma^3} f'(\frac{x-\mu}{\sigma}) \right).$$

From here it follows, by calculating the corresponding integrals, that the Fisher information matrix is

$$\mathsf{I}(\theta) = \frac{1}{\sigma^2} \begin{pmatrix} \int [f'(x)]^2/f(x)\boldsymbol{\lambda}(dx) & \int x[f'(x)]^2/f(x)\boldsymbol{\lambda}(dx) \\ \int x[f'(x)]^2/f(x)\boldsymbol{\lambda}(dx) & \int x^2[f'(x)]^2/f(x)\boldsymbol{\lambda}(dx) - 1 \end{pmatrix}.$$

Obviously, $I(\theta)$ is continuous, so that $(P_\theta)_{\theta\in\Delta}$, in view of Theorem 1.117, is $\mathbb{L}_2$-differentiable at any $\theta_0 \in \Delta$. The considered location-scale model is even $\mathbb{L}_2$-differentiable under weaker assumptions on f. For details we refer to Strasser (1985).

Example 1.120. Let $(P_\theta)_{\theta\in\Delta}$ be an exponential family with natural parameter θ and $\boldsymbol{\mu}$-densities $f_\theta(x) = \exp\{\langle\theta, T(x)\rangle - K(\theta)\}$, $\theta \in \Delta$. By Proposition 1.17 the function $K(\theta)$ is continuously differentiable in Δ^0. Furthermore, by Corollary 1.19 it holds $\mathsf{E}_\theta T = \nabla K(\theta)$ and $\mathsf{C}_\theta(T) = \nabla\nabla^T K(\theta)$. Hence

$$\int (\nabla \ln f_\theta)(\nabla \ln f_\theta)^T dP_\theta = \nabla\nabla^T K(\theta),$$

in view of Proposition 1.17, is a continuous function. Consequently, all assumptions in Theorem 1.117 are fulfilled so that $(P_\theta)_{\theta\in\Delta}$ is $\mathbb{L}_2$-differentiable at any $\theta_0 \in \Delta^0$with $\mathbb{L}_2$-derivative $\dot{L}_{\theta_0} = T - \nabla K(\theta_0)$ and Fisher information matrix $\mathsf{I}(\theta_0) = \nabla\nabla^T K(\theta_0)$.

In (A6) we have assumed that $\theta \mapsto f_\theta$ is continuously differentiable. This condition, however, is not satisfied in some special, but important, models. Thus we ask for weaker conditions, but confine ourselves to the location model. For the definition of absolutely continuity of a function we refer to the discussion before Theorem A.24. Absolute continuous functions are $\boldsymbol{\lambda}$-a.e. differentiable. The subsequent criterion goes back to Hájek (1972).

Lemma 1.121. *Let f be a Lebesgue density on $\mathbb{R}$, and let P_θ be the distribution with Lebesgue density $f(x-\theta)$. If $f(x)>0$ for every $x\in\mathbb{R}$, f is absolutely continuous, and $\mathsf{I}:=\int((f')^2/f)d\boldsymbol{\lambda}<\infty$, then the family $(P_\theta)_{\theta\in\mathbb{R}}$ is $\mathbb{L}_2$-differentiable at every $\theta_0\in\mathbb{R}$ with derivative $\dot{L}_{\theta_0}(x)=-f'(x-\theta_0)/f(x-\theta_0)$ and Fisher information I.*

Proof. As f is absolutely continuous it is also continuous. As f is positive $\inf_{a\le x\le b} f(x)>0$, say $f(x)\ge C>0$, $a\le x\le b$. Then $g(x):=f^{1/2}(x)$ satisfies

$$|g(x)-g(y)|=\frac{|f(x)-f(y)|}{g(x)+g(y)}\le\frac{1}{2C}\,|f(x)-f(y)|\,,$$

so that g is absolutely continuous, and it holds $g(x+h)-g(x)=\int_x^{x+h}g'(t)dt$; see Theorem A.24. To show that it holds

$$\int(g(x-(\theta_0+h))-g(x-\theta_0)+hg'(x-\theta_0))^2dx=o(h^2),$$

it is, in view of Vitali's theorem, and the fact that the integral on the left does not depend on θ_0, enough to show that

$$\limsup_{h\to0}\frac{1}{h^2}\int(g(x-h)-g(x))^2dx\le\int(g'(x))^2dx.$$

The Schwarz inequality gives for $h>0$

$$\begin{aligned}\frac{1}{h^2}\int(g(x-h)-g(x))^2dx&=\frac{1}{h^2}\int(\int I_{(x-h,x)}(t)g'(t)dt)^2dx\\&\le\frac{1}{h}\int(\int I_{(x-h,x)}(t)(g'(t))^2dt)dx=\int(g'(t))^2dt,\end{aligned}$$

which completes the proof. ■

The next example is from LeCam and Yang (2000).

Example 1.122. Suppose $f(x)=C(\beta)\exp\{-|x|^\beta\}$, $\beta>0$. Then f is $\boldsymbol{\lambda}$-a.e. differentiable with derivative

$$f'(x)=C(\beta)\,\mathrm{sgn}(x)\beta|x|^{\beta-1}\exp\{-|x|^\beta\},$$

and it holds $\int|f'(x)|dx<\infty$. Hence f is absolutely continuous. Apparently, $\int((f')^2/f)d\boldsymbol{\lambda}<\infty$ if and only if $\beta>1/2$. Hence in this case the location model $(P_\theta)_{\theta\in\mathbb{R}}$ with parent density f is $\mathbb{L}_2$-differentiable with derivative $\dot{L}_{\theta_0}(x)=-\,\mathrm{sgn}(x-\theta_0)\beta|x-\theta_0|^{\beta-1}$.

In Proposition 1.110 we have shown that the expectation of the $\mathbb{L}_2$-derivative $\dot{L}_{\theta_0}$ is zero at θ_0. For any $P\in\mathcal{P}(\mathfrak{A})$, and with $\mathbb{L}_{2,d}(P)$ analogously to (1.129), we set

$$\mathbb{L}_{2,d}^0(P)=\{T:T\in\mathbb{L}_{2,d}(P),\ \ \mathsf{E}_PT=0\}.\tag{1.140}$$

The question arises as to whether every $T\in\mathbb{L}_{2,d}^0(P)$ may appear as the $\mathbb{L}_2$-derivative of a differentiable model. That this is in fact true is the content of the next example. The construction is taken from Janssen (2004).

Example 1.123. For a fixed $T \in \mathbb{L}_{2,d}^0(P)$ and $\theta \in \Delta = \mathbb{R}^d$ we set

$$C(\theta) = \mathsf{E}_P(1 + \frac{1}{2}\langle\theta, T\rangle)^2, \quad f_\theta = C(\theta)^{-1}(1 + \frac{1}{2}\langle\theta, T\rangle)^2$$
$$P_\theta(A) = \int_A f_\theta dP, \quad \theta \in \Delta,\ A \in \mathfrak{A}.$$

We show that the family $(P_\theta)_{\theta\in\Delta}$ is $\mathbb{L}_2$-differentiable at $\theta_0 = 0$ and it holds $\dot{L}_0 = T$. To establish the $\mathbb{L}_2$-differentiability we apply Lemma 1.106. The condition (1.132) follows from $L_0(u) = f_u$,

$$C(u) = \mathsf{E}_P(1 + \frac{1}{2}\langle u, T\rangle)^2 = 1 + \frac{1}{4}\mathsf{E}_P(\langle u, T\rangle)^2 = 1 + o(\|u\|),$$
$$L_{\theta_0}(u) - 1 = (1 + o(\|u\|))^{-1}(1 + \langle u, T\rangle + \frac{1}{4}\langle u, T\rangle)^2) - 1 = \langle u, T\rangle + o_P(\|u\|).$$

To prove the uniform integrability of $\|u\|^{-2}(L_{\theta_0}^{1/2}(u) - 1)^2$ for $u \to 0$ we remark that

$$\|u\|^{-2}(L_{\theta_0}^{1/2}(u) - 1)^2 \le 2\|u\|^{-2}\left(\frac{|1 + \frac{1}{2}\langle u, T\rangle| - 1}{C(u)}\right)^2 + 2\|u\|^{-2}\left(\frac{C(u) - 1}{C(u)}\right)^2.$$

Applying $||a| - |b|| \le |a - b|$ and $|\langle u, T\rangle| \le \|u\|\,\|T\|$ to the first term we arrive at

$$\|u\|^{-2}(L_{\theta_0}^{1/2}(u) - 1)^2 \le \frac{1}{2(1 + o(\|u\|))}\|T\|^2 + O(\|u\|).$$

Hence by taking for u a sequence $u_n \to 0$ the left-hand side is dominated by a nonnegative random variable with finite expectation. This proves the uniform integrability of $\|u\|^{-2}(L_{\theta_0}^{1/2}(u) - 1)^2$.

1.6 Solutions to Selected Problems

Solution to Problem 1.2: Let M be a $d \times d$ matrix of full rank, $a, b \in \mathbb{R}^d$, and consider $\widetilde{\theta} = M^T(\theta - b)$ and $\widetilde{T} = M^{-1}(T - a)$. Then with $\widetilde{\theta} = M^T(\theta - b)$ we get

$$\langle\theta, T\rangle = \langle\widetilde{\theta}, \widetilde{T}\rangle + b^T M\widetilde{T} + \widetilde{\theta}^T M^{-1}a + b^T a.$$

By absorbing the last three summands into $\widetilde{K}(\widetilde{\theta})$ and $\widetilde{\boldsymbol{\mu}}$, say, we see that (1.5) still holds if $\theta, T, K, \boldsymbol{\mu}$ are replaced by $\widetilde{\theta}, \widetilde{T}, \widetilde{K}, \widetilde{\boldsymbol{\mu}}$, respectively. □

Solution to Problem 1.3: Without loss of generality we may assume $\mathbb{E}Y_i = 0$ and $a_0 = 0$. Then for some vector $a = (a_1, ..., a_d)$

$$\mathbb{E}(\sum\nolimits_{i=1}^d a_i Y_i)^2 = \sum\nolimits_{i,j=1}^d a_i a_j \mathsf{cov}(Y_i, Y_j) = 0$$

if and only if $a = 0$ or the matrix $(\mathsf{cov}(Y_i, Y_j))_{1\le i,j\le d}$ is singular. □

Solution to Problem 1.8: The extended family fails e.g. to satisfy (1.8). □

Solution to Problem 1.9: For $k \in \mathbb{N}$ and $\lambda > 0$, the measure $\boldsymbol{\mu}$ with point masses $\boldsymbol{\mu}(\{k\}) = 1/k!$, and $\theta = \ln \lambda$, $T(k) = k$, and $K(\theta) = \exp\{\theta\}$, it holds $(d\mathsf{Po}_\lambda / d\boldsymbol{\mu})(k) = \exp\{\theta T(k) - K(\theta)\}$. □

Solution to Problem 1.12: $(\mathsf{N}^{\otimes n}(\mu, \sigma^2)) \circ T_{\oplus n}^{-1}$ is the distribution of the random vector $(\sum_{i=1}^n X_i, \sum_{i=1}^n X_i^2) = (n\overline{X}_n, (n-1)S_n^2 + n\overline{X}_n^2)$. $n\overline{X}_n$ and $(n-1)S_n^2$ are independent, where $\mathcal{L}(n\overline{X}_n) = \mathsf{N}(n\mu, n\sigma^2)$ and the distribution of $(n-1)S_n^2/\sigma^2$ has the Lebesgue density h_{n-1}. For every $h : \mathbb{R}^2 \to_m \mathbb{R}_+$

$$\begin{aligned}
&\mathbb{E}h(n\overline{X}_n, (n-1)S_n^2 + n\overline{X}_n^2) \\
&= \int [\int h(t_1, t_2\sigma^2 + t_1^2/n)\varphi_{n\mu, n\sigma^2}(t_1)\mathsf{h}_{n-1}(t_2)dt_1]dt_2 \\
&= \int [\int h(s_1, s_2)\sigma^{-2}\varphi_{n\mu, n\sigma^2}(s_1)\mathsf{h}_{n-1}(s_2/\sigma^2 - s_1^2/(n\sigma^2))ds_1]ds_2. \quad \square
\end{aligned}$$

Solution to Problem 1.15: Set $\boldsymbol{\mu}(dx) = x^{-1}(1-x)^{-1}I_{(0,1)}(x)\boldsymbol{\lambda}(dx)$. For $\alpha, \beta > 0$ and $x \in \mathbb{R}$,

$$\begin{aligned}
\frac{d\mathsf{Be}_{\alpha,\beta}}{d\boldsymbol{\mu}}(x) &= \frac{\Gamma(\alpha+\beta)}{\Gamma(\alpha)\Gamma(\beta)} x^{\alpha-1}(1-x)^{\beta-1} x(1-x) I_{(0,1)}(x) \\
&= \exp\{\langle \theta, T(x)\rangle - K(\theta)\}, \quad \theta \in (0,\infty)^2,
\end{aligned}$$

where $\theta = (\alpha, \beta)$, $T(x) = (T_1(x), T_2(x)) = (\ln(x), \ln(1-x))$, and $K(\theta) = \ln(\Gamma(\theta_1)\Gamma(\theta_2)/\Gamma(\theta_1+\theta_2))$. □

Solution to Problem 1.31: If $A = \{x : \mathsf{m}(x) > 0\}$ and $B = \{\theta : \pi(\theta) > 0\}$, then

$$(\mathsf{P} \otimes \Pi)(A \times B) = \int_A [\int_B f_\theta(x)\pi(\theta)\boldsymbol{\tau}(d\theta)]\boldsymbol{\mu}(dx) = \int_A \mathsf{m}(x)\boldsymbol{\mu}(dx) = 1.$$

The proof follows from $(\boldsymbol{\mu} \otimes \boldsymbol{\tau})(\cdot \cap (A \times B)) \ll\gg (\mathsf{M} \otimes \Pi)(\cdot \cap (A \times B))$. □

Solution to Problem 1.32: Straightforward calculations give

$$\begin{aligned}
\mathsf{ga}_{\lambda,\beta}(x)\mathsf{ga}_{a,b}(\beta) &= \mathsf{ga}_{a+\lambda, b+x}(\beta)\frac{\Gamma(a+\lambda)}{\Gamma(a)\Gamma(\lambda)} b^a \frac{x^{\lambda-1}}{(b+x)^{a+\lambda}} I_{(0,\infty)}(x) \\
&= \pi(\beta|x)\mathsf{m}(x). \quad \square
\end{aligned}$$

Solution to Problem 1.42: First note that $D_d(1, a_2, ..., a_d) = \prod_{j=2}^d (a_j - 1)$. Evaluating the determinant according to the first column we get

$$D_d(a_1, ..., a_d) = a_1 D_{d-1}(a_2..., a_d) + b(a_2..., a_d),$$

where $b(a_2..., a_d)$ is a function of $a_2..., a_d$ only. Putting $a_1 = 1$ we see that

$$b(a_2..., a_d) = \prod_{j=2}^d (a_j - 1) - D_{d-1}(a_2..., a_d).$$

Hence $D_d(a_1,...,a_d) = (a_1-1)D_{d-1}(a_2...,a_d) + \prod_{j=2}^d (a_j - 1)$. The proof follows by induction. □

Solution to Problem 1.43: Put $p_d = 1 - \sum_{i=1}^{d-1} p_i$. Then

$$\begin{pmatrix} \frac{1}{p_1}+\frac{1}{p_d} & \frac{1}{p_d} & \cdots & \frac{1}{p_d} \\ \frac{1}{p_d} & \frac{1}{p_2}+\frac{1}{p_d} & & \frac{1}{p_d} \\ \cdot & & \cdot & \cdot \\ \cdot & & & \cdot \\ \frac{1}{p_d} & & \cdot & \frac{1}{p_d} \\ \frac{1}{p_d} & \cdot & \cdots \frac{1}{p_d} & \frac{1}{p_{d-1}}+\frac{1}{p_d} \end{pmatrix}.$$

Using the notation from the previous problem we get

$$\det(J(p)) = \frac{1}{p_d^d} D_{d-1}(1+\frac{p_d}{p_1},...,1+\frac{p_d}{p_{d-1}})$$
$$= \frac{1}{p_d^d}(\prod_{j=1}^{d-1}\frac{p_d}{p_j})(1+\sum_{j=1}^{d-1}\frac{p_j}{p_d}) = \frac{1}{p_d}(\prod_{j=1}^{d-1}\frac{1}{p_j}). \quad \square$$

Solution to Problem 1.50: The continuity is implied by the existence of the one-sided derivatives which we show now. The second inequality in (1.54) yields that the difference quotient is nondecreasing in the increment so that the derivative from the right exists and the inequality (1.55) is fulfilled. The proofs of the existence of the left-hand derivative and of (1.56) are similar. Add $\varepsilon_n \downarrow 0$ to x and y in (1.55) to get for $x < y$, $(x-y)^{-1}(\mathsf{v}(x+\varepsilon_n) - \mathsf{v}(y+\varepsilon_n)) \geq D^+\mathsf{v}(x+\varepsilon_n)$. If first $n \to \infty$ and then $y \downarrow x$ we get $\lim_{n\to\infty} D^+\mathsf{v}(x+\varepsilon_n) \leq D^+\mathsf{v}(x)$ which gives the right continuity as $D^+\mathsf{v}$ is nondecreasing. The left continuity of $D^-\mathsf{v}$ is similarly seen. (1.57) follows from (1.55) and the left and right continuity of $D^-\mathsf{v}$ and $D^+\mathsf{v}$, respectively. □

Solution to Problem 1.51: The inequality (1.55) implies

$$(v-t)D^+\mathsf{v}(t) \leq \mathsf{v}(v) - \mathsf{v}(t) \leq (v-t)D^+\mathsf{v}(v), \quad a < t < v < b.$$

For $h_n = (y-x)/n$ we get

$$\int_x^y D^+\mathsf{v}(s)ds = \sum_{i=1}^n \int_{x+(i-1)h_n}^{x+ih_n} D^+\mathsf{v}(s)ds \leq \sum_{i=1}^n h_n D^+\mathsf{v}(x+ih_n)$$
$$\leq \sum_{i=1}^n [\mathsf{v}(x+(i+1)h_n) - \mathsf{v}(x+ih_n)] \leq \mathsf{v}(y+h_n) - \mathsf{v}(x+h_n),$$

and analogously $\int_x^y D^+\mathsf{v}(s)ds \geq \mathsf{v}(y-h_n) - \mathsf{v}(x-h_n)$. The continuity of v completes the proof of the first statement for $D^+\mathsf{v}$. The statement for $D^-\mathsf{v}$ follows from the fact that $D^+\mathsf{v}$ and $D^-\mathsf{v}$ differ only on an at most countable set. As to the second statement, let $x < y$ and $\alpha \in (0,1)$. Set $z = \alpha x + (1-\alpha)y$, $A = \inf\{g(s) : z \leq s \leq y\}$, and $B = \sup\{g(s) : x \leq s \leq z\}$. Then

$$\alpha\mathsf{v}(x) + (1-\alpha)\mathsf{v}(y) - \mathsf{v}(z) = (1-\alpha)\int_z^y g(s)ds - \alpha\int_x^z g(s)ds$$
$$\geq \alpha(1-\alpha)(A-B)(y-x) \geq 0. \quad \square$$

Solution to Problem 1.53: As v is convex we have from (1.61) $\mathsf{v}(y) - \mathsf{v}(x) - D^+\mathsf{v}(x)(y-x) \geq 0$, so that v is strictly convex at $x_0 \in (a,b)$ if and only if for every $\varepsilon > 0$ it holds $\max(\mathsf{v}_l(\varepsilon), \mathsf{v}_r(\varepsilon)) > 0$, where

$$\mathsf{v}_r(\varepsilon) = \mathsf{v}(x_0+\varepsilon) - \mathsf{v}(x_0) - D^+\mathsf{v}(x_0)\varepsilon, \;\; \mathsf{v}_l(\varepsilon) = \mathsf{v}(x_0-\varepsilon) - \mathsf{v}(x_0) + D^+\mathsf{v}(x_0)\varepsilon.$$

As both v_r and v_l are nondecreasing the last statement is equivalent to the fact that at least one, say v_r, is positive for every $\varepsilon > 0$. Then by (1.61) $0 < \mathsf{v}_r(\varepsilon) = \int (x_0+\varepsilon-t) I_{(x_0,x_0+\varepsilon]}(t)\gamma_{\mathsf{v}}(dt)$ and consequently $\gamma_{\mathsf{v}}((x_0-\varepsilon, x_0+\varepsilon)) > 0$ for every $\varepsilon > 0$. Conversely, if the latter condition holds, then by (1.61) at least one of the functions v_r or v_l is strictly positive. If v is not strictly convex at every $a < x < b$, then there is some interval where v is linear, Taking x, y from this interval we see that $\mathsf{v}(\alpha x + (1-\alpha)y) = \alpha\mathsf{v}(x) + (1-\alpha)\mathsf{v}(y)$. Conversely, if v is strictly convex, $x_0 = \alpha x + (1-\alpha)y$, then for $0 < \alpha < 1, x < y$, it holds $\varepsilon_1 := x_0 - x > 0$, $\varepsilon_2 := y - x_0 > 0$ so that by the definition of $\mathsf{v}_r, \mathsf{v}_l$ and $\alpha(-\varepsilon_1) + (1-\alpha)\varepsilon_2 = 0$

$$\alpha\mathsf{v}(x) + (1-\alpha)\mathsf{v}(y) - \mathsf{v}(\alpha x + (1-\alpha)y) = \alpha\mathsf{v}_l(\varepsilon_1) + (1-\alpha)\mathsf{v}_r(\varepsilon_2) > 0. \quad \square$$

Solution to Problem 1.54: The statements $D^+\mathsf{v}_0(x) = D^+\mathsf{v}(x) - D^+\mathsf{v}(1)$ and $\mathsf{v}_0(1) = 0$ are clear from (1.62). $\mathsf{v}_0(x) \geq 0$ follows from (1.58) as $D^+\mathsf{v}_0$ is nondecreasing. From the Solution of Problem 1.53 we know that v is strictly convex at 1 if and only if one of the two functions $\mathsf{v}_r(\varepsilon) = \mathsf{v}_0(1+\varepsilon)$ or $\mathsf{v}_l(\varepsilon) = \mathsf{v}_0(1-\varepsilon)$ is positive for every $\varepsilon > 0$. $\square$

Solution to Problem 1.55: For $0 < \alpha < 1$, $0 < x_1 < x_2$, $x_0 = \alpha x_1 + (1-\alpha)x_2$,

$$\begin{aligned} \mathsf{v}^*(x_0) &= x_0\mathsf{v}(\frac{1}{x_0}) = x_0\mathsf{v}(\frac{\alpha x_1}{x_0}\frac{1}{x_1} + \frac{(1-\alpha)x_2}{x_0}\frac{1}{x_2}) \\ &\leq x_0\frac{\alpha x_1}{x_0}\mathsf{v}(\frac{1}{x_1}) + x_0\frac{(1-\alpha)x_2}{x_0}\mathsf{v}(\frac{1}{x_2}) = \alpha\mathsf{v}^*(x_1) + (1-\alpha)\mathsf{v}^*(x_2). \quad \square \end{aligned}$$

Solution to Problem 1.56: If $P_1 = P_{1,a} + P_{1,s} = \widetilde{P}_{1,a} + \widetilde{P}_{1,s}$ are two decompositions that satisfy (1.65), then there are P_0-null sets N and $\widetilde{N}$ such that $P_{1,s}(B \cap N) = P_{1,s}(B)$ and $\widetilde{P}_{1,s}(B \cap \widetilde{N}) = \widetilde{P}_{1,s}(B)$ for every $B \in \mathcal{A}$. Set $N_0 = N \cup \widetilde{N}$. Then $P_{1,s}(\mathcal{X}\backslash N_0) = \widetilde{P}_{1,s}(\mathcal{X}\backslash N_0) = 0$ and $P_{1,a}(A \cap N_0) = \widetilde{P}_{1,a}(A \cap N_0) = 0$ for every $A \in \mathcal{A}$. Hence $P_{1,s}(B) = P_{1,s}(B \cap N_0) = \widetilde{P}_{1,s}(B \cap N_0) = \widetilde{P}_{1,s}(B)$, which implies $P_{1,a} = \widetilde{P}_{1,a}$. $\square$

Solution to Problem 1.61: By $\gamma_{\mathsf{v}} = \gamma_{\mathsf{v}_0}$ it is enough to consider the function v_0 in (1.62). The relation (1.61) and the monotone convergence theorem give

$$\begin{aligned} \lim_{y\to\infty} \frac{1}{y}\mathsf{v}_0(y) &= \lim_{y\to\infty} \int \frac{1}{y}(y-t) I_{(1,y]}(t)\gamma_{\mathsf{v}_0}(dt) = \gamma_{\mathsf{v}_0}((1,\infty)), \\ \lim_{x\downarrow 0} \mathsf{v}_0(x) &= \lim_{x\downarrow 0} \int (t-x) I_{(x,1]}(t)\gamma_{\mathsf{v}_0}(dt) = \int t I_{(0,1]}(t)\gamma_{\mathsf{v}_0}(dt). \quad \square \end{aligned}$$

Solution to Problem 1.62: By the definition of $\mathsf{I}_{\mathsf{v}}(P_0, P_1)$ in (1.69)

$$\begin{aligned}\mathsf{I}_{\mathsf{w}}(P_0, P_1) &:= \int \mathsf{w}(f_0/f_1) f_1 I_{\{f_0>0, f_1>0\}} d\mu + \mathsf{w}(0) P_1(f_0 = 0) \\ &\quad + \mathsf{w}^*(0) P_0(f_1 = 0) \\ &= \int \mathsf{v}(f_0/f_1) f_1 I_{\{f_0>0, f_1>0\}} d\mu + \int a(f_0/f_1) f_1 I_{\{f_0>0, f_1>0\}} d\mu \\ &\quad + \int b f_1 I_{\{f_0>0, f_1>0\}} d\mu + (\mathsf{v}(0) + b) P_1(f_0 = 0) \\ &\quad + (\mathsf{v}^*(0) + a) P_0(f_1 = 0) = \mathsf{I}_{\mathsf{v}}(P_0, P_1) + a + b,\end{aligned}$$

so that $\mathsf{I}_{\mathsf{v}}(P_0, P_1) - \mathsf{v}(1) = \mathsf{I}_{\mathsf{w}}(P_0, P_1) - \mathsf{w}(1)$. □

Solution to Problem 1.67: Use $(\pi f_0) \wedge ((1-\pi) f_1) \le f_0 + f_1$ and Lebesgue's theorem. □

Solution to Problem 1.72: Apply (1.88) with $\mathsf{v}(x) = (\sqrt{x} - 1)^2$ and $\mathsf{v}(x) = |x - 1|$. □

Solution to Problem 1.74:

$$\begin{aligned}&\int I_C(x,y) L(x) (\mathsf{K} \otimes Q)(dx, dy) = \int [\int I_C(x,y) L(x) \mathsf{K}(dy|x)] Q(dx) \\ &= \int [\int I_C(x,y) \mathsf{K}(dy|x)] P(dx) = (\mathsf{K} \otimes P)(C).\end{aligned}$$

Similarly, by the definition of the conditional expectation and $(\mathsf{K} \otimes Q) \circ S^{-1} = \mathsf{K}Q$,

$$\begin{aligned}&\int I_B(y) \mathsf{E}_{\mathsf{K}\otimes Q}(L|S = y)(\mathsf{K}Q)(dy) = \int I_B(S) \mathsf{E}_{\mathsf{K}\otimes Q}(L|S)\, d(\mathsf{K} \otimes Q) \\ &= \int \mathsf{E}_{\mathsf{K}\otimes Q}(I_B(S) L|S)\, d(\mathsf{K} \otimes Q) = \int I_B(S) L\, d(\mathsf{K} \otimes Q) \\ &= \int [\int I_B(y) L(x) \mathsf{K}(dy|x)] Q(dx) = \int I_B(y)(\mathsf{K}P)(dy), \quad B \in \mathfrak{B}.\end{aligned}$$

Finally,

$$\begin{aligned}&\int I_B(y) \mathsf{E}_Q(L|T = y)(Q \circ T^{-1})(dy) = \mathsf{E}_Q I_B(T) \mathsf{E}_Q(L|T) \\ &= \mathsf{E}_Q(\mathsf{E}_Q(I_B(T) L|T)) = \int I_B(T) L\, dQ = \int I_B(t)(P \circ T^{-1})(dt). \qquad \square\end{aligned}$$

Solution to Problem 1.79: The first equation in the first statement follows from Corollary 1.78 with $\mathsf{v}(x) = |x - 1|$. It holds $\sum_{B\in\mathfrak{p}}(P_0(B) - P_1(B)) = 0$ and $\sum_{B\in\mathfrak{p}} |P_0(B) - P_1(B)| = \sum_{B:P_0(B)\ge P_1(B)}(P_0(B) - P_1(B)) - \sum_{B:P_0(B)<P_1(B)}(P_0(B) - P_1(B)) = 2(P_0(A) - P_1(A))$, $A = \cup_{B:P_0(B)\ge P_1(B)} B$, which implies the second equality. The second statement follows from Corollary 1.78 using the convex function $-x^s$, $0 < s < 1$. □

Solution to Problem 1.80: Using (1.76) we get

$$|\int hdP_0 - \int hdP_1| \;\leq\; |\int (hf_0 - hf_1)d\mu| \leq c\int |f_0 - f_1|\, d\mu = c\,\|P_0 - P_1\|\,.$$

The stated equality follows with $h_0 = I_A - I_{\overline{A}}$ where $A = \{f_0 > f_1\}$.

Now let $\mathcal{X}$ be a metric space. Without loss of generality we may assume that μ in (1.76) is a probability measure. It follows from Theorem 7.1.3 in Dudley (2002) that $\mu(B) = \sup\{\mu(F) : F \subseteq B,\ F \text{ closed}\}$. Hence there is a sequence of closed sets $F_1 \subseteq F_2 \subseteq \cdots \subseteq A$ such that $\mu(A \backslash F_n) \to 0$. $P_i \ll \mu$ implies that for every $\varepsilon > 0$ there is some n_0 with $|P_i(A) - P_i(F_{n_0})| < \varepsilon/4$. Let ρ be the metric of $\mathcal{X}$ and $\rho(x, F) = \inf\{\rho(x, y) : y \in F\}$. Set $\varphi(t) = 1$ if $t \leq 0$, $\varphi(t) = 1 - t$ if $0 \leq t \leq 1$, and $\varphi(t) = 0$ if $t \geq 1$. Put $h_{m,n}(x) = \varphi(m\rho(x, F_n))$. Then $0 \leq h_{m,n}(x) \leq 1$ is continuous and $h_{m,n_0}(x) \downarrow I_{F_{n_0}}(x)$ for every x. Hence $|\int h_{m_0,n_0} dP_i - P_i(F_{n_0})| < \varepsilon/4$ for some m_0. Put $h = 2h_{m_0,n_0} - 1$. Then $|h| \leq 1$, and with $\|P_0 - P_1\| = 2(P_0(A) - P_1(A))$ we get

$$\begin{aligned} &|\ \|P_0 - P_1\| - (\int hdP_0 - \int hdP_1)\ | \\ &\leq |\int (2I_A - 1 - h)dP_0| + |\int (2I_A - 1 - h)dP_1| < 2\varepsilon, \end{aligned}$$

and $\|P_0 - P_1\| \leq |\int hdP_0 - \int hdP_1\ | + 2\varepsilon$. $\square$

Solution to Problem 1.81: The statements (1.107) and (1.108) follow from (1.105), (1.106), $L_{0,1} = (f_1/f_0)I_{\{f_0>0\}} + \infty I_{\{f_0=0,f_1>0\}}$, $L_1 = 2f_1/(f_0 + f_1)$, and $L_0 = 2 - L_1$. (1.109) follows from (1.105) and Hölder's inequality as $\mathsf{H}_{s_2}(P_0, P_1) = \int (f_1/f_0)^{1-s_2} dP_0 \leq (\int (f_1/f_0)^{1-s_1} dP_0)^{(1-s_2)/(1-s_1)}$. (1.110) follows directly from the definitions of the terms that are involved. (1.111) follows from $sf_0 + (1 - s)f_1 - f_0^s f_1^{1-s} \geq 0$, $0 < s < 1$, with equality to 1 if and only if $f_0 = f_1$. Moreover, $P_0(f_0 > 0) = 1$ implies that $\mathsf{H}_s(P_0, P_1) = 0$ holds if and only if $P_0(f_1 > 0) = 0$, which is equivalent to the singularity of P_0 and P_1. This gives (1.112). The proof of (1.113) is similar to that of (1.109). The last statement follows from (1.106). $\square$

Solution to Problem 1.82: $f_0^s f_1^{1-s} \leq sf_0 + (1 - s)f_1 \leq f_0 + f_1$ and Lebesgue's theorem yield $\lim_{s\downarrow 0} \mathsf{H}_s(P_0, P_1) = \lim_{s\downarrow 0} \int f_0^s f_1^{1-s} d\mu = P_1(f_0 > 0) = P_1(L_{0,1} < \infty)$, which is one if and only if $P_1 \ll P_0$. $\square$

Solution to Problem 1.83: It holds

$$\mathsf{H}_{s+it}(P_0, P_1) = \int (\frac{f_0}{f_1})^{s+it} dP_1.$$

To complete the proof we have only to apply Lemma 1.16. $\square$

Solution to Problem 1.86: Dominate P_i, Q_i by the probability measure μ_i, $i = 1, ..., n$; use

$$\frac{d(\bigotimes_{i=1}^n P_i)}{d(\bigotimes_{i=1}^n \mu_i)}(x_1, ..., x_n) = \prod\nolimits_{i=1}^n \frac{dP_i}{d\mu_i}(x_i), \quad \bigotimes\nolimits_{i=1}^n \mu_i\text{-a.s.},$$

and a similar statement for $\bigotimes_{i=1}^n Q_i$. Then Fubini's theorem (see Theorem A.26) gives the first statement. Denote by $\mathfrak{G}_n$ the σ-algebra generated by the first n coordinates. Let $P^{\mathfrak{G}_n}$, $Q^{\mathfrak{G}_n}$, and $\mu^{\mathfrak{G}_n}$ be the restrictions of P, Q, and μ, respectively, to $\mathfrak{G}_n$. Then

$$\frac{dP^{\mathfrak{G}_n}}{d\mu^{\mathfrak{G}_n}}(x_1, x_2, \ldots) = \prod\nolimits_{i=1}^n \frac{dP_i}{d\mu_i}(x_i),$$

and a similar statement holds for $Q^{\mathfrak{G}_n}$. This gives

$$\mathsf{H}_s(P^{\mathfrak{G}_n}, Q^{\mathfrak{G}_n}) = \mathsf{H}_s(\bigotimes\nolimits_{i=1}^n P_i, \bigotimes\nolimits_{i=1}^n Q_i) = \prod\nolimits_{i=1}^n \mathsf{H}_s(P_i, Q_i).$$

An application of Theorem 1.77 completes the proof. To establish (1.118) we employ (1.110) to get

$$2[1 - \mathsf{H}_{1/2}(P_0, P_1)] = \mathsf{D}^2(P_0, P_1) \quad \text{and} \quad \mathsf{H}_{1/2}(P_0, P_1) = 1 - (1/2)\mathsf{D}^2(P_0, P_1).$$

Hence by the first statement,

$$\mathsf{D}^2(\bigotimes\nolimits_{i=1}^n P_i, \bigotimes\nolimits_{i=1}^n Q_i) = 2[1 - \prod_{i=1}^n (1 - (1/2)\mathsf{D}^2(P_i, Q_i))] \le \sum_{i=1}^n \mathsf{D}^2(P_i, Q_i),$$

by the well-known inequality $\left|1 - \prod_{i=1}^n (1 - a_i)\right| \le \sum_{i=1}^n a_i$, $0 \le a_i \le 1$. □

Solution to Problem 1.99: As $\mathsf{K}(\cdot|u) = \mathcal{L}(V|U = u)$ and $\mathsf{L}(\cdot|v) = \mathcal{L}(W|V = v)$,

$$\mathbb{P}(W \in dw, V \in dv, U \in du) = \mathsf{L}(dw|v)\mathsf{K}(dv|u)P_U(du). \tag{1.141}$$

If the stochastic kernel J is defined by $P_U(du)\mathsf{K}(dv|u) = \mathsf{J}(du|v)P_V(dv)$, then

$$\mathbb{P}(U \in du, V \in dv, W \in dw) = \mathsf{J}(du|v)\mathsf{L}(dw|v)P_V(dv),$$

which is the stated conditional independence. Conversely, if the last equation holds, then we introduce the kernel K by the requirement $P_U(du)\mathsf{K}(dv|u) = \mathsf{J}(du|v)P_V(dv)$, which implies the Markov property (1.141). As the condition of conditional independence of U and W, given V, is symmetric in U and W the sequence W, V, U is also a Markov chain. □

Solution to Problem 1.105: The condition (1.130) and the chain rule (see Proposition A.28) imply $L_{\theta_0}(u) = f_{\theta_0+u} f_{\theta_0}^{-1}$ so that

$$\mathsf{E}_{\theta_0}(L_{\theta_0}^{1/2}(u) - 1 - \frac{1}{2}\langle u, \dot{L}_{\theta_0}\rangle)^2 = \int (f_{\theta_0+u}^{1/2} - f_{\theta_0}^{1/2} - \frac{1}{2}\langle u, \dot{f}_{\theta_0}\rangle)^2 d\mu = o(\|u\|^2). \quad \square$$

Solution to Problem 1.107: We get from (1.107) that

$$1 - \mathsf{H}_{1-s}(P_{\theta_0}, P_{\theta_0+u}) = 1 - \mathsf{E}_{\theta_0} L_{\theta_0}^s(u) = \mathsf{E}_{\theta_0}(1 - s + sL_{\theta_0}(u) - L_{\theta_0}^s(u)).$$

$\psi(x) = (1 - s + sx - x^s)(1 - 1/2 + (1/2)x - x^{1/2})^{-1}$ is bounded on $0 \le x < \infty$, say by C. Let $u_n \to 0$ be a sequence which we may assume to be of the form $u_n = \varepsilon_n h_n$, where $h_n \to h$ and $\|h_n\| = 1$. Put $T_n =: \varepsilon_n^{-2}(1/2 + 1/2L_{\theta_0}(u_n) - L_{\theta_0}^{1/2}(u_n))$. Then $E_{\theta_0} T_n = \varepsilon_n^{-2}(1 - H_{1/2}(P_{\theta_0}, P_{\theta_0+u_n})) \to \frac{1}{8} h^T I(\theta_0) h^T$ by (1.134). It holds $T_n = \frac{1}{2}(\varepsilon_n^{-1}(L_{\theta_0}^{1/2}(u_n) - 1))^2$ and $T_n \to^{P_{\theta_0}} \frac{1}{8}(h^T \dot{L}_{\theta_0})^2$. As $E_{\theta_0} \frac{1}{8}(h^T \dot{L}_{\theta_0})^2 = \frac{1}{8} h^T I(\theta_0) h^T$ we get from the Theorem of Vitali (see Theorem A.21) that T_n is uniformly integrable. Then by $0 \le \varepsilon_n^{-2}(1 - s + sL_{\theta_0}(u_n) - L^s_{\theta_0}(u_n)) \le CT_n$ the sequence $\varepsilon_n^{-2}(1 - s + sL_{\theta_0}(u_n) - L^s_{\theta_0}(u_n))$ is also uniformly integrable and the statement follows from $\varepsilon_n^{-2}(1 - s + sL_{\theta_0}(u_n) - L^s_{\theta_0}(u_n)) \to^{P_{\theta_0}} \frac{1}{2}s(1-s)(h^T \dot{L}_{\theta_0})^2$. □

Solution to Problem 1.108: Set $\mathsf{v}_0(x) = \mathsf{v}(x) - \mathsf{v}(1) - \mathsf{v}'(1)(x-1)$. Then $\mathsf{v}(0) + \mathsf{v}^*(\infty) < \infty$ implies that $\psi(x) = \mathsf{v}_0(x)(\sqrt{x} - 1)^{-2}$ is bounded and $\lim_{x \to 1} \psi(x) = 4\mathsf{v}''(1)$. The rest is similar to the previous proof. □

Solution to Problem 1.109: It holds

$$|XY_n| I_{[N,\infty)}(|XY_n|) \le |XY_n| I_{[\sqrt{N},\infty)}(|X|) + |XY_n| I_{[\sqrt{N},\infty)}(|Y_n|),$$

and thus

$$\begin{aligned}&\mathbb{E}|XY_n| I_{[N,\infty)}(|XY_n|) \\ &\le \sup_n \left[\mathbb{E}Y_n^2\right]^{1/2} ([\mathbb{E}X^2 I_{[\sqrt{N},\infty)}(|X|)]^{1/2} + [\mathbb{E}X^2 I_{[\sqrt{N},\infty)}(|Y_n|)]^{1/2}).\end{aligned}$$

As $\mathbb{E}I_{[\sqrt{N},\infty)}(|Y_n|) \le N^{-1} \sup_n \mathbb{E}Y_n^2$ we get for every sequence $N_n \to \infty$ that $I_{[\sqrt{N_n},\infty)}(|Y_n|) \to^{\mathbb{P}} 0$ and $\mathbb{E}X^2 I_{[\sqrt{N_n},\infty)}(|Y_n|) \to 0$ by Lebesgue's theorem, which also implies $\mathbb{E}X^2 I_{[\sqrt{N_n},\infty)}(|X|) \to 0$. Hence XY_n, $n = 1, 2, ...$ is uniformly integrable. As

$$\mathbb{E}|XY_n - X_n Y_n| \le \left[\mathbb{E}(X - X_n)^2\right]^{1/2} \sup_n \left[\mathbb{E}Y_n^2\right]^{1/2} \to 0,$$

the uniform integrability follows from Vitali's theorem, A.21. □

Solution to Problem 1.113: It is enough to consider the case $n = 2$. The general case follows by mathematical induction. If u is sufficiently small, then by the definition of $\mathbb{L}_2$-differentiability $P_{i,\theta_0+u} \ll P_{i,\theta_0}$, and the likelihood in the product model is $L_{\otimes 2,\theta_0}(u)(x_1, x_2) = L_{1,\theta_0}(u)(x_1) L_{2,\theta_0}(u)(x_2)$, where

$$L_{i,\theta_0}(u)(x_i) = \frac{dP_{i,\theta_0+u}}{dP_{i,\theta_0}}(x_i).$$

The $\mathbb{L}_2$-differentiability implies $L_{i,\theta_0}(u) - 1 = o_{P_{i,\theta_0}}(1)$. The decomposition

$$L_{\otimes 2,\theta_0}(u) - 1 = (L_{1,\theta_0}(u) - 1) L_{2,\theta_0}(u) + (L_{2,\theta_0}(u) - 1)$$

shows that the first condition in Lemma 1.106 is satisfied. The second condition to be proved for $L_{\otimes 2,\theta_0}(u)$ follows from the corresponding condition for $L_{i,\theta_0}(u)$ and the relations (1.78) and Problem 1.86. □

Solution to Problem 1.116: Denote by $\rho_{\mathcal{S}}$ and $\rho_{\mathcal{T}}$ the metric in $\mathcal{S}$ and $\mathcal{T}$, respectively, and let D be a dense and at most countable subset of $\mathcal{S}$. Denote by $\rho_{\mathcal{T}}(t, C) = \inf\{\rho_{\mathcal{T}}(t, s), s \in C\}$ the distance of the point t to the set C. As the closed subsets C of $\mathcal{T}$ generate the σ-algebra $\mathfrak{T}$ it suffices to show that $\{(s, x) : \psi(s, x) \in C\} \in \mathfrak{S} \otimes \mathfrak{A}$ for every closed subset C. But this follows from

$$\begin{aligned}&\{(s, x) \; : \; \psi(s, x) \in C\} \\ &= \bigcap_{n=1}^{\infty} \bigcup_{u \in D} (\{s : \rho_{\mathcal{S}}(s, u) \le \frac{1}{n}\} \times \{(u, x) : \rho_{\mathcal{T}}(\psi(u, x), C) \le \frac{1}{n}\}). \quad \square\end{aligned}$$

2

Tests in Models with Monotonicity Properties

2.1 Stochastic Ordering and Monotone Likelihood Ratio

In this section we deal with problems where it is useful to extend the real line to $\overline{\mathbb{R}} = \mathbb{R} \cup \{\infty\} \cup \{-\infty\}$ and to admit the random variables to take values in $\overline{\mathbb{R}}$. If we equip $\overline{\mathbb{R}}$ with the metric $\overline{\rho}$ in Remark A.1, then $(\overline{\mathbb{R}}, \overline{\rho})$ becomes a compact metric space. By $\overline{\mathfrak{B}}$ we denote the σ-algebra of Borel sets and random variables with values in $\overline{\mathbb{R}}$ are $\mathfrak{F}$-$\overline{\mathfrak{B}}$ measurable mappings. We denote by $\mathcal{P}(\overline{\mathfrak{B}})$ the set of all distributions on $(\overline{\mathbb{R}}, \overline{\mathfrak{B}})$, whereas $\mathcal{P}(\mathfrak{B})$ is the set of all distributions on $(\mathbb{R}, \mathfrak{B})$. The distributions $P \in \mathcal{P}(\mathfrak{B})$ can be identified with the distributions Q from $\mathcal{P}(\overline{\mathfrak{B}})$ with $Q(\{\infty\} \cup \{-\infty\}) = 0$.

Now we introduce and study a semiorder, called *stochastic ordering*, in the space of distributions $\mathcal{P}(\overline{\mathfrak{B}})$. This ordering is used frequently in probability theory and statistical analysis. In particular it proves useful for the construction of level α tests for one-sided and two-sided testing problems and the study of their power functions. Although the concept of stochastic ordering is too weak to establish optimality, it leads to optimal level α tests whenever such tests exist. An important tool for the construction of level α tests is the *quantile function* of a distribution on $(\mathbb{R}, \mathfrak{B})$.

Definition 2.1. *Let F be the c.d.f. of a distribution $Q \in \mathcal{P}(\mathfrak{B})$; that is, $F(t) = Q((-\infty, t])$, $t \in \mathbb{R}$. The quantile function of Q, or generalized inverse of F, is defined by*

$$
\begin{aligned}
F^{-1}(u) &= \inf\{x : F(x) \geq u\}, \quad u \in (0,1], \quad \textit{and} \\
F^{-1}(0) &= \sup\{x : F(x) = 0\}.
\end{aligned} \tag{2.1}
$$

For $u \in [0,1]$ the point $c_u = F^{-1}(u)$ in $\mathbb{R}$ is called the u-quantile of F or the $(100u)$th percentile of F.

Problem 2.2. Show that for every c.d.f. F the following hold.

F. Liese, K.-J. Miescke, *Statistical Decision Theory*,
DOI: 10.1007/978-0-387-73194-0_2, © Springer Science+Business Media, LLC 2008

$$F(x) \geq y \quad \text{iff} \quad x \geq F^{-1}(y), \quad x \in \mathbb{R},\ y \in [0,1].$$
$$F(F^{-1}(y) - 0) \leq y \leq F(F^{-1}(y)), \quad y \in [0,1].$$
$$F^{-1}(F(x)) \leq x \leq F^{-1}(F(x) + 0), \quad x \in \{t : F(t) > 0,\ t \in \mathbb{R}\}.$$

Here we have used the notation $F(x-0) = \lim_{\varepsilon \downarrow 0} F(x - \varepsilon)$ and $F^{-1}(y+0) = \lim_{\varepsilon \downarrow 0} F^{-1}(y+\varepsilon)$.

Problem 2.3.* Let F be the c.d.f. of a distribution on $(\mathbb{R}, \mathfrak{B})$, and let U be a random variable that has a uniform distribution on $[0,1]$. Then $F^{-1}(U)$ has the c.d.f. F. On the other hand, if a random variable X with values in $\mathbb{R}$ has a c.d.f. $F : \mathbb{R} \to [0,1]$ that is continuous, then $F(X)$ has a uniform distribution on $[0,1]$.

Now we introduce the stochastic semiorder of distributions on $(\overline{\mathbb{R}}, \overline{\mathfrak{B}})$.

Definition 2.4. *For two distributions $Q_1, Q_2 \in \mathcal{P}(\overline{\mathfrak{B}})$ on $\overline{\mathbb{R}}$ with c.d.f.s $F_i(t) = Q_i([-\infty, t])$, $t \in \mathbb{R}$, $i = 1, 2$, we call Q_1 stochastically not larger than Q_2, denoted by $Q_1 \preceq Q_2$, if*

$$F_1(t) \geq F_2(t), \quad t \in \mathbb{R}.$$

A family of distributions $(P_\theta)_{\theta \in \Delta}$ on $(\overline{\mathbb{R}}, \overline{\mathfrak{B}})$ with $\Delta \subseteq \mathbb{R}$ is called stochastically nondecreasing if for every $\theta_1, \theta_2 \in \Delta$ with $\theta_1 < \theta_2$ it holds $P_{\theta_1} \preceq P_{\theta_2}$.

The relation $\preceq$ is obviously a semiorder in $\mathcal{P}(\overline{\mathfrak{B}})$ and is called the *stochastic semiorder*. This semiorder, restricted to $\mathcal{P}(\mathfrak{B})$, implies the pointwise semiorder for the quantile functions that were introduced in Definition 2.1.

Proposition 2.5. *Let $Q_1, Q_2 \in \mathcal{P}(\mathfrak{B})$ with c.d.f.s F_1, F_2, respectively. If $Q_1 \preceq Q_2$, then*

$$F_1^{-1}(\alpha) \leq F_2^{-1}(\alpha), \quad 0 \leq \alpha \leq 1.$$

Proof. If $Q_1 \preceq Q_2$, then $\{t : F_1(t) \geq \alpha\} \supseteq \{t : F_2(t) \geq \alpha\}$, which gives the statement. ■

If X_1 and X_2 are random variables with $X_1 \leq X_2$, $\mathbb{P}$-a.s., then obviously $\mathcal{L}(X_1) \preceq \mathcal{L}(X_2)$. On the other hand, it proves useful that for two distributions their comparison via $\preceq$ can be reduced to the pointwise comparison of random variables. This is the content of the next statement.

Proposition 2.6. *For $Q_1, Q_2 \in \mathcal{P}(\mathfrak{B})$ it holds $Q_1 \preceq Q_2$ if and only if there are random variables X_1 and X_2 with*

$$\mathcal{L}(X_i) = Q_i, \quad i = 1, 2, \quad \textit{and} \quad X_1 \leq X_2, \quad \mathbb{P}\text{-a.s.}$$

Proof. The statement that $X_1 \leq X_2$, $\mathbb{P}$-a.s., implies $Q_1 \preceq Q_2$ is trivial. Conversely, let $Q_1 \preceq Q_2$ with c.d.f.s F_1, F_2, respectively. For some U that is uniformly distributed on $[0,1]$, set $X_i = F_i^{-1}(U)$, $i = 1, 2$. By Proposition 2.5 we have $F_1^{-1}(U) \leq F_2^{-1}(U)$. Finally, by Problem 2.3, $\mathcal{L}(F_i^{-1}(U)) = Q_i$, $i = 1, 2$. ■

The stochastic semiorder can be characterized by a corresponding relation for expectations.

Proposition 2.7. *Suppose X_1, X_2 are random variables with values in $\overline{\mathbb{R}}$ and*

$$\mathcal{L}(X_1) \preceq \mathcal{L}(X_2). \tag{2.2}$$

If $\phi : \overline{\mathbb{R}} \to \mathbb{R}$ is a nondecreasing function with $\mathbb{E}|\phi(X_i)| < \infty$, $i = 1, 2$, then

$$\mathbb{E}\phi(X_1) \leq \mathbb{E}\phi(X_2). \tag{2.3}$$

Conversely, if (2.3) holds for every nonnegative, bounded, and nondecreasing function $\phi : \overline{\mathbb{R}} \to \mathbb{R}$, then the relation (2.2) holds.

Proof. If X_1 and X_2 are random variables with values in $\mathbb{R}$, then the first statement follows from Proposition 2.6. The extension to the case where X_1 or X_2 may also assume the values $-\infty$ and ∞ is straightforward. The converse statement of the proposition is obtained by choosing $\phi = I_{(t,\infty]}$, $t \in \overline{\mathbb{R}}$. ■

To illustrate the stochastic semiorder in $\mathcal{P}(\mathfrak{B})$ we give some examples.

Example 2.8. Let X be a random variable with c.d.f. F. For any $\theta \in \mathbb{R}$ the c.d.f. of $X + \theta$ is $F(\cdot - \theta)$. Thus, $(\mathcal{L}(X+\theta))_{\theta\in\mathbb{R}}$ is stochastically nondecreasing.

Problem 2.9. Let F be any c.d.f. and denote by Q_α, $\alpha > 0$, the distribution that has the c.d.f. $F_\alpha = 1 - (1-F)^\alpha$. Then Q_α is stochastically nonincreasing for $\alpha > 0$.

An important fact is that the distribution of likelihood ratio $L_{0,1}$ is under P_1 stochastically not smaller than under P_0. Somewhat more can be stated.

Theorem 2.10. *If for the binary model $(\mathcal{X}, \mathfrak{A}, \{P_0, P_1\})$ the likelihood ratio is of the form $L_{0,1} = h(T)$, where $T : \mathcal{X} \to_m \overline{\mathbb{R}}$ and $h : \overline{\mathbb{R}} \to_m \overline{\mathbb{R}}$ is nondecreasing, then $\mathcal{L}(T|P_0) \preceq \mathcal{L}(T|P_1)$.*

Proof. By the definition of the likelihood ratio $L_{0,1}$ (see Definition 1.57),

$$\int g dP_1 = \int g L_{0,1} dP_0 + \int g I_{\{\infty\}}(L_{0,1}) dP_1$$

holds for every $g : \mathcal{X} \to_m \mathbb{R}_+$. Let $\varphi : \overline{\mathbb{R}} \to \mathbb{R}$ be any nonnegative, bounded, and nondecreasing function. Put $A_0 = \{t : h(t) < 1\}$ and $A_1 = \{t : h(t) > 1\}$. As h is nondecreasing the sets A_i are some intervals, open or closed, and we have $a_0 := \sup_{t\in A_0} \varphi(t) \leq \inf_{t\in A_1} \varphi(t) =: a_1$. Hence,

$$\begin{aligned}
&\int \varphi(T) dP_1 - \int \varphi(T) dP_0 \\
&= \int I_{A_0 \cup A_1}(T)\varphi(T)(L_{0,1} - 1) dP_0 + \int \varphi(T) I_{\{\infty\}}(L_{0,1}) dP_1 \\
&\geq a_0 \int I_{A_0}(T)(L_{0,1} - 1) dP_0 + a_1 \int I_{A_1}(T)(L_{0,1} - 1) dP_0 \\
&\quad + a_1 \int I_{\{\infty\}}(L_{0,1}) dP_1 \geq a_0 \int (L_{0,1} - 1) dP_0 + a_0 \int I_{\{\infty\}}(L_{0,1}) dP_1 = 0.
\end{aligned}$$

■

Now we study families of distributions $(P_\theta)_{\theta\in\Delta}$ for which the likelihood ratio L_{θ_0,θ_1} of P_{θ_1} with respect to P_{θ_0}, $\theta_0, \theta_1 \in \Delta$, has a special structure.

Definition 2.11. *A model* $(\mathcal{X}, \mathfrak{A}, (P_\theta)_{\theta\in\Delta})$ *with* $\Delta \subseteq \mathbb{R}$ *is said to have a nondecreasing (increasing) likelihood ratio in the statistic* $T : \mathcal{X} \to_m \mathbb{R}$ *if for every* $\theta_0, \theta_1 \in \Delta$ *with* $\theta_0 < \theta_1$ *there is a nondecreasing (increasing) function* $h_{\theta_0,\theta_1} : \mathbb{R} \to_m \overline{\mathbb{R}}_+$ *such that* $L_{\theta_0,\theta_1} = h_{\theta_0,\theta_1}(T)$, $\{P_{\theta_0}, P_{\theta_1}\}$*-a.s. We also say in short that* $(P_\theta)_{\theta\in\Delta}$ *has (strict) MLR, i.e., (strict) monotone likelihood ratio, in* T.

Many of the families that are used in statistics have a nondecreasing likelihood ratio. Some examples follow below.

Example 2.12. Let $\mathsf{U}(0,\theta)$, $\theta \geq 0$, be the uniform distribution on $[0,\theta]$ with density $\mathsf{u}_{0,\theta}(x) = (1/\theta)I_{[0,\theta]}(x)$. Assume that $X_1, ..., X_n$ are i.i.d. with distribution $\mathsf{U}(0,\theta)$. Then the distribution $\mathsf{U}(0,\theta)^{\otimes n}$ of $(X_1, ..., X_n)$ has the Lebesgue density $g_\theta(M) = \theta^{-n} I_{[0,\theta]}(M)$, where $M(x_1, ..., x_n) = \max(x_1, ..., x_n)$. For $0 < \theta_0 < \theta_1 < \infty$ the likelihood ratio is

$$L_{\theta_0,\theta_1} = (\theta_0/\theta_1)^n \, I_{[0,\theta_0]}(M) + \infty I_{(\theta_0,\theta_1]}(M),$$

which is a nondecreasing function of M, $\{\mathsf{U}(0,\theta_0)^{\otimes n}, \mathsf{U}(0,\theta_1)^{\otimes n}\}$-a.s. Thus the family $(\mathsf{U}(0,\theta)^{\otimes n})_{\theta>0}$ has MLR in M.

Example 2.13. Consider the one-parameter exponential family $(P_\theta)_{\theta\in\Delta}$ from (1.5). The natural parameter set Δ is an interval and the likelihood ratio is

$$L_{\theta_0,\theta_1} = \exp\{(\theta_1 - \theta_0)T - K(\theta_1) + K(\theta_0)\},$$

which for $\theta_0 < \theta_1$, $\theta_0, \theta_1 \in \Delta$, is obviously an increasing function of T. Thus the family $(P_\theta)_{\theta\in\Delta}$ has strict MLR in T. Moreover, if Λ is an interval and $\kappa : \Lambda \to \Delta$ is a nondecreasing (increasing) function, then the family $(P_{\kappa(\eta)})_{\eta\in\Lambda}$ has (strict) MLR in T.

To study the structure of special location models that have the MLR property we need the concepts of unimodality and strong unimodality of a distribution. Furthermore we need the concepts of convexity and subconvexity. Convex functions of one variable have been introduced and studied already in Section 1.3. Let $C \subseteq \mathbb{R}^d$ be a convex set. A function $g : C \to \mathbb{R}$ is called *convex* if

$$g(\alpha x + (1-\alpha)y) \leq \alpha g(x) + (1-\alpha)g(y), \quad x, y \in C, \ 0 \leq \alpha \leq 1.$$

g is called *concave* if $-g$ is convex. A function $g : C \to \mathbb{R}$ is called *subconvex* if $\{t : g(t) \leq a, \, t \in C\}$ is a convex subset of $\mathbb{R}^d$ for every a. It is clear that every convex function is subconvex.

Problem 2.14. If $g : O \to \mathbb{R}$ is subconvex and $h : \mathbb{R} \to \mathbb{R}$ is nondecreasing, then $h(g)$ is again subconvex.

Problem 2.15.* If $g : \mathbb{R}^d \to \mathbb{R}$ is subconvex and $g(x) \geq g(x_0)$ for every $x \in \mathbb{R}^d$, then the function $s \mapsto g(x_0 + s(x - x_0))$ is nondecreasing on $[0, \infty)$ for every $x \in \mathbb{R}^d$.

Definition 2.16. *A distribution $P \in \mathcal{P}(\mathfrak{B})$ is called unimodal at mode m if $F(t) = P((-\infty, t])$, $t \in \mathbb{R}$, is convex in $(-\infty, m)$ and concave in (m, ∞). P is called strongly unimodal if $P \ll \boldsymbol{\lambda}$ and there is a $\boldsymbol{\lambda}$-density f and an open interval (a, b), finite or infinite, such that $f(t) = 0$ for $t \notin (a, b)$, $f(t) > 0$ for $t \in (a, b)$, and $g = -\ln f$ is convex in (a, b). Then we also say that P is log-concave.*

Next we characterize unimodal distributions.

Proposition 2.17. *A distribution Q is unimodal if and only if $Q = \alpha\boldsymbol{\delta}_m + (1-\alpha)P$, $0 \le \alpha \le 1$, where $P \ll \boldsymbol{\lambda}$ and there is a $\boldsymbol{\lambda}$-density f of P which is nondecreasing for $t < m$ and nonincreasing for $t > m$.*

Proof. Consider the representation $Q = \alpha\boldsymbol{\delta}_m + (1-\alpha)P$ with $P(\{m\}) = 0$, which is valid regardless of Q being unimodal or not. As the c.d.f. of $\boldsymbol{\delta}_m$ is constant for $t < m$ and $t > m$ we see that Q is unimodal if and only if P is unimodal. It suffices to consider the case $t < m$. If P is unimodal, then the c.d.f. F of P is convex, has in view of Problem 1.50 a nondecreasing derivative from the right, and (1.58) yields

$$F(b) - F(a) = \int_a^b D^+F(s)ds, \quad a < t < b < m.$$

Hence $f := D^+F$ is a version of the density and is nondecreasing for $t < m$. Conversely, if there exists a version f of the density that is nondecreasing for $t < m$, then by

$$F(b) - F(a) = \int_a^b f(s)ds, \quad a < b < m,$$

and Problem 1.51 the function F is convex for $t < m$. ■

Next we establish properties of strongly unimodal distributions.

Proposition 2.18. *If P is a strongly unimodal distribution with Lebesgue density f, then $\lim_{x\to\pm\infty} f(x) = 0$, there is some $m \in \mathbb{R}$ such that $f(x) \le f(m)$, and P is unimodal with mode m.*

Proof. If g is convex, then D^+g is nondecreasing and it follows from (1.58) that there are $x_0 < x_1$ such that g is monotone for $x < x_0$ and $x > x_1$. Hence $f = \exp\{-g\}$ is also monotone in this area. From $\int f dx = 1$ we get the first statement. This statement implies for the convex function g that $\lim_{x\to\pm\infty} g(x) = \infty$. As g is continuous we get that there is at least one minimum point, say m. Hence $f(x) \le f(m)$. The convex function g is subconvex. Hence $-f = -\exp\{-g\}$ is again subconvex; see Problem 2.14. It remains to apply Problem 2.15 to the function $-f$. ■

There is an important characterization of the strong unimodality which is due to Ibragimov (1956). For the definition of the generalized inverse of a c.d.f. that is used below we refer to Definition 2.1.

Proposition 2.19. *A distribution P with Lebesgue density f that is lower semicontinuous and satisfies $f(x) = 0$ on the complement of $(F^{-1}(0), F^{-1}(1))$ is strictly unimodal if and only if the distribution $P * Q$ is unimodal for every unimodal distribution Q.*

Examples of strongly unimodal distributions are the Laplace distribution with Lebesgue density $\mathsf{lp}(t)$, and the logistic distribution with Lebesgue density $\mathsf{lo}(t)$, where

$$\mathsf{lp}(t) = 2\exp\{-|t|\} \quad \text{and} \quad \mathsf{lo}(t) = \exp\{t\}(1 + \exp\{t\})^{-2}.$$

The convexity of $-\ln \mathsf{lp}(t)$ is clear, and the convexity of $-\ln \mathsf{lo}(t)$ follows from $(-\ln \mathsf{lo}(t))'' > 0$.

Now we show that a strongly unimodal distribution used as a parent distribution in a location model generates a model with MLR.

Proposition 2.20. *If the distribution $P = \mathcal{L}(X)$ is strongly unimodal with a density that is positive everywhere, then the location family $P_\theta := \mathcal{L}(X + \theta)$, $\theta \in \mathbb{R}$, has MLR in the identity.*

Proof. $f_\theta(t) := f(t-\theta)$ is a Lebesgue density of P_θ. As $f_\theta(t) > 0$ it suffices to show that

$$f(t_0 - \theta_1)f(t_1 - \theta_0) \leq f(t_0 - \theta_0)f(t_1 - \theta_1),$$

$t_0, t_1, \theta_0, \theta_1 \in \mathbb{R}$, $t_0 \leq t_1$, and $\theta_0 \leq \theta_1$. Put $r = t_0 - \theta_1$, $u = t_1 - \theta_0$, $s = \min(t_0 - \theta_0, t_1 - \theta_1)$, and $t = \max(t_0 - \theta_0, t_1 - \theta_1)$. Then $r \leq s \leq t \leq u$ and the above inequality is equivalent to

$$f(r)f(u) \leq f(s)f(t), \quad r \leq s \leq t \leq u, \ u - t = s - r.$$

For $g = -\ln f$ the last inequality is equivalent to $g(u) - g(t) \geq g(s) - g(r)$. As $u - t = s - r$ we obtain this inequality from the convexity of g and the inequality (1.54). ■

Suppose the sample space is $\mathbb{R}$, and that $(Q_\theta)_{\theta\in\Delta}$ is dominated by some $\mu \in \mathcal{M}^\sigma(\mathfrak{B})$ and has μ-densities $(g_\theta)_{\theta\in\Delta}$. Assume that it holds

$$g_{\theta_1}(x_0)g_{\theta_0}(x_1) \leq g_{\theta_0}(x_0)g_{\theta_1}(x_1), \quad \theta_0 < \theta_1, \ x_0 < x_1, \tag{2.4}$$

for $\theta_0, \theta_1 \in \Delta$ and $x_0, x_1 \in \mathbb{R}$.

Problem 2.21. Let $(Q_\theta)_{\theta\in\Delta}$ with $\Delta \subseteq \mathbb{R}$ be a family of distributions on $(\mathbb{R}, \mathfrak{B})$ that is dominated by some $\mu \in \mathcal{M}^\sigma(\mathfrak{B})$ and has μ-densities $(g_\theta)_{\theta\in\Delta}$, where $g_\theta(x) > 0$ for every $x \in \mathbb{R}$ and $\theta \in \Delta$. Then (2.4) holds if and only if $(Q_\theta)_{\theta\in\Delta}$ has MLR in the identity.

We get from Theorem 2.10 for the c.d.f.s $F_\theta(t) = P_\theta((-\infty, t])$, $t \in \mathbb{R}$, $\theta \in \Delta$, the inequality $F_{\theta_1}(t) \leq F_{\theta_0}(t)$ for $\theta_0 < \theta_1$. The following problem shows that an even stronger property holds.

Problem 2.22.* If (2.4) is satisfied, then for every x_0, x_1 and θ_0, θ_1 as specified there it holds $F_{\theta_0}(x_1)F_{\theta_1}(x_0) \leq F_{\theta_0}(x_0)F_{\theta_1}(x_1)$ as well as $(1 - F_{\theta_0}(x_1))(1 - F_{\theta_1}(x_0)) \leq (1 - F_{\theta_0}(x_0))(1 - F_{\theta_1}(x_1))$.

If the density g_θ is continuously differentiable with respect to θ, then (2.4) can be expressed with the help of derivatives. The condition (2.4) and the following criterion go back to Karlin (1957).

Problem 2.23. Let $(Q_\theta)_{\theta \in \Delta}$, where Δ is an open interval, be a family of distributions on $(\mathbb{R}, \mathfrak{B})$ with $Q_{\theta_0} \ll\gg Q_{\theta_1}$, $\theta_0, \theta_1 \in \Delta$. Suppose there is a $\mu \in \mathcal{M}^\sigma(\mathfrak{B})$ with $\mu \ll\gg Q_\theta$, $\theta \in \Delta$, such that $g_\theta(t) := dQ_\theta/d\mu$ is continuously differentiable with respect to $\theta \in \Delta$. Then (2.4) holds if and only if $t \mapsto ((\partial \ln g_\theta)/\partial\theta)(t)$ is nondecreasing for every fixed $\theta \in \Delta$.

For $x \in \mathbb{R}^n$ we introduce the *rank statistic* r and the *order statistic* s by

$$\begin{aligned}
r(x) &= (r_1(x), ..., r_n(x)) = (\sum\nolimits_{i=1}^n I_{[0,\infty)}(x_1 - x_i), ..., \sum\nolimits_{i=1}^n I_{[0,\infty)}(x_n - x_i)),\\
s(x) &= (s_1(x), ..., s_n(x)), \quad \text{where}\\
s_1(x) &= \min(x_1, ..., x_n), \quad \text{and}\\
s_i(x) &= \min(\{x_1, ..., x_n\} \backslash \{s_1(x), ..., s_{i-1}(x)\}), \quad i = 2, ..., n.
\end{aligned}$$

Following tradition, we use standard notation by setting for $x \in \mathbb{R}^n$, and for a random vector X with values in $\mathbb{R}^n$,

$$\begin{aligned}
(x_{[1]}, ..., x_{[n]}) &:= (s_1(x), ..., s_n(x)), \qquad (2.5)\\
X_{[\cdot]} = (X_{[1]}, ..., X_{[n]}) &:= (s_1(X), ..., s_n(X)),\\
X_{n,[\cdot]} = (X_{n,[1]}, ..., X_{n,[n]}) &:= (s_1(X), ..., s_n(X)),\\
R_n = (R_{n,1}, ..., R_{n,n}) &= (r_1(X), ..., r_n(X)).
\end{aligned}$$

$X_{n,[\cdot]}$ is used instead of $X_{[\cdot]}$ whenever the dependence of $X_{[\cdot]}$ on n is relevant and thus has to be indicated.

Problem 2.24.* Let $X_1, ..., X_n$ be i.i.d. random variables in $\mathbb{R}$ with a common distribution Q_θ that has the Lebesgue density g_θ, $\theta \in \Delta \subseteq \mathbb{R}$. Let $X_{[1]}, ..., X_{[n]}$ be the order statistics. If $(g_\theta)_{\theta \in \Delta}$ satisfies (2.4), then for every $i \in \{1, ..., n\}$ $(Q_{\theta,i})_{\theta \in \Delta}$ with $Q_{\theta,i} = \mathcal{L}(X_{[i]})$ has Lebesgue densities $(g_{\theta,i})_{\theta \in \Delta}$ that satisfy (2.4) as well.

We conclude this section by considering some families of distributions that are frequently used in statistics. The first are the noncentral chi-square distributions. Let $X_1, ..., X_n$ be independent random variables with $X_i \sim \mathsf{N}(\mu_i, 1)$, $i = 1, ..., n$. If $\mu_1 = \cdots = \mu_n = 0$, then $\sum_{i=1}^n X_i^2$ has a chi-square distribution $\mathsf{H}(n)$ with n degrees of freedom, which has the Lebesgue density

$$\mathsf{h}_n(t) = \frac{1}{2^{n/2}\Gamma(n/2)} t^{(n/2)-1} \exp\{-\frac{t}{2}\} I_{(0,\infty)}(t), \quad t \in \mathbb{R}.$$

To find the distribution of $\sum_{i=1}^{n} X_i^2$ for any $\mu_1, ..., \mu_n \in \mathbb{R}$ we make use of the following two facts. Since $\mathsf{H}(n) = \mathsf{Ga}(n/2, 1/2)$ the distribution of $w^{-1}\sum_{i=1}^{n} X_i^2$ is $\mathsf{Ga}(n/2, w/2)$ for any $w > 0$. The gamma distributions satisfy

$$\mathsf{Ga}(\alpha_1, \beta) * \mathsf{Ga}(\alpha_2, \beta) = \mathsf{Ga}(\alpha_1 + \alpha_2, \beta) \quad \alpha_1, \alpha_2, \beta > 0. \tag{2.6}$$

Problem 2.25.* Let $X_1, ..., X_n, W$ be independent, where $X_i \sim \mathsf{N}(\mu_i, 1)$, $i = 1, ..., n$, and W is positive with probability one. Consider the Poisson distributions $\mathsf{Po}(\lambda)$ with p.m.f. $\mathsf{po}_\lambda(k)$, $\lambda > 0$, and set in addition $\mathsf{Po}(0) = \boldsymbol{\delta}_0$. Then

$$\mathcal{L}(W^{-1}\sum\nolimits_{i=1}^{n} X_i^2)(B) = \sum\nolimits_{k=0}^{\infty} \mathsf{po}_{\delta^2/2}(k) \int_0^\infty \mathsf{Ga}(k + n/2, w/2)(B) P_W(dw),$$

where P_W is the distribution of W, B is a Borel set, and $\delta^2 = \sum_{i=1}^{n} \mu_i^2$.

For independent random variables $X_i \sim \mathsf{N}(\mu_i, 1)$, $i = 1, ..., n$, we call

$$\mathsf{H}(n, \delta^2) := \mathcal{L}(\sum\nolimits_{i=1}^{n} X_i^2), \quad \text{where } \delta^2 = \sum\nolimits_{i=1}^{n} \mu_i^2, \tag{2.7}$$

the noncentral χ^2-distribution with n degrees of freedom and noncentrality parameter δ^2. From Problem 2.25 we obtain, with the choice of $W \equiv 1$,

$$\mathsf{H}(n, \delta^2) = \sum\nolimits_{k=0}^{\infty} \mathsf{po}_{\delta^2/2}(k) \mathsf{H}(n + 2k). \tag{2.8}$$

Next we consider the noncentral F-distributions. Let $X_1, ..., X_{n_1}$ and $Y_1, ..., Y_{n_2}$ be independent random variables with $\mathcal{L}(X_i) = \mathsf{N}(\mu_i, 1)$, $i = 1, ..., n_1$, and $\mathcal{L}(Y_j) = \mathsf{N}(0, 1)$, $j = 1, ..., n_2$. Let $\delta^2 = \sum_{i=1}^{n} \mu_i^2$. We call

$$\mathsf{F}(n_1, n_2, \delta^2) := \mathcal{L}((n_2 \sum\nolimits_{i=1}^{n_1} X_i^2)/(n_1 \sum\nolimits_{i=1}^{n_2} Y_i^2))$$

the noncentral F-distribution with n_1 and n_2 *degrees of freedom and noncentrality parameter δ^2.* We could, of course, have also allowed a noncentral chi-square distribution for the denominator as well, but that more general setting would not be of any use later on. From Problem 2.25 we get

$$\mathsf{F}(n_1, n_2, \delta^2) = \sum\nolimits_{k=0}^{\infty} \mathsf{po}_{\delta^2/2}(k) \mathsf{F}(n_1 + 2k, n_2), \tag{2.9}$$

where $\mathsf{F}(n_1, n_2) = \mathsf{F}(n_1, n_2, 0)$ denotes the central F-distribution with n_1 and n_2 degrees of freedom.

Problem 2.26. The distributions $\mathsf{H}(m)$ and $\mathsf{F}(m, n)$ have MLR in the identity with respect to the parameter m.

Finally, we deal with the noncentral t-distributions. Let X and Y be independent random variables with $\mathcal{L}(X) = \mathsf{N}(\mu, 1)$ and $\mathcal{L}(Y) = \mathsf{H}(k)$. Then

$$\mathsf{T}(k, \mu) := \mathcal{L}((X\sqrt{k})/(\sqrt{Y})) \tag{2.10}$$

is called the *noncentral t-distribution with k degrees of freedom and noncentrality parameter μ.*

Theorem 2.27. *The families* $\mathsf{T}(k,\mu)$*,* $\mathsf{H}(m,\delta^2)$*, and* $\mathsf{F}(m,n,\delta^2)$ *have MLR in the identity with respect to the parameters* μ*,* δ^2*, and* δ^2*, respectively.*

Proof. For the proofs of the noncentral t- and F-distributions we refer to Lehmann (1986), p. 295 and p. 428. As to $\mathsf{H}(m,\delta^2)$, its Lebesgue density h_{m,δ^2} has, in view of (2.8), the representation

$$\mathsf{h}_{m,\delta^2} = \sum\nolimits_{k=0}^{\infty} \mathsf{po}_{\delta^2/2}(k)\mathsf{h}_{m+2k}.$$

Hence for $\delta_0^2 < \delta_1^2$ and $t_0 < t_1$,

$$\begin{aligned}
&\mathsf{h}_{m,\delta_0^2}(t_0)\mathsf{h}_{m,\delta_1^2}(t_1) - \mathsf{h}_{m,\delta_1^2}(t_0)\mathsf{h}_{m,\delta_0^2}(t_1) \\
&= \sum\nolimits_{k,l=0}^{\infty} \mathsf{po}_{\delta_0^2/2}(k)\mathsf{po}_{\delta_1^2/2}(l)[\mathsf{h}_{m+2k}(t_0)\mathsf{h}_{m+2l}(t_1) - \mathsf{h}_{m+2k}(t_1)\mathsf{h}_{m+2l}(t_0)] \\
&= \sum\nolimits_{k=0}^{\infty}\sum\nolimits_{l=k+1}^{\infty} [\mathsf{po}_{\delta_0^2/2}(k)\mathsf{po}_{\delta_1^2/2}(l) - \mathsf{po}_{\delta_0^2/2}(l)\mathsf{po}_{\delta_1^2/2}(k)] \\
&\qquad \times [\mathsf{h}_{m+2k}(t_0)\mathsf{h}_{m+2l}(t_1) - \mathsf{h}_{m+2k}(t_1)\mathsf{h}_{m+2l}(t_0)].
\end{aligned}$$

The first bracket is nonnegative as $\mathsf{Po}(\lambda)$, $\lambda > 0$, has MLR in the identity. Similarly, the second bracket is nonnegative as for fixed β_0 the family $\mathsf{Ga}(\alpha,\beta_0)$, $\alpha > 0$, is an exponential family with generating statistic $\ln t$ which has MLR in $\ln t$ and thus also in the identity. ■

2.2 Tests in Binary Models and Models with MLR

Often one has to decide, based on a sample, which of two different situations holds true. For instance, this could be the decision whether a new treatment is better than a standard, or whether a newly manufactured item has a longer expected lifetime than all other items of this type used so far. To deal with such a problem, we start with the statistical model

$$\mathcal{M} = (\mathcal{X}, \mathfrak{A}, (P_\theta)_{\theta\in\Delta}), \tag{2.11}$$

divide Δ into two disjoint subsets Δ_0 and Δ_A, and formulate the hypotheses

$$\mathsf{H}_0 : \theta \in \Delta_0 \quad \text{and} \quad \mathsf{H}_A : \theta \in \Delta_A. \tag{2.12}$$

H_0 is called the *null hypothesis*, H_A is called the *alternative hypothesis*, and we want to decide whether H_0 or H_A is true. In the classical Neyman–Pearson approach the tool for making decisions is a *test* $\varphi : \mathcal{X} \to_m [0,1]$, where $\varphi(x)$ is the probability of deciding in favor of H_A after $x \in \mathcal{X}$ has been observed.

Definition 2.28. *Every function* $\varphi : \mathcal{X} \to_m [0,1]$ *is called a test, and the function* $\theta \mapsto \mathsf{E}_\theta\varphi$ *is called the power function of the test* φ*. The set of all tests* $\varphi : \mathcal{X} \to_m [0,1]$ *is denoted by* $\mathcal{T}$*. A test* $\varphi \in \mathcal{T}$ *is called nonrandomized if* φ *takes on only the values* 0 *and* 1*, and it is called randomized otherwise.*

For a test φ, the set $\{\varphi = 0\}$ is its *acceptance area* of H_0, $\{\varphi = 1\}$ is its *acceptance area* of H_A, and $\{0 < \varphi < 1\}$ is its *randomization area* in $\mathcal{X}$. Deciding in favor of H_A when H_0 is true is called an *error of the first kind*. Deciding in favor of H_0 when H_A is true is called an *error of the second kind*. Consequently, for $\theta \in \Delta_0$, $\mathsf{E}_\theta\varphi$ is the probability of making an error of the first kind, and for $\theta \in \Delta_A$, $1 - \mathsf{E}_\theta\varphi$ is the probability of making an error of the second kind.

The simplest testing model consists of just two distributions P_0 and P_1. In such a so-called *binary model*,

$$\mathcal{M}_{0,1} = (\mathcal{X}, \mathfrak{A}, \{P_0, P_1\}), \tag{2.13}$$

each of the two hypotheses, expressed in terms of the associated distributions,

$$\mathsf{H}_0 : P_0 \quad \text{and} \quad \mathsf{H}_A : P_1 \tag{2.14}$$

consists of one single distribution. Such hypotheses are called *simple* and all others are called *composite*. Searching now for an optimal test leads to a dilemma due to the interrelation between the error probability of the first kind $\mathsf{E}_0\varphi$ and the error probability of the second kind $\mathsf{E}_1(1-\varphi)$. Minimizing the former calls for making φ small whereas minimizing the latter calls for making φ large. Indeed, as the following proposition shows, optimal tests exist only in trivial situations.

Proposition 2.29. *For the model $\mathcal{M}_{0,1}$ in (2.13) and the hypotheses (2.14) there exists a test ψ with minimum error probabilities, that is,*

$$\mathsf{E}_0\psi = \inf_{\varphi \in \mathcal{T}} \mathsf{E}_0\varphi \quad \textit{and} \quad \mathsf{E}_1(1-\psi) = \inf_{\varphi \in \mathcal{T}} \mathsf{E}_1(1-\varphi), \tag{2.15}$$

if and only if $P_0 \perp P_1$. In that case $\mathsf{E}_0\psi = \mathsf{E}_1(1-\psi) = 0$.

Proof. If $P_0 \perp P_1$, then there is some $A \in \mathfrak{A}$ with $P_0(A) = P_1(\mathcal{X}\backslash A) = 0$. Set $\psi = I_A$. Then $\mathsf{E}_0\psi = 1 - \mathsf{E}_1\psi = 0$, so that both conditions in (2.15) are satisfied. Conversely, suppose that an optimal test ψ in the sense of (2.15) exists. Then, in view of the trivial tests $\phi_0 \equiv 0$ and $\phi_1 \equiv 1$, $\mathsf{E}_0\psi = \mathsf{E}_1(1-\psi) = 0$. This implies $P_0(\psi = 0) = 1$ and $P_1(\psi = 1) = 1$ and thus $P_0 \perp P_1$. ■

Because there is no optimal test in the sense of (2.15), unless $P_0 \perp P_1$, we have to restrict ourselves to a suitable subclass of tests. This, of course, holds also for the general setting of (2.11) and (2.12). One approach is to control the error probabilities of the first kind and then to minimize the error probabilities of the second kind. This approach is suitable for situations where the consequences of the two errors are substantially different. Controlling the error probabilities of the first kind is appropriate if these errors are less acceptable than those of the second kind. The following two definitions are given in full generality. In the second "uniformly" may be dropped, of course, for binary models.

Definition 2.30. *For any test $\varphi : \mathcal{X} \to_m [0,1]$ its size is defined to be $\alpha(\varphi) := \sup_{\theta \in \Delta_0} \mathsf{E}_\theta \varphi$. For $\alpha \in (0,1)$ a test φ is called a level α test for H_0 if $\alpha(\varphi) \le \alpha$. Especially we say that a test φ attains the level α if $\alpha(\varphi) = \alpha$.*

Definition 2.31. *Let $\mathcal{T}_0 \subseteq \mathcal{T}$ be a subclass of tests. A test ψ is called a uniformly best test in $\mathcal{T}_0$ for H_0 versus H_A if $\psi \in \mathcal{T}_0$ and for every $\varphi \in \mathcal{T}_0$ it holds*

$$\mathsf{E}_\theta \varphi \le \mathsf{E}_\theta \psi, \quad \theta \in \Delta_A.$$

Especially, if $\mathcal{T}_0 = \{\varphi : \varphi \in \mathcal{T}, \ \alpha(\varphi) \le \alpha\}$, for some $\alpha \in (0,1)$, is the set of all level α tests, then ψ is called a uniformly best level α test.

Uniformly best is also called uniformly most powerful (UMP). In the above definitions we have deliberately excluded the cases of $\alpha = 0$ and $\alpha = 1$ because they are trivial from a statistical point of view. Nevertheless, they are somewhat intricate; see Remark 2.46. With the next problem it is shown that a uniformly best level α test exhausts the side condition under H_0 in the sense of $\alpha(\varphi) = \alpha$, unless it has power 1 throughout H_A.

Problem 2.32.* Let $\alpha \in (0,1)$ be fixed and ψ be a uniformly best level α test for the testing problem (2.12). If $\mathsf{E}_{\theta_1} \psi < 1$ for at least one $\theta_1 \in \Delta_A$, then $\alpha(\psi) = \alpha$. On the other hand, if $\mathsf{E}_\theta \psi = 1$ for all $\theta \in \Delta_A$, then $\alpha(\psi) < \alpha$ may occur, but there is also a test $\widetilde{\psi}$ with $\alpha(\widetilde{\psi}) = \alpha$ and $\mathsf{E}_\theta \widetilde{\psi} = 1$ for all $\theta \in \Delta_A$.

Thus, in the search for a uniformly best level α test for the testing problem (2.12) we may restrict ourselves to the class of tests that attain the level α and then search within that class for a uniformly best test.

For the sequel we assume that the parameter set Δ is a subset of the real line and that the hypotheses are one-sided.

$$\begin{aligned} &\mathcal{M} = (\mathcal{X}, \mathfrak{A}, (P_\theta)_{\theta \in \Delta}), \quad \Delta \subseteq \mathbb{R}, \\ &\mathsf{H}_0 \ : \ \Delta_0 = (-\infty, \theta_0] \cap \Delta, \quad \mathsf{H}_A : \Delta_A = (\theta_0, \infty) \cap \Delta, \end{aligned} \tag{2.16}$$

where $\theta_0 \in \Delta$ with $(\theta_0, \infty) \cap \Delta \neq \emptyset$.

Let us consider now tests that are based on a suitable statistic $T : \mathcal{X} \to_m \overline{\mathbb{R}}$. It is clear that a statistic that reflects the order structure of Δ should have on average larger values under larger parameters, which means that the family of distributions $Q_\theta = P_\theta \circ T^{-1}$, $\theta \in \Delta$, is stochastically nondecreasing. In this case it is reasonable to reject H_0 for large values of T. To formalize this idea, let $F_\theta(t) = P_\theta(T \le t)$, $t \in \mathbb{R}$, be the c.d.f. of T under P_θ, $\theta \in \Delta$. We assume that $P_{\theta_0}(T < \infty) = 1$. For a fixed $\alpha \in (0,1)$, let $c_{1-\alpha} = F_{\theta_0}^{-1}(1-\alpha)$ be the $(1-\alpha)$-quantile of T under P_{θ_0}, see Definition 2.1. Let the test $\psi_\alpha : \mathbb{R} \to_m [0,1]$ be given by

$$\psi_\alpha(t) = \begin{cases} 1 & \text{if} \quad t > c_{1-\alpha} \\ \gamma_\alpha & \text{if} \quad t = c_{1-\alpha} \\ 0 & \text{if} \quad t < c_{1-\alpha} \end{cases}, \quad t \in \overline{\mathbb{R}}, \tag{2.17}$$

where

$$\gamma_\alpha = [F_{\theta_0}(c_{1-\alpha}) - (1-\alpha)] \oslash P_{\theta_0}(T = c_{1-\alpha}), \quad \text{with} \tag{2.18}$$
$$b \oslash a = b/a \text{ if } a \neq 0 \text{ and } b \oslash a = 0 \text{ otherwise.}$$

Moreover, let the test $\varphi_{T,\alpha} : \mathcal{X} \to [0,1]$ be given by

$$\varphi_{T,\alpha}(x) = \psi_\alpha(T(x)), \quad x \in \mathcal{X}. \tag{2.19}$$

Obviously we have $\mathsf{E}_{\theta_0}\varphi_{T,\alpha} = \alpha$. Especially if F_{θ_0} is continuous at $c_{1-\alpha}$, then $P_{\theta_0}(T = c_{1-\alpha}) = 0$ and $\varphi_{T,\alpha}$ is a nonrandomized test.

Theorem 2.33. *Suppose that $\Delta = (a,b) \subseteq \mathbb{R}$ and that $\theta_0 \in (a,b)$ is fixed. Let $T : \mathcal{X} \to_m \overline{\mathbb{R}}$ be a statistic for which the family $Q_\theta = P_\theta \circ T^{-1}$, $\theta \in \Delta$, is stochastically nondecreasing. If $P_{\theta_0}(T < \infty) = 1$ and $\alpha \in (0,1)$, then $\varphi_{T,\alpha}$ from (2.19) is a level α test for the null hypothesis H_0 in (2.16) that attains the level. Moreover, the power function $\theta \mapsto \mathsf{E}_\theta\varphi_{T,\alpha}$ is nondecreasing on Δ.*

Proof. By construction of the test $\mathsf{E}_{\theta_0}\varphi_{T,\alpha} = \alpha$. As the function ψ_α is nondecreasing we get from Theorem 2.7 that the function

$$\theta \mapsto \mathsf{E}_\theta\varphi_{T,\alpha} = \int \psi_\alpha dQ_\theta$$

is nondecreasing on Δ. This also implies that $\varphi_{T,\alpha}$ is a level α test for H_0. ■

Let us now illustrate how such statistics T that satisfy the conditions of Theorem 2.33 can be found and utilized.

Problem 2.34. If the family P_θ, $\theta \in (a,b)$, is stochastically nondecreasing and $T : \mathbb{R}^n \to_m \overline{\mathbb{R}}$ is componentwise nondecreasing, then $Q_\theta = P_\theta^{\otimes n} \circ T^{-1}$, $\theta \in \Delta$, is again stochastically nondecreasing.

Example 2.35. There is a large variety of statistics $T : \mathbb{R}^n \to_m \mathbb{R}$ that are componentwise nondecreasing. To give a few examples, there is the arithmetic mean $\overline{x}_n = (1/n)\sum_{i=1}^n x_i$, the β-trimmed mean

$$T_\beta(x_1, ..., x_n) = \frac{1}{\beta n}\sum\nolimits_{i=[\beta n]}^{[(1-\beta)n]} x_{[i]}, \quad \beta \in (0, 1/2),$$

where $[\beta n]$ is the integer part of βn and $x_{[1]} \leq x_{[2]} \leq \cdots \leq x_{[n]}$ are the ordered values of $x_1, ..., x_n$, and the median

$$\mathsf{md}(x_1, ..., x_n) = \begin{cases} x_{[k+1]} & \text{if } n = 2k+1, \\ \frac{1}{2}(x_{[k]} + x_{[k+1]}) & \text{if } n = 2k. \end{cases}$$

Next we study Gaussian models, that is, normal distributions, where σ^2 may be known or unknown. For the latter case the following well-known fact is established first.

Problem 2.36.* Suppose $Z_1, ..., Z_n$ are i.i.d. $\mathsf{N}(0,1)$. Then $\overline{Z}_n := (1/n)\sum_{i=1}^n Z_i$ and $\sum_{i=1}^n (Z_i - \overline{Z}_n)^2$ are independent, where $\overline{Z}_n$ has the distribution $\mathsf{N}(0, 1/n)$ and $\sum_{i=1}^n (Z_i - \overline{Z}_n)^2$ has a χ^2-distribution with $n-1$ degrees of freedom.

Example 2.37. We consider the Gaussian model and the one-sided testing problem for μ when σ_0^2 is known; that is,

$$(\mathbb{R}^n, \mathfrak{B}_n, (\mathsf{N}^{\otimes n}(\mu, \sigma_0^2))_{\mu \in \mathbb{R}}), \quad \mathsf{H}_0 : \mu \leq \mu_0, \quad \mathsf{H}_A : \mu > \mu_0, \ \sigma_0^2 \text{ known.} \tag{2.20}$$

We denote by $X_1, ..., X_n$ the projections $\mathbb{R}^n \to \mathbb{R}$ and note that under $\mathsf{N}^{\otimes n}(\mu, \sigma_0^2)$ the $X_1, ..., X_n$ are i.i.d. random variables with common distribution $\mathsf{N}(\mu, \sigma_0^2)$. We set

$$U_0 = \frac{\sqrt{n}}{\sigma_0}(\overline{X}_n - \mu_0), \quad \text{where} \quad \overline{X}_n = \frac{1}{n}\sum\nolimits_{i=1}^n X_i. \tag{2.21}$$

Then

$$\mathcal{L}(U_0 | \mathsf{N}^{\otimes n}(\mu, \sigma_0^2)) = \mathsf{N}(\mu - \mu_0, 1).$$

For a fixed $0 < \alpha < 1$ we denote by $u_{1-\alpha}$ the α-quantile of $\mathsf{N}(0,1)$, that is, $u_{1-\alpha} = \Phi^{-1}(1-\alpha)$, and set $\psi_I(t) = I_{(u_{1-\alpha},\infty)}(t)$. The test

$$\varphi_{U_0,\alpha} = \psi_I(U_0) \tag{2.22}$$

is called the Gauss test, or U-test. As the family $(\mathsf{N}(\mu - \mu_0, 1))_{\mu \in \mathbb{R}}$ is stochastically nondecreasing in μ it is a level α-test for the hypotheses in (2.20). The power function is

$$\mathsf{E}_\mu \varphi_{U_0,\alpha} = 1 - \Phi(u_{1-\alpha} - \frac{\sqrt{n}(\mu - \mu_0)}{\sigma_0}), \quad \mu \in \mathbb{R}.$$

Now we consider the Gaussian model and the one-sided testing problem for μ when σ^2 is unknown, that is,

$$(\mathbb{R}^n, \mathfrak{B}_n, (\mathsf{N}^{\otimes n}(\mu, \sigma^2))_{\mu \in \mathbb{R}, \sigma^2 > 0}), \ \mathsf{H}_0 : \mu \leq \mu_0, \ \sigma^2 > 0, \ \mathsf{H}_A : \mu > \mu_0, \ \sigma^2 > 0. \tag{2.23}$$

We set

$$T = \frac{\sqrt{n}}{S_n}(\overline{X}_n - \mu_0), \quad \text{where} \quad S_n^2 = \frac{1}{n-1}\sum\nolimits_{i=1}^n (X_i - \overline{X}_n)^2.$$

Then by the definition of the noncentral t-distribution in (2.10) it holds

$$\mathcal{L}(T | \mathsf{N}^{\otimes n}(\mu, \sigma^2)) = \mathsf{T}(n-1, \sqrt{n}(\mu - \mu_0)/\sigma).$$

Denote by $\mathsf{T}_{n-1}(t) = \mathsf{T}(n-1, 0)((-\infty, t))$ the c.d.f. of the central t-distribution with $n-1$ degrees of freedom, and denote by $t_{1-\alpha,n-1} = \mathsf{T}_{n-1}^{-1}(1-\alpha)$ its $(1-\alpha)$ quantile. The test

$$\varphi_{T,\alpha} = \begin{cases} 1 & \text{if} \quad T > t_{1-\alpha,n-1} \\ 0 & \text{if} \quad T \leq t_{1-\alpha,n-1} \end{cases} \tag{2.24}$$

is called the t-test. As T_{n-1} is continuous, and $\mathsf{T}(n-1, \sqrt{n}(\mu - \mu_0)/\sigma)$ is stochastically nondecreasing, we see that $\varphi_{T,\alpha}$ is a level α test for the hypotheses in (2.23). The power function is

$$\mathsf{E}_\mu \varphi_{T,\alpha} = 1 - \mathsf{T}_{n-1,\sqrt{n}(\mu-\mu_0)/\sigma}(t_{1-\alpha,n-1}), \quad \mu \in \mathbb{R}.$$

Next we deal with a classical k sample problem which is called *analysis of variance* or ANOVA. Here we utilize the stochastic ordering of the noncentral F-distributions to construct a suitable level α test. Later, in Chapter 8, it is shown that this test is optimal in a certain class of invariant tests. Technical tools that are needed for the construction are prepared first.

A symmetric $d \times d$ matrix is called *idempotent* if $AA = A$. Let $\mathbb{L} \subseteq \mathbb{R}^d$ be the linear subspace spanned by the column vectors of A. Then for every $x \in \mathbb{R}^d$ the vector Ax belongs to $\mathbb{L}$ and $x - Ax \perp \mathbb{L}$ since $A(x - Ax) = Ax - AAx = 0$. In other words, A is the matrix of the projection onto the subspace $\mathbb{L}$.

Problem 2.38.* Show that the eigenvalues of a symmetric and idempotent matrix A are either 0 or 1, where the number of 1s is the rank of the matrix A.

Problem 2.39.* Assume that $X_1, ..., X_k$ are independent and $\mathcal{L}(X_i) = \mathsf{N}(\mu_i, \sigma^2)$, $i = 1, ..., k$. Set $X = (X_1, ..., X_k)^T$ and $\mu = (\mu_1, ..., \mu_k)^T$. If $d \le k$ and A is a $k \times k$ idempotent matrix of rank d, then $\mathcal{L}(X^T AX/\sigma^2)$ is a noncentral χ^2-distribution with d degrees of freedom and noncentrality parameter $\delta^2 = \mu^T A\mu/\sigma^2$.

Example 2.40. Let $X_{i,1}, ..., X_{i,n_i}$, $i = 1, ..., k$, be independent i.i.d. samples from k normal populations. More specifically, we assume that $\mathcal{L}(X_{i,j}) = \mathsf{N}(\theta_i, \sigma^2)$, $i = 1, ..., k$, where $\sigma^2 > 0$ is known. Suppose we want to test whether there are any differences between the k expectations. For this purpose we set $n = \sum_{i=1}^k n_i$, $\theta. = (1/n)\sum_{i=1}^k n_i\theta_i$,

$$\eta^2 = \sum\nolimits_{i=1}^k n_i(\theta_i - \theta.)^2, \tag{2.25}$$

and consider testing the null hypothesis $\mathsf{H}_0 : \eta^2 = 0$ against the alternative $\mathsf{H}_A : \eta^2 > 0$. To set up a suitable test statistic we introduce some standard notation. Let

$$X_{i,\cdot} = \frac{1}{n_i}\sum\nolimits_{j=1}^{n_i} X_{i,j} \quad \text{and} \quad X_{\cdot,\cdot} = \frac{1}{n}\sum\nolimits_{i=1}^k n_i X_{i,\cdot}\ . \tag{2.26}$$

A suitable test statistic may be obtained by substituting the parameters θ_i and $\theta.$ in (2.25) by estimators, and by attaching an appropriate normalizing factor. We consider the test statistic

$$\chi^2 = \frac{1}{\sigma^2}\sum\nolimits_{l=1}^k n_i(X_{i,\cdot} - X_{\cdot,\cdot})^2.$$

Set $\gamma_i = \sqrt{n_i/n}$, $\gamma = (\gamma_1, ..., \gamma_k)^T$and $A = \gamma\gamma^T$. Then $A = A^T$ and $AA = A$, so that A is an idempotent matrix of rank 1. Consequently, $I - A$ is an idempotent matrix of rank $k - 1$. It holds,

$$\chi^2 = \sum\nolimits_{i=1}^k (\sqrt{n_i}X_{i,\cdot}/\sigma)^2 - (\sum\nolimits_{i=1}^k \gamma_i\sqrt{n_i}X_{i,\cdot}/\sigma)^2 = Z^T(I - A)Z,$$

where $Z = (\sqrt{n_1}X_{1,\cdot}/\sigma, ..., \sqrt{n_k}X_{k,\cdot}/\sigma)^T$ is a vector with independent components that are distributed according to $\mathsf{N}(\delta_i, 1)$, where $\delta_i = \sqrt{n_i}\theta_i/\sigma$, $i = 1, ..., k$. From Problem 2.39 we see that χ^2 has a chi-square distribution with $k - 1$ degrees of freedom and noncentrality parameter

$$\begin{aligned}\delta^2 &= (\delta_1, ..., \delta_k)^T(I - A)(\delta_1, ..., \delta_k)\\ &= \sum\nolimits_{i=1}^k \delta_i^2 - (\sum\nolimits_{i=1}^k \gamma_i\delta_i)^2 = \frac{1}{\sigma^2}\sum\nolimits_{i=1}^k n_i(\theta_i - \theta.)^2.\end{aligned} \tag{2.27}$$

This leads to the χ^2-test given by

$$\varphi_{\chi^2,\alpha} = \begin{cases} 1 & \text{if} \quad \chi^2 > \chi^2_{1-\alpha,k-1} \\ 0 & \text{if} \quad \chi^2 \le \chi^2_{1-\alpha,k-1} \end{cases},$$

where $\chi^2_{1-\alpha,k-1}$ is the $(1-\alpha)$-quantile of the central chi-square distribution with $k-1$ degrees of freedom. The distribution of χ^2 at $\theta = (\theta_1, ..., \theta_k) \in \mathbb{R}^k$ is $\mathsf{H}(k-1, \delta^2)$, which by Theorems 2.27 and 2.10 is a family that is stochastically nondecreasing in the parameter $\delta^2 \geq 0$. Thus, by Theorem 2.33, the power function of the test φ_{χ^2} is a nondecreasing function in $\delta^2 \geq 0$. Also here, the power function can be evaluated explicitly. Indeed, from (2.8) we have

$$\mathsf{E}_\theta \varphi_{\chi^2,\alpha} = \sum\nolimits_{l=0}^{\infty} \mathsf{po}_{\delta^2/2}(l)(1 - \mathsf{H}_{k-1+2l}(\chi^2_{1-\alpha,k-1})).$$

The case of an unknown variance σ^2 is more involved.

Example 2.41. Under the assumptions of the previous example, but where $\sigma^2 > 0$ is now unknown, we use the following estimator of σ^2,

$$\widehat{\sigma^2} = \frac{1}{n-k} \sum\nolimits_{i=1}^{k} \sum\nolimits_{j=1}^{n_i} (X_{i,j} - X_{i,\cdot})^2.$$

Because $X_{1,1}, ..., X_{k,n_k}$ are independent normal random variables $(X_{1,1}, ..., X_{k,n_k})$ is multivariate normal. The relations $X_{\cdot,\cdot} = \sum_{i=1}^{k} (n_i/n) X_{i,\cdot}$ and

$$\begin{aligned}
&\mathsf{cov}((X_{i,\cdot} - X_{\cdot,\cdot}), (X_{l,j} - X_{l,\cdot})) \\
&= \mathsf{cov}(X_{i,\cdot}, X_{l,j}) - \mathsf{cov}(X_{i,\cdot}, X_{l,\cdot}) - \mathsf{cov}(X_{\cdot,\cdot}, X_{l,j}) + \mathsf{cov}(X_{\cdot,\cdot}, X_{l,\cdot}) \\
&= \frac{\sigma^2}{n_i}\delta_{i,l} - \frac{\sigma^2}{n_i}\delta_{i,l} - \sum\nolimits_{i=1}^{k} \frac{n_i}{n}\frac{\sigma^2}{n_i}\delta_{i,l} + \sum\nolimits_{i=1}^{k} \frac{n_i}{n}\frac{\sigma^2}{n_i}\delta_{i,l} = 0
\end{aligned}$$

show that χ^2 and $\widehat{\sigma^2}$ depend on random vectors that are independent, so that χ^2/σ^2 and $\widetilde{\chi}^2 = \widehat{\sigma^2}/\sigma^2$ are independent. Similarly as in the previous example one can show that $\widetilde{\chi}^2 = \widehat{\sigma^2}/\sigma^2$ has a chi-square distribution with $n-k$ degrees of freedom. The statistic $F = ((n-k)\chi^2)/((k-1)\widetilde{\chi}^2)$ has the noncentral F-distribution $\mathsf{F}(k-1, n-k, \delta^2)$ with δ^2 from (2.27). This can be utilized to construct a level α test. Let $z_{1-\alpha,k-1,n-k}$ be the $(1-\alpha)$-quantile of the central F-distribution $\mathsf{F}(k-1, n-k, 0)$. Then by Theorems 2.27 and 2.33, the test

$$\varphi_{F,\alpha} = \begin{cases} 1 & \text{if } F > z_{1-\alpha,k-1,n-k} \\ 0 & \text{if } F \leq z_{1-\alpha,k-1,n-k} \end{cases}$$

is a level α test for H_0, and its power function is a nondecreasing function of $\delta^2 \geq 0$. Also here, we have an explicit representation of the power function. It holds

$$\mathsf{E}_\theta \varphi_{F,\alpha} = \int_{z_{1-\alpha,k-1,n-k}}^{\infty} \mathsf{f}_{k-1,n-k,\delta^2}(t)dt,$$

where $\mathsf{f}_{k-1,n-k,\delta^2}$ is the Lebesgue density of $\mathsf{F}(k-1, n-k, \delta^2)$. This test $\varphi_{F,\alpha}$ is the well-known Fisher's F-test for the analysis of variance one-way layout. We conclude the example with the remark that the noncentral F-distribution $\mathsf{F}(k-1, n-k, \delta^2)$ can be, similarly as the noncentral χ^2-distribution, expanded as a Poisson mixture of central F-distributions; see (2.9).

The crucial question that has been left open so far is whether the tests in the previous examples have good power properties, and especially if they are uniformly best level α tests, at least in a suitable class of tests. To study this problem in detail we have to return to the very beginning, i.e., to a binary model $\mathcal{M}_{0,1}$ from (2.13). Likewise as in the previous examples the construction of suitable tests is usually based on a statistic, called the *test statistic*, whose values indicate somehow which of the two hypotheses is more likely to be true. When we are dealing with a binary model, then the likelihood ratio $L_{0,1}$ seems to be the appropriate test statistic because by Theorem 2.10 $L_{0,1}$ is under under P_1 stochastically not smaller than under P_0.

Set $\overline{P} = \frac{1}{2}(P_0 + P_1)$ and $L_i = dP_i/d\overline{P}$, $i = 0, 1$. Then $L_1 = 2 - L_0$ and according to (1.68) the likelihood ratio $L_{0,1}$ has the representation

$$L_{0,1} = ((2 - L_0)/L_0)I_{\{L_0>0\}} + \infty I_{\{L_0=0\}}, \quad \text{and} \tag{2.28}$$

$$L_0 = \frac{2}{1 + L_{0,1}} \quad \text{and} \quad L_1 = \frac{2L_{0,1}}{1 + L_{0,1}},$$

with the convention of $2/\infty = 0$ and $(2\infty)/(1 + \infty) = 2$.

Definition 2.42. *For the model $\mathcal{M}_{0,1}$ in (2.13) and the hypotheses in (2.14), a test $\varphi_c \in \mathcal{T}$ is called a likelihood ratio test at $c \in [0, \infty)$ if $\varphi_c = 1$ on $\{L_{0,1} > c\}$, $\{P_0, P_1\}$-a.s., and $\varphi_c = 0$ on $\{L_{0,1} < c\}$, $\{P_0, P_1\}$-a.s. It is called a likelihood ratio test at $c = \infty$ if $\varphi_\infty = 0$ on $\{L_{0,1} < \infty\}$, $\{P_0, P_1\}$-a.s. The set of all likelihood ratio tests for any $c \in [0, \infty)$ is denoted by $\mathcal{R}_{0,1}$.*

Set

$$F_0(t) := P_0(L_{0,1} \le t), \quad t \ge 0, \tag{2.29}$$

and recall that $P_0(L_{0,1} < \infty) = 1$ by $P_0(L_0 = 0) = 0$ and (2.28). The existence of a likelihood ratio test with size α for any given $\alpha \in (0, 1)$ is shown next. The construction follows along the lines of (2.17), (2.18), and (2.19). Here it is based on $c_{1-\alpha} := F_0^{-1}(1 - \alpha)$, the $(1 - \alpha)$-quantile of F_0.

Proposition 2.43. *Let $\alpha \in (0, 1)$ be fixed. If*

$$\psi_\alpha(t) = \begin{cases} 1 & \text{if } t > c_{1-\alpha} \\ \gamma_\alpha & \text{if } t = c_{1-\alpha} \\ 0 & \text{if } t < c_{1-\alpha} \end{cases}, \quad t \in \overline{\mathbb{R}}, \tag{2.30}$$

where

$$c_{1-\alpha} = F_0^{-1}(1 - \alpha), \quad \gamma_\alpha = [F_0(c_{1-\alpha}) - (1 - \alpha)] \oslash P_0(L_{0,1} = c_{1-\alpha}),$$

and F_0 is given by (2.29), then $\mathsf{E}_0\psi_\alpha(L_{0,1}) = \alpha$.

Proof. We have $\mathsf{E}_0\psi_\alpha(L_{0,1}) = (1 - F_0(c_{1-\alpha})) + \gamma_\alpha P_0(L_{0,1} = c_{1-\alpha}) = \alpha$. ■

The test $\psi_\alpha(L_{0,1})$ with ψ_α in (2.30) is of course a likelihood ratio test as $\psi_\alpha(L_{0,1})$ meets the conditions of Definition 2.42. In the testing problem (2.14) for the statistical model (2.13) let $\alpha \in (0, 1)$ be a fixed given level. In practice, the values of α are usually small; e.g. $\alpha = 0.05$ or $\alpha = 0.01$. Of minor interest are the excluded trivial cases of $\alpha = 0$ and $\alpha = 1$ which are, however, considered briefly in Remark 2.46.

The likelihood ratio tests introduced by Definition 2.42 have important properties.

Proposition 2.44. *Every $\varphi_c \in \mathcal{R}_{0,1}$ minimizes the second kind error probability in the class of all tests $\phi \in \mathcal{T}$ with size $\alpha(\phi) \le \alpha(\varphi_c)$. On the other hand, let $\varphi \in \mathcal{T}$ be any test that minimizes the second kind error probability in the class of all tests $\phi \in \mathcal{T}$ with size $\alpha(\phi) \le \alpha(\varphi)$. If $\alpha := \alpha(\varphi) \in (0,1)$, then φ and the test $\psi_\alpha(L_{0,1})$ with ψ_α from (2.30) are $\{P_0, P_1\}$-a.s. identical outside of $\{L_{0,1} = c_{1-\alpha}\}$.*

Proof. Let $\varphi_c \in \mathcal{R}_{0,1}$. We use the notation from (2.28). By Definition 2.42 it holds for every $\phi \in \mathcal{T}$,

$$0 \le (\varphi_c - \phi)(L_1 - cL_0), \quad \{P_0, P_1\}\text{-a.s.} \tag{2.31}$$

Hence for every $\phi \in \mathcal{T}$ with $\alpha(\phi) \le \alpha(\varphi_c)$ we get

$$0 \le \mathsf{E}_{\overline{P}}(\varphi_c - \phi)(L_1 - cL_0) \le \mathsf{E}_{\overline{P}}(\varphi_c - \phi)L_1 = \mathsf{E}_1\varphi_c - \mathsf{E}_1\phi.$$

To prove the second statement, assume that $0 < \alpha(\varphi) < 1$ and put $\alpha = \alpha(\varphi)$. Then $\psi_\alpha(L_{0,1})$ with ψ_α from (2.30) is another best test with size α. Hence

$$\mathsf{E}_{\overline{P}}(\psi_\alpha - \varphi)(L_1 - c_{1-\alpha}L_0) = \mathsf{E}_{\overline{P}}(\psi_\alpha - \varphi)L_1 = 0,$$

which, in view of (2.31), completes the proof. ■

We have seen in Proposition 2.29 that under the model $\mathcal{M}_{0,1}$ from (2.13) for the hypotheses (2.14) in general we cannot find a test that minimizes $\mathsf{E}_0\varphi$ and $\mathsf{E}_1(1-\varphi)$ simultaneously. We have then adopted the optimality criterion of minimizing $1 - \mathsf{E}_1\varphi$ subject to $\mathsf{E}_0\varphi \le \alpha$, where $\alpha \in (0,1)$ is given. This approach, where the two hypotheses are no longer treated symmetrically, has proved to be very useful in many real-life situations. It reflects a strong desire of accepting H_A, but with the insurance that under H_0 the probability of a wrong decision is at most α. Under the general model $\mathcal{M}$ from (2.11) for the hypotheses (2.12) this approach has been generalized to the concept of a uniformly best level α test for H_0 versus H_A in (2.12). The next result, combined with the previous two propositions, is called the fundamental *Neyman–Pearson lemma*, which characterizes completely the best level α test for H_0 versus H_A in (2.14). The first comprehensive presentation was presumably provided by Lehmann (1959).

Theorem 2.45. ***(Neyman–Pearson Lemma)*** *For testing $\mathsf{H}_0 : P_0$ versus $\mathsf{H}_A : P_1$, every $\varphi_c \in \mathcal{R}_{0,1}$ is a best level $\alpha(\varphi_c)$ test. For $\alpha \in (0,1)$ every best level α test φ for H_0 versus H_A with $\alpha(\varphi) = \alpha$ and the test $\psi_\alpha(L_{0,1})$ with ψ_α from (2.30) are $\{P_0, P_1\}$-a.s. identical outside of $\{L_{0,1} = c_{1-\alpha_0}\}$.*

Proof. The statements follow from Proposition 2.44. ∎

It should be noted that each test $\varphi_c \in \mathcal{R}_{0,1}$ has been left arbitrary on the set $\{L_{0,1} = c\}$. On that set we cannot distinguish the two distributions by means of the threshold c for $L_{0,1}$. All we know is that for every measurable set $B \subseteq \{L_{0,1} = c\}$ it holds $P_0(B) = cP_1(B)$. An important consequence of Theorem 2.45 and Proposition 2.43 is that every test $\varphi_c \in \mathcal{R}_{0,1}$ can be modified on $\{L_{0,1} = c\}$ to be constant there and still have the same power function.

Remark 2.46. Under the binary model (2.13) let us briefly consider some trivial tests for (2.14). Here we use the notation from (2.28). If φ is a level α test, then $\mathsf{E}_0\varphi = \mathsf{E}_{\overline{P}}L_0\varphi = 0$ implies $\overline{P}(\varphi > 0, L_0 > 0) = 0$. Such a test φ is a best test if and only if $\varphi = 1$ P_1-a.s. on $\{L_0 = 0\}$, which is equivalent with $\varphi = 1$ $\overline{P}$-a.s. on $\{L_0 = 0\}$ as $P_0(L_0 = 0) = 0$. Hence a test is a best level 0 test if and only if $\varphi = I_{\{L_0=0\}}$ $\overline{P}$-a.s. For $\alpha = 1$ the test $\phi \equiv 1$ is a best level 1 test. However, $\varphi_\infty = I_{\{L_1>0\}}$ is also a best level 1 test, but with size $\mathsf{E}_0\varphi_\infty$ that may be less than 1. A similar fact has been established in Problem 2.32 where it has been shown that a level $\alpha \in (0,1)$ test φ with $\mathsf{E}_1\varphi = 1$ may have $\alpha(\varphi) < \alpha$.

Problem 2.47.* For $\alpha \in (0,1)$ the best level α test $\psi_\alpha(L_{0,1})$ satisfies $\mathsf{E}_1\psi_\alpha(L_{0,1}) \geq \alpha$. Moreover, $\mathsf{E}_1\psi_\alpha(L_{0,1}) = \alpha$ holds if and only if $P_0 = P_1$.

Problem 2.48.* For fixed $0 < a < 1 < b$, consider the following model of two uniform distributions and the testing problem

$$\mathcal{M}_{0,1} = (\mathbb{R}, \mathfrak{B}, \{\mathsf{U}(0,1), \mathsf{U}(a,b)\}), \quad \mathsf{H}_0 : \mathsf{U}(0,1), \quad \mathsf{H}_A : \mathsf{U}(a,b).$$

Find a best level α test for H_0 versus H_A for a fixed given $\alpha \in (0,1)$.

The Neyman–Pearson lemma can be used to establish uniformly best level α tests for one-sided testing problems in one-parameter families. The next theorem is for families with MLR and due to Karlin and Rubin (1956). It was established earlier for exponential families by Blackwell and Girshick (1954).

Theorem 2.49. *Suppose that in the model $(\mathcal{X}, \mathfrak{A}, (P_\theta)_{\theta\in\Delta})$ the family of distributions has a nondecreasing likelihood ratio in the statistic $T : \mathcal{X} \to_m \mathbb{R}$. Let $\mathsf{H}_0 : \Delta_0 = (-\infty, \theta_0] \cap \Delta$ and $\mathsf{H}_A : \Delta_A = (\theta_0, \infty) \cap \Delta$ for some $\theta_0 \in \Delta$, where $\Delta_0, \Delta_A \neq \emptyset$. For $\alpha \in (0,1)$ the test $\varphi_{T,\alpha}$ from (2.19) is a uniformly best level α test for H_0 versus H_A. Furthermore, the power function $\mathsf{E}_\theta\varphi_{T,\alpha}$ is nondecreasing in $\theta \in \Delta$, and it satisfies*

$$\mathsf{E}_{\underline{\theta}}\varphi_{T,\alpha} \leq \alpha = \mathsf{E}_{\theta_0}\varphi_{T,\alpha} \leq \mathsf{E}_{\overline{\theta}}\varphi_{T,\alpha}, \quad \underline{\theta} \leq \theta_0 \leq \overline{\theta}, \quad \underline{\theta}, \overline{\theta} \in \Delta. \tag{2.32}$$

Moreover, for every test φ with $\mathsf{E}_{\theta_0}\varphi = \alpha$ it holds

$$\mathsf{E}_{\underline{\theta}}\varphi_{T,\alpha} \leq \mathsf{E}_{\underline{\theta}}\varphi, \quad \underline{\theta} < \theta_0, \quad \text{and} \quad \mathsf{E}_{\overline{\theta}}\varphi_{T,\alpha} \geq \mathsf{E}_{\overline{\theta}}\varphi, \quad \overline{\theta} > \theta_0. \tag{2.33}$$

Proof. Let $\overline{\theta} > \theta_0$ be fixed. By Definition 2.11, $L_{\theta_0,\overline{\theta}} = h_{\theta_0,\overline{\theta}}(T)$, $\{P_{\theta_0}, P_{\overline{\theta}}\}$-a.s., where $h_{\theta_0,\overline{\theta}}$ is a nondecreasing function. For $c := h_{\theta_0,\overline{\theta}}(c_{1-\alpha})$ it holds, $\{P_{\theta_0}, P_{\overline{\theta}}\}$-a.s.,

$$\begin{aligned} h_{\theta_0,\overline{\theta}}(T) > c \quad &\text{implies} \quad T > c_{1-\alpha} \quad \text{and} \\ h_{\theta_0,\overline{\theta}}(T) < c \quad &\text{implies} \quad T < c_{1-\alpha}. \end{aligned} \tag{2.34}$$

This means that the test $\varphi_{T,\alpha}$ is a likelihood ratio test at c. Hence by Theorem 2.45, $\varphi_{T,\alpha}$ is a best level α test at the level of $\alpha = \mathsf{E}_{\theta_0}\varphi_{T,\alpha}$ for testing P_{θ_0} versus $P_{\overline{\theta}}$. This proves the first statement and the second inequality of (2.33). The inequalities in (2.32) follow from Theorem 2.10, Proposition 2.7, and the fact that $\varphi_{T,\alpha}$ is a nondecreasing function of T. To prove the first inequality of (2.33) let $\underline{\theta} < \theta_0$ be fixed. We remark that the tests $1-\varphi$ and $1-\varphi_{T,\alpha}$ are level $(1-\alpha)$ tests for P_{θ_0} versus $P_{\underline{\theta}}$. Set $h_{\theta_0,\underline{\theta}} = (h_{\underline{\theta},\theta_0})^{-1}$ with the standard conventions of $1/0 = \infty$ and $1/\infty = 0$. Then $L_{\theta_0,\underline{\theta}} = h_{\theta_0,\underline{\theta}}(T)$, so that the likelihood ratio is a nonincreasing function of T. As in (2.34) one can see that $1-\varphi_{T,\alpha}$ is a level $(1-\alpha)$ likelihood ratio test for testing P_{θ_0} versus $P_{\underline{\theta}}$. An application of Theorem 2.45 now shows that

$$\mathsf{E}_{\underline{\theta}}\varphi_{T,\alpha} = 1 - \mathsf{E}_{\underline{\theta}}(1-\varphi_{T,\alpha}) \leq 1 - \mathsf{E}_{\underline{\theta}}(1-\varphi) = \mathsf{E}_{\underline{\theta}}\varphi,$$

which completes the proof. ■

For a symmetric and unimodal parent distribution the power of any test in the associated location model can be explicitly bounded by the power of the best test.

Problem 2.50.* Suppose X has a symmetric and strongly unimodal distribution P with a density f that is everywhere positive and c.d.f. F. For $0 < \alpha < 1$ let $u_{1-\alpha}$ be a $(1-\alpha)$–quantile. If $P_\theta = \mathcal{L}(X+\theta)$ and $\varphi : \mathbb{R} \to_m [0,1]$ is a test, then

$$\begin{aligned} \int \varphi dP_{\theta_1} \leq \alpha \quad &\text{implies} \quad \int \varphi dP_{\theta_2} \leq F(u_\alpha + |\theta_2 - \theta_1|), \\ \int \varphi dP_{\theta_1} \geq \alpha \quad &\text{implies} \quad \int \varphi dP_{\theta_2} \geq F(u_\alpha - |\theta_2 - \theta_1|). \end{aligned}$$

Now we apply the last theorem to a reparametrized one-parameter exponential family $(P_{\kappa(\eta)})_{\eta\in(a,b)}$ from (1.11) with $\boldsymbol{\mu}$-densities

$$\frac{dP_{\kappa(\eta)}}{d\boldsymbol{\mu}}(x) = \exp\{\kappa(\eta)T(x) - K(\kappa(\eta))\}, \quad x \in \mathcal{X}, \tag{2.35}$$

where $\kappa : (a,b) \to \Delta^0$ is increasing and continuous. Then $(P_{\kappa(\eta)})_{\eta\in(a,b)}$ has an increasing likelihood ratio in the generating statistic T. Denote by F_η the c.d.f. of T under $P_{\kappa(\eta)}$.

Proposition 2.51. *Suppose that $(P_{\kappa(\eta)})_{\eta\in(a,b)}$ is a one parameter exponential family on $(\mathcal{X},\mathfrak{A})$ with $\boldsymbol{\mu}$-densities (2.35) and generating statistic T, where κ is increasing. For a fixed $\eta_0 \in (a,b)$ let*

$$\mathsf{H}_0 : \eta \leq \eta_0 \quad \textit{versus} \quad \mathsf{H}_A : \eta > \eta_0, \quad \eta \in (a,b).$$

Then for $\alpha \in (0,1)$ the test $\varphi_{T,\alpha} = I_{(c_{1-\alpha},\infty)}(T) + \gamma_\alpha I_{\{c_{1-\alpha}\}}(T)$ with

$$c_{1-\alpha} = F_{\eta_0}^{-1}(1-\alpha), \quad \gamma_\alpha = [F_{\eta_0}(c_{1-\alpha}) - (1-\alpha)] \oslash P_{\kappa(\eta_0)}(T = c_{1-\alpha})$$

is a uniformly best level α test for H_0 versus H_A and has an increasing power function $\eta \mapsto \mathsf{E}_\eta \varphi_{T,\alpha}$. Moreover, for every other test ϕ with $\mathsf{E}_{\eta_0}\phi = \alpha$ it holds

$$\mathsf{E}_\eta \phi > \mathsf{E}_\eta \varphi_{T,\alpha}, \quad \eta < \eta_0,$$
$$\mathsf{E}_\eta \phi < \mathsf{E}_\eta \varphi_{T,\alpha}, \quad \eta > \eta_0.$$

Proof. In view of Theorem 2.49 it only remains to prove the strict monotonicity of the power function $\eta \mapsto \mathsf{E}_\eta \varphi_{T,\alpha}$. Assume that there are $\eta_1, \eta_2 \in (a,b)$ with $\eta_1 < \eta_2$ and $\mathsf{E}_{\eta_1}\varphi_{T,\alpha} = \mathsf{E}_{\eta_2}\varphi_{T,\alpha}$. We set $\alpha_0 = \mathsf{E}_{\eta_1}\varphi_{T,\alpha}$. As all distributions are equivalent we have $0 < \alpha_0 < 1$. Then for the testing problem $\mathsf{H}_0 : P_{\kappa(\eta_1)}$ versus $\mathsf{H}_A : P_{\kappa(\eta_2)}$ the test $\varphi_{T,\alpha}$ is a likelihood ratio test and thus by Theorem 2.45 a best level α_0-test. Because of $\mathsf{E}_{\eta_1}\varphi_{T,\alpha} = \mathsf{E}_{\eta_2}\varphi_{T,\alpha}$ the test $\psi \equiv \alpha_0$ is also a best level α_0 test for $\mathsf{H}_0 : P_{\kappa(\eta_1)}$ versus $\mathsf{H}_A : P_{\kappa(\eta_2)}$, and thus by Theorem 2.45 that $\varphi_{T,\alpha} = \alpha_0$, $P_{\kappa(\eta_1)}$-a.s., outside of $\{T = c_{1-\alpha}\}$. As $P_{\kappa(\eta_1)}(\varphi_{T,\alpha} = \alpha_0, T \neq c_{1-\alpha}) = 0$ we get $P_{\kappa(\eta_1)}(T = c_{1-\alpha}) = 1$. But this is impossible according to the assumption (A1) made in Section 1.1. ■

Although Proposition 2.51 gives the explicit structure of the UMP test the remaining problem is to determine the quantile $c_{1-\alpha}$, which leads to numerical methods unless the distribution of T under P_{θ_0} is known. The next example presents such a situation.

Example 2.52. Consider the testing problem (2.20). We know from the second part of Example 1.11 that $(\mathsf{N}^{\otimes n}(\mu, \sigma_0^2))_{\mu\in\mathbb{R}}$ is a reparametrized exponential family with generating statistic $U(x_1, ..., x_n) = \sum_{i=1}^n x_i$ and thus has MLR in U. But then this family has also MLR in U_0 from (2.21). Hence we see that the Gauss test (2.22) is a uniformly best level α test for the testing problem (2.20).

Example 2.53. Suppose $X_1, ..., X_n$ are i.i.d. with distribution $\mathcal{L}(X_i) = (1-p)\boldsymbol{\delta}_0 + p\boldsymbol{\delta}_1$, $p \in (0,1)$. As we have seen in Example 1.7, $((1-p)\boldsymbol{\delta}_0 + p\boldsymbol{\delta}_1)^{\otimes n}$ is an exponential family with generating statistic $T_{\oplus n}(x_1, ..., x_n) = \sum_{i=1}^n x_i$, where $x = (x_1, ..., x_n) \in \mathcal{X} = \{0,1\}^n$. We want to construct a uniformly best level α test for

$$\mathsf{H}_0 : 0 < p \leq p_0 \quad \text{versus} \quad \mathsf{H}_A : p_0 < p < 1. \tag{2.36}$$

Under $((1-p_0)\boldsymbol{\delta}_0 + p_0\boldsymbol{\delta}_1)^{\otimes n}$ the statistic $T_{\oplus n}$ has a binomial distribution $\mathsf{B}(n, p_0)$. Let $c_{1-\alpha}$ and $\gamma_{1-\alpha}$ be determined by

$$\sum_{l=0}^{c_{1-\alpha}-1} \mathsf{b}_{n,p_0}(l) < 1-\alpha \leq \sum_{l=0}^{c_{1-\alpha}} \mathsf{b}_{n,p_0}(l), \quad \text{and} \tag{2.37}$$
$$\gamma_\alpha = \frac{1}{\mathsf{b}_{n,p_0}(c_{1-\alpha})}\left[\sum_{l=0}^{c_{1-\alpha}} \mathsf{b}_{n,p_0}(l) - (1-\alpha)\right].$$

Then, according to Proposition 2.51, the test

$$\varphi_{T,\alpha}(x) = \begin{cases} 1 & \text{if } T_{\oplus n}(x) > c_{1-\alpha} \\ \gamma_\alpha & \text{if } T_{\oplus n}(x) = c_{1-\alpha} \\ 0 & \text{if } T_{\oplus n}(x) < c_{1-\alpha} \end{cases} \tag{2.38}$$

is a uniformly best level α test for H_0 versus H_A.

The next problem deals with the Poisson distribution where the natural parameter λ of the exponential family is the parameter of interest.

Problem 2.54. Let $\mathsf{Po}(\lambda)$ be a Poisson distribution with parameter $\lambda > 0$ and p.m.f. $\mathsf{po}_\lambda(x) = (1/k!)\lambda^k \exp\{-\lambda\}$, $k \in \mathbb{N}$. Similarly to the previous example, for $\lambda_0 > 0$ and $\alpha \in (0,1)$ fixed find a uniformly best level α test for $\mathsf{H}_0 : \lambda \le \lambda_0$ versus $\mathsf{H}_A : \lambda_0 < \lambda$.

Example 2.55. We know from Example 1.13 that the family of distributions $(\mathsf{Ga}^{\otimes n}(\lambda,\beta))_{\lambda,\beta>0}$ of an i.i.d. sample of size n from a gamma distribution is a two-parameter exponential family with generating statistic

$$T(x_1,...,x_n) = (T_1(x_1,...,x_n), T_2(x_1,...,x_n)) = (\sum\nolimits_{i=1}^n \ln x_i, -\sum\nolimits_{i=1}^n x_i).$$

By Problem 1.14 for a fixed $\lambda_0 > 0$ the family $\mathsf{Ga}^{\otimes n}(\lambda_0,\theta)$, $\theta > 0$, is a one-parameter exponential family with natural parameter θ and generating statistic $T_2(x_1,...,x_n) = -\sum_{i=1}^n x_i$. From (2.6) we know that under $\mathsf{Ga}^{\otimes n}(\lambda_0,\theta)$, $S := -T_2$ has the distribution $\mathsf{Ga}(n\lambda_0,\theta)$. Let $\theta_0 > 0$ and $\alpha \in (0,1)$ be fixed. We consider the testing problem $\mathsf{H}_0 : 0 < \theta \le \theta_0$ versus $\mathsf{H}_A : \theta_0 < \theta < \infty$. Let c_α be determined by

$$\int_0^{c_\alpha} \mathsf{ga}_{n\lambda_0,\theta_0}(s)ds = \alpha,$$

i.e., c_α is the α-quantile of $\mathsf{Ga}(n\lambda_0,\theta_0)$. Then by Proposition 2.51 the test

$$\varphi(x_1,...,x_n) = \begin{cases} 1 & \text{if} \quad \sum_{i=1}^n x_i < c_\alpha \\ 0 & \text{if} \quad \sum_{i=1}^n x_i \ge c_\alpha \end{cases}$$

is a uniformly best level α test for H_0 versus H_A. In this testing problem we have assumed that λ_0 is known. On the other hand, in many situations the nuisance parameter is unknown. In Chapter 8 it is shown how to find optimal tests in such more complicated settings.

Next we consider families with MLR that are not exponential families.

Problem 2.56.* Let $X_1,...,X_n$ be i.i.d. random variables with $\mathcal{L}(X_1) = \mathsf{U}(0,\theta)$, $\theta > 0$. For $\theta_0 > 0$ and $\alpha \in (0,1)$ fixed find a uniformly best level α test for $\mathsf{H}_0 : \theta \le \theta_0$ versus $\mathsf{H}_A : \theta > \theta_0$.

Problem 2.57.* Let $X_1,...,X_n$ be i.i.d. random variables with $\mathcal{L}(X_1) = \mathsf{U}(\theta,\theta+1)$, $\theta \in \mathbb{R}$. Let $\theta_0 \in \mathbb{R}$ and $\alpha \in (0,1)$ be fixed. Show that for $n = 1$ there exists a uniformly best level α test for $\mathsf{H}_0 : \theta \le \theta_0$ versus $\mathsf{H}_A : \theta > \theta_0$. Show also that the concept of MLR cannot be utilized in the case of $n > 1$.

Problem 2.56, and the first part of Problem 2.57 with $n = 1$, give examples of families with MLR that are not exponential families. This can be seen easily because the supports of the distributions depend on $\theta \in \Delta$. A converse of Proposition 2.51 is due to Pfanzagl (1968). Under mild conditions, the existence of a uniformly best level α test for one-sided alternatives, for one $\alpha \in (0,1)$ and all sample sizes $n = 1,2,...$, implies that the underlying family of distributions is an exponential family.

For a family with MLR in T the uniformly best level α test for one-sided hypotheses is nonrandomized whenever the statistic T has a continuous c.d.f. under P_{θ_0}. This is reflected by several of the previous examples and problems. In contrast to that situation the testing problem (2.36) for binomial distributions leads to a uniformly best level α test that is randomized, except for a few ("natural") choices of α for which $\gamma_a = 0$, that is, when

$$\sum\nolimits_{l=0}^{c_{1-\alpha}} \mathsf{b}_{n,p_0}(l) = 1 - \alpha$$

occurs. $\gamma_a = 1$ is not possible because of the definition of $c_{1-\alpha}$. If now $\gamma_a \in (0,1)$ and $T_{\oplus n}(x) = c_{1-\alpha}$, then we have to generate a random variable U that is uniformly distributed in $(0,1)$, reject H_0 if $U \leq \gamma_\alpha$, and accept H_0 if $U > \gamma_\alpha$.

Sometimes it is more convenient to incorporate a randomization right from the beginning, with the effect that the distributions in the new model have continuous c.d.f.s. More specifically, let X be a random variable with values in $\mathbb{N}$ and distribution P_θ, $\theta \in \Delta$. Let $p_\theta(k) = P_\theta(\{k\})$, $k \in \mathbb{N}$, $\theta \in \Delta$, denote the associated p.m.f.s, i.e., densities with respect to the counting measure $\boldsymbol{\kappa}$ on $\mathbb{N}$. Let U be uniformly distributed in $(0,1)$, independent of X, and set $Y = X + U$. It is important to know that there is a one-to-one relation between Y and the pair (X, U). Indeed, if $[x]$ denotes the largest integer less than or equal to x, then

$$(X, U) = ([Y], Y - [Y]).$$

Thus the random variable Y contains the same information as (X, U). For $\theta \in \Delta$ the distribution of $Y = X + U$ is $Q_\theta = P_\theta * \mathsf{U}(0,1)$, which has the piecewise constant Lebesgue density

$$f_\theta(t) = \sum\nolimits_{l=0}^{\infty} p_\theta(l) I_{[l,l+1)}(t). \tag{2.39}$$

Suppose now that $\Delta \subseteq \mathbb{R}$. The family $(P_\theta)_{\theta \in \Delta}$ of distributions on $\mathbb{N}$ has MLR in the identity if and only if the p.m.f.s p_θ satisfy

$$p_{\theta_0}(k_1)p_{\theta_1}(k_0) \leq p_{\theta_0}(k_0)p_{\theta_1}(k_1), \quad 0 \leq k_0 < k_1,\ k_i \in \mathbb{N},\ \theta_0 < \theta_1.$$

It is easy to verify that in this case

$$f_{\theta_0}(t_1)f_{\theta_1}(t_0) \leq f_{\theta_0}(t_0)f_{\theta_1}(t_1), \quad 0 \leq t_0 < t_1,\ \theta_0 < \theta_1,$$

so that the family $(Q_\theta)_{\theta \in \Delta}$ has MLR in the identity as well.

For $\theta_0 \in \Delta^0$ and $\alpha \in (0,1)$, consider now the testing problem

$$\mathsf{H}_0 : \theta \leq \theta_0 \quad \text{versus} \quad \mathsf{H}_A : \theta > \theta_0. \tag{2.40}$$

For the family $(P_\theta)_{\theta \in \Delta}$ a uniformly best level α test based on X is $\varphi_{X,\alpha}(k)$, $k \in \mathbb{N}$, with $c_{1-\alpha}$ and γ_α determined by (2.18), analogously to (2.38) with (2.37). On the other hand, for the family $(Q_\theta)_{\theta \in \Delta}$ the uniformly best level α test based on Y is $\varphi_{Y,\alpha}(y)$, $y > 0$, given by

$$\varphi_{Y,\alpha}(y) = \begin{cases} 1 & \text{if} \quad y \geq d_{1-\alpha} \\ 0 & \text{if} \quad y < d_{1-\alpha} \end{cases} \quad \text{where} \quad \int_{d_{1-\alpha}}^{\infty} f_{\theta_0}(t)dt = \alpha.$$

To summarize with the proposition below, by using the interpolation (2.39) of discrete distributions a randomized test for discrete distributions $(P_\theta)_{\theta\in\Delta}$ can be replaced by a nonrandomized test for distributions $(Q_\theta)_{\theta\in\Delta}$ that have Lebesgue densities.

Proposition 2.58. *Let $(P_\theta)_{\theta\in\Delta}$ be a family of distributions on $\mathbb{N}$ that has MLR in the identity. Let $Q_\theta = P_\theta * \mathsf{U}(0,1)$, $\theta \in \Delta$. Then $(Q_\theta)_{\theta\in\Delta}$ has MLR in the identity as well. For fixed $\alpha \in (0,1)$ the uniformly best level α test $\varphi_{X,\alpha}$ based on X with distribution P_θ, $\theta \in \Delta$, and the uniformly best level α test $\varphi_{Y,\alpha}$ based on Y with distribution P_θ, $\theta \in \Delta$, have the same power function; that is,*

$$\mathsf{E}_{P_\theta}\varphi_{X,\alpha} = \mathsf{E}_{Q_\theta}\varphi_{Y,\alpha}, \quad \theta \in \Delta.$$

Proof. It holds $d_{1-\alpha} = c_{1-\alpha} + \gamma_\alpha$, and thus

$$\mathsf{E}_{Q_\theta}\varphi_{Y,\alpha} = \int_{d_{1-\alpha}}^{\infty} f_\theta(t)dt = \gamma_a p_\theta(c_{1-\alpha}) + \sum\nolimits_{l=c_{1-\alpha}+1}^{\infty} p_\theta(l) = \mathsf{E}_{P_\theta}\varphi_{X,\alpha}.$$

■

We now return to the basic setting of a binary model $\mathcal{M}_{0,1}$ from (2.13). As has been pointed out already, a best level α test, in the sense of Neyman–Pearson, is set up for situations where an error of the first kind has more serious consequences than an error of the second kind. Protecting ourselves against an error of the first kind by means of the level α has a price, however. For smaller choices of α this may lead to a larger value of the probability of an error of the second kind. The same effect, of course, is caused by P_1 getting closer to P_0.

The Neyman–Pearson approach is not appropriate for all situations where tests in binary models are under concern. As an alternative approach one could utilize weights $c_0, c_1 > 0$ that reflect the relative importance of the two hypotheses and require to minimize the function $\varphi \mapsto c_0\mathsf{E}_0\varphi + c_1\mathsf{E}_1(1-\varphi)$. Apparently, in this approach only the relative weights $c_i/(c_0+c_1)$, $i = 0,1$, are relevant, and thus we may as well consider for any prior distribution $\Pi = \pi\boldsymbol{\delta}_0 + (1-\pi)\boldsymbol{\delta}_1$, $\pi \in (0,1)$, the Bayes risk of a test φ,

$$\mathsf{r}(\Pi,\varphi) = \pi\mathsf{E}_0\varphi + (1-\pi)\mathsf{E}_1(1-\varphi).$$

Definition 2.59. *Under the prior $\Pi = \pi\boldsymbol{\delta}_0 + (1-\pi)\boldsymbol{\delta}_1$ with $\pi \in [0,1]$, the value $\mathsf{b}_\pi(P_0,P_1) = \inf_{\varphi\in\mathcal{T}} \mathsf{r}(\Pi,\varphi)$ is called the Bayes risk of the testing problem $\mathsf{H}_0 : P_0$ versus $\mathsf{H}_1 : P_1$. The function $\pi \to \mathsf{b}_\pi(P_0,P_1)$ is called the error function. Every test φ_B with $\mathsf{r}(\Pi,\varphi_B) = \mathsf{b}_\pi(P_0,P_1)$ is called a Bayes test.*

In the above definition we have included the cases of $\pi = 0$ and $\pi = 1$ although they are trivial from a statistical point of view. Obviously, $\varphi_B \equiv 1$

is a Bayes test for $\pi = 0$, $\varphi_B \equiv 0$ is a Bayes test for $\pi = 1$, and the Bayes risk is 0 in both cases.

We dominate P_0, P_1 by $\mu \in \mathcal{M}^\sigma(\mathfrak{A})$ and set $f_i = dP_i/d\mu$, $i = 0, 1$. Then

$$L_{0,1} = I_{\{f_0>0\}} f_1/f_0 + \infty I_{\{f_1>0, f_0=0\}}$$

is a likelihood ratio of P_1 with respect to P_0; see (1.68). Recall that for any $c \geq 0$ every test φ_c that is 1 for $f_1 > cf_0$ and 0 for $f_1 < cf_0$ is called a likelihood ratio at c, see Definition 2.42. The next theorem extends Lemma 1.66.

Theorem 2.60. *If $\Pi = \pi\boldsymbol{\delta}_0 + (1-\pi)\boldsymbol{\delta}_1$ with $\pi \in (0,1)$, then a test φ_B is a Bayes test if and only if it is $\{P_0, P_1\}$-a.s. a likelihood ratio test at $c = \pi/(1-\pi)$; that is,*

$$\begin{aligned} \varphi_B &= 1, \quad \{P_0, P_1\}\text{-a.s. on } \{(1-\pi)f_1 > \pi f_0\}, \\ \varphi_B &= 0, \quad \{P_0, P_1\}\text{-a.s. on } \{(1-\pi)f_1 < \pi f_0\}. \end{aligned} \tag{2.41}$$

Moreover, the Bayes risk is given by

$$\begin{aligned} \mathsf{b}_\pi(P_0, P_1) &= \int (\pi\varphi_B f_0 + (1-\pi)(1-\varphi_B)f_1)d\mu \\ &= \int (\pi f_0) \wedge ((1-\pi)f_1)d\mu. \end{aligned} \tag{2.42}$$

Proof. Put $g_0 = \pi f_0$, $g_1 = (1-\pi)f_1$. For any test φ it holds

$$\begin{aligned} &\mathsf{r}(\Pi, \varphi) - \mathsf{r}(\Pi, \varphi_B) \\ &= \int [\pi\varphi f_0 + (1-\pi)(1-\varphi)f_1 - (\pi f_0) \wedge ((1-\pi)f_1)]d\mu \\ &= \int [\varphi(g_0 - g_0 \wedge g_1) + (1-\varphi)(g_1 - g_0 \wedge g_1)]d\mu \geq 0, \end{aligned}$$

and equality holds if and only if $\varphi = 0$ on $\{g_0 > g_1\}$ and $\varphi = 1$ on $\{g_0 < g_1\}$, μ-a.e., which is equivalent to (2.41). ■

It is important to know that any Bayes test φ_B from (2.41) can be arbitrarily modified on the set $\{f_1 = cf_0\}$. Thus in particular there always exists a nonrandomized Bayes test.

Problem 2.61.* The error function $\pi \mapsto \mathsf{b}_\pi(P_0, P_1)$, $\pi \in [0,1]$, has the following properties.

$$\mathsf{b}_0(P_0, P_1) = \mathsf{b}_1(P_0, P_1) = 0. \tag{2.43}$$

$$0 \leq \mathsf{b}_\pi(P_0, P_1) \leq \pi^s(1-\pi)^{1-s}\mathsf{H}_s(P_0, P_1), \quad s > 0,\ s \neq 1. \tag{2.44}$$

$$\mathsf{b}_\pi(P_0, P_1) = \pi \wedge (1-\pi) \quad \Leftrightarrow \quad P_0 = P_1. \tag{2.45}$$

$$\mathsf{b}_\pi(P_0, P_1) = 0 \quad \Leftrightarrow \quad P_0 \perp P_1. \tag{2.46}$$

$$\pi \mapsto \mathsf{b}_\pi(P_0, P_1) \text{ is continuous and concave.}$$

$$\|P_0 - P_1\| = 2 - 4\mathsf{b}_{1/2}(P_0, P_1). \tag{2.47}$$

For the binary model $\mathcal{M}_{0,1}$ from (2.13) the testing problem (2.14) admits an interesting geometric interpretation. For this purpose we introduce the *risk set of tests*,

$$\mathfrak{R} = \{(\mathsf{E}_0\varphi, 1 - \mathsf{E}_1\varphi) : \varphi \in \mathcal{T}\}. \tag{2.48}$$

As $\mathcal{T}$ is a convex set the set $\mathfrak{R}$ is also convex. To describe the lower bound of the set $\mathfrak{R}$ we put

$$\mathsf{g}_\alpha(P_0, P_1) := \inf\{\mathsf{E}_1(1 - \varphi) : \mathsf{E}_0\varphi \le \alpha, \ \ \varphi \in \mathcal{T}\}, \quad \alpha \in [0, 1]. \tag{2.49}$$

By Theorem 2.45, for $\alpha \in (0, 1)$ the point $(\alpha, \mathsf{g}_\alpha(P_0, P_1))$ corresponds to a best level α test $\psi_\alpha(L_{0,1})$. Therefore $\alpha \mapsto \mathsf{g}_\alpha(P_0, P_1)$ is a nonincreasing function of α. As the lower boundary of the convex set $\mathfrak{R}$ the function $\alpha \mapsto \mathsf{g}_\alpha(P_0, P_1)$ is also convex; that is,

$$\mathsf{g}_{q\alpha_1+(1-q)\alpha_2}(P_0, P_1) \le q\mathsf{g}_{\alpha_1}(P_0, P_1) + (1 - q)\mathsf{g}_{\alpha_2}(P_0, P_1), \quad q \in [0, 1]. \tag{2.50}$$

Furthermore, because $\psi \equiv \alpha$ is a level α test we get

$$\mathsf{g}_\alpha(P_0, P_1) \le 1 - \alpha, \quad \alpha \in (0, 1).$$

If $\mathsf{g}_\alpha(P_0, P_1) = 1 - \alpha$ for some $\alpha \in (0, 1)$, then by Theorem 2.45 $\psi \equiv \alpha$ is $\{P_0, P_1\}$-a.s. identical with ψ_α outside of $\{L_{0,1} = c_{1-\alpha}\}$, which, however, is impossible as $\alpha \in (0, 1)$. Consequently,

$$\mathsf{g}_\alpha(P_0, P_1) < 1 - \alpha, \quad \alpha \in (0, 1).$$

Next we study interrelations between $\mathsf{g}_\alpha(P_0, P_1)$ and $\mathsf{b}_\pi(P_0, P_1)$. The proof of the next theorem follows along the lines of Torgersen (1991), pp. 590–591.

Theorem 2.62. *For the binary model $\mathcal{M}_{0,1}$ from (2.13) it holds*

$$\mathsf{b}_\pi(P_0, P_1) = \min_{0<\alpha<1} [\pi\alpha + (1 - \pi)\mathsf{g}_\alpha(P_0, P_1)], \quad \pi \in (0, 1), \tag{2.51}$$

$$\mathsf{g}_\alpha(P_0, P_1) = \max_{0<\pi<1} \frac{1}{1 - \pi}[\mathsf{b}_\pi(P_0, P_1) - \pi\alpha], \quad \alpha \in (0, 1). \tag{2.52}$$

Proof. For any fixed $\pi \in (0, 1)$ the inequality

$$\mathsf{b}_\pi(P_0, P_1) \le \pi\alpha + (1 - \pi)\mathsf{g}_\alpha(P_0, P_1), \quad \alpha \in (0, 1),$$

follows from the definition of $\mathsf{g}_\alpha(P_0, P_1)$. It implies

$$\mathsf{b}_\pi(P_0, P_1) \le \inf_{0<\alpha<1} [\pi\alpha + (1 - \pi)\mathsf{g}_\alpha(P_0, P_1)], \quad \pi \in (0, 1),$$

$$\mathsf{g}_\alpha(P_0, P_1) \ge \sup_{0<\pi<1} \frac{1}{1 - \pi}[\mathsf{b}_\pi(P_0, P_1) - \pi\alpha], \quad \alpha \in (0, 1).$$

If $\alpha_0 \in (0, 1)$ is fixed, then let ψ_{α_0} be the likelihood ratio test $\psi_{\alpha_0}(L_{0,1})$ which by Theorem 2.45 is a best level α_0-test. Hence ψ_{α_0} is according to Theorem 2.60 also a Bayes test for $\pi = c_{1-\alpha_0}/(1 + c_{1-\alpha_0})$. This yields

$$\pi\alpha_0 + (1 - \pi)\mathsf{E}_1(1 - \psi_{\alpha_0}) = \mathsf{b}_{\pi(\alpha_0)}(P_0, P_1),$$

and we have established the equalities in (2.51) and (2.52). ■

Remark 2.63. The representation (2.51) shows that $\pi \mapsto \mathsf{b}_\pi(P_0, P_1)$ is a concave function, whereas (2.52) shows that $\alpha \mapsto \mathsf{g}_\alpha(P_0, P_1)$ is a convex function. The latter has been established already in (2.50).

Problem 2.64.* Given are two coins C_0 and C_1 with probability of heads equal to p_0 and p_1, respectively. Assume that $0 < p_0 < p_1 < 1$. One of these coins is tossed once and $x \in \{0, 1\}$ is observed, where 0 stands for tails and 1 stands for heads. Set $P_i = (1 - p_i)\boldsymbol{\delta}_0 + p_i\boldsymbol{\delta}_1$, $i = 0, 1$, and $\Delta = \{0, 1\}$. For a fixed $\alpha \in (0, 1)$ find a best level α test for $H_0 : P_0$ versus $H_1 : P_1$. For a fixed $\pi \in (0, 1)$ find a Bayes test for the prior $\Pi = \pi\boldsymbol{\delta}_0 + (1 - \pi)\boldsymbol{\delta}_1$.

2.3 Solutions to Selected Problems

Solution to Problem 2.3: From the first statement of Problem 2.2 we get $\mathbb{P}(F^{-1}(U) \le x) = \mathbb{P}(U \le F(x)) = F(x)$, $x \in \mathbb{R}$. On the other hand, if F is continuous, then by the second statement of Problem 2.2 $F(F^{-1}(u)) = u$, $u \in [0, 1]$. Thus $\mathbb{P}(F(F^{-1}(U)) \le y) = \mathbb{P}(U \le y) = y$, $y \in [0, 1]$. $F^{-1}(U)$ has the same c.d.f. F as X. □

Solution to Problem 2.15 Let $x, x_0 \in \mathbb{R}^d$ and $s_2 > 0$ be fixed. By assumption, $g(x_0) \le g(x_0 + s_2(x - x_0)) = t$, say. As g is subconvex $\{y : g(y) \le t\}$ is a convex subset of $\mathbb{R}^d$. Thus for every $\alpha \in [0, 1]$ it holds $\alpha x_0 + (1 - \alpha)(x_0 + s_2(x - x_0)) \in \{y : g(y) \le t\}$.

Let now $s_1 \in (0, s_2)$. For $\alpha = 1 - s_1/s_2$ we get

$$\alpha x_0 + (1 - \alpha)(x_0 + s_2(x - x_0)) = x_0 + s_1(x - x_0),$$

and thus $g(x_0 + s_1(x - x_0)) \le t$. □

Solution to Problem 2.22: With $x_0 < x_1$ and $\theta_0 < \theta_1$ (2.4) yields

$$\begin{aligned}
& F_{\theta_0}(x_1)F_{\theta_1}(x_0) \\
&= F_{\theta_0}(x_0)F_{\theta_1}(x_0) + [F_{\theta_0}(x_1) - F_{\theta_0}(x_0)]F_{\theta_1}(x_0) \\
&= F_{\theta_0}(x_0)F_{\theta_1}(x_0) + \int I_{(x_0,x_1]}(v)[\int I_{(-\infty,x_0]}(u)g_{\theta_1}(u)g_{\theta_0}(v)\mu(du)]\mu(dv) \\
&\le F_{\theta_0}(x_0)F_{\theta_1}(x_0) + \int I_{(x_0,x_1]}(v)[\int I_{(-\infty,x_0]}(u)g_{\theta_0}(u)g_{\theta_1}(v)\mu(du)]\mu(dv) \\
&= F_{\theta_0}(x_0)F_{\theta_1}(x_0) + [F_{\theta_1}(x_1) - F_{\theta_1}(x_0)]F_{\theta_0}(x_0) \\
&= F_{\theta_0}(x_0)F_{\theta_1}(x_1).
\end{aligned}$$

The second statement can be shown analogously. □

Solution to Problem 2.24: For any fixed $i \in \{1, ..., n\}$ the Lebesgue density of $X_{[i]}$ is given by $f_{\theta,i}(x) = i\binom{n}{i}F_\theta(x)^{i-1}[1 - F_\theta(x)]^{n-i} f_\theta(x), \quad x \in \mathbb{R}$. The condition

$$f_{\theta_1,i}(x_0)f_{\theta_0,i}(x_1) \le f_{\theta_0,i}(x_0)f_{\theta_1,i}(x_1), \quad \theta_0 < \theta_1,\ x_0 < x_1,$$

follows from (2.4) and Problem 2.22. □

Solution to Problem 2.25: Let $h_0 : \mathbb{R} \to_m \mathbb{R}_+$ and $\mathcal{L}(X) = \mathsf{N}(\mu, 1)$. Then

$$\mathbb{E}h_0(X^2) = \frac{1}{2}\mathbb{E}[h_0(X^2) + h_0((-X)^2)]$$
$$= \frac{1}{2\sqrt{2\pi}} \int h_0(s^2)[\exp\{-(s-\mu)^2/2\} + \exp\{-(s+\mu)^2/2\}]ds$$
$$= \frac{1}{2\sqrt{2\pi}} \exp\{-\mu^2/2\} \int_0^\infty h_0(s)s^{-(1/2)} \exp\{-s/2\} [2\sum\nolimits_{k=0}^\infty (\mu\sqrt{s})^{2k}/(2k)!]ds$$
$$= \sum\nolimits_{k=0}^\infty \mathsf{po}_{\mu^2/2}(k) \int_0^\infty h_0(s)\mathsf{h}_{2k+1}(s)ds,$$

where we have used $\Gamma(2k) = 2^{2k-1}\Gamma(k)\Gamma(k+1/2)/\sqrt{\pi}$. Hence for $w > 0$,

$$\mathbb{P}(X^2/w \le t) = \mathbb{E}I_{(-\infty,tw]}(X^2) = \sum\nolimits_{k=0}^\infty \mathsf{po}_{\mu^2/2}(k) \int_0^{tw} \mathsf{h}_{2k+1}(s)ds$$
$$= \sum\nolimits_{k=0}^\infty \mathsf{po}_{\mu^2/2}(k)\mathsf{Ga}(k+1/2, w/2)((0,t]),$$

and therefore

$$\mathcal{L}(X^2/w) = \sum\nolimits_{k=0}^\infty \mathsf{po}_{\mu^2/2}(k)\mathsf{Ga}(k+1/2, w/2).$$

Using (2.6) and $\mathsf{Po}(\lambda_1) * \mathsf{Po}(\lambda_2) = \mathsf{Po}(\lambda_1 + \lambda_2)$ we get

$$\mathcal{L}(\frac{1}{w}(X_1^2 + \cdots X_n^2)) = \mathcal{L}(X_1^2/w)^{*n} = \sum\nolimits_{l=0}^\infty \mathsf{po}_{\delta^2/2}(l)\mathsf{Ga}(l+n/2, w/2)$$

The random variables $W^{-1}X_1^2, ..., W^{-1}X_n^2$ are conditionally, given $W = w$, independent. Hence for every $h : \mathbb{R} \to_m \mathbb{R}_+$,

$$\mathbb{E}h(W^{-1}(\sum\nolimits_{i=1}^n X_i^2)) = \int [\mathbb{E}h(W^{-1}(\sum\nolimits_{i=1}^n X_i^2))|W = w)]P_W(dw)$$
$$= \int [\int h(t) \sum\nolimits_{l=0}^\infty \mathsf{po}_{\delta^2/2}(l)\mathsf{Ga}(l+n/2, w/2)(dt)]P_W(dw),$$

which completes the proof. □

Solution to Problem 2.32: Suppose that $\alpha(\psi) < \alpha < 1$. Then we take $\varepsilon = (\alpha-\alpha(\psi))/(1-\alpha(\psi))$, which satisfies $\varepsilon \in (0,1)$, and consider the test $\widetilde{\psi} := (1-\varepsilon)\psi+\varepsilon$. It holds $\alpha(\widetilde{\psi}) = \alpha$ and $\mathsf{E}_\theta\widetilde{\psi} = (1-\varepsilon)\mathsf{E}_\theta\psi + \varepsilon$, $\theta \in \Delta$. If now $\mathsf{E}_{\theta_1}\psi < 1$ for some $\theta_1 \in \Delta_A$, then $\mathsf{E}_{\theta_1}\widetilde{\psi} > \mathsf{E}_{\theta_1}\psi$. This is a contradiction to ψ being a uniformly best level α test, and thus $\alpha(\psi) = \alpha$ must hold. On the other hand, if $\mathsf{E}_\theta\psi = 1$ for all $\theta \in \Delta_A$, then the same test $\widetilde{\psi}$ has $\mathsf{E}_\theta\psi = 1$ for all $\theta \in \Delta_A$. □

Solution to Problem 2.36: Let O be an orthogonal $n \times n$ matrix with the last row equal to $(1/\sqrt{n}, ..., 1/\sqrt{n})$, and set $Y = OZ$ where $Z = (Z_1, ..., Z_n)^T$. Then $Y_1, ..., Y_n$ are i.i.d. $\mathsf{N}(0,1)$ and $Y_n = \sqrt{n}\overline{Z}_n$. Furthermore $\sum_{i=1}^n (Z_i - \overline{Z}_n)^2 = \|Z\|^2 - Y_n^2 = \|Y\|^2 - Y_n^2 = \sum_{i=1}^{n-1} Y_i^2$. Obviously, $\overline{Z}_n = Y_n/\sqrt{n}$ and $\sum_{i=1}^{n-1} Y_i^2$ are independent. The statements regarding the distributions are clear. □

Solution to Problem 2.38: There exists an orthogonal matrix O such that $O^TAO = \Lambda$ is a diagonal matrix consisting of the eigenvalues. The assumption

$AA = A$ yields $\Lambda\Lambda = O^T AOO^T AO = O^T AO = \Lambda$ which implies the first statement. The second statement follows from the fact that multiplying A with an orthogonal matrix does not change its rank. □

Solution to Problem 2.39: Let O be an orthogonal matrix with $O^T AO = \Lambda$, where Λ is a diagonal matrix. As A is an idempotent matrix of rank d, the diagonal matrix Λ has d entries 1 and the remaining are zero. Without loss of generality we assume that the first d entries in the diagonal are 1. The random vector $Y = \sigma^{-1}O^T X$ has a normal distribution with expectation $\nu = \sigma^{-1}O^T \mu$ and the unit matrix $\mathbf{I}$ as covariance matrix. It holds $\sigma^{-2}(X^T AX) = Y^T O^T AOY = Y^T \Lambda Y = \sum_{i=1}^d Y_i^2$ which has by the definition of the noncentral χ^2-distribution in (2.7) a noncentral χ^2-distribution with d degrees of freedom and noncentrality parameter $\sum_{i=1}^d \nu_i^2 = \nu^T \Lambda \nu = \sigma^{-2}\mu^T O\Lambda O^T \mu = \sigma^{-2}\mu^T A\mu$. □

Solution to Problem 2.47: The power of $\psi_\alpha(L_{0,1})$ is not smaller than the power of the no-data test $\phi \equiv \alpha$, and thus $\mathsf{E}_1\psi_\alpha(L_{0,1}) \geq \alpha$. If $\mathsf{E}_1\psi_\alpha(L_{0,1}) = \alpha$, then ϕ is a best test and by $\mathsf{E}_0\phi = \alpha \in (0,1)$ and the uniqueness statement in Theorem 2.45 we have $\phi = \psi_\alpha(L_{0,1})$, $\{P_0, P_1\}$-a.s., outside of $\{L_{0,1} = c_{1-\alpha}\}$. That means that

$$P_i(L_{0,1} < c_{1-\alpha}) = P_i(L_{0,1} > c_{1-\alpha}) = 0, \quad i = 0, 1.$$

From (1.68) we get $P_1(B) = c_{1-\alpha}P_0(B)$ for every Borel set B. This yields $c_{1-\alpha} = 1$ and $P_0 = P_1$. □

Solution to Problem 2.48: The Lebesgue densities of $\mathsf{U}(0,1)$ and $\mathsf{U}(a,b)$ are $f_0(x) = I_{[0,1]}(x)$ and $(b-a)^{-1}I_{[a,b]}(x)$, $x \in \mathbb{R}$, respectively. We get from (1.68) $L_{0,1}(x) = 0$ for $x \in [0,a)$, $L_{0,1}(x) = 1/(b-a)$ for $x \in [a,1]$, and $L_{0,1}(x) = \infty$ for $x \in (1,b]$. For $\alpha \in [1-a, 1]$ a best level α test is given by $\varphi(x) = I_{[a,b]}(x) + \frac{\alpha-1+a}{a} I_{[0,a]}(x)$, and for $\alpha \in [0, 1-a]$ a best level α test is given by $\psi(x) = I_{[1,b]}(x) + \frac{\alpha}{1-a} I_{[a,1]}(x)$. □

Solution to Problem 2.50: Let $\mathsf{H}_0 : P_{\theta_1}$ and $\mathsf{H}_A : P_{\theta_2}$. Suppose $\theta_1 < \theta_2$. As the c.d.f. is continuous for $t \neq 0$ and the family $(P_\theta)_{\theta\in\mathbb{R}}$ has MLR in the identity we get from Theorem 2.49 that the test $\psi = I_{(\theta_1+u_{1-\alpha},\infty)}$ is a best level α test. Hence by $u_{1-\alpha} = -u_\alpha$

$$\begin{aligned}\int \varphi dP_{\theta_2} \leq \int \psi dP_{\theta_2} &= \mathbb{P}(X + \theta_2 > \theta_1 + u_{1-\alpha}) = 1 - F(\theta_1 - \theta_2 + u_{1-\alpha}) \\ &= F(u_\alpha + (\theta_2 - \theta_1)).\end{aligned}$$

If $\theta_2 < \theta_1$ we switch the roles of θ_1 and θ_2. The second statement follows by turning to $1-\varphi$ and to $1-\alpha$ and using $1 - F(u_{1-\alpha} + |\theta_2 - \theta_1|) = 1 - F(-u_\alpha + |\theta_2 - \theta_1|) = F(u_\alpha - |\theta_2 - \theta_1|)$. □

Solution to Problem 2.56: By Example 2.12, for $0 < \theta_1 < \theta_2 < \infty$ the likelihood ratio is

$$L_{\theta_1,\theta_2} = (\theta_1/\theta_2)^n \, I_{[0,\theta_1]}(\max(x_1, ..., x_n)) + \infty I_{(\theta_1,\theta_2]}(\max(x_1, ..., x_n)),$$

which is a nondecreasing function of $\max(x_1, ..., x_n)$. $\max(X_1, ..., X_n)$ has the $1-\alpha$ quantile $\theta_0(1-\alpha)^{1/n}$ under $\mathsf{U}(0, \theta_0)$ so that the uniformly best level α test is

$$\varphi(x) = \begin{cases} 1 & \text{if } \max(x_1, ..., x_n) > \theta_0(1-\alpha)^{1/n}, \\ 0 & \text{if } \max(x_1, ..., x_n) \le \theta_0(1-\alpha)^{1/n}. \end{cases} \quad \square$$

Solution to Problem 2.57: For $n = 1$, $\theta_1 \in \mathbb{R}$, and $\theta_2 \in (\theta_1, \theta_1 + 1)$ the likelihood ratio is $L_{\theta_1,\theta_2}(x) = 0$ for $x \in [\theta_1, \theta_2)$, $L_{\theta_1,\theta_2}(x) = 1$ for $x \in [\theta_2, \theta_1 + 1]$, and $L_{\theta_1,\theta_2}(x) = \infty$ for $x \in (\theta_1 + 1, \theta_2 + 1]$. Thus we have a nondecreasing likelihood ratio. Without loss of generality we may assume that $\theta_0 = 0$, because we could subtract θ_0 from X_1 to get to that setting. The uniformly best level α test is now given by the best level α test in Problem 2.48 for $a = \theta_2$ and $b = \theta_2 + 1$.

For $n > 1$, the Lebesgue density of $X = (X_1, ..., X_n)$ is

$$\begin{aligned} f_\theta(x) &= \prod\nolimits_{i=1}^{n} I_{[\theta,\theta+1]}(x_i) \\ &= I_{[\theta,\infty)}(\min\{x_1, ..., x_n\}) I_{[0,\theta+1]}(\max\{x_1, ..., x_n\}), \quad x \in \mathbb{R}^n. \end{aligned}$$

For $\theta_1 < \theta_2$, $\theta_1, \theta_2 \in \mathbb{R}$, the likelihood ratio $L_{\theta_1,\theta_2}(x)$ cannot be represented as a nondecreasing function of a statistic $T(x)$ and thus the family of distributions of X does not have MLR. For further details see Lehmann (1986), p. 115. $\square$

Solution to Problem 2.61: The statements (2.43), (2.45), and (2.46) follow directly from (2.42). (2.44) follows from $a \wedge b \le a^s b^{1-s}$ and (2.42). The continuity was already established in Problem 1.67. If $0 < \pi_1, \pi_2 < 1$ and $0 < q < 1$, then

$$\begin{aligned} &\mathsf{b}_{q\pi_1+(1-q)\pi_2}(P_0, P_1) \\ &= \inf_\varphi \int [(q\pi_1 + (1-q)\pi_2)\varphi f_0 + (1 - (q\pi_1 + (1-q)\pi_2))(1-\varphi) f_1] d\mu \\ &\ge q \inf_\varphi \int [\pi_1 \varphi f_0 + (1-\pi_1)(1-\varphi) f_1] d\mu \\ &\quad + (1-q) \inf_\varphi \int [\pi_2 \varphi f_0 + (1-\pi_2)(1-\varphi) f_1] d\mu, \end{aligned}$$

which proves the concavity. (2.47) follows from $\|P_0 - P_1\| = \int |f_0 - f_1| d\mu$ along with $|a - b| = a + b - 2(a \wedge b)$. $\square$

Solution to Problem 2.64: The likelihood ratio is

$$L_{0,1}(x) = I_{\{0\}}(x)(\frac{1-p_1}{1-p_0}) + I_{\{1\}}(x)\frac{p_1}{p_0}.$$

For $\alpha \in [0, p_0]$, $\varphi(0) = 0$, and $\varphi(1) = \gamma$, where $\gamma = \alpha/p_0$ follows from $\alpha = \mathsf{E}_0\varphi = \gamma p_0$. For $\alpha \in (p_0, 1]$, $\varphi(0) = \gamma$ and $\varphi(1) = 1$, where $\gamma = (\alpha - p_0)/(1 - p_0)$ follows from $\alpha = \mathsf{E}_0\varphi = \gamma(1 - p_0) + p_0$. A nonrandomized Bayes test is given by $\varphi_B = I_{(\pi/(1-\pi),\infty)}(L_{0,1}(x))$, which is equivalent to

$$\varphi_B(0) = \begin{cases} 1 & \text{if } \pi < \frac{1-p_1}{1-p_0+1-p_1} \\ 0 & \text{otherwise} \end{cases}, \quad \varphi_B(1) = \begin{cases} 1 & \text{if } \pi < \frac{p_1}{p_0+p_1} \\ 0 & \text{otherwise} \end{cases}.$$

$\square$

3

Statistical Decision Theory

3.1 Decisions in Statistical Models

The concept of a statistical model $\mathcal{M} = (\mathcal{X}, \mathfrak{A}, (P_\theta)_{\theta \in \Delta})$ has been introduced at the beginning of Chapter 1. The purpose of statistical inference is to draw conclusions on the true but unknown distribution P_θ of X after the experiment has been carried out and the observation x is available.

To create a mathematical frame that makes "conclusions" more precise we choose a nonempty set $\mathcal{D}$ and call it the *decision space*. To be able to utilize tools from probability theory we assume that $\mathcal{D}$ is equipped with a σ-algebra $\mathfrak{D}$. The simplest way of making a decision is to select a point $a \in \mathcal{D}$ after $x \in \mathcal{X}$ has been observed. Such a decision, called a *nonrandomized decision*, is a measurable mapping $\mathsf{d} : \mathcal{X} \to_m \mathcal{D}$, where $\mathsf{d}(x)$ is the decision that is made after $x \in \mathcal{X}$ has been observed. However, for many statistical problems this approach turns out to be too narrow. Several arguments, some given below and others later, can be made for utilizing a more general approach of *randomized decisions*. Roughly speaking, a randomized decision selects a point in $\mathcal{D}$ at random after $x \in \mathcal{X}$ has been observed. An appropriate mathematical structure for representing randomized decisions is the *stochastic kernel*. Here and in the following, a stochastic kernel is understood to be a mapping $\mathsf{D} : \mathfrak{D} \times \mathcal{X} \to_k [0,1]$ which has the following two properties. For every fixed $x \in \mathcal{X}$ the object $\mathsf{D}(\cdot|x)$ is a probability distribution on $(\mathcal{D}, \mathfrak{D})$. For every fixed $B \in \mathfrak{D}$ the mapping $x \mapsto \mathsf{D}(B|x)$ from $\mathcal{X}$ into $[0,1]$ is $\mathfrak{A}$-$\mathfrak{B}_{[0,1]}$ measurable. We call every stochastic kernel $\mathsf{D} : \mathfrak{D} \times \mathcal{X} \to_k [0,1]$ a *randomized decision rule* or simply a *decision*. The interpretation of $\mathsf{D}(A|x)$ is that after $x \in \mathcal{X}$ has been observed $\mathsf{D}(A|x)$ is the probability that a point in $A \in \mathfrak{D}$ is selected.

The nonrandomized decisions can be embedded in this general approach by simply setting $\mathsf{D}(\cdot|x) = \boldsymbol{\delta}_{\mathsf{d}(x)}(\cdot)$ for every $\mathsf{d} : \mathcal{X} \to_m \mathcal{D}$. As such a decision $\mathsf{D}(\cdot|x)$ is concentrated at $\mathsf{d}(x)$ for every $x \in \mathcal{X}$ it may be called a nonrandomized decision as well. Among the two representations of a nonrandomized decision (i.e., in terms of D and d), the latter is usually more convenient to use.

F. Liese, K.-J. Miescke, *Statistical Decision Theory*,
DOI: 10.1007/978-0-387-73194-0_3,

Some preliminary justifications of the extension to randomized decisions seem appropriate. From a purely mathematical point of view by admitting randomized decisions the chances of finding decisions that are optimal in some way can only improve or, in the worst case, just remain the same. Both scenarios may occur, depending on the type of statistical problem considered. We have seen already in Chapter 2 that randomized tests should be included in the search for best level α tests. In the search for best decisions one often has to impose constraints on the class of decisions under consideration in order to guarantee the existence of a best decision. Such constraints are usually linear in some way, and then the fact that the set of all randomized decisions is convex allows us to use arguments and techniques from convex optimization. Moreover, if the constraints are set in terms of inequalities, then we can expect that optimal decisions, whenever they exist, must exhaust the constraints by establishing equalities in the inequalities. Without a convex structure of decisions it may not be possible to find a decision that exhausts the constraints.

By summarizing all the components of the above decision process we arrive at the following general definition of a decision.

Definition 3.1. *Given a statistical model* $(\mathcal{X},\mathfrak{A},(P_\theta)_{\theta\in\Delta})$ *and a decision space* $(\mathcal{D},\mathfrak{D})$*, a decision* D *is a stochastic kernel* $\mathsf{D}:\mathfrak{D}\times\mathcal{X}\to_k[0,1]$*. The class of all decisions* D *is denoted by* $\mathbb{D}$*. A decision* D *is called a nonrandomized decision if* $\mathsf{D}(\cdot|x)=\boldsymbol{\delta}_{\mathsf{d}(x)}$ *for some* $\mathsf{d}:\mathcal{X}\to_m\mathcal{D}$*.*

The process of first observing the data and then making a decision can also be described by means of a random vector (A,X) that is defined on some probability space, say $(\Omega,\mathfrak{F},\mathbb{P}_\theta)$, $\theta\in\Delta$. Hereby the random variable $X:\Omega\to_m\mathcal{X}$ is the observation, and $A:\Omega\to_m\mathcal{D}$ is the statistician's action after observing X. Clearly, A depends on the outcome of $X=x$ and is governed by the decision $\mathsf{D}(\cdot|x)$, $x\in\mathcal{X}$. More precisely, $\mathsf{D}(\cdot|x)$ is the conditional distribution of A given $X=x$, and $\mathcal{L}(X)=\mathbb{P}_\theta\circ X^{-1}=:P_\theta$, $\theta\in\Delta$, is the marginal distribution of X. This means that by the definition of the conditional distribution (see Definition A.36 and (A.3)), for every set $C\in\mathfrak{D}\otimes\mathfrak{A}$ it holds

$$\begin{aligned}\mathcal{L}(A,X)\ &:=\mathbb{P}_\theta\circ(A,X)^{-1}=\mathsf{D}\otimes P_\theta,\quad\text{where}\qquad(3.1)\\(\mathsf{D}\otimes P_\theta)(C)&=\int[\int I_C(a,x)\mathsf{D}(da|x)]P_\theta(dx),\quad C\in\mathfrak{D}\otimes\mathfrak{A}.\end{aligned}$$

By the standard extension technique, via linear combinations of indicator functions and the approximation of nonnegative measurable functions by increasing sequences of such linear combinations, one obtains

$$\mathbb{E}_\theta h(A,X)=\int[\int h(a,x)\mathsf{D}(da|x)]P_\theta(dx),\qquad(3.2)$$

for every $h:\mathcal{D}\times\mathcal{X}\to_m\mathbb{R}_+$.

The natural question that arises now is which decision should be chosen. Here we recall that the observations are subject to unavoidable random errors and thus no decision based on such data can be perfect. This suggests our adopting the concept of a loss due to decisions. To be able to measure the loss numerically we assume that it is given by some values $L(\theta, a)$, $\theta \in \Delta$, $a \in \mathcal{D}$, where $L(\theta, a)$ is the loss when a decision is made in favor of a and the true parameter is θ.

Definition 3.2. *A loss function L is a function $L : \Delta \times \mathcal{D} \to \mathbb{R}$ such that for every fixed $\theta \in \Delta$ the function $L(\theta, \cdot)$ is $\mathfrak{D}$-$\mathfrak{B}$ measurable and it holds*

$$-\infty < \inf_{a \in \mathcal{D}} L(\theta, a), \quad \theta \in \Delta. \tag{3.3}$$

The condition (3.3) guarantees that for any probability measure μ the integral $\int L(\theta, a)\mu(da)$ is well defined despite the fact that it may be ∞. In most cases we consider only nonnegative loss functions. Occasionally, however, when dealing with bounded loss functions, it proves convenient to have a real-valued loss function which allows us to keep the formulations simple.

Under a decision D, after X has been observed, the statistician's action toward a final decision is the random variable A, where the joint distribution of X and A is given by (3.1). Therefore the loss under D is a random variable $L(\theta, A)$. In view of condition (3.3) the expected loss exists, but it may assume the value ∞. We now introduce the *risk* of a decision as its expected loss.

Definition 3.3. *Given a statistical model $(\mathcal{X}, \mathfrak{A}, (P_\theta)_{\theta \in \Delta})$ and a loss function $L : \Delta \times \mathcal{D} \to \mathbb{R}$, the risk function (in short, risk) of a decision $\mathsf{D} \in \mathbb{D}$ is given by*

$$\mathsf{R}(\theta, \mathsf{D}) = \mathbb{E}_\theta L(\theta, A), \quad \theta \in \Delta,$$

where $\mathcal{L}((A, X)|\mathbb{P}_\theta) = \mathsf{D} \otimes P_\theta$, $\theta \in \Delta$.

Usually it proves more convenient to evaluate the risk in terms of the corresponding distributions. From $\mathcal{L}((A, X)|\mathbb{P}_\theta) = \mathsf{D} \otimes P_\theta$ and (3.2) we get that

$$\mathsf{R}(\theta, \mathsf{D}) = \int [\int L(\theta, a)\mathsf{D}(da|x)] P_\theta(dx), \quad \theta \in \Delta. \tag{3.4}$$

If $\mathsf{D}(\cdot|x) = \boldsymbol{\delta}_{\mathsf{d}(x)}(\cdot)$, $x \in \mathcal{X}$, is a nonrandomized decision based on $\mathsf{d} : \mathcal{X} \to_m \mathcal{D}$, then we write, instead of $\mathsf{R}(\theta, \mathsf{D})$, just

$$\mathsf{R}(\theta, \mathsf{d}) = \mathbb{E}_\theta L(\theta, \mathsf{d}(X)) = \int L(\theta, \mathsf{d}(x)) P_\theta(dx), \quad \theta \in \Delta.$$

Now we have all components together to be able to say what a *statistical decision problem* is.

Definition 3.4. *A statistical decision problem (in short, decision problem) is a triple $(\mathcal{M}, (\mathcal{D}, \mathfrak{D}), L)$ that consists of a statistical model $\mathcal{M}$ from (1.1), a decision space $(\mathcal{D}, \mathfrak{D})$, and a loss function L.*

Special types of statistical decision problems are estimation of parameters, testing of hypotheses, selection of populations, and classification. In an *estimation problem* it is intended to approximate the true parameter θ, or at least a function $\kappa(\theta)$ of it, after an observation has been made. Often θ is a vector and one is only interested in some of its components. Then $\kappa(\theta)$ is the subvector that consists of the components of interest. If Δ is a function space, for instance the space of all distribution functions F, then we may focus on some $\kappa(F)$ that is a real-valued functional of F, which may be the median or a specific quantile of F.

Let the model $\mathcal{M}$ from (1.1) be given. Suppose that $\kappa : \Delta \to \mathcal{S}$ is fixed and that we want to estimate $\kappa(\theta)$. We suppose that $\mathcal{S}$ is equipped with a σ-algebra $\mathfrak{S}$. In most cases it holds $\mathcal{S} = \mathbb{R}^d$ and $\mathfrak{S} = \mathfrak{B}_d$.

Definition 3.5. *An estimation problem is a decision problem that consists of a model $(\mathcal{X}, \mathfrak{A}, (P_\theta)_{\theta\in\Delta})$, a function $\kappa : \Delta \to \mathcal{S}$, the decision space $(\mathcal{D}, \mathfrak{D}) = (\mathcal{S}, \mathfrak{S})$, and a loss function $L(\theta, a) = l(\kappa(\theta), a)$, where $l : \mathcal{S} \times \mathcal{S} \to_m \mathbb{R}$ with $-\infty < \inf_{a\in\mathcal{S}} l(s, a)$, $s \in \mathcal{S}$. A decision $\mathsf{D} : \mathfrak{S} \times \mathcal{X} \to_k [0, 1]$ is called a randomized estimator. Every nonrandomized decision $S : \mathcal{X} \to_m \mathcal{S}$ is called an estimator.*

In the case of $\mathcal{S} = \mathbb{R}^d$ typical examples for l are given by $l(t, a) = \varrho(\|t - a\|)$, where $\varrho : \mathbb{R}_+ \to \mathbb{R}_+$ is a nondecreasing function; for example, $\varrho(s) = s$ or $\rho(s) = s^2$. The risk of a randomized estimator is given by

$$\mathsf{R}(\theta, \mathsf{D}) = \int [\int l(\kappa(\theta), a)\mathsf{D}(da|x)]P_\theta(dx), \quad \theta \in \Delta.$$

If $\mathsf{D}(\cdot|x) = \boldsymbol{\delta}_{S(x)}(\cdot)$, $x \in \mathcal{X}$, is nonrandomized, then its risk $\mathsf{R}(\theta, S)$, say, is

$$\mathsf{R}(\theta, S) = \int l(\kappa(\theta), S(x))P_\theta(dx) = \mathsf{E}_\theta l(\kappa(\theta), S), \quad \theta \in \Delta,$$

which can be interpreted as the average distance, measured by l, between S and $\kappa(\theta)$ that is to be estimated.

Suppose $\mathcal{S} = \mathbb{R}^d$ and $l(\kappa(\theta), a)$ is a convex function of a for every $\theta \in \Delta$. If

$$\int [\int \|a\| \, \mathsf{D}(da|x)]P_\theta(dx) < \infty, \quad \theta \in \Delta,$$

then the estimator $S(x) := \int a\mathsf{D}(da|x)$, $x \in \mathcal{X}$, satisfies, by Jensen's inequality (see Lemma A.33) for every $\theta \in \Delta$,

$$\mathsf{R}(\theta, S) = \int l(\kappa(\theta), S(x))P_\theta(dx) \le \int [\int l(\kappa(\theta), a)\mathsf{D}(da|x)]P_\theta(dx) = \mathsf{R}(\theta, \mathsf{D}). \tag{3.5}$$

This is one of the reasons why for convex loss functions only estimators, and not randomized estimators, are considered.

If $\kappa(\theta) = \theta$, $\theta \in \Delta$, then we want to estimate the parameter θ. Despite this it proves useful to allow the decision space to be a set $\mathcal{S}$ that contains Δ, but not necessarily being equal to Δ. For example, let $\Delta \subseteq \mathbb{R}^d$ where Δ is not closed. If the estimator is introduced by some minimization or maximization procedure, then typically the extreme points belong to the closure of Δ and the concrete value $S(x)$ of the estimator may be a boundary point. A similar situation is met when $\Delta = \mathbb{R}$ and due to compactness reasons we have to use the extended real line $\overline{\mathbb{R}} = [-\infty, \infty]$ as the decision space.

A *multiple decision problem* is a decision problem where the decision space $\mathcal{D}$ is finite. In this case $\mathfrak{D}$ is taken as the power set of $\mathcal{D}$. Let $\mathcal{D} = \{a_1, ..., a_k\}$. Put $\psi_i(x) := \mathsf{D}(\{a_i\}|x)$, $x \in \mathcal{X}$, $i = 1, ..., k$. Then

$$\mathsf{D}(A|x) = \sum\nolimits_{i=1}^{k} \psi_i(x)\boldsymbol{\delta}_{a_i}(A), \quad A \subseteq \mathcal{D}, \tag{3.6}$$

$$\psi_i(x) \geq 0, \ i = 1, ..., k, \ \text{ and } \sum\nolimits_{i=1}^{k} \psi_i(x) = 1, \quad x \in \mathcal{X}. \tag{3.7}$$

This allows us to represent every decision D by a vector $\psi = (\psi_1, ..., \psi_k)$, where $\psi_i(x)$ is the probability of deciding in favor of a_i after x has been observed, $i = 1, ..., k$. Conversely, every $\psi = (\psi_1, ..., \psi_k)$ that satisfies (3.7) leads to a decision D via (3.6). We may consider such a ψ as a measurable mapping $\psi : \mathcal{X} \to_m \mathbf{S}_k^c$, where

$$\mathbf{S}_k^c = \{(p_1, ..., p_k) : p_i \geq 0, \ i = 1, ..., k, \ \sum\nolimits_{i=1}^{k} p_i = 1\}$$

is the unit simplex, equipped with the σ-algebra of Borel sets $\mathfrak{S}_k^c$. If we consider $(\mathbf{S}_k^c, \mathfrak{S}_k^c)$ as a new decision space, then every randomized decision D for the decision space $(\mathcal{D}, \mathfrak{D})$ can be identified with a nonrandomized decision ψ for the new decision space $(\mathbf{S}_k^c, \mathfrak{S}_k^c)$.

The loss function $L(\theta, a)$ consists of k functions $L(\cdot, a_i) : \Delta \to_m \mathbb{R}$, $i = 1, ..., k$. The risk of a decision D is given by

$$\mathsf{R}(\theta, \mathsf{D}) = \sum\nolimits_{i=1}^{k} L(\theta, a_i) q_i(\theta), \quad \theta \in \Delta, \quad \text{where}$$

$$q_i(\theta) = \int \psi_i(x) P_\theta(dx)$$

is the probability of deciding in favor of a_i when θ is the true parameter, $i = 1, ..., k$.

Remark 3.6. Quite often we deal with ψ that represents D in (3.6) rather than with D itself, and then we use the notation $\mathsf{R}(\theta, \psi)$ rather than $\mathsf{R}(\theta, \mathsf{D})$ for the risk of D.

If the model is also finite, say $\Delta = \{1, ..., m\}$, then the loss function L can be represented by the k vectors

$$v_1 = \begin{pmatrix} L(1, a_1) \\ \vdots \\ L(m, a_1) \end{pmatrix}, ..., v_k = \begin{pmatrix} L(1, a_k) \\ \vdots \\ L(m, a_k) \end{pmatrix}, \tag{3.8}$$

and the risk function $\mathsf{R}(\theta, \mathsf{D})$ can be identified with the points

$$v = (\sum\nolimits_{i=1}^{k} L(1, a_i)q_i(1), ..., \sum\nolimits_{i=1}^{k} L(m, a_i)q_i(m))^T,$$

which belong to the polyhedron that is spanned by the vectors $v_1, ..., v_k$.

In multiple decision problems it is typical that the parameter set Δ is decomposed in k disjoint subsets, say $\Delta_1, ..., \Delta_k$, and for $\theta \in \Delta_i$ the decision a_i is correct whereas every other decision a_j, $j \neq i$ is false. To reflect this the so-called zero–one loss function $L_{0,1}$ is used. It is defined by

$$L_{0,1}(\theta, a_i) = 1 - I_{\Delta_i}(\theta), \quad \theta \in \Delta, \ i = 1, ..., k. \tag{3.9}$$

In this case $\mathsf{R}(\theta, q) = 1 - \sum_{i=1}^{k} I_{\Delta_i}(\theta)q_i(\theta)$, where $\sum_{i=1}^{k} I_{\Delta_i}(\theta)q_i(\theta)$ is the probability of making a correct decision and $\mathsf{R}(\theta, q)$ is the probability of making a false decision.

The special case of a multiple decision problem with $k = 2$ is called a *testing problem.* The subsequent definition integrates the concept of testing of hypotheses into the decision-theoretic framework.

Definition 3.7. *A testing problem is a decision problem that consists of a model* $(\mathcal{X}, \mathfrak{A}, (P_\theta)_{\theta\in\Delta})$, *a decomposition of* Δ *into two disjoint subsets* Δ_0 *and* Δ_A *that represent the null hypothesis* H_0 *and the alternative hypothesis* H_A, *the decision space* $(\mathcal{D}, \mathfrak{D}) = (\{0, 1\}, \mathfrak{P}(\{0, 1\}))$, *and a loss function* $L : \Delta \times \{0, 1\} \to \mathbb{R}$.

To link the decision-theoretical concept with the concept of a statistical test we set $\varphi(x) := \mathsf{D}(\{1\}|x)$ and note that $\varphi(x)$ is the probability of deciding in favor of H_A after x has been observed, $x \in \mathcal{X}$. This means that $\varphi(x)$ is the probability of rejecting the null hypothesis and therefore is a statistical test in the sense of Definition 2.28. Clearly then $1 - \varphi(x) = \mathsf{D}(\{0\}|x)$ is the probability of accepting H_0 after x has been observed. Conversely, every test $\varphi(x)$ defines by

$$\mathsf{D}(\cdot|x) = (1 - \varphi(x))\boldsymbol{\delta}_0(\cdot) + \varphi(x)\boldsymbol{\delta}_1(\cdot), \quad x \in \mathcal{X},$$

a decision for the decision space $\mathcal{D} = \{0, 1\}$.

In testing problems the loss is usually taken as the zero–one loss function in (3.9), which may be written as

$$L_{0,1}(\theta, a) = aI_{\Delta_0}(\theta) + (1 - a)I_{\Delta_A}(\theta), \quad \theta \in \Delta, \ a \in \{0, 1\}.$$

Then for $\theta \in \Delta_0$, $\mathsf{R}(\theta, \varphi) = \mathsf{E}_\theta\varphi$ is the probability of making an error of the first kind, and for $\theta \in \Delta_A$, $\mathsf{R}(\theta, \varphi) = 1 - \mathsf{E}_\theta\varphi$ is the probability of making an error of the second kind.

From Section 2.2 we know already that in the framework of the Neyman–Pearson theory tests (i.e., randomized decisions) have to be considered to find optimal decisions. This is necessary to be able to exhaust the constraints set by the concept of a level α test; see also Problem 2.32.

The simplest statistical model consists of just two distributions P_0 and P_1. In such a binary model $\mathcal{M}_{0,1} = (\mathcal{X}, \mathfrak{A}, \{P_0, P_1\})$ the hypotheses are $\mathsf{H}_0 : P_0$ and $\mathsf{H}_A : P_1$. The vectors in (3.8) are, adjusted to $\mathcal{D} = \{0,1\}$, $v_0 = (0,1)^T$ and $v_1 = (1,0)^T$, and $\mathfrak{R} = \{(\mathsf{E}_0\varphi, 1 - \mathsf{E}_1\varphi) : \varphi : \mathcal{X} \to_m [0,1]\}$ is the risk set of tests in (2.48) which is convex.

A *selection problem* is a special case of a multiple decision problem where the model consists of k populations, usually of a similar type. Let X_i, $i = 1, ..., k$, be observations with values in $\mathcal{X}_i$ and distributions P_{i,θ_i}, $\theta_i \in \Delta$, from k independent populations. More specifically, let the statistical model be

$$\mathcal{M} = (\times_{i=1}^k \mathcal{X}_i, \bigotimes_{i=1}^k \mathfrak{A}_i, (P_\theta)_{\theta \in \Delta^k}), \quad \theta = (\theta_1, ..., \theta_k), \tag{3.10}$$

where the P_{i,θ_i} are the marginal distributions of the distribution

$$P_\theta = \bigotimes_{i=1}^k P_{i,\theta_i} .$$

Typically, we have one common sample space $\mathcal{X}$ for all observations, but the sample sizes may not be the same for all populations. Then $\mathcal{X}_i = \mathcal{X}^{n_i}$ and $P_{i,\theta_i} = P_{\theta_i}^{\otimes n_i}$.

Suppose that $\kappa : \Delta \to \mathbb{R}$ is a real-valued function and we want to find a *best population*, which is meant to be a population with index i_0 for which $\kappa(\theta_{i_0}) = \max_{1 \le i \le k} \kappa(\theta_i)$, i.e., $i_0 \in \arg\max_{1 \le i \le k} \kappa(\theta_i)$. As we select one population this type of decision is called a *point selection*, in contrast to a *subset selection* where a subset of populations is selected.

Definition 3.8. *A point selection problem is a decision problem that consists of the model (3.10), a function $\kappa : \Delta \to \mathbb{R}$, the decision space $(\mathcal{D}_{pt}, \mathfrak{D}) = (\{1, ..., k\}, \mathfrak{P}(\{1, ..., k\}))$, and a loss function $L : \Delta^k \times \{1, ..., k\} \to \mathbb{R}$. Every $\psi = (\psi_1, ..., \psi_k) : \times_{i=1}^k \mathcal{X}_i \to_m \mathbf{S}_k^c$ is called a point selection rule, or in short a selection rule, and the associated decision kernel is defined by (3.6).*

Let the loss be given by the zero–one loss function

$$L_{0,1}(\theta, i) = 1 - I_{\mathsf{M}(\theta)}(i), \quad \theta = (\theta_1, ..., \theta_k) \in \Delta^k, \; i = 1, ..., k, \text{ where } \tag{3.11}$$
$$\mathsf{M}(\theta) = \arg\max_{1 \le i \le k} \kappa(\theta_i) = \{i : \kappa(\theta_i) = \max_{1 \le j \le k} \kappa(\theta_j)\}, \quad \theta \in \Delta^k.$$

Then the risk of a decision D, in terms of the associated selection rule ψ (see Remark 3.6) is

$$\mathsf{R}(\theta, \psi) = 1 - \sum\nolimits_{i=1}^k I_{\mathsf{M}(\theta)}(\theta_i) \mathsf{E}_\theta \psi_i, \quad \theta \in \Delta^k. \tag{3.12}$$

Representing the decision process by the random vector (A, X) defined on $(\Omega, \mathfrak{F}, \mathbb{P}_\theta)$, where $X = (X_1, ..., X_k)$, it can be seen from (3.1) that the risk

$\mathsf{R}(\theta,\psi) = \mathbb{P}_\theta(A \notin \mathsf{M}(\theta))$ is the probability that the action A does not fall into the set $\mathsf{M}(\theta)$ where the function κ attains its maximum. Therefore

$$\mathbb{P}_\theta(A \in \mathsf{M}(\theta)) = \sum\nolimits_{i=1}^{k} I_{\mathsf{M}(\theta)}(\theta_i)\mathsf{E}_\theta\psi_i, \tag{3.13}$$

is called the *probability of a correct selection*, which we also denote by

$$P_{cs}(\theta,\psi) := \sum\nolimits_{i=1}^{k} I_{\mathsf{M}(\theta)}(\theta_i)\mathsf{E}_\theta\psi_i.$$

In preparation for the next example some simple properties of the maximum of independent random variables are established.

Problem 3.9.* Let $X_1,...,X_k$ be independent random variables with distributions $P_1,...,P_k$ and continuous c.d.f.s $F_1,...,F_k$, respectively. Then

$$\mathbb{P}(X_i > \max_{j\neq i} X_j) = \int \prod\nolimits_{j\neq i} F_j(t)P_i(dt), \quad i = 1,...,k.$$

Problem 3.10.* Let $X_i \sim \mathsf{N}(\mu_i,\sigma^2)$, $\mu_i \in \mathbb{R}$, $\sigma^2 > 0$, $i = 1,...,k$, be independent, and $Z \sim \mathsf{N}(0,1)$ be a generic random variable. Then

$$\mathbb{P}(X_i > \max_{j\neq i} X_j) = \mathbb{E}\prod\nolimits_{j\neq i} \Phi(Z + (\mu_i - \mu_j)/\sigma) \quad \text{and}$$
$$\mathbb{P}(X_{i_0} > \max_{j\neq i_0} X_j) = \max_{1\le i\le k} \mathbb{P}(X_i > \max_{j\neq i} X_j) \quad \Leftrightarrow \quad \mu_{i_0} = \mu_{[k]},$$

where $\mu_{[k]} = \max\{\mu_1,...,\mu_k\}$.

Example 3.11. Let $X_i \sim P_{\theta_i}$, $i = 1,...,k$, be independent random variables in $\mathbb{R}$, where $(\theta_1,...,\theta_k) \in \Delta^k = \mathbb{R}^k$ is unknown. Suppose we want to find a population that has the largest parameter. Here $\kappa(\theta_i) = \theta_i$, $i = 1,...,k$, and $\mathcal{D}_{pt} = \{1,...,k\}$. We use the zero–one loss from (3.11), and thus the risk of a selection rule ψ is given by (3.12). If $(P_\theta)_{\theta\in\Delta}$ has MLR in the identity, then it seems natural to select a population with the largest value of the observations, and to break ties at random. This selection rule D_{nat} is called the *natural selection rule*. It is a randomized decision, as for every $x \in \mathbb{R}^k$ it is the uniform distribution on the set

$$\mathsf{M}(x) = \arg\max_{i\in\{1,...,k\}} x_i = \{i : x_i = \max_{1\le j\le k} x_j\}, \quad x = (x_1,...,x_k) \in \mathbb{R}^k.$$

Thus, with $|\mathsf{M}(x)|$ denoting the number of elements of $\mathsf{M}(x)$,

$$\varphi^{nat}(x) = (\varphi_1^{nat}(x),...,\varphi_k^{nat}(x)) := \frac{1}{|\mathsf{M}(x)|}(I_{\mathsf{M}(x)}(1),...,I_{\mathsf{M}(x)}(k)), \tag{3.14}$$
$$\mathsf{D}_{nat}(A|x) = \sum\nolimits_{i=1}^{k} \varphi_i^{nat}(x)\boldsymbol{\delta}_i(A), \quad A \subseteq \{1,...,k\}, \quad x \in \mathbb{R}^k.$$

Especially, let $X_i \sim \mathsf{N}(\mu_i,\sigma^2)$, $i = 1,...,k$, be independent with $\mu \in \Delta^k = \mathbb{R}^k$, where $\sigma^2 > 0$ may be known or unknown. In this case the natural selection rule turns out to be nonrandomized. If $\mu \in \mathbb{R}^k$ with $\mu_i > \max_{j\neq i}\mu_j$, then in view of Problem 3.10,

$$\mathbb{P}_\mu(A \in \mathsf{M}(\mu)) = \mathbb{P}_\mu(X_i > \max_{,j\neq i} X_j) = \int \prod\nolimits_{j\neq i} \Phi(t + (\mu_i - \mu_j)/\sigma)\varphi(t)dt,$$

$$\mathsf{R}(\mu, \varphi^{nat}) = 1 - \int \prod\nolimits_{j\neq i} \Phi(t + (\mu_i - \mu_j)/\sigma)\varphi(t)dt \in (0, 1 - \frac{1}{k}), \qquad (3.15)$$

with A in (3.13), where Φ and φ are the c.d.f. and density of $\mathsf{N}(0,1)$, respectively.

Consider now the no-data selection rule $\psi_{nd} \equiv (1, 0, ..., 0)$ which selects the first population, regardless of the outcome of the observations. If the first population has the largest mean, then ψ_{nd} has risk 0 and outperforms φ^{nat} because of (3.15). On the other hand, if the first population does not have the largest mean, then ψ_{nd} has risk 1 and is outperformed by φ^{nat} because of (3.15). We conclude that a uniformly best selection rule does not exist, unless we turn to a subclass of selection rules that excludes exotic rules such as ψ_{nd}. This can be achieved by imposing invariance requirements on the selection rules. For details we refer to Chapter 9.

Finally we introduce a type of statistical decision problem that is called a *classification problem*. Let $X, X_1, ..., X_k$ be independent random variables with values in $\mathcal{X}$, where X_i has the distribution P_{θ_i}, $\theta_i \in \Delta$, $i = 1, ..., k$ and X has the distribution P_θ, $\theta \in \{\theta_1, ..., \theta_k\}$. The statistical model is

$$\mathcal{M} = (\mathcal{X}^{k+1}, \mathfrak{A}^{\otimes(k+1)}, (P_\theta \otimes (\bigotimes\nolimits_{i=1}^k P_{\theta_i}))_{(\theta,\theta_1,...,\theta_k)\in\Gamma}), \quad \text{where} \quad (3.16)$$
$$\Gamma = \{(\theta, \theta_1, ..., \theta_k) : (\theta_1, ..., \theta_k) \in \Delta^k, \ \theta \in \{\theta_1, ..., \theta_k\}\}.$$

The model specification implies that $P_\theta \in \{P_{\theta_1}, ..., P_{\theta_k}\}$. Here we want to find a population with index i_0 for which $P_{\theta_{i_0}} = P_\theta$. If the θ_is are all different, and the parameter θ in $(P_\theta)_{\theta\in\Delta}$ is identifiable, then this index i_0 is uniquely determined. We take $\mathcal{D} = \{1, ..., k\}$ as the decision space.

Definition 3.12. *A classification problem is a decision problem that consists of the model (3.16), the decision space* $(\mathcal{D}, \mathfrak{D}) = (\{1, ..., k\}, \mathfrak{P}(\{1, ..., k\}))$*, and a loss function* $L : \Gamma \times \{1, ..., k\} \to \mathbb{R}$*. Every* $\psi = (\psi_1, ..., \psi_k) : \mathsf{X}_{i=1}^k \mathcal{X}_i \to_m \mathbf{S}_k^c$ *is called a classification rule, and the associated decision kernel is defined by (3.6).*

Let the loss be given by the zero–one loss function $L_{0,1}(\vartheta, i) = 1 - I_{G(\vartheta)}(i)$, $i = 1, ..., k$, where $G(\vartheta) = \{j : \theta = \theta_j\}$, $\vartheta = (\theta, \theta_1, ..., \theta_k) \in \Gamma$. Then the risk of a classification rule ψ is

$$\mathsf{R}(\vartheta, \psi) = 1 - \sum\nolimits_{i=1}^k I_{G(\vartheta)}(\theta_i)\mathsf{E}_\vartheta \psi_i, \quad \vartheta = (\theta, \theta_1, ..., \theta_k) \in \Gamma.$$

Similarly as in the previous example we see from (3.1) that the risk $\mathsf{R}(\vartheta, \psi) = \mathbb{P}_\vartheta(A \notin G(\vartheta))$ is the probability of an incorrect classification.

The purpose of statistical decision theory is to find decisions that are optimal within a specific framework that consists of three components: a statistical model, a decision space, and a loss function. The optimality of decisions is then determined by means of their risk functions.

The performance of a decision $\mathsf{D} \in \mathbb{D}$ is measured, pointwise at every $\theta \in \Delta$, by its expected loss, i.e., its risk $\mathsf{R}(\theta, \mathsf{D})$, which is given by (3.4). Now we want to compare two decisions $\mathsf{D}_1, \mathsf{D}_2 \in \mathbb{D}$ in terms of their risks. As the parameter θ is unknown we have to compare their risk functions at every $\theta \in \Delta$. The pointwise semiorder of the risk functions leads to a semiorder in the space of all decisions. If $\mathsf{R}(\theta, \mathsf{D}_2) \leq \mathsf{R}(\theta, \mathsf{D}_1)$ for all $\theta \in \Delta$, then D_2 is called *as good as* D_1. If in addition $\mathsf{R}(\theta_0, \mathsf{D}_2) < \mathsf{R}(\theta_0, \mathsf{D}_1)$ for some $\theta_0 \in \Delta$, then D_2 is called *better than* D_1. Ideally, we would like to find a decision $\mathsf{D}_* \in \mathbb{D}$ that is as good as all other decisions in $\mathbb{D}$, i.e., that is *uniformly best* in $\mathbb{D}$ in terms of the risk. Although a decision with such a strong optimality property does not exist in general, it may be possible to achieve this goal at least within a suitable subclass of decisions $\mathbb{D}_0 \subset \mathbb{D}$. In an estimation problem this could be the class of all unbiased estimators, in a testing problem the class of all level α tests, and in a selection problem the class of all permutation-invariant selection rules.

A decision that performs optimally in terms of the risk at every $\theta \in \Delta$ may not exist, as the pointwise comparison of the risk functions provides only a semiorder in the class of all decisions. Unfortunately, this situation may even prevail after a restriction to some suitable subclass $\mathbb{D}_0 \subset \mathbb{D}$ with a structure based on some widely accepted principle, such as the principle of invariance. Thus, working at the "good end" can be challenging. It can also be so at the "bad end", where one would like to discard any decision coming across for which another better decision exists. Obviously, one should ignore a decision once a better one has been found. The idea of discarding decisions for which better decisions can be found leads to the concept of *admissibility*.

Definition 3.13. *A decision $\mathsf{D}_a \in \mathbb{D}_0 \subseteq \mathbb{D}$ is called admissible in $\mathbb{D}_0$ if there does not exist a decision $\mathsf{D}_b \in \mathbb{D}_0$ that is better than D_a. A decision $\mathsf{D}_i \in \mathbb{D}_0 \subseteq \mathbb{D}$ is called inadmissible in $\mathbb{D}_0$ if it is not admissible in $\mathbb{D}_0$. Whenever $\mathbb{D}_0 = \mathbb{D}$, "in $\mathbb{D}_0$" is omitted for brevity.*

Obviously, a decision $\mathsf{D} \in \mathbb{D}_0 \subseteq \mathbb{D}$ that has a minimum risk in $\mathbb{D}_0$, uniformly in $\theta \in \Delta$, is admissible in $\mathbb{D}_0$. Proving the admissibility of other decisions can be difficult. On the other hand, proving that a particular decision D is inadmissible in $\mathbb{D}_0$ consists of presenting a better decision D_b from $\mathbb{D}_0$. In concrete situations, this can be a difficult task as well. Some useful techniques for such purposes are presented later. We conclude this section with a prominent example for an inadmissible decision.

Example 3.14. Let $X_i \sim \mathsf{N}(\theta_i, \sigma^2)$, $i = 1, ..., d$, be independent, where $\theta = (\theta_1, ..., \theta_d) \in \Delta = \mathbb{R}^d$ is unknown, but $\sigma^2 > 0$ is known. Without loss of generality let $\sigma^2 = 1$. Suppose that $\theta \in \Delta$ has to be estimated under the squared error loss $L(\theta, a) = \|\theta - a\|^2$. We consider the natural estimator $T_{nat}(x) = x$, $x \in \mathbb{R}^d$. It was a breakthrough in statistics when James and Stein (1960) showed that $T_{nat}(x)$ is inadmissible for $d \geq 3$ by proving that

$$
\begin{aligned}
\mathsf{R}(\theta, T_{nat}) - \mathsf{R}(\theta, S_{JS}) &= \mathsf{E}_\theta \parallel T_{nat} - \theta \parallel^2 - \mathsf{E}_\theta \parallel S_{JS} - \theta \parallel^2 \\
&= (d-2)^2 \mathsf{E}_\theta \parallel T_{nat} \parallel^{-2},
\end{aligned}
$$

where $S_{JS}(x) = (1-(d-2)\|x\|^{-2})x$, $x \in \mathbb{R}^d$, is the celebrated James–Stein estimator. It is a remarkable fact that for $d = 1$ (see Example 3.45) and $d = 2$ the estimator T_{nat} is admissible. For further details we refer to Chapter 7. It should be noted that S_{JS} belongs to the class of shrinkage estimators that has been studied in many papers. References can be found in Berger (1985) and Hoffmann (1992).

3.2 Convergence of Decisions

In this section we introduce a concept of convergence of decisions that is utilized in many situations later on. One such situation occurs in the search for optimal decisions in the Bayes or the minimax approach where the Bayes or the maximum risk, respectively, has to be minimized as a function of the decision. A natural question that arises here is whether a sequence of decisions converges to some decision if their associated risks converge to the minimum value. Other situations that require the use of convergence concepts occur when, for increasing sample sizes, asymptotically optimal decisions are desired.

For some purposes it proves useful to extend a stochastic kernel to a bilinear form on special function spaces. More precisely, let the decision space $\mathcal{D}$ be a metric space and $\mathfrak{D}$ be the σ-algebra of Borel sets. Furthermore, let Q be a distribution on the sample space $(\mathcal{X}, \mathfrak{A})$. Later on Q is specified as a probability measure that dominates a given model. Let $\mathsf{C}_b(\mathcal{D})$ be the space of all bounded continuous functions and $\mathbb{L}_1(Q)$ be the space of all Q-integrable real functions where we identify Q-a.s. identical functions. Denote by

$$\|f\|_u = \sup_{a \in \mathcal{D}} |f(a)| \quad \text{and} \quad \|g\|_1 := \int |g| dQ$$

the norm in $\mathsf{C}_b(\mathcal{D})$ and $\mathbb{L}_1(Q)$, respectively. Let $\mathsf{K} : \mathfrak{D} \times \mathcal{X} \to_k [0,1]$ be a stochastic kernel. Then for every $f \in \mathsf{C}_b(\mathcal{D})$ the function $x \mapsto \int f(a)\mathsf{K}(da|x)$ is bounded and measurable, and thus

$$B(f,g) = \int [\int f(a)\mathsf{K}(da|x)]g(x)Q(dx) \tag{3.17}$$

is well defined for every $g \in \mathbb{L}_1(Q)$. The mapping $B : \mathsf{C}_b(\mathcal{D}) \times \mathbb{L}_1(Q) \to \mathbb{R}$ has, for $f, f_1, f_2 \in \mathsf{C}_b(\mathcal{D})$ and $g, g_1, g_2 \in \mathbb{L}_1(Q)$, the following properties.

$$\begin{aligned} B(a_1 f_1 + a_2 f_2, g) &= a_1 B(f_1, g) + a_2 B(f_2, g), \quad a_1, a_2 \in \mathbb{R}, \\ B(f, b_1 g_1 + b_2 g_2) &= b_1 B(f, g_1) + b_2 B(f, g_2), \quad b_1, b_2 \in \mathbb{R}, \\ B(f,g) &\geq 0, \quad f, g \geq 0, \quad \text{with} \quad B(1,1) = 1, \\ |B(f,g)| &\leq \|f\|_u \|g\|_1 . \end{aligned} \tag{3.18}$$

Thus B is bilinear, nonnegative, and normalized. A mapping $B : \mathsf{C}_b(\mathcal{D}) \times \mathbb{L}_1(Q) \to \mathbb{R}$ that satisfies the conditions in (3.18) is called a *positive normed bilinear form*. The question of whether each positive normed bilinear form can be represented by (3.17) with a suitable stochastic kernel turns out to be a crucial point. It is clear that this question is closely related to the Riesz representation theorem for positive linear forms on the space of continuous functions on a compact space. The following measure-theoretic result is a special case of Theorem 6.11 in Strasser (1985).

Theorem 3.15. *Let* $(\mathcal{D}, \rho_{\mathcal{D}})$ *be a compact metric space,* $\mathfrak{D}$ *the* σ*-algebra of Borel sets, and* Q *a probability measure on* $(\mathcal{X}, \mathfrak{A})$*. If* $B : \mathsf{C}_b(\mathcal{D}) \times \mathbb{L}_1(Q) \to \mathbb{R}$ *is a mapping that satisfies the conditions in (3.18), then there exists a stochastic kernel* $\mathsf{K} : \mathfrak{D} \times \mathcal{X} \to_k [0,1]$ *such that*

$$B(f,g) = \int [\int f(a)\mathsf{K}(da|x)]g(x)Q(dx), \quad f \in \mathsf{C}_b(\mathcal{D}), \ g \in \mathbb{L}_1(Q).$$

It is clear that the stochastic kernel K in the last theorem can be modified on Q-null sets without changing the bilinear form B.

In the course of investigating the existence of an optimal decision we usually have to select a convergent subsequence from a given sequence. This, however, is only possible under additional conditions. As a preparation in this regard we have to deal with the separability of $\mathbb{L}_1(Q)$.

Problem 3.16.* Let $\mathfrak{A}_0 \subseteq \mathfrak{A}$ be a sub-σ-algebra that is countably generated; i.e., there exists a sequence $A_1, A_2, ... \in \mathfrak{A}$ which generates $\mathfrak{A}_0$. Let Q_0 be the restriction of Q to $\mathfrak{A}_0$. Then the space $\mathbb{L}_1(Q_0)$ is separable.

The next statement is essentially Theorem 20.4 in Heyer (1982).

Theorem 3.17. *Let* $(\mathcal{D}, \rho_{\mathcal{D}})$ *be a compact metric space,* $\mathfrak{D}$ *the* σ*-algebra of Borel sets, and* Q *a probability measure on* $(\mathcal{X}, \mathfrak{A})$*. If* $\mathsf{K}_n : \mathfrak{D} \times \mathcal{X} \to_k [0,1]$*,* $n = 1, 2, ...$*, is a sequence of stochastic kernels, then there is a subsequence* K_{n_k} *and a stochastic kernel* $\mathsf{K} : \mathfrak{D} \times \mathcal{X} \to_k [0,1]$ *such that*

$$\lim_{k\to\infty} \int [\int f(a)\mathsf{K}_{n_k}(da|x)]g(x)Q(dx)$$
$$= \int [\int f(a)\mathsf{K}(da|x)]g(x)Q(dx), \quad f \in \mathsf{C}_b(\mathcal{D}), \ g \in \mathbb{L}_1(Q).$$

Proof. The space $\mathsf{C}_b(\mathcal{D})$ is separable; see Proposition A.4. Let $\{f_1, f_2, ...\} \subseteq \mathsf{C}_b(\mathcal{D})$ be an at most countable dense subset. Denote by $\mathfrak{A}_0$ the smallest σ-algebra with respect to which all mappings $x \mapsto \int f_k(a)\mathsf{K}_n(da|x)$ are measurable. Then $\mathfrak{A}_0$ is generated by the countable system of sets $\{x : \int f_k(a)\mathsf{K}_n(da|x) < t\}$, where $t \in \mathbb{Q}$ and $k, n = 1, 2, ...$ Denote by Q_0 the restriction of Q to $\mathfrak{A}_0$. Then according to Problem 3.16 there exists an at most countable and dense subset $L = \{g_1, g_2, ...\}$ of $\mathbb{L}_1(Q_0)$. Define $B_n(f,g)$ by (3.17) where K is replaced with K_n. Consider the array $B_n(f_k, g_l)$, $n, k, l = 1, 2, ...,$ where $|B_n(f_k, g_l)| \le \|f_k\|_u \|g_l\|_1$ in view of (3.18). This array can be rearranged as a double array with $B_n(f_1, g_1)$, $B_n(f_1, g_2)$, $B_n(f_2, g_1)$,

$B_n(f_2, g_2), \ldots$ as its nth row, $n = 1, 2, \ldots$. By utilizing the diagonal technique we get a subsequence B_{n_m} such that $B_{n_m}(f_k, g_l)$ converges to some real number, say $B(f_k, g_l)$. To define B for every $f \in \mathsf{C}_b(\mathcal{D})$ and $g \in \mathbb{L}_1(Q_0)$ we choose f_{k_i} and g_{l_i} such that $\|f - f_{k_i}\|_u \to 0$ and $\|g - g_{l_i}\|_1 \to 0$. Then $D_1 = \sup_i \|f_{k_i}\|_u < \infty$, $D_2 = \sup_i \|g_{l_i}\|_1 < \infty$, and

$$\begin{aligned}
&|B(f_{k_i}, g_{l_i}) - B(f_{k_j}, g_{l_j})| \\
&\leq |\lim_{m\to\infty} B_{n_m}(f_{k_i} - f_{k_j}, g_{l_i})| + |\lim_{n\to\infty} B_{n_m}(f_{k_j}, g_{l_i} - g_{l_j}) \\
&\leq D_1 \left\|f_{k_i} - f_{k_j}\right\|_u + D_2 \left\|g_{l_i} - g_{l_j}\right\|_1,
\end{aligned}$$

so that $B(f_{k_i}, g_{l_i})$ is a Cauchy sequence and thus converges to some value, say $B(f, g)$. If $\widetilde{f}_i, \widetilde{g}_i$ are any other approximating sequences, then as above, with (f_{k_j}, g_{l_j}) replaced with $(\widetilde{f}_i, \widetilde{g}_i)$ and some constant D,

$$|B(f_{k_i}, g_{l_i}) - B(\widetilde{f}_i, \widetilde{g}_i)| \leq D \parallel f_{k_i} - \widetilde{f}_i \parallel_u + D \left\|g_{k_i} - \widetilde{g}_i\right\|_1 \to 0, \quad \text{as } i \to \infty.$$

This means that the definition of B is independent of the approximating sequence. As every B_{n_m} satisfies the conditions in (3.18) we see that $B(f, g) = \lim_{i\to\infty} \lim_{m\to\infty} B_{n_m}(f_{k_i}, g_{l_i})$ satisfies the conditions in (3.18), too. Hence by Theorem 3.15 with $\mathfrak{A}$ replaced with $\mathfrak{A}_0$ there exists a stochastic kernel $\mathsf{K} : \mathfrak{D} \times \mathcal{X} \to_k [0, 1]$ such that $\mathsf{K}(A|\cdot)$ is $\mathfrak{A}_0$ measurable and

$$B(f, g) = \int [\int f(a)\mathsf{K}(da|x)]g(x)Q(dx), \quad f \in \mathsf{C}_b(\mathcal{D}),\ g \in \mathbb{L}_1(Q_0).$$

Thus by construction, for every $f \in \mathsf{C}_b(\mathcal{D})$ and $g \in \mathbb{L}_1(Q_0)$

$$\lim_{m\to\infty} \int [\int f(a)\mathsf{K}_{n_m}(da|x)]g(x)Q(dx) = \int [\int f(a)\mathsf{K}(da|x)]g(x)Q(dx).$$

It remains to show that this statement holds for every $g \in \mathbb{L}_1(Q)$. For any $g \in \mathbb{L}_1(Q)$ we denote by $\mathsf{E}_Q(g|\mathfrak{A}_0)$ the conditional expectation of g under the condition $\mathfrak{A}_0$. Then

$$\mathsf{E}_Q(hg|\mathfrak{A}_0) = h\mathsf{E}_Q(g|\mathfrak{A}_0) \ \text{ and } \ \mathsf{E}_Q(hg) = \mathsf{E}_Q(h\mathsf{E}_Q(g|\mathfrak{A}_0))$$

for every bounded and $\mathfrak{A}_0$-measurable function h. Approximating any $f \in \mathsf{C}_b(\mathcal{D})$ uniformly by a sequence from the dense set $\{f_1, f_2, \ldots\} \subseteq \mathsf{C}_b(\mathcal{D})$ we see that $h_{n_k}(x) = \int f(a)\mathsf{K}_{n_k}(da|x)$ and $h(x) = \int f(a)\mathsf{K}(da|x)$ are $\mathfrak{A}_0$-measurable functions for every $f \in \mathsf{C}_b(\mathcal{D})$ by the definition of $\mathfrak{A}_0$. Hence

$$\begin{aligned}
&\lim_{k\to\infty} \int [\int f(a)\mathsf{K}_{n_k}(da|x)]g(x)Q(dx) \\
&= \lim_{k\to\infty} \int [\int f(a)\mathsf{K}_{n_k}(da|x)](\mathsf{E}_Q(g|\mathfrak{A}_0)(x))Q(dx) \\
&= \int [\int f(a)\mathsf{K}(da|x)](\mathsf{E}_Q(g|\mathfrak{A}_0)(x))Q(dx) = \int [\int f(a)\mathsf{K}(da|x)]g(x)Q(dx),
\end{aligned}$$

as $h(x) = \int f(a)\mathsf{K}(da|x)$ is $\mathfrak{A}_0$-measurable. ■

Remark 3.18. The above theorem has been proved in Heyer (1982), Theorem 20.4, with different techniques. To break down the considerations into suitable subsequences, instead of nets as in topology, Heyer has used in Lemma 20.5 Eberlein's theorem which states that bounded sets in $\mathbb{L}_1(Q)$ are w^*-sequentially compact. Our reduction step, using conditional expectations, is taken from Witting (1985), Theorem 2.14, which deals with statistical tests.

The concept of weak convergence of distributions is fundamental to all areas of statistics that deal with large sample sizes. Let $\mathcal{S}$ be a metric space and $\mathfrak{S}$ be the σ-algebra of Borel sets. According to Definition A.42 a sequence of distributions $Q_n \in \mathcal{P}(\mathfrak{S})$ is called weakly convergent to the distribution $Q \in \mathcal{P}(\mathfrak{S})$ if

$$\lim_{n\to\infty} \int f(s) Q_n(ds) = \int f(s) Q(ds)$$

for every bounded and continuous function f. Now we introduce, in a similar way, the concept of weakly convergent sequences of decisions for a fixed statistical model.

Definition 3.19. *Let $\mathcal{M} = (\mathcal{X}, \mathfrak{A}, (P_\theta)_{\theta\in\Delta})$ be a statistical model. Suppose that the decision space $\mathcal{D}$ is a metric space and let $\mathfrak{D}$ be the σ-algebra of Borel sets. A sequence of decisions $\mathsf{D}_n : \mathfrak{D} \times \mathcal{X} \to_k [0,1]$ is called weakly convergent to $\mathsf{D} : \mathfrak{D} \times \mathcal{X} \to_k [0,1]$ if for every $f \in \mathsf{C}_b(\mathcal{D})$ it holds*

$$\lim_{n\to\infty} \int [\int f(a)\mathsf{D}_n(da|x)] P_\theta(dx) = \int [\int f(a)\mathsf{D}(da|x)] P_\theta(dx), \ \theta \in \Delta. \quad (3.19)$$

In this case we write $\mathsf{D}_n \Rightarrow \mathsf{D}$. We call a set $\mathbb{D}_0$ of kernels weakly closed if for every sequence $\mathsf{D}_n \in \mathbb{D}_0$ with $\mathsf{D}_n \Rightarrow \mathsf{D}$ it follows that $\mathsf{D} \in \mathbb{D}_0$. Finally, we call such a set $\mathbb{D}_0$ weakly sequentially compact if for every sequence $\mathsf{D}_n : \mathfrak{D}\times\mathcal{X} \to_k [0,1]$ with $\mathsf{D}_n \in \mathbb{D}_0$ there exists a subsequence D_{n_k} and some $\mathsf{D} \in \mathbb{D}_0$ such that $\mathsf{D}_{n_k} \Rightarrow \mathsf{D}$.

To discuss the concept of weak convergence in terms of random variables let (A_n, X_n), $n = 1, 2, ...$, and (A, X) be random vectors that are defined on some probability space $(\Omega, \mathfrak{F}, \mathbb{P}_\theta)$, taking on values in $\mathcal{D} \times \mathcal{X}$, and have the distributions $\mathcal{L}(A_n, X_n) = \mathsf{D}_n \otimes P_\theta$ and $\mathcal{L}(A, X) = \mathsf{D} \otimes P_\theta$, respectively. Then for every $f \in \mathsf{C}_b(\mathcal{D})$ it holds

$$\int [\int f(a)\mathsf{D}_n(da|x)] P_\theta(dx) = \mathbb{E}_\theta f(A_n), \quad n = 1, 2, ...,$$

$$\int [\int f(a)\mathsf{D}(da|x)] P_\theta(dx) = \mathbb{E}_\theta f(A).$$

This shows that

$$\mathsf{D}_n \Rightarrow \mathsf{D} \quad \text{if and only if} \quad \mathcal{L}(A_n|\mathbb{P}_\theta) \Rightarrow \mathcal{L}(A|\mathbb{P}_\theta), \ \ \theta \in \Delta, \quad (3.20)$$

so that the weak convergence of decisions is nothing else than the convergence in distribution of the associated random variables A_n under each of the distributions in the model.

Lemma 3.20. *For every $\rho \in \mathcal{M}^\sigma(\mathfrak{B}_\Delta)$ there exists some $\Pi \in \mathcal{P}(\mathfrak{B}_\Delta)$ with $\rho \ll\gg \Pi$.*

Proof. There are pairwise disjoint sets $C_i \in \mathfrak{B}_\Delta$ with $\cup_{i=1}^\infty C_i = \Delta$ and $0 < \rho(C_i) < \infty$. Set $g(\theta) = \sum_{i=1}^\infty (2^i \rho(C_i))^{-1} I_{C_i}(\theta)$ and $\Pi(d\theta) = g(\theta)\rho(d\theta)$. Then Π is a probability measure which is equivalent to ρ as g is positive. ■

Although the following theorem is a direct consequence of Theorem 3.17 we present it here separately because it is a crucial tool in later chapters.

Theorem 3.21. *If the decision space $\mathcal{D}$ is a compact metric space and the model $\mathcal{M} = (\mathcal{X}, \mathfrak{A}, (P_\theta)_{\theta\in\Delta})$ is dominated, then the set of all decisions is weakly sequentially compact.*

Proof. Denote by Q a distribution that dominates the family of distributions $(P_\theta)_{\theta\in\Delta}$. Given any sequence of decisions D_n we get from Theorem 3.17 that there exists a subsequence D_{n_k} such that (3.19) holds. To complete the proof we have only to note that $dP_\theta/dQ \in \mathbb{L}_1(Q)$. ■

To illustrate the concept of weak convergence we turn to some special cases.

Example 3.22. If K_n and K are nonrandomized decisions, i.e., there are mappings $T_n : \mathcal{X}_n \to_m \mathcal{D}$ and $T : \mathcal{X} \to_m \mathcal{D}$ such that $\mathsf{K}_n = \delta_{T_n}$ and $\mathsf{K} = \delta_T$, then

$$\int [\int f(a)\mathsf{K}_n(da|x)]P_\theta(dx) = \int f(T_n(x))P_\theta(dx) = \mathsf{E}_{P_\theta} f(T_n),$$
$$\int [\int f(a)\mathsf{K}(da|x)]P_\theta(dx) = \int f(T(x))P_\theta(dx) = \mathsf{E}_{P_\theta} f(T).$$

Hence we see that $\mathsf{K}_n \Rightarrow \mathsf{K}$ holds if and only if the sequence of distributions $\mathcal{L}(T_n|P_\theta)$ converges weakly to the distribution $\mathcal{L}(T|P_\theta)$.

Example 3.23. Suppose the decision space is finite and considered as a metric space with the discrete metric. Set $q_{a,n}(x) = \mathsf{K}_n(\{a\}|x)$ and $q_a(x) = \mathsf{K}(\{a\}|x)$, $a \in \mathcal{D}$. Then $0 \le q_a, q_{a,n} \le 1$, $\sum_{a\in\mathcal{D}} q_{a,n} = 1$, and $\sum_{a\in\mathcal{D}} q_a = 1$. Instead of all functions f in (3.19) we have to consider only indicator functions of one point sets so that (3.19) is equivalent to

$$\lim_{n\to\infty} \int q_{a,n}(x)P_\theta(dx) = \int q_a(x)P_\theta(dx), \quad \theta \in \Delta,\ a \in \mathcal{D}.$$

If $\mathcal{D} = \{0, 1\}$, then we consider statistical tests and weak convergence of decisions is equivalent to the convergence of the power functions.

3.3 Continuity Properties of the Risk

According to its definition the risk function $\mathsf{R}(\theta, \mathsf{D})$ depends on θ and D. If θ belongs to a metric space, then we may investigate if $\mathsf{R}(\theta, \mathsf{D})$ as a function of θ is lower semicontinuous, or even continuous, for every fixed D. For a metric space $\mathcal{S}$ a function $f : \mathcal{S} \to (-\infty, \infty]$ is called *lower semicontinuous* if for every convergent sequence in $\mathcal{S}$, say $s_n \to s$, it holds $\liminf_{n\to\infty} f(s_n) \geq f(s)$. Analogously, f is called *upper semicontinuous* if $g = -f$ is lower semicontinuous. Obviously, if f is lower and upper semicontinuous, then it is continuous.

It is clear that such continuity properties, in connection with compactness arguments, are extremely helpful in the search for decisions that minimize the risk in compact subsets of decisions. To establish statements in this direction we need a suitable concept of continuity of the model $(\mathcal{X}, \mathfrak{A}, (P_\theta)_{\theta\in\Delta})$. For this purpose we make the following assumption.

(A7) In the model $(\mathcal{X}, \mathfrak{A}, (P_\theta)_{\theta\in\Delta})$ the parameter set Δ is a metric space, and $\theta \mapsto P_\theta$ is continuous in variational distance.

We call a model that satisfies (A7) a *continuous model*. Assumption (A7) seems to be relatively strong. However, if the model is dominated, say $P_\theta \ll \boldsymbol{\mu}$ with $f_\theta = dP_\theta/d\boldsymbol{\mu}$, and if $f_{\theta_n} \to f_\theta$, $\boldsymbol{\mu}$-a.e., for a sequence $\theta_n \to \theta$ in Δ, then by the lemma of Scheffé (see Lemma A.19),

$$\|P_{\theta_n} - P_\theta\| = \int |f_{\theta_n} - f_\theta| d\boldsymbol{\mu} \to 0.$$

Note that under (A7), if Δ is a separable metric space, then the model is dominated.

Problem 3.24.* Every continuous model with a separable parameter set is dominated.

For further results on dominance and separability we refer to Strasser (1985).

Proposition 3.25. *Let $(\mathcal{X}, \mathfrak{A}, (P_\theta)_{\theta\in\Delta})$ be a statistical model where Δ is a metric space. Let $L : \Delta \times \mathcal{D} \to \mathbb{R}_+$ be lower semicontinuous in $\theta \in \Delta$ for every fixed $a \in \mathcal{D}$, and measurable in $a \in \mathcal{D}$ for every fixed $\theta \in \Delta$. If (A7) is satisfied, then for every decision D the risk $\mathsf{R}(\theta, \mathsf{D})$ is a lower semicontinuous function of $\theta \in \Delta$. If $L : \Delta \times \mathcal{D} \to \mathbb{R}$ is bounded and continuous in $\theta \in \Delta$ for every fixed $a \in \mathcal{D}$, then for every fixed D the risk $\mathsf{R}(\theta, \mathsf{D})$ is a continuous function of $\theta \in \Delta$.*

Proof. Let $M(\theta, x, \mathsf{D}) := \int L(\theta, a)\mathsf{D}(da|x)$ and θ_n be any sequence in Δ with $\theta_n \to \theta$. Then by Fatou's lemma (see Lemma A.17) and the lower semicontinuity of $L(\cdot, a)$,

$$\liminf_{n\to\infty} M(\theta_n, x, \mathsf{D}) \geq M(\theta, x, \mathsf{D}), \quad x \in \mathcal{X}.$$

Hence by the inequality in Problem 1.80, for every $N > 0$,

$$\begin{aligned}
\liminf_{n\to\infty} \mathsf{R}(\theta_n, \mathsf{D}) &\geq \liminf_{n\to\infty} \int [M(\theta_n, x, \mathsf{D}) \wedge N] P_{\theta_n}(dx) \\
&\geq \liminf_{n\to\infty} \int [M(\theta_n, x, \mathsf{D}) \wedge N] P_\theta(dx) - N \limsup_{n\to\infty} \|P_{\theta_n} - P_\theta\| \\
&\geq \int [M(\theta, x, \mathsf{D}) \wedge N] P_\theta(dx).
\end{aligned}$$

As $M(\theta, x, \mathsf{D}) \geq 0$ we get from the monotone convergence theorem (see Theorem A.16) the first statement by taking $N \to \infty$ on the right-hand side. If $|L|$ is bounded by C and continuous in θ, then instead of Fatou's Lemma we may apply Lebesgue's theorem (see Theorem A.18) and get that $|M(\theta, x, \mathsf{D})|$ is bounded by C and continuous in θ for every fixed x. This gives

$$|\mathsf{R}(\theta_n, \mathsf{D}) - \mathsf{R}(\theta, \mathsf{D})| \leq |\int [M(\theta_n, x, \mathsf{D}) - M(\theta, x, \mathsf{D})] P_\theta(dx)| + 2C \|P_{\theta_n} - P_\theta\|.$$

Taking the limit as $n \to \infty$ we get the statement. ■

To prepare for the next statement we need a simple result on the equicontinuity of families of continuous functions.

Problem 3.26.* Let Δ be a metric space with metric ρ_Δ, $\mathcal{D}$ a compact metric space, and $L : \Delta \times \mathcal{D} \to \mathbb{R}$ be continuous. Then $\theta \mapsto L(\theta, a)$ is equicontinuous in a in the sense that for every fixed $\theta_0 \in \Delta$ it holds

$$\lim_{\delta\to 0} \sup_{\theta:\rho_\Delta(\theta,\theta_0)\leq\delta} \sup_{a\in\mathcal{D}} |L(\theta, a) - L(\theta_0, a)| = 0.$$

Proposition 3.27. *Let $(\mathcal{X}, \mathfrak{A}, (P_\theta)_{\theta\in\Delta})$ be a statistical model that is continuous in the sense of (A7). If $\mathcal{D}$ is a compact metric space, $L : \Delta \times \mathcal{D} \to \mathbb{R}_+$ is continuous, and $L(\theta, a) \leq C$ for some constant C, then the family of risk functions $\mathsf{R}(\theta, \mathsf{D})$, $\mathsf{D} \in \mathbb{D}$, is equicontinuous at every $\theta \in \Delta$ in the sense that*

$$\lim_{n\to\infty} \sup_{\mathsf{D}\in\mathbb{D}} |\mathsf{R}(\theta_n, \mathsf{D}) - \mathsf{R}(\theta, \mathsf{D})| = 0$$

for every sequence $\theta_n \to \theta$.

Proof. It holds

$$\begin{aligned}
\sup_{\mathsf{D}\in\mathbb{D}} |\mathsf{R}(\theta_n, \mathsf{D}) - \mathsf{R}(\theta, \mathsf{D})| &\leq \sup_{\mathsf{D}\in\mathbb{D}} |\int [\int (L(\theta_n, a) - L(\theta, a)) \mathsf{D}(da|x)] P_{\theta_n}(dx)| \\
&\quad + \sup_{\mathsf{D}\in\mathbb{D}} |\int [\int L(\theta, a) \mathsf{D}(da|x)](P_{\theta_n} - P_\theta)(dx)| \\
&\leq \sup_{a\in\mathcal{D}} |L(\theta_n, a) - L(\theta, a)| + C \|P_{\theta_n} - P_\theta\|,
\end{aligned}$$

where the last inequality follows from the inequality in Problem 1.80. The proof is completed by the application of the statement in Problem 3.26 and the continuity of the model. ■

Now we fix $\theta \in \Delta$ and consider $\mathsf{R}(\theta, \mathsf{D})$ as function of D. According to the concept of weak convergence, for every fixed θ the mapping $\mathsf{D} \mapsto \mathsf{R}(\theta, \mathsf{D})$ is continuous for bounded continuous loss functions. Next we study this continuity property in more details.

Proposition 3.28. *Let $(\mathcal{X}, \mathfrak{A}, (P_\theta)_{\theta\in\Delta})$ be a statistical model and $\mathcal{D}$ a metric space. Assume that the loss function $L : \Delta \times \mathcal{D} \to \mathbb{R}_+$ is lower semicontinuous in $a \in \mathcal{D}$ for every fixed $\theta \in \Delta$. Then for every sequence D_n with $\mathsf{D}_n \Rightarrow \mathsf{D}$ for some D it holds*

$$\liminf_{n\to\infty} \mathsf{R}(\theta, \mathsf{D}_n) \geq \mathsf{R}(\theta, \mathsf{D}).$$

Moreover, if in addition Δ is a compact metric space, $L : \Delta \times \mathcal{D} \to \mathbb{R}$ is continuous, $|L(\theta, a)| \leq C$ for some constant C, and (A7) holds, then

$$\lim_{n\to\infty} \sup_{\theta\in\Delta} |\mathsf{R}(\theta, \mathsf{D}_n) - \mathsf{R}(\theta, \mathsf{D})| = 0.$$

Proof. We know from (3.20) that $\mathsf{D}_n \Rightarrow \mathsf{D}$ if and only if $Q_{n,\theta} := \mathcal{L}(A_n|\mathbb{P}_\theta) \Rightarrow Q_\theta := \mathcal{L}(A|\mathbb{P}_\theta)$, $\theta \in \Delta$. The first statement follows from point (*G*) in Theorem A.49. To prove the second statement we introduce the family of continuous functions $\mathbb{F} = \{L(\theta, \cdot) : \theta \in \Delta\}$ on $\mathcal{D}$. Then by $|L(\theta, a)| \leq C$ and the compactness of Δ the family $\mathbb{F}$ satisfies the conditions in Proposition A.45 and the proof is complete. ■

3.4 Minimum Average Risk, Bayes Risk, Posterior Risk

In many situations the search for a decision that has, uniformly on the parameter set Δ, minimum risk within $\mathbb{D}$, or at least within a given class $\mathbb{D}_0 \subset \mathbb{D}$, remains unsuccessful. As we have pointed out already, such an optimal decision may not even exist. To deal with such difficulties one approach is to restrict the search even further to some suitable subclass $\mathbb{D}_s \subset \mathbb{D}_0$ where this task becomes feasible. Another approach is to replace the optimality that is based on uniform risk comparisons by some weaker optimality criterion. Rather than comparing decisions by their risks, pointwise throughout Δ, comparisons are now made by means of some suitably chosen functional of the risk function. Several functionals that reflect, in one way or another, the extent of risk can be chosen for this purpose. One choice is the supremum of the risk on Δ, which leads to *minimax decisions*. Another choice is an average of the risk, where some weight that depends on $\theta \in \Delta$ is associated with the risk function. This leads to *minimum average risk decisions*, and in particular to *Bayes decisions*.

It should be pointed out that in real-life situations an appropriate optimality criterion is not just simply given. It has to be adopted based on a thorough understanding of the task at hand, including the choice of the statistical model and the loss function. In the minimax approach, the supremum risk of some decisions may be attained only at parameter points $\theta \in \Delta$ that are very unlikely to occur in reality, and the risk function may be peaked there. In the minimum average risk approach, the weight used for averaging the risk may be much larger at some points, where less care of a loss is taken, than at other points. This interrelation, or redundancy, between loss and weight may cause conflicts in an attempt to choose a suitable optimality criterion.

In this section we study the structures of the average, Bayes, and posterior risk. Here we establish general concepts and results that provide a systematic way of finding decisions that minimize the average risk or the Bayes risk.

Let the parameter set Δ be equipped with a σ-algebra $\mathfrak{B}_\Delta$ of subsets, and denote by $\mathcal{M}^\sigma(\mathfrak{B}_\Delta)$ the set of all σ-finite measures on $(\Delta, \mathfrak{B}_\Delta)$. We assume that the family $(P_\theta)_{\theta\in\Delta}$ satisfies assumption (A3), i.e., that the mapping $\theta \mapsto P_\theta(B)$ is measurable for every $B \in \mathfrak{A}$. In addition we make the following assumption.

(A8) $(\Delta, \mathfrak{B}_\Delta)$ is a measurable space and $L : \Delta \times \mathcal{D} \to \mathbb{R}_+$ is $(\mathfrak{B}_\Delta \otimes \mathfrak{D})$-$\mathfrak{B}_+$ measurable.

To indicate that this condition holds we also write $L : \Delta \times \mathcal{D} \to_m \mathbb{R}_+$. It follows from Fubini's theorem for stochastic kernels (see Proposition A.40) that for every $\mathsf{D} \in \mathbb{D}$ the risk

$$\mathsf{R}(\theta, \mathsf{D}) = \int [\int L(\theta, a)\mathsf{D}(da|x)]P_\theta(dx)$$

is a nonnegative measurable function of θ that can be integrated with respect to every $\rho \in \mathcal{M}^\sigma(\mathfrak{B}_\Delta)$. We define the *average risk of* D *with respect to the weight measure* ρ by

$$\mathsf{r}(\rho, \mathsf{D}) = \int \mathsf{R}(\theta, \mathsf{D})\rho(d\theta).$$

Fubini's theorem and (3.4) yield

$$\mathsf{r}(\rho, \mathsf{D}) = \int (\int [\int L(\theta, a)\mathsf{D}(da|x)]\mathsf{P}(dx|\theta))\rho(d\theta). \tag{3.21}$$

If $\rho = \Pi$ is a probability measure, then we call Π the prior and $\mathsf{r}(\Pi, \mathsf{D})$ the *Bayes risk of* D *under the prior* Π.

Remark 3.29. (Convention). In view of (3.21), the assumptions (A3) and (A8) are essential, and thus tacitly assumed to hold, whenever we are dealing with the average risk or the Bayes risk. Occasionally, however, they are mentioned again as a reminder of this convention.

By a particular choice of ρ we emphasize special areas of the parameter set with the intention of getting decisions that perform well at least in the areas with high weights of ρ. If ρ is finite, but not a probability measure, then a normalization shows that minimizing $\mathsf{r}(\rho, \mathsf{D})$ is equivalent to minimizing $\mathsf{r}(\Pi, \mathsf{D})$ with $\Pi = \rho(\Delta)^{-1}\rho$. A similar reduction can be made if ρ is σ-finite.

Remark 3.30. Using the ρ-density of the distribution Π in Lemma 3.20 and the chain rule we may represent the average risk as

$$\begin{aligned}\mathsf{r}(\rho,\mathsf{D}) &= \int(\int[\int L(\theta,a)\mathsf{D}(da|x)]\mathsf{P}(dx|\theta))\rho(d\theta)\\ &= \int(\int[\int \widetilde{L}(\theta,a)\mathsf{D}(da|x)]\mathsf{P}(dx|\theta))\Pi(d\theta),\\ \widetilde{L}(\theta,a) &= L(\theta,a)\frac{d\rho}{d\Pi}(\theta).\end{aligned}$$

This means that every average risk can be written as a Bayes risk with a modified loss function. Thus, in principle, we could have operated with probability measures as averaging measures right from the beginning. However, the new loss function $\widetilde{L}$ may be less intuitive than the original L, and it may complicate the minimization of the posterior risk that is introduced later. As L is nonnegative $\widetilde{L}$ is bounded from below. The above representation reveals the duality between the averaging measure ρ and the loss function L. Especially in Bayes analysis, where Π is interpreted as a prior, one should be aware that the choice of the prior and the choice of the loss function are intimately linked.

To create an approach which simultaneously covers the case of finite and infinite averaging measures we need some technical results. For $\rho \in \mathcal{M}^{\sigma}(\mathfrak{B}_{\Delta})$ we set

$$(\mathsf{P}\rho)(A) = \int P_{\theta}(A)\rho(d\theta), \quad A \in \mathfrak{A}.$$

The measure $\mathsf{P}\rho$ is not necessarily σ-finite even though ρ is σ-finite. Indeed, if $\rho(\Delta) = \infty$ and $\mathcal{X}$ is finite, then $(\mathsf{P}\rho)(\{x\}) = \infty$ must hold for at least one point $x \in \mathcal{X}$ which rules out σ-finiteness. For dominated models the σ-finiteness of $\mathsf{P}\rho$ is equivalent with the fact that the normalizing factor $\mathsf{m}(x)$ in (1.35) is finite.

Problem 3.31.* If condition (A5) holds, then

$$\mathsf{m}(x) = \int f_{\theta}(x)\rho(d\theta) < \infty, \quad \boldsymbol{\mu}\text{-a.e. } x \in \mathcal{X} \quad \Leftrightarrow \quad \mathsf{P}\rho \text{ is } \sigma\text{-finite.} \tag{3.22}$$

$$\frac{d(\mathsf{P}\rho)}{d\boldsymbol{\mu}} = \mathsf{m}.$$

A crucial point in the subsequent considerations is a disintegration of the measure $\mathsf{P}\otimes\rho$, i.e., a representation $(\mathsf{P}\otimes\rho)(dx,d\theta) = (\boldsymbol{\Pi}\otimes\mathsf{P}\rho)(d\theta,dx)$ with the help of a stochastic kernel $\boldsymbol{\Pi}$. Such a decomposition of $\mathsf{P}\otimes\rho$ exists for a σ-finite ρ under weak conditions, as the next proposition shows. Let m be the marginal density in (3.22).

Proposition 3.32. *Let $\rho \in \mathcal{M}^{\sigma}(\mathfrak{B}_{\Delta})$ and assume that at least one of the following two conditions holds.*

$$\begin{aligned}&\textit{(A4) is satisfied and } \mathsf{P}\rho \textit{ is } \sigma\textit{-finite.}\\ &\textit{(A5) is satisfied and } \mathsf{m} < \infty, \quad \boldsymbol{\mu}\textit{-a.e.}\end{aligned} \tag{3.23}$$

Then there exists a stochastic kernel $\boldsymbol{\Pi} : \mathfrak{B}_\Delta \times \mathcal{X} \to_k [0,1]$ *such that*

$$\int [\int h(x,\theta) P_\theta(dx)] \rho(d\theta) = \int [\int h(x,\theta) \boldsymbol{\Pi}(d\theta|x)] \mathsf{P}\rho(dx), \tag{3.24}$$

for every $h : \mathcal{X} \times \Delta \to_m \mathbb{R}_+$. *If (3.23) is satisfied,* $\boldsymbol{\tau} \in \mathcal{M}^\sigma(\mathfrak{B}_\Delta)$, *and* $\pi = d\rho/d\boldsymbol{\tau}$, *then a version of the posterior distribution is given by*

$$\begin{aligned} \boldsymbol{\Pi}(d\theta|x) &= \pi(\theta|x)\boldsymbol{\tau}(d\theta), \quad \textit{where} \\ \pi(\theta|x) &= \frac{1}{\mathsf{m}(x)}\pi(\theta) f_\theta(x) I_{(0,\infty)}(\mathsf{m}(x)) + \pi(\theta) I_{\{0\}}(\mathsf{m}(x)). \end{aligned} \tag{3.25}$$

Proof. As ρ is σ-finite we get from Lemma 3.20 that there is a distribution Π with $\Pi \ll\gg \rho$. Condition (A4) and Theorem A.37 imply that there is a stochastic kernel $\boldsymbol{\Pi}_0 : \mathfrak{B}_\Delta \times \mathcal{X} \to_k [0,1]$ such that

$$\int [\int g(x,\theta) P_\theta(dx)] \Pi(d\theta) = \int [\int g(x,\theta) \boldsymbol{\Pi}_0(d\theta|x)] \mathsf{P}\Pi(dx),$$

for every $g : \mathcal{X} \times \Delta \to_m \mathbb{R}_+$. We note that $\Pi \ll\gg \rho$ implies $\mathsf{P}\Pi \ll\gg \mathsf{P}\rho$. As $\mathsf{P}\rho$ is σ-finite the theorem of Radon–Nikodym (see Theorem A.27) provides the existence of a density $\frac{d(\mathsf{P}\Pi)}{d(\mathsf{P}\rho)}$. Set

$$\boldsymbol{\Pi}_1(B|x) = \frac{d(\mathsf{P}\Pi)}{d(\mathsf{P}\rho)}(x) \int I_B(\theta) \frac{d\rho}{d\Pi}(\theta) \boldsymbol{\Pi}_0(d\theta|x).$$

Then $\boldsymbol{\Pi}_1(\cdot|x)$ is a measure on $\mathfrak{B}_\Delta$ for every fixed x and $\boldsymbol{\Pi}_1(B|x)$ is a measurable function of x for every fixed $B \in \mathfrak{B}_\Delta$. Moreover, for every $h : \mathcal{X} \times \Delta \to_m \mathbb{R}_+$, by the chain rule and $g(x,\theta) = h(x,\theta)\frac{d\rho}{d\Pi}(\theta)$,

$$\begin{aligned} &\int [\int h(x,\theta) \boldsymbol{\Pi}_1(d\theta|x)] \mathsf{P}\rho(dx) = \int [\int h(x,\theta) \frac{d\rho}{d\Pi}(\theta) \boldsymbol{\Pi}_0(d\theta|x)] \mathsf{P}\Pi(dx) \\ &= \int [\int (h(x,\theta) \frac{d\rho}{d\Pi}(\theta)) P_\theta(dx)] \Pi(d\theta) = \int [\int h(x,\theta) P_\theta(dx)] \rho(d\theta). \end{aligned} \tag{3.26}$$

Putting $h(x,\theta) = I_A(x)$, $A \in \mathfrak{A}$, we arrive at

$$\int I_A(x) \boldsymbol{\Pi}_1(\Delta|x) \mathsf{P}\rho(dx) = \int [\int I_A(x) P_\theta(dx)] \rho(d\theta) = \mathsf{P}\rho(A).$$

As $\mathsf{P}\rho$ is σ-finite we may conclude that

$$(\mathsf{P}\rho)(\{x : \boldsymbol{\Pi}_1(\Delta|x) \neq 1\}) = 0.$$

Set $\boldsymbol{\Pi}(\cdot|x) = \boldsymbol{\Pi}_1(\cdot|x)$ if $\boldsymbol{\Pi}_1(\Delta|x) = 1$ and $\boldsymbol{\Pi}(\cdot|x) = \boldsymbol{\Pi}_2(\cdot|x)$ if $\boldsymbol{\Pi}_1(\Delta|x) \neq 1$, where $\boldsymbol{\Pi}_2$ is any fixed stochastic kernel. Then $\boldsymbol{\Pi}$ is a stochastic kernel that satisfies (3.24) in view of (3.26). To prove the second statement we note that $\boldsymbol{\Pi}$ in (3.25) satisfies (3.24) by the definition of $\pi(\theta|x)$ and Fubini's theorem. ■

Definition 3.33. *If $\rho \in \mathcal{M}^\sigma(\mathfrak{B}_\Delta)$, then every stochastic kernel $\boldsymbol{\Pi} : \mathfrak{B}_\Delta \times \mathcal{X} \to_k [0,1]$ that satisfies (3.24) is called a posterior distribution. For any such stochastic kernel $\boldsymbol{\Pi}$,*

$$r(\rho, a|x) := \int L(\theta, a)\boldsymbol{\Pi}(d\theta|x), \quad \text{and} \tag{3.27}$$

$$\mathsf{r}(\rho, \mathsf{D}|x) := \int r(\rho, a|x)\mathsf{D}(da|x),$$

are called the posterior risk at x of making a decision $a \in \mathcal{D}$ and the posterior risk at x using the decision D, respectively.

For later reference we give the representations of the posterior risk in the dominated case. That the statement below holds for every $a \in \mathcal{D}$ follows immediately from Definition 3.33 and (3.25).

$$r(\rho, a|x) = \frac{1}{\mathsf{m}(x)} \int L(\theta, a) f_\theta(x)\pi(\theta)\boldsymbol{\tau}(d\theta), \quad \mathsf{P}\rho\text{-a.e.} \tag{3.28}$$

From (3.21) and (3.24) it follows that

$$\mathsf{r}(\rho, \mathsf{D}) = \int \mathsf{r}(\rho, \mathsf{D}|x)\mathsf{P}\rho(dx). \tag{3.29}$$

Now we study the posterior risk for exponential families under their conjugate priors. The next result is an immediate consequence of the definition of the densities of conjugate priors in (1.39) and the posterior densities in (1.41).

Proposition 3.34. *Let $(P_\theta)_{\theta \in \Delta}$ be an exponential family with generating statistic T and natural parameter θ. Let $\rho = \Pi_{u,v}$, $(u,v) \in \Upsilon$, be a conjugate prior with the $\boldsymbol{\tau}$-density from (1.39). Then*

$$r(\rho, a|x) = \int L(\theta, a)\pi_{u+1, v+T(x)}(\theta)\boldsymbol{\tau}(d\theta).$$

The next example deals with binomial distributions.

Example 3.35. Let $\mathsf{B}(n,p)$, $p \in (0,1)$, be the family of binomial distributions on $\mathcal{X} = \{0, 1, ..., n\}$. For an inference on p we utilize the conjugate prior $\mathsf{Be}(\alpha, \beta)$ from Example 1.45 and set $\rho = \mathsf{Be}(\alpha, \beta)$. The posterior at $x \in \mathcal{X}$ is $\mathsf{Be}(\alpha + x, \beta + n - x)$. Moreover,

$$r(\rho, a|x) = \int_0^1 L(\varkappa(p), a)\mathsf{be}_{\alpha+x, \beta+n-x}(p)dp, \quad \mathsf{r}(\rho, \mathsf{D}|x) = \int r(\rho, a|x)\mathsf{D}(da|x),$$

and $\varkappa(p) = \ln(p/(1-p)) = \theta$.

We introduce the concepts of minimum average risk and Bayes decisions. We recall that assumptions (A3) and (A8) are assumed to hold.

Definition 3.36. *Given a model* $(\mathcal{X}, \mathfrak{A}, (P_\theta)_{\theta \in \Delta})$, $\mathbb{D}_0 \subseteq \mathbb{D}$, *and* $\rho \in \mathcal{M}^\sigma(\mathfrak{B}_\Delta)$ *a decision* $\mathsf{D}_0 \in \mathbb{D}_0$ *with* $\mathsf{r}(\rho, \mathsf{D}_0) = \inf_{\mathsf{D} \in \mathbb{D}_0} \mathsf{r}(\rho, \mathsf{D})$ *is called a minimum average risk decision in* $\mathbb{D}_0$ *under the average measure* ρ. *If* $\rho = \Pi \in \mathcal{P}(\mathfrak{B}_\Delta)$, *then* D_0 *is called a Bayes decision in* $\mathbb{D}_0$ *under the prior* Π. *Whenever* $\mathbb{D}_0 = \mathbb{D}$, *"in* $\mathbb{D}_0$*" is not mentioned.*

Now we characterize the minimum average risk and Bayes decisions in $\mathbb{D}_0$.

Theorem 3.37. *Let* $\rho \in \mathcal{M}^\sigma(\mathfrak{B}_\Delta)$, $(\mathcal{X}, \mathfrak{A}, (P_\theta)_{\theta \in \Delta})$ *be a statistical model. Assume that a posterior distribution exists. Let* $\mathbb{D}_0 \subseteq \mathbb{D}$ *be a class of decisions. If* $\inf_{\mathsf{D} \in \mathbb{D}_0} \mathsf{r}(\rho, \mathsf{D}) < \infty$, *then a decision* $\mathsf{D}_0 \in \mathbb{D}_0$ *is a minimum average risk decision in* $\mathbb{D}_0$ *if and only if*

$$(\mathsf{P}\rho)(\{x : \mathsf{r}(\rho, \mathsf{D}_0|x) > \mathsf{r}(\rho, \mathsf{D}|x)\}) = 0, \quad \mathsf{D} \in \mathbb{D}_0.$$

Proof. The sufficiency follows from (3.29). Conversely, if for some $\mathsf{D}_1 \in \mathbb{D}_0$ we have $(\mathsf{P}\rho)(A) > 0$ for $A = \{x : \mathsf{r}(\rho, \mathsf{D}_0|x) > \mathsf{r}(\rho, \mathsf{D}_1|x)\}$, then

$$\mathsf{D}_2(\cdot|x) := \mathsf{D}_1(\cdot|x) I_A(x) + \mathsf{D}_0(\cdot|x) I_{\overline{A}}(x).$$

is a better decision in terms of the average risk. ■

Condition (3.31) says that a minimum average decision can be obtained by finding for every fixed x a distribution $\mathsf{D}(\cdot|x)$ that is concentrated on $\arg\min_{a \in \mathcal{D}} r(\rho, a|x)$, i.e., the set of all $a \in \mathcal{D}$ that minimizes $r(\rho, a|x)$.

Now we consider Bayes decisions in a normal model.

Example 3.38. Consider the family $(\mathsf{N}^{\otimes n}(\mu, \sigma^2))_{\mu \in \mathbb{R}}$, where $\sigma^2 > 0$ is known. We take $\rho = \mathsf{N}(\nu, \delta^2)$. Then by Lemma 1.37 it holds for every loss function $L : \mathbb{R} \times \mathcal{D} \to_m \mathbb{R}_+$,

$$r(\rho, a|x) = \int L(t, a) \varphi_{\mu(x), \tau^2}(t) dt, \quad \text{where}$$

$$\mu(x) = \frac{(n/\sigma^2)}{(n/\sigma^2) + (1/\delta^2)} \overline{x}_n + \frac{(1/\delta^2)}{(n/\sigma^2) + (1/\delta^2)} \nu, \quad \tau^2 = \frac{1}{(n/\sigma^2) + (1/\delta^2)}.$$

If $\mathcal{D} = \mathbb{R}$, then we have the problem of estimating the parameter μ. If the loss function is given by $L(\mu, a) = (\mu - a)^2$, then the pointwise minimization of

$$r(\mathsf{N}(\nu, \delta^2), a|x) = \int (t - a)^2 \varphi_{\mu(x), \tau^2}(t) dt$$

provides the Bayes estimator

$$T_B(x_1, ..., x_n) = \frac{(n/\sigma^2)}{(n/\sigma^2) + (1/\delta^2)} \overline{x}_n + \frac{(1/\delta^2)}{(n/\sigma^2) + (1/\delta^2)} \nu. \tag{3.30}$$

For infinite averaging measures ρ it often occurs that the risk $\mathsf{r}(\rho, \mathsf{D})$ is not finite, but the posterior risk $\mathsf{r}(\rho, \mathsf{D}|x)$ is $\mathsf{P}\rho$-a.e. finite. This occurs typically in invariant models if D is an invariant decision for which the risk $\mathsf{R}(\theta, \mathsf{D})$ is independent of θ and ρ is an invariant measure on the parameter space with infinite total mass. Such situations are met later on in Chapter 7 when we deal with the Pitman estimator.

Definition 3.39. *Let $\rho \in \mathcal{M}^\sigma(\mathfrak{B}_\Delta)$. Suppose that $(\mathcal{X}, \mathfrak{A}, (P_\theta)_{\theta\in\Delta})$ is a statistical model for which a posterior distribution exists. Let $\mathbb{D}_0 \subseteq \mathbb{D}$ be a class of decisions. A decision $\mathsf{D}_0 \in \mathbb{D}_0$ is called a generalized Bayes decision in $\mathbb{D}_0$ if it holds*

$$(\mathsf{P}\rho)(\{x : \mathsf{r}(\rho, \mathsf{D}_0|x) > \mathsf{r}(\rho, \mathsf{D}|x)\}) = 0, \quad \mathsf{D} \in \mathbb{D}_0.$$

It is clear from the definition that every minimum average risk decision with finite risk is a generalized Bayes decision.

Corollary 3.40. *Assume that for $r(\rho, a|x)$ from (3.27) $\inf_{a\in\mathcal{D}} r(\rho, a|x)$ is a measurable function of x. Then every decision $\mathsf{D}_0 \in \mathbb{D}_0$ that satisfies*

$$\mathsf{P}\rho(\{x : \mathsf{D}_0(\{a : r(\rho, a|x) > \inf_{b\in\mathcal{D}} r(\rho, b|x)\}|x) > 0\}) = 0 \tag{3.31}$$

is a generalized Bayes decision and a minimum average risk decision in $\mathbb{D}_0$.

Proof. Condition (3.31) implies $\mathsf{P}\rho$-*a.e.*

$$\mathsf{r}(\rho, \mathsf{D}_0|x) \le \inf_{b\in\mathcal{D}} r(\rho, b|x) \le \int r(\rho, a|x)\mathsf{D}(da|x) = \mathsf{r}(\rho, \mathsf{D}|x).$$

■

Subsequently we consider situations where $r(\rho, a|x)$ can be evaluated explicitly. This is especially the case when we are dealing with an exponential family and conjugate priors. On the other hand, there are many situations where the integrals associated with $r(\rho, a|x)$ in (3.27) and (3.28) cannot be evaluated explicitly. In this case one has to have recourse to numerical methods. Several techniques for that purpose have been developed in the last decades. A special role is hereby played by Monte Carlo methods that run fast and provide powerful tools for Bayes analyses. For details we refer to Chen, Shao, and Ibrahim (2000).

Example 3.41. Suppose that in the model $(\mathcal{X}, \mathfrak{A}, (P_\theta)_{\theta\in\Delta})$ the parameter set $\Delta = \{1, ..., m\}$ is finite. The task of deciding which of the values in $\mathcal{D} = \{1, ..., m\}$ is associated with the true distribution is called a classification problem. In the context of decision theory this can also be interpreted as the problem of estimating a parameter when the parameter set Δ is finite. For a given prior Π on Δ we set $\pi_i = \Pi(\{i\})$, $i = 1, ..., m$. Let f_i be the density of P_i with respect to a dominating σ-finite measure $\boldsymbol{\mu}$, $i = 1, ..., m$. Then the posterior risk $r(\Pi, a|x)$ from (3.27) has the form

$$r(\Pi, a|x) = \frac{1}{\mathsf{m}(x)} \sum\nolimits_{i=1}^{m} L(i, a) f_i(x)\pi_i, \quad a = 1, ..., m, \ x \in \mathcal{X}.$$

As the factor $1/\mathsf{m}(x)$ is irrelevant for minimizing $r(\Pi, a|x)$ as a function of a an application of Corollary 3.40 yields that a classification rule D_Π is Bayes with respect to the prior Π if and only if

$$\begin{aligned} \mathsf{D}_\Pi(A(x)|x) &= 1, \quad \boldsymbol{\mu}\text{-a.e.}, \quad \text{where} \\ A(x) &= \arg\min_{1\le a\le m} \sum\nolimits_{i=1}^{m} L(i, a) f_i(x)\pi_i. \end{aligned}$$

Especially, under the zero–one loss function $L_{0,1}(\theta, a) = 1 - I_{\{a\}}(\theta)$, we have

$$r(\Pi, a|x) = \frac{1}{\mathsf{m}(x)} \sum\nolimits_{i \neq a} f_i(x)\pi_i \quad \text{and} \quad A(x) = \underset{1 \leq a \leq m}{\arg\max} f_a(x)\pi_a.$$

We conclude the series of examples with a model where the posterior risk is finite whereas the risk is infinite for all invariant decisions.

Example 3.42. Let $P_\theta = \mathsf{N}^{\otimes n}(\theta, 1)$, $\theta \in \mathbb{R}$, and $\rho = \boldsymbol{\lambda}$. Then at $x = (x_1, ..., x_n) \in \mathbb{R}^n$, and with $\overline{x}_n = (1/n)\sum_{i=1}^n x_i$,

$$\begin{aligned} f_\theta(x) &= (2\pi)^{-n/2} \exp\{-\frac{1}{2}\sum\nolimits_{i=1}^n (x_i - \theta)^2\}, \\ \mathsf{m}(x) &= \int (2\pi)^{-n/2} \exp\{-\frac{1}{2}\sum\nolimits_{i=1}^n (x_i - \theta)^2\} d\theta \\ &= (2\pi)^{-(n-1)/2} n^{-1/2} \exp\{-\frac{1}{2}\sum\nolimits_{i=1}^n (x_i - \overline{x}_n)^2\} < \infty, \\ \pi(\theta|x) &= (2\pi)^{-1/2} n^{1/2} \exp\{-\frac{n}{2}(\theta - \overline{x}_n)^2\}. \end{aligned}$$

If $L(\theta, a) = (\theta - a)^2$ (i.e., the squared error loss function), then

$$r(\rho, a|x) = \int (\theta - a)^2 \pi(\theta|x) d\theta < \infty.$$

To find a generalized Bayes estimator we have to minimize the function $a \mapsto r(\rho, a|x)$ for every fixed x. This gives the estimator $T_\varrho(x) = \overline{x}_n$. The posterior risk of this estimator is

$$\mathsf{r}(\rho, T_\varrho|x) := \int r(\rho, a|x) \boldsymbol{\delta}_{\overline{x}_n}(da) = r(\rho, \overline{x}_n|x) = n^{-1}$$

and thus finite. In contrast to that the average risk of $T_\varrho(x) = \overline{x}_n$ is infinite. Indeed, the relation (3.29) implies

$$\mathsf{r}(\rho, T) = \int [\int \mathsf{r}(\rho, T_\varrho|x) P_\theta(dx)] d\theta = \int (\int n^{-1} P_\theta(dx)) d\theta = \infty.$$

Later on in Chapter 7 we show that $T_\varrho(x) = \overline{x}_n$ is the Pitman estimator, which is then introduced for a more general class of models.

Minimum average decisions reflect the risk according to the averaging measure ρ. If this measure is sufficiently well spread out across Δ, then a minimum average decision cannot be outperformed by any other decision uniformly in $\theta \in \Delta$. Indeed minimum average risk decisions are admissible under weak assumptions. To pursue this idea in a more flexible way, Stein (1955b) and LeCam (1955) were the first to consider sequences of priors to establish admissibility. These ideas have been used in many papers: see e.g. Diaconis and Stein (1983), Farrell (1968), Brown (1971), and Rukhin (1986). Here we present only the classical sufficient condition for admissibility, due to Blyth (1951), which is based on the continuity of the risk function. Conditions under which this continuity holds have been investigated in Section 3.3.

Theorem 3.43. *Assume that Δ is a metric space and that the risk function $\mathsf{R}(\theta, \mathsf{D})$ is continuous in θ for every $\mathsf{D} \in \mathbb{D}_0$. Let $\mathsf{D}_0 \in \mathbb{D}_0$. If there exists a sequence $\rho_n \in \mathcal{M}^\sigma(\mathfrak{B}_\Delta)$, and associated minimum average decisions $\mathsf{D}_n \in \mathbb{D}_0$ with $\mathsf{r}(\rho_n, \mathsf{D}_n) < \infty$, such that*

$$\lim_{n\to\infty} [\mathsf{r}(\rho_n, \mathsf{D}_0) - \mathsf{r}(\rho_n, \mathsf{D}_n)] = 0 \tag{3.32}$$

and

$$\liminf_{n\to\infty} \rho_n(B) > 0 \text{ for every open set } B \subseteq \Delta, \tag{3.33}$$

then D_0 is admissible in $\mathbb{D}_0$.

Corollary 3.44. *If the risk function $\mathsf{R}(\theta, \mathsf{D})$ is continuous in θ for every $\mathsf{D} \in \mathbb{D}_0$ and $\rho(B) > 0$ for every open set $B \subseteq \Delta$, then a minimum risk decision $\mathsf{D}_0 \in \mathbb{D}_0$ with finite risk is admissible.*

Proof. Suppose D_0 is not admissible. Then there are some $\mathsf{D} \in \mathbb{D}_0$ and $\theta_0 \in \Delta$ such that

$$\mathsf{R}(\theta, \mathsf{D}) \le \mathsf{R}(\theta, \mathsf{D}_0), \quad \theta \in \Delta, \quad \text{and} \quad \mathsf{R}(\theta_0, \mathsf{D}) < \mathsf{R}(\theta_0, \mathsf{D}_0).$$

The continuity of both $\mathsf{R}(\theta, \mathsf{D}_0)$ and $\mathsf{R}(\theta, \mathsf{D})$ implies that for every sufficiently small $\varepsilon > 0$ the set $U(\varepsilon, \theta_0) = \{\theta : \mathsf{R}(\theta, \mathsf{D}) + \varepsilon < \mathsf{R}(\theta, \mathsf{D}_0)\}$ is open and nonempty. By (3.33) we get

$$\begin{aligned}\lim_{n\to\infty} [\mathsf{r}(\rho_n, \mathsf{D}_0) - \mathsf{r}(\rho_n, \mathsf{D}_n)] &\ge \liminf_{n\to\infty} [\mathsf{r}(\rho_n, \mathsf{D}_0) - \mathsf{r}(\rho_n, \mathsf{D})] \\ = \liminf_{n\to\infty} \int [\mathsf{R}(\theta, \mathsf{D}_0) - \mathsf{R}(\theta, \mathsf{D})]\rho_n(d\theta) &\ge \liminf_{n\to\infty} \varepsilon \rho_n(U(\varepsilon, \theta_0)) > 0,\end{aligned}$$

which is a contradiction to (3.32). The corollary follows with $\rho_n = \rho$ and $\mathsf{D}_n = \mathsf{D}_0$. ■

In the above corollary the condition that the risk of D_0 is finite is essential. As we have seen already generalized Bayes decisions may have an infinite risk. Thus their possible admissibility cannot be concluded from the corollary.

We illustrate the last theorem with a classical result.

Example 3.45. Let X follow a normal distribution $\mathsf{N}(\mu, 1)$, $\mu \in \mathbb{R}$. Suppose that we want to estimate μ under the squared error loss $L(\mu, a) = (a - \mu)^2$. The natural estimator $T_{nat}(x) = x$ has the risk $\mathsf{R}(\mu, T_{nat}) = \int s^2 \varphi_{0,1}(s) ds = 1$. We compare this estimator with the Bayes estimator for the prior $\Pi = \mathsf{N}(0, \delta^2)$. According to (3.30) the Bayes estimator is

$$T_B(x) = \varrho_1^2 x, \quad \varrho_1^2 = \delta^2/(1 + \delta^2),$$

which has the risk $\mathsf{R}(\mu, T_B) = \varrho_1^4 + (1 - \varrho_1^2)^2 \mu^2$ and the Bayes risk $\mathsf{r}(\Pi, T_B) = \varrho_1^2$. Now we take $\rho_n = \sqrt{n}\mathsf{N}(0, n)$. The factor $\sqrt{n}$ does not change the minimum average risk estimator which is $m_n(x) = (n/(n + 1))x$. The density of ρ_n is $(2\pi)^{-1/2} \exp\{-s^2/(2n)\}$. As

$$\lim_{n\to\infty} \sqrt{n} \int_a^b \varphi_{0,n}(s)ds = \frac{b-a}{\sqrt{2\pi}} > 0, \quad a < b,$$

the condition (3.33) is satisfied. It remains to show that the condition (3.32) is met, too. Indeed,

$$\lim_{n\to\infty} \sqrt{n}[\mathsf{r}(\rho_n, T_{nat}) - \mathsf{r}(\rho_n, m_n)] = \lim_{n\to\infty} \sqrt{n}(1 - \frac{n}{1+n}) = 0.$$

Therefore the natural estimator $T_{nat}(x) = x$ is admissible.

Now we focus on decisions in Bayes models; i.e., we consider (X, Θ) as a pair of random variables where X is observable and takes on values in $(\mathcal{X}, \mathfrak{A})$, whereas Θ is not observable and takes on values in $(\Delta, \mathfrak{B}_\Delta)$. Θ is the object of interest and the inference. We assume that (A3) is satisfied which means that $\mathsf{P}(\cdot|\theta) = P_\theta(\cdot)$ is a stochastic kernel. The distribution Π of Θ is called the *prior distribution*, and the stochastic kernel $\mathsf{P}(\cdot|\theta) = P_\theta(\cdot)$ is the conditional distribution of X, given $\Theta = \theta$. Hence (X, Θ) has the joint distribution $\mathsf{P} \otimes \Pi$. When making an inference we have to deal with the random action A, where A is a random variable whose conditional distribution, given $X = x$ and $\Theta = \theta$, depends only on x and is specified by $\mathsf{D}(\cdot|x)$. Altogether, we model the observation and decision process by the probability space

$$(\Omega, \mathfrak{F}, \mathbb{P}) = (\mathcal{D} \times \mathcal{X} \times \Delta, \mathfrak{D} \otimes \mathfrak{A} \otimes \mathfrak{B}_\Delta, \mathsf{D} \otimes \mathsf{P} \otimes \Pi), \tag{3.34}$$

where $\mathbb{P} = \mathsf{D} \otimes \mathsf{P} \otimes \Pi$ is defined by

$$\mathbb{P}(E) = \int (\int [\int I_E(a, x, \theta) \mathsf{D}(da|x)] \mathsf{P}(dx|\theta)) \Pi(d\theta);$$

see Proposition A.40. If A, X, and Θ denote the projections on $\mathcal{D}$, $\mathcal{X}$, and Δ, respectively, then (A, X, Θ) is a random vector with values in $\mathcal{D} \times \mathcal{X} \times \Delta$ such that

$$\mathbb{E}h(A, X, \Theta) = \int (\int [\int h(a, x, \theta) \mathsf{D}(da|x)] \mathsf{P}(dx|\theta)) \Pi(d\theta) \tag{3.35}$$

holds for every $h : \mathcal{D} \times \mathcal{X} \times \Delta \to_m \mathbb{R}_+$. From here we see that Θ, X, A is a Markov chain; see (1.126). Suppose there exists a posterior distribution $\boldsymbol{\Pi}$ in the sense of Definition 3.33. Then for every $g : \mathcal{X} \times \Delta \to_m \mathbb{R}_+$,

$$\int [\int g(x, \theta) P_\theta(dx)] \Pi(d\theta) = \int [\int g(x, \theta) \boldsymbol{\Pi}(d\theta|x)] (\mathsf{P}\Pi)(dx).$$

This means that in the Bayes model the posterior distribution is just a regular conditional distribution of Θ, given $X = x$. The representation (3.35) implies that for every $h : \mathcal{D} \times \mathcal{X} \times \Delta \to_m \mathbb{R}_+$,

$$\mathbb{E}h(A, X, \Theta) = \int [\int [\int h(a, x, \theta) \mathsf{D}(da|x)] \boldsymbol{\Pi}(d\theta|x)] (\mathsf{P}\Pi)(dx).$$

We see from here that the conditional distribution of (A, Θ), given $X = x$, is $\mathsf{D}(\cdot|x) \otimes \boldsymbol{\Pi}(\cdot|x)$. This means that the decision A and the random parameter Θ are conditionally independent, given $X = x$, which is a property that holds for every Markov chain; see Problem 1.99. Moreover, we may write the posterior risk as a conditional expectation.

$$r(\Pi, a|x) = \int L(\theta, a)\boldsymbol{\Pi}(d\theta|x) = \mathbb{E}(L(\Theta, a)|X = x), \quad \mathsf{P}\Pi\text{-a.s.}$$

It also holds

$$r(\Pi, \mathsf{D}|x) = \mathbb{E}(L(\Theta, A)|X = x), \quad \mathsf{P}\Pi\text{-a.s.}$$

Now we consider examples of Bayes decision problems where explicit solutions can be obtained.

Example 3.46. Consider the Bayes decision model (3.34). Let $\kappa : \Delta \to_m \mathbb{R}$ be a function and suppose we want to predict the random variable $\kappa(\Theta)$. Each such prediction $T : \mathcal{X} \to_m \mathbb{R}$ is called a Bayes estimator of Θ. We use the squared error loss $L(\theta, a) = [\kappa(\theta) - a]^2$ and assume that $\mathbb{E}\kappa^2(\Theta) < \infty$. Then

$$\begin{aligned} r(\Pi, a|x) &= \int [\kappa(\theta) - a]^2 \boldsymbol{\Pi}(d\theta|x) \\ &= \int [\kappa(\theta) - \int \kappa(\tilde{\theta})\boldsymbol{\Pi}(d\tilde{\theta}|x)]^2 \boldsymbol{\Pi}(d\theta|x) + \int [a - \int \kappa(\tilde{\theta})\boldsymbol{\Pi}(d\tilde{\theta}|x)]^2 \boldsymbol{\Pi}(d\theta|x). \end{aligned}$$

Therefore,

$$T_\Pi(x) = \int \kappa(\theta)\boldsymbol{\Pi}(d\theta|x) = \mathbb{E}(\kappa(\Theta)|X = x)$$

is the $\mathsf{P}\Pi$-a.s. uniquely determined Bayes estimator.

Example 3.47. Let $(\mathcal{X}, \mathfrak{A}, (P_\theta)_{\theta\in\Delta})$ be a statistical model that satisfies condition (A5). We consider the problem of testing the hypotheses $\mathsf{H}_0 : \Delta_0$ versus $\mathsf{H}_A : \Delta_A$, where Δ_0 and Δ_A is a decomposition of Δ. Let Π be a prior on $(\Delta, \mathfrak{B}_\Delta)$. We assume that $v := \Pi(\Delta_0)$ satisfies $0 < v < 1$, and decompose Π into

$$\Pi = v\Pi_0 + (1 - v)\Pi_A, \quad \text{where} \quad \Pi_0 = \Pi(\cdot|\Delta_0) \quad \text{and} \quad \Pi_A = \Pi(\cdot|\Delta_A).$$

If $\mathsf{m}_0, \mathsf{m}_A$, and m denote the marginal densities

$$\mathsf{m}_0(x) = \int f_\theta(x)\Pi_0(d\theta), \ \mathsf{m}_A(x) = \int f_\theta(x)\Pi_A(d\theta), \ \mathsf{m}(x) = \int f_\theta(x)\Pi(d\theta), \quad (3.36)$$

then $\mathsf{m}(x) = v\mathsf{m}_0(x) + (1 - v)\mathsf{m}_A(x)$. The decision space is $\mathcal{D} = \{0, 1\}$. For any test φ the associated decision rule is $\mathsf{D}(\cdot|x) = (1 - \varphi(x))\boldsymbol{\delta}_0(\cdot) + \varphi(x)\boldsymbol{\delta}_1(\cdot)$. Assuming that there is no loss for making a correct decision, we adopt a loss function of the form

$$L(\theta, a) = al_0(\theta)I_{\Delta_0}(\theta) + (1 - a)l_1(\theta)I_{\Delta_A}(\theta),$$

where $l_0, l_1 : \Delta \to_m \mathbb{R}_+$. Then by (3.28) with $\boldsymbol{\tau} = \Pi$ and $\pi = 1$,

$$r(\Pi, 0|x) = \frac{v}{\mathsf{m}(x)} \int l_0(\theta) f_\theta(x)\Pi_0(d\theta), \quad r(\Pi, 1|x) = \frac{1 - v}{\mathsf{m}(x)} \int l_1(\theta) f_\theta(x)\Pi_A(d\theta).$$

According to Corollary 3.40, we get a Bayes decision by choosing any test φ_Π with

$$\varphi_\Pi(x) = \begin{cases} 1 & \text{if } r(\Pi,1|x) < r(\Pi,0|x), \\ 0 & \text{if } r(\Pi,1|x) > r(\Pi,0|x). \end{cases} \tag{3.37}$$

Especially, under the zero–one loss, i.e., $L(\theta,a) = aI_{\Delta_0}(\theta) + (1-a)I_{\Delta_A}(\theta)$, $a = 0,1$, we get

$$\varphi_\Pi(x) = \begin{cases} 1 & \text{if } (1-v)\mathsf{m}_A(x) < v\mathsf{m}_0(x), \\ 0 & \text{if } (1-v)\mathsf{m}_A(x) > v\mathsf{m}_0(x). \end{cases} \tag{3.38}$$

The test φ_Π in (3.37) and (3.38) can be chosen arbitrarily, but of course measurable, for observations x for which $r(\Pi,1|x) = r(\Pi,0|x)$ and $(1-v)\mathsf{m}_A(x) = v\mathsf{m}_0(x)$, respectively. Thus in particular, every Bayes test may be chosen to be a nonrandomized test. The test in (3.38) admits an interesting interpretation. The densities of the marginal distributions $\mathsf{P}\Pi_0$ and $\mathsf{P}\Pi_A$ with respect to the dominating measure $\boldsymbol{\mu}$ of the family $(P_\theta)_{\theta\in\Delta}$ are given by

$$\frac{d(\mathsf{P}\Pi_0)}{d\boldsymbol{\mu}}(x) = \mathsf{m}_0(x) \quad \text{and} \quad \frac{d(\mathsf{P}\Pi_A)}{d\boldsymbol{\mu}}(x) = \mathsf{m}_A(x).$$

This means that in view of Theorem 2.60 φ_Π in (3.38) is a Bayes test for the simple hypotheses $\mathsf{H}_0 : \mathsf{P}\Pi_0$ versus $\mathsf{H}_A : \mathsf{P}\Pi_A$ under the prior $(v, 1-v)$.

$$\begin{aligned} \varphi_\Pi(x) = 1 \quad &\text{if} \quad v\frac{d(\mathsf{P}\Pi_0)}{d\boldsymbol{\mu}}(x) < (1-v)\frac{d(\mathsf{P}\Pi_A)}{d\boldsymbol{\mu}}(x), \\ \varphi_\Pi(x) = 0 \quad &\text{if} \quad v\frac{d(\mathsf{P}\Pi_0)}{d\boldsymbol{\mu}}(x) > (1-v)\frac{d(\mathsf{P}\Pi_A)}{d\boldsymbol{\mu}}(x). \end{aligned}$$

Now we assume that the data are from the Bayes model (3.34). We consider the posterior probabilities of $\{\Theta \in \Delta_0\}$ and $\{\Theta \in \Delta_A\}$ at $X = x$. As $\boldsymbol{\tau} = \Pi$ and $\pi = 1$ the posterior density in (3.25) is $\pi(\theta|x) = (f_\theta(x)/\mathsf{m}(x))I_{(0,\infty)}(\mathsf{m}(x)) + I_{\{0\}}(\mathsf{m}(x))$ and

$$\begin{aligned} \mathbb{P}(\Theta \in \Delta_0|X = x) &= \int I_{\Delta_0}(\theta)\pi(\theta|x)\boldsymbol{\tau}(d\theta) = v\frac{\mathsf{m}_0(x)}{\mathsf{m}(x)}, \\ \mathbb{P}(\Theta \in \Delta_A|X = x) &= \int I_{\Delta_A}(\theta)\pi(\theta|x)\boldsymbol{\tau}(d\theta) = (1-v)\frac{\mathsf{m}_A(x)}{\mathsf{m}(x)}. \end{aligned}$$

Therefore

$$\begin{aligned} \varphi_\Pi(x) = 1 \quad &\text{if} \quad \mathbb{P}(\Theta \in \Delta_0|X = x) < \mathbb{P}(\Theta \in \Delta_A|X = x), \\ \varphi_\Pi(x) = 0 \quad &\text{if} \quad \mathbb{P}(\Theta \in \Delta_0|X = x) > \mathbb{P}(\Theta \in \Delta_A|X = x). \end{aligned}$$

The ratio of the posterior odds ratio and the prior odds ratio,

$$B(x) = \frac{\mathbb{P}(\Theta \in \Delta_0|X = x)/\mathbb{P}(\Theta \in \Delta_A|X = x)}{v/(1-v)}$$

is called the *Bayes factor* in favor of Δ_0.

3.5 Bayes and Minimax Decisions

As pointed out previously, in general we cannot find a decision D that minimizes $\mathsf{R}(\theta, D)$ uniformly in $\theta \in \Delta$. We also have already considered concepts for dealing with the risk function by means of a functional. The common idea behind these concepts is to assign a numerical value to the risk function that has to be minimized in order to get an optimal decision. One of these concepts is to average the risk function which leads to the concept of a minimum average risk decision and especially to Bayes decisions; see Definition 3.36. Another approach is to look at the worst case which is the maximum of the risk function.

Definition 3.48. *Given a statistical model $(\mathcal{X}, \mathfrak{A}, (P_\theta)_{\theta\in\Delta})$ and a class of decisions $\mathbb{D}_0 \subseteq \mathbb{D}$, a decision $\mathsf{D}_0 \in \mathbb{D}_0$ is called a minimax decision in $\mathbb{D}_0$ if $\sup_{\theta\in\Delta} \mathsf{R}(\theta, \mathsf{D}_0) = \inf_{\mathsf{D}\in\mathbb{D}_0} (\sup_{\theta\in\Delta} \mathsf{R}(\theta, \mathsf{D}))$. Whenever $\mathbb{D}_0 = \mathbb{D}$, "in $\mathbb{D}_0$" is not mentioned.*

In this section we study the relations between minimax and minimum average risk decisions. First we deal with the problem of the existence of minimax and minimum average risk decisions. The crucial point here is that under weak conditions the risk and the Bayes risk are lower semicontinuous functions of the decisions and the space of all decision is compact.

Proposition 3.49. *Suppose that in $(\mathcal{X}, \mathfrak{A}, (P_\theta)_{\theta\in\Delta})$ the family $(P_\theta)_{\theta\in\Delta}$ is dominated, the decision space $\mathcal{D}$ is a compact metric space, and the loss function $L(\theta, a)$ is lower semicontinuous in a for every $\theta \in \Delta$. If $\mathbb{D}_0$ is closed under the weak convergence of decisions, then there exists at least one decision that is minimax in $\mathbb{D}_0$.*

Proof. We know from Proposition 3.28 that for every fixed θ the function $\mathsf{D} \mapsto \mathsf{R}(\theta, \mathsf{D})$ is a lower semicontinuous function with respect to the weak convergence of decisions. As the model is dominated and the decision space is a compact metric space, the space of all decisions $\mathbb{D}$ is weakly sequentially compact; see Theorem 3.21. Hence, without loss of generality, we may choose a sequence D_n such that $\sup_{\theta\in\Delta} \mathsf{R}(\theta, \mathsf{D}_n) \to \inf_{\mathsf{D}\in\mathbb{D}_0} \sup_{\theta\in\Delta} \mathsf{R}(\theta, \mathsf{D})$ where D_n converges weakly to some D_0, and where $\mathsf{D}_0 \in \mathbb{D}_0$ by assumption. By the lower semicontinuity of $\mathsf{R}(\theta, \mathsf{D})$ it follows $\mathsf{R}(\theta, \mathsf{D}_0) \le \liminf_{n\to\infty} \mathsf{R}(\theta, \mathsf{D}_n)$ and thus

$$\sup_{\theta\in\Delta} \mathsf{R}(\theta, \mathsf{D}_0) \le \liminf_{n\to\infty} \sup_{\theta\in\Delta} \mathsf{R}(\theta, \mathsf{D}_n) = \lim_{n\to\infty} \sup_{\theta\in\Delta} \mathsf{R}(\theta, \mathsf{D}_n) = \inf_{\mathsf{D}\in\mathbb{D}_0} \sup_{\theta\in\Delta} \mathsf{R}(\theta, \mathsf{D}).$$

■

The existence of a minimum average decision can be guaranteed similarly. For the next proposition we assume that (A3) and (A8) are satisfied.

Proposition 3.50. *Suppose that in $(\mathcal{X}, \mathfrak{A}, (P_\theta)_{\theta\in\Delta})$ the family $(P_\theta)_{\theta\in\Delta}$ is dominated, the decision space $\mathcal{D}$ is a compact metric space, and the loss function $L(\theta, a)$ is nonnegative and lower semicontinuous in a for every $\theta \in \Delta$. If $\mathbb{D}_0$ is closed under the weak convergence of decisions, then for every $\rho \in \mathcal{M}^\sigma(\mathfrak{B}_\Delta)$ there exists at least one minimum average risk decision in $\mathbb{D}_0$.*

Proof. We again use Proposition 3.28 to see that for every fixed θ the function $\mathsf{D} \mapsto \mathsf{R}(\theta, \mathsf{D})$ is a lower semicontinuous function. By Fatou's lemma we get that $\mathsf{D} \mapsto \mathsf{r}(\rho, \mathsf{D})$ is a lower semicontinuous function on $\mathbb{D}$. The rest of the proof is analogous to the proof of Proposition 3.49. ■

Finding minimax decisions, provided there exists at least one, is a difficult problem which can be solved directly only in special situations. Therefore we study the minimax concept under various conditions, and also the relations to other concepts. To prepare for that, we establish for a class $\mathbb{D}_0 \subseteq \mathbb{D}$ a lower bound for the *minimax value* in $\mathbb{D}_0$, i.e., for $\inf_{\mathsf{D}\in\mathbb{D}_0} \sup_{\theta\in\Delta} \mathsf{R}(\theta, \mathsf{D})$. For any $\theta_0 \in \Delta$ and $\mathsf{D}_0 \in \mathbb{D}_0$ it holds $\inf_{\mathsf{D}\in\mathbb{D}_0} \mathsf{R}(\theta_0, \mathsf{D}) \le \sup_{\theta\in\Delta} \mathsf{R}(\theta, \mathsf{D}_0)$, which implies

$$\sup_{\theta\in\Delta} \inf_{\mathsf{D}\in\mathbb{D}_0} \mathsf{R}(\theta, \mathsf{D}) \le \inf_{\mathsf{D}\in\mathbb{D}_0} \sup_{\theta\in\Delta} \mathsf{R}(\theta, \mathsf{D}). \tag{3.39}$$

If there exists a decision $\mathsf{D}_0 \in \mathbb{D}_0$ that is minimax in $\mathbb{D}_0$, then the search for a minimax decision can be simplified considerably if there exists some $\theta_0 \in \Delta$ which is *least favorable* for D_0 in the sense of $\mathsf{R}(\theta_0, \mathsf{D}_0) = \sup_{\theta\in\Delta} \mathsf{R}(\theta, \mathsf{D}_0)$. The combination of these two ideas leads to the concept of a *saddle point*.

Definition 3.51. *Given a model $(\mathcal{X}, \mathfrak{A}, (P_\theta)_{\theta\in\Delta})$, a decision space $(\mathcal{D}, \mathfrak{D})$, and a class of decisions $\mathbb{D}_0 \subseteq \mathbb{D}$, a pair $(\theta_0, \mathsf{D}_0) \in \Delta \times \mathbb{D}_0$ is called a saddle point in $\Delta \times \mathbb{D}_0$ if*

$$\mathsf{R}(\theta, \mathsf{D}_0) \le \mathsf{R}(\theta_0, \mathsf{D}_0) \le \mathsf{R}(\theta_0, \mathsf{D}), \quad \theta \in \Delta,\ \mathsf{D} \in \mathbb{D}_0.$$

Whenever $\mathbb{D}_0 = \mathbb{D}$, "in $\Delta \times \mathbb{D}_0$" is not mentioned.

It is easy to see that the existence of a saddle point in $\Delta \times \mathbb{D}_0$ implies equality in (3.39).

Proposition 3.52. *If (θ_0, D_0) is a saddle point in $\Delta \times \mathbb{D}_0$, then*

$$\sup_{\theta\in\Delta} \inf_{D\in\mathbb{D}_0} \mathsf{R}(\theta, D) = \mathsf{R}(\theta_0, \mathsf{D}_0) = \sup_{\theta\in\Delta} \mathsf{R}(\theta, \mathsf{D}_0) = \inf_{D\in\mathbb{D}_0} \sup_{\theta\in\Delta} \mathsf{R}(\theta, \mathsf{D}),$$

so that in particular D_0 is minimax in $\mathbb{D}_0$ and θ_0 is least favorable for D_0.

Proof. It holds

$$\inf_{\mathsf{D}\in\mathbb{D}_0} \sup_{\theta\in\Delta} \mathsf{R}(\theta, \mathsf{D}) \le \mathsf{R}(\theta_0, \mathsf{D}_0) \le \inf_{\mathsf{D}\in\mathbb{D}_0} \mathsf{R}(\theta_0, \mathsf{D}) \le \sup_{\theta\in\Delta} \inf_{\mathsf{D}\in\mathbb{D}_0} \mathsf{R}(\theta, \mathsf{D}),$$

so that the statement follows from inequality (3.39). ■

The inequality $\mathsf{R}(\theta,\mathsf{D}_0) \le \mathsf{R}(\theta_0,\mathsf{D}_0)$ says that θ_0 is a least favorable parameter configuration for decision D_0, whereas $\mathsf{R}(\theta_0,\mathsf{D}_0) \le \mathsf{R}(\theta_0,\mathsf{D})$ says that D_0 is the best decision at parameter point θ_0. These can also be viewed as monotonicity properties of the risk function at the point (θ_0,D_0), and thus it is not surprising that in testing problems the MLR property is crucial in this regard.

Example 3.53. As before in Theorem 2.49, consider testing $\mathsf{H}_0 : (b_0,\theta_0]$ versus $\mathsf{H}_A : (\theta_0,b_1)$ as a decision problem with $\Delta = (b_0,b_1)$, where $\mathsf{D}(\cdot|x) = (1-\varphi(x))\boldsymbol{\delta}_0(\cdot)+\varphi(x)\boldsymbol{\delta}_1(\cdot)$ and $L(\theta,a) = aI_{(b_0,\theta_0]}(\theta) + (1-a)I_{(\theta_0,b_1)}(\theta)$, $\theta \in \Delta$, $a \in \mathcal{D} = \{0,1\}$. Here we have

$$\mathsf{R}(\theta,\mathsf{D}) = I_{(b_0,\theta_0]}(\theta)\mathsf{E}_\theta\varphi + I_{(\theta_0,b_1)}(\theta)(1-\mathsf{E}_\theta\varphi), \quad \theta \in \Delta.$$

For a fixed $\alpha \in (0,1/2]$ let $\mathbb{D}_0$ be the class of decisions associated with tests φ that satisfy $\mathsf{E}_{\theta_0}\varphi = \alpha$. Suppose that $(P_\theta)_{\theta\in(b_0,b_1)}$, has MLR in T. Let $\varphi_{T,\alpha}$ be the tests from (2.19) and

$$\mathsf{D}_0 = (1-\varphi_{T,\alpha})\boldsymbol{\delta}_0 + \varphi_{T,\alpha}\boldsymbol{\delta}_1.$$

The relation (2.32) in Theorem 2.49 gives $\mathsf{R}(\theta,\mathsf{D}_0) \le \mathsf{R}(\theta_0,\mathsf{D}_0) = \alpha$ for $\theta \in \Delta$. Furthermore, by the construction of $\mathbb{D}_0$ it holds $\mathsf{R}(\theta_0,\mathsf{D}_0) = \mathsf{R}(\theta_0,\mathsf{D})$ for every $\mathsf{D} \in \mathbb{D}_0$. Therefore (θ_0,D_0) is a saddle point in $\Delta \times \mathbb{D}_0$.

In the definition of a saddle point (θ_0,D_0) the point θ_0 plays the role of a least favorable parameter configuration for the decision D_0. A similar role of a least favorable prior can be assigned when we are dealing with Bayes decisions. For the remainder of this section it is assumed that conditions (A3) and (A8) are satisfied.

Definition 3.54. *For a decision problem under the model $(\mathcal{X},\mathfrak{A},(P_\theta)_{\theta\in\Delta})$, and a class of decisions $\mathbb{D}_0 \subseteq \mathbb{D}$, a distribution $\Pi_0 \in \mathcal{P}(\mathfrak{B}_\Delta)$ is called a least favorable prior for $\mathbb{D}_0$ if*

$$\sup_{\Pi\in\mathcal{P}(\mathfrak{B}_\Delta)} \inf_{\mathsf{D}\in\mathbb{D}_0} \mathsf{r}(\Pi,\mathsf{D}) = \inf_{\mathsf{D}\in\mathbb{D}_0} \mathsf{r}(\Pi_0,\mathsf{D}).$$

Whenever $\mathbb{D}_0 = \mathbb{D}$, "for $\mathbb{D}_0$" is not mentioned.

There is a close connection between the minimax approach and the concept of a least favorable prior which is studied next. In Bayes analysis, after a prior Π and a class $\mathbb{D}_0 \subseteq \mathbb{D}$ have been chosen, the goal is to find a decision D_0 for which $\mathsf{r}(\Pi,\mathsf{D}_0) \le \mathsf{r}(\Pi,\mathsf{D})$, $\mathsf{D} \in \mathbb{D}_0$. Regarding the choice of the prior, the concept of a least favorable prior for $\mathbb{D}_0$ targets a prior Π_0 that maximizes the minimal Bayes risks in $\mathbb{D}_0$ for all priors $\Pi \in \mathcal{P}(\mathfrak{B}_\Delta)$. To examine how this is related to the minimax approach, we start with some obvious facts. The following hold.

$$\inf_{\mathsf{D}\in\mathbb{D}_0} \sup_{\theta\in\Delta} \mathsf{R}(\theta,\mathsf{D}) = \inf_{\mathsf{D}\in\mathbb{D}_0} \sup_{\Pi} \mathsf{r}(\Pi,\mathsf{D}),$$

$$\underline{w} := \sup_{\Pi} \inf_{\mathsf{D}\in\mathbb{D}_0} \mathsf{r}(\Pi,\mathsf{D}) \le \inf_{\mathsf{D}\in\mathbb{D}_0} \sup_{\theta\in\Delta} \mathsf{R}(\theta,\mathsf{D}) = \inf_{\mathsf{D}\in\mathbb{D}_0} \sup_{\Pi} \mathsf{r}(\Pi,\mathsf{D}) =: \overline{w}. \quad (3.40)$$

If Π_0 is least favorable for $\mathbb{D}_0$ and D_0 is Bayes in $\mathbb{D}_0$ under Π_0, then

$$\sup_{\Pi} \inf_{\mathsf{D}\in\mathbb{D}_0} \mathsf{r}(\Pi, \mathsf{D}) = \mathsf{r}(\Pi_0, \mathsf{D}_0).$$

The question that remains open is whether equality holds in (3.40), i.e., whether

$$\sup_{\Pi} \inf_{\mathsf{D}\in\mathbb{D}_0} \mathsf{r}(\Pi, \mathsf{D}) = \inf_{\mathsf{D}\in\mathbb{D}_0} \sup_{\theta\in\Delta} \mathsf{R}(\theta, \mathsf{D}). \tag{3.41}$$

Any statement that establishes (3.41) is called a *minimax theorem*. If (3.41) holds, then, at least approximately, every minimax decision is a Bayes decision. This is a completeness statement to which we come back later on. The question regarding the validity of a minimax theorem is also related to another concept.

Definition 3.55. *Given a model $(\mathcal{X}, \mathfrak{A}, (P_\theta)_{\theta\in\Delta})$ and a decision space $(\mathcal{D}, \mathfrak{D})$, a pair $(\Pi_0, \mathsf{D}_0) \in \mathcal{P}(\mathfrak{B}_\Delta) \times \mathbb{D}_0$ is called a saddle point in $\mathcal{P}(\mathfrak{B}_\Delta) \times \mathbb{D}_0$ if*

$$\mathsf{r}(\Pi, \mathsf{D}_0) \le \mathsf{r}(\Pi_0, \mathsf{D}_0) \le \mathsf{r}(\Pi_0, \mathsf{D}), \quad \mathsf{D} \in \mathbb{D}_0,\ \Pi \in \mathcal{P}(\mathfrak{B}_\Delta). \tag{3.42}$$

The existence of a saddle point is of great importance for establishing (3.41), as the next result shows.

Proposition 3.56. *If (Π_0, D_0) is a saddle point in $\mathcal{P}(\mathfrak{B}_\Delta) \times \mathbb{D}_0$, then (3.41) holds, Π_0 is a least favorable prior for $\mathbb{D}_0$, D_0 is Bayes in $\mathbb{D}_0$ under Π_0, and D_0 is minimax in $\mathbb{D}_0$.*

Proof. To establish (3.41) we remark that (3.42) yields

$$\overline{w} = \inf_{\mathsf{D}\in\mathbb{D}_0} \sup_{\Pi} \mathsf{r}(\Pi, \mathsf{D}) \le \mathsf{r}(\Pi_0, \mathsf{D}_0) \le \sup_{\Pi} \inf_{\mathsf{D}\in\mathbb{D}_0} \mathsf{r}(\Pi, \mathsf{D}) = \underline{w}.$$

In view of (3.40) we have equality. The statements that Π_0 is a least favorable prior for $\mathbb{D}_0$, and that D_0 is Bayes in $\mathbb{D}_0$ under Π_0, are direct consequences of (3.42). It remains to show that D_0 is minimax in $\mathbb{D}_0$. As a prior can also be chosen to be a δ-distribution we get

$$\sup_{\theta\in\Delta} \mathsf{R}(\theta, \mathsf{D}_0) = \sup_{\Pi} \mathsf{r}(\Pi, \mathsf{D}_0) \le \mathsf{r}(\Pi_0, \mathsf{D}_0) \le \mathsf{r}(\Pi_0, \mathsf{D}) \le \sup_{\theta\in\Delta} \mathsf{R}(\theta, \mathsf{D}),$$

which proves the minimaxity. ■

We also consider the following condition.

$$\sup_{\tilde{\theta}\in\Delta} \mathsf{R}(\tilde{\theta}, \mathsf{D}_0) = \mathsf{R}(\theta, \mathsf{D}_0), \quad \Pi_0\text{-a.s. } \theta \in \Delta, \quad \mathsf{D}_0 \text{ is Bayes under } \Pi_0. \tag{3.43}$$

Decisions that satisfy the first part in (3.43) have constant risk and are called *equalizer decisions*.

Theorem 3.57. *The conditions (3.42) and (3.43) are equivalent, and they imply that D_0 is minimax in $\mathbb{D}_0$.*

Proof. The conditions on the right-hand sides of (3.42) and (3.43) are identical. Then the left-hand condition in (3.42) is equivalent to

$$\sup_{\Pi} \mathsf{r}(\Pi, \mathsf{D}_0) = \sup_{\theta\in\Delta} \mathsf{R}(\theta, \mathsf{D}_0) = \mathsf{r}(\Pi_0, \mathsf{D}_0) = \int \mathsf{R}(\theta, \mathsf{D}_0)\Pi_0(d\theta).$$

But this in turn is equivalent to

$$\int (\sup_{\widetilde{\theta}\in\Delta} \mathsf{R}(\widetilde{\theta}, \mathsf{D}_0) - \mathsf{R}(\theta, \mathsf{D}_0))\Pi_0(d\theta) = 0,$$

which is equivalent to the first part in (3.43). The second statement follows from Proposition 3.56. ■

The next proposition is based on a more general version of condition (3.43).

Proposition 3.58. *If there exist a sequence of priors Π_n, and an associated sequence D_n of Bayes decisions in $\mathbb{D}_0$ under Π_n, $n = 1, 2, ...,$ such that*

$$\sup_{\theta\in\Delta} \mathsf{R}(\theta, \mathsf{D}_0) = \lim_{n\to\infty} \mathsf{r}(\Pi_n, \mathsf{D}_n), \tag{3.44}$$

then D_0 is a minimax decision in $\mathbb{D}_0$.

Proof. The assumption and (3.40) imply

$$\sup_{\theta\in\Delta} \mathsf{R}(\theta, \mathsf{D}_0) \le \sup_{\Pi\in\mathcal{P}(\mathfrak{B}_\Delta)} \inf_{\mathsf{D}\in\mathbb{D}_0} \mathsf{r}(\Pi, \mathsf{D}) \le \inf_{\mathsf{D}\in\mathbb{D}_0} \sup_{\theta\in\Delta} \mathsf{R}(\theta, \mathsf{D}).$$

■

Every sequence Π_n that satisfies (3.44) is called a *least favorable sequence of priors* for $\mathbb{D}_0$, and D_0 is called an *approximate Bayes decision* in $\mathbb{D}_0$.

Example 3.59. We consider the problem of estimating the parameter p in a binomial distribution $\mathsf{B}(n,p)$, $p \in \Delta = (0,1)$. Here $\mathcal{X} = \{0, 1, ..., n\}$ and the decision space is $\mathcal{D} = \Delta = (0,1)$. We use the squared error loss $L(p,a) = (a-p)^2$. Let the prior Π on $\Delta = (0,1)$ be the beta distribution $\mathsf{Be}(\alpha,\beta)$ with parameters $\alpha, \beta > 0$. From Example 3.35, with the loss $L(\kappa(p), a) = (a-p)^2$, we get

$$r(\Pi, a|x) = \int_0^1 (p-a)^2 \mathsf{be}_{\alpha+x,\beta+n-x}(p)dp, \quad a \in \Delta,\ x \in \mathcal{X}.$$

According to Corollary 3.40 we have to minimize $r(\Pi, a|x)$ for every fixed x. Hence

$$T_{\alpha,\beta}(x) := \int_0^1 p\mathsf{be}_{\alpha+x,\beta+n-x}(p)dp = \frac{\alpha+x}{\alpha+\beta+n}$$

is a Bayes estimator for p under the prior $\mathsf{Be}(\alpha,\beta)$, and

$$\begin{aligned}\mathsf{R}(p, T_{\alpha,\beta}) &= \sum\nolimits_{l=0}^{n} \mathsf{b}_{n,p}(l)(\frac{\alpha+l}{\alpha+\beta+n} - p)^2 \\ &= \frac{1}{(\alpha+\beta+n)^2}\left[p^2((\alpha+\beta)^2 - n) + p(n - 2\alpha(\alpha+\beta)) + \alpha^2\right].\end{aligned}$$

We see that $\mathsf{R}(p, T_{\alpha,\beta})$ is constant in p if and only if $\alpha = \beta = \sqrt{n}/2$. By Theorem 3.57

$$T_{\sqrt{n}/2,\sqrt{n}/2}(x) = \frac{1}{\sqrt{n}+n}(x + \sqrt{n}/2)$$

is a minimax estimator. To better understand which areas of the parameter space are emphasized by the prior, we recall that $\mathsf{Be}(\alpha, \beta)$ has the expectation μ and variance σ^2, respectively,

$$\mu = \frac{\alpha}{\alpha+\beta} \quad \text{and} \quad \sigma^2 = \frac{\alpha\beta}{(\alpha+\beta)^2(\alpha+\beta+1)},$$

which for $\alpha = \beta = \sqrt{n}/2$ turn out to be $\mu_n = 1/2$ and $\sigma_n^2 = [4(\sqrt{n}+1)]^{-1}$. Although n is fixed here, we see that σ_n^2 is small for large n so that the interpretation of the least favorable prior is that parameter values close to 0 and 1 are easier to estimate than those in the middle, an effect that is increasing with n.

Next we consider the problem of estimating the mean θ of a d-variate normal distribution. We use a loss function that depends on the difference of the estimate and the unknown parameter, similarly as it has been done in some of the previous examples. Especially we use subconvex functions l that are an extension of the class of convex functions. We recall from Section 2.1 that a function $l : \mathbb{R}^d \to \mathbb{R}$ is subconvex if $\{x : l(x) \leq c,\ x \in \mathbb{R}^d\}$ is a convex set for every $c \in \mathbb{R}$, and convex if $l(\alpha x + (1-\alpha)y) \leq \alpha l(x) + (1-\alpha)l(y)$, $\alpha \in [0,1]$, $x, y \in \mathbb{R}^d$. The next problem shows that subconvexity can be expressed in terms of some less restrictive inequalities.

Problem 3.60. $l : \mathbb{R}^d \to \mathbb{R}$ is subconvex if and only if

$$l(\alpha x + (1-\alpha)y) \leq \max(l(x), l(y)), \quad \alpha \in [0,1],\ x, y \in \mathbb{R}^d.$$

A function l is called *centrally symmetric* if $l(-x) = l(x)$. We denote by $\mathfrak{L}_d$ the class of all nonnegative centrally symmetric subconvex Borel measurable functions on $\mathbb{R}^d$ with $l(0) = 0$. Examples of subconvex functions on $\mathbb{R}$ are all convex functions, such as $|x|^p$, $p \geq 1$. However, the class of subconvex functions is larger, and in particular it contains functions that are used in *robust statistics* to reduce the influence of outliers. Examples are

$$l^*(x) = x^2 I_{[-1,1]}(x) + |x| I_{\mathbb{R}\setminus[-1,1]}(x) \quad \text{and} \quad l^{**}(x) = x^2 I_{[-1,1]}(x) + I_{\mathbb{R}\setminus[-1,1]}(x).$$

The function l^* is a linear continuation of the quadratic function, whereas l^{**} is bounded.

Problem 3.61. It holds $l \in \mathfrak{L}_1$ if l is symmetric and $l : [0,\infty) \to [0,\infty)$ is nondecreasing. If $l \in \mathfrak{L}_1$, then for $l_d(x) := l(\|x\|)$ it holds $l_d \in \mathfrak{L}_d$.

The following result is the famous Anderson's lemma. For a proof we refer to Strasser (1985), Lemma 38.21.

Proposition 3.62. *If P is a distribution on the Borel sets of $\mathbb{R}^d$ that has a centrally symmetric Lebesgue density f for which $-f$ is subconvex and $l \in \mathfrak{L}_d$, then*

$$\int l(x)P(dx) \leq \int l(x+a)P(dx), \quad a \in \mathbb{R}^d.$$

If $P = \mathsf{N}(0, \Sigma)$ is a normal distribution with regular covariance matrix Σ and expectation zero, then the Lebesgue density is

$$\varphi_{0,\Sigma}(x) = (2\pi)^{-n/2}(\det(\Sigma))^{-1/2}\exp\{-\frac{1}{2}x^T\Sigma^{-1}x\}.$$

Obviously, $-\varphi_{0,\Sigma}$ is centrally symmetric and subconvex. Hence by Anderson's lemma for every $l \in \mathfrak{L}_d$,

$$\int l(x)\varphi_{0,\Sigma}(x)\boldsymbol{\lambda}_d(dx) \leq \int l(x+a)\varphi_{0,\Sigma}(x)\boldsymbol{\lambda}_d(dx), \quad a \in \mathbb{R}^d. \tag{3.45}$$

Problem 3.63.* Let $Y_i \sim \mathsf{N}(\mu_i, \Sigma_i)$, where $\mu_i \in \mathbb{R}^d$ and Σ_i is a nonsingular $d \times d$ matrix, $i = 1, 2$. If Y_1 and Y_2 are independent, then the conditional distribution of Y_2 given $X := Y_1 + Y_2 = x$ is a normal distribution with expectation $\mu_2 + \Sigma_2(\Sigma_1 + \Sigma_2)^{-1}(x - (\mu_1 + \mu_2))$ and covariance matrix $\Sigma_2 - \Sigma_2(\Sigma_1 + \Sigma_2)^{-1}\Sigma_2$.

Example 3.64. We consider again the estimation problem of Example 3.14 in a somewhat more general setting. We want to estimate the parameter θ in the family $(\mathsf{N}(\theta, \Sigma_0))_{\theta \in \mathbb{R}^d}$ where Σ_0 is nonsingular and known. The decision space is $\mathcal{D} = \mathbb{R}^d$. Let the prior be $\Pi_n = \mathsf{N}(0, n\Sigma_0)$. Then by Problem 3.63 the posterior distribution is the normal distribution $\mathsf{N}(\mu_n(x), \Sigma_n)$ with

$$\mu_n(x) = \frac{n}{n+1}x \quad \text{and} \quad \Sigma_n = \frac{n}{n+1}\Sigma_0.$$

If $L(\theta, a) = l(\theta - a)$ for some $l \in \mathfrak{L}_d$, then the function $r(\Pi_n, a|x)$ in (3.28) turns out to be

$$\begin{aligned} r(\Pi_n, a|x) &= \int l(\theta - a)\varphi_{\mu_n(x),\Sigma_n}(\theta)\boldsymbol{\lambda}_d(d\theta) \\ &= \int l(\theta - [a - \mu_n(x)])\varphi_{0,\Sigma_n}(\theta)\boldsymbol{\lambda}_d(d\theta). \end{aligned}$$

We get from (3.45) that

$$\int l(\theta - [a - \mu_n(x)])\varphi_{0,\Sigma_n}(\theta)\boldsymbol{\lambda}_d(d\theta) \geq \int l(\theta)\varphi_{0,\Sigma_n}(\theta)\boldsymbol{\lambda}_d(d\theta).$$

As

$$\inf_{a \in \mathcal{D}} r(\Pi_n, a|x) = \int l(\theta)\varphi_{0,\Sigma_n}(\theta)\boldsymbol{\lambda}_d(d\theta)$$

is independent of x and therefore a measurable function of x, we get from Corollary 3.40 that $T_n(x) = (n/(n+1))x$ is a Bayes estimator for the prior Π_n under the loss function l. The Bayes risk of this estimator is given by

$$\begin{aligned} \mathsf{r}(\Pi_n, T_n) &= \int [\int l(\frac{n}{n+1}x - \theta)\mathsf{N}(\theta, \Sigma_0)(dx)]\mathsf{N}(0, n\Sigma_0)(d\theta) \\ &= \int l(\tau_n w)\mathsf{N}(0, \Sigma_0)(dw), \quad \text{where } \tau_n^2 = \frac{n}{n+1}. \end{aligned}$$

Theorem 3.65. *For the family of normal distributions* $(\mathsf{N}(\theta, \Sigma_0))_{\theta \in \mathbb{R}^d}$, *with a known nonsingular* Σ_0, *and the loss function* $L(\theta, a) = l(a - \theta)$, $l \in \mathfrak{L}_d$, $\theta \in \mathbb{R}^d$, $a \in \mathbb{R}^d$, *the estimator* $T_{nat}(x) = x$ *is a minimax estimator in the class of all randomized estimators; that is,*

$$\begin{aligned} &\inf_{\mathsf{D}} \sup_{\theta \in \mathbb{R}^d} \int [\int l(a - \theta)\mathsf{D}(da|x)]\mathsf{N}(\theta, \Sigma_0)(dx) \\ &= \sup_{\theta \in \mathbb{R}^d} \int l(T_{nat}(x) - \theta)\mathsf{N}(\theta, \Sigma_0)(dx) = \int l(t)\mathsf{N}(0, \Sigma_0)(dt). \end{aligned}$$

Moreover, the minimax theorem

$$\begin{aligned} &\sup_{\Pi} \inf_{\mathsf{D}} \int (\int [\int l(a - \theta)\mathsf{D}(da|x)]\mathsf{N}(\theta, \Sigma_0)(dx))\Pi(d\theta) \qquad (3.46) \\ &= \int l(t)\mathsf{N}(0, \Sigma_0)(dt) \end{aligned}$$

holds. If l *is bounded, then it holds*

$$\begin{aligned} &\lim_{m \to \infty} \inf_{\mathsf{D}} \sup_{\theta \in \mathbb{R}^d, \|\theta\| \le m} \int (\int l(a - \theta)\mathsf{D}(da|x))\mathsf{N}(\theta, \Sigma_0)(dx) \qquad (3.47) \\ &= \int l(t)\mathsf{N}(0, \Sigma_0)(dt). \end{aligned}$$

Proof. We use the notations from Example 3.64. It holds

$$\begin{aligned} \mathsf{r}(\Pi_n, T_n) &= C \int l(\tau_n w) \exp\{-w^T \Sigma_0^{-1} w/2\} \boldsymbol{\lambda}_d(dw) \\ &= \tau_n^{-d} C \int \exp\{-\frac{1}{2n} t^T \Sigma_0^{-1} t\} l(t) \exp\{-t^T \Sigma_0^{-1} t/2\} \boldsymbol{\lambda}_d(dt), \end{aligned}$$

where $C = (2\pi)^{-d/2}(\det(\Sigma_0))^{-1/2}$. An application of the monotone convergence theorem yields

$$\lim_{n \to \infty} \mathsf{r}(\Pi_n, T_n) = C \int l(t) \exp\{-t^T \Sigma_0^{-1} t/2\} \boldsymbol{\lambda}_d(dt) = \mathsf{R}(0, T_{nat}) = \mathsf{R}(\theta, T_{nat}).$$

Hence

$$\lim_{n \to \infty} \mathsf{r}(\Pi_n, T_n) = \sup_{\theta \in \mathbb{R}^d} \mathsf{R}(\theta, T_{nat}),$$

and Proposition 3.58 yields the minimaxity of T_{nat}. As T_n is a Bayes estimator under the prior Π_n we get with $\underline{w}$ and $\overline{w}$ in (3.40) that

$$\overline{w} \le \lim_{n \to \infty} \mathsf{r}(\Pi_n, T_n) = \lim_{n \to \infty} \inf_{\mathsf{D}} \mathsf{r}(\Pi_n, \mathsf{D}) \le \sup_{\Pi} \inf_{\mathsf{D}} \mathsf{r}(\Pi, \mathsf{D}) = \underline{w},$$

and the minimax theorem (3.46) is established. If l is bounded, then for some $C > 0$ we have $0 \le l \le C$ and $0 \le \mathsf{R}(\theta, \mathsf{D}) \le C$. For every prior Π we set $B_m = \{\theta : \|\theta\| \leq m\}$ and $\Pi^{(m)}(B) = \Pi(B|B_m)$, where the conditional probability is well defined for all sufficiently large m. Then

$$\begin{aligned}
&| \ \mathsf{r}(\Pi, \mathsf{D}) - \mathsf{r}(\Pi^{(m)}, \mathsf{D}) \ | \\
&\leq | \ (1 - \frac{1}{\Pi(\Delta_m)}) \int I_{\Delta_m}(\theta) \mathsf{R}(\theta, \mathsf{D}) \Pi(d\theta) \ | + \int I_{\Delta \backslash \Delta_m}(\theta) \mathsf{R}(\theta, \mathsf{D}) \Pi(d\theta) \\
&\leq C \ | \ 1 - \frac{1}{\Pi(\Delta_m)} \ | + C(1 - \Pi(\Delta_m)).
\end{aligned}$$

Furthermore,

$$\begin{aligned}
&\sup_{\Pi : \Pi(B_m) = 1} \inf_{\mathsf{D}} \mathsf{r}(\Pi, \mathsf{D}) \leq \inf_{\mathsf{D}} \sup_{\Pi : \Pi(B_m) = 1} \mathsf{r}(\Pi, \mathsf{D}) \\
&\leq \inf_{\mathsf{D}} \sup_{\|\theta\| \leq m} \mathsf{R}(\theta, \mathsf{D}) \leq \inf_{\mathsf{D}} \sup_{\theta \in \mathbb{R}^d} \mathsf{R}(\theta, \mathsf{D}) = \sup_{\Pi} \inf_{\mathsf{D}} \mathsf{r}(\Pi, \mathsf{D}) \\
&\leq \sup_{\Pi : \Pi(B_m) = 1} \inf_{\mathsf{D}} \mathsf{r}(\Pi, \mathsf{D}) + C \ | \ 1 - \frac{1}{\Pi(\Delta_m)} \ | + C(1 - \Pi(\Delta_m)).
\end{aligned}$$

Taking $m \to \infty$ we get (3.47). ■

3.6 Γ-Minimax Decisions

Sometimes the prior information on the unknown parameter is rather vague, consisting only of a set of priors known to contain the true prior. In this case the Bayes approach cannot be utilized due to the incomplete knowledge of the prior. To overcome this shortcoming we choose here a minimax approach that protects against the effects of priors that are causing the worst Bayes risks. Let $(\mathcal{X}, \mathfrak{A}, (P_\theta)_{\theta \in \Delta})$ be the statistical model. We recall that whenever we are dealing with Bayes risks, assumptions (A3) and (A8) are assumed to hold; see Remark 3.29. Let now $\Gamma \subseteq \mathcal{P}(\mathfrak{B}_\Delta)$ be a given set of priors.

Definition 3.66. *For given classes $\mathbb{D}_0 \subseteq \mathbb{D}$ and $\Gamma \subseteq \mathcal{P}(\mathfrak{B}_\Delta)$, we call $\mathsf{D}_0 \in \mathbb{D}_0$ a Γ-minimax decision in $\mathbb{D}_0$ if*

$$\sup_{\Pi \in \Gamma} \mathsf{r}(\Pi, \mathsf{D}_0) = \inf_{\mathsf{D} \in \mathbb{D}_0} \sup_{\Pi \in \Gamma} \mathsf{r}(\Pi, \mathsf{D}).$$

Whenever $\mathbb{D}_0 = \mathbb{D}$, "in $\mathbb{D}_0$" is not mentioned.

The Γ-minimax approach is set up somewhere in between the Bayes and the minimax approaches which have been studied already in the previous section. It is equivalent to the Bayes approach if $\Gamma = \{\Pi_0\}$ for some $\Pi_0 \in \mathcal{P}(\mathfrak{B}_\Delta)$, and it is equivalent to the minimax approach if $\Gamma = \mathcal{P}(\mathfrak{B}_\Delta)$. As an analogue to (3.39) we have

$$\sup_{\Pi \in \Gamma} \inf_{\mathsf{D} \in \mathbb{D}_0} \mathsf{r}(\Pi, \mathsf{D}) \leq \inf_{\mathsf{D} \in \mathbb{D}_0} \sup_{\Pi \in \Gamma} \mathsf{r}(\Pi, \mathsf{D}). \tag{3.48}$$

The converse inequality, which would imply the minimax statement

$$\sup_{\Pi\in\Gamma}\inf_{\mathsf{D}\in\mathbb{D}_0}\mathsf{r}(\Pi,\mathsf{D})=\inf_{\mathsf{D}\in\mathbb{D}_0}\sup_{\Pi\in\Gamma}\mathsf{r}(\Pi,\mathsf{D}), \tag{3.49}$$

holds only under some additional conditions. A first simple result concerns the case of a saddle point.

Problem 3.67. If (Π_0,D_0) is a saddle point in $\Gamma\times\mathbb{D}_0$ in the sense of (3.42), then the statement (3.49) holds and the decision D_0 is Γ-minimax in $\mathbb{D}_0$.

A typical application concerns the testing of one-sided hypotheses for a one-dimensional parameter where the power function is monotone. This is certainly true for families with MLR; see Theorem 2.49. To be more precise, we assume that $\Delta\subseteq\mathbb{R}$ is an interval and $(P_\theta)_{\theta\in\Delta}$ is a family of distributions on $(\mathcal{X},\mathfrak{A})$ where (A3) is satisfied. For $\theta_0\in\Delta$ and some $d>0$ with $\theta_0+d\in\Delta$ we consider the testing problem

$$\mathsf{H}_0:\theta\in(-\infty,\theta_0]\cap\Delta \quad\text{versus}\quad \mathsf{H}_A:\theta\in[\theta_0+d,\infty)\cap\Delta. \tag{3.50}$$

Instead of the zero–one loss function, which gives the same weight to the error of the first and the second kind, we use the loss function

$$L(\theta,a)=L_1 I_{\{0\}}(a)I_{[\theta_0+d,\infty)}(\theta)+L_2 I_{\{1\}}(a)I_{(-\infty,\theta_0]}(\theta),\quad a\in\{0,1\}, \tag{3.51}$$

where $L_1,L_2\geq 0$ are fixed. Given $\pi,\pi'>0$ with $\pi+\pi'\leq 1$, let Γ be the set of all priors Π on the Borel sets of Δ that satisfy

$$\Pi([\theta_0+d,\infty)\cap\Delta)\leq\pi \quad\text{and}\quad \Pi((-\infty,\theta_0]\cap\Delta)\leq\pi'. \tag{3.52}$$

These conditions guarantee that not too much mass of the prior is on either side. We denote by $\varphi_{\rho,B}$ a Bayes test for the zero–one loss function, where the hypotheses and the prior are given by

$$\begin{aligned}&\mathsf{H}_0 : P_{\theta_0} \quad\text{versus}\quad \mathsf{H}_A:P_{\theta_0+d},\quad\text{and}\\ &\Pi_\rho=\rho\boldsymbol{\delta}_{\theta_0}+(1-\rho)\boldsymbol{\delta}_{\theta_0+d},\quad \rho=L_2\pi'/(L_1\pi+L_2\pi').\end{aligned} \tag{3.53}$$

According to Theorem 2.60 every Bayes test for the Bayes testing problem (3.53) under the zero–one loss function is a likelihood ratio test and thus of the form $\varphi_{\rho,B}=I_{(c,\infty)}(T)+\gamma I_{\{c\}}(T)$ for some $c\in\mathbb{R}$ and $\gamma\in[0,1]$.

Proposition 3.68. *Suppose that $(P_\theta)_{\theta\in\Delta}$ has MLR in T, and that c and γ are chosen such that, under the zero–one loss function, the test*

$$\varphi_{\rho,B}=I_{(c,\infty)}(T)+\gamma I_{\{c\}}(T)$$

is a Bayes test for the Bayes testing problem (3.53). If Π_0 satisfies $\Pi_0(\{\theta_0+d\})=\pi$, $\Pi_0(\{\theta_0\})=\pi'$, and $\Pi_0((\theta_0,\theta_0+d))=1-\pi-\pi'$, then the pair $(\varphi_{\rho,B},\Pi_0)$ is a saddle point for the testing problem (3.50) under the loss function (3.51).

Proof. Using the fact that $\varphi_{\rho,B}$ is a nondecreasing function of T, so that $\theta \mapsto \mathsf{E}_\theta \varphi_{\rho,B}$ is a nondecreasing function of θ by Theorem 2.10 and Proposition 2.7, we get for any prior on Δ,

$$\begin{aligned}
\mathsf{r}(\varphi_{\rho,B}, \Pi) &= \int [L_1 I_{[\theta_0+d,\infty)}(\theta)[1 - \mathsf{E}_\theta \varphi_{\rho,B}] + L_2 I_{(-\infty,\theta_0]}(\theta) \mathsf{E}_\theta \varphi_{\rho,B}] \Pi(d\theta) \\
&\le \int L_1 I_{[\theta_0+d,\infty)}(\theta)[1 - \mathsf{E}_{\theta_0+d} \varphi_{\rho,B}] \Pi(d\theta) + \int L_2 I_{(-\infty,\theta_0]}(\theta) \mathsf{E}_{\theta_0} \varphi_{\rho,B} \Pi(d\theta) \\
&\le L_1 \pi [1 - \mathsf{E}_{\theta_0+d} \varphi_{\rho,B}] + L_2 \pi' \mathsf{E}_{\theta_0} \varphi_{\rho,B} = \mathsf{r}(\varphi_{\rho,B}, \Pi_0).
\end{aligned}$$

The remaining inequality $\mathsf{r}(\varphi_{\rho,B}, \Pi_0) \le \mathsf{r}(\varphi, \Pi_0)$, for every test φ, follows from the fact that $\varphi_{\rho,B}$ is a Bayes test for (3.53) under the zero–one loss. ■

To construct the Bayes test $\varphi_{\rho,B}$ in concrete situations we use the fact that by Theorem 2.60 $\varphi_{\rho,B}$ is a likelihood ratio test.

Example 3.69. Suppose $(P_\theta)_{\theta \in \Delta}$ is a one-parameter exponential family on $(\mathcal{X}, \mathfrak{A})$ with generating statistic T, where the conditions (A1) and (A2) are fulfilled. If the sample size is n, then $(P_\theta^{\otimes n})_{\theta \in \Delta}$ is again an exponential family with generating statistic $T_{\oplus n}$ and $\boldsymbol{\mu}^{\otimes n}$-density

$$\frac{dP_\theta^{\otimes n}}{d\boldsymbol{\mu}^{\otimes n}}(x) = \exp\{\theta T_{\oplus n}(x) - nK(\theta)\}, \quad x \in \mathcal{X}^n.$$

Let $\theta_0 \in \Delta$ and $d > 0$ with $\theta_0 + d \in \Delta$ be fixed. Then according to Theorem 2.60 the test $\varphi_{\rho,B}$ in Proposition 3.68 can be written as

$$\varphi_{\rho,B}(x) = I_{(c,\infty)}(T_{\oplus n}) + \gamma I_{\{c\}}(T_{\oplus n}),$$

where

$$c = \frac{1}{d}[\ln \frac{L_2 \pi'}{L_1 \pi} + nK(\theta_0 + d) - nK(\theta_0)],$$

and $\gamma \in [0, 1]$ is arbitrary. It follows from Proposition 3.68 that $\varphi_{\rho,B}$ is a Γ-minimax test for the testing problem 3.50 under the loss function (3.51) for the class Γ of all priors Π that satisfy (3.52).

Another example concerns location models that are generated by a log-concave, i.e., strongly unimodal, density; see Definition 2.16.

Example 3.70. Let f be a Lebesgue density on $\mathbb{R}$ that is positive, where $\ln f$ is a concave function. Let P_θ be defined by $P_\theta(dx) = f(x - \theta)dx$.The family $(P_\theta)_{\theta \in \mathbb{R}}$ has, in view of Proposition 2.20, MLR in the identity $T(x) = x$. Let ϑ_0 and $d > 0$ be fixed. Then $L_{\theta_0,\theta_0+d}(x) = f(x - \theta_0 - d)/f(x - \theta_0)$ is a nondecreasing function of x. Hence, for

$$c_0 = \inf\{x : L_{\theta_0,\theta_0+d}(x) > L_2 \pi'/(L_1 \pi)\}$$

it holds that $x \le c_0$ implies $L_{\theta_0,\theta_0+d}(x) \le L_2 \pi'/(L_1 \pi)$ and $x > c_0$ implies $L_{\theta_0,\theta_0+d}(x) \ge L_2 \pi'/(L_1 \pi)$. Then $L_{\theta_0,\theta_0+d}(x) > L_{\theta_0,\theta_0+d}(c_0)$ implies $x > c_0$ and $L_{\theta_0,\theta_0+d}(x) < L_{\theta_0,\theta_0+d}(c_0)$ implies $x < c_0$ so that every test that satisfies

$$\varphi_{\rho,B}(x) = \begin{cases} 1 & \text{if } x > c_0 \\ 0 & \text{if } x < c_0 \end{cases}$$

is a likelihood ratio test and thus a Bayes test for P_{θ_0} versus P_{θ_0+d} under the zero–one loss and the prior $(\rho, 1-\rho)$ with $\rho = L_2\pi'/(L_1\pi + L_2\pi')$. It follows from Proposition 3.68 that $\varphi_{\rho,B}$ is a Γ-minimax test for the testing problem 3.50 under the loss function (3.51) for the class Γ of all priors Π that satisfy (3.52).

Another condition that establishes Γ-minimaxity is similar to Proposition *3.58* and employs suitable sequences of priors.

Proposition 3.71. *If for* $\mathsf{D}_0 \in \mathbb{D}_0$ *there exists a sequence of priors* $\Pi_n \in \Gamma$, *and a sequence* D_n *of associated Bayes decisions in* $\mathbb{D}_0$, *such that*

$$\sup_{\Pi\in\Gamma} \mathsf{r}(\Pi, \mathsf{D}_0) \leq \liminf_{n\to\infty} \mathsf{r}(\Pi_n, \mathsf{D}_n),$$

then D_0 *is* Γ*-minimax in* $\mathbb{D}_0$.

Proof. We have

$$\sup_{\Pi\in\Gamma} \mathsf{r}(\Pi, \mathsf{D}_0) \leq \liminf_{n\to\infty} \mathsf{r}(\Pi_n, \mathsf{D}_n) \leq \sup_{\Pi\in\Gamma} \inf_{\mathsf{D}\in\mathbb{D}_0} \mathsf{r}(\Pi, \mathsf{D}).$$

The rest follows from inequality (3.48). ■

The usefulness of the above proposition is demonstrated below by some examples.

Example 3.72. We consider the problem of estimating the parameter μ in the model $(\mathbb{R}, \mathfrak{B}, (\mathsf{N}(\mu,1))_{\mu\in\mathbb{R}})$ under the squared error loss $L(\mu, a) = (\mu - a)^2$. Let $\nu \in \mathbb{R}$ and $\delta^2 > 0$ be fixed, and the family of priors be given by

$$\Gamma = \{\Pi : \Pi \in \mathcal{P}(\mathfrak{B}_1), \int t\Pi(dt) = \nu, \int (t-\nu)^2\Pi(dt) = \delta^2\}. \tag{3.54}$$

As we know already from (3.30) in Example 3.38, for the prior $\Pi_0 = \mathsf{N}(\nu, \delta^2)$ the Bayes estimator is given by

$$\mathsf{d}_0(x) = \frac{\delta^2}{1+\delta^2}x + \frac{1}{1+\delta^2}\nu. \tag{3.55}$$

Its risk function is

$$\mathsf{R}(\mu, \mathsf{d}_0) = \int [\frac{\delta^2}{1+\delta^2}x + \frac{1}{1+\delta^2}\nu - \mu]^2 \varphi(x)(dx) = \frac{\delta^4}{(1+\delta^2)^2} + \frac{(\nu-\mu)^2}{(1+\delta^2)^2}.$$

Thus,

$$\mathsf{r}(\Pi, \mathsf{d}_0) = \frac{\delta^4}{(1+\delta^2)^2} + \frac{\delta^2}{(1+\delta^2)^2} = \frac{\delta^2}{1+\delta^2}$$

for every $\Pi \in \Gamma$. This means that

$$\sup_{\Pi\in\Gamma} \mathsf{r}(\Pi, \mathsf{d}_0) = \mathsf{r}(\Pi_0, \mathsf{d}_0) = \inf_{\mathsf{d}} \mathsf{r}(\Pi_0, \mathsf{d}),$$

as d_0 is a Bayes estimator under the prior Π_0. Hence d_0 from (3.55) is Γ-minimax for Γ given by (3.54).

Next we apply the Γ-minimax approach to a type of estimation problem where it is known that the parameter belongs to a given bounded interval. Let the family be given by $\mathsf{N}(\mu, 1)$, $\mu \in [0, c]$, where $c > 0$ is known. The problem of constructing a Γ-minimax estimator in this setting has been studied in Zinzius (1981). Here we use the prior $\Pi_c = \frac{1}{2}(\boldsymbol{\delta}_0 + \boldsymbol{\delta}_c)$.

Problem 3.73.* For the family $\mathsf{N}(\mu, 1)$, $\mu \in [0, c]$, where $c > 0$ is fixed,

$$T_c(x) = c - \frac{c}{1 + \exp\{cx - c^2/2\}}$$

is the Bayes estimator for the prior $\Pi_c = \frac{1}{2}(\boldsymbol{\delta}_0 + \boldsymbol{\delta}_c)$ under the squared error loss.

The risk of T_c is given by

$$\begin{aligned} \mathsf{R}(\mu, T_c) &= (2\pi)^{-1/2} \int (T_c(x) - \mu)^2 \exp\{-(x-\mu)^2/2\} dx \\ &= (2\pi)^{-1/2} \int (T_c(x+\mu) - \mu)^2 \exp\{-x^2/2\} dx, \quad \mu \in [0, c]. \end{aligned}$$

$\mathsf{R}(\mu, T_c)$ as a function of μ is twice continuously differentiable and it holds for every fixed $a > 0$

$$\lim_{c \downarrow 0} \sup_{0 \le \mu \le a} |\mathsf{R}''(\mu, T_c) - \mathsf{R}''(\mu, T_0)| = 0.$$

As $T_0 = 0$ it holds $\mathsf{R}(\mu, T_0) = \mu^2$ and $\mathsf{R}''(\mu, T_0) = 2$. We see that there is a sufficiently small $c_0 > 0$ such that for $c \le c_0$ the function $\mathsf{R}(\mu, T_c)$ is convex.

As $T_c(c - x) = c - T_c(x)$ we get $\mathsf{R}(\mu, T_c) = \mathsf{R}(c - \mu, T_c)$, $\mu \in [0, c]$. This together with the convexity of $\mathsf{R}(\mu, T_c)$ yields for $c \le c_0$

$$\max_{0 \le \mu \le c} \mathsf{R}(\mu, T_c) = \mathsf{R}(0, T_c) = \mathsf{R}(c, T_c). \tag{3.56}$$

Proposition 3.74. *There exists a $c_0 > 0$ such that for any fixed $c \le c_0$ the following statement holds. For the family $\mathsf{N}(\mu, 1)$, $\mu \in [0, c]$, under the squared error loss, the estimator $T_c(x) = c(1 - [1 + \exp\{cx - c^2/2\}]^{-1})$ is Γ-minimax for the class $\Gamma = \{\Pi : \Pi \in \mathcal{P}(\mathfrak{B}), \Pi([0, c]) = 1\}$ and minimax.*

Proof. The relation (3.56) allows us to apply Proposition 3.71 with $\Pi_n = \Pi_c$. As the class Γ contains all one-point distributions on the interval $[0, c]$ the concepts Γ-minimax and minimax are identical. ■

It has been shown in Zinzius (1981) that the largest possible c_0 is the unique solution of the equation

$$2 - c^2 - \frac{2c^2}{\sqrt{2\pi e}}(1 + \exp\{-c - c^2/2\})^2 = 0.$$

It holds $1.2 < c_0 < 1.21$. Numerical calculations show that for $c > 1.21$ the two-point prior Π_c is no longer least favorable. Instead, a Bayes estimator based on a discrete distribution on three points becomes a minimax estimator.

Finally, we consider Γ-minimax estimators for a nonsymmetric loss function. A popular choice for the loss is the LINEX loss function, which is defined by

$$L_\alpha(\theta, a) = \exp\{\alpha(a - \theta)\} - \alpha(a - \theta) - 1,$$

where $\alpha \neq 0$ is any real number. The family $\mathsf{N}(\mu, 1)$, $\mu \in [-c, c]$, where $c > 0$ is given, along with the class of priors $\Gamma = \{\Pi : \Pi \in \mathcal{P}(\mathfrak{B}), \Pi([-c, c]) = 1\}$, has been studied in Bischoff, Fieger, and Wulfert (1995).

Problem 3.75. The Bayes estimator for the prior $\Pi_{\beta,c} = \beta\boldsymbol{\delta}_{-c} + (1-\beta)\boldsymbol{\delta}_c$, under the LINEX loss function, is given by

$$T_{\beta,c}(x) = \frac{1}{\alpha}\ln\Big(\frac{\beta\exp\{-xc\} + (1-\beta)\exp\{xc\}}{\beta\exp\{-xc + \alpha c\} + (1-\beta)\exp\{xc - ac\}}\Big).$$

Proposition 3.76. *Let $\alpha > 0$, $c_0 = \min((\sqrt{3}/2) - 1)\alpha, (\ln 3)/(2\alpha))$, and $c \leq c_0$. Then there exists a $\beta^* \in (0,1)$ such that, under the LINEX loss function, the prior $\Pi_{\beta^*,c} = \frac{1}{2}(\beta^*\boldsymbol{\delta}_{-c} + (1-\beta^*)\boldsymbol{\delta}_c)$ is least favorable, and the Bayes estimator $T_{\beta^*,c}$ for $\Pi_{\beta^*,c}$ is Γ-minimax and minimax.*

The proof of the proposition in Bischoff et al. (1995) consists mainly of two steps. First it is shown that the risk function $\mathsf{R}(\mu, T_{\beta,c})$ is convex for $c \leq c_0$. Then, secondly, it is shown that there exists some $\beta^* \in (0,1)$ such that $\mathsf{R}(-c, T_{\beta^*,c}) = \mathsf{R}(c, T_{\beta^*,c})$.

3.7 Minimax Theorem

As we have mentioned already the minimax condition (3.41) relates approximately minimax decisions to Bayes decisions under a least favorable prior. If the stronger condition of the existence of a saddle point is satisfied, then, as we have seen in Proposition 3.56, this statement is true not only approximately but in the strict sense. The minimax statement (3.41) also plays an important role in later chapters.

Now we are ready to prove the *minimax theorem of decision theory.* In a first step we consider a parameter set $\digamma$ that is a finite subset of Δ. By $\mathfrak{F}(\Delta)$ we denote the system of finite subsets of Δ. For a finite subset $\digamma \subseteq \Delta$ let $\mathcal{P}(\digamma)$ denote the set of all distributions on $\digamma$. Suppose that (A3) is satisfied.

Theorem 3.77. *Let $(\mathcal{X}, \mathfrak{A}, (P_\theta)_{\theta\in\digamma})$ be a finite model. Assume that $\mathbb{D}_0$ is a convex set of decisions that is closed under the weak convergence. If the decision space is a compact metric space and $a \mapsto L(\theta, a)$ is continuous, then it holds*

$$\sup_{\Pi\in\mathcal{P}(\digamma)} \inf_{\mathsf{D}\in\mathbb{D}_0} \mathsf{r}(\Pi, \mathsf{D}) = \inf_{\mathsf{D}\in\mathbb{D}_0} \sup_{\theta\in\digamma} \mathsf{R}(\theta, \mathsf{D}).$$

Proof. As the decision space $\mathcal{D}$ is compact and $L(\theta, a)$ is continuous it holds $L(\theta, a) \geq C$, $a \in \mathcal{D}$, $\theta \in \digamma$, for some constant C. Hence we may assume that $L(\theta, a) \geq 0$. In view of (3.40) we have only to show that

$$\inf_{\mathsf{D}\in\mathbb{D}_0} \sup_{\theta\in\digamma} \mathsf{R}(\theta, \mathsf{D}) \leq \sup_{\Pi\in\mathcal{P}(\digamma)} \inf_{\mathsf{D}\in\mathbb{D}_0} \mathsf{r}(\Pi, \mathsf{D}).$$

Let $\theta_1, ..., \theta_n$ be the elements of $\digamma$. Due to the compactness theorem (see Theorem 3.21) and the continuity of the loss function, the set

$$K = \{(\mathsf{R}(\theta_1, \mathsf{D}), ..., \mathsf{R}(\theta_n, \mathsf{D})), \quad \mathsf{D} \in \mathbb{D}\},$$

is compact and convex. Thus the set $A(K)$ of all vectors which are, in the componentwise semiorder, larger than or equal to some vector from K, by Problem 3.81, is closed and convex. Set

$$C := \sup_{\Pi\in\mathcal{P}(\digamma)} \inf_{\mathsf{D}\in\mathbb{D}_0} \sum\nolimits_{\theta\in\digamma} \mathsf{R}(\theta, \mathsf{D})\Pi(\{\theta\}).$$

Then for every $\Pi \in \mathcal{P}(\digamma)$,

$$\int C\Pi(d\theta) \geq \inf_{\mathsf{D}\in\mathbb{D}_0} \sum\nolimits_{\theta\in\digamma} \mathsf{R}(\theta, \mathsf{D})\Pi(\{\theta\}).$$

From Problem 3.81 we get that there exists a decision D_0 such that the vector $(C, ..., C)$ is bounded from below by the vector $(R(\theta_1, \mathsf{D}_0), ..., R(\theta_n, \mathsf{D}_0))$ in the componentwise semiorder. This means $C \geq R(\theta, \mathsf{D}_0)$ for $\theta \in \digamma$ and therefore

$$\begin{aligned}\inf_{\mathsf{D}\in\mathbb{D}_0} \sup_{\theta\in\digamma} R(\theta, \mathsf{D}) &\leq \sup_{\theta\in\digamma} R(\theta, \mathsf{D}_0) \leq \sup_{\Pi\in\mathcal{P}(\digamma)} \inf_{\mathsf{D}\in\mathbb{D}_0} \sum\nolimits_{\theta\in\digamma} \mathsf{R}(\theta, \mathsf{D})\Pi(\{\theta\}) \\ &\leq \sup_{\Pi\in\mathcal{P}(\digamma)} \inf_{\mathsf{D}\in\mathbb{D}_0} \sum\nolimits_{\theta\in\digamma} \mathsf{R}(\theta, \mathsf{D})\Pi(\{\theta\}).\end{aligned}$$

■

In most decision problems, of course, the parameter set is not finite. Therefore we look for minimax statements without this restrictive condition. A simple consequence of the previous theorem is that

$$\sup_{\Pi\in\mathcal{P}(\Delta)} \inf_{\mathsf{D}\in\mathbb{D}_0} \mathsf{r}(\Pi, \mathsf{D}) \geq \sup_{\digamma\in\mathfrak{F}(\Delta)} \inf_{\mathsf{D}\in\mathbb{D}_0} \sup_{\theta\in\digamma} \mathsf{R}(\theta, \mathsf{D}).$$

Instead of the right-hand term it is desired to have $\inf_{\mathsf{D}\in\mathbb{D}_0} \sup_{\theta\in\Delta} \mathsf{R}(\theta, \mathsf{D})$, of course, which is equivalent to having

$$\sup_{\digamma\in\mathfrak{F}(\Delta)} \inf_{\mathsf{D}\in\mathbb{D}_0} \sup_{\theta\in\digamma} \mathsf{R}(\theta, \mathsf{D}) = \inf_{\mathsf{D}\in\mathbb{D}_0} \sup_{\theta\in\Delta} \mathsf{R}(\theta, \mathsf{D}).$$

The next theorem gives conditions under which this statement holds.

Theorem 3.78. *Suppose that both the parameter set Δ and the decision space $\mathcal{D}$ are compact metric spaces, and that $L(\theta, a)$ is a continuous function of (θ, a). If the model $(\mathcal{X}, \mathfrak{A}, (P_\theta)_{\theta\in\Delta})$ is continuous in the sense of (A7), and $\mathbb{D}_0$ is convex and closed with respect to the weak convergence, then*

$$\sup_{\Pi\in\mathcal{P}(\Delta)} \inf_{\mathsf{D}\in\mathbb{D}_0} \mathsf{r}(\Pi, \mathsf{D}) = \inf_{\mathsf{D}\in\mathbb{D}_0} \sup_{\theta\in\Delta} \mathsf{R}(\theta, \mathsf{D}). \tag{3.57}$$

Proof. Denote by ρ_Δ the metric on Δ. Proposition 3.27 implies that the family of functions $\theta \mapsto \mathsf{R}(\theta, \mathsf{D})$, $\mathsf{D} \in \mathbb{D}_0$, is equicontinuous at every $\theta \in \Delta$. As Δ is compact this equicontinuity holds uniformly in θ. This means that for every $\varepsilon > 0$ there exists a $\delta_\varepsilon > 0$ such that

$$\sup_{\rho_\Delta(\theta_1,\theta_2)<\delta_\varepsilon} \sup_{\mathsf{D}\in\mathbb{D}_0} |\mathsf{R}(\theta_1, \mathsf{D}) - \mathsf{R}(\theta_2, \mathsf{D})| < \varepsilon.$$

The compactness of Δ yields that there exists a finite set $\digamma$ such that the open balls with radius δ_ε and center at a point in $\digamma$ cover Δ. This yields

$$|\sup_{\Pi\in\mathcal{P}(\digamma)} \inf_{\mathsf{D}\in\mathbb{D}_0} \mathsf{r}(\Pi, \mathsf{D}) - \sup_{\Pi} \inf_{\mathsf{D}\in\mathbb{D}_0} \mathsf{r}(\Pi, \mathsf{D})| \le \varepsilon \quad \text{and}$$
$$|\inf_{\mathsf{D}\in\mathbb{D}_0} \sup_{\theta\in\Delta} \mathsf{R}(\theta, \mathsf{D}) - \inf_{\mathsf{D}\in\mathbb{D}_0} \sup_{\theta\in\digamma} \mathsf{R}(\theta, \mathsf{D})| \le \varepsilon.$$

According to Theorem 3.77, for every finite subset $\digamma \subseteq \Delta$, it holds

$$\sup_{\Pi\in\mathcal{P}(\digamma)} \inf_{\mathsf{D}\in\mathbb{D}_0} \mathsf{r}(\Pi, \mathsf{D}) = \inf_{\mathsf{D}\in\mathbb{D}_0} \sup_{\theta\in\digamma} \mathsf{R}(\theta, \mathsf{D}).$$

This implies

$$|\sup_{\Pi\in\mathcal{P}(\Delta)} \inf_{\mathsf{D}\in\mathbb{D}_0} \mathsf{r}(\Pi, \mathsf{D}) - \inf_{\mathsf{D}\in\mathbb{D}_0} \sup_{\theta\in\Delta} \mathsf{R}(\theta, \mathsf{D})| \le 2\varepsilon.$$

■

We conclude this section with the remark that more general minimax theorems can be found in LeCam (1986), Strasser (1985), and Torgersen (1991). Theorem 3.78 corresponds to Theorem 46.5 in Strasser (1985), where the assumptions on the model are weaker than those we have made here. The techniques that are used in the above books go beyond the scope of this book.

The subsequent problems concern technical results that have been used above.

Problem 3.79.* If $K \subseteq \mathbb{R}^n$ is a closed convex set, then every minimal sequence $x_n \in K$, i.e., every sequence with $\lim_{n\to\infty} \|x_n\| = \inf\{\|x\|, x \in K\}$, converges to some x_K with $\|x_K\| = \inf\{\|x\|, x \in K\}$, and the point $x_K \in K$ is unique.

Problem 3.80.* Let $K \subseteq \mathbb{R}^n$ be a closed convex set. If $z \notin K$ and $x_K \in K$ satisfies $\|x_K - z\| = \inf\{\|x - z\|, x \in K\}$, then $\inf_{x\in K} \langle x, b\rangle > \langle z, b\rangle$ for $b = x_K - z$.

Problem 3.81.* For every bounded, closed, and convex set $B \subseteq \mathbb{R}_+^n$ the set

$$A(B) := \{(y_1, ..., y_n) : y_i \geq x_i,\ i = 1, ..., n, \text{ for some } (x_1, ..., x_n) \in B\} \tag{3.58}$$

is a closed convex set. A point $z \in \mathbb{R}_+^n$ is contained in $A(B)$ if and only if

$$\langle \alpha, z \rangle \geq \inf_{y \in A(B)} \langle \alpha, y \rangle, \quad \alpha = (\alpha_1, ..., \alpha_n),\ \alpha_1, ..., \alpha_n \geq 0,\ \sum\nolimits_{i=1}^{n} \alpha_i = 1.$$

3.8 Complete Classes

In the previous sections we have seen that Bayes decisions are also minimax decisions if, roughly speaking, the risk is constant; see Theorem 3.57 and Proposition 3.58. This means that in the search for minimax decisions one should focus on the class of Bayes decisions with constant risk. In this section we deal with the important question of whether every decision can be outperformed by a Bayes decision, where of course the priors have to be suitably constructed. Statements of this type are referred to in the literature as *complete class theorems.* Assume that (A3) is satisfied.

Theorem 3.82. *Suppose that Δ is a separable metric space, $\mathcal{D}$ is a compact metric space, and $L(\theta, a)$ is bounded and continuous in (θ, a). Assume that the model is continuous in the sense of (A7). Then for every $\mathsf{D} \in \mathbb{D}$ there exists a $\mathsf{D}_0 \in \mathbb{D}$ and a sequence of priors Π_k with respective Bayes decisions $\mathsf{D}_{\Pi_k} \in \mathbb{D}$ such that*

$$\mathsf{D}_{\Pi_k} \Rightarrow \mathsf{D}_0 \quad \text{and} \quad \mathsf{R}(\theta, \mathsf{D}_0) \leq \mathsf{R}(\theta, \mathsf{D}), \quad \theta \in \Delta.$$

Corollary 3.83. *If in addition the parameter set Δ is a compact metric space, then for every $\mathsf{D} \in \mathbb{D}$ there exists a prior Π_0 and a Bayes decision $\mathsf{D}_0 \in \mathbb{D}$ with respect to Π_0 such that*

$$\mathsf{R}(\theta, \mathsf{D}_0) \leq \mathsf{R}(\theta, \mathsf{D}), \quad \theta \in \Delta.$$

Proof. We follow the ideas of the proof of Theorem 47.7 in Strasser (1985) and start with a finite set $\Delta_n = \{\theta_1, ..., \theta_n\} \subseteq \Delta$. By Proposition 3.50, for every $\Pi \in \mathcal{P}(\Delta_n)$ there exists a Bayes decision D_Π with respect to Π. Due to the compactness theorem (see Theorem 3.21) the set $B_n = \{(\mathsf{R}(\theta_1, \mathsf{D}), ..., \mathsf{R}(\theta_n, \mathsf{D})) : \mathsf{D} \in \mathbb{D}\}$ is compact and convex so that by Problem 3.81 the set $A(B_n)$ from (3.58) is closed and convex. Set

$$\varepsilon = \inf_{\Pi \in \mathcal{P}(\Delta_n)} [\sum\nolimits_{\theta \in \Delta_n} \mathsf{R}(\theta, \mathsf{D}) \Pi(\{\theta\}) - \inf_{\widetilde{\mathsf{D}} \in \mathbb{D}} \mathsf{r}(\Pi, \widetilde{\mathsf{D}})].$$

Then $\varepsilon \geq 0$, and for every $\Pi \in \mathcal{P}(\Delta_n)$,

$$\int (R(\theta, \mathsf{D}) - \varepsilon) \Pi(d\theta) \geq \inf_{\widetilde{\mathsf{D}} \in \mathbb{D}} \mathsf{r}(\Pi, \widetilde{\mathsf{D}}) = \inf_{\widetilde{\mathsf{D}} \in \mathbb{D}} \sum\nolimits_{\theta \in \Delta_n} \mathsf{R}(\theta, \widetilde{\mathsf{D}}) \Pi(\{\theta\}).$$

From Problem 3.81 we get that there exists a decision $\mathsf{D}_{0,n}$ such that the vector $(R(\theta_1, \mathsf{D}) - \varepsilon, ..., R(\theta_n, \mathsf{D}) - \varepsilon)$ is bounded from below by the vector $(R(\theta_1, \mathsf{D}_{0,n}), ..., R(\theta_n, \mathsf{D}_{0,n}))$ in the componentwise semiorder. This means that $R(\theta, \mathsf{D}) - \varepsilon \geq R(\theta, \mathsf{D}_{0,n})$, $\theta \in \Delta_n$, and thus

$$\begin{aligned} 0 &\leq \inf_{\Pi \in \mathcal{P}(\Delta_n)} [\sum\nolimits_{\theta \in \Delta_n} \mathsf{R}(\theta, \mathsf{D}_{0,n}) \Pi(\{\theta\}) - \inf_{\widetilde{\mathsf{D}} \in \mathbb{D}} \mathsf{r}(\Pi, \widetilde{\mathsf{D}})] \\ &\leq \inf_{\Pi \in \mathcal{P}(\Delta_n)} [\sum\nolimits_{\theta \in \Delta_n} \mathsf{R}(\theta, \mathsf{D}) \Pi(\{\theta\}) - \inf_{\widetilde{\mathsf{D}} \in \mathbb{D}} \mathsf{r}(\Pi, \widetilde{\mathsf{D}})] - \varepsilon = 0. \end{aligned}$$

This yields $\varepsilon = 0$, and therefore

$$R(\theta, \mathsf{D}) \geq R(\theta, \mathsf{D}_{0,n}), \quad \theta \in \Delta_n. \tag{3.59}$$

As Δ_n is finite, the function $\Pi \mapsto \int R(\theta, \mathsf{D}) \Pi(d\theta)$, $\Pi \in \mathcal{P}(\Delta_n)$, is continuous, so that the function $\Pi \mapsto [\sum_{\theta \in \Delta_n} \mathsf{R}(\theta, \mathsf{D}) \Pi(\{\theta\}) - \inf_{\mathsf{D}} \mathsf{r}(\Pi, \mathsf{D})]$ is lower semicontinuous with respect to the pointwise convergence of the probability mass function and attains the minimum on the closed simplex $\mathcal{P}(\Delta_n)$. Hence for some $\widetilde{\Pi}_n$

$$\sum\nolimits_{\theta \in \Delta_n} \mathsf{R}(\theta, \mathsf{D}) \widetilde{\Pi}_n(\{\theta\}) \leq \inf_{\mathsf{D}} \mathsf{r}(\widetilde{\Pi}_n, \mathsf{D}),$$

which means that $\mathsf{D}_{0,n}$ is Bayes with respect to $\widetilde{\Pi}_n$.

Now we turn to the general case. Let $\{\theta_1, \theta_2, ...\} \subseteq \Delta$ be an at most countable subset of Δ that is dense in Δ. As $\theta \mapsto P_\theta$ is continuous we get from Problem 3.24 that the family $(P_\theta)_{\theta \in \Delta}$ is dominated. Hence by Theorem 3.21 and the continuity of L we find a subsequence Π_k of $\widetilde{\Pi}_n$, decisions D_{Π_k} that are Bayes with respect to Π_k, and a decision D_0 such that $\mathsf{D}_{\Pi_k} \Rightarrow \mathsf{D}_0$. Hence

$$\lim_{k \to \infty} \mathsf{R}(\theta, \mathsf{D}_{\Pi_k}) = \mathsf{R}(\theta, \mathsf{D}_0) \leq \mathsf{R}(\theta, \mathsf{D}), \quad \theta \in \{\theta_1, \theta_2, ...\},$$

where the inequality follows from (3.59). We already know from Proposition 3.25 that $\mathsf{R}(\theta, \mathsf{D})$ is continuous in θ for every D. This gives $\mathsf{R}(\theta, \mathsf{D}_0) \leq \mathsf{R}(\theta, \mathsf{D})$ for every $\theta \in \Delta$.

To prove the corollary, we remark that the compactness of Δ and Prohorov's theorem (see Theorem A.48) imply that Π_k contains a subsequence Π_{k_l} which converges weakly to some prior Π_0. Hence we have $\mathsf{D}_l := \mathsf{D}_{\Pi_{k_l}} \Rightarrow \mathsf{D}_0$ and $\Pi_{k_l} \Rightarrow \Pi$. It remains to prove that D_0 is Bayes with respect to Π_0. The weak convergence $\mathsf{D}_l \Rightarrow \mathsf{D}_0$ implies the pointwise convergence of the risk functions. But as Δ is compact we get from Proposition 3.28 that $\sup_{\theta \in \Delta} |\mathsf{R}(\theta, \mathsf{D}_l) - \mathsf{R}(\theta, \mathsf{D}_0)| \to 0$ which, together with $\Pi_{k_l} \Rightarrow \Pi_0$, implies

$$\inf_{\mathsf{D}} \sum\nolimits_{\theta \in \Delta} \mathsf{R}(\theta, \mathsf{D}) \Pi_{k_l}(\{\theta\}) = \sum\nolimits_{\theta \in \Delta} \mathsf{R}(\theta, \mathsf{D}_l) \Pi_{k_l}(\{\theta\}) \to \int \mathsf{R}(\theta, \mathsf{D}_0) \Pi(d\theta).$$

On the other hand,

$$\sum_{\theta \in \Delta} \mathsf{R}(\theta, \mathsf{D}) \Pi_{k_l}(\{\theta\}) \to \int \mathsf{R}(\theta, \mathsf{D}) \Pi(d\theta)$$

for every fixed D, which proves $\int \mathsf{R}(\theta, \mathsf{D})\Pi(d\theta) \geq \int \mathsf{R}(\theta, \mathsf{D}_0)\Pi(d\theta)$, so that D_0 is Bayes with respect to Π_0. ■

If Δ is not compact, then the limit of Bayes decisions is not necessarily a Bayes decision. Thus the assumption of Δ being a compact metric space is indispensable in general. We illustrate this by an example.

Example 3.84. We consider the problem of estimating, under the squared error loss, the parameter μ in the family $(\mathsf{N}(\mu, \sigma^2))_{\mu\in\mathbb{R}}$, where $\sigma^2 > 0$ is known. If we use the same prior $\mathsf{N}(\nu, \delta^2)$ for μ as in Problem 3.38, then by (3.30) the Bayes estimator is given by

$$T_{\nu,\delta}(x) = \frac{(1/\sigma^2)}{(1/\sigma^2) + (1/\delta^2)} x + \frac{(1/\delta^2)}{(1/\sigma^2) + (1/\delta^2)} \nu.$$

Set $T_{nat}(x) = x$. We recall that by Example 3.22 for $\delta \to \infty$ the decisions $\boldsymbol{\delta}_{T_\delta}$ converge weakly to $\boldsymbol{\delta}_{T_{nat}}$ if and only if $\mathcal{L}(\boldsymbol{\delta}_{T_{\nu,\delta}}|\mathsf{N}(\mu, \sigma^2)) \Rightarrow \mathcal{L}(\boldsymbol{\delta}_{T_{nat}}|\mathsf{N}(\mu, \sigma^2))$. But this is true as $T_{\nu,\delta}(x) \to T_{nat}(x)$ for every x and the pointwise convergence of random variables implies the weak convergence of the distributions.

However, as shown in the next problem, the estimator $T_{nat}(x)$ is not a Bayes estimator for any prior with a finite second moment.

Problem 3.85.* For the family $(\mathsf{N}(\mu, 1))_{\mu\in\mathbb{R}}$, under the squared error loss $L(\theta, a) = (\theta - a)^2$, the estimator $T_{nat}(x) = x$ is not a Bayes estimator for any prior with a finite second moment.

Now we use the above complete class theorem to characterize minimax classification rules and minimax tests. First we consider minimax classification rules under the zero–one loss function. Classification rules have been introduced in Example 3.41. From Corollary 3.83 we already know that there exist a prior Π_0 and a Bayes classification rule D_{Π_0} with respect to Π_0 that is minimax.

Proposition 3.86. *Let $(P_\theta)_{\theta\in\Delta}$, $\Delta = \{1, ..., m\}$, be a finite family of distributions. For the classification problem with decision space $\mathcal{D} = \{1, ..., m\}$, under the zero–one loss $L(\theta, a) = 1 - I_{\{\theta\}}(a)$, there exists at least one minimax classification rule. For every minimax classification rule D there exists a prior Π such that D is Bayes with respect to Π. If a classification rule D satisfies*

$$\int [1 - \mathsf{D}(\{1\}|x)]P_1(dx) = \cdots = \int [1 - \mathsf{D}(\{m\}|x)]P_m(dx), \tag{3.60}$$

and is a Bayes decision for some prior Π, then D is minimax.

Proof. The existence of a minimax decision D follows from Proposition 3.49. The second statement follows from Corollary 3.83, and the third follows from Proposition 3.58. ■

Condition (3.60) can be utilized to find a least favorable prior for which a minimax classification rule is Bayes.

Example 3.87. Suppose that the distributions P_{θ_i}, $i = 1, ..., m$, are from an exponential family $(P_\theta)_{a<\theta<b}$ with natural parameter θ and generating statistic T; see Definition 1.1. As in Example 3.41 we set $\Delta = \mathcal{D} = \{1, ..., m\}$, use the zero–one loss, and get the following. D_π is a Bayes classification rule with respect to a prior $\pi = (\pi_1, ..., \pi_m)$ if and only if $\mathsf{D}_\pi(A(x)|x) = 1$, $\boldsymbol{\mu}$-a.s. $x \in \mathcal{X}$, where

$$\begin{aligned} A(x) &= \{a : f_{\theta_a}(x)\pi_a = \max_{1\le j\le m} f_{\theta_j}(x)\pi_j\} \\ &= \{a : \theta_a T(x) + \gamma_a = \max_{1\le j\le m} (\theta_j T(x) + \gamma_j)\}, \end{aligned}$$

and $\gamma_j = \ln \pi_j - K(\theta_j)$, $j = 1, ..., m$. Such a Bayes classification rule D_π is minimax if the prior π satisfies $\mathsf{R}(1, \mathsf{D}_\pi) = \cdots = \mathsf{R}(m, \mathsf{D}_\pi)$. Especially if $P_{\theta_i}(T \le t)$, $t \in \mathbb{R}$, is continuous for every $i = 1, ..., m$, then

$$\mathsf{R}(i, \mathsf{D}_\pi) = P_{\theta_i}((\theta_i - \theta_j)T < (\gamma_j - \gamma_i),\ j \ne i), \quad i = 1, ..., m.$$

In the special case of $m = 2$ the classification problem reduces to a testing problem. Instead of using the, use the zero–one loss let us adopt here the more general loss function

$$L(i, a) = a\rho I_{\{0\}}(i) + (1 - a)(1 - \rho)I_{\{1\}}(i), \quad \rho \in (0, 1), \quad i, a \in \{0, 1\}. \tag{3.61}$$

Instead of the factors ρ and $1 - \rho$ in the loss function L any other pair of numbers $l_0 \ge 0$ and $l_1 \ge 0$ could have been chosen. However, then dividing the loss by $l_0 + l_1$ would not change the decision problem, except for a factor $1/(l_0 + l_1)$ on the risk, and the modified loss would be of the type (3.61). The quantity

$$\mathsf{m}_\rho(P_0, P_1) = \inf\{\max(\rho \mathsf{E}_0\varphi,\ (1 - \rho)\mathsf{E}_1(1 - \varphi)), \quad \varphi \in \mathcal{T}\} \tag{3.62}$$

is called the *minimax value* of the testing problem.

Problem 3.88.* For $0 < \rho < 1$ it holds $0 \le \mathsf{m}_\rho(P_0, P_1) \le \rho(1 - \rho)$, where

$$\begin{aligned} \mathsf{m}_\rho(P_0, P_1) = 0 \quad &\Leftrightarrow \quad P_0 \perp P_1, \\ \mathsf{m}_\rho(P_0, P_1) = \rho(1 - \rho) \quad &\Leftrightarrow \quad P_0 = P_1. \end{aligned}$$

We consider now the nontrivial case where P_0 and P_1 are neither identical nor mutually singular. Let $L_{0,1}$ be the likelihood ratio of P_1 with respect to P_0; set $F_i(t) = P_i(L_{0,1} \le t)$, $i = 0, 1$, and $G = \rho F_0 + (1 - \rho)F_1$. As P_0 and P_1 are not mutually singular it holds $\lim_{t\uparrow\infty} F_1(t) > 0$ so that $\lim_{t\uparrow\infty} F_0(t) = 1$ implies $\lim_{t\uparrow\infty} G(t) > \rho$. Hence $\{t : G(t) > \rho\} \ne \emptyset$ and we may put $c = G^{-1}(\rho)$ (see Definition 2.1) and $\gamma = (G(c) - \rho) \oslash (G(c) - G(c - 0))$. Let

$$\psi(t) = I_{(c,\infty)}(t) + \gamma I_{\{c\}}(t). \tag{3.63}$$

It holds

$$\begin{aligned} \rho(1 - F_0(c) + \gamma[F_0(c) - F_0(c - 0)]) - (1 - \rho)(F_1(c) - \gamma[F_1(c) - F_1(c - 0)]) \\ = \rho - G(c) + \gamma[G(c) - G(c - 0)] = 0, \end{aligned}$$

$$\rho\mathsf{E}_0\psi(L_{0,1}) = (1-\rho)\mathsf{E}_1\psi(L_{0,1}).$$

As customary, tests that provide minimax decisions are called, in reference to their power properties, *maximin tests.*

Theorem 3.89. *Let P_0 and P_1 be the two distributions of a binary model. If the loss function is defined as in (3.61), then there exists at least one maximin test for $\mathsf{H}_0 : P_0$ versus $\mathsf{H}_A : P_1$. Every maximin test φ satisfies*

$$\rho\mathsf{E}_0\varphi = (1-\rho)(1-\mathsf{E}_1\varphi). \tag{3.64}$$

Every likelihood ratio test that satisfies (3.64) is a maximin test. If P_0 and P_1 are not mutually singular, then the test $\psi(L_{0,1})$ with ψ from (3.63) is a maximin test. If $0 < \mathsf{m}_\rho(P_0, P_1) < \rho(1-\rho)$, then every maximin test φ and $\psi(L_{0,1})$ are $\{P_0, P_1\}$-a.s. identical outside of $\{L_{0,1} = c\}$.

Proof. The existence follows from Proposition 3.49. Let φ_0 be any maximin test. If $\rho\mathsf{E}_0\varphi_0 \neq (1-\rho)(1-\mathsf{E}_1\varphi_0)$, say $\rho\mathsf{E}_0\varphi_0 < (1-\rho)\mathsf{E}_1(1-\varphi_0)$, then we set $\varphi_\varepsilon = (1-\varepsilon)\varphi_0 + \varepsilon$ for a sufficiently small $\varepsilon > 0$ and get

$$\max(\rho\mathsf{E}_0\varphi_\varepsilon, (1-\rho)(1-\mathsf{E}_1\varphi_\varepsilon)) < \max(\rho\mathsf{E}_0\varphi_0, (1-\rho)(1-\mathsf{E}_1\varphi_0)),$$

which contradicts the assumption that φ_0 is a maximin test. The likelihood ratio test $\psi(L_{0,1})$, is according to Theorem 2.60, a Bayes test for a suitable prior. Hence $\psi(L_{0,1})$ is maximin by Proposition 3.58. The condition $\mathsf{m}_\rho(P_0, P_1) > 0$ implies $0 < \rho < 1$, $\mathsf{E}_0\varphi = \mathsf{E}_0\psi(L_{0,1}) > 0$, and $\mathsf{E}_1\varphi = \mathsf{E}_1\psi(L_{0,1}) < 1$. Hence the test φ has the same size and the same power as the likelihood ratio test $\psi(L_{0,1})$. An application of Theorem 2.45 yields the stated uniqueness. ■

Problem 3.90. For $\alpha \in (0,1)$, construct a maximin level α test for $\mathsf{H}_0 : \mathsf{N}(\mu_1, \sigma_1^2)$ versus $\mathsf{H}_A : \mathsf{N}(\mu_2, \sigma_2^2)$, where $\mu_1, \mu_2 \in \mathbb{R}$ and $\sigma_1^2, \sigma_2^2 > 0$ are fixed given.

3.9 Solutions to Selected Problems

Solution to Problem 3.9: It holds

$$\begin{aligned}\mathbb{P}(X_i > \max_{j\neq i} X_j) &= \int [\int \prod\nolimits_{j\neq i} I_{(-\infty,t)}(t_j) \bigotimes\nolimits_{j\neq i} P_j(dt_j)] P_i(dt) \\ &= \int \prod\nolimits_{j\neq i} F_j(t) P_i(dt), \quad i = 1, ..., k.\end{aligned}$$

The last equation follows from the continuity of $F_1, ..., F_k$. □

Solution to Problem 3.10: Using Problem 3.9 we have

$$\begin{aligned}\mathbb{P}(X_i > \max_{j\neq i} X_j) &= \int \prod\nolimits_{j\neq i} \Phi(\frac{t-\mu_j}{\sigma}) \frac{1}{\sigma} \varphi(\frac{t-\mu_i}{\sigma}) dt \\ &= \int \prod\nolimits_{j\neq i} \Phi(s + \frac{\mu_i - \mu_j}{\sigma}) \varphi(s) ds.\end{aligned}$$

As $\prod_{j\neq i}\Phi(s+(\mu_i-\mu_j)/\sigma)$ is an increasing function of $\mu_i-\mu_j$ the probability $\mathbb{P}(X_{i_0}>\max_{j\neq i_0}X_j)$ is maximal if $\mu_{i_0}-\mu_j$ is maximal for each $j\neq i_0$. This is true if and only if $\mu_{i_0}=\mu_{[k]}$. □

Solution to Problem 3.16: As $\mathfrak{A}_0$ is countably generated there exists an increasing sequence of finite partitions $\mathfrak{z}_n=\{A_{1,n},...,A_{m_n,n}\}$ so that $\mathfrak{A}_0$ is generated by $\mathfrak{z}_1,\mathfrak{z}_2,...$. Then for $\mathfrak{A}_{0,n}=\sigma(\mathfrak{z}_n)$ and $f_n=\mathsf{E}_{Q_0}(f|\mathfrak{A}_{0,n})$ it holds $\int|f_n-f|dQ_0\to 0$ by Levy's martingale theorem; see Theorem A.34. Modify the finite number of values of f_n in such a way that the new function $\widetilde{f}_n$ has rational values and the sequence satisfies again $\int|\widetilde{f}_n-f|dQ_0\to 0$. Hence the family of functions $f:\mathcal{X}\to_m\mathbb{Q}$ that are $\mathfrak{A}_{0,n}$-measurable for some n rational values is at most countable and dense in $\mathbb{L}_1(Q_0)$. □

Solution to Problem 3.24: Let $\Delta_0=\{\theta_1,\theta_2,...\}$ be a subset that is dense in Δ. $Q=\sum_{k=1}^{\infty}2^{-k}P_{\theta_k}$ is a probability measure. For a fixed $\theta\in\Delta$ choose k_l such that $\sup_{B\in\mathfrak{A}}2|P_{\theta_{k_l}}(B)-P_\theta(B)|=\|P_{\theta_{k_l}}-P_\theta\|\to 0$ as $l\to\infty$. If $Q(A)=0$, then $P_{\theta_k}(A)=0$ for every k and $P_\theta(A)=0$. □

Solution to Problem 3.26: If the statement is not true, then there is a subsequence $\theta_n\to\theta_0$, and a sequence a_n which may be assumed to be convergent in view of the compactness of $\mathcal{D}$, say $a_n\to a$, such that for some $\varepsilon>0$ it holds $|L(\theta_n,a_n)-L(\theta_0,a_n)|\geq\varepsilon$, which contradicts the continuity of L. □

Solution to Problem 3.31: If $\mathsf{m}(x)<\infty$, $\boldsymbol{\mu}$-a.e., then for every $A_n\in\mathfrak{A}$ with $\boldsymbol{\mu}(A_n)<\infty$, $\cup_{n=1}^{\infty}A_n=\mathcal{X}$, and $B_N=\{x:\mathsf{m}(x)<N\}$ it holds $(\mathsf{P}\rho)(A_n\cap B_N)=\int_{A_n\cap B_N}\mathsf{m}(x)\boldsymbol{\mu}(dx)<\infty$. On the other hand, $C_n\uparrow\mathcal{X}$ with $(\mathsf{P}\rho)(C_n)=\int_{C_n}\mathsf{m}(x)\boldsymbol{\mu}(dx)<\infty$ yields $\mathsf{m}(x)<\infty,\boldsymbol{\mu}$-a.e. on C_n for every n, and therefore $\mathsf{m}(x)<\infty,\boldsymbol{\mu}$-a.e. The statement $d(\mathsf{P}\rho)/d\boldsymbol{\mu}=\mathsf{m}$ follows from $\int_A\mathsf{m}(x)\boldsymbol{\mu}(dx)=\int[\int I_A(x)f_\theta(x)\rho(d\theta)]\boldsymbol{\mu}(dx)=\int P_\theta(A)\rho(d\theta)$. □

Solution to Problem 3.63: The joint density of (X,Y_2) is $\varphi_{\mu_1,\Sigma_1}(x-y_2)\varphi_{\mu_2,\Sigma_2}(y_2)$. The marginal density of X is $\varphi_{\mu_1+\mu_2,\Sigma_1+\Sigma_2}(x)$. Consider the ratio to get the result. □

Solution to Problem 3.73: We have to minimize $r(\rho,a|x)$ in (3.28). Hence we have to minimize $g(a)=\frac{1}{2}a^2\varphi_{0,1}(x)+\frac{1}{2}(a-c)^2\varphi_{0,c}(x)$ which is a convex and differentiable function. The zero of $g'(a)$ gives T_c. □

Solution to Problem 3.79: If x_n satisfies $\|x_n\|\to\inf_{x\in K}\|x\|$, then it follows from $\|\frac{1}{2}(x_n+x_m)\|\leq\frac{1}{2}\|x_n\|+\frac{1}{2}\|x_m\|$ that $\lim_{m,n\to\infty}\|\frac{1}{2}(x_n+x_m)\|=\inf_{x\in K}\|x\|$. Hence

$$\frac{1}{4}\|x_n-x_m\|^2=\frac{1}{2}\|x_n\|^2+\frac{1}{2}\|x_m\|^2-\|\frac{1}{2}(x_n+x_m)\|^2$$

implies $\lim_{m,n\to\infty}\|x_n-x_m\|=0$, so that by the completeness of $\mathbb{R}^d$ it holds $x_n\to x_K$, where $x_K\in K$ as K is closed. □

Solution to Problem 3.80: For every $x \in K$ and every $\alpha \in [0,1]$,

$$0 \leq f(\alpha) = \|\alpha(x-z) + (1-\alpha)(x_K - z)\|^2 - \|x_K - z\|^2 ,$$

so that $f(0) = 0$ implies

$$f'(0) = 2(\langle x - z, x_K - z\rangle - \|x_K - z\|^2) \geq 0$$

and $\inf_{x \in K} \langle x - z, x_K - z\rangle \geq \|x_K - z\|^2 > 0$. □

Solution to Problem 3.81: The convexity of $A(B)$ follows from the convexity of B. Let $y \succeq x$ be the coordinatewise semiorder and $y_n \to y$, $y_n \in A(B)$. Then there exist $x_n \in B$ with $y_n \succeq x_n$, and by the compactness of B a convergent subsequence $x_{n_k} \to x \in B$, so that $y \succeq x$ and thus $y \in A(B)$. If $z \notin A(B)$, then by Problem 3.80 there is some $b \in \mathbb{R}^n$ with $\langle b, z\rangle < \inf_{y \in A(B)} \langle b, y\rangle$. By construction of $A(B)$ and $z \in \mathbb{R}^n_+$ each coordinate of $b \neq 0$ must be nonnegative as otherwise $\inf_{y \in A(B)} \langle b, y\rangle = -\infty$. The vector b with nonnegative components can be normalized to be a distribution. □

Solution to Problem 3.85: The posterior density is

$$\pi(\theta|x) = (\mathsf{m}(x)\sqrt{2\pi})^{-1} \exp\{-\frac{1}{2}(\theta - x)^2\}, \quad \text{where}$$
$$\mathsf{m}(x) = \int \frac{1}{\sqrt{2\pi}} \exp\{-\frac{1}{2}(\theta - x)^2\} \Pi(d\theta).$$

If $T_{nat}(x) = x$ is a Bayes estimator for the prior Π, then by $\mathbb{E}(\Theta|X = x) = x$,

$$\int \frac{1}{\sqrt{2\pi}} \theta \exp\{-\frac{1}{2}(\theta - x)^2\}\Pi(d\theta) = x\mathsf{m}(x), \quad \boldsymbol{\lambda}\text{-a.e.}$$
$$\int (\theta - x) \exp\{-\frac{1}{2}(\theta - x)^2\}\Pi(d\theta) = 0, \quad \boldsymbol{\lambda}\text{-a.e.}$$

The function $g(x) = \int \exp\{-\frac{1}{2}(\theta - x)^2\}\Pi(d\theta)$ is continuously differentiable and $g' = 0$ $\boldsymbol{\lambda}$-a.e. Hence g is a constant. As $g(0) > 0$ and $\lim_{x\to\infty} g(x) = 0$ we get a contradiction. □

Solution to Problem 3.88: Let φ be a test with

$$\mathsf{m}_\rho(P_0, P_1) = \max(\rho \mathsf{E}_0\varphi,\ (1-\rho)\mathsf{E}_1(1-\varphi)).$$

If $m_\rho(P_0, P_1) = 0$, then $E_0\varphi = E_1(1-\varphi) = 0$, and $P_0(\varphi = 0) = 1$, $P_1(\varphi = 0) = 0$ implies $P_0 \perp P_1$. If $P_0 \perp P_1$, then for some A, $P_0(A) = 0 = 1 - P_1(A)$. $\varphi = I_A$ yields $m_\rho(P_0, P_1) = 0$. Let $\psi_{1-\rho}(L_{0,1})$ be the best level $1-\rho$ test with $E_0\varphi = 1-\rho$. Then by Problem 2.47 $E_1(1 - \psi_{1-\rho}(L_{0,1})) \leq \rho$ with equality if and only if $P_0 = P_1$. □

4

Comparison of Models, Reduction by Sufficiency

4.1 Comparison and Randomization of Models

A statistical inference on an unknown distribution parameter is made after observations have been drawn that contain information about it. Hereby it is clear that samples of larger sizes would produce more information. Large sets of data, however, may become too unwieldy. Then summary data, such as the sample mean or sample variance, may be utilized instead to get some insight into the situation and to draw first conclusions. It is clear that there may be some loss of information if we take into account only such summary data for our inference, which in turn may have a negative effect on the quality of the inference. Some framework that provides a precise formulation of the loss of information due to the reduction of data becomes necessary. We start with a statistical model $\mathcal{M} = (\mathcal{X}, \mathfrak{A}, (P_\theta)_{\theta\in\Delta})$, another measurable space $(\mathcal{Y}, \mathfrak{B})$, and a statistic $T : \mathcal{X} \to_m \mathcal{Y}$ that provides the reduction of the data from $x \in \mathcal{X}$ to $T(x) \in \mathcal{Y}$. Here we call

$$\mathcal{N} = (\mathcal{Y}, \mathfrak{B}, (Q_\theta)_{\theta\in\Delta}), \quad \text{where } Q_\theta = P_\theta \circ T^{-1},\ \theta \in \Delta, \tag{4.1}$$

the *reduced model.*

Example 4.1. Let $(\mathcal{X}, \mathfrak{A}) = (\mathbb{R}^n, \mathfrak{B}_n)$. For a sample $X_1, ..., X_n$ we consider the rank statistic $R_n = (R_{n,1}, ..., R_{n,n})$ that has been introduced in (2.5). R_n takes on values in $\{1, ..., n\}^n$ and gives information on the order relations between the observations, but does not say anything about their values. On the other hand, together with the order statistic $X_{[\cdot]} = (X_{[1]}, ..., X_{[n]})$ in (2.5) the sample $X_1, ..., X_n$ can be reconstructed from $(R_n, X_{[\cdot]})$, and therefore $(R_n, X_{[\cdot]})$ carries the complete information that is contained in the data. However, if we use only R_n, or only $X_{[\cdot]}$, then the full information cannot be exploited.

In some situations the reduction of data is not made by the statistician. Instead, the observation process is incomplete and even possibly disturbed by random errors.

F. Liese, K.-J. Miescke, *Statistical Decision Theory*,
DOI: 10.1007/978-0-387-73194-0_4, © Springer Science+Business Media, LLC 2008

Example 4.2. In reliability theory, when random lifetimes are considered, it is typical that the lifetimes can only be observed up to certain time points. The simplest case is where from an i.i.d. sample $X_1, ..., X_n$ of lifetimes we can only observe $X_1 \wedge c, ..., X_n \wedge c$, where $c > 0$ is a given constant. More generally, let $C_1, ..., C_n$ be i.i.d. censoring times. Suppose that we can observe only $(X_1 \wedge C_1, D_1), ..., (X_n \wedge C_n, D_n)$, where $D_i = I_{[C_i,\infty)}(X_i)$, $i = 1, ..., n$. In this model, which is called a *random censorship model,* we may observe only the minimum of the lifetime and the censoring time, but in addition we get the information whether this value was a lifetime or a censoring time.

Occasionally, in interviews additional errors are included on purpose by the statistician in order to protect the identities of interviewed people.

Example 4.3. Here we consider a randomized response technique that is used, among others, in sampling surveys when a sensitive question about a personality property $\mathfrak{C}$, say, is asked that a person may perhaps not want to answer truthfully. $\mathfrak{C}$ may, for example, stand for using drugs. Let $X = 1$ if the interviewed person has property $\mathfrak{C}$, and let $X = 0$ otherwise. Suppose we want to estimate the proportion of people that have property $\mathfrak{C}$. In an interview two questions are presented to a respondent. The first is whether the respondent's social security number is even, and the second is whether the respondent has the property $\mathfrak{C}$. Instead of answering both questions the respondent is now asked to flip a coin and not to reveal the result to the interviewer. Then the respondent is asked to answer the first question if the coin had turned up "tails", and to answer the second question if the coin had turned up "heads". Here it can be assumed that the respondent will answer truthfully.

Let $Z = 0$ if the coin turns up "tails", and let $Z = 1$ otherwise. Moreover, let $S = 1$ if the respondent's social security number is even, and let $S = 0$ otherwise. What the interviewer observes in this setting is $Y := (1-Z)S+ZX$. The random variable X remains hidden from the interviewer, and he cannot find out whether the respondent has the property $\mathfrak{C}$, as long as he does not know whether the respondent's social security number is even or odd. There are several other randomization techniques available for estimating proportions that circumvent possibly biased answers. For further details we refer to Chaudhuri and Mukerjee (1988).

The examples 4.2 and 4.3 are covered by the following more general setup. Let X be a random variable that takes on values in $(\mathcal{X}, \mathfrak{A})$. Let V be another random variable with values in $(\mathcal{V}, \mathfrak{V})$ that is independent of X. Let $(\mathcal{Y}, \mathfrak{B})$ be a third measurable space, $g : \mathcal{X} \times \mathcal{V} \to_m \mathcal{Y}$, and suppose that $Y = g(X, V)$ is the random variable that we observe. Then by the independence of X and V for any $B \in \mathfrak{B}$ it holds

$$\mathsf{K}(B|x) := \mathbb{E}(I_B(Y)|X = x) = \mathbb{E}I_B(g(x, V)), \quad x \in \mathcal{X},$$

where the right-hand side determines the version of the conditional expectation used in the middle. It is easy to see that K is a stochastic kernel that satisfies

$$(\mathcal{L}(Y, X))(C) = (\mathsf{K} \otimes P)(C) = \int [\int I_C(x, y)\mathsf{K}(dy|x)]P(dx) \quad \text{and}$$

$$(\mathcal{L}(Y))(B) = (\mathsf{K}P)(B) = \int \mathsf{K}(B|x)P(dx), \quad P = \mathcal{L}(X). \tag{4.2}$$

The two equations in (4.2) describe a fairly general situation of restrictedly observable data and lead to the following definition.

Definition 4.4. *Given a model* $\mathcal{M} = (\mathcal{X}, \mathfrak{A}, (P_\theta)_{\theta\in\Delta})$, *a measurable space* $(\mathcal{Y}, \mathfrak{B})$, *and a stochastic kernel* $\mathsf{K} : \mathfrak{B} \times \mathcal{X} \to_k [0,1]$, *we call the model* $\mathcal{N} = (\mathcal{Y}, \mathfrak{B}, (\mathsf{K}P_\theta)_{\theta\in\Delta})$ *the randomization of* $\mathcal{M}$ *by the kernel* K, *and we write* $\mathcal{N} = \mathsf{K}\mathcal{M}$.

For $T : \mathcal{X} \to_m \mathcal{Y}$ we introduce the special kernel K by setting $\mathsf{K}(\cdot|x) = \boldsymbol{\delta}_{T(x)}$. Then $\mathsf{K}P_\theta = P_\theta \circ T^{-1}$ so that the reduced model in (4.1) is a special case of randomization.

Up to this point we have studied situations where a possible random influence on the data leads to a new model which we know is a randomization of the original model. Now we look at the more general situation where any two models $\mathcal{M} = (\mathcal{X}, \mathfrak{A}, (P_\theta)_{\theta\in\Delta})$ and $\mathcal{N} = (\mathcal{Y}, \mathfrak{B}, (Q_\theta)_{\theta\in\Delta})$ are given and we want to decide, say, if $\mathcal{N}$ is a randomization of $\mathcal{M}$. This allows us to clarify if $\mathcal{M}$ contains more information on the parameter than $\mathcal{N}$. To this end we formulate first necessary conditions. We already know that the application of a stochastic kernel to distributions is a contraction in the sense that the distance between the new distributions does not exceed the original distance if the distance is measured by means of v-divergences. Indeed, the monotonicity theorem for v divergences (see Theorem 1.70) and the corresponding statement for Hellinger transforms in Proposition 1.93 show that for $(Q_\theta)_{\theta\in\Delta} = (\mathsf{K}P_\theta)_{\theta\in\Delta}$ it holds

$$\begin{aligned} \mathsf{I}_\mathsf{v}(P_{\theta_1}, P_{\theta_2}) &\geq \mathsf{I}_\mathsf{v}(Q_{\theta_1}, Q_{\theta_2}), \quad \theta_1, \theta_2 \in \Delta, \\ \mathsf{H}_s(P_{\theta_1}, ..., P_{\theta_k}) &\leq \mathsf{H}_s(Q_{\theta_1}, ..., Q_{\theta_k}), \quad \theta_1, ..., \theta_k \in \Delta, \ s \in \mathbf{S}_k^o. \end{aligned} \tag{4.3}$$

Although the inequalities in (4.3) are only necessary conditions for a randomization they are also sufficient if we restrict ourselves to special families of distributions.

Example 4.5. For fixed positive σ^2 and τ^2 we consider the models

$$\mathcal{M} = (\mathbb{R}, \mathfrak{B}, (\mathsf{N}(\mu, \sigma^2)_{\mu\in\mathbb{R}}) \quad \text{and} \quad \mathcal{N} = (\mathbb{R}, \mathfrak{B}, (\mathsf{N}(\mu, \tau^2)_{\mu\in\mathbb{R}}).$$

If $\sigma^2 \leq \tau^2$, then

$$\mathsf{N}(\mu, \tau^2) = \mathsf{N}(\mu, \sigma^2) * \mathsf{N}(0, \tau^2 - \sigma^2),$$

so that $\mathcal{N} = \mathsf{K}\mathcal{M}$, where K is the convolution kernel $\mathsf{K}(A|x) = (\mathsf{N}(0, \tau^2 - \sigma^2))(A - x)$.

Conversely, suppose that $\mathcal{N} = \mathsf{K}\mathcal{M}$ for some kernel K. Then

$$\mathsf{H}_s(\mathsf{N}(\mu_1, \sigma^2), \mathsf{N}(\mu_2, \sigma^2)) \leq \mathsf{H}_s(\mathsf{N}(\mu_1, \tau^2), \mathsf{N}(\mu_2, \tau^2)), \quad 0 < s < 1,$$

and by (1.79) and (4.3),

$$\begin{aligned} \mathsf{H}_{1/2}(\mathsf{N}(\mu_1, \sigma^2), \mathsf{N}(\mu_2, \sigma^2)) &= \exp\{-\frac{1}{8}\frac{(\mu_1 - \mu_2)^2}{\sigma^2}\} \\ &\leq \exp\{-\frac{1}{8}\frac{(\mu_1 - \mu_2)^2}{\tau^2}\} = \mathsf{H}_{1/2}(\mathsf{N}(\mu_1, \tau^2), \mathsf{N}(\mu_2, \tau^2)), \end{aligned}$$

which means $\sigma^2 \leq \tau^2$. Altogether we have obtained that $\mathcal{N} = \mathsf{K}\mathcal{M}$ if and only if $\sigma^2 \leq \tau^2$.

In Example 4.5 we used a normally distributed Z to randomize the model $\mathcal{M}$. The question that remains open is whether another random variable could have been used instead.

Problem 4.6.* If X and Z are independent with $\mathcal{L}(X) = \mathsf{N}(\mu_1, \sigma_1^2)$ and $\mathcal{L}(Y) = \mathsf{N}(\mu_2, \sigma_2^2)$, then it holds $\mathcal{L}(Y) = \mathcal{L}(X + Z)$ if and only if $\sigma_1^2 \le \sigma_2^2$ and $\mathcal{L}(Z) = \mathsf{N}(\mu_2 - \mu_1, \sigma_2^2 - \sigma_1^2)$.

Problem 4.7.* Let Σ_1 be nonsingular. $(\mathsf{N}(\mu, \Sigma_2))_{\mu \in \mathbb{R}^d}$ is a randomization of $(\mathsf{N}(\mu, \Sigma_1))_{\mu \in \mathbb{R}^d}$ if and only if $\Sigma_1 \preceq \Sigma_2$ in the Löwner semiorder, i.e., $u^T(\Sigma_2 - \Sigma_1)u \ge 0$, $u \in \mathbb{R}^d$.

In the above problem multivariate normal distributions are compared. The more general situation of linear models with normally distributed errors is studied in Lehmann (1988), Torgersen (1991), and Luschgy (1992b).

For fixed variances the two families compared in Example 4.5 are location models with parent normal distributions. The question arises as to whether similar statements can be made for any location model. This problem was first studied by Boll (1955). His result has been generalized by many authors and is called the *convolution theorem.* Here we present only a special version of the general convolution theorem. We refer to Strasser (1985), Theorem 55.12, for the general case and its proof.

Theorem 4.8. *Let P and Q be distributions on the Borel sets of $\mathbb{R}^d$, and set $\mathcal{M} = (\mathbb{R}^d, \mathfrak{B}_d, (P(\cdot - \theta))_{\theta \in \mathbb{R}})$ and $\mathcal{N} = (\mathbb{R}^d, \mathfrak{B}_d, (Q(\cdot - \theta))_{\theta \in \mathbb{R}})$. Then the following holds. Model $\mathcal{N}$ is a randomization of $\mathcal{M}$ if and only if there exists a distribution R such that $Q = R * P$.*

To see that $Q = R * P$ leads to a randomization, we set $P_\theta(B) = P(B - \theta)$, $Q_\theta(B) = Q(B - \theta)$ and introduce K as convolution kernel $\mathsf{K}(B|x) = R(B - x)$. Then

$$Q_\theta(B) = \int P_\theta(B - x)R(dx) = \int R(B - x)P_\theta(x) = (\mathsf{K}P_\theta)(B).$$

Now we consider decision-theoretic consequences of the concept of randomization. Whenever two models $\mathcal{M}$ and $\mathcal{N}$ are simultaneously under consideration, then we indicate by a subscript $\mathcal{M}$ or $\mathcal{N}$ at a decision D to which model it belongs. Let $\mathcal{M} = (\mathcal{X}, \mathfrak{A}, (P_\theta)_{\theta \in \Delta})$ be a model and $\mathcal{N} = \mathsf{K}\mathcal{M} = (\mathcal{Y}, \mathfrak{B}, (\mathsf{K}P_\theta)_{\theta \in \Delta})$ be a randomization of $\mathcal{M}$. Let $L(\theta, \cdot) : \mathcal{D} \to_m \mathbb{R}$ be a bounded loss function and $\mathsf{D}_\mathcal{N} : \mathfrak{D} \times \mathcal{Y} \to_k [0, 1]$ a decision for the model $\mathcal{N}$. Using Fubini's theorem for stochastic kernels (see Proposition A.40) and the definition of $\mathsf{K}P_\theta$ in (4.2), the risk function may be written as

$$\begin{aligned}
\mathsf{R}(\theta, \mathsf{D}_\mathcal{N}) &= \int [\int L(\theta, a)\mathsf{D}_\mathcal{N}(da|y)](\mathsf{K}P_\theta)(dy) \\
&= \int [\int L(\theta, a)(\mathsf{D}_\mathcal{N}\mathsf{K})(da|x)]P_\theta(dx) = \mathsf{R}(\theta, (\mathsf{D}_\mathcal{N}\mathsf{K})_\mathcal{M}), \quad \text{where}
\end{aligned}$$

$$(\mathsf{D}_\mathcal{N}\mathsf{K})_\mathcal{M}(A|x) = \int \mathsf{D}_\mathcal{N}(A|y)\mathsf{K}(dy|x), \quad A \in \mathfrak{D},\ x \in \mathcal{X},$$

is again a stochastic kernel, and $(\mathsf{D}_{\mathcal{N}}\mathsf{K})_{\mathcal{M}}$ is a decision for the model $\mathcal{M}$. This means that for every decision $\mathsf{D}_{\mathcal{N}}$ for the model $\mathcal{N}$ there is a decision $(\mathsf{D}_{\mathcal{N}}\mathsf{K})_{\mathcal{M}}$ for the model $\mathcal{M}$ such that the associated risk functions are the same. Hence a minimization over a set of decisions will lead for the model $\mathcal{M}$ to a smaller risk than the minimization for the model $\mathcal{N}$.

Now we consider the case where the risk function of the model $\mathcal{N}$ is only up to some $\varepsilon \geq 0$ larger than the risk function of the model $\mathcal{M}$. For every loss function $L : \Delta \times \mathcal{D} \to \mathbb{R}$ we set $\|L\|_u = \sup_{\theta\in\Delta, a\in\mathcal{D}} |L(\theta, a)|$. The following definition is taken from Torgersen (1991), Section 6.2.

Definition 4.9. *For $\varepsilon \geq 0$ a model $\mathcal{M} = (\mathcal{X}, \mathfrak{A}, (P_\theta)_{\theta\in\Delta})$ is called ε-deficient with respect to $\mathcal{N} = (\mathcal{Y}, \mathfrak{B}, (Q_\theta)_{\theta\in\Delta})$ if for every finite subset $F \subseteq \Delta$, every finite decision space $\mathcal{D}$, every loss function L with $\|L\|_u \leq 1$, and every $\mathsf{D}_{\mathcal{N}} : \mathfrak{D} \times \mathcal{Y} \to_k [0, 1]$, there exists a $\mathsf{D}_{\mathcal{M}} : \mathfrak{D} \times \mathcal{X} \to_k [0, 1]$ such that*

$$\mathsf{R}(\theta, \mathsf{D}_{\mathcal{M}}) \leq \mathsf{R}(\theta, \mathsf{D}_{\mathcal{N}}) + \varepsilon, \quad \theta \in F. \tag{4.4}$$

In this case we write $\mathcal{M} \succeq^{\varepsilon} \mathcal{N}$. If $\mathcal{M}$ is 0-deficient with respect to $\mathcal{N}$, then we call $\mathcal{M}$ at least as informative as $\mathcal{N}$ and write $\mathcal{M} \succeq \mathcal{N}$ instead of $\mathcal{M} \succeq^0 \mathcal{N}$. If $\mathcal{M} \succeq \mathcal{N}$ and $\mathcal{N} \succeq \mathcal{M}$, then we call $\mathcal{M}$ and $\mathcal{N}$ equivalent and write $\mathcal{M} \sim \mathcal{N}$.

If L is any bounded loss function, with absolute value not necessarily bounded by 1, then we may switch from L to $\widetilde{L} = (\sup_{a\in\mathcal{D}} |L(\theta, a)|)^{-1} L$ and see that (4.4) is equivalent to

$$\mathsf{R}(\theta, \mathsf{D}_{\mathcal{M}}) \leq \mathsf{R}(\theta, \mathsf{D}_{\mathcal{N}}) + \varepsilon \sup_{a\in\mathcal{D}} |L(\theta, a)|, \quad \theta \in F.$$

Similarly, in terms of nonnegative and bounded loss functions we can say that (4.4) is equivalent to

$$\mathsf{R}(\theta, \mathsf{D}_{\mathcal{M}}) \leq \mathsf{R}(\theta, \mathsf{D}_{\mathcal{N}}) + \frac{\varepsilon}{2} \sup_{a\in\mathcal{D}} L(\theta, a), \quad \theta \in F,$$

for every nonnegative and bounded loss function L.

Let $\mathfrak{F}(\Delta)$ denote the system of all finite subsets of Δ, and let $\mathcal{M}_F$ be the finite submodel of $\mathcal{M}$ that is obtained from $\mathcal{M}$ by restricting the parameter set to some $F \in \mathfrak{F}(\Delta)$. It is immediately clear from the definition that

$$\mathcal{M} \succeq^{\varepsilon} \mathcal{N} \quad \text{if and only if} \quad \mathcal{M}_F \succeq^{\varepsilon} \mathcal{N}_F \quad \text{for every } F \in \mathfrak{F}(\Delta).$$

Corollary 4.10. *Let $\mathcal{M} = (\mathcal{X}, \mathfrak{A}, (P_\theta)_{\theta\in\Delta})$ and $\mathcal{N} = (\mathcal{Y}, \mathfrak{B}, (Q_\theta)_{\theta\in\Delta})$ be two statistical models. If there is a kernel $\mathsf{K} : \mathfrak{B} \times \mathcal{X} \to_k [0, 1]$ such that*

$$\|\mathsf{K}P_\theta - Q_\theta\| \leq \varepsilon, \quad \theta \in \Delta, \tag{4.5}$$

then $\mathcal{M} \succeq^{\varepsilon} \mathcal{N}$, and especially

$$\mathcal{N} = \mathsf{K}\mathcal{M} \quad \textit{implies} \quad \mathcal{M} \succeq \mathcal{N}.$$

Proof. If $\|L\|_u \leq 1$, then

$$\begin{aligned}
&\mathsf{R}(\theta, (\mathsf{D}_{\mathcal{N}}\mathsf{K})_{\mathcal{M}}) = \mathsf{R}(\theta, \mathsf{D}_{\mathcal{N}}) + \int [\int L(\theta, a)\mathsf{D}_{\mathcal{N}}(da|y)](\mathsf{K}P_\theta - Q_\theta)(dy) \\
&\leq \mathsf{R}(\theta, \mathsf{D}_{\mathcal{N}}) + \|\mathsf{K}P_\theta - Q_\theta\| \leq \mathsf{R}(\theta, \mathsf{D}_{\mathcal{N}}) + \varepsilon,
\end{aligned}$$

where the first inequality follows from the inequality in Problem 1.80. ■

Problem 4.11.* Let $\mathcal{M} = (\mathcal{X}, \mathfrak{A}, (P_\theta)_{\theta\in\Delta})$ be a model. Assume that there is a $T : \mathcal{X} \to_m \mathcal{Y}$ where T is a mapping onto $B \in \mathfrak{B}$ and the inverse mapping $S : B \to \mathcal{X}$ is measurable with respect to $\mathfrak{B}_B = \{C : C \in \mathfrak{B},\ C \subseteq B\}$. If $\mathcal{N} = (\mathcal{Y}, \mathfrak{B}, (Q_\theta)_{\theta\in\Delta})$ with $Q_\theta = P_\theta \circ T^{-1}$, then $\mathcal{M}$ and $\mathcal{N}$ are mutual randomizations of each other and therefore equivalent.

Problem 4.12. Suppose that for the model $\mathcal{M} = (\mathcal{X}, \mathfrak{A}, (P_\theta)_{\theta\in\Delta})$ the σ-algebra $\mathfrak{A}$ is finite with the atoms $A_1, ..., A_m$. Let $\mathcal{Y} = \{y_1, ..., y_m\}$ be any set. If Q_θ is defined by $Q_\theta(\{y_i\}) = P_\theta(A_i)$, $i = 1, ..., m$, then $\mathcal{M}$ and $\mathcal{N}$ are mutual randomizations of each other and therefore equivalent.

Next we establish a criterion which states that under certain conditions (4.5) is not only sufficient but also necessary for the relation $\mathcal{M} \succeq^\varepsilon \mathcal{N}$. This criterion, called the *randomization criterion*, is shown to play a crucial role when we deal with the convergence of models in the next section. Setting up this randomization criterion requires some technical preparations. One step hereby is a simple special version of the classical minimax theorem. As we need only a special result we prove this statement directly. For a general form of the minimax theorem we refer to LeCam and Yang (1990) and to Torgersen (1991) for the relations to game theory. Recall that a subset $\mathbb{L} \subseteq \mathbb{R}^d$ is called a *polytope* if it is the convex hull of a finite number of vectors from $\mathbb{R}^d$.

Lemma 4.13. *Let $\mathbb{K} \subseteq \mathbb{R}^k$ be a convex and compact set and $\mathbb{L} \subseteq \mathbb{R}^d$ be a polytope. If $\Psi : \mathbb{K} \times \mathbb{L} \to \mathbb{R}$ is continuous and satisfies*

$$\begin{aligned}
\Psi(\alpha y_1 + (1-\alpha)y_2, z) &= \alpha\Psi(y_1, z) + (1-\alpha)\Psi(y_2, z), \qquad (4.6)\\
\Psi(y, \alpha z_1 + (1-\alpha)z_2) &= \alpha\Psi(y, z_1) + (1-\alpha)\Psi(y, z_2),
\end{aligned}$$

for $\alpha \in [0,1]$, $y, y_1, y_2 \in \mathbb{K}$, and $z, z_1, z_2 \in \mathbb{L}$, then there is some y_0 with

$$\sup_{z\in\mathbb{L}} \Psi(y_0, z) = \inf_{y\in\mathbb{K}} \sup_{z\in\mathbb{L}} \Psi(y, z) = \sup_{z\in\mathbb{L}} \inf_{y\in\mathbb{K}} \Psi(y, z).$$

Proof. The inequality

$$\underline{v} := \sup_{z\in\mathbb{L}} \inf_{y\in\mathbb{K}} \Psi(y, z) \leq \inf_{y\in\mathbb{K}} \sup_{z\in\mathbb{L}} \Psi(y, z) =: \overline{v}. \qquad (4.7)$$

holds in any case. To prove the converse inequality we note that we may assume that Ψ is nonnegative and $\mathbb{L}$ is the convex hull of $z_1, ..., z_N$. By the assumption on Ψ the set

$$\widetilde{K} = \{(\Psi(y, z_1), ..., \Psi(y, z_N)), \quad y \in \mathbb{K}\}$$

is a compact and convex subset of $\mathbb{R}_+^N$. For any $\alpha_i \geq 0$ with $\sum_{i=1}^N \alpha_i = 1$,

$$\inf_{y \in \mathbb{K}} \sum\nolimits_{i=1}^N \alpha_i \Psi(y, z_i) = \inf_{y \in \mathbb{K}} \Psi(y, \sum\nolimits_{i=1}^N \alpha_i z_i) \leq \underline{v}.$$

By Problem 3.81 there is some $y_0 \in \mathbb{K}$ such that $(\Psi(y_0, z_1), ..., \Psi(y_0, z_N))$ is, with respect to the componentwise semiorder, not larger than the vector $(\underline{v}, ..., \underline{v})$, i.e., $\Psi(y_0, z_i) \leq \underline{v}$, $i = 1, ..., N$. As every $z \in \mathbb{L}$ is a convex linear combination of the z_i we get $\sup_{z \in \mathbb{L}} \Psi(y_0, z) \leq \underline{v}$ and thus

$$\overline{v} = \inf_{y \in \mathbb{K}} \sup_{z \in \mathbb{L}} \Psi(y, z) \leq \sup_{z \in \mathbb{L}} \Psi(y_0, z) \leq \underline{v},$$

which together with (4.7) completes the proof. ■

In the definition of ε-deficiency the risk functions are compared pointwise. For later purpose, when we deal with the convergence of models, we need also characterizations in terms of the Bayes risk and in terms of the performance function. The latter assigns to each θ the distribution of the decision, i.e., this function is defined by $\theta \mapsto \mathsf{D}P_\theta$. The following theorem is Corollary 6.3.2 in Torgersen (1991).

Theorem 4.14. *For any two statistical models $\mathcal{M} = (\mathcal{X}, \mathfrak{A}, (P_i)_{i \in \Delta})$ and $\mathcal{N} = (\mathcal{Y}, \mathfrak{B}, (Q_i)_{i \in \Delta})$ with a common finite parameter set Δ, any finite decision space $\mathcal{D}$, and any $\varepsilon \geq 0$, the following statements are equivalent.*

(A) Pointwise comparison of risk:

For every loss function L with $\|L\|_u \leq 1$ and every decision $\mathsf{D}_\mathcal{N}$ there exists a decision $\mathsf{D}_\mathcal{M}$ with

$$\mathsf{R}(\theta, \mathsf{D}_\mathcal{M}) \leq \mathsf{R}(\theta, \mathsf{D}_\mathcal{N}) + \varepsilon, \quad \theta \in \Delta.$$

(B) Comparison of Bayes risks:

For every loss function L with $\|L\|_u \leq 1$, every prior Π on Δ, and every decision $\mathsf{D}_\mathcal{N}$, there exists a decision $\mathsf{D}_\mathcal{M}$ with

$$\mathsf{r}(\Pi, \mathsf{D}_\mathcal{M}) \leq \mathsf{r}(\Pi, \mathsf{D}_\mathcal{N}) + \varepsilon.$$

(C) Comparison of performance functions:

For every decision $\mathsf{D}_\mathcal{N}$ there exists a decision $\mathsf{D}_\mathcal{M}$ with

$$\sum\nolimits_{a \in \mathcal{D}} \left| \int \mathsf{D}_\mathcal{M}(\{a\}|x) P_\theta(dx) - \int \mathsf{D}_\mathcal{N}(\{a\}|x) Q_\theta(dx) \right| \leq \varepsilon, \quad \theta \in \Delta.$$

Proof. $(A) \to (B)$ is a consequence of $\mathsf{r}(\Pi, \mathsf{D}_\mathcal{M}) = \sum_{\theta \in \Delta} \mathsf{R}(\theta, \mathsf{D}_\mathcal{M}) \Pi(\{\theta\})$ and a similar relation for $\mathsf{r}(\Pi, \mathsf{D}_\mathcal{N})$.

To show $(B) \to (C)$, let $\Delta = \{\theta_1, ..., \theta_m\}$ and $\mathcal{D} = \{a_1, ..., a_N\}$. We set for any prior Π with $\pi_i := \Pi(\{\theta_i\}) > 0$ for every $i = 1, ..., m$,

$$\begin{aligned} M(\mathsf{D}_{\mathcal{M}}, L) &:= (\mathsf{r}(\Pi, \mathsf{D}_{\mathcal{M}}) - \mathsf{r}(\Pi, \mathsf{D}_{\mathcal{N}})) - \varepsilon \qquad (4.8)\\ &= (\sum_{i=1}^{m} \sum_{j=1}^{N} [\pi_i \int L(\theta_i, a_j) \mathsf{D}_{\mathcal{M}}(\{a_j\}|x) P_{\theta_i}(dx) - \pi_i \gamma_{i,j}]) - \varepsilon, \\ \gamma_{i,j} &= \int L(\theta_i, a_j) \mathsf{D}_{\mathcal{N}}(\{a_j\}|y) Q_{\theta_i}(dy), \quad i = 1, ..., m, \ j = 1, ..., N. \end{aligned}$$

Denote by $\mathbb{D}$ the set of all decisions $\mathsf{D}_{\mathcal{M}}$ for the model $\mathcal{M}$. Set

$$\begin{aligned} \mathbb{K} &= \{(\int \mathsf{D}_{\mathcal{M}}(\{a_1\}|x) P_{\theta_1}(dx), ..., \int \mathsf{D}_{\mathcal{M}}(\{a_N\}|x) P_{\theta_m}(dx)) : \mathsf{D}_{\mathcal{M}} \in \mathbb{D}\}, \\ \mathbb{L} &= \mathsf{X}_{i=1}^{m} [-\pi_i, \pi_i]^N. \end{aligned}$$

The set $\mathbb{K}$ is convex and by the compactness theorem (see Theorem 3.21) is a compact subset of $\mathbb{R}^{mN}$, and $\mathbb{L}$ is a polytope. Put

$$\begin{aligned} y &= (\int \mathsf{D}_{\mathcal{M}}(\{a_j\}|x) P_{\theta_i}(dx))_{1 \le i \le m, 1 \le j \le N} \in \mathbb{K}, \\ z &= (\pi_i L(\theta_i, a_j))_{1 \le i \le m, 1 \le j \le N} \in \mathbb{L}, \end{aligned}$$

and $\Psi(y, z) = M(\mathsf{D}_{\mathcal{M}}, L)$. The function $\Psi(y, z)$ is continuous and satisfies the conditions in (4.6). Thus by Lemma 4.13 there is some decision $\mathsf{D}_{\mathcal{M}}^{(0)}$ with

$$\sup_{L \in \mathbb{L}} M(\mathsf{D}_{\mathcal{M}}^{(0)}, L) = \inf_{\mathsf{D}_{\mathcal{M}} \in \mathbb{D}} \sup_{L \in \mathbb{L}} M(\mathsf{D}_{\mathcal{M}}, L) = \sup_{L \in \mathbb{L}} \inf_{\mathsf{D}_{\mathcal{M}} \in \mathbb{D}} M(\mathsf{D}_{\mathcal{M}}, L) \le 0,$$

where the inequality on the right-hand side follows from condition (B) and the representation (4.8). Hence,

$$\begin{aligned} \sup_{\|L\|_u \le 1} & \sum_{i=1}^{m} \sum_{j=1}^{N} \pi_i L(\theta_i, a_j) \\ & \times [\int \mathsf{D}_{\mathcal{M}}^{(0)}(\{a_j\}|x) P_{\theta_i}(dx) - \int \mathsf{D}_{\mathcal{N}}(\{a_j\}|y) Q_{\theta_i}(dy)] \le \varepsilon. \end{aligned}$$

Now we fix $i_0 \in \{1, ..., m\}$ and a function $g : \mathcal{D} \to [-1, 1]$. We set $L(\theta_{i_0}, a_j) = g(a_j)$ and $L(\theta_i, a_j) = 0$ for $i \ne i_0$. Then by $\pi_{i_0} > 0$,

$$\sup_{\|g\|_u \le 1} \sum_{j=1}^{N} g(a_j) [\int \mathsf{D}_{\mathcal{M}}^{(0)}(\{a_j\}|x) P_{\theta_{i_0}}(dx) - \int \mathsf{D}_{\mathcal{N}}(\{a_j\}|y) Q_{\theta_{i_0}}(dy)] \le \frac{\varepsilon}{\pi_{i_0}}.$$

To get (C) we let $\pi_{i_0} \uparrow 1$ and set $g(a_j) = 1$ if

$$\int \mathsf{D}_{\mathcal{M}}^{(0)}(\{a_j\}|x) P_{\theta_{i_0}}(dx) \ge \int L(i_0, a_j) \mathsf{D}_{\mathcal{N}}(\{a_j\}|y) Q_{\theta_{i_0}}(dy).$$

and $g(a_j) = -1$ otherwise. To prove $(C) \to (A)$, for $i = 1, ..., m$ it holds

$$
\begin{aligned}
&|\mathsf{R}(\theta_i, \mathsf{D}_{\mathcal{M}}) - \mathsf{R}(\theta_i, \mathsf{D}_{\mathcal{N}})| \\
&\leq |\sum\nolimits_{j=1}^{N} L(\theta_i, a_j)[\int \mathsf{D}_{\mathcal{M}}(\{a_j\}|x)P_{\theta_i}(dx) - \int \mathsf{D}_{\mathcal{N}}(\{a_j\}|y)Q_{\theta_i}(dy)]\,| \\
&\leq \sum\nolimits_{j=1}^{N} |\int \mathsf{D}_{\mathcal{M}}(\{a_j\}|x)P_{\theta_i}(dx) - \int \mathsf{D}_{\mathcal{N}}(\{a_j\}|y)Q_{\theta_i}(dy)\,| \leq \varepsilon.
\end{aligned}
$$

■

A simple fact, with far-reaching consequences, is that in the comparison of two finite models we have to consider only the minimal Bayes risks, i.e., the Bayes risks of the two problems. For a fixed decision space $\mathcal{D}$ denote by $\mathbb{D}_{\mathcal{M}}$ and $\mathbb{D}_{\mathcal{N}}$ the set of all decisions for model $\mathcal{M}$ and $\mathcal{N}$, respectively.

Corollary 4.15. *Under the assumptions of Theorem 4.14, if the conditions $(A) - (C)$ are satisfied, the following holds. $\mathcal{M} \succeq^{\varepsilon} \mathcal{N}$ if and only if for every finite decision space $\mathcal{D}$, for every loss function L with $\|L\|_u \leq 1$, and for every prior Π*

$$\inf_{\mathsf{D}\in\mathbb{D}_{\mathcal{M}}} \mathsf{r}(\Pi, \mathsf{D}) \leq \inf_{\mathsf{D}\in\mathbb{D}_{\mathcal{N}}} \mathsf{r}(\Pi, \mathsf{D}) + \varepsilon. \tag{4.9}$$

Proof. As the models and the decision space are finite we know from Proposition 3.50 that the infima are attained at some decisions which are the Bayes decisions $\mathsf{D}_{\Pi,\mathcal{M}}$ and $\mathsf{D}_{\Pi,\mathcal{N}}$, respectively. If $\mathcal{M} \succeq^{\varepsilon} \mathcal{N}$, then by condition (B) in Theorem 4.14 for $\mathsf{D}_{\Pi,\mathcal{N}}$ there exists a $\mathsf{D}_{\mathcal{M}}$ with $\mathsf{r}(\Pi, \mathsf{D}_{\mathcal{M}}) \leq \mathsf{r}(\Pi, \mathsf{D}_{\Pi,\mathcal{N}}) + \varepsilon$. The inequality (4.9) follows from $\mathsf{r}(\Pi, \mathsf{D}_{\Pi,\mathcal{M}}) \leq \mathsf{r}(\Pi, \mathsf{D}_{\mathcal{M}})$. The proof of the opposite direction is similar. ■

In Corollary 4.10 we have seen that $\mathcal{M} \succeq^{\varepsilon} \mathcal{N}$ if $\mathcal{N}$ is a randomization of $\mathcal{M}$ "up to ε". The natural question arises whether $\mathcal{M} \succeq^{\varepsilon} \mathcal{N}$ implies (4.5). This means that we ask whether a model $\mathcal{M}$ is approximately as informative as $\mathcal{N}$ if and only if $\mathcal{N}$ is an approximate randomization of $\mathcal{M}$. This problem was studied by Blackwell (1951, 1953) for finite decision spaces. For general results we refer to Strasser (1985), LeCam (1986), and Torgersen (1991). We establish here the randomization criterion only for Borel spaces $(\mathcal{Y}, \mathfrak{B})$ and finite models. This is sufficient for our purposes. Some remarks on more general results in the literature are made after the following theorem which is a special case of Theorem 6.4.1 in Torgersen (1991).

Theorem 4.16. ***(Randomization Criterion)*** *Let $\mathcal{M} = (\mathcal{X}, \mathfrak{A}, (P_\theta)_{\theta\in\Delta})$ and $\mathcal{N} = (\mathcal{Y}, \mathfrak{B}, (Q_\theta)_{\theta\in\Delta})$ with $\Delta = \{\theta_1, ..., \theta_m\}$ be two finite models. If $(\mathcal{Y}, \mathfrak{B})$ is a Borel space, then for every fixed $\varepsilon \geq 0$ the model $\mathcal{M}$ is ε-deficient with respect to $\mathcal{N}$ if and only if there exists a kernel $\mathsf{K} : \mathfrak{B} \times \mathcal{X} \to_k [0, 1]$ such that $\|\mathsf{K}P_\theta - Q_\theta\| \leq \varepsilon$, $\theta \in \Delta$.*

Corollary 4.17. *Under the assumptions of the theorem it holds $\mathcal{N} \preceq \mathcal{M}$ if and only if $\mathcal{N}$ is a randomization of $\mathcal{M}$.*

Proof. One direction is clear from Corollary 4.10. To prove the opposite direction we note that by the definition of a Borel space and Problem 4.11

we may restrict ourselves to the case where $\mathcal{Y}$ is a compact metric space with metric ρ and $\mathfrak{B}$ is the σ-algebra of Borel sets. From the compactness of $\mathcal{Y}$ we get the existence of an increasing sequence of partitions $\mathfrak{p}_1 \subseteq \mathfrak{p}_2 \subseteq ...$, $\mathfrak{p}_n = \{A_{1,n}, ..., A_{N_n,n}\}$ where $A_{i,n}$ has a diameter not exceeding $1/n$. It holds for every continuous function φ and points $y_{i,n} \in A_{i,n}$,

$$\sup_{y\in\mathcal{Y}} |\varphi(y) - \sum\nolimits_{i=1}^{N_n} I_{A_{i,n}}(y)\varphi(y_{i,n})| \le \omega_{1/n}(\varphi),$$

where $\omega_\delta(\varphi) = \sup_{\rho(s,t)\le\delta} |\varphi(s) - \varphi(t)|$ is the modulus of continuity of φ. Introduce the finite decision space $\mathcal{D}_n$ by $\mathcal{D}_n = \{y_{1,n}, ..., y_{N_n,n}\}$ and set $\mathsf{D}_n(\cdot|y) = \boldsymbol{\delta}_{T(y)}(\cdot)$, where $T(y) = y_{j,n}$, $y \in A_{j,n}$. Let $\mathbb{D}$ be the set of all stochastic kernels $\mathsf{L}_n : \mathfrak{P}(\mathcal{D}_n) \times \mathcal{X} \to_k [0,1]$. By (C) in Theorem 4.14 we find a decision L_n^0 such that

$$\max_{\theta\in\Delta} \sum\nolimits_{j=1}^{N_n} |\int \mathsf{L}_n^0(\{y_{j,n}\}|x)P_\theta(dx) - \int \mathsf{D}_n(\{y_{j,n}\}|y)Q_\theta(dy)| \le \varepsilon.$$

Introduce the kernel $\mathsf{K}_n^0 : \mathfrak{B} \times \mathcal{X} \to_k [0,1]$ by

$$\mathsf{K}_n^0(B|x) = \sum\nolimits_{j=1}^{N_n} \mathsf{L}_n^0(\{y_{j,n}\}|x)\boldsymbol{\delta}_{y_{j,n}}(B).$$

The relations

$$\int \varphi(y)(\mathsf{K}_n^0 P_\theta)(dy) = \sum\nolimits_{i=1}^{N_n} \varphi(y_{i,n}) \int \mathsf{L}_n^0(\{y_{i,n}\}|x)P_\theta(dx),$$

$$\left|\sum\nolimits_{j=1}^{N} \int \varphi(y)\mathsf{D}_n(\{y_{j,n}\}|y)Q_\theta(dy) - \int \varphi(y)Q_\theta(dy)\right| \le \omega_{1/n}(\varphi),$$

yield for a continuous function φ with values in $[-1,1]$,

$$|\int \varphi(y)(\mathsf{K}_n^0 P_\theta)(dy) - \int \varphi(y)Q_\theta(dy)| \le \varepsilon + \omega_{1/n}(\varphi), \quad \theta \in \Delta.$$

In view of the compactness theorem (see Theorem 3.21) and the fact that Δ is finite there is a subsequence n_m and a kernel $\mathsf{K} : \mathfrak{B} \times \mathcal{X} \to_k [0,1]$ such that

$$\lim_{m\to\infty} \int \varphi(y)(\mathsf{K}_{n_m}^0 P_\theta)(dy) = \int \varphi(y)(\mathsf{K}P_\theta)(dy), \quad \theta \in \Delta,$$

for every continuous function φ. Hence $|\int \varphi(y)(\mathsf{K}P_\theta)(dy) - \int \varphi(y)Q_\theta(dy)| \le \varepsilon$. The supremum over all continuous functions φ with $|\varphi(x)| \le 1$ provides the variational distance; see Problem 1.80. This completes the proof. The corollary follows for $\varepsilon = 0$. ■

We conclude this section with some remarks on the concept of deficiency and the randomization theorem. Our definition of deficiency in Definition 4.9 is taken from Torgersen (1991), Section 6.2. Other authors have relaxed the assumption of a finite decision space by assuming only that it is a topological space, and they have assumed only that the loss function is continuous; see Strasser (1985) and LeCam (1986). But this approach is equivalent to that given in Definition 4.9; see Corollary 6.4.4 in Torgersen (1991). We have confined ourselves here to Borel spaces and to a finite parameter set in order to deal with kernels and make use of the compactness theorem which provides the sequential compactness of decisions. By using topological arguments that are more far-reaching one may drop the assumption that the parameter set is finite. However, one would then have to deal with randomizations in a more general sense that requires additional topological tools. For details we refer to LeCam (1986), Strasser (1985), and Torgersen (1991).

4.2 Comparison of Finite Models by Standard Distributions

It is clear that models with the same sample space are easier to compare than models with different sample spaces. Therefore we investigate how one can switch from a given model to another equivalent model for which the sample space is universal, that is, common to all models. First we formulate a sufficient condition that guarantees that a reduced model is equivalent to the original model.

To find conditions on a statistic T that lead to an equivalent model we start with the decision-theoretic framework that was introduced in Chapter 3; see (3.1). Given a model $(\mathcal{X}, \mathfrak{A}, (P_\theta)_{\theta\in\Delta})$, a decision space $(\mathcal{D}, \mathfrak{D})$, and a decision $\mathsf{D} : \mathfrak{D} \times \mathcal{X} \to_k [0,1]$, we assume that (A, X) is a random vector that is defined on the probability space, say $(\Omega, \mathfrak{F}, \mathbb{P}_\theta)$, $\theta \in \Delta$, such that $\mathcal{L}(X|\mathbb{P}_\theta) = P_\theta$ and $\mathsf{D}(\cdot|x)$ is the conditional distribution of the decision A given $X = x$. This means that

$$\begin{aligned} \mathcal{L}((A, X)|\mathbb{P}_\theta) &= \mathsf{D} \otimes P_\theta, \quad \text{where} \\ (\mathsf{D} \otimes P_\theta)(C) &= \int [\int I_C(a, x)\mathsf{D}(da|x)]P_\theta(dx), \quad C \in \mathfrak{D} \otimes \mathfrak{A}. \end{aligned}$$

Suppose now that we have another measurable space $(\mathcal{Y}, \mathfrak{B})$ and a statistic $T : \mathcal{X} \to_m \mathcal{Y}$. If $\mathsf{C} : \mathfrak{D} \times \mathcal{Y} \to_k [0,1]$ is a decision for the model $(\mathcal{Y}, \mathfrak{B}, (P_\theta \circ T^{-1})_{\theta\in\Delta})$, then we call

$$\mathsf{D}(\cdot|x) = \mathsf{C}(\cdot|T(x)), \quad x \in \mathcal{X},$$

a decision *factorized by the statistic* T and write in short $\mathsf{D} = \mathsf{C} \circ T$. Such decisions depend on the observations only through the statistic T. The next theorem provides a sufficient criterion for the existence of factorized decisions. Later on, when we deal with the concept of sufficiency, it is shown that this condition is also necessary.

Theorem 4.18. *Suppose that the model $\mathcal{M} = (\mathcal{X}, \mathfrak{A}, (P_\theta)_{\theta\in\Delta})$ is dominated and the decision space $(\mathcal{D}, \mathfrak{D})$ is a Borel space. Let $(\mathcal{Y}, \mathfrak{B})$ be a measurable space and $T : \mathcal{X} \to_m \mathcal{Y}$. If there exists a dominating probability measure Q such that the densities dP_θ/dQ are $\sigma(T)$-measurable, then for every decision D there exists a decision factorized by T, say $\mathsf{C} \circ T$, such that for any loss function $L(\theta, \cdot) : \mathcal{D} \to_m \mathbb{R}_+$,*

$$\mathsf{R}(\theta, \mathsf{D}) = \mathsf{R}(\theta, \mathsf{C} \circ T), \quad \theta \in \Delta. \tag{4.10}$$

In particular, for $\mathcal{N} = (\mathcal{Y}, \mathfrak{B}, (P_\theta \circ T^{-1})_{\theta\in\Delta})$ it holds that $\mathcal{M} \sim \mathcal{N}$.

Proof. Let (A, X) be a random vector, defined on $(\Omega, \mathfrak{F}, \mathbb{P})$, such that $\mathcal{L}((A, X)|\mathbb{P}) = \mathsf{D} \otimes Q$. As $(\mathcal{D}, \mathfrak{D})$ is a Borel space there exists a stochastic kernel $\mathsf{C} : \mathfrak{D} \times \mathcal{Y} \to_k [0, 1]$ that is the conditional distribution of A given $T = y$, i.e., $\mathcal{L}((A, T(X))|\mathbb{P}) = \mathsf{C} \otimes (Q \circ T^{-1})$. For every $h : \mathcal{D} \times \mathcal{Y} \to \mathbb{R}_+$ it holds that

$$\begin{aligned}\int [\int h(a, T(x))\mathsf{D}(da|x)]Q(dx) &= \mathbb{E}h(A, T(X)) \\ &= \int [\int h(a, T(x))\mathsf{C}(da|T(x))]Q(dx).\end{aligned}$$

As dP_θ/dQ is $\sigma(T)$-measurable it holds $(dP_\theta/dQ)(x) = g_\theta(T(x))$ for some $g_\theta : \mathcal{Y} \to_m \mathbb{R}_+$. Hence with $h(a, T(x)) = L(\theta, a)g_\theta(T(x))$ we get

$$\begin{aligned}\mathsf{R}(\theta, \mathsf{C} \circ T) &= \int [\int L(\theta, a)\mathsf{C}(da|T(x))]g_\theta(T(x))Q(dx) \\ &= \int [\int L(\theta, a)\mathsf{D}(da|x)]g_\theta(T(x))Q(dx) = \mathsf{R}(\theta, \mathsf{D}),\end{aligned}$$

which proves the first statement. To prove the second statement, we remark that $\mathcal{M} \succeq \mathcal{N}$ follows from Corollary 4.10. The converse statement $\mathcal{N} \succeq \mathcal{M}$ follows from (4.10) as each finite decision space $\mathcal{D}$ is a Borel space. ■

The above decision D is the conditional distribution of A given X. The kernel C that is used to factorize decision D is nothing else than the conditional distribution of A given T. The crucial point has been to find a version of the conditional distribution that is independent of the parameter.

Problem 4.19.* Under the assumptions of Theorem 4.18, C is a conditional distribution of A given T that is independent of the parameter, which is equivalent to

$$\int [\int h(a, t)\mathsf{C}(da|t)](P_\theta \circ T^{-1})(dt) = \int [\int h(a, T(x))\mathsf{D}(da|x)]P_\theta(dx)$$

for every $h : \mathcal{D} \times \mathcal{Y} \to_m \mathbb{R}_+$ and $\theta \in \Delta$.

Now we consider a special situation where the conditions of the previous theorem are satisfied. Every finite model (i.e., every model that consists only of a finite number of different distributions) is dominated. For such a model we can use the arithmetic mean as the dominating distribution. Let

$$\begin{aligned}&\mathcal{M}=(\mathcal{X},\mathfrak{A},\{P_1,...,P_m\}),\\&\overline{P}=\tfrac{1}{m}\textstyle\sum_{i=1}^m P_i,\quad M=(M_1,...,M_m)=(\tfrac{dP_1}{d\overline{P}},...,\tfrac{dP_m}{d\overline{P}}).\end{aligned}\tag{4.11}$$

We recall $\mathcal{S}_m$ to be the simplex

$$\mathcal{S}_m=\{x:x=(x_1,...,x_m),\ x_i\geq 0,\ \sum\nolimits_{i=1}^m x_i=m\},\tag{4.12}$$

and $\mathfrak{S}_m$ the Borel σ-algebra of subsets of $\mathcal{S}_m$. By construction the vector M from (4.11) is a measurable mapping $M:\mathcal{X}\to_m\mathcal{S}_m$. We call

$$\mu:=\overline{P}\circ M^{-1},\tag{4.13}$$

the *standard distribution of the model* $\mathcal{M}$. Denote by $Z_i(x)=x_i$ the projection of $\mathcal{S}_m$ onto the ith coordinate. It holds for any $h:\mathcal{S}_m\to_m\mathbb{R}_+$,

$$\begin{aligned}&\int h(x)(P_i\circ M^{-1})(dx)=\int h(M(x))P_i(dx)=\int h(M(x))\frac{dP_i}{d\overline{P}}(x)\overline{P}(dx)\\&=\int h(M(x))Z_i(M(x))\overline{P}(dx)=\int hZ_i d\mu,\quad\text{so that}\\&\frac{dQ_i}{d\mu}=Z_i,\quad\text{where}\quad Q_i:=P_i\circ M^{-1},\quad i=1,...,m.\end{aligned}\tag{4.14}$$

Definition 4.20. *A distribution μ on $(\mathcal{S}_m,\mathfrak{S}_m)$ is called a standard distribution if $\int Z_i d\mu=1$, $i=1,...,m$, where $Z_i(x)=x_i$ is the projection of $\mathcal{S}_m$ onto the ith coordinate. The model*

$$\mathcal{N}=(\mathcal{S}_m,\mathfrak{S}_m,\{Q_1,...,Q_m\}),\quad\textit{with}\ dQ_i=Z_i d\mu,\tag{4.15}$$

is called a standard model. If μ is defined by (4.13), then we call $\mathcal{N}$ the standard model of $\mathcal{M}$.

Remark 4.21. The finite model $(\mathcal{X},\mathfrak{A},\{P_1,...,P_m\})$ can also be dominated by the finite measure $\overline{\mu}=\sum_{i=1}^m P_i$ which differs from $\overline{P}$ only by the factor m. Then the vector $(dP_1/d\overline{\mu},,...,dP_m/d\overline{\mu})$ takes on values in the unit simplex, and instead of dealing with a standard distribution one deals with a standard measure with total mass m.

An essential step toward the comparison of models is the next proposition. Its statement is an immediate consequence of Theorem 4.18 for $T=M$.

Proposition 4.22. *The models $\mathcal{M}$ in (4.11) and $\mathcal{N}$ in (4.15) are equivalent in the sense of Definition 4.9.*

We recall the Hellinger transform; see Definition 1.87. Using the notations in (4.11) we have

$$\begin{aligned} \mathsf{H}_s(P_1,...,P_m) &= \int M_1^{s_1}\cdots M_m^{s_m}\, d\overline{P} \\ &= \int t_1^{s_1}\cdots t_m^{s_m}\,\mu(dt_1,...dt_m), \quad s\in \mathbf{S}_m^o, \qquad (4.16)\\ \mathbf{S}_m^o &= \{s : s=(s_1,..,s_m), s_i>0, \sum\nolimits_{i=1}^m s_i = 1\}, \end{aligned}$$

where $\mu = \mathcal{L}((M_1,...,M_m)|\overline{P})$ is the standard distribution, which is defined on the σ-algebra $\mathfrak{S}_m$ of Borel sets in $\mathcal{S}_m$ from (4.12). $\mathsf{H}_s(P_1,...,P_m)$ may also be viewed as a transformation of the standard distribution μ. If $m=2$, then by (1.77) $\mathsf{H}_{(s_1,s_2)}(P_1,P_2)$ is related to $\mathsf{H}_s(P_1,P_2)$ in (1.105) by

$$\mathsf{H}_s(P_1,P_2) = \mathsf{H}_{(s_1,s_2)}(P_1,P_2), \quad s_1 = s,\ s_2 = 1-s.$$

If the model is not homogeneous (i.e., if not all P_i and P_j are mutually absolutely continuous) then there is the difficulty that $\mathsf{H}_s(P_1,...,P_m)$, for every $s\in\mathbf{S}_m^o$, does not provide the Hellinger transforms for submodels, say for the model $\{P_1,...,P_{m-1}\}$. Indeed, if, for example, P_m is mutually singular to $P_1,...,P_{m-1}$, then $\mathsf{H}_s(P_1,...,P_m)=0$ for every $s\in\mathbf{S}_m^o$. Nevertheless, $\mathsf{H}_s(P_1,...,P_m)$, for every $s\in\mathbf{S}_m^o$, determines uniquely the amount of μ on

$$\mathcal{S}_m^o = \{(x_1,...,x_m) : x_i>0,\ \sum\nolimits_{i=1}^m x_i = m\}.$$

Lemma 4.23. *For two models $(\mathcal{X}_i,\mathfrak{A}_i,\{P_{i,1},...,P_{i,m}\})$, $i=1,2$, it holds*

$$\mathsf{H}_s(P_{1,1},...,P_{1,m}) = \mathsf{H}_s(P_{2,1},...,P_{2,m}), \quad s\in\mathbf{S}_m^o,$$

if and only if

$$\mu_1(B)=\mu_2(B), \quad B\in\mathfrak{S}_m^o,$$

where $\mathfrak{S}_m^o = \{B : B\in\mathfrak{S}_m,\ B\subseteq\mathcal{S}_m^o\}$.

Proof. The fact that $\mu_1=\mu_2$ on $\mathfrak{S}_m^o$ implies the equality of the Hellinger transforms follows from (4.16). To prove the opposite statement we denote by $Z_1,...,Z_m$ the projections of $\mathcal{S}_m^o$ on the coordinates. Set $\widetilde{\mu}_i(B)=\mu_i(B\cap\mathcal{S}_m^o)$. Then (4.16) yields

$$\int \exp\{\sum\nolimits_{i=2}^m s_i\ln(Z_i/Z_1)\}Z_1 d\widetilde{\mu}_1 = \int \exp\{\sum\nolimits_{i=2}^m s_i\ln(Z_i/Z_1)\}Z_1 d\widetilde{\mu}_2 \qquad (4.17)$$

for every $s_2,...,s_m\in(0,1)$ with $\sum_{i=2}^m s_i<1$. Set

$$d\nu_i = Z_1 d\widetilde{\mu}_i \quad\text{and}\quad T=(\ln(Z_2/Z_1),...,\ln(Z_m/Z_1)).$$

As the set of all vectors $(s_2,...,s_m)$ that satisfy (4.17) contains an open rectangle $\mathsf{X}_{j=2}^m(a_j,b_j)$, we get from Proposition 1.25 $\nu_1\circ T^{-1}=\nu_2\circ T^{-1}$. The mapping $T : \mathcal{S}_m^o \to \mathbb{R}^{m-1}$ is one-to-one, and obviously both T and the inverse mapping are measurable. Hence $\nu_1 \circ T^{-1} = \nu_2 \circ T^{-1}$ implies $\nu_1 = \nu_2$ or $Z_1 d\widetilde{\mu}_1 = Z_1 d\widetilde{\mu}_2$. Similar considerations show that $Z_i d\widetilde{\mu}_1 = Z_i d\widetilde{\mu}_2$, $i = 2, ..., m$. Taking the sum on both sides the relation $\sum_{k=1}^m Z_k = m$ yields $\widetilde{\mu}_1 = \widetilde{\mu}_2$. ■

If the model $\mathcal{M}$ is homogeneous, then the standard distribution is concentrated on $\mathcal{S}_m^o$ so that in the class of homogeneous models the Hellinger transform determines the standard distribution uniquely. To deal with the general case there are different possibilities. One way is to extend H_s to $s \in \mathbf{S}_m^c$, where $\mathbf{S}_m^c \supset \mathbf{S}_m^o$ is the unit simplex where some of the coordinates may vanish; see (1.13). Using the convention $0^0 = 1$ all Hellinger transforms of submodels can be concluded from $\mathsf{H}_s(P_1, ..., P_m)$. A technical disadvantage is that additional formulations are necessary to distinguish between the cases of $s \in \mathbf{S}_m^o$ and $s \in \mathbf{S}_m^c \backslash \mathbf{S}_m^o$. For our purposes the way of smoothing the model seems to be more appropriate. More precisely, given $\mathcal{M} = (\mathcal{X}, \mathfrak{A}, \{P_1, ..., P_m\})$, we set

$$\begin{aligned} \mathcal{M}_\alpha &= (\mathcal{X}, \mathfrak{A}, \{P_{1,\alpha}, ..., P_{m,\alpha}\}), \quad 0 \le \alpha \le 1, \\ P_{i,\alpha} &= (1-\alpha)P_i + \alpha \overline{P}, \quad 0 \le \alpha \le 1, \ i = 1, ..., m. \end{aligned} \tag{4.18}$$

Obviously$\mathcal{M} = \mathcal{M}_0$. But $\mathcal{M}_\alpha$ is homogeneous for $0 < \alpha \le 1$ and $\|P_i - P_{i,\alpha}\| \le 2\alpha$, so that $\mathcal{M}$ can be approximated by homogeneous models $\mathcal{M}_\alpha$ in the sense of the variational distance. Moreover,

$$\begin{aligned} \overline{P}_\alpha &= \frac{1}{m}\sum\nolimits_{i=1}^m P_{i,\alpha} = \overline{P} \\ M_{i,\alpha} &= \frac{dP_{i,\alpha}}{d\overline{P}_\alpha} = \frac{dP_{i,\alpha}}{d\overline{P}} = (1-\alpha)M_i + \alpha. \end{aligned}$$

Denote by $\mu_\alpha = \mathcal{L}((M_{1,\alpha}, ..., M_{m,\alpha})|\overline{P})$ the standard distribution of $\mathcal{M}_\alpha$. Let $\varphi : \mathcal{S}_m \to \mathbb{R}$ be a continuous function on $\mathcal{S}_m$ which is bounded due to the compactness of $\mathcal{S}_m$. Then

$$\begin{aligned} &\int \varphi(u_1, ..., u_m)\mu_\alpha(du_1, ..., du_m) \\ &= \int \varphi((1-\alpha)u_1 + \alpha, ..., (1-\alpha)u_m + \alpha)\mu(du_1, ..., du_m). \end{aligned}$$

For two models $\mathcal{M}_i$, $i = 1, 2$, with the parameter set $\{1, ..., m\}$ we denote by μ_i and $\mu_{i,\alpha}$ the standard distributions of the models $\mathcal{M}_i$ and $\mathcal{M}_{i,\alpha}$, respectively. Taking the limit $\alpha \downarrow 0$ in the above equality we get from the fact that φ was any continuous function the following statement.

$$\begin{aligned} &\mu_1 = \mu_2 \text{ implies } \mu_{1,\alpha} = \mu_{2,\alpha} \text{ for every } 0 \le \alpha \le 1, \\ &\mu_{1,\alpha} = \mu_{2,\alpha} \text{ for every } 0 < \alpha < 1 \text{ implies } \mu_1 = \mu_2. \end{aligned} \tag{4.19}$$

Problem 4.24.* If $\mathcal{M}_i$, $i = 1, 2$, are models with the parameter set $\{1, ..., m\}$, and the $\mathcal{M}_{i,\alpha}, i = 1, 2$ are defined as in (4.11), then $\mathcal{M}_1 \sim \mathcal{M}_2$ implies $\mathcal{M}_{1,\alpha} \sim \mathcal{M}_{2,\alpha}$ for $0 \le \alpha \le 1$.

Because for any finite model the standard distribution is defined on the simplex $\mathcal{S}_m$, no matter what the original sample spaces may have been, we now have the opportunity to compare models with possibly different sample spaces. Moreover, if the distributions in the model were given by means of densities, then by turning to $dP_i/d\overline{P}$ the new statistic $M = (M_1, ..., M_m)$ is independent of the original dominating measure. The choice of $\overline{P}$ as dominating measure is some type of self-normalization.

Theorem 4.25. *Suppose that $\mathcal{M}_i = (\mathcal{X}_i, \mathfrak{A}_i, \{P_{i,1}, ..., P_{i,m}\})$ is a finite model with the standard distribution μ_i, $i = 1, 2$. Then the following conditions are equivalent.*

(A) $\mathcal{M}_1 \sim \mathcal{M}_2$.
(B) $\mu_1 = \mu_2$.
(C) $\mathsf{H}_s(P_{1,1,\alpha}, ..., P_{1,m,\alpha}) = \mathsf{H}_s(P_{2,1,\alpha}, ..., P_{2,m,\alpha}), \quad s \in \mathbf{S}_m^o, \ 0 \le \alpha \le 1.$

Corollary 4.26. *If one of the models $\mathcal{M}_1$ or $\mathcal{M}_2$ is homogeneous, then condition (C) can be replaced by the weaker condition*

(D) $\mathsf{H}_s(P_{1,1}, ..., P_{1,m}) = \mathsf{H}_s(P_{2,1}, ..., P_{2,m}), \quad s \in \mathbf{S}_m^o.$

Corollary 4.27. *For $m = 2$ the equivalent conditions (A) and (B) hold if and only if*

(E) $\mathsf{H}_s(P_{1,1}, P_{1,2}) = \mathsf{H}_s(P_{2,1}, P_{2,2}), \quad 0 < s < 1.$

Proof. (A) implies $\mathcal{M}_{1,\alpha} \sim \mathcal{M}_{2,\alpha}$ for every $0 \le \alpha \le 1$ in view of Problem 4.24. Proposition 4.22 yields that the corresponding standard models $\mathcal{N}_{i,\alpha}$ are equivalent. As the simplex $\mathcal{S}_m$ is a compact metric space we get from the randomization criterion (see Theorem 4.16) that $\mathcal{N}_{1,\alpha}$ and $\mathcal{N}_{2,\alpha}$ are mutual randomizations. A twofold application of Proposition 1.93 implies (C) for every $0 \le \alpha \le 1$. Assume now that (C) holds. The models $\mathcal{M}_{i,\alpha}$ are homogeneous for $0 < \alpha < 1$. Hence the $\mu_{i,\alpha}$ are concentrated on $\mathcal{S}_m^o$ and we get from Lemma 4.23 that $\mu_{1,\alpha} = \mu_{2,\alpha}$. Thus (4.19) provides (B). If (B) holds, then $\mathcal{N}_1 = \mathcal{N}_2$, and thus $\mathcal{M}_1 \sim \mathcal{M}_2$ by Proposition 4.22.

To prove the first corollary, we have only to note that (D) implies that μ_1 and μ_2 are identical for the Borel subsets of $\mathcal{S}_m^o$ and therefore identical on $\mathfrak{S}_m$, as at least one of the two probability measures is concentrated on $\mathcal{S}_m^o$. To prove the second corollary, we note that

$$\mathcal{S}_2 = \{(x_1, x_2) : x_i \ge 0, x_1 + x_2 = 2\},$$

and Lemma 4.23 shows that $\mu_1(B) = \mu_2(B)$ for every Borel set B with

$$B \subseteq \mathcal{S}_2^o = \{(x_1, x_2) : x_i > 0, \ x_1 + x_2 = 2\}.$$

Thus it remains to show that $\mu_1(\{(2,0)\}) = \mu_2(\{(2,0)\})$ and $\mu_1(\{(0,2)\}) = \mu_2(\{(0,2)\})$. But this follows from $\int t_j \mu_i(dt_1, dt_2) = 1$,

$$\begin{aligned}\lim_{s\downarrow 0}\mathsf{H}_s(P_{i,1},P_{i,2}) &= \lim_{s\downarrow 0}\int t_1^s t_2^{1-s}\mu_i(dt_1,dt_2)\\ &= \int I_{(0,2]}(t_1)t_2\mu_i(dt_1,dt_2) = 1-2\mu_i(\{(0,2)\}),\end{aligned}$$

and by a similar consideration for $s\uparrow 1$. ■

We consider the situation in Example 1.88. Let $(P_{i,\theta})_{\theta\in\Delta}$ be a natural exponential family on $(\mathcal{X}_i,\mathfrak{A}_i)$ with $\boldsymbol{\mu}_i$-density $f_{i,\theta}(x)=\exp\{\langle\theta,T_i(x)\rangle-K_i(\theta)\}$, $i=1,2$, where the parameter θ belongs to the same parameter set $\Delta\subseteq\mathbb{R}^d$. Then we know from Example 1.120 that for $\theta\in\Delta^0$ the Fisher information matrices for the models

$$\mathcal{M}_i=(\mathcal{X}_i,\mathfrak{A}_i,(P_{i,\theta})_{\theta\in\Delta^0}),\quad \frac{dP_{i,\theta}}{d\boldsymbol{\mu}_i}=\exp\{\langle\theta,T_i\rangle-K_i(\theta)\},\ i=1,2,\tag{4.20}$$

are given by

$$\mathsf{I}_i(\theta)=\nabla\nabla^T K_i(\theta),\quad \theta\in\Delta^0,\ \ i=1,2.\tag{4.21}$$

Proposition 4.28. *For the exponential family models in (4.20) the following statements are equivalent.*

(A) $\mathcal{M}_1\sim\mathcal{M}_2$.
(B) $(\mathcal{X}_1,\mathfrak{A}_1,\{P_{1,\theta_1},P_{1,\theta_2}\})\sim(\mathcal{X}_2,\mathfrak{A}_2,\{P_{2,\theta_1},P_{2,\theta_2}\}),\ \ \theta_1,\theta_2\in\Delta^0$.
(C) $\mathsf{I}_1(\theta)=\mathsf{I}_2(\theta),\quad \theta\in\Delta^0$.

Proof. According to Definition 4.9 condition (A) is equivalent to the statement that for every finite subset $\digamma=\{\theta_1,...,\theta_m\}\subseteq\Delta$ the finite models $\mathcal{M}_{1,\digamma}$ and $\mathcal{M}_{2,\digamma}$ are equivalent. As the exponential families provide homogeneous models we get from condition (D) in Corollary 4.26 and from Example 1.88 the equivalent condition

$$\sum\nolimits_{j=1}^m s_jK_1(\theta_j)-K_1(\sum\nolimits_{j=1}^m s_j\theta_j)=\sum\nolimits_{j=1}^m s_jK_2(\theta_j)-K_2(\sum\nolimits_{j=1}^m s_j\theta_j)\tag{4.22}$$

for every $(s_1,...,s_m)\in\mathbf{S}_m^o$ and every finite subset $\digamma=\{\theta_1,...,\theta_m\}\subseteq\Delta$. Using mathematical induction it is easy to see that (4.22) is equivalent to the requirement that (4.22) holds for $m=2$, i.e., for $0<s<1$,

$$\begin{aligned}&sK_1(\theta_1)+(1-s)K_1(\theta_2)-K_1(s\theta_1+(1-s)\theta_2)\\ &=sK_2(\theta_1)+(1-s)K_2(\theta_2)-K_2(s\theta_1+(1-s)\theta_2),\quad \theta_1,\theta_2\in\Delta^0.\end{aligned}\tag{4.23}$$

But this is, according to condition (A) in Theorem 4.25 and Corollary 4.26, equivalent to condition (B).

It remains to show that condition (C) is equivalent to (4.23). To this end we fix $\theta_2\in\Delta^0$ and consider the function

$$\begin{aligned}G(\theta_1)=&\ s(K_1(\theta_1)-K_2(\theta_1))+(1-s)(K_1(\theta_2)-K_2(\theta_2))\\ &-K_1(s\theta_1+(1-s)\theta_2)+K_2(s\theta_1+(1-s)\theta_2).\end{aligned}$$

Then

$$G(\theta_2) = 0 \quad \text{and} \quad \nabla G(\theta_1)|_{\theta_1=\theta_2} = 0,$$
$$\nabla\nabla^T G(\theta_1)|_{\theta_1=\theta_2} = s(1-s)[\nabla\nabla^T K_1(\theta_2) - \nabla\nabla^T K_2(\theta_2)].$$

From here we see that the condition (4.23) implies that for every $\theta_2 \in \Delta^0$ condition C) is satisfied in view of (4.21). Conversely, if (4.21) and thus $\nabla\nabla^T K_1(\theta_2) = \nabla\nabla^T K_2(\theta_2)$ holds for every $\theta_2 \in \Delta^0$, then a Taylor expansion of $G(\theta_1)$ at θ_2 shows that $G(\theta_1) = 0$. As $\theta_1 \in \Delta^0$ was arbitrary we get (4.23). ■

The importance of the standard distribution is also reflected by its relation to the minimal Bayes risk for a finite model. More precisely, consider the finite model in (4.11) and suppose that $\mathcal{D}$ is a compact metric space that is equipped with the Borel sets $\mathfrak{D}$. Let Π be a prior on $\Delta = \{1, ..., m\}$ and set $\pi_i = \Pi(\{i\})$, $i = 1, ..., m$. Then for any loss function $L(\theta, \cdot) : \mathcal{D} \to_m \mathbb{R}_+$ and any decision D the Bayes risk is given by

$$\mathsf{r}(\Pi, \mathsf{D}) = \sum\nolimits_{i=1}^{m} \int L(i, a)\mathsf{D}(da|x)P_i(dx)\pi_i.$$

We know from Theorem 4.18 that we only have to consider decisions that are factorized by the statistic M in (4.11). Hence for some $\mathsf{C} : \mathfrak{D} \times \mathcal{S}_m \to_k [0, 1]$ it holds

$$\mathsf{r}(\Pi, \mathsf{D}) = \sum\nolimits_{i=1}^{m} \int \pi_i L(i, a)\mathsf{C}(da|t)t_i\mu(dt),$$

where we have utilized the fact that by (4.14),

$$\frac{d(P_i \circ M^{-1})}{d(\overline{P} \circ M^{-1})}(t) = Z_i(t) = t_i, \quad t \in \mathcal{S}_m.$$

We consider the vector $(\int \pi_1 L(1, a)\mathsf{C}(da|t), ..., \int \pi_m L(m, a)\mathsf{C}(da|t))$. If C runs through all possible kernels and t takes on all possible points in the simplex $\mathcal{S}_m$, then we get a set of vectors which is obviously given by

$$D_{L,\Pi} = \{(\int \pi_1 L(1, a)P(da), ..., \int \pi_m L(m, a)P(da)) : P \in \mathcal{P}(\mathfrak{A})\}. \quad (4.24)$$

Problem 4.29. If $\mathcal{D}$ is a compact metric space and $L(i, a)$ is continuous in a for every $i = 1, ..., m$, then $D_{L,\Pi}$ is a compact and convex subset of $\mathbb{R}^m$.

If $\mathsf{c} : \mathcal{S}_m \to_m D_{L,\Pi}$ denotes the measurable mapping

$$\mathsf{c}(t) = (\int \pi_1 L(1, a)\mathsf{C}(da|t), ..., \int \pi_m L(m, a)\mathsf{C}(da|t)),$$

then we get the following representation of the Bayes risk.

$$\mathsf{r}(\Pi, \mathsf{D}) = \int \langle \mathsf{c}(t), t\rangle \, \mu(dt). \quad (4.25)$$

Therefore the problem of finding a Bayes decision is reduced to the minimization of $\mathsf{c} \mapsto \int \langle \mathsf{c}(t), t \rangle \, \mu(dt)$, where the minimum is taken over all measurable mappings $\mathsf{c} : \mathcal{S}_m \to_m D_{L,\Pi}$. No matter how the convex and compact set D has been defined, the minimization of $\mathsf{c} \mapsto \int \langle \mathsf{c}(t), t \rangle \, \mu(dt)$ over $\mathcal{C} := \{\mathsf{c} : \mathsf{c} : \mathcal{S}_m \to_m D\}$ is considered to be a *standard decision problem.* To solve the minimization problem we call

$$\psi_D(t) := \inf_{y \in D} \langle y, t \rangle \,, \quad t \in \mathbb{R}_+^m, \tag{4.26}$$

the *lower envelope function* of a convex set D. Subsequently we use, instead of the Euclidean metric, the maximum norm which is defined by $\|x\|_u = \max_{1 \le i \le m} |x_i|$.

Problem 4.30.* If D is a convex and compact set, then $\psi_D(t) = \inf_{y \in D} \langle y, t \rangle$ is a concave and Lipschitz-continuous function with a Lipschitz constant not exceeding $m \sup_{y \in D} \|y\|_u$, i.e., it holds

$$|\psi_D(t_1) - \psi_D(t_2)| \le m \left(\sup_{y \in D} \|y\|_u \right) \|t_1 - t_2\|_u \,. \tag{4.27}$$

We recall from (A.6) that the Dudley metric $\|\mu_1 - \mu_2\|_{\mathsf{D}}$ of two distributions on a metric space $(\mathcal{S}, \rho_{\mathcal{S}})$ is defined by

$$\|\mu_1 - \mu_2\|_{\mathsf{D}} = \sup_{\varphi} | \int \varphi d\mu_1 - \int \varphi d\mu_2 |,$$

where the supremum is taken over all functions φ with $|\varphi(s)| \le 1$ and $|\varphi(t) - \varphi(s)| \le \rho_{\mathcal{S}}(s,t)$, $t, s \in \mathcal{S}$. If $\mathcal{S} = \mathcal{S}_m$, then we use the metric $\rho_{\mathcal{S}_m}(s,t) = \|s - t\|_u = \max_{1 \le i \le m} |s_i - t_i|$.

Theorem 4.31. *Let $\mathcal{M} = (\mathcal{X}, \mathfrak{A}, \{P_1, ..., P_m\})$ be a finite model with the standard distribution μ on $(\mathcal{S}_m, \mathfrak{S}_m)$ given by (4.13). For every compact and convex set $D \subseteq \mathbb{R}^m$ there exists a decision $\mathsf{c}_D : \mathcal{S}_m \to_m D$ for the standard decision problem which minimizes the risk, i.e.,*

$$\int \langle \mathsf{c}_D(t), t \rangle \, \mu(dt) = \inf_{\mathsf{c} \in \mathcal{C}} \int \langle \mathsf{c}(t), t \rangle \, \mu(dt). \tag{4.28}$$

Moreover,

$$\inf_{\mathsf{c} \in \mathcal{C}} \int \langle \mathsf{c}(t), t \rangle \, \mu(dt) = \int \psi_D(t) \mu(dt), \tag{4.29}$$

where ψ_D is the envelope function in (4.26).

Corollary 4.32. *If $\mathcal{D}$ is a compact metric space, $L(\theta_i, \cdot) : \mathcal{D} \to \mathbb{R}$ is a continuous function for every $1 \le i \le m$, and Π is a prior on $\Delta = \{1, ..., m\}$, then the minimal Bayes risk is given by*

$$\inf_{\mathsf{D}} \mathsf{r}(\Pi, \mathsf{D}) = \int \psi_{D_{L,\Pi}}(t) \mu(dt), \tag{4.30}$$

where $D_{L,\Pi}$ is defined in (4.24).

Corollary 4.33. *Let $\mathcal{M}_i = (\mathcal{X}_i, \mathfrak{A}_i, (P_{i,\theta})_{\theta\in\Delta})$, $i = 1, 2$, be finite models with $|\Delta| = m$, and with the respective standard distributions μ_i on $(\mathcal{S}_m, \mathfrak{S}_m)$ given by (4.13), $i = 1, 2$. If $\mathcal{D}$ and L satisfy the conditions in Corollary 4.32, then*

$$| \inf_{\mathsf{D}_{\mathcal{M}_1}} \mathsf{r}(\Pi, \mathsf{D}_{\mathcal{M}_1}) - \inf_{\mathsf{D}_{\mathcal{M}_2}} \mathsf{r}(\Pi, \mathsf{D}_{\mathcal{M}_2}) | \le \|L\|_u m \|\mu_1 - \mu_2\|_{\mathsf{D}},$$

where $\|L\|_u = \sup_{a\in\mathcal{D},\theta\in\Delta} |L(\theta, a)|$.

Proof. Set $\Psi(x,t) = \langle x, t\rangle$, $x \in D$, $t \in \mathcal{S}_m$. As D is compact for every t there is some $\mathsf{c}_D(t)$ such that $\Psi(\mathsf{c}_D(t), t) = \inf_{x\in D} \Psi(x,t)$. As Ψ satisfies the assumptions of the measurable selection theorem (see Theorem A.10) we may choose c_D to be measurable. This proves (4.28). The statement (4.29) results from the definition of the lower envelope function. The statement (4.30) is a consequence of the representation of the Bayes risk in (4.25). To prove the statement of the second corollary we remark that for $t \in \mathcal{S}_m$ and $m = |\Delta|$,

$$\begin{aligned}
&|\psi_{D_{L,\Pi}}(t)| \\
&\le |\inf_{y\in D_{L,\Pi}} \langle y, t\rangle | \le \sup_{y\in D_{L,\Pi}} | \langle y, t\rangle | = \sup_{y\in D_{L,\Pi}} |\textstyle\sum_{i=1}^{m} t_i y_i| \\
&\le (\sup_{y\in D_{L,\Pi}} \|y\|_u)(\textstyle\sum_{i=1}^{m} t_i) \le \|L\|_u m, \\
&|\psi_{D_{L,\Pi}}(t_1) - \psi_{D_{L,\Pi}}(t_2)| \\
&\le (\sup_{y\in D_{L,\Pi}} \|y\|_u) m \|t_1 - t_2\|_u \le \|L\|_u m \|t_1 - t_2\|_u,
\end{aligned}$$

where the first inequality in the last line follows from (4.27). Hence by the definition of the Dudley metric,

$$\begin{aligned}
| \inf_{\mathsf{D}_{\mathcal{M}_1}} \mathsf{r}(\Pi, \mathsf{D}_{\mathcal{M}_1}) - \inf_{\mathsf{D}_{\mathcal{M}_2}} \mathsf{r}(\Pi, \mathsf{D}_{\mathcal{M}_2})| &= | \int \psi_{D_{L,\Pi}}(t)(\mu_1 - \mu_2)(dt)| \\
&\le \|L\|_u m \|\mu_1 - \mu_2\|_{\mathsf{D}}.
\end{aligned}$$

■

In Chapter 1 we have introduced and studied v-divergences as a concept to measure the distance between two distributions. We have also interpreted this distance as a measure of informativeness of a binary model. The idea hereby is that it is easier to distinguish between two distributions if there is a large distance between them. A statement that makes this interpretation precise is Theorem 1.68 where v-divergences are expressed in terms of the Bayes risk. Now we show that v-divergences are the minimal risks for a standard decision problem in a binary model where the convex function v is the lower envelope of the compact convex set that appears in the definition of the standard decision problem. More precisely, let $D \subseteq \mathbb{R}^m$ be a compact and convex set and μ be a standard measure that generates the standard model $\mathcal{N} = (\mathcal{S}_m, \mathfrak{S}_m, \{Q_1, ..., Q_m\})$, where $dQ_i = Z_i d\mu$, $i = 1, ..., m$. We use $\mathcal{D} := D$ as the decision space and define the loss function L by $L(i, a) = a_i$, $a = (a_1, ..., a_m) \in D$. For every decision $\mathsf{D} : \mathfrak{D} \times \mathcal{S}_m \to_k [0, 1]$ the risk is

$$\mathsf{R}(i,\mathsf{D}) = \int [\int a_i \mathsf{D}(da_1,...,da_m|t_1,...,t_m)] t_i \mu(dt_1,...,dt_m),$$

$i = 1,...,m$. As D is closed and convex it holds for $t = (t_1,...,t_m)$ that

$$\mathsf{d}(t) := (\int a_1 \mathsf{D}(da_1,...,da_m|t_1,...,t_m),..., \int a_m \mathsf{D}(da_1,...,da_m|t_1,...,t_m)) \in D,$$

and d is a nonrandomized decision that has the same risk function as D. Hence by (4.28) and (4.29),

$$\inf_{\mathsf{D}} \sum\nolimits_{i=1}^{m} \mathsf{R}(i,\mathsf{D}) = \int \psi_D(t)\mu(dt),$$

where ψ_D is defined in (4.26). Let D be a compact and convex subset of $\mathbb{R}^2$ and ψ_D be the associated lower envelope function in (4.26). We have already seen in Problem 4.30 that for a compact and convex set $D \subseteq \mathbb{R}^2$ the lower envelope function

$$\psi_D(t_0,t_1) = \inf\{t_0x_0 + t_1x_1, (x_0,x_1) \in D\}, \quad t_0,t_1 \geq 0,$$

is Lipschitz-continuous and concave. The function ψ_D is homogeneous in the sense that it satisfies $\psi_D(at_0,at_1) = a\psi_D(t_0,t_1)$ for every $a > 0$. We introduce the convex function v by

$$\mathsf{v}(x) = -\psi_D(x,1).$$

Then the conjugate convex function $\mathsf{v}^*(x) = x\mathsf{v}(1/x)$ is given by $\mathsf{v}^*(x) = -\psi_D(1,x)$.

We recall that according to (4.11), $M = (M_0, M_1)$, $M_i = dP_i/d\overline{P}$, $i = 0,1$, where $\overline{P} = \frac{1}{2}(P_0 + P_1)$, and the standard distribution is defined by $\mu = \overline{P} \circ M^{-1}$. As $M_1 = 2 - M_0$, $\overline{P}$-a.s., we get

$$\begin{aligned}
\int \psi_D(t_0,t_1)\mu(dt_0,dt_1) &= \int \psi_D(M_0,M_1)d\overline{P} = \int \psi_D(M_0, 2-M_0)d\overline{P} \\
&= \psi_D(2,0)\overline{P}(M_1=0) + \psi_D(0,2)\overline{P}(M_0=0) \\
&\quad + \int I_{(0,2)}(M_0)\psi_D(\frac{M_0}{M_1},1)M_1 d\overline{P}.
\end{aligned}$$

It holds $\overline{P}(M_1 = 0) = \frac{1}{2}P_0(M_1 = 0)$ and $\overline{P}(M_0 = 0) = \frac{1}{2}P_1(M_0 = 0)$. Hence by $\psi_D(2,0) = -2\mathsf{v}^*(0)$, $\psi_D(0,2) = -2\mathsf{v}(0)$, and the definition of the v-divergence in Definition 1.60,

$$\int \psi_D(t_0,t_1)\mu(dt_0,dt_1) = -\mathsf{I}_{\mathsf{v}}(P_0,P_1). \tag{4.31}$$

The following theorem establishes the relation between the minimal risk in standard decision problems and v-divergences. It explains why v-divergences are not only distances, but also characterize the informativeness of a model in the sense that it is harder to distinguish between distributions with smaller distances than between those with larger distances. The theorem shows that $-\mathsf{I}_{\mathsf{v}}(P_0, P_1)$ is the minimal Bayes risk of a standard decision problem, where $-\mathsf{v}$ is the lower envelope of the convex set that appears in the standard decision problem. Thus we can say that each standard decision problem defines in an inherent manner a special distance.

We recall $\mathsf{b}_\pi(P_0, P_1)$, the minimal Bayes risk for testing $\mathsf{H}_0 : P_0$ versus $\mathsf{H}_A : P_1$ with prior $(\pi, 1-\pi)$ that was introduced in (1.82), and ρ_{v}, the curvature measure in (1.84). The next theorem and its corollary show that for binary models one model is already more informative than the other if it is more informative than the other under all testing problems, which means that for every $0 < \pi < 1$ the minimum Bayes risk is not larger than the other one.

Theorem 4.34. *Let $\mathcal{M} = (\mathcal{X}, \mathfrak{A}, \{P_0, P_1\})$ be a binary model with the standard distribution μ, and let D be a convex and compact subset of $\mathbb{R}^2$. If ψ_D is the lower envelope and $\mathsf{v} = -\psi_D(x, 1)$, then*

$$\inf_{\mathsf{D}}(\mathsf{R}(0, \mathsf{D}) + \mathsf{R}(1, \mathsf{D})) = -\mathsf{I}_{\mathsf{v}}(P_0, P_1),$$

$$\inf_{\mathsf{D}}(\mathsf{R}(0, \mathsf{D}) + \mathsf{R}(1, \mathsf{D})) = \psi_D(1, 1) - \int \mathsf{B}_\pi(P_0, P_1)\rho_{\mathsf{v}}(d\pi), \tag{4.32}$$

where $\mathsf{B}_\pi(P_0, P_1) = \pi \wedge (1-\pi) - \mathsf{b}_\pi(P_0, P_1)$.

Corollary 4.35. *If $\mathcal{M} = (\mathcal{X}, \mathfrak{A}, \{P_0, P_1\})$ and $\mathcal{N} = (\mathcal{Y}, \mathfrak{B}, \{Q_0, Q_1\})$ are two binary models, then $\mathcal{M} \succeq \mathcal{N}$ holds if and only if $\mathsf{b}_\pi(P_0, P_1) \le \mathsf{b}_\pi(Q_0, Q_1)$, $0 < \pi < 1$.*

Proof. The first statement follows from (4.31) and (4.29). The second statement follows from the first statement and Theorem 1.68.

To prove the corollary, let $\mathcal{M} \succeq \mathcal{N}$. Then $\mathsf{b}_\pi(P_0, P_1) \le \mathsf{b}_\pi(Q_0, Q_1)$, $0 < \pi < 1$, by condition (B) in Theorem 4.14. Conversely, if the last inequality holds, then by (4.30) and (4.32),

$$\begin{aligned}\inf_{\mathsf{D}_{\mathcal{M}}} \mathsf{r}(\Pi, \mathsf{D}_{\mathcal{M}}) &= \psi_{D_{L,\Pi}}(1, 1) - \int \mathsf{B}_\pi(P_0, P_1)\rho_{\mathsf{v}}(d\pi) \\ &\le \psi_{D_{L,\Pi}}(1, 1) - \int \mathsf{B}_\pi(Q_0, Q_1)\rho_{\mathsf{v}}(d\pi) = \inf_{\mathsf{D}_{\mathcal{N}}} \mathsf{r}(\Pi, \mathsf{D}_{\mathcal{N}}).\end{aligned}$$

To complete the proof we have only to apply (B) from Theorem 4.14 for $\varepsilon = 0$. ■

4.3 Sufficiency in Dominated Models

In this section we introduce and discuss the classical concepts of sufficiency and show how these concepts are related to the decision-theoretic concept of equivalence of models. Roughly speaking, it is shown that a model induced by a sufficient statistic contains the same information as the original model in the sense that the two models are equivalent. As dominated models play a crucial role hereby their structure is studied in the first part of this section.

For any model $(\mathcal{X}, \mathfrak{A}, (P_\theta)_{\theta\in\Delta})$ we call

$$[(P_\theta)_{\theta\in\Delta}] = \{\overline{P} : \overline{P} = \sum\nolimits_{j=1}^{\infty} c_j P_{\theta_j},\ c_j \geq 0,\ \sum\nolimits_{j=1}^{\infty} c_j = 1,\ \theta_j \in \Delta\} \quad (4.33)$$

the *convex hull* of $(P_\theta)_{\theta\in\Delta}$.

If the family $(P_\theta)_{\theta\in\Delta}$ is separable, in the sense that there is an at most countable family $\{P_0, P_1, ...\}$ that is dense in $(P_\theta)_{\theta\in\Delta}$ with respect to the variational distance, then it is easy to see that for any positive numbers α_i with $\sum_{i=0}^{\infty} \alpha_i = 1$ the distribution $\sum_{i=0}^{\infty} \alpha_i P_i$ dominates the model. The general case, where $(P_\theta)_{\theta\in\Delta}$ is any dominated family, is covered by the following famous lemma due to Halmos and Savage (1949).

Lemma 4.36. *If $(\mathcal{X}, \mathfrak{A}, (P_\theta)_{\theta\in\Delta})$ is a model where the family $(P_\theta)_{\theta\in\Delta}$ is dominated by some $\mu \in \mathcal{M}^\sigma(\mathfrak{A})$, then there exists a $\overline{P} \in [(P_\theta)_{\theta\in\Delta}]$ that dominates the family $(P_\theta)_{\theta\in\Delta}$.*

Proof. By Lemma 3.20 we may assume without loss of generality that the dominating measure is a probability measure, say Q. For every P from the convex hull $[(P_\theta)_{\theta\in\Delta}]$ (see (4.33)) we set $S_P = \{x : (dP/dQ)(x) > 0\}$. It holds for any $P_1, P_2 \in [(P_\theta)_{\theta\in\Delta}]$ and any $0 < \alpha < 1$,

$$Q(S_{\alpha P_1 + (1-\alpha)P_2}) = Q(S_{P_1} \cup S_{P_2}) \geq \max(Q(S_{P_1}), Q(S_{P_2})). \quad (4.34)$$

If $P_n \in [(P_\theta)_{\theta\in\Delta}]$ is a sequence with

$$Q(S_{P_n}) \to \sup_{P\in[(P_\theta)_{\theta\in\Delta}]} Q(S_P) =: s,$$

and $\overline{P} = \sum_{n=1}^{\infty} 2^{-n} P_n$, then $Q(S_{\overline{P}}) = s$ by (4.34). Hence for any $P \in [(P_\theta)_{\theta\in\Delta}]$

$$s \geq Q(S_{(\overline{P}+P)/2}) = Q(S_{\overline{P}}) + Q(S_P \backslash S_{\overline{P}}) = s + Q(S_P \backslash S_{\overline{P}}),$$

which implies $Q(S_P \backslash S_{\overline{P}}) = 0$, and by $P \ll Q$ also $P(S_P \backslash S_{\overline{P}}) = 0$. This yields for any $A \in \mathfrak{A}$,

$$P(A \backslash S_{\overline{P}}) = P((A \cap S_P) \backslash S_{\overline{P}}) = 0. \quad (4.35)$$

If $A \in \mathfrak{A}$ and $\overline{P}(A) = 0$, then $\overline{P}(A \cap S_{\overline{P}}) = 0$. As the density of $\overline{P}$ is positive on $S_{\overline{P}}$ we get $Q(A \cap S_{\overline{P}}) = 0$ and $P(A \cap S_{\overline{P}}) = 0$ by $P \ll Q$. Hence by (4.35)

$$P(A) = P(A \cap S_{\overline{P}}) + P(A \backslash S_{\overline{P}}) = 0.$$

■

Now we introduce several types of sufficiency.

Definition 4.37. *Let $(\mathcal{X}, \mathfrak{A}, (P_\theta)_{\theta\in\Delta})$ be a model, $(\mathcal{Y}, \mathfrak{B})$ a measurable space, and $T : \mathcal{X} \to_m \mathcal{Y}$ a statistic.*
T is called sufficient for $(P_\theta)_{\theta\in\Delta}$ if for every $A \in \mathfrak{A}$ there is a function $k_A : \mathcal{Y} \to_m \mathbb{R}$ such that

$$\mathsf{E}_\theta(I_A|T) = k_A(T), \quad P_\theta\text{-a.s.}, \theta \in \Delta. \tag{4.36}$$

T is called regular sufficient for $(P_\theta)_{\theta\in\Delta}$ if there exists a stochastic kernel $\mathsf{M} : \mathfrak{A} \times \mathcal{Y} \to_k [0,1]$ such that

$$\mathsf{E}_\theta(I_A|T) = \mathsf{M}(A|T), \quad P_\theta\text{-a.s.}, \theta \in \Delta,\ A \in \mathfrak{A}. \tag{4.37}$$

T is called Blackwell sufficient for $(P_\theta)_{\theta\in\Delta}$ if there exists a stochastic kernel $\mathsf{M} : \mathfrak{A} \times \mathcal{Y} \to_k [0,1]$ such that

$$P_\theta = \mathsf{M}(P_\theta \circ T^{-1}), \quad \theta \in \Delta.$$

T is called pairwise sufficient for $(P_\theta)_{\theta\in\Delta}$ if T is sufficient for every binary submodel $\{P_{\theta_1}, P_{\theta_2}\}$, $\theta_1, \theta_2 \in \Delta$.

A sub-σ-algebra $\mathfrak{G} \subseteq \mathfrak{A}$ is called (regular, pairwise) sufficient for $(P_\theta)_{\theta\in\Delta}$ if $(\mathcal{Y}, \mathfrak{B}) = (\mathcal{X}, \mathfrak{A})$ and the identical mapping T is (regular, pairwise) sufficient.

Historically, the independence of the conditional probabilities, given T, of the parameter was the starting point of the concept of sufficiency. This goes back to Fisher (1920, 1934) who considered a statistic T to be sufficient if the conditional distribution of any other statistic S, given $T = t$, is independent of the parameter, so that T contains the complete information. This means that given $T(x) = t$, the additional information on which values x have led to $T(x) = t$ contains no additional information on the parameter.

The difference between sufficiency and pairwise sufficiency is that for a pairwise sufficient statistic T the function $k_A(T)$, defined by $k_A(T) = \mathsf{E}_\theta(I_A|T)$, depends on the particular pair θ_1 and θ_2. This difficulty disappears in dominated models.

Problem 4.38.* If $(\mathcal{X}, \mathfrak{A}, (P_\theta)_{\theta\in\Delta})$ is dominated by some $\mu \in \mathcal{M}^\sigma(\mathfrak{A})$, then $T : \mathcal{X} \to_m \mathcal{Y}$ is sufficient for $(P_\theta)_{\theta\in\Delta}$ if and only if T is pairwise sufficient for $(P_\theta)_{\theta\in\Delta}$.

Example 1.5.7 from Torgersen (1991), which follows below, shows that the assumption of the family to be dominated is indispensable.

Example 4.39. Assume that $\mathfrak{A}_0 \subseteq \mathfrak{A}$ is a sub-σ-algebra that separates the distributions in the family $(P_\theta)_{\theta\in\Delta}$ in the sense that for every $\theta_1, \theta_2 \in \Delta$ with $\theta_1 \neq \theta_2$ there exists a set $B \in \mathfrak{A}_0$ such that $P_{\theta_1}(B) = 1$ and $P_{\theta_2}(\overline{B}) = 1$. Then

$$\int_B \mathsf{E}_{\theta_1}(I_A|\mathfrak{A}_0)dP_{\theta_2} = \int_{\overline{B}} \mathsf{E}_{\theta_2}(I_A|\mathfrak{A}_0)dP_{\theta_1} = 0,$$

which shows that $k_A = \mathsf{E}_{\theta_1}(I_A|\mathfrak{A}_0) + \mathsf{E}_{\theta_2}(I_A|\mathfrak{A}_0)$ satisfies $k_A = \mathsf{E}_{\theta_i}(I_A|\mathfrak{A}_0)$, $i = 1, 2$. Hence $\mathfrak{A}_0$ is pairwise sufficient for $(P_\theta)_{\theta\in\Delta}$. The model $([0,1], \mathfrak{B}_{[0,1]}, (\delta_\theta)_{\theta\in[0,1]})$ is an example for which the separation condition holds. This model cannot be dominated by a σ-finite measure μ as the set of all points x with $\mu(\{x\}) > 0$ is at most countable. Let $\mathfrak{A}_0$ be the sub-σ-algebra of $\mathfrak{B}_{[0,1]}$ that consists of all sets A for which either A or $\overline{A}$ is at most countable. $\mathfrak{A}_0$ separates the model $(\boldsymbol{\delta}_\theta)_{\theta\in[0,1]}$ and is therefore pairwise sufficient for $(P_\theta)_{\theta\in\Delta}$. However, $\mathfrak{A}_0$ is not sufficient for $(P_\theta)_{\theta\in\Delta}$. Indeed, $k_{[0,1/2]}$ from the definition of sufficiency would have to satisfy

$$k_{[0,1/2]}(\theta) = \int_{\{\theta\}} k_{[0,1/2]}(x)\boldsymbol{\delta}_\theta(dx) = \boldsymbol{\delta}_\theta(\{\theta\} \cap [0,1/2]) = I_{[0,1/2]}(\theta),$$

which contradicts the requirement that $k_{[0,1/2]}(\theta)$ is $\mathfrak{A}_0$-measurable.

The independence of the parameter of the conditional probability in the definition of sufficiency extends easily to the conditional expectation of any nonnegative random variable and thus also to any random variable with existing expectation.

Problem 4.40.* Suppose that $S : \mathcal{X} \to_m \mathbb{R}_+$. If T is sufficient for $(P_\theta)_{\theta\in\Delta}$, then there exists some $k_S : \mathcal{Y} \to_m \mathbb{R}$ such that $\mathsf{E}_\theta(S|T) = k_S(T)$, P_θ-a.s., $\theta \in \Delta$.

A simple fact is that one-to-one measurable mappings are sufficient.

Problem 4.41.* Suppose that $T : \mathcal{X} \to \mathcal{Y}$ is a one-to-one mapping of $\mathcal{X}$ onto $\mathcal{Y}$ and that both, T and the inverse mapping U, are measurable. Then T is sufficient.

For discrete distributions the sufficiency of a statistic can often be verified directly from the definition.

Problem 4.42. The statistic $T : \mathbb{N}^n \to \mathbb{N}$, defined by $T(k_1, ..., k_n) = k_1 + \cdots + k_n$, is sufficient for the family $(\mathsf{Po}^{\otimes n}(\lambda))_{\lambda>0}$.

It is an interesting fact, which is used on several occasions, that the sufficiency of a statistic continues to hold if we turn to the convex hull of the model.

Problem 4.43.* If T is sufficient for $(P_\theta)_{\theta\in\Delta}$, then T is also sufficient for $[(P_\theta)_{\theta\in\Delta}]$.

Condition (4.37) is equivalent to

$$P_\theta(A \cap \{T \in B\}) = \int I_B(T(x))\mathsf{M}(A|T(x))P_\theta(dx), \quad A \in \mathfrak{A},\ B \in \mathfrak{B}. \tag{4.38}$$

It is easy to see that the set of all $C \in \mathfrak{A} \otimes \mathfrak{B}$ for which

$$\int I_C(x, T(x))P_\theta(dx) = \int [\int I_C(x,t)\mathsf{M}(dx|t)](P_\theta \circ T^{-1})(dt) \tag{4.39}$$

holds is a σ-algebra that contains, in view of (4.38), all sets $A \times B$, $A \in \mathfrak{A}$, $B \in \mathfrak{B}$. As these product sets generate $\mathfrak{A} \otimes \mathfrak{B}$ we see that (4.39) holds for every $C \in \mathfrak{A} \otimes \mathfrak{B}$. If X is a random variable with $\mathcal{L}(X) = P_\theta$, then (4.39) can be written as

$$\mathcal{L}(X, T(X)) = \mathsf{M} \otimes (P_\theta \circ T^{-1}).$$

Turning to the marginal distribution we get $P_\theta = \mathsf{M}(P_\theta \circ T^{-1})$ which is the Blackwell sufficiency. Blackwell sufficiency, sometimes called *exhaustivity*, is a condition that refers only to the marginal distribution, whereas sufficiency and regular sufficiency refer to the joint distribution of X and $T(X)$. Thus we cannot expect that the Blackwell sufficiency implies the regular sufficiency in general. However, under additional assumptions it is shown later that the two concepts are equivalent.

For $\mathsf{E}_Q(I_A|\sigma(T))$, the conditional probability of A, given the σ-algebra $\sigma(T)$, under Q, an interesting question is if it can also be the conditional probability under another distribution, say P. A sufficient condition on the relation of P and Q for this to be true is given in the next problem.

Problem 4.44.* Let P and Q be distributions on $(\mathcal{X}, \mathfrak{A})$ with $P \ll Q$. If $(\mathcal{Y}, \mathfrak{B})$ is another measurable space, $T : \mathcal{X} \to_m \mathcal{Y}$ a statistic, and dP/dQ is $\sigma(T)$-measurable, then for every $h : \mathcal{X} \to_m \mathbb{R}_+$

$$\mathsf{E}_P(h|\sigma(T)) = \mathsf{E}_Q(h|\sigma(T)), \quad P\text{-}a.s.$$

Later on, in the proof of the factorization lemma, it is shown that criteria for sufficiency of the binary submodels are essential. Thus, let us consider a binary model $\mathcal{M} = (\mathcal{X}, \mathfrak{A}, \{P_0, P_1\})$ and the reduced model $\mathcal{N} = (\mathcal{Y}, \mathfrak{B}, \{Q_0, Q_1\})$ with $Q_i = P_i \circ T^{-1}$, $i = 0, 1$, for some $T : \mathcal{X} \to_m \mathcal{Y}$. We set

$$\overline{P} = \frac{1}{2}(P_0 + P_1), \quad \overline{Q} = \frac{1}{2}(Q_0 + Q_1), \quad L_i := \frac{dP_i}{d\overline{P}}, \quad M_i := \frac{dQ_i}{d\overline{Q}}, \quad i = 0, 1.$$

Let $\mathsf{D}(P_0, P_1)$ be the Hellinger distance in (1.75) and $\mathsf{b}_\pi(P_0, P_1)$ from (1.82).

Theorem 4.45. *Given $\mathcal{M} = (\mathcal{X}, \mathfrak{A}, \{P_0, P_1\})$, a statistic $T : \mathcal{X} \to_m \mathcal{Y}$, and $\mathcal{N} = (\mathcal{Y}, \mathfrak{B}, \{Q_0, Q_1\})$ with $Q_i = P_i \circ T^{-1}$, $i = 0, 1$, the following statements are equivalent.*

(A) T *is sufficient for* $\{P_0, P_1\}$.
(B) $\mathsf{E}_{\overline{P}}(L_0|T) = L_0, \quad \overline{P}\text{-}a.s.$
(C) $\mathsf{D}(Q_0, Q_1) = \mathsf{D}(P_0, P_1)$.
(D) $\mathsf{b}_\pi(Q_0, Q_1) = \mathsf{b}_\pi(P_0, P_1), \quad 0 < \pi < 1.$

Corollary 4.46. *If T is sufficient for $\{P_0, P_1\}$, then*

$$\mathsf{I}_\mathsf{v}(Q_0, Q_1) = \mathsf{I}_\mathsf{v}(P_0, P_1) \tag{4.40}$$

for every convex function $\mathsf{v} : (0, \infty) \to \mathbb{R}$. Conversely, if (4.40) holds for one strictly convex function $\mathsf{v} : (0, \infty) \to \mathbb{R}$ with $\mathsf{I}_\mathsf{v}(P_0, P_1) < \infty$, then T is sufficient for $\{P_0, P_1\}$.

Corollary 4.47. *If T is Blackwell sufficient for $\{P_0, P_1\}$, then T is sufficient for $\{P_0, P_1\}$.*

Proof. The proof follows the scheme $(A) \to (D) \to (C) \to (B) \to (A)$. To show $(A) \to (D)$, let φ_B be a Bayes test for $\mathsf{H}_0 : P_0$ versus $\mathsf{H}_A : P_1$. Then by Problem 4.40 there is a test ψ with $\psi(T) = \mathsf{E}_{P_i}(\varphi_B|T)$, P_i-a.s. Hence $\mathsf{E}_{P_i \circ T^{-1}}\psi = \mathsf{E}_{P_i}\varphi_B$ and $\mathsf{b}_\pi(Q_0, Q_1) \le \mathsf{b}_\pi(P_0, P_1)$. The opposite inequality is trivial as the set of the tests $\varphi : \mathcal{X} \to_m [0,1]$ that are functions of T is a subset of all tests. To establish $(D) \to (C)$, we set $\mathsf{v}(x) = (\sqrt{x} - 1)^2$. Then $\mathsf{D}^2(P_0, P_1) = \mathsf{I}_\mathsf{v}(P_0, P_1)$ and (C) follows from Theorem 1.68. $(C) \to (B)$ follows from Proposition 1.75. Finally, $(B) \to (A)$ follows from Problem 4.44 with $Q = \overline{P}$ and the fact that L_0 and $L_1 = 2 - L_0$ are $\sigma(T)$-measurable.

Corollary 4.46 follows from Theorem 1.70 and condition (D). To prove Corollary 4.47, we fix any strictly convex function v with $\mathsf{I}_\mathsf{v}(P_0, P_1) < \infty$. For example, we may choose $\mathsf{v}(x) = (\sqrt{x} - 1)^2$. A twofold application of Theorem 1.70 yields $\mathsf{I}_\mathsf{v}(P_0, P_1) \ge \mathsf{I}_\mathsf{v}(P_0 \circ T^{-1}, P_1 \circ T^{-1}) \ge \mathsf{I}_\mathsf{v}(\mathsf{M}(P_0 \circ T^{-1}), \mathsf{M}(P_1 \circ T^{-1})) = \mathsf{I}_\mathsf{v}(P_0, P_1)$, so that the statement follows from Corollary 4.46. ■

Problem 4.48.* $(\mathcal{X}, \mathfrak{A}, \{P_0, P_1\}) \sim (\mathcal{Y}, \mathfrak{B}, \{Q_0, Q_1\})$ holds if and only if we have $\mathsf{g}_\alpha(P_0, P_1) = \mathsf{g}_\alpha(Q_0, Q_1)$, $0 < \alpha < 1$, where g_α is the minimal probability of an error of the second kind of a level α test; see (2.49).

Remark 4.49. The statement of Corollary 4.46 is an information-theoretic characterization of sufficiency that presumably goes back to Csiszár (1963). For the Hellinger distance this relation is condition (C) in Theorem 4.45 and can also be found in LeCam (1986). The equivalence of conditions (A) and (D) in Theorem 4.45 is a testing-theoretic characterization of sufficiency which is due to Pfanzagl (1974), where the equivalent condition in Problem 4.48 is used. For further references and historical remarks we refer to Pfanzagl (1994) and Torgersen (1991). Sverdrup (1966) is an exposition of the central ideas of classical papers on sufficiency.

Now we extend the above results for binary models to dominated models. Here we utilize the statement in Problem 4.44 where the $\sigma(T)$-measurability of the density has been used. It turns out that this condition works also for dominated models, and is also a necessary condition. This is the content of the famous factorization criterion for sufficiency due to Neyman (1935).

Theorem 4.50. ***(Neyman Criterion)*** *In $(\mathcal{X}, \mathfrak{A}, (P_\theta)_{\theta\in\Delta})$, let $(P_\theta)_{\theta\in\Delta}$ be dominated by some $\mu \in \mathcal{M}^\sigma(\mathfrak{A})$, and $f_\theta = dP_\theta/d\mu$, $\theta \in \Delta$. Then for any statistic $T : \mathcal{X} \to_m \mathcal{Y}$ the following holds. T is sufficient for $(P_\theta)_{\theta\in\Delta}$ if and only if there are functions $g_\theta : \mathcal{Y} \to_m \mathbb{R}$, $\theta \in \Delta$, and $h : \mathcal{X} \to_m \mathbb{R}$, such that*

$$f_\theta = g_\theta(T)h, \quad \mu\text{-a.e.} \tag{4.41}$$

Proof. Suppose that T is sufficient and $\overline{P}$ is from Lemma 4.36. Then according to Problem 4.43 the statistic T is sufficient for $\{P_\theta, \frac{1}{2}(P_\theta + \overline{P})\}$. Hence we get from (B) in Theorem 4.45 that $dP_\theta/d(\frac{1}{2}(P_\theta + \overline{P}))$ is a measurable function of T. Then

$$\frac{dP_\theta}{d(\frac{1}{2}(P_\theta + \overline{P}))} = \frac{2dP_\theta/d\overline{P}}{1 + dP_\theta/d\overline{P}}$$

shows that $dP_\theta/d\overline{P}$ is a measurable function of T, say $g_\theta(T)$. Putting $h = d\overline{P}/d\mu$ we get (4.41). Conversely, if (4.41) is satisfied, then by Problem 4.44 $\mathsf{E}_{\overline{P}}(I_A|T)$ is a version of the conditional probability that is independent of the parameter. ■

Example 4.51. Let $X_1, ..., X_n$ be an i.i.d. sample from a distribution that belongs to an exponential family $(P_\theta)_{\theta\in\Delta}$ in natural form with $\Delta \subseteq \mathbb{R}^d$ and generating statistic $T : \mathcal{X} \to_m \mathbb{R}^d$. Then $\mathcal{L}((X_1, ..., X_n)) = P_\theta^{\otimes n}$, and by Proposition 1.4

$$\frac{dP_\theta^{\otimes n}}{d\boldsymbol{\mu}^{\otimes n}} = \exp\left\{\langle\theta, T_{\oplus n}\rangle - nK(\theta)\right\}.$$

Consequently, by Theorem 4.50 the generating statistic $T_{\oplus n}(x_1, ..., x_n) = \sum_{i=1}^n T(x_i)$ is sufficient for the family $(P_\theta^{\otimes n})_{\theta\in\Delta}$.

In a finite model the likelihood carries the complete information.

Example 4.52. For every finite model $(\mathcal{X}, \mathfrak{A}, \{P_1, ..., P_m\})$ the statistic M in (4.11) is sufficient for $\{P_1, ..., P_m\}$. This follows directly from the factorization criterion Theorem 4.50.

If we turn from one statistic by a one-to-one bimeasurable mapping to another statistic, then by Problem 4.41 the new statistic is again sufficient. The next problem illustrates this for normal distributions.

Problem 4.53.* For a model there are in general several different sufficient statistics available. Consider the model

$$(\mathbb{R}^n, \mathfrak{B}_n, (\mathsf{N}^{\otimes n}(\mu, \sigma^2))_{\mu\in\mathbb{R}, \sigma^2>0}).$$

Let $\overline{X}_n = (1/n)\sum_{i=1}^n X_i$ and $S_n^2 = (1/(n-1))\sum_{i=1}^n (X_i - \overline{X}_n)^2$, where $X_i : \mathbb{R}^n \to \mathbb{R}$, $i = 1, ..., k$, are the projections on the coordinates. The statistics $(\sum_{i=1}^n X_i, \sum_{i=1}^n X_i^2)$ and $(\overline{X}_n, S_n^2)$ are both sufficient for $(\mathsf{N}^{\otimes n}(\mu, \sigma^2))_{\mu\in\mathbb{R}\ \sigma^2>0}$.

The next problem shows that there is a close relationship between the concepts of a monotone likelihood ratio (MLR) and sufficiency.

Problem 4.54.* If $\mathcal{M} = (\mathcal{X}, \mathfrak{A}, (P_\theta)_{\theta\in\Delta})$, $\Delta \subseteq \mathbb{R}$, is dominated by some $\mu \in \mathcal{M}^\sigma(\mathfrak{A})$ and has a monotone likelihood ratio in $T : \mathcal{X} \to_m \overline{\mathbb{R}}$ (see Definition 2.11), then T is sufficient for $(P_\theta)_{\theta\in\Delta}$.

We have seen in Problem 4.41 that a one-to-one statistic is sufficient. For dominated models we can say more.

Problem 4.55. Let $(\mathcal{X}, \mathfrak{A}, (P_\theta)_{\theta\in\Delta})$ be dominated by some $\mu \in \mathcal{M}^\sigma(\mathfrak{A})$, $T : \mathcal{X} \to_m \mathcal{Y}$, and $g : \mathcal{Y} \to_m \mathcal{Y}$. If $S = g(T)$ is sufficient for $(P_\theta)_{\theta\in\Delta}$, then T is sufficient for $(P_\theta)_{\theta\in\Delta}$ as well.

In dominated models sufficiency can be characterized by the fact that the distances between the distributions in the reduced model are the same as in the original model.

Problem 4.56.* Let $(\mathcal{X}, \mathfrak{A}, (P_\theta)_{\theta\in\Delta})$ be dominated by some $\mu \in \mathcal{M}^\sigma(\mathfrak{A})$, and let $T : \mathcal{X} \to_m \mathcal{Y}$ be a statistic. T is sufficient for $(P_\theta)_{\theta\in\Delta}$ if and only if $\mathsf{D}(P_{\theta_1}, P_{\theta_2}) = \mathsf{D}(P_{\theta_1} \circ T^{-1}, P_{\theta_2} \circ T^{-1})$ for every $\theta_1, \theta_2 \in \Delta$.

Problem 4.57.* Suppose that $(\mathcal{X}, \mathfrak{A}, (P_\theta)_{\theta\in\Delta})$, $\Delta \subseteq \mathbb{R}^d$, is a statistical model which is $\mathbb{L}_2$-differentiable at $\theta_0 \in \Delta^0$ with Fisher information matrix $\mathsf{I}(\theta_0)$. Let $T : \mathcal{X} \to_m \mathcal{Y}$ be a statistic and $\mathsf{I}_T(\theta_0)$ be the Fisher information matrix in the family $Q_\theta = P_\theta \circ T^{-1}$, $\theta \in \Delta$; see Theorem 1.114. If T is sufficient for $(P_\theta)_{\theta\in\Delta}$, then $\mathsf{I}_T(\theta_0) = \mathsf{I}(\theta_0)$.

By Corollary 4.47 Blackwell sufficiency implies pairwise sufficiency which for dominated models is equivalent to sufficiency. Moreover, regular sufficiency is stronger than sufficiency and Blackwell sufficiency. This leads to the following relation between the different types of sufficiency.

Proposition 4.58. *Let $(\mathcal{X}, \mathfrak{A}, (P_\theta)_{\theta\in\Delta})$ be dominated by some $\mu \in \mathcal{M}^\sigma(\mathfrak{A})$ and $(\mathcal{X}, \mathfrak{A})$ be a Borel space. Then for a statistic $T : \mathcal{X} \to_m \mathcal{Y}$ the conditions pairwise sufficient, sufficient, Blackwell sufficient, and regular sufficient for $(P_\theta)_{\theta\in\Delta}$ are equivalent.*

Proof. We have only to show that the sufficiency implies the regular sufficiency. Let $\overline{P}$ be from Lemma 4.36 and $\mathsf{M} : \mathfrak{A} \times \mathcal{Y} \to_k [0,1]$ be a stochastic kernel such that $\overline{P}(A|\sigma(T)) = \mathsf{M}(A|T)$, $\overline{P}$-a.s. By the factorization criterion (see Theorem 4.50) the density $dP_\theta/d\overline{P}$ is $\sigma(T)$-measurable and therefore by Problem 4.44 it holds $P_\theta(A|\sigma(T)) = \mathsf{M}(A|T)$, P_θ-a.s., which gives the regular sufficiency. ■

Remark 4.59. If $\mathfrak{A}$ is finite, then the model $(\mathcal{X}, \mathfrak{A}, (P_\theta)_{\theta\in\Delta})$ is readily dominated. If $\mathfrak{A}$ has the atoms $A_1, ..., A_n$, then we may turn to the new finite sample space $\widetilde{\mathcal{X}} = \{\{A_1\}, ..., \{A_n\}\}$. As every finite set is a Borel space we get from Proposition 4.58 that for a finite σ-algebra $\mathfrak{A}$ the conditions pairwise sufficient, sufficient, Blackwell sufficient, and regular sufficient are equivalent.

The concept of sufficiency is also meaningful in the Bayes framework. Let $(\mathcal{X}, \mathfrak{A}, (P_\theta)_{\theta\in\Delta})$ be a given statistical model. We assume that the condition (A3) in Section 1.2 is fulfilled so that (X, Θ) has the distribution $\mathsf{P} \otimes \Pi$.

Definition 4.60. $T : \mathcal{X} \to_m \mathcal{Y}$ *is called Bayes sufficient for* $(P_\theta)_{\theta\in\Delta}$ *if for every prior* Π *and every* $B \in \mathfrak{B}_\Delta$ *there is a function* $b_B : \mathcal{Y} \to_m \mathbb{R}_+$ *such that*

$$\mathbb{P}(\Theta \in B|X) = b_B(T(X)), \quad \mathbb{P}\text{-}a.s.$$

The next proposition clarifies how Bayes sufficiency is related to the other types of sufficiency. It is due to Blackwell and Ramamoorthi (1982); see also Schervish (1995), Theorem 2.14.

Proposition 4.61. *If $(\Delta, \mathfrak{B}_\Delta)$ is a Borel space, $(P_\theta)_{\theta\in\Delta}$ satisfies (A3), and $T : \mathcal{X} \to_m \mathcal{Y}$ is regular sufficient for $(P_\theta)_{\theta\in\Delta}$, then T is Bayes sufficient for $(P_\theta)_{\theta\in\Delta}$. Conversely, if $(\Delta, \mathfrak{B}_\Delta)$ is arbitrary but $\{\theta\} \in \mathfrak{B}_\Delta$, $\theta \in \Delta$, $(P_\theta)_{\theta\in\Delta}$ is dominated by some $\mu \in \mathcal{M}^\sigma(\mathfrak{A})$, and $T : \mathcal{X} \to_m \mathcal{Y}$ is Bayes sufficient for $(P_\theta)_{\theta\in\Delta}$, then T is sufficient for $(P_\theta)_{\theta\in\Delta}$.*

Proof. If T is regular sufficient, then by the standard extension technique we get from (4.39) that for every $h : \mathcal{X} \times \mathcal{T} \to_m \mathbb{R}_+$,

$$\int [\int h(x, T(y))\mathsf{M}(dx|T(y))]P_\theta(dy) = \int h(x, T(x))P_\theta(dx). \tag{4.42}$$

The family of distributions $Q_\theta = P_\theta \circ T^{-1}$ is a stochastic kernel, say Q. As $(\Delta, \mathfrak{B}_\Delta)$ is a Borel space we can find a stochastic kernel L such that

$$Q_\theta(dt)\Pi(d\theta) = \mathsf{L}(d\theta|t)(\mathsf{Q}\Pi)(dt). \tag{4.43}$$

It follows from (4.42) that for every $g : \mathcal{X} \times \Delta \to_m \mathbb{R}_+$ and $h(x,t) = g(x, \theta)$ for fixed θ

$$\int g(x, \theta)P_\theta(dx) = \int [\int g(x, \theta)\mathsf{M}(dx|T(y))]P_\theta(dy).$$

Integration with respect to θ yields

$$\begin{aligned}\int [\int g(x, \theta)P_\theta(dx)]\Pi(d\theta) &= \int [\int [\int g(x, \theta)\mathsf{M}(dx|T(y))]P_\theta(dy)]\Pi(d\theta) \\ &= \int [\int [\int g(x, \theta)\mathsf{M}(dx|t)]Q_\theta(dt)]\Pi(d\theta) \\ &= \int [\int [\int g(x, \theta)\mathsf{M}(dx|t)]\mathsf{L}(d\theta|t)](\mathsf{Q}\Pi)(dt).\end{aligned}$$

Hence with $h(x,t) = \int g(x, \theta)\mathsf{L}(d\theta|t)$ and (4.42),

$$\begin{aligned}&\int [\int g(x, \theta)P_\theta(dx)]\Pi(d\theta) = \int [\int [\int h(x, T(y))\mathsf{M}(dx|T(y))]P_{\widetilde{\theta}}(dy)]\Pi(d\widetilde{\theta}) \\ &= \int [\int h(x, T(x))P_{\widetilde{\theta}}(dx)]\Pi(d\widetilde{\theta}) = \int [\int [\int g(x, \theta)\mathsf{L}(d\theta|T(x))]P_{\widetilde{\theta}}(dx)]\Pi(d\widetilde{\theta}) \\ &= \int [\int g(x, \theta)\mathsf{L}(d\theta|T(x))](\mathsf{P}\Pi)(dx).\end{aligned}$$

This means that $\mathbb{P}(\Theta \in B|X = x) = \mathsf{L}(B|T(x))$, so that the Bayes sufficiency is established.

Conversely suppose that T is Bayes sufficient. Choose the prior $\Pi = \frac{1}{2}(\delta_{\theta_1} + \delta_{\theta_2})$ and put $f_{\theta_i} = 2dP_{\theta_i}/d(P_{\theta_1} + P_{\theta_2})$. Then $f_{\theta_1} + f_{\theta_2} = 2$ and

$$\mathbb{P}(\Theta = \theta_i|X = x) = \frac{2f_{\theta_i}(x)}{f_{\theta_1}(x) + f_{\theta_2}(x)} = f_{\theta_i}(x)$$

is a measurable function of T. From the factorization criterion in Theorem 4.50 we get that T is sufficient for $\{P_{\theta_1}, P_{\theta_2}\}$. Hence T is pairwise sufficient for $(P_\theta)_{\theta\in\Delta}$. The rest follows from Problem 4.38. ∎

Now we study the case where T is sufficient, not necessarily regular sufficient, and the model $(\mathcal{X}, \mathfrak{A}, (P_\theta)_{\theta\in\Delta})$ is dominated by some $\mu \in \mathcal{M}^\sigma(\mathfrak{A})$. We use the notations in Section 1.2, and take $\mu = \overline{P}$ from Lemma 4.36 as the dominating measure. Then $g_\theta(T) = dP_\theta/d\overline{P}$ by Theorem 4.50. Suppose that condition (A5) is satisfied for $\boldsymbol{\mu} = \overline{P}$. If $T : \mathcal{X} \to_m \mathcal{Y}$ is sufficient for $(P_\theta)_{\theta\in\Delta}$, then we obtain the marginal and the posterior density, respectively,

$$\mathsf{m}(x) = \mathsf{n}(T(x)), \quad \mathsf{n}(t) = \int g_\theta(t)\pi(\theta)\boldsymbol{\tau}(d\theta),$$

$$\pi(\theta|x) = \xi(\theta|T(x)), \quad \xi(\theta|t) = \begin{cases} g_\theta(t)\pi(\theta)/\mathsf{n}(t) & \text{if } \mathsf{n}(t) > 0, \\ \pi(\theta) & \text{if } \mathsf{n}(t) = 0, \end{cases}$$

with g_θ in (4.41). Thus we have obtained the following result.

Proposition 4.62. *If the model $(\mathcal{X}, \mathfrak{A}, (P_\theta)_{\theta\in\Delta})$ is dominated by some $\mu \in \mathcal{M}^\sigma(\mathfrak{A})$, $T : \mathcal{X} \to_m \mathcal{Y}$ is sufficient for $(P_\theta)_{\theta\in\Delta}$, and $g_\theta(t)$ is a measurable function of (θ, t), then T is Bayes sufficient for $(P_\theta)_{\theta\in\Delta}$ and*

$$\mathbb{P}(\Theta \in B|X = x) = \int_B \xi(\theta|T(x))\boldsymbol{\tau}(d\theta), \quad \mathcal{L}(X|\mathbb{P})\text{-}a.s.$$

Example 4.63. Let $(P_\theta)_{\theta\in\Delta}$ be a natural exponential family with dominating measure $\boldsymbol{\mu}$ and generating statistic T. We consider the model $(\mathcal{X}^n, \mathfrak{A}^{\otimes n}, (P_\theta^{\otimes n})_{\theta\in\Delta})$, which by Proposition 1.4 is again a natural exponential family, but now with the $\boldsymbol{\mu}^{\otimes n}$-density $\exp\{\langle\theta, T_{\oplus n}\rangle - nK(\theta)\}$. For $(a, b) \in \Upsilon$, which was introduced in (1.36), we use the conjugate prior with the $\boldsymbol{\tau}$-density $\pi_{a,b} = \exp\{\langle b, \theta\rangle - aK(\theta) - L(a, b)\}$. From (1.43) in Lemma 1.35 we get that the posterior distribution of Θ, given $X = x$, is

$$\boldsymbol{\Pi}_{n,a,b}(B|x) = \int_B \pi_{a+n, b+T_{\oplus n}(x)}(\theta)\boldsymbol{\tau}(d\theta), \quad x \in \mathcal{X}^n, \ B \in \mathfrak{B}_\Delta.$$

Hence the posterior distribution depends on the data x only through the statistic $T_{\oplus n}(x)$. This is in accordance with the fact that $T_{\oplus n}$ is sufficient; see Example 4.51.

The display below gives relations among different types of sufficiency.

Relations between types of sufficiency

Regular sufficient	$\longrightarrow$	Blackwell sufficient
	$\searrow$	
$\downarrow$	Sufficient	$\downarrow$
	$\searrow$	
Bayes sufficient	$\longrightarrow$	Pairwise sufficient

For further relations among different concepts of sufficiency under additional assumptions on the model see Problem 4.38 and Propositions 4.58 and 4.61.

For an i.i.d. sample from a member of a family of distributions that are equivalent to the Lebesgue measure on $\mathbb{R}$ the reduction by a one-dimensional sufficient statistic works only for exponential families. This is a very important fact, as it clarifies the role of reduction by sufficiency as well as the role of exponential families in statistical analysis. Following up on previous results by Koopman (1936), Pitman (1936), Dynkin (1951), Borges and Pfanzagl (1963), Brown (1964), Denny (1970), and Pfanzagl (1972), the next theorem, due to Hipp (1974), establishes this fact for statistics that satisfy the following condition. A statistic $T : \mathbb{R}^n \to_m \mathbb{R}$ is called *locally Lipschitz* if for every $x \in \mathbb{R}^n$, there exists a $c \geq 0$ and an open set $U \subseteq \mathbb{R}^n$ with $x \in U$ such that $(T(y) - T(z))^2 \leq c\sum_{i=1}^n (y_i - z_i)^2$ for all $y, z \in U$.

Theorem 4.64. *Let $(P_\theta)_{\theta\in\Delta}$ be a family of distributions on the Borel sets of $\mathbb{R}$, where each P_θ is equivalent to the Lebesgue measure on $\mathbb{R}$. If for some $n \geq 2$ there exists a statistic $T : \mathbb{R}^n \to_m \mathbb{R}$ that is sufficient for $(P_\theta^{\otimes n})_{\theta\in\Delta}$ and locally Lipschitz, then $(P_\theta)_{\theta\in\Delta}$ is a one-parameter exponential family.*

An essential assumption in the last theorem is the equivalence of the distributions in the family $(P_\theta)_{\theta\in\Delta}$ to the Lebesgue measure on $\mathbb{R}$. This means that examples for dominated families which have a one-dimensional sufficient statistic but do not form an exponential family are those for which the supports of the distributions depend on the parameter. An example follows below; see also Problem 2.56.

Example 4.65. The statistic $M_n(x_1, ..., x_n) = \max\{x_1, ..., x_n\}$ is sufficient for $(\mathsf{U}^{\otimes n}(0,\theta))_{\theta>0}$. This follows from the fact that $\mathsf{U}^{\otimes n}(0,\theta)$ has the Lebesgue density $\theta^{-n} I_{[0,\theta]}(M_n)$ so that M_n is a sufficient statistic by the Neyman criterion.

Now we study how the different concepts of sufficiency are related to the decision-theoretically motivated concept of equivalence of models. Let the model $\mathcal{M} = (\mathcal{X}, \mathfrak{A}, (P_\theta)_{\theta\in\Delta})$ be given, $(\mathcal{Y}, \mathfrak{B})$ be another measurable space, and $T : \mathcal{X} \to_m \mathcal{Y}$. The reduced model $\mathcal{N} = (\mathcal{Y}, \mathfrak{B}, (Q_\theta)_{\theta\in\Delta})$ with $Q_\theta = P_\theta \circ T^{-1}$, $\theta \in \Delta$, is a randomization and thus at most as informative as $\mathcal{M}$, i.e., $\mathcal{N} \preceq \mathcal{M}$. If T is Blackwell sufficient, then we can find a kernel $\mathsf{M} : \mathfrak{A} \times \mathcal{Y} \to_k [0,1]$ such that $P_\theta = \mathsf{M}Q_\theta$ and thus $\mathcal{M} = \mathsf{M}\mathcal{N}$. Consequently, the two models $\mathcal{M}$ and $\mathcal{N}$ are mutual randomizations of each other and thus, by Corollary 4.10, they are equivalent. Beyond this equivalence we have even more. For every decision $\mathsf{D}_\mathcal{M}$ for the model $\mathcal{M}$ the decision $\mathsf{D}_\mathcal{N} := \mathsf{D}_\mathcal{M}\mathsf{M}$ for the model $\mathcal{N}$ has the same risk function,

$$\mathsf{R}(\theta, \mathsf{D}_\mathcal{M}) = \mathsf{R}(\theta, \mathsf{D}_\mathcal{N}), \quad \theta \in \Delta,$$

which follows from the fact that for every $\theta \in \Delta$,

$$\begin{aligned} \mathsf{R}(\theta, \mathsf{D}_\mathcal{M}) &= \int [\int L(\theta, a)\mathsf{D}_\mathcal{M}(da|x)]P_\theta(dx) \\ &= \int [\int L(\theta, a)\mathsf{D}_\mathcal{N}(da|y)]Q_\theta(dy) = \mathsf{R}(\theta, \mathsf{D}_\mathcal{N}). \end{aligned} \tag{4.44}$$

An analogous statement holds if T is sufficient, the model is dominated, and the decision space is a Borel space. The next theorem extends the results of

Theorem 4.18 and relates the different types of sufficiency to the decision-theoretically based concept of equivalence of models.

Theorem 4.66. *Assume that either T is Blackwell sufficient or that the model $(\mathcal{X}, \mathfrak{A}, (P_\theta)_{\theta\in\Delta})$ is dominated by some $\mu \in \mathcal{M}^\sigma(\mathfrak{A})$, $T : \mathcal{X} \to_m \mathcal{Y}$ is sufficient, and $(\mathcal{D}, \mathfrak{D})$ is a Borel space. Then for every decision $\mathsf{D} : \mathfrak{D} \times \mathcal{X} \to_k [0,1]$ there is a decision factorized by T which has the same risk function; i.e., there is a decision $\mathsf{C} : \mathfrak{D} \times \mathcal{Y} \to_k [0,1]$ such that*

$$\mathsf{R}(\theta, \mathsf{D}) = \mathsf{R}(\theta, \mathsf{C} \circ T), \quad \theta \in \Delta. \tag{4.45}$$

Corollary 4.67. *If $T : \mathcal{X} \to_m \mathcal{Y}$ is a statistic, then the following holds. The models $\mathcal{M} = (\mathcal{X}, \mathfrak{A}, (P_\theta)_{\theta\in\Delta})$ and $\mathcal{N} = (\mathcal{Y}, \mathfrak{B}, (Q_\theta)_{\theta\in\Delta})$ with $Q_\theta = P_\theta \circ T^{-1}$, $\theta \in \Delta$, are equivalent if and only if T is pairwise sufficient.*

Proof. In Theorem 4.66 the statement under the first assumption follows from (4.44) by setting $\mathsf{C} = \mathsf{D}_\mathcal{N}$. The statement under the second assumption follows from Theorems 4.18 and 4.50.

To prove Corollary 4.67, suppose $\mathcal{M}$ and $\mathcal{N}$ are equivalent. Then for every fixed $\theta_1, \theta_2 \in \Delta$ the models $(\mathcal{X}, \mathfrak{A}, \{P_{\theta_1}, P_{\theta_2}\})$ and $(\mathcal{Y}, \mathfrak{B}, \{Q_{\theta_1}, Q_{\theta_2}\})$ are also equivalent, and thus $\mathsf{H}_{1/2}(P_{\theta_1}, P_{\theta_2}) = \mathsf{H}_{1/2}(Q_{\theta_1}, Q_{\theta_2})$ by Theorem 4.25. Theorem 4.45 provides that T is sufficient for $\{P_{\theta_1}, P_{\theta_2}\}$ so that T is pairwise sufficient. Conversely, if T is pairwise sufficient, then it is sufficient for every finite submodel $(P_\theta)_{\theta\in F}$, see Problem 4.38. As $(P_\theta)_{\theta\in F}$ is dominated, and every finite decision space is a Borel space, we get the stated equivalence from $\mathsf{R}(\theta, \mathsf{D}) = \mathsf{R}(\theta, \mathsf{C}_F \circ T)$, $\theta \in F$, and the definition of equivalence in Definition 4.9. ■

From (4.45) we see the decision theoretic difference between dominated and undominated models. If we have a pairwise sufficient statistic T, then we can find for every decision D and every finite set $F \subseteq \Delta$ a decision C_F that gives the same risk for $\theta \in F$. In the dominated case we can put all these C_F together, i.e., find one universal C that satisfies (4.45).

Remark 4.68. (Reduction by Sufficiency) If either T is Blackwell sufficient or the model $\mathcal{M} = (\mathcal{X}, \mathfrak{A}, (P_\theta)_{\theta\in\Delta})$ is dominated, T is sufficient, and the decision space is a Borel space, then by Theorem 4.66 we may restrict all risk considerations to decisions that are factorized by T, i.e., that depend on $x \in \mathcal{X}$ only through $T(x)$. Following such a restriction is called a reduction by sufficiency.

4.4 Completeness, Ancillarity, and Minimal Sufficiency

On several occasions we have already been faced with the necessity of the family $(P_\theta \circ T^{-1})_{\theta\in\Delta}$ being sufficiently large. This becomes an important issue later on in Chapter 7 when we characterize uniformly best unbiased estimators. However, we have to deal with it already in this section for studying the interrelation of sufficiency and ancillarity. To give a precise formulation we introduce the concept of *completeness* which is due to Lehmann and Scheffé (1947, 1950). Before, it has been used implicitly already in Scheffé (1943) and Halmos (1946).

Definition 4.69. *For any model $(\mathcal{X}, \mathfrak{A}, (P_\theta)_{\theta\in\Delta})$ a statistic $T : \mathcal{X} \to_m \mathcal{Y}$ is called (boundedly) complete if for every (bounded) $h \in \bigcap_{\theta\in\Delta} \mathbb{L}_1(P_\theta \circ T^{-1})$ with $\int h(T)dP_\theta = 0$ for every $\theta \in \Delta$ it follows that $h(T) = 0$, P_θ-a.s., for every $\theta \in \Delta$. If T is complete, then we also say that the family $(P_\theta \circ T^{-1})_{\theta\in\Delta}$ is complete.*

Remark 4.70. If $g : \mathcal{Y} \to_m \mathcal{S}$ is one-to-one and the inverse mapping is also measurable, then $S = g(T)$ is (boundedly) complete if and only if T is (boundedly) complete.

It is clear that completeness implies boundedly completeness. An example which shows that the converse statement is not true in general can be found in Lehmann and Scheffé (1950). For recent results on this topic we refer to Mattner (1993).

Roughly speaking the completeness of a family of distributions means that it is sufficiently large. The next problem illustrates completeness in the dominated case.

Problem 4.71. If $(P_\theta)_{\theta\in\Delta}$ is dominated by $\mu \in \mathcal{M}^\sigma(\mathfrak{A})$ and the set of finite linear combinations of $\{dP_\theta/d\mu : \theta \in \Delta\}$ is dense in $\mathbb{L}_1(\mu)$, then $(P_\theta)_{\theta\in\Delta}$ is complete.

For families of discrete distributions completeness can be often directly verified.

Problem 4.72.* The family of distributions $\mathsf{B}(n,p)$, $p \in (0,1)$, on $\{0,1,...,n\}$ is complete.

The binomial distributions in the last problem are a special exponential family. Now we show that the generating statistic of any exponential family in natural form, in the sense of Definition 1.1, is complete. This fact, which appeared first in Sverdrup (1953), was established explicitly in Lehmann and Scheffé (1955).

Theorem 4.73. *Let $(\mathcal{X}, \mathfrak{A}, (P_\theta)_{\theta\in\Delta})$ be a d-parameter exponential family in natural form, with natural parameter $\theta \in \Delta$ and generating statistic $T : \mathcal{X} \to_m \mathbb{R}^d$, that satisfies conditions (A1) and (A2). If Δ_0 is a subset of Δ so that the interior Δ_0^0 of Δ_0 is nonempty, then for the model $(\mathcal{X}, \mathfrak{A}, (P_\theta)_{\theta\in\Delta_0})$ the statistic T is complete.*

Proof. Suppose that $h : \mathbb{R}^d \to_m \mathbb{R}$ satisfies $\mathsf{E}_\theta|h(T)| < \infty$, and $\mathsf{E}_\theta h(T) = 0$, $\theta \in \Delta_0$. Then with $d\mu_1 = h^+(T)d\boldsymbol{\mu}$ and $d\mu_2 = h^-(T)d\boldsymbol{\mu}$ it holds

$$\int \exp\{\langle\theta, T\rangle - K(\theta)\}d\mu_1 = \int \exp\{\langle\theta, T\rangle - K(\theta)\}d\mu_2$$

for every θ in some open rectangle. An application of Proposition 1.25 yields $\mu_1 = \mu_2$, which implies $h^+(T) = h^-(T)$, $\boldsymbol{\mu}$-a.e., which is equivalent to $h(T) = 0$, $\boldsymbol{\mu}$-a.e., and thus we get $h(T) = 0$, P_θ-a.s., as $P_\theta \ll \boldsymbol{\mu}$. ■

Example 4.74. The model $(\mathbb{R}^d, \mathfrak{B}_d, \mathsf{N}^{\otimes n}(\mu, \sigma^2)_{\mu\in\mathbb{R},\sigma^2>0})$ is, according to Example 1.11, an exponential family with natural parameter $\theta = (\mu/\sigma^2, -1/\sigma^2) \in \mathbb{R}\times(-\infty, 0)$ and generating statistic $T_{\oplus n}(x_1, ..., x_n) = (\sum_{i=1}^n x_i, \sum_{i=1}^n x_i^2)$. The latter is complete according to Theorem 4.73. It should be noted that the statistic $(\overline{X}_n, S_n^2)$ is also complete, which follows from Remark 4.70.

The next problem establishes the completeness of a family that is not an exponential family.

Problem 4.75.* The family of uniform distributions $\mathsf{U}(0, \theta)$, $\theta > 0$, is complete.

Now we consider a type of statistic that plays a role opposite to that of a sufficient statistic, in the sense that we cannot get any information about the parameter if we observe only this statistic.

Definition 4.76. *Given the model $(\mathcal{X}, \mathfrak{A}, (P_\theta)_{\theta\in\Delta})$ and a statistic $V : \mathcal{X} \to_m \mathcal{V}$, V is called ancillary if $P_\theta \circ V^{-1}$ is independent of θ for every $\theta \in \Delta$.*

To discuss jointly the concepts of ancillarity and sufficiency let us consider the following situation where $(\mathcal{X}, \mathfrak{A})$, $(\mathcal{U}, \mathfrak{U})$, and $(\mathcal{V}, \mathfrak{V})$ are measurable spaces. Suppose that $U : \mathcal{X} \to_m \mathcal{U}$ and $V : \mathcal{X} \to_m \mathcal{V}$ are statistics which carry the complete information in the sense that there exists a mapping

$$\psi : \mathcal{U} \times \mathcal{V} \to_m \mathcal{X} \quad \text{with} \quad \psi(U(x), V(x)) = x, \quad x \in \mathcal{X}. \tag{4.46}$$

The next theorem shows that for independent statistics U and V, whenever V is ancillary, the ancillary component V can be cancelled and the remaining component U is sufficient.

Theorem 4.77. *Assume that there exists a mapping ψ that satisfies (4.46) for two statistics $U : \mathcal{X} \to_m \mathcal{U}$ and $V : \mathcal{X} \to_m \mathcal{V}$ that are independent with respect to P_θ for every $\theta \in \Delta$. If now V is ancillary, then U is regular sufficient.*

Proof. Set $Q = P_\theta \circ V^{-1}$, which by assumption is independent of $\theta \in \Delta$. Introduce the stochastic kernel M by $\mathsf{M}(A|u) = \int I_A(\psi(u, v))Q(dv)$, $A \in \mathfrak{A}$, $u \in \mathcal{U}$. Then for every $A \in \mathfrak{A}$ and $B \in \mathfrak{U}$,

$$\begin{aligned}
&\int \mathsf{M}(A|u)\, I_B(u)\, (P_\theta \circ U^{-1})(du) \\
&= \int I_B(u)\, [\int I_A(\psi(u, v))Q(dv)](P_\theta \circ U^{-1})(du) \\
&= \int I_B(u)\, I_A(\psi(u, v))(P_\theta \circ (U, V)^{-1})(du, dv) = \int I_A(x)\, I_B(U(x))\, P_\theta(dx).
\end{aligned}$$

Consequently $\mathsf{E}_\theta(I_A|U = u) = \mathsf{M}(A|u)$ is independent of $\theta \in \Delta$ which implies the regular sufficiency. ■

The concept of ancillarity is somewhat subtle and deserves further explanation. If $P_\theta \circ V^{-1}$ is independent of θ, then an observation of V only does not provide any information about the unknown parameter. However, this does not mean that for any other statistic $W : \mathcal{X} \to_m \mathcal{W}$ the observation (W, V) contains the same information as W. Indeed, if W and V are not independent, then V may contribute some information about the unknown parameter via (W, V). To illustrate this issue we use an example from Pfanzagl (1994). First, a problem is given that serves as a preparation.

Problem 4.78.* Let $X_i \sim \mathsf{N}(0, \sigma_i^2)$ for $i = 1, 2$, and $L : \mathbb{R} \to \mathbb{R}_+$ be symmetric and strictly increasing on $\mathbb{R}_+$. If $\sigma_1^2 < \sigma_2^2$ and $\mathbb{E}L(X_1) < \infty$, then $\mathbb{E}L(X_1) < \mathbb{E}L(X_2)$.

Example 4.79. We consider the model $\mathcal{M} = \left(\mathbb{R}^n, \mathfrak{B}_n, (\mathsf{N}^{\otimes n}(\theta, a^2\theta^2)\right)_{\theta\in\mathbb{R}})$, where $a > 0$ is fixed, with the statistics $\overline{X}_n$ and S_n^2. Obviously, $\overline{X}_n / S_n$ is ancillary. Let us focus on two types of estimators of θ. The first is $\overline{X}_n$, and here we have $\mathcal{L}(\sqrt{n}\,(\overline{X}_n - \theta)) = \mathsf{N}(0, a^2\theta^2)$. The other is the maximum likelihood estimator (MLE) which is

$$T_n = \frac{1}{2a^2}[4a^2 S_n^2 + \left(1 + 4a^2\right)(\overline{X}_n)^2]^{1/2} - \overline{X}_n.$$

Then by Pfanzagl (1994), p. 305, the distribution of $\sqrt{n}\,(T_n - \theta)$ converges to $\mathsf{N}(0, a^2\theta^2/(1+2a^2))$ as $n \to \infty$. Hence by Problem 4.78, for any symmetric function L which is increasing, continuous, and bounded on $\mathbb{R}_+$ we have for all $\theta \in \mathbb{R}$ and sufficiently large n,

$$\mathsf{E}_\theta L\left(\sqrt{n}(\overline{X}_n - \theta)\right) > \mathsf{E}_\theta L\left(\sqrt{n}(T_n - \theta)\right)$$

which implies that $\left(\overline{X}_n, \overline{X}_n/S_n\right)$ contains more information about θ than $\overline{X}_n$ alone. Here, of course, $\overline{X}_n$ and $\overline{X}_n/S_n$ are not independent.

Problem 4.80.* Suppose that $X = (X_1, ..., X_n) \sim \mathsf{N}^{\otimes n}(\mu, \sigma^2)$. Then $\overline{X}_n$ and $V = \left(X_1 - \overline{X}_n, ..., X_n - \overline{X}_n\right)$ are independent.

Example 4.81. For $\mathcal{M} = \left(\mathbb{R}^n, \mathfrak{B}_n, (\mathsf{N}^{\otimes n}(\mu, \sigma^2))_{\mu\in\mathbb{R}}\right)$, where σ^2 is known, we set $U(x_1, ..., x_n) = \overline{x}_n$ and $V(x_1, ..., x_n) = (x_1 - \overline{x}_n, ..., x_n - \overline{x}_n)$. For $\psi(u, v_1, ..., v_n) = (v_1 + u, ..., v_n + u)$ we have $\psi(U, V_1, ..., V_n)(x_1, ..., x_n) = (x_1, ..., x_n)$. Since V is ancillary and independent of U it follows that U is sufficient for $(\mathsf{N}^{\otimes n}(\mu, \sigma^2))_{\mu\in\mathbb{R}}$.

In Theorem 4.77 we have decomposed the observation in two independent components U and V and have shown that ancillarity of V implies that U is sufficient. The next theorem, due to Basu (1955, 1958), deals with the converse direction as it shows that independence is necessary under weak additional assumptions.

Theorem 4.82. *Let $(\mathcal{U}, \mathfrak{U})$ and $(\mathcal{V}, \mathfrak{V})$ be measurable spaces, $U : \mathcal{X} \to_m \mathcal{U}$ a sufficient statistic, and $V : \mathcal{X} \to_m \mathcal{V}$ another arbitrary statistic. Then the following hold.*
(A) If $(P_\theta)_{\theta\in\Delta} \ll P_{\theta_0}$ for some $\theta_0 \in \Delta$ and U and V are independent with respect to P_{θ_0}, then V is ancillary.
(B) If V is ancillary and $\left(P_\theta \circ U^{-1}\right)_{\theta\in\Delta}$ is boundedly complete, then U and V are independent with respect to P_θ for every $\theta \in \Delta$.

Proof. First we prove (A). Using the function $k_A : \mathcal{U} \to_m \mathbb{R}$ that appears in Definition 4.37, where T corresponds to U, we get

$$\int I_B(U) I_C(V)\, dP_\theta = \int I_B(U) k_{V^{-1}(C)}(U)\, dP_\theta, \quad \theta \in \Delta, \tag{4.47}$$

for $B \in \mathfrak{U}$ and $C \in \mathfrak{V}$. The independence of U and V under P_{θ_0} yields

$$0 = \int \left[P_{\theta_0}(V \in C) - k_{V^{-1}(C)}(U)\right] I_B(U)\, dP_{\theta_0}.$$

As B was arbitrary and $P_\theta \ll P_{\theta_0}$ we get $P_{\theta_0}(V \in C) = k_{V^{-1}(C)}(U)$, P_θ-a.s., $\theta \in \Delta$. Hence by (4.47) with $B = \mathcal{U}$,

$$P_{\theta_0}(V \in C) = \int k_{V^{-1}(C)}(U)\, dP_\theta = P_\theta(V \in C).$$

To prove (B), we fix $\theta_1 \in \Delta$. Then by $P_{\theta_1}(V \in C) = P_\theta(V \in C)$ and (4.47) with $B = \mathcal{U}$,

$$\int \left[P_{\theta_1}(V \in C) - k_{V^{-1}(C)}(U)\right] dP_\theta = 0, \quad \theta \in \Delta.$$

The boundedly completeness of $\left(P_\theta \circ U^{-1}\right)_{\theta \in \Delta}$ yields

$$P_{\theta_1}(V \in C) = k_{V^{-1}(C)}(U), \quad P_\theta\text{-a.s.}, \theta \in \Delta.$$

Hence by (4.47)

$$P_\theta(U \in B,\ V \in C) = \int I_B(U) I_C(V)\, dP_\theta = \int I_B(U) k_{V^{-1}(C)}(U)\, dP_\theta$$
$$= \int I_B(U) P_{\theta_1}(V \in C)\, dP_\theta = P_\theta(U \in B) P_\theta(V \in C).$$

■

Example 4.83. Let $X_1, ..., X_n$ be an i.i.d. sample from a gamma distribution $\mathsf{Ga}(\lambda, \beta)$ with Lebesgue density $\mathsf{ga}_{\lambda,\beta}(x) = I_{(0,\infty)}(x) \beta^\lambda \Gamma(\lambda)^{-1} x^{\lambda-1} \exp\{-\beta x\}$, $\lambda, \beta > 0$. We consider the statistics $\overline{X}_n$ and

$$T_n(X_1, ..., X_n) = \overline{X}_n \left[\prod\nolimits_{i=1}^n X_i\right]^{-n}.$$

Fix $\lambda > 0$. As $\{(\mathsf{Ga}(\lambda, \beta))^{\otimes n} : \beta > 0\}$ is an exponential family generated by $\overline{X}_n$ we see that the statistic $\overline{X}_n$ is sufficient and complete for $\{(\mathsf{Ga}(\lambda, \beta))^{\otimes n} : \beta > 0\}$. The distribution of T_n does not depend on β. Hence T_n and $\overline{X}_n$ are stochastically independent under $(\mathsf{Ga}(\lambda, \beta))^{\otimes n}$ for every $\lambda, \beta > 0$.

Example 4.84. Suppose that $X = (X_1, ..., X_n) \sim \mathsf{N}^{\otimes n}(\mu, \sigma^2)$. We consider the statistics $\overline{X}_n$ and S_n^2. S_n^2 is a measurable function of $V = \left(X_1 - \overline{X}_n, ..., X_n - \overline{X}_n\right)$ and thus by Problem 4.80 independent of $\overline{X}_n$.

If $\sigma^2 > 0$ is fixed, then $\overline{X}_n$ is a generating statistic for the exponential family $(\mathsf{N}^{\otimes n}(\mu, \sigma^2))_{\mu \in \mathbb{R}}$ and therefore complete. As S_n^2 is ancillary, by Theorem 4.82 $\overline{X}_n$ and S_n^2 are independent, which provides an alternative proof to that in Problem 4.80.

Obviously, the identical mapping is always a sufficient statistic, but it does not provide any reduction of the data. For a given model there is usually more than one sufficient statistic. The question arises as to whether there are sufficient statistics that reduce the data by mapping multiple points into single points, and if so, how far this reduction can be pushed without losing sufficiency. Clearly, a sufficient statistic S reduces the data more than a sufficient statistic T if S is a function of T, but T is not a function of S. In this case the range of S is in a way smaller than the range of T. This leads to the concept of a *minimal sufficient statistic.*

Definition 4.85. *Given the model $(\mathcal{X}, \mathfrak{A}, (P_\theta)_{\theta\in\Delta})$ and a sufficient statistic $T : \mathcal{X} \to_m \mathcal{Y}$, T is called minimal sufficient if for any other sufficient statistic S that takes values in some $(\mathcal{S}, \mathfrak{S})$ there exists a mapping $h : \mathcal{S} \to_m \mathcal{Y}$ such that $T = h(S)$, P_θ-a.s., for every $\theta \in \Delta$.*

Example 4.86. Given the model $(\mathcal{X}, \mathfrak{A}, (P_\theta)_{\theta\in\Delta})$, let $\mathbb{R}^\Delta$ be the set of all mappings $\psi : \Delta \to \mathbb{R}$ and let Z_θ, $\theta \in \Delta$, be the family of evaluation maps, i.e., $Z_\theta(\psi) = \psi(\theta)$. Let $\mathfrak{B}^{\otimes\Delta}$ be the σ-algebra of subsets of $\mathbb{R}^\Delta$ generated by the evaluation maps. Assume that the family $\mathcal{P} = (P_\theta)_{\theta\in\Delta}$ is dominated and take $\overline{P}$ from Lemma 4.36 as the dominating measure. Set $f_\theta = dP_\theta/d\overline{P}$ and introduce $T : \mathcal{X} \to \mathbb{R}^\Delta$ by $T : x \mapsto (f_\theta(x))_{\theta\in\Delta}$. The sufficiency of T follows from Theorem 4.50. However, T is even minimal sufficient. Indeed, if $S : \mathcal{X} \to_m \mathcal{S}$ is a sufficient statistic, then by Theorem 4.50 there are functions $g_\theta : \mathcal{S} \to_m \mathbb{R}$ so that $dP_\theta/d\overline{P} = g_\theta(S)$, $\mathcal{P}$-a.s. Introduce $h : \mathcal{S} \to_m \mathbb{R}^\Delta$ by $h(s) = (g_\theta(s))_{\theta\in\Delta}$. Then $T = h(S)$ and therefore T is minimal sufficient. If $\Delta = \{1, ..., m\}$, then M in (4.11) is minimal sufficient.

Next a sufficient condition is given for a statistic to be minimal sufficient.

Theorem 4.87. *Assume that the model $(\mathcal{X}, \mathfrak{A}, (P_\theta)_{\theta\in\Delta})$ is dominated and $(\mathcal{Y}, \mathfrak{B})$ is a Borel space. If $T : \mathcal{X} \to_m \mathcal{Y}$ is sufficient and boundedly complete, then T is minimal sufficient.*

Proof. We follow Pfanzagl (1994). Suppose $S : \mathcal{X} \to_m \mathcal{S}$ is any sufficient statistic. Let the dominating measure be chosen to be $\overline{P}$ from Lemma 4.36. Sufficiency of S implies that for every $A \in \mathfrak{A}$ there exists a function $k_A : \mathcal{S} \to_m [0, 1]$ such that $k_A(S) = \mathsf{E}_\theta(I_A|S)$, P_θ-a.s. for every $\theta \in \Delta$. Sufficiency of T implies that there is a function l_A such that $l_A(T) = \mathsf{E}_\theta(k_A(S)|T)$, P_θ-a.s. for every $\theta \in \Delta$. Hence for $A = T^{-1}(B)$, $B \in \mathfrak{B}$, it holds

$$\int l_{T^{-1}(B)}(T)dP_\theta = \int I_{T^{-1}(B)}dP_\theta = \int I_B(T)dP_\theta, \quad P_\theta\text{-a.s. for every } \theta \in \Delta.$$

Boundedly completeness of T yields $l_{T^{-1}(B)}(T) = I_B(T)$, P_θ-a.s. for every $\theta \in \Delta$, and by the definition of $k_{T^{-1}(B)}(S)$ and $l_{T^{-1}(B)}(T)$ we have

$$\begin{aligned}&\int I_C(T)(I_B(T) - k_{T^{-1}(B)}(S))dP_\theta \\ &= \int I_C(T)(l_{T^{-1}(B)}(T) - k_{T^{-1}(B)}(S))dP_\theta = 0, \quad C \in \mathfrak{B}.\end{aligned}$$

As $\overline{P}$ is a convex linear combination of some distributions P_θ we get

$$\int I_C(T)[I_B(T) - k_{T^{-1}(B)}(S)]d\overline{P} = 0, \quad C \in \mathfrak{B},$$
$$\int I_B(T)[I_B(T) - k_{T^{-1}(B)}(S)]d\overline{P} = 0, \text{ and } \int I_{\overline{B}}(T)k_{T^{-1}(B)}(S)d\overline{P} = 0.$$

Hence by $0 \le k_{T^{-1}(B)}(S) \le 1$,

$$\int |I_B(T) - k_{T^{-1}(B)}(S)|d\overline{P}$$
$$\le \int I_B(T)[I_B(T) - k_{T^{-1}(B)}(S)]d\overline{P} + \int I_{\overline{B}}(T)k_{T^{-1}(B)}(S)d\overline{P} = 0,$$

and $I_B(T) = k_{T^{-1}(B)}(S)$, $\overline{P}$-a.s. This means that $I_B(T)$ is $\overline{P}$-a.s. identical with some $\sigma(S)$-measurable function. It remains to show that there is a $g : \mathcal{Y} \to_m \mathcal{S}$ such that $T = g(S)$, $\overline{P}$-a.s. As the Borel space $\mathcal{Y}$ is Borel isomorphic to a Borel set of $[0,1]$ this statement has to be proved only for a real-valued T. But then the statement follows via a pointwise approximation by linear combinations of indicator functions. ■

An application of the last theorem, in combination with Theorem 4.73, to exponential families gives the following statement.

Proposition 4.88. *Under the assumptions of Theorem 4.73 the generating statistic T in an exponential family with natural parameter $\theta \in \Delta$ is minimal sufficient for the model $(\mathcal{X}, \mathfrak{A}, (P_\theta)_{\theta \in \Delta_0})$.*

4.5 Solutions to Selected Problems

Solution to Problem 4.6: The inequality $|Z| \le |Y| + |X|$ shows that Z has a finite second moment. Hence $\sigma_2^2 = \mathbb{V}(X + Z) = \sigma_1^2 + \mathbb{V}(Z) \ge \sigma_1^2$. Denote by $\varphi_X, \varphi_X, \varphi_Z$ the characteristic functions of X, Y, Z. Then $\exp\{it\mu_2 - \frac{1}{2}\sigma_2^2 t^2\} = \exp\{it\mu_1 - \frac{1}{2}\sigma_1^2 t^2\}\varphi_Z(t)$, which proves $\mathcal{L}(Z) = \mathsf{N}(\mu_2 - \mu_1, \sigma_2^2 - \sigma_1^2)$. □

Solution to Problem 4.7: If $\Sigma_1 \preceq \Sigma_2$, then $\Sigma_2 - \Sigma_1$ is positive semidefinite and $\mathsf{N}(\mu, \Sigma_2) = \mathsf{N}(\mu, \Sigma_1) * \mathsf{N}(\mu, \Sigma_2 - \Sigma_1)$. Conversely, there exists a nonsingular matrix A such that $A^T \Sigma_1 A = \mathbf{I}$ is the unit matrix and $A^T \Sigma_2 A = \Lambda$ is a diagonal matrix with diagonal elements $\lambda_1 \ge \cdots \ge \lambda_d \ge 0$; see Anderson (1984), p. 589. Hence we may assume that $\Sigma_1 = \mathbf{I}$ and $\Sigma_2 = \Lambda$. Then $\mathsf{N}(\mu, \Sigma_1) = \bigotimes_{i=1}^d \mathsf{N}(\mu_i, 1)$ and $\mathsf{N}(\mu, \Sigma_2) = \bigotimes_{i=1}^d \mathsf{N}(\mu_i, \lambda_i)$, $\mu \in \mathbb{R}^d$. Hence

$$\mathsf{H}_s(\mathsf{N}(0, \Sigma_1), \mathsf{N}(\mu, \Sigma_1) = \exp\{-s(1-s)\sum\nolimits_{i=1}^d \mu_i^2\}$$
$$\ge \mathsf{H}_s(\mathsf{N}(0, \Sigma_2)\mathsf{N}(\mu, \Sigma_2)) = \exp\{-s(1-s)\sum\nolimits_{i=1}^d \mu_i^2/\lambda_i\}.$$

As the μ_i are arbitrary it follows $\lambda_i \le 1$, $i = 1, ..., d$. □

Solution to Problem 4.11: We have only to show that $\mathcal{M}$ is a randomization of $\mathcal{N}$. We extend S by setting $S(y) = x_0$ if $y \notin B$, where x_0 is any fixed point. Fix $A \in \mathfrak{A}$. Then for $x_0 \notin A$ it holds $\{y : S(y) \in A\} = \{y : S(y) \in A\} \cap B \in \mathfrak{B}$. If $x_0 \in A$, then $\{y : S(y) \in A\} = (\{y : S(y) \in A\} \cap B) \cup \overline{B} \in \mathfrak{B}$ so that $S : \mathcal{Y} \to \mathcal{X}$ is measurable. As $(P_\theta \circ T^{-1})(\overline{B}) = 0$ and $T : \mathcal{X} \longleftrightarrow B$ we get $(P_\theta \circ T^{-1}) \circ S^{-1} = P_\theta$. Then $P_\theta = \mathsf{L}(P_\theta \circ T^{-1})$ where $\mathsf{L} = \boldsymbol{\delta}_S$. □

Solution to Problem 4.19: This follows from Lemma A.15. □

Solution to Problem 4.24: Let $\mathsf{D}_{\mathcal{M}_{2,\alpha}}$ be a decision for the finite decision space $\mathcal{D}$ and the model $\mathcal{M}_{2,\alpha}$. As $\mathcal{M}_2$ has the same sample space it is also a decision for $\mathcal{M}_2$. Then by (C) in Theorem 4.14 there is a decision $D_{\mathcal{M}_1}$ such that for $j = 1, ..., m$

$$\int \mathsf{D}_{\mathcal{M}_1}(\{a\}|x_1)P_{1,j}(dx_1) = \int \mathsf{D}_{\mathcal{M}_{2,\alpha}}(\{a\}|x_2)P_{2,j}(dx_2), \quad \text{and thus}$$
$$\int \mathsf{D}_{\mathcal{M}_1}(\{a\}|x_1)\overline{P}_1(dx_1) = \int \mathsf{D}_{\mathcal{M}_{2,\alpha}}(\{a\}|x_2)\overline{P}_2(dx_2).$$

Taking the convex linear combination with weights $1-\alpha$ and α we get for $j = 1, ..., m$

$$\int \mathsf{D}_{\mathcal{M}_1}(\{a\}|x_1)P_{1,j,\alpha}(dx_1) = \int \mathsf{D}_{\mathcal{M}_{2,\alpha}}(\{a\}|x_2)P_{2,j,\alpha}(dx_2).$$

Putting $\mathsf{D}_{\mathcal{M}_{1,\alpha}} = \mathsf{D}_{\mathcal{M}_1}$ we see from (C) in Theorem 4.14 that $\mathcal{M}_{1,\alpha} \succeq \mathcal{M}_{2,\alpha}$. Interchanging the role of $\mathcal{M}_{1,\alpha}$ and $\mathcal{M}_{2,\alpha}$ we get the statement. □

Solution to Problem 4.30: The concavity is obvious. If t_1 and t_2 are fixed, then

$$\inf_{y\in D} \langle y, t_1\rangle \le \inf_{y\in D} \langle y, t_2\rangle + \sup_{y\in D} |\langle y, t_1 - t_2\rangle| \quad \text{and}$$
$$\inf_{y\in D} \langle y, t_2\rangle \le \inf_{y\in D} \langle y, t_1\rangle + \sup_{y\in D} |\langle y, t_1 - t_2\rangle|.$$

The proof follows from

$$|\langle y, t_1 - t_2\rangle| = |\sum_{i=1}^m y_i(t_{1,i} - t_{2,i})| \le \|y\|_u \sum_{i=1}^m |t_{1,i} - t_{2,i}|$$
$$\le m\,\|y\|_u\,\|t_2 - t_1\|_u\,. \qquad \square$$

Solution to Problem 4.38: We must show that pairwise sufficiency implies sufficiency. Let $\overline{P} = \sum_{j=1}^\infty c_j P_{\theta_j}$ be from Lemma 4.36. Set $f_\theta = dP_\theta/d\overline{P}$ and $g_\theta = E_{\overline{P}}(f_\theta|T)$. As $E_{\overline{P}} I_A(\sum_{j=1}^\infty c_j f_{\theta_j}) = \overline{P}(A)$ for every $A \in \mathfrak{A}$ the function $\sum_{j=1}^\infty c_j f_{\theta_j}$ is a version of the density of $\overline{P}$ with respect to $\overline{P}$. Hence

$$\sum\nolimits_{j=1}^\infty c_j f_{\theta_j} = 1 \text{ and } \sum\nolimits_{j=1}^\infty c_j g_{\theta_j} = 1, \quad \overline{P}\text{-a.s.} \tag{4.48}$$

Denote by Q_θ and $\overline{Q}$ the restrictions of P_θ and $\overline{P}$ on $\sigma(T)$. Then as in Problem 1.74 $Q_\theta \ll \overline{Q}$ and $g_\theta = dQ_\theta/d\overline{Q}$. Hence for every $B \in \sigma(T)$,

$$\mathsf{E}_{\overline{P}}(I_B I_A f_\theta) = \mathsf{E}_\theta(I_B \mathsf{E}_\theta(I_A|T)) = \mathsf{E}_{\overline{P}}(I_B g_\theta \mathsf{E}_\theta(I_A|T))$$

for every $A \in \mathfrak{A}$, $B \in \sigma(T)$, which implies

$$\mathsf{E}_{\overline{P}}(I_A f_\theta|T) = g_\theta \mathsf{E}_\theta(I_A|T), \quad \overline{P}\text{-a.s.} \tag{4.49}$$

By the pairwise sufficiency, for every $A \in \mathfrak{A}$ there are $\sigma(T)$-measurable functions $k_{A,\theta,j} : \mathcal{X} \to_m \mathbb{R}$ such that

$$k_{A,\theta,j} = \mathsf{E}_\theta(I_A|T), \quad P_\theta\text{-}a.s. \quad \text{and} \quad k_{A,\theta,j} = \mathsf{E}_{\theta_j}(I_A|T), \quad P_{\theta_j}\text{-a.s.} \tag{4.50}$$

Set $A_j = \{g_{\theta_j} > 0\}$. Then $\overline{Q}(\cdot \cap A_j) \sim Q_{\theta_j}(\cdot \cap A_j)$ and thus $Q_\theta(\cdot \cap A_j) \ll Q_{\theta_j}(\cdot \cap A_j)$. As the functions that appear in (4.50) are $\sigma(T)$-measurable we may replace P_θ with Q_θ and P_{θ_j} with Q_{θ_j}. This together with $Q_\theta(\cdot \cap A_j) \ll Q_{\theta_j}(\cdot \cap A_j)$ yields

$$\begin{aligned} I_{A_j}\mathsf{E}_\theta(I_A|T) &= I_{A_j}\mathsf{E}_{\theta_j}(I_A|T), \quad Q_\theta\text{-a.s.,} \quad \text{and} \\ g_{\theta_j}\mathsf{E}_\theta(I_A|T) &= g_{\theta_j}\mathsf{E}_{\theta_j}(I_A|T), \quad P_\theta\text{-a.s.} \end{aligned}$$

By (4.48), (4.49), and (4.50),

$$\begin{aligned} \mathsf{E}_{\overline{P}}(I_A|T) &= \sum\nolimits_{j=1}^{\infty} c_j \mathsf{E}_{\overline{P}}(I_A f_{\theta_j}|T) = \sum\nolimits_{j=1}^{\infty} c_j g_{\theta_j} \mathsf{E}_{\theta_j}(I_A|T) \\ &= \sum\nolimits_{j=1}^{\infty} c_j g_{\theta_j} \mathsf{E}_\theta(I_A|T) = \mathsf{E}_\theta(I_A|T), \quad P_\theta\text{-a.s.,} \end{aligned}$$

is a version of the conditional expectation that is independent of θ. □

Solution to Problem 4.40: Define $k_S(T) = \sum_{i=1}^n c_i k_{A_i}$ for $S = \sum_{i=1}^n c_i I_{A_i}$. Approximate any nonnegative S by an increasing sequence of step functions S_n and set for the nondecreasing sequence k_{S_n} $k_S = \lim_{n\to\infty} k_{S_n}$. □

Solution to Problem 4.41: By assumption we find for every $A \in \mathfrak{A}$ a set $B \in \mathfrak{B}$ with $A = T^{-1}(B)$. Then $k_A(T) = I_B(T)$ is $\sigma(T)$-measurable, and for every $C \in \mathfrak{B}$ it holds that $\mathsf{E}_\theta I_A I_C(T) = P_\theta(T \in B \cap C) = \mathsf{E}_\theta k_A(T) I_C(T)$. □

Solution to Problem 4.43: Fix $A \in \mathfrak{A}$ and $B \in \mathfrak{A}$. Then

$$\int I_A(x) I_B(T(x)) P_{\theta_j}(dx) = \int k_A(T(x)) I_B(T(x)) P_{\theta_j}(dx)$$

by (4.36) for every j. Multiplying this equation by c_j and taking the sum we see that $k_A(T)$ is also a version of $\mathsf{E}_{\overline{P}}(I_A|T)$ where $\overline{P} = \sum_{j=1}^{\infty} c_j P_{\theta_j}$. □

Solution to Problem 4.44: By assumption $dP/dQ = g(T)$ for some measurable function g of T. Then for any fixed $B \in B$,

$$\begin{aligned}
&\int I_B(T)\mathsf{E}_Q(h|\sigma(T))dP \\
&= \int I_B(T)\mathsf{E}_Q(h|\sigma(T))dP = \int \mathsf{E}_Q(I_B(T)h|\sigma(T))g(T)dQ \\
&= \int \mathsf{E}_Q(I_B(T)hg(T)|\sigma(T))dQ = \int I_B(T)g(T)hdQ \\
&= \int I_B(T)hdP. \qquad \square
\end{aligned}$$

Solution to Problem 4.48: Use Theorem 2.62. □

Solution to Problem 4.53: The sufficiency of $(\sum_{i=1}^n X_i, \sum_{i=1}^n X_i^2)$ follows from Example 4.51. To prove the sufficiency of $(\overline{X}_n, S_n^2)$ use Problem 4.41. □

Solution to Problem 4.54: By the definition of MLR the likelihood ratio L_{θ_0,θ_1} of P_{θ_1} with respect to P_{θ_0} is a function of T. Hence $dP_{\theta_i}/d(\frac{1}{2}(P_{\theta_0}+P_{\theta_1}))$, $i=0,1$, are also measurable functions of T. The factorization criterion yields the statement. □

Solution to Problem 4.56: Combine Problem 4.38 with Theorem 4.45. □

Solution to Problem 4.57: Use Problem 4.56 and Lemma 1.106. □

Solution to Problem 4.72: Put $p = \eta/(1+\eta)$. Then

$$0 = (1+\eta)^{-n} \sum\nolimits_{k=0}^{n} h(k)\tbinom{n}{k}\eta^k$$

for every $0 < \eta < \infty$, which implies $h(k) = 0$ for $k = 0, 1, ..., n$. □

Solution to Problem 4.75: $\mathsf{E}_\theta|h| < \infty$ and $\mathsf{E}_\theta h = 0$ for every θ imply that $d\mu^+ = h^+ d\lambda$ and $d\mu^- = h^- d\lambda$ are locally finite measures that are identical on $[0,\theta]$. Hence $\mu^+([a,b)) = \mu^-([a,b))$ for every $0 \le a < b < \infty$ so that $\mu^+ = \mu^-$ by the uniqueness theorem for σ-finite measures. As the densities h^+, h^- are $\boldsymbol{\lambda}$-a.e. uniquely determined it follows that $h = 0$, $\boldsymbol{\lambda}$-a.e., and thus $\mathsf{U}(0,\theta)$-a.s. □

Solution to Problem 4.78: It holds,

$$\mathbb{E}L(X_i) = 2\int_0^\infty L(t)\varphi_{0,\sigma_i^2}(t)dt = 2\int_0^\infty L(\sigma_i s)\varphi_{0,1}(s)ds, \quad i = 1,2.$$

Therefore,

$$\mathbb{E}L(X_2) - \mathbb{E}L(X_1) = 2\int_0^\infty [L(\sigma_2 s) - L(\sigma_1 s)]\varphi_{0,1}(s)ds > 0,$$

as L is strictly increasing and $\sigma_2 > \sigma_1$. □

Solution to Problem 4.80: $\left(X_1 - \overline{X}_n, ..., X_n - \overline{X}_n, \overline{X}_n\right)$ is a linear function of X and thus again normal. The independence follows from $\mathsf{cov}(X_1 - \overline{X}_n, \overline{X}_n) = 0$. □

5

Invariant Statistical Decision Models

5.1 Invariant Models and Invariant Statistics

In this section we study statistical models and decision problems that have special invariance properties. By adopting the principle of invariance the search for best decisions may then be restricted to the class of invariant decisions. To introduce the concept of invariance, and to get results on optimal invariant decisions, a suitable mathematical framework has to be established.

Given a set $\mathcal{X}$ we denote by $\mathcal{S}_{\mathcal{X}}$ the set of all one-to-one mappings $u : \mathcal{X} \to \mathcal{X}$ of $\mathcal{X}$ onto $\mathcal{X}$. With the composition $(u \circ v)(x) := u(v(x))$ the set $\mathcal{S}_{\mathcal{X}}$ becomes a group. This group is known as the *symmetric group*. The identical mapping is the unit element in this group. For a measurable space $(\mathcal{X}, \mathfrak{A})$ we denote by $\mathcal{S}_{m,\mathcal{X}} \subseteq \mathcal{S}_{\mathcal{X}}$ the set of all $u \in \mathcal{S}_{\mathcal{X}}$ for which u and the inverse mapping u^{-1} are measurable with respect to the σ-algebra $\mathfrak{A}$. Such mappings are called *bimeasurable bijections*. Instead of $u \in \mathcal{S}_{m,\mathcal{X}}$ we also write $u : \mathcal{X} \leftrightarrow_m \mathcal{X}$.

Problem 5.1. If $u \in \mathcal{S}_{\mathcal{X}}$ and $B \subseteq \mathcal{X}$, then the inverse image of a set is the image of the set under the inverse mapping, i.e., $\{x : u(x) \in B\} = \{u^{-1}(y) : y \in B\}$. Moreover, for $u \in \mathcal{S}_{m,\mathcal{X}}$ and $A \in \mathfrak{A}$ it holds $u(A) \in \mathfrak{A}$.

For a given statistical model $(\mathcal{X}, \mathfrak{A}, (P_\theta)_{\theta \in \Delta})$ we study subgroups of $\mathcal{S}_{m,\mathcal{X}}$ that act on the sample space and leave the model unchanged. Often such subgroups are parametrized in a natural way by some parameter γ that belongs to a multiplicative group $\mathcal{G}$.

Definition 5.2. *Let $\mathcal{G}$ be a group that satisfies (5.1) and let $\mathcal{U} = \{u_\gamma : \gamma \in \mathcal{G}\}$ be a subgroup of $\mathcal{S}_{m,\mathcal{X}}$. We call $\mathcal{U}$ a group of measurable transformations if $\gamma \mapsto u_\gamma$ is a homomorphism; that is, $u_{\gamma_1}(u_{\gamma_2}(x)) = u_{\gamma_1\gamma_2}(x)$ for $\gamma_1, \gamma_2 \in \mathcal{G}$ and $x \in \mathcal{X}$.*

Often $\mathcal{G}$ is endowed with a σ-algebra $\mathfrak{G}$ so that the following holds.

$$\begin{aligned} &(\gamma_1, \gamma_2) \mapsto \gamma_1\gamma_2 \quad \text{is } (\mathfrak{G} \otimes \mathfrak{G})\text{-}\mathfrak{G} \text{ measurable}, \\ &\gamma \mapsto \gamma^{-1} \quad \text{is } \mathfrak{G}\text{-}\mathfrak{G} \text{ measurable}. \end{aligned} \tag{5.1}$$

F. Liese, K.-J. Miescke, *Statistical Decision Theory*,
DOI: 10.1007/978-0-387-73194-0_5,

To average over γ we need $u_\gamma(x)$ to be a measurable function of (x, γ), i.e.,

$$(x, \gamma) \mapsto u_\gamma(x), \quad \text{is } (\mathfrak{A} \otimes \mathfrak{G})\text{-}\mathfrak{A} \text{ measurable.} \tag{5.2}$$

Groups of measurable transformations that act on statistical models are not studied here in full generality. We rather consider only such groups $\mathcal{G}$ that are either finite or Borel subsets of an Euclidean space $\mathbb{R}^m$. In the latter case the sample space $\mathcal{X}$ is also an Euclidean space, say $\mathcal{X} = \mathbb{R}^n$, and the group of operations as well as $(x, \gamma) \mapsto u_\gamma(x)$ are seen to be continuous mappings. The measurability conditions in (5.1) and (5.2) are then fulfilled if we use

$$\mathfrak{G} := \mathfrak{B}_\mathcal{G} = \{B \cap \mathcal{G} : B \in \mathfrak{B}_m\} \quad \text{and} \quad \mathfrak{A} = \mathfrak{B}_n. \tag{5.3}$$

We also need the concept of an invariant measure. Let $A\gamma = \{\tilde{\gamma}\gamma : \tilde{\gamma} \in A\}$, $\gamma \in \mathcal{G}$, $A \in \mathfrak{G}$. A measure μ on $\mathfrak{G}$ is called *right invariant* if $\mu(A\gamma) = \mu(A)$, $\gamma \in \mathcal{G}$, $A \in \mathfrak{G}$. If the group is at most countable, then the counting measure is, up to a factor, the only invariant measure, and especially there exists an invariant distribution, namely the uniform distribution, if and only if the group is finite. More generally, right invariant measures exist on locally compact groups and are uniquely determined up to a factor; see Nachbin (1976) or Wijsman (1990). Such right invariant measures are called Haar measures. Right invariant distributions exist only if the locally compact group is compact. The special groups that are considered here have right invariant measures that are absolutely continuous with respect to the Lebesgue measure, and the special form of their densities can be verified directly without having recourse to the general theory of Haar measures.

Now we collect some transformation groups that are used systematically in the sequel. In each of the following cases the sample space is $(\mathcal{X}, \mathfrak{A}) = (\mathbb{R}^n, \mathfrak{B}_n)$ and the set $\mathcal{G}$ can be considered as a Borel set of $\mathbb{R}^m$ for a suitable m. We use $\mathfrak{G} = \mathfrak{B}_\mathcal{G}$ from (5.3) as the σ-algebra of subsets of $\mathcal{G}$.

$\mathcal{G}$	Combination Rule	Group Notation (description)	$u_\gamma(x)$	Right Invariant Measure
$\mathcal{M}^r_{n\times n}$	B_1B_2	$\mathcal{U}_{gl}$ (general linear)	Bx	$\frac{1}{\lvert\det(B)\rvert^n}\boldsymbol{\lambda}_{n^2}(dB)$
$\mathbb{L}_d$	$a_1 + a_2$	$\mathcal{U}_t$ (translation)	$a + x$	$\boldsymbol{\lambda}_n$
$\mathcal{O}_{n\times n}$	O_1O_2	$\mathcal{U}_{rot}$ (rotation)	Ox	R_n see Problem 5.4
$\mathbb{R}_\oplus$	$a_1 + a_2$	$\mathcal{U}_l$ (location)	$x + a\mathbf{1}$	$\boldsymbol{\lambda}$
$\mathbb{R}^+_\bullet$	b_1b_2	$\mathcal{U}_s$ (scale)	bx	$\frac{1}{b}\boldsymbol{\lambda}(db)$
Π_k	$\gamma_1 \circ \gamma_2$	$\mathcal{U}_{per}$ (permutation)	$(x_{\gamma(1)}, ..., x_{\gamma(k)})$	$\boldsymbol{\lambda}(\gamma) = \frac{1}{k!}$, $\gamma \in \Pi_k$

(5.4)

where $\mathbf{1} = (1, ..., 1)^T$ and the following notations have been used.

$\mathbb{R}_{\oplus n} = \mathbb{R}^n$ as additive group, especially $\mathbb{R}_\oplus = \mathbb{R}_{\oplus 1}$,
$\mathbb{R}_\bullet^+ = (0, \infty)$ as multiplicative group,
$\mathcal{M}_{n\times n}^r$ as multiplicative group of nonsingular $n \times n$ matrices,
$\mathcal{O}_{n\times n}$ as multiplicative group of orthogonal $n \times n$ matrices,
$\mathbb{L}_d \subseteq \mathbb{R}^n$ linear subspace as group of translations of $\mathbb{R}^n$.

That all of these groups are groups of measurable transformations follows from the obvious fact that the mappings that appear in (5.1) and (5.2) are continuous and thus measurable. As to the statement on the right invariant measure for the general linear group $\mathcal{U}_{gl}$ we refer to Wijsman (1990). The statements for $\mathcal{U}_t$ and $\mathcal{U}_l$ follow directly from the translation invariance of the Lebesgue measure. The statements for $\mathcal{U}_s$ and $\mathcal{U}_{rot}$ are settled with the next two problems.

Problem 5.3. Consider the multiplicative group $\mathbb{R}_\bullet^+$ of the positive numbers. Set $\mu(B) := \int t^{-1} I_B(t) \boldsymbol{\lambda}(dt)$, $B \in \mathfrak{B}_+$. Then $\mu(B) = \mu(Bs)$, $B \in \mathfrak{B}_+$, $s > 0$.

Problem 5.4.* Let $Z_1, ..., Z_n$ be i.i.d. random (column) vectors with common distribution $\mathsf{N}(0, \mathbf{I})$. We apply the Gram–Schmidt procedure and arrange the obtained vectors as column vectors in a matrix, say M. Then the distribution R_n of M^T on $(\mathcal{G}, \mathfrak{G})$ is a right invariant distribution on the group of all orthogonal matrices; that is, on the goup of all rotations.

The right invariance of a measure is equivalent to the following invariance of integrals.

Problem 5.5. μ is right invariant if and only if for every $g : \mathcal{G} \to_m \mathbb{R}_+$

$$\int g(\gamma\gamma_0^{-1})\mu(d\gamma) = \int g(\gamma)\mu(d\gamma), \quad \gamma_0 \in \mathcal{G}.$$

Often, two transformations from the above groups have to be applied one after the other. To get again a transformation group one has to set up the rule of combination in a suitable manner. Furthermore, when the product of two groups from (5.4) is formed, where both have invariant measures that are absolutely continuous with respect to the Lebesgue measure, an invariant measure on the product space can be found by applying the transformation rules for the Lebesgue measure.

Problem 5.6.* $\mathcal{G} = \mathbb{R}_\oplus \times \mathbb{R}_\bullet^+$ with

$$(\alpha_1, \beta_1) \odot (\alpha_2, \beta_2) = (\alpha_1 + \beta_1\alpha_2, \beta_1\beta_2), \quad (\alpha_i, \beta_i) \in \mathbb{R}_\oplus \times \mathbb{R}_\bullet^+,$$

where $\odot$ stands for the combination, is a group. Moreover,

$$\mathcal{U}_{ls} = \{u_{\alpha,\beta} : u_{\alpha,\beta}(x_1, ..., x_n) = (\beta x_1 + \alpha, ..., \beta x_n + \alpha), \quad (\alpha, \beta) \in \mathbb{R}_\oplus \times \mathbb{R}_\bullet^+\}, \tag{5.5}$$

is a group of measurable transformations which is called the location-scale group. The measure

$$\mu_{ls}(C) = \int [\int I_C(\alpha, \beta) \frac{1}{\beta} \boldsymbol{\lambda}(d\alpha)] \boldsymbol{\lambda}(d\beta)$$

is right invariant.

The following example is relevant for nonparametric statistical models.

Example 5.7. Let $\mathcal{F}_m$ be the set of all strictly monotone increasing and continuous functions γ that map $\mathbb{R}$ onto $\mathbb{R}$. Equipped with the composition of functions the set $\mathcal{F}_m$ is a group. Introduce $u_\gamma : (\mathbb{R}^n, \mathfrak{B}_n) \leftrightarrow_m (\mathbb{R}^n, \mathfrak{B}_n)$ by

$$u_\gamma(x_1, ..., x_n) = (\gamma(x_1), ..., \gamma(x_n)) . \tag{5.6}$$

Then $\mathcal{U} = \{u_\gamma : \gamma \in \mathcal{F}_m\}$ is a group of measurable bijections.

The invariance of a statistical model is introduced next.

Definition 5.8. *For a group of measurable transformations* $\mathcal{U} = \{u_\gamma : \gamma \in \mathcal{G}\}$ *we call the statistical model* $(\mathcal{X}, \mathfrak{A}, (P_\theta)_{\theta \in \Delta})$, *and likewise the family* $(P_\theta)_{\theta \in \Delta}$, $\mathcal{U}$*-invariant if*

$$P_\vartheta \circ u_\gamma^{-1} \in \{P_\theta : \theta \in \Delta\}, \quad \vartheta \in \Delta, \gamma \in \mathcal{G}. \tag{5.7}$$

Whenever it is clear which group $\mathcal{U}$ *is involved we simply call them invariant.*

Establishing invariance becomes a trivial task when the parametrized model is generated by the application of a transformation group on one fixed distribution.

Example 5.9. Let $\mathcal{U} = \{u_\gamma : \gamma \in \mathcal{G}\} \subseteq \mathcal{S}_{m,\mathcal{X}}$ be given, and a distribution P on $(\mathcal{X}, \mathfrak{A})$ be fixed. Set $\Delta = \mathcal{G}$ and $(P_\gamma)_{\gamma \in \mathcal{G}} = (P \circ u_\gamma^{-1})_{\gamma \in \mathcal{G}}$. It is obvious that $\mathcal{M} = (\mathcal{X}, \mathfrak{A}, (P_\theta)_{\theta \in \Delta})$ is $\mathcal{U}$*-invariant.* This model is called a *group model.* For the special subgroups collected in (5.4) we obtain the following models for any fixed $P \in \mathcal{P}(\mathfrak{B}_n)$.

$$\mathcal{P}_l = (P \circ u_\alpha^{-1})_{u_\alpha \in \mathcal{U}_l} \quad \text{location model.} \tag{5.8}$$

$$\mathcal{P}_s = (P \circ u_\beta^{-1})_{u_\beta \in \mathcal{U}_s} \quad \text{scale model.} \tag{5.9}$$

$$\mathcal{P}_{ls} = (P \circ u_{\alpha,\beta}^{-1})_{u_{\alpha,\beta} \in \mathcal{U}_{ls}} \quad \text{location-scale model.} \tag{5.10}$$

Problem 5.10. Let $\mathcal{G} = \mathbb{R}^n_\oplus \times \mathcal{M}^r_{n \times n}$ be the affine group with the rule of combination $(a_1, B_1) \odot (a_2, B_2) = (a_1 + B_1 a_2, B_1 B_2)$. Then the invariant model that is generated by the standard normal distribution $\mathsf{N}(0, \mathbf{I})$ is the set of all normal distributions with nonsingular covariance matrices Σ, say $\Sigma \in \mathcal{M}^r_{s,n \times n}$.

$$\{\mathsf{N}(0, \mathbf{I}) \circ u_\gamma^{-1} : \gamma = (a,B) \in \mathbb{R}^n_\oplus \times \mathcal{M}^r_{n \times n}\} = \{\mathsf{N}(\mu, \Sigma) : \mu \in \mathbb{R}^n, \ \Sigma \in \mathcal{M}^r_{s,n \times n}\}.$$

The next example is relevant for the k-sample problem.

Example 5.11. Suppose that k experiments are performed where $(\mathcal{X},\mathfrak{A})$ is the common sample space. Then the sample space for all k observations is $(\mathcal{X}^k,\mathfrak{A}^{\otimes k})$. Denote by Π_k the group of all permutations γ of $(1,...,k)$ and define $u_\gamma : \mathcal{X}^n \to_m \mathcal{X}^n$ by $u_\gamma(x_1,...,x_k) = (x_{\gamma(1)},...,x_{\gamma(k)})$. Then

$$\mathcal{U}_{per} = \{u_\gamma : \gamma \in \Pi_k\} \tag{5.11}$$

is a group of measurable transformations. Moreover, for any family $(P_\theta)_{\theta\in\Delta}$ the model $(\mathcal{X}^k,\mathfrak{A}^{\otimes k},(\bigotimes_{j=1}^k P_{\theta_j})_{(\theta_1,...,\theta_k)\in\Delta^k})$ is an $\mathcal{U}_{per}$-invariant statistical model.

Next we introduce the concept of an invariant statistic.

Definition 5.12. *Let $(\mathcal{X},\mathfrak{A})$ be a sample space, $\mathcal{U} = \{u_\gamma : \gamma \in \mathcal{G}\}$ a group of measurable transformations of $(\mathcal{X},\mathfrak{A})$, and $(\mathcal{T},\mathfrak{T})$ another measurable space. A statistic $T : \mathcal{X} \to_m \mathcal{T}$ is called $\mathcal{U}$-invariant if*

$$T(u_\gamma(x)) = T(x), \quad x \in \mathcal{X},\ \gamma \in \mathcal{G}. \tag{5.12}$$

Whenever it is clear which group $\mathcal{U}$ is involved we simply say that T is invariant. A subset $A \subseteq \mathcal{X}$ is called invariant if the indicator function I_A is invariant.

For every $x \in \mathcal{X}$ we call the set $\{u_\gamma(x) : \gamma \in \mathcal{G}\}$ the *orbit* of x in $\mathcal{X}$. Then we can say that a statistic is invariant if and only if it is constant on every orbit in $\mathcal{X}$.

In the class of all $\mathcal{U}$-invariant statistics the so-called maximal invariant statistics are of special importance.

Definition 5.13. *An invariant statistic $T : \mathcal{X} \to_m \mathcal{T}$ is called maximal invariant if for every $x, y \in \mathcal{X}$ the equality $T(x) = T(y)$ implies that there exists some $\gamma \in \mathcal{G}$ with $y = u_\gamma(x)$.*

Apparently, a maximal invariant statistic separates and thus identifies the orbits in $\mathcal{X}$.

Problem 5.14. A statistic $T : \mathcal{X} \to_m \mathcal{T}$ is maximal invariant if and only if for every $t \in \mathcal{T}$ the set $\{x : T(x) = t\}$ is either empty or an orbit in $\mathcal{X}$.

Problem 5.15.* The system $\mathfrak{J}$ of all measurable and invariant subsets is a sub-σ-algebra of $\mathfrak{A}$ and is called the σ-algebra of invariant sets. Every invariant statistic $T : \mathcal{X} \to_m \mathcal{T}$ is measurable with respect to $\mathfrak{J}$.

Problem 5.16. If $S : \mathbb{R}^n \to_m \mathbb{R}$ is equivariant with respect to $\mathcal{U}_l$ from (5.8) in the sense that $S(x_1+\alpha,...,x_n+\alpha) = \alpha + S(x_1,...,x_n)$, $x \in \mathbb{R}^n$, $\alpha \in \mathbb{R}$, then $T_S(x_1,...,x_n) = (x_1 - S(x_1,...,x_n),...,x_n - S(x_1,...,x_n))$ is maximal invariant with respect to $\mathcal{U}_l$. Especially for the equivariant statistic $S(x_1,...,x_n) = x_1$ we get

$$T_S(x_1,...,x_n) = (0, x_2 - x_1,...,x_n - x_1).$$

Problem 5.17. If $S : (\mathbb{R}_{\neq 0})^n \to_m \mathbb{R}_{\neq 0}$ is equivariant with respect to $\mathcal{U}_s$ from (5.4) in the sense that $S(\beta x_1, ..., \beta x_n) = \beta S(x_1, ..., x_n)$, $x \in (\mathbb{R}_{\neq 0})^n$, $\beta > 0$, then each of the two statistics,

$$T_S(x_1, ..., x_n) = \frac{1}{S(x_1, ..., x_n)}(x_1, ..., x_n), \; T_s(x_1, ..., x_n) = \frac{1}{|x_1|}(x_1, x_2, ..., x_n),$$

is maximal invariant with respect to $\mathcal{U}_s$ in (5.9). T_s is a special version of T_S.

Problem 5.18.* Set $\mathbb{R}^n_{\neq} = \{x : x_i \neq x_j \text{ for } i \neq j, \; x \in \mathbb{R}^n\}$. The statistics $T_{ls} : \mathbb{R}^n_{\neq} \to_m \{-1, 1\} \times \mathbb{R}^{n-1}$, defined by

$$T_{ls}(x_1, ..., x_n) = (\text{sgn}(x_2 - x_1), \frac{x_3 - x_1}{x_2 - x_1}, ..., \frac{x_n - x_1}{x_2 - x_1}), \tag{5.13}$$

is maximal invariant with respect to $\mathcal{U}_{ls}$ in (5.10).

Example 5.19. We reconsider the group $\mathcal{F}_m$ of Example 5.7 along with the transformations $u_\gamma : \mathbb{R}^n_{\neq} \leftrightarrow_m \mathbb{R}^n_{\neq}$ from (5.6). Let $\mathcal{P}_c = \{P : P \in \mathcal{P}(\mathfrak{B}), P(\{x\}) = 0, x \in \mathbb{R}\}$ be the set of all atomless distributions on the Borel sets $\mathfrak{B}$ of $\mathbb{R}$. Then $P^{\otimes n}(\mathbb{R}^n_{\neq}) = 1$ for $P \in \mathcal{P}_c$, and the nonparametric model $\mathcal{M}_{np} = (\mathbb{R}^n_{\neq}, \mathfrak{B}_{n,\neq}, (P^{\otimes n})_{P \in \mathcal{P}_c})$ is invariant with respect to $\mathcal{U} = \{u_\gamma : \gamma \in \mathcal{F}_m\}$. Let $R = (R_1, ..., R_n)$ be the vector of ranks (see Example 4.1) which is here a random permutation of $(1, ..., n)$. The corresponding inverse permutation $S = (S_1, ..., S_n)$ is called the vector of antiranks. Both vectors, R and S, are obviously $\mathcal{F}_m$-invariant. To see that they are even maximal invariant it suffices to consider the antiranks. If $S(x_1, ..., x_n) = S(y_1, ..., y_n) = (s_1, ..., s_n)$, say, then by the definition of the antiranks $x_{s_1} < \cdots < x_{s_n}$ and $y_{s_1} < \cdots < y_{s_n}$. Let γ be a strictly increasing piecewise linear function with $\gamma(x_{s_i}) = y_{s_i}$, $i = 1, ..., n$. Then $(\gamma(x_1), ..., \gamma(x_n)) = (y_1, ..., y_n)$ proves that S is maximal invariant.

It is clear that every function of an invariant statistic is again an invariant statistic. A maximal invariant statistic T is maximal in the sense that every invariant statistic S is a function of T.

Proposition 5.20. *Let $(\mathcal{T}, \mathfrak{T})$ and $(\mathcal{S}, \mathfrak{S})$ be measurable spaces. If $T : \mathcal{X} \to_m \mathcal{T}$ is maximal invariant and $S : \mathcal{X} \to_m \mathcal{S}$ is invariant, then there exists a function $h : \mathcal{T} \to \mathcal{S}$ with $S = h(T)$.*

Proof. For $t \in T(\mathcal{X})$ we choose any $x \in T^{-1}(\{t\})$ and set $h(t) = S(x)$. Due to the maximal invariance of T and the invariance of S the definition of $h(t)$ is independent of the actual choice of x. To complete the proof we have only to fix any $y_0 \in \mathcal{S}$ and to define $h(t) = y_0$ for $t \in \mathcal{T} \backslash T(\mathcal{X})$. ■

Although Proposition 5.20 clarifies the concept of maximal invariance it leaves the question open as to whether h is measurable. In the subsequent models such a measurable representation is given directly. Therefore we do not formulate general conditions that guarantee that such a measurable choice of h is possible. We only give a simple but useful sufficient condition for the measurability of h.

Problem 5.21.* Suppose that $T : \mathcal{X} \to_m \mathcal{T}$ is maximal invariant and $S : \mathcal{X} \to_m \mathcal{S}$ is invariant. If $(\mathcal{S}, \mathfrak{S})$ is a Borel space and T generates the σ-algebra $\mathfrak{I}$ of invariant sets, then there exists a function $h : \mathcal{T} \to_m \mathcal{S}$ with $S = h(T)$.

5.2 Invariant Decision Problems

From now on we assume that in the statistical model $(\mathcal{X}, \mathfrak{A}, (P_\theta)_{\theta\in\Delta})$ the parameter $\theta \in \Delta$ is *identifiable*, which means that

$$P_{\theta_1} = P_{\theta_2} \quad \text{implies} \quad \theta_1 = \theta_2, \quad \theta_1, \theta_2 \in \Delta. \tag{5.14}$$

Suppose that the model is invariant under a group of measurable transformations $\mathcal{U} = \{u_\gamma : \gamma \in \mathcal{G}\} \subseteq \mathcal{S}_{m,\mathcal{X}}$. Then we see that for every $\theta \in \Delta$ and $\gamma \in \mathcal{G}$ there is a uniquely determined, say, $v_\gamma(\theta) \in \Delta$, such that

$$P_\theta \circ u_\gamma^{-1} = P_{v_\gamma(\theta)}. \tag{5.15}$$

Problem 5.22. $\mathcal{V} = \{v_\gamma : \gamma \in \mathcal{G}\}$ is a subgroup of $\mathcal{S}_\Delta$, i.e., $v_\gamma : \Delta \to \Delta$ is one-to-one and it holds $v_{\gamma_1}(v_{\gamma_2}(\theta)) = v_{\gamma_1\gamma_2}(\theta)$ for every $\theta \in \Delta$ and every $\gamma_1, \gamma_2 \in \mathcal{G}$.

The group $\mathcal{V} = \{v_\gamma : \gamma \in \mathcal{G}\}$ is called the *induced group*. A simple consequence is that the distribution of an invariant statistic is also invariant in the following sense.

Problem 5.23.* If (5.12) and (5.15) are satisfied, then $P_\theta \circ T^{-1} = P_{v_\gamma(\theta)} \circ T^{-1}$ for every $\theta \in \Delta$ and $\gamma \in \mathcal{G}$.

The following transformation rule follows by the standard extension technique via linear combinations of indicator functions and monotone increasing approximating sequences of nonnegative measurable functions.

Problem 5.24. The condition (5.15) is equivalent to

$$\int g(x) P_{v_\gamma(\theta)}(dx) = \int g(u_\gamma(x)) P_\theta(dx), \quad g : \mathcal{X} \to_m \mathbb{R}_+, \gamma \in \mathcal{G}. \tag{5.16}$$

Suppose now that in addition a decision space $(\mathcal{D}, \mathfrak{D})$ and a loss function $L(\theta, \cdot) : \mathcal{D} \to_m \mathbb{R}$ are given. The loss function L is called *invariant* if there is a subgroup $\mathcal{W} = \{w_\gamma : \gamma \in \mathcal{G}\} \subseteq \mathcal{S}_{m,\mathcal{D}}$ such that

$$L(\theta, a) = L(v_\gamma(\theta), w_\gamma(a)), \quad \theta \in \Delta, \ a \in \mathcal{D}, \ \gamma \in \mathcal{G}. \tag{5.17}$$

Definition 5.25. *Given a statistical model $(\mathcal{X}, \mathfrak{A}, (P_\theta)_{\theta\in\Delta})$, a decision space $(\mathcal{D}, \mathfrak{D})$, and a loss function L, the decision problem is called invariant if the family $(P_\theta)_{\theta\in\Delta}$ of distributions is invariant in the sense of (5.7), the parameter $\theta \in \Delta$ is identifiable (i.e., (5.14) is fulfilled), and the loss function L satisfies (5.17).*

For an invariant decision problem the group $\mathcal{W} = \{w_\gamma : \gamma \in \mathcal{G}\} \subseteq \mathcal{S}_{m,\mathcal{D}}$ comes into consideration via the invariant loss function. Sometimes, however, the transformation group $\mathcal{W}$ is given directly. Independently of how the group $\mathcal{W}$ has been constructed we define invariant decisions in the following way.

Definition 5.26. *Given an invariant decision problem and a group of measurable transformations* $\mathcal{W} = \{w_\gamma : \gamma \in \mathcal{G}\} \subseteq \mathcal{S}_{m,\mathcal{D}}$ *we call a decision* $\mathsf{D} \in \mathbb{D}$ *invariant if*

$$\mathsf{D}(A|x) = \mathsf{D}(w_\gamma(A)|u_\gamma(x)), \quad A \in \mathfrak{D},\ x \in \mathcal{X},\ \gamma \in \mathcal{G}. \tag{5.18}$$

The subclass of all decisions from $\mathbb{D}$ *that are invariant is denoted by* $\mathbb{D}_{inv}$. *A uniformly best invariant decision is a decision in* $\mathbb{D}_{inv}$ *that is uniformly best in* $\mathbb{D}_{inv}$.

If $\mathsf{D} \in \mathbb{D}$ is a nonrandomized decision, then it has the representation $\mathsf{D}(A|x) = \boldsymbol{\delta}_{d(x)}(A)$ for some $d : \mathcal{X} \to_m \mathcal{D}$. In this case D is invariant if and only if

$$\boldsymbol{\delta}_{d(x)}(A) = \boldsymbol{\delta}_{d(u_\gamma(x))}(w_\gamma(A))$$

holds for every $A \in \mathfrak{D}$, $x \in \mathcal{X}$, and $\gamma \in \mathcal{G}$. If now $\{a\} \in \mathfrak{D}$ for every $a \in \mathcal{D}$, then the invariance of a nonrandomized decision D is equivalent with the fact that the associated $d : \mathcal{X} \to_m \mathcal{D}$ is *equivariant* in the following sense.

$$d(u_\gamma(x)) = w_\gamma(d(x)), \quad x \in \mathcal{X},\ \gamma \in \mathcal{G}. \tag{5.19}$$

The following is a transformation rule for integrals under an application of $\mathcal{G}$.

Lemma 5.27. *If (5.15) and (5.18) hold, then for every* $\theta \in \Delta$, $\gamma \in \mathcal{G}$, *and* $L(\theta, \cdot) : \mathcal{D} \to_m \mathbb{R}$,

$$\int \Big[\int L(v_\gamma(\theta), a)\, \mathsf{D}(da|x)\Big] P_{v_\gamma(\theta)}(dx) = \int \Big[\int L(v_\gamma(\theta), w_\gamma(a)) \mathsf{D}(da|x)\Big] P_\theta(dx).$$

If in addition (5.17) is satisfied, then

$$\mathsf{R}(\theta, \mathsf{D}) = \mathsf{R}(v_\gamma(\theta), \mathsf{D}), \qquad \theta \in \Delta,\ \gamma \in \mathcal{G}. \tag{5.20}$$

Proof. As $L(\theta, a)$ is lower bounded for fixed θ we may assume that L is nonnegative. Replace $w_\gamma(A)$ in (5.18) with $w_\gamma(w_{\gamma^{-1}}(B))$ to see that

$$\mathsf{D}(B|u_\gamma(x)) = \mathsf{D}(w_{\gamma^{-1}}(B)|x) = \int I_B(w_\gamma(a)) \mathsf{D}(da|x).$$

Applying the standard extension technique via linear combinations of indicator functions and monotone increasing approximating sequences the invariance of D implies

$$\int h(a) \mathsf{D}(da|u_\gamma(x)) = \int h(w_\gamma(a)) \mathsf{D}(da|x), \quad h : \mathcal{X} \to_m \mathbb{R}_+.$$

Combined with (5.16) this yields

$$\int [\int L(v_\gamma(\theta), w_\gamma(a))\, \mathsf{D}(da|x)] P_\theta(dx) = \int [\int L(v_\gamma(\theta), a)\, \mathsf{D}(da|u_\gamma(x))] P_\theta(dx)$$
$$= \int [\int L(v_\gamma(\theta), a)\, \mathsf{D}(da|x))] P_{v_\gamma(\theta)}(dx).$$

The second statement of the lemma is obvious. ■

For every $\theta \in \Delta$ we call the set $\{v_\gamma(\theta) : \gamma \in \mathcal{G}\}$ the orbit of θ in Δ. Then we can say that for an invariant decision problem the risk function of any invariant decision is constant on every orbit of Δ.

Let us consider invariant tests for the model $(\mathcal{X}, \mathfrak{A}, (P_\theta)_{\theta\in\Delta})$ and the testing problem $\mathsf{H}_0 : \theta \in \Delta_0$ versus $\mathsf{H}_A : \theta \in \Delta_A$ from (2.12). The decision space is $(\mathcal{D}, \mathfrak{D}) = (\{0,1\}, \mathfrak{P}(\{0,1\}))$. Only the identical mapping of the decision space makes sense here and thus the transformation group $\mathcal{W}$ becomes trivial. Let L be the zero–one loss function $L(\theta, a) = aI_{\Delta_0}(\theta) + (1-a)I_{\Delta_A}(\theta)$, $\theta \in \Delta$, $a = 0, 1$. The invariance requirement (5.17) imposed on L means $L(\theta, a) = L(v_\gamma(\theta), a)$, $\theta \in \Delta$, $a \in \{0,1\}$, $\gamma \in \mathcal{G}$, which implies that

$$v_\gamma(\Delta_0) = \Delta_0, \quad v_\gamma(\Delta_A) = \Delta_A, \quad \gamma \in \mathcal{G}. \tag{5.21}$$

For an invariant model $(\mathcal{X}, \mathfrak{A}, (P_\theta)_{\theta\in\Delta})$ we call the testing problem $\mathsf{H}_0 : \theta \in \Delta_0$ versus $\mathsf{H}_A : \theta \in \Delta_A$ under the zero–one loss function *invariant* if (5.21) holds. Because of $w_\gamma(0) = 0$ and $w_\gamma(1) = 1$ a decision

$$\mathsf{D}(A|x) = \varphi(x)\boldsymbol{\delta}_1(A) + (1-\varphi(x))\boldsymbol{\delta}_0(A), \quad x \in \mathcal{X},\ A \in \mathfrak{D},$$

is invariant, i.e., satisfies (5.18), if and only if the test $\varphi(x) = \mathsf{D}(\{1\}|x)$ is invariant in the sense that

$$\varphi(u_\gamma(x)) = \varphi(x), \quad x \in \mathcal{X},\ \gamma \in \mathcal{G}. \tag{5.22}$$

Such tests are called *invariant tests.* Some examples for invariant models and invariant decision problems follow.

Example 5.28. Consider the model

$$(\mathbb{R}^n, \mathfrak{B}_n, (P \circ u_\theta^{-1})_{\theta\in\mathbb{R}}),$$

where $x = (x_1, ..., x_n)$, $\mathbf{1} = (1, ..., 1)$, and $u_\theta(x) = x + \theta\mathbf{1}$, which is the location model with the family $\mathcal{P}_l$ in (5.8) that has the parent distribution $P \in \mathcal{P}(\mathfrak{B}_n)$. Denote by $(X_1, ..., X_n)$ the projections on the coordinates. The parameter θ is identifiable and the associated group $\mathcal{V} = \{v_\gamma : \gamma \in \mathbb{R}_\oplus\}$ from (5.15) is given by $v_\gamma(\theta) = \theta + \gamma$. If we want to estimate the unknown location parameter θ the decision space is $(\mathcal{D}, \mathfrak{D}) = (\mathbb{R}, \mathfrak{B})$. For $l : \mathbb{R} \to_m \mathbb{R}_+$ we introduce the loss function by $L(\theta, a) = l(a-\theta)$. Taking $\mathcal{W} = \mathcal{V}$ we see that the decision problem of estimating $\theta \in \mathbb{R}$ is invariant. For any estimator $T : \mathbb{R}^n \to_m \mathbb{R}$ the equivariance (5.19) means that $T(x_1 + \alpha, ..., x_n + \alpha) = \alpha + T(x_1, ..., x_n)$. Every such estimator provides an invariant decision in the sense of Definition 5.26. Examples of equivariant estimators are $\overline{X}_n = (1/n)\sum_{i=1}^n X_i$, $\wedge_{i=1}^n X_i$, and $\vee_{i=1}^n X_i$.

Problem 5.29. $(1/n)\sum_{i=1}^{n} X_i$, $\wedge_{i=1}^{n} X_i$, and $\vee_{i=1}^{n} X_i$ are equivariant in the scale model (5.9).

The next example is a combination of the last two cases.

Example 5.30. Consider the model

$$(\mathbb{R}^n, \mathfrak{B}_n, (P \circ u_\theta^{-1})_{\theta \in \mathbb{R}}),$$

where $x = (x_1, ..., x_n)$, $\mathbf{1} = (1, ..., 1)$, and $u_\theta(x) = \beta x + \alpha \mathbf{1}$, which is the multivariate location-scale model that has the parent distribution $P \in \mathcal{P}(\mathfrak{B}_n)$. The parameter $\theta = (\alpha, \beta)$ is identifiable unless $P = \boldsymbol{\delta}_0$ a.s., which we assume to be excluded. An example for an equivariant estimator of $\theta = (\alpha, \beta)$ is $\left(\overline{X}_n, S_n\right)$.

Example 5.31. Consider the group of measurable transformations $\mathcal{U}_{rot}$ in (5.4) such that $u_O : \mathbb{R}^n \leftrightarrow_m \mathbb{R}^n$ is defined by $u_O(x) = Ox$, $O \in \mathcal{O}_{n\times n}$. As the normal distribution $\mathsf{N}(\mu, \mathbf{I})$, $\mu \in \mathbb{R}$, satisfies $\mathsf{N}(\mu, \mathbf{I}) \circ u_O^{-1} = \mathsf{N}(O\mu, \mathbf{I})$ we see that the model

$$(\mathbb{R}^n, \mathfrak{B}_n, (\mathsf{N}(\mu, \mathbf{I}))_{\mu \in \mathbb{R}^n})$$

is $\mathcal{O}_{n\times n}$-invariant. The testing problem $\mathsf{H}_0 : \mu = 0$ versus $\mathsf{H}_A : \mu \neq 0$ is invariant in the sense of (5.21). An intuitively appealing test is based on the $\mathcal{U}_{rot}$-invariant statistic $\chi^2(x) = \sum_{i=1}^{n} x_i^2$ and given by $\varphi(x) = I_{(c,\infty)}(\chi^2(x))$, $x \in \mathbb{R}^n$. This test is obviously invariant in the sense of (5.22). Later on in Theorem 5.33 we show that for a proper choice of c this test is a uniformly best level α test for $\mathsf{H}_0 : \mu = 0$ versus $\mathsf{H}_A : \mu \neq 0$ in the class of all rotation invariant level α tests.

Now we look at a typical selection problem.

Example 5.32. According to Definition 3.8 for the model

$$(\mathcal{X}^k, \mathfrak{A}^{\otimes k}, (\textstyle\bigotimes_{i=1}^{k} P_{\theta_i})_{(\theta_1,...,\theta_k)\in\Delta^k})$$

every $\varphi = (\varphi_1, ..., \varphi_k) : \mathsf{X}_{i=1}^{k} \mathcal{X}_i \to_m \mathbf{S}_k^c$ is called a point selection rule, and the associated decision is given by

$$\mathsf{D}(A|x) = \sum\nolimits_{i=1}^{k} \varphi_i(x)\, \boldsymbol{\delta}_i(A), \quad A \subseteq \mathcal{D} = \{1, ..., k\}.$$

This is a special case of the model in (3.10). It is permutation invariant, i.e., $\mathcal{U}_{per}$-invariant, where $\mathcal{U}_{per}$ is defined in (5.11). By assumption the parameter θ is identifiable in the family $(P_\theta)_{\theta\in\Delta}$ and thus the transformation group $\mathcal{V}$ is given by $v_\gamma(\theta_1, ..., \theta_k) = (\theta_{\gamma(1)}, ..., \theta_{\gamma(k)})$, $\gamma \in \Pi_k$. Let $\kappa : \Delta \to \mathbb{R}$ be a functional and suppose that we want to select a population which is associated with the largest value of $\kappa(\theta_1), ..., \kappa(\theta_k)$. Denote by γ^{-1} the inverse of permutation $\gamma \in \Pi_k$ and introduce the group $\mathcal{W} = \{w_\gamma : \gamma \in \Pi_k\}$ by $w_\gamma(a) = \gamma^{-1}(a)$, $a \in \mathcal{D}$. As

$$\mathsf{D}(w_\gamma(A)\,|u_\gamma(x)) = \sum\nolimits_{i=1}^{k} \varphi_i(u_\gamma(x))\, \boldsymbol{\delta}_i(\gamma^{-1}(A)),$$

a decision D is invariant in the sense of (5.18) if and only if

$$\varphi_i\left(x_{\gamma(1)}, ..., x_{\gamma(k)}\right) = \varphi_{\gamma(i)}\left(x_1, ..., x_k\right), \quad x \in \mathcal{X}^k,\ \gamma \in \Pi_k,\ i = 1, ..., k.$$

After a permutation γ of the k populations any particular population i now appears in $(P_{\theta_{\gamma(1)}}, ..., P_{\theta_{\gamma(k)}})$ at position $\gamma^{-1}(i)$. Let now $S : \mathcal{X}^k \to_m \mathbb{R}^k$ be an estimator of $(\kappa(\theta_1), ..., \kappa(\theta_k))$. We call

$$\varphi_S^{nat} = (\varphi_{S,1}^{nat}, ..., \varphi_{S,k}^{nat}) \text{ with } \varphi_{S,i}^{nat}(x) = \frac{1}{|\mathsf{M}(S(x))|} I_{\mathsf{M}(S(x))}(i), \ i = 1, ..., k,$$

and $\mathsf{M}(S(x)) = \{i : S_i(x) = \max_{1 \le j \le k} S_j(x)\}$, the *natural selection rule* based on S. Obviously, φ_S^{nat} is invariant in the sense of Definition 5.25. Especially, if the components $S_1, ..., S_k$ are all different, then φ_S^{nat} selects the population which is associated with the largest value of $S_1, ..., S_k$.

We return to the topic of invariant tests. Let $T : \mathcal{X} \to_m \mathcal{T}$, be an invariant statistic that generates the σ-algebra $\mathfrak{J}$. Then every invariant test is a function of T, see Problem 5.21. This means that in the search for an optimal invariant test we may switch to the reduced model

$$(\mathcal{T}, \mathfrak{T}, (P_\theta \circ T^{-1})_{\theta \in \Delta}). \tag{5.23}$$

Quite often this model has additional useful properties, such as monotone likelihood ratio, which facilitates the search for optimal tests. A switch to the model (5.23), which may also be made in any other invariant decision problem, is called *reduction by invariance*. Often one can find an optimal decision for the smaller model (5.23). By representing any invariant decision for the original model $(\mathcal{X}, \mathfrak{A}, (P_\theta)_{\theta \in \Delta})$ as a function of the maximal invariant statistic T we can find an optimal invariant decision for the model $(\mathcal{X}, \mathfrak{A}, (P_\theta)_{\theta \in \Delta})$. The following testing problem illustrates this approach.

Suppose we want to test $\mathsf{H}_0 : \mu = 0$ versus $\mathsf{H}_A : \mu \neq 0$ in the statistical model

$$\mathcal{M} = (\mathbb{R}^n, \mathfrak{B}_n, (\mathsf{N}(\mu, \mathbf{I}))_{\mu \in \mathbb{R}^n}), \tag{5.24}$$

which obviously is rotation invariant. Using the invariant statistic $\chi^2(x) = \|x\|^2$ we switch to the reduced model

$$(\mathbb{R}_+, \mathfrak{B}_+, (\mathsf{N}(\mu, \mathbf{I}) \circ (\chi^2)^{-1})_{\mu \in \mathbb{R}^n}), \tag{5.25}$$

where $\mathsf{N}(\mu, \mathbf{I}) \circ (\chi^2)^{-1} = \mathsf{H}(n, \delta^2(\mu))$, the χ^2-distribution with n degrees of freedom and noncentrality parameter $\delta^2(\mu) = \sum_{i=1}^n \mu_i^2$. Denote by $S_r = \{x : \|x\| = r\}$ the sphere with radius $r \geq 0$ and fix a unit vector e_0. If the test φ is rotation invariant, then φ is constant on S_r. Hence

$$\varphi(x) = h(\|x\|), \quad \text{where} \quad h(r) = \varphi(re_0), \quad x \neq 0. \tag{5.26}$$

Theorem 5.33. *For the model $\mathcal{M}$ in (5.24) and $\delta_0^2 \geq 0$ the χ^2-test*

$$\varphi_{\chi^2,\alpha}(x) = I_{(\chi^2_{1-\alpha,n}, \infty)}(\chi^2(x))$$

is a uniformly best invariant level α test for $\mathsf{H}_0 : \delta^2(\mu) = 0$ versus $\mathsf{H}_A : \delta^2(\mu) > \delta_0^2$; that is,

$$\mathsf{E}_\mu \varphi_{\chi^2,\alpha} \geq \mathsf{E}_\mu \varphi, \quad \delta^2(\mu) > \delta_0^2,$$

holds for every test φ *that satisfies* $\mathsf{E}_0\varphi \leq \alpha$ *and* $\varphi(Ox) = \varphi(x)$, $x \in \mathbb{R}^n$, *for every orthogonal matrix* O.

Proof. We know from Theorem 2.27 that the family $(\mathsf{H}(n, \delta^2))_{\delta^2>0}$ has a nondecreasing likelihood ratio in the identity. Hence we get from Theorem 2.49 that $\psi_\alpha(t) = I_{(\chi^2_{1-\alpha,n},\infty)}(t)$ is a uniformly best level α test for $\mathsf{H}_0 : \delta^2 = 0$ versus $\mathsf{H}_A : \delta^2 > 0$ in the reduced model (5.25). Hence ψ_α is also a uniformly best level α test for $\mathsf{H}_0 : \delta^2 = 0$ versus $\mathsf{H}_A : \delta^2 > \delta_0^2$. As any rotation invariant test φ can in view of (5.26) be represented as a measurable function of $\chi^2(x)$ we get the statement. ■

That the χ^2-test in the last theorem is also a maximin test for $\mathsf{H}_0 : \delta^2(\mu) = 0$ versus $\mathsf{H}_{A,\delta_0^2} : \delta^2(\mu) > \delta_0^2$ for every fixed $\delta_0^2 > 0$ is shown in Theorem 5.43.

Now we consider a decision problem where permutation, reflection, and location invariance are natural requirements. Suppose we have independent samples of common size n from three normal populations $\mathsf{N}(\mu_1, \sigma^2)$, $\mathsf{N}(\mu_2, \sigma^2)$, $\mathsf{N}(\mu, \sigma^2)$, where $\mu_1, \mu_2, \mu \in \mathbb{R}$ are unknown and $\sigma^2 > 0$ is known. By a reduction by sufficiency (see Remark 4.68) we may assume that $n = 1$. Let X_1, X_2, X be the respective observations. We consider the following *classification problem.* Let it be known that $\mu \in \{\mu_1, \mu_2\}$ holds, and suppose we want to decide whether $\mu = \mu_1$ or $\mu = \mu_2$ is true. The statistical model is a special case of (3.16),

$$\begin{aligned}
&(\mathbb{R}^3, \mathfrak{B}_3, (P_\theta)_{\theta\in\Delta}), \quad \text{where} \qquad (5.27)\\
&P_\theta = \mathsf{N}(\mu_1, \sigma^2) \otimes \mathsf{N}(\mu_2, \sigma^2) \otimes \mathsf{N}(\mu, \sigma^2),\\
&\Delta = \{\theta : \theta = (\mu_1, \mu_2, \mu),\ (\mu_1, \mu_2) \in \mathbb{R}^2,\ \mu \in \{\mu_1, \mu_2\}\},
\end{aligned}$$

and the decision space is $\mathcal{D} = \{1, 2\}$. It is clear that every decision D, called a *classification rule* here, may be written as

$$\mathsf{D}(A|x_1, x_2, y) = \varphi(x_1, x_2, y)\boldsymbol{\delta}_1(A) + [1 - \varphi(x_1, x_2, y)]\boldsymbol{\delta}_2(A), \quad A \subseteq \mathcal{D},$$

where $\varphi : \mathbb{R}^3 \to_m [0, 1]$ is a test and $\varphi(x_1, x_2, y)$ is the probability of deciding in favor of $\mu = \mu_1$ after $(x_1, x_2, y) \in \mathbb{R}^3$ has been observed. We adopt the zero–one loss function

$$L((\mu_1, \mu_2, \mu), a) = I_{\{(\mu_1,2)\}}(\mu, a) + I_{\{(\mu_2,1)\}}(\mu, a). \qquad (5.28)$$

The risk $\mathsf{R}(\theta, \varphi) := \mathsf{R}(\theta, \mathsf{D})$, $\theta = (\mu_1, \mu_2, \mu) \in \Delta$, is given by

$$\begin{aligned}
&\mathsf{R}((\mu_1, \mu_2, \mu), \varphi) \qquad (5.29)\\
&= \int (\varphi(x_1, x_2, y)I_{\{\mu_2\}}(\mu) + [1 - \varphi(x_1, x_2, y)]I_{\{\mu_1\}}(\mu))P_\theta(dx_1, dx_2, dy).
\end{aligned}$$

Set $\mathcal{G} = \{1, 2, 3, 4\} \times \mathbb{R}$ and consider the set of transformations $u_{(i,b)} :\to \mathbb{R}^3$, $(i, b) \in \mathcal{G}$, defined by

$$\begin{aligned} u_{(1,b)}(x_1,x_2,y) &= (x_1+b, x_2+b, y+b), \\ u_{(2,b)}(x_1,x_2,y) &= (-x_1+b, -x_2+b, -y+b), \\ u_{(3,b)}(x_1,x_2,y) &= (x_2+b, x_1+b, y+b), \\ u_{(4,b)}(x_1,x_2,y) &= (-x_2+b, -x_1+b, -y+b). \end{aligned} \tag{5.30}$$

It is clear that a suitable multiplication can be introduced in $\mathcal{G}$ so that $u_{\gamma_1} \circ u_{\gamma_2} = u_{\gamma_1\gamma_2}$ holds. For example, $(1, b_1)(2, b_2) = (2, b_2 - b_1)$. We do not need the explicit structure of the multiplication in the sequel. The induced group on the parameter set then consists of the four families of transformations $v_{(i,b)}(\mu_1, \mu_2, \mu) = u_{(i,b)}(\mu_1, \mu_2, \mu)$, $i = 1, ..., 4$, with common parameter $b \in \mathbb{R}$. Finally we introduce the family of transformations on the decision space by $w_{(1,b)}(1) = w_{(2,b)}(1) = 1$, $w_{(1,b)}(2) = w_{(2,b)}(2) = 2$, $w_{(3,b)}(1) = w_{(4,b)}(1) = 2$, $w_{(3,b)}(2) = w_{(4,b)}(2) = 1$. Then with the loss function from (5.28) we get an invariant decision problem.

Problem 5.34. A classification rule $\varphi(x_1, x_2, y)$ is invariant if and only there is some $\psi : \mathbb{R}^2 \to_m [0, 1]$ such that

$$\begin{aligned} \varphi(x_1,x_2,y) &= \psi(x_1-y, x_2-y), \\ \psi(t_1,t_2) &= \psi(-t_1,-t_2), \quad \text{and} \quad \psi(t_1,t_2) = 1 - \psi(t_2,t_1). \end{aligned} \tag{5.31}$$

The following theorem is due to Kudo (1959).

Theorem 5.35. *In the above invariant decision problem for model (5.27) and loss function (5.28) the classification rule*

$$\varphi_0(x_1,x_2,y) = \begin{cases} 1 & \text{if} \quad |x_1 - y| \le |x_2 - y|, \\ 0 & \text{if} \quad |x_1 - y| > |x_2 - y|, \end{cases}$$

is a uniformly best rule in the class of all classification rules that are invariant under the group of transformations given by (5.30).

Proof. We use (5.29) to get

$$\begin{aligned} \mathsf{R}((\mu_1,\mu_2,\mu_2),\varphi) &= \mathbb{E}\varphi(X_1,X_2,X), \quad \text{if } X \sim \mathsf{N}(\mu_2,\sigma^2), \\ \mathsf{R}((\mu_1,\mu_2,\mu_1),\varphi) &= \mathbb{E}(1-\varphi(X_1,X_2,X)), \quad \text{if } X \sim \mathsf{N}(\mu_1,\sigma^2). \end{aligned}$$

We know from (5.20) that the risk of an invariant classification rule φ is constant on orbits $\theta \mapsto v_\gamma(\theta)$, $\gamma = (i, b) \in \mathcal{G}$, $\theta = (\mu_1, \mu_2, \mu) \in \Delta$. This means that

$$\begin{aligned} \mathsf{R}((\mu_1,\mu_2,\mu_1),\varphi) &= \mathsf{R}((0,\mu_2-\mu_1,0),\varphi) = \mathsf{R}((\mu_2-\mu_1,0,0),\varphi) \\ &= \mathsf{R}((\mu_1-\mu_2,0,0),\varphi) = \mathsf{R}((\mu_1,\mu_2,\mu_2),\varphi). \end{aligned}$$

Hence for any invariant classification rule φ, the test ψ in (5.31), and $X \sim \mathsf{N}(\mu_2, \sigma^2)$,

$$\mathsf{R}(\theta,\varphi) = \frac{1}{4}\mathbb{E}[\psi(X_1 - X, X_2 - X) + \psi(X - X_1, X - X_2) \\ +(1 - \psi(X_2 - X, X_1 - X)) + (1 - \psi(X - X_2, X - X_1))],$$

where $\theta = (\mu_1, \mu_2, \mu) \in \Delta$. Therefore we have to minimize

$$\mathbb{E}\psi(X_1 - X, X_2 - X) - \mathbb{E}\psi(X_2 - X, X_1 - X) \\ +\mathbb{E}\psi(X - X_1, X - X_2) - \mathbb{E}\psi(X - X_2, X - X_1) \tag{5.32}$$

as a function of ψ. The vector $(X_1 - X, X_2 - X)$ has a normal distribution with mean vector $(\delta, 0) = (\mu_1 - \mu_2, 0)$ and covariance matrix

$$\Sigma = \sigma^2 \begin{pmatrix} 2 & 1 \\ 1 & 2 \end{pmatrix},$$

so that the Lebesgue density of $(X_1 - X, X_2 - X)$ is given by

$$\varphi_{(\delta,0),\Sigma}(t_1, t_2) = \frac{1}{\sqrt{3}2\pi\sigma^2}\exp\{-\frac{1}{3\sigma^2}[(t_1 - \delta)^2 - (t_1 - \delta)t_2 + t_2^2]\}.$$

The expectation in (5.32) is given by

$$\frac{1}{\sqrt{3}2\pi\sigma^2}\int[\int \psi(t_1,t_2)(\exp\{-\frac{1}{3\sigma^2}[(t_1-\delta)^2 - (t_1-\delta)t_2 + t_2^2]\} \\ - \exp\{-\frac{1}{3\sigma^2}[t_1^2 - t_1(t_2 - \delta) + (t_2 - \delta)^2]\} \\ + \exp\{-\frac{1}{3\sigma^2}[(t_1+\delta)^2 - (t_1+\delta)t_2 + t_2^2]\} \\ - \exp\{-\frac{1}{3\sigma^2}[t_1^2 - t_1(t_2 + \delta) + (t_2 + \delta)^2]\})dt_1]dt_2 \\ = \int[\int \psi(t_1,t_2)g(t_1,t_2)h(t_1,t_2)dt_1]dt_2,$$

where for $\tilde{\delta} = \delta/(3\sigma^2)$,

$$g(t_1,t_2) = \frac{1}{\sqrt{3}2\pi\sigma^2}\exp\{-\frac{\delta^2}{3\sigma^2}\}\exp\{-\frac{1}{3\sigma^2}(t_1^2 - t_1t_2 + t_2^2)\},$$
$$h(t_1,t_2) = \exp\{2\tilde{\delta}t_1 - \tilde{\delta}t_2\} - \exp\{2\tilde{\delta}t_2 - \tilde{\delta}t_1\} + \exp\{\tilde{\delta}t_2 - 2\tilde{\delta}t_1\} \\ - \exp\{\tilde{\delta}t_1 - 2\tilde{\delta}t_2\} \\ = [\exp\{\frac{\tilde{\delta}}{2}(t_1+t_2)\} - \exp\{-\frac{\tilde{\delta}}{2}(t_1+t_2)\}] \\ \times [\exp\{\frac{3\tilde{\delta}}{2}(t_1-t_2)\} - \exp\{-\frac{3\tilde{\delta}}{2}(t_1-t_2)\}].$$

Now, $h(t_1,t_2) \le 0$ holds if and only if

$$\begin{aligned} \text{either} \quad & \delta(t_1+t_2) \geq 0 \quad \text{and} \quad \delta(t_1-t_2) \leq 0, \\ \text{or} \quad & \delta(t_1+t_2) \leq 0 \quad \text{and} \quad \delta(t_1-t_2) \geq 0, \end{aligned}$$

which is equivalent to

$$\delta^2(t_1+t_2)(t_1-t_2) = \delta^2(t_1^2 - t_2^2) \leq 0.$$

This means that $\psi(t_1, t_2) = I_{[0,|t_2|]}(|t_1|)$ provides a classification rule that minimizes the expression in (5.32), and the proof is thus completed. ■

Because φ_0 does not depend on σ^2, Theorem 5.35 holds also in the case where σ^2 is unknown. It should be noted that in Kudo (1959) the case of unequal sample sizes has also been considered. In that case the invariance conditions on the decisions have to be changed and depend now on the sample sizes. Because the main ideas in Kudo (1959) have become clear already with a common sample size we omit the general case for brevity.

We conclude this section with some remarks on the relation between the concepts of sufficiency and invariance. The concept of sufficiency is closely related to possible invariance properties of a model under consideration. Below we look at the simple situation where the group $\mathcal{G}$ is finite and not only the model, but all distributions in the model, are invariant.

Proposition 5.36. *Let $(\mathcal{X}, \mathfrak{A}, (P_\theta)_{\theta\in\Delta})$ be a model and $\mathcal{U} = \{u_\gamma : \gamma \in \mathcal{G}\} \subseteq \mathcal{S}_{m,\mathcal{X}}$ a finite group. If $P_\theta \circ u_\gamma^{-1} = P_\theta$ for every $\gamma \in \mathcal{G}$ and $\theta \in \Delta$, then the σ-algebra of invariant sets is sufficient.*

Proof. Denote by $|\mathcal{G}|$ the cardinality of $\mathcal{G}$. Let $A \in \mathfrak{A}$. The measurable function $H(x) = |\mathcal{G}|^{-1}\sum_{\gamma\in\mathcal{G}} I_A(u_\gamma(x))$ is invariant and thus $\mathfrak{J}$-measurable. For $C \in \mathfrak{J}$ it holds $I_C(x) = I_C(u_\gamma(x))$, $\gamma \in \mathcal{G}$. This yields

$$\begin{aligned} &\int I_C(x)H(x)P_\theta(dx) = |\mathcal{G}|^{-1}\sum\nolimits_{\gamma\in\mathcal{G}} \int I_C(u_\gamma(x))I_A(u_\gamma(x))P_\theta(dx) \\ &= |\mathcal{G}|^{-1}\sum\nolimits_{\gamma\in\mathcal{G}} \int I_C(y)I_A(y)(P_\theta \circ u_\gamma^{-1})(dy) \\ &= |\mathcal{G}|^{-1}\sum\nolimits_{\gamma\in\mathcal{G}} \int I_C(y)I_A(y)P_\theta(dy) = \int I_C(y)I_A(y)P_\theta(dy). \end{aligned}$$

Consequently, $\mathsf{E}_\theta(I_A|\mathfrak{J}) = H$, P_θ-*a.s.*, so that H is a version of the conditional probability that is independent of $\theta \in \Delta$. ■

Example 5.37. Assume $X_1, ..., X_n$ are i.i.d. real-valued random variables with a common distribution P. We consider the full nonparametric model

$$\mathcal{M} = (\mathbb{R}^n, \mathfrak{B}_n, (P^{\otimes n})_{P\in\mathcal{P}(\mathfrak{B})}). \tag{5.33}$$

Let $T_\uparrow : \mathbb{R}^n \to \mathbb{R}^n$ be the order statistic $T_\uparrow(x_1, ..., x_n) = (x_{[1]}, ..., x_{[n]})$ that was introduced in Example 4.1. $T_\uparrow$ is obviously continuous and thus $\mathfrak{B}_n$-$\mathfrak{B}_n$ measurable. Let $\mathcal{G}$ be the transformation group induced by the permutations of the coordinates, i.e., for a permutation γ the mapping u_γ is defined by $u_\gamma(x) = (x_{\gamma(1)}, ..., x_{\gamma(n)})$.

The mapping $u_\gamma : \mathbb{R}^n \to \mathbb{R}^n$ is continuous and consequently $\mathfrak{B}_n$-$\mathfrak{B}_n$ measurable. As $\sigma(T_\uparrow) = \mathfrak{I}$ we obtain from Proposition 5.36 that the order statistic $T_\uparrow$ is sufficient for the model (5.33).

In conclusion, we remark that the requirement of a statistic to be sufficient as well as equivariant is restrictive and can be fulfilled only in certain situations. The following has been shown in Dynkin (1951). For a statistical model on the real line with a continuous, sufficient, and equivariant statistic the family of distributions must either be a location family with a normal parent distribution, or a scale family with a gamma parent distribution. For more details and further references we refer to Pfanzagl (1972, 1994).

5.3 Hunt–Stein Theorem

In the examples of the previous sections we have constructed invariant decisions mainly by using suitably constructed invariant or equivariant statistics. However, if there exists an invariant distribution on the group $\mathcal{G}$, then we may construct invariant decisions also by averaging. To this end we suppose that that the conditions (5.1) and (5.2) are satisfied and that $L : \Delta \times \mathcal{D} \to_m \mathbb{R}_+$ is a measurable function of (θ, a). Let Q be any distribution on $(\mathcal{G}, \mathfrak{G})$. Given any decision D we denote by D_Q the decision

$$\mathsf{D}_Q(A|x) = \int \mathsf{D}(w_\gamma(A)|u_\gamma(x))Q(d\gamma). \tag{5.34}$$

By the standard extension technique we get for any $g : \mathcal{D} \to_m \mathbb{R}_+$,

$$\int g(a)\mathsf{D}_Q(da|x) = \int [\int g(w_{\gamma^{-1}}(a))\mathsf{D}(da|u_\gamma(x))]Q(d\gamma). \tag{5.35}$$

For any $Q \in \mathcal{P}(\mathfrak{G})$ we set $Q_{\gamma_0}(G) = Q(G\gamma_0^{-1})$, $G \in \mathfrak{G}$, $\gamma_0 \in \mathcal{G}$. Then

$$\int h(\gamma)Q_{\gamma_0}(d\gamma) = \int h(\gamma\gamma_0)Q(d\gamma), \quad h : \mathcal{G} \to_m \mathbb{R}_+. \tag{5.36}$$

Recall that Q is right invariant if $Q(G) = Q(G\gamma)$, $G \in \mathfrak{G}$, $\gamma \in \mathcal{G}$.

Lemma 5.38. *If Q is right invariant, then D_Q is invariant in the sense of Definition 5.26, and for every invariant loss function $L : \Delta \times \mathcal{D} \to_m \mathbb{R}_+$,*

$$\mathsf{R}(\theta, \mathsf{D}_Q) = \int \mathsf{R}(v_\gamma(\theta), \mathsf{D})Q(d\gamma), \quad \theta \in \Delta. \tag{5.37}$$

Proof. The relation (5.36) and $Q = Q_{\gamma_0}$ yield

$$\begin{aligned}\mathsf{D}_Q(w_{\gamma_0}(A)|u_{\gamma_0}(x)) &= \int \mathsf{D}(w_{\gamma\gamma_0}(A)|u_{\gamma\gamma_0}(x))Q(d\gamma)\\ &= \int \mathsf{D}(w_\gamma(A)|u_\gamma(x))Q_{\gamma_0}(d\gamma) = \mathsf{D}_Q(A|x).\end{aligned}$$

Using (5.35) and the invariance of L (see (5.17)) we get for $\theta \in \Delta$,

$$\begin{aligned}\int L(\theta, a)\mathsf{D}_Q(da|x) &= \int [\int L(\theta, w_{\gamma^{-1}}(a))\mathsf{D}(da|u_\gamma(x))]Q(d\gamma)\\ &= \int [\int L(v_\gamma(\theta), a)\mathsf{D}(da|u_\gamma(x))]Q(d\gamma).\end{aligned}$$

Hence by (5.16),

$$\begin{aligned}\mathsf{R}(\theta, \mathsf{D}_Q) &= \int [\int L(\theta, a)\mathsf{D}_Q(da|x)]P_\theta(dx)\\ &= \int [\int [\int L(v_\gamma(\theta), a)\mathsf{D}(da|u_\gamma(x))]P_\theta(dx)]Q(d\gamma)\\ &= \int [\int [\int L(v_\gamma(\theta), a)\mathsf{D}(da|x)]P_{v_\gamma(\theta)}(dx)]Q(d\gamma)\\ &= \int \mathsf{R}(v_\gamma(\theta), \mathsf{D})Q(d\gamma), \quad \theta \in \Delta.\end{aligned}$$

■

Now we study the relation between the concept of invariance and the Bayes approach. Despite the fact that more general results can be achieved we restrict ourselves here to results on finite groups $\mathcal{G}$ that are used later on. A prior Π is called invariant if $\Pi \circ v_\gamma^{-1} = \Pi$. If Π is any prior, then it is obvious that

$$\overline{\Pi}(B) := \frac{1}{|\mathcal{G}|}\sum\nolimits_{\gamma \in \mathcal{G}} (\Pi \circ v_\gamma^{-1})(B)$$

is an invariant prior.

Proposition 5.39. *Assume that $\mathcal{G}$ is finite. Given an invariant decision problem specified by the invariant model $(\mathcal{X}, \mathfrak{A}, (P_\theta)_{\theta\in\Delta})$, the invariant loss function L, and the decision space $(\mathcal{D}, \mathfrak{D})$, an invariant decision D_0 is a uniformly best invariant decision if and only if it is a Bayes decision with respect to every discrete invariant prior.*

Proof. For every $\theta \in \Delta$ we denote by $\Pi_\theta = |\mathcal{G}|^{-1}\sum_{\gamma\in\mathcal{G}} \boldsymbol{\delta}_{v_\gamma(\theta)}$ the uniform distribution on the finite orbit $\{v_\gamma(\theta) : \gamma \in \mathcal{G}\}$. Every invariant discrete prior Π satisfies $\Pi(\{v_{\gamma_1}(\theta)\}) = \Pi(\{v_{\gamma_2}(\theta)\})$, $\gamma_1, \gamma_2 \in \mathcal{G}$. Hence Π is the mixture of some Π_{θ_i}, i.e., $\Pi = \sum_{i=1}^m \alpha_i \Pi_{\theta_i}$.

From here we see that a decision is a Bayes decision with respect to every discrete invariant prior if and only if it is a Bayes decision with respect to every Π_θ. Suppose that D_0 is a uniformly best invariant decision and D is any decision. Let Q be the uniform distribution on $\mathcal{G}$. As D_Q is invariant we have $\mathsf{R}(\theta, \mathsf{D}_0) \le \mathsf{R}(\theta, \mathsf{D}_Q)$, and in view of (5.37), we get

$$\frac{1}{|\mathcal{G}|}\sum\nolimits_{\gamma\in\mathcal{G}} \mathsf{R}(v_\gamma(\theta),\mathsf{D}_0) \le \frac{1}{|\mathcal{G}|}\sum\nolimits_{\gamma\in\mathcal{G}} \mathsf{R}(v_\gamma(\theta),\mathsf{D}_Q)$$
$$= \frac{1}{|\mathcal{G}|}\sum\nolimits_{\gamma\in\mathcal{G}} \frac{1}{|\mathcal{G}|}\sum\nolimits_{\widetilde{\gamma}\in\mathcal{G}} \mathsf{R}(v_{\widetilde{\gamma}}(v_\gamma(\theta)),\mathsf{D}) = \frac{1}{|\mathcal{G}|}\sum\nolimits_{\gamma\in\mathcal{G}} \mathsf{R}(v_\gamma(\theta),\mathsf{D}),$$

which shows that D_0 is Bayes with respect to the prior Π_θ. Conversely, if D_0 and D are invariant, then by (5.20) for any fixed $\theta_0 \in \Delta$,

$$\mathsf{R}(\theta_0,\mathsf{D}_0) = \mathsf{R}(v_\gamma(\theta_0),\mathsf{D}_0) \quad \text{and} \quad \mathsf{R}(\theta_0,\mathsf{D}) = \mathsf{R}(v_\gamma(\theta_0),\mathsf{D}), \quad \gamma\in\mathcal{G}.$$

Averaging over $\gamma \in \mathcal{G}$ by using the uniform distribution on the orbit $\{v_\gamma(\theta) : \gamma \in \mathcal{G}\}$ we get

$$\mathsf{R}(\theta,\mathsf{D}_0) = \frac{1}{|\mathcal{G}|}\sum\nolimits_{\gamma\in\mathcal{G}} \mathsf{R}(v_\gamma(\theta),\mathsf{D}_0) = \mathsf{r}(\Pi_\theta,\mathsf{D}_0),$$
$$\mathsf{R}(\theta,\mathsf{D}) = \frac{1}{|\mathcal{G}|}\sum\nolimits_{\gamma\in\mathcal{G}} \mathsf{R}(v_\gamma(\theta),\mathsf{D}) = \mathsf{r}(\Pi_\theta,\mathsf{D}).$$

If now D_0 is a Bayes decision with respect to Π_θ, then $\mathsf{R}(\theta,\mathsf{D}_0) \le \mathsf{R}(\theta,\mathsf{D})$. ■

Another relation exists between admissibility and the property of a decision to be uniformly best invariant.

Proposition 5.40. *Assume that $\mathcal{G}$ is finite. Then every uniformly best invariant decision D_0 is admissible.*

Proof. Assume D_0 is not admissible. Then there is some D with $\mathsf{R}(\theta,\mathsf{D}) \le \mathsf{R}(\theta,\mathsf{D}_0)$, $\theta \in \Delta$, and $\mathsf{R}(\theta_0,\mathsf{D}) < \mathsf{R}(\theta_0,\mathsf{D}_0)$ for some $\theta_0 \in \Delta$. If Π_{θ_0} is the uniform distribution on the orbit $\{v_\gamma(\theta_0) : \gamma \in \mathcal{G}\}$, then by the invariance of D and D_0 it follows $\mathsf{r}(\Pi_{\theta_0},\mathsf{D}) < \mathsf{r}(\Pi_{\theta_0},\mathsf{D}_0)$. Hence D_0 is not Bayes with respect to Π_{θ_0} and thus not uniformly best invariant by Proposition 5.39. ■

Next we consider the relation between invariant decisions and minimax decisions. The main idea here is the following. If there exists an invariant distribution Q on the group $\mathcal{G}$, then for any given decision D the decision D_Q is invariant, and its maximum risk satisfies in view of (5.37)

$$\sup_\theta \mathsf{R}(\theta,\mathsf{D}_Q) \le \sup_\theta \mathsf{R}(\theta,\mathsf{D}).$$

This gives the following statement.

Proposition 5.41. *Assume that there exists an invariant distribution Q on $(\mathcal{G},\mathfrak{G})$. Suppose that $\mathbb{D}_0$ is a subset of decisions for a decision problem that is invariant in the sense of Definition 5.25. If $\mathbb{D}_0$ is closed under averaging with Q (i.e., if for every $\mathsf{D} \in \mathbb{D}_0$ the decision D_Q belongs to $\mathbb{D}_0$), then*

$$\inf_{\mathsf{D}\in\mathbb{D}_0} \sup_\theta \mathsf{R}(\theta,\mathsf{D}) = \inf_{\mathsf{D}\in\mathbb{D}_0\cap\mathbb{D}_{inv}} \sup_\theta \mathsf{R}(\theta,\mathsf{D}) = \inf_{\mathsf{D}\in\mathbb{D}_0} \sup_\theta \mathsf{R}(\theta,\mathsf{D}_Q),$$

where $\mathbb{D}_{inv}$ is the set of all invariant decisions.

Remark 5.42. In an invariant testing problem all transformations w_γ, $\gamma \in \mathcal{G}$, coincide with the identical mapping. Moreover, the group $\mathcal{V} = \{v_\gamma : \gamma \in \mathcal{G}\}$ leaves Δ_0 and Δ_A invariant, see (5.21). Suppose φ is a test that defines the decision $\mathsf{D} = \varphi\boldsymbol{\delta}_1 + (1-\varphi)\,\boldsymbol{\delta}_0$. Then

$$\begin{aligned}\mathsf{D}_Q &= \varphi_Q\boldsymbol{\delta}_1 + (1-\varphi_Q)\,\boldsymbol{\delta}_0, \quad \text{where}\\ \varphi_Q(x) &= \int \varphi(u_\gamma(x))Q(d\gamma).\end{aligned}$$

If φ is a level α test for $\mathsf{H}_0 : \theta \in \Delta_0$, then φ_Q is also a level α test for H_0.

Now we consider again the χ^2-test for the testing problem that was treated in Theorem 5.33. From there we know already that this test is a uniformly best invariant level α test for $\mathsf{H}_0 : \delta^2(\mu) = 0$ versus $\mathsf{H}_A : \delta^2(\mu) > 0$, where $\delta^2(\mu) = \sum_{i=1}^n \mu_i^2$. Now we establish the maximin property. For $\delta_0^2 \ge 0$ we set $\widetilde{\Delta} = \{0\} \cup \{\mu : \delta^2(\mu) > \delta_0^2\}$ and consider the restricted model and the restricted testing problem given by

$$\begin{aligned}&(\mathbb{R}^n, \mathfrak{B}_n, (\mathsf{N}(\mu,\mathbf{I}))_{\mu\in\widetilde{\Delta}})\\ &\mathsf{H}_0 : \delta^2(\mu) = 0 \quad \text{versus} \quad \mathsf{H}_{A,\delta_0^2} : \delta^2(\mu) > \delta_0^2.\end{aligned} \tag{5.38}$$

Theorem 5.43. *For the statistical model and the testing problem in (5.38) the χ^2-test $\varphi_{\chi^2,\alpha}(x) = I_{(\chi^2_{1-\alpha,n},\infty)}(\chi^2(x))$ is a maximin level α test; that is, it holds*

$$\inf_{\delta^2(\mu)>\delta_0^2} \mathsf{E}_\mu\varphi \le \inf_{\delta^2(\mu)>\delta_0^2} \mathsf{E}_\mu\varphi_{\chi^2,\alpha} \tag{5.39}$$

for every test φ that satisfies $\mathsf{E}_0\varphi \le \alpha$.

Proof. The restricted model $(\mathbb{R}^n, \mathfrak{B}_n, (\mathsf{N}(\mu,\mathbf{I}))_{\mu\in\widetilde{\Delta}})$ is rotation invariant. Let $(\mathcal{G}, \mathfrak{G}) = (\mathcal{O}_{n\times n}, \mathfrak{B}_{\mathcal{O}_{n\times n}})$ be the group of orthogonal matrices equipped with the Borel sets. Then R_n in Problem 5.4 is a right invariant distribution on $(\mathcal{G}, \mathfrak{G})$. If we use the zero–one loss function so that $\mathsf{R}(\theta,\mathsf{D}) = 1 - \mathsf{E}_\mu\varphi$, and $\Delta = \{\mu : \delta^2(\mu) > \delta_0^2\}$, we get from Proposition 5.41 and Remark 5.42 that it holds

$$\sup_{\varphi\in\mathcal{T}_\alpha} \inf_{\delta^2(\mu)>\delta_0^2} \mathsf{E}_\mu\varphi = \sup_{\varphi\in\mathcal{T}_{\alpha,inv}} \inf_{\delta^2(\mu)>\delta_0^2} \mathsf{E}_\mu\varphi,$$

where $\mathcal{T}_\alpha$ is the class of all level α tests and $\mathcal{T}_{\alpha,inv}$ is the class of all rotation invariant level α tests. It remains to prove that inequality (5.39) holds for every rotation invariant level α test φ. This, however, follows from Theorem 5.33. ■

The crucial point in Proposition 5.41 and in Theorem 5.43 has been that there exists a right invariant distribution on the group $\mathcal{G}$. For compact topological groups such a distribution always exists and is called the Haar measure; see Nachbin (1965).

For noncompact groups that are at least locally compact there exist only right invariant measures, also called Haar measures, which have infinite total mass; see Nachbin (1965). Examples of such measures are $|\det(B)|^{-n}\boldsymbol{\lambda}_{n^2}(dB)$ for the general linear group, $b^{-1}\boldsymbol{\lambda}(db)$ for the scale group, and $\boldsymbol{\lambda}_d$ for the translation group; see (5.4). Although for noncompact groups such as $\mathbb{R}^d$ and $\mathbb{R}_\bullet^+$, right invariant distributions fail to exist there are often sequences of distributions available that behave as uniform distributions. For a measurable group $(\mathcal{G}, \mathfrak{G})$ we call a sequence $Q_n \in \mathcal{P}(\mathfrak{G})$, $n = 1, 2, ...$, *asymptotically right invariant* if

$$\lim_{n\to\infty} |Q_n(B) - Q_n(B\gamma_0)| = 0, \quad B \in \mathfrak{G}, \gamma_0 \in \mathcal{G}. \tag{5.40}$$

As every bounded measurable function $h : \mathcal{G} \to_m \mathbb{R}$ can be approximated uniformly by finite linear combinations of indicator functions we get for such a function h and an asymptotically right invariant sequence Q_n,

$$\lim_{n\to\infty} \int (h(\gamma) - h(\gamma\gamma_0))Q_n(d\gamma) = 0, \quad \gamma_0 \in \mathcal{G}. \tag{5.41}$$

Due to this property such a sequence Q_n is also called an *invariant mean.* Groups that have invariant means are called *amenable.* Such groups have been studied in many papers. A classical reference is Bondar and Milnes (1981). Often one can construct sequences Q_n that satisfy (5.40) by using an invariant σ-finite measure λ, a suitably chosen sequence A_n with $0 < \lambda(A_n) < \infty$, and setting

$$Q_n(\cdot) = \frac{1}{\lambda(A_n)}\lambda(\cdot \cap A_n). \tag{5.42}$$

This technique of constructing sequences Q_n can be applied to locally compact groups under weak additional assumptions. If $\mathcal{G}$ is a locally compact and σ-compact Abelian group, $\mathfrak{G}$ the σ-algebra of Borel sets, and λ a Haar measure, then there exists a sequence $A_1 \subseteq A_2 \subseteq \cdots$ of elements in $\mathfrak{G}$ such that Q_n from (5.42) satisfies (5.40). For a proof we refer to Kerstan and Matthes (1969). Here we consider only some special cases where the sets A_n can be constructed in a straightforward manner.

Problem 5.44. Set $A_n = [-n, n]$. Then the uniform distribution Q_n on A_n satisfies $\lim_{n\to\infty} |Q_n(A) - Q_n(A+t)| = 0$ for every $A \in \mathfrak{B}$ and $t \in \mathbb{R}$. Similarly, for the multiplicative measurable group $(\mathbb{R}_\bullet^+, \mathfrak{B}_+)$ we introduce the sequence $B_n = [n^{-1}, n]$ and set $Q_n(B) = \mu(B_n)^{-1}\mu(B \cap B_n)$, $B \in \mathfrak{B}_+$, with μ from Example 5.3. Then $|Q_n(B) - Q_n(Bs)| \to 0$ for every $B \in \mathfrak{B}_+$ and $s > 0$.

In the remainder of this section we consider only testing problems. Lemma 5.38 provides an averaging technique, based on a right invariant distribution Q on $(\mathcal{G}, \mathfrak{G})$, to get an invariant decision from a decision, and thus also an invariant test from a test. If, instead of such a Q, we have only a sequence Q_n that satisfies (5.40), then we may consider the sequence of tests $\int \varphi(u_\gamma(x))Q_n(d\gamma)$ with the idea that perhaps the limit, if such exists, or at least an accumulation point is invariant. However, as we show in the course of the proof of the next lemma, a limiting test is only almost invariant in the sense of the subsequent lemma. This leads to the question of whether for an almost invariant statistic there exists an invariant statistic that differs from it only on a set of probability zero. A precise formulation of this question and conditions that lead to a positive answer are given below.

Lemma 5.45. *Suppose that the model $(\mathcal{X}, \mathfrak{A}, (P_\theta)_{\theta\in\Delta})$ is invariant. Let $T : \mathcal{X} \to_m \mathbb{R}$ be a statistic such that for every $\gamma \in \mathcal{G}$ there exists an $N(\gamma) \in \mathfrak{A}$ with $\{x : T(u_\gamma(x)) \neq T(x)\} \subseteq N(\gamma)$ and $P_\theta(N(\gamma)) = 0$ for every $\theta \in \Delta$. Suppose that there exists a σ-finite measure μ on $(\mathcal{G}, \mathfrak{G})$ that satisfies the following condition.*

$$\mu(A) = 0 \Rightarrow \mu(A\gamma) = 0, \quad \gamma \in \mathcal{G},\ A \in \mathfrak{G}. \tag{5.43}$$

Then there exists an invariant statistic $\overline{T} : \mathcal{X} \to_m \mathbb{R}$ and an $N \in \mathfrak{A}$ with $P_\theta(N) = 0$, $\theta \in \Delta$, such that $T(x) = \overline{T}(x)$ for every $x \notin N$.

For a proof we refer to Pfanzagl (1994), Theorem 1.9.8. It should be noted that the condition (5.43) is satisfied if there exists a σ-finite invariant measure, which then can be taken for μ. In (5.34) we have constructed an invariant decision by averaging with an invariant distribution on $(\mathcal{G}, \mathfrak{G})$. The next lemma provides a similar technique for cases where only sequences of distributions on $(\mathcal{G}, \mathfrak{G})$ exist that satisfy (5.40).

Lemma 5.46. *Let $(\mathcal{X}, \mathfrak{A}, (P_\theta)_{\theta\in\Delta})$ be an invariant and dominated model. Suppose there exists a σ-finite measure μ on $(\mathcal{G}, \mathfrak{G})$ that satisfies (5.43). If there exists a sequence of distributions Q_n on $(\mathcal{G}, \mathfrak{G})$ that satisfies (5.40), then for each test φ there exists an invariant test $\overline{\varphi}$ and a subsequence $\int \varphi(u_\gamma(x))Q_{n_k}(d\gamma)$ such that*

$$\lim_{k\to\infty} \int [\int \varphi(u_\gamma(x))Q_{n_k}(d\gamma)]P_\theta(dx) = \int \overline{\varphi}(x)P_\theta(dx), \quad \theta \in \Delta.$$

Proof. As the decision space $\{0, 1\}$ is finite, and the model is dominated, from Theorem 3.21 it follows that there exists a subsequence and a test, say ψ, such that

$$\lim_{k\to\infty} \int [\int \varphi(u_\gamma(x))Q_{n_k}(d\gamma)]g(x)\overline{P}(dx) = \int \psi(x)g(x)\overline{P}(dx) \tag{5.44}$$

for every $g : \mathcal{X} \to_m \mathbb{R}$ with $\int |g|d\overline{P} < \infty$, where $\overline{P} = \sum_{i=0}^{\infty} c_i P_{\theta_i}$ is taken from Lemma 4.36.

Let h be any bounded and measurable function and set $g = h(dP_\theta/d\overline{P})$. Then by (5.16), for any $\gamma_0 \in \mathcal{G}$,

$$\lim_{k\to\infty} \int [\int \varphi(u_{\gamma\gamma_0}(x)) Q_{n_k}(d\gamma)] h(x) P_{\theta_i}(dx)$$
$$= \lim_{k\to\infty} \int [\int \varphi(u_\gamma(u_{\gamma_0}(x))) Q_{n_k}(d\gamma)] h(u_{\gamma_0^{-1}}(u_{\gamma_0}(x))) P_{\theta_i}(dx)$$
$$= \lim_{k\to\infty} \int [\int \varphi(u_\gamma(x)) Q_{n_k}(d\gamma)] h(u_{\gamma_0^{-1}}(x)) P_{v_{\gamma_0}(\theta_i)}(dx)$$
$$= \int \psi(x) h(u_{\gamma_0^{-1}}(x)) P_{v_{\gamma_0}(\theta_i)}(dx) = \int \psi(u_{\gamma_0}(x)) h(x) P_{\theta_i}(dx).$$

For every fixed $\gamma_0 \in \mathcal{G}$ the sequence

$$\int [\varphi(u_\gamma(x)) - \varphi(u_{\gamma\gamma_0}(x))] Q_{n_k}(d\gamma)$$

is bounded by 2 and tends, in view of (5.41), to zero for every fixed $x \in \mathcal{X}$. Hence by Lebesgue's theorem for every bounded and measurable function h,

$$\int [\psi(x) - \psi(u_{\gamma_0}(x))] h(x) P_{\theta_i}(dx) = 0,$$

which implies

$$P_{\theta_i}(\{x : \psi(x) = \psi(u_{\gamma_0}(x))\}) = 1 \quad \text{and} \quad \overline{P}(\{x : \psi(x) = \psi(u_{\gamma_0}(x))\}) = 1.$$

Hence $P_\theta(\{x : \psi(x) = \psi(u_{\gamma_0}(x))\}) = 1$, $\theta \in \Delta$, so that $T = \psi$ satisfies the condition in Lemma 5.45. Thus there exists an invariant test $\overline{\varphi}$ with $P_\theta(\{x : \psi(x) = \overline{\varphi}(x)\}) = 1$. ■

The next statement is a version of a famous theorem by Hunt and Stein who did not publish it. Lehmann (1986), p. 536, refers to an unpublished paper of Hunt and Stein (1946). We establish here only the classical version for testing hypotheses, which says that under weak assumptions in the search for a maximin test we may restrict ourselves, without loss in terms of the risk, to considering only the invariant tests. A more general Hunt–Stein type theorem can be found in Strasser (1985) and LeCam (1986). For references to general results of the Hunt–Stein type we refer to Lehmann (1998), pp. 421–422.

Theorem 5.47. ***(Hunt–Stein)*** *Suppose that $(\mathcal{X}, \mathfrak{A}, (P_\theta)_{\theta\in\Delta})$ is an invariant and dominated model, that there exists a σ-finite measure μ that satisfies (5.43), and that there exist distributions Q_n on $(\mathcal{G}, \mathfrak{G})$ that satisfy (5.40). If the testing problem for $\mathsf{H}_0 : \theta \in \Delta_0$ versus $\mathsf{H}_A : \theta \in \Delta_A$ is invariant, then for every test φ there exists an invariant test $\overline{\varphi}$ such that*

$$\sup_{\theta\in\Delta_0} \mathsf{E}_\theta \overline{\varphi} \le \sup_{\theta\in\Delta_0} \mathsf{E}_\theta \varphi \quad \textit{and} \quad \inf_{\theta\in\Delta_A} \mathsf{E}_\theta \varphi \le \inf_{\theta\in\Delta_A} \mathsf{E}_\theta \overline{\varphi}.$$

Corollary 5.48. *Under the assumptions of the last theorem for every level α test φ for H_0 there exists an invariant level α test $\overline{\varphi}$ such that $\inf_{\theta\in\Delta_A} \mathsf{E}_\theta \varphi \le \inf_{\theta\in\Delta_A} \mathsf{E}_\theta \overline{\varphi}$.*

Proof. By Lemma 5.46, (5.16), (5.21), and Fubini's theorem,

$$\sup_{\theta\in\Delta_0}\int\overline{\varphi}(x)P_\theta(dx) = \sup_{\theta\in\Delta_0}\lim_{k\to\infty}\int[\int\varphi(u_\gamma(x))Q_{n_k}(d\gamma)]P_\theta(dx)$$
$$= \sup_{\theta\in\Delta_0}\lim_{k\to\infty}\int[\int\varphi(x)P_{v_\gamma(\theta)}(dx)]Q_{n_k}(d\gamma) \le \sup_{\theta\in\Delta_0}\mathsf{E}_\theta\varphi.$$

Analogously, $\inf_{\theta\in\Delta_A}\int\overline{\varphi}(x)P_\theta(dx) \ge \inf_{\theta\in\Delta_A}\mathsf{E}_\theta\varphi$. ■

Now we consider again the model $(\mathbb{R}^k,\mathfrak{B}_k,(\mathsf{N}(\mu,\mathbf{I}))_{\mu\in\mathbb{R}^k})$ and set

$$\chi^2_*(x) = \sum\nolimits_{i=1}^k(x_i-\overline{x}_k)^2 \quad\text{and}\quad \delta^2_*(\mu) = \sum\nolimits_{i=1}^k(\mu_i-\overline{\mu}_k)^2,$$

where $\overline{x}_k = k^{-1}\sum_{i=1}^k x_i$ and $\overline{\mu}_k = k^{-1}\sum_{i=1}^k\mu_i$. Here we want to test if all components of μ are equal. This testing problem can be formulated as

$$\mathsf{H}_0^*:\delta^2_*(\mu)=0 \quad\text{versus}\quad \mathsf{H}_A^*:\delta^2_*(\mu)>0. \tag{5.45}$$

Let $\mathbb{L}_1$ be the one-dimensional subspace of $\mathbb{R}^k$ that consists of vectors $x = (a,...,a) = a\mathbf{1}$, $a\in\mathbb{R}$. Then we have

$$\mathsf{H}_0^*:\mu\in\mathbb{L}_1 \quad\text{versus}\quad \mathsf{H}_A^*:\mu\notin\mathbb{L}_1.$$

Let $\mathcal{O}$ be the set of all orthogonal matrices O that leave $\mathbb{L}_1$ invariant, and thus also leave the orthogonal complement $\mathbb{L}_1^\perp$ invariant. We introduce the group $\mathcal{G}=\mathbb{L}_1\times\mathcal{O}$ by setting

$$(a_1,O_1)\circ(a_2,O_2) = (a_1+a_2,O_1O_2),$$
$$u_\gamma(x) = O(x+a),\quad \gamma=(a,O)\in\mathbb{L}_1\times\mathcal{O}.$$

Then $u_{\gamma_1}(u_{\gamma_2}(x)) = u_{\gamma_1\gamma_2}(x)$. We consider $\mathcal{G}$ as a subset of $\mathbb{R}\times\mathbb{R}^{(k-1)^2}$ and use the Borel subsets as the σ-algebra $\mathfrak{G}$.

Problem 5.49.* For the group $(\mathcal{G},\mathfrak{G}) = (\mathbb{L}_1\times\mathcal{O},\mathfrak{B}_{\mathbb{L}_1\times\mathcal{O}})$ there exists a σ-finite right invariant measure μ and a sequence of distributions Q_n that satisfies (5.40).

The next proposition establishes the maximin property of the χ^2-test for comparing the means of k independent normal distributions with a common known variance. The latter, without loss of generality, is put to 1.

Proposition 5.50. *For the model* $(\mathbb{R}^k,\mathfrak{B}_k,(\mathsf{N}(\mu,\mathbf{I})_{\mu\in\mathbb{R}^k})$ *the* χ^2*-test*

$$\varphi_{\chi^2_*,\alpha}(x) = I_{(\chi^2_{1-\alpha,k-1},\infty)}(\chi^2_*(x))$$

is a uniformly best $(\mathbb{L}_1\times\mathcal{O})$*-invariant level* α *test for the testing problem (5.45). Furthermore, for every fixed* $\delta_0^2\ge 0$ *it holds*

$$\inf_{\delta^2_*(\mu)>\delta^2_0}\mathsf{E}_\mu\varphi_{\chi^2_*,\alpha} \ge \inf_{\delta^2_*(\mu)>\delta^2_0}\mathsf{E}_\mu\varphi$$

for every test φ *that satisfies* $\mathsf{E}_0\varphi \le \alpha$, *so that the test* $\varphi_{\chi^2_*,\alpha}$ *is a maximin level* α *test for*

$$\mathsf{H}_0^* : \delta_*^2(\mu) = 0 \quad \textit{versus} \quad \mathsf{H}_{A,\delta_0^2}^* : \delta_*^2(\mu) > \delta_0^2.$$

Proof. The proof of the first statement is similar to the proof of Theorem 5.33. First of all we note that for $\delta_*^2(\mu) = 0$ the distribution $\mathsf{N}(\mu, \mathbf{I}) \circ (\chi_*^2)^{-1}$ is a χ^2-distribution with $k-1$ degrees of freedom. Next we show that every $(\mathbb{L}_1 \times \mathcal{O})$-invariant test φ can be represented as a measurable function of $\chi_*^2(x)$. To this end we fix a unit vector e_0 and note that

$$\begin{aligned}\varphi(x) &= h(\|x - \overline{x}_k\mathbf{1}\|), \quad x - \overline{x}_k\mathbf{1} \neq 0, \quad \text{where } h(r) = \varphi(re_0), \\ \varphi(x) &= h(\|x - \overline{x}_k\mathbf{1}\|)I_{(0,\infty)}(\|x - \overline{x}_k\mathbf{1}\|) + \varphi(0)I_{\{0\}}(\|x - \overline{x}_k\mathbf{1}\|).\end{aligned}$$

As in the proof of Theorem 5.33 it suffices to find a uniformly best level α test for the reduced model

$$(\mathbb{R}_+, \mathfrak{B}_+, (\mathsf{N}(\mu, \mathbf{I}) \circ (\chi_*^2)^{-1})_{\mu \in \mathbb{R}^k}).$$

But Theorem 2.49 yields that $\psi_\alpha(t) = I_{(\chi^2_{1-\alpha,k-1},\infty)}(t)$ is a uniformly best level α test for $\mathsf{H}_0 : \delta^2 = 0$ versus $\mathsf{H}_A : \delta^2 > 0$ in the reduced model. Hence

$$\psi_\alpha(\chi_*^2(x)) = \varphi_{\chi_*^2,\alpha}(x)$$

is a uniformly best invariant level α test. To prove the second statement we note that also the reduced model with $\mathsf{N}(\mu, \mathbf{I})$, $\mu \in \Delta_0^* \cup \Delta_A^* = \mathbb{L}_1 \cup (\{\mu : \delta_*^2(\mu) > \delta_0^2\})$, as well as the hypotheses are invariant.

In view of Problem 5.49 there is a σ-finite invariant measure μ that satisfies (5.43) and a sequence Q_n that satisfies (5.40). As $\varphi_{\chi_*^2,\alpha}$ is a test of size α we get the statement from Corollary 5.48. ∎

It is known that asymptotically right invariant sequences of distributions exist for every Abelian locally compact group; see Bondar and Milnes (1981). If a group is not Abelian, then in relatively simple situations already condition (5.40) may not be satisfied. The following example, due to Stein, is taken from Lehmann (1986), Example 9 in Chapter 9.

Example 5.51. Let $X = (X_1, ..., X_n)$ and $Y = (Y_1, ..., Y_n)$ be independent and normally distributed random vectors with expectation zero and covariance matrices Σ and $\sigma\Sigma$, respectively, where Σ is nonsingular, $\sigma > 0$, and σ and Σ are both unknown. Suppose we want to test $\mathsf{H}_0 : \sigma \le 1$ versus $\mathsf{H}_A : \sigma > 1 + \delta$, where $\delta > 0$ is fixed given. The statistical model is

$$\begin{aligned}&(\mathbb{R}^{2n}, \mathfrak{B}_{2n}, \mathsf{N}(0, \Sigma) \otimes \mathsf{N}(0, \sigma\Sigma)), \quad \text{where} \\ &\theta = (\sigma, \Sigma) \in \Delta = \{(\sigma, \Sigma) : \sigma \in (0, 1] \cup (1 + \delta, \infty],\ \Sigma \text{ symmetric positive definite}\}.\end{aligned}$$

We use the general linear group $\mathcal{M}_{n\times n}^r$ from (5.4) and equip $\mathcal{M}_{n\times n}^r$ with the σ-algebra of Borel sets. For $\gamma = B \in \mathcal{M}_{n\times n}^r$ we set $u_\gamma(x, y) = (Bx, By)$ and

$v_\gamma(\theta) = (\sigma, B\Sigma B^T)$. This shows that the model and the testing problem are invariant. According to (5.4) the measure $|\det(B)|^{-n}\boldsymbol{\lambda}_{n^2}(dB)$ is σ-finite and invariant, so that (5.43) is fulfilled. Now we show that for $n \geq 2$ the statement of the Hunt–Stein theorem is not true, i.e., that the group $\mathcal{M}^r_{n\times n}$ is not amenable. Indeed, if a test φ is invariant, then $\varphi(x) = \varphi(Bx)$ for every $B \in \mathcal{M}^r_{n\times n}$. As for $n \geq 2$ and any $x_1, x_2 \in \mathbb{R}^n$ there exists some $B \in \mathcal{M}^r_{n\times n}$ with $x_2 = Bx_1$, we see that every invariant test is constant. Thus every best invariant level α test is constant and takes on the value α. But this test cannot be a maximin test. Note that $F = \sigma^2 X_1^2/Y_1^2$ has a F distribution with $(1,1)$ degrees of freedom. If $f_{1-\alpha,1,1}$ denotes the $1-a$ quantile and we reject H_0 whenever $X_1^2/Y_1^2 > f_{1-\alpha,1,1}$, then we get a level α test with a strictly increasing power function. This means that the constant test cannot be a maximin level α test.

5.4 Equivariant Estimators, Girshick–Savage Theorem

In this section we study equivariant estimators in an invariant location model, establish a minimax theorem, and prove the minimax property of the Pitman estimator. In Chapter 7 we continue the study of the Pitman estimator. There we emphasize more the fact that the Pitman estimator is a generalized Bayes estimator, and establish conditions under which the Pitman estimator is admissible.

The results on the minimaxity of the Pitman estimator go back to Girshick and Savage (1951). These results have been generalized by other authors to more general models that are generated by a fixed distribution and the application of a transformation group. Here in this section we concentrate on the location model and mainly use the quadratic loss function. The presentation below follows Witting (1985). Results related to general transformation groups can be found in Strasser (1985).

We use the additive group $\mathbb{R}_\oplus$ and the corresponding transformation group $\mathcal{U}_l$ from (5.4). Hence

$$u_\theta(x) = x + \theta\mathbf{1}, \quad x \in \mathbb{R}^n,\ \theta \in \mathbb{R}, \tag{5.46}$$

where $\mathbf{1} = (1,...,1)^T \in \mathbb{R}^n$. We denote by $\mathfrak{J} \subseteq \mathfrak{B}_n$ the sub-σ-algebra of *invariant Borel sets*, which are the sets $B \in \mathfrak{B}_n$ with $I_B(x+\theta\mathbf{1}) = I_B(x)$ for every $x \in \mathbb{R}^n$ and $\theta \in \mathbb{R}$. According to Definition 5.13 an invariant statistic $T : \mathbb{R}^n \to_m \mathbb{R}^n$ is *maximal invariant* if $T(x) = T(y)$ implies $x = y + \theta\mathbf{1}$ for some $\theta \in \mathbb{R}$. A statistic $\mathcal{E} : \mathbb{R}^n \to_m \mathbb{R}$ is called *equivariant* if $\mathcal{E}(x + \alpha\mathbf{1}) = \mathcal{E}(x) + \alpha$ holds for every $x \in \mathbb{R}^n$ and $\alpha \in \mathbb{R}$. By $\mathfrak{E}$ we denote the set of all equivariant statistics.

Example 5.52. The following statistics are equivariant.

$$(x_1,...,x_n) \mapsto \overline{x}_n, \quad (x_1,...,x_n) \mapsto \min_{1\leq i\leq n} x_i, \quad (x_1,...,x_n) \mapsto x_1.$$

If $\mathcal{E} \in \mathfrak{E}$, then $T_{\mathcal{E}}(x) := x - \mathcal{E}(x)\mathbf{1}$ is invariant. If $T_{\mathcal{E}}(x) = T_{\mathcal{E}}(y)$, then $x - y = (\mathcal{E}(x) - \mathcal{E}(y))\mathbf{1}$ so that $T_{\mathcal{E}}$ is maximal invariant. Moreover, if $S : \mathbb{R}^n \to_m \mathbb{R}^n$ is any invariant statistic, then $S(x - \mathcal{E}(x)\mathbf{1}) = S(x)$. Hence every invariant statistic $S : \mathbb{R}^n \to_m \mathbb{R}^n$ satisfies $S = S(T_{\mathcal{E}})$ and can therefore be written as a measurable function of the equivariant statistic $T_{\mathcal{E}}$.

We study the *location model*

$$\mathcal{M}_{lo} = \left(\mathbb{R}^n, \mathfrak{B}_n, (P \circ u_\theta^{-1})_{\theta \in \mathbb{R}}\right), \tag{5.47}$$

where by the definition of u_θ in (5.46) $P_\theta := P \circ u_\theta^{-1}$ is given by

$$P_\theta(B) = \int I_B(x + \theta\mathbf{1})P(dx),$$

and especially $P = P_0$. We use the loss function $L(\theta, \alpha) = l(\theta - \alpha)$, where $l : \mathbb{R} \to_m \mathbb{R}_+$. Defining $\mathcal{V} = \mathcal{W}$ as in Example 5.28 we get a decision problem that is invariant in the sense of Definition 5.25. According to (5.19) a non-randomized decision D is invariant if and only if the associated statistic d is equivariant.

Under the loss function $L(\theta, a) = l(\theta - a)$ the risk of any estimator S is given by $R(\theta, S) = \mathsf{E}_\theta l(S - \theta)$, $\theta \in \mathbb{R}$. If now S is equivariant, then

$$\begin{aligned}
\mathsf{R}(\theta, S) &= \int l(S(x) - \theta)\,(P \circ u_\theta^{-1})(dx) \\
&= \int l(S(x + \theta\mathbf{1}) - \theta)\,P_0(dx) = \mathsf{R}(0, S) = \mathsf{E}_0 l(S), \quad \theta \in \mathbb{R}.
\end{aligned}$$

It should be noted that this is a special case of (5.20).

Definition 5.53. *Let $L(\theta, a) = l(\theta - a)$, $\theta, a \in \mathbb{R}$. An equivariant estimator $\mathcal{P}$ that satisfies $\mathsf{E}_0 l(\mathcal{P}) < \infty$ and $\mathsf{E}_0 l(\mathcal{P}) \le \mathsf{E}_0 l(S)$, $S \in \mathfrak{E}$, is called a Pitman estimator under the loss function L. In the special case of $l(t) = t^2$, $t \in \mathbb{R}$, it is simply called a Pitman estimator.*

For a strictly convex l the Pitman estimator under the loss L is unique.

Problem 5.54.* Let $\mathfrak{C}$ be a convex subset of estimators and l be strictly convex and nonnegative. Then for $S_0, S_1 \in \mathfrak{C}$ the condition

$$\mathsf{E}_0 l(S_0) = \mathsf{E}_0 l(S_1) = \inf_{S \in \mathfrak{C}} \mathsf{E}_0 l(S) < \infty$$

implies that $S_0 = S_1$, P_0-a.s.

Problem 5.55.* Given the probability space $(\mathbb{R}^n, \mathfrak{B}_n, P)$ and the sub-σ-algebra of invariant Borel sets, we fix a regular conditional distribution given $\mathfrak{I}$, i.e., some $\mathsf{K} : \mathfrak{B}_n \times \mathbb{R}^n \to_k [0, 1]$ for which $x \mapsto \mathsf{K}(A|x)$ is $\mathfrak{I}$-measurable for every $A \in \mathfrak{B}_n$ and that satisfies

$$P(A \cap B) = \int_B \mathsf{K}(A|x)P(dx), \quad A \in \mathfrak{B}_n,\ B \in \mathfrak{I}.$$

Such a stochastic kernel exists; see Theorem A.37. Then for every $T : \mathbb{R}^n \to \mathbb{R}$ that is $\mathfrak{I}$-$\mathfrak{B}_n$ measurable, and every $h : \mathbb{R}^n \times \mathbb{R} \to_m \mathbb{R}_+$, it holds

$$\int h(x, T(x))P(dx) = \int [\int h(y, T(x))\mathsf{K}(dy|x)]P(dx). \tag{5.48}$$

For $S = \mathcal{E}_0 - T$ with $\mathcal{E}_0 \in \mathfrak{E}$ it follows from (5.48) that

$$\begin{aligned} \mathsf{E}_0 l\,(S) &= \mathsf{E}_0(\mathsf{E}_0(l(\mathcal{E}_0 - T)|\mathfrak{I})) \\ &= \int [\int l(\mathcal{E}_0(y) - T(x))\mathsf{K}(dy|x)]P(dx). \end{aligned} \tag{5.49}$$

If K is known, then we may fix any $\mathcal{E}_0 \in \mathfrak{E}$ and find a T that minimizes $\mathsf{E}_0 l(S)$ by minimizing $t \mapsto \int l(\mathcal{E}_0(y) - t)\mathsf{K}(dy|x)$ for every fixed x. Then T becomes automatically equivariant, provided the minimization point is unique. If $l(t) = t^2$ and $\mathsf{E}_0\mathcal{E}_0^2 < \infty$, then $\int \mathcal{E}_0^2(y)\mathsf{K}(dy|x) < \infty$, P-a.s., and

$$\arg\min_{t\in\mathbb{R}} \int (\mathcal{E}_0(y) - t)^2\mathsf{K}(dy|x) = \{\int \mathcal{E}_0(y)\mathsf{K}(dy|x)\}, \quad P\text{-a.s.} \tag{5.50}$$

Theorem 5.56. *If $\mathcal{E}_0$ is equivariant and satisfies $\mathsf{E}_0\mathcal{E}_0^2 < \infty$, then the Pitman estimator exists, is P-a.s. uniquely determined, and it holds*

$$\mathcal{P}(x) = \mathcal{E}_0(x) - \int \mathcal{E}_0(y)\mathsf{K}(dy|x), \quad P\text{-a.s.}, \tag{5.51}$$

where K is a regular conditional distribution given the σ-algebra of invariant Borel sets $\mathfrak{I}$.

Proof. Combine (5.50) and Problem 5.54. ∎

To calculate the Pitman estimator we need the distribution $\mathsf{K}_{\mathcal{E}_0}(\cdot|x) := \mathsf{K}(\mathcal{E}_0^{-1}(\cdot)|x)$ for only one $\mathcal{E}_0$. This means that we have to find a stochastic kernel $\mathsf{K}_{\mathcal{E}_0}$ that satisfies

$$\begin{aligned} &\mathsf{K}_{\mathcal{E}_0}(A|x + a\mathbf{1}) = \mathsf{K}_{\mathcal{E}_0}(A|x), \quad A \in \mathfrak{B}_n,\ x \in \mathbb{R}^n,\ a \in \mathbb{R} \\ &\textstyle\int I_B(x)\mathsf{K}_{\mathcal{E}_0}(A|x)P(dx) = P(\mathcal{E}_0^{-1}(A) \cap B), \quad B \in \mathfrak{I}. \end{aligned} \tag{5.52}$$

We use $\mathcal{E}_0(x_1, ..., x_n) = x_1$. The conditional distribution $\mathsf{K}_{\mathcal{E}_0}(\cdot|x)$ can be represented by means of the conditional density if P has a Lebesgue density.

Lemma 5.57. *If P has the Lebesgue density f, then $\int f\,(x + s\mathbf{1})\,ds > 0$, P-a.s. Let $\mathcal{E}_0(y_1, ..., y_n) = y_1$, $y \in \mathbb{R}^n$, and*

$$f(y_1|x) = \frac{f\,(y_1, x_2 - x_1 + y_1, ..., x_n - x_1 + y_1)}{\int f\,(s, x_2 - x_1 + s, ..., x_n - x_1 + s)\,ds}$$

if $\int f\,(s, x_2 - x_1 + s, ..., x_n - x_1 + s)\,ds > 0$, and $f(y_1|x) = g\,(y_1)$ otherwise, where g is any Lebesgue density. Then

$$\mathsf{K}_{\mathcal{E}_0}(C|x) = \int_C f(y_1|x)dy_1$$

is a stochastic kernel that satisfies the conditions in (5.52).

Proof. The set $A_0 = \{x : \int f(x + s\mathbf{1})\, ds = 0\}$ is invariant; that is, $I_{A_0}(x) = I_{A_0}(x + t\mathbf{1})$ for every $t \in \mathbb{R}$. Hence by Fubini's theorem and the definition of A_0,

$$0 = \int [\int I_{A_0}(x) f(x + s\mathbf{1})\, ds] \boldsymbol{\lambda}_n(dx) = \int [\int I_{A_0}(x - s\mathbf{1}) f(x)\, \boldsymbol{\lambda}_n(dx)] ds$$
$$= \int (\int I_{A_0}(x) P(dx)) ds,$$

which implies $P(A_0) = 0$. For every $D \in \mathfrak{B}_{n-1}$ and $\widetilde{T}(x_1, ..., x_n) = (x_2 - x_1, ..., x_n - x_1)$ it holds

$$P(\widetilde{T} \in B) = \int I_D(y_2 - y_1, ..., y_n - y_1) P(dy)$$
$$= \int I_D(s) [\int f(y_1, s_2 + y_1, ..., s_n + y_1)\, dy_1] \boldsymbol{\lambda}_{n-1}(ds),$$

where $s = (s_2, ..., s_n)$. Hence

$$\frac{d(P \circ \widetilde{T}^{-1})}{d\boldsymbol{\lambda}_{n-1}}(s) = \int f(y_1, s_2 + y_1, ..., s_n + y_1)\, dy_1, \quad \boldsymbol{\lambda}_{n-1}\text{-a.e.} \tag{5.53}$$

Suppose that $A \in \mathfrak{B}$ and $B \in \mathfrak{J}$. Put $D = \{(s_2, ..., s_n) : I_B(0, s_2, ..., s_n) = 1\}$. Then $D \in \mathfrak{B}_{n-1}$ and by the invariance of B it holds $I_B(x_1, ..., x_n) = I_D(x_2 - x_1, ..., x_n - x_1)$.

$$\int_B [\int_A f(y_1|x) dy_1] P(dx)$$
$$= \int [\int I_A(y_1) I_D(x_2 - x_1, ..., x_n - x_1)$$
$$\times \frac{f(y_1, x_2 - x_1 + y_1, ..., x_n - x_1 + y_1)}{\int f(t, x_2 - x_1 + t, ..., x_n - x_1 + t)\, dt} dy_1] P(dx)$$
$$= \int [\int I_A(y_1) I_D(s_2, ..., s_n)$$
$$\times \frac{f(y_1, s_2 + y_1, ..., s_n + y_1)}{\int f(t, s_2 + t, ..., s_n + t)\, dt} dy_1] (P \circ \widetilde{T}^{-1})(ds_2, ..., ds_n).$$

Using (5.53) we get

$$\int_B [\int_A f(y_1|x) dy_1] P(dx)$$
$$= \int [\int I_A(y_1) I_D(s_2, ..., s_n) f(y_1, s_2 + y_1, ..., s_n + y_1)\, dy_1] \boldsymbol{\lambda}_{n-1}(ds_2, ..., ds_n)$$
$$= \int [\int I_A(y_1) I_D(y_2 - y_1, ..., y_n - y_1) f(y_1, ..., y_n)\, dy_1] \boldsymbol{\lambda}_{n-1}(dy_2, ..., dy_n)$$
$$= \int I_A(y_1) I_B(y_1, ..., y_n) P(dy_1, ..., dy_n) = P(\mathcal{E}_0^{-1}(A) \cap B).$$

■

Using the conditional density in Lemma 5.57 we find an explicit representation of the Pitman estimator.

Proposition 5.58. *If $l(t) = t^2$, $t \in \mathbb{R}$, $\int x_1^2 P(dx_1, ..., dx_n) < \infty$, and P has the Lebesgue density f, then the Pitman estimator $\mathcal{P}$ in (5.51) is given by*

$$\mathcal{P}(x_1, ..., x_n) = \frac{\int s f(x_1 - s, ..., x_n - s)\, ds}{\int f(x_1 - s, ..., x_n - s)\, ds}, \quad P\text{-a.s.} \tag{5.54}$$

Proof. We use $\mathcal{E}_0(x_1, ..., x_n) = x_1$ in Theorem 5.56. Then by Lemma 5.57,

$$\begin{aligned}\mathcal{P}(x) &= \mathcal{E}_0(x) - \int \mathcal{E}_0(y)\mathsf{K}(dy|x) \\ &= x_1 - \frac{\int y_1 f(y_1, x_2 - x_1 + y_1, ..., x_n - x_1 + y_1)\, dy_1}{\int f(s, x_2 - x_1 + s, ..., x_n - x_1 + s)\, ds} \\ &= \frac{\int s f(x_1 - s, ..., x_n - s)\, ds}{\int f(x_1 - s, ..., x_n - s)\, ds}, \quad P\text{-a.s.}\end{aligned}$$

■

Example 5.59. Assume that we have observed n independent random variables with a common normal distribution $\mathsf{N}_{\theta,\sigma^2}$. Then P in (5.47) is given by $P = \mathsf{N}_{0,\sigma^2}^{\otimes n}$. Hence $f(x) = (2\pi)^{-n/2}\sigma^{-n}\exp\{-(2\sigma^2)^{-1}\|x\|^2\}$, $x = (x_1, ..., x_n) \in \mathbb{R}^n$, and by (5.54),

$$\begin{aligned}\mathcal{P}(x) &= \left[\int \prod\nolimits_{i=1}^n \varphi_{s,\sigma^2}(x_i) ds\right]^{-1} \left[\int s \prod\nolimits_{i=1}^n \varphi_{s,\sigma^2}(x_i) ds\right] \\ &= \frac{\exp\{-\frac{1}{2\sigma^2}\sum_{i=1}^n (x_i - \overline{x}_n)^2\} \int s \exp\{-\frac{n}{2\sigma^2}(s - \overline{x}_n)^2\} ds}{\exp\{-\frac{1}{2\sigma^2}\sum_{i=1}^n (x_i - \overline{x}_n)^2\} \int \exp\{-\frac{n}{2\sigma^2}(s - \overline{x}_n)^2\} ds} = \overline{x}_n.\end{aligned}$$

Example 5.60. Assume that we have observed n independent random variables with a common exponential distribution $\mathsf{Ex}(1)$. Then the Lebesgue density of P in (5.47) is given by $f(x) = I_{(0,\infty)}(x_{[1]})\exp\{-\sum_{i=1}^n x_i\}$, where $x_{[1]} = \min\{x_1, ..., x_n\}$, $x = (x_1, ..., x_n) \in \mathbb{R}^n$. Then by (5.54),

$$\begin{aligned}\mathcal{P}(x) &= \frac{\int s I_{(0,\infty)}(x_{[1]} - s)\exp\{-\sum_{i=1}^n (x_i - s)\} ds}{\int I_{(0,\infty)}(x_{[1]} - s)\exp\{-\sum_{i=1}^n (x_i - s)\} ds} \\ &= \frac{\int s I_{(-\infty, x_{[1]})}(s)\exp\{ns\} ds}{\int I_{(-\infty, x_{[1]})}(s)\exp\{ns\} ds} = x_{[1]} - \frac{1}{n}.\end{aligned}$$

Pitman estimators can also be considered for more general models where the family of distributions is generated by any group of measurable transformations. We do not go into more detail but refer to Strasser (1985) and Witting (1985).

Now we study the problem of whether optimal equivariant estimators are also optimal in the minimax sense in the class of all estimators. The answer is given by a theorem due to Girshick and Savage (1951). We prove, in a first step, a minimax result that is similar to Theorem 3.57 and corresponds to the Hunt–Stein theorem 5.47 for testing problems. In contrast to Theorem 3.57 we have here a special model, but the decision space is not compact. For any $l : \mathbb{R} \to_m \mathbb{R}_+$ and an estimator S for θ in the location model (5.47) we set

$$\begin{aligned} \mathsf{R}(\theta, S) &= \int l(S(x) - \theta) P_\theta(dx) \qquad (5.55) \\ &= \int l(S(x + \theta\mathbf{1}) - \theta) P_0(dx), \quad \theta \in \mathbb{R}, \\ \mathsf{r}(\Pi, S) &= \int \mathsf{R}(\theta, S)\Pi(d\theta), \end{aligned}$$

where $\Pi \in \mathcal{P}(\mathfrak{B})$ is any prior. For an equivariant estimator S it holds

$$\mathsf{R}(\theta, S) = \mathsf{R}(0, S) = \int l(S(x)) P_0(dx), \quad \theta \in \mathbb{R}.$$

Theorem 5.61. ***(Girshick–Savage)*** *If $l : \mathbb{R} \to_m \mathbb{R}_+$ is bounded, then it holds in the location model (5.47),*

$$\sup_\Pi \inf_S \mathsf{r}(\Pi, S) = \inf_S \sup_\theta \mathsf{R}(\theta, S) = \inf_{S \in \mathfrak{E}} \mathsf{R}(0, S),$$

where $\inf_S$ *is the infimum over all estimators S.*

Proof. It holds,

$$\sup_\Pi \inf_S \mathsf{r}(\Pi, S) \le \inf_S \sup_\theta \mathsf{R}(\theta, S) \le \inf_{S \in \mathfrak{E}} \sup_\theta \mathsf{R}(\theta, S) = \inf_{S \in \mathfrak{E}} \mathsf{R}(0, S). \qquad (5.56)$$

Therefore we have to show that $\inf_{T \in \mathfrak{E}} \mathsf{R}(0, T) \le \sup_\Pi \inf_S \mathsf{r}(\Pi, S)$. We use the prior $\mathsf{N}(0, \sigma^2)$ and fix an estimator $S : \mathbb{R}^n \to_m \mathbb{R}$. Let $\mathcal{E}$ be equivariant; e.g., $\mathcal{E}(x_1, ..., x_n) = x_1$. It follows from (5.55) and Fubini's theorem that

$$\begin{aligned} \mathsf{r}(\mathsf{N}(0, \sigma^2), S) &= \int [\int l(S(x + \theta\mathbf{1}) - \theta)\varphi_{0,\sigma^2}(\theta) d\theta] P_0(dx) \\ &= \int [\int l(S(x + \theta\mathbf{1}) - \mathcal{E}(x + \theta\mathbf{1}) + \mathcal{E}(x))\varphi_{0,\sigma^2}(\theta) d\theta] P_0(dx) \\ &= \int [\int l(S_\eta(x))\varphi_{\mathcal{E}(x),\sigma^2}(\eta) d\eta] P_0(dx), \end{aligned}$$

where

$$S_\eta(x) = S(x + (\eta - \mathcal{E}(x))\mathbf{1}) - \mathcal{E}(x + (\eta - \mathcal{E}(x))\mathbf{1}) + \mathcal{E}(x),$$

is an equivariant estimator for every η. The risk of S_η under P_0 satisfies

$$\begin{aligned} \int [\int l(S_\eta(x))\varphi_{0,\sigma^2}(\eta) d\eta] P_0(dx) &= \int [\int l(S_\eta(x)) P_0(dx)]\varphi_{0,\sigma^2}(\eta) d\eta \\ &\ge \inf_{T \in \mathfrak{E}} \int l(T(x)) P_0(dx). \end{aligned}$$

Because l is bounded, say by C, then by the inequality in Problem 5.64,

$$\begin{aligned}
&\mathsf{r}(\mathsf{N}(0,\sigma^2),S) = \int [\int l(S_\eta(x))\varphi_{\mathcal{E}(x),\sigma^2}(\eta)d\eta]P_0(dx) \\
&\geq \int [\int l(S_\eta(x))\varphi_{0,\sigma^2}(\eta)d\eta]P_0(dx) \\
&\qquad - 2\sqrt{2}C\int [1-\exp\{-\mathcal{E}^2(x)/(8\sigma^2)\}]^{1/2}P_0(dx) \\
&\geq \inf_{T\in\mathfrak{E}}\int l(T(x))P_0(dx) - 2\sqrt{2}C\int [1-\exp\{-\mathcal{E}^2(x)/(8\sigma^2)\}]^{1/2}P_0(dx).
\end{aligned}$$

Taking the infimum over all estimators S on the the left-hand side and letting $\sigma^2\to\infty$ we get from Lebesgue's theorem that

$$\begin{aligned}
\sup_{\Pi}\inf_{S}\mathsf{r}(\Pi,S) &\geq \limsup_{\sigma^2\to\infty}\inf_{S}\mathsf{r}(\mathsf{N}(0,\sigma^2),S) \\
&\geq \inf_{T\in\mathfrak{E}}\int l(T(x))P_0(dx) = \inf_{T\in\mathfrak{E}}\mathsf{R}(0,T).
\end{aligned}$$

■

Now we consider the important case of a quadratic loss function.

Theorem 5.62. ***(Girshick–Savage)*** *If for the location model (5.47) there exists an equivariant statistic $\mathcal{E}_0$ with $\mathsf{E}_0\mathcal{E}_0^2<\infty$, then*

$$\sup_{\Pi}\inf_{S}\int \mathsf{E}_\theta(S-\theta)^2\Pi(d\theta) = \inf_{S}\sup_{\theta}\mathsf{E}_\theta(S-\theta)^2 = \inf_{S\in\mathfrak{E}}\mathsf{E}_0 S^2 = \mathsf{E}_0\mathcal{P}^2,$$

where $\mathcal{P}$ is the Pitman estimator in (5.51).

Proof. As the last equality in the statement follows from Theorem 5.56 we have in view of (5.56) only to show

$$\mathsf{E}_0\mathcal{P}^2 \leq \sup_{\Pi}\inf_{S}\int \mathsf{E}_\theta(S-\theta)^2\Pi(d\theta). \tag{5.57}$$

To this end we set $A_N=\{x:|\mathcal{E}_0(x)|\leq N\}$, $P_N(B)=P_0(B|A_N)$, and note that $P_0(A_N)\to 1$ as $N\to\infty$. Furthermore, we put $\mathfrak{E}_N=\{\mathcal{E}:P_N(|\mathcal{E}|\leq 2N)=1\}$ and $l_N(t)=\min(t^2,4N^2)$. Let $\mathcal{P}_N$ be the Pitman estimator for the model (5.47) according to Theorem 5.61 when P is replaced by P_N. The application of Theorem 5.61 to the loss function l_N yields

$$\begin{aligned}
&\inf_{S\in\mathfrak{E}}\int \min(S^2,4N^2)dP_N \\
&= \sup_{\Pi}\inf_{S}\int [\int \min((S(x-\theta\mathbf{1}))^2,4N^2)P_N(dx)]\Pi(d\theta) \\
&\leq \frac{1}{P(A_N)}\sup_{\Pi}\inf_{S}\int \mathsf{E}_\theta(S-\theta)^2\Pi(d\theta).
\end{aligned} \tag{5.58}$$

For $\varepsilon > 0$ we find $S_{N,0} \in \mathfrak{E}$ with

$$\int \min(S_{N,0}^2, 4N^2)dP_N \le \inf_{S\in\mathfrak{E}} \int \min(S^2, 4N^2)dP_N + \varepsilon. \tag{5.59}$$

Note that the set $B_N = \{|S_{N,0} - \mathcal{E}_0| \le N\}$ is invariant. Hence $\widetilde{S}_{N,0} = S_{N,0}I_{B_N} + \mathcal{E}_0 I_{\overline{B}_N}$ is equivariant. It holds

$$\begin{aligned}
&\{\widetilde{S}_{N,0}^2 > \min(S_{N,0}^2, 4N^2)\} \cap B_N \\
&\quad = (\{\widetilde{S}_{N,0}^2 > S_{N,0}^2\} \cap B_N) \cup (\{\widetilde{S}_{N,0}^2 > 4N^2\} \cap B_N) \\
&\quad = \{|S_{N,0}| > 2N\} \cap \{|S_{N,0} - \mathcal{E}_0| \le N\} \subseteq \{|\mathcal{E}_0| > N\} \\
&\{\widetilde{S}_{N,0}^2 > \min(S_{N,0}^2, 4N^2\} \cap \overline{B}_N = (\{\mathcal{E}_0^2 > S_{N,0}^2\} \cap \overline{B}_N) \cup (\{\mathcal{E}_0^2 > 4N^2\} \cap \overline{B}_N) \\
&\quad \subseteq \{\mathcal{E}_0^2 > S_{N,0}^2\} \cap \{|S_{N,0}| + |\mathcal{E}_0| > N\} \cup \{|\mathcal{E}_0| > 2N\} \subseteq \{|\mathcal{E}_0| > N/2\}.
\end{aligned}$$

Hence $\{\widetilde{S}_{N,0}^2 > \min(S_{N,0}^2, 4N^2)\} \subseteq \{|\mathcal{E}_0| > N/2\}$, and by $|\mathcal{E}_0| \le N$ P_N-a.s.,

$$\int \widetilde{S}_{N,0}^2 dP_N \le \int \min(S_{N,0}^2, 4N^2)dP_N + \int I_{[N/2,\infty)}(|\mathcal{E}_0|)\mathcal{E}_0^2 dP_N.$$

From $\mathcal{P}_N = \mathcal{E}_0 - \mathsf{E}_{P_N}(\mathcal{E}_0|\mathfrak{I})$, P_N-a.s., we get

$$\begin{aligned}
&\mathsf{E}_{P_N}(\mathcal{E}_0 - \mathsf{E}_{P_N}(\mathcal{E}_0|\mathfrak{I}))^2 = \int \mathcal{P}_N^2 dP_N = \inf_{S\in\mathfrak{E}} \int \min S^2 dP_N \le \int \widetilde{S}_{N,0}^2 dP_N \\
&\le \int \min(S_{N,0}^2, 4N^2)dP_N + \int I_{[N/2,\infty)}(|\mathcal{E}_0|)\mathcal{E}_0^2 dP_N \\
&\le \inf_{S\in\mathfrak{E}} \int \min(S^2, 4N^2)dP_N + \varepsilon + \int I_{[N/2,\infty)}(|\mathcal{E}_0|)\mathcal{E}_0^2 dP_N \\
&\le \frac{1}{P(A_N)} \sup_{\Pi} \inf_{S} \int \mathsf{E}_\theta(S-\theta)^2 \Pi(d\theta) + \varepsilon + \int I_{[N/2,\infty)}(|\mathcal{E}_0|)\mathcal{E}_0^2 dP_N,
\end{aligned}$$

where the last two inequalities follow from (5.59) and (5.58). Taking $N \to \infty$, by Problem 5.66 with $P = P_0$, $P(A_N) \to 1$, and $\int I_{[N/2,\infty)}(|\mathcal{E}_0|)\mathcal{E}_0^2 dP_N \to 0$, we get that

$$\mathsf{E}_0\mathcal{P}^2 = \mathsf{E}_0(\mathcal{E}_0 - \mathsf{E}_0(\mathcal{E}_0|\mathfrak{I}))^2 \le \sup_{\Pi} \inf_{S} \int \mathsf{E}_\theta(S-\theta)^2 \Pi(d\theta) + \varepsilon.$$

Letting $\varepsilon \to 0$ we get (5.57). ■

We conclude this section with some remarks on Pitman estimators in the scale model. To this end we assume that P is a distribution on $(\mathbb{R}^n, \mathfrak{B}_n)$ and consider the scale model given by (5.9). For some function $l : (0,\infty) \to_m \mathbb{R}_+$ we introduce the loss function by $L(\theta, a) = l(a/\theta)$, where $\theta > 0$ is the unknown parameter in the model

$$(\mathbb{R}^n, \mathfrak{B}_n, (P_\theta)_{\theta\in(0,\infty)}), \quad P_\theta = P \circ u_\theta^{-1}, \; u_\theta(x) = \theta x, \; x \in \mathbb{R}^n. \tag{5.60}$$

Hence $P_\theta(B) = \int I_B(\theta x)P(dx)$. By construction the problem of estimating the parameter θ, under the loss function $L(\theta, a) = l(a/\theta)$, is an invariant decision problem. An estimator $T : \mathbb{R}^n \to_m (0, \infty)$ is equivariant if $T(\theta x) = \theta T(x)$, $\theta > 0$, $x \in \mathbb{R}^n$. Denote by $\mathfrak{I}$ the σ-algebra of Borel sets that are invariant under the group of measurable transformations $\mathcal{U}_s = \{u_\theta : \theta > 0\}$. Then for every equivariant estimator T

$$\mathsf{E}_\theta L(\theta, T) = \mathsf{E}_\theta l(\frac{T}{\theta}) = \mathsf{E}_1 l(T), \tag{5.61}$$

and for any equivariant estimator $\mathcal{E}_1$, similarly as in (5.49),

$$\mathsf{E}_1 l\,(S) = \mathsf{E}_1(\mathsf{E}_1(l(\mathcal{E}_1 \frac{S}{\mathcal{E}_1})|\mathfrak{I})) = \int [\int l(\frac{\mathcal{E}_1(y)}{T(x)})\mathsf{K}(dy|x)]P(dx),$$

where $T = \mathcal{E}_1/S$ is invariant, and K is a regular conditional distribution, given the σ-algebra of scale invariant Borel sets. We may fix any $\mathcal{E}_1$ and find the equivariant estimator with minimum risk, $T(x)$ say, by minimizing $t \mapsto \int l(\mathcal{E}_1(y)/t)\mathsf{K}(dy|x)$. If $l(t) = (t-1)^2$, then the Pitman estimator $\mathcal{P}$ is defined to be that equivariant estimator $\mathcal{P}$ which minimizes the risk $\mathsf{E}_1\,(S-1)^2$. To find $\mathcal{P}$ we start with an equivariant estimator $\mathcal{E}_1$ with $\mathsf{E}_1(\mathcal{E}_1 - 1)^2 < \infty$. Then $\int \mathcal{E}_1^2(y)\mathsf{K}(dy|x) < \infty$, P-a.s. The function $\int (\mathcal{E}_1(y)/t - 1)^2\mathsf{K}(dy|x)$ attains the minimum at

$$T(x) = \left[\int \mathcal{E}_1(y)\mathsf{K}(dy|x)\right]^{-1} \int \mathcal{E}_1^2(y)\mathsf{K}(dy|x).$$

Hence

$$\mathcal{P}(x) = \frac{\mathcal{E}_1(x) \int \mathcal{E}_1(y)\mathsf{K}(dy|x)}{\int \mathcal{E}_1^2(y)\mathsf{K}(dy|x)}.$$

Again we denote by $\mathfrak{E}$ the set of all equivariant estimators.

Proposition 5.63. *If $l : (0,\infty) \to_m \mathbb{R}_+$ is bounded, then it holds in the model (5.60),*

$$\sup_\Pi \inf_S \mathsf{r}(\Pi, S) = \inf_S \sup_\theta \mathsf{R}(\theta, S) = \inf_{S \in \mathfrak{E}} \mathsf{R}(1, S).$$

Proof. Similarly as in the proof of Theorem 5.61 we have to show that $\inf_{S\in\mathfrak{E}} \mathsf{R}(1, S) \le \sup_\Pi \inf_S \mathsf{r}(\Pi, S)$. We use the prior $\mathsf{Ga}(\alpha, 1)$ and fix an estimator $S : \mathbb{R}^n \to_m (0, \infty)$. It follows from (5.61) and Fubini's theorem with $\eta = \theta\mathcal{E}(x)$,

$$\begin{aligned} \mathsf{r}(\mathsf{Ga}(\alpha,1), S) &= \int [\int l(S(x)/\theta)P_\theta(dx)]\mathsf{ga}_{\alpha,1}(\theta)d\theta \\ &= \int [\int l(S(\theta x)/\theta)\mathsf{ga}_{\alpha,1}(\theta)d\theta]P_1(dx) \\ &= \int [\int l\left(S(\frac{\eta x}{\mathcal{E}(x)})\frac{\mathcal{E}(x)}{\eta}\right) \mathsf{ga}_{\alpha,1/\mathcal{E}(x)}(\eta)d\eta]P_1(dx). \end{aligned}$$

Note that $S_\eta(x) = S(\eta x/\mathcal{E}(x))\mathcal{E}(x)/\eta$ is an equivariant estimator. Hence

$$\mathsf{r}(\mathsf{Ga}(\alpha,\beta),S) \geq \inf_{T\in\mathfrak{E}} \int l(T(x))P_1(dx).$$

By assumption l is bounded, say by C. Then by the inequality in Problem 5.67 with $\beta_1 = 1$ and $\beta_2 = 1/\mathcal{E}(x)$,

$$\begin{aligned}
\mathsf{r}(\mathsf{Ga}(\alpha,1),S) &= \int [\int l(S_\eta(x))\mathsf{ga}_{\alpha,1/\mathcal{E}(x)}(\eta)d\eta]P_1(dx)\\
&\geq \int [\int l(S_\eta(x))\mathsf{ga}_{\alpha,1}(\eta)d\eta]P_1(dx)\\
&\quad - 2\sqrt{2}C\int (1-2^\alpha[\sqrt{\mathcal{E}(x)}+1/\sqrt{\mathcal{E}(x)}]^{-\alpha})^{1/2}P_1(dx)\\
&\geq \inf_{T\in\mathfrak{E}} \int l(T(x))P_1(dx)\\
&\quad - 2\sqrt{2}C\int (1-2^\alpha[\sqrt{\mathcal{E}(x)}+1/\sqrt{\mathcal{E}(x)}]^{-\alpha})^{1/2}P_0(dx).
\end{aligned}$$

Taking the infimum over all estimators S on the the left-hand side and letting $\alpha \to 0$ we get from Lebesgue's theorem that

$$\begin{aligned}
\sup_{\Pi}\inf_{S} \mathsf{r}(\Pi,S) &\geq \limsup_{\alpha\to 0}\inf_{S} \mathsf{r}(\mathsf{Ga}(\alpha,1),S)\\
&\geq \inf_{T\in\mathfrak{E}} \int l(T(x))P_1(dx) = \inf_{T\in\mathfrak{E}} \mathsf{R}(1,T).
\end{aligned}$$

■

Subsequently we collect some problems that have been used in the previous proofs.

Problem 5.64.* If $g : \mathbb{R} \to_m \mathbb{R}$ is bounded, say $|g| \leq C$, then

$$|\int g(t)\mathsf{N}(0,\sigma^2)(dt) - \int g(t)\mathsf{N}(\mu,\sigma^2)(dt)| \leq 2\sqrt{2}C[1-\exp\{-\mu^2/(8\sigma^2)\}]^{1/2}.$$

Problem 5.65.* Let P and Q be distributions on $(\mathcal{X},\mathfrak{A})$ with $Q \ll P$. Let $\mathfrak{G} \subseteq \mathfrak{A}$ be a sub-σ-algebra of $\mathfrak{A}$. If $T : \mathcal{X} \to_m \mathbb{R}$ is a statistic with $\mathsf{E}_P|T| < \infty$ and $\mathsf{E}_Q|T| < \infty$, then

$$\begin{aligned}
&\mathsf{E}_P(T\tfrac{dQ}{dP}|\mathfrak{G}) = \mathsf{E}_P(\tfrac{dQ}{dP}|\mathfrak{G})\mathsf{E}_Q(T|\mathfrak{G}), \quad P\text{-a.s.}\\
&\mathsf{E}_Q(T|\mathfrak{G}) = \left[\mathsf{E}_P(\tfrac{dQ}{dP}|\mathfrak{G})\right]^{-1}\mathsf{E}_P(T\tfrac{dQ}{dP}|\mathfrak{G}), \quad Q\text{-a.s.}
\end{aligned} \tag{5.62}$$

Problem 5.66.* Let $A_n \in \mathfrak{A}$ with $\lim_{n\to\infty} P(A_n) = 1$ and set $P_n(B) = P(B|A_n)$, $B \in \mathfrak{A}$. Let $\mathfrak{G} \subseteq \mathfrak{A}$ be a sub-σ-algebra and $T : \mathcal{X} \to_m \mathbb{R}$ with $\mathsf{E}_P T^2 < \infty$. Then

$$\lim_{n\to\infty} \mathsf{E}_{P_n}(T - \mathsf{E}_{P_n}(T|\mathfrak{G}))^2 = \mathsf{E}_P(T - \mathsf{E}_P(T|\mathfrak{G}))^2.$$

Problem 5.67.* If $g : \mathbb{R} \to_m \mathbb{R}$ is bounded, say $|g| \leq C$, then

$$|\int g(t)\mathsf{Ga}(\alpha,\beta_1)(dt) - \int g(t)\mathsf{Ga}(\alpha,\beta_2)(dt)| \leq 2\sqrt{2}C\left(1-\left(\frac{2\sqrt{\beta_1\beta_2}}{\beta_1+\beta_2}\right)^\alpha\right)^{1/2}.$$

5.5 Solutions to Selected Problems

Solution to Problem 5.4: For column vectors $x_1, ..., x_n$ we use the Gram–Schmidt orthogonalization procedure by setting $y_1 = x_1/\|x_1\|$ and

$$y_i = (x_i - \sum\nolimits_{j=1}^{i-1}(x_i^T y_j)y_j) \parallel x_i - \sum\nolimits_{j=1}^{i-1}(x_i^T y_j)y_j \parallel^{-1}, \quad i = 2, ..., n,$$

to define the continuous mapping $M : \mathcal{D}(M) \to \mathcal{G}$ by $M((x_1, ..., x_n)) = (y_1, ..., y_n)^T$, where $\mathcal{D}(M) \subseteq \mathbb{R}^{n^2}$ is the set of all nonsingular matrices $(x_1, ..., x_n)$. Let O be an orthogonal matrix. We replace x_i with $\widetilde{x}_i = Ox_i$ and denote the resulting vectors after the mapping M by $\widetilde{y}_i$. Then $\widetilde{y}_1 = (\|\widetilde{x}_1\|)^{-1}\widetilde{x}_1 = Oy_1$ and

$$\begin{aligned}\widetilde{y}_i &= (Ox_i - \sum\nolimits_{j=1}^{i-1}(Ox_i)^T(Oy_j)Oy_j)) \parallel Ox_i - \sum\nolimits_{j=1}^{i-1}(Ox_i)^T(Oy_j)Oy_j) \parallel^{-1} \\ &= O(x_i - \sum\nolimits_{j=1}^{i-1}(x_i^T y_j)y_j) \parallel x_i - \sum\nolimits_{j=1}^{i-1}(x_i^T y_j)y_j) \parallel^{-1}, \quad i = 2, ..., n.\end{aligned}$$

Hence $OM((x_1, ..., x_n)) = M(O(x_1, ..., x_n))$ for every orthogonal matrix O. If $X_{i,j}$, $1 \le i, j \le n$, are i.i.d. $\mathsf{N}(0,1)$, then $(X_1, ..., X_n) = (X_{i,j})_{1\le i,j\le n} \in \mathcal{D}(M)$ a.s. As $\mathcal{L}(OM(X_1, ..., X_n)) = \mathcal{L}(M(X_1, ..., X_n))$ it follows that the distribution of $M^T(X_1, ..., X_n)$ is a right invariant distribution. □

Solution to Problem 5.6: The first statement is clear. For the second apply the transformation rule for the Lebesgue measure; see Theorem A.23. □

Solution to Problem 5.15: A is invariant if and only if $u_\gamma^{-1}(A) = A$ for every γ. As the intersection, the union, and the complement can be interchanged with the inverse image operation $\mathfrak{I}$ is a sub-σ-algebra of $\mathfrak{A}$. If T is invariant, then $T \circ u_\gamma = T$. Therefore $u_\gamma^{-1}(T^{-1}(B)) = T^{-1}(B)$. As $u_\gamma^{-1}(T^{-1}(B) = u_{\gamma^{-1}}(T^{-1}(B))$ and γ is arbitrary the invariance follows. □

Solution to Problem 5.18: If

$$\left(\operatorname{sgn}(x_2 - x_1), \frac{x_3 - x_1}{x_2 - x_1}, ..., \frac{x_n - x_1}{x_2 - x_1}\right) = \left(\operatorname{sgn}(y_2 - y_1), \frac{y_3 - y_1}{y_2 - y_1}, ..., \frac{y_n - y_1}{y_2 - y_1}\right),$$

then for $\alpha = x_1 - \beta y_1$, and $\beta = (x_2 - x_1)/(y_2 - y_1) > 0$ it holds $x_i = \beta y_i + \alpha$. □

Solution to Problem 5.21: Apply the factorization lemma; see Lemma A.9. □

Solution to Problem 5.23:

$$\begin{aligned}(P_{v_\gamma(\theta)} \circ T^{-1})(B) &= P_{v_\gamma(\theta)}(T^{-1}(B)) = P_\theta(u_\gamma^{-1}(T^{-1}(B)) \\ &= P_\theta(T(u_\gamma) \in B) = P_\theta(T \in B).\end{aligned} \qquad \square$$

Solution to Problem 5.49: If $(\mathcal{G}, \mathfrak{G})$ and $(\widetilde{\mathcal{G}}, \widetilde{\mathfrak{G}})$ are two measurable groups, where between them there is a bimeasurable isomorphism, then for one group there is a σ-finite invariant measure if and only if the same holds for the other group. The same statement holds true for the existence of distributions Q_n that have the property

(5.40). Consider now the group $(\widetilde{\mathcal{G}}, \widetilde{\mathfrak{G}}) = (R \times O_{(k-1)\times(k-1)}, \mathfrak{B}_{\mathbb{R}\times O_{(k-1)\times(k-1)}})$. There is bimeasurable isomorphism between $(\mathcal{G}, \mathfrak{G})$ and $(\widetilde{\mathcal{G}}, \widetilde{\mathfrak{G}})$. Hence we have only to consider $(\widetilde{\mathcal{G}}, \widetilde{\mathfrak{G}})$. Using the invariant distribution R_{k-1} on the group $O_{(k-1)\times(k-1)}$ of all rotations of R^{k-1}, whose existence has been proved in Problem 5.4, we set $\mu = \boldsymbol{\lambda} \otimes R_{k-1}$ to get a σ-finite invariant measure. If $\mathsf{U}(-n,n)$ is the uniform distribution, then $Q_n = \mathsf{U}(-n,n) \otimes R_{k-1}$ satisfies (5.40). □

Solution to Problem 5.54: The inequalities $l(\frac{1}{2}(S_0 + S_1)) \leq \frac{1}{2}l(S_0) + \frac{1}{2}l(S_1)$ and $\mathsf{E}_0 l(S_0) = \mathsf{E}_0 l(S_1) \leq \mathsf{E}_0 l(\frac{1}{2}(S_0 + S_1))$ imply

$$\mathsf{E}_0 \left[\frac{1}{2}l(S_0) + \frac{1}{2}l(S_1) - l(\frac{1}{2}(S_0 + S_1))\right] = 0.$$

As the bracket is nonnegative it must vanish P_0-a.s. The strict convexity of l yields $S_0 = S_1$, P_0-a.s. □

Solution to Problem 5.55: If $h(y,t) = I_A(y)I_C(t)$, $A \in \mathfrak{B}_n$, $C \in \mathfrak{B}$, then $B = T^{-1}(C) \in \mathfrak{I}$ and $\int h(x, T(x))P(dx) = \int [\int h(y, T(x))\mathsf{K}(dy|x)]P(dx)$ by the definition of K. To complete the proof one has only to apply the standard extension technique. □

Solution to Problem 5.64: The first inequality in Problem 1.80 yields

$$\left|\int g(t)\mathsf{N}(0,\sigma^2)(dt) - \int g(t)\mathsf{N}(\mu,\sigma^2)(dt)\right| \leq C \left\|\mathsf{N}(0,\sigma^2) - \mathsf{N}(\mu,\sigma^2)\right\|.$$

As to the variational distance, we combine $\|P_0 - P_1\| \leq 2\mathsf{D}(P_0, P_1)$ from Proposition 1.84, $\mathsf{D}^2(P_0, P_1) = 2(1 - \mathsf{H}_{1/2}(P_0, P_1))$ from (1.110), and $\mathsf{H}_{1/2}(\mathsf{N}(0,\sigma^2), \mathsf{N}(\mu,\sigma^2)) = \exp\{-\mu^2/(8\sigma^2)\}$ from (1.79). □

Solution to Problem 5.65: Denote by $P^{\mathfrak{G}}$ and $Q^{\mathfrak{G}}$ the restrictions of P and Q, respectively, to $\mathfrak{G}$. It follows from

$$\int_B \mathsf{E}_P(\frac{dQ}{dP}|\mathfrak{G})dP = \int_B \frac{dQ}{dP}dP = Q(B), \quad B \in \mathfrak{G},$$

that

$$\mathsf{E}_P(\frac{dQ}{dP}|\mathfrak{G}) = \frac{dQ^{\mathfrak{G}}}{dP^{\mathfrak{G}}}.$$

Hence for $B \in \mathfrak{G}$,

$$\int_B \mathsf{E}_P(\frac{dQ}{dP}|\mathfrak{G})\mathsf{E}_Q(T|\mathfrak{G})dP = \int_B \frac{dQ^{\mathfrak{G}}}{dP^{\mathfrak{G}}}\mathsf{E}_Q(T|\mathfrak{G})dP^{\mathfrak{G}}.$$

The $\mathfrak{G}$-measurability of $\mathsf{E}_Q(T|\mathfrak{G})$ yields

$$\int_B \frac{dQ^{\mathfrak{G}}}{dP^{\mathfrak{G}}}\mathsf{E}_Q(T|\mathfrak{G})dP^{\mathfrak{G}} = \int_B \mathsf{E}_Q(T|\mathfrak{G})dQ^{\mathfrak{G}} = \int_B T dQ = \int_B T\frac{dQ}{dP}dP.$$

To prove the second statement we set $A = \{\mathsf{E}_P(dQ/dP|\mathfrak{G}) = 0\}$. Then $A \in \mathfrak{G}$ and

$$Q(A) = \int_A \frac{dQ}{dP} dP = \int_A \mathsf{E}_P(\frac{dQ}{dP}|\mathfrak{G}) dP = 0. \qquad \square$$

Solution to Problem 5.66: Set $Q = P_n$. Then $P_n \ll P$ and $dQ/dP = L_n := P(A_n)^{-1} I_{A_n}$. Then by (5.62)

$$\mathsf{E}_P(L_n|\mathfrak{G})\mathsf{E}_{P_n}(T|\mathfrak{G}) = \mathsf{E}_P(TL_n|\mathfrak{G}). \tag{5.63}$$

By Lemma A.33 and $P(A_n) \to 1$

$$\begin{aligned} &\mathsf{E}_P(\mathsf{E}_P(L_n|\mathfrak{G}) - 1)^2 \le \mathsf{E}_P(\mathsf{E}_P(L_n|\mathfrak{G}) - 1)^2 \le \mathsf{E}_P(L_n - 1)^2 \to 0, \\ &\mathsf{E}_P(\mathsf{E}_P(TL_n - T|\mathfrak{G}))^2 \le \mathsf{E}_P(TL_n - T)^2 \to 0, \end{aligned}$$

where the last statement follows from Lebesgue's theorem. Hence $\mathsf{E}_{P_n}(T|\mathfrak{G}) \to^P \mathsf{E}_P(T|\mathfrak{G})$. Then by Proposition A.12 for every subsequence n_k there is a new subsequence, say n_{k_l}, so that $TI_{A_{n_{k_l}}} - \mathsf{E}_{P_{n_{k_l}}}(T|\mathfrak{G})I_{A_{n_{k_l}}} \to T - \mathsf{E}_P(T|\mathfrak{G})$, P-a.s. Hence by Fatou's lemma,

$$\begin{aligned} \liminf_{l\to\infty} \mathsf{E}_{P_{n_{k_l}}}(T - \mathsf{E}_{P_{n_{k_l}}}(T|\mathfrak{G}))^2 &= \liminf_{l\to\infty} \mathsf{E}_P(T - \mathsf{E}_{P_{n_{k_l}}}(T|\mathfrak{G}))^2 I_{A_{n_{k_l}}} \\ &\ge \mathsf{E}_P(T - \mathsf{E}_P(T|\mathfrak{G}))^2. \end{aligned}$$

As the subsequence n_k was arbitrary we get

$$\liminf_{n\to\infty} \mathsf{E}_{P_n}(T - \mathsf{E}_{P_n}(T|\mathfrak{G}))^2 \ge \mathsf{E}_P(T - \mathsf{E}_P(T|\mathfrak{G}))^2. \tag{5.64}$$

The assumption $P(A_n) \to 1$ implies $\mathsf{E}_{P_n}(T|\mathfrak{G})I_{A_n} \to^P \mathsf{E}_P(T|\mathfrak{G})$ and $TI_{A_n} \to^P T$. As

$$\mathsf{E}_{P_n}(\mathsf{E}_P(T|\mathfrak{G}))^2 \le \frac{1}{P(A_n)}\mathsf{E}_P(\mathsf{E}_P(T|\mathfrak{G}))^2 \le \frac{1}{P(A_n)}\mathsf{E}_P T^2 < \infty$$

and the conditional expectation provides the best approximation in mean square it holds

$$\begin{aligned} &\limsup_{n\to\infty} \mathsf{E}_{P_n}(T - \mathsf{E}_{P_n}(T|\mathfrak{G}))^2 = \limsup_{n\to\infty} \mathsf{E}_P(T - \mathsf{E}_{P_n}(T|\mathfrak{G}))^2 \frac{1}{P(A_n)} I_{A_n} \\ &\le \limsup_{n\to\infty} \mathsf{E}_P(T - \mathsf{E}_P(T|\mathfrak{G}))^2 \frac{1}{P(A_n)} = \mathsf{E}_P(T - \mathsf{E}_P(T|\mathfrak{G}))^2, \end{aligned}$$

where the last equality follows from $P(A_n) \to 1$. In view of (5.64) the proof is completed. $\square$

Solution to Problem 5.67: The solution is similar to that of Problem 5.64 if one uses

$$\mathsf{H}_{1/2}(\mathsf{Ga}(\alpha, \beta_1), \mathsf{Ga}(\alpha, \beta_2)) = \left(\frac{\sqrt{\beta_1 \beta_2}}{\frac{1}{2}(\beta_1 + \beta_2)} \right)^{\alpha}. \qquad \square$$

6

Large Sample Approximations of Models and Decisions

6.1 Distances of Statistical Models

Let us be given two models $\mathcal{M}_i = (\mathcal{X}_i, \mathfrak{A}_i, (P_{i,\theta})_{\theta\in\Delta}), i = 1, 2$. According to Definition 4.9, for every $\varepsilon \geq 0$ the model $\mathcal{M}_1$ is ε-deficient with respect to model $\mathcal{M}_2$, denoted by $\mathcal{M}_1 \succeq^{\varepsilon} \mathcal{M}_2$, if the following holds. For every finite subset $\mathcal{F} \subseteq \Delta$, every finite decision space $\mathcal{D}$, every real-valued loss function L with $\|L\|_u \leq 1$, and every decision $\mathsf{D}_{\mathcal{M}_2} : \mathfrak{D} \times \mathcal{X}_2 \to_k [0,1]$, there exists a decision $\mathsf{D}_{\mathcal{M}_1} : \mathfrak{D} \times \mathcal{X}_1 \to_k [0,1]$ with

$$\mathsf{R}(\theta, \mathsf{D}_{\mathcal{M}_1}) \leq \mathsf{R}(\theta, \mathsf{D}_{\mathcal{M}_2}) + \varepsilon, \quad \theta \in \mathcal{F}. \tag{6.1}$$

We set

$$\begin{aligned} d(\mathcal{M}_1, \mathcal{M}_2) &= \inf\{\varepsilon : \mathcal{M}_1 \succeq^{\varepsilon} \mathcal{M}_2, \ \varepsilon \geq 0\}, \\ \delta(\mathcal{M}_1, \mathcal{M}_2) &= \max\{d(\mathcal{M}_1, \mathcal{M}_2), d(\mathcal{M}_2, \mathcal{M}_1)\}, \end{aligned} \tag{6.2}$$

and call $\delta(\mathcal{M}_1, \mathcal{M}_2)$ the *deficiency* of the models $\mathcal{M}_1$ and $\mathcal{M}_2$. $\delta(\mathcal{M}_1, \mathcal{M}_2)$ is nonnegative and symmetric. Moreover, if $\mathcal{M}_1, \mathcal{M}_2, \mathcal{M}_3$ are any models with the same parameter set, then

$$d(\mathcal{M}_1, \mathcal{M}_3) \leq d(\mathcal{M}_1, \mathcal{M}_2) + d(\mathcal{M}_2, \mathcal{M}_3),$$

and thus δ satisfies the triangular inequality

$$\delta(\mathcal{M}_1, \mathcal{M}_3) \leq \delta(\mathcal{M}_1, \mathcal{M}_2) + \delta(\mathcal{M}_2, \mathcal{M}_3).$$

This means that $\delta(\mathcal{M}_1, \mathcal{M}_2)$ is a pseudometric in the space $\mathfrak{M}(\Delta)$, say, of all models with parameter set Δ. If $\mathcal{F}$ is any subset of Δ and $\mathcal{M}_{i,\mathcal{F}}$ is the model extracted from $\mathcal{M}_i$ with the restricted parameter set $\mathcal{F}$, then by the definition of d and δ it holds

$$\delta(\mathcal{M}_{1,\mathcal{F}}, \mathcal{M}_{2,\mathcal{F}}) \leq \delta(\mathcal{M}_1, \mathcal{M}_2). \tag{6.3}$$

F. Liese, K.-J. Miescke, *Statistical Decision Theory*,
DOI: 10.1007/978-0-387-73194-0_6, © Springer Science+Business Media, LLC 2008

Problem 6.1.* For any models $\mathcal{M}_1$ and $\mathcal{M}_2$ with the same parameter set Δ it holds

$$\delta(\mathcal{M}_1, \mathcal{M}_2) = \sup_{F \subseteq \Delta, |F| < \infty} \delta(\mathcal{M}_{1,F}, \mathcal{M}_{2,F}). \tag{6.4}$$

The pseudometric δ is consistent with the equivalence of models in the sense of Definition 4.9.

Lemma 6.2. *For two models $\mathcal{M}_1$ $\mathcal{M}_2$ the condition $\delta(\mathcal{M}_1, \mathcal{M}_2) = 0$ holds if and only if $\mathcal{M}_1 \sim \mathcal{M}_2$; that is, $\mathcal{M}_1 \succeq \mathcal{M}_2$ and $\mathcal{M}_2 \succeq \mathcal{M}_1$.*

Proof. As $\mathcal{M}_1 \sim \mathcal{M}_2$ if and only if $\mathcal{M}_{1,F} \sim \mathcal{M}_{2,F}$ for every finite subset $F \subseteq \Delta$, and in view of (6.4) $\delta(\mathcal{M}_1, \mathcal{M}_2) = 0$ holds if and only if $\delta(\mathcal{M}_{1,F}, \mathcal{M}_{2,F}) = 0$ for every finite subset $F \subseteq \Delta$, we may restrict ourselves to the finite models $\mathcal{M}_{1,F}$ and $\mathcal{M}_{2,F}$. If $\delta(\mathcal{M}_{1,F}, \mathcal{M}_{2,F}) = 0$, then there is a sequence $\varepsilon_n \geq 0$ with $\varepsilon_n \to 0$ such that $\mathcal{M}_{1,F} \succeq^{\varepsilon_n} \mathcal{M}_{2,F}$. We have to show that this implies $\mathcal{M}_{1,F} \succeq^0 \mathcal{M}_{2,F}$. For any finite decision space $\mathcal{D}$, any loss function L with $\|L\|_u \leq 1$, and every decision $\mathsf{D}_{\mathcal{M}_2} : \mathfrak{D} \times \mathcal{X}_2 \to_k [0,1]$, there exist decisions $\mathsf{D}_{\mathcal{M}_1,n} : \mathfrak{D} \times \mathcal{X}_1 \to_k [0,1]$ with

$$\mathsf{R}(\theta, \mathsf{D}_{\mathcal{M}_1,n}) \leq \mathsf{R}(\theta, \mathsf{D}_{\mathcal{M}_2}) + \varepsilon_n, \quad \theta \in F.$$

As the finite model $\mathcal{M}_{1,F}$ is dominated and the decision space is finite, and therefore compact, Theorem 3.17 provides the existence of a subsequence n_k and a decision $\mathsf{D}_{\mathcal{M}_1,0}$ such that

$$\lim_{k\to\infty} \mathsf{R}(\theta, \mathsf{D}_{\mathcal{M}_1,n_k}) = \mathsf{R}(\theta, \mathsf{D}_{\mathcal{M}_1,0}), \quad \theta \in F.$$

Hence $\mathsf{R}(\theta, \mathsf{D}_{\mathcal{M}_1,0}) \leq \mathsf{R}(\theta, \mathsf{D}_{\mathcal{M}_2})$, $\theta \in F$, which yields $\mathcal{M}_{1,F} \succeq^0 \mathcal{M}_{2,F}$. Interchanging the roles of $\mathcal{M}_{1,F}$ and $\mathcal{M}_{2,F}$ completes the proof. ■

There is another way of expressing that one model is up to ε as informative as another model. If $\mathcal{M}_2$ is a randomization of $\mathcal{M}_1$ up to ε, then we have shown already in Corollary 4.10 that $\mathcal{M}_2$ is ε-deficient with respect to $\mathcal{M}_1$. For any finite models $\mathcal{M}_i = (\mathcal{X}_i, \mathfrak{A}_i, \{P_{i,1}, ..., P_{i,m}\})$, $i = 1, 2$, we set

$$\begin{aligned} D(\mathcal{M}_1, \mathcal{M}_2) &= \inf_{\mathsf{K}} \max_{j=1,...,m} \|P_{2,j} - \mathsf{K}P_{1,j}\|, \\ \Delta(\mathcal{M}_1, \mathcal{M}_2) &= \max\{D(\mathcal{M}_1, \mathcal{M}_2), D(\mathcal{M}_2, \mathcal{M}_1)\}, \end{aligned} \tag{6.5}$$

where the infimum is taken over all stochastic kernels $\mathsf{K} : \mathfrak{A}_2 \times \mathcal{X}_1 \to_k [0,1]$. The definition of $\Delta(\mathcal{M}_1, \mathcal{M}_2)$ gives

$$\Delta(\mathcal{M}_1, \mathcal{M}_2) \geq 0 \quad \text{and} \quad \Delta(\mathcal{M}_1, \mathcal{M}_2) = \Delta(\mathcal{M}_2, \mathcal{M}_1). \tag{6.6}$$

Let $\mathcal{M}_1, \mathcal{M}_2, \mathcal{M}_3$ be finite models with the same parameter set $\{1, ..., m\}$. Suppose there are kernels $\mathsf{K} : \mathfrak{A}_2 \times \mathcal{X}_1 \to_k [0,1]$ and $\mathsf{L} : \mathfrak{A}_3 \times \mathcal{X}_2 \to_k [0,1]$ such that for some $\varepsilon_1, \varepsilon_2 > 0$,

$$\|P_{2,j} - \mathsf{K}P_{1,j}\| < \varepsilon_1 \quad \text{and} \quad \|P_{3,j} - \mathsf{L}P_{2,j}\| < \varepsilon_2, \ \ j = 1, ..., m.$$

Then the kernel $(\mathsf{LK})(dx_3|x_1) := \int \mathsf{L}(dx_3|x_2)\mathsf{K}(dx_2|x_1)$ satisfies

$$\begin{aligned}\|P_{3,j} - \mathsf{LK}P_{1,j}\| &\leq \|P_{3,j} - \mathsf{L}P_{2,j}\| + \|\mathsf{L}P_{2,j} - \mathsf{LK}P_{1,j}\| \\ &\leq \|P_{3,j} - \mathsf{L}P_{2,j}\| + \|P_{2,j} - \mathsf{K}P_{1,j}\| \leq \varepsilon_1 + \varepsilon_2,\end{aligned}$$

where the second inequality follows from (1.94). Hence we get $D(\mathcal{M}_1, \mathcal{M}_3) \leq D(\mathcal{M}_1, \mathcal{M}_2) + D(\mathcal{M}_2, \mathcal{M}_3)$, which together with (6.5) gives the triangular inequality

$$\Delta(\mathcal{M}_1, \mathcal{M}_3) \leq \Delta(\mathcal{M}_1, \mathcal{M}_2) + \Delta(\mathcal{M}_2, \mathcal{M}_3). \tag{6.7}$$

Remark 6.3. The definition of $\Delta(\mathcal{M}_1, \mathcal{M}_2)$ is soon extended to all models $\mathcal{M}_1$ and $\mathcal{M}_2$ with a common, but not necessarily finite, parameter set. It follows then from (6.6) and (6.7) that $\Delta(\mathcal{M}_1, \mathcal{M}_2)$ is also a pseudometric on $\mathfrak{M}(\Delta)$. The relations between the two pseudometrics $\delta(\mathcal{M}_1, \mathcal{M}_2)$ and $\Delta(\mathcal{M}_1, \mathcal{M}_2)$ are fundamental to decision theory.

For $\mathcal{M}_i = (\mathcal{X}_i, \mathfrak{A}_i, \{P_{i,1}, ..., P_{i,m}\})$, $i = 1, 2$, with $(\mathcal{X}_1, \mathfrak{A}_1) = (\mathcal{X}_2, \mathfrak{A}_2)$, we may use the kernel generated by the identical mapping to get an upper bound for $\Delta(\mathcal{M}_1, \mathcal{M}_2)$.

$$\Delta(\mathcal{M}_1, \mathcal{M}_2) \leq \max_{j=1,\ldots,m} \|P_{1,j} - P_{2,j}\|. \tag{6.8}$$

Another conclusion of Corollary 4.10 is the following statement.

Lemma 6.4. *For any finite models* $\mathcal{M}_i = (\mathcal{X}_i, \mathfrak{A}_i, \{P_{i,1}, ..., P_{i,m}\})$, $i = 1, 2$, *it holds*

$$\delta(\mathcal{M}_1, \mathcal{M}_2) \leq \Delta(\mathcal{M}_1, \mathcal{M}_2). \tag{6.9}$$

Proof. For every $\alpha > \Delta(\mathcal{M}_1, \mathcal{M}_2)$ there is a kernel $\mathsf{K} : \mathfrak{A}_2 \times \mathcal{X}_1 \to_k [0, 1]$ such that $\max_{j=1,\ldots,m} \|P_{2,j} - \mathsf{K}P_{1,j}\| < \alpha$. For any decision $\mathsf{D}_{\mathcal{M}_2}$ we set

$$(\mathsf{D}_{\mathcal{M}_2}\mathsf{K})(da|x_1) = \int \mathsf{D}_{\mathcal{M}_2}(da|x_2)\mathsf{K}(dx_2|x_1).$$

Hence

$$\begin{aligned}&\mathsf{R}(j, \mathsf{D}_{\mathcal{M}_2}) - \mathsf{R}(j, \mathsf{D}_{\mathcal{M}_2}\mathsf{K}) \\ &= \int [\int L(j, a)\mathsf{D}_{\mathcal{M}_2}(da|x_2)]P_{2,j}(dx_2) - \int [\int L(j, a)(\mathsf{D}_{\mathcal{M}_2}\mathsf{K})(da|x_1)]P_{1,j}(dx_1) \\ &= \int g_j(x_2)[P_{2,j}(dx_2) - (\mathsf{K}P_{1,j})(dx_2)], \quad \text{where} \\ &\qquad g_j(x_2) = \int L(j, a)\mathsf{D}_{\mathcal{M}_2}(da|x_2).\end{aligned}$$

As $|g_j| \leq 1$ we obtain $|\mathsf{R}(j, \mathsf{D}_{\mathcal{M}_2}) - \mathsf{R}(j, \mathsf{D}_{\mathcal{M}_2}\mathsf{K})| \leq \|P_{2,j} - \mathsf{K}P_{1,j}\|$ and $\mathsf{R}(j, \mathsf{D}_{\mathcal{M}_1}) \leq \mathsf{R}(j, \mathsf{D}_{\mathcal{M}_2}) + \alpha$ with the decision $\mathsf{D}_{\mathcal{M}_1} := \mathsf{D}_{\mathcal{M}_2}\mathsf{K}$ for the model $\mathcal{M}_1$. Hence $\mathcal{M}_1 \succeq^{\alpha} \mathcal{M}_2$. Similarly $\mathcal{M}_2 \succeq^{\alpha} \mathcal{M}_1$ and therefore $\delta(\mathcal{M}_1, \mathcal{M}_2) \leq \alpha$. Taking $\alpha \downarrow \Delta(\mathcal{M}_1, \mathcal{M}_2)$ we get the statement. ■

Our goal is now to show that equality holds in (6.9), and that it can be generalized to any, not necessarily finite, models $\mathcal{M}_1$ and $\mathcal{M}_2$. A first step in this direction concerns models $\mathcal{N}_i = (\mathcal{Y}_i, \mathfrak{B}_i, (Q_{i,\theta})_{\theta\in\digamma})$ with finite sample spaces $\mathcal{Y}_i$. Then the randomization criterion Theorem 4.16 applies and we get

$$\delta(\mathcal{N}_1, \mathcal{N}_2) = \Delta(\mathcal{N}_1, \mathcal{N}_2), \quad \text{for } |\mathcal{Y}_i| < \infty. \tag{6.10}$$

The technique used in the following is to approximate any model by models with a finite sample space. Here we often use the simple fact that

$$\delta(\mathcal{M}, \mathcal{N}) = \Delta(\mathcal{M}, \mathcal{N}) = 0 \quad \text{if } \mathcal{M} \text{ and } \mathcal{N} \text{ are mutual randomizations.} \tag{6.11}$$

The crucial point for the subsequent considerations is that models with a finite parameter set can be approximated by models with a finite sample space. For a model $\mathcal{M} = (\mathcal{X}, \mathfrak{A}, \{P_1, ..., P_m\})$ we set

$$\begin{aligned} \overline{P} &= \tfrac{1}{m}\textstyle\sum_{j=1}^{m} P_j, \quad M_j = \tfrac{dP_j}{d\overline{P}}, \quad j = 1, ..., m, \\ L_j &= \tfrac{M_j}{M_1} I_{(0,\infty)}(M_1) + \infty I_{\{0\}}(M_1), \quad j = 2, ..., m. \end{aligned} \tag{6.12}$$

Lemma 6.5. *Let $\mathcal{M} = (\mathcal{X}, \mathfrak{A}, \{P_1, ..., P_m\})$ be a finite model. Then for every $\varepsilon > 0$ there exists a model $\mathcal{N} = (\mathcal{Y}, \mathfrak{B}, \{Q_1, ..., Q_m\})$ with $|\mathcal{Y}| < \infty$ and*

$$\Delta(\mathcal{M}, \mathcal{N}) \le \varepsilon.$$

Proof. Approximating M_j by nonnegative step functions in the sense of $\mathbb{L}_1(\overline{P})$ we see that there is a partition $A_1, ..., A_N$ of $\mathcal{X}$ with $A_i \in \mathfrak{A}$ and probability densities $\widetilde{M}_j = \sum_{i=1}^{N} c_{i,j} I_{A_i}(x)$ such that the distributions

$$\widetilde{P}_j(A) = \int I_A(x)\widetilde{M}_j(x)\overline{P}(dx), \quad A \in \mathfrak{A}, \ j \in \digamma,$$

satisfy $\| \widetilde{P}_j - P_j \| \le \varepsilon$. Set $\widetilde{\mathcal{M}} = (\mathcal{X}, \mathfrak{A}, \{\widetilde{P}_1, ..., \widetilde{P}_m\})$. The inequality (6.8) implies $\Delta(\widetilde{\mathcal{M}}, \mathcal{M}) \le \varepsilon$. From each A_i we select a point $x_i \in A_i$ and set $\mathcal{Y} = \{x_1, ..., x_N\}$, $\mathfrak{B} = \mathfrak{P}(\mathcal{Y})$, $B \in \mathfrak{B}$,

$$Q_j(B) = \sum\nolimits_{i=1}^{N} \widetilde{P}_j(A_i)\boldsymbol{\delta}_{x_i}(B), \quad \text{and} \quad \mathcal{N} = (\mathcal{Y}, \mathfrak{B}, \{Q_1, ..., Q_m\}).$$

Then obviously $Q_j = \widetilde{P}_j \circ T^{-1}$, where $T : \mathcal{X} \to \mathcal{Y}$ is the mapping defined by $T(x) = x_i$ for $x \in A_i$. Hence $\mathcal{N}$ is a randomization of $\widetilde{\mathcal{M}}$. To show that $\widetilde{\mathcal{M}}$ is also a randomization of $\mathcal{N}$ we introduce the kernel $\mathsf{K} : \mathfrak{A} \times \mathcal{Y} \to_k [0, 1]$ by

$$\mathsf{K}(A|y) = \sum\nolimits_{i=1}^{N} \frac{\overline{P}(A \cap A_i)}{\overline{P}(A_i)} \boldsymbol{\delta}_y(A_i).$$

Then with $c_{i,j} = \widetilde{P}_j(A_i)/\overline{P}(A_i)$,

$$
\begin{aligned}
(\mathsf{K}Q_j)(A) &= \sum\nolimits_{i=1}^{N} \frac{\overline{P}(A \cap A_i)}{\overline{P}(A_i)} Q_j(\{x_i\}) \\
&= \sum\nolimits_{i=1}^{N} \frac{\overline{P}(A \cap A_i)}{\overline{P}(A_i)} \widetilde{P}_j(A_i) = \int I_A \sum\nolimits_{i=1}^{N} I_{A_i} c_{i,j} d\overline{P} \\
&= \int I_A \widetilde{M}_j d\overline{P} = \widetilde{P}_j(A).
\end{aligned}
$$

Hence $\widetilde{\mathcal{M}}$ is a randomization of $\mathcal{N}$ and the proof is completed in view of (6.11). ■

Let $F_* \subseteq F \subseteq \Delta$ be finite subsets of Δ. If $\mathcal{M}_{i,F_*}$ and $\mathcal{M}_{i,F}$ are the models with the restricted parameter sets F_* and F, respectively, $i = 1, 2$, then by (6.5)

$$\Delta(\mathcal{M}_{1,F_*}, \mathcal{M}_{2,F_*}) \leq \Delta(\mathcal{M}_{1,F}, \mathcal{M}_{2,F}). \tag{6.13}$$

Suppose now that that $\mathcal{M}_i = (\mathcal{X}_i, \mathfrak{A}_i, (P_{i,\theta})_{\theta \in \Lambda})$, $i = 1, 2$ are two models with any parameter set Λ, say. Then we set

$$\Delta(\mathcal{M}_1, \mathcal{M}_2) = \sup_{F \subseteq \Lambda, |F| < \infty} \Delta(\mathcal{M}_{1,F}, \mathcal{M}_{2,F}) \tag{6.14}$$

and call $\Delta(\mathcal{M}_1, \mathcal{M}_2)$ the Δ-*distance* of $\mathcal{M}_1$ and $\mathcal{M}_2$. For every finite subset $F \subseteq \Lambda$ with $F = \{\theta_1, ..., \theta_m\}$ we introduce the standard distribution $\mu_{i,F}$ as in (4.13), the standard model as in (4.15), and denote the latter by

$$\mathcal{N}_{i,F} = (\mathcal{S}_{|F|}, \mathfrak{S}_{|F|}, \{Q_{i,1}, ..., Q_{i,|F|}\}), \quad i = 1, 2.$$

The next result, in a different formulation, is due to LeCam (1964); see also Torgersen (1991) and Strasser (1985).

Theorem 6.6. *For any models $\mathcal{M}_i = (\mathcal{X}_i, \mathfrak{A}_i, (P_{i,\theta})_{\theta \in \Lambda})$, $i = 1, 2$, it holds*

$$\Delta(\mathcal{M}_1, \mathcal{M}_2) = \delta(\mathcal{M}_1, \mathcal{M}_2).$$

Moreover, $\Delta(\mathcal{M}_1, \mathcal{M}_2) = 0$ holds if and only if for every finite subset $F \subseteq \Lambda$ the standard distributions $\mu_{i,F}$, $i = 1, 2$, are identical.

Proof. In view of (6.4) and (6.14) it suffices to deal with finite models. By (6.9) we have only to show that

$$\Delta(\mathcal{M}_1, \mathcal{M}_2) \leq \delta(\mathcal{M}_1, \mathcal{M}_2).$$

By Lemma 6.5 and (6.10), for $\varepsilon > 0$ there are models $\mathcal{N}_i$ with finite samples spaces such that

$$\Delta(\mathcal{M}_i, \mathcal{N}_i) \leq \varepsilon, \quad i = 1, 2, \quad \text{and} \quad \Delta(\mathcal{N}_1, \mathcal{N}_2) = \delta(\mathcal{N}_1, \mathcal{N}_2).$$

Hence by the triangular inequalities for δ and Δ, together with (6.9),

$$\begin{aligned}
&\Delta(\mathcal{M}_1,\mathcal{M}_2) \le \Delta(\mathcal{N}_1,\mathcal{N}_2) + 2\varepsilon = \delta(\mathcal{N}_1,\mathcal{N}_2) + 2\varepsilon \\
&\le \delta(\mathcal{M}_1,\mathcal{M}_2) + 2\varepsilon + \delta(\mathcal{M}_1,\mathcal{N}_1) + \delta(\mathcal{M}_2,\mathcal{N}_2) \\
&\le \delta(\mathcal{M}_1,\mathcal{M}_2) + 2\varepsilon + \Delta(\mathcal{M}_1,\mathcal{N}_1) + \Delta(\mathcal{M}_2,\mathcal{N}_2) \le \delta(\mathcal{M}_1,\mathcal{M}_2) + 4\varepsilon,
\end{aligned}$$

which completes the proof of the first statement. As to the second, we note that by (6.14), the first statement, and Lemma 6.2 it holds $\Delta(\mathcal{M}_1,\mathcal{M}_2) = 0$ if $\mathcal{M}_{1,\mathsf{F}} \sim \mathcal{M}_{2,\mathsf{F}}$ for every finite F. By Proposition 4.22 this is equivalent to $\mathcal{N}_{1,\mathsf{F}} \sim \mathcal{N}_{2,\mathsf{F}}$. An application of Theorem 4.25 completes the proof. ■

For the next section, which deals with the convergence of models, it is important to have an upper bound for $\Delta(\mathcal{M}_1,\mathcal{M}_2)$ in terms of the Dudley metric (see (A.6)) of the associated standard distributions. The bridge is the equality of $\delta(\mathcal{M}_1,\mathcal{M}_2)$ and $\Delta(\mathcal{M}_1,\mathcal{M}_2)$, which has been established in Theorem 6.6, and the relation of $\delta(\mathcal{M}_1,\mathcal{M}_2)$ to the Bayes risks in $\mathcal{M}_1$,$\mathcal{M}_2$.

Theorem 6.7. *For any finite models $\mathcal{M}_i = (\mathcal{X}_i, \mathfrak{A}_i, \{P_{i,1},...,P_{i,m}\})$, $i = 1,2$,*

$$\Delta(\mathcal{M}_1,\mathcal{M}_2) = \sup_{\Pi, L:\|L\|_u \le 1, \mathcal{D}} |\inf\nolimits_{\mathsf{D}_{\mathcal{M}_1}} \mathsf{r}(\Pi, \mathsf{D}_{\mathcal{M}_1}) - \inf\nolimits_{\mathsf{D}_{\mathcal{M}_2}} \mathsf{r}(\Pi, \mathsf{D}_{\mathcal{M}_2})|,$$

where the supremum is taken over all priors Π, all loss functions L with $\|L\|_u \le 1$, and all finite decision spaces $\mathcal{D}$. Furthermore,

$$\Delta(\mathcal{M}_1,\mathcal{M}_2) \le m \|\mu_1 - \mu_2\|_{\mathsf{D}}\,, \tag{6.15}$$

where $\|\mu_1 - \mu_2\|_{\mathsf{D}}$ is the Dudley distance of the standard distributions μ_1 and μ_2 that belong to the models $\mathcal{M}_1$ and $\mathcal{M}_2$, respectively.

Proof. The expression $\inf_{\mathsf{D}_{\mathcal{M}_i}} \mathsf{r}(\Pi, \mathsf{D}_{\mathcal{M}_i})$ is a function of Π, L, and $\mathcal{D}$, say $\psi_i(\Pi, L, \mathcal{D})$, $i = 1,2$. Corollary 4.15 implies that $\mathcal{M}_1 \succeq^{\varepsilon} \mathcal{M}_2$ if and only if $\psi_1(\Pi, L, \mathcal{D}) - \psi_2(\Pi, L, \mathcal{D}) \le \varepsilon$ for every prior Π, every loss function L with $\|L\|_u \le 1$, and every finite decision space $\mathcal{D}$. This yields

$$\inf\{\varepsilon : \mathcal{M}_1 \succeq^{\varepsilon} \mathcal{M}_2\} = \sup\nolimits_{\Pi,L,\mathcal{D}}\{\psi_1(\Pi, L, \mathcal{D}) - \psi_2(\Pi, L, \mathcal{D})\}.$$

By switching the roles of $\mathcal{M}_1$ and $\mathcal{M}_2$ and using the fact that

$$\delta(\mathcal{M}_1,\mathcal{M}_2) = \max\left(\inf\{\varepsilon : \mathcal{M}_1 \succeq^{\varepsilon} \mathcal{M}_2\}, \inf\{\varepsilon : \mathcal{M}_2 \succeq^{\varepsilon} \mathcal{M}_1\}\right),$$

we get the first statement. To prove the stated inequality we note that $\delta(\mathcal{M}_1,\mathcal{M}_2) = \Delta(\mathcal{M}_1,\mathcal{M}_2)$ by Theorem 6.6. To complete the proof we apply Corollary 4.33. ■

We conclude this section with the remark that the explicit evaluation of $\Delta(\mathcal{M}_1,\mathcal{M}_2)$ is a challenging problem that can be completed only in special cases. The deficiency of binary models was obtained by Torgersen (1970); see also LeCam (1986) and Strasser (1985). We also refer to Luschgy (1992b), Lehmann (1988), and Torgersen (1991), where the deficiency of two linear models is evaluated. Finally we remark that in many cases one needs only tractable upper bounds for $\Delta(\mathcal{M}_1,\mathcal{M}_2)$ for two models with the same sample space. Then the inequality (6.8) can be used as long as the variational distance, or a suitable bound for it, can be evaluated. Results in this direction can be found, for example, in Shiryaev and Spokoiny (2000).

6.2 Convergence of Models

In this section we introduce and study convergence concepts for statistical models by using the deficiency of models. As weak convergence is concerned with the finite submodels, we first consider relations among standard distributions, Hellinger transforms, and deficiencies of the finite submodels to prepare for the results on the convergence of models.

When dealing with the convergence of not necessarily homogeneous models it proves useful to turn to a smoothed model which is homogenous. More precisely, given $\mathcal{M} = (\mathcal{X}, \mathfrak{A}, \{P_1, ..., P_m\})$, we set $\overline{P} = (1/m)\sum_{i=1}^m P_i$ and

$$\begin{aligned} \mathcal{M}_\alpha &= (\mathcal{X}, \mathfrak{A}, \{P_{1,\alpha}, ..., P_{m,\alpha}\}), \quad 0 \le \alpha \le 1, \quad \text{where} \\ P_{i,\alpha} &= (1-\alpha)P_i + \alpha\overline{P}, \quad 0 \le \alpha \le 1, \ i = 1, ..., m. \end{aligned} \tag{6.16}$$

If $\mathcal{N} = (\mathcal{Y}, \mathfrak{B}, \{Q_1, ..., Q_m\})$ is another finite model and $\mathcal{N}_\alpha$ is introduced in the same way as $\mathcal{M}_\alpha$, then for kernels $\mathsf{K} : \mathfrak{B} \times \mathcal{X} \to_k [0,1]$ and $\mathsf{L} : \mathfrak{A} \times \mathcal{Y} \to_k [0,1]$ it holds

$$\begin{aligned} &\left\| (1-\alpha)Q_i + \alpha\overline{Q} - \mathsf{K}((1-\alpha)P_i + \alpha\overline{P}) \right\| \\ &\le (1-\alpha)\left\| Q_i - \mathsf{K}P_i \right\| + \alpha\frac{1}{m}\sum_{i=1}^m \left\| Q_i - \mathsf{K}P_i \right\| \le \max_{1\le i\le m} \left\| Q_i - \mathsf{K}P_i \right\|. \end{aligned}$$

Similarly,

$$\left\| (1-\alpha)P_i + \alpha\overline{P} - \mathsf{L}((1-\alpha)Q_i + \alpha\overline{Q}) \right\| \le \max_{1\le i\le m} \left\| P_i - \mathsf{L}Q_i \right\|.$$

Hence by the definition of the Δ-distance in (6.5),

$$\Delta(\mathcal{M}_\alpha, \mathcal{N}_\alpha) \le \Delta(\mathcal{M}, \mathcal{N}), \quad 0 \le \alpha \le 1. \tag{6.17}$$

Problem 6.8.* If $P_1, ..., P_m, Q_1, ..., Q_m$ are distributions on $(\mathcal{X}, \mathfrak{A})$, then

$$\begin{aligned} &|\mathsf{H}_s(P_1, ..., P_m) - \mathsf{H}_s(Q_1, ..., Q_m)| \le 2\sum_{i=1}^m \|P_i - Q_i\|^{m(s)}, \\ &s \in \mathbf{S}_m^o = \{s : s = (s_1, ..., s_m), \ s_i > 0, \ \sum_{i=1}^m s_i = 1\}, \ m(s) = \min\{s_1, ..., s_m\}. \end{aligned}$$

The distance $\Delta(\mathcal{M}, \mathcal{N})$ gives upper bounds for the differences of the Hellinger transforms.

Proposition 6.9. *For any statistical models* $\mathcal{M} = (\mathcal{X}, \mathfrak{A}, \{P_1, ..., P_m\})$ *and* $\mathcal{N} = (\mathcal{Y}, \mathfrak{B}, \{Q_1, ..., Q_m\})$ *and* $0 \le \alpha \le 1$ *it holds*

$$|\mathsf{H}_s(P_{1,\alpha}, ..., P_{m,\alpha}) - \mathsf{H}_s(Q_{1,\alpha}, ..., Q_{m,\alpha})| \le 2m(\Delta(\mathcal{M}, \mathcal{N}))^{m(s)}, \quad s \in \mathbf{S}_m^o.$$

Proof. Inequality 1.123 yields for any stochastic kernels $\mathsf{K} : \mathfrak{B} \times \mathcal{X} \to_k [0,1]$ and $\mathsf{L} : \mathfrak{A} \times \mathcal{Y} \to_k [0,1]$,

$$\begin{aligned} \mathsf{H}_s(P_{1,\alpha}, ..., P_{m,\alpha}) &\le \mathsf{H}_s(\mathsf{K}P_{1,\alpha}, ..., \mathsf{K}P_{m,\alpha}), \\ \mathsf{H}_s(Q_{1,\alpha}, ..., Q_{m,\alpha}) &\le \mathsf{H}_s(\mathsf{L}Q_{1,\alpha}, ..., \mathsf{L}Q_{m,\alpha}). \end{aligned}$$

Hence by Problem 6.8, applied to $P_{i,\alpha}, Q_{i,\alpha}$ instead of P_i, Q_i,

$$\begin{aligned}
&\mathsf{H}_s(P_{1,\alpha},...,P_{m,\alpha}) \le \mathsf{H}_s(\mathsf{K}P_{1,\alpha},...,\mathsf{K}P_{m,\alpha}) \\
&\le 2\sum\nolimits_{i=1}^{m} \|\mathsf{K}P_{i,\alpha} - Q_{i,\alpha}\|^{m(s)} + \mathsf{H}_s(Q_{1,\alpha},...,Q_{m,\alpha}), \\
&\mathsf{H}_s(Q_{1,\alpha},...,Q_{m,\alpha}) \le \mathsf{H}_s(\mathsf{L}Q_{1,\alpha},...,\mathsf{L}Q_{m,\alpha}) \\
&\le 2\sum\nolimits_{i=1}^{m} \|\mathsf{L}Q_{i,\alpha} - P_{i,\alpha}\|^{m(s)} + \mathsf{H}_s(P_{1,\alpha},...,P_{m,\alpha}), \\
&|\mathsf{H}_s(P_{1,\alpha},...,P_{m,\alpha}) - \mathsf{H}_s(Q_{1,\alpha},...,Q_{m,\alpha})| \\
&\le 2m \max_{1\le i\le m} \max\{\|\mathsf{K}P_{i,\alpha} - Q_{i,\alpha}\|^{m(s)}, \|\mathsf{L}Q_{i,\alpha} - P_{i,\alpha}\|^{m(s)}\}.
\end{aligned}$$

Taking the infimum over K and L we get

$$|\mathsf{H}_s(P_{1,\alpha},...,P_{m,\alpha}) - \mathsf{H}_s(Q_{1,\alpha},...,Q_{m,\alpha})| \le 2m(\Delta(\mathcal{M}_\alpha,\mathcal{N}_a))^{m(s)}.$$

To complete the proof we apply inequality (6.17). ∎

Now we use the concept of deficiency to introduce convergence concepts for any, not necessarily finite, statistical models. We recall that for two models the Δ-distance has been defined by (6.14).

Definition 6.10. *If $\mathcal{M}_n = (\mathcal{X}_n, \mathfrak{A}_n, (P_{n,\theta})_{\theta\in\Lambda})$ and $\mathcal{M} = (\mathcal{X}, \mathfrak{A}, (P_\theta)_{\theta\in\Lambda})$ are models with the same parameter set Λ, then the sequence of models $\mathcal{M}_n$ is said to be convergent to the model $\mathcal{M}$ if $\lim_{n\to\infty} \Delta(\mathcal{M}_n, \mathcal{M}) = 0$. If $\Lambda_n \uparrow \Lambda$, then the sequence of models $\mathcal{M}_n = (\mathcal{X}_n, \mathfrak{A}_n, (P_{n,\theta})_{\theta\in\Lambda_n})$ is called weakly convergent to the model $\mathcal{M} = (\mathcal{X}, \mathfrak{A}, (P_\theta)_{\theta\in\Lambda})$ if for every finite subset $F \subseteq \Lambda$ it holds $\lim_{n\to\infty} \Delta(\mathcal{M}_{n,F}, \mathcal{M}_F) = 0$. In this case we write $\mathcal{M}_n \Rightarrow \mathcal{M}$. We call $\mathcal{M}_0$ an accumulation point of the sequence $\mathcal{M}_n$ if there exists a subsequence $\mathcal{M}_{n_k}$ with $\mathcal{M}_{n_k} \Rightarrow \mathcal{M}_0$.*

Remark 6.11. If the parameter set Λ is finite, then it follows from the previous definition and (6.13) that the convergence and the weak convergence are identical.

Remark 6.12. It should be noted that the limiting model is not uniquely determined as Δ is only a pseudometric. Indeed, Lemma 6.2 and Theorem 6.6 imply that

$$\Delta(\mathcal{M}_1, \mathcal{M}_2) = 0 \text{ if and only if } \mathcal{M}_1 \sim \mathcal{M}_2. \tag{6.18}$$

This means that the convergence of models is, more precisely, a convergence of classes of equivalent models to an equivalence class of models.

The relation between the convergence and the weak convergence becomes clear if we take into account (6.14). The distance $\Delta(\mathcal{M}_{n,F}, \mathcal{M}_F)$ is small if there are kernels K_F and L_F depending on F such that $\max_{\theta\in F} \|\mathsf{K}_F P_{n,\theta} - P_\theta\|$ and $\max_{\theta\in F} \|P_{n,\theta} - \mathsf{L}_F P_\theta\|$ are small. On the other hand, $\Delta(\mathcal{M}_n, \mathcal{M})$ is small if and only if the supremum over all such maxima is small.

As it turns out, the concept of weak convergence is flexible enough to cover typical situations in both parametric and nonparametric statistics, and strong enough to create a rich and fruitful asymptotic theory of statistical models. A key role in this regard is played by manageable criteria for weak convergence which are criteria for the convergence of finite models. Let

$$\mathcal{M}_n = (\mathcal{X}_n, \mathfrak{A}_n, \{P_{n,1}, ..., P_{n,m}\}), \quad \mathcal{M} = (\mathcal{X}, \mathfrak{A}, \{P_1, ..., P_m\}),$$

$n = 1, 2, ...,$ be models. We set

$$\begin{aligned}
\overline{P}_n &= \frac{1}{m}\sum\nolimits_{i=1}^m P_{n,i}, \quad \overline{P} = \frac{1}{m}\sum\nolimits_{i=1}^m P_i, \quad M_{n,i} = \frac{dP_{n,i}}{d\overline{P}_n}, \quad M_i = \frac{dP_i}{d\overline{P}},\\
L_{n,j} &= \frac{M_{n,j}}{M_{n,1}} I_{(0,\infty)}(M_{n,1}) + \infty I_{\{0\}}(M_{n,1}),\\
L_j &= \frac{M_j}{M_1} I_{(0,\infty)}(M_1) + \infty I_{\{0\}}(M_1), \quad j = 2, ..., m.
\end{aligned}$$

We recall that $L_{n,j}$ and $\ln L_{n,j}$ are random variables with values in $\overline{\mathbb{R}} = [-\infty, \infty]$ which, together with the metric $\overline{\rho}(x, y)$, form the compact metric space $(\overline{\mathbb{R}}, \overline{\rho})$; see Remark A.1. The weak convergence of distributions refers to this metric. Furthermore, we denote by

$$\mu_n = \mathcal{L}((M_{n,1}, ..., M_{n,m})|\overline{P}_n) \quad \text{and} \quad \mu = \mathcal{L}((M_1, ..., M_m)|\overline{P}) \tag{6.19}$$

the standard distributions of the models $\mathcal{M}_n$ and $\mathcal{M}$, respectively. Note that μ_n and μ are defined on the Borel sets of the simplex

$$\mathcal{S}_m = \{(t_1, ..., t_m) : t_i \geq 0, \ \frac{1}{m}\sum\nolimits_{i=1}^m t_i = 1\}.$$

Introduce $P_{n,i,\alpha}$ and $P_{i,\alpha}$ as in (6.16). Then by the definition of the Hellinger transforms (see Definition 1.87),

$$\begin{aligned}
\mathsf{H}_s(P_{n,1,\alpha}, ..., P_{n,m,\alpha}) &= \int \prod\nolimits_{i=1}^m ((1-\alpha)t_i + \alpha)^{s_i} \mu_n(dt),\\
\mathsf{H}_s(P_{1,\alpha}, ..., P_{m,\alpha}) &= \int \prod\nolimits_{i=1}^m ((1-\alpha)t_i + \alpha)^{s_i} \mu(dt).
\end{aligned}$$

Theorem 6.13. *For any models $\mathcal{M}_n = (\mathcal{X}_n, \mathfrak{A}_n, \{P_{n,1}, ..., P_{n,m}\})$ and $\mathcal{M} = (\mathcal{X}, \mathfrak{A}, \{P_1, ..., P_m\})$, the following statements are equivalent.*

(A) $\lim_{n\to\infty} \Delta(\mathcal{M}_n, \mathcal{M}) = 0$.

(B) $\mu_n \Rightarrow \mu$.

(C) $\lim_{n\to\infty} \mathsf{H}_s(P_{n,1,\alpha}, ..., P_{n,m,\alpha}) = \mathsf{H}_s(P_{1,\alpha}, ..., P_{m,\alpha}), \quad \alpha \in [0,1], \ s \in \mathbf{S}_m^0$.

Corollary 6.14. *If the model $\mathcal{M}$ is homogeneous, then condition (C) can be replaced by the weaker condition*

(D) $\lim_{n\to\infty} \mathsf{H}_s(P_{n,1}, ..., P_{n,m}) = \mathsf{H}_s(P_1, ..., P_m), \quad s \in \mathbf{S}_m^0$.

Corollary 6.15. *If the model $\mathcal{M}$ is homogeneous, then each of the conditions (A) through (D) is equivalent to*

(E) $\mathcal{L}((\ln L_{n,2},...,\ln L_{n,m})|P_{n,1}) \Rightarrow \mathcal{L}((\ln L_2,...,\ln L_m)|P_1)$.

Proof. $(A) \to (C)$: the stated convergence follows from Proposition 6.9. $(C) \to (B)$: as the simplex $\mathcal{S}_m$ is a compact metric space according to Prohorov's theorem (see Theorem A.48), every subsequence of $\{\mu_n\}$ contains a weakly convergent subsequence. It remains to show that all accumulation points are identical. But this follows from the fact that

$$t \mapsto \prod_{i=1}^{m}((1-\alpha)t_i + \alpha)^{s_i}$$

is a continuous and bounded function on $\mathcal{S}_m$ and the uniqueness theorem 4.25. $(B) \to (A)$: as the Dudley metric metricizes the weak convergence (see Theorem A.50) condition (B) implies $\|\mu_n - \mu\|_{\mathsf{D}} \to 0$. Inequality (6.15) implies (A). The first corollary follows from Corollary 4.26. To prove the second corollary we note that, as $\ln x$ is a homeomorphism between $\overline{\mathbb{R}}_+$ and $\overline{\mathbb{R}}$, the statement (E) is equivalent to

$$\mathcal{L}((L_{n,2},...,L_{n,m})|P_{n,1}) \Rightarrow \mathcal{L}((L_2,...,L_m)|P_1). \tag{6.20}$$

It follows from the definition of the Hellinger transform in Definition 1.87 and the definition of the likelihood ratios $L_{n,i}$ and L_i, respectively,

$$\mathsf{H}_s(P_{n,1},...,P_{n,m}) = \int M_{n,1}^{s_1} \cdots M_{n,m}^{s_m} d\overline{P}_n = \int L_{n,2}^{s_2} \cdots L_{n,m}^{s_m} dP_{n,1},$$
$$\mathsf{H}_s(P_1,...,P_m) = \int L_2^{s_2} \cdots L_m^{s_m} dP_1.$$

Set $f_n = L_{n,2}^{s_2} \cdots L_{n,m}^{s_m}$. Then

$$\int f_n^{1/(1-s_1)} dP_{n,1} = \int L_{n,2}^{s_2/(1-s_1)} \cdots L_{n,m}^{s_m/(1-s_1)} dP_{n,1}$$
$$\leq \prod_{i=2}^{m} (\int L_{n,2} dP_{n,1})^{s_i/(1-s_1)} \leq 1.$$

Hence,

$$\limsup_{c\to\infty} \limsup_{n\to\infty} \int f_n I_{[c,\infty)}(f_n) dP_{n,1} \leq \limsup_{c\to\infty} c^{-1/(1-s_1)} = 0,$$

and we get from (6.20) and Proposition A.44 that $\mathsf{H}_s(P_{n,1},...,P_{n,m}) \to \mathsf{H}_s(P_1,...,P_m)$. The statement follows from the previous corollary. ■

The next proposition shows that every sequence of finite models is relatively compact, so that the convergence of the sequence of models depends only on whether there is just one or perhaps several accumulation points.

Proposition 6.16. *Every sequence* $\mathcal{M}_n = (\mathcal{X}_n, \mathfrak{A}_n, \{P_{n,1}, ..., P_{n,m}\})$, $n = 1, 2, ...,$ *of finite models is relatively sequentially compact in the sense that for every subsequence* n_k *there is a subsequence* n_{k_l} *and a model* $\mathcal{M} = (\mathcal{X}, \mathfrak{A}, \{P_1, ..., P_m\})$ *such that*

$$\lim_{l\to\infty} \Delta(\mathcal{M}_{n_{k_l}}, \mathcal{M}) = 0.$$

Proof. We recall that by Theorem 4.25 every finite model $\mathcal{M}$ and the associated standard model $\mathcal{N}$ are equivalent and therefore $\Delta(\mathcal{M}, \mathcal{N}) = 0$ by (6.18). Hence it suffices to consider standard models

$$\mathcal{N}_n = (\mathcal{S}_m, \mathfrak{S}_m, \{P_{n,1}, ..., P_{n,m}\}), \quad \text{where}$$
$$P_{n,i}(B) = \int I_B(x_1, ..., x_m) x_i \mu_n(dx_1, ..., dx_m), \quad B \in \mathfrak{S}_m.$$

$\mathcal{S}_m$ is compact, and by Prohorov's theorem (see Theorem A.48) the sequence μ_{n_k} contains a subsequence n_{k_l} that converges weakly to some distribution μ. The latter is again a standard distribution as the mapping $(x_1, ..., x_m) \to x_i$ is a bounded and continuous function on $\mathcal{S}_m$. Put $\mathcal{N} = (\mathcal{S}_m, \mathfrak{S}_m, \{P_1, ..., P_m\})$, where $dP_i = x_i \mu(dx_1, ..., dx_m)$. Then $\lim_{l\to\infty} \Delta(\mathcal{N}_{n_{k_l}}, \mathcal{N}) = 0$ by Theorem 6.13. To complete the proof we set $\mathcal{M} = (\mathcal{X}, \mathfrak{A}, \{P_1, ..., P_m\}) = (\mathcal{S}_m, \mathfrak{S}_m, \{P_1, ..., P_m\})$. ■

Example 6.17. A model $\mathcal{N} = (\mathcal{Y}, \mathfrak{B}, (Q_\theta)_{\theta\in\Delta})$ is called *totally informative* if

$$Q_{\theta_1} \perp Q_{\theta_2}, \quad \theta_1 \neq \theta_2, \quad \theta_1, \theta_2 \in \Delta.$$

If Δ is finite, say $\Delta = \{1, ..., m\}$, then one may decompose $\mathcal{Y}$ into m disjoint sets A_i with $Q_i(A_i) = 1$. This means that the true distribution can be identified without any error by a sample of size $n = 1$. Consider now a sequence of models

$$\mathcal{M}_n = (\mathcal{X}_n, \mathfrak{A}_n, (P_{n,\theta})_{\theta\in\Delta_n}).$$

If $\mathcal{N}$ is totally informative, then for every finite set $F \subseteq \Delta$ the model $\mathcal{N}_F$ is *totally informative as well. By* Problem 1.81 it holds $Q_{\theta_1} \perp Q_{\theta_2}$ if and only if $\mathsf{H}_{1/2}(Q_{\theta_1}, Q_{\theta_2}) = 0$. *If* $\mathcal{M}_n \Rightarrow \mathcal{N}$, then by (C) in Theorem 6.13

$$\lim_{n\to\infty} \mathsf{H}_{1/2}(P_{n,\theta_1}, P_{n,\theta_2}) = \mathsf{H}_{1/2}(Q_{\theta_1}, Q_{\theta_2}) = 0, \quad \theta_1 \neq \theta_2. \tag{6.21}$$

But this condition is also sufficient for $\mathcal{M}_n \Rightarrow \mathcal{N}$. To see this we have, in view of the sequential compactness in Proposition 6.16, only to show that every accumulation point $(\mathcal{Y}, \mathfrak{B}, (Q_\theta)_{\theta\in F})$ of $\mathcal{M}_{n,F}$ is totally informative. But this follows from

$$\mathsf{H}_{1/2}(Q_{\theta_1}, Q_{\theta_2}) = \lim_{n\to\infty} \mathsf{H}_{1/2}(P_{n,\theta_1}, P_{n,\theta_2}) = 0$$

and the fact that $\mathsf{H}_{1/2}(Q_{\theta_1}, Q_{\theta_2}) = 0$ implies $Q_{\theta_1} \perp Q_{\theta_2}$. Hence we have obtained that (6.21) is necessary and sufficient for the weak convergence of $\mathcal{M}_n$ to the totally informative model $\mathcal{N} = (\mathcal{Y}, \mathfrak{B}, (Q_\theta)_{\theta\in\Delta})$.

Suppose we are given i.i.d. observations $X_1, ..., X_n$ with common distribution P_θ. Then we have the sequence of models

$$\mathcal{M}_n = (\mathcal{X}^n, \mathfrak{A}^{\otimes n}, (P_\theta^{\otimes n})_{\theta \in \Delta}).$$

If the parameter θ is identifiable, then $\theta_1 \neq \theta_2$ implies $P_{\theta_1} \neq P_{\theta_2}$ and thus $\mathsf{H}_{1/2}(P_{\theta_1}, P_{\theta_2}) < 1$; see Problem 1.81. Then by Problem 1.86,

$$\lim_{n\to\infty} \mathsf{H}_{1/2}(P_{\theta_1}^{\otimes n}, P_{\theta_2}^{\otimes n}) = 0,$$
$$\lim_{n\to\infty} \mathsf{H}_{1/2}(P_{\theta_1}^{\otimes n}, P_{\theta_2}^{\otimes n}) = \mathsf{H}_{1/2}(P_{\theta_1}^{\otimes \infty}, P_{\theta_2}^{\otimes \infty}).$$

Hence we get that the sequence of models $(\mathcal{X}^n, \mathfrak{A}^{\otimes n}, (P_\theta^{\otimes n})_{\theta \in \Delta})$ tends weakly to the totally informative model $(\mathcal{X}^\infty, \mathfrak{A}^{\otimes \infty}, (P_\theta^{\otimes \infty})_{\theta \in \Delta})$.

The exponential families constitute another class of models for which the weak convergence of models can be reduced to the weak convergence of all binary submodels. Suppose we are given the following exponential models.

$$\mathcal{M} = (\mathcal{X}, \mathfrak{A}, (P_\theta)_{\theta \in \Delta}), \qquad \frac{dP_\theta}{d\boldsymbol{\mu}}(x) = \exp\{\langle \theta, T(x)\rangle - K(\theta)\},$$
$$\mathcal{M}_n = (\mathcal{X}_n, \mathfrak{A}_n, (P_{n,\theta})_{\theta \in \Delta_n}), \ \frac{dP_{n,\theta}}{d\boldsymbol{\mu}_n}(x_n) = \exp\{\langle \theta, T_n(x_n)\rangle - K_n(\theta)\}.$$

Proposition 6.18. *Let $\mathcal{M}_n$ and $\mathcal{M}$ be exponential models where the parameter sets satisfy $\Delta_n \subseteq \Delta \subseteq \mathbb{R}^d$ with $\Delta_n \uparrow \Delta$, and $\mathcal{M}$ satisfies the conditions (A1) and (A2). Then $\mathcal{M}_n \Rightarrow \mathcal{M}$ if and only if for every $\theta_1, \theta_2 \in \Delta$, and every $0 < s < 1$,*

$$\begin{aligned} &\lim\nolimits_{n\to\infty}[sK_n(\theta_1) + (1-s)K_n(\theta_2) - K_n(s\theta_1 + (1-s)\theta_2)] \\ &\quad = sK(\theta_1) + (1-s)K(\theta_2) - K(s\theta_1 + (1-s)\theta_2). \end{aligned} \tag{6.22}$$

The latter condition is equivalent to

$$\lim_{n\to\infty} \mathsf{H}_s(P_{n,\theta_1}, P_{n,\theta_2}) = \mathsf{H}_s(P_{\theta_1}, P_{\theta_2}), \quad \theta_1, \theta_2 \in \Delta, \quad s \in (0,1). \tag{6.23}$$

Proof. As $\mathcal{M}$ is homogeneous Theorem 6.13 and Example 1.88 show that $\mathcal{M}_n \Rightarrow \mathcal{M}$ holds if and only if

$$K_n(\sum\nolimits_{i=1}^m s_i\theta_i) - \sum\nolimits_{i=1}^m s_i K_n(\theta_i) \to K(\sum\nolimits_{i=1}^m s_i\theta_i) - \sum\nolimits_{i=1}^m s_i K(\theta_i) \tag{6.24}$$

for every $s_i > 0$, $\sum_{i=1}^m s_i = 1$, every θ_i and every m, so that (6.22) is necessary. Set

$$\widetilde{\theta}_1 = \theta_m, \quad \widetilde{\theta}_2 = \sum\nolimits_{i=1}^{m-1} \frac{s_i}{1-s_m}\theta_i, \quad s = s_m.$$

Then

$$\begin{aligned} &K_n(\sum\nolimits_{i=1}^m s_i\theta_i) - \sum\nolimits_{i=1}^m s_i K_n(\theta_i) \\ &= K_n(s\widetilde{\theta}_1 + (1-s)\widetilde{\theta}_2) - sK_n(\widetilde{\theta}_1) - (1-s)K_n(\widetilde{\theta}_2) \\ &\qquad + (1-s)(K_n(\sum\nolimits_{i=1}^{m-1} \frac{s_i}{1-s}\theta_i) - \sum\nolimits_{i=1}^{m-1} \frac{s_i}{1-s}K(\theta_i)). \end{aligned}$$

This relation and mathematical induction show that (6.22) and (6.23) are equivalent. ■

The question arises of how we can model situations for i.i.d. samples if nontrivial limit models are desired. The idea is to turn to a double array $X_{n,i}$ with $\mathcal{L}(X_{n,i}) = P_{n,\theta}$, $i = 1, ..., n$, $n = 1, 2, ...$ If for large n one more observation $X_{n,i}$ provides little additional information (i.e., if the distributions P_{n,θ_1} and P_{n,θ_2} get closer and closer for large n), then we may expect to get a nontrivial limit model for $(P_{n,\theta}^{\otimes n})_{\theta\in\Delta}$. One way to achieve this goal is the localization of the parameter. We fix some $\theta_0 \in \Delta$, introduce a local parameter h by setting $\theta = \theta_0 + h/\sqrt{n}$, and study the sequence of models

$$\mathcal{M}_n = (\mathcal{X}^n, \mathfrak{A}^{\otimes n}, (P_{\theta_0+h/\sqrt{n}}^{\otimes n})_{h\in\Delta_n}), \quad \Delta_n = \{h : \theta_0 + h/\sqrt{n} \in \Delta\}.$$

Example 6.19. Suppose $\mathcal{M} = (\mathcal{X}, \mathfrak{A}, (P_\theta)_{\theta\in\Delta})$, $\Delta \subseteq \mathbb{R}$, is a one-parameter exponential family with density $dP_\theta/d\boldsymbol{\mu} = \exp\{\theta T - K(\theta)\}$. Then

$$P_{n,h} = P_{\theta_0+h/\sqrt{n}}^{\otimes n}$$

leads again to an exponential family, but now with the parameter h. It follows from Example 1.88 and Problem 1.86 that

$$\begin{aligned}\mathsf{H}_s(P_{n,h_1}, P_{n,h_2}) &= (\mathsf{H}_s(P_{\theta_0+h_1/\sqrt{n}}, P_{\theta_0+h_2/\sqrt{n}}))^n \\ &= \exp\{-n[sK(\theta_0 + \frac{h_1}{\sqrt{n}}) + (1-s)K(\theta_0 + \frac{h_2}{\sqrt{n}}) - K(\theta_0 + \frac{sh_1 + (1-s)h_2}{\sqrt{n}})]\}.\end{aligned}$$

Consider the function

$$f(t) = sK(\theta_0 + th_1) + (1-s)K(\theta_0 + th_2) - K(\theta_0 + t(sh_1 + (1-s)h_2)).$$

It holds $f(0) = 0$, $f'(0) = 0$, and

$$\begin{aligned}f''(0) &= K''(\theta_0)(sh_1^2 + (1-s)h_2^2 - (sh_1 + (1-s)h_2)^2) \\ &= -s(1-s)(h_1 - h_2)^2 K''(\theta_0).\end{aligned}$$

We know from Example 1.120 that $K''(\theta_0)$ is the Fisher information $\mathsf{I}(\theta_0)$. Hence

$$\mathsf{H}_s(P_{n,h_1}, P_{n,h_2}) \to \exp\{-\frac{1}{2}s(1-s)(h_2 - h_1)^2 \mathsf{I}(\theta_0)\}.$$

The relation 1.79 yields

$$\mathsf{H}_s(\mathsf{N}(\mathsf{I}(\theta_0)h_1, \mathsf{I}(\theta_0)), \mathsf{N}(\mathsf{I}(\theta_0)h_2, \mathsf{I}(\theta_0))) = \exp\{-\frac{1}{2}s(1-s)\mathsf{I}(\theta_0)(h_2 - h_1)^2\}.$$

Using the criterion for the weak convergence of models given by exponential families, which has been established in Proposition 6.18, we arrive at

$$(\mathcal{X}^n, \mathfrak{A}^{\otimes n}, (P_{\theta_0+h/\sqrt{n}}^{\otimes n})_{h\in\Delta_n}) \Rightarrow \mathcal{G} = (\mathbb{R}, \mathfrak{B}, (\mathsf{N}(\mathsf{I}(\theta_0)h, \mathsf{I}(\theta_0)))_{h\in\mathbb{R}}).$$

Later we show that such a convergence to the Gaussian model $\mathcal{G}$ holds for any $\mathbb{L}_2$-differentiable family of distributions $(P_\theta)_{\theta\in\Delta}$.

Another example of a situation where for large n one more observation provides little additional information is met in the Poisson limit theorem considered below.

Example 6.20. For every fixed n let $\varepsilon_{n,1},...,\varepsilon_{n,n}$ be i.i.d. Bernoulli variables with success probability $p_n = \lambda/n$. Set $P_{n,\lambda} = ((1-(\lambda/n))\boldsymbol{\delta}_0 + (\lambda/n)\boldsymbol{\delta}_1)^{\otimes n}$, $\lambda \in \Delta_n = (0,n)$. We consider the sequence of models

$$\begin{aligned}\mathcal{M}_n &= (\{0,1\}^n, \mathfrak{P}(\{0,1\}^n), ((1-\frac{\lambda}{n})\boldsymbol{\delta}_0 + \frac{\lambda}{n}\boldsymbol{\delta}_1)^{\otimes n}_{\lambda\in\Delta_n}),\\ \mathcal{M} &= (\mathbb{N}, \mathfrak{P}(\mathbb{N}), (\mathsf{Po}(\lambda))_{\lambda\in\Lambda}), \quad \Lambda = (0,\infty).\end{aligned}$$

Both families of distributions are exponential families. Using

$$\mathsf{H}_s((1-p_1)\boldsymbol{\delta}_0 + p_1\boldsymbol{\delta}_1, (1-p_2)\boldsymbol{\delta}_0 + p_2\boldsymbol{\delta}_1) = (1-p_1)^s(1-p_2)^{1-s} + p_1^s p_2^{1-s}$$

we get

$$\begin{aligned}&\mathsf{H}_s((1-\frac{\lambda_1}{n})\boldsymbol{\delta}_0 + \frac{\lambda_1}{n}\boldsymbol{\delta}_1)^{\otimes n}, (1-\frac{\lambda_2}{n})\boldsymbol{\delta}_0 + \frac{\lambda_2}{n}\boldsymbol{\delta}_1)^{\otimes n})\\ &= \left((1-\frac{\lambda_1}{n})^s(1-\frac{\lambda_2}{n})^{1-s} + \frac{1}{n}\lambda_1^s\lambda_2^{1-s}\right)^n \to \exp\left\{-s\lambda_1 - (1-s)\lambda_2 + \lambda_1^s\lambda_2^{1-s}\right\}.\end{aligned}$$

As $\mathsf{H}_s(\mathsf{Po}(\lambda_1), \mathsf{Po}(\lambda_2)) = \exp\left\{-s\lambda_1 - (1-s)\lambda_2 + \lambda_1^s\lambda_2^{1-s}\right\}$ we get from Proposition 6.18 that the sequence of Bernoulli models $\mathcal{M}_n$ converges weakly to the Poisson model $\mathcal{M}$. This is, roughly speaking, the Poisson limit theorem in the language of convergence of models.

6.3 Weak Convergence of Binary Models

To prepare the criteria for the convergence of products of models we study binary models in this section. Compared with multivariate limit theorems, one could say here that the binary models are comparable with the one-dimensional marginal distributions, whose convergence provides already important properties of the sequence of multivariate distributions. For example, one may conclude the tightness of a sequence of multivariate distributions from the tightness of all marginal distributions. We start with a binary model $\mathcal{M}$ and set

$$\begin{aligned}&\mathcal{M} = (\mathcal{X}, \mathfrak{A}, \{P,Q\}), \qquad R = \tfrac{1}{2}(P+Q), \quad M = \tfrac{dP}{dR},\\ &L = \tfrac{2-M}{M} I_{(0,2)}(M) + \infty I_{\{0\}}(M), \ \mu = \mathcal{L}((M, 2-M)|R).\end{aligned} \tag{6.25}$$

It holds that $dQ/dR = 2-M$, and L is the likelihood ratio of Q with respect to P. We note that L is a random variable with values in $\overline{\mathbb{R}}_+ = [0,\infty]$. The latter is a compact metric space with the metric that is induced by the one-to-one mapping $\psi(x) = x/(1+|x|)$ of $\overline{\mathbb{R}}_+$ on $[0,1]$; see Remark A.1. Let $\overline{\mathfrak{B}}_+$ denote the σ-algebra of Borel sets of $\overline{\mathbb{R}}_+$. Then L is a measurable mapping with values in $\overline{\mathbb{R}}_+$, and by the distribution of L we mean $\mathcal{L}(L|P) = P\circ L^{-1}$. We also consider the log-likelihood and note that the extension of the logarithm, by setting $\ln 0 = -\infty$ and $\ln \infty = \infty$, is a homeomorphism of the compact metric spaces $\overline{\mathbb{R}}_+$ and $\overline{\mathbb{R}}$. If $\overline{\mathfrak{B}}$ is the σ-algebra of Borel sets of $\overline{\mathbb{R}}$, then $\ln L$ is a random variable with value in $(\overline{\mathbb{R}}, \overline{\mathfrak{B}})$. Finally, $\chi(x) = (2x/(1+x), 2/(1+x))$ is a homeomorphism of the compact metric spaces $\overline{\mathbb{R}}_+ = [0, \infty]$ and

$$\mathcal{S}_2 = \{(t_1, t_2) : t_1, t_2 \geq 0,\ t_1 + t_2 = 2\}.$$

We have already characterized the weak convergence of models in terms of the standard measures. For binary models we show that this is equivalent to the convergence of the distributions of likelihood ratios. Suppose we are given a sequence of binary models $\mathcal{M}_n$. We set

$$\begin{aligned} &\mathcal{M}_n = (\mathcal{X}_n, \mathfrak{A}_n, \{P_n, Q_n\}), \quad M_n = \tfrac{dP_n}{dR_n}, \quad R_n = \tfrac{1}{2}(P_n + Q_n), \\ &L_n = \tfrac{2-M_n}{M_n} I_{(0,\infty)}(M_n) + \infty I_{\{0\}}(M_n), \quad \mu_n = \mathcal{L}((M_n, 2-M_n)|R_n). \end{aligned} \tag{6.26}$$

Theorem 6.21. *For binary models* $\mathcal{M}_n = (\mathcal{X}_n, \mathfrak{A}_n, \{P_n, Q_n\})$, $n = 1, 2, ...,$ *and* $\mathcal{M} = (\mathcal{X}, \mathfrak{A}, \{P, Q\})$ *with the standard distributions* μ_n *and* μ, *respectively, the following statements are equivalent.*

(A) $\lim_{n\to\infty} \Delta(\mathcal{M}_n, \mathcal{M}) = 0$.

(B) $\mu_n \Rightarrow \mu$.

(C) $\lim_{n\to\infty} \mathsf{H}_s(P_n, Q_n) = \mathsf{H}_s(P, Q),\ 0 < s < 1$.

(D) $\mathcal{L}(L_n|P_n) \Rightarrow \mathcal{L}(L|P)$.

Proof. Theorem 6.13 implies the equivalence of (A) and (B) and the necessity of (C). If (C) is fulfilled and $\widetilde{\mu}$ is an accumulation point of the sequence μ_n, then the fact that $x_1^s x_2^{1-s}$ is a continuous function on $\mathcal{S}_2$ implies

$$\int x_1^s x_2^{1-s} \widetilde{\mu}(dx_1, dx_2) = \mathsf{H}_s(P, Q) = \int x_1^s x_2^{1-s} \mu(dx_1, dx_2), \quad 0 < s < 1,$$

so that by Corollary 4.27 all accumulation points of the sequence μ_n are identical with μ. As $\mathcal{S}_2$ is compact the sequence μ_n is sequentially compact with respect to the weak convergence by Prohorov's theorem (see Theorem A.48). Hence we have the weak convergence of μ_n to μ which is (B). Now we show $(D) \to (C)$. It holds for $0 < s < 1$,

$$\lim_{N\to\infty} \sup_n \int I_{[N,\infty)}(L_n) L_n^{1-s} dP_n \leq \lim_{N\to\infty} \sup_n \frac{1}{N^s} \int L_n dP_n \leq \lim_{N\to\infty} \frac{1}{N^s} = 0.$$

Then (D) and Proposition A.44 give

$$\mathsf{H}_s(P_n, Q_n) = \int M_n^s (2 - M_n)^{1-s} dR_n = \int L_n^{1-s} dP_n \to \int L^{1-s} dP = \mathsf{H}_s(P, Q).$$

Finally we show $(B) \to (D)$. Suppose $\varphi : [0,\infty] \to \mathbb{R}$ is continuous. Then $\psi(x_1,x_2) = \varphi(x_2/x_1)x_1$, $0 < x_1 \le 2$, $x_2 = 2 - x_1$, with $\psi(0, 2-0) = 0$, is continuous on $\mathcal{S}_2$. Hence

$$\lim_{n\to\infty} \int \varphi(L_n)dP_n = \lim_{n\to\infty} \int \varphi(\frac{2-M_n}{M_n})M_n dR_n = \lim_{n\to\infty} \int \psi d\mu_n$$
$$= \int \psi d\mu = \int \varphi(L)dP.$$

■

If μ is the standard distribution of $\mathcal{M} = (\mathcal{X}, \mathfrak{A}, \{P, Q\})$, then the properties of P and Q to be absolutely continuous, or to be mutually singular, can be expressed with the help of the standard distribution. Indeed,

$$\begin{aligned} Q \ll P \quad &\Leftrightarrow \quad R(M=0) = \mu(\{(0,2)\}) = 0, \\ P \ll Q \quad &\Leftrightarrow \quad R(M=2) = \mu(\{(2,0)\}) = 0, \\ P \perp Q \quad &\Leftrightarrow \quad \begin{cases} R(M=2) = \mu(\{(2,0)\}) = 1/2, \\ R(M=0) = \mu(\{(0,2)\}) = 1/2. \end{cases} \end{aligned} \tag{6.27}$$

Now we study sequences $\mathcal{M}_n = (\mathcal{X}_n, \mathfrak{A}_n, \{P_n, Q_n\})$ for which each accumulation point $\mathcal{M} = (\mathcal{X}, \mathfrak{A}, \{P, Q\})$ has the property $Q \ll P$, which is equivalent to $\mu(\{(0,2)\}) = 0$.

Definition 6.22. *For a sequence of binary models $\mathcal{M}_n = (\mathcal{X}_n, \mathfrak{A}_n, \{P_n, Q_n\})$ we call the sequence $\{Q_n\}$ contiguous with respect to the sequence $\{P_n\}$ if $Q \ll P$ for each accumulation point $\mathcal{M} = (\mathcal{X}, \mathfrak{A}, \{P, Q\})$. In this case we write $\{Q_n\} \lhd \{P_n\}$. If $\{P_n\} \lhd \{Q_n\}$ holds as well, then we write $\{P_n\} \lhd\rhd \{Q_n\}$.*

The definition can be reformulated by saying that $\{Q_n\}$ is contiguous with respect to $\{P_n\}$ if and only if each accumulation point ν of the sequence of standard distributions satisfies $Q(L = \infty) = \nu(\{(0,2)\}) = 0$. It is clear from its definition that contiguity is in a way an asymptotic version of the concept of absolute continuity. Indeed, if $(\mathcal{X}_n, \mathfrak{A}_n, \{P_n, Q_n\}) = (\mathcal{X}, \mathfrak{A}, \{P, Q\})$, then $\{Q_n\} \lhd \{P_n\}$ if and only if $Q \ll P$. However, if we have only $\mathcal{M}_n \Rightarrow \mathcal{M}$, then from $Q_n \ll P_n$ for every n one cannot conclude that the limit model satisfies $Q \ll P$. This is due to the simple fact that for a convergent sequence of standard measures μ_n with $\mu_n(\{(0,2)\}) = 0$ the limit μ does not necessarily satisfy $\mu(\{(0,2)\}) = 0$.

Example 6.23. This is a special case of Example 6.17. Let P_0 and P_1 be two different distributions on $(\mathcal{X}, \mathfrak{A})$ that are mutually absolutely continuous. The relation $P_0 \neq P_1$ yields $\mathsf{H}_s(P_0, P_1) < 1$. We consider the sequence of models $\mathcal{M}_n = (\mathcal{X}^n, \mathfrak{A}^{\otimes n}, \{P_0^{\otimes n}, P_1^{\otimes n}\})$. It follows from Problem 1.86 that

$$\mathsf{H}_s(P_0^{\otimes n}, P_1^{\otimes n}) = (\mathsf{H}_s(P_0, P_1))^n \to 0, \quad s \in (0,1).$$

Hence by Theorem 6.21 $\Delta(\mathcal{M}_n, \mathcal{M}) \to 0$, where $\mathcal{M} = (\mathcal{X}, \mathfrak{A}, \{P, Q\})$ is the total informative model that consists of two mutually singular distributions P and Q. Hence the associated standard distributions satisfy $\mu_n(\{(0,2),(2,0)\}) = 0$ and $\mu(\{(0,2),(2,0)\}) = 1$.

We see from the previous example that absolute continuity in the limit model will occur only if the absolute continuity is uniform along the sequence. This is reflected in the next theorem by the condition that the likelihood ratios are uniformly integrable. Sometimes it is not advisable to try to verify this condition directly. Then it may be better to deal with Hellinger integrals that are often directly available and reflect uniform integrability by their behavior under $s \to 0$. This conjecture is based on the fact that

$$Q \ll P \quad \Leftrightarrow \quad Q(L = \infty) = \lim_{s \downarrow 0}(1 - \mathsf{H}_s(P, Q)) = 0, \tag{6.28}$$

which has been established in Problem 1.82.

To study $\mathsf{H}_s(P_n, Q_n)$ as a function of both variables s and n we need suitable inequalities for the family of functions

$$\mathsf{u}_s(x) = sx + (1-s)(2-x) - x^s(2-x)^{1-s}, \quad 0 \le x \le 2,\ 0 < s < 1.$$

Note in particular that $\mathsf{u}_{1/2}(x) = \frac{1}{2}(\sqrt{x} - \sqrt{2-x})^2$. The importance of this family originates from the relations

$$\begin{aligned} 1 - \mathsf{H}_s(P, Q) &= \int \mathsf{u}_s(M)dR, \\ 2[1 - \mathsf{H}_{1/2}(P, Q)] &= 2\int \mathsf{u}_{1/2}(M)dR = \mathsf{D}^2(P, Q). \end{aligned}$$

Problem 6.24.* For every $0 < s < 1$ there are positive constants c_s and d_s such that for $0 \le x \le 2$,

$$c_s \mathsf{u}_s(x) \le \mathsf{u}_{1/2}(x) \le d_s \mathsf{u}_s(x). \tag{6.29}$$

It holds for every $0 < \varepsilon < 1$,

$$\alpha(\varepsilon) := \sup_{0<s<1, |x-1| \le \varepsilon} \left| \frac{\mathsf{u}_s(x) - 4s(1-s)\mathsf{u}_{1/2}(x)}{\mathsf{u}_{1/2}(x)} \right| \to 0, \ \text{ as } \varepsilon \downarrow 0, \tag{6.30}$$

and

$$\beta(\varepsilon) := \sup_{0<s\le 1/2} \sup_{\varepsilon \le x \le 2} \frac{\mathsf{u}_s(x)}{4s(1-s)\mathsf{u}_{1/2}(x)} < \infty, \quad \varepsilon > 0. \tag{6.31}$$

It holds for every $0 \le x, y \le 1$ with $y \ge cx$, $c > 0$, and $0 < s < 1$,

$$1 - x^s y^{1-s} - (1-x)^s (1-y)^{1-s} \ge ((1-s) - s/c - c^{-s})y. \tag{6.32}$$

Lemma 6.25. *It holds, in the settings of Problem 6.24, for every $c > 1$,*

$$1 - \mathsf{H}_s(P, Q) \ge [(1-s) - s/c - c^{-s}]Q(L \ge c), \tag{6.33}$$

$$1 - \mathsf{H}_s(P, Q) \le 4\beta(2/(1+c))s(1-s)\mathsf{D}^2(P, Q) + Q(L > c). \tag{6.34}$$

Proof. Set $A = \{2 - M \ge cM\}$ for $c > 0$. An application of Hölder's inequality yields

$$\begin{aligned} \mathsf{H}_s(P, Q) &= \mathsf{E}_R I_A M^s (2-M)^{1-s} + \mathsf{E}_R I_{\overline{A}} M^s (2-M)^{1-s} \\ &\le P^s(A) Q^{1-s}(A) + P^s(\overline{A}) Q^{1-s}(\overline{A}). \end{aligned} \tag{6.35}$$

Hence by $Q(2 - M \geq cM) \geq cP(2 - M \geq cM)$ and inequality (6.32),

$$1 - \mathsf{H}_s(P, Q) \geq [(1-s) - s/c - c^{-s}]Q(2 - M \geq cM).$$

We have for $0 < \varepsilon < 1$

$$\begin{aligned}
1 - \mathsf{H}_s(P, Q) &= \int \mathsf{u}_s(M)dR = \int \mathsf{u}_s(M)I_{[0,\varepsilon)}(M)dR + \int \mathsf{u}_s(M)I_{[\varepsilon,2]}(M)dR \\
&\leq \beta(\varepsilon)4s(1-s)\int \mathsf{u}_{1/2}(M)I_{[\varepsilon,2]}(M)dR \\
&+ \int [sM + (1-s)(2-M) - M^s(2-M)^{1-s}]I_{[0,\varepsilon)}(M)dR \\
&\leq \beta(\varepsilon)4s(1-s)\int \mathsf{u}_{1/2}(M)dR + \int (2-M)(s\frac{\varepsilon}{2-\varepsilon} + (1-s))I_{[0,\varepsilon)}(M)dR \\
&\leq \beta(\varepsilon)4s(1-s)\mathsf{D}^2(P,Q) + Q(M < \varepsilon).
\end{aligned}$$

It remains to set $\varepsilon = 2/(1+c)$ and to use the definition of L in (6.25). ■

The next theorem presents necessary and sufficient conditions for contiguity. One condition is the uniform integrability of the likelihood ratios, whereas the other is an asymptotic version of (6.28).

Theorem 6.26. *For a sequence of binary models* $(\mathcal{X}_n, \mathfrak{A}_n, \{P_n, Q_n\})$, $n = 1, 2, ...$, *the condition* $\{Q_n\} \lhd \{P_n\}$ *is equivalent to any one of the following conditions.*

(A) *For any* $A_n \in \mathfrak{A}_n$, $\lim_{n\to\infty} P_n(A_n) = 0$ *implies* $\lim_{n\to\infty} Q_n(A_n) = 0$.

(B) *It holds* $\lim_{n\to\infty} Q_n(L_n = \infty) = 0$ *and* $\lim_{c\to\infty} \limsup_{n\to\infty} \int I_{(c,\infty)}(L_n)L_n dP_n = 0$.

(C) $\lim_{c\to\infty} \limsup_{n\to\infty} Q_n(L_n > c) = 0$.

(D) $\liminf_{s\downarrow 0} \liminf_{n\to\infty} \mathsf{H}_s(P_n, Q_n) = 1$.

Corollary 6.27. *If* $(\mathcal{X}, \mathfrak{A}, \{P, Q\})$ *is a model, and* L *in (6.25) and* L_n *in (6.26) satisfy* $\mathcal{L}(L_n|P_n) \Rightarrow \mathcal{L}(L|P)$, *then* $\{Q_n\} \lhd \{P_n\}$ *holds if and only if* $Q \ll P$.

Proof. Conditions (B) and (C) are equivalent by the definition of the likelihood ratio. $(A) \to (C)$: if (C) is not valid, then there is a sequence $c_n \to \infty$ with $\limsup_{n\to\infty} Q_n(L_n > c_n) > 0$. Put $A_n = \{L_n > c_n\}$. Then

$$\limsup_{n\to\infty} P_n(L_n > c_n) \leq \limsup_{n\to\infty} \frac{1}{c_n} \int L_n dP_n \leq \limsup_{n\to\infty} \frac{1}{c_n} = 0,$$

which contradicts (A). $(C) \to (D)$: it follows from (6.34), $\mathsf{D}^2(P_n, Q_n) \leq 2$, and the second condition in (B) that

$$\begin{aligned}&\limsup_{s\downarrow 0}\limsup_{n\to\infty}(1-\mathsf{H}_s(P_n,Q_n))\\&\le \limsup_{c\to\infty}\limsup_{s\downarrow 0}8\beta(2/(1+c))s(1-s)+\limsup_{c\to\infty}\limsup_{n\to\infty}Q_n(L_n>c)=0.\end{aligned}$$

$(D)\to(A)$: an application of Hölder's inequality yields, as in (6.35),

$$\mathsf{H}_s(P_n,Q_n)\le P_n^s(A_n)Q_n^{1-s}(A_n)+P_n^s(\overline{A}_n)Q_n^{1-s}(\overline{A}_n).$$

If (A) is not fulfilled, then there is a sequence A_{n_k} with $\lim_{k\to\infty}P_{n_k}(A_{n_k})=0$, where $\alpha=\lim_{k\to\infty}Q_{n_k}(A_{n_k})>0$. Then

$$\begin{aligned}&\liminf_{s\downarrow 0}\liminf_{n\to\infty}\mathsf{H}_s(P_n,Q_n) &(6.36)\\&\le \liminf_{s\downarrow 0}\limsup_{k\to\infty}[P_n^s(A_{n_k})Q_n^{1-s}(A_{n_k})+P_n^s(\overline{A}_{n_k})Q_n^{1-s}(\overline{A}_{n_k})]\\&=\liminf_{s\downarrow 0}\limsup_{k\to\infty}P_n^s(\overline{A}_{n_k})Q_{n_k}^{1-s}(\overline{A}_{n_k})=\liminf_{s\downarrow 0}(1-\alpha)^{1-s}=1-\alpha<1,\end{aligned}$$

which contradicts (D).

It remains to show that the conditions (A) through (D) are equivalent to $\{Q_n\}\lhd\{P_n\}$. Suppose m_n is a subsequence with $\mathcal{M}_{m_n}\Rightarrow\mathcal{M}=(\mathcal{X},\mathfrak{A},\{P,Q\})$. Then by Theorem 6.13 $\lim_{n\to\infty}\mathsf{H}_s(P_{m_n},Q_{m_n})=\mathsf{H}_s(P,Q)$. If (D) is satisfied, then

$$\lim_{s\downarrow 0}\mathsf{H}_s(P,Q)=\lim_{s\downarrow 0}\lim_{n\to\infty}\mathsf{H}_s(P_{m_n},Q_{m_n})\ge\lim_{s\downarrow 0}\liminf_{n\to\infty}\mathsf{H}_s(P_n,Q_n)=1,$$

which together with (6.28) gives $Q\ll P$ and thus $\{Q_n\}\lhd\{P_n\}$.

If (C) is not fulfilled, then there is a subsequence n_k and $c_k\to\infty$ such that $\alpha=\liminf_{k\to\infty}Q_{n_k}(L_{n_k}>c_k)>0$. Set $A_{n_k}=\{L_{n_k}>c_k\}$. Then

$$P_{n_k}(A_{n_k})=\int\frac{1}{L_{n_k}}I_{A_{n_k}}\,dQ_{n_k}\le\frac{1}{c_k}\to 0,$$

which, as in (6.36), implies

$$\limsup_{k\to\infty}\mathsf{H}_s(P_{n_k},Q_{n_k})\le(1-\alpha)^{1-s}.$$

If $\mathcal{M}_{n_{k_l}}\Rightarrow\mathcal{M}=(\mathcal{X},\mathfrak{A},\{P,Q\})$, then

$$\mathsf{H}_s(P,Q)=\lim_{l\to\infty}\mathsf{H}_s(P_{n_{k_l}},Q_{n_{kl}})\le(1-\alpha)^{1-s}$$

by Theorem 6.13. Hence $\lim_{s\downarrow 0}\mathsf{H}_s(P,Q)<1$ and we have an accumulation point $\mathcal{M}$ for which $Q\ll P$ is not satisfied, so that $\{Q_n\}\lhd\{P_n\}$ is not true.

To prove the corollary we remark that the assumption and Theorem 6.21 yield that $\mathcal{M}_n\Rightarrow\mathcal{M}=(\mathcal{X},\mathfrak{A},\{P,Q\})$, so that all accumulation points of $\mathcal{M}_n$ are identical with $\mathcal{M}$. The statement then follows directly by the definition of the contiguity. ∎

A simple sufficient condition for contiguity is given in the next problem.

Problem 6.28.* If $\mathsf{v} : (0,\infty) \to \mathbb{R}$ is a convex function with $\lim_{x\downarrow 0} \mathsf{v}(x) = \infty$, then

$$\sup_n \mathsf{I}_\mathsf{v}(P_n, Q_n) < \infty \quad \text{implies} \quad \{Q_n\} \lhd \{P_n\}.$$

In the sequel we often make use of the fact that stochastic convergence is preserved under switching from a sequence of distributions to a contiguous sequence. A precise formulation of this fact is as follows.

Problem 6.29.* Let $(\mathcal{X}_n, \mathfrak{A}_n, \{P_n, Q_n\})$, $n = 1, 2, ...$, be a sequence of binary models and $X_n : \mathcal{X}_n \to_m \mathbb{R}$, $n = 1, 2, ...$ If $\{Q_n\} \lhd \{P_n\}$, then $X_n \to^{P_n} 0$ implies that $X_n \to^{Q_n} 0$.

Problem 6.30.* If $T_n : \mathcal{X}_n \to_m \mathbb{R}^d$ satisfies $T_n = O_{P_n}(1)$ and $\{Q_n\} \lhd \{P_n\}$, then $T_n = O_{Q_n}(1)$.

Remark 6.31. The concept of contiguity was introduced by LeCam (1960); see LeCam and Yang (1990), p. 29. If $Q_n \ll P_n$, then condition (B) in Theorem 6.26 means that the likelihood ratios are uniformly integrable. Condition (D) is due to Jacod and Shiryaev (1987) and Liese (1986, 1987a,b) and has been used to study the contiguity of sequences of stochastic processes. A comprehensive discussion of topics related to the concept of contiguity can be found in Roussas (1972).

Now we consider situations where the log-likelihood is asymptotically normal.

Proposition 6.32. ***(First Lemma of LeCam)*** *For any sequence of models $\mathcal{M}_n = (\mathcal{X}_n, \mathfrak{A}_n, \{P_n, Q_n\})$ the following statements are equivalent.*

(A) $\mathcal{L}(\ln L_n | P_n) \Rightarrow \mathsf{N}(-\sigma^2/2, \sigma^2)$.

(B) $\mathcal{L}(\ln L_n | Q_n) \Rightarrow \mathsf{N}(\sigma^2/2, \sigma^2)$.

(C) $\mathcal{M}_n \Rightarrow (\mathbb{R}, \mathfrak{B}, \{\mathsf{N}(0,1), \mathsf{N}(\sigma,1)\})$.

Proof. We set $P = \mathsf{N}(0,1)$, $Q = \mathsf{N}(\sigma,1)$, and $L = dQ/dP$. Then $\ln L(x) = \sigma x - \sigma^2/2$. As $\ln x$ is a homeomorphism between $\overline{\mathbb{R}}$ and $\mathbb{R}_+$ the weak convergence of $\mathcal{L}(\ln L_n|Q_n)$ to $\mathcal{L}(\ln L|Q)$ is equivalent to the weak convergence of $\mathcal{L}(L_n|Q_n)$ to $\mathcal{L}(L|Q)$. To prove the equivalence of (A) and (C) we apply Theorem 6.21 and note that

$$\mathcal{L}(\ln L|P) = \mathsf{N}(-\sigma^2/2, \sigma^2).$$

The proof of the equivalence of (B) and (C) is similar if we exchange the roles of P_n and Q_n, note that $-\ln L_n$ is the log-likelihood of P_n with respect to Q_n, and do that analogously with P and Q. ■

The following situation is typically met in the area of testing hypotheses for large sample sizes. Given a sequence of models $\mathcal{M}_n$ and a sequence of statistics $S_n : \mathcal{X}_n \to_m \mathcal{S}$, where we know the limit distribution of the S_n under the null hypothesis, the question of what can be said about the distributions under the alternative is answered by the so-called third lemma of LeCam. We remark at this point that the notation of the first, second, and third lemma of LeCam is due to Hájek (1962).

Let $\mathcal{M}_n = (\mathcal{X}_n, \mathfrak{A}_n, \{P_n, Q_n\})$, $n = 1, 2, ...$, and $\mathcal{M} = (\mathcal{X}, \mathfrak{A}, \{P, Q\})$ be models, $\mathcal{S}$ a metric space with the σ-algebra of Borel sets $\mathfrak{B}$, and $S_n : \mathcal{X}_n \to_m \mathcal{S}$ and $S : \mathcal{X} \to_m \mathcal{S}$ statistics. With L_n and L from (6.26) and (6.25), respectively, we set

$$\begin{aligned} P^* &= \mathcal{L}((S, L)|P), & Q^* &= \mathcal{L}((S, L)|Q), \\ P_n^* &= \mathcal{L}((S_n, L_n)|P_n), & Q_n^* &= \mathcal{L}((S_n, L_n)|Q_n). \end{aligned} \tag{6.37}$$

If $Q \ll P$, then for any $h : \mathcal{S} \times \mathbb{R} \to_m \mathbb{R}_+$,

$$\begin{aligned} \int h(s,t) Q^*(ds, dt) &= \int h(S(x), L(x)) Q(dx) \\ &= \int h(S(x), L(x)) L(x) P(dx) = \int h(s,t) t P^*(ds, dt). \end{aligned}$$

Hence,

$$\frac{dQ^*}{dP^*}(s,t) = t, \quad P^*\text{-a.s.} \tag{6.38}$$

Theorem 6.33. *If P_n^* and Q_n^* are defined by (6.37), then $P_n^* \Rightarrow P^*$ and $Q \ll P$ imply $Q_n^* \Rightarrow Q^*$.*

Proof. As the weak convergence of distributions of random vectors implies the weak convergence of the marginal distributions we get $\mathcal{L}(L_n|P_n) \Rightarrow \mathcal{L}(L|P)$ and thus $\mathcal{M}_n \Rightarrow \mathcal{M}$ from Theorem 6.21. Then $\{Q_n\} \lhd \{P_n\}$ by Corollary 6.27, and the weight of the part of Q_n which is singular to P_n tends to zero. Hence we may assume that $Q_n \ll P_n$. Let $\varphi : \mathcal{S} \times \mathbb{R} \to \mathbb{R}_+$ be a continuous and bounded function, say $|\varphi| \leq b$. Similarly as in (6.38) it holds $Q_n^*(ds, dt) = t P_n^*(ds, dt)$. Set $f(s,t) = \varphi(s,t)t$. Then

$$\int I_{[c,\infty)}(|f(s,t)|) P_n^*(ds, dt) \leq \int I_{[c/b,\infty)}(t) P_n^*(ds, dt) \leq \int I_{[c/b,\infty)}(\frac{dQ_n}{dP_n}) dP_n.$$

Hence $\{Q_n\} \lhd \{P_n\}$ and condition (B) in Theorem 6.26 yield

$$\limsup_{c\to\infty} \limsup_{n\to\infty} \int I_{[c,\infty)}(|f(s,t)|) P_n^*(ds, dt) = 0.$$

Hence by Proposition A.44,

$$\lim_{n\to\infty} \int \varphi dQ_n^* = \lim_{n\to\infty} \int \varphi(s,t) t P_n^*(ds, dt) = \int \varphi(s,t) t P^*(ds, dt) = \int \varphi dQ^*,$$

which yields the statement. ■

We consider now the situation where the joint distribution of the log-likelihood $\ln L_n$ and a statistic S_n is asymptotically normal.

Proposition 6.34. ***(Third Lemma of LeCam for Binary Models)*** *Let $\mathcal{M}_n = (\mathcal{X}_n, \mathfrak{A}_n, \{P_n, Q_n\})$, $n = 1, 2, ...,$ be any sequence of models and $S_n : \mathcal{X}_n \to_m \mathbb{R}$. Then for any fixed $\mu \in \mathbb{R}$, $\sigma^2 \geq 0$, and $\tau \geq 0$,*

$$\mathcal{L}\left((S_n, \ln L_n)^T | P_n\right) \Rightarrow \mathsf{N}\left(\begin{pmatrix} \mu \\ -\sigma^2/2 \end{pmatrix}, \begin{pmatrix} \tau^2 & \rho\sigma\tau \\ \rho\sigma\tau & \sigma^2 \end{pmatrix}\right) \tag{6.39}$$

implies

$$\mathcal{L}\left((S_n, \ln L_n)^T | Q_n\right) \Rightarrow \mathsf{N}\left(\begin{pmatrix} \mu + \rho\sigma\tau \\ \sigma^2/2 \end{pmatrix}, \begin{pmatrix} \tau^2 & \rho\sigma\tau \\ \rho\sigma\tau & \sigma^2 \end{pmatrix}\right). \tag{6.40}$$

Proof. As $\ln x$ is a homeomorphism between $\overline{\mathbb{R}}_+$ and $\overline{\mathbb{R}}$ we see that under the assumptions of Theorem 6.33 the relation $\mathcal{L}((S_n, \ln L_n)|P_n) \Rightarrow \mathcal{L}((S, \ln L)|P)$ implies $\mathcal{L}((S_n, L_n)|Q_n) \Rightarrow \mathcal{L}((S, L)|Q)$. We introduce the model $\mathcal{M}$ and S such that $\mathcal{L}((S, \ln L)|P)$ becomes the normal distribution on the right-hand side of (6.39). Put

$$\begin{aligned} &\mathcal{M} = (\mathbb{R}_2, \mathfrak{B}_2, \{P, Q\}), \quad \text{where} \\ &P = \mathsf{N}(\mu, 1) \otimes \mathsf{N}(0, 1) \quad \text{and} \quad Q = \mathsf{N}(\mu, 1) \otimes \mathsf{N}(\sigma, 1). \end{aligned}$$

Denote by X and Y the projections $\mathbb{R}_2 \to \mathbb{R}$ onto the coordinates. Then

$$\ln L = \ln \frac{dQ}{dP} = \sigma Y - \frac{\sigma^2}{2}.$$

Define $S : \mathbb{R}_2 \to \mathbb{R}$ for $\sigma^2 = 0$ by $S = X$ and for $\sigma^2 > 0$ by

$$S = a(X - \mu) + \frac{\rho\tau}{\sigma}(\ln L + \frac{\sigma^2}{2}) + \mu, \quad \text{where} \quad a = \tau(1 - \rho^2)^{1/2}.$$

Then the vector $(S, \ln L)$ is a linear image of (X, Y) and has thus a normal distribution under both, P and Q. It holds

$$\begin{aligned} &\mathsf{E}_P \ln L = -\sigma^2/2, \quad \mathsf{E}_P S = \mu, \quad \mathsf{V}_P(S) = (1 - \rho^2)\tau^2 + (\frac{\rho\tau}{\sigma})^2\sigma^2 = \tau^2, \\ &\mathsf{V}_P(\ln L) = \sigma^2, \quad \mathsf{cov}_P(S, \ln L) = \frac{\rho\tau}{\sigma}\mathsf{V}_P(\ln L) = \rho\sigma\tau. \end{aligned}$$

Hence $\mathcal{L}((S, \ln L)|P)$ is the normal distribution on the right-hand side of (6.39). To apply Theorem 6.33 we have to calculate $\mathcal{L}((S, \ln L)|Q)$. As $(S, \ln L)$ is a linear image of (X, Y) the distribution $\mathcal{L}((S, \ln L)|Q)$ is normal, and it holds

$$\begin{aligned} &\mathsf{E}_Q \ln L = \sigma^2/2, \quad \mathsf{E}_Q S = \mu + \frac{\rho\tau}{\sigma}\sigma^2 = \mu + \rho\sigma\tau, \\ &\mathsf{V}_Q(S) = (1 - \rho^2)\tau^2 + (\frac{\rho\tau}{\sigma})^2\sigma^2 = \tau^2, \\ &\mathsf{V}_Q(\ln L) = \sigma^2, \quad \mathsf{cov}_Q(S, \ln L) = \frac{\rho\tau}{\sigma}\mathsf{V}_Q(\ln L) = \rho\sigma\tau. \end{aligned}$$

■

The third lemma of LeCam can be used to calculate the asymptotic power of tests under the alternative.

Example 6.35. Consider the sequence of testing problems $\mathsf{H}_0 : P_n$ versus $\mathsf{H}_A : Q_n$, $n = 1, 2, ...$ Let $\mu \in \mathbb{R}$, $\tau^2 > 0$, and $\alpha \in (0,1)$ be fixed. For a sequence of test statistics $S_n : \mathcal{X} \to_m \mathbb{R}$ that satisfies $\mathcal{L}(S_n|P_n) \Rightarrow \mathsf{N}(\mu, \tau^2)$ we set

$$\varphi_n = I_{[\mu+\tau u_{1-\alpha},\infty)}(S_n),$$

where $u_{1-\alpha} = \Phi^{-1}(1-\alpha)$. Then $\mathsf{E}_{P_n}\varphi_n \to \alpha$, so that the sequence is an asymptotic level α test. To study its power under the alternatives let us assume that (6.39) holds. Then by (6.40),

$$\begin{aligned} \mathcal{L}(S_n|Q_n) &\Rightarrow \mathsf{N}(\mu + \rho\sigma\tau, \tau^2), \\ \lim_{n\to\infty} \mathsf{E}_{Q_n}\varphi_n &= 1 - \Phi(u_{1-\alpha} - \rho\sigma). \end{aligned}$$

This means that the limiting power of the tests φ_n based on the test statistics S_n depends on the asymptotic correlation ρ, so that test statistics that have a high positive correlation with the log-likelihood provide a large asymptotic power. The maximum asymptotic power is attained with $S_n = \ln L_n$ because in this case $\rho = 1$. Later on in Chapter 8 we make systematic use of the third lemma of LeCam to find the asymptotic power of given tests, and to characterize best tests under local alternatives.

Above we studied sequences of models for which the limiting model $(\mathcal{X}, \mathfrak{A}, \{P, Q\})$ satisfies the condition $Q \ll P$. Now we consider the case of $P \perp Q$.

Definition 6.36. *For a sequence of binary models $\mathcal{M}_n = (\mathcal{X}_n, \mathfrak{A}_n, \{P_n, Q_n\})$, $n = 1, 2, ...$, the sequences $\{P_n\}$ and $\{Q_n\}$ are called entirely separated if there exists at least one accumulation point $\mathcal{M} = (\mathcal{X}, \mathfrak{A}, \{P, Q\})$ for which $P \perp Q$. In this case we write $\{P_n\} \vartriangle \{Q_n\}$.*

Entire separation is closely related to sequences of models for which sequences of tests exist with their error probabilities of the first and second kind tending to zero.

Theorem 6.37. *For a sequence of binary models $(\mathcal{X}_n, \mathfrak{A}_n, \{P_n, Q_n\})$, $n = 1, 2, ..$, the following statements are equivalent.*

(A) $\{P_n\} \vartriangle \{Q_n\}$.

(B) $\liminf_{n\to\infty} \mathsf{H}_s(P_n, Q_n) = 0, \quad 0 < s < 1.$

(C) $\liminf_{n\to\infty} \mathsf{b}_\pi(P_n, Q_n) = 0, \quad 0 < \pi < 1.$

Corollary 6.38. *Condition (B) is satisfied if and only if there exists an $s_0 \in (0,1)$ with $\liminf_{n\to\infty} \mathsf{H}_{s_0}(P_n, Q_n) = 0$. Condition (C) is satisfied if and only if there exists a $\pi_0 \in (0,1)$ with $\liminf_{n\to\infty} \mathsf{b}_{\pi_0}(P_n, Q_n) = 0$.*

Corollary 6.39. *$\{P_n\} \vartriangle \{Q_n\}$ holds if and only if there exists a sequence $A_{n_k} \in \mathfrak{A}_{n_k}$ with*

$$\lim_{k\to\infty} P_{n_k}(A_{n_k}) = 0 \quad \textit{and} \quad \lim_{k\to\infty} Q_{n_k}(A_{n_k}) = 1. \tag{6.41}$$

Proof. $(B) \to (C)$ follows from inequality (2.44). As $\mathsf{b}_\pi(P_n, Q_n) \le \pi \wedge (1-\pi)$ we may carry out the limit in the integral representation of $\mathsf{H}_s(P_n, Q_n)$ in Corollary 1.69 and obtain that (C) implies (B). The equivalence of (A) and (B) follows from the relative compactness of the sequence $\mathcal{M}_n$ and Theorem 6.21. The sufficiency of the first condition of Corollary 6.38 follows from inequality (2.44) for $s = s_0$ and the equivalence of (C) and (B). The sufficiency of the second condition can be obtained from the inequalities

$$\mathsf{b}_\pi(P, Q) \le 2\mathsf{b}_{1/2}(P_n, Q_n), \quad 2(\pi \wedge (1-\pi))\mathsf{b}_{1/2}(P_n, Q_n) \le \mathsf{b}_\pi(P, Q).$$

If (6.41) holds, then the tests $\varphi_{n_k} = I_{A_{n_k}}$ satisfy $\mathsf{b}_\pi(P_{n_k}, Q_{n_k}) \le \pi P_{n_k}(A_{n_k}) + (1-\pi)Q_{n_k}(\overline{A}_{n_k}) \to 0$. Conversely, if $\mathsf{b}_\pi(P_{n_k}, Q_{n_k}) \to 0$, then let ψ_{n_k} be a nonrandomized Bayes test. Putting $A_{n_k} = \{\psi_{n_k} = 1\}$ we get (6.41). ■

Subsequently we need an elementary inequality for the Hellinger distance.

Problem 6.40.* It holds

$$\mathsf{D}^2(P, Q) \le 2\mathsf{D}^2(P, \frac{1}{2}(P+Q)) + 2\mathsf{D}^2(Q, \frac{1}{2}(P+Q)) \le 2\mathsf{D}^2(P, Q).$$

In preparation for the next theorem we collect some properties of Hellinger integrals of product measures. We set $\mathsf{G}_s(P, Q) = -\ln \mathsf{H}_s(P, Q)$, where $\mathsf{G}_s(P, Q) := \infty$ if $\mathsf{H}_s(P, Q) = 0$. Then by (1.109),

$$\begin{aligned} &0 \le \mathsf{G}_s(P, Q) \le \infty, \\ &\tfrac{1}{1-s_1}\mathsf{G}_{s_1}(P, Q) \le \tfrac{1}{1-s_2}\mathsf{G}_{s_2}(P, Q), \quad 0 < s_1 < s_2 < 1, \\ &\mathsf{G}_s(\textstyle\bigotimes_{i=1}^n P_i, \bigotimes_{i=1}^n Q_i) = \sum_{i=1}^n \mathsf{G}_s(P_i, Q_i). \end{aligned} \tag{6.42}$$

Lemma 6.41. *It holds for* $0 < s < 1$,

$$\begin{aligned} &\textstyle\sum_{i=1}^n (1 - \mathsf{H}_s(P_i, Q_i)) \le \sum_{i=1}^n \mathsf{G}_s(P_i, Q_i) \\ &\le \tfrac{3}{2}(\min_{1\le i\le n} \mathsf{H}_s(P_i, Q_i))^{-2} \textstyle\sum_{i=1}^n (1 - \mathsf{H}_s(P_i, Q_i)), \end{aligned} \tag{6.43}$$

and

$$\begin{aligned} &1 - \exp\{-3[2 - \max_{1\le i\le n} \mathsf{D}^2(P_i, Q_i)]^{-2} \textstyle\sum_{i=1}^n \mathsf{D}^2(P_i, Q_i)\} \\ &\le \tfrac{1}{2}\mathsf{D}^2(\textstyle\bigotimes_{i=1}^n P_i, \bigotimes_{i=1}^n Q_i) \le 1 - \exp\{-\tfrac{1}{2}\sum_{i=1}^n \mathsf{D}^2(P_i, Q_i)\}. \end{aligned} \tag{6.44}$$

Proof. The Taylor expansion up to the second order gives for $0 < x < 1$,

$$x \le x + \frac{1}{2}x^2 \le -\ln(1-x) \le x + \frac{x^2}{2(1-x)^2} \le \frac{3}{2}\frac{x}{(1-x)^2}.$$

Putting $x = 1 - \mathsf{H}_s(P_i, Q_i)$ and taking the sum we get (6.43) from the third statement in (6.42). Furthermore, with $x = 1 - \mathsf{H}_{1/2}(P_i, Q_i) = \frac{1}{2}\mathsf{D}^2(P_i, Q_i)$,

$$\exp\{-\sum_{i=1}^{n} \mathsf{G}_{1/2}(P_i, Q_i)\} = \prod_{i=1}^{n} \mathsf{H}_{1/2}(P_i, Q_i) = 1 - \frac{1}{2}\mathsf{D}^2(\bigotimes_{i=1}^n P_i, \bigotimes_{i=1}^n Q_i),$$

$$\begin{aligned} \mathsf{D}^2(\bigotimes_{i=1}^n P_i, \bigotimes_{i=1}^n Q_i) &= 2[1 - \exp\{-\textstyle\sum_{i=1}^n \mathsf{G}_{1/2}(P_i, Q_i)\}] \\ &\le 2[1 - \exp\{-\frac{1}{2}\textstyle\sum_{i=1}^n \mathsf{D}^2(P_i, Q_i)\}], \end{aligned}$$

$$\mathsf{D}^2(\bigotimes_{i=1}^n P_i, \bigotimes_{i=1}^n Q_i) \ge 2[1 - \exp\{-\frac{3}{2}\textstyle\sum_{i=1}^n \frac{\mathsf{D}^2(P_i,Q_i)/2}{(1-\mathsf{D}^2(P_i,Q_i)/2)^2}\}].$$

■

Now we look at sequences of statistical models that correspond to independent observations and establish conditions for the contiguity and the entire separation. For $n = 1, 2, ...$ let

$$\begin{aligned} &\mathcal{M}_n = (\mathsf{X}_{i=1}^n \mathcal{X}_{n,i}, \bigotimes_{i=1}^n \mathfrak{A}_{n,i}, \{P_n, Q_n\}), \\ &P_n = \bigotimes_{i=1}^n P_{n,i}, \quad Q_n = \bigotimes_{i=1}^n Q_{n,i}, \quad \text{and} \quad R_n = \bigotimes_{i=1}^n R_{n,i}, \\ &R_{n,i} = \tfrac{1}{2}(P_{n,i} + Q_{n,i}), \\ &M_{n,i} = \frac{dP_{n,i}}{dR_{n,i}} \quad \text{and} \quad L_{n,i} = \frac{2-M_{n,i}}{M_{n,i}} I_{(0,\infty)}(M_{n,i}) + \infty I_{\{0\}}(M_{n,i}), \end{aligned} \tag{6.45}$$

where $M_{n,i}$ and $L_{n,i}$ are considered to be random variables on $\mathsf{X}_{i=1}^n \mathcal{X}_{n,i}$.

Theorem 6.42. *For the models $\mathcal{M}_n$ in (6.45) the following statements are equivalent.*

(A) $\{\bigotimes_{i=1}^n Q_{n,i}\} \lhd \{\bigotimes_{i=1}^n P_{n,i}\}$.

(B) $\lim_{s\downarrow 0} \limsup_{n\to\infty} \sum_{i=1}^n \mathsf{G}_s(P_{n,i}, Q_{n,i}) = 0$.

(C) $\lim_{s\downarrow 0} \limsup_{n\to\infty} \sum_{i=1}^n (1 - \mathsf{H}_s(P_{n,i}, Q_{n,i})) = 0$.

(D) $\begin{cases} (D1) \ \limsup_{n\to\infty} \sum_{i=1}^n \mathsf{D}^2(P_{n,i}, Q_{n,i}) < \infty, \ \textit{and} \\ (D2) \ \lim_{c\to\infty} \limsup_{n\to\infty} \sum_{i=1}^n Q_n(L_{n,i} > c) = 0. \end{cases}$

Proof. The equivalence of (A) and (B) follows from Theorem 6.26. The equivalence of (B) and (C) follows from (6.43). Hence (A), (B), and (C) are equivalent. If condition (C) is satisfied, then the inequality (6.29) and

$$1 - \mathsf{H}_s(P_{n,i}, Q_{n,i}) = \int \mathsf{u}_s(M_{n,i}) dR_n$$

provide $(D1)$. The inequality (6.33) gives

$$\sum_{i=1}^n (1 - \mathsf{H}_s(P_{n,i}, Q_{n,i})) \ge [(1-s) - s/c - c^{-s}] \sum_{i=1}^n Q_n(L_{n,i} \ge c).$$

Taking first $n \to \infty$, then $c \to \infty$, and finally $s \downarrow 0$, we get $(D2)$. Similarly, if (D) is satisfied, then

$$\begin{aligned}&\sum\nolimits_{i=1}^{n}(1-\mathsf{H}_s(P_{n,i},Q_{n,i}))\\ &\le \sum\nolimits_{i=1}^{n} 4\beta(2/(1+c))s(1-s)\mathsf{D}^2(P_{n,i},Q_{n,i})+Q_n(L_{n,i}>c)\end{aligned}$$

by (6.33). Taking first $n \to \infty$, then $c \to \infty$, and finally $s \downarrow 0$, we get (C). ■

Remark 6.43. The equivalence of (A) and (D) in Theorem 6.42 is a well-known result by Oosterhoff and van Zwet (1979). Our proof is taken from Liese and Vajda (1987).

We apply Theorem 6.42 to the special case where the double array of distributions in (6.45) is in fact a sequence. Then the question of contiguity reduces to that of the absolute continuity of the corresponding product measures. The next theorem is due to Kakutani (1948). The subsequent proof is taken from Kühn and Liese (1978).

Theorem 6.44. *It holds $\bigotimes_{i=1}^{\infty} Q_i \ll \bigotimes_{i=1}^{\infty} P_i$ if and only if*

$$Q_i \ll P_i, \quad i=1,2,\ldots \quad \textit{and} \tag{6.46}$$

$$\sum\nolimits_{i=1}^{\infty} \mathsf{D}^2(P_i,Q_i)<\infty. \tag{6.47}$$

Under the condition (6.46) $\bigotimes_{i=1}^{\infty} Q_i \ll \bigotimes_{i=1}^{\infty} P_i$ or $\bigotimes_{i=1}^{\infty} Q_i \perp \bigotimes_{i=1}^{\infty} P_i$ holds, depending on whether (6.47) is satisfied.

Proof. The necessity of (6.46) is clear, and the necessity of (6.47) follows from condition $(D1)$ in Theorem 6.42. Conversely, (1.117) and the inequality (6.42) imply for $0<s<1/2$ that

$$\begin{aligned}&-\ln \mathsf{H}_s(\textstyle\bigotimes_{i=1}^{\infty} P_i,\bigotimes_{i=1}^{\infty} Q_i)=\sum_{i=1}^{\infty}\mathsf{G}_s(P_i,Q_i)\\ &\le \sum\nolimits_{i=1}^{N}\mathsf{G}_s(P_i,Q_i)+2(1-s)\sum\nolimits_{i=N+1}^{\infty}\mathsf{G}_{1/2}(P_i,Q_i).\end{aligned}$$

If (6.46) is fulfilled, then by (6.28) and $\mathsf{G}_s(P_i,Q_i)=-\ln(1-\mathsf{H}_s(P_i,Q_i))$,

$$\lim_{s\downarrow 0}(-\ln \mathsf{H}_s(\textstyle\bigotimes_{i=1}^{\infty} P_i,\bigotimes_{i=1}^{\infty} Q_i))\le 2\sum_{i=N+1}^{\infty}\mathsf{G}_{1/2}(P_i,Q_i). \tag{6.48}$$

That inequality $\mathsf{H}_{1/2}(P_i,Q_i)>0$ holds for every i follows from (6.46), because $\mathsf{H}_{1/2}(P_i,Q_i)=0$ would imply $P_i \perp Q_i$. If in addition (6.47) holds, then $\inf_{1\le i<\infty}\mathsf{H}_{1/2}(P_i,Q_i)>0$ by

$$\sum\nolimits_{i=1}^{\infty}\mathsf{D}^2(P_i,Q_i)=\sum\nolimits_{i=1}^{\infty}2(1-\mathsf{H}_{1/2}(P_i,Q_i))<\infty.$$

Hence $\sum_{i=1}^{\infty}\mathsf{G}_{1/2}(P_i,Q_i)<\infty$ by inequality (6.43). To complete the proof we let $N\to\infty$ in (6.48) and apply (6.28) to $P=\bigotimes_{i=1}^{\infty} P_i$ and $Q=\bigotimes_{i=1}^{\infty} Q_i$.

To prove the second statement it is enough to note that by $\ln(1-x)\le x$,

$$\mathsf{H}_{1/2}(\bigotimes_{i=1}^{\infty} P_i, \bigotimes_{i=1}^{\infty} Q_i) = \prod_{i=1}^{\infty} \mathsf{H}_{1/2}(P_i, Q_i)$$
$$\leq \exp\{-\sum\nolimits_{i=1}^{\infty}(1 - \mathsf{H}_{1/2}(P_i, Q_i))\} = \exp\{-\frac{1}{2}\sum\nolimits_{i=1}^{\infty} \mathsf{D}^2(P_i, Q_i)\},$$

so that $\sum_{i=1}^{\infty} \mathsf{D}^2(P_i, Q_i) = \infty$ implies $\bigotimes_{i=1}^{\infty} P_i \perp \bigotimes_{i=1}^{\infty} Q_i$. ■

The second statement of Theorem 6.44 is referred to as a *dichotomy* as only two extreme situations are possible.

Example 6.45. For normal distributions the Hellinger integral can be explicitly calculated. Indeed, (1.79) yields

$$\mathsf{H}_{1/2}(\mathsf{N}(\mu_1, \sigma^2), \mathsf{N}(\mu_2, \sigma^2)) = \exp\{-\frac{1}{8}\frac{(\mu_1 - \mu_2)^2}{\sigma^2}\},$$
$$\mathsf{D}^2(\mathsf{N}(\mu_1, \sigma^2), \mathsf{N}(\mu_2, \sigma^2)) = 2(1 - \mathsf{H}_{1/2}(\mathsf{N}(\mu_1, \sigma^2), \mathsf{N}(\mu_2, \sigma^2))).$$

Using the inequality $x - x^2/2 \leq 1 - \exp\{-x\} \leq x$, $x \geq 0$ it follows that

$$\sum\nolimits_{i=1}^{\infty}(1 - \mathsf{D}^2(\mathsf{N}(\mu_i, \sigma^2), \mathsf{N}(0, \sigma^2))) < \infty \quad \Leftrightarrow \quad \sum\nolimits_{i=1}^{\infty} \mu_i^2 < \infty.$$

Hence by Theorem 6.44 it holds

$$\bigotimes_{i=1}^{\infty} \mathsf{N}(\mu_i, 1) \ll\gg \bigotimes_{i=1}^{\infty} \mathsf{N}(0, 1) \quad \text{or} \quad \bigotimes_{i=1}^{\infty} \mathsf{N}(\mu_i, 1) \perp \bigotimes_{i=1}^{\infty} \mathsf{N}(0, 1),$$

depending on whether $\sum_{i=1}^{\infty} \mu_i^2$ is finite or infinite. A more general situation for exponential families is studied in Problem 6.54.

In order to establish limit theorems for the log-likelihood of independent observations we use concepts from the classical field of limit theorems of sums of independent random variables. We recall the notations in (6.45) and set $Y_{n,i} = L_{n,i}^{1/2} - 1$. The first two moments of $Y_{n,i}$ are closely related to the Hellinger distance, and also to each other. Indeed, the following statements are direct consequences of (1.110).

$$\begin{aligned}
&\mathsf{H}_{1/2}(P_{n,i}, Q_{n,i}) = \mathsf{E}_{P_n} L_{n,i}^{1/2}, \quad \mathsf{D}^2(P_{n,i}, Q_{n,i}) = 2[1 - \mathsf{H}_{1/2}(P_{n,i}, Q_{n,i})],\\
&\mathsf{E}_{P_n} Y_{n,i} = -\tfrac{1}{2}\mathsf{D}^2(P_{n,i}, Q_{n,i}), \quad \mathsf{E}_{P_n} Y_{n,i}^2 + Q_n(Y_{n,i} = \infty) = \mathsf{D}^2(P_{n,i}, Q_{n,i}),\\
&\mathsf{V}_{P_n}(Y_{n,i}) = \mathsf{D}^2(P_{n,i}, Q_{n,i}) - \tfrac{1}{4}(\mathsf{D}^2(P_{n,i}, Q_{n,i}))^2 - Q_n(Y_{n,i} = \infty).
\end{aligned} \tag{6.49}$$

Definition 6.46. *A double array of models* $(\mathcal{X}_{n,i}, \mathfrak{A}_{n,i}, \{P_{n,i}, Q_{n,i}\})$, $i = 1, ..., n$, $n = 1, 2, ...$, *is called bounded if*

$$\limsup_{n\to\infty} \mathsf{D}^2(\bigotimes_{i=1}^{n} P_{n,i}, \bigotimes_{i=1}^{n} Q_{n,i}) < 2. \tag{6.50}$$

It is called infinitesimal if

$$\lim_{n\to\infty} \max_{1\leq i\leq n} \mathsf{D}^2(P_{n,i}, Q_{n,i}) = 0. \tag{6.51}$$

It is called to satisfy the Lindeberg condition if for every $\varepsilon > 0$,

$$\lim_{n\to\infty} \sum\nolimits_{i=1}^{n}(\mathsf{E}_{P_{n,i}} I_{(\varepsilon,\infty)}(|Y_{n,i}|)Y_{n,i}^2 + Q_n(Y_{n,i} = \infty)) = 0. \tag{6.52}$$

Condition (6.50) means that the models $\mathcal{M}_n$ in (6.45) stay away from each totally informative model $\mathcal{M} = (\mathcal{X}, \mathfrak{A}, \{P, Q\})$ which is characterized by $\mathsf{D}^2(P,Q) = 2$, or equivalently by $P \perp Q$. Condition (6.51) says that the individual models $(\mathcal{X}_{n,i}, \mathfrak{A}_{n,i}, \{P_{n,i}, Q_{n,i}\})$ contribute very little information to $\mathcal{M}_n$ as the Hellinger distance $\mathsf{D}^2(P_{n,i}, Q_{n,i})$ is, uniformly in $i = 1, ..., n$, small for large n.

Proposition 6.47. *Condition (6.52) implies (6.51). If (6.51) holds, then (6.50) is satisfied if and only if*

$$\limsup_{n\to\infty} \sum\nolimits_{i=1}^{n} \mathsf{D}^2(P_{n,i}, Q_{n,i}) < \infty. \tag{6.53}$$

Proof. The relation (6.52) yields

$$\begin{aligned}
&\max_{1\le i\le n} \mathsf{D}^2(P_{n,i}, Q_{n,i}) = \max_{1\le i\le n} [\mathsf{E}_{P_n} Y_{n,i}^2 + Q_n(Y_{n,i} = \infty)] \\
&\le \varepsilon^2 + \max_{1\le i\le n} [\mathsf{E}_{P_n} I_{(\varepsilon,\infty)}(|Y_{n,i}|) Y_{n,i}^2 + Q_n(Y_{n,i} = \infty)] \\
&\le \varepsilon^2 + \sum\nolimits_{i=1}^{n} [\mathsf{E}_{P_n} I_{(\varepsilon,\infty)}(|Y_{n,i}|) Y_{n,i}^2 + Q_n(Y_{n,i} = \infty)].
\end{aligned}$$

Taking first $n \to \infty$, and then $\varepsilon \to 0$, we get the first statement. The second statement follows from inequality (6.44). ■

The Lindeberg condition (6.52) implies that the likelihood ratios $L_{n,i}$ are uniformly close to 1, so that $\ln L_{n,i}$ is uniformly small. The Problems 6.55 and 6.56 formulate other conditions to express this property in terms of the $Y_{n,i}$.

Lemma 6.48. *If the conditions (6.53) and (6.52) are satisfied and* $Y_{n,i} = L_{n,i}^{1/2} - 1$, *then*

$$\sum\nolimits_{i=1}^{n} Y_{n,i}^2 - \sum\nolimits_{i=1}^{n} \mathsf{D}^2(P_{n,i}, Q_{n,i}) = o_{P_n}(1). \tag{6.54}$$

Proof. We set $C_{n,i} = \{|Y_{n,i}| \le \frac{1}{2}\}$. Then

$$\begin{aligned}
&\sum\nolimits_{i=1}^{n} (Y_{n,i}^2 - \mathsf{E}_{P_n} Y_{n,i}^2) = \sum\nolimits_{i=1}^{n} W_{n,i} + S_n, \quad \text{where} \\
&W_{n,i} = I_{C_{n,i}} Y_{n,i}^2 - \mathsf{E}_{P_n} I_{C_{n,i}} Y_{n,i}^2 \quad \text{and} \quad \mathsf{E}_{P_n}|S_n| \le 2\sum\nolimits_{i=1}^{n} \mathsf{E}_{P_n} I_{\overline{C}_{n,i}} Y_{n,i}^2.
\end{aligned}$$

The Lindeberg condition yields $\mathsf{E}_{P_n}|S_n| \to 0$. The independence of the $W_{n,i}$ and $\mathsf{E}_{P_n} W_{n,i} = 0$ give

$$\begin{aligned}
&\mathsf{E}_{P_n}(\sum\nolimits_{i=1}^{n} W_{n,i})^2 = \sum\nolimits_{i=1}^{n} \mathsf{E}_{P_n} W_{n,i}^2 \\
&\le \sum\nolimits_{i=1}^{n} \mathsf{E}_{P_n} I_{C_{n,i}} Y_{n,i}^4 \le \sum\nolimits_{i=1}^{n} \mathsf{E}_{P_n}(Y_{n,i}^2 \wedge |Y_{n,i}|^3) \to 0,
\end{aligned}$$

where the last statement follows from Problem 6.55. To complete the proof we use (6.49). ■

Now we prove the asymptotic normality of a double array of models that correspond to a double array of independent observations. More precisely, we consider the models $\mathcal{M}_n$ in (6.45). Let L_n be the likelihood ratio of $Q_n = \bigotimes_{i=1}^n Q_{n,i}$ with respect to $P_n = \bigotimes_{i=1}^n P_{n,i}$. With the likelihood ratios $L_{n,i}$ of $Q_{n,i}$ with respect to $P_{n,i}$ that have been introduced in (6.45) we have

$$L_n = \prod\nolimits_{i=1}^n L_{n,i}, \quad P_n\text{-a.s.},$$

where we use the convention $0 \cdot \infty = 0$. The next theorem is similar to Theorem 6.3 in Janssen, Milbrodt, and Strasser (1985), and closely related to Proposition 3 in LeCam and Yang (1990).

Theorem 6.49. *Suppose that the sequence of binary models $\mathcal{M}_n$ in (6.45) satisfies (6.52). If*

$$\lim_{n\to\infty} \sum\nolimits_{i=1}^n \mathsf{D}^2(P_{n,i}, Q_{n,i}) = \sigma^2/4 > 0, \tag{6.55}$$

then

$$\begin{aligned} \ln L_n &= 2\sum\nolimits_{i=1}^n Y_{n,i} - \sum\nolimits_{i=1}^n Y_{n,i}^2 + o_{P_n}(1) && (6.56)\\ &= 2\sum\nolimits_{i=1}^n Y_{n,i} - \sigma^2/4 + o_{P_n}(1), && (6.57)\end{aligned}$$

and

$$\mathcal{M}_n \Rightarrow \mathcal{M} = (\mathbb{R}, \mathfrak{B}, \{\mathsf{N}(0,1), \mathsf{N}(\sigma,1)\}). \tag{6.58}$$

Corollary 6.50. *Under the assumptions of the above theorem it holds that $\{\bigotimes_{i=1}^n P_{n,i}\} \lhd\rhd \{\bigotimes_{i=1}^n Q_{n,i}\}$.*

Proof. The Taylor expansion of $\ln(1+x)$ up to the third order gives

$$|\ln(1+x) - x + \frac{1}{2}x^2| \le \frac{1}{3}\frac{|x|^3}{(1+x)^3} \le \frac{8}{3}(|x|^3 \wedge x^2), \quad |x| \le \frac{1}{2}.$$

To establish the expansion of $\ln L_n$ we note that, by the definition of $Y_{n,i}$,

$$\ln L_n = \sum\nolimits_{i=1}^n \ln L_{n,i} = \sum\nolimits_{i=1}^n 2\ln(1+Y_{n,i}), \quad P_n\text{-a.s.}$$

Set $A_n = \{\max_{1\le i\le n} |Y_{n,i}| \le 1/2\}$. As the double array $\{Y_{n,i}\}$ satisfies the Lindeberg condition (6.52) it follows from (6.60) that $P_n(A_n) \to 1$. Hence

$$\begin{aligned} \ln L_n &= \sum\nolimits_{i=1}^n (2Y_{n,i} - Y_{n,i}^2) + \sum\nolimits_{i=1}^n [2\ln(1+Y_{n,i}) - (2Y_{n,i} - Y_{n,i}^2)]I_{A_n} \\ &\quad + \sum\nolimits_{i=1}^n [2\ln(1+Y_{n,i}) - (2Y_{n,i} - Y_{n,i}^2)]I_{\overline{A}_n} \\ &= \sum\nolimits_{i=1}^n (2Y_{n,i} - Y_{n,i}^2) + T_n + o_{P_n}(1),\end{aligned}$$

where we used $Z_n I_{\overline{A}_n} = o_{P_n}(1)$ for any sequence Z_n. As

$$\mathsf{E}_{P_n}|T_n| \leq \frac{16}{3}\sum\nolimits_{i=1}^{n} \mathsf{E}_{P_n}(|Y_{n,i}|^3 \wedge Y_{n,i}^2) \to 0$$

by Problem 6.55 the proof of (6.56) is completed. The statement (6.57) follows from (6.54). Set $\widetilde{X}_{n,i} = Y_{n,i} - \mathsf{E}_{P_n} Y_{n,i}$. Then by (6.49)

$$\sum\nolimits_{i=1}^{n} \mathsf{E}_{P_n}(Y_{n,i} - \widetilde{X}_{n,i})^2 = \sum\nolimits_{i=1}^{n} (\mathsf{E}_{P_n} Y_{n,i})^2 = \frac{1}{4}\sum\nolimits_{i=1}^{n} (\mathsf{D}^2(P_{n,i}, Q_{n,i}))^2.$$

As $\max_{1 \leq i \leq n} \mathsf{D}^2(P_{n,i}, Q_{n,i}) \to 0$ by Proposition 6.47, and $\sum_{i=1}^{n} \mathsf{D}^2(P_{n,i}, Q_{n,i})$ is bounded by assumption (6.55), we get from Problem 6.59 that the double array $\widetilde{X}_{n,i}$ satisfies the Lindeberg condition (6.52). We know from Problem 6.57 and (6.55) that $\lim_{n\to\infty} \sigma_n^2 = \lim_{n\to\infty} \sum_{i=1}^{n} \mathsf{D}^2(P_{n,i}, Q_{n,i}) = \sigma^2/4$. Hence the double array $X_{n,i} = (1/\sigma_n)\widetilde{X}_{n,i}$ satisfies the Lindeberg condition in Theorem A.54, which implies that

$$\mathcal{L}(\sum\nolimits_{i=1}^{n} X_{n,i}|P_n) \Rightarrow \mathsf{N}(0,1),$$

$$\lim_{n\to\infty} \mu_n = -\frac{\sigma^2}{8}, \quad \lim_{n\to\infty} \sigma_n = \frac{\sigma}{2}, \quad \text{and}$$

$$\mathcal{L}(\sum\nolimits_{i=1}^{n} 2Y_{n,i} - \frac{\sigma^2}{4}|P_n) \Rightarrow \mathsf{N}(-\frac{1}{2}\sigma^2, \sigma^2),$$

in view of Problem 6.57. Hence (6.58) follows from (6.57) and Proposition 6.32. ■

Remark 6.51. Theorem 6.49 can be found, for example, in LeCam and Yang (1990), Witting and Müller-Funk (1995), Rieder (1994), and Bickel, Klaassen, Ritov, and Wellner (1993). A general theory for double arrays of binary models can be found in Janssen, Milbrodt, and Strasser (1985) and LeCam and Yang (1990). In both books a general theory of infinitely divisible models is created. The possible limit models are then not necessarily Gaussian models.

Subsequently we collect additional and some technical problems.

Problem 6.52.* If the condition $(D1)$ in Theorem 6.42 is not satisfied, then $\{\bigotimes_{i=1}^{n} Q_{n,i}\} \vartriangle \{\bigotimes_{i=1}^{n} P_{n,i}\}$.

Problem 6.53. Let $\mathsf{K}(Q,P)$ be the Kullback–Leibler distance introduced in (1.81); that is, $\mathsf{K}(Q,P) = \int (\ln(dQ/dP))dQ$ if $Q \ll P$, and $\mathsf{K}(Q,P) = \infty$ otherwise. Then $\limsup_{n\to\infty} \sum_{i=1}^{n} \mathsf{K}(Q_{n,i}, P_{n,i}) < \infty$ implies $\{\bigotimes_{i=1}^{n} Q_{n,i}\} \vartriangleleft \{\bigotimes_{i=1}^{n} P_{n,i}\}$.

Problem 6.54. Let $(P_\theta)_{\theta\in\Delta}$, $\Delta \subseteq \mathbb{R}$, be a one-parameter exponential family, and $\theta_0 \in \Delta^0$ be fixed. By Example 1.88

$$\mathsf{H}_s(P_{\theta_1}, P_{\theta_2}) = \exp\{K(s\theta_1 + (1-s)\theta_2) - sK(\theta_1) - (1-s)K(\theta_2)\}.$$

If $C := \limsup_{n\to\infty} \max_{1\leq i\leq n} |c_{n,i}| < \infty$ and $[\theta_0 - C, \theta_0 + C] \subseteq \Delta^0$, then it holds either $\{\bigotimes_{i=1}^{n} P_{\theta_0 + c_{n,i}}\} \vartriangleleft \{P_{\theta_0}^{\otimes n}\}$ or $\{\bigotimes_{i=1}^{n} P_{\theta_0 + c_{n,i}}\} \vartriangle \{P_{\theta_0}^{\otimes n}\}$, depending on whether $\limsup_{n\to\infty} \sum_{i=1}^{n} c_{n,i}^2 < \infty$ or $\limsup_{n\to\infty} \sum_{i=1}^{n} c_{n,i}^2 = \infty$.

Problem 6.55.* If the condition (6.53) holds, then (6.52) is equivalent to

$$\lim_{n\to\infty} \sum\nolimits_{i=1}^{n} [\mathsf{E}_{P_n}(Y_{n,i}^2 \wedge |Y_{n,i}|^3) + Q_n(Y_{n,i} = \infty)] = 0.$$

Problem 6.56.* The condition (6.52) implies

$$\limsup_{n\to\infty} \max_{1\le i\le n} \mathsf{E}_{P_n} Y_{n,i}^2 = 0 \tag{6.59}$$

as well as

$$\lim_{n\to\infty} P_n(\max_{1\le i\le n} |Y_{n,i}| > \varepsilon) = 0, \quad \varepsilon > 0. \tag{6.60}$$

Problem 6.57.* If the conditions (6.53) and (6.52) are satisfied, then $\mu_n := \sum_{i=1}^n \mathsf{E}_{P_n} Y_{n,i}$ and $\sigma_n^2 := \sum_{i=1}^n \mathsf{V}_{P_n}(Y_{n,i})$ satisfy for $n \to \infty$,

$$\mu_n = \sum\nolimits_{i=1}^{n} (-\frac{1}{2}\mathsf{D}^2(P_{n,i}, Q_{n,i})) \quad \text{and} \quad \sigma_n^2 - \sum\nolimits_{i=1}^{n} \mathsf{D}^2(P_{n,i}, Q_{n,i}) \to 0.$$

Problem 6.58. *The Lindeberg condition (6.52) holds if and only if for every* $\varepsilon > 0$, $\lim_{n\to\infty} \sum_{i=1}^n \int g_\varepsilon(M_{n,i})dR_n = 0$, where $g_\varepsilon(x) = I_{(\varepsilon,\infty)}(|x-1|)(\sqrt{2-x} - \sqrt{x})^2$.

Problem 6.59.* Suppose $Y_{n,i}$ and $Z_{n,i}$ are random variables. Then the following holds. $\lim_{n\to\infty} \sum_{i=1}^n \mathsf{E}_{\theta_0}(Y_{n,i} - Z_{n,i})^2 = 0$ and $\lim_{n\to\infty} \sum_{i=1}^n \mathbb{E}Z_{n,i}^2 I_{(\varepsilon,\infty)}(|Z_{n,i}|) = 0$ for every $\varepsilon > 0$ implies

$$\lim_{n\to\infty} \sum\nolimits_{i=1}^{n} \mathbb{E}Y_{n,i}^2 I_{(\varepsilon,\infty)}(|Y_{n,i}|) = 0 \quad \text{for every } \varepsilon > 0.$$

6.4 Asymptotically Normal Models

In this section we study the convergence to a Gaussian shift model and establish conditions under which this convergence holds. In particular we investigate models that are associated with independent observations and a localized parameter. This leads to the concept of local asymptotic normality of models which has been introduced by LeCam (1960). It isolates the central idea of proving optimality of sequences of decisions that were obtained in Wald (1943) and LeCam (1953) by exponential approximation. The concept of convergence of models allows us to break up the proof of asymptotic optimality of decisions into three subproblems which can be treated separately. The first is to establish the convergence of the models. The second is to find optimal solutions in the limit model. Once optimal decisions have been found in the limit model, the third is to determine whether these decisions, applied to the sequence of models, provide asymptotically optimal solutions of the decision problem under consideration. To establish a relation between the members in sequences of decisions for the sequence of models it is not enough to have convergence of the models. Moreover, one has to relate the risks of the sequence of decisions to the risk of the optimal decision in the limit model. This is the topic of the next section. Here in this section we prove the convergence to a Gaussian model, a property that is called the *asymptotic normality of a sequence of models.* This concept, with all of its consequences, is of similar fundamental importance as the central limit theorem is to probability theory and classical areas of mathematical statistics. The reason is that for Gaussian shift models explicit optimal solutions for large classes of decisions are known. This covers estimation problems under a wide class of loss functions, one- and two-sided testing problems, and selection problems. The results of this section have been taken from Strasser (1985), LeCam and Yang (1990), Rieder (1994), and Witting and Müller-Funk (1995).

6.4.1 Gaussian Models

To start with we collect some well-known facts on the multivariate normal distribution and introduce Gaussian models that appear as limit models in subsequent chapters. For the following considerations we use moment generating functions. If X is any random vector with values in $\mathbb{R}^d$, then

$$\varphi_X(t) = \mathbb{E}\exp\{t^T X\}, \quad t \in \mathbb{R}^d,$$

is the moment generating function which, according to Theorem A.51, determines the distribution $\mathcal{L}(X)$ of X uniquely, provided that $\{t : \varphi_X(t) < \infty\}$ contains an open neighborhood of 0, see also Proposition 1.25. A random vector $Z = (Z_1, ..., Z_d)^T$ that consists of i.i.d. standard normally distributed components Z_i is called a *standard normal vector*. Its moment generating function is given by

$$\varphi_Z(t) = \mathbb{E}\exp\{t^T Z\} = \exp\{\|t\|^2/2\}.$$

Recall that any random (column) vector X with values in $\mathbb{R}^d$ has a multivariate normal distribution if it can be written as $X = AZ + b$, where A is a possibly singular $d \times d$ matrix and $b \in \mathbb{R}^d$. It is easy to see that

$$\begin{aligned} \mu &: = \mathbb{E}X = b \quad \text{and} \quad \Sigma := \mathsf{C}(X) = AA^T, \\ \varphi_X(t) &= \mathbb{E}\exp\{t^T(AZ+b)\} = \exp\{t^T b\}\varphi_Z(A^T t) \\ &= \exp\{t^T\mu + \frac{1}{2}t^T\Sigma t\}, \quad t \in \mathbb{R}^d. \end{aligned} \tag{6.61}$$

This representation, in conjunction with the uniqueness statement in Proposition 1.25, shows that the distribution of $X = AZ + b$ depends only on μ and Σ and is independent of the concrete representation of X, and the following well-known transformation rule holds. If $\mathcal{L}(X) = \mathsf{N}(\mu, \Sigma)$ then for a matrix A and a vector b

$$\mathcal{L}(AX + b) = \mathsf{N}(A\mu + b, A\Sigma A^T). \tag{6.62}$$

It is well known that $\mathcal{L}(X) = \mathsf{N}(\mu, \Sigma)$ is absolutely continuous with respect to the Lebesgue measure if and only if $\det(\Sigma) \neq 0$. If Σ is nonsingular, then the Lebesgue density is

$$\varphi_{\mu,\Sigma}(x) \tag{6.63}$$
$$= (2\pi)^{-n/2}(\det(\Sigma))^{-1/2}\exp\{-\frac{1}{2}(x-\mu)^T\Sigma^{-1}(x-\mu)\}, \quad x \in \mathbb{R}^d,$$

and as $\varphi_{\mu,\Sigma}$ is positive all distributions $\mathsf{N}(\mu,\Sigma)$, $\mu \in \mathbb{R}^d$, are equivalent (i.e., mutually absolutely continuous). This equivalence is no longer true if Σ is singular. To study this case in more detail we use a suitable rotation. As Σ is symmetric and positive semidefinite there are nonnegative numbers λ_i, the eigenvalues, and orthonormal column vectors e_i such that $\Sigma e_i = \lambda_i e_i$, $i = 1,...,d$. Let r be the rank of Σ, and assume without loss of generality that $\lambda_i > 0$, $1 \le i \le r$, and $\lambda_i = 0$, $r+1 \le i \le d$. It follows from $x = \sum_{i=1}^d (x^T e_i)e_i$ that

$$\mathbb{L} := \{\Sigma x : \ x \in \mathbb{R}^d\} = \{\sum\nolimits_{i=1}^r t_i e_i : \ t_1,...,t_r \in \mathbb{R}\}. \tag{6.64}$$

Let Λ be the diagonal matrix with entries $\lambda_1,...,\lambda_d$, and O the orthogonal matrix with columns $e_1,...,e_d$. The rotation to principal axes and the inverse transformation are given by

$$O^T\Sigma O = \Lambda \quad \text{and} \quad \Sigma = O\Lambda O^T,$$
$$O^T\mathbb{L} = \{x : x = (x_1,...,x_d) \in \mathbb{R}^d, \ x_i = 0, \ i = r+1,...,d\}.$$

Lemma 6.60. *If X has a normal distribution with expectation zero and covariance matrix Σ, then $X \in \mathbb{L}$, $\mathbb{P}$-a.s. For every $\theta \in \mathbb{R}^d$ it holds either $\mathsf{N}(\theta,\Sigma) \ll\gg \mathsf{N}(0,\Sigma)$ or $\mathsf{N}(\theta,\Sigma) \perp \mathsf{N}(0,\Sigma)$, where the latter case holds if and only if $\theta \notin \mathbb{L}$. If $\theta \in \mathbb{L}$, $\theta = \Sigma h$, then $\mathsf{N}(\theta,\Sigma) \ll\gg \mathsf{N}(0,\Sigma)$ and*

$$\frac{d\mathsf{N}(\Sigma h,\Sigma)}{d\mathsf{N}(0,\Sigma)}(x) = \exp\{h^T x - \frac{1}{2}h^T\Sigma h\}, \quad \mathsf{N}(0,\Sigma)\text{-}a.s.$$

Proof. Put $X_i = e_i^T X$. Then $\mathbb{E}X_i^2 = e_i^T\Sigma e_i = \lambda_i$. Hence $\mathbb{E}X_i^2 = 0$, $i = r+1,...,d$,

$$X = \sum\nolimits_{i=1}^d X_i e_i = \sum\nolimits_{i=1}^r X_i e_i, \quad \mathbb{P}\text{-a.s.},$$

and thus $X \in \mathbb{L}$, $\mathbb{P}$-a.s., by (6.64). If $\theta \notin \mathbb{L}$, then $\mathsf{N}(0,\Sigma)(\mathbb{L}) = 1$ and $\mathsf{N}(\theta,\Sigma)(\mathbb{L}+\theta) = 1$. As $\theta \notin \mathbb{L}$ implies $\mathbb{L} \cap (\mathbb{L}+\theta) = \emptyset$ we get $\mathsf{N}(\theta,\Sigma) \perp \mathsf{N}(0,\Sigma)$. If $\theta \in \mathbb{L}$ then by 6.64 it holds $\theta = \Sigma h$ for some $h \in \mathbb{R}^d$. It remains to show $\mathsf{N}(\Sigma h,\Sigma) \ll\gg \mathsf{N}(0,\Sigma)$ and to establish the density formula. To this end we consider the measure

$$Q(B) = \int I_B(x)\exp\{h^T x - \frac{1}{2}h^T\Sigma h\}\mathsf{N}(0,\Sigma)(dx), \quad B \in \mathfrak{B}_d,$$

and calculate the moment generating function. If X has the distribution $\mathsf{N}(0,\Sigma)$, then by (6.61) with $\mu = 0$ it follows

$$\int \exp\{t^T x\} Q(dx) = \mathbb{E} \exp\{t^T X + h^T X - \frac{1}{2} h^T \Sigma h\}$$
$$= \exp\{-\frac{1}{2} h^T \Sigma h\} \exp\{\frac{1}{2}(t+h)\Sigma(t+h)\} = \exp\{\frac{1}{2} t^T \Sigma t + t^T \Sigma h\}$$
$$= \int \exp\{t^T x\} \mathsf{N}(\Sigma h, \Sigma)(dx),$$

where the last equation follows from (6.61) with $\mu = \Sigma h$. The uniqueness statement for the moment generating function (see Proposition 1.25) gives $Q = \mathsf{N}(\Sigma h, \Sigma)$ and the proof is completed. ■

For a known symmetric and positive semidefinite matrix I_0 we consider the Gaussian model

$$\mathcal{G} = (\mathbb{R}^d, \mathfrak{B}_d, (\mathsf{N}(\mathsf{I}_0 h, \mathsf{I}_0))_{h \in \mathbb{R}^d}). \tag{6.65}$$

Regardless of whether I_0 is invertible, Lemma 6.60 implies that $\mathcal{G}$ is homogeneous and it holds

$$\begin{aligned} \frac{d\mathsf{N}(\mathsf{I}_0 h, \mathsf{I}_0)}{d\mathsf{N}(0, \mathsf{I}_0)} &= \exp\{h^T Z - \tfrac{1}{2} h^T \mathsf{I}_0 h\}, \\ \ln\left(\frac{d\mathsf{N}(\mathsf{I}_0 h, \mathsf{I}_0)}{d\mathsf{N}(0, \mathsf{I}_0)}\right) &= h^T Z - \tfrac{1}{2} h^T \mathsf{I}_0 h, \quad Z(x) = x. \end{aligned} \tag{6.66}$$

Hence $(\mathsf{N}(\mathsf{I}_0 h, \mathsf{I}_0))_{h \in \mathbb{R}^d}$ is an exponential family. The generating statistic is $Z(x) = x$ and is called the *central variable* of $\mathcal{G}$. Note that the exponential family $(\mathsf{N}(\mathsf{I}_0 h, \mathsf{I}_0))_{h \in \mathbb{R}^d}$ does not satisfy the condition (A1) unless I_0 is invertible.

For a known covariance matrix Σ_0 we consider the Gaussian model

$$\mathcal{G}_0 = (\mathbb{R}^d, \mathfrak{B}_d, (\mathsf{N}(h, \Sigma_0))_{h \in \mathbb{R}^d}).$$

We get from Lemma 6.60 that this model is homogenous, i.e. $\mathsf{N}(h_1, \Sigma_0) \ll\gg \mathsf{N}(h_2, \Sigma_0)$ for every $h_1, h_2 \in \mathbb{R}^d$ if and only if Σ_0 is nonsingular. In this case

$$\frac{d\mathsf{N}(h, \Sigma_0)}{d\mathsf{N}(0, \Sigma_0)}(x) = \exp\{h^T \Sigma_0^{-1} x - \frac{1}{2} h^T \Sigma_0^{-1} h\}, \tag{6.67}$$

and $(\mathsf{N}(h, \Sigma_0))_{h \in \mathbb{R}^d}$ is an exponential family with generating statistic $T = \Sigma_0^{-1} Z$. If I_0 is nonsingular then for $\Sigma_0 := \mathsf{I}_0^{-1}$, $S(x) = \Sigma_0 x$, and $T(x) = \mathsf{I}_0 x$ it holds

$$\begin{aligned} \mathsf{N}(\mathsf{I}_0 h, \mathsf{I}_0) &= \mathsf{N}(h, \Sigma_0) \circ T^{-1} \\ \mathsf{N}(h, \Sigma_0) &= \mathsf{N}(\mathsf{I}_0 h, \mathsf{I}_0) \circ S^{-1}. \end{aligned} \tag{6.68}$$

Thus the two models $\mathcal{G}_0$ and $\mathcal{G}$ are mutual randomizations and we get

$$\Delta(\mathcal{G}_0, \mathcal{G}) = 0, \tag{6.69}$$

which implies that $\mathcal{G}_0$ and $\mathcal{G}$ belong to the same equivalence class of models. This is also reflected by the Hellinger transforms.

Problem 6.61. For every $s \in \mathbf{S}_m^o$ and $h_1, ..., h_m \in \mathbb{R}^d$ it holds

$$\begin{aligned}&\mathsf{H}_s(\mathsf{N}(\mathsf{I}_0 h_1, \mathsf{I}_0), ..., \mathsf{N}(\mathsf{I}_0 h_m, \mathsf{I}_0))\\ &= \exp\{\frac{1}{2}(\sum_{i=1}^m s_i h_i)^T \mathsf{I}_0 (\sum_{i=1}^m s_i h_i) - \frac{1}{2}\sum_{i=1}^m s_i h_i^T \mathsf{I}_0 h_i\}.\end{aligned}$$

If I_0 is invertible, then we may replace I_0 with I_0^{-1} and h_i with $\mathsf{I}_0^{-1} h_i$. As the right-hand term remains unchanged after this transformation we get

$$\mathsf{H}_s(\mathsf{N}(\mathsf{I}_0 h_1, \mathsf{I}_0), ..., \mathsf{N}(\mathsf{I}_0 h_m, \mathsf{I}_0)) = \mathsf{H}_s(\mathsf{N}(h_1, \mathsf{I}_0^{-1}), ..., \mathsf{N}(h_m, \mathsf{I}_0^{-1})). \qquad (6.70)$$

Problem 6.62.* It holds for every $h_1, h_2 \in \mathbb{R}^d$,

$$\begin{aligned}\mathsf{H}_{1/2}(\mathsf{N}(\mathsf{I}_0 h_1, \mathsf{I}_0), \mathsf{N}(\mathsf{I}_0 h_2, \mathsf{I}_0)) &= \exp\{-\frac{1}{8}(h_1 - h_2)^T \mathsf{I}_0 (h_1 - h_2)\},\\ \mathsf{D}^2(\mathsf{N}(\mathsf{I}_0 h_1, \mathsf{I}_0), \mathsf{N}(\mathsf{I}_0 h_2, \mathsf{I}_0)) &= 2[1 - \exp\{-\frac{1}{8}(h_1 - h_2)^T \mathsf{I}_0 (h_1 - h_2)\}],\\ \|\mathsf{N}(\mathsf{I}_0 h_1, \mathsf{I}_0) - \mathsf{N}(\mathsf{I}_0 h_2, \mathsf{I}_0)\|^2 &\le (h_1 - h_2)^T \mathsf{I}_0 (h_1 - h_2).\end{aligned}$$

If $\det(\mathsf{I}_0) \neq 0$ and $\Sigma_0 = \mathsf{I}_0^{-1}$, then $\mathsf{N}(\mathsf{I}_0 h_i, \mathsf{I}_0)$ can be replaced with $\mathsf{N}(h_i, \Sigma_0)$, $i = 1, 2$.

The models $\mathcal{G}_0$ and $\mathcal{G}$ are special cases of a more general concept, the concept of a *Gaussian shift model.* Let $\mathbb{H}$ be a Hilbert space with scalar product $\langle \cdot, \cdot \rangle$. $\mathbb{H}$ serves as the parameter space. A statistical model $\mathcal{G} = (\mathcal{X}, \mathfrak{A}, (P_h)_{h \in \mathbb{H}})$ is called a Gaussian shift model on $\mathbb{H}$ if $P_h \ll P_0$, $h \in \mathbb{H}$, and there is a family $\mathsf{L}(h) : \mathcal{X} \to_m \mathbb{R}$, $h \in \mathbb{H}$, called the *central process*, such that

$$\begin{aligned}\mathsf{L}(ah + bg) &= a\mathsf{L}(h) + b\mathsf{L}(g) \quad h, g \in \mathbb{H},\ a, b \in \mathbb{R},\\ \frac{dP_h}{dP_0} &= \exp\{\mathsf{L}(h) - \frac{1}{2}\|h\|^2\}, \quad \text{and}\\ \mathcal{L}(\mathsf{L}(h)|P_0) &= \mathsf{N}(0, \|h\|^2), \quad h \in \mathbb{H}.\end{aligned}$$

It is not hard to show (see, e.g., Strasser (1985)) that for a finite-dimensional Hilbert space $\mathbb{H}$ there is a random variable T, called the *central variable*, such that $\mathsf{L}(h) = \langle T, h \rangle$. If we use the Euclidean scalar product, then we see from (6.66) that $\mathcal{G}$ in (6.65) is a Gaussian shift with central variable $Z(x) = x$. For $\mathcal{G}_0$ we introduce the new scalar product $\langle a, b \rangle = a^T \Sigma_0^{-1} b$. Then $\mathcal{G}_0$ is a Gaussian shift with central variable Z. As we deal mainly with parametric models the general theory of Gaussian shift models is not needed. We refer to Strasser (1985) for details on general Gaussian shift models.

6.4.2 The LAN and ULAN Property

The concept of asymptotic normality of models means, roughly speaking, that for a sequence of models the log-likelihood admits asymptotically a linearization as in (6.66). This leads to an asymptotic linearization of the log-likelihood with the help of a central sequence that plays the role of the central variable Z in (6.66). This property, called the LAN property, is the backbone of modern asymptotic statistics.

Definition 6.63. *A sequence of models* $\mathcal{M}_n = (\mathcal{X}_n, \mathfrak{A}_n, (P_{n,h})_{h\in\Delta_n})$ *with* $\Delta_n \uparrow \mathbb{R}^d$ *is called locally asymptotically normal (LAN) if there exist a sequence* $Z_n : \mathcal{X}_n \to_m \mathbb{R}^d$, *called the central sequence, a positive semidefinite and symmetric* $d \times d$ *matrix* I_0, *called the Fisher information matrix, and* $r_n(h) : \mathcal{X}_n \to_m \overline{\mathbb{R}}$ *such that the log-likelihood* $\ln L_{n,h}$ *of* $P_{n,h}$ *with respect to* $P_{n,0}$ *admits the expansion*

$$\ln L_{n,h} = h^T Z_n - \frac{1}{2} h^T \mathsf{I}_0 h + r_n(h), \tag{6.71}$$

$$r_n(h) \to^{P_{n,0}} 0, \quad h \in \mathbb{R}^d, \tag{6.72}$$

$$\mathcal{L}(Z_n | P_{n,0}) \Rightarrow \mathsf{N}(0, \mathsf{I}_0). \tag{6.73}$$

The sequence $\mathcal{M}_n$ *is called uniformly locally asymptotically normal (ULAN) if instead of (6.72) the stronger condition*

$$\lim_{n\to\infty} \sup_{h\in C} P_{n,0}(|r_n(h)| > \varepsilon) = 0, \quad \varepsilon > 0,$$

holds for every compact subset $C \subseteq \mathbb{R}^d$.

Remark 6.64. Local means in the LAN and ULAN conditions an approximation of the log-likelihood in (6.71) if a localized parameter is used and the sample size is large. For example, we show later that the sequence of localized models with $P_{n,h} = P_{\theta_0+h/\sqrt{n}}^{\otimes n}$ satisfies the ULAN condition, provided that $(P_\theta)_{\theta\in\Delta}$ is $\mathbb{L}_2$-differentiable at θ_0. As a short notation we also use LAN(Z_n, I_0) and ULAN(Z_n, I_0) to indicate the central sequence and the information matrix.

The next theorem connects the LAN property with the convergence of models. We show that the LAN property is equivalent to the weak convergence of the models $\mathcal{M}_n$ to the model $\mathcal{G}$ in (6.65). This relation has far-reaching consequences as it connects the convergence of distributions of the log-likelihoods with the decision-theoretic convergence of models concept that is based on the Δ-distance.

We recall that the weak convergence of models is denoted by $\Rightarrow$; see Definition 6.10.

Theorem 6.65. *If a sequence of models* $\mathcal{M}_n = (\mathcal{X}_n, \mathfrak{A}_n, (P_{n,h})_{h\in\Delta_n})$ *satisfies the LAN*(Z_n, I_0) *condition, then*

$$\mathcal{M}_n = (\mathcal{X}_n, \mathfrak{A}_n, (P_{n,h})_{h\in\Delta_n}) \Rightarrow \mathcal{G} = (\mathbb{R}^d, \mathfrak{B}_d, (\mathsf{N}(\mathsf{I}_0 h, \mathsf{I}_0))_{h\in\mathbb{R}^d}). \tag{6.74}$$

Conversely, if (6.74) holds, then there is a sequence $Z_n : \mathcal{X}_n \to_m \mathbb{R}^d$ *such that* $\mathcal{M}_n$ *satisfies the LAN*(Z_n, I_0) *condition.*

Corollary 6.66. *If* Σ_0 *is a symmetric positive definite* $d \times d$ *matrix, then the LAN*(Z_n, I_0) *condition for* $\mathsf{I}_0 = \Sigma_0^{-1}$ *implies*

$$\mathcal{M}_n = (\mathcal{X}_n, \mathfrak{A}_n, (P_{n,h})_{h\in\Delta_n}) \Rightarrow \mathcal{G}_0 = (\mathbb{R}^d, \mathfrak{B}_d, (\mathsf{N}(h, \Sigma_0))_{h\in\mathbb{R}^d}). \tag{6.75}$$

Corollary 6.67. *If the sequence of models $\mathcal{M}_n$ satisfies the LAN(Z_n, I_0) condition, then for every $h_1, h_2 \in \mathbb{R}^d$ the sequences $\{P_{n,h_1}\}$ and $\{P_{n,h_2}\}$ are mutually contiguous.*

Proof. We know from Lemma 6.60 that for $Z(x) = x$ the log-likelihood of $\mathsf{N}(\mathsf{I}_0 h, \mathsf{I}_0)$ with respect to $\mathsf{N}(0, \mathsf{I}_0)$ satisfies

$$\ln L_h = h^T Z - \frac{1}{2} h^T \mathsf{I}_0 h \quad \text{and} \quad \mathcal{L}(Z|\mathsf{N}(0, \mathsf{I}_0)) = \mathsf{N}(0, \mathsf{I}_0). \tag{6.76}$$

Hence the condition LAN(Z_n, I_0) implies that for every $h_1, ..., h_m \in \mathbb{R}^d$

$$\mathcal{L}((\ln L_{n,h_1}, ..., \ln L_{n,h_m})|P_{n,0}) \Rightarrow \mathcal{L}((\ln L_{h_1}, ..., \ln L_{h_m})|\mathsf{N}(0, \mathsf{I}_0)). \tag{6.77}$$

This implies $\mathcal{M}_n \Rightarrow \mathcal{G}$ in view of Corollary 6.15 as $\mathcal{G}$ is homogeneous. Conversely, fix $h \in \mathbb{R}^d$ and an orthonormal system $e_1, ..., e_d$ and set for $A_n = \{\max_{1 \le i \le d} |\ln L_{n,e_i}| + |\ln L_{n,h}| < \infty\}$,

$$Z_n = \sum\nolimits_{i=1}^{d} (\ln L_{n,e_i} + \frac{1}{2} e_i^T \mathsf{I}_0 e_i) I_{A_n} e_i.$$

Suppose that $\mathcal{M}_n \Rightarrow \mathcal{G}$. As the limit model of the sequence of binary submodels $(\mathcal{X}_n, \mathfrak{A}_n, \{P_{n,g}, P_{n,h}\})$ is

$$(\mathbb{R}^d, \mathfrak{B}_d, \{\mathsf{N}(\mathsf{I}_0 g, \mathsf{I}_0), \mathsf{N}(\mathsf{I}_0 h, \mathsf{I}_0)\}),$$

and $\mathsf{N}(\mathsf{I}_0 g, \mathsf{I}_0) \ll\gg \mathsf{N}(\mathsf{I}_0 h, \mathsf{I}_0)$, we get

$$\{P_{n,g}\} \lhd\rhd \{P_{n,h}\}, \quad g, h \in \mathbb{R}^d, \tag{6.78}$$

and $\lim_{n\to\infty} P_{n,0}(A_n) = 1$ and $\lim_{n\to\infty} P_{n,0}(|\ln L_{n,h}| = \infty) = 0$ from Theorem 6.26. Hence $\mathcal{M}_n \Rightarrow \mathcal{G}$ and Corollary 6.15 show that the distribution under $P_{n,0}$ of every linear function of $(\ln L_{n,e_1}) I_{A_n}, ..., (\ln L_{n,e_d}) I_{A_n}, (\ln L_{n,h}) I_{A_n}$ tends weakly to the distribution under $\mathsf{N}(0, \mathsf{I}_0)$ of the same linear function of $\ln L_{e_1}, ..., \ln L_{e_d}, \ln L_h$. For $h_i = e_i^T h$ it holds

$$\mathcal{L}(\ln L_h + \frac{1}{2} h^T \mathsf{I}_0 h - \sum\nolimits_{i=1}^{d} h_i (\ln L_{e_i} + \frac{1}{2} e_i^T \mathsf{I}_0 e_i) | \mathsf{N}(0, \mathsf{I}_0)) = \boldsymbol{\delta}_0.$$

The weak convergence of distributions to $\boldsymbol{\delta}_0$ is equivalent to the stochastic convergence of the associated random variables. Taking into account $\lim_{n\to\infty} P_{n,0}(A_n) = 1$ we get

$$\ln L_{n,h} + \frac{1}{2} h^T \mathsf{I}_0 h - \sum\nolimits_{i=1}^{d} h_i (\ln L_{n,e_i} + \frac{1}{2} e_i^T \mathsf{I}_0 e_i) \to^{P_{n,0}} 0.$$

The first corollary is only a reformulation of the theorem as $\mathcal{G}$ and $\mathcal{G}_0$ are equivalent; see (6.69). The second corollary follows from (6.78). ■

Now we study sequences of models for which the stochastic convergence to zero of the remainder term $r_n(h)$ in (6.72) is uniform in $h \in C$ for compact subsets $C \subseteq \mathbb{R}^d$. It turns out that this condition is equivalent to an equicontinuity property of the sequence of models. The next statement is a modification of Theorem 80.13 in Strasser (1985).

Proposition 6.68. *Let* $\mathcal{M}_n = (\mathcal{X}_n, \mathfrak{A}_n, (P_{n,h})_{h\in\Delta_n})$, $\Delta_n \uparrow \mathbb{R}^d$ *be a sequence of models that satisfies the condition LAN*(Z_n, I_0) *in Definition 6.63. Then for every compact set* $C \subseteq \mathbb{R}^d$ *the following conditions are equivalent.*

(A) *ULAN*(Z_n, I_0).

(B) $r_n(h_n) \to^{P_{n,0}} 0$ *for every convergent sequence* h_n. (6.79)

(C) $\lim_{\delta\downarrow 0} \limsup_{n\to\infty} \sup_{\|h_1-h_2\|\le\delta,\, h_1,h_2\in C} \|P_{n,h_1} - P_{n,h_2}\| = 0.$ (6.80)

Proof. (A) $\to$ (B) is clear. (B) $\to$ (A): set $\varphi_n(h) = P_{n,0}(|r_n(h)| > \varepsilon)$ and $\varphi(h) = 0$ and use Problem 6.79. (C) $\to$ (B): suppose $h_n \to h$. Set $Q_n = (1/3)(P_{n,h_n} + P_{n,h} + P_{n,0})$ and $f_{n,h} = dP_{n,h}/dQ_n$. Then

$$\begin{aligned}\int |L_{n,h_n} - L_{n,h}|dP_{n,0} &= \int I_{(0,\infty)}(f_{n,0}) \mid \frac{f_{n,h_n}}{f_{n,0}} - \frac{f_{n,h}}{f_{n,0}} \mid f_{n,0}\, dQ_n \\ &\le \int |f_{n,h_n} - f_{n,h}|\, dQ_n = \|P_{n,h_n} - P_{n,h}\|\,.\end{aligned}$$

If condition (6.80) is satisfied, then

$$\lim_{n\to\infty} \int |L_{n,h_n} - L_{n,h}|dP_{n,0} = 0. \tag{6.81}$$

The conditions (6.71), (6.72), and (6.73) imply

$$\liminf_{\delta\downarrow 0} \liminf_{n\to\infty} P_{n,0}(L_{n,h} > \delta) = 1,$$

which together with (6.81) provide $\ln L_{n,h_n} - \ln L_{n,h} \to^{P_{n,0}} 0$. As (6.73) implies that the sequence Z_n is stochastically bounded with respect to $P_{n,0}$ we get

$$\begin{aligned}&r_n(h_n) - r_n(h) \\ &= \ln L_{n,h_n} - \ln L_{n,h} - \langle Z_n, h_n - h\rangle + \frac{1}{2}h_n^T \mathsf{I}_0 h_n - \frac{1}{2}h^T \mathsf{I}_0 h \to^{P_{n,0}} 0.\end{aligned}$$

(B) $\to$ (C): put $\varphi_n(h) = P_{n,h}$, and denote by $\mathcal{T}_n$ the space of all distributions on $(\mathcal{X}_n, \mathfrak{A}_n)$. Use the variational distance as metric $\rho_{\mathcal{T}_n}$ to conclude from condition (6.100) in Problem 6.81 that in order to prove (6.80) it is enough to show that $\lim_{n\to\infty} \|P_{n,h_n} - P_{n,h}\| = 0$ for every h and every sequence h_n with $h_n \to h$. The conditions (6.71) and (6.73) imply $P_{n,0}(L_{n,h} > 0) \to 1$. This implies for the likelihood ratio of P_{n,h_n} with respect to $P_{n,h}$, say M_n, that

$$\begin{aligned}\ln M_n &= \ln L_{n,h_n} - \ln L_{n,h} \\ &= \langle Z_n, h_n - h\rangle - \frac{1}{2}(h_n^T \mathsf{I}_0 h_n - h^T \mathsf{I}_0 h) + r_n(h_n) - r_n(h) + o_{P_{n,0}}(1).\end{aligned}$$

Corollary 6.67 yields $\{P_{n,h}\} \lhd\rhd \{P_{n,0}\}$ and therefore

$$r_n(h_n) = o_{P_{n,0}}(1) = o_{P_{n,h}}(1) \quad \text{and} \quad Z_n = O_{P_{n,0}}(1) = O_{P_{n,h}}(1),$$

where the last statement follows from Problem 6.30. Hence $\mathcal{L}(\ln M_n|P_{n,h}) \Rightarrow \boldsymbol{\delta}_0$. By Theorem 6.21 the weak convergence of the binary models $\{P_{n,h_n}, P_{n,h}\}$ to a binary model $\{P, P\}$ follows, where P is any distribution on, $(\mathcal{X}, \mathfrak{A})$, say. From condition (B) in Theorem 6.21 we get $\lim_{n\to\infty} \mathsf{D}(P_{n,h_n}, P_{n,h}) = \mathsf{D}(P, P) = 0$ and $\lim_{n\to\infty} \|P_{n,h_n} - P_{n,h}\| = 0$ by Proposition 1.84. ■

Now we study sequences of models that satisfy the LAN condition. We start with a parametrized family of distributions $(P_\theta)_{\theta\in\Delta}$ on $(\mathcal{X}, \mathfrak{A})$, where $\Delta \subseteq \mathbb{R}^d$, and $\theta_0 \in \Delta^0$ is fixed. As we have explained already previously (see Examples 6.17, 6.19, and 6.20) one can only arrive at nontrivial limit models by turning away from the standard i.i.d. case to a double array of observations, where for increasing sample sizes each individual observation contributes less and less information to the whole model. With this in mind, let us introduce a local parameter by setting

$$\Delta_n = \{h : h \in \mathbb{R}^d,\ \theta_0 + h/\sqrt{n} \in \Delta\},\ n = 1, 2, ...,\quad \Delta_n \uparrow \mathbb{R}^d,$$

and study the sequence of models

$$\mathcal{M}_n = (\mathcal{X}^n, \mathfrak{A}^{\otimes n}, (P^{\otimes n}_{\theta_0+h/\sqrt{n}})_{h\in\Delta_n}). \tag{6.82}$$

To establish the ULAN condition we use the linearization of the log-likelihood for binary models. Let $\{c_{n,i}\}$ be a double array of regression coefficients that satisfy

$$\lim_{n\to\infty} \max_{1\le i\le n} c^2_{n,i} = 0 \quad \text{and} \quad \lim_{n\to\infty} \sum\nolimits_{i=1}^n c^2_{n,i} = 1. \tag{6.83}$$

For a fixed $\theta_0 \in \Delta^0$ and a convergent sequence $h_n \in \mathbb{R}^d$ we consider the sequence of binary models

$$\mathcal{M}_n = (\mathcal{X}^n, \mathfrak{A}^{\otimes n}, \{P^{\otimes n}_{\theta_0}, \textstyle\bigotimes_{i=1}^n P_{\theta_0+c_{n,i}h_n}\}).$$

We denote by $L_{\theta_0}(h)$ the likelihood ratio of P_{θ_0+h} with respect to P_{θ_0}. Set

$$\begin{aligned} X_{n,i}(\mathbf{x}_n) &= x_{n,i}, \quad \mathbf{x}_n = (x_{n,1}, ..., x_{n,n}) \in \mathcal{X}^n, \\ L_{n,i} &= L_{\theta_0+c_{n,i}h_n}(X_{n,i}), \quad \text{and} \quad Y_{n,i} = L^{1/2}_{n,i} - 1. \end{aligned}$$

To apply the results on binary models we assume that $(P_\theta)_{\theta\in\Delta}$ is $\mathbb{L}_2$-differentiable at θ_0 with derivative $\dot{L}_{\theta_0}$ and Fisher information $\mathsf{I}(\theta_0) = \mathsf{E}_{\theta_0}\dot{L}_{\theta_0}\dot{L}^T_{\theta_0}$; see Definition 1.103. Especially, we get from this definition that for all sufficiently large n it holds $Q_{n,i} := P_{\theta_0+c_{n,i}h_n} \ll P_{n,i} := P_{\theta_0}$, and therefore $Q_{n,i}(L_{n,i} = \infty) = 0$. Set

$$\begin{aligned} R(\delta) &:= \sup_{\|h\|\le\delta} \|h\|^{-2}\, \mathsf{E}_{\theta_0}\left|[L^{1/2}_{\theta_0}(h) - 1] - \frac{1}{2}h^T\dot{L}_{\theta_0}\right|^2, \\ \gamma(\delta) &:= \sup_{\|h\|\le\delta} \|h\|^{-2}\left|\mathsf{E}_{\theta_0}[L^{1/2}_{\theta_0}(h) - 1]^2 - \frac{1}{4}h^T\mathsf{I}(\theta_0)h\right|. \end{aligned}$$

If $(P_\theta)_{\theta\in\Delta}$ is $\mathbb{L}_2$-differentiable at θ_0, then $R(\delta)$ and $\gamma(\delta)$ tend to zero as $\delta \to 0$. An approximation of the first two moments of $Y_{n,i}$ is considered next.

Problem 6.69.* It holds

$$|\,\mathsf{E}_{\theta_0} Y_{n,i} - (-\frac{1}{8}c_{n,i}^2 h_n^T \mathsf{I}(\theta_0) h_n)\,| \le \gamma(\delta_n)\,\|c_{n,i}h_n\|^2\,, \tag{6.84}$$

$$|\,\mathsf{E}_{\theta_0} Y_{n,i}^2 - \frac{1}{4}c_{n,i}^2 h_n^T \mathsf{I}(\theta_0) h_n\,| \le \gamma(\delta_n)\,\|c_{n,i}h_n\|^2\,, \tag{6.85}$$

$$\mathsf{E}_{\theta_0}(Y_{n,i} - \mathsf{E}_{\theta_0} Y_{n,i} - \frac{1}{2}c_{n,i}h_n^T \dot{L}_{\theta_0}(X_{n,i}))^2 \le \rho_{n,i}, \quad \text{where} \tag{6.86}$$

$$\rho_{n,i} = 2R(\delta_n)\,\|c_{n,i}h_n\|^2 + 2(\frac{1}{8}c_{n,i}^2 h_n^T \mathsf{I}(\theta_0) h_n + \gamma(\delta_n)\,\|c_{n,i}h_n\|^2)^2,$$

for every sufficiently large n, and where $\delta_n = \max_{1\le i\le n}\|c_{n,i}h_n\|$.

The following version of the so-called second lemma of LeCam is taken from Witting and Müller-Funk (1995).

Theorem 6.70. ***(Second Lemma of LeCam)*** *Suppose that $(P_\theta)_{\theta\in\Delta}$ is $\mathbb{L}_2$-differentiable at $\theta_0 \in \Delta^0$ with derivative $\dot{L}_{\theta_0}$ and Fisher information matrix $\mathsf{I}(\theta_0) = \mathsf{E}_{\theta_0}\dot{L}_{\theta_0}\dot{L}_{\theta_0}^T$. Suppose that the conditions in (6.83) are satisfied. Then the sequence of models*

$$(\mathcal{X}^n, \mathfrak{A}^{\otimes n}, (\textstyle\bigotimes_{i=1}^n P_{\theta_0+c_{n,i}h})_{h\in\Delta_n}), \quad \Delta_n = \{h : \theta_0 + c_{n,i}h \in \Delta,\ i = 1, ..., n\},$$

satisfies the $ULAN(Z_n, \mathsf{I}_0)$ condition with $Z_n = \sum_{i=1}^n c_{n,i}\dot{L}_{\theta_0}(X_{n,i})$ and $\mathsf{I}_0 = \mathsf{I}(\theta_0)$.

Corollary 6.71. *Under the assumptions of the theorem the sequence*

$$(\mathcal{X}^n, \mathfrak{A}^{\otimes n}, (P_{\theta_0+h/\sqrt{n}}^{\otimes n})_{h\in\Delta_n}), \quad \Delta_n = \{h : \theta_0 + h/\sqrt{n} \in \Delta\},$$

satisfies the $ULAN(Z_n, \mathsf{I}_0)$ condition with $Z_n = (1/\sqrt{n})\sum_{i=1}^n \dot{L}_{\theta_0}(X_{n,i})$ and $\mathsf{I}_0 = \mathsf{I}(\theta_0)$.

Proof. We apply Theorem 6.49 with $P_{n,i} = P_{\theta_0}$ and $Q_{n,i} = P_{\theta_0+c_{n,i}h_n}$, where $h_n \to h$. To establish (6.52) we remark that by the definition of the $\mathbb{L}_2$-differentiability (see Definition 1.103) it holds $Q_{n,i}(Y_{n,i} = \infty) = 0$ for all sufficiently large n. It follows from Problem 6.59 and (6.86) that we have to establish the Lindeberg condition only for the double array $\mathsf{E}_{\theta_0}Y_{n,i} + \frac{1}{2}c_{n,i}h_n^T\dot{L}_{\theta_0}(X_{n,i})$. Utilizing Problem 6.59 once more (6.84) shows that we have to prove the Lindeberg condition for $c_{n,i}h_n^T\dot{L}_{\theta_0}(X_{n,i})$ only. As the $\dot{L}_{\theta_0}(X_{n,i})$ are i.i.d. under $P_{\theta_0}^{\otimes n}$, and $\mathsf{E}_{\theta_0}\parallel \dot{L}_{\theta_0}(X_{1,1})\parallel^2 < \infty$, the Lindeberg condition for $c_{n,i}h_n^T\dot{L}_{\theta_0}(X_{n,i})$ follows from Problem 6.82.

Condition (6.55) with $\sigma^2 = h^T\mathsf{I}(\theta_0)h$ follows from $\mathsf{E}_{\theta_0}Y_{n,i}^2 = \mathsf{D}^2(P_{n,i}, Q_{n,i})$, (6.85), $h_n \to h$, and (6.83). Hence we get from Theorem 6.49,

$$\ln L_{n,h_n} = 2\sum\nolimits_{i=1}^n Y_{n,i} - \sigma^2/4 + o_{P_{\theta_0}^{\otimes n}}(1), \quad \sigma^2 = h^T\mathsf{I}(\theta_0)h.$$

The random variables $V_{n,i} = Y_{n,i} - \mathsf{E}_{\theta_0} Y_{n,i} - \frac{1}{2} c_{n,i} h_n^T \dot{L}_{\theta_0}(X_{n,i})$ are independent and have expectation zero. Hence,

$$\mathsf{E}_{\theta_0}(\sum\nolimits_{i=1}^n V_{n,i})^2 = \sum\nolimits_{i=1}^n \mathsf{E}_{\theta_0} V_{n,i}^2 \le \sum\nolimits_{i=1}^n \rho_{n,i} \to 0,$$

by (6.86). Thus by (6.84) and (6.83),

$$\begin{aligned}\ln L_{n,h_n} &= \sum\nolimits_{i=1}^n [c_{n,i} h_n^T \dot{L}_{\theta_0}(X_{n,i}) + 2\mathsf{E}_{\theta_0} Y_{n,i}] - \sigma^2/4 + o_{P_{\theta_0}^{\otimes n}}(1) \\ &= h_n^T Z_n - \frac{1}{2} h_n^T \mathsf{I}_0 h_n + o_{P_{\theta_0}^{\otimes n}}(1),\end{aligned}$$

which is the expansion (6.71) with $\mathsf{I}_0 = \mathsf{I}(\theta_0)$. It remains to show that $\mathcal{L}(Z_n | P_{\theta_0}^{\otimes n}) \Rightarrow \mathsf{N}(0, \mathsf{I}_0)$. To this end we fix h and consider the random variables $Z_{n,i} = c_{n,i} h^T \dot{L}_{\theta_0}(X_{n,i})$, which satisfy the Lindeberg condition by Problem 6.82. As $\mathsf{E}_{\theta_0} Z_{n,i} = 0$ by Proposition 1.110, and

$$\sum\nolimits_{i=1}^n \mathsf{V}_{\theta_0}(Z_{n,i}) = \sum\nolimits_{i=1}^n c_{n,i}^2 h^T \mathsf{I}(\theta_0) h \to h^T \mathsf{I}(\theta_0) h,$$

we get $\mathcal{L}(Z_n | P_{\theta_0}^{\otimes n}) \Rightarrow \mathsf{N}(0, \mathsf{I}_0)$ from Theorem A.54 and the Cramér–Wold device; see Criterion A.52. The proof of the corollary follows by setting $c_{n,i} = 1/\sqrt{n}$. ■

In the following theorem we formulate the third lemma of LeCam in the language of the LAN condition. In the next three chapters we systematically use the statements of the subsequent theorem and its corollaries. They allow us to evaluate the risk of estimators, tests, and selection rules asymptotically under local alternatives. Such statements are the main tools to evaluate the efficiency of decisions, and thus they are the backbone of the entire LeCam theory. The concept of weak convergence of distributions relies on the expectation of bounded and continuous functions of the random variables under consideration. Sometimes one has to deal with functions that are discontinuous at a few points. Such functions are indicator functions of intervals that appear in the theory of testing statistical hypotheses, or convex loss functions when estimating parameters. A function $\varphi : \mathbb{R}^d \to_m \mathbb{R}$ is called $\boldsymbol{\lambda}_d$-a.e. continuous if the set of all points of discontinuity has the Lebesgue measure zero.

Theorem 6.72. ***(Third Lemma of LeCam Under LAN)*** *Suppose that the sequence of models $\mathcal{M}_n = (\mathcal{X}_n, \mathfrak{A}_n, (P_{n,h})_{h \in \Delta_n})$ with $\Delta_n \uparrow \mathbb{R}^d$ fulfils the LAN(Z_n, I_0) condition, and let $S_n : \mathcal{X}_n \to \mathbb{R}^m$. Then*

$$\mathcal{L}\left((S_n^T, Z_n^T)^T | P_{n,0}\right) \Rightarrow \mathsf{N}(0, \Sigma) \tag{6.87}$$

implies

$$\mathcal{L}\left((S_n^T, Z_n^T)^T | P_{n,h}\right) \Rightarrow \mathsf{N}\left(\begin{pmatrix}\Sigma_{1,2} h \\ \Sigma_{2,2} h\end{pmatrix}, \Sigma\right), \quad h \in \mathbb{R}^d, \tag{6.88}$$

where $(\Sigma_{i,j})_{1 \le i,j \le 2}$ *is the partition of* Σ *into submatrices and* $\Sigma_{2,2} = \mathsf{I}_0$*. Furthermore, for every bounded and continuous* $\varphi : \mathbb{R}^{m+d} \to \mathbb{R}$ *and every* $h \in \mathbb{R}^d$,

$$\lim_{n\to\infty} \int \varphi(S_n(x_n), Z_n(x_n)) P_{n,h}(dx_n) \tag{6.89}$$
$$= \int \varphi(x + \Sigma_{1,2}h, y + \Sigma_{2,2}h) \mathsf{N}(0, \Sigma)(dx, dy).$$

If $\varphi : \mathbb{R}^d \to_m \mathbb{R}$ *is bounded and* $\boldsymbol{\lambda}_d$*-a.e. continuous, and* I_0 *is nonsingular, then*

$$\lim_{n\to\infty} \int \varphi(Z_n(x_n)) P_{n,h}(dx_n) = \int \varphi(x + \mathsf{I}_0 h) \mathsf{N}(0, \mathsf{I}_0)(dx). \tag{6.90}$$

If $\varphi : \mathbb{R}^m \to_m \mathbb{R}$ *is bounded and* $\boldsymbol{\lambda}_m$*-a.e. continuous, and* $\Sigma_{1,1}$ *is nonsingular, then*

$$\lim_{n\to\infty} \int \varphi(S_n(x_n)) P_{n,h}(dx_n) = \int \varphi(x + \Sigma_{1,2}h) \mathsf{N}(0, \Sigma_{1,1})(dx). \tag{6.91}$$

Corollary 6.73. *If in addition the ULAN*(Z_n, I_0) *condition holds, then the convergence in (6.89), (6.90), and (6.91) is locally uniform in* h.

Proof. Fix $g_0 \in \mathbb{R}^m$, $h_0 \in \mathbb{R}^d$ and set $\widetilde{S}_n = g_0^T S_n + h_0^T Z_n$. Then by (6.71) and (6.87),

$$\mathcal{L}((\widetilde{S}_n, \ln L_{n,h})^T | P_{n,0}) \Rightarrow \mathsf{N}\left(\begin{pmatrix} 0 \\ -\frac{1}{2}\sigma^2 \end{pmatrix}, \begin{pmatrix} \tau^2 & \rho\tau\sigma \\ \rho\tau\sigma & \sigma^2 \end{pmatrix}\right),$$

where

$$\tau^2 = g_0^T \Sigma_{1,1} g_0 + h_0^T \Sigma_{2,2} h_0 + g_0^T \Sigma_{1,2} h_0 + h_0^T \Sigma_{2,1} g_0, \ \sigma^2 = h^T \Sigma_{2,2} h,$$
$$\rho\tau\sigma = g_0^T \Sigma_{1,2} h + h_0^T \Sigma_{2,2} h.$$

Hence by Proposition 6.34

$$\mathcal{L}(\widetilde{S}_n | P_{n,h}) \Rightarrow \mathsf{N}\left(\rho\tau\sigma, \tau^2\right).$$

As g_0 and h_0 are arbitrary we get the statement (6.88) from the Cramér–Wold device; see Criterion A.52. The statement (6.88) implies (6.89) for every bounded and continuous $\varphi : \mathbb{R}^{m+d} \to \mathbb{R}$. Suppose now $\varphi : \mathbb{R}^d \to \mathbb{R}$ is bounded and $\boldsymbol{\lambda}_d$-a.e. continuous. If I_0 is nonsingular, then φ is also $\mathsf{N}(\mathsf{I}_0 h, \mathsf{I}_0)$-a.s. continuous. Hence (6.90) follows from the fact that (6.88) implies the weak convergence of the marginal distributions and (F) in Theorem A.49. The proof of (6.91) is similar. To prove the corollary, we set

$$\varphi_n(h) = \int \varphi(S_n(x_n), Z_n(x_n)) P_{n,h}(dx_n).$$

If the ULAN(Z_n, I_0) condition holds, then in view of (6.80) and the inequality in Problem 1.80,

$$\lim_{\delta\downarrow 0}\limsup_{n\to\infty}\sup_{\|h_1-h_2\|\le\delta,\ h_1,h_2\in C}|\varphi_n(h_1)-\varphi_n(h_2)|=0.$$

An application of the statement in Problem 6.80 completes the proof. ■

For i.i.d. observations the sequence of models has a product structure., i.e.

$$\mathcal{M}_n=(\mathcal{X}^n,\mathfrak{A}^{\otimes n},(P^{\otimes n}_{\theta_0+h/\sqrt{n}})_{h\in\Delta_n}),\quad \Delta_n\uparrow\mathbb{R}^d. \tag{6.92}$$

Denote by $X_1,...,X_n$ the projections of $\mathcal{X}^n$ on $\mathcal{X}$. The sequence of statistics $S_n:\mathcal{X}^n\to_m\mathbb{R}^m$ admits in many cases a so-called stochastic Taylor expansion; that is,

$$S_n=\frac{1}{\sqrt{n}}\sum\nolimits_{i=1}^n\Psi(X_i)+o_{P^{\otimes n}_{\theta_0}}(1), \tag{6.93}$$

where Ψ belongs to the space $\mathbb{L}^0_{2,m}(P_{\theta_0})$ of functions $\Psi:\mathcal{X}\to_m\mathbb{R}^d$ with $\mathsf{E}_{\theta_0}\Psi=0$ and $\mathsf{E}_{\theta_0}\|\Psi\|^2<\infty$. For such sequences we get the following version of the third lemma of LeCam.

Corollary 6.74. *Assume that for the model* $(\mathcal{X},\mathfrak{A},(P_\theta)_{\theta\in\Delta})$, $\Delta\subseteq\mathbb{R}^d$, *the family* $(P_\theta)_{\theta\in\Delta}$ *is* $\mathbb{L}_2$*-differentiable at* $\theta_0\in\Delta^0$ *with derivative* $\dot{L}_{\theta_0}$. *Suppose that* $S_n:\mathcal{X}^n\to_m\mathbb{R}^m$ *is a sequence of statistics that admits the representation (6.93). Then it holds*

$$\mathcal{L}(S_n|P^{\otimes n}_{\theta_0+h/\sqrt{n}})\Rightarrow\mathsf{N}(\mathsf{C}_{\theta_0}(\Psi,\dot{L}_{\theta_0})h,\mathsf{C}_{\theta_0}(\Psi)),\quad h\in\mathbb{R}^d. \tag{6.94}$$

For every bounded and continuous function $\varphi:\mathbb{R}^m\to\mathbb{R}$ *it holds locally uniform in* h,

$$\lim_{n\to\infty}\int\varphi(S_n(x_n))P^{\otimes n}_{\theta_0+h/\sqrt{n}}(dx_n)=\int\varphi(x+\mathsf{C}_{\theta_0}(\Psi,\dot{L}_{\theta_0})h)\mathsf{N}(0,\mathsf{C}_{\theta_0}(\Psi))(dx). \tag{6.95}$$

If $\mathsf{C}_{\theta_0}(\Psi)$ *is nonsingular, then (6.95) holds locally uniform in* h *for every* $\varphi:\mathbb{R}^m\to_m\mathbb{R}$ *that is bounded and* $\boldsymbol{\lambda}_d$*-a.e. continuous.*

Proof. We note that by the second lemma of LeCam (see Corollary 6.71) the sequence of models $(\mathcal{X}^n,\mathfrak{A}^{\otimes n},(P^{\otimes n}_{\theta_0+h/\sqrt{n}})_{h\in\Delta_n})$, $\Delta_n=\{h:\theta_0+h/\sqrt{n}\in\Delta\}$, satisfies the ULAN($Z_n,\mathsf{I}(\theta_0)$) condition with central sequence $Z_n=n^{-1/2}\sum_{i=1}^n\dot{L}_{\theta_0}(X_i)$ and Fisher information matrix $\mathsf{I}(\theta_0)=\mathsf{C}_{\theta_0}(\dot{L}_{\theta_0})$. The central limit theorem for i.i.d. random vectors and Slutsky's lemma yield that the distribution of

$$\begin{pmatrix}S_n\\Z_n\end{pmatrix}=\frac{1}{\sqrt{n}}\sum_{i=1}^n\begin{pmatrix}\Psi(X_i)\\\dot{L}_{\theta_0}(X_i)\end{pmatrix}+o_{P^{\otimes n}_{\theta_0}}(1)$$

tends to a normal distribution with expectation zero and covariance matrix Σ given by

$$\Sigma = \begin{pmatrix} \mathsf{C}_{\theta_0}(\Psi) & \mathsf{C}_{\theta_0}(\Psi, \dot{L}_{\theta_0}) \\ \mathsf{C}_{\theta_0}(\dot{L}_{\theta_0}, \Psi) & \mathsf{I}(\theta_0) \end{pmatrix}.$$

To complete the proof we have only to apply the already proved statements in Theorem 6.72 and Corollary 6.73. ■

The shift $\mathsf{C}_{\theta_0}(\Psi, \dot{L}_{\theta_0})h$ that appears in (6.95) indicates how the statistic S_n measures the deviation from θ_0 when the data are from a local alternative. Suppose for simplicity that $d = 1$. Then $\mathsf{C}_{\theta_0}(\Psi)$ is the variance $\mathsf{V}_{\theta_0}(\Psi)$. If we turn to the normalized statistic

$$\widetilde{S}_n = (\mathsf{V}_{\theta_0}(\Psi))^{-1/2} S_n,$$

and denote by ρ the correlation between Ψ and $\dot{L}_{\theta_0}$, and by $\mathsf{I}(\theta_0)$ the Fisher information, then

$$\mathcal{L}(\widetilde{S}_n | P^{\otimes n}_{\theta_0 + h/\sqrt{n}}) \Rightarrow \mathsf{N}(\rho \mathsf{I}^{1/2}(\theta_0)h, 1).$$

Thus, to maximize the shift means to maximize the correlation coefficient. If we are interested in a positive shift, then the maximum shift is produced by taking $\Psi = \dot{L}_{\theta_0}$. This is easy to understand as the tangent $\dot{L}_{\theta_0}$ of the model reflects best the behavior of the model if we operate with first-order linearizations.

The simplest situations where the ULAN condition is satisfied are those with exponential families.

Example 6.75. Let $(P_\theta)_{\theta \in \Delta}$ be an exponential family on $(\mathcal{X}, \mathfrak{A})$ with generating statistic $T : \mathcal{X} \to \mathbb{R}^d$ and natural parameter $\theta \in \Delta$. We assume that the conditions (A1) and (A2) are satisfied. Then by (1.6)

$$\frac{dP_\theta}{d\mu}(x) = \exp\{\langle \theta, T(x) \rangle - K(\theta)\}, \quad x \in \mathcal{X}.$$

By Example 1.120 the family is $\mathbb{L}_2$-differentiable with derivative $\dot{L}_{\theta_0} = T - \nabla K(\theta_0)$, $\theta_0 \in \Delta^0$. Hence the sequence $P^{\otimes n}_{\theta_0 + h_n/\sqrt{n}}$ satisfies the ULAN(Z_n, I_0) condition with

$$Z_n = \frac{1}{\sqrt{n}} \sum\nolimits_{i=1}^n \dot{L}_{\theta_0}(X_i) = \frac{1}{\sqrt{n}} \sum\nolimits_{i=1}^n (T(X_i) - \nabla K(\theta_0))$$

and $\mathsf{I}_0 = \nabla\nabla^T K(\theta_0)$. But this can also be obtained directly. Indeed, for $h_n \to h$,

$$\begin{aligned} \ln L_{n,h_n} &:= \ln(dP^{\otimes n}_{\theta_0 + h_n/\sqrt{n}} / dP^{\otimes n}_{\theta_0}) \\ &= \sum\nolimits_{i=1}^n (h_n/\sqrt{n})^T [T(X_i) - \nabla K(\theta_0)] + \frac{1}{2} h^T \mathsf{I}_0 h + r_n(\theta_0, h), \quad \text{where} \end{aligned}$$

$$r_n(\theta_0, h_n) = n[K(\theta_0 + h_n/\sqrt{n}) - K(\theta_0) - (h_n/\sqrt{n})^T \nabla K(\theta_0)] - \frac{1}{2} h_n^T \mathsf{I}_0 h_n \to 0.$$

If a sequence of models satisfies the LAN condition, then by (6.71),

$$L_{n,h} \approx \exp\{h^T Z_n - \frac{1}{2} h^T \mathsf{I}_0 h\},$$

which means that $(P_{n,h})_{h\in\Delta_n}$ is an approximate exponential family. Two problems arise in an attempt to bring this statement into a mathematically rigorous form. First, the LAN(Z_n, I_0) condition does not say anything about possible moments of Z_n. Especially we cannot expect that $\int \exp\{h^T Z_n\} dP_{n,0} < \infty$, and thus we cannot turn $\exp\{h^T Z_n\}$ into a density by a proper normalization. To overcome this difficulty we apply a truncation technique by setting, for a sequence c_n with $c_n \to \infty$,

$$\begin{aligned} Z_n^* &= I_{[0,c_n]}(\|Z_n\|)Z_n, \\ K_n(h) &= \ln(\int \exp\{h^T Z_n^*\} dP_{n,0}), \\ \frac{dQ_{n,h}}{dP_{n,0}} &= \exp\{h^T Z_n^* - K_n(h)\}. \end{aligned} \tag{6.96}$$

Second, to make the phrase "approximate exponential family" precise, we measure the difference between $P_{n,h}$ and $Q_{n,h}$ with the variational distance. The next theorem is due to LeCam (1960). For a proof we refer to Strasser (1985), Theorem 81.1.

Theorem 6.76. *Let $\mathcal{M}_n = (\mathcal{X}_n, \mathfrak{A}_n, (P_{n,h})_{h\in\Delta_n})$ be a sequence of models with $\Delta_n \uparrow \mathbb{R}^d$ that fulfils the LAN(Z_n, I_0) condition. Then there exists a sequence c_n with $c_n \to \infty$ such that*

$$\lim_{n\to\infty} \|P_{n,h} - Q_{n,h}\| = 0, \quad h \in \mathbb{R}^d,$$

$$\lim_{n\to\infty} K_n(h) = \frac{1}{2} h_0^T \mathsf{I}_0 h, \quad h \in \mathbb{R}^d,$$

for Q_n defined in (6.96).

The above theorem has far-reaching consequences. As the sequence of generating statistics Z_n^* is sufficient for the family $Q_{n,h}$, which approximates $P_{n,h}$ in terms of the variational distance, Z_n^* and thus Z_n are approximately sufficient for the family $P_{n,h}$ in the following sense.

Theorem 6.77. *Let the assumptions of Theorem 6.76 be satisfied. If the decision space $(\mathcal{D}, \mathfrak{D})$ is a Borel space, and $\mathsf{D}_{\mathcal{M}_n} : \mathfrak{D} \times \mathcal{X}_n \to_k [0,1]$ is a sequence of decisions for the models $\mathcal{M}_n$, then there are decisions $\mathsf{D}^*_{\mathcal{M}_n}$ factorized by Z_n; that is, there are $\mathsf{D}_n : \mathfrak{D} \times \mathbb{R}^d \to_k [0,1]$ with $\mathsf{D}^*_{\mathcal{M}_n}(A|x) = \mathsf{D}_n(A|Z_n(x))$, such that for every bounded loss function L,*

$$\lim_{n\to\infty} (\mathsf{R}(h, \mathsf{D}_{\mathcal{M}_n}) - \mathsf{R}(h, \mathsf{D}^*_{\mathcal{M}_n})) = 0, \quad h \in \mathbb{R}^d.$$

Proof. By Theorem 4.18 there are decisions $\mathsf{D}^*_{\mathcal{M}_n}$ factorized by Z_n^*, and thus factorized by Z_n, such that

$$\int [\int L(h,a) \mathsf{D}_{\mathcal{M}_n}(da|x)] Q_{n,h}(dx) = \int [\int L(h,a) \mathsf{D}_n(da|Z_n(x))] Q_{n,h}(dx).$$

If $\|L\|_u \leq c$, then $|\int L(h,a)\mathsf{D}_n(da|Z_n(x))| \leq c$. Hence by the inequality $|\int fd(P_0 - P_1)| \leq c\|P_0 - P_1\|$ for $\|f\|_u \leq c$ (see Problem 1.80) we get

$$\begin{aligned}
&|\mathsf{R}(h,\mathsf{D}_{\mathcal{M}_n}) - \mathsf{R}(h,\mathsf{D}^*_{\mathcal{M}_n})| \\
&= |\int [\int L(h,a)(\mathsf{D}_{\mathcal{M}_n}(da|x) - \mathsf{D}_n(da|Z_n(x)))]P_{n,h}(dx)| \\
&\leq |\int [\int L(h,a)(\mathsf{D}_{\mathcal{M}_n}(da|x) - \mathsf{D}_n(da|Z_n(x)))]Q_{n,h}(dx)| \\
&\quad + |\int [\int L(h,a)(\mathsf{D}_{\mathcal{M}_n}(da|x) - \mathsf{D}_n(da|Z_n(x)))](P_{n,h} - Q_{n,h})(dx)| \\
&\leq 2c\,\|P_{n,h} - Q_{n,h}\|\,.
\end{aligned}$$

■

Remark 6.78. The LAN condition was introduced by LeCam (1960). The more general concept of local asymptotic mixed normality (LAMN) is due to Jeganathan (1980a,b). A somewhat more general concept is that of locally asymptotically quadratic families; see LeCam and Yang (1990). Shiryaev and Spokoiny (2000) introduced another concept for asymptotic normality. The starting point for this is the fact that weak convergence of distributions can be replaced by the a.s. convergence of random variables with the same distributions if one defines the new random variables on a suitably constructed probability space.

We have established the LAN property only for i.i.d. observations. This fundamental condition can be proved in a more general setting. For independent but not necessarily identically distributed observations we refer to Strasser (1985) and Rieder (1994). There are many papers dealing with the LAN condition for stochastic processes. First results are contained in the books by Basawa and Prakasa Rao (1980) and Basawa and Scott (1983). The LAN condition appears then as a special case of the LAMN condition when the considered process is ergodic. Without being complete, for diffusion processes we refer to Jeganathan (1980a,b), Basawa and Prakasa Rao (1980), Basawa and Scott (1983), Davies (1985), and Luschgy (1992a). Ergodic diffusion processes are considered in Kutoyanc (2004), and Markov statistical models are studied in Höpfner, Jacod, and Ladell (1990). The LAN condition for spatial Poisson processes has been established in Liese and Lorz (1999). Thinned point processes are considered in Falk and Liese (1998).

For further details of the history and development of the concept of asymptotic normality of sequences of models and related concepts we refer to Strasser (1985), Chapter 13, LeCam and Yang (1990), and Shiryaev and Spokoiny (2000).

Finally, we collect some technical problems that concern well-known results on the uniform convergence of functions.

Problem 6.79.* Let $\mathcal{S}$ and $\mathcal{T}$ be any metric spaces with metrics $\rho_{\mathcal{S}}$ and $\rho_{\mathcal{T}}$, respectively, and let $\varphi, \varphi_1, ...$ be functions on $\mathcal{S}$ with values in $\mathcal{T}$. If $\mathcal{S}$ is compact and φ is continuous, then

$$\lim_{n\to\infty} \sup_{s\in\mathcal{S}} \rho_{\mathcal{T}}(\varphi_n(s), \varphi(s)) = 0 \quad \Longleftrightarrow \quad \lim_{n\to\infty} \rho_{\mathcal{T}}(\varphi_n(s_n), \varphi(s)) = 0, \tag{6.97}$$

for every $s_n, s \in \mathcal{S}$ with $s_n \to s$.

Problem 6.80.* Suppose $\mathcal{S}$ and $\mathcal{T}$ are metric spaces and $\varphi, \varphi_n : \mathcal{S} \to \mathcal{T}$. If $\mathcal{S}$ is compact, φ is continuous, and the sequence φ_n satisfies

$$\lim_{\delta \downarrow 0} \limsup_{n\to\infty} \sup_{s_1,s_2:\rho_{\mathcal{S}}(s_1,s_2)\le\delta} \rho_{\mathcal{T}}(\varphi_n(s_1), \varphi_n(s_2)) = 0,$$

then the pointwise convergence $\varphi_n(s) \to \varphi(s)$ for all s from a dense subset implies the uniform convergence of φ_n to φ.

Problem 6.81.* Suppose $\mathcal{S}$ and $\mathcal{T}_n$ are metric spaces and $\varphi_n : \mathcal{S} \to \mathcal{T}_n$. If $\mathcal{S}$ is compact, then the following conditions are equivalent.

$$\lim_{\delta \downarrow 0} \limsup_{n\to\infty} \sup_{s_1,s_2:\rho_{\mathcal{S}}(s_1,s_2)\le\delta} \rho_{\mathcal{T}_n}(\varphi_n(s_1), \varphi_n(s_2)) = 0. \tag{6.98}$$

$$\lim_{\delta \downarrow 0} \limsup_{n\to\infty} \sup_{s:\rho_{\mathcal{S}}(s,s_0)\le\delta} \rho_{\mathcal{T}_n}(\varphi_n(s), \varphi_n(s_0)) = 0, \quad s_0 \in \mathcal{S}. \tag{6.99}$$

$$\lim_{n\to\infty} \rho_{\mathcal{T}_n}(\varphi_n(s_n), \varphi_n(s_0)) = 0, \quad s_n, s_0 \in \mathcal{S}, \; s_n \to s_0. \tag{6.100}$$

Problem 6.82.* Let $V_1, V_2, ...$ be i.i.d. random vectors with $\mathbb{E}\,\|V_1\|^2 < \infty$. If $h_{n,i}$, $i = 1, ..., n$, $n = 1, 2, ...$, is a double array of vectors with $\sup_n \sum_{i=1}^n \|h_{n,i}\|^2 < \infty$ and $\lim_{n\to\infty} \max_{1\le i\le n} \|h_{n,i}\|^2 = 0$, then $Z_{n,i} = h_{n,i}^T V_i$ satisfies the Lindeberg condition

$$\lim_{n\to\infty} \sum\nolimits_{i=1}^{n} \mathbb{E} Z_{n,i}^2 I_{(\varepsilon,\infty)}(|Z_{n,i}|) = 0, \quad \varepsilon > 0.$$

6.5 Asymptotic Lower Risk Bounds, Hájek–LeCam Bound

In the previous sections we have studied the dependence of the risk and the Bayes risk on their relevant components. In Chapter 3 it has been shown that under mild assumptions the risk function $\mathsf{R}(\theta, \mathsf{D})$ is continuous in ϑ. Moreover, for a bounded and continuous loss function L the risk $\mathsf{R}(\theta, \mathsf{D})$ has been shown to depend continuously on D. Now we study the dependence of the risk on the model; that is, we consider the behavior of the risk under the weak convergence of models. This investigation is motivated by the following fact. We have seen that under mild assumptions the limit model is a Gaussian model. In such models one can often find optimal decisions under weak restrictions on the class of decisions under consideration. Consider, for example, the problem of testing the hypotheses $\mathsf{H}_0 : \mu \le \mu_0$ versus $\mathsf{H}_A : \mu > \mu_0$ in the model $(\mathbb{R}^n, \mathfrak{B}_n, (\mathsf{N}(\mu, \sigma^2))_{\mu\in\mathbb{R}})$ where σ^2 is known. In Example 2.52 we have seen that the Gauss test φ that rejects the null hypothesis for large values of $\overline{X}_n$ is a uniformly best level α test for $\alpha = \mathsf{E}_{\mu_0}\varphi$. In the next chapter it is shown that $\overline{X}_n$ as an estimator of μ has several optimality properties under the squared error loss. Another example is testing $\mathsf{H}_0 : \mu_1 = \cdots = \mu_k$ versus $\mathsf{H}_A : \delta^2 > 0$, where $\delta^2 = \sum_{i=1}^k \mu_i^2$, in the model $(\mathbb{R}^n, \mathfrak{B}_n, (\bigotimes_{i=1}^n \mathsf{N}(\mu_i, \sigma^2))_{(\mu_1,\ldots,\mu_n)\in\mathbb{R}^n})$ with a known σ^2. Here the maximin property of the χ^2-test has been established in Theorem 5.43. These are just a few examples. In the next three chapters we study estimation, testing, and selection problems systematically with the aim of finding optimal decisions. Especially optimal decisions for Gaussian models are found there.

Suppose we are given a sequence of models $\mathcal{M}_n$ and consider a sequence of associated decisions $\mathsf{D}_{\mathcal{M}_n}$. Let the models $\mathcal{M}_n$ tend to a model $\mathcal{M}$ for which we have found an optimal decision $\mathsf{D}_{\mathcal{M}}$. Then we could call the sequence $\mathsf{D}_{\mathcal{M}_n}$ optimal for the sequence $\mathcal{M}_n$ if the risk functions of $\mathsf{D}_{\mathcal{M}_n}$ in $\mathcal{M}_n$ approximate in some specific way the risk function of $\mathsf{D}_{\mathcal{M}}$ in $\mathcal{M}$. For example, this could mean that the maximum risks of $\mathsf{D}_{\mathcal{M}_n}$ converge to the maximum risk of a minimax decision in the limit model, if such a decision exists. Or it could mean that the risk functions of $\mathsf{D}_{\mathcal{M}_n}$ converge pointwise to the risk function of a uniformly best decision in the limit model, if such a decision exists. With these ideas in mind we set up the following program for finding asymptotic optimal decisions.

(A) Establish the convergence $\mathcal{M}_n \Rightarrow \mathcal{G}$.

(B) Establish an asymptotic lower bound for $\mathsf{R}(\theta, \mathsf{D}_{\mathcal{M}_n})$ in terms of the limit model.

(C) Find an optimal decision $\mathsf{D}_{\mathcal{G}}(\cdot|Z)$ for $\mathcal{G}$ that attains the lower bound and is factorized by the central variable Z.

(D) Replace Z by Z_n and show that for $\mathsf{D}_{\mathcal{M}_n^*} := \mathsf{D}_{\mathcal{G}}(\cdot|Z_n)$ the risks $\mathsf{R}(\theta, \mathsf{D}_{\mathcal{M}_n^*})$ tend to the lower bound of the risks. (6.101)

Point (A) has been the topic of the previous section on asymptotic normality of models and the LAN and ULAN condition. As to point (C), some results on finding optimal decisions have been established already in the previous chapters. The motivation of (D) is the following. As the central variable in a Gaussian model that appears as a limit model is a sufficient statistic each optimal decision can be written as a function of the central variable. As we know from Theorem 6.77 that the central sequence is asymptotically sufficient it is obvious to replace the central variable by the central sequence. To show that the sequence of decisions $\mathsf{D}_{\mathcal{M}_n^*}$ obtained in this way is asymptotically optimal, one has to prove the convergence of the risks. This is not a difficult task as the LAN condition implies the weak convergence of the distributions of the central sequences. It just remains to find conditions under which the expectations involved in the risks converge as well. Here in this section we focus on point (B).

The crucial point in the subsequent considerations is a compactness property for sequences of decisions that belong to weakly convergent sequences of models. To make this property precise we need a concept of convergence of decisions, similar to that which has been introduced in Chapter 3 for a fixed model. Below we introduce a similar concept for a weakly convergent sequence of models.

Definition 6.83. *Let $\mathcal{M}_n = (\mathcal{X}_n, \mathfrak{A}_n, (P_{n,\theta})_{\theta\in\Delta_n})$ with $\Delta_n \uparrow \Delta$ be a sequence of statistical models that converge weakly to the model $\mathcal{M} = (\mathcal{X}, \mathfrak{A}, (P_\theta)_{\theta\in\Delta})$. Suppose that the decision space $\mathcal{D}$ is a metric space and $\mathfrak{D}$ is the σ-algebra of Borel sets. A sequence of decisions $\mathsf{D}_n : \mathfrak{D} \times \mathcal{X}_n \to_k [0,1]$ is called weakly convergent to $\mathsf{D} : \mathfrak{D} \times \mathcal{X} \to_k [0,1]$ if for every $f \in \mathsf{C}_b(\mathcal{D})$ and $\theta \in \Delta$,*

$$\lim_{n\to\infty} \int [\int f(a)\mathsf{D}_n(da|x)]P_{n,\theta}(dx) = \int [\int f(a)\mathsf{D}(da|x)]P_\theta(dx). \tag{6.102}$$

In this case we write $\mathsf{D}_n \Rightarrow \mathsf{D}$. We call $\mathsf{D}_0 : \mathfrak{D} \times \mathcal{X} \to_k [0,1]$ an accumulation point of the sequence D_n if there exists a subsequence D_{n_k} with $\mathsf{D}_{n_k} \Rightarrow \mathsf{D}_0$.

Turning to a special case, let K_n and K be nonrandomized decisions. Then there are mappings $T_n : \mathcal{X}_n \to_m \mathcal{D}$ and $T : \mathcal{X} \to_m \mathcal{D}$ such that $\mathsf{K}_n = \boldsymbol{\delta}_{T_n}$ and $\mathsf{K} = \boldsymbol{\delta}_T$, and thus

$$\int [\int f(a)\mathsf{K}_n(da|x)]P_{n,\theta}(dx) = \int f(T_n(x))P_{n,\theta}(dx) = \mathsf{E}_{n,\theta} f(T_n),$$
$$\int [\int f(a)\mathsf{K}(da|x)]P_\theta(dx) = \int f(T(x))P_\theta(dx) = \mathsf{E}_\theta f(T).$$

Hence we see that $\mathsf{K}_n \Rightarrow \mathsf{K}$ holds if and only if the sequence of distributions $\mathcal{L}(T_n|P_{n,\theta})$ converges weakly to the distribution $\mathcal{L}(T|P_\theta)$ for every $\theta \in \Delta$.

To prepare for the next consideration we study the convergence (6.102) in more detail.

Problem 6.84.* If (6.102) holds and $\mathcal{D}$ is a metric space, then

$$\liminf_{n\to\infty} \int [\int f(a)\mathsf{D}_n(da|x)]P_{n,\theta}(dx) \geq \int [\int f(a)\mathsf{D}(da|x)]P_\theta(dx), \quad \theta \in \Delta,$$

for every lower bounded and lower semicontinuous function $f : \mathcal{D} \to (-\infty, \infty]$.

Problem 6.85.* If Δ is a metric space and $\mathcal{M}$ is continuous in the sense of (A7), then for a fixed $f \in \mathsf{C}_b(\mathcal{D})$ the convergence in (6.102) for every θ from a dense subset of Δ implies the convergence for every $\theta \in \Delta$.

Now we establish the existence of an accumulation point, or the compactness, of any sequence of decisions that belong to a weakly convergent sequence of models.

Proposition 6.86. *Let $\mathcal{M}_n = (\mathcal{X}_n, \mathfrak{A}_n, (P_{n,\theta})_{\theta\in\Delta_n})$, $n = 1, 2, ...,$ and $\mathcal{M} = (\mathcal{X}, \mathfrak{A}, (P_\theta)_{\theta\in\Delta})$ be models with $\Delta_n \uparrow \Delta$. Assume that $\mathcal{D}$ is a compact metric space. If $\mathcal{M}_n \Rightarrow \mathcal{M}$, then for every sequence of decisions $\mathsf{D}_{\mathcal{M}_n}$ for $\mathcal{M}_n$ and every at most countable subset $\Delta_0 \subseteq \Delta$ there exists a subsequence $\mathsf{D}_{\mathcal{M}_{n_k}}$ and a decision $\mathsf{D}_\mathcal{M}$ for $\mathcal{M}$ such that for every $f \in \mathsf{C}_b(\mathcal{D})$ and every $\theta \in \Delta_0$,*

$$\lim_{l\to\infty} \int [\int f(a)\mathsf{D}_{\mathcal{M}_{n_l}}(da|x)]P_{n_l,\theta}(dx) = \int [\int f(a)\mathsf{D}_\mathcal{M}(da|x)]P_\theta(dx). \tag{6.103}$$

If Δ is a separable metric space and the model $\mathcal{M}$ is continuous in the sense of (A7), then (6.103) holds for every $\theta \in \Delta$.

Proof. Let $\Delta_0 = \{\theta_1, \theta_2, ...\} \subseteq \Delta$ and set $F_m = \{\theta_1, ..., \theta_m\}$. According to Definition 6.10 we have $\Delta(\mathcal{M}_{n,F_m}, \mathcal{M}_{F_m}) \to 0$. It follows from (6.5) that for every m there is a sequence of kernels $\mathsf{K}_{n,m}$ with $\lim_{n\to\infty} \|P_{\theta_i,n} - \mathsf{K}_{n,m}P_{\theta_i}\| = 0$, $i = 1, ..., m$. We consider the decision

$$(\mathsf{D}_{\mathcal{M}_n}\mathsf{K}_{n,m})(B|x) := \int \mathsf{D}_{\mathcal{M}_n}(B|x_n)\mathsf{K}_{n,m}(dx_n|x)$$

for the model $\mathcal{M}_{F_m}$. Then by $|\int f(a)\mathsf{D}_{\mathcal{M}_n}(da|x_n)| \le \|f\|_u$ and Problem 1.80,

$$\begin{aligned}
&\left|\int [\int f(a)\mathsf{D}_{\mathcal{M}_n}(da|x_n)]P_{n,\theta_i}(dx_n) - \int [\int f(a)(\mathsf{D}_{\mathcal{M}_n}\mathsf{K}_{n,m})(da|x)]P_{\theta_i}(dx)\right| \\
&= \left|\int [\int f(a)\mathsf{D}_{\mathcal{M}_n}(da|x_n)]P_{n,\theta_i}(dx_n)\right. \\
&\left. - \int [\int f(a)\mathsf{D}_{\mathcal{M}_n}(da|x)](\mathsf{K}_{n,m}P_{\theta_i})(dx_n)\right| \le \|f\|_u \|P_{n,\theta_i} - \mathsf{K}_{n,m}P_{\theta_i}\| \to 0
\end{aligned}$$

as $n \to \infty$, $i = 1, ..., m$. From the compactness theorem (see Theorem 3.21) we get that for every m there exists a subsequence $n_{m,k}$ and a decision $\mathsf{D}_{0,m}$ such that $\mathsf{D}_{\mathcal{M}_{n_{m,k}}}\mathsf{K}_{n_{m,k},m} \Rightarrow \mathsf{D}_{0,m}$, which implies that for $k \to \infty$,

$$\begin{aligned}
&\lim_{k\to\infty} \int [\int f(a)\mathsf{D}_{\mathcal{M}_{n_{m,k}}}(da|x)]P_{n_{m,k},\theta_i}(dx) \qquad (6.104)\\
&= \int [\int f(a)\mathsf{D}_{0,m}(da|x)]P_{\theta_i}(dx),
\end{aligned}$$

for every $f \in \mathsf{C}_b(\mathcal{D})$ and $i = 1, ..., m$. It is clear that we may choose the subsequences to satisfy $\{n_{m,k}\} \subseteq \{n_{m-1,k}\}$. Due to Theorem 3.21 again we may assume that for some $\mathsf{D}_{\mathcal{M}}$ it holds $\mathsf{D}_{0,m} \Rightarrow \mathsf{D}_{\mathcal{M}}$. Otherwise we turn to a subsequence. Let n_l be the diagonal sequence if we arrange $\{n_{m,k}\}$, $m = 1, 2, ...$ in a double array. Then by (6.104), for every $f \in \mathsf{C}_b(\mathcal{D})$ and every $\theta \in \Delta_0$,

$$\lim_{l\to\infty} \int [\int f(a)\mathsf{D}_{\mathcal{M}_{n_l}}(da|x)]P_{n,\theta}(dx) = \int [\int f(a)\mathsf{D}_{\mathcal{M}}(da|x)]P_\theta(dx).$$

The additional statement follows from Problem 6.85. ■

Recall that a real-valued function f on a metric space $\mathcal{T}$ is lower semicontinuous at t_0 if $\liminf_{n\to\infty} f(t_n) \ge f(t_0)$ for every sequence $t_n \to t_0$. Now we formulate and prove the famous asymptotic Hájek–LeCam bound for the maximum risk in a sequence of weakly convergent models. This statement corresponds in some sense to the Cramér–Rao inequality and other lower bounds for the risk for a finite sample size, and it takes care of point (B) in (6.101).

Theorem 6.87. *Let $\mathcal{M}_n = (\mathcal{X}_n, \mathfrak{A}_n, (P_{n,\theta})_{\theta\in\Delta_n})$, $n = 1, 2, ...,$ and $\mathcal{M} = (\mathcal{X}, \mathfrak{A}, (P_\theta)_{\theta\in\Delta})$ be models with $\Delta_n \uparrow \Delta$ that satisfy $\mathcal{M}_n \Rightarrow \mathcal{M}$. Assume that the parameter space Δ is a separable metric space, and that the model $\mathcal{M}$ is* *continuous in the sense of (A7). Assume that the decision space is a compact metric space and* $\mathbb{D}_0$ *is a set of decisions that contains all accumulation points of the sequence* $\mathsf{D}_{\mathcal{M}_n}$. *Let for every fixed* $\theta \in \Delta$ *the loss function* $a \mapsto L(\theta, a)$ *be nonnegative and lower semicontinuous at every* $a \in \mathcal{D}$. *Then*

$$\liminf_{n\to\infty} \mathsf{R}(\theta, \mathsf{D}_{\mathcal{M}_n}) \geq \inf_{\mathsf{D}_{\mathcal{M}} \in \mathbb{D}_0} \mathsf{R}(\theta, \mathsf{D}_{\mathcal{M}}), \quad \theta \in \Delta. \tag{6.105}$$

If $\widetilde{\Delta} \subseteq \Delta$, *and in addition for every fixed* $a \in \mathcal{D}$ *the function* $\theta \mapsto L(\theta, a)$ *is lower semicontinuous at every* $\theta \in \widetilde{\Delta}$, *then*

$$\liminf_{n\to\infty} \sup_{\theta \in \widetilde{\Delta}} \mathsf{R}(\theta, \mathsf{D}_{\mathcal{M}_n}) \geq \inf_{\mathsf{D}_{\mathcal{M}} \in \mathbb{D}_0} \sup_{\theta \in \widetilde{\Delta}} \mathsf{R}(\theta, \mathsf{D}_{\mathcal{M}}). \tag{6.106}$$

Proof. To prove (6.105) we fix θ_0 and choose a subsequence n_l such that

$$\liminf_{n\to\infty} \mathsf{R}(\theta_0, \mathsf{D}_{\mathcal{M}_n}) = \lim_{l\to\infty} \mathsf{R}(\theta_0, \mathsf{D}_{\mathcal{M}_{n_l}}).$$

In view of Proposition 6.86 we may assume that $\mathsf{D}_{\mathcal{M}_{n_l}}$ converges weakly to some decision $\mathsf{D}_{0,\mathcal{M}}$ for the model $\mathcal{M}$. Otherwise we turn to a subsequence. Hence by Problem 6.84,

$$\liminf_{n\to\infty} \mathsf{R}(\theta_0, \mathsf{D}_{\mathcal{M}_n}) = \lim_{l\to\infty} \mathsf{R}(\theta_0, \mathsf{D}_{\mathcal{M}_{n_l}}) \geq \mathsf{R}(\theta_0, \mathsf{D}_{0,\mathcal{M}}) \geq \inf_{\mathsf{D}_{\mathcal{M}} \in \mathbb{D}_0} \mathsf{R}(\theta_0, \mathsf{D}_{\mathcal{M}}).$$

Now let Δ_0 be a countable and dense subset of $\widetilde{\Delta}$ and set

$$A = \liminf_{n\to\infty} \sup_{\theta \in \Delta_0} \mathsf{R}(\theta, \mathsf{D}_{\mathcal{M}_n}).$$

Let n_k be a subsequence such that $\lim_{k\to\infty} \sup_{\theta\in\Delta_0} \mathsf{R}(\theta, \mathsf{D}_{\mathcal{M}_{n_k}}) = A$. Similarly to the first part of the proof we may assume that there exists a decision $\mathsf{D}_{\mathcal{M}} \in \mathbb{D}_0$ such that $\mathsf{D}_{\mathcal{M}_{n_k}} \Rightarrow \mathsf{D}_{\mathcal{M}}$ as $k \to \infty$. Hence,

$$\liminf_{n\to\infty} \sup_{\theta \in \Delta_0} \mathsf{R}(\theta, \mathsf{D}_{\mathcal{M}_n}) \geq \mathsf{R}(\theta_0, \mathsf{D}_{\mathcal{M}})$$

for every fixed $\theta_0 \in \Delta_0$. As θ_0 is arbitrary we may take the supremum on the right-hand side over $\theta_0 \in \Delta_0$, and then the infimum over all accumulation points. This gives

$$\liminf_{n\to\infty} \sup_{\theta \in \widetilde{\Delta}} \mathsf{R}(\theta, \mathsf{D}_{\mathcal{M}_n}) \geq \inf_{\mathsf{D}_{\mathcal{M}} \in \mathbb{D}_0} \sup_{\theta \in \Delta_0} \mathsf{R}(\theta, \mathsf{D}_{\mathcal{M}}).$$

To complete the proof we have only to note that the continuity of the model and the lower semicontinuity of L imply, in view of Proposition 3.25, that $\mathsf{R}(\theta, \mathsf{D}_{\mathcal{M}})$ is lower semicontinuous. Hence $\sup_{\theta\in\Delta_0} \mathsf{R}(\theta, \mathsf{D}_{\mathcal{M}}) = \sup_{\theta\in\widetilde{\Delta}} \mathsf{R}(\theta, \mathsf{D}_{\mathcal{M}})$ and the proof is completed. ■

Remark 6.88. The asymptotic lower bound that has been established in (6.106) is called the lower Hájek–LeCam bound. It is only a special case of more general results established in LeCam (1972, 1979), and for estimators in Hájek (1972). See also Millar (1983) and Strasser (1985).

Suppose now that the $\mathrm{LAN}(Z_n, \mathsf{I}_0)$ condition in Definition 6.63 holds; that is, that the sequence of models $\mathcal{M}_n$ converges in view of Theorem 6.65 weakly to

$$\mathcal{G} = (\mathbb{R}^d, \mathfrak{B}_d, (\mathsf{N}(\mathsf{I}_0 h, \mathsf{I}_0))_{h\in\mathbb{R}^d}) \quad \text{or} \tag{6.107}$$

$$\mathcal{G}_0 = (\mathbb{R}^d, \mathfrak{B}_d, (\mathsf{N}(h, \Sigma_0))_{h\in\mathbb{R}^d}), \quad \text{for } \det(\mathsf{I}_0) \neq 0,\ \Sigma_0 = \mathsf{I}_0^{-1}. \tag{6.108}$$

We now specify the two lower bounds, for the risk and the maximum risk, that have been established in Theorem 6.87 for the case where the limit model is given by (6.107) or (6.108).

Proposition 6.89. *Let $\mathcal{M}_n = (\mathcal{X}_n, \mathfrak{A}_n, (P_{n,h})_{h\in\Delta_n})$ be a sequence of models with $\Delta_n \uparrow \mathbb{R}^d$ that fulfills the LAN(Z_n, I_0) condition in Definition 6.63. If the decision space $\mathcal{D}$ is a compact metric space, $C \subseteq \mathbb{R}^d$, $L : C \times \mathcal{D} \to \mathbb{R}_+$ is lower semicontinuous, and $\mathbb{D}_\mathcal{G}$ contains all accumulation points of the sequence $\mathsf{D}_{\mathcal{M}_n}$, then*

$$\liminf_{n\to\infty} \sup_{h\in C} \mathsf{R}(h, \mathsf{D}_{\mathcal{M}_n}) \geq \inf_{\mathsf{D}_\mathcal{G}\in\mathbb{D}_\mathcal{G}} \sup_{h\in C} \int \Big[\int L(h,a)\mathsf{D}_\mathcal{G}(da|x)\Big]\mathsf{N}(\mathsf{I}_0 h, \mathsf{I}_0)(dx),$$

$$\liminf_{n\to\infty} \mathsf{R}(h, \mathsf{D}_{\mathcal{M}_n}) \geq \inf_{\mathsf{D}_\mathcal{G}\in\mathbb{D}_\mathcal{G}} \int \Big[\int L(h,a)\mathsf{D}_\mathcal{G}(da|x)\Big]\mathsf{N}(\mathsf{I}_0 h, \mathsf{I}_0)(dx), \tag{6.109}$$

for all $h \in \mathbb{R}^d$. If $\mathbb{D}_{\mathcal{G}_0}$ contains all accumulation points of the sequence $\mathsf{D}_{\mathcal{M}_n}$ and $\det(\mathsf{I}_0) \neq 0$, then with $\Sigma_0 = \mathsf{I}_0^{-1}$ it holds

$$\liminf_{n\to\infty} \sup_{h\in C} \mathsf{R}(h, \mathsf{D}_{\mathcal{M}_n}) \geq \inf_{\mathsf{D}_{\mathcal{G}_0}\in\mathbb{D}_{\mathcal{G}_0}} \sup_{h\in C} \int \Big[\int L(h,a)\mathsf{D}_{\mathcal{G}_0}(da|x)\Big]\mathsf{N}(h, \Sigma_0)(dx),$$

$$\liminf_{n\to\infty} \mathsf{R}(h, \mathsf{D}_{\mathcal{M}_n}) \geq \inf_{\mathsf{D}_{\mathcal{G}_0}\in\mathbb{D}_{\mathcal{G}_0}} \int \Big[\int L(h,a)\mathsf{D}_{\mathcal{G}_0}(da|x)\Big]\mathsf{N}(h, \Sigma_0)(dx), \tag{6.110}$$

for all $h \in \mathbb{R}^d$.

Proof. Problem 6.62 shows that the models with $(\mathsf{N}(\mathsf{I}_0 h, \mathsf{I}_0))_{h\in\mathbb{R}^d}$, and with $(\mathsf{N}(h, \Sigma_0))_{h\in\mathbb{R}^d}$ for $\det(\mathsf{I}_0) \neq 0$ and $\Sigma_0 = \mathsf{I}_0^{-1}$, are continuous in the sense of condition (A7). Therefore (6.109) and (6.110) follow from (6.106), where the corresponding second statement follows from the former one by setting $C = \{h\}$. ■

Although the statements (6.109) and (6.110) are equivalent due to the equivalence of $\mathcal{G}$ and $\mathcal{G}_0$, it is convenient to have separate formulations, as it depends on the concrete situation which model is easier to deal with.

The term on the right-hand side of (6.109) is the lower Hájek–LeCam bound if the limit model is a finite-dimensional Gaussian shift model. As we have pointed out already such bounds become important when characterizing the asymptotic optimality of decisions in a minimax sense. In a first step, of course, one has to evaluate such a bound. The lower bound in (6.109) refers to the limit model which is the model (6.107). If there exists a uniformly best decision in $\mathbb{D}_0$ for the limit model $\mathcal{G}$, say D_0, then we call every sequence of decisions $\mathsf{D}_{\mathcal{M}_n}$ *asymptotically uniformly best* with respect to $\mathbb{D}_0$ that satisfies

$$\lim_{n\to\infty} \mathsf{R}(h, \mathsf{D}_{\mathcal{M}_n}) = \mathsf{R}(h, \mathsf{D}_0), \quad h \in \Delta,$$

where Δ is either $\mathbb{R}^d$ or a subset of $\mathbb{R}^d$, depending on the problem under consideration. In testing problems, $\mathbb{D}_0$ is typically chosen to be the class of all level α tests, or the class of all unbiased level α tests. Likewise, we call a sequence of decisions $D_{\mathcal{M}_n}$ that attains the lower Hájek–LeCam minimax bounds in Proposition 6.89 *asymptotically minimax.* Sequences cf estimators, tests, and selection rules that are asymptotically uniformly best, or asymptotically minimax, are studied systematically in the next three chapters.

6.6 Solutions to Selected Problems

Solution to Problem 6.1: In view of the inequality (6.3) and the definition of $\delta(\mathcal{M}_1, \mathcal{M}_2)$ in (6.2) it is enough to show that

$$d(\mathcal{M}_1, \mathcal{M}_2) \le \sup_{F \subseteq \Delta, |F| < \infty} d(\mathcal{M}_{1,F}, \mathcal{M}_{2,F}),$$

and a corresponding statement holds for $d(\mathcal{M}_2, \mathcal{M}_1)$. It suffices to prove the first one. If the right-hand term is smaller than A, then $d(\mathcal{M}_{1,F}, \mathcal{M}_{2,F}) < A$ for every finite F. Hence for every finite decision space $\mathcal{D}$ and every loss function L with $\|L\|_u \le 1$ and every decision $\mathsf{D}_{\mathcal{M}_2} : \mathfrak{D} \times \mathcal{X}_2 \to_k [0,1]$, there exists a decision $\mathsf{D}_{\mathcal{M}_1} : \mathfrak{D} \times \mathcal{X}_1 \to_k [0,1]$ with

$$\mathsf{R}(\theta, \mathsf{D}_{\mathcal{M}_1}) \le \mathsf{R}(\theta, \mathsf{D}_{\mathcal{M}_2}) + A, \quad \theta \in F.$$

But this implies $d(\mathcal{M}_1, \mathcal{M}_2) \le A$ and the proof is complete. □

Solution to Problem 6.8: Let μ dominate P_i and Q_i with densities f_i and g_i, respectively, $i = 1, ..., m$. Then $|a^s - b^s| \le |a - b|^s$ for $a, b \ge 0$ and $0 < s < 1$ implies

$$\begin{aligned} &|\mathsf{H}_s(P_1, ..., P_m) - \mathsf{H}_s(Q_1, ..., Q_m)| \\ &\le \int |f_1 - g_1|^{s_1} f_2^{s_2} \cdots f_m^{s_m} d\mu + \cdots + \int g_1^{s_1} \cdots g_{m-1}^{s_{m-1}} |f_m - g_m|^{s_m} d\mu \\ &\le \sum\nolimits_{i=1}^m \left(\int |f_i - g_i| d\mu\right)^{s_i} = \sum\nolimits_{i=1}^m \|P_i - Q_i\|^{s_i}. \end{aligned}$$

The last inequality follows from the generalized Hölder inequality $\int h_1^{\alpha_1} \cdots h_m^{\alpha_m} d\mu \le (\int h_1 d\mu)^{\alpha_1} \cdots (\int h_m d\mu)^{\alpha_m}$, $h_i \ge 0$, $\alpha_i > 0$, $\sum_{i=1}^m \alpha_i = 1$. Hence by $\|P_i - Q_i\| \le 2$,

$$\begin{aligned} |\mathsf{H}_s(P_1, ..., P_m) - \mathsf{H}_s(Q_1, ..., Q_m)| &\le \sum\nolimits_{i=1}^m 2^{s_i} (\|P_i - Q_i\| /2)^{s_i} \\ &\le 2 \sum\nolimits_{i=1}^m \|P_i - Q_i\|^{m(s)}. \qquad \square \end{aligned}$$

Solution to Problem 6.24: The functions $f_s(x) = \mathsf{u}_s(x)/\mathsf{u}_{1/2}(x)$ and $1/f_s(x)$ are bounded and continuous on $[0, 2]$. This proves (6.29). The function

$$\mathsf{w}_s(z) = sz + (1-s) - z^s - 4s(1-s)(\frac{1}{2}z + \frac{1}{2} - z^{1/2})$$

satisfies $\mathsf{w}_s(1) = \mathsf{w}_s'(1) = 0$ and $\mathsf{w}_s''(z) = s(1-s)(z^{s-2} - z^{-3/2})$. Hence

$$\sup\{\frac{|\mathsf{w}_s(z)|}{(z-1)^2} : \frac{1-\varepsilon}{1+\varepsilon} \le z \le \frac{1+\varepsilon}{1-\varepsilon},\ 0 < s < 1\} \to 0, \quad \varepsilon \to 0.$$

Otherwise $\lim_{z\to 1}(z-1)^{-2}(\frac{1}{2}z + \frac{1}{2} - z^{1/2}) = \frac{1}{8}$. To complete the proof of (6.30) we remark that

$$\mathsf{u}_s(x) - 4s(1-s)\mathsf{u}_{1/2}(x) = \mathsf{w}_s(\frac{x}{2-x})(2-x).$$

To prove (6.31), we set

$$\mathsf{v}_s(z) = \frac{1}{s(1-s)}(sz + (1-s) - z^s).$$

Then $\mathsf{v}_s(1) = \mathsf{v}_s'(1) = \mathsf{v}_{1/2}(1) = \mathsf{v}_{1/2}'(1) = 0$. If $z \ge \varepsilon/(2-\varepsilon)$, $0 < \varepsilon < 1$, and $0 < s \le 1/2$, then

$$\begin{aligned}\mathsf{v}_s''(z) &= z^{s-2} = z^{s-1/2}\mathsf{v}_{1/2}''(z) \le (\frac{2-\varepsilon}{\varepsilon})^{1/2}\mathsf{v}_{1/2}''(z),\\ \mathsf{v}_s(z) &= \int_1^z (z-t)\mathsf{v}_s''(t)dt\\ &\le (\frac{2-\varepsilon}{\varepsilon})^{1/2}\int_1^z (z-t)\mathsf{v}_{1/2}''(t)dt = (\frac{2-\varepsilon}{\varepsilon})^{1/2}\mathsf{v}_{1/2}(z).\end{aligned}$$

Replacing z with $x/(2-x)$ we get

$$\sup_{0<s<1/2,\,\varepsilon\le x<2}\frac{\mathsf{v}_s(x/(2-x))(2-x)}{\mathsf{v}_{1/2}(x/(2-x))(2-x)} = \sup_{0<s<1/2,\,\varepsilon\le x<2}\frac{\mathsf{u}_s(x)}{\mathsf{u}_{1/2}(x)} \le (\frac{2-\varepsilon}{\varepsilon})^{1/2}.$$

To prove (6.32) we remark that

$$\begin{aligned}&1 - x^s y^{1-s} - (1-x)^s(1-y)^{1-s} \ge 1 - x^s y^{1-s} - s(1-x) - (1-s)(1-y)\\ &= sx + (1-s)y - x^s y^{1-s} \ge sx + (1-s)y - c^{-s}y \ge (1-s-c^{-s})y. \quad \square\end{aligned}$$

Solution to Problem 6.28: It holds $\mathsf{v}^*(x) = x\mathsf{v}(1/x)$ and thus $\lim_{x\to\infty} x^{-1}\mathsf{v}^*(x) = \infty$. Furthermore, by (1.73) it holds $\sup_n \mathsf{I}_{\mathsf{v}^*}(Q_n, P_n) = \sup_n \mathsf{I}_{\mathsf{v}}(P_n, Q_n) < \infty$. Without loss of generality we may assume that $\mathsf{v}^* \ge 0$ and that v^* is nondecreasing for $x \ge 1$. Otherwise we could turn to v_0^* in (1.62). We get from (1.69) that $Q_n \ll P_n$ for every n. Put $\mathsf{w}(c) = \sup_{x\ge c}(x/\mathsf{v}^*(x))$. Then $\lim_{c\to\infty}\mathsf{w}(c) = 0$ and by $\mathsf{v}^* \ge 0$,

$$\begin{aligned}Q_n(L_n > c) &= \int I_{(c,\infty)}(L_n)L_n dP_n \le \int \frac{L_n}{\mathsf{v}^*(L_n)} I_{(c,\infty)}(L_n)\mathsf{v}^*(L_n)dP_n\\ &\le \mathsf{w}(c)\int \mathsf{v}^*(L_n)dP_n = \mathsf{w}(c)(\sup_n \mathsf{I}_{\mathsf{v}}(P_n, Q_n)).\end{aligned}$$

Hence $\lim_{c\to\infty}\sup_n Q_n(L_n > c) = 0$ and we get the condition (C) in Theorem 6.26.
$\square$

Solution to Problem 6.29: Put $A_n = \{|X_n| > \varepsilon\}$ and apply (A) in Theorem 6.26. □

Solution to Problem 6.30: $T_n = O_{P_n}(1)$ means $\lim_{c\to\infty} \limsup_{n\to\infty} P_n(\|T_n\| > c) = 0$, which implies $\lim_{n\to\infty} P_n(\|T_n\| > c_n) = 0$ for every sequence $c_n \to \infty$. Conversely if $\lim_{c\to\infty} \limsup_{n\to\infty} P_n(\|T_n\| > c) > 0$, then there is a sequence $c_n \to \infty$ with $\liminf_{n\to\infty} P_n(\|T_n\| > c_n) > 0$. Hence $T_n = O_{P_n}(1)$ if and only if $\lim_{n\to\infty} P_n(\|T_n\| > c_n) = 0$ for every sequence $c_n \to \infty$. Hence $\lim_{n\to\infty} Q_n(\|T_n\| > c_n) = 0$ by $\{Q_n\} \lhd \{P_n\}$ (see Theorem 6.26) and thus $T_n = O_{Q_n}(1)$. □

Solution to Problem 6.40: Set $R = \frac{1}{2}(P+Q)$, $M = dP/dR$. Then $\mathsf{D}^2(P,Q) = \mathsf{E}_R(M^{1/2} - (2-M)^{1/2})^2$, $\mathsf{D}^2(P, \frac{1}{2}(P+Q)) = \mathsf{E}_R(M^{1/2}-1)^2$, $\mathsf{D}^2(Q, \frac{1}{2}(P+Q)) = \mathsf{E}_R((2-M)^{1/2}-1)^2$. Then by $(a+b)^2 \le 2(a^2+b^2)$ the left-hand side inequality follows with $a = M^{1/2} - 1$ and $b = 1-(2-M)^{1/2}$. On the other hand, by the concavity of $\sqrt{x}$,

$$\begin{aligned}
&\mathsf{E}_R(M^{1/2}-1)^2 = 2(1-\mathsf{E}_R M^{1/2}) = 2 - 2\mathsf{E}_R(\frac{1}{2}[M + (2-M)]M)^{1/2} \\
&\le 2 - \mathsf{E}_R\,(MM)^{1/2} - \mathsf{E}_R\,((2-M)M)^{1/2} = 1 - \mathsf{H}_{1/2}(P,Q) = \frac{1}{2}\mathsf{D}^2(P,Q).
\end{aligned}$$

Hence $2\mathsf{D}^2(P, \frac{1}{2}(P+Q)) \le \mathsf{D}^2(P,Q)$ and $2\mathsf{D}^2(Q, \frac{1}{2}(P+Q)) \le \mathsf{D}^2(Q,P) = \mathsf{D}^2(P,Q)$ by exchanging the roles of P and Q. □

Solution to Problem 6.52: Using (6.43) it holds

$$\begin{aligned}
&\frac{1}{2}\sum\nolimits_{i=1}^n \mathsf{D}^2(P_{n,i}, Q_{n,i}) = \sum\nolimits_{i=1}^n (1 - \mathsf{H}_{1/2}(P_{n,i}, Q_{n,i})) \le \sum\nolimits_{i=1}^n \mathsf{G}_{1/2}(P_{n,i}, Q_{n,i}) \\
&\quad = -\ln \prod\nolimits_{i=1}^n \mathsf{H}_{1/2}(P_{n,i}, Q_{n,i}) = -\ln \mathsf{H}_{1/2}(\textstyle\bigotimes_{i=1}^n P_{n,i}, \bigotimes_{i=1}^n Q_{n,i}).
\end{aligned}$$

It remains to apply Theorem 6.37. □

Solution to Problem 6.55: It holds for $0 < \varepsilon < 1$,

$$\begin{aligned}
&\sum\nolimits_{i=1}^n \mathsf{E}_{P_n}(Y_{n,i}^2 \wedge |Y_{n,i}|^3) \\
&\le \sum\nolimits_{i=1}^n \mathsf{E}_{P_n} I_{[0,\varepsilon)}(|Y_{n,i}|)|Y_{n,i}|^3 + \sum\nolimits_{i=1}^n \mathsf{E}_{P_n} I_{(\varepsilon,\infty)}(|Y_{n,i}|)Y_{n,i}^2 \\
&\le \varepsilon \sum\nolimits_{i=1}^n \mathsf{E}_{P_n} Y_{n,i}^2 + \sum\nolimits_{i=1}^n \mathsf{E}_{P_n} I_{(\varepsilon,\infty)}(|Y_{n,i}|)Y_{n,i}^2.
\end{aligned}$$

By taking first $n \to \infty$, and then $\varepsilon \to 0$, we get the statement from *(6.52)*. Conversely,

$$\begin{aligned}
&\sum\nolimits_{i=1}^n \mathsf{E}_{P_n} I_{(\varepsilon,\infty)}(|Y_{n,i}|)Y_{n,i}^2 \le \varepsilon^2 \sum\nolimits_{i=1}^n \mathsf{E}_{P_n}(\frac{Y_{n,i}}{\varepsilon})^2 \wedge (\frac{|Y_{n,i}|}{\varepsilon})^3 \le \\
&\varepsilon^2 \sum\nolimits_{i=1}^n \mathsf{E}_{P_n} \frac{Y_{n,i}^2}{\varepsilon^2 \wedge \varepsilon^3} \wedge \frac{|Y_{n,i}|^3}{\varepsilon^2 \wedge \varepsilon^3} \le \frac{\varepsilon^2}{\varepsilon^2 \wedge \varepsilon^3} \sum\nolimits_{i=1}^n \mathsf{E}_{P_n}(Y_{n,i}^2 \wedge |Y_{n,i}|^3). \quad \square
\end{aligned}$$

Solution to Problem 6.56:

$$\max_{1\le i\le n} \mathsf{E}_{P_n} Y_{n,i}^2 \le \varepsilon^2 + \max_{1\le i\le n} \mathsf{E}_{P_n} I_{[\varepsilon,\infty)}(|Y_{n,i}|)Y_{n,i}^2 \le \varepsilon^2 + \sum\nolimits_{i=1}^{n} \mathsf{E}_{P_n} I_{[\varepsilon,\infty)}(|Y_{n,i}|)Y_{n,i}^2.$$

By taking first $n \to \infty$, and then $\varepsilon \to 0$, we get the statement (6.59). The second statement follows from Problem 6.55 and

$$\begin{aligned} P_n(\max_{1\le i\le n} |Y_{n,i}| > \varepsilon) &\le \sum\nolimits_{i=1}^{n} P_n(|Y_{n,i}| > \varepsilon) \\ &\le \frac{1}{\varepsilon^2 \wedge \varepsilon^3} \sum\nolimits_{i=1}^{n} \mathsf{E}_{P_n}(Y_{n,i}^2 \wedge |Y_{n,i}|^3). \quad \square \end{aligned}$$

Solution to Problem 6.57: The first statement follows from (6.49). The second statement follows from (6.49), (6.51), and (6.53). □

Solution to Problem 6.59: The inequalities $(a+b)^2 \le 2a^2 + 2b^2$ and

$$Z_{n,i}^2 I_{(0,\varepsilon/2]}(|Z_{n,i}|) I_{(\varepsilon,\infty)}(|Y_{n,i}|) \le (Y_{n,i} - Z_{n,i})^2$$

imply

$$\begin{aligned} &\sum\nolimits_{i=1}^{n} \mathbb{E} Y_{n,i}^2 I_{(\varepsilon,\infty)}(|Y_{n,i}|) \\ &\le 2\sum\nolimits_{i=1}^{n} \mathbb{E} Z_{n,i}^2 I_{(\varepsilon,\infty)}(|Y_{n,i}|) + 2\sum\nolimits_{i=1}^{n} \mathbb{E}(Y_{n,i} - Z_{n,i})^2, \\ &\sum\nolimits_{i=1}^{n} \mathbb{E} Z_{n,i}^2 I_{(\varepsilon,\infty)}(|Y_{n,i}|) \\ &\le \sum\nolimits_{i=1}^{n} \mathbb{E} Z_{n,i}^2 I_{(\varepsilon/2,\infty)}(|Z_{n,i}|) + \sum\nolimits_{i=1}^{n} \mathbb{E} Z_{n,i}^2 I_{(0,\varepsilon/2]}(|Z_{n,i}|) I_{(\varepsilon,\infty)}(|Y_{n,i}|) \\ &\le \sum\nolimits_{i=1}^{n} \mathbb{E} Z_{n,i}^2 I_{(\varepsilon/2,\infty)}(|Z_{n,i}|) + \sum\nolimits_{i=1}^{n} \mathbb{E}(Y_{n,i} - Z_{n,i})^2. \quad \square \end{aligned}$$

Solution to Problem 6.62: We get from (6.66), $\mathcal{L}(Z|N(0, I_0)) = N(0, I_0)$, and (6.61),

$$\begin{aligned} &H_{1/2}(N(I_0 h_1, I_0), N(I_0 h_2, I_0)) \\ &= \mathsf{E}_0 \exp\{\frac{1}{2}(h_1 + h_2)^T Z - \frac{1}{4} h_1^T I_0 h_1 - \frac{1}{4} h_2^T I_0 h_2\} \\ &= \exp\{\frac{1}{8}(h_1 + h_2)^T I_0 (h_1 + h_2) - \frac{1}{4} h_1^T I_0 h_1 - \frac{1}{4} h_2^T I_0 h_2\} \\ &= \exp\{-\frac{1}{8}(h_1 - h_2)^T I_0 (h_1 - h_2)\}. \end{aligned}$$

The relation $D^2(P_{h_1}, P_{h_2}) = 2(1 - H_{1/2}(P_{h_1}, P_{h_2}))$ yields the second statement. The third statement follows from the inequality $\|P_0 - P_1\| \le 2D(P_0, P_1)$; see Proposition 1.84. □

Solution to Problem 6.69: The statement (6.85) follows from the definition of $\gamma(\delta)$. It holds $Q_{n,i}(Y_{n,i} = \infty) = 0$ for all sufficiently large n. The relation (6.49) yields $\mathsf{E}_{P_n} Y_{n,i}^2 = -2\mathsf{E}_{P_n} Y_{n,i}$ for such n. This gives (6.84). The statement (6.86) follows from $(a+b)^2 \le 2a^2 + 2b^2$, the definition of $R(\delta)$, and (6.84). □

Solution to Problem 6.79: The direction $\Rightarrow$ is clear. Conversely, if $A := \limsup_{n\to\infty} \sup_{s\in\mathcal{S}} \rho_{\mathcal{T}}(\varphi_n(s), \varphi(s)) > 0$, then there exist subsequences φ_{n_k} and s_{n_k}, where the sequence s_{n_k} by the compactness of $\mathcal{S}$ may be assumed to be convergent, say to s, such that $A = \lim_{k\to\infty} \rho_{\mathcal{T}}(\varphi_{n_k}(s_{n_k}), \varphi(s_{n_k}))$. The continuity of φ yields

$$\begin{aligned}
0 = &\lim_{k\to\infty} \rho_{\mathcal{T}}(\varphi_{n_k}(s_{n_k}), \varphi(s)) \geq \\
&\lim_{k\to\infty} \rho_{\mathcal{T}}(\varphi_{n_k}(s_{n_k}), \varphi(s_{n_k})) - \lim_{k\to\infty} \rho_{\mathcal{T}}(\varphi(s_{n_k}), \varphi(s)) = A > 0. \quad \square
\end{aligned}$$

Solution to Problem 6.80: Cover $\mathcal{S}$ with a finite number of balls $B_\delta(s_i)$, $i = 1, ..., N$, with a diameter not exceeding δ and the centers s_i from the dense subset on which the pointwise convergence holds. Then

$$\begin{aligned}
&\sup_{s\in\mathcal{S}} \rho_{\mathcal{T}}(\varphi_n(s), \varphi(s)) \leq \max_{1\leq i\leq N} \rho_{\mathcal{T}}(\varphi_n(s_i), \varphi(s_i)) \\
&+ \sup_{s_1,s_2:\rho_{\mathcal{S}}(s_1,s_2)\leq\delta} \rho_{\mathcal{T}}(\varphi_n(s_1), \varphi_n(s_2)) + \sup_{s_1,s_2:\rho_{\mathcal{S}}(s_1,s_2)\leq\delta} \rho_{\mathcal{T}}(\varphi(s_1), \varphi(s_2)).
\end{aligned}$$

Take first $n \to \infty$, and then $\delta \to 0$, to get the statement. $\square$

Solution to Problem 6.81: The direction (6.98) $\Rightarrow$ (6.99) is clear. If (6.98) is not fulfilled, then there is a subsequence n_k, a sequence $\delta_k \to 0$, and sequences $s_{k,i}$ with $\rho_{\mathcal{S}}(s_{k,1}, s_{k,2}) \leq \delta_k$ and $\rho_{\mathcal{T}_{n_k}}(\varphi_{n_k}(s_{k,1}), \varphi_{n_k}(s_{k,2})) \geq \varepsilon > 0$ for some $\varepsilon > 0$. Due to the compactness of $\mathcal{S}$ we may assume that the sequences $s_{k,i}$ converge. They converge to the same point, say s_0, as $\rho_{\mathcal{S}}(s_{k,1}, s_{k,2}) \leq \delta_k \to 0$. The inequality

$$\rho_{\mathcal{T}_{n_k}}(\varphi_{n_k}(s_{k,1}), \varphi_{n_k}(s_{k,2})) \leq \rho_{\mathcal{T}_{n_k}}(\varphi_{n_k}(s_{k,1}), \varphi_{n_k}(s_0)) + \rho_{\mathcal{T}_{n_k}}(\varphi_{n_k}(s_0), \varphi_{n_k}(s_{k,2}))$$

contradicts (6.99).

The implication (6.98) $\Rightarrow$ (6.100) again is clear. If (6.98) is not fulfilled there is a subsequence s_{n_k} which may assumed to be convergent to s_0, say, such that $\lim_{k\to\infty} \rho_{\mathcal{T}_{n_k}}(\varphi_{n_k}(s_{n_k}), \varphi_{n_k}(s_0)) > 0$, which contradicts (6.100). Hence the equivalence of (6.98), (6.99), and (6.100) is established. $\square$

Solution to Problem 6.82: Put $\alpha_n = \max_{1\leq i\leq n} \|h_{n,i}\|$. Then for $\varepsilon > 0$,

$$\begin{aligned}
&\sum\nolimits_{i=1}^n \mathbb{E}I_{(\varepsilon,\infty)}(|Z_{n,i}|)Z_{n,i}^2 \\
&\leq \sum\nolimits_{i=1}^n \|h_{n,i}\|^2 \mathbb{E}I_{(\varepsilon,\infty)}(\alpha_n \|V_i\|) \|V_i\|^2 \\
&\leq [\mathbb{E}\|V_1\|^2 I_{[\varepsilon/\alpha_n,\infty)}(\|V_1\|)] \sum\nolimits_{i=1}^n \|h_{n,i}\|^2 .
\end{aligned}$$

Lebesgue's theorem, $\alpha_n \to 0$, and $\sup_n \sum_{i=1}^n \|h_{n,i}\|^2 < \infty$ yield the statement for $n \to \infty$. $\square$

Solution to Problem 6.84: Use (6.102) and Theorem A.49. $\square$

Solution to Problem 6.85: As the weak convergence of models implies the convergence of the Hellinger distances (see Theorem 6.13) we get from the first inequality in Proposition 1.84,

$$\begin{aligned}
&\limsup_{n\to\infty} \|P_{n,\theta_1} - P_{n,\theta_2}\| \\
&\le 2 \lim_{n\to\infty} \mathsf{D}(P_{n,\theta_1}, P_{n,\theta_2}) = 2\mathsf{D}(P_{\theta_1}, P_{\theta_2}) \le 2 \|P_{\theta_1} - P_{\theta_2}\|^{1/2}.
\end{aligned}$$

Let $\theta \in \Delta$ be fixed and θ_m be from the dense set with $\theta_m \to \theta$. Then

$$\begin{aligned}
&\limsup_{n\to\infty} \left| \int [\int f(a)\mathsf{D}_n(da|x)] P_{n,\theta}(dx) - \int [\int f(a)\mathsf{D}(da|x)] P_\theta(dx) \right| \\
&\le \limsup_{n\to\infty} \left| \int [\int f(a)\mathsf{D}_n(da|x)] (P_{n,\theta} - P_{n,\theta_m})(dx) \right| \\
&\quad + \limsup_{n\to\infty} \left| \int [\int f(a)\mathsf{D}_n(da|x)] P_{n,\theta_m}(dx) - \int [\int f(a)\mathsf{D}(da|x)] P_{\theta_m}(dx) \right| \\
&\quad + \limsup_{n\to\infty} \left| \int [\int f(a)\mathsf{D}(da|x)] (P_{\theta_m} - P_\theta)(dx) \right| \\
&\le 2 \|f\|_u \|P_{\theta_m} - P_\theta\|^{1/2} + \|f\|_u \|P_{\theta_m} - P_\theta\|,
\end{aligned}$$

as the middle term on the right-hand side of the first inequality vanishes by assumption. Take $m \to \infty$ to complete the proof. □

7

Estimation

7.1 Lower Information Bounds in Estimation Problems

The theory of estimation is a fundamental part of mathematical statistics with a long history. A complete presentation is not given in this section. Instead, we concentrate on basic ideas and establish selected results, which, admittedly, reflect the personal taste of the authors. However, the main ideas are presented in a way that allows the reader to get connected with the classical monographs, such as Lehmann and Casella (1998), Pfanzagl (1994), Witting (1985), and Witting and Müller-Funk (1995). Readers who are mainly interested in asymptotic results are referred to the fundamental monograph by Bickel, Klaassen, Ritov, and Wellner (1993) and to the books by Lehmann (1998) and Serfling (1980).

As we have seen already in the previous sections, a statistical inference for a model $(\mathcal{X}, \mathfrak{A}, (P_\theta)_{\theta\in\Delta})$ can be made under different points of view, depending on the goals that are pursued. The goals are then specified by the choice of the decision space. In this section our goal is to estimate the parameter θ or a function $\kappa(\theta)$ of it. For this purpose we want to construct a mapping $S : \mathcal{X} \to_m \Delta$ that approximates the unknown θ, or $\kappa(\theta)$, as well as possible. According to Definition 3.5 the estimation problem consists of a statistical model $(\mathcal{X}, \mathfrak{A}, (P_\theta)_{\theta\in\Delta})$, a function $\kappa : \Delta \to \mathcal{S}$, where $\mathcal{S}$ is equipped with the σ-algebra $\mathfrak{S}$, and a function $l : \mathcal{S} \times \mathcal{S} \to_m \mathbb{R}$ that is bounded from below and defines the loss by $L(\theta, a) = l(\kappa(\theta), a)$. If $\Delta \subseteq \mathbb{R}^d$ and $\kappa(\theta) = \theta$, so that $\mathcal{S} = \Delta$, then we assume that $\Delta \in \mathfrak{B}_d$ and $\mathfrak{S} = \mathfrak{B}_\Delta$. A decision (i.e., a randomized estimator) is then a stochastic kernel $\mathsf{D} : \mathfrak{S} \times \mathcal{X} \to_k [0, 1]$. If $\mathsf{D}(\cdot|x) = \boldsymbol{\delta}_{S(x)}$, then S is called an estimator; see Definition 3.5. The risk of both types of estimators is given by

$$\mathsf{R}(\theta, \mathsf{D}) = \int [\int l(\kappa(\theta), a)\mathsf{D}(da|x)]P_\theta(dx),$$

$$\mathsf{R}(\theta, S) = \int l(\kappa(\theta), S(x))P_\theta(dx) = \mathsf{E}_\theta l(\kappa(\theta), S).$$

F. Liese, K.-J. Miescke, *Statistical Decision Theory*,
DOI: 10.1007/978-0-387-73194-0_7,

To exclude exotic or meaningless estimators, such as no-data estimators, one often has to restrict the class of estimators considered. One commonly used restriction is the requirement of unbiasedness.

Definition 7.1. *Given a model* $(\mathcal{X}, \mathfrak{A}, (P_\theta)_{\theta\in\Delta})$ *and a function* $\kappa : \Delta \to \mathbb{R}^d$, *we call an estimator* $S : \mathcal{X} \to_m \mathbb{R}^d$ *unbiased if* $\mathsf{E}_\theta \|S\| < \infty$, *and* $\mathsf{E}_\theta S = \kappa(\theta)$, $\theta \in \Delta$.

Depending on the structure of the model and the function κ unbiased estimators may or may not exist, as the following example shows.

Example 7.2. Consider the problem of estimating the success probability in the binomial distribution, so that the model is $(\{0,...,n\}, \mathfrak{P}(\{0,...,n\}), (\mathsf{B}_{n,p})_{0<p<1})$. If there exists an unbiased estimator of $\kappa : (0,1) \to \mathbb{R}$, then $\kappa(p)$ must be a polynomial with a degree not exceeding n, because every unbiased estimator must satisfy

$$\sum\nolimits_{k=0}^{n} S(k)\mathsf{b}_{n,p}(k) = \sum\nolimits_{k=0}^{n} S(k)\binom{n}{k} p^k(1-p)^{n-k} = \kappa(p), \quad 0 < p < 1.$$

Consider the case of some real-valued function $\kappa : \Delta \to \mathbb{R}$ and use the squared error loss $l(t,a) = (t-a)^2$. Then the risk of an estimator $S : \mathcal{X} \to_m \mathbb{R}$ is given by $\mathsf{R}(\theta, S) = \mathsf{E}_\theta(S - \kappa(\theta))^2$. If the estimator is unbiased, then $\mathsf{R}(\theta, S) = \mathsf{V}_\theta(S) = \mathsf{E}_\theta(S - \mathsf{E}_\theta S)^2$, whereas in general the *bias* $\mathsf{E}_\theta S - \kappa(\theta)$ leads to an additional term; that is, $\mathsf{R}(\theta, S) = \mathsf{V}_\theta(S) + (\mathsf{E}_\theta S - \kappa(\theta))^2$.

We now construct lower bounds for the variance, and also for the more general risk $\mathsf{E}_\theta|S - \kappa(\theta)|^\beta$, $\beta > 1$, of an unbiased estimator S. These lower bounds include special v-divergences that have been introduced in (1.74). Especially, we recall the Hellinger distance introduced in (1.75) and the χ^s-divergence in (1.74). We also recall that by (1.69) and the definition of $\chi^s(P_0, P_1)$ in (1.74) for $s > 1$ it holds $\chi^s(P_0, P_1) = \infty$ if P_0 is not absolutely continuous with respect to P_1.

Theorem 7.3. *If S is an unbiased estimator of $\kappa : \Delta \to \mathbb{R}$ with finite second moment, then for every point $\theta_1, \theta_2 \in \Delta$ it holds*

$$\mathsf{V}_{\theta_1}(S) + \mathsf{V}_{\theta_2}(S) \ge \frac{1 - \mathsf{D}^2(P_{\theta_1}, P_{\theta_2})}{2\mathsf{D}^2(P_{\theta_1}, P_{\theta_2})}[\kappa(\theta_1) - \kappa(\theta_2)]^2. \tag{7.1}$$

If $P_{\theta_2} \ll P_{\theta_1}$, then

$$\mathsf{E}_{\theta_1}|S - \kappa(\theta_1)|^\beta \ge \frac{|\kappa(\theta_2) - \kappa(\theta_1)|^\beta}{[\chi^\alpha(P_{\theta_2}, P_{\theta_1})]^{\beta/\alpha}}, \quad \alpha > 1, \quad \frac{1}{\alpha} + \frac{1}{\beta} = 1. \tag{7.2}$$

Corollary 7.4. *Let $(P_\theta)_{\theta\in\Delta}$, $\Delta \subseteq \mathbb{R}$, be a one-parameter family that is $\mathbb{L}_2$-differentiable at $\theta_0 \in \Delta^0$ with Fisher information $\mathsf{I}(\theta_0) > 0$. Suppose that $\kappa : \Delta \to \mathbb{R}$ is differentiable at θ_0. If S is an unbiased estimator of $\kappa(\theta)$ with finite second moment, and $\theta \mapsto \mathsf{V}_\theta(S)$ is continuous at θ_0, then*

$$\mathsf{V}_{\theta_0}(S) \ge \frac{[\kappa'(\theta_0)]^2}{\mathsf{I}(\theta_0)}. \tag{7.3}$$

Proof. Set $\overline{P} = \frac{1}{2}(P_{\theta_1} + P_{\theta_2})$, $\overline{\kappa} = \frac{1}{2}(\kappa(\theta_1) - \kappa(\theta_2))$, and $f_i = dP_{\theta_i}/d\overline{P}$, $i = 1, 2$. Then

$$\kappa(\theta_1) - \kappa(\theta_2) = \int (S - \overline{\kappa})(f_1^{1/2} + f_2^{1/2})(f_1^{1/2} - f_2^{1/2})d\overline{P}.$$

By Schwarz' inequality, and $(f_1^{1/2} + f_2^{1/2})^2 \leq 2(f_2 + f_2)$,

$$(\kappa(\theta_1) - \kappa(\theta_2))^2 \leq 2[\int ((S - \overline{\kappa})^2 (f_1 + f_2)d\overline{P}][\int (f_1^{1/2} - f_2^{1/2})^2 d\overline{P}].$$

Note that the second factor on the right-hand side is the square of the Hellinger distance $\mathsf{D}(P_{\theta_1}, P_{\theta_2})$. The term in the first brackets is

$$\mathsf{E}_{\theta_1}(S - \overline{\kappa})^2 + \mathsf{E}_{\theta_2}(S - \overline{\kappa})^2 = \mathsf{V}_{\theta_1}(S) + \mathsf{V}_{\theta_2}(S) + \frac{1}{2}(\kappa(\theta_1) - \kappa(\theta_2))^2,$$

which proves (7.1). Set $L_{\theta_1,\theta_2} = dP_{\theta_2}/dP_{\theta_1}$. Then by Hölder's inequality, with $\beta > 1$ and $\alpha^{-1} + \beta^{-1} = 1$,

$$\begin{aligned} |\kappa(\theta_2) - \kappa(\theta_1)| &= |\mathsf{E}_{\theta_1}(S - \kappa(\theta_1))(L_{\theta_1,\theta_2} - 1)| \\ &\leq [\mathsf{E}_{\theta_1}|S - \kappa(\theta_1)|^{\beta}]^{1/\beta}[\mathsf{E}_{\theta_1}|L_{\theta_1,\theta_2} - 1|^{\alpha}]^{1/\alpha} \\ &= [\mathsf{E}_{\theta_1}|S - \kappa(\theta_1)|^{\beta}]^{1/\beta}[\chi^{\alpha}(P_{\theta_2}, P_{\theta_1})]^{1/\alpha}. \end{aligned}$$

To prove the corollary, we note that by Lemma 1.106

$$\mathsf{D}^2(P_\theta, P_{\theta_0}) = \frac{1}{4}\mathsf{I}(\theta_0)(\theta - \theta_0)^2 + o((\theta - \theta_0)^2).$$

Hence by the inequality (7.1) and the continuity of $\theta \mapsto \mathsf{V}_\theta(S)$,

$$2\mathsf{V}_{\theta_0}(S) \geq \lim_{\theta \to \theta_0} \frac{[1 - \mathsf{D}^2(P_{\theta_0}, P_\theta)](\theta_0 - \theta)^2}{2\mathsf{D}^2(P_{\theta_0}, P_\theta)} \frac{[\kappa(\theta_0) - \kappa(\theta)]^2}{(\theta_0 - \theta)^2} = \frac{2[\kappa'(\theta_0)]^2}{\mathsf{I}(\theta_0)}.$$

■

The subsequent problems show that the bounds that appear in the above theorem can be explicitly evaluated in many cases.

Problem 7.5. Consider a one-parameter exponential family $(P_\theta)_{\theta \in \Delta}$, $\Delta \subseteq \mathbb{R}$. Recall that $\gamma_{\mathsf{m}}(\theta) = K'(\theta)$, $\theta \in \Delta^0$, is the expectation $\mathsf{E}_\theta T$ of the generating statistic T; see (1.24). If $\kappa(\theta) = \gamma_{\mathsf{m}}(\theta)$, then T is an unbiased estimator of κ. To evaluate the right-hand side note that by (1.110) and Example 1.88 it holds

$$\mathsf{D}^2(P_{\theta_1}, P_{\theta_2}) = 2[1 - \exp\{K(\frac{1}{2}(\theta_1 + \theta_2)) - \frac{1}{2}K(\theta_1) - \frac{1}{2}K(\theta_2)\}].$$

Plug in this expression to get an explicit lower bound for $\mathsf{V}_{\theta_1}(T) + \mathsf{V}_{\theta_2}(T)$.

Problem 7.6. For an exponential family the χ^2-distance that appears in (7.2) is

$$\chi^2(P_{\theta_2}, P_{\theta_1}) = \mathsf{E}_{\theta_1} L^2_{\theta_2,\theta_1} - 1 = \exp\{K(2\theta_2 - \theta_1) - 2K(\theta_2) + K(\theta_1)\} - 1.$$

Inequality (7.1) is taken from Ibragimov and Has'minskii (1981). Inequality (7.2) is due to Vajda (1973), where the special case of $\alpha = \beta = 2$ was studied in Chapman and Robbins (1951). Although both inequalities are similar, the two-point inequality (7.1) has in comparison to inequality (7.2) for $\alpha = \beta = 2$ the advantage that in any case the right-hand term is not degenerate because $0 \le \mathsf{D}^2(P_{\theta_1}, P_{\theta_2}) \le 2$, where equality on the left-hand side appears only for $P_{\theta_1} = P_{\theta_2}$, which means for $\theta_1 \neq \theta_2$ that the parameter is not identifiable. The χ^α-distance $\chi^\alpha(P_{\theta_2}, P_{\theta_1})$ that appears in (7.2) is finite only if the likelihood ratio is integrable to the power of $\alpha > 1$. Moreover, we see from the inequalities in Theorem 7.3 that the lower bound for the risks connected with the estimation of the function depends on the distance between the corresponding distributions, where the type of the distance used depends on the chosen loss function. Roughly speaking, to every loss function there is a specific distance that gives a lower bound for the risk. Instead of discussing this question in full generality we refer to Vajda (1973), Kozek (1977a,b), and Liese (1988). Inequality (7.3) is a version of the Cramér–Rao inequality that is due to Cramér (1946) and Rao (1945). Forerunners are Fréchet (1943) and Darmois (1945).

Classical approaches to the Cramér–Rao bound do not use the concept of $\mathbb{L}_2$-differentiability and the continuity condition of the variance function. Instead, direct sufficient conditions are established that guarantee that the derivative with respect to the parameter can be carried out under the integral sign. The Cramér–Rao inequality is also called the "information inequality" by some authors, a notation introduced by Savage (1954).

Next we establish the multivariate Cramér–Rao inequality. Hereby we utilize the Löwner semiorder of matrices, which is defined by $A \succeq B$ if and only if $A - B$ is positive semidefinite. The subsequent version of the Cramér–Rao inequality, and the discussion of its stability, is taken from Witting (1985) and Ibragimov and Has'minskii (1981). Recall that for a differentiable function $g = (g_1, ..., g_k)^T : \mathbb{R}^d \to \mathbb{R}^k$ the Jacobian $J_g(\theta)$ is the $k \times d$ matrix of partial derivatives; that is, $J_g(\theta) = ((\partial g_i / \partial \theta_j))_{1 \le i \le k, 1 \le j \le d}$. The transposed matrix $J_g^T(\theta)$ will be denoted by $\dot{g}(\theta)$ and called the derivative of g. The columns of $\dot{g}(\theta)$ are ∇g_l; that is, the gradient of g_l, $l = 1, ..., k$. Especially if $g : \mathbb{R}^d \to \mathbb{R}$, then $\dot{g}(\theta) = J_g^T(\theta) = \nabla g(\theta)$.

Theorem 7.7. ***(Cramér–Rao Inequality)*** *Suppose* $(P_\theta)_{\theta \in \Delta}$, $\Delta \subseteq \mathbb{R}^d$, *is* $\mathbb{L}_2$*-differentiable at* $\theta_0 \in \Delta^0$ *with derivative* $\dot{L}_{\theta_0}$ *and Fisher information matrix* $\mathsf{I}(\theta_0)$, *where* $\det(\mathsf{I}(\theta_0)) \neq 0$. *Suppose* $S = (S_1, ..., S_k)^T : \mathcal{X} \to_m \mathbb{R}^k$ *satisfies* $\sup_{\theta \in U(\theta_0)} \mathsf{E}_\theta \|S\|^2 < \infty$ *for some neighborhood* $U(\theta_0)$ *of* θ_0. *Then* $g(\theta) := \mathsf{E}_\theta S$ *is differentiable at* θ_0 *with derivative*

$$\dot{g}(\theta_0) = \mathsf{E}_{\theta_0} \dot{L}_{\theta_0} S^T, \tag{7.4}$$

and the covariance matrix $\mathsf{C}_\theta(S)$ *satisfies the Cramér–Rao inequality*

$$\mathsf{C}_{\theta_0}(S) \succeq \dot{g}^T(\theta_0)(\mathsf{I}(\theta_0))^{-1}\dot{g}(\theta_0). \tag{7.5}$$

Equality holds in (7.5) if and only if

$$S - \mathsf{E}_{\theta_0} S = \dot{g}^T(\theta_0)(\mathsf{I}(\theta_0))^{-1}\dot{L}_{\theta_0}, \quad P_{\theta_0}\text{-a.s.} \tag{7.6}$$

Proof. That $g(\theta)$ is differentiable and (7.4) holds follows from a componentwise application of Proposition 1.111. As θ_0 is fixed, we omit θ_0 in $\mathsf{I}(\theta_0)$ and $\dot{g}(\theta_0)$ to simplify the formulas. The relation (7.4) and $\mathsf{E}_{\theta_0}\dot{L}_{\theta_0} = 0$ (see Proposition 1.110) imply that for every $u \in \mathbb{R}^k$ and $v \in \mathbb{R}^d$,

$$\mathsf{E}_{\theta_0} u^T(S - g)(\dot{L}_{\theta_0}^T v) = u^T \dot{g}^T v.$$

Hence by the Schwarz inequality $(u^T\dot{g}^Tv)^2 \le (u^T\mathsf{C}_{\theta_0}(S)u)(v^T\mathsf{I}v)$. If we set $v = \mathsf{I}^{-1}\dot{g}u$, then

$$\begin{aligned}(u^T\dot{g}^T\mathsf{I}^{-1}\dot{g}u)^2 &\le (u^T\mathsf{C}_{\theta_0}(S)u)(u^T\dot{g}^T\mathsf{I}^{-1}\mathsf{I}\mathsf{I}^{-1}\dot{g}u)\\ &= (u^T\mathsf{C}_{\theta_0}(S)u)(u^T\dot{g}^T\mathsf{I}^{-1}\dot{g}u),\end{aligned}$$

which implies (7.5). To study the case when equality holds in (7.5), we note that by $\mathsf{E}_{\theta_0}\dot{L}_{\theta_0} = 0$ and relation (7.4) it holds

$$\mathsf{E}_{\theta_0}(S - \mathsf{E}_{\theta_0}S)(\dot{g}^T\mathsf{I}^{-1}\dot{L}_{\theta_0})^T = \dot{g}^T\mathsf{I}^{-1}\dot{g}.$$

As the matrix on the right-hand side is symmetric we get

$$\begin{aligned}&\mathsf{E}_{\theta_0}\left[S - \mathsf{E}_{\theta_0}S - \dot{g}^T\mathsf{I}^{-1}\dot{L}_{\theta_0}\right]\left[S - \mathsf{E}_{\theta_0}S - \dot{g}^T\mathsf{I}^{-1}\dot{L}_{\theta_0}\right]^T\\ &= \mathsf{C}_{\theta_0}(S) - 2\dot{g}^T\mathsf{I}^{-1}\dot{g} + \dot{g}^T\mathsf{I}^{-1}\dot{g} = 0,\end{aligned}$$

which proves (7.6) provided that equality holds in (7.5). ■

Remark 7.8. The lower bound in (7.5) does not depend on the concrete type of parametrization. To see this let Δ and Λ be open subsets of $\mathbb{R}^d$ and $\kappa : \Lambda \leftrightarrow \Delta$ be a diffeomorphism. If $(P_\theta)_{\theta\in\Delta}$, $\Delta \subseteq \mathbb{R}^d$, is $\mathbb{L}_2$-differentiable at $\theta_0 = \kappa(\eta_0) \in \Delta^0$ with Fisher information $\mathsf{I}(\theta_0)$, then by Proposition 1.112 $(P_{\kappa(\eta)})_{\eta\in\Lambda}$ is $\mathbb{L}_2$-differentiable at η_0 with Fisher information

$$\dot{\kappa}^T(\eta_0)\mathsf{I}(\kappa(\eta_0))\dot{\kappa}(\eta_0) = J_\kappa(\eta_0)\mathsf{I}(\kappa(\eta_0))J_\kappa^T(\eta_0).$$

Moreover, by the chain rule the Jacobian of $g(\kappa(\eta))$ is $J_g(\kappa(\eta))J_\kappa(\eta)$. Hence the lower bound (7.5) for the model $(P_{\kappa(\eta)})_{\eta\in\Lambda}$ at $\theta_0 = \kappa(\eta_0)$ is

$$\begin{aligned}&J_g(\kappa(\eta_0))J_\kappa(\eta_0)[J_\kappa(\eta_0)\mathsf{I}(\kappa(\eta_0))J_\kappa^T(\eta_0)]^{-1}(J_g(\kappa(\eta_0))J_\kappa(\eta_0))^T\\ &= J_g(\theta_0)\mathsf{I}^{-1}(\theta_0)J_g^T(\theta_0).\end{aligned}$$

Remark 7.9. If we turn from the model $(\mathcal{X}, \mathfrak{A}, (P_\theta)_{\theta\in\Delta})$, $\Delta \subseteq \mathbb{R}^d$, to the model $(\mathcal{X}^n, \mathfrak{A}^{\otimes n}, (P_\theta^{\otimes n})_{\theta\in\Delta})$, i.e., to the case of a sample of size n, then according to Problem 1.113 the Fisher information matrix now is $n\mathsf{I}(\theta_0)$. For any unbiased estimator $S : \mathcal{X}^n \to_m \Delta$ that satisfies $\sup_{\theta\in U(\theta_0)} \mathsf{E}_\theta \|S\|^2 < \infty$ the Cramér–Rao inequality becomes

$$\mathsf{C}_{\theta_0}(S) \succeq \frac{1}{n}\dot{g}^T(\theta_0)(\mathsf{I}(\theta_0))^{-1}\dot{g}(\theta_0). \tag{7.7}$$

Example 7.10. We consider an exponential family $(P_\theta)_{\theta\in\Delta}$ with generating statistic T and a density given by (1.6). A sample of size n has then a distribution from the family $(P_\theta^{\otimes n})_{\theta\in\Delta}$, which again is an exponential family but now with generating statistic $T_{\oplus n}$; see Proposition 1.4. The Fisher information matrix for the model $(P_\theta^{\otimes n})_{\theta\in\Delta}$ is, according to Example 1.120 and Problem 1.113, given by $\mathsf{I}_{\otimes n}(\theta_0) = n\nabla\nabla^T K(\theta_0)$. Let $S : \mathcal{X}^n \to_m \Delta$ be an estimator with $\mathsf{E}_\theta \|S\|^2 < \infty$, $\theta \in \Delta$. According to Lemma 1.16 the function $\theta \mapsto \mathsf{E}_\theta \|S\|^2$ is continuous in Δ^0, which implies that for some neighborhood of θ_0 the condition $\sup_{\theta\in U(\theta_0)} \mathsf{E}_\theta \|S\|^2 < \infty$ is fulfilled. If S is unbiased, then (7.5) for $g(\theta) = \theta$ implies

$$\mathsf{C}_{\theta_0}(S) \succeq (n\nabla\nabla^T K(\theta_0))^{-1}.$$

If we want to estimate the expectation $\mu = \nabla K(\theta)$, then we have to put $g(\theta) = \mathsf{E}_\theta T = \nabla K(\theta)$ to make T an unbiased estimator of $g(\theta)$. It holds $\dot{g}(\theta_0) = \nabla\nabla^T K(\theta_0)$. The symmetry of $\nabla\nabla^T K(\theta_0)$ and (7.5) yield

$$\mathsf{C}_{\theta_0}(S) \succeq (\nabla\nabla^T K(\theta_0))(n\nabla\nabla^T K(\theta_0))^{-1}(\nabla\nabla^T K(\theta_0)) = \frac{1}{n}\nabla\nabla^T K(\theta_0).$$

The estimator $(1/n)T_{\oplus n}$ is the arithmetic mean of i.i.d. random vectors with covariance matrix $\mathsf{C}_{\theta_0}(n^{-1}T_{\oplus n}) = n^{-1}\nabla\nabla^T K(\theta_0)$. Hence

$$\frac{1}{n}\dot{g}^T(\theta_0)(\mathsf{I}(\theta_0))^{-1}\dot{g}(\theta_0) = \frac{1}{n}\nabla\nabla^T K(\theta_0),$$

so that $(1/n)T_{\oplus n}$ attains the lower Cramér–Rao bound.

Example 7.11. We consider a sample of size n from a normal distribution, so that $P_\theta = \mathsf{N}^{\otimes n}(\mu, \sigma^2)$, $\theta = (\mu, \sigma^2) \in \mathbb{R} \times (0, \infty)$. We apply Theorem 1.117 to calculate the Fisher information matrix. It holds

$$\begin{aligned}\nabla \ln \varphi_{\mu,\sigma^2}(x) &= \left(\frac{x-\mu}{\sigma^2}, \frac{(x-\mu)^2}{2\sigma^4} - \frac{1}{2\sigma^2}\right)^T,\\ \mathsf{I}(\mu,\sigma^2) &= \int (\nabla \ln \varphi_{\mu,\sigma^2}(x))(\nabla \ln \varphi_{\mu,\sigma^2}(x))^T \varphi_{\mu,\sigma^2}(x)dx = \begin{pmatrix} \frac{1}{\sigma^2} & 0 \\ 0 & \frac{1}{2\sigma^4}\end{pmatrix},\end{aligned}$$

and $\mathsf{N}^{\otimes n}(\mu, \sigma^2)$ has the Fisher information matrix $n\mathsf{I}(\mu, \sigma^2)$. Let $S = (S_1, S_2)$ be any unbiased estimator of the parameter $\theta = (\mu, \sigma^2)$ with $\mathsf{E}_{\mu,\sigma^2} \|S\|^2 < \infty$, $\mu \in \mathbb{R}$, $\sigma^2 > 0$. As in Example 7.10 one can use Lemma 1.16 to see that the condition $\sup_{(\mu,\sigma^2)\in U(\mu_0,\sigma_0^2)} \mathsf{E}_{\mu,\sigma^2} \|S\|^2 < \infty$ is satisfied. Hence by (7.7) with $g(\theta) = \theta$, and $\dot{g}$ thus being the unit matrix,

$$\mathsf{C}_{\mu,\sigma^2}(S) \succeq \begin{pmatrix} \sigma^2/n & 0 \\ 0 & 2\sigma^4/n \end{pmatrix}.$$

Set $\overline{X}_n = (1/n)\sum_{i=1}^n X_i$ and $S_n^2 = (1/(n-1))\sum_{i=1}^n (X_i - \overline{X}_n)^2$. Then $\mathsf{E}_\theta \overline{X}_n = \mu$ and $\mathsf{E}_\theta S_n^2 = \sigma^2$. Furthermore, $\mathsf{V}_\theta(\overline{X}_n) = \sigma^2/n$, $\mathsf{V}_{\mu,\sigma^2}(S_n^2) = 2\sigma^4/(n-1)$, and $\mathsf{cov}_{\mu,\sigma^2}(\overline{X}_n, S_n^2) = 0$ due to the independence of $\overline{X}_n$ and S_n^2. Hence the relation (7.5) with $\kappa(\theta) = \theta$ reads

$$\mathsf{C}_{\mu,\sigma^2}(\overline{X}_n, S_n^2) = \begin{pmatrix} \sigma^2/n & 0 \\ 0 & 2\sigma^4/(n-1) \end{pmatrix} \succeq \begin{pmatrix} \sigma^2/n & 0 \\ 0 & 2\sigma^4/n \end{pmatrix},$$

so that the arithmetic mean attains the lower Cramér–Rao bound whereas S_n^2 does not attain the lower bound despite the fact that S_n^2 is a uniformly best unbiased estimator, as we show in Example 7.20.

There are several improvements of the Cramér–Rao inequality that remove, at least up to a certain extent, the unsatisfactory fact that some uniformly best unbiased estimators do not attain the lower bound in the Cramér–Rao inequality. To explain the basic idea let P_θ, $\theta \in \Delta \subseteq \mathbb{R}$, be dominated with density f_θ that is positive and k times differentiable. If $\kappa(\theta) = \int T(x) f_\theta(x)\mu(dx)$ is also k times differentiable, and the derivative and the integral can be exchanged, then one has for any $c_1, ..., c_k \in \mathbb{R}$,

$$\int ([(T(x) - \kappa(\theta)) f_\theta^{1/2}(x)] \sum\nolimits_{l=1}^k c_l f_\theta^{(l)}(x) f_\theta^{-1/2}(x))\mu(dx) = \sum\nolimits_{l=1}^k c_l \kappa^{(l)}(\theta).$$

An application of Schwarz' inequality gives

$$\left[\mathsf{E}_\theta(\sum\nolimits_{l=1}^k c_l f_\theta^{(l)} f_\theta^{-1})^2\right]^{-1} \left[\sum\nolimits_{l=1}^k c_l \kappa^{(l)}(\theta)\right]^2 \leq \mathsf{V}_\theta(T).$$

As the c_l can be chosen freely one can try to maximize the left-hand expression as a function of $c_1, ..., c_k$. This leads to the improvement of the Cramér–Rao bound due to Bhattacharyya (1946, 1947a), to which we refer for further details.

Examples 7.10 and 7.11 lead to the question if there are there other families of distribution, besides exponential families, for which equality is attained in the Cramér–Rao inequality. This problem has been studied in several papers which differ in the differentiability concepts applied to the family of distributions under consideration; see Wijsman (1973), Barankin (1949), Fabian and Hannan (1977), Joshi (1976), Čenvoc (1982), and Müller-Funk, Pukelsheim, and Witting (1989). Subsequently we follow the representation in Müller-Funk, Pukelsheim, and Witting (1989) and use the concept of continuous $\mathbb{L}_2$-differentiability. To avoid technical difficulties and to simplify the proof we consider only homogeneous models. Let Δ be a open subset of $\mathbb{R}^d$. A homogeneous model $(\mathcal{X}, \mathfrak{A}, (P_\theta)_{\theta\in\Delta})$ is called *continuously* $\mathbb{L}_2$*-differentiable* on Δ if $(P_\theta)_{\theta\in\Delta}$ is $\mathbb{L}_2$-differentiable at every $\theta_0 \in \Delta$, say with derivative $\dot{L}_{\theta_0}$, and it holds

$$\lim_{\theta\to\theta_0} \mathsf{E}_{\theta_0} \parallel L_{\theta,\theta_0}^{1/2} \dot{L}_\theta - \dot{L}_{\theta_0} \parallel^2 = 0, \quad \theta_0 \in \Delta.$$

An important feature is that continuous differentiability implies continuity and even differentiability of other functions that are derived from the model.

Problem 7.12.* If $(P_\theta)_{\theta\in\Delta}$ is continuously $\mathbb{L}_2$-differentiable on Δ, then $\theta \mapsto \mathsf{I}(\theta)$ is continuous. For every $B \in \mathfrak{A}$ the function $\theta \mapsto P_\theta(B)$ is continuously differentiable with derivative

$$\nabla P_\theta(B) = \mathsf{E}_\theta I_B \dot{L}_\theta. \tag{7.8}$$

Problem 7.13.* Suppose $(P_\theta)_{\theta\in\Delta}$ is continuously $\mathbb{L}_2$-differentiable on Δ. If $\theta \mapsto \mathsf{E}_\theta \|S\|^2$ is continuous, then $\dot{g}(\theta) = \mathsf{E}_\theta \dot{L}_\theta S^T$ is continuous in $\theta \in \Delta$.

Problem 7.14.* Let $T : \mathcal{X} \to_m \mathbb{R}^d$ be a statistic with $\mathsf{E}_P \|T\|^2 < \infty$ and $\det(\mathsf{C}_P(T)) \neq 0$. If for some $a \in \mathbb{R}^d$ and $b \in \mathbb{R}$ $f(x) = \exp\{\langle a, T(x)\rangle + b\}$ is a probability density with respect to P, then $\exp\{\langle a, T(x)\rangle + b\} = \exp\{\langle c, T(x)\rangle + d\}$, P-a.s., implies $a = c$ and $b = d$.

The following result is a special case of Theorem 1 in Müller-Funk, Pukelsheim, and Witting (1989).

Theorem 7.15. *Let* $(\mathcal{X}, \mathfrak{A}, (P_\theta)_{\theta\in\Delta})$ *be a model where* $\Delta \subseteq \mathbb{R}^d$ *is an open and connected set, the family* $(P_\theta)_{\theta\in\Delta}$ *is continuously* $\mathbb{L}_2$*-differentiable on* Δ*, and it holds* $\det(\mathsf{I}(\theta)) \neq 0$, $\theta \in \Delta$. *Suppose* $S : \mathcal{X} \to_m \mathbb{R}^d$ *is a statistic for which* $\mathsf{E}_\theta \|S\|^2 < \infty$, $\theta \in \Delta$, $\det(\mathsf{C}_\theta(S)) \neq 0$, *and* $\theta \mapsto \mathsf{C}_\theta(S)$ *is continuous. If the equality in the Cramér–Rao inequality (7.5) holds for* $g(\theta) = \mathsf{E}_\theta S$ *at every* $\theta \in \Delta$*, then for a fixed* $\theta_0 \in \Delta$ *there are functions* $b, c : \Delta \to \mathbb{R}^k$ *such that*

$$\frac{dP_\theta}{dP_{\theta_0}} = \exp\{\langle b(\theta), S\rangle + c(\theta)\}, \quad P_{\theta_0}\text{-}a.s.$$

Proof. The continuity of $\theta \mapsto \mathsf{C}_\theta(S)$ implies $\sup_{\theta\in U(\theta_0)} \mathsf{E}_\theta \|S\|^2 < \infty$ for some neighborhood of θ_0. Hence by Theorem 7.7 the function $g(\theta) = \mathsf{E}_\theta S$ is differentiable and the Jacobian $\dot{g}(\theta)^T = \mathsf{E}_\theta S\dot{L}_\theta^T$ is continuous in view of Problem 7.13. The assumption $\det(\mathsf{C}_\theta(S)) \neq 0$ and the equality in (7.5) show that $\dot{g}^T(\theta)(\mathsf{I}(\theta))^{-1}$ has the rank d. Hence the inverse matrix of $\dot{g}^T(\theta)(\mathsf{I}(\theta))^{-1}$ exists and we get from (7.6) that P_θ-a.s.,

$$\dot{L}_\theta = A(\theta)S + a(\theta), \quad A(\theta) = \mathsf{I}(\theta)(\dot{g}^T(\theta))^{-1}, \quad a(\theta) = -\mathsf{I}(\theta)(\dot{g}^T(\theta))^{-1}(\mathsf{E}_\theta S),$$

where $A(\theta)$ and $a(\theta)$ depend continuously on θ as $\dot{g}(\theta)$, $\mathsf{I}(\theta)$, and $g(\theta) = \mathsf{E}_\theta S$ do. We fix θ_0 and connect θ_0 and θ with a continuously differentiable path $\theta(s)$, $0 \le s \le 1$. This is possible as Δ is connected and open. Denote by $\dot{\theta}(s)$ the derivative and set

$$\beta(s) = A^T(\theta(s))\dot{\theta}(s), \quad b(\theta) = \int_0^1 \beta(s)ds, \quad c(\theta) = \int_0^1 a^T(\theta(s))\dot{\theta}(s)ds$$

$$f(x) = \exp\{\int_0^1 \dot{\theta}^T(s)\dot{L}_{\theta(s)}(x)ds\} = \exp\{\int_0^1 \langle\beta(s), S(x)\rangle\, ds + c(\theta)\}.$$

We show that $f = dP_\theta/dP_{\theta_0}$. Then by Problem 7.14 the functions $b(\theta)$ and $c(\theta)$ are independent of the chosen path and the statement is established. To prove $f = dP_\theta/dP_{\theta_0}$ we fix $\varepsilon > 0$ and a partition of $\mathbb{R}^d$ into rectangles R_i, $i = 1, 2, ...$ with a diameter not exceeding ε. We fix $B \in \mathfrak{A}$, set $B_i = B \cap \{S \in R_i\}$, and let $p(s) = P_{\theta(s)}(B_i)$. We suppose $P_{\theta_0}(B_i) > 0$. Then by the homogeneity of the model $P_{\theta(s)}(B_i) > 0$ for every $0 \le s \le 1$. Hence by (7.8),

$$\begin{aligned}
\ln\frac{P_\theta(B_i)}{P_{\theta_0}(B_i)} &= \int_0^1 \frac{d}{ds}\ln p(s)ds = \int_0^1 \frac{1}{P_{\theta(s)}(B_i)}\mathsf{E}_{\theta(s)} I_{B_i}\langle\dot{\theta}(s), \dot{L}_{\theta(s)}\rangle ds \\
&= \int_0^1 \frac{1}{P_{\theta(s)}(B_i)}\mathsf{E}_{\theta(s)} I_{B_i}\dot{\theta}^T(s)(A(\theta(s))S + a(\theta(s)))ds \\
&= \int_0^1 \frac{1}{P_{\theta(s)}(B_i)}\mathsf{E}_{\theta(s)} I_{B_i}\,\langle\beta(s), S\rangle\, ds + c(\theta), \quad \text{and}
\end{aligned}$$

$$\frac{P_\theta(B_i)}{P_{\theta_0}(B_i)} \exp\{-\int_0^1 \frac{1}{P_{\theta(s)}(B_i)} \mathsf{E}_{\theta(s)} I_{B_i} \langle \beta(s), S \rangle \, ds - c(\theta)\} = 1.$$

Hence,

$$\int_{B_i} f dP_{\theta_0} = \frac{P_\theta(B_i)}{P_{\theta_0}(B_i)} \int_{B_i} \exp\{\int_0^1 \langle \beta(s), S - \frac{1}{P_{\theta(s)}(B_i)} \mathsf{E}_{\theta(s)} I_{B_i} S \rangle ds\} dP_{\theta_0}.$$

For $x \in B_i$ the vector $S(x)$ belongs to some R_i. Hence $(P_{\theta(s)}(B_i))^{-1} \mathsf{E}_{\theta(s)} I_{B_i} S$ also belongs to R_i so that $\|S(x) - (P_{\theta(s)}(B_i))^{-1} \mathsf{E}_{\theta(s)} I_{B_i} S\| \le 2\varepsilon$. The continuity of $\beta(s)$ yields $c := \sup_{0 \le s \le 1} \|\beta(s)\| < \infty$. Hence

$$\exp\{-2c\varepsilon\} P_\theta(B_i) \le \int_{B_i} f dP_{\theta_0} \le \exp\{2c\varepsilon\} P_\theta(B_i),$$

and by taking the sum over all i,

$$\exp\{-2c\varepsilon\} P_\theta(B) \le \int_B f dP_{\theta_0} \le \exp\{2c\varepsilon\} P_\theta(B),$$

which completes the proof if we let ε tend to zero. ∎

7.2 Unbiased Estimators with Minimal Risk

On several occasions we have used already a reduction by sufficiency, as in the presence of a sufficient statistic we have to make our decisions only based on the sufficient statistic; see Theorem 4.66.

Suppose $T : \mathcal{X} \to_m \mathcal{T}$ is sufficient and $S : \mathcal{X} \to_m \mathbb{R}^m$ satisfies $\mathsf{E}_\theta \|S\| < \infty$, $\theta \in \Delta$. Then a componentwise application of Problem 4.40 provides the existence of some $g : \mathcal{T} \to_m \mathbb{R}^m$ with

$$\mathsf{E}_\theta(S|T) = g(T), \quad P_\theta\text{-a.s.}, \ \theta \in \Delta.$$

Theorem 7.16. ***(Rao–Blackwell)*** *Suppose $T : \mathcal{X} \to_m \mathcal{T}$ is sufficient for the model $(\mathcal{X}, \mathfrak{A}, (P_\theta)_{\theta \in \Delta})$ and $S : \mathcal{X} \to_m \mathbb{R}^m$ is an estimator of the function $\kappa : \Delta \to \mathbb{R}^m$ that satisfies $\mathsf{E}_\theta \|S\| < \infty$, $\theta \in \Delta$. If the loss function $L(\theta, a)$ is convex in $a \in \mathbb{R}^m$, then $\mathsf{R}(\theta, g(T)) \le \mathsf{R}(\theta, S)$. Moreover, if S is unbiased, then $g(T)$ is also unbiased.*

Proof. By the iterated expectation rule (see (a) in A.31) and Jensen's inequality for the conditional expectation (see Lemma A.33),

$$\begin{aligned} \mathsf{R}(\theta, S) &= \mathsf{E}_\theta L(\theta, S) = \mathsf{E}_\theta(\mathsf{E}_\theta(L(\theta, S)|T)) \ge \mathsf{E}_\theta(L(\theta, \mathsf{E}_\theta(S|T))) \\ &= \mathsf{E}_\theta(L(\theta, g(T))) = \mathsf{R}(\theta, g(T)). \end{aligned}$$

The unbiasedness of $g(T)$ follows from the iterated expectation rule. ∎

As the set of all unbiased estimators is convex we get from Problem 5.54 that a uniformly best unbiased estimator is uniquely determined for strictly convex loss functions. The subsequent theorem of Lehmann and Scheffé shows that this estimator is a function of the complete sufficient statistic whenever such a statistic exists.

Theorem 7.17. *(Lehmann–Scheffé)* *Suppose $T : \mathcal{X} \to_m \mathcal{T}$ is sufficient and complete for the model $(\mathcal{X}, \mathfrak{A}, (P_\theta)_{\theta\in\Delta})$, there exists at least one unbiased estimator S for the function $\kappa : \Delta \to \mathbb{R}^m$, and the loss function $L(\theta, a)$ is convex in $a \in \mathbb{R}^m$.*

(A) There exists at least one unbiased estimator of the form $h(T)$, where $h : \mathcal{T} \to_m \mathbb{R}^m$.
(B) If $g : \mathcal{T} \to_m \mathbb{R}^m$ and $g(T)$ is unbiased, then $g(T)$ is a uniformly best unbiased estimator and it is P_θ-a.s., $\theta \in \Delta$, uniquely determined in the class of all estimators that are functions of T.
(C) If $L(\theta, a)$ is strictly convex for every θ, then a uniformly best unbiased estimator with finite risk is, P_θ-a.s., uniquely determined for every $\theta \in \Delta$.

Proof. To prove (A) we note that the sufficiency of T implies that by Problem 4.40 there is some $h : \mathcal{T} \to_m \mathbb{R}^m$ such that $\mathsf{E}_\theta(S|T) = h(T)$, P_θ-a.s., $\theta \in \Delta$. The iterated expectation rule, see (a) in A.31, provides the unbiasedness of $h(T)$. To prove (B) let S_0 be any unbiased estimator. Applying Rao–Blackwell's theorem we get functions h and h_0 such that

$$\mathsf{E}_\theta L(\theta, h(T)) \le \mathsf{E}_\theta L(\theta, S) \quad \text{and} \quad \mathsf{E}_\theta L(\theta, h_0(T)) \le \mathsf{E}_\theta L(\theta, S_0), \quad \theta \in \Delta.$$

As $h(T)$ and $h_0(T)$ are unbiased we may conclude that

$$\mathsf{E}_\theta(h(T) - h_0(T)) = \int (h(t) - h_0(t))(P_\theta \circ T^{-1})(dt) = 0.$$

The completeness of T yields $h(T) = h_0(T)$, P_θ-a.s., and therefore

$$\mathsf{E}_\theta L(\theta, h(T)) = \mathsf{E}_\theta L(\theta, h_0(T)) \le \mathsf{E}_\theta L(\theta, S_0).$$

This shows that $h(T)$ is uniformly best and unbiased which follows from the iterated expectation rule. The uniqueness follows as above. If $h_1(T)$ and $h_2(T)$ are any unbiased estimators, then $\mathsf{E}_\theta(h_1(T) - h_2(T)) = 0$ and the completeness of T give the uniqueness. The statement (C) follows from Problem 5.54. ■

Definition 7.18. *Given a model $(\mathcal{X}, \mathfrak{A}, (P_\theta)_{\theta\in\Delta})$ and a function $\kappa : \Delta \to \mathbb{R}^m$, let $\mathfrak{U}$ be the family of all unbiased estimators $U : \mathcal{X} \to_m \mathbb{R}^m$, that is, estimators with $\mathsf{E}_\theta U = \kappa(\theta)$, that satisfy $\mathsf{E}_\theta \|U\|^2 < \infty$, $\theta \in \Delta$. We call $S \in \mathfrak{U}$ a UMVU estimator if $\mathsf{C}_\theta(S) \preceq \mathsf{C}_\theta(U)$ in the Löwner semiorder for every $\theta \in \Delta$ and $U \in \mathfrak{U}$.*

UMVU stands for "uniformly minimum variance unbiased". If the conditions of the Lehmann–Scheffé theorem are satisfied, then, to find a uniformly best unbiased estimator, we have to find a function h that satisfies the system of equations

$$\int h(T(x))P_\theta(dx) = \kappa(\theta), \quad \theta \in \Delta. \tag{7.9}$$

On the other hand, if h in (7.9) runs through the class of all functions that are integrable with respect to $P_\theta \circ T^{-1}$, then we obtain all possible unbiasedly estimable functions κ. The next examples present special exponential families for which the system of equations (7.9) can be solved for h when κ is given. The first example is taken from Pfanzagl (1994).

Example 7.19. If X has a binomial distribution with parameters n and p, then

$$\begin{aligned}
&\mathbb{E}X(X-1)\cdots(X-m+1) = \sum\nolimits_{k=0}^{n} k(k-1)\cdots(k-m+1)\tbinom{n}{k}p^k(1-p)^{n-k} \\
&= \sum\nolimits_{k=m}^{n} \frac{k!}{(k-m)!}\tbinom{n}{k}p^k(1-p)^{n-k} \\
&= p^m \frac{n!}{(n-m)!}\sum\nolimits_{l=0}^{n-m} \frac{(n-m)!}{((n-m)-l)!l!}p^l(1-p)^{n-m-l} = p^m \frac{n!}{(n-m)!}.
\end{aligned}$$

For $\kappa(p) = \sum_{m=0}^{n} a_m p^m$ the estimator $S(k) = \sum_{m=0}^{n} a_m S_m(k)$ with

$$S_m(k) = \frac{k(k-1)\cdots(k-m+1)}{n(n-1)\cdots(n-m+1)},$$

is an unbiased estimator that is a function of the complete and sufficient statistic $T(k) = k$. As we are considering the reduced model every estimator is a function of T. Hence by (B) in Theorem 7.17 S is the uniquely determined UMVU estimator.

Example 7.20. We consider the model $(\mathbb{R}^n, \mathfrak{B}_n, (\mathsf{N}^{\otimes n}(\mu,\sigma^2))_{\mu\in\mathbb{R}.,\sigma^2>0})$ which, by Proposition 1.4 and Example 1.11, is an exponential family with generating statistic $T_{\oplus n} = (\sum_{i=1}^n X_i, \sum_{i=1}^n X_i^2)$ and natural parameter $\theta = (\theta_1,\theta_2) = (\mu/\sigma^2, -1/(2\sigma^2))$. Put $\kappa(\theta) = (\mu,\sigma^2)$ and consider the statistic $(\overline{X}_n, S_n^2)$ in Example 7.11. Then $\mathsf{E}_\theta(\overline{X}_n, S_n^2) = (\mu,\sigma^2)$. As $(\overline{X}_n, S_n^2)$ is a function of $T_{\oplus n}$ we get that $(\overline{X}_n, S_n^2)$ is the $\mathsf{N}^{\otimes n}(\mu,\sigma^2)$-a.s. uniquely determined UMVU estimator. That $\overline{X}_n$ is the UMVU estimator of μ was already established in 7.11. Now the gap regarding S_n^2, that has been left open there, is also closed.

Problem 7.21. In the model of the previous example $\frac{\sqrt{2}\Gamma((n-1)/2)}{\Gamma((n-2)/2)}\frac{\overline{X}_n}{S_n}$ is a UMVU estimator of μ/σ.

Problem 7.22. Consider the model $(\mathbb{R}^n, \mathfrak{B}_n, (\mathsf{Ga}^{\otimes n}(\lambda,\beta))_{\lambda,\beta>0})$. Then $\overline{X}_n$ is a UMVU estimator of λ/β.

Problem 7.23.* For the model $(\mathbb{R}^n, \mathfrak{B}_n, (\mathsf{N}(0,\sigma^2\mathbf{I}))_{\sigma^2>0})$, $n > 2$, the estimator $\widehat{1/\sigma^2}(x) = (n-2)\,\|x\|^{-2}$ is a UMVU estimator of $1/\sigma^2$.

We now consider linear models and show that for normal errors the least squares estimator is a UMVU estimator.

Let $\mathbb{L}$ be a linear subspace of $\mathbb{R}^n$ that has dimension $d < n$. Denote by $\mathbb{L}^\perp = \{x : x^T y = 0,\ y \in \mathbb{L}\}$ the orthogonal complement of $\mathbb{L}$. Recall that every $x \in \mathbb{R}^n$ can be uniquely represented as $x = x_{\mathbb{L}} + x_{\mathbb{L}^\perp}$, where $x_{\mathbb{L}} \in \mathbb{L}$ and $x_{\mathbb{L}^\perp} \in \mathbb{L}^\perp$. The mapping $x \mapsto x_{\mathbb{L}}$ is called the *projection* of x on $\mathbb{L}$ and is denoted by $\Pi_{\mathbb{L}}$. The definition of the projection shows that

$$z = \Pi_{\mathbb{L}}(x) \quad \text{if and only if } z \in \mathbb{L} \text{ and } x - z \perp \mathbb{L}. \tag{7.10}$$

The next problems collect some well-known properties of the projection.

Problem 7.24.* It holds $\|x - x_{\mathbb{L}}\|^2 = \inf_{y\in\mathbb{L}} \|x - y\|^2$.

As the mapping $\Pi_{\mathbb{L}}$ is linear there is a matrix $C_{\mathbb{L}}$ with $\Pi_{\mathbb{L}}(x) = C_{\mathbb{L}} x$. We call $C_{\mathbb{L}}$ a *projection matrix*. Such matrices have been briefly considered already in Section 2.2; see Problem 2.38.

Problem 7.25.* A matrix C is a projection matrix if and only if $C^T = C$ and $CC = C$. In this case Cx is a projection onto the linear subspace that is generated by the column vectors of C.

Problem 7.26.* If $\mathbb{L}$ is generated by the d linearly independent column vectors of the $n \times d$ matrix B, then the $d \times d$ matrix $B^T B$ is nonsingular and it holds $\Pi_{\mathbb{L}}(x) = B(B^T B)^{-1} B^T x$.

If the column vectors of the matrix B generate the linear subspace $\mathbb{L}$, then $\mathbb{L} = \{Bw,\ w \in \mathbb{R}^d\}$. This means that for every $x \in \mathbb{R}^n$ the projection $\Pi_{\mathbb{L}}(x)$ of x onto $\mathbb{L}$ can be written as $\Pi_{\mathbb{L}}(x) = Bv$ with some $v \in \mathbb{R}^d$. Putting $z = Bv$ in (7.10) we see that $(Bw)^T(x - Bv) = 0$ for every $w \in \mathbb{R}^d$. Hence we obtain $\Pi_{\mathbb{L}}(x) = Bv$ from the following so-called *normal equations*

$$\Pi_{\mathbb{L}}(x) = Bv \quad \text{and} \quad B^T Bv = B^T x. \tag{7.11}$$

If B has full rank, then $(B^T B)^{-1}$ exists and $\Pi_{\mathbb{L}}(x) = B(B^T B)^{-1} B^T x$ as in Problem 7.26.

If X and Y are random (column) vectors of dimensions m and n, respectively, and have finite second moments, then we set

$$\mathsf{C}(X, Y) = \mathbb{E}(X - \mathbb{E}X)(Y - \mathbb{E}Y)^T,$$

so that $\mathsf{C}(X, Y)$ is the matrix of all covariances of components of the vectors X and Y. It is called the *covariance matrix* of X and Y. An immediate consequence is that

$$\mathsf{C}((AX), (BY)) = A\mathsf{C}(X, Y)B^T. \tag{7.12}$$

The random vectors X and Y are called *uncorrelated* if $\mathsf{C}(X, Y) = 0$.

Problem 7.27. If X and Y are random vectors of dimension d with finite second moments and expectation zero, then $\mathbb{E}\|X+Y\|^2 = \mathbb{E}\|X\|^2 + \mathbb{E}\|Y\|^2 + 2tr(\mathsf{C}(X,Y))$, where $tr(A)$ is the trace of the matrix A.

Suppose $\mathbb{L}$ and $\mathbb{M}$ are linear subspaces and the matrices $C_{\mathbb{L}}$ and $C_{\mathbb{M}}$ provide the projections onto $\mathbb{L}$ and $\mathbb{M}$, respectively. If $\mathbb{L} \perp \mathbb{M}$, then $(C_{\mathbb{L}}x)^T(C_{\mathbb{M}}y) = 0$ for every $x, y \in \mathbb{R}^n$, which is equivalent to $C_{\mathbb{L}}C_{\mathbb{M}} = 0$. An often-used fact is that measurable functions of independent random vectors are independent. A similar statement holds if we restrict the functions to be linear and require only that the vectors are uncorrelated.

Problem 7.28.* Suppose Z is a random vector with $\mathsf{C}(Z) = \sigma^2\mathbf{I}$. Let $\mathbb{L}$ and $\mathbb{M}$ be linear subspaces of $\mathbb{R}^n$. Then $\mathbb{L} \perp \mathbb{M}$, or equivalently $C_{\mathbb{L}}C_{\mathbb{M}} = 0$, implies

$$\mathsf{C}(A\Pi_{\mathbb{L}}(Z), B\Pi_{\mathbb{M}}(Z)) = 0,$$

for any $k \times n$ and $m \times n$ matrices A and B, respectively.

We consider the problem of estimating μ and σ in the *linear model*

$$X = \mu + \sigma Z, \quad \mu \in \mathbb{L},\ \sigma > 0, \quad \text{with} \quad \mathbb{E}Z = 0,\ \mathsf{C}(Z) = \mathbf{I}, \tag{7.13}$$

where $\mathbb{L}$ is a d-dimensional subspace of $\mathbb{R}^n$. The method of least squares goes back to Gauss and Legendre. The estimator $\widehat{\mu}$ is defined by the requirement that the distance between vectors $\mu \in \mathbb{L}$ and the observation $x \in \mathbb{R}^n$ is to be minimized , i.e.,

$$\widehat{\mu}(x) \in \arg\min_{\mu \in \mathbb{L}} \|x - \mu\|^2 .$$

Hence $\widehat{\mu}(x) = \Pi_{\mathbb{L}}(x)$ by Problem 7.24. Often the subspace $\mathbb{L}$ is specified by d linearly independent column vectors $b_1, ..., b_d \in \mathbb{R}^n$ that generate $\mathbb{L}$. If B is the $n \times d$ matrix with the columns $b_1, ..., b_d$, then $\widehat{\mu}(x) = B(B^TB)^{-1}B^Tx$ by Problem 7.26. Every $\mu \in \mathbb{L}$ can be written in a unique way as $\mu = B\theta$, $\theta \in \mathbb{R}^d$. Then we define $\widehat{\theta}$ by the relation $B(B^TB)^{-1}B^Tx = B\widehat{\theta}$, which implies

$$\widehat{\theta} = (B^TB)^{-1}B^Tx.$$

We estimate the variance σ^2 in the model (7.13) by

$$\widehat{\sigma^2}(x) = \frac{1}{n-d}\|x - \widehat{\mu}(x)\|^2 .$$

Problem 7.29.* The estimator $\widehat{\sigma^2}$ is unbiased.

Let $\mathfrak{L}_u$ be the class of all linear unbiased estimators $T(x) = Dx$ in the model (7.13). We call an estimator $S \in \mathfrak{L}_u$ UMVU in $\mathfrak{L}_u$ if $\mathsf{C}_\theta(S) \preceq \mathsf{C}_\theta(U)$ in the Löwner semiorder for every $\theta \in \Delta$ and $U \in \mathfrak{L}_u$.

Theorem 7.30. ***(Gauss–Markov)*** *The least squares estimator for μ in the model (7.13) (i.e., $\Pi_{\mathbb{L}}(x) = \widehat{\mu}(x) = B(B^TB)^{-1}B^Tx$) is UMVU in $\mathfrak{L}_u$. If $\mathcal{L}(Z) = \mathsf{N}(0, \mathbf{I})$, then $\Pi_{\mathbb{L}}(x)$ is a UMVU estimator of μ and $\widehat{\sigma^2}(x)$ is a UMVU estimator of σ^2.*

Proof. It holds

$$\mathbb{E}\Pi_{\mathbb{L}}(X) = \mathbb{E}\Pi_{\mathbb{L}}(\mu + \sigma Z) = \mu.$$

If DX is any linear and unbiased estimator, then $\mu = \mathbb{E}DX = D\mathbb{E}X = D\mu$ for every $\mu \in \mathbb{L}$. Hence $D\mu = \Pi_{\mathbb{L}}\mu$ and $D\Pi_{\mathbb{L}}(Z) = \Pi_{\mathbb{L}}(Z)$. Thus we arrive at

$$\begin{aligned}\mathbb{E}\left\|DX - \mu\right\|^2 &= \mathbb{E}\left\|D(X - \mu)\right\|^2 = \mathbb{E}\left\|D(\Pi_{\mathbb{L}} + \Pi_{\mathbb{L}^\perp})(X - \mu)\right\|^2 \\ &= \sigma^2\mathbb{E}\left\|D\Pi_{\mathbb{L}}(Z) + D\Pi_{\mathbb{L}^\perp}(Z)\right\|^2 = \sigma^2\mathbb{E}\left\|\Pi_{\mathbb{L}}(Z)\right\|^2 + \sigma^2\mathbb{E}\left\|D\Pi_{\mathbb{L}^\perp}(Z)\right\|^2,\end{aligned}$$

where the last equality follows from Problems 7.28 and 7.27. Hence

$$\mathbb{E}\left\|DX - \mu\right\|^2 \geq \sigma^2\mathbb{E}\left\|\Pi_{\mathbb{L}}(Z)\right\|^2 = \mathbb{E}\left\|\Pi_{\mathbb{L}}(X - \mu)\right\|^2 = \mathbb{E}\left\|\Pi_{\mathbb{L}}(X) - \mu\right\|^2,$$

and the first statement is established. To prove the second statement we note that the Lebesgue density of $\mathcal{L}(X) = \mathsf{N}(\mu, \sigma^2\mathbf{I})$ is given by

$$\varphi_{\mu,\sigma^2\mathbf{I}}(x) = (2\pi\sigma^2)^{-n/2}\exp\{-\frac{1}{2\sigma^2}(T_1(x) - 2\langle T_2(x), \mu\rangle + \|\mu\|^2)\},$$

where $T_1(x) = \|x\|^2$ and $T_2(x) = \Pi_{\mathbb{L}}(x)$.

It follows from (1.6) that $(\mathsf{N}(\mu, \sigma^2\mathbf{I}))_{\mu\in\mathbb{L},\sigma^2>0}$ is an exponential family with generating statistic (T_1, T_2) and natural parameter $\theta = (-1/(2\sigma^2), \mu/\sigma^2) \in \Delta = (-\infty, 0) \times \mathbb{L}$. By Example 4.51 with $n = 1$ and $T = (T_1, T_2)$ we see that T is sufficient. Moreover by Theorem 4.73 the statistic T is complete. As

$$(\widehat{\mu}(x), \widehat{\sigma^2}(x)) = (T_2(x), \frac{1}{n-d}(T_1(x) - \|T_2(x)\|^2))$$

is a function of (T_1, T_2), and $\widehat{\mu}$ and $\widehat{\sigma^2}$ are unbiased, the statement follows from Lehmann–Scheffé's theorem; see Theorem 7.17. ■

Example 7.31. Consider the linear regression model

$$X_i = \alpha + \beta t_i + \sigma\varepsilon_i, \quad i = 1, ..., n,$$

with uncorrelated ε_i that have expectation zero and variance one. $t_1, ..., t_n$ are fixed given values at which the measurements have been made. The model space $\mathbb{L}$ is generated by the vectors $\mathbf{1} = (1, ..., 1)^T$ and $t = (t_1, ..., t_n)^T$ which are linearly independent by assuming that $\sum_{i=1}^n (t_i - \overline{t}_n)^2 > 0$. The normal equations in (7.11) are here two linear equations for the parameters α and β,

$$\begin{pmatrix} n & \sum_{i=1}^n t_i \\ \sum_{i=1}^n t_i & \sum_{i=1}^n t_i^2 \end{pmatrix} \begin{pmatrix} \alpha \\ \beta \end{pmatrix} = \begin{pmatrix} \sum_{i=1}^n x_i \\ \sum_{i=1}^n t_i x_i \end{pmatrix},$$

which leads to the well-known estimators

$$\begin{aligned}\widehat{\alpha}_n(x) &= \overline{x}_n - \widehat{\beta}(x)\overline{t}_n, \quad \widehat{\beta}_n(x) = \frac{\sum_{i=1}^n (x_i - \overline{x}_n)(t_i - \overline{t}_n)}{\sum_{i=1}^n (t_i - \overline{t}_n)^2}, \\ \widehat{\sigma^2}(x) &= \frac{1}{n-2}\sum\nolimits_{i=1}^n [x_i - \widehat{\alpha}_n(x) - \widehat{\beta}_n(x)t_i]^2,\end{aligned}$$

where $x = (x_1, ..., x_n)$.

Problem 7.32. Show directly that the estimators $\widehat{\alpha}_n$, $\widehat{\beta}_n$, and $\widehat{\sigma^2}$ are unbiased, and that

$$\mathsf{V}_{\alpha,\beta}(\widehat{\alpha}_n) = \frac{\sigma^2 \sum_{i=1}^n t_i^2}{n \sum_{i=1}^n (t_i - \bar{t}_n)^2}, \quad \mathsf{V}_{\alpha,\beta}(\widehat{\beta}_n) = \frac{\sigma^2}{\sum_{i=1}^n (t_i - \bar{t}_n)^2}.$$

We refer to Christensen (1987) for a comprehensive representation of linear models.

In the above considerations we have restricted the class of estimators to unbiased estimators in order to find estimators that are optimal under the squared error loss. The requirement that the estimator be unbiased means that the true parameter value is in the center of the distribution of the estimator. Another possibility of making the notion of "center" precise is to use the median of the distribution of the estimator. This leads to the theory of *median unbiased estimators.* Recall that $\overline{\mathbb{R}} = [-\infty, \infty]$ is the extended real line and $\overline{\mathfrak{B}}$ the σ-algebra of Borel sets of $\overline{\mathbb{R}}$.

Definition 7.33. *Given a model* $(\mathcal{X}, \mathfrak{A}, (P_\theta)_{\theta\in\Delta})$ *and a function* $\kappa : \Delta \to \overline{\mathbb{R}}$, *the randomized estimator* $\mathsf{D} : \overline{\mathfrak{B}} \times \mathcal{X} \to_k [0,1]$ *is called median unbiased if*

$$\int \mathsf{D}([\kappa(\theta), \infty]|x) P_\theta(dx) \geq \frac{1}{2} \quad \text{and} \quad \int \mathsf{D}([-\infty, \kappa(\theta)]|x) P_\theta(dx) \geq \frac{1}{2}, \quad \theta \in \Delta.$$

The concept of median unbiasedness was already used by Laplace (1774) who showed that $\mathbb{L}_1$-estimators are median unbiased in the location model. Brown (1947) seems to have been the first who reestablished this concept. For discussions on the concept of unbiasedness we refer to Lehmann (1951), Birnbaum (1964), and van der Vaart (1961).

Problem 7.34.* The set $\mathbb{D}_0$ of all median unbiased randomized estimators for the model $(\mathcal{X}, \mathfrak{A}, (P_\theta)_{\theta\in\Delta})$ with values in $\overline{\mathbb{R}}$ and the function $\kappa : \Delta \to \mathbb{R}$ is closed with respect to the weak convergence of decisions in the sense of Definition 3.19.

Suppose $\varrho : \mathbb{R} \to \mathbb{R}$ is a continuous function, where ϱ is nonincreasing for $x < 0$, nondecreasing for $x > 0$, and $\varrho(0) = 0$. Let μ_ϱ be the measure on the Borel sets of $\mathbb{R}$ with

$$\begin{aligned} \mu_\varrho((a,b]) &= \varrho(b) - \varrho(a), \quad 0 \leq a < b < \infty, \\ \mu_\varrho((a,b]) &= \varrho(a) - \varrho(b), \quad -\infty < a < b \leq 0. \end{aligned} \tag{7.14}$$

Subsequently we apply a well-known integration by parts technique.

Problem 7.35.* If $\varrho : \mathbb{R} \to \mathbb{R}$ is a continuous function, nonincreasing on $(-\infty, 0)$, nondecreasing on $(0, \infty)$, with $\varrho(0) = 0$, and Q is a distribution on $(\mathbb{R}, \mathfrak{B})$, then

$$\int I_{(\theta,\infty)}(s)\varrho(s-\theta)Q(ds) = \int I_{(0,\infty)}(t)Q([t+\theta, \infty))\mu_\varrho(dt) \quad \text{and}$$

$$\int I_{(-\infty,\theta]}(s)\varrho(s-\theta)Q(ds) = \int I_{(-\infty,0)}(t)Q((-\infty, t+\theta])\mu_\varrho(dt), \quad \theta \in \mathbb{R}.$$

If Q and ϱ are symmetric, i.e., $Q(-B) = Q(B)$, $B \in \mathfrak{B}$, and $\varrho(-s) = \varrho(s)$, $s \in \mathbb{R}$, then

$$\int \varrho(s-\theta)Q(ds) = \int I_{(0,\infty)}(t)(Q([t+\theta,\infty)) + Q([t-\theta,\infty)))\mu_\varrho(dt). \tag{7.15}$$

The next theorem is due to Pfanzagl (1970, 1971). It gives the optimal unbiased estimator in a location model with parent distribution P that is symmetric in the sense of $P(-B) = P(B)$, $B \in \mathfrak{B}$, and strongly unimodal.

Theorem 7.36. *Let $\varrho : \mathbb{R} \to [0,\infty]$ be a symmetric continuous function that is nondecreasing on $[0,\infty]$. Suppose $P = \mathcal{L}(Z)$ is a symmetric and strongly unimodal distribution on $(\mathbb{R}, \mathfrak{B})$ with a Lebesgue density positive everywhere. Let $P_\theta = \mathcal{L}(Z+\theta)$, $\theta \in \mathbb{R}$. If $\mathsf{D} : \overline{\mathfrak{B}} \times \mathcal{X} \to_k [0,1]$ is a median unbiased estimator for the model $(\mathcal{X}, \mathfrak{A}, (P_\theta)_{\theta\in\Delta})$, and $\kappa(\theta) = \theta$, then*

$$\mathsf{R}(\theta,\mathsf{D}) := \int [\int \varrho(y-\theta)\mathsf{D}(dy|x)]P_\theta(dx) \geq \int \varrho(s)P(ds).$$

The natural estimator $T_{nat}(x) = x$ is a uniformly best estimator in the class of all median unbiased estimators under the loss function $L(\theta,a) = \varrho(a-\theta)$.

Proof. Without loss of generality we may assume that $\varrho(0) = 0$. As $\varrho(\infty) \geq 0$ it is sufficient to consider randomized estimators D with $\mathsf{D}(\{\infty\}|x) = 0$ for every x. Using the integration by parts in Problem 7.35 with $Q = \mathsf{D}$, and then $Q = P$ for $\theta = 0$, we see that the statement to be proved is equivalent to

$$\begin{aligned}&\int [\int I_{[0,\infty)}(t)(\mathsf{D}([t+\theta,\infty)|x) + \mathsf{D}([t-\theta,\infty)|x))P_\theta(dx)]\mu_\varrho(dt)\\ &\geq 2\int I_{[0,\infty)}(t)P([t,\infty))\mu_\varrho(dt).\end{aligned}$$

As the density of P is symmetric it holds $P_{-\theta} = P_\theta$ so that it is sufficient to show that

$$\int \mathsf{D}([t+\theta,\infty)|x)P_\theta(dx) \geq 1 - F(t), \quad t > 0, \tag{7.16}$$

where F is the c.d.f. of P. We set $\varphi(x) = \mathsf{D}([t+\theta,\infty)|x)$ and $\alpha = 1/2$. The median unbiasedness of D and $F(-t) = 1 - F(t)$ yield

$$\int \mathsf{D}([t+\theta,\infty)|x)P_{\theta+t}(dx) \geq 1/2.$$

An application of the second inequality in Problem 2.50 with $\theta_1 = \theta + t$ and $\theta_2 = \theta$ yields (7.16). The statement that T_{nat} is a uniformly best unbiased estimator follows from the fact that, due to the symmetry of P, the estimator T_{nat} is median unbiased and

$$2\int I_{[0,\infty)}(t)P((t,\infty))\mu_\varrho(dt) = 2\int I_{[0,\infty)}(t)\varrho(t)P(dt) = \int \varrho(t)P(dt)$$

is the risk of T_{nat} under the loss function $L(\theta,a) = \varrho(a-\theta)$. ■

7.3 Bayes and Generalized Bayes Estimators

We start with the model $(\mathcal{X}, \mathfrak{A}, (P_\theta)_{\theta\in\Delta})$ and assume that condition (A3) is satisfied. Suppose we want to estimate a function $\kappa : \Delta \to_m \mathbb{R}^d$. Then $(\mathcal{D}, \mathfrak{D}) = (\mathbb{R}^d, \mathfrak{B}_d)$ is the decision space. We fix $l : \mathbb{R}^d \times \mathbb{R}^d \to_m \mathbb{R}_+$ and introduce the loss function by setting $L(\theta, a) = l(\kappa(\theta), a)$. Recall that a randomized estimator is a stochastic kernel $\mathsf{D} : \mathfrak{B}_d \times \mathcal{X} \to_k [0,1]$. Especially if $T : \mathcal{X} \to_m \mathbb{R}^d$, then the kernel $\mathsf{D} = \boldsymbol{\delta}_T$ is an estimator. The risk is given by

$$\mathsf{R}(\theta, \mathsf{D}) = \int [\int l(\kappa(\theta), a)\mathsf{D}(da|x)]P_\theta(dx).$$

We recall that for $\rho \in \mathcal{M}^\sigma(\mathfrak{B}_\Delta)$ a posterior distribution is a stochastic kernel $\boldsymbol{\Pi}$ such that $P_\theta(dx)\rho(d\theta) = \boldsymbol{\Pi}(d\theta|x)(\mathsf{P}\rho)(dx)$, where $(\mathsf{P}\rho)(dx) = \int P_\theta(dx)\rho(d\theta)$. Conditions for the existence of a posterior distribution have been established in Proposition 3.32. If we turn to the location or the scale model, then for an invariant averaging measure the posterior distribution turns out to be closely related to the conditional distribution with respect to the σ-algebra of invariant Borel sets of some equivariant statistics.

Example 7.37. We consider the model

$$\mathcal{M}_{lo} = \left(\mathbb{R}^n, \mathfrak{B}_n, (P \circ u_\theta^{-1})_{\theta\in\mathbb{R}}\right), \tag{7.17}$$

where $(P \circ u_\theta^{-1})(B) = P(B - \theta\mathbf{1})$, and assume that P has the Lebesgue density f. Set $\rho = \boldsymbol{\lambda}$ and $\mathsf{m}(x) = \int f(x - \theta\mathbf{1})\boldsymbol{\lambda}(d\theta)$. It holds for $a < b$,

$$\int I_{[a,b]}(x_1)\mathsf{m}(x_1, ..., x_n)\boldsymbol{\lambda}_n(dx_1, ..., dx_n) = b - a.$$

Hence $\mathsf{m} < \infty$, $\boldsymbol{\lambda}_n$-a.e., and we see from (3.25) that $\boldsymbol{\Pi}(d\theta|x) = \pi(\theta|x)\boldsymbol{\lambda}(d\theta)$, where

$$\pi(\theta|x) = \begin{cases} \frac{f(x-\theta\mathbf{1})}{\int f(x-s\mathbf{1})ds} & \text{if } \mathsf{m}(x) > 0, \\ g(\theta) & \text{if } \mathsf{m}(x) = 0, \end{cases} \tag{7.18}$$

and where g is a fixed Lebesgue density. Hence $f(x_1 - \theta|x) = \pi(\theta|x)$, where $f(y_1|x)$ has been defined in Lemma 5.57. This means that the posterior distribution $\boldsymbol{\Pi}(\cdot|x)$ and the conditional distribution $\mathsf{K}_{\mathcal{E}_0}$, defined in Lemma 5.57, of the equivariant statistic $\mathcal{E}_0(x) = x_1$ given the σ-algebra of invariant Borel sets, are related by

$$\mathsf{K}_{\mathcal{E}_0}(x_1 - C|x) = \boldsymbol{\Pi}(C|x). \tag{7.19}$$

Definition 7.38. *Given a weight measure $\rho \in \mathcal{M}^\sigma(\mathfrak{B}_\Delta)$, a randomized estimator D_0 is called a minimum average estimator if $\mathsf{r}(\rho, \mathsf{D}_0) \le \mathsf{r}(\rho, \mathsf{D})$ for every randomized estimator D. If $\rho = \Pi$ is a prior, then D_0 is called a Bayes estimator. For any $\rho \in \mathcal{M}^\sigma(\mathfrak{B}_\Delta)$ for which a posterior distribution exists the decision D_0 is called a generalized Bayes estimator if the posterior risk in Definition 3.33 satisfies*

$$\mathsf{r}(\rho, \mathsf{D}_0|x) \le \mathsf{r}(\rho, \mathsf{D}|x), \quad \mathsf{P}\rho\text{-a.e.}$$

If a posterior distribution exists it follows from Theorem 3.37 that a randomized estimator D_0 with $\mathsf{r}(\rho, \mathsf{D}_0) < \infty$ is a minimum average estimator if and only if it is a generalized Bayes estimator. Especially every Bayes estimator with finite risk is a generalized Bayes estimator. Another feature is that, according to Corollary 3.40, a Bayes estimator can be obtained by a pointwise minimization of the function $a \mapsto r(\rho, a|x)$ introduced in Definition 3.33, i.e., by finding a mapping $S : \mathcal{X} \to_m \mathbb{R}^d$ such that

$$r(\rho, S(x)|x) \leq r(\rho, b|x), \quad b \in \mathbb{R}^d.$$

Proposition 7.39. *Assume that condition (A3) is satisfied for the model* $(\mathcal{X}, \mathfrak{A}, (P_\theta)_{\theta\in\Delta})$, $\rho \in \mathcal{M}^\sigma(\mathfrak{B}_\Delta)$, *a posterior distribution exists, and the loss function is* $L(\theta, a) = \|\kappa(\theta) - a\|^2$. *If* $\int \|\kappa(\theta)\|^2 \boldsymbol{\Pi}(d\theta|x) < \infty$, $\mathsf{P}\Pi$*-a.e., then*

$$S(x) = \int \kappa(\theta)\boldsymbol{\Pi}(d\theta|x),$$

is a generalized Bayes estimator, and S *is,* $\mathsf{P}\Pi$*-a.e., uniquely determined in the class of all nonrandomized estimators.*

Proof. The posterior risk in Definition 3.33 satisfies

$$r(\rho, a|x) = \int \|\kappa(\theta) - a\|^2 \boldsymbol{\Pi}(d\theta|x) \geq \int \|\kappa(\theta) - S(x)\|^2 \boldsymbol{\Pi}(d\theta|x).$$

If $T : \mathcal{X} \to_m \mathbb{R}^d$ is another generalized Bayes estimator, then by the convexity of L the estimator $\frac{1}{2}(S+T)$ is again a generalized Bayes estimator and it holds, $\mathsf{P}\Pi$-a.e.,

$$\begin{aligned} 0 &= \frac{1}{2}\int (\|\kappa(\theta) - S(x)\|^2 + \|\kappa(\theta) - T(x)\|^2)\boldsymbol{\Pi}(d\theta|x) \\ &\quad - \int \| \kappa(\theta) - \frac{1}{2}(S(x) + T(x)) \|^2 \boldsymbol{\Pi}(d\theta|x) = \frac{1}{4}\|S(x) - T(x)\|^2 . \end{aligned}$$

■

Now we consider the Bayes model

$$(\Omega, \mathfrak{F}, \mathbb{P}) = (\mathcal{X} \times \Delta, \mathfrak{A} \otimes \mathfrak{B}_\Delta, \mathsf{P} \otimes \Pi),$$

where Π is a prior and X and Θ are the projections of $\mathcal{X} \times \Delta$ on $\mathcal{X}$ and Δ, respectively. In this Bayes model we may operate directly with the conditional expectation to get the Bayes estimator. Let $L(\theta, a) = \|\kappa(\theta) - a\|^2$. Then for any randomized estimator D it holds

$$\mathsf{r}(\Pi, \mathsf{D}) = \int (\int \|\kappa(\theta) - a\|^2 \mathsf{D}(da|x))(\mathsf{P}\Pi)(dx).$$

If $\mathbb{E}\|\kappa(\Theta)\|^2 < \infty$ and $\mathsf{r}(\Pi, \mathsf{D}) < \infty$, then

$$\mathsf{r}(\Pi, \mathsf{D}) \geq \int \|\kappa(\theta) - S_{\mathsf{D}}(x)\|^2 (\mathsf{P}\Pi)(dx) = \mathsf{r}(\Pi, S_{\mathsf{D}}), \quad \text{where} \tag{7.20}$$
$$S_{\mathsf{D}}(x) = \int a\mathsf{D}(da|x).$$

This means that it suffices to deal with nonrandomized estimators if we use the squared error loss; see also the discussion of (3.5).

Proposition 7.40. *If* $L(\theta, a) = \|\kappa(\theta) - a\|^2$ *and* $\mathbb{E}\|\kappa(\Theta)\|^2 < \infty$, *then* $S(x) = \mathbb{E}(\kappa(\Theta)|X = x))$ *is a Bayes estimator. Moreover* S *is,* $\mathsf{P}\Pi$*-a.e., uniquely determined in the class of all nonrandomized estimators.*

Proof. If $S_0(X)$ is any estimator with finite risk $\mathbb{E}\|S_0 - \kappa(\Theta)\|^2 < \infty$, then the relation

$$\begin{aligned}&\mathbb{E}(\,(S(X) - S_0(X))^T(S(X) - \kappa(\Theta))|X) \\ &= (S(X) - S_0(X))^T\mathbb{E}\,(S(X) - \kappa(\Theta)|X) = 0\end{aligned}$$

gives

$$\mathbb{E}\|S(X) - \kappa(\Theta)\|^2 = \mathbb{E}\|S(X) - S_0(X)\|^2 + \mathbb{E}\|S(X) - \kappa(\Theta)\|^2$$

and the proof is completed. ■

In the Bayes model, under the weak assumption that at least one of the conditions (A4) or (A5) is satisfied, a regular conditional distribution of Θ, given X, (i.e., a posterior distribution) exists and it holds

$$\mathbb{E}(\kappa(\Theta)|X = x) = \int \kappa(\theta)\boldsymbol{\Pi}(d\theta|x), \quad \mathsf{P}\Pi\text{-a.e.}, \tag{7.21}$$

provided that $\mathbb{E}\|\kappa(\Theta)\| < \infty$; see (1.32). Moreover, if $\mathbb{E}\|\kappa(\Theta)\|^2 < \infty$, then

$$\mathbb{E}\|\kappa(\Theta)\|^2 = \int (\int \|\kappa(\theta)\|^2 \boldsymbol{\Pi}(d\theta|x))(\mathsf{P}\Pi)(dx) < \infty,$$

which implies $\int \|\kappa(\theta)\|^2 \boldsymbol{\Pi}(d\theta|x) < \infty$, $\mathsf{P}\Pi$-a.e. Hence the generalized Bayes estimator is identical with the Bayes estimator, and the latter can be obtained via the posterior distribution. In exponential families there is the opportunity to give more explicit expressions. Under a conjugate prior the conditional expectation is directly available from the posterior, which is again an exponential family.

Example 7.41. Let $(P_\theta)_{\theta\in\Delta}$ be a one-parameter exponential family with natural parameter θ and generating statistic T. Let $\Pi_{a,b}$, $(a, b) \in \Upsilon$, be the family of conjugate priors. Then the posterior distributions are $\Pi_{a+1,b+T(x)}$, $(a, b) \in \Upsilon$. If $(a, b) \in \Upsilon^0$, then by Theorem 1.17 $\int \|\theta\|^2 \Pi_{a,b}(d\theta) < \infty$, so that at $x \in \mathcal{X}$,

$$S_{a,b}(x) = \int \theta\Pi_{a+1,b+T(x)}(d\theta) \tag{7.22}$$

is the Bayes estimator of θ under the prior $\Pi_{a,b}$. Similar results hold under reparametrization. The Bayes estimator of the success probability in a binomial distribution, under the squared error loss and a $\mathsf{Be}(\alpha, \beta)$ prior, is a special case that has been considered in Example 3.59.

Example 7.42. We consider the location model (7.17), assume that P has a Lebesgue density, and choose ρ to be the Lebesgue measure. For the equivariant statistic $\mathcal{E}_0(x) = x_1$ let $\mathsf{K}_{\mathcal{E}_0}$ be given by Lemma 5.57. Then by (7.19) the posterior distribution exists and is given by $\boldsymbol{\Pi}(C|x) = \mathsf{K}_{\mathcal{E}_0}(x_1 - C|x)$. From here we see that under the assumption $\int x_1^2 P(dx) < \infty$ it holds

$$\int \theta^2 \boldsymbol{\Pi}(d\theta|x) = \int (x_1 - t)^2 \mathsf{K}_{\mathcal{E}_0}(dt|x) < \infty, \quad P\text{-a.s.},$$

and the generalized Bayes estimator

$$\int \theta \boldsymbol{\Pi}(d\theta|x) = x_1 - \int t \mathsf{K}_{\mathcal{E}_0}(dt|x) = \mathcal{P}(x)$$

is the Pitman estimator in (5.51) for $\mathcal{E}_0(x) = x_1$.

Problem 7.43.* Let $0 < \alpha < 1$ and $\tau_\alpha(t) = (1-\alpha)|t| I_{(-\infty,0]}(t) + \alpha t I_{(0,\infty)}(t)$. Let X be a random variable with c.d.f. F, and u_α be an α-quantile, i.e., $F(u_\alpha - 0) \le \alpha \le F(u_\alpha)$. Then

$$\mathbb{E}(\tau_\alpha(X-\theta) - \tau_\alpha(X - u_\alpha)) = \int I_{[u_\alpha,\theta]}(s)(F(s) - \alpha)ds, \quad \theta > u_\alpha,$$

$$\mathbb{E}(\tau_\alpha(X-\theta) - \tau_\alpha(X - u_\alpha)) = \int I_{[\theta,u_\alpha]}(s)(\alpha - F(s))ds, \quad \theta < u_\alpha.$$

Example 7.44. Let $(P_\theta)_{\theta\in\Delta}$ be a one-parameter exponential family with natural parameter θ and generating statistic T. Let $\Pi_{a,b}$, $(a,b) \in \Upsilon^0$, be a conjugate prior, so that the posterior is $\Pi_{a+1,b+T(x)}$ and by Theorem 1.17 $\int |\theta| \Pi_{a,b}(d\theta) < \infty$. If we use the loss function $L(\theta, a) = \tau_\alpha(\theta - a)$, $0 < \alpha < 1$, then by Problem 7.43 every α-quantile $q_\alpha(x)$ of the distribution $\Pi_{a+1,b+T(x)}$ is a Bayes estimator of $\theta \in \Delta$.

Example 7.45. We consider the situation in Example 7.42 and assume that P has a Lebesgue density and $\int |x_1| P(dx) < \infty$. The posterior distribution is given by (7.18). If we use the loss function $L(\theta, a) = \tau_\alpha(\theta - a)$, then every solution of the equation

$$\int_{-\infty}^{S(x)} \pi(\theta|x)d\theta = \alpha$$

that depends measurably on x is a generalized Bayes estimator.

The construction of a Bayes or a generalized Bayes estimator depends on the given model, a fixed loss function, and an appropriately chosen prior or averaging measure. After the loss function has been fixed the remaining task is to adjust a prior. Now we use the *hierarchical Bayes approach* to construct estimators. This means that instead of one prior Π we use a family of priors Π_ξ that depend on a *hyperparameter* ξ, the outcome of a random variable Ξ, where the model is completed by a choice of the distribution Γ of Ξ.

Often, the conditional distributions involved in the hierarchical Bayes model are given by conditional densities. Assume that $\boldsymbol{\mu}, \boldsymbol{\tau}$, and $\boldsymbol{\rho}$ are σ-finite measures on $(\mathcal{X}, \mathfrak{A})$, $(\Delta, \mathfrak{B}_\Delta)$, and $(\Upsilon, \mathfrak{B}_\Upsilon)$, respectively. Suppose $(x,\theta) \mapsto f_\theta(x)$, $(\theta,\xi) \mapsto p(\theta|\xi)$, and $\xi \mapsto r(\xi)$ are nonnegative and measurable with respect to the corresponding σ-algebras, and assume that

$$\int f_\theta(x)\boldsymbol{\mu}(dx) = 1, \quad \int p(\theta|\xi)\boldsymbol{\tau}(d\theta) = 1, \quad \int r(\xi)\boldsymbol{\rho}(d\xi) = 1.$$

The distribution $\mathcal{L}(X, \Theta, \Xi)$ is specified by the requirement that it has the density

$$\frac{d\mathcal{L}(X,\Theta,\Xi)}{d(\boldsymbol{\mu}\otimes\boldsymbol{\tau}\otimes\boldsymbol{\rho})}(x,\theta,\xi) = f_\theta(x)p(\theta|\xi)r(\xi). \tag{7.23}$$

Let

$$\mathsf{m}(x) = \int [\int f_\theta(x)p(\theta|\xi)r(\xi)\boldsymbol{\rho}(d\xi)]\boldsymbol{\tau}(d\theta)$$

be the marginal density of X. Then the posterior density $\pi(\theta|x)$ and the posterior distribution $\boldsymbol{\Pi}(d\theta|x)$ are given by

$$\pi(\theta|x) = \begin{cases} \frac{1}{\mathsf{m}(x)} \int f_\theta(x)p(\theta|\xi)r(\xi)\boldsymbol{\rho}(d\xi) & \text{if } \mathsf{m}(x) > 0, \\ \pi(\theta) & \text{if } \mathsf{m}(x) = 0, \end{cases}$$

$$\boldsymbol{\Pi}(B|x) = \int_B \pi(\theta|x)\boldsymbol{\tau}(d\theta), \quad B \in \mathfrak{B}_\Delta,$$

and according to (7.21) the Bayes estimator can be represented as

$$\mathbb{E}(\kappa(\Theta)|X = x) = \int \kappa(\theta)\pi(\theta|x)\boldsymbol{\tau}(d\theta).$$

In the hierarchical Bayes approach the choice of a prior is only completed after the hyperparameter has been implemented and its distribution has been specified. According to (7.23) the random variables X, Θ, Ξ form a Markov chain, and by Problem 1.100 the random variables Ξ, Θ, X form a Markov chain again. Hence by Proposition 1.101,

$$\mathsf{I}_\mathsf{v}(X||\Xi) \leq \mathsf{I}_\mathsf{v}(X||\Theta),$$

so that the hyperparameter Ξ has less influence on the inference than the parameter Θ.

The next example presents special Bayes hierachical models. For more models we refer to Lehmann and Casella (1998).

Example 7.46. The normal Bayes hierarchical model is given by the following specification of the conditional distributions.

$$\mathcal{L}(X|\Theta = \theta) = \mathsf{N}^{\otimes n}(\theta, \sigma^2), \quad \mathcal{L}(\Theta|\Xi = \xi) = \mathsf{N}(0, \xi), \quad \mathcal{L}(\Xi) = \mathsf{Ig}(\lambda, \beta),$$

where σ^2, λ, and β are known. Similarly, the Poisson Bayes hierarchical model is given by

$$\mathcal{L}(X|\Theta = \theta) = \mathsf{Po}(\theta), \quad \mathcal{L}(\Theta|\Xi = \xi) = \mathsf{Ga}(\lambda, \xi), \quad \mathcal{L}(\Xi) = \mathsf{Ga}(\kappa, \tau),$$

where again λ, κ, and τ are known. The calculation of the conditional density of Θ, given $X = x$, is facilitated by the choice of the prior to be a conjugate prior. For further details we refer to Lehmann and Casella (1998).

After the implementation of a hyperparameter in the prior the model choice is not complete until the distribution of the hyperparameter has been specified. The *empirical Bayes approach* solves the problem of not knowing the hyperparameter in the prior by estimating it based on the marginal distribution of the data. This means that instead of averaging on the hyperparameter in the prior distribution we estimate this parameter using the data. We illustrate this approach by an example.

Example 7.47. Let $P_\theta = \mathsf{N}(\theta, \mathbf{I})$, $\theta \in \mathbb{R}^d$, and suppose we want to estimate θ under the loss $L(\theta, a) = \|\theta - a\|^2$. We use the prior $\mathsf{N}(0, \tau^2\mathbf{I})$ for Θ. Then

$$\begin{aligned} f_\theta(x)\pi(\theta) &= (2\pi)^{-d} \exp\{-\frac{1}{2}(\|x-\theta\|^2 + \tau^{-2}\|\theta\|^2)\} \\ &= (2\pi)^{-d} \exp\{-\frac{1+\tau^2}{2\tau^2}\|\theta - \tau^2 x/(1+\tau^2)\|^2 - \|x\|^2/(2(1+\tau^2))\}. \end{aligned}$$

Hence $\boldsymbol{\Pi}(\cdot|x) = \mathsf{N}(\frac{\tau^2}{1+\tau^2}x, \tau^2\mathbf{I})$ is the posterior distribution, and the Bayes estimator is

$$\mathbb{E}(\Theta|X = x) = \frac{\tau^2}{1+\tau^2}x = (1 - \frac{1}{1+\tau^2})x. \tag{7.24}$$

The marginal distribution of X is $\mathsf{N}(0, (1+\tau^2)\mathbf{I})$. According to Problem 7.23 the UMVU estimator of $1/\sigma^2 = (1+\tau^2)^{-1}$ is $(n-2)/\|x\|^2$. Plugging in this estimator we get the empirical Bayes estimator

$$S_{JS}(x) = (1 - (d-2)\|x\|^{-2})x, \tag{7.25}$$

which is nothing else than the famous James–Stein estimator that has been discussed already in Example 3.14. We may also estimate $1/\sigma^2 = (1+\tau^2)^{-1}$ by the maximum likelihood estimator that is obtained by maximizing the function

$$\tau^2 \mapsto (2\pi)^{-d}(1+\tau^2)^{-d/2} \exp\{-\frac{1}{2(1+\tau^2)}\|x\|^2\}.$$

This gives the estimator

$$\widehat{(1+\tau^2)^{-1}} = \begin{cases} d/\|x\|^2 & \text{if } \|x\|^2 > d, \\ 0 & \text{if } \|x\|^2 \le d. \end{cases}$$

Plugging this estimator into the Bayes estimator we get a truncated James–Stein estimator given by

$$S_{JS}^{+}(x) = \max((1 - d\|x\|^{-2}), 0)x.$$

For empirical Bayes estimators in other models, and evaluations of the risks, we refer to Chapter 4 in Lehmann and Casella (1998).

7.4 Admissibility of Estimators, Shrinkage Estimators

In Theorem 3.43 we have studied the problem of admissibility of any decision by comparisons with minimum average and especially Bayes decisions, which is the method by Blyth (1951). In this section we study the admissibility of some special estimators. We start with a simple but useful result for Bayes estimators. Suppose $(\mathcal{X}, \mathfrak{A}, (P_\theta)_{\theta\in\Delta})$ satisfies condition (A3) and Π is a prior on $(\Delta, \mathfrak{B}_\Delta)$ which is a Borel space. Suppose we want to estimate a function $\kappa : \Delta \to_m \mathbb{R}^d$. We fix $l : \mathbb{R}^d \times \mathbb{R}^d \to_m \mathbb{R}_+$ and introduce the loss function by setting $L(\theta, a) = l(\kappa(\theta), a)$.

Proposition 7.48. *Assume that for the prior Π there exists a nonrandomized Bayes estimator $T : \mathcal{X} \to_m \mathbb{R}^d$ that is uniquely determined in the sense that for any further nonrandomized estimator $S : \mathcal{X} \to_m \mathbb{R}^d$ the relation $\mathsf{r}(\Pi, S) = \mathsf{r}(\Pi, T)$ implies $S = T$, P_θ-a.s., for every $\theta \in \Delta$. Then the Bayes estimator T is admissible is the class of all nonrandomized estimators.*

Proof. If T_0 dominates T in the sense that $\mathsf{R}(\theta, T_0) \le \mathsf{R}(\theta, T)$ for every θ, with strict inequality for at least one θ_0, then by the assumed uniqueness it holds $T_0 = T$, P_{θ_0}-a.s., which contradicts $\mathsf{R}(\theta_0, T_0) < \mathsf{R}(\theta_0, T)$. ■

Problem 7.49.* If $(P_\theta)_{\theta\in\Delta}$ satisfies (A3) and the family $(P_\theta)_{\theta\in\Delta}$ is homogeneous, then for every prior Π it holds $\mathsf{P}\Pi \ll\gg P_\theta$, $\theta \in \Delta$.

It follows from (7.20) that for the squared error loss function every estimator that is admissible in the class of all nonrandomized estimators is automatically admissible in the class of all estimators. Combining this fact with Propositions 7.40 and 7.48 we get the following statement.

Proposition 7.50. *Suppose $(P_\theta)_{\theta\in\Delta}$ satisfies (A3) and is homogeneous. Assume that $\int \|\kappa(\theta)\|^2\, \Pi(d\theta) < \infty$. Then under the squared error loss $L(\theta, a) = \|\kappa(\theta) - a\|^2$ the Bayes estimator $T(x) = \mathbb{E}(\kappa(\Theta)|X = x)$ is admissible.*

Example 7.51. Under the squared error loss, in (7.22) the Bayes estimator $S_{a,b}$ for the parameter in an exponential family is admissible.

Another criterion for admissibility was provided by Theorem 3.43 and its corollary. Since the Pitman estimator is a generalized Bayes estimator under an infinite average measure (see Example 7.42) the simplified version of Blyth's method in the corollary of Theorem 3.43 is not applicable as it would be in the case of Bayes decisions with finite risk. Stein succeeded in establishing the admissibility of the Pitman estimator under the squared loss by using suitable sequences of priors. For details we refer to Stein (1959), Lehmann and Casella (1998), p. 342, and Perng (1970).

Now we investigate the case of a Gaussian location model in higher dimensions. Here we study especially the so-called *Stein effect* which says that for dimensions larger than two the natural estimator, which is here the identity, is no longer admissible. This result, due to Stein (1955a) and James and Stein (1960), was a breakthrough in mathematical statistics. It was the starting point of a new branch of mathematical statistics called *shrinkage estimators.*

Let $X = (X_1, ..., X_d)^T$ be a d-dimensional normal vector with distribution $\mathsf{N}(\theta, \Sigma)$ and known covariance matrix Σ. Then we may assume $\Sigma = \mathbf{I}$ and have the model

$$\mathcal{M} = (\mathbb{R}^d, \mathfrak{B}_d, (\mathsf{N}(\theta, \mathbf{I}))_{\theta\in\mathbb{R}^d}).$$

Using the squared error loss $L(\theta, a) = \parallel \theta - a \parallel^2$ we want to estimate θ. We already know from Theorem 3.65 that the natural estimator $T_{nat}(x) = x$ is minimax, not only for the squared error loss but also for all subconvex and symmetric loss functions. Moreover, we know from Proposition 7.50 that the Bayes estimator $\mathbb{E}(\Theta|X = x) = (\tau^2/(1+\tau^2))x$ in (7.24) is admissible. By letting $\tau^2 \to \infty$ we would obtain the natural estimator T_{nat}. However, the pointwise limit of admissible estimators is not necessarily admissible. To see this, recall that we have estimated the hyperparameter τ^2 to get the James–Stein estimator $S_{JS}(x) = (1 - (d-2)\,\|x\|^{-2})x$ in (7.25). Now we show that this estimator has for $d \geq 3$ a smaller risk than T_{nat}. Our starting point is a suitable integration by parts formula. The subsequent result is known as *Stein's identity* for the normal distribution. A similar result, not needed in our context, holds also for any exponential family, see Lehmann and Casella (1998), p. 31.

Lemma 7.52. *Let* $X = (X_1, ..., X_d)$, $\mathcal{L}(X) = \mathsf{N}(\theta, \mathbf{I})$, $\theta \in \mathbb{R}^d$, *and assume that* $g : \mathbb{R}^d \to \mathbb{R}$ *is continuously differentiable. If* $\mathbb{E}|g(X)(X_i - \theta_i)| < \infty$ *for some* $i \in \{1, ..., d\}$, *and* $\mathbb{E}\,\|\nabla g(X)\| < \infty$, *then*

$$\mathbb{E}g(X)(X_i - \theta) = \mathbb{E}\frac{\partial g}{\partial x_i}(X), \quad i = 1, ..., d. \tag{7.26}$$

Proof. First we assume that $g(x) = 0$ for $|x_1| > N$. The density $\varphi_{\theta_1,1}(s)$ of X_1 satisfies $\varphi'_{\theta_1,1}(s) = (\theta_1 - s)\varphi_{\theta_1,1}(s)$. Hence we get

$$\begin{aligned}
\mathbb{E}\frac{\partial g}{\partial x_1}(X) &= \mathbb{E}\int_{-\infty}^{\infty} \frac{\partial g}{\partial x_1}(t, X_2, ..., X_d)\varphi_{\theta_1,1}(t)dt \\
&= \mathbb{E}\int_{-\infty}^{\infty} \frac{\partial g}{\partial x_1}(t, X_2, ..., X_d)[\int_{-\infty}^{t} (\theta_1 - s)\varphi_{\theta_1,1}(s)ds]dt \\
&= \mathbb{E}\int_{-\infty}^{\infty} [\int_{s}^{\infty} \frac{\partial g}{\partial x_1}(t, X_2, ..., X_d)dt](\theta_1 - s)\varphi_{\theta_1,1}(s)ds \\
&= -\mathbb{E}\int_{-\infty}^{\infty} g(s, X_2, ..., X_d)(\theta_1 - s)\varphi_{\theta_1,1}(s)ds = \mathbb{E}g(X)(X_1 - \theta_1).
\end{aligned}$$

To deal with the general case, we denote by $h_N(t)$ a sequence of continuously differentiable functions with $|h_N| \leq 1$ which is 1 for $|t| \leq N$, zero for $|t| \geq N+1$, and for which $C := \sup_{t,N} |h'_N(t)| < \infty$. Set $g_N(x_1, ..., x_d) = g(x_1, ..., x_d)h_N(x_1)$. Then we have

$$\mathbb{E}g_N(X)(X_1-\theta_1)=\mathbb{E}\frac{\partial g_N}{\partial x_1}(X). \tag{7.27}$$

Because of $|g_N(X)(X_1-\theta_1)|\le|g(X)(X_1-\theta_1)|$ and

$$\left|\frac{\partial g_N}{\partial x_1}(X)\right|\le\left|\frac{\partial g}{\partial x_1}(X)\right|+C\,|g(X)|,$$

we may apply Lebesgue's theorem to (7.27) and obtain the statement for the first coordinate in (7.26). The statement for the other coordinates can be treated analogously. ■

Problem 7.53.* If $X=(X_1,...,X_d)$ is a random vector with $\mathcal{L}(X)=\mathsf{N}(0,\mathbf{I})$, then $\mathbb{E}\,\|X-\theta\|^{-2}<\infty$ for $d\ge 3$ and $\theta\in\mathbb{R}^d$.

We employ a function $g=(g_1,...,g_d)^T:\mathbb{R}^d\to\mathbb{R}^d$ that is continuously differentiable to introduce a class of estimators by

$$T_g(X)=X-g(X).$$

Using (7.26) for the functions g_i and the squared error loss function we get, with $T_0(x)=x$, for the risk

$$\begin{aligned}\mathsf{R}(\theta,T_g)&=\mathbb{E}(X-g(X)-\theta)^T(X-g(X)-\theta)\\&=\mathbb{E}(X-\theta)^T(X-\theta)-2\mathbb{E}(X-\theta)^Tg(X)+\mathbb{E}g(X)^Tg(X)\\&=\mathsf{R}(\theta,T_0)-2\sum\nolimits_{i=1}^d\mathbb{E}\frac{\partial g_i}{\partial x_i}(X)+\mathbb{E}\,\|g(X)\|^2.\end{aligned} \tag{7.28}$$

In an attempt to improve the estimator T_0, one may look for functions g that satisfy

$$\mathbb{E}\,\|g(X)\|^2-2\sum\nolimits_{i=1}^d\mathbb{E}\frac{\partial g_i(X)}{\partial x_i}\le 0.$$

The next theorem shows that the idea of correcting the natural estimator by a term $g(X)$ works well for a large class of functions g. The estimators

$$S_r(x)=\left(1-\frac{r(\|x\|)}{\|x\|^2}\right)x \tag{7.29}$$

with $r>0$ are called *shrinkage estimators*. The construction is based on the idea to shrink samples x according to the length $\|x\|$ of x. The special case of $(1-(d-2)\,\|x\|^{-2})x$ is the James–Stein estimator that was discovered by Stein (1955a). It was used by James and Stein (1960) to prove the inadmissibility of the natural estimator for $d\ge 3$. The following result is due to Baranchik (1970).

Theorem 7.54. *Let* $r : [0, \infty) \to [0, 2(d-2)]$ *be a nondecreasing function that is continuously differentiable in* $(0, \infty)$. *Then for* $d \geq 3$, *under the squared error loss, the risk of the estimator* S_r *in (7.29) for* θ *in the model* $(\mathbb{R}^d, \mathfrak{B}_d, (\mathsf{N}(\theta, \mathbf{I}))_{\theta \in \mathbb{R}^d})$ *satisfies*

$$\mathsf{R}(\theta, S_r) = \mathbb{E}\,\|S_r(X) - \theta\|^2 \leq R(\theta, T_0), \quad \theta \in \mathbb{R}^d,$$

where $\mathcal{L}(X) = \mathsf{N}(\theta, \mathbf{I})$ *and* $T_0(x) = x$. *Moreover,* S_r *is a minimax estimator.*

Corollary 7.55. *For* $d \geq 3$ *the risk of the estimator* $T_c = (1 - c\frac{d-2}{\|x\|^2})x$ *is*

$$\mathsf{R}(\theta, T_c) = d - (d-2)^2 c(2-c)\mathbb{E}\frac{1}{\|X\|^2}.$$

T_1 *is the James–Stein estimator, which has minimum risk in the class of the estimators* T_c. *For* $c < 2$ *it holds*

$$\mathsf{R}(\theta, T_c) < \mathsf{R}(\theta, T_0), \quad \theta \in \mathbb{R}^d. \tag{7.30}$$

Proof. Suppose that r is continuously differentiable, and set $g_i(x) = \|x\|^{-2} r(\|x\|)x_i$. Using $\frac{\partial \|x\|}{\partial x_i} = \frac{x_i}{\|x\|}$ we get

$$\sum\nolimits_{i=1}^{d} \frac{\partial g_i(x)}{\partial x_i} = \frac{dr(\|x\|)}{\|x\|^2} + \frac{r'(\|x\|)}{\|x\|} - \frac{2r(\|x\|)}{\|x\|^2}.$$

Hence,

$$\begin{aligned}&\mathbb{E}\,\|g(X)\|^2 - 2\sum\nolimits_{i=1}^{d} \mathbb{E}\frac{\partial g_i(X)}{\partial x_i}\\ &= \mathbb{E}\left[\frac{r(\|X\|)(4 - 2d + r(\|X\|))}{\|X\|^2} - 2\frac{r'(\|X\|)}{\|X\|}\right] \leq 0.\end{aligned}$$

Thus by (7.28) the maximum risk of S_r does not exceed the maximum risk of the estimator T_0, which is a minimax estimator, as it has been established in Theorem 3.65. To prove the corollary we set $r = c(d-2)$. Then

$$\begin{aligned}\mathsf{R}(\theta, T_c) &= \mathsf{R}(\theta, T_0) - \mathbb{E}\,\|g(X)\|^2 + 2\sum\nolimits_{i=1}^{d} \mathbb{E}\frac{\partial g_i(X)}{\partial x_i}\\ &= d - c(2-c)(d-2)^2\mathbb{E}\frac{1}{\|X\|^2}.\end{aligned}$$

■

The next theorem summarizes results on the admissibility of the natural estimator for multivariate normal distributions.

Theorem 7.56. *For the model* $(\mathbb{R}^d, \mathfrak{B}_d, (\mathsf{N}(\theta, \mathbf{I}))_{\theta\in\mathbb{R}^d})$*, under the squared error loss, the natural estimator* $T_{nat}(x) = x$ *is admissible for* $d = 1$ *and* $d = 2$*.* T_{nat} *is inadmissible for* $d \geq 3$*.*

Proof. The admissibility for $d = 1$ was proved in Theorem 3.65. The proof of the admissibility for $d = 2$ is more convoluted as the conjugate priors fail to work, see Lehmann and Casella (1998), p. 398. Another sequence of prior was used in James and Stein (1960). These priors were modified in Brown and Hwang (1982), who also proved a general result for the admissibility of estimators that includes the admissibility of T_{nat} for $d = 2$. According to (7.30) for $d \geq 3$ and $c < 2$ each estimator T_c, and especially the James–Stein estimator, has a smaller risk. ■

We conclude this section with the remark that the so-called *Stein phenomenon*, i.e., the fact that well established estimators, say maximum likelihood estimators or UMVU estimators, are admissible for low dimensions and inadmissible for higher dimensions. The Stein effect is not restricted to normal distributions or to distributions that have a Lebesgue density. An example is the Clevenson–Zidek estimator for estimating the parameters of independent Poisson random variables under the loss function $(\theta - a)^2/\theta$. For details and a comprehensive overview of the Stein effect, shrinkage estimators, and different types of improvements of standard estimators, we refer to Lehmann and Casella (1998), where it is also shown that the James–Stein estimator is inadmissible.

7.5 Consistency of Estimators

7.5.1 Consistency of M-Estimators and MLEs

Consistency of M-Estimators, Argmin Theorem

So far we have constructed estimators for a fixed sample size n, and they are tainted with uncertainty due to the randomness of the observations. The question arises as to whether this uncertainty can be reduced by increasing the sample size. Answers can be found in the area of consistency of decisions. Here we deal with the consistency of estimators, which is a first but important step toward an asymptotic investigation of estimators. If a sequence of estimators is consistent (i.e., tends in a suitable manner to the unknown parameter), then in smooth models all further considerations can be based on a locally linear approximation of the estimator and the model. This is an important fact that is used systematically later on.

When we consider estimators under increasing sample sizes, then we have to deal with a sequence of models $\mathcal{M}_n = (\mathcal{X}_n, \mathfrak{A}_n, (P_{n,\theta})_{\theta\in\Delta})$. We assume that the parameter set Δ is a separable metric space endowed with the metric ρ_Δ. If $\mathbf{X}_n$ is the observation with values in $\mathcal{X}_n$, then we envision that the sequence $\mathbf{X}_n$ is defined on a probability space $(\Omega, \mathfrak{F}, \mathbb{P}_{\theta_0})$, where $\mathcal{L}(\mathbf{X}_n|\mathbb{P}_{\theta_0}) = \mathbb{P}_{\theta_0} \circ \mathbf{X}_n = P_{n,\theta_0}$ for some $\theta_0 \in \Delta$. For i.i.d. observations we may use the infinite product space. More precisely, we consider the product space $(\mathcal{X}^\infty, \mathfrak{A}^{\otimes\infty})$ where $\mathcal{X}^\infty = \{(x_1, x_2, ...) : x_i \in \mathcal{X}\}$ and $\mathfrak{A}^{\otimes\infty}$ is the smallest σ-algebra for which all coordinate mappings $X_i : \mathcal{X}^\infty \to \mathcal{X}$, defined by $X_i((x_1, x_2, ...)) = x_i$, $i = 1, 2, ...,$ are measurable. Let $P_\theta^{\otimes\infty}$ be the infinite product measure on $(\mathcal{X}^\infty, \mathfrak{A}^{\otimes\infty})$, which is defined by the condition

$$P_\theta^{\otimes\infty}(B \times \mathcal{X} \times \mathcal{X} \times \cdots) = P_\theta^{\otimes n}(B), \quad B \in \mathfrak{A}^{\otimes n}, \ \theta \in \Delta, \ n = 1, 2, ...$$

For the existence of the infinite product measure we refer to Kallenberg (1997). Altogether we arrive at the probability space

$$(\Omega, \mathfrak{F}, (\mathbb{P}_\theta)_{\theta\in\Delta}) = (\mathcal{X}^\infty, \mathfrak{A}^{\otimes\infty}, (P_\theta^{\otimes\infty})_{\theta\in\Delta})$$

with the coordinate mappings $X_1, X_2, ...$ as observations. For a finite sample size n the observation $\mathbf{X}_n = (X_1, ..., X_n)$ is modeled by

$$\mathcal{M}_n = (\mathcal{X}_n, \mathfrak{A}_n, (P_{n,\theta})_{\theta\in\Delta}) = (\mathcal{X}^n, \mathfrak{A}^{\otimes n}, (P_\theta^{\otimes n})_{\theta\in\Delta}),$$

where again the X_i are the coordinate mappings that formally depend on n. We suppress this dependence in the notation by writing X_i only. Furthermore, to avoid unnecessarily convoluted formulations it is not always mentioned explicitly whether the X_i are defined on $(\Omega, \mathfrak{F}, \mathbb{P}_{\theta_0})$ or on $(\mathcal{X}^n, \mathfrak{A}^{\otimes n}, P_{\theta_0}^{\otimes n})$. It is, however, indicated by $\mathbb{E}_{\theta_0}$ and $\mathbb{P}_{\theta_0}$ if we use any $(\Omega, \mathfrak{F}, \mathbb{P}_{\theta_0})$, and by E_{n,θ_0} and $P_{n,\theta_0} = P_{\theta_0}^{\otimes n}$ if we use $\mathcal{M}_n$.

Definition 7.57. *Given the sequence of models $(\mathcal{X}_n, \mathfrak{A}_n, (P_{n,\theta})_{\theta\in\Delta})$, a sequence of estimators $\widehat{\theta}_n : \mathcal{X}_n \to_m \Delta$ is called consistent at θ_0 if $\widehat{\theta}_n(\mathbf{X}_n) \to^{\mathbb{P}_{\theta_0}} \theta_0$, and strongly consistent at θ_0 if $\widehat{\theta}_n(\mathbf{X}_n) \to \theta_0$, $\mathbb{P}_{\theta_0}$-a.s.. If a type of consistency holds for every θ_0, then "at θ_0" is omitted.*

The consistency of estimators can be directly expressed by the $P_{n,\theta}$. Indeed, the sequence $\widehat{\theta}_n$ is consistent if and only if

$$\lim_{n\to\infty} P_{n,\theta_0}(\rho_\Delta(\widehat{\theta}_n, \theta_0) > \varepsilon) = 0, \quad \varepsilon > 0.$$

A stronger version is the uniform consistency on a subset of the parameter space. A sequence $\widehat{\theta}_n$ of estimators is called *uniformly consistent* on $K \subseteq \Delta$ if

$$\lim_{n\to\infty} \sup_{\theta\in K} P_{n,\theta}(\rho_\Delta(\widehat{\theta}_n, \theta) > \varepsilon) = 0, \quad \varepsilon > 0.$$

There are different methods for constructing consistent estimators. One of these, and probably one of the most important methods, is the concept of *M-estimators* which includes as special cases the maximum likelihood principle, the method of least squares, and the concept of $\mathbb{L}_1$*-estimators*. Moreover, this approach can be used to construct *robust estimators* and is applicable to semiparametric situations where the likelihood approach fails to apply.

The starting point is to find an estimator by minimizing a suitably constructed sequence of *criterion functions* $M_n(\theta, x_1, ..., x_n)$ about θ. Here the criterion function is constructed in such a way that if θ_0 is the true parameter, then these random functions tend to a deterministic function that attains its minimum at θ_0. To guarantee the convergence of the criterion functions they are often defined to be an arithmetic mean of independent terms that include both the observations and the unknown parameter. More specifically, for the models $\mathcal{M}_n = (\mathcal{X}^n, \mathfrak{A}^{\otimes n}, (P_\theta^{\otimes n})_{\theta\in\Delta})$ we consider the criterion functions

$$M_n(\theta, \mathbf{x}_n) = \frac{1}{n}\sum\nolimits_{i=1}^n \varrho_\theta(x_i), \quad \theta \in \Delta, \ \mathbf{x}_n = (x_1, ..., x_n) \in \mathcal{X}^n,$$

where ϱ_θ has to be adapted to the model $(\mathcal{X}, \mathfrak{A}, (P_\theta)_{\theta\in\Delta})$. We also use $M_n(\theta) = (1/n)\sum_{i=1}^n \varrho_\theta(X_i)$ for brevity.

Definition 7.58. *Let $\mathcal{M} = (\mathcal{X}, \mathfrak{A}, (P_\theta)_{\theta\in\Delta})$ be a model, where Δ is a separable metric space with Borel σ-algebra $\mathfrak{B}_\Delta$. A function $\varrho : \Delta \times \mathcal{X} \to \mathbb{R}$ is called a contrast function for $\mathcal{M}$ if for every fixed $\theta \in \Delta$ the function $x \mapsto \varrho_\theta(x)$ is measurable, for every $x \in \mathcal{X}$ the function $\theta \mapsto \varrho_\theta(x)$ is continuous, it holds $\mathsf{E}_{\theta_0} |\varrho_\theta - \varrho_{\theta_0}| < \infty$ for every $\theta, \theta_0 \in \Delta$, and ϱ satisfies the contrast condition*

$$M(\theta, \theta_0) := \mathsf{E}_{\theta_0}(\varrho_\theta - \varrho_{\theta_0}) > 0, \quad \theta \neq \theta_0. \tag{7.31}$$

Remark 7.59. It follows from Example 1.116 that the function $(\theta, x) \mapsto \varrho_\theta(x)$ is a measurable function of (θ, x).

Example 7.60. Consider the location model $(\mathcal{X}, \mathfrak{A}, (P_\theta)_{\theta\in\mathbb{R}})$ where $P_\theta = \mathcal{L}(X+\theta)$, $\int |t| P_\theta(dt) < \infty$, and $\int t P_0(dt) = 0$. Then $\mathsf{E}_{\theta_0} \left|(X-\theta)^2 - (X-\theta_0)^2\right| < \infty$ and

$$\mathsf{E}_{\theta_0}((X-\theta)^2 - (X-\theta_0)^2) = (\theta - \theta_0)^2,$$

so that the contrast condition is satisfied for $\varrho_\theta(x) = (x-\theta)^2$.

It should be noted that the condition $\mathsf{E}_{\theta_0}\varrho_\theta > \mathsf{E}_{\theta_0}\varrho_{\theta_0}$, $\theta \neq \theta_0$, which seems to be more intuitive is equivalent to (7.31) if $\mathsf{E}_{\theta_0}|\varrho_{\theta_0}| < \infty$. However, by working with (7.31) unnecessary additional conditions can be avoided. In the above example, if $\varrho_\theta(x) = (x-\theta)^2$, $\mathsf{E}_{\theta_0}|X| < \infty$, and $\mathsf{E}_{\theta_0}X^2 = \infty$, then $\mathsf{E}_{\theta_0}\varrho_\theta = \infty$ for every θ so that $\mathsf{E}_{\theta_0}\varrho_\theta > \mathsf{E}_{\theta_0}\varrho_{\theta_0}$ is not fulfilled.

Example 7.61. Consider the location model $(\mathcal{X}, \mathfrak{A}, (P_\theta)_{\theta\in\mathbb{R}})$ with $P_\theta = \mathcal{L}(Z+\theta)$, where Z has the c.d.f. G which has the α-quantile 0, so that $G(0-0) \leq \alpha \leq G(0)$. Let $X = Z + \theta_0$ be the observation, which has the c.d.f. $F(t) = G(t - \theta_0)$. Put $\tau_\alpha(t) = (1-\alpha)|t| I_{(-\infty,0]}(t) + \alpha t I_{(0,\infty)}(t)$ and $\varrho_\theta(x) = \tau_\alpha(x-\theta)$. From $||a+b|-|a|| \leq |b|$ we then get $\mathsf{E}_{\theta_0} |\varrho_\theta(X) - \varrho_{\theta_0}(X)| < \infty$. As $u_\alpha = \theta_0$ is an α-quantile of F, Problem 7.43 yields

$$\mathsf{E}_{\theta_0}(\varrho_\theta(X) - \varrho_{\theta_0}(X)) = \mathsf{E}_0(\tau_\alpha(X-\theta) - \tau_\alpha(X-\theta_0))$$
$$= \begin{cases} \int I_{[\theta_0,\theta]}(s)(F(s)-\alpha)ds & \text{if } \theta > \theta_0, \\ \int I_{[\theta,\theta_0]}(s)(\alpha - F(s))ds & \text{if } \theta < \theta_0. \end{cases}$$

Using $F(t) = G(t-\theta_0)$ we see that the contrast condition (7.31) is satisfied if and only if $G(-\varepsilon) < \alpha < G(\varepsilon)$, $\varepsilon > 0$, which is equivalent to the α-quantile of being unique.

Definition 7.62. *Given a contrast function* $\varrho : \Delta \times \mathcal{X} \to \mathbb{R}$ *an estimator* $\widehat{\theta}_n : \mathcal{X}^n \to_m \Delta$ *is called an* M*-estimator at* $\mathbf{x}_n \in \mathcal{X}^n$ *if* $M_n(\widehat{\theta}_n(\mathbf{x}_n), \mathbf{x}_n) = \inf_{\theta\in\Delta} M_n(\theta, \mathbf{x}_n)$. *If* $\widehat{\theta}_n$ *is an* M*-estimator at every* $\mathbf{x}_n \in \mathcal{X}^n$, *then we simply call it an* M*-estimator.*

The existence of an M-estimator can be easily established if the parameter space is compact.

Proposition 7.63. *If* Δ *is a compact metric space and* $\varrho : \Delta \times \mathcal{X} \to \mathbb{R}$ *is a contrast function, then for every* n *there exists at least one* M*-estimator.*

Proof. The compactness of Δ and the continuity of $M_n(\theta, \mathbf{x}_n)$ as a function of θ imply that for every $\mathbf{x}_n \in \mathcal{X}^n$ there exists at least one $\widehat{\theta}_n(\mathbf{x}_n)$ such that $\widehat{\theta}_n(\mathbf{x}_n)$ is a minimum point. As Δ is a Polish space, Theorem A.10 yields the existence of a measurable version of the minimizer $\widehat{\theta}_n$. ■

The minimization of $M_n(\theta, \mathbf{x}_n)$ over θ is equivalent to the minimization of $M_n(\theta, \mathbf{x}_n) - M_n(\theta_0, \mathbf{x}_n)$ over θ. Due to the law of large numbers it holds $M_n(\theta) - M_n(\theta_0) \to M(\theta, \theta_0)$, $\mathbb{P}$-a.s. The contrast condition (7.31) suggests that each sequence of minimizers of M_n converges to θ_0. To explain this idea we remark that for $\mathbf{X}_n = (X_1, ..., X_n)$,

$$\begin{aligned} &\rho_\Delta(\widehat{\theta}_n(\mathbf{X}_n), \theta_0) > \varepsilon \quad \text{implies} \\ &M_n(\theta_0, \mathbf{X}_n) \geq \inf_{\theta\in B_\varepsilon} M_n(\theta, \mathbf{X}_n) \geq \frac{1}{n}\sum\nolimits_{i=1}^n \varrho_{B_\varepsilon}(X_i), \quad \text{where} \\ &\varrho_{B_\varepsilon}(x_i) = \inf_{\theta\in B_\varepsilon} \varrho_\theta(x_i) \text{ and } B_\varepsilon = \{\theta : \rho_\Delta(\theta, \theta_0) > \varepsilon\}. \end{aligned} \tag{7.32}$$

Note that due to the continuity of ϱ and the separability of Δ the function ϱ_{B_ε} is measurable and takes on values in $[-\infty, \infty)$. The term $(1/n)\sum_{i=1}^n \varrho_{B_\varepsilon}(X_i)$ is approximately $\mathbb{E}_{\theta_0}\varrho_{B_\varepsilon}(X_1)$. If $\mathbb{E}_{\theta_0}\varrho_{B_\varepsilon}(X_1) > \mathbb{E}_{\theta_0}\varrho_{\theta_0}(X_1)$, then $\widehat{\theta}_n(\mathbf{X}_n)$ cannot have a distance of more than ε from θ_0 for large n. This basic idea for proving consistency appeared already in Wald (1949), and it has been used since then by many authors in various situations. Our representation follows Perlman (1972), Pfanzagl (1969, 1994), Liese and Vajda (1994, 1995), and Berlinet, Liese, and Vajda (2000). We note that by $\varrho_A - \varrho_\eta \leq 0$ for $\eta \in A$ and $\mathsf{E}_{\theta_0}|\varrho_\eta - \varrho_{\theta_0}| < \infty$ the expectation $\mathsf{E}_{\theta_0}(\varrho_A - \varrho_{\theta_0})$ is well defined.

Condition 7.64 *Let Δ be a separable metric space and $\varrho : \Delta \times \mathcal{X} \to \mathbb{R}$ a contrast function for the model $\mathcal{M} = (\mathcal{X}, \mathfrak{A}, (P_\theta)_{\theta\in\Delta})$. We say that ϱ and $\mathcal{M}$ satisfy the finite covering property at θ_0 if for every $\varepsilon > 0$ there are a finite number of sets $\Delta_1, ..., \Delta_N$ that satisfy $B_\varepsilon = \{\theta : \rho_\Delta(\theta, \theta_0) \geq \varepsilon\} \subseteq \cup_{i=1}^N \Delta_i$, and*

$$\mathsf{E}_{\theta_0}(\varrho_{\Delta_i} - \varrho_{\theta_0}) > 0, \quad i = 1, ..., N,$$

where $\varrho_{\Delta_i}(x) = \inf_{\theta\in\Delta_i} \varrho_\theta(x)$.

The finite covering property can easily be verified for compact parameter spaces.

Lemma 7.65. *If Δ is a compact metric space and*

$$\mathsf{E}_{\theta_0}(\varrho_{\theta_0} - \inf\nolimits_{\theta\in\Delta} \varrho_\theta) < \infty, \tag{7.33}$$

then the finite covering property at θ_0 is satisfied.

Proof. Let $U(\theta, \delta)$ be an open ball with diameter δ and center θ. By the continuity of $\eta \mapsto \varrho_\eta(x)$, condition (7.33) and Lebesgue's theorem yield

$$\lim_{\delta\downarrow 0} \mathsf{E}_{\theta_0} \inf_{\eta\in U(\theta,\delta)} (\varrho_\eta - \varrho_{\theta_0}) = \mathsf{E}_{\theta_0}(\varrho_\theta - \varrho_{\theta_0}) > 0,$$

for every $\theta \neq \theta_0$ by the contrast condition. Hence for every $\theta \in B_\varepsilon$ there is some $\delta(\varepsilon, \theta) > 0$ such that $\mathsf{E}_{\theta_0}(\inf_{\eta\in U(\theta,\delta(\varepsilon,\theta))} \varrho_\eta - \varrho_{\theta_0}) > 0$. As Δ is compact we may cover B_ε already by finitely many $\Delta_i = U(\theta, \delta(\varepsilon, \theta_i))$, $i = 1, ...N$. ∎

We also consider estimators $\widehat{\theta}_n$ that are only approximate M-estimators. To be more precise let $\mathcal{L}(\mathbf{X}_n|\mathbb{P}_{\theta_0}) = P_{\theta_0}^{\otimes n}$ and set

$$D_n(\mathbf{X}_n) = M_n(\widehat{\theta}_n(\mathbf{X}_n), \mathbf{X}_n) - \inf_{\theta\in\Delta} M_n(\theta, \mathbf{X}_n),$$

for any estimator $\widehat{\theta}_n : \mathcal{X}^n \to_m \Delta$. Depending on in which sense $D_n(\mathbf{X}_n)$ becomes small for large n we distinguish between different types of estimators.

For subsequent purposes we introduce some new notation for a sequence Y_n of random variables. We say that

$$Y_n = \mathbf{o}_{\mathbb{P}_{\theta_0}}(0) \quad \text{if} \quad \mathbb{P}_{\theta_0}(Y_n \neq 0) \to 0. \tag{7.34}$$

Note that such sequences Y_n converge stochastically to zero, but in a special way. For any sequence C_n of numbers, or even random variables, we see that

$$Y_n = \mathbf{o}_{\mathbb{P}_{\theta_0}}(0) \quad \text{implies} \quad C_n Y_n = \mathbf{o}_{\mathbb{P}_{\theta_0}}(0).$$

Definition 7.66. *A sequence of estimators $\{\widehat{\theta}_n\}$ is called a strongly approximate M-estimator at θ_0 if $D_n(\mathbf{X}_n) \to 0$, $\mathbb{P}_{\theta_0}$-a.s. Similarly, $\{\widehat{\theta}_n\}$ is called an approximate M-estimator if $D_n(\mathbf{X}_n) \to^{\mathbb{P}_{\theta_0}} 0$. We call the sequence $\{\widehat{\theta}_n\}$ an asymptotic M-estimator at θ_0 if $D_n(\mathbf{X}_n) = \mathbf{o}_{\mathbb{P}_{\theta_0}}(0)$.*

Obviously every asymptotic M-estimator is an approximate M-estimator.

Theorem 7.67. *Let Δ be a separable metric space and $\varrho : \Delta \times \mathcal{X} \to \mathbb{R}$ be a contrast function for $\mathcal{M} = (\mathcal{X}, \mathfrak{A}, (P_\theta)_{\theta\in\Delta})$. If the finite covering property is satisfied at θ_0, and the estimator $\widehat{\theta}_n : \mathcal{X}^n \to_m \Delta$ is a strongly approximate M-estimator at θ_0, then $\widehat{\theta}_n$ is strongly consistent at θ_0. Similarly an approximate M-estimator at θ_0 is consistent at θ_0. Especially every asymptotic M-estimator is consistent.*

Corollary 7.68. *The statement of the theorem is valid if the finite covering condition is replaced by the assumption that Δ is compact and (7.33) holds.*

Proof. For B_ε in Condition 7.64 and every $m = 1, 2, ...$ there are $\Delta_{m,1}, ..., \Delta_{m,N_m}$ such that

$$B_{1/m} \subseteq \bigcup\nolimits_{i=1}^{N_m} \Delta_{m,i} \quad \text{and} \quad \mathsf{E}_{\theta_0}(\varrho_{\Delta_{m,i}}(X_1) - \varrho_{\theta_0}(X_1)) > 0, \ i = 1, ..., N_m.$$

Set

$$A_{m,i} = \{\lim_{n\to\infty} \frac{1}{n} \sum\nolimits_{j=1}^{n} (\varrho_{\Delta_{m,i}}(X_j) - \varrho_{\theta_0}(X_j)) = \mathsf{E}_{\theta_0}(\varrho_{\Delta_{m,i}}(X_1) - \varrho_{\theta_0}(X_1))\}.$$

The strong law of large numbers yields $\mathbb{P}_{\theta_0}(A_{m,i}) = 1$. Suppose that $D_n \to 0$, $\mathbb{P}_{\theta_0}$-a.s., and set

$$A = \{\lim_{n\to\infty} D_n(\mathbf{X}_n) = 0\} \cap \bigcap\nolimits_{m=1}^{\infty} \bigcap\nolimits_{i=1}^{N_m} A_{m,i}.$$

Then $\mathbb{P}_{\theta_0}(A) = 1$ and for fixed $\omega \in A$ and $\varepsilon > 0$ we choose m such that $\varepsilon > 1/m$. If $\rho_\Delta(\widehat{\theta}_n, \theta_0) > 1/m$, then by the definition of D_n

$$\inf_{\theta\in B_{1/m}} M_n(\theta, \mathbf{X}_n) \le M_n(\widehat{\theta}_n, \mathbf{X}_n) = \inf_{\theta\in\Delta} M_n(\theta, \mathbf{X}_n) + D_n \le M_n(\theta_0, \mathbf{X}_n) + D_n.$$

Hence,

$$\begin{aligned}
0 &\ge \inf_{\theta\in B_{1/m}} M_n(\theta, \mathbf{X}_n(\omega)) - M_n(\theta_0, \mathbf{X}_n(\omega)) - D_n(\mathbf{X}_n(\omega)) \\
&\ge \min_{1\le i\le N_m} \frac{1}{n} \sum\nolimits_{j=1}^{n} [\varrho_{\Delta_{m,i}}(X_j(\omega)) - \varrho_{\theta_0}(X_j(\omega))] - D_n(\mathbf{X}_n(\omega)) \\
&\to \min_{1\le i\le N_m} \mathsf{E}_{\theta_0}(\varrho_{\Delta_{m,i}} - \varrho_{\theta_0}) > 0.
\end{aligned}$$

This shows that the inequality $\rho_\Delta(\widehat{\theta}_n(\mathbf{X}_n(\omega)), \theta_0) > 1/m$ may hold only for a finite number of n. Hence $\widehat{\theta}_n \to \theta_0$, $\mathbb{P}_{\theta_0}$-a.s. The proof of the stochastic convergence follows from a subsequence argument and Proposition A.12. The corollary follows from Lemma 7.65. ∎

Remark 7.69. There are situations where Condition 7.64 is not fulfilled, but it can be made valid if we replace $\varrho_\theta(x)$ with $\widetilde{\varrho}_\theta(x_1,...,x_k) = (1/k)\sum_{j=1}^k \varrho_\theta(x_j)$. Using this idea Condition 7.64 can be weakened. For details we refer to Perlman (1972) and Pfanzagl (1994).

The statement in the corollary is closely related to proofs of consistency that are based on uniform laws of large numbers. Then the special structure of M_n of being an arithmetic mean is irrelevant in this respect.

Theorem 7.70. ***(Argmin Theorem)*** *Let Δ be a compact metric space. Suppose $(W(\theta))_{\theta\in\Delta}$ and $(W_n(\theta))_{\theta\in\Delta}$ are continuous stochastic processes, defined on $(\Omega,\mathfrak{F},\mathbb{P})$, and let $\|W_n - W\|_u = \sup_{\theta\in\Delta}|W_n(\theta) - W(\theta)|$. Suppose W has a unique minimizer $\widehat{\theta} : \Omega \to_m \Delta$ and $\widehat{\theta}_n : \Omega \to_m \Delta$. Then for $D_n = W_n(\widehat{\theta}_n) - \inf_{\theta\in\Delta} W_n(\theta)$ the following hold.*

(A) If $D_n \to 0$, $\mathbb{P}$-a.s., and $\|W_n - W\|_u \to 0$, $\mathbb{P}$-a.s., then $\widehat{\theta}_n \to \widehat{\theta}$, $\mathbb{P}$-a.s.

(B) If $D_n \to^{\mathbb{P}} 0$ and $\|W_n - W\|_u \to^{\mathbb{P}} 0$, then $\widehat{\theta}_n \to^{\mathbb{P}} \widehat{\theta}$.

Proof. The process W is continuous on the compact set Δ and has a unique minimum at $\widehat{\theta}$. Hence $\delta_\varepsilon = \inf_{\rho_\Delta(\theta,\widehat{\theta})\geq\varepsilon}(W(\theta) - W(\widehat{\theta})) > 0$, $\mathbb{P}$-a.s. If $\rho_\Delta(\widehat{\theta}_n,\widehat{\theta}) \geq \varepsilon$, then

$$\inf_{\rho_\Delta(\theta,\widehat{\theta})\geq\varepsilon} W_n(\theta) \leq W_n(\widehat{\theta}_n) = \inf_{\theta\in\Delta} W_n(\theta) + D_n,$$

and

$$\begin{aligned} W(\widehat{\theta}) + \delta_\varepsilon = \inf_{\rho_\Delta(\theta,\widehat{\theta})\geq\varepsilon} W(\theta) &\leq \sup_{\theta\in\Delta}|W_n(\theta) - W(\theta)| + \inf_{\rho_\Delta(\theta,\widehat{\theta})\geq\varepsilon} W_n(\theta) \\ &\leq \sup_{\theta\in\Delta}|W_n(\theta) - W(\theta)| + \inf_{\theta\in\Delta} W_n(\theta) + D_n \\ &\leq 2\sup_{\theta\in\Delta}|W_n(\theta) - W(\theta)| + W(\widehat{\theta}) + D_n. \end{aligned}$$

$D_n \to 0$ and $\|W_n - W\|_u \to 0$, $\mathbb{P}$-a.s., imply that $\rho_\Delta(\widehat{\theta}_n,\widehat{\theta}) \geq \varepsilon$ is possible only for a finite number of n. The statement (B) follows from a subsequence argument and Proposition A.12. ■

The uniform law of large numbers has been assumed to hold in Theorem 7.70. For the sequence of arithmetic means it holds under a condition that is a bit stronger than the conditions of Corollary 7.68.

Proposition 7.71. *Suppose $X_1, X_2, ...$ are defined on $(\Omega,\mathfrak{F},\mathbb{P})$ and are i.i.d. If Δ is a compact metric space and*

$$M_n(\theta) = \frac{1}{n}\sum\nolimits_{i=1}^n \varrho_\theta(X_i),$$

then $\mathbb{E}\sup_{\theta\in\Delta}|\varrho_\theta(X_1) - \varrho_{\theta_0}(X_1)| < \infty$ implies

$$\sup_{\theta\in\Delta}|(M_n(\theta) - M_n(\theta_0)) - \mathbb{E}(\varrho_\theta(X_1) - \varrho_{\theta_0}(X_1))| \to 0, \ \mathbb{P}\text{-}a.s.$$

Proof. Set

$$\pi_\delta(X_1) = \sup_{\rho_\Delta(\theta_1,\theta_2)\le\delta} |\varrho_{\theta_1}(X_1) - \varrho_{\theta_2}(X_1)|.$$

Then $\pi_\delta(X_1)$ is nondecreasing in δ. Hence $\lim_{\delta\downarrow 0}\pi_\delta(X_1)$ exists, and by $\mathbb{E}\sup_{\theta\in\Delta}|\varrho_\theta(X_1) - \varrho_{\theta_0}(X_1)| < \infty$ and Lebesgue's theorem it holds

$$\lim_{\delta\downarrow 0}\mathbb{E}\pi_\delta(X_1) = \mathbb{E}\lim_{\delta\downarrow 0}\pi_\delta(X_1) = 0.$$

Set $\omega_\delta(M_n) = \sup_{\rho_\Delta(\theta_1,\theta_2)\le\delta}|M_n(\theta_1) - M_n(\theta_2)|$. Then

$$\omega_\delta(M_n) \le \frac{1}{n}\sum\nolimits_{i=1}^{n}\pi_\delta(X_i).$$

Cover Δ by a finite number of balls $B_\delta(\theta_i)$ with center in θ_i and a radius that does not exceed δ, $i = 1, ..., m$. Then with $W_n(\theta) = M_n(\theta) - M_n(\theta_0)$, $W(\theta) = \mathbb{E}(\varrho_\theta(X_1) - \varrho_{\theta_0}(X_1))$, and $\omega_\delta(M_n) = \omega_\delta(W_n)$,

$$\sup\nolimits_{\theta\in\Delta}|W_n(\theta) - W(\theta)| \le \max_{1\le i\le m}|W_n(\theta_i) - W(\theta_i)| + \omega_\delta(M_n) + \omega_\delta(W).$$

$W_n(\theta_i) \to W(\theta_i)$, P-a.s., holds by the strong law of large numbers. We apply the latter to the $\pi_\delta(X_i)$ to get

$$\limsup_{n\to\infty}\sup\nolimits_{\theta\in\Delta}|W_n(\theta)-W(\theta)| \le \limsup_{n\to\infty}\omega_\delta(M_n)+\omega_\delta(W) \le \mathbb{E}\pi_\delta(X_1)+\omega_\delta(W).$$

$\lim_{\delta\downarrow 0}\omega_\delta(W) = 0$ holds as the continuous function W on the compact set Δ is uniformly continuous. The statement $\lim_{\delta\downarrow 0}\mathbb{E}\pi_\delta(X_1) = 0$ has been already established. ■

Statements on the convergence of minimizers (maximizers) of convergent sequences of stochastic processes M_n, are called *argmin (argmax) theorems.* General forms of argmax theorems can be found in van der Vaart and Wellner (1996). We note that such statements are also used in stochastic optimization; see, for example, Wets (1989).

Theorem 7.70 used the compactness of Δ. If the parameter space is not compact, say $\Delta = \mathbb{R}^d$, then one needs additional conditions which, roughly speaking, prevent the minimizer from running out of every compact set. There is one important class of criterion functions, namely convex functions, for which this property is automatically satisfied. This is one of the reasons why convex criterion functions are so popular. The other point is the equicontinuity of sequences that are pointwise convergent. This property is needed to conclude local uniform convergence from pointwise convergence that is provided by the classical laws of large numbers.

Let $O \subseteq \mathbb{R}^d$ be an open convex set and $f : O \to \mathbb{R}$ a convex function. Then for $x = (x_1, ..., x_d) \in O$ and fixed $x_1, ..., x_{i-1}, x_{i+1}, ..., x_d$ the function $x_i \mapsto f(x_1, ..., x_d)$ is a convex function of x_i. The left- and right-hand derivatives $\partial^- f/\partial x_i$ and $\partial^+ f/\partial x_i$ exist according to Problem 1.50.

Problem 7.72.* If $x, c \in O$, then for any $\varepsilon_i \in \{+, -\}$,

$$g(x) := f(x) - f(c) - \sum_{i=1}^{d} (\partial^{\varepsilon_i} f(c)/\partial x_i)(x_i - c_i) \geq 0.$$

The next problem establishes the well-known Lipschitz property of convex functions; see, for example, Rockafellar (1970).

Problem 7.73.* Suppose $O \subseteq \mathbb{R}^d$ is an open convex set, $K \subseteq O$ is a compact set, and $f : O \to \mathbb{R}$ is convex. Then for every $\varepsilon > 0$ with

$$K_\varepsilon := \{y : y = x + z,\ x \in K,\ z \in \mathbb{R}^d,\ \|z\| \leq \varepsilon\} \subseteq O,$$

it holds

$$|f(x) - f(y)| \leq \frac{1}{\varepsilon}(\sup_{x \in K_\varepsilon} f(x) - \inf_{x \in K_\varepsilon} f(x)) \|y - x\|.$$

We modify this result to get a Lipschitz constant that is better suited for our purposes.

Problem 7.74.* Suppose $O \subseteq \mathbb{R}^d$ is an open convex set and $K \subseteq O$ is a compact set. Then there are a constant C and points $b_1, ..., b_m \in O$ such that for every convex function $f : O \to \mathbb{R}$ it holds

$$|f(x) - f(y)| \leq C \sum_{i=1}^{m} |f(b_i)| \|y - x\|, \quad x, y \in K. \tag{7.35}$$

A consequence of the inequality (7.35) is that every convex function is uniformly continuous on compact subsets. Moreover, if the sequence f_n converges pointwise to f on O and K is a fixed compact set, then $L = \sup_n C \sum_{i=1}^{m} |f_n(b_i)|$ is a common Lipschitz constant for the f_n. Hence this sequence is equicontinuous on K and the pointwise convergence implies the uniform convergence on K. This is a classical result of convex analysis; see Theorem 10.8 in Rockafellar (1970) for a different proof. The stochastic counterpart is known as the *convexity lemma* which was established and reestablished by several authors; see, e.g., Haberman (1989), Niemiro (1992), Jurečková (1977, 1992), Anderson and Gill (1982), Pollard (1990, 1991), and Hjort and Pollard (1993). If $O \subseteq \mathbb{R}^d$ is an open convex set, we call a stochastic process $(M(\theta))_{\theta \in O}$, convex if $M(\cdot, \omega)$ is a convex function for every $\omega \in \Omega$. As a convex function on O is Lipschitz on compact subsets (see Problem 7.74), we get that every convex process is continuous.

Lemma 7.75. ***(Convexity Lemma)*** *Let $O \subseteq \mathbb{R}^d$ be open and convex, $M(\theta)$ and $M_n(\theta)$, $\theta \in O$, be convex stochastic processes, and D be a dense subset of O. Then for any compact subset K of O the following hold.*

(A) $M_n(\theta) \to M(\theta)$, $\mathbb{P}$-a.s., $\theta \in D$, implies $\sup_{\theta \in K} |M_n(\theta) - M(\theta)| \to 0$, $\mathbb{P}$-a.s.

(B) $M_n(\theta) \to^{\mathbb{P}} M(\theta)$, $\theta \in D$, implies $\sup_{\theta \in K} |M_n(\theta) - M(\theta)| \to^{\mathbb{P}} 0$.

Proof. Let K be fixed. Without loss of generality we may assume that $b_1, ..., b_m$ in Problem 7.74 belong to D. Set $A = \{\omega : M_n(\theta, \omega) \to M(\theta, \omega), \theta \in D\}$. Then $\mathbb{P}(A) = 1$, and for every fixed $\omega \in A$ the sequence of functions $\theta \mapsto M_n(\theta, \omega)$ converges pointwise on a dense subset of K to the function $\theta \mapsto M(\theta, \omega)$ and satisfies

$$\sup_{\theta_i \in K, \|\theta_1 - \theta_2\| \leq \delta} |M_n(\theta_1, \omega) - M_n(\theta_2, \omega)| \leq L(\omega)\delta,$$

for $L(\omega) = \sup_n C \sum_{i=1}^{m} |M_n(b_i, \omega)| < \infty$. The statement follows from the fact that an equicontinuous sequence of functions that converges pointwise on a dense subset converges uniformly; see Problem 6.80. To prove statement (B) it suffices to show that for every subsequence M_{n_k} there is again a subsequence that converges uniformly in θ, $\mathbb{P}$-a.s. By a diagonal technique there is a subsequence $M_{n_{k_l}}$ such $M_{n_{k_l}}(\theta) \to M(\theta)$, $\mathbb{P}$-a.s., for every θ from the countable set D in the first part of the proof. With the same argument as in that part we get $\sup_{\theta \in K} |M_{n_{k_l}}(\theta) - M(\theta)| \to 0$, $\mathbb{P}$-a.s. ■

If we combine the convexity lemma with Theorem 7.70 we can say that for a pointwise converging sequence of criterion functions, and a compact parameter space, the minimizers are consistent estimators, unless the minimizer in the limiting function is not unique. The compactness condition could be removed if we would know that minimizers of sequences of convex criterion functions cannot run out of compact sets. This can be concluded from the following lemma by Hjort and Pollard (1993), which gives a bound for the distance of minimizers of two convex functions.

Lemma 7.76. *Suppose $O \subseteq \mathbb{R}^d$ is open and convex, $f : O \to \mathbb{R}$ is a convex function, and $g : O \to \mathbb{R}$ is continuous. If $x_0 \in O$ and $y_0 \in O$ are minimizers of f and g, respectively, then*

$$\sup_{y \in O, \|y - y_0\| \leq \varepsilon} 2\,|f(y) - g(y)| < \inf_{y \in O, \|y - y_0\| = \varepsilon} [g(y) - g(y_0)] \tag{7.36}$$

implies $\|x_0 - y_0\| \leq \varepsilon$.

Proof. Suppose $x_0, y_0 \in O$ satisfy $a := \|x_0 - y_0\| > \varepsilon > 0$. Set $u = a^{-1}(x_0 - y_0)$. Then $\|u\| = 1$, $x_0 = y_0 + au$, and the convexity of f implies $(1 - \varepsilon/a)f(y_0) + (\varepsilon/a)f(x_0) \geq f(y_0 + \varepsilon u)$. Hence

$$\begin{aligned}
&\frac{\varepsilon}{a}(f(x_0) - f(y_0)) \geq f(y_0 + \varepsilon u) - f(y_0) \\
&= g(y_0 + \varepsilon u) - g(y_0) + [f(y_0 + \varepsilon u) - g(y_0 + \varepsilon u)] + [g(y_0) - f(y_0)] \\
&\geq \inf_{y \in O, \|y - y_0\| = \varepsilon} [g(y) - g(y_0)] - \sup_{y \in O, \|y - y_0\| \leq \varepsilon} 2|f(y) - g(y)|.
\end{aligned}$$

If (7.36) is satisfied, then the left-hand term is positive. This contradicts the fact that x_0 is a minimizer of f. Hence $\|x_0 - y_0\| > \varepsilon$ is impossible. ■

Now we combine the Hjort–Pollard lemma with the convexity lemma to establish an argmin theorem for convex processes.

Theorem 7.77. ***(Argmin Theorem for Convex Processes)*** *Suppose $O \subseteq \mathbb{R}^d$ is open and convex and assume that $M_n(\theta)$ and $M(\theta)$, $\theta \in O$, are convex stochastic processes, where M has a unique minimizer $\widehat{\theta} : \Omega \to_m O$. If $\widehat{\theta}_n : \Omega \to_m O$ are respective minimizers of M_n, then the following hold.*

(A) $M_n(\theta) \to M(\theta)$, $\mathbb{P}$-a.s., $\theta \in O$, implies $\widehat{\theta}_n \to \widehat{\theta}$, $\mathbb{P}$-a.s.
(B) $M_n(\theta) \to^{\mathbb{P}} M(\theta)$, $\theta \in O$, implies $\widehat{\theta}_n \to^{\mathbb{P}} \widehat{\theta}$.

Proof. To prove (A) let $M_n(\theta) \to M(\theta)$, $\theta \in O$, $\mathbb{P}$-a.s. Let K_m be an increasing sequence of compact sets with $K_m \uparrow O$. By Lemma 7.75 we find some $A \in \mathfrak{F}$ with $\mathbb{P}(A) = 1$ and

$$\sup_{\theta \in K_m} |M_n(\theta,\omega) - M(\theta,\omega)| \to 0, \quad \omega \in A, \quad m = 1, 2, \ldots .$$

Let $\varepsilon > 0$ and $\omega \in A$ be fixed. As $\widehat{\theta}(\omega)$ is the unique minimizer of $M(\cdot,\omega)$ it holds

$$D(\omega) = \inf_{\theta \in O, \|\theta - \widehat{\theta}(\omega)\| = \varepsilon} \mid M(\theta,\omega) - M(\widehat{\theta}(\omega),\omega) \mid > 0.$$

As O is open and $K_m \uparrow O$ we find some m_0 with

$$\{\theta : \theta \in O, \ \| \theta - \widehat{\theta}(\omega) \| \leq \varepsilon\} \subseteq K_{m_0}.$$

Hence,

$$\sup_{\theta \in O, \|\theta - \widehat{\theta}(\omega)\| \leq \varepsilon} 2\,|M_n(\theta,\omega) - M(\theta,\omega)| < D(\omega),$$

for all sufficiently large n, say $n \geq n_0$, and therefore $\| \widehat{\theta}_n(\omega) - \widehat{\vartheta}(\omega) \| \leq \varepsilon$ for $n \geq n_0$ by Lemma 7.76. To prove (B) we fix a countable subset D of O that is dense in O. Then by the diagonal technique we find a subsequence $M_{n_k}(\theta)$ of $M_n(\theta)$ that converges to $M(\theta)$, $\mathbb{P}$-a.s., for every $\theta \in D$. Hence by the convexity lemma we have $M_{n_k}(\theta) \to M(\theta)$, $\mathbb{P}$-a.s., for every $\theta \in O$. To complete the proof it suffices to apply the first statement to this subsequence and to use the subsequence characterization of the stochastic convergence; see Proposition A.12. ■

Consistency of MLEs

Here in this section we apply the general results on M-estimators to the problem of consistency of maximum likelihood estimators.

Suppose $\mathcal{M} = (\mathcal{X}, \mathfrak{A}, (P_\theta)_{\theta \in \Delta})$ is a model for which the family $(P_\theta)_{\theta \in \Delta}$ is dominated by a σ-finite measure μ. Set $f_\theta(x) = (dP_\theta/d\mu)(x)$. Then the functions $\theta \mapsto f_\theta(x)$ and $\Lambda(\theta) : \theta \mapsto \ln f_\theta(x)$ are the *likelihood* and the *log-likelihood* functions at x, respectively. Analogously, for i.i.d. observations the likelihood and log-likelihood functions are, according to Proposition A.29, given by

$$f_{n,\theta}(\mathbf{x}_n) = \prod\nolimits_{i=1}^{n} f_\theta(x_i), \quad \Lambda_n(\theta, \mathbf{x}_n) = \sum\nolimits_{i=1}^{n} \ln f_\theta(x_i), \tag{7.37}$$

where $\mathbf{x}_n = (x_1, ..., x_n) \in \mathcal{X}^n$. We also use $\Lambda_n(\theta) = \sum_{i=1}^n \ln f_\theta(X_i)$ for brevity.

Let Δ be a separable metric space with the σ-algebra of Borel sets $\mathfrak{B}_\Delta$. The estimator $\widehat{\theta}_n : \mathcal{X}^n \to_m \Delta$ is called a *maximum likelihood estimator* (MLE) at $\mathbf{x}_n \in \mathcal{X}^n$ if $f_{n,\theta}(\mathbf{x}_n) \leq f_{n,\widehat{\theta}_n(\mathbf{x}_n)}(\mathbf{x}_n)$, $\theta \in \Delta$. If $\widehat{\theta}_n$ is an MLE at every $\mathbf{x}_n \in \mathcal{X}^n$, then we omit "at every $\mathbf{x}_n \in \mathcal{X}^n$". Equivalently, we may require for the log-likelihood function Λ_n that

$$\Lambda_n(\theta, \mathbf{x}_n) \leq \Lambda_n(\widehat{\theta}_n(\mathbf{x}_n), \mathbf{x}_n), \quad \theta \in \Delta, \ \mathbf{x}_n \in \mathcal{X}^n, \tag{7.38}$$

where we use the convention $\ln 0 = -\infty$.

Let $\mathcal{M} = (\mathcal{X}, \mathfrak{A}, (P_\theta)_{\theta\in\Delta})$ be a dominated model, where Δ is a separable metric space. We assume that the sample space and the dominating σ-finite measure μ can be chosen in such a way that for some version of the density $f_\theta(x)$ the following conditions hold.

$$\begin{aligned} &\theta \mapsto f_\theta(x) \quad \text{is continuous for every } x \in \mathcal{X}, \\ &f_\theta(x) > 0, \quad x \in \mathcal{X}, \ \theta \in \Delta. \end{aligned} \tag{7.39}$$

As $x \mapsto f_\theta(x)$ is measurable for every $\theta \in \Delta$ it follows from Problem 1.116 that $(\theta, x) \mapsto f_\theta(x)$ is measurable.

We recall the Kullback–Leibler distance (see (1.81)),

$$\mathsf{K}(P_{\theta_0}, P_\theta) = \mathsf{E}_{\theta_0} \ln \frac{f_{\theta_0}}{f_\theta}, \quad \text{if } P_{\theta_0} \ll P_\theta, \quad \text{and} \quad \mathsf{K}(P_{\theta_0}, P_\theta) = \infty, \text{ else.}$$

As the convex function $\mathsf{v}(x) = x\ln x - x + 1$, on which the introduction of $\mathsf{K}(P_{\theta_0}, P_\theta)$ in (1.74) has been based, is strictly convex at $x_0 = 1$ we get from Proposition 1.63 that $\mathsf{K}(P_{\theta_0}, P_\theta) \geq 0$, with equality holding if and only if $P_{\theta_0} = P_\theta$. Moreover, if $P_{\theta_0} \ll P_\theta$, then

$$\mathsf{K}(P_{\theta_0}, P_\theta) = \int \mathsf{v}(f_{\theta_0}/f_\theta) f_\theta d\mu.$$

As $\mathsf{v}(x) \geq 0$ and $\int |f_{\theta_0}/f_\theta - 1| f_\theta d\mu < \infty$ we see that $\mathsf{E}_{\theta_0} |\ln f_\theta - \ln f_{\theta_0}| < \infty$ holds if and only if $\mathsf{K}(P_{\theta_0}, P_\theta) < \infty$. Hence $\varrho_\theta = -\ln f_\theta$ satisfies the contrast condition (7.31), provided that $\mathsf{K}(P_{\theta_0}, P_\theta) < \infty$, $\theta_0, \theta \in \Delta$, and the parameter is identifiable. We call $\varrho_\theta = -\ln f_\theta$ the *likelihood contrast function.* By turning from the likelihood to the log-likelihood function we may directly apply the results for M-estimators to the criterion function $M_n(\theta) = -\Lambda_n(\theta)$.

We also consider estimators $\widehat{\theta}_n$ that are only an approximate MLE in the sense that the value of the likelihood function at $\widehat{\theta}_n$ is only an approximate maximum. To be more precise let

$$D_n(\mathbf{X}_n) = \sup_{\theta\in\Delta} \Lambda_n(\theta, \mathbf{X}_n) - \Lambda_n(\widehat{\theta}_n(\mathbf{X}_n), \mathbf{X}_n)$$

for the estimator $\widehat{\theta}_n : \mathcal{X}^n \to_m \Delta$. Similarly as for M-estimators we say that $\{\widehat{\theta}_n\}$ is a *strongly approximate MLE* at θ_0 if $D_n(\mathbf{X}_n) \to 0$, $\mathbb{P}_{\theta_0}$-a.s. Similarly we say that$\{\widehat{\theta}_n\}$ is an *approximate MLE* at θ_0 if $D_n(\mathbf{X}_n) \to^{\mathbb{P}_{\theta_0}} 0$. We say that $\{\widehat{\theta}_n\}$ is an *asymptotic MLE* at θ_0 if

$$D_n(\mathbf{X}_n) = \mathbf{o}_{\mathbb{P}_{\theta_0}}(0), \tag{7.40}$$

with $\mathbf{o}_{\mathbb{P}_{\theta_0}}(0)$ as defined in (7.34).

Theorem 7.78. *Suppose Δ is a separable metric space, $(P_\theta)_{\theta\in\Delta}$ is dominated, and (7.39) is satisfied. Suppose $\theta_1 \neq \theta_2$ implies $P_{\theta_1} \neq P_{\theta_2}$, and $\mathsf{K}(P_{\theta_0}, P_\theta) < \infty$ for every $\theta \in \Delta$. If the finite covering condition 7.64 at θ_0 is fulfilled for the likelihood contrast function $\varrho_\theta(x) = -\ln f_\theta(x)$, then every strongly approximate MLE is strongly consistent, and every approximate MLE is consistent. Especially every asymptotic MLE is consistent.*

Corollary 7.79. *The statement of the theorem remains valid if the finite covering condition is replaced by the assumption that Δ is compact and it holds $\mathsf{E}_{\theta_0}(\sup_{\theta\in\Delta} \ln f_\theta - \ln f_{\theta_0}) < \infty$.*

Proof. Apply Theorem 7.67 and its corollary. ■

Remark 7.80. If Δ is compact and instead of $\mathsf{E}_{\theta_0}(\sup_{\theta\in\Delta} \ln f_\theta - \ln f_{\theta_0}) < \infty$ the stronger condition $\mathsf{E}_{\theta_0} \sup_{\theta\in\Delta} |\ln f_\theta - \ln f_{\theta_0}| < \infty$ holds, then the consistency can be also concluded from a combination of Theorem 7.70 and Proposition 7.71.

The following example for a location model is taken from Pfanzagl (1994).

Example 7.81. Let f be a positive and continuous density on $\mathbb{R}$ and consider the location model $(\mathcal{X}, \mathfrak{A}, (P_\theta)_{\theta\in\mathbb{R}})$ where P_θ has the Lebesgue density $f(x-\theta)$. Assume in addition that

$$\lim_{x\to\pm\infty} f(x) = 0, \quad \mathsf{K}(P, P_a) < \infty,\ a \in \mathbb{R}, \quad \text{and} \quad -\int f(x)\ln f(x)dx < \infty. \tag{7.41}$$

Using characteristic functions one can see that $P_{\theta_1} = P_{\theta_2}$ if and only if $\theta_1 = \theta_2$.Then $\varrho_\theta(x) = -\ln f(x-\theta)$ is a contrast function. We establish the finite covering property for the likelihood contrast function $\varrho_\theta(x) = -\ln f(x-\theta)$. The condition $\lim_{x\to\pm\infty} f(x) = 0$ and the continuity of f imply that f is bounded, say $f \le C$. Furthermore $\lim_{c\to\infty} \sup_{|\theta|>c} \ln f(x-\theta) = -\infty$. As $\ln f(x-\theta) \le \ln C$ we get from the monotone convergence theorem

$$\lim_{c\to\infty} \int (\sup_{|\theta|>c} \ln f(x-\theta)) f(x)dx = -\infty.$$

Consequently, if θ_0 is fixed, then there exists some c_0 such that

$$\int [\inf_{\theta:|\theta-\theta_0|>c_0}(-\ln f(x-\theta)) + \ln f(x-\theta_0)] f(x-\theta_0)dx > 0.$$

For the finite covering property it remains to consider $\Delta_0 = \{\theta : |\theta-\theta_0| \le c_0\}$. As $f \le C$ we have $[-\ln f(x-\theta_0) - \inf_{\theta\in\Delta_0}(-\ln f(x-\theta))] \le \ln C - \ln f(x-\theta_0)$ and by (7.41)

$$\int [-\ln f(x-\theta_0) - \inf_{\theta\in\Delta_0}(-\ln f(x-\theta))] f(x-\theta_0)dx < \infty,$$

so that we may apply Lemma 7.65 and obtain the finite covering property. Then the previous theorem shows that every sequence of MLEs is strongly consistent. Below are two typical densities that satisfy the conditions in (7.41).

Distribution	f	ϱ
Normal	$(2\pi)^{-1/2}\exp\{-t^2/2\}$	$\frac{1}{2}\ln(2\pi)+\frac{1}{2}t^2$
Laplace	$\frac{1}{2}\exp\{-\lvert t\rvert\}$	$\ln 2+\lvert t\rvert$

In the first case the MLE is the arithmetic mean $\overline{X}_n$, so that the strong consistency deduced from Proposition 7.101 is just the strong law of large numbers. We show below that the sample median is the M-estimator that belongs to the contrast function $|t|$ in a location model. Hence we get that the median, which is the MLE for the Laplace parent distribution, is strongly consistent; see Example 7.108.

Next we present another approach to the consistency of an MLE. It provides bounds for the rate of convergence of the stochastic convergence of an MLE. More precisely, let $\mathcal{M} = (\mathcal{X}, \mathfrak{A}, (P_\theta)_{\theta\in\Delta})$ be a dominated model, where Δ is a separable metric space. Set $f_\theta = dP_\theta/d\mu$ and suppose that $\theta \mapsto f_\theta(x)$ is continuous for every $x \in \mathcal{X}$. For any Borel set $A \subseteq \Delta$ we set $f_A(x) = \sup_{\theta\in A} f_\theta(x)$. The separability of Δ and the continuity of $f_\theta(x)$ guarantees that f_A is a measurable function. If $B_\varepsilon = \{\theta : \rho_\Delta(\theta,\theta_0) > \varepsilon\}$ and $\widehat{\theta}_n$ is a maximum likelihood estimator, then, similarly to (7.32),

$$\{x : \rho_\Delta(\widehat{\theta}(x), \theta_0) > \varepsilon\} \subseteq \{x : f_{B_\varepsilon}(x) \geq f_{\theta_0}(x)\}. \tag{7.42}$$

We set for any Borel set $A \subseteq \Delta$ and $0 < s < 1$,

$$\mathcal{H}_s(\theta_0, A) = \int (f_A(x))^s\, f_{\theta_0}^{1-s}(x)\mu(dx). \tag{7.43}$$

Then $P_{\theta_0}(\{x : f_A(x) \geq f_{\theta_0}(x)\}) \leq \mathcal{H}_s(\theta_0, A)$. If $A \subseteq \cup_{i=1}^N A_i$ for some Borel sets $A_1, ..., A_N$, then

$$\begin{aligned} P_{\theta_0}(\{x\ :\ f_A(x) \geq f_{\theta_0}(x)\}) &\leq P_{\theta_0}(\{x : \max_{1\leq i\leq m} f_{A_i}(x) \geq f_{\theta_0}(x)\}) \qquad (7.44)\\ &\leq \sum\nolimits_{i=1}^N P_{\theta_0}(\{x : f_{A_i}(x) \geq f_{\theta_0}(x)\}) \leq \sum\nolimits_{i=1}^N \mathcal{H}_s(\theta_0, A_i). \end{aligned}$$

The following property of $\mathcal{H}_s(\theta_0, A)$ corresponds to the product property of Hellinger integrals in Problem 1.86. Let $\mathcal{M}_i = (\mathcal{X}_i, \mathfrak{A}_i, (P_{i,\theta})_{\theta\in\Delta})$, $i = 1, 2$, be two dominated models with densities $f_{i,\theta}$, $i = 1, 2$, depending continuously on θ. Set $\mathcal{M} = (\mathcal{X}_1 \times \mathcal{X}_2, \mathfrak{A}_1 \otimes \mathfrak{A}_2, (P_{1,\theta} \otimes P_{2,\theta})_{\theta\in\Delta})$. Then the inequality

$$\sup_{\theta\in A}(f_{1,\theta}(x_1) f_{2,\theta}(x_1)) \leq (\sup_{\theta\in A} f_{1,\theta}(x_1))(\sup_{\theta\in A} f_{2,\theta}(x_1))$$

implies

$$\mathcal{H}_{\otimes,s}(\theta, A) \le \mathcal{H}_{1,s}(\theta, A)\mathcal{H}_{2,s}(\theta, A), \tag{7.45}$$

where $\mathcal{H}_{\otimes,s}$ and $\mathcal{H}_{i,s}$ refer to $\mathcal{M}$ and $\mathcal{M}_i$, $i = 1, 2$, respectively. The next condition is similar to the finite covering condition in Condition 7.64.

Condition 7.82 *We say that the finite Hellinger covering property at θ_0 is fulfilled if there exists some $0 < s < 1$ such that for every $\varepsilon > 0$ there is a finite number of sets $C_{\varepsilon,i}$, $i = 1, ..., N(\varepsilon)$, with*

$$B_\varepsilon = \{\theta : \rho_\Delta(\theta, \theta_0) \ge \varepsilon\} \subseteq \bigcup\nolimits_{i=1}^{N(\varepsilon)} C_{\varepsilon,i} \text{ and } \overline{\mathcal{H}}_s(\varepsilon) := \max_{1 \le i \le N(\epsilon)} \mathcal{H}_s(\theta_0, C_{\varepsilon,i}) < 1.$$

Similarly as for Condition 7.64 the Hellinger covering property is satisfied for compact parameter spaces.

Problem 7.83.* Suppose that $\theta \mapsto f_\theta(x)$ is continuous for every $x \in \mathcal{X}$, the parameter θ is identifiable, Δ is a compact metric space, and it holds

$$\int (\sup\nolimits_{\theta \in \Delta} f_\theta(x))^s f_{\theta_0}^{1-s}(x)\mu(dx) < \infty. \tag{7.46}$$

Then the finite Hellinger covering property at θ_0 is fulfilled.

The subsequent theorem is taken from Rüschendorf (1988).

Theorem 7.84. *Let $\mathcal{M} = (\mathcal{X}, \mathfrak{A}, (P_\theta)_{\theta \in \Delta})$ be dominated, where Δ is a separable metric space, and suppose that $\theta \mapsto f_\theta(x)$ is continuous for every $x \in \mathcal{X}$. If the finite Hellinger covering property is satisfied, then every sequence of MLEs $\widehat{\theta}_n : \mathcal{X}^n \to_m \Delta$ satisfies*

$$P_{\theta_0}^{\otimes n}(\rho_\Delta(\widehat{\theta}_n, \theta_0) \ge \varepsilon) \le N(\varepsilon)(\overline{\mathcal{H}}_s(\varepsilon))^n \tag{7.47}$$

and is strongly consistent at θ_0.

Corollary 7.85. *The statement of the theorem remains valid if the finite Hellinger covering property is replaced by the assumption that Δ is compact and (7.46) holds.*

Proof. Set $f_{n,\theta}(x_1, ..., x_n) = \prod_{i=1}^n f_\theta(x_i)$ and introduce $\mathcal{H}_{n,s}(\theta_0, A)$ as $\mathcal{H}_s(\theta_0, A)$ in (7.43) with f_θ replaced by $f_{n,\theta}$. Then $\mathcal{H}_{n,s}(\theta_0, A) \le (\mathcal{H}_s(\theta_0, A))^n$ by (7.45). Hence (7.42) and (7.44) yield

$$P_{\theta_0}^{\otimes n}(\rho_\Delta(\widehat{\theta}_n, \theta_0) \ge \varepsilon) \le \sum\nolimits_{i=1}^{N(\varepsilon)} \mathcal{H}_{n,s}(\theta_0, C_{\varepsilon,i}) \le N(\varepsilon)(\overline{\mathcal{H}}_s(\varepsilon))^n.$$

Thus $\sum_{n=1}^\infty P_{n,\theta_0}(\rho_\Delta(\widehat{\theta}_n, \theta_0) \ge \varepsilon) < \infty$ for every $\varepsilon > 0$. The Borel–Cantelli lemma implies that $\mathbb{P}_{\theta_0}$-a.s. for every fixed $\varepsilon > 0$ the inequality $\rho_\Delta(\widehat{\theta}_n, \theta_0) \ge \varepsilon$ holds only for a finite number of n, and thus the proof is completed. The corollary follows from Problem 7.83. ■

The inequality (7.47) gives an exponential rate for the stochastic convergence of the MLE. For further results and references in this regard we refer to Pfaff (1982).

Remark 7.86. There are numerous papers that deal with the consistency of maximum likelihood estimators. Cramér (1946) proved the existence of consistent solutions of the likelihood equations. The consistency of maximum likelihood estimators, defined by maximizing the likelihood function, was established in Wald (1949). This result was generalized and refined by several authors; see, e.g., LeCam (1953), Kraft (1955), Bahadur (1958), and Huber (1967). Perlman (1972) introduced the weaker form of the finite covering property, discussed in Remark 7.69, and avoids the compactness conditions in earlier papers. We also refer to Strasser (1981a) who studied the relation between the consistency of MLEs and Bayes estimators. For further details in the history of maximum likelihood estimation the reader is referred to Pratt (1976), LeCam (1990), and Pfanzagl (1994). The paper by Pfanzagl (1969) contains fundamental results on the consistency of M-estimators.

There are several classical examples of inconsistent maximum likelihood estimators, which show that the sufficient conditions set in Theorems 7.78 and 7.84 are in general indispensable. The first example concerns the case where $\theta \mapsto f_\theta$ is not continuous.

Example 7.87. Let $P_\theta = \mathsf{N}(\theta, 1)$, $\theta \in \mathbb{R}\backslash\{-1,1\}$, $P_{-1} = \mathsf{N}(1,1)$, and $P_1 = \mathsf{N}(-1,1)$. Then the density f_θ of P_θ is not a continuous function. The maximum likelihood estimator is given by

$$\widehat{\theta}_n(x_1, ..., x_n) = \begin{cases} \overline{x}_n & \text{if } \overline{x}_n \notin \{-1,1\} \\ -\overline{x}_n & \text{if } \overline{x}_n \in \{-1,1\} \end{cases}.$$

Hence $\widehat{\theta}_n = \overline{x}_n$, $P_\theta^{\otimes n}$-a.s., and $\widehat{\theta}_n \to^{P_1^{\otimes n}} -1$, so that $\widehat{\theta}_n$ is not consistent.

We refer to Bahadur (1958) for other examples of models where the MLE is inconsistent. The next example shows that maximum likelihood estimators may run to a boundary point of the parameter set that may not belong to it. It is taken from Pfanzagl (1994), to which we refer for a proof.

Example 7.88. Let $\mathcal{X} = (0,1)$ and $\mathfrak{A}$ be the σ-algebra of Borel sets in $(0,1)$. Suppose the family $(P_\theta)_{\theta\in\Delta}$, $\Delta = (0,1]$, is given by Lebesgue densities f_θ that satisfy the following conditions.

$$\begin{aligned} \theta &\mapsto f_\theta(x) \text{ is continuous for } x \in (0,1), \\ \frac{1}{2} &\leq f_\theta(x) \leq f_\theta(\theta), \quad (\theta, x) \in (0,1] \times (0,1], \\ f_\theta(\theta) &= 2^{\theta^{-3}-1}, \quad \theta \in (0,1], \quad \text{and} \quad f_1(x) = 1. \end{aligned}$$

Then every sequence of MLEs converges, $\mathbb{P}_\theta$-a.s., to 0, $\theta \in \Delta$.

Although we have established conditions that guarantee the consistency of MLEs, these conditions say nothing about the existence of an MLE for a finite sample size or whether an MLE can be found as the solution of the likelihood equation. We now study these questions regarding the existence of an MLE for exponential families. We recall the notations from (7.37) and (7.38). If $\Delta \subseteq \mathbb{R}^d$, then a necessary condition for $\widehat{\theta}_n(\mathbf{x}_n)$ to be an MLE is to satisfy the *likelihood equation*

$$\dot{\Lambda}_n(\widehat{\theta}_n(\mathbf{x}_n), \mathbf{x}_n) = 0, \tag{7.48}$$

provided that $\theta \mapsto \Lambda_n(\theta, \mathbf{x}_n)$ is differentiable and $\widehat{\theta}_n(\mathbf{x}_n)$ is an inner point of Δ. The dot on top of $\dot{\Lambda}_n$ denotes the gradient with respect to the parameter vector θ, i.e., $\dot{\Lambda}_n(\theta, \mathbf{x}_n) = \nabla \Lambda_n(\theta, \mathbf{x}_n)$. It should be mentioned that a solution of (7.48) is not necessarily a maximizer of the likelihood function, or equivalently, of the log-likelihood function. On the other hand, if $\widehat{\theta}_n(\mathbf{x}_n)$ is a boundary point of Δ, then (7.48) may not be satisfied. Moreover, this equation, as a necessary condition, is based on the differentiability of the likelihood function which may not be given. Nevertheless, in such cases an MLE can often be calculated directly.

Example 7.89. Let $X_1, ..., X_n$ be an i.i.d. sample from a uniform distribution $\mathsf{U}(0, \theta)$, $\theta > 0$. The likelihood function is then given by

$$f_{n,\theta}(x_1, ..., x_n) = \frac{1}{\theta^n} I_{[0,\theta]}(\max_{1 \le i \le n} x_i).$$

To maximize $f_{n,\theta}$ over $\theta \in (0, \infty)$ we have only to consider the case $\max_{1 \le i \le n} x_i \le \theta$. In this area the function θ^{-n} attains its maximum at $\widehat{\theta}_n(x_1, ..., x_n) = \max_{1 \le i \le n} x_i$.

The existence of an MLE can be concluded from the compactness of the parameter space and the continuity of the likelihood function, as has been done in Proposition 7.63. Without the compactness an MLE may fail to exist for special outcomes of a sample.

Example 7.90. Let $\mathcal{X} = \{0, 1, .., n\}$ and consider the binomial distribution $\mathsf{B}(n, p)$, $p \in (0, 1)$, as an exponential family parametrized by the natural parameter $\theta = \ln(p/(1-p)) \in \Delta = \mathbb{R}$. Using the measure with point masses $\binom{n}{k}$ as dominating measure the density is given by $f_\theta(k) = \exp\{\theta k - n \ln(1 + e^\theta)\}$, $k \in \mathcal{X}$, $\theta \in \Delta$; see Problem 1.7. The maximum likelihood estimator, provided it exists, maximizes the function $f_\theta(k)$ as a function of θ for any observation $k \in \mathcal{X}$. For $k = 0$ the function $f_\theta(0) = (1 + \exp\{\theta\})^{-n}$ has no maximum point.

Now we study systematically the existence of maximum likelihood estimators in exponential families. As $T_{\oplus n}$ is a sufficient statistic it is enough to consider the case $n = 1$. Let $(P_\theta)_{\theta \in \Delta}$ be a d-parameter exponential family in natural form, with natural parameter θ and generating statistic T, and $\boldsymbol{\mu}$-density $f_\theta(x) = \exp\{\langle \theta, T(x) \rangle - K(\theta)\}$, $x \in \mathcal{X}$, $\theta \in \Delta$; see (1.6). The log-likelihood function $\Lambda(\theta, x) = \langle \theta, T(x) \rangle - K(\theta)$ is a concave function of θ. It is strictly concave on Δ^0 as the Hessian $\nabla \nabla^T K(\theta)$ is positive definite there.

Problem 7.91.* If conditions (A1) and (A2) are fulfilled, then for every fixed $x \in \mathcal{X}$ the following statements for Λ are equivalent.

(A) $\Lambda(\cdot, x)$ has a global maximum at $\widehat{\theta}(x) \in \Delta^0$.

(B) $\Lambda(\cdot, x)$ has a local maximum at $\widehat{\theta}(x) \in \Delta^0$.

(C) $\widehat{\theta}(x) \in \Delta^0$ is a solution of $\dot{\Lambda}(\theta, x) = 0$.

In each of the equivalent cases (A), (B), and (C), $\widehat{\theta}(x)$ is uniquely determined.

Let $\Delta \subseteq \mathbb{R}^d$ be a convex set with $\Delta^0 \neq \emptyset$ and boundary $\partial\Delta$, and let $K : \Delta \to \mathbb{R}$ be a convex function which is differentiable in Δ^0. K is called *steep* if for every $\theta \in \Delta^0$ and every $\eta \in \partial\Delta$ the function $\kappa(\lambda) := K(\lambda\theta + (1-\lambda)\eta)$ satisfies the condition

$$\lim_{\lambda \downarrow 0} \frac{d\kappa(\lambda)}{d\lambda} = -\infty. \tag{7.49}$$

If $d = 1$, then Δ is some interval. If this is a finite and open interval, say $\Delta = (a, b)$, then for $\eta = b$ it holds $\kappa(\lambda) = K(\lambda\theta + (1-\lambda)b)$, so that (7.49) holds if and only if $\lim_{\theta\uparrow b} K'(\theta) = \infty$. The case of $\eta = a$ can be treated similarly.

Problem 7.92.* Let $K : \Delta \to \mathbb{R}$ be a steep convex function and $t \in \mathbb{R}^d$ be fixed. If $\langle\theta, t\rangle - K(\theta)$, as a function of θ, has a global maximum over Δ at the point θ_0, then $\theta_0 \in \Delta^0$.

Now we are ready to present the main result on the existence of an MLE in exponential families.

Proposition 7.93. *Let $(P_\theta)_{\theta\in\Delta}$ be an exponential family in natural form, with natural parameter θ, that satisfies (A1) and (A2). Let $\gamma_{\mathsf{m}}(\theta) = \mathsf{E}_\theta T$. Then for every x with $T(x) \in \gamma_{\mathsf{m}}(\Delta^0)$, there is a uniquely determined solution $\widehat{\theta}(x)$ of the equation*

$$\nabla K(\theta) = T(x), \tag{7.50}$$

and $\widehat{\theta}(x)$ is the unique point at which the log-likelihood function $\Lambda(\theta, x) = \langle\theta, T(x)\rangle - K(\theta)$ attains its maximum. On the other hand, if $T(x) \notin \gamma_{\mathsf{m}}(\Delta^0)$, and K is steep, then the function $\theta \mapsto \Lambda(\theta, x)$ does not attain a maximum at any point of Δ.

Proof. If $T(x) \in \gamma_{\mathsf{m}}(\Delta^0)$, then (7.50) yields condition (C) in Problem 7.91 which gives the statement. From Problem 7.92 and the steepness condition, we see that the function $\Lambda(\theta, x) = \langle\theta, T(x)\rangle - K(\theta)$ cannot attain its maximum at any boundary point. Therefore, if $\Lambda(\theta, x)$ would attain the maximum at some $\widehat{\theta}(x) \in \Delta$, then $\widehat{\theta}(x) \in \Delta^0$, and consequently $\widehat{\theta}(x)$ would be a point of a local maximum and thus a solution of equation (7.50). But this would mean that $T(x) \in \gamma_{\mathsf{m}}(\Delta^0)$, in contradiction to the assumption. ■

Proposition 7.93 gives a complete characterization of the existence of an MLE in an exponential family under (A1) and (A2). However, the set $\gamma_{\mathsf{m}}(\Delta^0)$ may not be easy to characterize, which could be a drawback. An insight of the structure of $\gamma_{\mathsf{m}}(\Delta^0)$ gives the following proposition. The convex support $\mathsf{CS}(\boldsymbol{\nu})$ of $\boldsymbol{\nu} = \boldsymbol{\mu} \circ T^{-1}$ is the closure of the convex hull of the support of $\boldsymbol{\nu}$, where the support of $\boldsymbol{\nu}$ is the minimal closed subset C of $\mathbb{R}^d$ with $\boldsymbol{\nu}(\mathbb{R}^d \backslash C) = 0$. For a proof of the next statement we refer to Brown (1986), pp. 73–75.

Proposition 7.94. *Let $(P_\theta)_{\theta\in\Delta}$ be an exponential family with natural parameter $\theta \in \Delta$, where (A1) and (A2) are fulfilled. Let $\gamma_{\mathsf{m}}(\theta) = \mathsf{E}_\theta T$. If the function K satisfies the steepness condition (7.49), then $\gamma_{\mathsf{m}}(\Delta^0) = (\mathsf{CS}(\boldsymbol{\nu}))^0$, the interior of $\mathsf{CS}(\boldsymbol{\nu})$.*

The measures $Q_\theta = P_\theta \circ T^{-1}$ and $\boldsymbol{\nu} = \boldsymbol{\mu} \circ T^{-1}$ are equivalent and thus have the same support. The statistic T takes, Q_θ-a.s., only values in the support of Q_θ which is a subset of $\mathsf{CS}(Q_\theta)$. But since in general $Q_\theta(\mathsf{CS}(Q_\theta)\backslash(\mathsf{CS}(Q_\theta))^0) > 0$, it may occur that the statistic T takes on values outside of $\gamma_{\mathsf{m}}(\Delta^0)$ with positive probability. However, there is one important special case where this situation cannot occur. For the Lebesgue measure $\boldsymbol{\lambda}_d$ on $\mathbb{R}^d$ it is well known that the boundary ∂C of a convex set C satisfies $\boldsymbol{\lambda}_d(\partial C) = 0$; see Lang (1986). Hence, if Q_θ is absolutely continuous with respect to $\boldsymbol{\lambda}_d$, then $Q_\theta(\mathsf{CS}(Q_\theta)\backslash(\mathsf{CS}(Q_\theta))^0) = 0$, and consequently $Q_\theta((\mathsf{CS}(Q_\theta))^0) = 1$. Thus in this case T takes on values in $(\mathsf{CS}(Q_\theta))^0 = \gamma_{\mathsf{m}}(\Delta^0)$ with probability one. To summarize, the following can be stated.

Proposition 7.95. *Suppose $(P_\theta)_{\theta\in\Delta}$ is an exponential family in natural form where (A1) and (A2) are fulfilled. Let $Q_\theta = P_\theta \circ T^{-1}$ be absolutely continuous with respect to the Lebesgue measure. If the function $K(\theta)$ is steep, then for every $\theta \in \Delta$ the maximum likelihood estimator $\widehat{\theta}(x)$ exists for P_θ-almost all x, it holds $\widehat{\theta}(x) \in \Delta^0$, and $\widehat{\theta}(x)$ is the unique solution of the likelihood equation $\nabla K(\theta) = T(x)$.*

The setting of Proposition 7.95 covers many families of distributions that are relevant in statistics, e.g., normal, exponential, and gamma distributions.

Example 7.96. Consider the exponential family $(\mathsf{N}^{\otimes n}(\mu, \sigma^2))_{\mu\in\mathbb{R},\sigma^2>0}$. Represented in natural form (see Example 1.11) it has $K(\theta_1, \theta_2) = -(n/2)(\theta_1^2/(2\theta_2) + \ln(-\theta_2/\pi))$. Hence,

$$\gamma_{\mathsf{m}}(\theta_1, \theta_2) = \nabla K(\theta_1, \theta_2) = \frac{n}{2}(-\frac{2\theta_1}{\theta_2}, \frac{\theta_1^2}{2\theta_2^2} - \frac{1}{\theta_2}),$$

so that $\gamma_{\mathsf{m}}(\Delta^0) = \gamma_{\mathsf{m}}(\Delta) = \gamma_{\mathsf{m}}(\mathbb{R} \times (-\infty, 0)) = \mathbb{R} \times (0, \infty)$. Fix $\eta \in \partial\Delta = \mathbb{R} \times \{0\}$, say $\eta = (\eta_1, 0)$. Then $\kappa(\lambda) = K(\lambda\theta + (1-\lambda)\eta)$ satisfies

$$\frac{d\kappa(\lambda)}{d\lambda} = -\frac{n}{2}\left(\frac{2(\lambda\theta_1 + (1-\lambda)\eta_1)}{2\theta_2}(\theta_1 - \eta_1) + \frac{1}{\lambda}\right) \underset{\lambda\to 0}{\to} -\infty,$$

and the steepness condition (7.49) is fulfilled. We can use Proposition 7.95 to conclude that the maximum likelihood estimator exists P_θ-a.s. But this route is not necessary as $T_{\oplus n}$ takes on only values in $\gamma_{\mathsf{m}}(\Delta^0)$ so that we may apply Problem 7.91. As there is a one-to-one relationship between (θ_1, θ_2) and (μ, σ^2) we get that the maximum likelihood estimator of (μ, σ^2) also exists P_θ-a.s., that it is the unique solution of the likelihood equations

$$\frac{\partial}{\partial\mu}\sum\nolimits_{i=1}^{n} \ln \varphi_{\mu,\sigma^2}(x_i) = 0 \quad \text{and} \quad \frac{\partial}{\partial\sigma^2}\sum\nolimits_{i=1}^{n} \ln \varphi_{\mu,\sigma^2}(x_i) = 0,$$

and that it is given by

$$\widehat{\mu}(x_1, ..., x_n) = \overline{x}_n \quad \text{and} \quad \widehat{\sigma^2}(x_1, ..., x_n) = \frac{1}{n}\sum\nolimits_{i=1}^{n}(x_i - \overline{x}_n)^2.$$

The next example shows that the assumption $Q_\theta \ll \boldsymbol{\lambda}_d$ is essential.

Example 7.97. As in Example 1.7, let X follow a binomial distribution $\mathsf{B}(n,p)$ with parameters $n \in \{1,2,...\}$ and $p \in (0,1)$. The sample space is $\mathcal{X} = \{0,1,...,n\}$. The density of $\mathsf{B}(n,p)$ with respect to the measure $\boldsymbol{\mu}$ that has the point masses $\boldsymbol{\mu}(\{x\}) = \binom{n}{x}$, $x \in \mathcal{X}$, is given by

$$f_\theta(x) = \exp\{\theta x - K(\theta)\}, \quad x \in \mathcal{X},\ \theta \in \Delta = \mathbb{R},$$

where $K(\theta) = n\ln(1+e^\theta)$, and $\theta(p) = \ln(p/(1-p))$, $p \in (0,1)$. As $\partial\Delta = \partial\mathbb{R} = \emptyset$ the steepness condition is satisfied. Furthermore,

$$\gamma_{\mathsf{m}}(\theta) = n\frac{e^\theta}{1+e^\theta}, \quad \gamma_{\mathsf{m}}(\Delta^0) = \gamma_{\mathsf{m}}(\Delta) = (0,n). \tag{7.51}$$

Hence $T(x) = x \in \gamma_{\mathsf{m}}(\Delta^0)$ if and only if $x \in \{1,...,n-1\}$. At every $x \in \{1,2,...,n-1\}$ an MLE $\widehat{\theta}(x)$ is readily found by maximizing $f_\theta(x) = \exp\{\theta x\}(1+\exp\{\theta\})^{-n}$, which gives $\widehat{\theta}(x) = \ln(x/(n-x))$. However, for $x \in \{0,n\}$ MLEs $\widehat{\theta}(0)$ and $\widehat{\theta}(n)$ do not exist, which is in accordance with (7.51). If we extend the parameter space in the original parametrization, i.e., we allow $p \in [0,1]$, with the convention $\mathsf{b}_{n,0}(0) = \mathsf{b}_{n,1}(n) = 1$, then there always exists an MLE of p, and it is unique. It is $\widehat{p}(x) = x/n$, $x \in \{0,1,...,n\}$. But $\widehat{p}(0)$ and $\widehat{p}(n)$ are not solutions of the likelihood equation, which fail to exist for $x = 0$ and $x = n$.

We conclude this section with the asymptotic of the MLE in exponential families. Let $X_1,...,X_n$ be a sample from an exponential family with generating statistic T and natural parameter θ. The log-likelihood function is then given by $\Lambda_n(\theta,\mathbf{x}_n) = \langle\theta, T_{\oplus n}(\mathbf{x}_n)\rangle - nK(\theta)$, $\mathbf{x}_n = (x_1,...,x_n) \in \mathcal{X}^n$, so that the likelihood equation reads $T_{\oplus n}(\mathbf{x}_n) = n\nabla K(\theta)$. Set $\gamma_{\mathsf{m}}(\theta) = \mathsf{E}_\theta T = \nabla K(\theta)$ and denote by $\kappa_{\mathsf{m}} : \gamma_{\mathsf{m}}(\Delta^0) \to \Delta^0$ the inverse mapping, which exists according to Theorem 1.22. Set $\overline{T}_n = (1/n)T_{\oplus n}$ and

$$\widetilde{\theta}_n(\mathbf{x}_n) = \begin{cases} \kappa_{\mathsf{m}}(\overline{T}_n(\mathbf{x}_n)), & \text{if } \overline{T}_n(\mathbf{x}_n) \in \gamma_{\mathsf{m}}(\Delta^0), \\ \theta^*, & \text{if } \overline{T}_n(\mathbf{x}_n) \notin \gamma_{\mathsf{m}}(\Delta^0), \end{cases} \tag{7.52}$$

where $\theta^* \in \Delta$ is any fixed parameter value.

Proposition 7.98. *If the conditions (A1) and (A2) are satisfied, then for the sequence of models* $(\mathcal{X}^n, \mathfrak{A}^{\otimes n}, (P_\theta)_{\theta\in\Delta})$ *the sequence* $\widetilde{\theta}_n$ *is a strongly approximate and asymptotic MLE.* $\widetilde{\theta}_n$ *is strongly consistent at every* $\theta_0 \in \Delta^0$.

Proof. If $D_n := \sup_{\theta\in\Delta}\Lambda_n(\theta) - \Lambda_n(\widetilde{\theta}_n)$, then by Problem 7.91 it holds $D_n = 0$ for $\overline{T}_n \in \gamma_{\mathsf{m}}(\Delta^0)$. By the strong law of large numbers there is some $A \in \mathfrak{F}$ with $\mathbb{P}_{\theta_0}(A) = 1$ such that $\overline{T}_n(X_1(\omega),...,X_n(\omega)) \to \gamma_{\mathsf{m}}(\theta_0)$, $\omega \in A$. Hence $D_n(\omega) = 0$ for all sufficiently large n. This means that $\widetilde{\theta}_n$ is a strongly approximate and asymptotic MLE. As $\theta_0 \in \Delta^0$, and $\gamma_{\mathsf{m}}(\Delta^0)$ is open by Theorem 1.22, there is some $n_0(\omega)$ such that

$$\widetilde{\theta}_n(X_1(\omega),...,X_n(\omega)) = \kappa_{\mathsf{m}}(\overline{T}_n(X_1(\omega),...,X_n(\omega))), \quad n \geq n_0(\omega).$$

The continuity of κ_{m} (see Theorem 1.22) implies

$$\kappa_{\mathsf{m}}(\overline{T}_n(X_1(\omega),...,X_n(\omega))) \to \kappa_{\mathsf{m}}(\gamma_{\mathsf{m}}(\theta_0)) = \theta_0, \quad \omega \in A,$$

which proves the strong consistency. ■

Consistency in Location and Regression Models

To prepare for the results on the consistency in regression models we deal with the location model and construct contrast functions. We assume that $\varrho : \mathbb{R} \to \mathbb{R}_+$ satisfies the following conditions.

$$\begin{aligned}&\varrho \text{ is continuous, nonincreasing in } (-\infty,0), \text{ nondecreasing in } (0,\infty),\\ &\varrho(0)=0,\ \lim_{t\to-\infty}\varrho(t)>0, \text{ and } \lim_{t\to\infty}\varrho(t)>0.\end{aligned} \tag{7.53}$$

Let $P_\theta = \mathcal{L}(Z+\theta)$ be a location model with parent distribution $P_0 = P = \mathcal{L}(Z)$. Suppose that

$$\mathbb{E}\,|\varrho(Z+a)-\varrho(Z)| < \infty, \quad a \in \mathbb{R}. \tag{7.54}$$

We set $\varrho_\theta(x) = \varrho(x-\theta)$ and want to impose conditions on ϱ and P such that ϱ_θ becomes a contrast condition, that is, satisfies

$$\begin{aligned}&\mathsf{E}_{\theta_0}(\varrho_\theta - \varrho_{\theta_0}) = \int(\varrho_\theta(t)-\varrho_{\theta_0}(t))P_{\theta_0}(dt)\\ &= \int(\varrho(t-(\theta-\theta_0))-\varrho(t))P(dt) \ge 0 \quad \text{and} \quad = 0 \Leftrightarrow \theta=\theta_0.\end{aligned} \tag{7.55}$$

It is shown that this condition is satisfied for symmetric unimodal distributions and a symmetric ϱ. To this end we need the following statement.

Problem 7.99.* If Z has a distribution that is symmetric about 0 and unimodal with mode 0, then the function $a \mapsto \mathbb{P}(|Z-a|<t)$ is nonincreasing for $a>0$.

We recall the measure μ_ϱ in (7.14), defined by a continuous function ϱ that satisfies (7.53). If ϱ is symmetric, then $\mu_\varrho(-B) = \mu_\varrho(B)$ for every Borel subset of $(0,\infty)$. This symmetry is used in the subsequent lemma that presents conditions for a function ϱ to generate a contrast function in the location model.

Lemma 7.100. *Suppose $P = \mathcal{L}(Z)$ and ϱ are symmetric, and (7.53) and (7.54) are fulfilled. Then*

$$\begin{aligned}\mathbb{E}(\varrho(Z+a)-\varrho(Z)) &= \mathbb{E}[\frac{1}{2}(\varrho(Z+a)+\varrho(Z-a))-\varrho(Z)]\\ &= \int(\mathbb{P}(|Z|<t)-\mathbb{P}(|Z+a|<t))I_{(0,\infty)}(t)\mu_\varrho(dt).\end{aligned}$$

Proof. The first statement follows from $\mathcal{L}(\varrho(Z-a)) = \mathcal{L}(\varrho(-Z+a)) = \mathcal{L}(\varrho(Z+a))$. If ϱ is bounded, then by (7.15)

$$\begin{aligned}
&\int (\varrho(s+a) - \varrho(s))P(ds) \\
&= \int (P([t+a,\infty)) + P([t-a,\infty)) - 2P([t,\infty)))I_{(0,\infty)}(t)\mu_\varrho(dt) \\
&= \int (\mathbb{P}(|Z| < t) - \mathbb{P}(|Z+a| < t))I_{(0,\infty)}(t)\mu_\varrho(dt),
\end{aligned}$$

where we have used $\mu_\varrho(\{x\}) = 0$ for every x, which comes from the continuity of ϱ. To complete the proof we approximate ϱ by $\varrho_N = \min(\varrho, N)$. ■

We see from the above lemma that $\mathbb{E}(\varrho(Z+a) - \varrho(Z)) \geq 0$ holds if ϱ is convex or $\mathcal{L}(Z)$ is unimodal. According to Proposition 2.17 every unimodal, symmetric about 0, distribution P with $P(\{0\}) = 0$ has a Lebesgue density. Conditions for the consistency of M-estimators in location models are now given.

Proposition 7.101. *Suppose $\mathcal{L}(Z)$ is unimodal with mode $m = 0$ and $\mathbb{P}(Z = 0) = 0$. Assume in addition that $\mathcal{L}(Z)$ and ϱ are symmetric, and (7.53) and (7.54) are fulfilled. Suppose that at least one of the following conditions is satisfied.*

(A) The density f of $\mathcal{L}(Z)$ is decreasing on $(0,\infty)$.
(B) ϱ is increasing in $(0,\infty)$.

Then for $P_\theta = \mathcal{L}(Z+\theta)$ and $\varrho_\theta(t) = \varrho(t-\theta)$ the contrast condition (7.55) holds, and every strongly approximate sequence of M-estimators for the sequence of location models $(\mathbb{R}^n, \mathfrak{B}_n, (P_\theta^{\otimes n})_{\theta\in\mathbb{R}})$ is strongly consistent.

Proof. If (A) is fulfilled, then for every $a, t > 0$,

$$\mathbb{P}(|Z| < t) - \mathbb{P}(|Z+a| < t) = \int_{t-a}^{t} f(s)ds - \int_{t}^{t+a} f(s)ds > 0.$$

As $\mu_\varrho((0,\infty)) = \varrho(\infty) > 0$ we get from the previous lemma $\mathsf{v}_\varrho(a) = \mathbb{E}(\varrho(Z+a) - \varrho(Z)) > 0$. The case of $a < 0$ follows from $\mathsf{v}_\varrho(-a) = \mathsf{v}_\varrho(a)$. Suppose (B) is fulfilled. The set

$$A_a = \{t : t > 0, \int_{t-a}^{t} f(s)ds - \int_{t}^{t+a} f(s)ds > 0\}$$

is nonempty and open. As ϱ is increasing it holds $\mu_\varrho((b_1, b_2)) > 0$ for every $0 < b_1 < b_2$, and thus $\mu_\varrho(A_a) > 0$. Hence $\mathsf{v}_\varrho(a) > 0$ for $a > 0$, and $\mathsf{v}_\varrho(a) > 0$ for $a < 0$ by $\mathsf{v}_\varrho(-a) = \mathsf{v}_\varrho(a)$. Thus the contrast condition is proved under (A) and (B).

We establish the finite covering property. We have to show that for every $\varepsilon > 0$ the set $(-\infty, -\varepsilon) \cup (\varepsilon, \infty)$ can be covered by finitely many Δ_i with $\mathbb{E}(\inf_{a\in\Delta_i} \varrho(Z+a) - \varrho(Z)) > 0$. As $\varrho \geq 0$ and is nondecreasing in $|t|$ we get for $|a| \geq 1$

$$|\inf_{|a|\geq 1} \varrho(Z+|a|) - \varrho(Z)| \leq |\varrho(Z+1) - \varrho(Z)| + |\varrho(Z-1) - \varrho(Z)|.$$

This means that the nondecreasing sequence $\inf_{|a|\geq N} \varrho(Z+a) - \varrho(Z)$ is bounded from below by a random variable with finite expectation. Hence by the monotone convergence theorem

$$\lim_{N\to\infty} \mathbb{E}(\inf_{|a|\geq N} \varrho(Z+a) - \varrho(Z)) = \mathbb{E}(\varrho(\infty) - \varrho(Z)) > 0,$$

and thus $\mathbb{E}(\inf_{|a|\geq N_0} \varrho(Z+a) - \varrho(Z)) > 0$ for some $N_0 > 0$. Put $\Delta_0 = (-\infty, -N_0) \cup (N_0, \infty)$. As for $|t-a| < \delta$

$$|\varrho(Z+t) - \varrho(Z)| \leq |\varrho(Z+a+\delta) - \varrho(Z)| + |\varrho(Z+a-\delta) - \varrho(Z)|,$$

we get from condition (7.54) and Lebesgue's theorem that

$$\lim_{\delta\downarrow 0} \mathbb{E}(\inf_{|t-a|<\delta} \varrho(Z+t) - \varrho(Z)) = \mathbb{E}(\varrho(Z+a) - \varrho(Z)) > 0.$$

Hence we may cover the compact set $[-N_0, -\varepsilon] \cup [\varepsilon, N_0]$ by a finite number of sets $\Delta_1, ..., \Delta_N$ that satisfy $\mathbb{E}(\inf_{a\in\Delta_i} \varrho(Z+a) - \varrho(Z)) > 0$. To complete the proof we have only to apply Theorem 7.67. ■

Example 7.102. A contrast function that originated from robust statistics is $\varrho(t) = \frac{1}{2}t^2 I_{[-c,c]}(t) + (c|t| - c^2/2)I_{(c,\infty)}(|t|)$. $\dot{\varrho}(t) = tI_{[-c,c]}(t) + \text{sgn}(t)cI_{(c,\infty)}(|t|)$, its derivative, is nondecreasing. Hence ϱ is convex with a curvature measure $\gamma_\varrho = \lambda(\cdot \cap [-c,c])$, so that ϱ is strictly convex in $[-c,c]$. The linear continuation guarantees that possible gross errors have less influence than under a quadratic contrast function. If the influence of outliers is reduced one more step, then one arrives at nonconvex contrast functions. For a collection of such functions and their statistical properties we refer to Andrews et al. (1972). Some of these functions have special names. The M-estimator that belongs to $\varrho(t) = \frac{1}{2}t^2 I_{[-c,c]}(t) + cI_{(c,\infty)}(|t|)$ is called the skipped mean, and to $\varrho(t) = \frac{1}{2}|t| I_{[-c,c]}(t) + cI_{(c,\infty)}(|t|)$ the skipped median.

Now we construct estimators by minimizing convex criterion functions. It is clear that for a convex function $\mathsf{v} : \mathbb{R} \to \mathbb{R}$ the set $\arg\min_{t\in\mathbb{R}} \mathsf{v}(t)$ is either an interval or empty. To avoid the interval consisting of more than one point one has to require that v be strictly convex in a neighborhood of the minimum point.

Problem 7.103.* Let $\mathsf{v} : \mathbb{R} \to \mathbb{R}$ be convex. Then θ_0 is a minimizer if and only if $D^-\mathsf{v}(\theta_0) \leq 0 \leq D^+\mathsf{v}(\theta_0)$. The minimizer is uniquely determined if and only if $D^-\mathsf{v}(\theta_0 - \varepsilon) < 0 < D^+\mathsf{v}(\theta_0 + \varepsilon)$, $\varepsilon > 0$. If v is strictly convex in a neighborhood of θ_0, then the minimizer θ_0 is unique.

If ϱ is convex and (7.54) is satisfied, then

$$\mathsf{v}_\varrho(a) := \mathbb{E}(\varrho(Z+a) - \varrho(Z)) \tag{7.56}$$

is obviously a convex function on $\mathbb{R}$. The next problem gives the relation between the one-sided derivatives and the curvature measures γ_ϱ and $\gamma_{\mathsf{v}_\varrho}$ of the convex functions ϱ and v_ϱ. We recall that according to (1.59),

$$\gamma_\varrho((a,b]) = D^+\varrho(b) - D^+\varrho(a), \quad \gamma_{\mathsf{v}_\varrho}((a,b]) = D^+\mathsf{v}_\varrho(b) - D^+\mathsf{v}_\varrho(a). \tag{7.57}$$

Problem 7.104.* If ϱ is a convex function, then (7.54) implies that for every $a \in \mathbb{R}$ it holds $\mathbb{E}\left|D^{\pm}\varrho(Z+a)\right| < \infty$,

$$\begin{aligned} D^+\mathsf{v}_\varrho(a) &= \mathbb{E}D^+\varrho(Z+a), \quad D^-\mathsf{v}_\varrho(a) = \mathbb{E}D^-\varrho(Z+a), \quad \text{and} \\ \gamma_{\mathsf{v}_\varrho}(B) &= \int \gamma_\varrho(B+s)P(ds), \end{aligned} \tag{7.58}$$

where $P = \mathcal{L}(Z)$.

By the definition of v_ϱ we see that the question of whether the contrast condition (7.31) is satisfied is here whether $a = 0$ is the unique minimum point of v_ϱ. The uniqueness is clear as long as v_ϱ is strictly convex in some interval $(-\delta_0, \delta_0)$. From Problem 1.53 we know that this is equivalent with

$$\gamma_{\mathsf{v}_\varrho}((a,b)) > 0, \quad \text{for every } (a,b) \subseteq (-\delta_0, \delta_0),\ a < b. \tag{7.59}$$

Remark 7.105. In view of (7.58) condition (7.59) is certainly fulfilled if ϱ is strictly convex on $\mathbb{R}$, as in this case $\gamma_\varrho((a,b)) > 0$ for every $a < b$. This holds, for example, for $\varrho(t) = |t|^p$ with $p > 1$. If $\varrho(x) = |x|$, then ϱ is strictly convex at $x_0 = 0$ but not in $(-\delta_0, \delta_0)$. To make v_ϱ strictly convex in $(-\delta_0, \delta_0)$ it is, for example, enough to assume that P has a Lebesgue density that is positive and continuous at $x_0 = 0$. Indeed, we get from $\gamma_\varrho = 2\boldsymbol{\delta}_0$ and (7.58) that (7.59) is fulfilled.

The next proposition presents conditions for a function ϱ to generate a contrast function in the location model.

Proposition 7.106. *Let Z be a random variable, and let $\varrho : \mathbb{R} \to \mathbb{R}$ be a convex function that satisfies (7.54). Then $a_0 = 0$ is the unique minimum point of v_ϱ in (7.56) if and only if*

$$\mathbb{E}D^-\varrho(Z-\varepsilon) < 0 < \mathbb{E}D^+\varrho(Z+\varepsilon), \quad \varepsilon > 0. \tag{7.60}$$

If (7.59) is fulfilled, then $\mathbb{E}D^-\varrho(Z) \le 0 \le \mathbb{E}D^+\varrho(Z)$ implies that $a_0 = 0$ is the unique minimum point of v_ϱ.

Proof. The proof follows from the Problems 7.103 and 7.104, and the fact that (7.59) implies that v_ϱ is strictly convex in a neighborhood of 0. ■

Example 7.107. If $\varrho(t) = t^2$, then $\mathsf{v}_\varrho(a)$ in (7.56) satisfies the condition (7.54) if and only if $\mathbb{E}|X| < \infty$. As $D^+\mathsf{v}_\varrho(a) = D^-\mathsf{v}_\varrho(a) = \mathbb{E}X + a$ we see that $-\mathbb{E}X$ is the unique minimum point of $\mathsf{v}_\varrho(a)$. Set

$$\tau_\alpha(t) = (1-\alpha)|t|I_{(-\infty,0]}(t) + \alpha t I_{(0,\infty)}(t), \quad \alpha \in (0,1).$$

Then

$$\begin{aligned} D^+\tau_\alpha(t) &= -(1-\alpha)I_{(-\infty,0)}(t) + \alpha I_{[0,\infty)}(t) = \alpha - I_{(-\infty,0)}(t) \\ D^-\tau_\alpha(t) &= -(1-\alpha)I_{(-\infty,0]}(t) + \alpha I_{(0,\infty)}(t) = \alpha - I_{(-\infty,0]}(t). \end{aligned} \tag{7.61}$$

If $\varrho(t) = \tau_\alpha(t)$, then (7.54) is satisfied for every random variable, and it holds

$$\begin{aligned} D^+\mathsf{v}_\varrho(a) &= \mathbb{E}D^+\mathsf{v}_\alpha(X+a) = \alpha - F(-a-0), \\ D^-\mathsf{v}_\varrho(a) &= \mathbb{E}D^-\mathsf{v}_\alpha(X+a) = \alpha - F(-a). \end{aligned}$$

Hence $D^-\mathsf{v}_\varrho(a) \le 0 \le D^+\mathsf{v}_\varrho(a)$ holds if and only if $u_\alpha = -a$ is an α-quantile of the distribution of X. According to (7.60) the minimum point u_α is unique if and only if

$$F(u_\alpha - \varepsilon) < \alpha < F(u_\alpha + \varepsilon), \quad \varepsilon > 0. \tag{7.62}$$

If now $x_1, ..., x_n \in \mathbb{R}$, $\widehat{F}_n(t) = n^{-1}\sum_{i=1}^n I_{(-\infty,t]}(x_i)$, and $M_n(\theta) = n^{-1}\sum_{i=1}^n \tau_\alpha(x_i - \theta)$, then $\widehat{\theta}_n \in \arg\min_{\theta\in\mathbb{R}} M_n$ if and only if

$$\widehat{F}_n(\widehat{\theta}_n - 0) \le \alpha \le \widehat{F}_n(\widehat{\theta}_n), \tag{7.63}$$

which means that $\widehat{\theta}_n$ is a sample α-quantile.

Example 7.108. Suppose ϱ is convex and (7.54) is satisfied. Suppose further that $\mathbb{E}D^+\varrho(Z-\varepsilon) < 0 < \mathbb{E}D^+\varrho(Z+\varepsilon)$, $\varepsilon > 0$. Then $\mathsf{v}_\varrho(a) = \mathbb{E}(\varrho(Z+a) - \varrho(X))$ has a unique minimum at $a = 0$ (see Proposition 7.106) so that the contrast condition (7.55) is satisfied. Let $X_1, ..., X_n$ be independent replications of $X = Z + \theta_0$, and $\widehat{\theta}_n$ be a minimizer of

$$M_n(\theta) = \frac{1}{n}\sum\nolimits_{i=1}^n \varrho(X_i - \theta).$$

As $M_n(\theta) \to \mathsf{v}_\varrho(\theta_0 - \theta)$, $\mathbb{P}_{\theta_0}$-a.s., we get $\widehat{\theta}_n \to \theta_0$, $\mathbb{P}_{\theta_0}$-a.s., from Theorem 7.77. Let τ_α be as in Example 7.107, F be the c.d.f. of X, and u_α be an α-quantile. Then we know from Example 7.107 that under the assumption (7.62) the function

$$M(\theta, \theta_0) = \mathbb{E}_{\theta_0}(\tau_\alpha(X_1 - \theta) - \tau_\alpha(X_1))$$

is convex and has a unique minimum at u_α. Hence we obtain from Theorem 7.77 that every sequence of sample α-quantiles $\widehat{\theta}_n$ tends, $\mathbb{P}_{\theta_0}$-a.s., to the α-quantile of F. This is a well-known result on the strong consistency of the sample quantiles.

Now we study the problem of consistency in *regression models*. The starting point for constructing a regression model is a one-parameter parent family of distributions $(P_\eta)_{\eta\in(a,b)}$, defined on the sample space $(\mathcal{Y}, \mathfrak{B})$, where we assume that this family is a stochastic kernel, i.e., $\eta \mapsto P_\eta(B)$ is a measurable mapping for every $B \in \mathfrak{B}$. The sample is influenced by *covariables*, also called *regressors*, that take on values in $\mathcal{X}$, with $(\mathcal{X}, \mathfrak{A})$ as the measurable space.

The distribution of the observation and its covariable are connected by the regression function $g : \Delta \times \mathcal{X} \to (a,b)$ in such a way that for a fixed value x of the covariable the observation Y has the distribution $P_{g_\theta(x)}$. We assume that Δ is a separable metric space, let $\mathsf{C}_m(\Delta, \mathcal{X})$ denote the class of functions that are continuous in θ and measurable in x, and assume that the regression function satisfies

$$g \in \mathsf{C}_m(\Delta, \mathcal{X}). \tag{7.64}$$

One has to distinguish between two types of regression models, depending on whether the covariable is random. A regression model with random regressor is given by a pair of random variables (Y, X), where the conditional distribution of Y, given $X = x$, is given by $\mathsf{K}_\theta(\cdot|x) = P_{g_\theta(x)}$. Denote the marginal distribution of X on $(\mathcal{X}, \mathfrak{A})$ by μ. A sample of size n consists of independent pairs (Y_i, X_i), $i = 1, ..., n$, with common distribution

$$\mathcal{L}(Y_i, X_i) = \mathsf{K}_\theta \otimes \mu, \quad i = 1, ..., n. \tag{7.65}$$

In a regression model with nonrandom regressors we fix an experimental *design*, i.e., values $x_1, ..., x_n \in \mathcal{X}$ at which the n measurements are taken. Then the observations $Y_1, ..., Y_n$ are independent and have the distributions $P_{g_\theta(x_1)}, ..., P_{g_\theta(x_n)}$, respectively. To make an asymptotic treatment of this regression model possible it is generally assumed that $\mathcal{X}$ is a separable metric space, and that the sequence of empirical measures $\mu_n = (1/n)\sum_{i=1}^n \boldsymbol{\delta}_{x_i}$ tends weakly to some limiting experimental design. Although both regression models are different their technical treatments are similar. Although regression models with random regressors admit the application of limit theorems for i.i.d. random variables, regression models with nonrandom regressors require additional conditions that, roughly speaking, guarantee that the distributions $P_{g_\theta(x_i)}$ do not fluctuate too much. These conditions increase the technical details, and because of this we mainly deal with regression models with random regressors.

Depending on the structures of the regression function and the parent model that are combined, different types of regression models are obtained.

Linear regression:	$g_\theta(x) = \theta^T x,$	$\theta, x \in \mathbb{R}^d.$
Generalized linear regression:	$g_\theta(x) = \varphi(\theta^T x),$	$\theta, x \in \mathbb{R}^d.$
Nonlinear regression:	$g_\theta(x),$	$\theta \in \Delta \subseteq \mathbb{R}^k, x \in \mathbb{R}^d.$

The function φ in generalized linear regression is assumed to be continuous and is called the *link function.* Often-used parent models are based on the following.

Location family:	$P_\eta = \mathcal{L}(\varepsilon + \eta), \quad \eta \in \mathbb{R}.$
Exponential family:	$dP_\eta/d\boldsymbol{\mu} = \exp\{\eta T - K(\eta)\}, \quad \eta \in (a,b).$

In general, the likelihood approach with the construction of maximum likelihood estimators cannot be applied to regression models as these models are not completely specified and thus a likelihood function is not available. This occurs, for example, if the location family is used as parent model, and about the distribution of the errors $\varepsilon_i = Y_i - g(X_i, \theta)$ it is only known that its expectation, median, or some other location invariant functional is zero.

Lemma 7.109. *If ϱ_η is a contrast function for the parent family $(P_\eta)_{\eta\in(a,b)}$, and*

$$\int [\int \mid \varrho_{g_\theta(x)}(y) - \varrho_{g_{\theta_0}(x)}(y) \mid P_{g_{\theta_0}(x)}(dy)]\mu(dx) < \infty \tag{7.66}$$

is fulfilled, then the condition

$$\mu(\{x : g_\theta(x) = g_{\theta_0}(x)\}) = 1 \quad \Rightarrow \quad \theta = \theta_0 \tag{7.67}$$

implies that $\widetilde{\varrho}_\theta(x, y) := \varrho_{g_\theta(x)}(y)$ is a contrast function for the regression model $(\mathcal{Y} \times \mathcal{X}, \mathfrak{B} \otimes \mathfrak{A}, (\mathsf{K}_\theta \otimes \mu)_{\theta\in\Delta})$, where $\mathsf{K}_\theta \otimes \mu$ is from (7.65).

Proof. It holds

$$\mathsf{E}_{\theta_0}(\varrho_{g_\theta(X)}(Y) - \varrho_{g_{\theta_0}(X)}(Y)) = \int [\int (\varrho_{g_\theta(x)}(y) - \varrho_{g_{\theta_0}(x)}(y))P_{g_\theta(x)}(dy)]\mu(dx).$$

As

$$\int (\varrho_\eta(y) - \varrho_{\eta_0}(y))P_{\eta_0}(dy) > 0, \quad \eta \neq \eta_0,$$

the statement follows from (7.67). ■

The next example illustrates the identifiability condition (7.67).

Example 7.110. In the linear regression model with random regressor it holds $Y = \theta^T X + \varepsilon$ and $g_\theta(x) = \theta^T x$. Set $\varrho_\eta(t) = (t - \eta)^2$. Then ϱ_η is a contrast function for the location model $P_\eta = \mathcal{L}(\varepsilon + \eta)$, provided that $\mathbb{E}\,|\varepsilon| < \infty$ and $\mathbb{E}\varepsilon = 0$. If X is a random covariable with $\mu = \mathcal{L}(X)$, then the condition (7.67) may be written as

$$\mathbb{P}((\theta - \theta_0)^T X = 0) = 1 \quad \text{implies} \quad \theta = \theta_0. \tag{7.68}$$

If $\mathbb{E}\,\|X\|^2 < \infty$, then the latter condition holds if and only if the matrix $\mathbb{E}XX^T$ is nonsingular.

If the location family is used as parent family, then regardless of whether the regression function is linear, we make the assumption that ϱ is a contrast function in the location model generated by the i.i.d. errors ε_i, i.e.,

$$\begin{aligned} &\varrho \text{ satisfies (7.53)}, \\ &\mathbb{E}|\varrho(\varepsilon_1 + a) - \varrho(\varepsilon_1)| < \infty, \\ &\mathbb{E}(\varrho(\varepsilon_1 + a) - \varrho(\varepsilon_1)) > 0, \quad a \neq 0. \end{aligned} \tag{7.69}$$

We impose the following conditions on the regression model.

$$\begin{aligned} &X_1, ..., X_n, \varepsilon_1, ..., \varepsilon_n, \text{ are independent}, \ \mathcal{L}(X_i) = \mu, \ \mathcal{L}(\varepsilon_i) = P, \\ &Y_i = g_{\theta_0}(X_i) + \varepsilon_i, \quad \theta_0 \in \Delta, \ i = 1, ..., n, \quad g \in \mathsf{C}_m(\Delta, \mathcal{X}). \end{aligned} \tag{7.70}$$

Theorem 7.111. ***(Nonlinear Regression)*** *Assume that Δ is a compact metric space and the conditions (7.69) and (7.70) are satisfied. If in addition*

$$\mathbb{E} \sup_{\theta \in \Delta} |\varrho(\varepsilon_1 - [g_\theta(X_1) - g_{\theta_0}(X_1)]) - \varrho(\varepsilon_1)| < \infty$$

and (7.67) hold, then every sequence of minimizers $\widehat{\theta}_n : \mathcal{X}^n \times \mathcal{Y}^n \to_m \Delta$ of

$$M_n(\theta) = \frac{1}{n} \sum\nolimits_{i=1}^n \varrho(Y_i - g_\theta(X_i))$$

is strongly consistent.

Proof. The statement follows from Theorem 7.70, Proposition 7.71, and Lemma 7.109 if we use $\widetilde{\varrho}_\theta(x, y) = \varrho(y - g_\theta(x))$ as the contrast function, and set $W_n(\theta) = M_n(\theta) - M_n(\theta_0)$ and $W(\theta) = \mathbb{E}(\varrho(Y_1 - g_\theta(X_1)) - \varrho(\varepsilon_1))$. ■

If $\mathbb{E}|\varepsilon_1| < \infty$, then $\varrho(t-\eta) = (t-\eta)^2$ is a contrast function for the location model $P_\eta = \mathcal{L}(\varepsilon_1 + \eta)$. On the other hand, assume that the common c.d.f. F of the ε_i satisfies $F(-\varepsilon) < 1/2 < F(\varepsilon)$, $\varepsilon > 0$, and that $\varrho(t) = |t|$. Then $\varrho(\cdot - \eta)$ is a contrast function for the location model $P_\eta = \mathcal{L}(\varepsilon_1 + \eta)$, see Example 7.107. By combining these contrast functions for the location model with Theorem 7.111 one can easily establish sufficient conditions for the strong consistency of least squares and $\mathbb{L}_1$-estimators for nonlinear regression models. There are a large number of papers that deal with the consistency of M-estimators in nonlinear regression models, where in some papers random regressors and in others nonrandom regressors are considered. See, e.g., Jurečková and Sen (1996) and Liese and Vajda (1999, 2003a,b, 2004) for references. If $\varrho = \tau_\alpha$, then the estimators are called *regression quantiles*, which have been introduced by Koenker and Bassat (1978) and studied by other authors. See Jurečková and Sen (1996) for references.

Now we study the linear regression model and use convex criterion functions.

Theorem 7.112. ***(Linear Regression)*** *Suppose X_i, ε_i satisfy (7.70), and $\varrho : \mathbb{R} \to \mathbb{R}$ is a convex function that satisfies*

$$\mathbb{E}|\varrho(\varepsilon_1 + a) - \varrho(\varepsilon_1)| < \infty \quad \textit{and} \quad \mathbb{E}(\varrho(\varepsilon_1 + a) - \varrho(\varepsilon_1)) > 0, \ a \neq 0. \tag{7.71}$$

If (7.68) holds, then for the linear regression model $Y_i = \theta_0^T X_i + \varepsilon_i$, $i = 1, ..., n$, every sequence of minimizers $\widehat{\theta}_n$ that minimize

$$\frac{1}{n} \sum\nolimits_{i=1}^n \varrho(Y_i - \theta^T X_i)$$

over $\theta \in \mathbb{R}^d$ is strongly consistent.

Proof. $\widehat{\theta}_n$ minimizes

$$M_n(\theta) = \frac{1}{n}\sum\nolimits_{i=1}^{n}(\varrho(Y_i - \theta^T X_i) - \varrho(\varepsilon_i)),$$

and

$$M(\theta) = \mathbb{E}(\varrho(Y_1 - \theta^T X_1) - \varrho(\varepsilon_1))$$

has in view of (7.68) and (7.71) a unique minimum at θ_0. It remains to apply the strong law of large numbers to $M_n(\theta)$ for fixed θ and the argmin theorem for convex processes; see Theorem 7.77. ■

Example 7.113. For $p \geq 1$ it holds $|s+t|^p \leq 2^{p-1}(|s|^p+|t|^p)$. As $||x+a|^p - |x|^p| \leq p(|x|^{p-1} + |x+a|^{p-1})|a|$, we get

$$||x+a|^p - |x|^p| \leq p((1+2^{p-1})|a||x|^{p-1} + |a|^p).$$

Hence $\mathbb{E}|\varepsilon_i|^{p-1} < \infty$ implies the first condition in (7.71) for $\varrho = |x|^p$. To verify the second condition we set $\mathsf{v}_\varrho(a) = \mathbb{E}(|\varepsilon_1 + a|^p - |\varepsilon_1|^p)$. This function is strictly convex for $p > 1$ and $D^-\mathsf{v}_\varrho(0-0) = D^+\mathsf{v}_\varrho(0+0) = \mathbb{E}(\operatorname{sgn}(\varepsilon_1)\,|\varepsilon_1|^{p-1})$. Hence we get from Proposition 7.106 that

$$\mathbb{E}(\operatorname{sgn}(\varepsilon_1)\,|\varepsilon_1|^{p-1}) = 0$$

implies that the second condition in (7.71) holds for $p > 1$. In the case of least squares estimators (i.e., $p = 2$), the last condition means $\mathbb{E}\varepsilon_1 = 0$. If $p = 1$, then instead of $\varrho(x) = |x|$ we may also use $\tau_{1/2}(x) = \frac{1}{2}|x|$ and get from (7.62) in Example 7.107 that $\mathsf{v}_\varrho(a) = \mathbb{E}(|\varepsilon_1 + a| - |\varepsilon_1|)$ has a unique minimum at $a_0 = 0$ if and only if $a_0 = 0$ is the unique median of the c.d.f. of ε_1.

7.5.2 Consistency in Bayes Models

In this section we study the behavior of the posterior distribution for large sample sizes. There are several reasons for doing this. One is to investigate the influence of the prior on the posterior distribution, and thus on the posterior risk introduced in Definition 3.33. Results in this direction are referred to as Bayes robustness. Another is a frequentist verification of Bayes procedures with the aim to show that Bayes estimators are consistent and even asymptotically efficient.

We start with the Bayes model introduced in Chapter 1 and suppose that $(P_\theta)_{\theta\in\Delta}$ is a family of distributions on $(\mathcal{X}, \mathfrak{A})$ where $(\Delta, \mathfrak{B}_\Delta)$ is a measurable space. We assume that the condition (A3) is satisfied so that $\mathsf{P}(A|\theta) = P_\theta(A)$, $A \in \mathfrak{A}$, is a stochastic kernel. Then by Proposition A.39

$$\mathsf{P}_n(B|\theta) := P_\theta^{\otimes n}(B), \quad B \in \mathfrak{A}^{\otimes n},\ \theta \in \Delta,$$

is also a stochastic kernel. We consider the product space $(\mathcal{X}^n \times \Delta, \mathfrak{A}^{\otimes n} \otimes \mathfrak{B}_\Delta)$ and denote by $X_1, ..., X_n, \Theta$ the projections on the coordinates. For a prior distribution Π on $(\Delta, \mathfrak{B}_\Delta)$ we consider the probability space

$$(\mathcal{X}^n \times \Delta, \mathfrak{A}^{\otimes n} \otimes \mathfrak{B}_\Delta, \mathsf{P}_n \otimes \Pi).$$

Then $\mathcal{L}(X_1, ..., X_n, \Theta) = \mathsf{P}_n \otimes \Pi$, so that $X_1, ..., X_n$ are conditionally i.i.d., given $\Theta = \theta$, with regular conditional distribution P_n. We now assume that $(\Delta, \mathfrak{B}_\Delta)$ is a Borel space. Then (see Theorem A.37) there exists a stochastic kernel $\boldsymbol{\Pi}_n : \mathfrak{B}_\Delta \times \mathcal{X}^n \to_k [0,1]$ such that for every $h : \mathcal{X}^n \times \Delta \to_m \mathbb{R}_+$

$$\int [\int h(x,\theta)\mathsf{P}_n(dx|\theta)]\Pi(d\theta) = \int [\int h(x,\theta)\boldsymbol{\Pi}_n(d\theta|x)](\mathsf{P}_n\Pi)(dx). \quad (7.72)$$

This is equivalent to the fact that $\boldsymbol{\Pi}_n$ is a regular conditional distribution of Θ, given $(X_1, ..., X_n) = x$, and $\mathsf{P}_n\Pi$ is the marginal distribution of $(X_1, ..., X_n)$. Our aim is to study the sequence of posterior distributions $\boldsymbol{\Pi}_n$ and to show that under mild conditions these distributions concentrate themselves more and more around the "true parameter". To make such a statement precise we consider the product space $(\mathcal{X}^\infty, \mathfrak{A}^{\otimes\infty})$ where $\mathfrak{A}^{\otimes\infty}$ is the smallest σ-algebra for which all projections X_i, $i = 1, 2, ...$, are measurable. The family of subsets of $\mathcal{X}^\infty$,

$$\mathfrak{C} = \bigcup\nolimits_{n=1}^{\infty} \{(X_1, ..., X_n)^{-1}(B) : B \in \mathfrak{A}^{\otimes n}\},$$

called the family of cylinder sets, is an algebra that generates $\mathfrak{A}^{\otimes\infty}$. As in Section 7.5.1, let $P_\theta^{\otimes\infty}$ be the infinite product measure on $(\mathcal{X}^\infty, \mathfrak{A}^{\otimes\infty})$, defined by the condition

$$P_\theta^{\otimes\infty}(B \times \mathcal{X} \times \mathcal{X} \times \cdots) = P_\theta^{\otimes n}(B), \quad B \in \mathfrak{A}^{\otimes n}, \ \theta \in \Delta, \ n = 1, 2, ...$$

Hence the mapping $\theta \mapsto P_\theta^{\otimes\infty}(A)$ is measurable for every $A \in \mathfrak{C}$. As the class of all $A \in \mathfrak{A}^{\otimes\infty}$ for which $\theta \mapsto P_\theta^{\otimes\infty}(A)$ is measurable is a monotone class that contains the algebra $\mathfrak{C}$ it follows from the monotone class theorem (see e.g. Kallenberg (1997)) that

$$\mathsf{P}_\infty(A|\theta) = P_\theta^{\otimes\infty}(A), \quad A \in \mathfrak{A}^{\otimes\infty}, \ \theta \in \Delta, \quad (7.73)$$

is again a stochastic kernel. To summarize, we arrive at the probability space

$$(\Omega, \mathfrak{F}, \mathbb{P}) = (\mathcal{X}^\infty \times \Delta, \mathfrak{A}^{\otimes\infty} \otimes \mathfrak{B}_\Delta, \mathsf{P}_\infty \otimes \Pi), \quad (7.74)$$

on which the random variables X_i, $i = 1, 2, ...$, and Θ are defined to be the coordinate mappings. In a first step of the consistency considerations we study the posterior distributions of Θ, given $X_1, ..., X_n$, and ask for conditions under which the sequences of posterior distributions concentrate themselves more and more on a neighborhood of Θ. Let us first consider this problem in a special case.

Example 7.114. Consider the special case of i.i.d. Bernoulli variables $X_1, ..., X_n$ with distribution $(1-p)\boldsymbol{\delta}_0 + p\boldsymbol{\delta}_1$, $p \in \Delta = (0,1)$. Then $\sum_{i=1}^n X_i$ is a sufficient statistic for the model $(((1-p)\boldsymbol{\delta}_0 + p\boldsymbol{\delta}_1)^{\otimes n})_{p\in(0,1)}$. Hence by Proposition 4.62 the posterior distribution depends on $(X_1, ..., X_n)$ only through $T = \sum_{i=1}^n X_i$. If $\Pi = \mathsf{Be}(\alpha, \beta)$ is the beta distribution with parameters $\alpha, \beta > 0$, then by Example 1.45 it holds for the posterior $\boldsymbol{\Pi}_n$ in (7.72) that

$$\boldsymbol{\Pi}_n(\cdot|x_1,...,x_n) = \mathsf{Be}(\alpha + \sum\nolimits_{i=1}^n x_i, \beta + n - \sum\nolimits_{i=1}^n x_i).$$

As the beta distribution $\mathsf{Be}(\alpha,\beta)$ has the expectation $\alpha/(\alpha+\beta)$ we get

$$\mathbb{E}(\Theta|X_1 = x_1,...,X_n = x_n) = \frac{\alpha+\beta}{\alpha+\beta+n}\frac{\alpha}{\alpha+\beta} + \frac{n}{\alpha+\beta+n}\frac{1}{n}\sum\nolimits_{i=1}^n x_i.$$

Moreover, the variance of the beta distribution $\mathsf{Be}(\alpha,\beta)$ is $\frac{\alpha\beta}{(\alpha+\beta)^2(\alpha+\beta+1)}$. Hence the variance of the posterior distribution is

$$\frac{(\alpha+\sum_{i=1}^n X_i)(\beta+n-\sum_{i=1}^n X_i)}{(\alpha+\beta+n)^2(\alpha+\beta+n+1)} \le \frac{(\alpha+n)(\beta+n)}{(\alpha+\beta+n)^2(\alpha+\beta+n+1)} \to 0.$$

Therefore,

$$\begin{aligned}\mathbb{E}(\Theta - \mathbb{E}(\Theta|X_1,...,X_n))^2 &= \mathbb{E}(\mathbb{E}(\Theta - \mathbb{E}(\Theta|X_1,...,X_n))^2|X_1,...,X_n)\\ &\le \frac{(\alpha+n)(\beta+n)}{(\alpha+\beta+n)^2(\alpha+\beta+n+1)} \to 0,\end{aligned}$$

so that for large n the posterior $\boldsymbol{\Pi}_n$ concentrates more and more around the unobservable Θ. Taking $T_n = \mathbb{E}(\Theta|X_1,...,X_n)$ or the arithmetic mean $(1/n)\sum_{i=1}^n X_i$ we get that T_n consistently approximates Θ, or is a consistent estimator in the Bayes model (7.74).

Now we establish a first result on the consistency of the posterior distribution in the Bayes model (7.74). The subsequent result is due to Doob (1949). Stronger results can be found in Schwartz (1965). The version of Doob's result presented here is taken from Schervish (1995); see also Schervish and Seidenfels (1990).

Theorem 7.115. *Suppose $(\mathcal{X},\mathfrak{A})$ and $(\Delta,\mathfrak{B}_\Delta)$ are Polish spaces. Let Π be a prior on $(\Delta,\mathfrak{B}_\Delta)$ and $(P_\theta)_{\theta\in\Delta}$ be a family that satisfies the condition (A3). Let $\boldsymbol{\Pi}_n$ be a regular version of the posterior distribution specified by (7.72). If there exists a sequence $T_n : \mathcal{X}^n \to_m \Delta$ with $T_n(X_1,...,X_n) \to^{\mathbb{P}} \Theta$, then for every $B \in \mathfrak{B}_\Delta$ it holds*

$$\lim_{n\to\infty} \boldsymbol{\Pi}_n(B|X_1,...,X_n) = I_B(\Theta), \quad \mathbb{P}\text{-a.s.}$$

Proof. Recall that X_i and Θ are the projection mappings on Ω in (7.74). We introduce a sequence of sub-σ-algebras of $\mathfrak{A}^{\otimes\infty} \otimes \mathfrak{B}_\Delta$ by $\mathfrak{F}_n = \sigma(X_1,...,X_n)$ and set $\mathfrak{F}_\infty = \sigma(X_1,X_2,...)$. Then by Levy's martingale theorem (see Theorem A.34) it holds

$$\lim_{n\to\infty} \boldsymbol{\Pi}_n(B|X_1,...,X_n) = \lim_{n\to\infty} \mathbb{E}(I_B(\Theta)|\mathfrak{F}_n) = \mathbb{E}(I_B(\Theta)|\mathfrak{F}_\infty), \quad \mathbb{P}\text{-a.s.}$$

The condition imposed on T_n and Proposition A.12 yield the existence of a subsequence n_k such that $T_{n_k}(X_1,...,X_{n_k})$ converges to Θ, $\mathbb{P}$-a.s. For every $\omega = (x,\theta) \in \mathcal{X}^\infty \times \Delta$ and any fixed $\theta_0 \in \Delta$ we set

$$T(\omega) = \lim_{k\to\infty} T_{n_k}(X_1(\omega),...,X_{n_k}(\omega)),$$

if the limit exists, and $T = \theta_0$ else. Then T is measurable with respect to $\mathfrak{F}_\infty$ and it holds $I_B(T) = I_B(\Theta)$, $\mathbb{P}$-a.s. Hence $\mathbb{E}(I_B(\Theta)|\mathfrak{F}_\infty) = I_B(\Theta)$, $\mathbb{P}$-a.s., and the proof is completed. ∎

Another focus of an asymptotic investigation of the posterior distribution concerns the dependence of the predictive distribution on the prior distribution. More precisely, let $U_n = (X_1, ..., X_n)$ and $V_n = (X_{n+1}, X_{n+2}, ...)$. Consider the random variables to be defined on the probability space (7.74). Then for every $h : \mathcal{X}^n \times \mathcal{X} \times \cdots \times \Delta \to_m \mathbb{R}_+$ it holds

$$\mathbb{E}h(U_n, V_n, \Theta) = \int [\int [\int h(u, v, \theta)(\bigotimes_{i=1}^n P_\theta)(du)](\bigotimes_{i=n+1}^\infty P_\theta)(dv)]\Pi(d\theta).$$

Especially, by taking into account (7.72), it holds

$$\begin{aligned}\mathbb{E}g(U_n, V_n) &= \int [\int [\int g(u, v)(\bigotimes_{i=1}^n P_\theta)(du)](\bigotimes_{i=n+1}^\infty P_\theta)(dv)]\Pi(d\theta) \\ &= \int [\int [\int g(u, v)(\bigotimes_{i=n+1}^\infty P_\theta)(dv)]\boldsymbol{\Pi}_n(d\theta|u)](\mathsf{P}_n\Pi)(du),\end{aligned}$$

for every $g : \mathcal{X}^n \times \mathsf{X}_{i=n+1}^\infty \mathcal{X} \to_m \mathbb{R}_+$. We see from here that

$$\boldsymbol{\Lambda}_n(B|u) = \int (\bigotimes_{i=n+1}^\infty P_\theta(B))\boldsymbol{\Pi}_n(d\theta|u), \quad B \in \bigotimes_{i=n+1}^\infty \mathfrak{A},$$

is a regular version of the conditional distribution of $V_n = (X_{n+1}, X_{n+2}, ...)$, given $U_n = (X_1, ..., X_n)$. $\boldsymbol{\Lambda}_n$ is called the *predictive distribution* of V_n, given U_n. The conditional distribution $\boldsymbol{\Lambda}_n$ depends, of course, on the prior Π. To indicate this dependence we also write $\boldsymbol{\Lambda}_{n,\Pi}$. The next theorem shows that for two priors Π_1 and Π_2 which assign the same parameters a positive weight, i.e., are measure-theoretic equivalent, the predictive distributions $\boldsymbol{\Lambda}_{n,\Pi_1}$ and $\boldsymbol{\Lambda}_{n,\Pi_2}$ are close for large n. We recall that by the definition of $\boldsymbol{\Lambda}_{n,\Pi_i}$ it holds

$$\mathcal{L}((U_n, V_n)|\mathsf{P}_\infty\Pi_i) = \boldsymbol{\Lambda}_{n,\Pi_i} \otimes (\mathsf{P}_n\Pi_i), \quad i = 1, 2.$$

One way to formulate the agreement of the predictive distributions is to replace $\mathsf{P}_n\Pi_2$ with $\mathsf{P}_n\Pi_1$ and to compare the distributions $\boldsymbol{\Lambda}_{n,\Pi_1} \otimes (\mathsf{P}_n\Pi_1)$ and $\boldsymbol{\Lambda}_{n,\Pi_2} \otimes (\mathsf{P}_n\Pi_1)$. Suppose $\mathfrak{A}$ and consequently $\mathfrak{A}^{\otimes\infty}$ are countably generated. Then by (1.76) and Proposition 1.95 it follows that

$$\begin{aligned}&\|\boldsymbol{\Lambda}_{n,\Pi_1} \otimes (\mathsf{P}_n\Pi_1) - \boldsymbol{\Lambda}_{n,\Pi_2} \otimes (\mathsf{P}_n\Pi_1)\| \\ &= \int \|\boldsymbol{\Lambda}_{n,\Pi_1}(\cdot|u_n) - \boldsymbol{\Lambda}_{n,\Pi_2}(\cdot|u_n)\| \, (\mathsf{P}_n\Pi_1)(du_n).\end{aligned} \tag{7.75}$$

For every prior Π and P_∞ from (7.73) we introduce the distribution $\mathbb{P}_\Pi$ on $(\mathcal{X}^\infty, \mathfrak{A}^{\otimes\infty})$ by $\mathbb{P}_\Pi = \mathsf{P}_\infty\Pi$ and denote by $\mathbb{E}_\Pi$ the expectation with respect to $\mathbb{P}_\Pi$. If $\Pi_1 \ll \Pi_2$, then

$$\mathbb{P}_{\Pi_2}(B) = \int P_\theta^{\otimes\infty}(B)\Pi_2(d\theta) = 0$$

implies $\int P_\theta^{\otimes\infty}(B)\Pi_1(d\theta) = 0$ and therefore $\mathbb{P}_{\Pi_1} \ll \mathbb{P}_{\Pi_2}$. Put

$$Z = d\mathbb{P}_{\Pi_1}/d\mathbb{P}_{\Pi_2}, \quad \mathfrak{F}_n = \sigma(X_1, ..., X_n), \quad \text{and} \quad Z_n = \mathbb{E}_{\Pi_2}(Z|\mathfrak{F}_n).$$

Then for every $B \in \mathfrak{F}_n$ it holds

$$\int I_B \mathbb{E}_{\Pi_2}(Z|\mathfrak{F}_n) d\mathbb{P}_{\Pi_2} = \int I_B Z d\mathbb{P}_{\Pi_2} = \mathbb{P}_{\Pi_1}(B),$$
$$\frac{d\mathbb{P}_{n,\Pi_1}}{d\mathbb{P}_{n,\Pi_2}} = Z_n,$$

where $\mathbb{P}_{n,\Pi_i}$ is the restriction of $\mathbb{P}_{\Pi_i}$ on $\mathfrak{F}_n$. Hence $\mathbb{E}_{\Pi_2}(Z|\mathfrak{F}_n)$ is a martingale, so that Levy's martingale convergence theorem (see Theorem A.34) implies

$$\lim_{n\to\infty} \mathbb{E}_{\Pi_2}|Z_n - Z| = 0. \tag{7.76}$$

Theorem 7.116. *Let $(\mathcal{X}, \mathfrak{A})$ and $(\Delta, \mathfrak{B}_\Delta)$ be standard Borel spaces, and Π_1 and Π_2 be priors on $(\Delta, \mathfrak{B}_\Delta)$ with $\Pi_1 \ll \Pi_2$. If the family $(P_\theta)_{\theta\in\Delta}$ satisfies the condition (A3), and $\boldsymbol{\Lambda}_{n,\Pi_i}$, $i = 1, 2$, are any regular versions of the predictive distributions, then*

$$\lim_{n\to\infty} \|\boldsymbol{\Lambda}_{n,\Pi_1} \otimes (\mathsf{P}_n\Pi_1) - \boldsymbol{\Lambda}_{n,\Pi_2} \otimes (\mathsf{P}_n\Pi_1)\| = 0,$$
$$\|\boldsymbol{\Lambda}_{n,\Pi_1}(\cdot|X_1, ..., X_n) - \boldsymbol{\Lambda}_{n,\Pi_2}(\cdot|X_1, ..., X_n)\| \to^{\mathbb{P}_{\Pi_1}} 0. \tag{7.77}$$

Proof. The variational distance of two distributions is nonnegative and does not exceed 2. This, in conjunction with (7.75), yields that the two statements are equivalent. To show the first one we use $Z_n = d\mathbb{P}_{n,\Pi_1}/d\mathbb{P}_{n,\Pi_2}$ and get for every measurable function h that

$$\sup_{\|h\|_u\le 1} \left| \int [\int h(u_n, v_n)(\boldsymbol{\Lambda}_{n,\Pi_1}(dv_n|u_n) - \boldsymbol{\Lambda}_{n,\Pi_2}(dv_n|u_n))](\mathsf{P}_n\Pi_1)(du_n) \right|$$
$$= \sup_{\|h\|_u\le 1} \left| \int [\int h(u_n, v_n)Z(u_n, v_n)\boldsymbol{\Lambda}_{n,\Pi_2}(dv_n|u_n)](\mathsf{P}_n\Pi_2)(du_n) \right.$$
$$\left. - \int [\int h(u_n, v_n)Z_n(u_n)\boldsymbol{\Lambda}_{n,\Pi_2}(dv_n|u_n)](\mathsf{P}_n\Pi_2)(du_n) \right| \le \mathbb{E}_{\Pi_2}|Z - Z_n| \to 0,$$

where the last statement follows from (7.76). To complete the proof it suffices to note that in view of Problem 1.80 the term on the left-hand side of the equality is $\|\boldsymbol{\Lambda}_{n,\Pi_1} \otimes (\mathsf{P}_n\Pi_1) - \boldsymbol{\Lambda}_{n,\Pi_2} \otimes (\mathsf{P}_n\Pi_1)\|$. ■

There are other versions of the agreement of predictive distributions. For example, in Schervish (1995), p. 456, it is proved that the convergence in (7.77) holds true also for the $\mathbb{P}_{\Pi_1}$-a.s. convergence if a version for the predictive distribution is used that is defined via the densities Z and Z_n.

Up to this point we have studied the posterior and the predictive distribution in the Bayes model (7.74). Now we fix a parameter value, say θ_0, and assume that the data $X_1, X_2, ...$ are from the frequentist model

$$(\Omega, \mathfrak{F}, \mathbb{P}) := (\mathcal{X}^\infty, \mathfrak{A}^{\otimes\infty}, P_{\theta_0}^{\otimes\infty}). \tag{7.78}$$

We ask whether the sequence of posteriors concentrates more and more around the true value θ_0.

Definition 7.117. *Let Δ be a Polish space, $\mathfrak{B}_\Delta$ the σ-algebra of Borel sets, and $(\mathcal{X}, \mathfrak{A})$ a standard Borel space. Suppose (A3) is satisfied. For a prior Π on $(\Delta, \mathfrak{B}_\Delta)$ the sequence of posteriors $\boldsymbol{\Pi}_n$ is called strongly consistent at θ_0 if for every open set O that contains the point θ_0 it holds*

$$\boldsymbol{\Pi}_n(O|X_1, ..., X_n) \to 1, \quad P_{\theta_0}^{\otimes\infty}\text{-a.s.}$$

We assume that the prior Π has a density π with respect to some $\boldsymbol{\tau} \in \mathcal{M}^\sigma(\mathfrak{B}_\Delta)$, the model $(P_\theta)_{\theta\in\Delta}$ is dominated, and the family $(f_\theta)_{\theta\in\Delta}$ satisfies the condition (A5). Set $\mathbf{x}_n = (x_1, ..., x_n)$,

$$\mathsf{m}_n(\mathbf{x}_n) = \int \prod_{i=1}^n f_\theta(x_i)\pi(\theta)\boldsymbol{\tau}(d\theta),$$

$$\pi_n(\theta|\mathbf{x}_n) = \begin{cases} \frac{1}{\mathsf{m}_n(\mathbf{x}_n)} \prod_{i=1}^n f_\theta(x_i)\pi(\theta) & \text{if } \mathsf{m}_n(\mathbf{x}_n) > 0, \\ \pi(\theta) & \text{if } \mathsf{m}_n(\mathbf{x}_n) = 0, \end{cases} \tag{7.79}$$

$$\boldsymbol{\Pi}_n(B|\mathbf{x}_n) = \int_B \pi_n(\theta|\mathbf{x}_n)\boldsymbol{\tau}(d\theta), \quad B \in \mathfrak{B}_\Delta, \ \mathsf{P}_n\Pi\text{-a.s.} \tag{7.80}$$

Then $\boldsymbol{\Pi}_n$ is a version of the posterior distribution.

The subsequent result on the posterior robustness can be found in Ghosh and Ramamoorthi (2003), from where the proof has been taken. Suppose the priors Π_i, $i = 1, 2$, are absolutely continuous with respect to $\boldsymbol{\tau}$ with densities $\pi_i, i = 1, 2$. Suppose that the posterior densities $\pi_{i,n}$ are defined by (7.79), and the posterior distributions $\boldsymbol{\Pi}_{i,n}$, $i = 1, 2$, are defined by (7.80) with π_n replaced with $\pi_{i,n}$, $i = 1, 2$.

Theorem 7.118. *Assume that $\mathcal{X}$ and Δ are Polish spaces, endowed with the Borel σ-algebras $\mathfrak{A}$ and $\mathfrak{B}_\Delta$, and that (A5) is satisfied. Suppose $\Pi_i \ll \boldsymbol{\tau}$, $\pi_i = d\Pi_i/d\boldsymbol{\tau}$, $i = 1, 2$, are continuous and positive at θ_0. If the posterior distributions $\boldsymbol{\Pi}_{i,n}$, $i = 1, 2$, are both strongly consistent at θ_0, then*

$$\|\boldsymbol{\Pi}_{1,n}(\cdot|X_1, ..., X_n) - \boldsymbol{\Pi}_{2,n}(\cdot|X_1, ..., X_n)\| \to 0, \quad P_{\theta_0}^{\otimes\infty}\text{-a.s.}$$

Proof. Set $h_n(\theta, \omega) = \prod_{i=1}^n f_\theta(X_i(\omega))$ and put, for $A \in \mathfrak{B}_\Delta$,

$$\mu_{i,n}(A, \omega) = \int I_A(\theta)h_n(\theta, \omega)\Pi_i(d\theta).$$

Let $O_1 \supseteq O_2 \supseteq \cdots$ be a sequence of open balls with centers θ_0 and diameters tending to zero. Then there is a set $A \in \mathfrak{F}$ with $P_{\theta_0}^{\otimes\infty}(A) = 1$ so that for every $\omega \in A$ and a given $\eta > 1$ there exists some $n_0(\eta, m)$ with

$$\boldsymbol{\Pi}_{i,n}(O_m|X_1(\omega),...,X_n(\omega)) = \frac{\mu_{i,n}(O_m,\omega)}{\mu_{i,n}(\Delta,\omega)} \geq \frac{1}{\eta}$$

for every $n \geq n_0(\eta, m)$. The continuity of π_1 and π_2 yields the existence of m_0 such that

$$\frac{\alpha}{\eta} \leq \frac{\pi_1(\theta)}{\pi_2(\theta)} \leq \eta\alpha, \quad \theta \in O_{m_0},$$

where $\alpha = \pi_1(\theta_0)/\pi_2(\theta_0)$. Hence for every $n \geq n_0(\eta, m_0)$ and $A \in \mathfrak{B}_\Delta$

$$\frac{\alpha}{\eta} \leq \frac{\mu_{1,n}(A \cap O_{m_0},\omega)}{\mu_{2,n}(A \cap O_{m_0},\omega)} \leq \eta\alpha,$$

$$\frac{\mu_{1,n}(\Delta,\omega)}{\mu_{2,n}(\Delta,\omega)} = \frac{\mu_{1,n}(\Delta,\omega)}{\mu_{1,n}(O_{m_0},\omega)} \frac{\mu_{2,n}(O_{m_0},\omega)}{\mu_{2,n}(\Delta,\omega)} \frac{\mu_{1,n}(O_{m_0},\omega)}{\mu_{2,n}(O_{m_0},\omega)},$$

$$\frac{\alpha}{\eta^2} \leq \frac{\mu_{1,n}(\Delta,\omega)}{\mu_{2,n}(\Delta,\omega)} \leq \eta^2\alpha.$$

Therefore, and with $\mu_{i,n}(\Delta,\omega) = \mathsf{m}_{i,n}(\mathbf{X}_n(\omega))$,

$$\begin{aligned}
&\boldsymbol{\Pi}_{1,n}(A|X_1(\omega),...,X_n(\omega)) - \boldsymbol{\Pi}_{2,n}(A|X_1(\omega),...,X_n(\omega)) \\
&\leq \frac{\mu_{1,n}(A \cap O_{m_0},\omega)}{\mu_{1,n}(\Delta,\omega)} - \frac{\mu_{2,n}(A \cap O_{m_0},\omega)}{\mu_{2,n}(\Delta,\omega)} + \frac{\mu_{1,n}(\Delta \backslash O_{m_0},\omega)}{\mu_{1,n}(\Delta,\omega)} \\
&\leq (\eta^3 - 1)\frac{\mu_{2,n}(O_{m_0},\omega)}{\mu_{2,n}(\Delta,\omega)} + \frac{\mu_{1,n}(\Delta \backslash O_{m_0},\omega)}{\mu_{1,n}(\Delta,\omega)}.
\end{aligned}$$

Hence by the consistency of $\boldsymbol{\Pi}_{i,n}$, $i = 1, 2$,

$$\limsup_{n\to\infty} \sup_{A\in\mathfrak{B}_\Delta} (\boldsymbol{\Pi}_{1,n}(A|X_1(\omega),...,X_n(\omega)) - \boldsymbol{\Pi}_{2,n}(A|X_1(\omega),...,X_n(\omega))) \leq \eta^3 - 1.$$

To complete the proof we note that $\eta > 1$ was arbitrary and

$$\|\boldsymbol{\Pi}_{1,n} - \boldsymbol{\Pi}_{2,n}\| = 2 \sup_{A\in\mathfrak{B}_\Delta} (\boldsymbol{\Pi}_{1,n}(A) - \boldsymbol{\Pi}_{2,n}(A)).$$

■

Now we establish conditions that guarantee consistency. A first result in this direction goes back to Doob (1949). It says that for any prior Π the posterior distributions are consistent at Π-almost all $\theta \in \Delta$. Although this results says that consistency holds in almost all cases, the proof ensures only the existence of a subset Δ_0 of the parameter space with $\Pi(\Delta_0) = 1$ so that we have consistency for $\theta \in \Delta_0$. For details of the proof we refer to Ghosh and Ramamoorthi (2003). As Doob's result guarantees only the existence of Δ_0, but does not specify Δ_0, for a given θ one cannot decide whether the sequence of priors is consistent at θ. This problem was studied by Schwartz (1965). To present this result we recall the Kullback–Leibler distance, i.e., $\mathsf{K}(P,Q) = \int \ln(dP/dQ)dP$ if $P \ll Q$, and $\mathsf{K}(P,Q) = \infty$ else; see (1.81). Applying this to the product measures we get from Proposition A.29 that

$$\mathsf{K}(P_{\theta_0}^{\otimes n}, P_{\theta}^{\otimes n}) = n\mathsf{K}(P_{\theta_0}, P_\theta). \tag{7.81}$$

Subsequently, we often have some functions that are well defined, $\mathsf{P}_n\Pi$-a.s., but we want them to be well defined, $P_{\theta_0}^{\otimes n}$-a.s. This is certainly true if $P_{\theta_0}^{\otimes n} \ll \mathsf{P}_n\Pi$. The next problem gives a sufficient condition.

Problem 7.119.* If (A5) is satisfied and $\Pi(\{\theta : \mathsf{K}(P_{\theta_0}, P_\theta) < \infty\}) > 0$, then $P_{\theta_0}^{\otimes n} \ll \mathsf{P}_n\Pi$ for every n.

Given a family $(P_\theta)_{\theta\in\Delta}$ we introduce the *Kullback–Leibler neighborhood* of θ_0 by

$$\mathcal{K}_\varepsilon(\theta_0) = \{\theta : \mathsf{K}(P_{\theta_0}, P_\theta) < \varepsilon\}.$$

If condition (A5) is satisfied, then $\theta \mapsto \mathsf{K}(P_{\theta_0}, P_\theta) = \int \ln(f_{\theta_0}/f_\theta)dP_{\theta_0}$ is a measurable function so that $\mathcal{K}_\varepsilon(\theta_0) \in \mathfrak{B}_\Delta$. If Π is a prior, we say that θ_0 *belongs to the Kullback–Leibler support of* Π if $\Pi(\mathcal{K}_\varepsilon(\theta_0)) > 0$ for every $\varepsilon > 0$.

We express the condition for consistency with the help of the conditional densities π_n in (7.79). If θ_0 belongs to the Kullback–Leibler support of the prior Π, then by $\mathcal{L}(X_1, ..., X_n|P_{\theta_0}^{\otimes\infty}) = P_{\theta_0}^{\otimes n}$ and Problem 7.119 it holds

$$\boldsymbol{\Pi}_n(\Delta\backslash O|X_1, ..., X_n) = \frac{\int I_{\Delta\backslash O}(\theta)\prod_{i=1}^n f_\theta(X_i)\Pi(d\theta)}{\int \prod_{i=1}^n f_\theta(X_i)\Pi(d\theta)}, \quad P_{\theta_0}^{\otimes\infty}\text{-}a.s.$$

Thus it suffices to show that for every open set O that contains θ_0,

$$\frac{\int I_{\Delta\backslash O}(\theta)\prod_{i=1}^n (f_\theta(X_i)/f_{\theta_0}(X_i))\Pi(d\theta)}{\int \prod_{i=1}^n (f_\theta(X_i)/f_{\theta_0}(X_i))\Pi(d\theta)} \to 0 \quad P_{\theta_0}^{\otimes\infty}\text{-}a.s.$$

The idea is to find conditions so that, $P_{\theta_0}^{\otimes\infty}$-a.s., for every $\beta > 0$, and for some $\beta_0 > 0$,

$$\liminf_{n\to\infty} \exp\{n\beta\} \int \prod\nolimits_{i=1}^n \frac{f_\theta(X_i)}{f_{\theta_0}(X_i)}\Pi(d\theta) = \infty, \tag{7.82}$$

$$\limsup_{n\to\infty} \exp\{n\beta_0\} \int I_{\Delta\backslash O}(\theta) \prod\nolimits_{i=1}^n \frac{f_\theta(X_i)}{f_{\theta_0}(X_i)}\Pi(d\theta) = 0. \tag{7.83}$$

Lemma 7.120. *Suppose condition (A5) is fulfilled. If θ_0 is in the Kullback–Leibler support of the prior Π, then (7.82) holds.*

Proof. Suppose $\Pi(\mathcal{K}_\varepsilon(\theta_0)) > 0$, and set $\Pi_\varepsilon(B) = \Pi(B|\mathcal{K}_\varepsilon(\theta_0))$. For every $\theta \in \mathcal{K}_\varepsilon(\theta_0)$ it holds $\mathsf{K}(P_{\theta_0}, P_\theta) < \infty$ and thus $P_{\theta_0} \ll P_\theta$. Hence for $\mathsf{w}_1(t) = t\ln t - t + 1$,

$$\mathsf{K}(P_{\theta_0}, P_\theta) = \int \ln(\frac{f_{\theta_0}(x)}{f_\theta(x)})P_{\theta_0}(dx) = \int \mathsf{w}_1(\frac{f_{\theta_0}(x)}{f_\theta(x)})f_\theta(x)\boldsymbol{\mu}(dx).$$

As $\mathsf{w}_1 \geq 0$ the subsequent integrals are well defined, and by Fubini's theorem

$$\int [\int \mathsf{w}_1(\frac{f_{\theta_0}(x)}{f_\theta(x)}) f_\theta(x) \Pi_\varepsilon(d\theta)] \boldsymbol{\mu}(dx) = \int \mathsf{K}(P_{\theta_0}, P_\theta) \Pi_\varepsilon(d\theta) \le \varepsilon.$$

Hence,

$$-\infty < \int [\int \ln \frac{f_{\theta_0}(x)}{f_\theta(x)} \Pi_\varepsilon(d\theta)] P_{\theta_0}(dx) \le \varepsilon. \tag{7.84}$$

We consider the i.i.d. random variables

$$Y_i = \int (\ln \frac{f_{\theta_0}(X_i)}{f_\theta(X_i)}) \Pi_\varepsilon(d\theta), \quad i = 1, 2, ...$$

on the probability space (7.78). Then $\mathbb{E}Y_1 \le \varepsilon$ by (7.84). The strong law of large numbers provides the existence of an $\Omega_0 \in \mathfrak{F}$ with $P_{\theta_0}^{\otimes\infty}(\Omega_0) = 1$ and

$$\frac{1}{n} \sum\nolimits_{i=1}^{n} Y_i(\omega) \to \mathbb{E}Y_1 \le \varepsilon, \quad \omega \in \Omega_0.$$

Hence by the convexity of $\exp\{x\}$, Jensen's inequality, and Fatou's lemma,

$$\begin{aligned}
&\liminf_{n\to\infty} \exp\{2n\varepsilon\} \int \prod\nolimits_{i=1}^{n} \frac{f_\theta(X_i(\omega))}{f_{\theta_0}(X_i(\omega))} \Pi(d\theta) \\
&\ge \liminf_{n\to\infty} \int I_{\mathcal{K}_\varepsilon(\theta_0)}(\theta) \exp\{n[2\varepsilon - \frac{1}{n}\sum\nolimits_{i=1}^{n} \ln \frac{f_\theta(X_i(\omega))}{f_{\theta_0}(X_i(\omega))}]\} \Pi(d\theta) \\
&\ge \Pi(\mathcal{K}_\varepsilon(\theta_0)) \liminf_{n\to\infty} \exp\{n(2\varepsilon - \frac{1}{n}\sum\nolimits_{i=1}^{n} Y_i(\omega))\} = \infty.
\end{aligned}$$

■

For $\Pi(\Delta\backslash O) > 0$ we set $\Pi_{\Delta\backslash O}(B) = \Pi(B|\Delta\backslash O)$ and assume that for some $D_O > 0$ and $q_O < 1$,

$$\mathsf{H}_{1/2}(P_{\theta_0}^{\otimes n}, \mathsf{P}_n \Pi_{\Delta\backslash O}) \le D_O q_O^n. \tag{7.85}$$

Lemma 7.121. *If (A5) is satisfied, then the condition (7.85) implies the existence of some $\beta_0 > 0$ such that (7.83) holds.*

Proof. We have only to consider the case of $\Pi(\Delta\backslash O) > 0$. Then

$$\begin{aligned}
&\mathsf{H}_{1/2}(P_{\theta_0}^{\otimes n}, \mathsf{P}_n \Pi_{\Delta\backslash O}) \\
&= \int [\prod\nolimits_{i=1}^{n} f_{\theta_0}(x_i)]^{1/2} [\int \prod\nolimits_{i=1}^{n} f_\theta(x_i) \Pi_{\Delta\backslash O}(d\theta)]^{1/2} \boldsymbol{\mu}^{\otimes n}(dx_1, ..., dx_n) \\
&= \mathbb{E}g^{1/2}(X_1, ..., X_n) \le D_O q_O^n,
\end{aligned}$$

where $g(X_1, ..., X_n) = \int \prod_{i=1}^{n} (f_\theta(X_i)/f_{\theta_0}(X_i)) \Pi_{\Delta\backslash O}(d\theta)$. If $Z \ge 0$, then $\mathbb{P}(Z > 1) \le \mathbb{E}Z$. Hence for $r > 0$,

$$\mathbb{P}(r^{2n} g(X_1, ..., X_n) \ge 1) = \mathbb{P}(r^n g^{1/2}(X_1, ..., X_n) \ge 1) \le D_O (r q_O)^n.$$

Choose $1 < r < 1/q_O$. Then by Borel–Cantelli's lemma, with $P_{\theta_0}^{\otimes\infty}$-probability one, the inequality $r^{2n} g(X_1, ..., X_n) \geq 1$ holds only for a finite number of n. Hence we get (7.83) for every $\beta_0 > 0$ that satisfies $\exp\{\beta_0\} < r$. ■

The condition (7.85) implies, in view of Lemma 1.66 and the inequality $\min(a, b) \leq \sqrt{ab}$, that the minimal Bayes risk $\mathsf{b}_{1/2}(P_{\theta_0}^{\otimes n}, \mathsf{P}_n \Pi_{\Delta\backslash O})$ for testing $\mathsf{H}_0 : P_{\theta_0}^{\otimes n}$ versus $\mathsf{H}_A : \mathsf{P}_n \Pi_{\Delta\backslash O}$ tends to zero at an exponential rate.

Now we establish equivalent conditions that are sufficient for (7.85). In the proof we make use of a special case of Hoeffding's inequality. If $Z_1, ..., Z_n$ are independent and $|Z_i| \leq 1$, then

$$\mathbb{P}(\sum\nolimits_{i=1}^{n} (Z_i - \mathbb{E}Z_i) \geq t) \leq \exp\{-\frac{2t^2}{n}\}, \quad t \geq 0. \tag{7.86}$$

For a proof we refer to Hoeffding (1963). The next proposition is taken from Ghosh and Ramamoorthi (2003). It studies the behavior of error probabilities for the sequence of testing problems $\mathsf{H}_0 : P_{\theta_0}^{\otimes n}$ versus $\mathsf{H}_A : P_{\theta}^{\otimes n}$, $\theta \in \Delta\backslash O$.

Proposition 7.122. *Let O be an open set that contains θ_0. Then the following conditions are equivalent and imply (7.85).*

(A) There exists a sequence of tests φ_n that is uniformly consistent in the sense that

$$\int \varphi_n dP_{\theta_0}^{\otimes n} + \sup_{\theta \in \Delta\backslash O} \int (1 - \varphi_n) dP_{\theta}^{\otimes n} \to 0.$$

(B) For some m there exists a strictly unbiased test φ in the sense that

$$\int \varphi dP_{\theta_0}^{\otimes m} < \inf_{\theta \in \Delta\backslash O} \int \varphi dP_{\theta}^{\otimes m}.$$

(C) There exists a sequence of nonrandomized tests φ_n and $C, \beta > 0$ with

$$\int \varphi_n dP_{\theta_0}^{\otimes n} + \sup_{\theta \in \Delta\backslash O} \int (1 - \varphi_n) dP_{\theta}^{\otimes n} \leq C \exp\{-n\beta\}.$$

Proof. It is clear that (A) implies (B), and (C) implies (A). We show that (B) implies (C) and consider first the case $m = 1$. It holds $\alpha = \int \varphi dP_{\theta_0} < \gamma = \inf_{\theta \in \Delta\backslash O} \int \varphi dP_{\theta}$. Set

$$A_n = \{(x_1, ..., x_n) : \sum\nolimits_{i=1}^{n} (\varphi(x_i) - \alpha) > \frac{n}{2}(\gamma - \alpha)\}.$$

Then by Hoeffding's inequality (7.86) it follows that

$$P_{\theta_0}^{\otimes n}(A_n) \leq \exp\{-\frac{n^2(\gamma - \alpha)^2}{4n}\}.$$

On the other hand, for $\theta \in \Delta\backslash O$ by $\alpha < \int \varphi dP_{\theta}$ it follows that

$$P_\theta^{\otimes n}(A_n) \geq P_\theta^{\otimes n}(\sum_{i=1}^{n}[\varphi(X_i) - \int \varphi dP_\theta] > \frac{n}{2}(\alpha - \gamma)) \geq 1 - \exp\{-\frac{n^2(\gamma-\alpha)^2}{4n}\},$$

where the second inequality follows from an application of Hoeffding's inequality (7.86) to $\varphi(X_i)$. Thus $\varphi_n = I_{A_n}$ is the required sequence of tests.

Now we consider the case where m is arbitrary. Then we replace P_θ with $P_\theta^{\otimes m}$ to get a sequence of tests ψ_k, some $\widetilde{\beta} > 0$, and $\widetilde{C} > 0$ so that

$$\int \psi_k dP_{\theta_0}^{\otimes km} + \sup_{\theta \in \Delta \backslash O} \int (1-\psi_k) dP_\theta^{\otimes km} \leq \widetilde{C} \exp\{-k\widetilde{\beta}\}.$$

It remains to set $\varphi_n = \psi_k$ if $(k-1)m \leq n < km$. Indeed, as φ_n depends only on $x_1, ..., x_{(k-1)m}$ it follows that

$$\int \varphi_n dP_{\theta_0}^{\otimes n} + \sup_{\theta \in \Delta \backslash O} \int (1-\varphi_n) dP_\theta^{\otimes n} \leq \widetilde{C} \exp\{-(k-1)\widetilde{\beta}\} \leq C \exp\{-n\beta\},$$

where $C = \widetilde{C} \exp\{\widetilde{\beta}\}$ and $\beta = \widetilde{\beta}/m$. Finally we show that (C) implies the condition (7.85). Set $A_n = \{\varphi_n = 1\}$. Then by Schwarz' inequality

$$\begin{aligned}
&\mathsf{H}_{1/2}(P_{\theta_0}^{\otimes n}, \mathsf{P}_n \Pi_{\Delta \backslash O}) \\
&= \int (\textstyle\prod_{i=1}^{n} f_{\theta_0}(x_i))^{1/2} [\int \prod_{i=1}^{n} f_\theta(x_i) \Pi_{\Delta \backslash O}(d\theta)]^{1/2} \boldsymbol{\mu}^{\otimes n}(dx_1, ..., dx_n) \\
&\leq (P_{\theta_0}^{\otimes n}(A_n))^{1/2} (\mathsf{P}_n \Pi_{\Delta \backslash O}(A_n))^{1/2} \\
&\quad + (P_{\theta_0}^{\otimes n}(\mathcal{X}^n \backslash A_n))^{1/2} (\mathsf{P}_n \Pi_{\Delta \backslash O}(\mathcal{X}^n \backslash A_n))^{1/2}.
\end{aligned}$$

The inequality (7.85) follows from $P_{\theta_0}^{\otimes n}(A_n) \leq C \exp\{-n\beta\}$ and

$$\begin{aligned}
&(\mathsf{P}_n \Pi_{\Delta \backslash O})(\mathcal{X}^n \backslash A_n) \\
&\leq \frac{1}{\Pi(\Delta \backslash O)} \int I_{\Delta \backslash O}(\theta) P_\theta^{\otimes n}(A_n) \Pi(d\theta) \leq \frac{C}{\Pi(\Delta \backslash O)} \exp\{-n\beta\}.
\end{aligned}$$

■

Now we are ready to formulate the result of Schwartz (1965) on the consistency of posterior distributions.

Theorem 7.123. ***(Schwartz)*** *Let $\mathcal{X}$ and Δ be Polish spaces, endowed with the Borel σ-algebras $\mathfrak{A}$ and $\mathfrak{B}_\Delta$, respectively, and assume that condition (A5) is satisfied. Let Π be a prior on $(\Delta, \mathfrak{B}_\Delta)$. If θ_0 belongs to the Kullback–Leibler support of Π, and the condition (7.85) holds for every open set O that contains θ_0, then the sequence of posterior $\boldsymbol{\Pi}_n$ in (7.72) is strongly consistent.*

Proof. Apply Lemmas 7.120 and 7.121 with $\beta = \beta_0$. ■

Now we discuss the condition that θ_0 belongs to the Kullback–Leibler support of Π and present examples where condition (7.85), or one of the conditions in Lemma 7.122, is satisfied. Suppose that θ_0 belongs to the support of the prior Π, i.e., $\Pi(U_{\theta_0}) > 0$ for every open neighborhood of θ_0. If the function $\theta \mapsto \mathsf{K}(P_{\theta_0}, P_\theta)$ is continuous at θ_0, then $\mathsf{K}(P_{\theta_0}, P_{\theta_0}) = 0$ implies that θ_0 belongs to the Kullback–Leibler support of Π. If $\theta \mapsto \mathsf{K}(P_{\theta_0}, P_\theta)$ is not continuous, and instead of the condition that θ_0 belongs to the support of Π the stronger condition $\Pi(\{\theta_0\}) > 0$ holds, then θ_0 belongs again to the Kullback–Leibler support of Π. We give examples where $\theta \mapsto \mathsf{K}(P_{\theta_0}, P_\theta)$ is continuous, and one where it is not.

Example 7.124. Suppose Δ is a compact metric space and $\mathfrak{B}_\Delta$ is the σ-algebra of Borel sets. Suppose that $(P_\theta)_{\theta\in\Delta}$ is dominated, say by $\boldsymbol{\mu}$, the densities satisfy $f_\theta(x) > 0$, $x \in \mathcal{X}$, $\theta \in \Delta$, and $\theta \mapsto f_\theta(x)$ is continuous for every x. If

$$\int \sup_{\theta\in\Delta} |\ln f_{\theta_0}(x) - \ln f_\theta(x)|\, P_{\theta_0}(dx) < \infty,$$

then Lebesgue's theorem implies the continuity of $\theta \mapsto \mathsf{K}(P_{\theta_0}, P_\theta)$. If in addition θ_0 belongs to the support of the prior Π, then under (7.85) we have a situation where the sequence of posteriors is strongly consistent, and in view of Corollary 7.79 we have at the same time that every sequence of maximum likelihood estimators is strongly consistent.

Example 7.125. If $(P_\theta)_{\theta\in\Delta}$ is an exponential family with natural parameter $\theta \in \Delta$, then $\mathsf{K}(P_{\theta_0}, P_\theta)$ can be explicitly evaluated. By (1.6) and (1.23), for $\theta_0 \in \Delta^0$,

$$\begin{aligned}\mathsf{K}(P_{\theta_0}, P_\theta) &= \int \ln(dP_{\theta_0}/dP_\theta)dP_{\theta_0} = \int (\langle\theta_0 - \theta, T(x)\rangle - K(\theta_0) + K(\theta))P_{\theta_0}(dx)\\ &= K(\theta) - K(\theta_0) - \langle\theta - \theta_0, \nabla K(\theta_0)\rangle\,.\end{aligned}$$

For $\theta_0 \in \Delta^0$ the continuity of $\theta \mapsto \mathsf{K}(P_{\theta_0}, P_\theta)$ at θ_0 follows from Theorem 1.17.

Example 7.126. Consider the family of uniform distributions $\mathsf{U}(0,\theta)$ on $(0,\theta)$, where $\theta \in \Delta = (0,\infty)$. Fix $\theta_0 \in (0,\infty)$. For $\theta < \theta_0$ the distribution $\mathsf{U}(0,\theta_0)$ is not absolutely continuous with respect $\mathsf{U}(0,\theta)$ so that $\mathsf{K}(\mathsf{U}(0,\theta_0), \mathsf{U}(0,\theta)) = \infty$ and $\theta \mapsto \mathsf{K}(\mathsf{U}(0,\theta_0), \mathsf{U}(0,\theta))$ is not continuous at θ_0.

Now we give examples where the assumptions of Schwartz' theorem hold.

Example 7.127. We assume that Condition 7.82 is fulfilled. Using the notation from there we get with $O = \{\theta : \rho_\Delta(\theta,\theta_0) < \varepsilon\}$, $f_{n,\theta}(x) = \prod_{i=1}^n f_\theta(x_i)$, $x = (x_1,...,x_n)$, and $f_{n,C_{\varepsilon,i}}(x) = \sup_{\theta\in C_{\varepsilon,i}} f_{n,\theta}(x)$,

$$\begin{aligned}\mathsf{H}_{1/2}(P_{\theta_0}^{\otimes n}, \mathsf{P}_n\Pi_{\Delta\backslash O}) &= \int [f_{n,\theta_0}(x)]^{1/2} [\int f_{n,\theta}(x)\Pi_{\Delta\backslash O}(d\theta)]^{1/2}\boldsymbol{\mu}^{\otimes n}(dx)\\ &\le \int [f_{n,\theta_0}(x)]^{1/2} [\sum\nolimits_{i=1}^{N_\varepsilon} \Pi_{\Delta\backslash O}(C_{\varepsilon,i}) f_{n,C_{\varepsilon,i}}(x)]^{1/2}\boldsymbol{\mu}^{\otimes n}(dx).\end{aligned}$$

The inequality $(\sum_{i=1}^{N_\varepsilon} a_i)^{1/2} \le \sum_{i=1}^{N_\varepsilon} a_i^{1/2}$, $\Pi_{\Delta\backslash O}(C_{\varepsilon,i}) \le 1$, and (7.45) yield

$$\begin{aligned}\mathsf{H}_{1/2}(P_{\theta_0}^{\otimes n}, \mathsf{P}_n\Pi_{\Delta\backslash O}) &\le \sum\nolimits_{i=1}^{N_\varepsilon} \int [f_{n,\theta_0}(x)]^{1/2} [f_{n,C_{\varepsilon,i}}(x)]^{1/2}\boldsymbol{\mu}^{\otimes n}(dx)\\ &\le N_\varepsilon(\max_{1\le i\le N_\varepsilon} \int [f_{\theta_0}(x)]^{1/2} [\sup\nolimits_{\theta\in C_{\varepsilon,i}} f_\theta(x)]^{1/2}\boldsymbol{\mu}(dx))^n.\end{aligned}$$

Hence Condition 7.82 implies (7.85). If in addition θ_0 belongs to the Kullback–Leibler support of the prior Π, then we have a situation where the sequence of posteriors is strongly consistent, and in view of Theorem 7.84 we have at the same time that every sequence of maximum likelihood estimators is strongly consistent.

Now we give an example where the condition (B) in Proposition 7.122 is fulfilled.

Example 7.128. Let $(P_\theta)_{\theta\in(a,b)}$ be a one-parameter exponential family with natural parameter θ, and let $\theta_0 \in (a,b)$ and $\alpha \in (0,1)$ be fixed. Under the standard conditions (A1) and (A2) we show in Theorem 8.11 that the power function of the uniformly best unbiased level α test φ for testing $\mathsf{H}_0 : \theta = \theta_0$ versus $\mathsf{H}_A : \theta \neq \theta_0$ is strictly decreasing for $\theta < \theta_0$, strictly increasing for $\theta > \theta_0$, and equal to α at θ_0. This means that the condition (B) in Proposition 7.122 is satisfied.

Above we have given examples where both the strong consistency of the posterior distributions and the strong consistency of every sequence of MLEs are satisfied. One could get here the impression that one has in any case either consistency or inconsistency of both the posteriors and the MLE. This, however, is not true. There are examples where the MLE is consistent but the posteriors are inconsistent, and vice versa. For the detailed construction of these examples, and for references to other examples in literature, we refer to Ghosh and Ramamoorthi (2003) and Diaconis and Freedman (1986a,b). Finally, we refer to Strasser (1981a), who refined Perlman's necessary and sufficient conditions for the consistency of the MLE to show that under weak regularity conditions the consistency of the MLE implies the consistency of the Bayes estimator.

7.6 Asymptotic Distributions of Estimators

7.6.1 Asymptotic Distributions of M-Estimators

Approximation of the Criterion Function

After the consistency of a special type of estimators has been established the question is now what are good sequences of estimators. Consistency is a qualitative property that does not say anything about the rate of convergence. One way of specifying what good estimators are is based on the rate of convergence of $\mathbb{P}(\rho_\Delta(\widehat{\theta}_n, \theta) > \varepsilon)$ to zero. This sequence typically tends to zero at an exponential rate, so that $\limsup_{n\to\infty} \ln \mathbb{P}(\rho_\Delta(\widehat{\theta}_n, \theta) > \varepsilon)$ is an appropriate measure of the quality of estimators. The proofs of the corresponding results require techniques from the area of large deviations. We do not go into detail for results of this type which can be found in Bahadur (1960) and Kester and Kallenberg (1986). Instead of following this way we use a sequence of constants $c_n \to \infty$ and study the limit distribution of $c_n(\widehat{\theta}_n - \theta)$, provided it exists. A typical form of the constants c_n is $c_n = \sqrt{n}$, and the limit distribution is usually a normal distribution, say $\mathsf{N}(0, \Sigma)$, where Σ depends on the concrete sequence of estimators under consideration. As different types of estimators have different covariance matrices in the limit distribution comparisons of estimators can be made based on their associated covariance matrices of the limit distributions.

It proves useful to reformulate the weak convergence of distributions of statistics $S_n : \mathcal{X}_n \to_m \mathbb{R}^d$ directly with the help of the models $(\mathcal{X}_n, \mathfrak{A}_n, P_n)$. If Q is a distribution on $(\mathbb{R}^d, \mathfrak{B}_d)$, then we write $\mathcal{L}(S_n|P_n) \Rightarrow Q$ if

$$\lim_{n\to\infty} \mathsf{E}_{P_n}\varphi(S_n) = \int \varphi(x)Q(dx),$$

for every $\varphi \in \mathsf{C}_b(\mathbb{R}^d)$. We say that S_n converges P_n-stochastically to $a \in \mathbb{R}^d$ if $P_n(\|S_n - a\| > \varepsilon) \to 0$, $\varepsilon > 0$, and denote this by $S_n \to^{P_n} a$.

We recall the useful Landau and stochastic Landau symbols. If a_n and b_n are any sequences of real numbers, then we write $a_n = O(b_n)$ if $b_n \neq 0$ for all $n \geq n_0$ and a_n/b_n is bounded for $n \geq n_0$. Similarly $a_n = o(b_n)$ if $b_n \neq 0$ for all $n \geq n_0$ and a_n/b_n tends to zero as $n \to \infty$.

We call $S_n : \mathcal{X}_n \to_m \mathbb{R}^m$ stochastically bounded, if

$$\lim_{c\to\infty} \limsup_{n\to\infty} P_n(\|S_n\| > c) = 0. \tag{7.87}$$

We write $S_n = O_{P_n}(c_n)$ if $(1/c_n)S_n$ is stochastically bounded. We write $S_n = o_{P_n}(c_n)$ if $(1/c_n)S_n$ tends P_n-stochastically to zero. If S_n are random matrices, then we write $S_n = O_{P_n}(c_n)$ if all entries of S_n are $O_{P_n}(c_n)$ and use a similar convention for the other cases. We write, similarly to (7.34),

$$S_n = \mathbf{o}_{P_n}(0), \quad \text{if} \quad P_n(S_n \neq 0) \to 0, \text{ as } n \to \infty. \tag{7.88}$$

We do not use separate symbols for scalars, vectors, and matrices. Whether $O_{P_n}, o_{P_n}, \mathbf{o}_{P_n}$ is a scalar, vector, or matrix should be clear from the context. The next problem collects well-known rules for dealing with $O_{P_n}, o_{P_n}, \mathbf{o}_{P_n}$ symbols.

Problem 7.129.* If $T_n : \mathcal{X}_n \to_m \mathbb{R}^d$, $T : \mathcal{X} \to_m \mathbb{R}^d$, and $\mathcal{L}(T_n|P_n) \Rightarrow \mathcal{L}(T|P)$, then $T_n = O_{P_n}(1)$. It holds $O_{P_n}(1) + o_{P_n}(1) = O_{P_n}(1)$, $O_{P_n}(1)o_{P_n}(1) = o_{P_n}(1)$. If $S_n : \mathcal{X}_n \to_m \mathbb{R}^d$ and $A_n : \mathcal{X}_n \to_m \mathbb{R}^{kd}$ are random matrices, then $S_n = \mathbf{o}_{P_n}(0)$ implies $A_n S_n = \mathbf{o}_{P_n}(0)$.

Problem 7.130.* If $U_n : \mathcal{X}_n \to \mathbb{R}^d$ is stochastically bounded and $U_n = (\Sigma + o_{P_n}(1))V_n + o_{P_n}(1)$, where Σ is a nonsingular matrix, then $V_n = \Sigma^{-1}U_n + o_{P_n}(1)$.

The first question, of course, is which constants should be used when studying the limit distribution of $T_n = c_n(\widehat{\theta}_n - \theta_0)$. If the distributions of $c_n(\widehat{\theta}_n - \theta_0)$ converge, then this sequence is stochastically bounded. Given a sequence of models $(\mathcal{X}_n, \mathfrak{A}_n, (P_{n,\theta})_{\theta\in\Delta})$, where $\Delta \subseteq \mathbb{R}^d$ is a Borel set, and $c_n \to \infty$, we say a sequence of estimators $\widehat{\theta}_n : \mathcal{X}_n \to_m \Delta$ is c_n-*consistent* at θ_0 if

$$\lim_{c\to\infty}\limsup_{n\to\infty} P_{n,\theta_0}(c_n \parallel \widehat{\theta}_n - \theta_0 \parallel > c) = 0,$$

i.e., $c_n(\widehat{\theta}_n - \theta_0)$ is stochastically bounded with respect to P_{n,θ_0}. If this condition is satisfied for every θ_0, then we say $\widehat{\theta}_n$ is *c_n-consistent.* We call $\widehat{\theta}_n$ *uniformly c_n-consistent on K* if

$$\lim_{c\to\infty}\limsup_{n\to\infty}\sup_{\theta\in K} P_{n,\theta}(c_n \parallel \widehat{\theta}_n - \theta \parallel > c) = 0.$$

For regular models and uniformly consistent estimators the sequence c_n cannot tend to infinity faster than $\sqrt{n}$.

Proposition 7.131. *Let $\Delta \subseteq \mathbb{R}^d$, and suppose $(P_\theta)_{\theta\in\Delta}$ is $\mathbb{L}_2$-differentiable at $\theta_0 \in \Delta^0$. Suppose $O \subseteq \Delta^0$ is an open set that contains θ_0. Then for the sequence of models $(\mathcal{X}^n, \mathfrak{A}^{\otimes n}, (P_\theta^{\otimes n})_{\theta\in\Delta})$ there is no sequence of estimators that is uniformly c_n-consistent on O for any sequence c_n with $c_n/\sqrt{n} \to \infty$.*

Proof. We get from (1.134) that

$$\mathsf{H}_{1/2}(P_{\theta_0+u}, P_{\theta_0}) = 1 - \frac{1}{8}u^T \mathsf{I}(\theta_0)u + R(u) \quad \text{and} \quad \lim_{u\to 0} \|u\|^{-2} R(u) = 0.$$

Hence by (1.117)

$$\begin{aligned}\mathsf{H}_{1/2}(P^{\otimes n}_{\theta_0+h/\sqrt{n}}, P^{\otimes n}_{\theta_0}) &= (1 - \frac{1}{8n}h^T \mathsf{I}(\theta_0)h + R(h/\sqrt{n}))^n \\ &\to \exp\{-\frac{1}{8}h^T \mathsf{I}(\theta_0)h\}.\end{aligned} \tag{7.89}$$

Set

$$A_n = \{\parallel c_n(\widehat{\theta}_n - \theta_0) \parallel \geq \parallel c_n(\widehat{\theta}_n - \theta_0) - c_n h/\sqrt{n} \parallel\}.$$

Suppose there are $\widehat{\theta}_n$ and c_n with $c_n/\sqrt{n} \to \infty$ such that the sequence $T_n = c_n(\widehat{\theta}_n - \theta_0)$ is stochastically bounded under $P^{\otimes n}_{\theta_0}$. Then

$$P^{\otimes n}_{\theta_0}(A_n) = P^{\otimes n}_{\theta_0}\left(\|T_n\| \geq \left\|T_n - c_n h/\sqrt{n}\right\|\right) \to 0, \quad h \neq 0.$$

On the other hand, for all sufficiently large n it holds $\theta_0 + h/\sqrt{n} \in O$, and $c_n(\widehat{\theta}_n - (\theta_0 + h/\sqrt{n})) = T_n - c_n h/\sqrt{n}$ is, in view of the uniform consistency for $\theta \in O$, stochastically bounded under $P^{\otimes n}_{\theta_0+h/\sqrt{n}}$. Otherwise, $\|T_n\|$ tends stochastically to ∞ with respect to $P^{\otimes n}_{\theta_0+h/\sqrt{n}}$. Hence $P^{\otimes n}_{\theta_0+h/\sqrt{n}}(A_n) \to 1$. Using the inequality in Problem 1.79 we get

$$\begin{aligned}\mathsf{H}_{1/2}(P^{\otimes n}_{\theta_0+h/\sqrt{n}}, P^{\otimes n}_{\theta_0}) \leq &[P^{\otimes n}_{\theta_0}(A_n) P^{\otimes n}_{\theta_0+h/\sqrt{n}}(A_n)]^{1/2} \\ &+ [(1 - P^{\otimes n}_{\theta_0}(A_n))(1 - P^{\otimes n}_{\theta_0+h/\sqrt{n}}(A_n))]^{1/2} \to 0,\end{aligned}$$

which contradicts (7.89). ■

In nonregular models there may exist estimators that are n^{α}-consistent with $\alpha > 1/2$. We demonstrate this by an example. For a general treatment of such models we refer to Akahira and Takeuchi (1981) and Janssen and Mason (1990).

Example 7.132. Suppose $X_1, ..., X_n$ is a sample from a uniform distribution $\mathsf{U}(0,\theta)$ with $\theta > 0$. For $\theta > 1$ we consider $1 - \mathsf{H}_{1/2}(\mathsf{U}(0,\theta), \mathsf{U}(0,1)) = 1 - \theta^{-1/2}$. Comparing this with (1.134) we see that $(\mathsf{U}(0,\theta))_{\theta>0}$ is not $\mathbb{L}_2$-differentiable at $\theta_0 = 1$. Similarly it follows that $(\mathsf{U}(0,\theta))_{\theta>0}$ is not $\mathbb{L}_2$-differentiable at any $\theta_0 > 0$. We consider the estimator $\widehat{\theta}_n = \max(X_1, ..., X_n)$. It holds $F_n(t) = \mathbb{P}(\widehat{\theta}_n \le t) = F^n(t)$, where $F(t) = \theta^{-1} t I_{[0,\theta)}(t) + I_{[\theta,\infty)}(t)$. Hence

$$\mathbb{P}(n(\widehat{\theta}_n - \theta) \le t) = I_{[0,\infty)}(t) + I_{[-n\theta,0)}(t)(1 + \frac{t}{n\theta})^n \to \exp\{\frac{t}{\theta}\} I_{(-\infty,0)}(t) + I_{[0,\infty)}(t)$$

as $n \to \infty$. As the weak convergence of the distributions of $n(\widehat{\theta}_n - \theta)$ implies the stochastic boundedness we see that $\widehat{\theta}_n$ is n-consistent.

Suppose $\varrho : \Delta \times \mathcal{X} \to \mathbb{R}$ is a contrast function in the sense of Definition 7.58 which yields the criterion function

$$M_n(\theta, (x_1, ..., x_n)) = \frac{1}{n} \sum\nolimits_{i=1}^{n} \varrho_\theta(x_i).$$

To establish the asymptotic distribution of a consistent minimizer there are several approaches. One way to get the asymptotic distribution of the minimizer is to study $(M_n(\theta))_{\theta\in\Delta}$ as a stochastic process. Then the technique of local parameters proves useful. We set

$$W_n(h) = \sum\nolimits_{i=1}^{n} (\varrho_{\theta_0 + h/\sqrt{n}}(X_i) - \varrho_{\theta_0}(X_i)), \tag{7.90}$$

and minimize $W_n(h)$ over h. The minimizers obtained by minimizing M_n and W_n are related by $\widehat{h}_n = \sqrt{n}(\widehat{\theta}_n - \theta_0)$. Let $\dot{\varrho}_{\theta_0} := \nabla \varrho_{\theta_0}$ be the gradient (a column vector) and $\ddot{\varrho}_{\theta_0} = \nabla\nabla^T \varrho_{\theta_0}$ be the matrix of the second derivatives, where the derivatives refer to the components of θ. A formal Taylor expansion gives

$$W_n(h) = \frac{1}{\sqrt{n}} \sum\nolimits_{i=1}^{n} h^T \dot{\varrho}_{\theta_0}(X_i) + \frac{1}{2n} \sum\nolimits_{i=1}^{n} h^T \ddot{\varrho}_{\theta_0}(X_i) h + R_n(h).$$

If we could neglect the remainder $R_n(h)$, then we would have to minimize only a quadratic function. The difficulty comes with the remainder term which we want to be uniformly small in $h \in \mathbb{R}^d$, but that is hard to prove. One needs maximal inequalities and techniques from the theory of large deviations. For details we refer to Ibragimov and Has'minskii (1981). The above-mentioned difficulties disappear when the W_n are convex stochastic processes. For such processes Hjort and Pollard (1993) obtained the following result.

Theorem 7.133. *Let $(W_n(h))_{h\in\mathbb{R}^d}$ be a sequence of convex stochastic processes on $(\Omega, \mathfrak{F}, \mathbb{P})$. If there exists a sequence of random vectors S_n with $\mathcal{L}(S_n) \Rightarrow \mathsf{N}(0, \Sigma_0)$, where $\det(\Sigma_0) \neq 0$, and a nonsingular symmetric matrix Σ_1 such that*

$$W_n(h) - h^T S_n \to^{\mathbb{P}} \frac{1}{2} h^T \Sigma_1 h, \quad h \in \mathbb{R}^d, \tag{7.91}$$

then for every sequence $\widehat{h}_n$ of minimizers of W_n it holds

$$\mathcal{L}(\widehat{h}_n|\mathbb{P}) \Rightarrow \mathsf{N}(0, \Sigma_1^{-1}\Sigma_0\Sigma_1^{-1}).$$

Corollary 7.134. *Suppose ϱ_θ is convex in $\theta \in \mathbb{R}^d$, $W_n(h)$, $h \in \mathbb{R}^d$, is defined in (7.90), and $\widehat{\theta}_n$ is a minimizer of $M_n(\theta) = (1/n)\sum_{i=1}^n \varrho_\theta(X_i)$. If W_n satisfies the conditions of the theorem, then*

$$\mathcal{L}(\sqrt{n}(\widehat{\theta}_n - \theta_0)|P_{\theta_0}^{\otimes n}) \Rightarrow \mathsf{N}(0, \Sigma_1^{-1}\Sigma_0\Sigma_1^{-1}).$$

Proof. The stochastic processes $\widetilde{W}_n(h) = W_n(h) - h^T S_n$, $h \in \mathbb{R}^d$, are convex, and by assumption $\widetilde{W}_n(h) \to^{\mathbb{P}} \frac{1}{2}h^T\Sigma_1 h$. Set $V_n(h) = h^T S_n + \frac{1}{2}h^T\Sigma_1 h$. Then by Lemma 7.75 it holds for every $c > 0$,

$$\sup_{\|h\|\leq c} |W_n(h) - V_n(h)| = \sup_{\|h\|\leq c} |\, \widetilde{W}_n(h) - \frac{1}{2}h^T\Sigma_1 h \,| \to^{\mathbb{P}} 0. \tag{7.92}$$

Note that $U_n = -\Sigma_1^{-1}S_n$ is the minimizer of $V_n(h)$ over $h \in \mathbb{R}^d$. Set

$$\Delta_n = \sup_{h\in\mathbb{R}^d, \|h-U_n\|\leq\varepsilon} |W_n(h) - V_n(h)|\,, \quad \delta_n = \inf_{h\in\mathbb{R}^d, \|h-U_n\|=\varepsilon} V_n(h) - V_n(U_n).$$

Then

$$\begin{aligned}\delta_n = \inf\nolimits_{h\in\mathbb{R}^d,\|h\|=\varepsilon}[(U_n+h)^T S_n + \frac{1}{2}(U_n+h)^T\Sigma_1(U_n+h) - U_n^T S_n \\ - \frac{1}{2}U_n^T\Sigma_1 U_n] = \frac{1}{2}\inf\nolimits_{h\in\mathbb{R}^d,\|h\|=\varepsilon} h^T\Sigma_1 h = \frac{1}{2}\lambda_{\min}\varepsilon^2,\end{aligned}$$

where $\lambda_{\min} > 0$ is the smallest eigenvalue of Σ_1. We apply Lemma 7.76 to the convex processes W_n and V_n with $f = W_n$ and $g = V_n$ to get

$$\mathbb{P}(\|\, \widehat{h}_n - U_n \,\| > \varepsilon) \leq \mathbb{P}(\Delta_n \geq \frac{1}{2}\delta_n) = \mathbb{P}(\Delta_n \geq \frac{1}{4}\lambda_{\min}\varepsilon^2),$$

where $\widehat{h}_n$ is a minimizer of $W_n(h)$. The statement follows from Slutsky's lemma (see Lemma A.46) and $\mathcal{L}(S_n) \Rightarrow \mathsf{N}(0, \Sigma_0)$. To prove the corollary we remark that for every minimizer $\widehat{\theta}_n$ of $M_n(\theta)$ the function $W_n(h)$ in (7.90) is minimized by $\widehat{h}_n = \sqrt{n}(\widehat{\theta}_n - \theta_0)$. ■

To apply Theorem 7.133 to linear regression models with a convex criterion function we study first the location model. Suppose $\varrho : \mathbb{R} \to \mathbb{R}$ is a convex function and assume that condition (7.54) is satisfied. We study the behavior of the function

$$\mathsf{w}_\varrho(a) = \mathbb{E}(\varrho(Z+a) - \varrho(Z) - a(D^+\varrho(Z))) = \mathsf{v}_\varrho(a) - a\mathbb{E}(D^+\varrho(Z))$$

in a neighborhood of $a = 0$. As in view of Problem 7.104 $D^+\mathsf{v}_\varrho(0) = \mathbb{E}D^+\varrho(Z)$ we have $D^+\mathsf{w}_\varrho(0) = 0$. The definition of $\gamma_{\mathsf{v}_\varrho}$ in (7.57) yields $\gamma_{\mathsf{v}_\varrho} = \gamma_{\mathsf{w}_\varrho}$. Using the generalized Taylor formula in Lemma 1.52 we arrive at

$$\mathsf{w}_\varrho(a) = \begin{cases} \int (a-u) I_{(0,a]}(u)\gamma_{\mathsf{v}_\varrho}(du), & a > 0, \\ \int (u-a) I_{(a,0]}(u)\gamma_{\mathsf{v}_\varrho}(du), & a < 0. \end{cases} \tag{7.93}$$

Lemma 7.135. *Let $\varrho : \mathbb{R} \to \mathbb{R}$ be a convex function, and assume that condition (7.54) is satisfied. If for some $\delta_0 > 0$,*

$$\gamma_{\mathsf{v}_\varrho}(\cdot \cap (-\delta_0, \delta_0)) \ll \boldsymbol{\lambda}, \quad g = \frac{d\gamma_{\mathsf{v}_\varrho}(\cdot \cap (-\delta_0, \delta_0))}{d\boldsymbol{\lambda}}, \tag{7.94}$$

and $g(t)$ is continuous at $t = 0$, then $\lim_{a\to 0} \mathsf{w}_\varrho(a)/a^2 = \frac{1}{2}g(0)$.

Proof. If $a > 0$, then by (7.93)

$$\frac{1}{a^2}\mathsf{w}_\varrho(a) - \frac{1}{2}g(0) = \frac{1}{a^2}\int (a-u)I_{(0,a]}(u)(g(u)-g(0))du \to \frac{1}{2}g(0),$$

where the last statement follows from the continuity of g at $u = 0$. The case $a < 0$ is similar. ■

Now we consider a linear regression model under the following conditions.

$$\begin{aligned} &Y_i = \theta_0^T x_i + \varepsilon_i, \quad i = 1, ..., n, \quad \text{where } \varepsilon_1, ..., \varepsilon_n \text{ are i.i.d.}, \\ &x_1, ..., x_n \in \mathbb{R}^d, \ \|x_i\| \le C, \quad \text{and } \Sigma_n := \tfrac{1}{n}\textstyle\sum_{i=1}^n x_i x_i^T \to \Sigma. \end{aligned} \tag{7.95}$$

Set $\mathbf{Y}_n = (Y_1, ..., Y_n)$. The subsequent result is closely related to the results in Jurečková and Sen (1996), Chapter 5, and Liese and Vajda (2003b, 2004).

Theorem 7.136. *Let $\varrho : \mathbb{R} \to \mathbb{R}$ be convex with $\mathbb{E}\,|\varrho(\varepsilon_i + a) - \varrho(\varepsilon_i)| < \infty$, $a \in \mathbb{R}$, $\mathbb{E}D^+\varrho(\varepsilon_i) = 0$, and $\sigma^2 = \mathbb{E}(D^+\varrho(\varepsilon_i))^2 < \infty$. Suppose that (7.94) is satisfied, g is continuous in $(-\delta_0, \delta_0)$, $g(0) > 0$, and*

$$R(a) := \mathbb{E}(\varrho(\varepsilon_1 - a) - \varrho(\varepsilon_1) + aD^+\varrho(\varepsilon_1))^2 = o(a^2). \tag{7.96}$$

Assume that (7.95) is fulfilled and Σ is nonsingular. If $\widehat{\theta}_n : (\mathbb{R}\times\mathbb{R}^d)^n \to_m \mathbb{R}^d$ minimizes $M_n(\theta) = (1/n)\sum_{i=1}^n \varrho(Y_i - \theta^T x_i)$, then

$$\mathcal{L}(\sqrt{n}(\widehat{\theta}_n(\mathbf{Y}_n) - \theta_0)) \Rightarrow \mathsf{N}(0, \frac{\sigma^2}{g^2(0)}\Sigma^{-1}).$$

Proof. Set

$$W_n(h) = \sum_{i=1}^{n} [\varrho(Y_i - (\theta_0 + n^{-1/2}h)^T x_i) - \varrho(Y_i - \theta_0^T x_i)]$$
$$= \sum_{i=1}^{n} [\varrho(\varepsilon_i - n^{-1/2}h^T x_i) - \varrho(\varepsilon_i)],$$

and define S_n in (7.91) to be $S_n = -n^{-1/2}\sum_{i=1}^{n} D^+\varrho(\varepsilon_i)x_i$. Note that $\mathbb{E}D^+\varrho(\varepsilon_i)x_i = 0$ follows from $\mathbb{E}D^+\varrho(\varepsilon_i) = 0$. Then

$$W_n(h) - h^T S_n = \sum_{i=1}^{n} V_i(h), \quad \text{where}$$
$$V_i(h) = \varrho(\varepsilon_i - n^{-1/2}h^T x_i) - \varrho(\varepsilon_i) + n^{-1/2}D^+\varrho(\varepsilon_i)h^T x_i.$$

By the independence of $\varepsilon_1, ..., \varepsilon_n$,

$$\mathbb{E}\sum_{i=1}^{n}(V_i(h) - \mathbb{E}V_i(h))^2 = \sum_{i=1}^{n} \mathbb{V}(V_i(h)) \leq \sum_{i=1}^{n} \mathbb{E}(V_i(h))^2$$
$$= \frac{1}{n}\sum_{i=1}^{n} nR(h^T x_i/\sqrt{n}).$$

As $\|x_i\| \leq C$ condition (7.96) implies $\lim_{n\to\infty} \max_{1\leq i\leq n} nR(h^T x_i/\sqrt{n}) = 0$. Hence $\sum_{i=1}^{n}(V_i(h) - \mathbb{E}V_i(h)) \to^{\mathbb{P}} 0$. Set

$$r(\delta) = \sup_{|a|\leq\delta} \frac{1}{a^2} \mid \mathsf{w}_\varrho(a) - \frac{a^2}{2}g(0) \mid .$$

Then

$$\mid \sum_{i=1}^{n} \mathbb{E}V_i(h) - \frac{1}{n}\sum_{i=1}^{n}(h^T x_i)^2 \frac{1}{2}g(0) \mid \leq \frac{1}{n}\sum_{i=1}^{n}(h^T x_i)^2 r(\|x_i\| \, \|h\| / \sqrt{n})$$
$$\leq r(C\|h\|/\sqrt{n})h^T(\frac{1}{n}\sum_{i=1}^{n}(x_i x_i^T))h \to 0,$$

as $(1/n)\sum_{i=1}^{n}(x_i x_i^T) \to \Sigma$. Hence we have obtained

$$W_n(h) + n^{-1/2}\sum_{i=1}^{n} D^+\varrho(\varepsilon_i)h^T x_i \to^{\mathbb{P}} \frac{1}{2}g(0)h^T \Sigma h.$$

The sequence of random variables $(1/\sqrt{n})D^+\varrho(\varepsilon_i)h^T x_i$ satisfies the Lindeberg condition, which follows from Problem 6.82. As $\mathbb{E}D^+\varrho(\varepsilon_i) = 0$, and the variance of $(1/\sqrt{n})\sum_{i=1}^{n} D^+\varrho(\varepsilon_i)h^T x_i$ is $\sigma^2 h^T(\Sigma_n)h \to \sigma^2 h^T \Sigma h$, we obtain

$$\mathcal{L}(n^{-1/2}\sum_{i=1}^{n} D^+\varrho(\varepsilon_i)h^T x_i) \Rightarrow \mathsf{N}(0, \sigma^2 h^T \Sigma h),$$

so that by the Cramér–Wold device

$$\mathcal{L}(n^{-1/2}\sum_{i=1}^{n} D^+\varrho(\varepsilon_i)x_i) \Rightarrow \mathsf{N}(0, \sigma^2 \Sigma).$$

It remains to apply Theorem 7.133 with $\Sigma_0 = \sigma^2\Sigma$ and $\Sigma_1 = g(0)\Sigma$. ■

Problem 7.137.* If the convex function ϱ is twice continuously differentiable, $\ddot{\varrho}(x) > 0, x \in \mathbb{R}$, and

$$\mathbb{E} \sup_{|a| \le \delta_0} |\ddot{\varrho}(\varepsilon_1 + a)|^2 < \infty \tag{7.97}$$

for some $\delta_0 > 0$, then the conditions (7.94) and (7.96) are fulfilled.

Problem 7.138.* If $\varrho(t) = \tau_\alpha(t)$ is from Example 7.107 and the distribution P of ε_1 has a Lebesgue density f that is continuous in a neighborhood of 0 with $f(0) > 0$, then the conditions (7.94) and (7.96) are fulfilled.

Example 7.139. This is a continuation of Example 7.108. Let $X_1, ..., X_n$ be i.i.d. with common distribution P and c.d.f. F. Let u_α be an α-quantile of F; that is, $F(u_\alpha - 0) \le \alpha \le F(u_\alpha)$. Assume that P has a Lebesgue density f which is positive at u_α and continuous in a neighborhood of u_α. Let $\widehat{\theta}_n$ be a sample α-quantile; that is, $\widehat{F}_n(\widehat{\theta}_n - 0) \le \alpha \le \widehat{F}_n(\widehat{\theta}_n)$, where $\widehat{F}_n(t) = (1/n)\sum_{i=1}^n I_{(-\infty,t]}(X_i)$. By Example 7.107 $\widehat{\theta}_n$ is a minimizer of

$$M_n(\theta) = \frac{1}{n}\sum\nolimits_{i=1}^n \tau_\alpha(X_i - \theta).$$

The assumptions on f imply (7.62). Hence $\widehat{\theta}_n \to u_\alpha$, $\mathbb{P}$-a.s., by Example 7.108. The c.d.f. F is continuous at u_α and it holds $F(u_\alpha) = \alpha$. Hence

$$\begin{aligned} \mathbb{E}D^+\tau_\alpha(X_1) &= -(1-\alpha)\mathbb{P}(X_1 < u_\alpha) + \alpha\mathbb{P}(X_1 \ge u_\alpha) = 0, \\ \mathbb{E}(D^+\tau_\alpha(X_1 - u_\alpha))^2 &= (1-\alpha)^2\alpha + \alpha^2(1-\alpha) = \alpha(1-\alpha). \end{aligned}$$

We set $\varepsilon_i = X_i - u_\alpha$. Then the ε_i have the density $g(t) = f(t + u_\alpha)$. As $D^+\tau_\alpha(t) = -(1-\alpha)I_{(-\infty,0)}(t) + \alpha I_{[0,\infty)}(t)$ the curvature measure γ_{τ_α} in (7.57) is given by $\gamma_{\tau_\alpha} = \delta_0$, so that by Problem 7.104 it holds $\gamma_{\mathsf{v}_{\tau_\alpha}} = Q$, where $Q = \mathcal{L}(\varepsilon_i)$. As Q has the Lebesgue density g we get from Problem 7.138 that (7.94) and (7.96) are satisfied. We consider the location model $Y_i = \theta + \varepsilon_i$ as a special regression model with $x_i = 1$ and obtain from Theorem 7.136, with $\sigma^2 = \alpha(1-\alpha)$, the well-known asymptotic normality of the sample α-quantiles,

$$\mathcal{L}(\sqrt{n}(\widehat{\theta}_n - u_\alpha)|P^{\otimes n}) \Rightarrow \mathsf{N}(0, \alpha(1-\alpha)/f^2(u_\alpha)).$$

We conclude this section with the remark that the technique for proving Theorem 7.136 also works if the function ϱ is not necessarily convex, but of locally bounded variation. The advantage of this approach is that typical functions of robust statistics which have at some points only one-sided derivatives are included. But in contrast to the case of a convex function ϱ one needs additional conditions to guarantee the consistency. For details we refer to Liese and Vajda (2003b, 2004).

Asymptotic Solution of M-Equation

As before at some places in previous chapters subsequently we need regularity conditions that guarantee a Taylor expansion. If A is a $k \times m$ matrix with entries $a_{i,j}$ we set $\|A\| = \sup_{\|x\| \le 1} \|Ax\|$. Then obviously $\|Ax\| \le \|A\| \, \|x\|$.

Suppose $O \subseteq \mathbb{R}^d$ is open. We denote by $\mathsf{C}^{(k)}(O)$ the class of all k times continuously differentiable functions $\psi : O \to \mathbb{R}$. For the space of continuous functions we simply write also $\mathsf{C}(O)$ instead of $\mathsf{C}^{(0)}(O)$. Let $(\mathcal{X}, \mathfrak{A})$ be a measurable space, and let $\mathsf{C}_m^{(k)}(O, \mathcal{X})$ be defined as follows.

$$\begin{aligned}\mathsf{C}_m^{(k)}(O, \mathcal{X}) = \{\psi : \psi : O \times \mathcal{X} \to \mathbb{R},\\ \theta \mapsto \psi_\theta(x) \text{ belongs to } \mathsf{C}^{(k)}(O), \text{ for every } x \in \mathcal{X},\\ x \mapsto \psi_\theta(x) \text{ is measurable for every } \theta \in O\}.\end{aligned} \tag{7.98}$$

We note that every $\psi \in \mathsf{C}_m^{(0)}(O, \mathcal{X})$ is a measurable function of (θ, x). This follows from Problem 1.116. If $\psi : O \to \mathbb{R}^l$, then $\psi \in \mathsf{C}_m^{(k)}(O, \mathcal{X})$ means that $\psi = (\psi_1, ..., \psi_l)^T$ and $\psi_i \in \mathsf{C}_m^{(k)}(O, \mathcal{X})$, $i = 1, ..., l$. Then $\dot{\psi} = J_\psi^T$, where J_ψ is the Jacobian $J_\psi(\theta) = ((\partial\psi_i/\partial\theta_j))_{1 \le i \le l, 1 \le j \le d}$, is the derivative of ψ. The columns of $\dot{\psi}_\theta$ are $\nabla\psi_i$, i.e., the gradient of ψ_i, $i = 1, ..., l$. All derivatives refer to the components of $\theta = (\theta_1, ..., \theta_d)$. With these notations we see that the first-order Taylor expansion in Theorem A.2 can be written as

$$\psi_{\theta+h} - \psi_\theta = \int_0^1 \dot{\psi}_{\theta+sh}^T h ds, \quad \theta, \theta + h \in O. \tag{7.99}$$

(A9) Given a model $(\mathcal{X}, \mathfrak{A}, (P_\theta)_{\theta\in\Delta})$ with $\Delta \subseteq \mathbb{R}^d$, we assume that $\psi_\theta(x)$, $\theta \in \Delta$, $x \in \mathcal{X}$, satisfies for some open neighborhood $U(\theta_0)$ of $\theta_0 \in \Delta^0$

(A) $\psi \in \mathsf{C}_m^{(1)}(U(\theta_0), \mathcal{X})$.

(B) $\mathsf{E}_{\theta_0}(\sup_{\theta\in U(\theta_0)} \| \dot{\psi}_\theta \|) < \infty$.

(C) $\mathsf{E}_{\theta_0} \| \psi_{\theta_0} \|^2 < \infty$.

A consequence of these regularity conditions is that differentiation and expectation can be exchanged.

Problem 7.140.* If $\psi \in \mathsf{C}_m^{(1)}(U(\theta_0), \mathcal{X})$ and condition (B) in (A9) is satisfied, then

$$\nabla(\mathsf{E}_{\theta_0}\psi_\theta) = \mathsf{E}_{\theta_0}\dot{\psi}_\theta, \quad \theta \in U(\theta_0).$$

The following statement on the remainder term in a stochastic Taylor expansion is used subsequently at several places. For a set $\Delta \subseteq \mathbb{R}^d$ we consider the sequence of models $(\mathcal{X}^n, \mathfrak{A}^{\otimes n}, (P_\theta^{\otimes n})_{\theta\in\Delta})$, and denote by $X_1, ..., X_n$ the projections which are i.i.d. with common distribution P_θ under $P_\theta^{\otimes n}$.

Lemma 7.141. *Let $\Delta \subseteq \mathbb{R}^d$ be a Borel set and $\widehat{\theta}_n : \mathcal{X}^n \to_m \Delta$ with $\widehat{\theta}_n \to^{P_{\theta_0}^{\otimes n}} \theta_0$. If for some open neighborhood $U(\theta_0)$ of $\theta_0 \in \Delta^0$ the function $\varphi : \Delta \times \mathcal{X} \to \mathbb{R}^k$ satisfies $\varphi \in \mathsf{C}_m(U(\theta_0), \mathcal{X})$, and it holds $\mathsf{E}_{\theta_0} \sup_{\theta\in U(\theta_0)} \|\varphi_\theta\| < \infty$, then*

$$S_n = \frac{1}{n}\sum\nolimits_{i=1}^n \varphi_{\widehat{\theta}_n}(X_i) \to^{P_{\theta_0}^{\otimes n}} \mathsf{E}_{\theta_0}\varphi_{\theta_0}.$$

If $\psi : \Delta \times \mathcal{X} \to \mathbb{R}^d$ satisfies (A) and (B) in (A9), then

$$T_n = \frac{1}{n}\sum\nolimits_{i=1}^{n} \int_0^1 (\dot{\psi}_{\theta_0+s(\widehat{\theta}_n-\theta_0)}(X_i) - \dot{\psi}_{\theta_0}(X_i))ds \to^{P_{\theta_0}^{\otimes n}} 0.$$

Proof. Let $\delta > 0$ be so small that $\{\|\theta - \theta_0\| < \delta\} \subseteq U(\theta_0)$ and set $A_n = \{\| \widehat{\theta}_n - \theta_0 \|< \delta\}$. Then

$$\begin{aligned}\|S_n - \mathsf{E}_{\theta_0}\varphi_{\theta_0}\|\, I_{A_n} &\le I_{A_n}\frac{1}{n}\sum\nolimits_{i=1}^{n} \sup\nolimits_{\|\theta-\theta_0\|<\delta} \|\varphi_\theta(X_i) - \varphi_{\theta_0}(X_i)\| \\ &+ \| \frac{1}{n}\sum\nolimits_{i=1}^{n} \varphi_{\theta_0}(X_i) - \mathsf{E}_{\theta_0}\varphi_{\theta_0} \|= S_{1,n}(\delta) + S_{2,n}.\end{aligned}$$

To deal with $S_{1,n}(\delta)$ we note that

$$\begin{aligned}&\lim_{\delta\downarrow 0}\limsup_{n\to\infty} \mathsf{E}_{P_{\theta_0}^{\otimes n}} S_{1,n}(\delta) \\ &\le \lim_{\delta\downarrow 0} \mathsf{E}_{P_{\theta_0}} \sup\nolimits_{\|\theta-\theta_0\|<\delta} \|\varphi_\theta(X_1) - \varphi_{\theta_0}(X_1)\| = 0\end{aligned} \tag{7.100}$$

by Lebesgue's theorem and the continuity of $\theta \mapsto \varphi_\theta$. Hence

$$\begin{aligned}P_{\theta_0}^{\otimes n}(\|S_n - \mathsf{E}_{\theta_0}\varphi_{\theta_0}\| > \varepsilon) &\le P_{\theta_0}^{\otimes n}(\overline{A}_n) + P_{\theta_0}^{\otimes n}(S_{1,n}(\delta) + S_{2,n} > \varepsilon) \\ &\le P_{\theta_0}^{\otimes n}(\overline{A}_n) + P_{\theta_0}^{\otimes n}(S_{2,n} > \frac{\varepsilon}{2}) + P_{\theta_0}^{\otimes n}(S_{1,n}(\delta) > \frac{\varepsilon}{2}) \\ &\le P_{\theta_0}^{\otimes n}(\overline{A}_n) + P_{\theta_0}^{\otimes n}(S_{2,n} > \frac{\varepsilon}{2}) + \frac{2}{\varepsilon}\mathsf{E}_{P_{\theta_0}^{\otimes n}} S_{1,n}(\delta).\end{aligned}$$

It holds $P_{\theta_0}^{\otimes n}(\overline{A}_n) \to 0$ by the consistency of $\widehat{\theta}_n$. The relation $P_{\theta_0}^{\otimes n}(S_{2,n} > \varepsilon/2) \to 0$ follows from the law of large numbers. Hence

$$\limsup_{n\to\infty} P_{\theta_0}^{\otimes n}(\|S_n - \mathsf{E}_{\theta_0}\varphi_{\theta_0}\| > \varepsilon) \le \frac{2}{\varepsilon}\mathsf{E}_{\theta_0} \sup\nolimits_{\|\theta-\theta_0\|<\delta} \|\varphi_\theta(X_1) - \varphi_{\theta_0}(X_1)\|\,.$$

Letting $\delta \to 0$ we get the first statement from (7.100). The proof of the second statement is similar. It holds

$$\|T_n\|\, I_{A_n} \le I_{A_n}\frac{1}{n}\sum\nolimits_{i=1}^{n} \sup_{\|\theta-\theta_0\|<\delta} \| \dot{\psi}_\theta(X_i) - \dot{\psi}_{\theta_0}(X_i) \|= T_n(\delta).$$

Condition (A), Lebesgue's theorem, and the continuity of $\theta \mapsto \dot{\psi}_\theta$ yield

$$\lim_{\delta\downarrow 0}\limsup_{n\to\infty} \mathsf{E}_{P_{\theta_0}^{\otimes n}} T_n(\delta) \le \lim_{\delta\downarrow 0} \mathsf{E}_{\theta_0} \sup_{\|\theta-\theta_0\|<\delta} \| \dot{\psi}_\theta(X_1) - \dot{\psi}_{\theta_0}(X_1) \|= 0.$$

The rest of the proof is similar to that of the first statement. ■

If $\widehat{\theta}_n : \mathcal{X}^n \to_m \Delta$ is an M-estimator that minimizes the criterion function $M_n(\theta, \mathbf{x}_n) = (1/n)\sum_{i=1}^{n} \varrho_\theta(x_i)$, $\mathbf{x}_n = (x_1, ..., x_n) \in \mathcal{X}^n$, then

$$\dot{M}_n(\widehat{\theta}_n(\mathbf{x}_n), \mathbf{x}_n) = \frac{1}{n}\sum\nolimits_{i=1}^{n} \dot{\varrho}_{\widehat{\theta}_n(\mathbf{x}_n)}(x_i) = 0, \tag{7.101}$$

provided that $\widehat{\theta}_n(\mathbf{x}_n)$ belongs to the interior of Δ and ϱ is differentiable with respect to θ. It should be pointed out that the notion of an M-estimator is not unique in the literature. Some authors call every solution of the equation (7.101) an M-estimator. It is clear that a minimum point may not solve (7.101) unless it is an interior point, and the solution of (7.101) may not be a minimizer of the criterion function.

We know that even in exponential families the special M-estimator, an MLE, may not exist for some outcomes of a sample. However, the probabilities of the events on which an MLE fails to exist typically tend to zero for increasing sample size. Let A_n be the event on which $\widehat{\theta}_n$ is a minimizer. Suppose ϱ_θ is differentiable with respect to θ, $\theta_0 \in \Delta^0$, and $\widehat{\theta}_n$ is consistent at θ_0. If $\varepsilon > 0$ is small enough, then $\{\theta : \|\theta - \theta_0\| < \varepsilon\} \subseteq \Delta^0$. Hence for $\mathbf{X}_n \in B_n = A_n \cap \{\| \widehat{\theta}_n - \theta_0 \| < \varepsilon\}$ the criterion function has a local minimum at the interior point $\widehat{\theta}_n(\mathbf{x}_n)$, so that

$$\dot{M}_n(\widehat{\theta}_n(\mathbf{X}_n), \mathbf{X}_n) = \frac{1}{n} \sum\nolimits_{i=1}^{n} \dot{\varrho}_{\widehat{\theta}_n(\mathbf{X}_n)}(X_i) = 0, \quad \mathbf{X}_n \in B_n, \quad P_{n,\theta_0}(B_n) \to 1.$$

It turns out that this property of $\widehat{\theta}_n$ only in conjunction with the consistency is enough to develop the complete asymptotic theory of M-estimators. As long as we know that a solution of (7.101) is consistent we don't need to check whether the criterion function $M_n(\theta, \mathbf{x}_n)$ really has a minimum at $\widehat{\theta}_n$. This is a fortunate situation as the sufficient conditions for a function of several variables to have a minimum at a given point are rather unwieldy. Now we adopt a more general point of view. Regardless of whether the function $\psi : \Delta \times \mathcal{X} \to_m \mathbb{R}^m$ is given by $\psi_\theta = \dot{\varrho}_\theta$ or has been motivated by other arguments, we say that a sequence of estimators $\widehat{\theta}_n : \mathcal{X}^n \to_m \Delta$ *solves the M-equation asymptotically* if

$$\frac{1}{n} \sum\nolimits_{i=1}^{n} \psi_{\widehat{\theta}_n(\mathbf{X}_n)}(X_i) = \mathbf{o}_{P_{n,\theta_0}}(0), \tag{7.102}$$

where $\mathbf{o}_{P_{n,\theta_0}}(0)$ is defined in (7.88). We study systematically sequences of estimators that are consistent and solve a given equation asymptotically in the sense of (7.102). We emphasize that equations of the above type are not necessarily motivated by the minimization of a criterion function. For example, the method of moments that expresses the moments by the parameters, and estimates these moments in a nonparametric way, is another way to establish equations of the type (7.101). For a systematic approach to estimators defined by equations we refer to Godambe (1991).

The function ψ_θ characterizes the influence of the observed data on the values of the estimators. In the so-called *gross error model* it is allowed that a few values of the sample take on extremely large values and the goal is to construct estimators that are robust to such occurrences. Roughly speaking we have this property if the influence function is either bounded or tends slowly to infinity for $|x| \to \infty$. A precise formulation can be given by considering an estimator as a functional of the empirical distribution and by characterizing the influence by suitable differentiability concepts for functionals. For details we refer to the monographs by Huber (1981) and Hampel, Ronchetti, Rousseeuw, and Stahel (1986), and the references given there.

We consider the sequence of models

$$(\mathcal{X}^n, \mathfrak{A}^{\otimes n}, (P_{n,\theta})_{\theta\in\Delta}), \quad P_{n,\theta} = P_\theta^{\otimes n}, \quad \Delta \text{ open subset of } \mathbb{R}^d,$$

denote again by $X_i : \mathcal{X}^n \to \mathcal{X}$ the projections, and study estimators that solve asymptotically the M-equation in the sense of (7.102).

Theorem 7.142. *Assume that $\widehat{\theta}_n : \mathcal{X}^n \to_m \Delta$ is a sequence of estimators that is consistent at θ_0 and solves the M-equation (7.102) asymptotically, condition (A9) is fulfilled, $\mathsf{E}_{\theta_0}\psi_{\theta_0} = 0$, $\Sigma_1 = \mathsf{E}_{\theta_0}\psi_{\theta_0}\psi_{\theta_0}^T$, $\Sigma_2 = \mathsf{E}_{\theta_0}\dot{\psi}_{\theta_0}^T$, and* $\det(\Sigma_2) \neq 0$. *Then $\sqrt{n}(\widehat{\theta}_n - \theta_0)$ admits the asymptotic linearization*

$$\sqrt{n}(\widehat{\theta}_n - \theta_0) = -\frac{1}{\sqrt{n}}\sum\nolimits_{i=1}^n \Sigma_2^{-1}\psi_{\theta_0}(X_i) + o_{P_{\theta_0}^{\otimes n}}(1), \tag{7.103}$$

and it holds

$$\mathcal{L}(\sqrt{n}(\widehat{\theta}_n - \theta_0)|P_{\theta_0}^{\otimes n}) \Rightarrow \mathsf{N}(0, \Sigma), \tag{7.104}$$

where $\Sigma = \Sigma_2^{-1}\Sigma_1\Sigma_2^{-T}$.

Proof. Set $G_n(\theta) = (1/n)\sum_{i=1}^n \psi_\theta(X_i)$. For $U(\theta_0)$ in (A9) we set

$$\widetilde{\theta}_n = \widehat{\theta}_n I_{U(\theta_0)}(\widehat{\theta}_n) + \theta_0 I_{\Delta\setminus U(\theta_0)}(\widehat{\theta}_n).$$

Then $\widetilde{\theta}_n$ is again an estimator that is consistent at θ_0 and satisfies $G_n(\widetilde{\theta}_n) = \mathbf{o}_{P_{n,\theta_0}}(0)$. We apply the first-order Taylor expansion (see (7.99)) to G_n and get

$$G_n(\widetilde{\theta}_n) - G_n(\theta_0) = [\int_0^1 \frac{1}{n}\sum\nolimits_{i=1}^n \dot{\psi}_{\theta_0}^T(\theta_0 + s(\widetilde{\theta}_n - \theta_0), X_i)ds](\widetilde{\theta}_n - \theta_0).$$

Note that $G_n(\widetilde{\theta}_n) = \mathbf{o}_{P_{n,\theta_0}}(0)$ implies

$$\sqrt{n}G_n(\widetilde{\theta}_n) = \sqrt{n}\mathbf{o}_{P_{n,\theta_0}}(0) = \mathbf{o}_{P_{n,\theta_0}}(0).$$

Hence by Lemma 7.141, the law of large numbers, and the fact that $\mathbf{o}_{P_{n,\theta_0}}(0)$ is a sequence $o_{P_{n,\theta_0}}(1)$, we get with $\Sigma_2 = \mathsf{E}_{\theta_0}\dot{\psi}_{\theta_0}^T$,

$$-\sqrt{n}G_n(\theta_0) = (\Sigma_2 + o_{P_{n,\theta_0}}(1))\sqrt{n}(\widehat{\theta}_n - \theta_0) + o_{P_{n,\theta_0}}(1).$$

In view of $\mathsf{E}_{\theta_0}\psi_{\theta_0} = 0$ and the assumption (C) in (A9) we may apply the central limit theorem for i.i.d. random vectors to

$$\sqrt{n}G_n(\theta_0) = n^{-1/2}\sum\nolimits_{i=1}^{n} \psi_{\theta_0}(X_i)$$

and get $\mathcal{L}(\sqrt{n}G_n(\theta_0)|P_{n,\theta_0}) \Rightarrow \mathsf{N}(0, \Sigma_1)$. Hence $\sqrt{n}G_n(\theta_0)$ is stochastically bounded. An application of Problem 7.130 yields the statement (7.103). The already established asymptotic normality of $G_n(\theta_0)$ yields (7.104). ■

In the previous theorem we have proved the asymptotic normality of consistent estimators that solve the M-equation asymptotically, where the M-equation was a necessary condition for M-estimators. Let us go back one more step to the contrast condition (7.31). If we set $\psi_\theta = \dot{\varrho}_\theta$ and formally carry out the derivative under the expectation in (7.31), then we arrive at $\mathsf{E}_{\theta_0}\psi_{\theta_0} = 0$ for every $\theta_0 \in \Delta$. If this condition is at least satisfied in a neighborhood of θ_0, and the model is differentiable, then we may express $\mathsf{E}_{\theta_0}\dot{\psi}_{\theta_0}$ with the help of the $\mathbb{L}_2$-derivative. The formal calculation is nothing but the product rule. Denote by $L_{\theta_0,\theta}$ the likelihood ratio of P_θ with respect to P_{θ_0}. Then

$$0 = \nabla \mathsf{E}_\theta \psi_\theta = \nabla \mathsf{E}_{\theta_0} L_{\theta_0,\theta}\psi_\theta = \mathsf{E}_{\theta_0} \dot{L}_{\theta_0,\theta}\psi_\theta^T + \mathsf{E}_{\theta_0} L_{\theta_0,\theta}\dot{\psi}_\theta.$$

The next lemma justifies this calculation.

Lemma 7.143. *Assume that the family $(P_\theta)_{\theta\in\Delta}$ is $\mathbb{L}_2$-differentiable at θ_0 with derivative $\dot{L}_{\theta_0}$. Suppose that $\psi : \Delta \times \mathcal{X} \to \mathbb{R}$ satisfies the conditions in (A9). If in addition the conditions*

$$\begin{aligned} &\lim\nolimits_{\theta\to\theta_0} \mathsf{E}_\theta \left\|\psi_\theta - \psi_{\theta_0}\right\|^2 = 0, \quad \lim\nolimits_{\theta\to\theta_0} \mathsf{E}_{\theta_0} \left\|\psi_\theta - \psi_{\theta_0}\right\|^2 = 0, \\ &\mathsf{E}_\theta\psi_\theta = 0, \quad \theta \in U(\theta_0), \end{aligned} \tag{7.105}$$

are fulfilled, then $\mathsf{E}_{\theta_0}\dot{\psi}_{\theta_0} = -\mathsf{E}_{\theta_0}\psi_{\theta_0}\dot{L}_{\theta_0}^T$.

Proof. By the definition of $\mathbb{L}_2$-differentiability it holds $P_\theta \ll P_{\theta_0}$ for all θ in some neighborhood $U_0 \subseteq U(\theta_0)$ of θ_0.Then $\mathsf{E}_{\theta_0+h}\psi_{\theta_0} - \mathsf{E}_{\theta_0}\psi_{\theta_0} = -\mathsf{E}_{\theta_0+h}(\psi_{\theta_0+h} - \psi_{\theta_0})$ by $\mathsf{E}_\theta\psi_\theta = 0$, $\theta \in U(\theta_0)$, for $\theta_0 + h \in U_0$. The first condition in (7.105) yields $\sup_{\theta\in U_1} \mathsf{E}_\theta \left\|\psi_\theta\right\|^2 < \infty$ for some neighborhood $U_1 \subseteq U_0$. Hence $\mathsf{E}_{\theta_0+h}\psi_{\theta_0} - \mathsf{E}_{\theta_0}\psi_{\theta_0} = (\mathsf{E}_{\theta_0}\psi_{\theta_0}\dot{L}_{\theta_0}^T)h + o(h)$ by Proposition 1.111. Furthermore by $(L_{\theta_0+h,\theta_0} - 1) = (L_{\theta_0+h,\theta_0}^{1/2} - 1)(L_{\theta_0+h,\theta_0}^{1/2} + 1)$ and Schwarz' inequality,

$$\begin{aligned} &\left\|\mathsf{E}_{\theta_0+h}(\psi_{\theta_0+h} - \psi_{\theta_0}) - \mathsf{E}_{\theta_0}(\psi_{\theta_0+h} - \psi_{\theta_0})\right\| \\ &= \left\|\mathsf{E}_{\theta_0}(L_{\theta_0+h,\theta_0} - 1)(\psi_{\theta_0+h} - \psi_{\theta_0})\right\| \\ &\le (\mathsf{E}_{\theta_0}(L_{\theta_0+h,\theta_0}^{1/2} - 1)^2)^{1/2}(\mathsf{E}_{\theta_0}(L_{\theta_0+h,\theta_0}^{1/2} + 1)^2 \left\|\psi_{\theta_0+h} - \psi_{\theta_0}\right\|^2)^{1/2}. \end{aligned}$$

The first factor is $O(\|h\|)$ by the definition of $\mathbb{L}_2$-differentiability; see Lemma 1.106. To deal with the second factor we remark that $(\sqrt{x}+1)^2 = x+2\sqrt{x}+1 \le 2x + 2$. Hence

$$\begin{aligned} &\mathsf{E}_{\theta_0}((L_{\theta_0+h,\theta_0}^{1/2} + 1)\left\|\psi_{\theta_0+h} - \psi_{\theta_0}\right\|)^2 \\ &\le 2\mathsf{E}_{\theta_0}\left\|\psi_{\theta_0+h} - \psi_{\theta_0}\right\|^2 + 2\mathsf{E}_{\theta_0+h}\left\|\psi_{\theta_0+h} - \psi_{\theta_0}\right\|^2, \end{aligned}$$

which tend to zero by assumption (7.105). Thus we have by $\mathsf{E}_\theta \psi_\theta = 0$

$$(\mathsf{E}_{\theta_0} \psi_{\theta_0} \dot{L}_{\theta_0}^T)h + o(h) = -\mathsf{E}_{\theta_0+h}(\psi_{\theta_0+h} - \psi_{\theta_0}) = -\mathsf{E}_{\theta_0}(\psi_{\theta_0+h} - \psi_{\theta_0}) + o(\|h\|).$$

To complete the proof we apply Problem 7.140 to $\mathsf{E}_{\theta_0} \psi_\theta$. ■

The next statement is a direct consequence of the last lemma and (7.103).

Proposition 7.144. *Suppose the family $(P_\theta)_{\theta \in \Delta}$ is $\mathbb{L}_2$-differentiable at θ_0 with derivative $\dot{L}_{\theta_0}$. Suppose condition (A9) is satisfied and (7.105) holds. Suppose $\widehat{\theta}_n : \mathcal{X}^n \to_m \Delta$ is a sequence of consistent estimators that is consistent at θ_0 and solves the M-equation (7.102) asymptotically. If $\Sigma_0 = \mathsf{E}_{\theta_0} \psi_{\theta_0} \dot{L}_{\theta_0}^T$ is nonsingular, then*

$$\sqrt{n}(\widehat{\theta}_n - \theta_0) = \frac{1}{\sqrt{n}} \sum\nolimits_{i=1}^n \Sigma_0^{-1} \psi_{\theta_0}(X_i) + o_{P_{\theta_0}^{\otimes n}}(1).$$

Theorem 7.142 is a basic result in the theory of M-estimators. It was established and reestablished by several authors; see, for example, Jurečková and Sen (1996), Chapter 5.2, and Liese and Vajda (2004). Later on we use this theorem to establish the limit distribution for consistent MLEs. In the present section we apply this theorem to regression models with random covariates. We consider the nonlinear regression model (7.70) and set $\mathbf{X}_n = (X_1, ..., X_n)$ and $\mathbf{Y}_n = (Y_1, ..., Y_n)$.

The M-equation in (7.102) that is associated with the minimization of

$$M_n(\theta) = \frac{1}{n} \sum\nolimits_{i=1}^n \varrho(Y_i - g_\theta(X_i))$$

is as follows.

$$\dot{M}_n(\widehat{\theta}_n) = \frac{1}{n} \sum\nolimits_{i=1}^n \psi_{\widehat{\theta}_n(\mathbf{X}_n, \mathbf{Y}_n)}(X_i, Y_i) = \mathbf{o}_{P_{n,\theta_0}}(0), \tag{7.106}$$

where $P_{n,\theta_0} = \mathcal{L}(\mathbf{X}_n, \mathbf{Y}_n)$, and ψ_θ and its derivative are given by

$$\begin{aligned}
\psi_\theta(X_i, Y_i) &= -\dot{\varrho}(Y_i - g_\theta(X_i))\dot{g}_\theta(X_i), \\
\dot{\psi}_\theta(X_i, Y_i) &= \ddot{\varrho}(Y_i - g_\theta(X_i))(\dot{g}_\theta(X_i))(\dot{g}_\theta(X_i))^T - \dot{\varrho}(Y_i - g_\theta(X_i))\ddot{g}_\theta(X_i).
\end{aligned}$$

As we assume that the estimators are already known to be consistent we operate with the corresponding M-equation and require that

$$\begin{aligned}
&\sigma^2 := \mathsf{E}_{\theta_0}(\dot{\varrho}(\varepsilon_1))^2 < \infty, \quad \mathsf{E}_{\theta_0}\dot{\varrho}(\varepsilon_1) = 0, \\
&\mathsf{E}_{\theta_0} |\ddot{\varrho}(\varepsilon_1)| < \infty, \quad \lambda := \mathsf{E}_{\theta_0}\ddot{\varrho}(\varepsilon_1) \neq 0 \\
&\mathsf{E}_{\theta_0} \|\dot{g}_{\theta_0}(X_1)\|^2 < \infty, \quad \mathsf{E}_{\theta_0} \|\ddot{g}_{\theta_0}(X_1)\|^2 < \infty.
\end{aligned} \tag{7.107}$$

We also suppose that

$$\begin{aligned}
&\Sigma_0 = \mathsf{E}_{\theta_0}(\dot{g}_{\theta_0}(X_1))(\dot{g}_{\theta_0}(X_1))^T \quad \text{is nonsingular,} \\
&\mathsf{E}_{\theta_0} \|\psi_{\theta_0}(X_1, Y_1)\|^2 < \infty, \quad \mathsf{E}_{\theta_0}(\sup\nolimits_{\theta \in U(\theta_0)} \| \dot{\psi}_\theta(X_1, Y_1) \|) < \infty.
\end{aligned} \tag{7.108}$$

We note that by the independence of the ε_1 and X_1,

$$\mathsf{E}_{\theta_0}\psi_{\theta_0}(X_1,Y_1) = -\mathsf{E}_{\theta_0}\dot{\varrho}(\varepsilon_1)\mathsf{E}_{\theta_0}\dot{g}_{\theta_0}(X_1) = 0,$$
$$\mathsf{E}_{\theta_0}(\psi_{\theta_0}(X_1,Y_1))(\psi_{\theta_0}(X_1,Y_1))^T = \sigma^2\Sigma_0, \quad \text{and} \quad \mathsf{E}_{\theta_0}\dot{\psi}_{\theta_0}(X_1,Y_1) = \lambda\Sigma_0.$$

Theorem 7.145. *Let $Y_1,...,Y_n$ be from the regression model (7.70). Suppose that $\varrho \in \mathsf{C}^{(2)}(\mathbb{R})$, $g \in \mathsf{C}_m^{(2)}(U(\theta_0),\mathcal{X})$, and (7.107) and (7.108) are satisfied. If $\widehat{\theta}_n : \mathbb{R}^n \times \mathbb{R}^n \to_m \Delta$ is consistent at θ_0 and satisfies (7.106), then*

$$\mathcal{L}(\sqrt{n}(\widehat{\theta}_n(\mathbf{X}_n,\mathbf{Y}_n) - \theta_0)|P_{n,\theta_0}) \Rightarrow \mathsf{N}(0,\Sigma),$$

where $\Sigma = (\sigma^2/\lambda^2)\Sigma_0^{-1}$ with Σ_0 from (7.108) and σ^2 and λ in (7.107).

Proof. We replace the observations X_i by (X_i,Y_i) and apply Theorem 7.142 to the function $\varrho_\theta(x,y) = \varrho(y - g_\theta(x))$. The conditions for $\psi_\theta = \dot{\varrho}_\theta$ in (A9) follow immediately from (7.107) and (7.108). The structure of Σ follows from $\Sigma_1 = \mathsf{E}_{\theta_0}\psi_{\theta_0}\psi_{\theta_0}^T = \sigma^2\Sigma_0$ and $\Sigma_2 = \mathsf{E}_{\theta_0}\dot{\psi}_{\theta_0}^T = \lambda\Sigma_0^T = \lambda\Sigma_0$. ■

Example 7.146. The conditions in the above theorem simplify if the function ϱ is more specific or the regression model is linear. Let us consider robust estimators for the linear regression model $Y_i = \theta_0^T X_i + \varepsilon_i$. Robust estimators belong to ϱ-functions that do not increase too rapidly if $|t| \to \infty$. This condition guarantees that possible gross errors have only a small influence on the estimator. Suppose that $g(\theta,x) = \theta^T x$, and that ϱ is twice continuously differentiable and $\varrho, \dot{\varrho}, \ddot{\varrho}$ are bounded. We suppose that $\mathbb{E}\dot{\varrho}(\varepsilon_1) = 0$, $\mathbb{E}\ddot{\varrho}(\varepsilon_1) \neq 0$, and $\mathbb{E}\|X_1\|^2 < \infty$. Let $\widehat{\theta}_n$ be a sequence of consistent estimators that satisfy

$$\frac{1}{n}\sum\nolimits_{i=1}^n \dot{\varrho}(Y_i - \widehat{\theta}_n^T X_i)X_i = \mathbf{o}_{P_{n,\theta_0}}(0).$$

To calculate the matrix Σ_0 we note that $\dot{g}_{\theta_0}(X_1) = X_1$. Hence $\Sigma_0 = \mathbb{E}(X_1X_1^T)$, which we assume to be nonsingular. Then all conditions of Theorem 7.145 are satisfied and $\mathcal{L}(\sqrt{n}(\widehat{\theta}_n - \theta_0)|P_{n,\theta_0}) \Rightarrow \mathsf{N}(0,\Sigma)$, where

$$\Sigma = \frac{\mathbb{E}(\dot{\varrho}(\varepsilon_1))^2}{(\mathbb{E}\ddot{\varrho}(\varepsilon_1))^2}(\mathbb{E}(X_1X_1^T))^{-1}.$$

A simple example is $\varrho(t) = t^2$. Then the above M-equation has an explicit solution. Indeed,

$$\frac{1}{n}\sum\nolimits_{i=1}^n \dot{\varrho}(Y_i - X_i^T\widehat{\theta}_n)X_i = \frac{2}{n}\sum\nolimits_{i=1}^n (Y_iX_i - X_iX_i^T\widehat{\theta}_n).$$

Hence we set

$$\widehat{\theta}_n = \begin{cases} (\frac{1}{n}\sum_{i=1}^n X_iX_i^T)^{-1}(\frac{1}{n}\sum_{i=1}^n Y_iX_i) & \text{if } \det(\frac{1}{n}\sum_{i=1}^n X_iX_i^T) \neq 0, \\ \widetilde{\theta} \quad \text{otherwise,} \end{cases}$$

where $\widetilde{\theta}$ is any fixed value from the parameter space. The strong law of large numbers yields

$$\frac{1}{n}\sum_{i=1}^{n} X_i X_i^T \to \mathbb{E}X_1 X_1^T, \quad \mathbb{P}\text{-a.s.}$$

If the matrix $\mathbb{E}X_1X_1^T$ is nonsingular, then the probabilities of the events $A_n = \{\det((1/n)\sum_{i=1}^n X_iX_i^T) \neq 0\}$ tend to one. Hence $\widehat{\theta}_n$ satisfies (7.106). To verify that $\widehat{\theta}_n$ is consistent we remark that

$$\frac{1}{n}\sum_{i=1}^{n} Y_i X_i \to \mathbb{E}Y_1 X_1 = \mathbb{E}(\theta_0^T X_1 + \varepsilon_1)X_1 = \mathbb{E}(X_1 X_1^T)\theta_0, \quad \mathbb{P}\text{-a.s.},$$

by the strong law of large numbers, provided that $\mathbb{E}\varepsilon_1 = 0$. Hence $\widehat{\theta}_n$ is strongly consistent. Suppose that $\mathbb{E}\varepsilon_i = 0$ and $\mathbb{E}\varepsilon_i^2 < \infty$. The constants σ^2 and λ in (7.107) are given by $\sigma^2 = \mathbb{E}(2\varepsilon_i)^2$ and $\lambda = \mathbb{E}2 = 2$. Hence we see that $\mathcal{L}(\sqrt{n}(\widehat{\theta}_n - \theta_0)|P_{n,\theta_0}) \Rightarrow \mathsf{N}(0,\Sigma)$, where $\Sigma = (\mathbb{E}\varepsilon_1^2)(\mathbb{E}(X_1X_1^T))^{-1}$.

7.6.2 Asymptotic Distributions of MLEs

Let $\mathcal{M} = (\mathcal{X}, \mathfrak{A}, (P_\theta)_{\theta\in\Delta})$ be a dominated model, where $\Delta \subseteq \mathbb{R}^d$ is a Borel set. We consider the sequence of models

$$(\mathcal{X}^n, \mathfrak{A}^{\otimes n}, (P_\theta^{\otimes n})_{\theta\in\Delta}),$$

where the projections $X_i : \mathcal{X}^n \to \mathcal{X}$, $i = 1, ..., n$, are i.i.d. under $P_\theta^{\otimes n}$. Set $f_\theta = (dP_\theta/d\mu)$ and recall that the log-likelihood is given by $\Lambda_n(\theta) = \sum_{i=1}^n \ln f_\theta(X_i)$, where $\Lambda_n(\theta) = \Lambda_n(\theta, \mathbf{X}_n)$ is used for brevity. The aim of this section is to study the limit distribution of $\sqrt{n}(\widehat{\theta}_n - \theta_0)$ when $\widehat{\theta}_n$ is an MLE. We apply the theory of M-estimators for the likelihood contrast function $\varrho_\theta(x) := -\ln f_\theta(x)$. Similar to the considerations of the asymptotic normality of M-estimators we suppose that the MLE $\widehat{\theta}_n$ is consistent. As a consequence of the considerations on M-estimators (see (7.102)) and the definition of an asymptotic MLE (see (7.40)), we get that every consistent sequence of MLEs is an *asymptotic solution of the likelihood equation*. This means that

$$\dot{\Lambda}_n(\widehat{\theta}_n) = \sum_{i=1}^{n} \nabla \ln f_{\widehat{\theta}_n}(X_i) = \mathbf{o}_{P_{\theta_0}^{\otimes n}}(0), \tag{7.109}$$

for the log-likelihood $\Lambda_n(\theta) = \sum_{i=1}^n \ln f_\theta(X_i)$, provided that $\ln f_\theta(x)$ is differentiable with respect to θ, $\widehat{\theta}_n$ is consistent, and θ_0 is an inner point of Δ. The sharper requirement

$$\dot{\Lambda}_n(\widehat{\theta}_n) = \sum_{i=1}^{n} \nabla \ln f_{\widehat{\theta}_n}(X_i) = 0, \tag{7.110}$$

is called the *likelihood equation*, which may not have a solution for every sample outcome. We have discussed this issue for exponential families in detail in Proposition 7.93. Similarly as for M-estimators we need some regularity conditions. The following assumptions may be considered as a tightening of the regularity condition (A6).

(A10) Let $\mathcal{M} = (\mathcal{X}, \mathfrak{A}, (P_\theta)_{\theta\in\Delta})$, where $\Delta \subseteq \mathbb{R}^d$ is open. Suppose that for some $\mu \in \mathfrak{M}^\sigma(\mathfrak{A})$ and $\theta_0 \in \Delta^0$ the following conditions hold.

(A) $P_\theta \ll \mu, \quad f_\theta(x) = \frac{dP_\theta}{d\mu}(x) > 0, \quad (\theta, x) \in \Delta \times \mathcal{X}$.

(B) $f \in \mathsf{C}_m^{(2)}(U(\theta_0), \mathcal{X})$, for some open neighborhood $U(\theta_0) \subseteq \Delta$.

(C) $\int \nabla\nabla^T f_\theta d\mu = 0, \quad \theta \in U(\theta_0)$.

(D) $\mathsf{E}_{\theta_0}(\sup_{\theta \in U(\theta_0)} \left\|\nabla\nabla^T \ln f_\theta\right\|) < \infty$.

(E) $\mathsf{E}_{\theta_0}(\sup_{\theta \in U(\theta_0)} \left\|\nabla \ln f_\theta\right\|^2) < \infty$.

The next lemma shows that under (A10) the function $\psi_\theta = -\nabla \ln f_\theta$ satisfies the regularity conditions in (A9).

Lemma 7.147. *If condition (A10) is satisfied, then the following hold.*

(A) The family $(P_\theta)_{\theta \in U(\theta_0)}$ is $\mathbb{L}_2$-differentiable with derivative $\dot{L}_\theta = \nabla \ln f_\theta$, and $\psi_\theta = -\dot{L}_\theta$ satisfies condition (A9).
(B) The Fisher information $\mathsf{I}(\theta) := \mathsf{E}_\theta \dot{L}_\theta \dot{L}_\theta^T$ is continuous in $U(\theta_0)$, and

$$\mathsf{I}(\theta) = -\mathsf{E}_\theta(\nabla\nabla^T \ln f_\theta), \tag{7.111}$$

$$\int \nabla f_\theta d\mu = \mathsf{E}_\theta \dot{L}_\theta = 0, \quad \theta \in U(\theta_0). \tag{7.112}$$

Proof. First of all we note that by $f_\theta > 0$ and condition (B) the function $\theta \mapsto \nabla \ln f_\theta(x)$ is twice continuously differentiable for every x and $\psi \in \mathsf{C}_m^{(1)}(U(\theta_0), \mathcal{X})$. The conditions (B) and (C) in (A9) follow from (D) and (E) in (A10). To prove the $\mathbb{L}_2$-differentiability we note that (A10) implies (A6). Thus we have only to prove the continuity of $\mathsf{I}(\theta)$ to get the $\mathbb{L}_2$-differentiability from Theorem 1.117. This continuity follows from the condition (E), the continuity of $\theta \mapsto \nabla \ln f_\theta(x)$ for every x, and Lebesgue's theorem. The first equality in (7.112) follows from Theorem 1.117. The second equality in (7.112) follows for every $\theta \in U(\theta_0)$ from the $\mathbb{L}_2$-differentiability and Proposition 1.110. Furthermore,

$$\begin{aligned}\nabla\nabla^T \ln f_\theta &= \nabla(\frac{1}{f_\theta}\nabla^T f_\theta) = \frac{1}{f_\theta}\nabla\nabla^T f_\theta - \frac{1}{f_\theta^2}(\nabla f_\theta)(\nabla^T f_\theta) \\ &= \frac{1}{f_\theta}\nabla\nabla^T f_\theta - (\nabla \ln f_\theta)(\nabla \ln f_\theta)^T.\end{aligned}$$

Integrating $f_\theta(\nabla\nabla^T \ln f_\theta) = \nabla\nabla^T f_\theta - (\nabla \ln f_\theta)(\nabla \ln f_\theta)^T f_\theta$ with respect to μ, and using (C), we get the statement (7.111). ■

Now we are ready to establish the asymptotic normality of MLEs. As in the case of M-estimators we assume that the estimators are already known to be consistent.

Theorem 7.148. ***(Asymptotic Normality of MLEs)*** *Let $(\mathcal{X}, \mathfrak{A}, (P_\theta)_{\theta \in \Delta})$ satisfy condition (A10), and $\widehat{\theta}_n : \mathcal{X}^n \to_m \Delta$ be consistent at θ_0 and fulfil (7.109). If the Fisher information matrix is nonsingular, then*

$$\sqrt{n}(\widehat{\theta}_n - \theta_0) = \frac{1}{\sqrt{n}} \sum\nolimits_{i=1}^{n} \mathsf{I}^{-1}(\theta_0)\dot{L}_{\theta_0}(X_i) + o_{P_{\theta_0}^{\otimes n}}(1), \tag{7.113}$$
$$\mathcal{L}(\sqrt{n}(\widehat{\theta}_n - \theta_0)|P_{\theta_0}^{\otimes n}) \Rightarrow \mathsf{N}(0, \mathsf{I}^{-1}(\theta_0)).$$

Proof. The function $\psi_\theta(x) = -\nabla \ln f_\theta(x)$ satisfies condition (A9). As

$$\Sigma_1 = \mathsf{E}_{\theta_0}(\psi_{\theta_0}(X_1))(\psi_{\theta_0}(X_1))^T = \mathsf{E}_{\theta_0}\dot{L}_{\theta_0}\dot{L}_{\theta_0}^T = \mathsf{I}(\theta_0),$$
$$\Sigma_2 = \mathsf{E}_{\theta_0}\dot{\psi}_{\theta_0}(X_1) = -\mathsf{E}_{\theta_0}(\nabla\nabla^T \ln f_{\theta_0}(x)) = \mathsf{I}(\theta_0),$$

and $\mathsf{E}_{\theta_0}\psi_{\theta_0} = \mathsf{E}_{\theta_0}\dot{L}_{\theta_0} = 0$, we get the statement from Theorem 7.142. ■

We apply the above theorem to an exponential family.

Example 7.149. Let $(P_\theta)_{\theta\in\Delta}$ be an exponential family with natural parameter θ and generating statistic T. Then $\ln f_\theta(x) = \langle \theta, T(x)\rangle - K(\theta)$,

$$\dot{L}_\theta = \nabla \ln f_\theta(x) = T(x) - \nabla K(\theta) \quad \text{and} \quad \nabla\nabla^T \ln f_\theta(x) = -\nabla\nabla^T K(\theta).$$

We see from here that all conditions in (A10) are satisfied, where (C) follows directly from Theorem 1.17. It holds $\mathsf{I}(\theta) = \nabla\nabla^T K(\theta)$. We consider the estimator $\widetilde{\theta}_n$ introduced in (7.52). We have shown in Proposition 7.98 that $\widetilde{\theta}_n$ is consistent and an asymptotic solution of the likelihood equation

$$\sum\nolimits_{i=1}^{n} T(X_i) - n\nabla K(\widetilde{\theta}_n) = \mathbf{o}_{P_{\theta_0}^{\otimes n}}(0).$$

If the standard assumptions (A1) and (A2) are satisfied, then $\det(\mathsf{I}(\theta)) \neq 0$ and we get from Theorem 7.148 that

$$\mathcal{L}(\sqrt{n}(\widetilde{\theta}_n - \theta_0)|P_{\theta_0}^{\otimes n}) \Rightarrow \mathsf{N}(0, \mathsf{I}^{-1}(\theta_0)).$$

Next we illustrate the limit distribution of an MLE with the location model.

Example 7.150. Suppose f is a Lebesgue density that is everywhere positive and twice continuously differentiable. We consider the location family $f_\theta(x) = f(x-\theta)$, $\theta \in \Delta = \mathbb{R}$. Then

$$\frac{\partial}{\partial\theta} \ln f_\theta(x) = -\frac{\dot{f}(x-\theta)}{f(x-\theta)}, \quad \frac{\partial^2}{\partial\theta^2} \ln f_\theta(x) = \frac{f(x-\theta)\ddot{f}(x-\theta) + (\dot{f}(x-\theta))^2}{f^2(x-\theta)},$$

and the Fisher information is independent of θ and given by $\mathsf{I} = \int (\dot{f}(t))^2/f(t)dt$, provided the integral is finite. If $X_1, ..., X_n$ have the density $f(x-\theta_0)$, then condition (A10) is satisfied if

$$\int \ddot{f}(x)dx = 0, \int \sup_{\theta:|\theta|<\varepsilon} (\frac{\dot{f}(x-\theta)}{f(x-\theta)})^2 f(x)dx < \infty, \int \sup_{\theta:|\theta|<\varepsilon} \frac{|\ddot{f}(x-\theta)|}{f(x-\theta)} f(x)dx < \infty. \tag{7.114}$$

The MLE $\widehat{\theta}_n$ satisfies the likelihood equation

$$\sum\nolimits_{i=1}^{n} \dot{f}(X_i - \widehat{\theta}_n)/f(X_i - \widehat{\theta}_n) = 0.$$

Suppose that f satisfies in addition the conditions in Example 7.81. Then from here we see that the MLE $\widehat{\theta}_n$ is strongly consistent. Hence under the condition (7.114) and $\mathsf{I} > 0$ Theorem 7.148 yields that

$$\begin{aligned}\sqrt{n}(\widehat{\theta}_n - \theta_0) &= -\sum\nolimits_{i=1}^{n} \mathsf{I}^{-1}\dot{f}(X_i - \theta_0)/f(X_i - \theta_0) + o_{P_{\theta_0}^{\otimes n}}(1), \\ \mathcal{L}(\sqrt{n}(\widehat{\theta}_n - \theta_0)|P_{\theta_0}^{\otimes n}) &\Rightarrow \mathsf{N}(0, \mathsf{I}^{-1}).\end{aligned}$$

The likelihood equation in (7.110) is in general a nonlinear system of equations. To solve it one has to use numerical methods, e.g. a Newton approximation or some suitable modification. For all these methods one needs some starting values. One idea that goes back to LeCam is to start with any $\sqrt{n}$-consistent estimator θ_n^* and then to make a one-step Newton approximation. Recall that by $\sqrt{n}$-consistency of θ_n^* we mean that $\sqrt{n}(\theta_n^* - \theta_0) = O_{P_{\theta_0}^{\otimes n}}(1)$. The somewhat surprising fact is that the new estimator $\widetilde{\theta}_n$ obtained from the one-step Newton approximation differs from the MLE $\widehat{\theta}_n$ only by terms $o_{P_{\theta_0}^{\otimes n}}(1/\sqrt{n})$, so that $\sqrt{n}(\widetilde{\theta}_n - \theta_0)$ and $\sqrt{n}(\widehat{\theta}_n - \theta_0)$ have the same limit distribution. To be more specific let us assume that condition (A10) is satisfied. Then the log-likelihood function $\Lambda_n(\theta) = \sum_{i=1}^{n} \ln f_\theta(X_i)$ is twice continuously differentiable and the one-step Newton approximation to the likelihood equation $\dot{\Lambda}_n(\theta) = 0$ is a solution of the equation

$$\dot{\Lambda}_n(\theta_n^*) + \ddot{\Lambda}_n(\theta_n^*)(\theta - \theta_n^*) = 0. \tag{7.115}$$

We see from Lemma 7.141 that $(1/n)\ddot{\Lambda}_n(\theta_n^*) \to^{P_{\theta_0}^{\otimes n}} \mathsf{E}_{\theta_0}(\nabla\nabla^T \ln f_{\theta_0}) = -\mathsf{I}(\theta_0)$. If the Fisher information matrix is nonsingular, then $P_{\theta_0}^{\otimes n}(A_n) \to 1$, where $A_n = \{\det(\ddot{\Lambda}_n(\theta_n^*)) \neq 0\}$. On A_n we find the one-step Newton approximation by solving the equation (7.115) and set

$$\widetilde{\theta}_n = [\theta_n^* - (\ddot{\Lambda}_n(\theta_n^*))^{-1}\dot{\Lambda}_n(\theta_n^*)]I_{A_n} + \theta_n^* I_{\mathcal{X}^n \setminus A_n}. \tag{7.116}$$

The first statement in Lemma 7.141 gives

$$\ddot{\Lambda}_n(\theta_n^*) = o_{P_{\theta_0}^{\otimes n}}(n) - n\mathsf{I}(\theta_0).$$

Using a first-order Taylor expansion we get from the second statement in Lemma 7.141,

$$\begin{aligned}\dot{\Lambda}_n(\theta_n^*) &= \dot{\Lambda}_n(\theta_0) + [\frac{1}{n}\int_0^1 \ddot{\Lambda}_n(\theta_0 + s(\theta_n^* - \theta_0))ds]n(\theta_n^* - \theta_0) \\ &= \dot{\Lambda}_n(\theta_0) + (o_{P_{\theta_0}^{\otimes n}}(n) - n\mathsf{I}(\theta_0))(\theta_n^* - \theta_0).\end{aligned} \tag{7.117}$$

Hence,

$$\begin{aligned}(\widetilde{\theta}_n - \theta_n^*)I_{A_n} &= -(o_{P_{\theta_0}^{\otimes n}}(n) - n\mathsf{I}(\theta_0))^{-1}\dot{\Lambda}_n(\theta_0)I_{A_n} \\ &\quad + (o_{P_{\theta_0}^{\otimes n}}(n) - n\mathsf{I}(\theta_0))^{-1}(o_{P_{\theta_0}^{\otimes n}}(n) - n\mathsf{I}(\theta_0))(\theta_n^* - \theta_0)I_{A_n}.\end{aligned}$$

Multiplying this equation with $\sqrt{n}$ and using $P_{\theta_0}^{\otimes n}(A_n) \to 0$ we get

$$\begin{aligned}\sqrt{n}(\widetilde{\theta}_n - \theta_n^*) = -(o_{P_{\theta_0}^{\otimes n}}(1) - \mathsf{I}(\theta_0))^{-1} n^{-1/2} \dot{\Lambda}_n(\theta_0) + (o_{P_{\theta_0}^{\otimes n}}(1) - \mathsf{I}(\theta_0))^{-1} \\ \times (o_{P_{\theta_0}^{\otimes n}}(1) - \mathsf{I}(\theta_0))\sqrt{n}(\theta_n^* - \theta_0) + o_{P_{\theta_0}^{\otimes n}}(1).\end{aligned}$$

As the distribution of $n^{-1/2}\dot{\Lambda}_n(\theta_0) = n^{-1/2}\sum_{i=1}^n \dot{L}_{\theta_0}(X_i)$ tends to a normal distribution we see that $n^{-1/2}\dot{\Lambda}_n(\theta_0)$ is stochastically bounded. $\sqrt{n}(\theta_n^* - \theta_0)$ is stochastically bounded by assumption. Hence,

$$\begin{aligned}\sqrt{n}(\widetilde{\theta}_n - \theta_n^*) &= \mathsf{I}^{-1}(\theta_0) n^{-1/2} \dot{\Lambda}_n(\theta_0) - \sqrt{n}(\theta_n^* - \theta_0) + o_{P_{\theta_0}^{\otimes n}}(1), \\ \sqrt{n}(\widetilde{\theta}_n - \theta_0) &= \mathsf{I}^{-1}(\theta_0) n^{-1/2} \dot{\Lambda}_n(\theta_0) + o_{P_{\theta_0}^{\otimes n}}(1). \qquad (7.118)\end{aligned}$$

Now we compare the one-step Newton approximation $\widetilde{\theta}_n$ with the MLE $\widehat{\theta}_n$.

Proposition 7.151. *If condition (A10) is satisfied, and θ_n^* is any estimator that is $\sqrt{n}$-consistent at θ_0, then the one-step Newton approximation $\widetilde{\theta}_n$ in (7.116) satisfies (7.118) and*

$$\mathcal{L}(\sqrt{n}(\widetilde{\theta}_n - \theta_0) | P_{\theta_0}^{\otimes n}) \Rightarrow \mathsf{N}(0, \mathsf{I}^{-1}(\theta_0)).$$

If $\widehat{\theta}_n$ is a consistent estimator that solves the likelihood equation asymptotically (i.e., if it holds (7.109)), then the one-step Newton approximation $\widetilde{\theta}_n$ in (7.116) is asymptotically equivalent to $\widehat{\theta}_n$ in the sense that $\sqrt{n}(\widehat{\theta}_n - \widetilde{\theta}_n) = o_{P_{\theta_0}^{\otimes n}}(1)$.

Proof. The first statement follows from the already established relation (7.118). To prove the second statement we have only to use (7.113). ■

In (7.117) we have approximated $n^{-1}\ddot{\Lambda}_n(\theta_n^*)$ by $-\mathsf{I}(\theta_0)$. Under assumption (A10) the Fisher information is a continuous function of θ in a neighborhood of θ_0. Hence $n^{-1}\ddot{\Lambda}_n(\theta_n^*) = -\mathsf{I}(\theta_n^*) + o_{P_{\theta_0}^{\otimes n}}(1)$. If we use this approximation instead of (7.117), then by almost the same arguments as above one can see that the modified one-step approximation $\widetilde{\widetilde{\theta}}_n = \theta_n^* + \mathsf{I}(\theta_n^*)^{-1}\dot{\Lambda}_n(\theta_n^*)$ satisfies

$$\sqrt{n}(\widehat{\theta}_n - \widetilde{\widetilde{\theta}}_n) = o_{P_{\theta_0}^{\otimes n}}(1), \quad \mathcal{L}(\sqrt{n}(\widetilde{\widetilde{\theta}}_n - \theta_0) | P_{\theta_0}^{\otimes n}) \Rightarrow \mathsf{N}(0, \mathsf{I}^{-1}(\theta_0)). \qquad (7.119)$$

Example 7.152. We consider the location model $f(x-\theta)$, $\theta \in \mathbb{R}$, where f is a positive and twice continuously differentiable Lebesgue density with Fisher information $\mathsf{I}(\theta) = \int [\dot{f}(x-\theta)]^2 [f(x-\theta)]^{-1} dx = \int [\dot{f}(x)]^2 [f(x)]^{-1} dx =: \mathsf{I} < \infty$. It holds

$$\dot{\Lambda}_n(\theta, \mathbf{x}_n) = -\sum\nolimits_{i=1}^n \dot{f}(x_i - \theta)/f(x_i - \theta),$$

and the modified one-step approximation to the likelihood equation is given by $\widetilde{\widetilde{\theta}}_n = \theta_n^* + \mathsf{I}^{-1}\dot{\Lambda}_n(\theta_n^*)$. If θ_n^* is a $\sqrt{n}$-consistent estimator and the condition (7.114) is satisfied, then (7.119) holds. If f has a finite second moment and $\int xf(x)dx = 0$, then we may estimate θ by $\overline{X}_n$, and the $\sqrt{n}$-consistency follows from the central limit theorem. However, if f is the Cauchy distribution, then the moment condition is not satisfied. But then we may estimate the location parameter θ by the median which is $\sqrt{n}$-consistent in view of Example 7.139.

We conclude this section with a reference to Searle, Casella, and McCulloch (1992) who have studied the performance of the one-step approximation.

7.6.3 Asymptotic Normality of the Posterior

In Section 7.5.2 we have studied the posterior distributions in the frequentist model and have given conditions that imply the consistency. It is clear that asymptotic normality can only be achieved by an appropriate centering and scaling. The famous Bernstein–von Mises theorem states that the right centering point is the MLE, and that after scaling with $\sqrt{n}$ the posterior distributions converge to a normal distribution. Under an additional moment condition on the prior the Bernstein–von Mises theorem implies the asymptotic equivalence of the MLE and the Bayes estimator. We start with an example to illustrate the situation.

Example 7.153. With $X_1, ..., X_n$ as the coordinate projections the MLE for the family $(((1-p)\boldsymbol{\delta}_0 + p\boldsymbol{\delta}_1)^{\otimes n})_{p\in(0,1)}$ is the relative frequency $\widehat{p}_n = (1/n)\sum_{i=1}^n X_i$. On the other hand, we have seen in Example 7.114 that the Bayes estimator for a Beta prior under the quadratic loss is the conditional expectation, given by

$$\mathbb{E}(\Theta|X_1, ..., X_n) = \frac{\alpha+\beta}{\alpha+\beta+n}\frac{\alpha}{\alpha+\beta} + \frac{n}{\alpha+\beta+n}\frac{1}{n}\sum\nolimits_{i=1}^n X_i.$$

Hence in the frequentist model (7.78), where the X_i are i.i.d., we get

$$\begin{aligned}
&\sqrt{n}(\mathbb{E}(\Theta|X_1, ..., X_n) - \widehat{p}_n)\\
&= \frac{\sqrt{n}(\alpha+\beta)}{\alpha+\beta+n}\frac{\alpha}{\alpha+\beta} + \sqrt{n}(\frac{n}{\alpha+\beta+n} - 1)\frac{1}{n}\sum\nolimits_{i=1}^n X_i\\
&= \frac{\sqrt{n}(\alpha+\beta)}{\alpha+\beta+n}\frac{\alpha}{\alpha+\beta} - \sqrt{n}\frac{\alpha+\beta}{\alpha+\beta+n}\frac{1}{n}\sum\nolimits_{i=1}^n X_i \to^{\mathbb{P}} 0.
\end{aligned}$$

This means especially that $\sqrt{n}(\mathbb{E}(\Theta|X_1, ..., X_n) - p)$ and $\sqrt{n}(\widehat{p}_n - p)$ have the same limit distribution and the Bayes estimator shares the asymptotic optimality of the MLE.

We assume now that the parameter set Δ is an open subset of $\mathbb{R}^d$, the prior Π has a density π with respect to the Lebesgue measure $\boldsymbol{\lambda}_d$, the model $(P_\theta)_{\theta\in\Delta}$ is dominated, and the family $(f_\theta)_{\theta\in\Delta}$ satisfies the condition (A5). We set $\pi(\theta) = 0$ for $\theta \in \mathbb{R}^d\backslash\Delta$. Let the density of the posterior distribution be defined as in (7.79); that is, for $\mathbf{x}_n = (x_1, ..., x_n) \in \mathcal{X}^n$ we have

$$\pi_n(\theta|\mathbf{x}_n) = \begin{cases} \frac{1}{\mathsf{m}_n(\mathbf{x}_n)} \prod_{i=1}^n f_\theta(x_i)\pi(\theta) & \text{if } \mathsf{m}_n(\mathbf{x}_n) > 0, \\ \pi(\theta) & \text{if } \mathsf{m}_n(\mathbf{x}_n) = 0, \quad \text{where} \end{cases} \tag{7.120}$$

$$\mathsf{m}_n(\mathbf{x}_n) = \int \prod_{i=1}^n f_\theta(x_i)\pi(\theta)\boldsymbol{\lambda}_d(d\theta).$$

We denote the log-likelihood function by $\Lambda_n(\theta, \mathbf{x}_n) = \sum_{i=1}^n \ln f_\theta(x_i)$. Suppose that $\widehat{\theta}_n : \mathcal{X}^n \to_m \Delta$ is consistent at θ_0 and an asymptotic solution of the likelihood equation, i.e., it holds (7.109). In the Bayes model $\pi_n(\theta|\mathbf{x}_n)$ is the posterior density of Θ under the prior Π. We center Θ at $\widehat{\theta}_n$ and consider the scaled variable $\sqrt{n}(\Theta - \widehat{\theta}_n)$ which has, for $\mathsf{m}_n(\mathbf{x}_n) > 0$, the posterior density

$$\pi_n^*(\eta|\mathbf{x}_n) = \frac{\exp\{U_n(\eta, \mathbf{x}_n)\}\pi(\widehat{\theta}_n(\mathbf{x}_n) + \eta/\sqrt{n})}{\int \exp\{U_n(\eta, \mathbf{x}_n)\}\pi(\widehat{\theta}_n(\mathbf{x}_n) + \eta/\sqrt{n})\boldsymbol{\lambda}_d(d\eta)}, \quad \text{where}$$

$$U_n(\eta, \mathbf{x}_n) = \Lambda_n(\widehat{\theta}_n(\mathbf{x}_n) + \eta/\sqrt{n}, \mathbf{x}_n) - \Lambda_n(\widehat{\theta}_n(\mathbf{x}_n), \mathbf{x}_n), \ \eta \in \mathbb{R}^d.$$

Although the density $\pi_n^*(\eta|\mathbf{x}_n)$ has been obtained from the Bayes model, where Θ is a random variable, we may consider $\pi_n^*(\eta|\mathbf{x}_n)$ as a statistic in the frequentist model $(\mathcal{X}^n, \mathfrak{A}^{\otimes n}, P_{\theta_0}^{\otimes n})$ and study the asymptotic as n tends to infinity. In a first step we expand $U_n(\eta, \mathbf{X}_n)$.

Lemma 7.154. *Suppose that condition (A10) is satisfied, where Δ is an open subset of $\mathbb{R}^d$. Suppose $\widehat{\theta}_n : \mathcal{X}^n \to_m \Delta$ is consistent at θ_0 and an asymptotic solution of the likelihood equation; that is, it holds (7.109). Then*

$$U_n(\eta, \mathbf{X}_n) \to^{P_{\theta_0}^{\otimes\infty}} -\frac{1}{2}\eta^T \mathsf{I}(\theta_0)\eta$$

for every fixed $\eta \in \mathbb{R}^d$, where $\mathsf{I}(\theta_0)$ is the Fisher information matrix.

Proof. Set $A_n = \{\mathbf{x}_n : \dot{\Lambda}_n(\widehat{\theta}_n(\mathbf{x}_n), \mathbf{x}_n) = 0\}$. Using a second-order Taylor expansion (see Theorem A.2) we get for

$$V_n = U_n(\eta, \mathbf{X}_n) - \frac{1}{2}\eta^T \ddot{\Lambda}_n(\widehat{\theta}_n(\mathbf{X}_n), \mathbf{X}_n)\eta,$$

the representation

$$V_n I_{A_n}(\mathbf{X}_n)$$
$$= \frac{1}{n}\eta^T [\int_0^1 (1-s)(\ddot{\Lambda}_n(\widehat{\theta}_n(\mathbf{X}_n) + \frac{s}{\sqrt{n}}\eta, \mathbf{X}_n) - \ddot{\Lambda}_n(\widehat{\theta}_n(\mathbf{X}_n), \mathbf{X}_n))ds]\eta I_{A_n}(\mathbf{X}_n).$$

Set $B_{n,\varepsilon} = \{\| \widehat{\theta}_n(\mathbf{X}_n) - \theta_0 \| \leq \varepsilon/2\}$ and denote by $\|A\|$ the matrix norm $\|A\| = \sup_{\|x\|\leq 1} \|Ax\|$. Then for $n^{-1/2} \leq \varepsilon/2$ it holds

$$\mathsf{E}_{n,\theta_0}|V_n| I_{A_n}(\mathbf{X}_n)$$
$$\leq \|\eta\|^2 \frac{1}{2n}\sum_{i=1}^n \mathsf{E}_{n,\theta_0} \sup_{\|\theta-\theta_0\|\leq\varepsilon} \left\|\nabla\nabla^T \ln f_\theta(X_i) - \nabla\nabla^T \ln f_{\theta_0}(X_i)\right\|$$
$$\leq \|\eta\|^2 \frac{1}{2}\mathsf{E}_{\theta_0} \sup_{\|\theta-\theta_0\|\leq\varepsilon} \left\|\nabla\nabla^T \ln f_\theta(X_1) - \nabla\nabla^T \ln f_{\theta_0}(X_1)\right\|.$$

Hence $\lim_{\varepsilon\to 0}\sup_n \mathsf{E}_{n,\theta_0} \mid V_n|I_{A_n}(\mathbf{X}_n) = 0$ and

$$P_{\theta_0}^{\otimes n}(|V_n| > \delta) \le \frac{1}{\delta}\sup_n \mathsf{E}_{n,\theta_0}|V_n|I_{A_n}(\mathbf{X}_n) + P_{\theta_0}^{\otimes n}(\overline{A}_n).$$

Taking first $n \to \infty$ and then $\varepsilon \to 0$ we get

$$V_n \to^{P_{\theta_0}^{\otimes n}} 0.$$

To complete the proof we apply Lemma 7.141 to $\varphi_\theta = \nabla\nabla^T \ln f_\theta$ to get

$$\begin{aligned}\ddot{\Lambda}_n(\widehat{\theta}_n(\mathbf{X}_n), \mathbf{X}_n) &= \frac{1}{n}\sum\nolimits_{i=1}^n \nabla\nabla^T \ln f_{\widehat{\theta}_n(\mathbf{X}_n)}(X_i)\\ &= \mathsf{E}_{\theta_0}\nabla\nabla^T \ln f_{\theta_0}(X_1) + o_{P_{\theta_0}^{\otimes n}}(1) = -\mathsf{I}(\theta_0) + o_{P_{\theta_0}^{\otimes n}}(1).\end{aligned}$$

Hence

$$U_n(\eta, \mathbf{X}_n) = V_n + \frac{1}{2}\eta^T \ddot{\Lambda}_n(\widehat{\theta}_n(\mathbf{X}_n), \mathbf{X}_n)\eta = -\frac{1}{2}\eta^T \mathsf{I}(\theta_0)\eta + o_{P_{\theta_0}^{\otimes n}}(1).$$

■

Theorem 7.155. ***(Bernstein–von Mises)*** *Assume that condition (A10) is satisfied, where Δ is an open subset of $\mathbb{R}^d$. Suppose $\widehat{\theta}_n : \mathcal{X}^n \to_m \Delta$ is consistent at θ_0 and an asymptotic solution of the likelihood equation; that is, (7.109) holds. If the prior Π has a density π with respect to the Lebesgue measure which is continuous at θ_0, then*

$$\exp\{U_n(\eta, \mathbf{X}_n)\}\pi(\widehat{\theta}_n(\mathbf{X}_n) + \eta/\sqrt{n}) \to^{P_{\theta_0}^{\otimes\infty}} \pi(\theta_0)\exp\{-\frac{1}{2}\eta^T \mathsf{I}(\theta_0)\eta\}. \quad (7.121)$$

If in addition the matrix $\mathsf{I}(\theta_0)$ is nonsingular, $\pi(\theta_0) > 0$, and the condition

$$\begin{aligned}&\int \exp\{U_n(\eta, \mathbf{X}_n)\}\pi(\widehat{\theta}_n(\mathbf{X}_n) + \eta/\sqrt{n})\boldsymbol{\lambda}_d(d\eta) \quad (7.122)\\ &\qquad \to^{P_{\theta_0}^{\otimes\infty}} \pi(\theta_0)(2\pi)^{d/2}(\det(\mathsf{I}(\theta_0)))^{-1/2}\end{aligned}$$

holds, then

$$\int |\pi_n^*(\eta|\mathbf{X}_n) - \varphi_{0,\mathsf{I}^{-1}(\theta_0)}(\eta)|\boldsymbol{\lambda}_d(d\eta) \to^{P_{\theta_0}^{\otimes\infty}} 0. \quad (7.123)$$

Corollary 7.156. *If instead of (7.122) the condition*

$$\begin{aligned}&\int (1 + \|\eta\|)\exp\{U_n(\eta, \mathbf{X}_n)\}\pi(\widehat{\theta}_n(\mathbf{X}_n) + \eta/\sqrt{n})\boldsymbol{\lambda}_d(d\eta) \quad (7.124)\\ &\to^{P_{\theta_0}^{\otimes\infty}} \pi(\theta_0)\int (1 + \|\eta\|)\exp\{-\frac{1}{2}\eta^T \mathsf{I}(\theta_0)\eta\}\boldsymbol{\lambda}_d(d\eta)\end{aligned}$$

holds, then

$$\int (1 + \|\eta\|)|\pi_n^*(\eta|\mathbf{X}_n) - \varphi_{0,\mathsf{I}^{-1}(\theta_0)}(\eta)|\boldsymbol{\lambda}_d(d\eta) \to^{P_{\theta_0}^{\otimes\infty}} 0. \quad (7.125)$$

Proof. The first statement follows from Lemma 7.154, the continuity of π, and the consistency of $\widehat{\theta}_n$. To prove the second statement, we introduce a probability measure μ that is equivalent to the Lebesgue measure by setting $\mu(d\eta) = h(\eta)\boldsymbol{\lambda}_d(d\eta)$, where h is a positive and bounded probability density. Furthermore, let

$$\begin{aligned} V_n(\eta, \mathbf{X}_n) &= \exp\{U_n(\eta, \mathbf{X}_n)\}\pi(\widehat{\theta}_n(\mathbf{X}_n) + \eta/\sqrt{n}), \\ W_n(\eta, \mathbf{X}_n) &= V_n(\eta, \mathbf{X}_n) - \pi(\theta_0)\exp\{-\frac{1}{2}\eta^T \mathsf{I}(\theta_0)\eta\}. \end{aligned}$$

Then the statement (7.121) may be written as

$$g_n(\eta) := \mathsf{E}_{n,\theta_0} \frac{|W_n(\eta, \mathbf{X}_n)|}{1 + |W_n(\eta, \mathbf{X}_n)|} \to 0.$$

As $0 \le g_n(\eta) \le 1$ we get $\int g_n(\eta)\mu(d\eta) \to 0$ and from Fubini's theorem that

$$\int \frac{|W_n(\eta, \mathbf{X}_n(\omega))|}{1 + |W_n(\eta, \mathbf{X}_n(\omega))|}(P_{\theta_0}^{\otimes\infty} \otimes \mu)(d\omega, d\eta) \to 0.$$

Hence,

$$W_n(\eta, \mathbf{X}_n) \to^{P_{\theta_0}^{\otimes\infty} \otimes \mu} 0. \tag{7.126}$$

The integrals in (7.123) are bounded by 2. If (7.123) is not true, then by (7.126) and by turning to a suitable subsequence n_k we find a set $A \in \mathfrak{A}^{\otimes\infty} \otimes \mathfrak{B}_d$ with $(P_{\theta_0}^{\otimes\infty} \otimes \mu)(A) = 1$ and a set $B \in \mathfrak{A}^{\otimes\infty}$ with $P_{\theta_0}^{\otimes\infty}(B) = 1$ such that

$$\liminf_{k\to\infty} \mathsf{E}_{n_k,\theta_0} \int |\pi_{n_k}^*(\eta|\mathbf{X}_n) - \varphi_{0,\mathsf{I}^{-1}(\theta_0)}(\eta)|\boldsymbol{\lambda}_d(d\eta) > 0, \tag{7.127}$$

and

$$\begin{aligned} &\lim\nolimits_{k\to\infty} V_{n_k}(\eta, \mathbf{X}_{n_k}(\omega)) = \pi(\theta_0)\exp\{-\tfrac{1}{2}\eta^T \mathsf{I}(\theta_0)\eta\}, \quad (\omega, \eta) \in A, \\ &\int V_{n_k}(\eta, \mathbf{X}_{n_k}(\omega))\boldsymbol{\lambda}_d(d\eta) \to \pi(\theta_0)(2\pi)^{d/2}\sqrt{\det(\mathsf{I}(\theta_0))}, \quad \omega \in B. \end{aligned} \tag{7.128}$$

Set $A_\omega = \{\eta : (\omega, \eta) \in A\}$. Then $A_\omega \in \mathfrak{B}_d$, and by Fubini's theorem

$$\int [\int I_{A_\omega}(\eta)\mu(d\eta)]P_{\theta_0}^{\otimes\infty}(d\omega) = \int I_A(\omega, \eta)(P_{\theta_0}^{\otimes\infty} \otimes \mu)(d\omega, d\eta) = 1.$$

Hence $P_{\theta_0}^{\otimes\infty}(\{\omega : \mu(A_\omega) = 1\}) = 1$, and by the equivalence of the measures μ and $\boldsymbol{\lambda}_d$ we get $P_{\theta_0}^{\otimes\infty}(\{\omega : \boldsymbol{\lambda}_d(\mathbb{R}_d \backslash A_\omega) = 0\}) = 1$. Put

$$C = \{\omega : \boldsymbol{\lambda}_d(\mathbb{R}_d \backslash A_\omega) = 0\} \cap B.$$

Then for every fixed $\omega \in C$ we obtain from

$$\pi_{n_k}^*(\eta|\mathbf{X}_{n_k}(\omega)) = (\int V_{n_k}(\eta, \mathbf{X}_{n_k}(\omega))\boldsymbol{\lambda}_d(d\eta))^{-1}V_{n_k}(\eta, \mathbf{X}_{n_k}(\omega))$$

and (7.128) that

$$\pi^*_{n_k}(\eta|\mathbf{X}_{n_k}(\omega)) \to \varphi_{0,\mathsf{I}^{-1}(\theta_0)}(\eta), \ \boldsymbol{\lambda}_d\text{-a.e., and } \int \pi^*_{n_k}(\eta|\mathbf{X}_n)\boldsymbol{\lambda}_d(d\eta) \to 1.$$

An application of Scheffé's lemma (see Lemma A.19) gives for every $\omega \in C$,

$$\lim_{k\to\infty} \int |\pi^*_{n_k}(\eta|\mathbf{X}_{n_k}(\omega)) - \varphi_{0,\mathsf{I}^{-1}(\theta_0)}(\eta)|\boldsymbol{\lambda}_d(d\eta) = 0.$$

As $P_{\theta_0}^{\otimes\infty}(C) = 1$ we may take the expectation, which we may exchange with the limit as the integrals are bounded by 2, and obtain a contradiction to (7.127). The proof of the corollary is similar. ■

Remark 7.157. The Bernstein–von Mises theorem has a long history. An early version is in Laplace (1820). This result was reobtained by Bernstein (1917) and von Mises (1931). More general versions are due to LeCam (1958), Bickel and Yahav (1969), and Ibragimov and Has'minskii (1972, 1981). Ghosh, Sinha, and Joshi (1981) give expansions of the posterior that refine posterior normality. For further references we refer to Ghosh and Ramamoorthi (2003). The version that we have presented here follows Ferguson (1986).

We finally note that the assumption that $\mathbf{X}_n$ consists of i.i.d. components is used only at a few places. Indeed, to establish Lemma 7.154 one needs only that the MLE is consistent and that the sequence $\ddot{\Lambda}_n(\theta)$ is equicontinuous at θ_0 in the sense that we may plug in $\widehat{\theta}_n(\mathbf{X}_n)$ without changing the asymptotic. For details in the more general situation where $X_1, X_2, \ldots$ are not necessarily i.i.d. we refer to Schervish (1995) and the references there.

In the above version of the Bernstein–von Mises theorem, besides a standard Taylor expansion, the crucial points are the conditions (7.122) and (7.124). These conditions guarantee a uniform integrability of the corresponding functions of η which, together with the pointwise convergence in (7.121), provide the $\mathbb{L}_1$-convergence. To find sufficient conditions for (7.122) and (7.124) one needs assumptions that allow one to control the log-likelihood and henceforth the posterior density for large η. For details we refer to Ghosh and Ramamoorthi (2003) and Lehmann (1998). Here we only show that (7.122) and (7.124) are satisfied for exponential families.

Example 7.158. We consider a one-parameter exponential family $(P_\theta)_{\theta\in\Delta}$ and recall that the density is given by $f_\theta(x) = \exp\{\theta T(x) - K(\theta)\}$. If $\theta_0 \in \Delta^0$, then by Proposition 7.93 and the arguments in the proof of Proposition 7.98 there is a set $A \in \mathfrak{A}^{\otimes\infty}$ with $P_{\theta_0}^{\otimes\infty}(A) = 1$ such that for every $\omega \in A$ the MLE $\widehat{\theta}_n(\mathbf{X}_n(\omega))$ exists for all sufficiently large n,

$$\dot{K}(\widehat{\theta}_n(\mathbf{X}_n(\omega))) = \frac{1}{n}\sum\nolimits_{i=1}^n T(X_i(\omega)) \quad \text{and} \quad \widehat{\theta}_n(\mathbf{X}_n(\omega)) \to \theta_0.$$

We fix $\varepsilon > 0$ so that $C = \{\theta : |\theta - \theta_0| \le 2\varepsilon\} \subseteq \Delta^0$. Put $D_0 = \inf_{\theta\in C} \ddot{K}(\theta)$. Then $D_0 > 0$, and for all sufficiently large n and $|\eta| < \varepsilon\sqrt{n}$,

$$
\begin{aligned}
&U_n(\eta, \mathbf{X}_n(\omega)) \\
&= n[\frac{\eta}{\sqrt{n}}\frac{1}{n}\sum\nolimits_{i=1}^{n} T(X_i(\omega)) - K(\widehat{\theta}_n(\mathbf{X}_n(\omega)) + \eta/\sqrt{n}) + K(\widehat{\theta}_n(\mathbf{X}_n(\omega)))] \\
&= n[\frac{\eta}{\sqrt{n}}\dot{K}(\widehat{\theta}_n(\mathbf{X}_n(\omega))) - K(\widehat{\theta}_n(\mathbf{X}_n(\omega)) + \eta/\sqrt{n}) + K(\widehat{\theta}_n(\mathbf{X}_n(\omega)))] \\
&= -\eta^2 \int_0^1 (1-s)\ddot{K}(\widehat{\theta}_n(\mathbf{X}_n(\omega)) + s\eta/\sqrt{n})\eta ds \le -\frac{\eta^2}{2} D_0.
\end{aligned}
$$

Hence with $D_1 = \sup_{\theta \in C} \pi(\theta)$, for fixed ω, and all sufficiently large n,

$$V_n(\eta, \mathbf{X}_n(\omega)) I_{[-\varepsilon\sqrt{n}, \varepsilon\sqrt{n}]}(\eta) \le D_1 \exp\{-\frac{\eta^2}{2} D_0\}.$$

To deal with the case $|\eta| \ge \varepsilon\sqrt{n}$ we note that by the strict convexity of K

$$\inf_{|\theta-\theta_0|\le\varepsilon, |h|>\varepsilon} -[h\dot{K}(\theta) - K(\theta+h) + K(\theta)] =: A > 0.$$

Hence,

$$
\begin{aligned}
&V_n(\eta, \mathbf{X}_n(\omega)) I_{[\varepsilon\sqrt{n},\infty)}(|\eta|) \le \exp\{-An\}\pi(\widehat{\theta}_n(\mathbf{X}_n(\omega)) + \eta/\sqrt{n}) I_{[\varepsilon\sqrt{n},\infty)}(|\eta|), \\
&\int V_n(\eta, \mathbf{X}_n(\omega)) I_{[\varepsilon\sqrt{n},\infty)}(|\eta|) d\eta \le \sqrt{n}\exp\{-An\} \to 0.
\end{aligned}
$$

Lebesgue's theorem yields

$$
\begin{aligned}
&\lim_{n\to\infty} \int V_n(\eta, \mathbf{X}_n(\omega)) d\eta = \lim_{n\to\infty} \int V_n(\eta, \mathbf{X}_n(\omega)) I_{[-\varepsilon\sqrt{n}, \varepsilon\sqrt{n}]}(\eta) d\eta \\
&= \int \lim_{n\to\infty} V_n(\eta, \mathbf{X}_n(\omega)) d\eta = \int \pi(\theta_0) \exp\{-\frac{1}{2}\eta^T \mathsf{I}(\theta_0)\eta\} d\eta \\
&= \pi(\theta_0)(2\pi)^{d/2}(\det(\mathsf{I}(\theta_0)))^{-1/2}.
\end{aligned}
$$

With $\pi_n(\theta|\mathbf{x}_n)$ in (7.120) it holds

$$\int \|\theta\| \,\Pi(d\theta) = \int [\int \|\theta\| \,\pi_n(\theta|x) \boldsymbol{\lambda}_d(d\theta)](\mathsf{P}_n\Pi)(dx).$$

If $\Pi(\{\theta : \mathsf{K}(P_{\theta_0}, P_\theta) < \infty\}) > 0$, then $P_{\theta_0}^{\otimes n} \ll \mathsf{P}_n\Pi$ by Problem 7.119, so that under the condition $\int \|\theta\| \,\Pi(d\theta) < \infty$,

$$\widetilde{\theta}_n(\mathbf{x}_n) := \int \theta\pi_n(\theta|\mathbf{x}_n)\boldsymbol{\lambda}_d(d\theta)$$

is $\mathsf{P}_n\Pi$-a.s. and $P_{\theta_0}^{\otimes n}$-a.s. well defined. If $\int \|\theta\|^2 \,\Pi(d\theta) < \infty$, then $\widetilde{\theta}_n$ is the Bayes estimator with respect to the quadratic loss function $L(\theta, a) = \|\theta - a\|^2$. If the weaker condition

$$\int \|\theta\|^2 \,\pi_n(\theta|\mathbf{x}_n)\boldsymbol{\lambda}_d(d\theta) < \infty, \quad \mathsf{P}_n\Pi\text{-a.s.},$$

holds, then $\widetilde{\theta}_n(\mathbf{x}_n)$ is a generalized Bayes estimator in the sense that $\widetilde{\theta}_n(\mathbf{x}_n)$ minimizes the posterior risk

$$r(\Pi, a|\mathbf{x}_n) = \int \|\theta - a\|^2 \, \pi_n(\theta|\mathbf{x}_n)\boldsymbol{\lambda}_d(d\theta)$$

of making the decision a after $\mathbf{x}_n$ has been observed, see Proposition 7.39.

Theorem 7.159. *Suppose condition (A10) is satisfied, where Δ is an open subset of $\mathbb{R}^d$. Suppose $\widehat{\theta}_n : \mathcal{X}^n \to_m \Delta$ is consistent at θ_0 and an asymptotic solution of the likelihood equation; that is, (7.109) holds. Assume that the prior Π has a density π with respect to the Lebesgue measure which is continuous at θ_0, $\pi(\theta_0) > 0$, and satisfies $\int \|\theta\| \, \pi(\theta)\boldsymbol{\lambda}_d(d\theta) < \infty$. If $\mathsf{I}(\theta_0)$ is nonsingular and the condition (7.124) holds, then*

$$\sqrt{n}(\widetilde{\theta}_n(\mathbf{X}_n) - \widehat{\theta}_n(\mathbf{X}_n)) \to^{P_{\theta_0}^{\otimes\infty}} 0 \quad \textit{and}$$
$$\mathcal{L}(\sqrt{n}(\widetilde{\theta}_n(\mathbf{X}_n) - \theta_0)|P_{\theta_0}^{\otimes\infty}) \Rightarrow \mathsf{N}(0, \mathsf{I}^{-1}(\theta_0)).$$

Proof. It holds

$$\sqrt{n}(\widetilde{\theta}_n(\mathbf{X}_n) - \widehat{\theta}_n(\mathbf{X}_n)) = \int \eta \pi_n^*(\eta|\mathbf{X}_n)\boldsymbol{\lambda}_d(d\eta), \quad \int \eta \varphi_{0,\mathsf{I}^{-1}(\theta_0)}(\eta)\boldsymbol{\lambda}_d(d\eta) = 0.$$

Hence

$$\| \sqrt{n}(\widetilde{\theta}_n(\mathbf{X}_n) - \widehat{\theta}_n(\mathbf{X}_n)) \| \leq \int \|\eta\| \, |\pi_n^*(\eta|\mathbf{X}_n) - \varphi_{0,\mathsf{I}^{-1}(\theta_0)}(\eta)|\boldsymbol{\lambda}_d(d\eta) \to^{P_{\theta_0}^{\otimes\infty}} 0$$

in view of (7.125). The second statement follows from Theorem 7.148 and Slutsky's lemma; see Lemma A.46. ■

We illustrate the Bernstein–von Mises theorem. A first example is Example 7.153 where we have established the equivalence of the MLE and the Bayes estimators for the binomial model with the conjugate beta prior. The next example concerns the gamma distribution with a gamma prior.

Example 7.160. By taking the derivative with respect to β of the log-likelihood function $\sum_{i=1}^n \ln \mathsf{ga}_{\lambda,\beta}(x_i)$, and then putting the corresponding expression equal to zero, one can see that for a fixed known λ_0 the solution of the likelihood equation for β in the model with $\mathsf{Ga}^{\otimes n}(\lambda_0, \beta)$ is

$$\widehat{\beta}_n(x_1, ..., x_n) = \frac{\lambda_0}{\overline{x}_n},$$

which according to Problem 7.91 is the MLE. If we take for P_θ the distribution $\mathsf{Ga}(\lambda_0, \beta)$, and $\Pi = \mathsf{Ga}(a, b)$ as the prior for the parameter β, then similar to Problem 1.32 the posterior density in π_n in (7.120) is given by

$$\pi_n(\beta|\mathbf{x}_n) = \mathsf{ga}_{n\lambda_0+a, n\overline{x}_n+b}(\beta),$$

where $\mathbf{x}_n = (x_1, ..., x_n)$. As for any $\beta, \lambda > 0$ the distribution $\mathsf{Ga}(\lambda, \beta)$ has the expectation λ/β we get that the Bayes estimator under the quadratic loss is

$$\widetilde{\beta}_n(x_1, ..., x_n) = \frac{\lambda_0 + a/n}{\overline{x}_n + b/n}.$$

Hence for $\mathbf{X}_n \sim \mathsf{Ga}^{\otimes n}(\lambda_0, \beta_0)$,

$$\begin{aligned}\sqrt{n}(\widetilde{\beta}_n(\mathbf{X}_n) - \widehat{\beta}_n(\mathbf{X}_n)) &= \sqrt{n}(\frac{\lambda_0 + a/n}{\overline{X}_n + b/n} - \frac{\lambda_0}{\overline{X}_n}) \\ &= \frac{1}{\sqrt{n}\,\overline{X}_n(\overline{X}_n + b/n)}\overline{X}_n(a - \lambda_0 b) \longrightarrow^{\mathsf{Ga}^{\otimes n}(\lambda_0,\beta_0)} 0,\end{aligned}$$

which is the equivalence stated in the previous theorem.

7.7 Local Asymptotic Optimality of MLEs

In the first part of this chapter we have established lower bounds for the risk of estimating parameters under finite sample sizes. One of them is given by the Cramér–Rao inequality, and unbiased estimators that attain this lower bound are automatically UMVU. The large sample counterpart of the Cramér–Rao inequality is the lower Hájek–LeCam bound for the risks in a sequence of convergent models, and the natural question arises as to which estimators attain this lower bound.

Another aspect is that in some introductory textbooks a sequence of estimators is called *efficient* if $\mathcal{L}(\sqrt{n}(\widehat{\theta}_n - \theta_0)|P_{\theta_0}^{\otimes n}) \Rightarrow \mathsf{N}(0, \mathsf{I}^{-1}(\theta_0))$. One can easily find various examples of estimators that have this property. Moreover, we already know that under weak regularity conditions the MLE also has this property. So one could conclude that the MLE is asymptotically efficient. However, this approach is not satisfactory from a rigorous mathematical point of view. The question that remains open is whether in special situations there are any other suitably constructed estimators that outperform the MLE. The goal of this section is to find additional conditions on the estimators such that within this class one can find asymptotically optimal estimators. Then it turns out that the MLE asymptotically outperforms all estimators from this class. These additional conditions correspond to the finite sample size concepts of unbiasedness, median unbiasedness, and equivariance. It has been a long way for statisticians to figure out such additional conditions that clarify in which sense the MLE is asymptotically the best estimator. For a detailed presentation of the history of asymptotic properties of the MLE we refer to Pfanzagl (1994). Here in this section we follow Pfanzagl (1994), Rieder (1994), and Witting and Müller-Funk (1995), from which the main results have been taken.

The starting point for the development of proving the asymptotic efficiency of the MLE is probably Fisher (1922). A breakthrough was an example by Hodges, Jr. which shows that even in the simple case of estimating the mean of $\mathsf{N}^{\otimes n}(\theta, 1)$, $\theta \in \mathbb{R}$, where the sample mean $\overline{X}_n$ is the UMVU estimator and also the MLE, this estimator is asymptotically not the best estimator for every

$\theta \in \mathbb{R}$. This fact is called the effect of *superefficiency*. The following example is due to Hodges, Jr. who did not publish this result. LeCam (1953) and LeCam and Yang (1990) refer to Hodges, Jr.

Example 7.161. We consider the model $(\mathbb{R}, \mathfrak{B}, (\mathsf{N}(\theta,1)_{\theta\in\mathbb{R}}))$, which at every $\theta_0 \in \mathbb{R}$ is $\mathbb{L}_2$-differentiable and has Fisher information $\mathsf{I}(\theta_0) = 1$. If we have an i.i.d. sample $X_1, ..., X_n$, then $\overline{X}_n$ is the UMVU estimator as well as the MLE, and we have $\mathcal{L}(\sqrt{n}(\overline{X}_n - \theta_0)|\mathsf{N}^{\otimes n}(\theta_0, 1)) = \mathsf{N}(0,1)$. As $\mathsf{I}(\theta_0) = 1$ one might consider, in view of the above discussion, $\overline{X}_n$ as asymptotically efficient. Now we modify $\overline{X}_n$ by setting

$$\widetilde{X}_n = \begin{cases} \overline{X}_n & \text{if } |\overline{X}_n| > n^{-1/4}, \\ \frac{1}{2}\overline{X}_n & \text{if } |\overline{X}_n| \le n^{-1/4}. \end{cases}$$

Put $P_{n,\theta_0} = \mathsf{N}^{\otimes n}(\theta_0, 1)$. Then

$$\begin{aligned} P_{n,\theta_0}(\widetilde{X}_n \neq \overline{X}_n) &= P_{n,\theta_0}(|\overline{X}_n| \le n^{-1/4}) \\ &= \Phi_{0,1}(-\sqrt{n}\theta_0 + n^{1/4}) - \Phi_{0,1}(-\sqrt{n}\theta_0 - n^{1/4}). \end{aligned}$$

For $\theta_0 \neq 0$ it follows $P_{n,\theta_0}(\widetilde{X}_n \neq \overline{X}_n) \to 0$, and Slutsky's lemma yields

$$\mathcal{L}(\sqrt{n}(\widetilde{X}_n - \theta_0)|\mathsf{N}^{\otimes n}(\theta_0, 1)) \Rightarrow \mathsf{N}(0,1).$$

For $\theta_0 = 0$ it holds

$$P_{n,\theta_0}(\widetilde{X}_n \neq \frac{1}{2}\overline{X}_n) = P_{n,0}(|\overline{X}_n| > n^{-1/4}) = 1 - \Phi_{0,1}(n^{1/4}) + \Phi_{0,1}(-n^{1/4}) \to 0,$$

and, again by Slutsky's lemma, $\mathcal{L}(\sqrt{n}\widetilde{X}_n|\mathsf{N}^{\otimes n}(0,1)) \Rightarrow \mathsf{N}(0, 1/4)$. Hence at $\theta_0 = 0$ we have the effect of superefficiency in the sense that the variance of the asymptotic normal distribution of the normalized estimator $\sqrt{n}(\widetilde{X}_n - \theta_0)$ is smaller than the inverse of the Fisher information, i.e., the lower bound in the Cramér–Rao inequality.

From the above example one can get the impression that a suitable modification of a good estimator with the aim of producing superefficiency is only possible for special points of the parameter space, i.e., that the set of such exceptional points is expected to be small. The subsequent theorem makes "small" precise by showing that this set has Lebesgue measure zero. This results was established by LeCam (1953) and Bahadur (1964).

Theorem 7.162. ***(LeCam–Bahadur)*** *Suppose $\Delta \subseteq \mathbb{R}^d$ is open, the model $(\mathcal{X}, \mathfrak{A}, (P_\theta)_{\theta\in\Delta})$ is $\mathbb{L}_2$-differentiable at every $\theta_0 \in \Delta$, and the Fisher information matrix $\mathsf{I}(\theta)$ is nonsingular for every $\theta \in \Delta$. Let $T_n : \mathcal{X}^n \to_m \Delta$ be a sequence of estimators such that $\mathcal{L}(\sqrt{n}(T_n - \theta)|P_\theta^{\otimes n}) \Rightarrow \mathsf{N}(0, \Sigma(\theta))$ for every $\theta \in \Delta$. Then there is a set N of Lebesgue measure zero such that, in terms of the Löwner semiorder,*

$$\mathsf{I}^{-1}(\theta) \preceq \Sigma(\theta), \quad \theta \in \Delta \backslash N.$$

The proof of this theorem is based on the asymptotic power of best tests. We deal with this topic in Chapter 8 and thus present a proof of the above theorem there on page 493.

In Theorem 7.162 the sequence of estimators was completely arbitrary. Now we consider estimators that admit similar representations as the M-estimators. Let $\mathcal{M}_n = (\mathcal{X}^n, \mathfrak{A}^{\otimes n}, (P_\theta^{\otimes n})_{\theta \in \Delta})$, with $\Delta \subseteq \mathbb{R}^d$, be a sequence of models, and for a fixed $\theta_0 \in \Delta^0$ set $P_{n,\theta_0} = P_{\theta_0}^{\otimes n}$. Given the sequence of models $\mathcal{M}_n$, we say that the sequence of estimators $\widehat{\theta}_n : \mathcal{X}^n \to_m \Delta$ is *asymptotically linear at* θ_0 *with influence function* Ψ_{θ_0} if $\Psi_{\theta_0} \in \mathbb{L}_{2,d}^0(P_{\theta_0})$ and

$$\sqrt{n}(\widehat{\theta}_n - \theta_0) = \frac{1}{\sqrt{n}} \sum\nolimits_{i=1}^n \Psi_{\theta_0}(X_i) + o_{P_{n,\theta_0}}(1) \quad \text{or} \tag{7.129}$$

$$\widehat{\theta}_n = \theta_0 + \frac{1}{n} \sum\nolimits_{i=1}^n \Psi_{\theta_0}(X_i) + o_{P_{n,\theta_0}}(n^{-1/2}).$$

The second statement is often referred to as a *first-order stochastic Taylor expansion.* The stochastic Taylor expansion (7.129) has been established in Proposition 7.144 for consistent estimators that solve the M-equation asymptotically. In that case the influence function had the special structure $\Psi_{\theta_0} = \Sigma_0^{-1} \psi_{\theta_0}$, where $\Sigma_0 = \mathsf{C}_{\theta_0}(\psi_{\theta_0}, \dot{L}_{\theta_0})$, and where $\mathsf{C}_{\theta_0}(X, Y)$ is the covariance matrix of the random vectors X and Y. If $\widehat{\theta}_{n,MLE}$ is a sequence of MLEs, then the expansion (7.129) is also valid for $\Psi_{\theta_0} = \mathsf{I}^{-1}(\theta_0)\dot{L}_{\theta_0}$ under the assumptions formulated in Theorem 7.148. A sequence of estimators $\widehat{\theta}_n : \mathcal{X}_n \to_m \Delta$ is called *regular at* $\theta_0 \in \Delta^0$ if there is a distribution Q_{θ_0} on $(\mathbb{R}^d, \mathfrak{B}_d)$ such that

$$\mathcal{L}(\sqrt{n}(\widehat{\theta}_n - (\theta_0 + h/\sqrt{n}))|P_{n,\theta_0+h/\sqrt{n}}) \Rightarrow Q_{\theta_0}, \quad h \in \mathbb{R}^d.$$

Proposition 7.163. *Suppose that* $(P_\theta)_{\theta \in \Delta}$ *is* $\mathbb{L}_2$*-differentiable with derivative* $\dot{L}_{\theta_0}$. *Then for the sequence of models* $(\mathcal{X}^n, \mathfrak{A}^{\otimes n}, (P_\theta^{\otimes n})_{\theta \in \Delta})$ *a sequence of estimators* $\widehat{\theta}_n$ *that satisfy (7.129) is regular at* θ_0 *if and only if* $\mathsf{C}_{\theta_0}(\Psi_{\theta_0}, \dot{L}_{\theta_0}) = \mathbf{I}$.

Proof. Corollary 6.74 shows that

$$\mathcal{L}(\sqrt{n}(\widehat{\theta}_n - (\theta_0 + h/\sqrt{n}))|P_{n,\theta_0+h/\sqrt{n}}) \Rightarrow \mathsf{N}(\mathsf{C}_{\theta_0}(\Psi_{\theta_0}, \dot{L}_{\theta_0})h - h, \mathsf{C}_{\theta_0}(\Psi_{\theta_0})).$$

The statement follows as h is arbitrary. ■

If the conditions in Proposition 7.144 are satisfied, then

$$\sqrt{n}(\widehat{\theta}_n - \theta_0) = \frac{1}{\sqrt{n}} \sum\nolimits_{i=1}^n (\mathsf{C}_{\theta_0}(\psi_{\theta_0}, \dot{L}_{\theta_0}))^{-1} \psi_{\theta_0}(X_i) + o_{P_{\theta_0}^{\otimes n}}(1),$$

so that $\Psi_{\theta_0} = (\mathsf{C}_{\theta_0}(\psi_{\theta_0}, \dot{L}_{\theta_0}))^{-1}\psi_{\theta_0}$ satisfies the condition $\mathsf{C}_{\theta_0}(\Psi_{\theta_0}, \dot{L}_{\theta_0}) = \mathbf{I}$ and $\widehat{\theta}_n$ is regular.

The motivation for introducing the concept of regularity is to exclude the effect of superefficiency; that is, the opportunity to get an estimator that is better than the MLE at isolated points. To see that the Hodges estimator $\widetilde{X}_n$ is not regular we note that by Problem 1.86 and (1.79) it holds

$$\begin{aligned}\mathsf{H}_s(\mathsf{N}^{\otimes n}(0,1),\mathsf{N}^{\otimes n}(h/\sqrt{n},1)) &= \left(\mathsf{H}_s(\mathsf{N}(0,1),\mathsf{N}(h/\sqrt{n},1))\right)^n \\ &= \exp\{-\frac{1}{2}s(1-s)h^2\}.\end{aligned}$$

Hence $P_{n,h} = \mathsf{N}^{\otimes n}(h/\sqrt{n},1) \lhd\rhd P_{n,0} = \mathsf{N}^{\otimes n}(0,1)$ by Theorem 6.26. Using the notation in Example 7.161 and the fact that $P_{n,h} \lhd\rhd P_{n,0}$ we get from (A) in Theorem 6.26 that $P_{n,h}(\widetilde{X}_n \neq \frac{1}{2}\overline{X}_n) \to 0$, and by Slutsky's lemma

$$\begin{aligned}&\lim_{n\to\infty} \mathcal{L}(\sqrt{n}(\widetilde{X}_n - h/\sqrt{n})|\mathsf{N}^{\otimes n}(h/\sqrt{n},1)) \\ &= \lim_{n\to\infty} \mathcal{L}(\sqrt{n}(\frac{1}{2}\overline{X}_n - h/\sqrt{n})|\mathsf{N}^{\otimes n}(h/\sqrt{n},1)) = \mathsf{N}(-h/2,1/4),\end{aligned}$$

which obviously depends on h, so that $\widetilde{X}_n$ is not regular.

Now we turn back to estimators that have the stochastic Taylor expansion (7.129), which implies the asymptotic normality

$$\mathcal{L}(\sqrt{n}(\widehat{\theta}_n - \theta_0)|P_{\theta_0}^{\otimes n}) \Rightarrow \mathsf{N}(0, \mathsf{C}_{\theta_0}(\Psi_{\theta_0})).$$

This allows a comparison, within the class of regular estimators that satisfy (7.129), of the asymptotic efficiency of the estimators by comparing the covariance matrices in the limit distributions.

Proposition 7.164. *Suppose $(P_\theta)_{\theta\in\Delta}$ is $\mathbb{L}_2$-differentiable at $\theta_0 \in \Delta^0$ with nonsingular Fisher information matrix $\mathsf{I}(\theta_0)$. Let $\widehat{\theta}_n$ be an asymptotic MLE that has the stochastic Taylor expansion (7.113). Then $\widehat{\theta}_n$ is asymptotically efficient, in the sense of $\mathsf{C}_{\theta_0}(\Psi_{\theta_0}) \succeq \mathsf{I}^{-1}(\theta_0)$, within the class of all regular estimators $\widetilde{\theta}_n$ that admit a stochastic Taylor expansion (7.129). It holds $\mathsf{C}_{\theta_0}(\Psi_{\theta_0}) = \mathsf{I}^{-1}(\theta_0)$ if and only if $\Psi_{\theta_0} = \mathsf{I}^{-1}(\theta_0)\dot{L}_{\theta_0}$, P_{θ_0}-a.s..*

Proof. First of all we note that by (7.113) the MLE has the influence function $\mathsf{I}^{-1}(\theta_0)\dot{L}_{\theta_0}$ and is thus regular by Proposition 7.163. If $\widetilde{\theta}_n$ has the expansion (7.129), then the assumed regularity implies, again by Proposition 7.163, that $\mathsf{C}_{\theta_0}(\Psi_{\theta_0}, \dot{L}_{\theta_0}) = \mathbf{I}$. Hence, in terms of the Löwner semiorder,

$$\begin{aligned}0 &\preceq \mathsf{E}_{\theta_0}(\Psi_{\theta_0} - \mathsf{I}^{-1}(\theta_0)\dot{L}_{\theta_0})(\Psi_{\theta_0} - \mathsf{I}^{-1}(\theta_0)\dot{L}_{\theta_0})^T \\ &= \mathsf{C}_{\theta_0}(\Psi_{\theta_0}) + \mathsf{E}_{\theta_0}\mathsf{I}^{-1}(\theta_0)\dot{L}_{\theta_0}\dot{L}_{\theta_0}^T\mathsf{I}^{-1}(\theta_0) - \mathsf{E}_{\theta_0}\mathsf{I}^{-1}(\theta_0)\dot{L}_{\theta_0}\Psi_{\theta_0}^T \\ &\quad - \mathsf{E}_{\theta_0}\Psi_{\theta_0}\dot{L}_{\theta_0}^T\mathsf{I}^{-1}(\theta_0) = \mathsf{C}_{\theta_0}(\Psi_{\theta_0}) - \mathsf{I}^{-1}(\theta_0).\end{aligned}$$

To complete the proof we note that for a random vector X with expectation zero it holds $\mathsf{C}_{\theta_0}(X) = 0$ if and only if $X = 0$, P_{θ_0}-a.s. ■

Remark 7.165. There is a simple interpretation for the inequality $\mathsf{C}_{\theta_0}(\Psi_{\theta_0}) \succeq \mathsf{I}^{-1}(\theta_0)$. In the class of all sequences of estimators that admit the representation

(7.129) the estimators for which the influence function Ψ_{θ_0} is constructed with the help of the "gradient" $\dot{L}_{\theta_0}$ of the model at θ_0 are most efficient. It is clear that after a linearization of the model such statistics reflect best the local behavior of the model. The influence function $\mathsf{I}^{-1}(\theta_0)\dot{L}_{\theta_0}$ that belongs to the MLE is, in view of $\mathsf{C}_{\theta_0}(\Psi_{\theta_0}) \succeq \mathsf{I}^{-1}(\theta_0) = \mathsf{C}_{\theta_0}(\mathsf{I}^{-1}(\theta_0)\dot{L}_{\theta_0})$, the most efficient influence function. The $\mathbb{L}_2$-derivative $\dot{L}_{\theta_0}$ is also called the score function. Thus the score function provides, after a normalization, the most efficient influence function.

We have studied regular estimators with a stochastic Taylor expansion in Proposition 7.164. Now we investigate the larger class of all regular estimators. The condition of regularity corresponds for a fixed sample size to the requirement of equivariance. The latter is a condition that refers to every sample. In contrast to this requirement of equivariance the regularity is formulated in terms of the distributions induced by the sequence of statistics and is adapted to convergent sequences of models. The convolution theorem of Hájek (1970) and Inagaki (1970) states that for sequences of asymptotic normal models every sequence of regular estimators is asymptotically more spread out than the central variable in the limiting Gaussian model. The finite sample counterpart is the fact that the identity is under the squared error loss the best equivariant estimator for the family $(\mathsf{N}(\mu,\sigma^2))_{\mu\in\mathbb{R}}$; see Example 5.59 and Theorem 5.56.

Theorem 7.166. ***(Hájek–Inagaki)*** *Suppose* $\mathcal{M}_n = (\mathcal{X}_n, \mathfrak{A}_n, (P_{n,h})_{h\in\Delta_n})$, $\Delta_n \uparrow \mathbb{R}^d$, *satisfies the LAN*$(Z_n, \mathsf{I}_0)$ *condition in Definition 6.63. If* I_0 *is nonsingular and* $S_n : \mathcal{X}_n \to_m \mathbb{R}^d$ *is a sequence such that* $\mathcal{L}(S_n - h|P_{n,h})$ *converges for every* $h \in \mathbb{R}$ *weakly to some distribution* Q *that is independent of* h*, then there is a distribution* R *on* $(\mathbb{R}^d, \mathfrak{B}_d)$ *such that the following hold.*

(A) $Q = \mathsf{N}(0, \mathsf{I}_0^{-1}) * R.$

(B) $\mathcal{L}(S_n - \mathsf{I}_0^{-1}Z_n|P_{n,h}) \Rightarrow R.$

(C) $S_n = \mathsf{I}_0^{-1}Z_n + o_{P_{n,0}}(1) \;\Leftrightarrow\; R = \boldsymbol{\delta}_0.$

Proof. The LAN condition implies $\mathcal{L}(Z_n|P_{n,0}) \Rightarrow \mathsf{N}(0,\mathsf{I}_0)$. As by assumption $\mathcal{L}(S_n|P_{n,0}) \Rightarrow Q$, we get that Z_n and S_n are stochastically bounded with respect to $P_{n,0}$. Then the vector (S_n, Z_n) with values in $\mathbb{R}^{2d}$ is stochastically bounded as well. As a subset of $\mathbb{R}^{2d}$ is compact if and only if it is bounded and closed, we get from Theorem A.48 that the sequence $\mathcal{L}((S_n, Z_n)|P_{n,0})$ is tight, so that there is a subsequence n_k with $\mathcal{L}((S_{n_k}, Z_{n_k})|P_{n,0}) \Rightarrow \mu$ for some distribution μ on $\mathfrak{B}_{2d}$. Denote by S and Z the projections of $\mathbb{R}^{2d} = \mathbb{R}^d \times \mathbb{R}^d$ on the first and second component of this product, respectively. Hence with this notation and the LAN property

$$
\begin{aligned}
&\mathcal{L}((S_{n_k}, Z_{n_k})|P_{n_k,0}) \Rightarrow \mathcal{L}((S,Z)|\mu) \quad \text{and} \quad \mathcal{L}(Z|\mu) = \mathsf{N}(0,\mathsf{I}_0),\\
&L_{n,h} = dP_{n,h}/dP_{n,0} = \exp\{h^T Z_n - \frac{1}{2}h^T \mathsf{I}_0 h + o_{P_{n,0}}(1)\}.
\end{aligned}
$$

Set

$$\begin{aligned}V_n &= \exp\{it^T S_n - it^T h\} L_{n,h} \\ &= \exp\{it^T S_n - it^T h + h^T Z_n - \frac{1}{2} h^T \mathsf{I}_0 h + o_{P_{n,0}}(1)\} \quad \text{and} \\ V &= \exp\{it^T S - it^T h + h^T Z - \frac{1}{2} h^T \mathsf{I}_0 h\}.\end{aligned}$$

The LAN condition implies that the sequence $P_{n,h}$ is contiguous with respect to $P_{n,0}$; see Corollary 6.67. This implies in view of Theorem 6.26 that the sequence $L_{n_k,h}$, and consequently also the sequence V_{n_k}, is uniformly integrable with respect to $P_{n_k,0}$ in the sense that

$$\lim_{N\to\infty} \limsup_{k\to\infty} \mathsf{E}_{n_k,0} |V_{n_k}| I_{[N,\infty)}(|V_n|) = 0.$$

Hence by Slutsky's lemma (see Lemma A.46) and Proposition A.44,

$$\begin{aligned}\lim_{k\to\infty} \mathsf{E}_{n_k,0} V_{n_k} &= \lim_{k\to\infty} \mathsf{E}_{n_k,h} \exp\{it^T S_{n_k} - it^T h\} \\ &= \mathsf{E}_\mu \exp\{it^T S - it^T h + h^T Z - \frac{1}{2} h^T \mathsf{I}_0 h\}.\end{aligned}$$

Otherwise, by the assumption $\lim_{n\to\infty} \mathcal{L}(S_n - h|P_{n,h}) = \lim_{n\to\infty} \mathcal{L}(S_n|P_{n,0})$,

$$\lim_{k\to\infty} \mathsf{E}_{n_k,h} \exp\{it^T S_{n_k} - it^T h\} = \lim_{k\to\infty} \mathsf{E}_{n_k,0} \exp\{it^T S_{n_k}\} = \mathsf{E}_\mu \exp\{it^T S\}.$$

Hence for every $t, h \in \mathbb{R}^d$,

$$\exp\{-it^T h - \frac{1}{2} h^T \mathsf{I}_0 h\} \mathsf{E}_\mu \exp\{it^T S + h^T Z\} = \mathsf{E}_\mu \exp\{it^T S\}. \qquad (7.130)$$

As $\mathcal{L}(Z|\mu) = \mathsf{N}(0, \mathsf{I}_0)$ it holds $\mathsf{E}_\mu \exp\{h^T Z\} < \infty$ for every h. If we represent $\exp\{it^T S\}$ as a linear combination of four nonnegative functions f_j, and set $d\boldsymbol{\mu}_j = f_j d\mu$, $j = 1, ..., 4$, then we get from Lemma 1.16 that $\mathsf{E}_\mu \exp\{it^T S + z^T Z\}$ is an analytic function of z. Hence (7.130) remains valid if we replace h with $z \in \mathbb{R}^d + i\mathbb{R}^d$. Thus we may set $h = -i\mathsf{I}_0^{-1} t$ and obtain

$$\exp\{-\frac{1}{2} t^T \mathsf{I}_0^{-1} t\} \mathsf{E}_\mu \exp\{it^T (S - \mathsf{I}_0^{-1} Z)\} = \mathsf{E}_\mu \exp\{it^T S\}.$$

The facts that the characteristic function of a convolution of distributions is the product of their characteristic functions, and that $\exp\{-\frac{1}{2} t^T \mathsf{I}_0^{-1} t\}$ is the characteristic function of $\mathsf{N}(0, \mathsf{I}_0^{-1})$, in conjunction with the uniqueness theorem for characteristic functions (see Theorem A.51) yield statement (A) for $Q = \mathcal{L}(S|\mu)$ and $R = \mathcal{L}(S - \mathsf{I}_0^{-1} Z|\mu)$.

To prove (B) we recall that by the proof of (A) the sequence (S_n, Z_n) is stochastically bounded with respect to $P_{n,0}$. Hence by the contiguity of $P_{n,h}$ with respect to $P_{n,0}$ the sequence (S_n, Z_n), and consequently the sequence $S_n - \mathsf{I}_0^{-1} Z_n$, is stochastically bounded under $P_{n,h}$, and thus the sequence of distributions $\mathcal{L}(S_n - \mathsf{I}_0^{-1} Z_n|P_{n,h})$ is tight. We have already shown in the proof of (A) that $R = \mathcal{L}(S - \mathsf{I}_0^{-1}Z|\mu)$ is an accumulation point of the sequence $\mathcal{L}(S_n - \mathsf{I}_0^{-1}Z_n|P_{n,h})$. If n_l is any subsequence so that $\mathcal{L}(S_{n_l} - \mathsf{I}_0^{-1}Z_{n_l}|P_{n_l,h})$ tends weakly to some distribution $\widetilde{R}$, say, then we apply the already established statement (A) to the sequence n_l to get $Q = \mathsf{N}(0, \mathsf{I}_0^{-1}) * \widetilde{R}$. On the other hand we already know that $Q = \mathsf{N}(0, \mathsf{I}_0^{-1}) * R$. Turning to characteristic functions and using the fact that the characteristic function of $\mathsf{N}(0, \mathsf{I}_0^{-1})$ is $\exp\{-\frac{1}{2}t^T\mathsf{I}_0^{-1}t\}$, which is nonzero for every t, we get from Theorem A.51 that $\widetilde{R} = R$. Hence all accumulation points, in the sense of weak convergence, of the tight sequence $\mathcal{L}(S_n - \mathsf{I}_0^{-1}Z_n|P_{n,h})$ are identical, so that (B) is established. (C) follows from (B) as the weak convergence of the distributions to $\boldsymbol{\delta}_0$ and the stochastic convergence to 0 are identical. ■

Remark 7.167. The statement in Theorem 7.166 was independently established in papers by Hájek (1970) and Inagaki (1970). The proof given above follows Witting and Müller-Funk (1995). For forerunners, other versions, and historical remarks we refer to Pfanzagl (1994) and Strasser (1985).

The essence of the Hájek–Inagaki convolution theorem is that the limit distribution Q_{θ_0} is more spread out than the limit distribution of the MLE. An easy way to see this is to fix independent random vectors S and U such that $\mathcal{L}(S) = \mathsf{N}(0, \mathsf{I}^{-1}(\theta_0))$ and $\mathcal{L}(U) = R$. Then $\mathcal{L}(S+U) = \mathsf{N}(0, \mathsf{I}^{-1}(\theta_0)) * R$. Moreover, if $\mathbb{E}\,\|U\|^2 < \infty$, then $\mathsf{C}(S+U) \succeq \mathsf{C}(S)$, in the Löwner semiorder, which clarifies in which sense $\mathsf{N}(0, \mathsf{I}^{-1}(\theta_0)) * R$ is more spread out than $\mathsf{N}(0, \mathsf{I}^{-1}(\theta_0))$. The consideration of the covariance matrices holds for any sum of independent random vectors. However, as we are in the specific situation where S has a normal distribution more can be said. It follows from Anderson's lemma (see Proposition 3.62) that for any nonnegative subconvex and centrally symmetric function $l : \mathbb{R}^d \to_m \mathbb{R}_+$ it holds

$$\begin{aligned} \mathbb{E}l(S+U) &= \int [\int l(s+u)\mathsf{N}(0, \mathsf{I}^{-1}(\theta_0))(ds)]R(du) \\ &\geq \int [\int l(s)\mathsf{N}(0, \mathsf{I}^{-1}(\theta_0))(ds)]R(du) = \mathbb{E}l(S), \end{aligned} \tag{7.131}$$

and the statement that $\mathcal{L}(S+U)$ is more spread out than $\mathcal{L}(S)$ is specified by this inequality.

We apply the convolution theorem to regular estimators that are based on i.i.d. observations. Recall that $\mathfrak{L}_d$ is the class of all nonnegative measurable subconvex and centrally symmetric functions l on $\mathbb{R}^d$.

Proposition 7.168. *Let $\Delta \subseteq \mathbb{R}^d$ be open. Assume that the family $(P_\theta)_{\theta\in\Delta}$ is $\mathbb{L}_2$-differentiable at θ_0, and that the Fisher information matrix $\mathsf{I}(\theta_0)$ is nonsingular. If $\widetilde{\theta}_n : \mathcal{X}^n \to_m \Delta$ is a sequence of regular estimators for which the distributions $\mathcal{L}(\sqrt{n}(\widetilde{\theta}_n - \theta_0)|P_{\theta_0}^{\otimes n})$ tend weakly to some limit distribution, then*

$$\liminf_{n\to\infty} \int l(\sqrt{n}(\widetilde{\theta}_n - \theta_0))dP_{\theta_0}^{\otimes n} \geq \int l(t)\mathsf{N}(0, \mathsf{I}^{-1}(\theta_0))(dt),$$

for every continuous $l \in \mathfrak{L}_d$. If condition (A10) is fulfilled and $\widehat{\theta}_n$ is a sequence of estimators that solves the likelihood equation asymptotically, i.e., satisfies (7.109), and is consistent at θ_0, then $\widehat{\theta}_n$ is a sequence of regular estimators that is asymptotically efficient in the sense that this sequence achieves the above lower bound for every bounded and continuous $l \in \mathfrak{L}_d$, i.e.,

$$\lim_{n\to\infty} \int l(\sqrt{n}(\widehat{\theta}_n - \theta_0))dP_{\theta_0}^{\otimes n} = \int l(t)\mathsf{N}(0, \mathsf{I}^{-1}(\theta_0))(dt).$$

Proof. The $\mathbb{L}_2$-differentiability implies that the sequence $P_{\theta_0+h/\sqrt{n}}^{\otimes n}$ satisfies the ULAN condition. If l is bounded, then the stated inequality follows from Theorem 7.166 and (7.131). To deal with the general case we remark that l is lower bounded and set $l_N = \min(l, N)$. Then

$$\liminf_{n\to\infty} \int l(\sqrt{n}(\widetilde{\theta}_n - \theta_0))dP_{\theta_0}^{\otimes n} \geq \int l_N(t)\varphi_{0,\mathsf{I}^{-1}(\theta_0)}(t)\boldsymbol{\lambda}(dt).$$

The monotone convergence theorem completes the proof of the first statement. The regularity of $\widehat{\theta}_n$ follows from Proposition 7.163 and (7.113). The fact that $\widehat{\theta}_n$ attains the lower bound for a bounded and continuous $l \in \mathfrak{L}_d$ follows from Theorem 7.148. ■

In Proposition 7.168 we have established a lower bound for the risk of all regular estimators for which the distributions $\mathcal{L}(\sqrt{n}(\widetilde{\theta}_n - \theta_0)|P_{\theta_0}^{\otimes n})$ converge. This is true for all regular estimators that admit a stochastic Taylor expansion (7.129). Now we operate directly with the family of localized models and find an asymptotic lower bound for the maximum risk by using the Hájek–LeCam bound in Proposition 6.89, where the concrete bound is provided by the optimal estimator in the Gaussian limit model. More precisely, assume that $\Delta \subseteq \mathbb{R}^d$ is open and that $(P_\theta)_{\theta\in\Delta}$ is $\mathbb{L}_2$-differentiable at $\theta_0 \in \Delta^0$. We remark that

$$\begin{aligned} &\mathcal{M}_n = (\mathcal{X}^n, \mathfrak{A}^{\otimes n}, (P_{n,h})_{h\in\Delta_n}), \quad \text{where} \\ &P_{n,h} = P_{\theta_0+h/\sqrt{n}}^{\otimes n}, \quad h \in \Delta_n = \{h : h \in \mathbb{R}^d,\ \theta_0 + h/\sqrt{n} \in \Delta\}, \end{aligned}$$

satisfies the ULAN condition; see Theorem 6.70. This implies especially that $\mathcal{M}_n$ converges weakly to the Gaussian model $\mathcal{G}_0$; that is,

$$\mathcal{M}_n \Rightarrow \mathcal{G}_0 = (\mathbb{R}^d, \mathfrak{B}_d, \mathsf{N}(h, \mathsf{I}^{-1}(\theta_0))_{h\in\mathbb{R}^d}) \tag{7.132}$$

(see Corollary 6.66). For any given sequence of estimators $\widehat{\theta}_n : \mathcal{X}^n \to \Delta$ we introduce estimators for the local parameter h by setting $\widehat{h}_n = \sqrt{n}(\widehat{\theta}_n - \theta_0)$. Given $l : \mathbb{R}^d \to_m \mathbb{R}_+$ we introduce the risk in the local model by setting it equal to $\int l(\widehat{h}_n - h)dP_{n,h}$. Turning back to the original model we get

$$\int l(\sqrt{n}(\widehat{\theta}_n - \theta_0) - h/\sqrt{n})dP_{\theta_0+h/\sqrt{n}}^{\otimes n} = \int l(\widehat{h}_n - h)dP_{n,h}.$$

Using this relation, an application of Proposition 6.89 gives for $h = 0$ asymptotic lower bounds for $\int l(\sqrt{n}(\widehat{\theta}_n - \theta_0))dP_{\theta_0}^{\otimes n}$, but we can also arrive at minimax results by letting h vary on compact sets.

Let us first consider the limiting model $\mathcal{G}_0$. In Theorem 3.65 it has been shown that

$$\begin{aligned}\inf_{\mathsf{D}} \sup_{h\in\mathbb{R}} \mathsf{R}(h,\mathsf{D}) &= \inf_{\mathsf{D}} \sup_{h\in\mathbb{R}} \int [\int l(a-h)\mathsf{D}(da|x)]\mathsf{N}(h,\mathsf{I}^{-1}(\theta_0))(dx) \\ &= \int l(t)\mathsf{N}(0,\mathsf{I}^{-1}(\theta_0))(dt) = \mathsf{R}(h,T_{nat}) = \mathsf{R}(0,T_{nat}),\end{aligned}$$

where $T_{nat}(x) = x$. Furthermore, by Theorem 3.65,

$$\begin{aligned}\lim_{m\to\infty} \inf_{\mathsf{D}} \sup_{\theta\in\mathbb{R}^d,\|\theta\|\le m} &\int [\int l(a-\theta)\mathsf{D}(da|x)]\mathsf{N}(\theta,\mathsf{I}^{-1}(\theta_0))(dx) \qquad (7.133)\\ &= \int l(t)\mathsf{N}(0,\mathsf{I}^{-1}(\theta_0))(dt),\end{aligned}$$

provided that, in addition, the above l is bounded.

Theorem 7.169. ***(Local Asymptotic Minimax Theorem)*** *Suppose that $\Delta \subseteq \mathbb{R}^d$ is open, $(P_\theta)_{\theta\in\Delta}$ is $\mathbb{L}_2$-differentiable at $\theta_0 \in \Delta$, $\widehat{\theta}_n : \mathcal{X}^n \to_m \Delta$ is a sequence of estimators, and $l \in \mathfrak{L}_d$ is continuous. Then*

$$\begin{aligned}\lim_{m\to\infty} \liminf_{n\to\infty} \sup_{\|h\|\le m} &\int l(\sqrt{n}(\widehat{\theta}_n - \theta_0) - h)dP_{\theta_0+h/\sqrt{n}}^{\otimes n} \qquad (7.134)\\ &\ge \int l(t)\mathsf{N}(0,\mathsf{I}^{-1}(\theta_0))(dt).\end{aligned}$$

If condition (A10) is fulfilled and $\widehat{\theta}_n$ is a sequence of estimators that solves the likelihood equation asymptotically (i.e., satisfies (7.109)) and is consistent at θ_0, then $\widehat{\theta}_n$ is asymptotically minimax in the sense that this sequence achieves the above lower bound for every bounded and continuous $l \in \mathfrak{L}_d$, i.e.,

$$\lim_{m\to\infty} \lim_{n\to\infty} \sup_{\|h\|\le m} \int l(\sqrt{n}(\widehat{\theta}_n - \theta_0) - h)dP_{\theta_0+h/\sqrt{n}}^{\otimes n} = \int l(t)\mathsf{N}(0,\mathsf{I}^{-1}(\theta_0))(dt).$$

Proof. By the subconvexity of l, $l \ge 0$, and $l(0) = 0$ the function $l(t_1,...,t_d)$ is nondecreasing in t_i for $t_i \ge 0$ and fixed t_j, $j \ne i$; see Problem 2.15. Furthermore this function is symmetric in $t_i \in \mathbb{R}$. From here it follows that every continuous $l \in \mathfrak{L}_d$ can be extended to a continuous function $l : \overline{\mathbb{R}}^d \to [0,\infty]$ in such a way that $l(t_1,...,t_i,...,t_d) \le l(t_1,...,\infty,...,t_d)$ for every $t_i \in \mathbb{R}$ and $t_j \in \overline{\mathbb{R}}$, $j \ne i$. We use the decision space $(\mathcal{D},\mathfrak{D}) = (\overline{\mathbb{R}}^d, \overline{\mathfrak{B}}_d)$ and note that $\mathcal{D}$ is a compact metric space. Let $\mathbb{D}_0$ be the set of all randomized estimators $\mathsf{D} : \overline{\mathfrak{B}}_d \times \mathbb{R}^d \to_k [0,1]$ for the model $\mathsf{N}(h,\mathsf{I}^{-1}(\theta_0))_{h\in\mathbb{R}^d}$. $\mathbb{D}_0$ is closed as the set of all decisions is sequentially compact; see Theorem 3.17. In view of (7.132) we may apply Proposition 6.89 and obtain

$$\liminf_{n\to\infty} \sup_{\|h\|\le m} \int l(\sqrt{n}(\widehat{\theta}_n - \theta_0) - h)dP^{\otimes n}_{\theta_0+h/\sqrt{n}}$$
$$\ge \inf_{\mathsf{D}\in\mathbb{D}_0} \sup_{\|h\|\le m} \int [\int l(t-h)\mathsf{D}(dt|x)]\mathsf{N}(h,\mathsf{I}^{-1}(\theta_0))(dx).$$

To carry out the minimization over $\mathsf{D} \in \mathbb{D}_0$ for the model $\mathcal{G}_0$ we note that we have to extend the infimum only over all estimators D with $\mathsf{D}(\overline{\mathbb{R}}^d\backslash\mathbb{R}^d|x) = 0$. Hence by (7.133),

$$\lim_{m\to\infty} \liminf_{n\to\infty} \sup_{\|h\|\le m} \int l(\sqrt{n}(\widehat{\theta}_n - \theta_0) - h)dP^{\otimes n}_{\theta_0+h/\sqrt{n}} \ge \int l(t)\mathsf{N}(0,\mathsf{I}^{-1}(\theta_0))(dt),$$

provided that l is bounded. If l is unbounded we set $l_N = \min(l, N)$. Then

$$\int l(\sqrt{n}(\widehat{\theta}_n - \theta_0) - h)dP^{\otimes n}_{\theta_0+h/\sqrt{n}} \ge \int l_N(\sqrt{n}(\widehat{\theta}_n - \theta_0) - h)dP^{\otimes n}_{\theta_0+h/\sqrt{n}},$$
$$\lim_{N\to\infty} \int l_N(t)\mathsf{N}(0,\mathsf{I}^{-1}(\theta_0))(dt) = \int l(t)\mathsf{N}(0,\mathsf{I}^{-1}(\theta_0))(dt),$$

completes the proof of the first statement.

If condition (A10) is fulfilled and $\widehat{\theta}_n$ is a sequence of estimators that solves the likelihood equation asymptotically, and is consistent at θ_0, then $\sqrt{n}(\widehat{\theta}_n - \theta_0) = \mathsf{I}^{-1}(\theta_0)Z_n + o_{P^{\otimes n}_{\theta_0}}(1)$; see (7.113). Set $\Psi = \mathsf{I}^{-1}(\theta_0)\dot{L}_{\theta_0}$. Then $\mathsf{C}_{\theta_0}(\Psi, \dot{L}_{\theta_0}) = \mathbf{I}$, $\mathsf{C}_{\theta_0}(\Psi) = \mathsf{I}^{-1}(\theta_0)$ and by Corollary 6.74 it follows for every bounded and continuous function φ

$$\lim_{n\to\infty} \int \varphi(\sqrt{n}(\widehat{\theta}_n - \theta_0))dP^{\otimes n}_{\theta_0+h_1/\sqrt{n}} = \int \varphi(t+h_1)\mathsf{N}(0,\mathsf{I}^{-1}(\theta_0))(dt)$$

for every fixed h_1. Putting $\varphi_{h_2}(t) = l(t-h_2)$ we get for every fixed h_2

$$\lim_{n\to\infty} \int l(\sqrt{n}(\widehat{\theta}_n - \theta_0) - h_2)dP^{\otimes n}_{\theta_0+h_1/\sqrt{n}} = \int l(t+h_1-h_2)\mathsf{N}(0,\mathsf{I}^{-1}(\theta_0))(dt).$$

The family $\varphi_{h_2}(t) = l(t-h_2)$, $||h_2|| \le m$, satisfies the conditions in Proposition A.45. Hence the above convergence is uniform in h_2 for every fixed h_1. As the family $P^{\otimes n}_{\theta_0+h_1/\sqrt{n}}$ satisfies the ULAN-condition we get from the asymptotic local equicontinuity of $P^{\otimes n}_{\theta_0+h_1/\sqrt{n}}$ in the sense of variational distance (see (6.80)) and the boundedness of l that for every $m > 0$ the above convergence is simultaneously uniform in h_1 and h_2 for $\|h_1\| \le m$ and $\|h_2\| \le m$. Putting $h_1 = h_2 = h$ we get

$$\lim_{n\to\infty} \int l(\sqrt{n}(\widehat{\theta}_n - \theta_0) - h)dP^{\otimes n}_{\theta_0+h/\sqrt{n}} = \int l(t)\mathsf{N}(0,\mathsf{I}^{-1}(\theta_0))(dt)$$

locally uniformly in h. Hence,

$$\lim_{n\to\infty} \sup_{\|h\|\le m} \int l(\sqrt{n}(\widehat{\theta}_n - \theta_0) - h) dP^{\otimes n}_{\theta_0+h/\sqrt{n}} = \int l(t)\mathsf{N}(0, \mathsf{I}^{-1}(\theta_0))(dt).$$

Taking $m \to \infty$ we get the second statement. ■

Remark 7.170. Instead of (7.134) one might also prove

$$\liminf_{n\to\infty} \sup_{h\in\Delta_n} \int l(\sqrt{n}(\widehat{\theta}_n - \theta_0) - h) dP^{\otimes n}_{\theta_0+h/\sqrt{n}} \ge \int l(t)\mathsf{N}(0, \mathsf{I}^{-1}(\theta_0))(dt),$$

where $\Delta_n = \{h : \theta_0 + \sqrt{n}h \in \Delta\}$. If $\Delta = \mathbb{R}^d$, then $\Delta_n = \mathbb{R}^d$ and the lower bound will not be attained in general as the ULAN property gives only the local uniform convergence.

Now we consider the one-dimensional case and asymptotically median unbiased estimators. We show that the MLE is within this class asymptotically uniformly best. To establish this result we apply the lower Hájek–LeCam bound which was established in Section 6.6, where we have assumed that the decision space is compact. Therefore we use the decision space $\overline{\mathbb{R}} = [-\infty, \infty]$. As every $l \in \mathfrak{L}_1$ is symmetric, nondecreasing for $t \ge 0$, and nonincreasing for $t \le 0$, we may extend every continuous $l \in \mathfrak{L}_1$ continuously on $\overline{\mathbb{R}}$ by setting $l(-\infty) = l(\infty) := \lim_{t\to\infty} l(t)$.

Let $\Delta \subseteq \mathbb{R}$ be open and $(P_\theta)_{\theta\in\Delta}$ be a one-parameter family of distributions on $(\mathcal{X}, \mathfrak{A})$. A sequence of estimators $\widehat{\theta}_n : \mathcal{X}^n \to_m \overline{\mathbb{R}}$ for the parameter $\theta \in \Delta$ is called *asymptotically median unbiased* at θ_0 if for every $h \in \mathbb{R}$ and $P_{n,h} = P^{\otimes n}_{\theta_0+h/\sqrt{n}}$ it holds

$$\liminf_{n\to\infty}([P_{n,h}(\sqrt{n}(\widehat{\theta}_n - \theta_0) \le h)] \wedge [P_{n,h}(\sqrt{n}(\widehat{\theta}_n - \theta_0) \ge h)]) \ge \frac{1}{2}. \tag{7.135}$$

The following result is due to Pfanzagl (1970).

Theorem 7.171. *Let $(P_\theta)_{\theta\in\Delta}$ be $\mathbb{L}_2$-differentiable at $\theta_0 \in \Delta^0$ with $\mathsf{I}(\theta_0) > 0$. Suppose $\widehat{\theta}_n : \mathcal{X}^n \to_m \overline{\mathbb{R}}$ is a sequence of asymptotically median unbiased estimators at θ_0. If $l \in \mathfrak{L}_1$ is continuous, then for every $h \in \mathbb{R}$,*

$$\liminf_{n\to\infty} \int l(\sqrt{n}(\widehat{\theta}_n - \theta_0) - h) dP^{\otimes n}_{\theta_0+h/\sqrt{n}} \ge \int l(t)\mathsf{N}(0, \mathsf{I}^{-1}(\theta_0))(dt).$$

If condition (A10) is fulfilled and $\widehat{\theta}_n$ is a sequence of estimators that solves the likelihood equation asymptotically (i.e., satisfies (7.109)), and is consistent at θ_0, then $\widehat{\theta}_n$ is a sequence of asymptotically median unbiased estimators at θ_0 that is asymptotically efficient in the sense that this sequence achieves the above lower bound for every bounded and continuous $l \in \mathfrak{L}_1$, i.e.,

$$\lim_{n\to\infty} \int l(\sqrt{n}(\widehat{\theta}_n - \theta_0) - h) dP^{\otimes n}_{\theta_0+h/\sqrt{n}} = \int l(t)\mathsf{N}(0, \mathsf{I}^{-1}(\theta_0))(dt),$$

where the convergence is locally uniform in h.

Proof. The decision space is $(\mathcal{D}, \mathfrak{D}) = (\overline{\mathbb{R}}, \overline{\mathfrak{B}})$ and $\mathcal{D}$ is a compact metric space. Let $\mathbb{D}_0$ be the set of all median unbiased randomized estimators $\mathsf{D} : \overline{\mathfrak{B}} \times \mathbb{R} \to_k [0,1]$ for the model $\mathsf{N}(h, \mathsf{I}^{-1}(\theta_0))_{h \in \mathbb{R}^d}$. As by Problem 7.34 $\mathbb{D}_0$ is closed and it holds (7.132) we may apply Proposition 6.89 to obtain

$$\begin{aligned}
&\liminf_{n\to\infty} \int l(\sqrt{n}(\widehat{\theta}_n - \theta_0) - h) dP^{\otimes n}_{\theta_0 + h/\sqrt{n}} \\
&\geq \inf_{\mathsf{D}\in\mathbb{D}_0} \int [\int l(t-h)\mathsf{D}(dt|x)]\mathsf{N}(h, \mathsf{I}^{-1}(\theta_0))(dx) \\
&= \int l(t)\mathsf{N}(0, \mathsf{I}^{-1}(\theta_0))(dt),
\end{aligned}$$

where the last equality follows from Theorem 7.36 with $\varrho = l$. If (A10) and (7.109) hold, then by Corollary 6.74 it follows that $\mathcal{L}(\sqrt{n}(\widehat{\theta}_n - \theta_0)|P^{\otimes n}_{\theta_0+h/\sqrt{n}}) \Rightarrow \mathsf{N}(h, \mathsf{I}^{-1}(\theta_0))$ locally uniformly in h, which gives the asymptotic median unbiasedness and the stated equality. ■

We illustrate the efficiency of estimators by considering location models.

Example 7.172. Suppose $f : \mathbb{R} \to (0,\infty)$ is a twice continuously differentiable Lebesgue density with finite Fisher information $\mathsf{I} = \int [\dot{f}(x)]^2 [f(x)]^{-1} dx < \infty$. The log-likelihood function for the location model is $\Lambda_n(\theta, \mathbf{x}_n) = \sum_{i=1}^n \ln f(x_i - \theta)$, and the likelihood equation reads

$$\dot{\Lambda}_n(\theta) = \sum\nolimits_{i=1}^n \dot{f}(X_i - \theta)/f(X_i - \theta) = 0.$$

Example 7.81 gives the strong consistency of the MLE $\widehat{\theta}_n$ under the conditions there which we assume to be fulfilled. If in addition 7.114 is fulfilled, then by Example 7.150 it holds

$$\sqrt{n}(\widehat{\theta}_n - \theta_0) = -\sum\nolimits_{i=1}^n \mathsf{I}^{-1}\dot{f}(X_i - \theta_0)/f(X_i - \theta_0) + o_{P^{\otimes n}_{\theta_0}}(1).$$

Hence $\widehat{\theta}_n$ is a regular and asymptotically median unbiased estimator which is efficient in the sense that it attains the lower bounds in 7.169. Suppose now that the second moment $\int t^2 f(t)dt$ is finite and it holds $\int tf(t)dt = 0$. Then the parameter θ is the expectation which we may estimate by the arithmetic mean $\overline{X}_n$. If σ^2 is the variance of the density f, then $\mathcal{L}(\sqrt{n}(\overline{X}_n - \theta_0)|P^{\otimes n}_{\theta_0}) \Rightarrow \mathsf{N}(0, \sigma^2)$, so that by Theorem 7.171 $\sigma^2 \geq \mathsf{I}^{-1}$. If $f = \varphi_{0,\sigma^2}$, then we have equality. The question arises as to whether there are other distributions with $\sigma^2 = \mathsf{I}^{-1}$. The answer is given by Proposition 7.164 which says that $\sigma^2 = \mathsf{I}^{-1}$ implies $\mathsf{I}^{-1}\dot{L}_{\theta_0} = \mathsf{I}^{-1}\dot{f}(x)/f(x) = \Psi_{\theta_0}(x) = x - \theta_0$. By integrating this differential equation we see that $f = \varphi_{\theta_0,\sigma^2}$. Hence in the class of all location models that satisfy the above regularity conditions the arithmetic mean is asymptotically efficient if and only if the location model is generated by a normal distribution.

Example 7.173. In the previous example we have used relatively strong assumptions to make the general results applicable to the location model. But in more specific location models the MLE is directly available and can be seen to be asymptotically normal. Suppose we are given a location model generated by a positive strongly unimodal and continuous density f. This is equivalent to the log-concavity so that $f(t) = \exp\{-\varrho(t)\}$ for some convex function ϱ. We know from Proposition 2.18 that $\lim_{x\to\pm\infty} f(x) = 0$. Suppose that in addition the remaining conditions in (7.41) are satisfied. Then we get from Example 7.81 that the MLE is strongly consistent.

A special strongly unimodal distribution is the Laplace distribution, i.e., $f(t) = \frac{1}{2}\exp\{-|t|\}$. The convex function $\varrho(t) = \frac{1}{2}|t|$ has derivatives from the right and from the left which differ only at zero. Hence ϱ and thus f are differentiable, $\boldsymbol{\lambda}$-a.e., and it holds $\dot{f}(t) = -\operatorname{sgn}(t)\exp\{-|t|\}$ for $t \neq 0$. As $\int_a^b |\dot{f}(t)|dt < \infty$ for every $a < b$ we get that f is absolutely continuous. Moreover,

$$\mathsf{I} = \int [\dot{f}(t)/f(t)]^2 f(t)dt = \int f(t)dt = 1.$$

We get from Lemma 1.121 that the location model is $\mathbb{L}_2$-differentiable with $\mathbb{L}_2$-derivative $\dot{L}_{\theta_0}(t) = -\operatorname{sgn}(t-\theta_0)$, $\boldsymbol{\lambda}$-a.e. As

$$\Lambda_n(\theta) = -\sum\nolimits_{i=1}^n |X_i - \theta| - n\ln 2,$$

we get from Example 7.107 that the MLE is the sample median. From Example 7.139 we get for $\alpha = 1/2$ the representation of the sample median $\widehat{\theta}_n$,

$$\begin{aligned}\sqrt{n}(\widehat{\theta}_n - u_{1/2}) &= -(f(u_{1/2}))^{-1} n^{-1/2} 2^{-1} \sum\nolimits_{i=1}^n \operatorname{sgn}(X_i - \theta_0) + o_{P^{\otimes n}}(1) \\ &= -n^{-1/2} \sum\nolimits_{i=1}^n \operatorname{sgn}(X_i - \theta_0) + o_{P^{\otimes n}}(1),\end{aligned}$$

where $X_1, ..., X_n$ are i.i.d. with common density $f(t-\theta_0)$. This is, in view of $\dot{L}_{\theta_0}(t) = -\operatorname{sgn}(t-\theta_0)$, exactly the expansion (7.113). As $\widehat{\theta}_n$ is consistent we see that the median is the asymptotically efficient estimator in the location model generated by the Laplace distribution, where "efficient" is specified by Proposition 7.164 and Theorems 7.171 and 7.169.

Sometimes the parameter vector $\theta = (\tau^T, \xi^T)^T$ consists of two parts, where τ is the k-dimensional parameter of interest and ξ is the $(d-k)$-dimensional nuisance parameter that is only used to fit the model to the data. We suppose that $(P_\theta)_{\theta\in\Delta}$ is $\mathbb{L}_2$-differentiable and split the $\mathbb{L}_2$-derivative $\dot{L}_\theta$ into U_θ and V_θ, respectively.

$$\dot{L}_\theta = \begin{pmatrix} U_\theta \\ V_\theta \end{pmatrix}, \quad \theta = \begin{pmatrix} \tau \\ \xi \end{pmatrix} \in \Delta^0.$$

The covariance matrix $\mathsf{C}_{\theta_0}(\dot{L}_{\theta_0})$ is the Fisher information matrix, and we partition it analogously.

$$\begin{aligned}\mathsf{C}_{\theta_0}(\dot{L}_{\theta_0}) = \mathsf{I}(\theta_0) &= \begin{pmatrix} \mathsf{I}_{1,1}(\theta_0) & \mathsf{I}_{1,2}(\theta_0) \\ \mathsf{I}_{2,1}(\theta_0) & \mathsf{I}_{2,2}(\theta_0) \end{pmatrix} \\ &= \begin{pmatrix} \mathsf{C}_{\theta_0}(U_{\theta_0}) & \mathsf{C}_{\theta_0}(U_{\theta_0}, V_{\theta_0}) \\ \mathsf{C}_{\theta_0}(V_{\theta_0}, U_{\theta_0}) & \mathsf{C}_{\theta_0}(V_{\theta_0}) \end{pmatrix}.\end{aligned}$$

Later on we use the inverse matrix $\mathsf{I}^{-1}(\theta_0)$ also represented by block matrices.

Problem 7.174.* Let I be a symmetric positive semidefinite $d \times d$ matrix consisting of the block matrices $(\mathsf{I}_{i,j})_{1 \le i,j \le 2}$. Then I is invertible if and only if $\mathsf{I}_{2,2}$ and $\mathsf{G} := \mathsf{I}_{1,1} - \mathsf{I}_{1,2}\mathsf{I}_{2,2}^{-1}\mathsf{I}_{2,1}$ are invertible. In this case I and G are positive definite and the inverse matrix $\mathsf{J} = \mathsf{I}^{-1}$ has the block matrices $(\mathsf{J}_{i,j})_{1 \le i,j \le 2}$ given by

$$\begin{pmatrix} \mathsf{J}_{1,1} & \mathsf{J}_{1,2} \\ \mathsf{J}_{2,1} & \mathsf{J}_{2,2} \end{pmatrix} = \begin{pmatrix} \mathsf{G}^{-1} & -\mathsf{G}^{-1}\mathsf{I}_{1,2}\mathsf{I}_{2,2}^{-1} \\ -(\mathsf{I}_{1,2}\mathsf{I}_{2,2}^{-1})^T\mathsf{G}^{-1} & \mathsf{I}_{2,2}^{-1} + (\mathsf{I}_{1,2}\mathsf{I}_{2,2}^{-1})^T\mathsf{G}^{-1}(\mathsf{I}_{1,2}\mathsf{I}_{2,2}^{-1}) \end{pmatrix}.$$

Problem 7.175.* Let I be symmetric positive definite, and $\mathsf{J}_{1,1}$ and G be the matrices introduced in Problem 7.174. Then $\mathsf{I}_{1,1} - \mathsf{G}$ is positive semidefinite, so that $\mathsf{I}_{1,1} \succeq \mathsf{G}$ in the Löwner semiorder.

Problem 7.176. Let A and B be positive definite symmetric matrices. If $A \preceq B$ in the Löwner semiorder, then $B^{-1} \preceq A^{-1}$.

In the presence of a nuisance parameter ξ_0 one has to distinguish two cases when the parameter of interest τ_0 is to be estimated. If ξ_0 is known, then one has the model $(P_{\tau,\xi_0})_{\tau \in \Upsilon}$. If the regularity conditions of Theorem 7.148 are satisfied, then the MLE $\widehat{\tau}_n$ satisfies $\mathcal{L}(\sqrt{n}(\widehat{\tau}_n - \tau_0)|P_{\tau_0,\xi_0}^{\otimes n}) \Rightarrow \mathsf{N}(0, \mathsf{I}_{1,1}^{-1}(\tau_0, \xi_0))$. On the other hand, if ξ_0 is unknown, then one has to estimate both τ_0 and ξ_0; that is, we have the model $(P_\theta)_{\theta \in \Delta}$, where $\theta = (\tau, \xi) \in \Delta$. If the regularity conditions of Theorem 7.148 are satisfied for this model, then the MLE $\widetilde{\theta}_n = (\widetilde{\tau}_n, \widetilde{\xi}_n)$ satisfies

$$\mathcal{L}\left(\sqrt{n}\left(\begin{pmatrix} \widetilde{\tau}_n \\ \widetilde{\xi}_n \end{pmatrix} - \begin{pmatrix} \tau_0 \\ \xi_0 \end{pmatrix}\right) \middle| P_{\tau_0,\xi_0}^{\otimes n}\right) \Rightarrow \mathsf{N}\left(0, \begin{pmatrix} \mathsf{J}_{1,1} & \mathsf{J}_{1,2} \\ \mathsf{J}_{2,1} & \mathsf{J}_{2,2} \end{pmatrix}\right),$$

and therefore $\mathcal{L}(\sqrt{n}(\widetilde{\tau}_n - \tau_0)|P_{\tau_0,\xi_0}^{\otimes n}) \Rightarrow \mathsf{N}(0, \mathsf{J}_{1,1})$. In view of Problem 7.175 we have $\mathsf{I}_{1,1} \succeq \mathsf{J}_{1,1}^{-1}$ and thus $\mathsf{I}_{1,1}^{-1} \preceq \mathsf{J}_{1,1}$, i.e., for large n the estimator $\widehat{\tau}_n$ is more concentrated around τ_0 than $\widetilde{\tau}_n$. This is not surprising since for the construction of $\widehat{\tau}_n$ we could use the known nuisance parameter ξ_0, but not for the construction of $\widetilde{\tau}_n$. Moreover, if $\mathsf{I}_{1,2}(\theta_0) = \mathsf{C}_{\theta_0}(U_{\theta_0}, V_{\theta_0}) = 0$, then $\mathsf{I}_{1,1}^{-1} = \mathsf{J}_{1,1}$ and both estimators have asymptotically the same precision. The condition $\mathsf{C}_{\theta_0}(U_{\theta_0}, V_{\theta_0}) = 0$ can be considered as an orthogonality of the tangent vectors that belong to the parameters τ_0 and ξ_0. In this case it does not matter if ξ_0 is known or unknown when we estimate τ by the MLE. Such situations are called *adaptive cases*. We return to this point in Chapter 8 when we are constructing tests in the presence of nuisance parameters.

Example 7.177. To illustrate the concept of adaptivity we consider the problem of estimating parameters in a location-scale model. Suppose $f : \mathbb{R} \to (0, \infty)$ is a continuously differentiable Lebesgue density with $\int [\dot{f}(x)]^2 [f(x)]^{-1} dx < \infty$ and $\int x^2 [\dot{f}(x)]^2 [f(x)]^{-1} dx < \infty$. Let P_θ be the distribution with the Lebesgue density

$$f_\theta(x) = \frac{1}{\sigma} f(\frac{x - \mu}{\sigma}), \quad \theta = (\mu, \sigma) \in \Delta = \mathbb{R} \times (0, \infty).$$

Then by Example 1.119 the family $(P_\theta)_{\theta \in \Delta}$ is $\mathbb{L}_2$-differentiable, and the Fisher information matrix is given by

$$\mathsf{I}(\theta) = \frac{1}{\sigma^2}\begin{pmatrix} \int[\dot{f}(x)]^2[f(x)]^{-1}dx & \int x[\dot{f}(x)]^2[f(x)]^{-1}dx \\ \int x[\dot{f}(x)]^2[f(x)]^{-1}dx & \int x^2[\dot{f}(x)]^2[f(x)]^{-1}dx - 1 \end{pmatrix}.$$

If now f is symmetric, then $\int x[\dot{f}(x)]^2[f(x)]^{-1}dx = 0$, so that adaptivity holds. This situation is especially met for the family of normal distributions.

7.8 Solutions to Selected Problems

Solution to Problem 7.12 The continuous differentiability means that the mapping $\theta \mapsto L_{\theta,\theta_0}^{1/2}\dot{L}_\theta$ is continuous in the sense of $\mathbb{L}_{2,d}(P_{\theta_0})$. The continuity of the scalar product yields the first statement. The $\mathbb{L}_1$-differentiability established in Proposition 1.110 implies $P_{\theta_0+u}(B) - P_{\theta_0}(B) = \mathsf{E}_{\theta_0} I_B\langle u, \dot{L}_{\theta_0}\rangle + o(\|u\|)$. Hence $\mathsf{E}_{\theta_0} I_B \dot{L}_{\theta_0}$ is the gradient. The continuity follows from

$$\begin{aligned}
&\| \mathsf{E}_\theta I_B \dot{L}_\theta - \mathsf{E}_{\theta_0} I_B \dot{L}_{\theta_0} \| = \| \mathsf{E}_{\theta_0} I_B (L_{\theta,\theta_0}\dot{L}_\theta - \dot{L}_{\theta_0}) \| \\
&\le \mathsf{E}_{\theta_0} \| L_{\theta,\theta_0}^{1/2}\dot{L}_\theta - \dot{L}_{\theta_0} \| + \mathsf{E}_{\theta_0} \| L_{\theta,\theta_0}^{1/2}(\dot{L}_\theta - L_{\theta,\theta_0}^{1/2}\dot{L}_\theta) \| \\
&\le (\mathsf{E}_{\theta_0} \| L_{\theta,\theta_0}^{1/2}\dot{L}_\theta - \dot{L}_{\theta_0} \|^2)^{1/2} + (\mathsf{E}_\theta \| \dot{L}_\theta \|^2)^{1/2}(\mathsf{E}_{\theta_0}(1 - L_{\theta,\theta_0}^{1/2})^2)^{1/2}.
\end{aligned}$$

The first term tends to zero by the continuous differentiability. $\mathsf{E}_{\theta_0}(1 - L_{\theta,\theta_0}^{1/2})^2 = o(\| \theta - \theta_0 \|)$ by (1.133). $\mathsf{E}_\theta \| \dot{L}_\theta \|^2$ is the trace of $\mathsf{I}(\theta)$ and therefore bounded as $\mathsf{I}(\theta)$ is continuous in θ. □

Solution to Problem 7.13: The mapping $\theta \mapsto L_{\theta,\theta_0}^{1/2} S$ is stochastically continuous with respect to P_{θ_0}. The assumed continuity of $\theta \mapsto \mathsf{E}_\theta \|S\|^2$ and Vitali's theorem imply the continuity of $\theta \mapsto L_{\theta,\theta_0}^{1/2} S$ in the sense of $\mathbb{L}_{2,d}(P_{\theta_0})$. As $\theta \mapsto L_{\theta,\theta_0}^{1/2}\dot{L}_\theta$ is continuous in the sense of $\mathbb{L}_{2,d}(P_{\theta_0})$ by the continuous differentiability, the statement follows from the continuity of the scalar product. □

Solution to Problem 7.14: The required equality yields $\langle c - a, T(x)\rangle = d - b$. $\det(\mathsf{C}_\theta(T)) \neq 0$ implies $c = a$ and thus $d = b$, see Problem 1.3. □

Solution to Problem 7.23: $\widehat{\sigma^2}$ is a function of the complete and sufficient statistic $T(x) = \|x\|^2$. T/σ^2 has a χ^2-distribution with n degrees of freedom. Hence $\mathbb{E}(\sigma^2/T) = (2^{n/2}\Gamma(n/2))^{-1}\sigma^2 \int_0^\infty (1/t)t^{(n/2)-1}\exp\{-t/2\}dt = 1/(n-2)$. □

Solution to Problem 7.24: For $y \in \mathbb{L}$ it holds $\|x - y\|^2 = \|x_{\mathbb{L}} + x_{\mathbb{L}^\perp} - y\|^2 = \|x_{\mathbb{L}} - y\|^2 + \|x_{\mathbb{L}^\perp}\|^2$, and $\|x_{\mathbb{L}} - y\|^2 = 0$ for $y = x_{\mathbb{L}}$. The statement follows now with $x_{\mathbb{L}^\perp} = x - x_{\mathbb{L}}$. □

Solution to Problem 7.25: If $Cx = \Pi_{\mathbb{L}}(x)$ for some subspace $\mathbb{L}$, then by (7.10) $x^T C y = x^T \Pi_{\mathbb{L}}(y) = (\Pi_{\mathbb{L}}(x))^T \Pi_{\mathbb{L}}(y) = (Cx)^T \Pi_{\mathbb{L}}(y) = x^T C^T y$ for every $x, y \in \mathbb{R}^n$. $CC = C$ follows from $\Pi_{\mathbb{L}}(\Pi_{\mathbb{L}}(x)) = x$. Conversely if $C^T = C$ and $CC = C$ hold, and $\mathbb{L}$ is generated by the column vectors of C, then $Cx \in \mathbb{L}$ and $(x - Cx)^T(Cy) = 0$, so that $x - Cx \perp \mathbb{L}$. □

Solution to Problem 7.26: If for some $x \neq 0$ it holds $B^T Bx = 0$, then $\|Bx\|^2 = x^T B^T Bx = 0$, which is impossible as the rank of B is d. $C = B(B^T B)^{-1} B^T$ satisfies $C = C^T$, $CC = C$, $Cx = x$ if $x \in \mathbb{L} = \{By : y \in \mathbb{R}^d\}$, and $Cx = 0$ if $x \perp \mathbb{L}$, as $x \perp \mathbb{L}$ holds if and only if $x^T B = 0$. □

Solution to Problem 7.28: Represent $\Pi_{\mathbb{L}}(Z)$ and $\Pi_{\mathbb{M}}(Z)$ by matrices $\Pi_{\mathbb{L}}(Z) = C_{\mathbb{L}} Z$ and $\Pi_{\mathbb{M}}(Z) = C_{\mathbb{M}} Z$. Then by (7.12) $\mathsf{C}(AC_{\mathbb{L}}Z, BC_{\mathbb{M}}Z) = A\mathsf{C}(C_{\mathbb{L}}Z, C_{\mathbb{M}}Z)B^T = AC_{\mathbb{L}}\mathsf{C}(Z,Z)C_{\mathbb{M}}B^T = \sigma^2 AC_{\mathbb{L}}\mathbf{I}C_{\mathbb{M}}B^T = 0$ if $C_{\mathbb{L}}C_{\mathbb{M}} = 0$, i.e., $\mathbb{L} \perp \mathbb{M}$. □

Solution to Problem 7.29: Let $b_1, ..., b_n$ be an orthonormal basis such that $b_1, ..., b_d$ is a basis for $\mathbb{L}$ and $b_{d+1}, ..., b_n$ for $\mathbb{L}^\perp$. Set $Y = (Y_1, ..., Y_n)^T$, $Y_i = X^T b_i$.The covariance matrix of Y is $\sigma^2 \mathbf{I}$. The claim follows from $\|\Pi_{\mathbb{L}^\perp}(X)\|^2 = \sum_{i=d+1}^n Y_i^2$. □

Solution to Problem 7.34: Let $f_m : \mathbb{R}$ be the piecewise linear function which is one for $t \geq \kappa(\theta)$, and zero for $t \leq \kappa(\theta) - \frac{1}{m}$. Then f_m is bounded and continuous on $\overline{\mathbb{R}}$ and $f_m \downarrow I_{[\kappa(\theta),\infty]}$. If the D_n are median unbiased and $\mathsf{D}_n \Rightarrow \mathsf{D}$, then

$$\frac{1}{2} \leq \lim_{n\to\infty} \int [\int f_m(a)\mathsf{D}_n(da|x)]P_\theta(dx) = \int [\int f_m(a)\mathsf{D}(da|x)]P_\theta(dx),$$
$$\frac{1}{2} \leq \lim_{m\to\infty} \int [\int f_m(a)\mathsf{D}(da|x)]P_\theta(dx) = \int \mathsf{D}([\kappa(\theta),\infty]|x)P_\theta(dx).$$

The case of $[-\infty, \kappa(\theta)]$ is similar. □

Solution to Problem 7.35: As $\varrho(0) = 0$ we get from Fubini's theorem

$$\int I_{(\theta,\infty)}(s)\varrho(s-\theta)Q(ds) = \int I_{(\theta,\infty)}(s)[\int I_{(0,s-\theta]}(t)\mu_\varrho(dt)]Q(ds)$$
$$= \int I_{(0,\infty)}(t)[\int I_{[t+\theta,\infty)}(s)Q(ds)]\mu_\varrho(dt) = \int I_{(0,\infty)}(t)Q([t+\theta,\infty))\mu_\varrho(dt).$$
$$\int I_{(-\infty,\theta]}(s)\varrho(s-\theta)Q(ds) = \int I_{(-\infty,\theta]}(s)[\int I_{(s-\theta,0]}(t)\mu_\varrho(dt)]Q(ds)$$
$$= \int I_{(-\infty,0]}(t)[\int I_{(-\infty,t+\theta)}(s)Q(ds)]\mu_\varrho(dt) = \int I_{(-\infty,0)}(t)Q((-\infty,t+\theta])\mu_\varrho(dt),$$

where the last equality follows from $\mu_\varrho(\{a\}) = 0$, $a \in \mathbb{R}$. If Q and μ_ϱ are symmetric, then

$$\int I_{(-\infty,0)}(t)Q((-\infty,t+\theta])\mu_\varrho(dt) = \int I_{(-\infty,0)}(t)Q([-t-\theta,\infty))\mu_\varrho(dt)$$
$$= \int I_{(0,\infty)}(t)Q([t-\theta,\infty))\mu_\varrho(dt). \quad \square$$

Solution to Problem 7.43: Without loss of generality we assume that $u_\alpha = 0$. For $\theta > 0$ and $P = \mathcal{L}(X)$ it holds

$$\tau_\alpha(X-\theta) - \tau_\alpha(X) = \theta(1-\alpha)I_{(-\infty,0]}(X) + [-X+\theta(1-\alpha)]I_{(0,\theta]}(X) - \alpha\theta I_{(\theta,\infty)}(X),$$

$$
\begin{aligned}
&\mathbb{E}(\tau_\alpha(X-\theta)-\tau_\alpha(X))\\
&=(1-\alpha)\theta F(0)-\mathbb{E}XI_{(0,\theta]}(X)+(1-\alpha)\theta(F(\theta)-F(0))-\alpha\theta(1-F(\theta))\\
&=-\int I_{(0,\theta]}(t)[\int I_{(0,t)}(s)ds)]P(dt)-\theta(\alpha-F(\theta))\\
&=-\int I_{(0,\theta]}(s)[\int I_{(s,\theta]}(t)P(dt)]ds-\theta(\alpha-F(\theta))\\
&=-\int I_{(0,\theta]}(s)(F(\theta)-F(s))ds-\theta(\alpha-F(\theta))=\int I_{[0,\theta]}(s)(F(s)-\alpha)ds.
\end{aligned}
$$

The case of $\theta < 0$ is similar. □

Solution to Problem 7.49: If $P_{\theta_0}(A) = 0$, then by the homogeneity of the model $P_\theta(A) = 0$, $\theta \in \Delta$, and thus $(\mathsf{P}\Pi)(A) = \int P_\theta(A)\Pi(d\theta) = 0$, $\theta \in \Delta$. Conversely, $(\mathsf{P}\Pi)(A) = 0$ implies $P_{\theta_0}(A) = 0$ for at least one θ_0 and thus $P_\theta(A) = 0$, $\theta \in \Delta$. □

Solution to Problem 7.53: $\mathcal{L}(\|X-\theta\|^2)$ is a χ^2-distribution with noncentrality $\delta^2 = \|\theta\|^2$ and d degrees of freedom. We know from Theorem 2.27 that $\mathcal{L}(\|X-\theta\|^2)$ is stochastically nondecreasing in δ^2. Hence by Proposition 2.7 $\mathbb{E}\|X-\theta\|^{-2} \le \mathbb{E}\|X\|^{-2} = 2^{-d/2}\Gamma(d/2)\int_0^\infty (1/t)t^{(d/2)-1}\exp\{-t/2\}dt < \infty$ if $d \ge 3$. □

Solution to Problem 7.72: As in the one-dimensional case one can see that the left- and right-hand partial derivatives exist and it holds $\partial^- f/\partial x_i \le \partial^+ f/\partial x_i$. Set $\varepsilon_i(x) = \operatorname{sgn}(x_i - c_i)$ for fixed $x = (x_1, ..., x_d)$ and $c = (c_1, ..., c_d)$. The function $v(t) = f(c+t(x-c))$ is convex and satisfies $D^+v(0) = \sum_{i=1}^d (\partial^{\varepsilon_i(x)} f(c)/\partial x_i)(x_i - c_i)$. Hence $f(c+(x-c)) - f(c) - \sum_{i=1}^d (\partial^{\varepsilon_i(x)} f(c)/\partial x_i)(x_i - c_i) \ge 0$ by (1.55). For any $\varepsilon_i \in \{+,-\}$ it holds $(\partial^{\varepsilon_i(x)} f(c)/\partial x_i)(x_i - c_i) \ge (\partial^{\varepsilon_i} f(c)/\partial x_i)(x_i - c_i)$ and the statement is established. □

Solution to Problem 7.73: Fix $x, y \in K$ and put $z = x + \varepsilon(y-x)/\|y-x\|$. Then $z \in K_\varepsilon$ and for $\alpha = (1/\varepsilon)\|y-x\| \le 1$ it holds $y = (1-\alpha)x + \alpha z$, and

$$
\begin{aligned}
f(y) &\le (1-\alpha)f(x) + \alpha f(z) = f(x) + \alpha(f(z) - f(x))\\
f(y) - f(x) &\le \alpha(f(z) - f(x)) \le \alpha(\sup\nolimits_{x\in K_\varepsilon} f(x) - \inf\nolimits_{x\in K_\varepsilon} f(x)).
\end{aligned}
$$

As x and y are arbitrary we get for $\|y-x\| \le \varepsilon$

$$
|f(x) - f(y)| \le \varepsilon^{-1}\|y-x\|(\sup\nolimits_{x\in K_\varepsilon} f(x) - \inf\nolimits_{x\in K_\varepsilon} f(x)).
$$

But for $\|y-x\| > \varepsilon$ this inequality is trivial. □

Solution to Problem 7.74: If v is a convex function on an open interval (a,b) of the real line, then by (1.54) the function $(v(t)-v(u))/(t-u)$ is nondecreasing in t and nonincreasing in u. Hence for $a+\varepsilon < x < b-\varepsilon$,

$$
\frac{\mathsf{v}(x) - \mathsf{v}(x-\varepsilon)}{\varepsilon} \le \mathsf{D}^+\mathsf{v}(x) \le \frac{\mathsf{v}(x+\varepsilon) - \mathsf{v}(x)}{\varepsilon}.
$$

Let e_i be the vector with 1 in the ith component and 0 in the others. Let $c \in K$ be fixed. Let $\varepsilon > 0$ be so small that $K_\varepsilon \subseteq O$. It holds

$$|\frac{\partial^+ f}{\partial x_i}(c)| \leq \frac{1}{\varepsilon}|f(c+\varepsilon e_i) - f(c)| \leq \frac{1}{\varepsilon}(\max(|f(c+\varepsilon e_i)|, |f(c-\varepsilon e_i)|) + |f(c)|).$$

Denote the vectors $c \pm \varepsilon e_i$ c by $b_1, ..., b_{2^d+1}$. Then $g_f(x) = f(c) + \sum_{i=1}^n \frac{\partial^+ f}{\partial x_i}(c)(x_i - c_i)$ satisfies

$$|g_f(x) - g_f(y)| \leq \|x-y\| \sum_{i=1}^{d} \left|\frac{\partial^+ f}{\partial x_i}(c)\right| \leq \|x-y\| \frac{1}{\varepsilon}\left(\sum_{i=1}^{2^d} |f(b_i)| + d|f(c)|\right)$$

$$|g_f(x)| \leq (\sup_{x,y \in K_\varepsilon} \|x-y\|)\frac{1}{\varepsilon}\sum_{i=1}^{2^d} |f(b_i)| + |f(c)| \leq M \sum_{i=1}^{2^d+1} |f(b_i)|,$$

where M is some constant. $\widetilde{f}(x) = f(x) - g_f(x)$ is a nonnegative convex function. Hence by Problem7.73 $|\widetilde{f}(x) - \widetilde{f}(y)| \leq \varepsilon^{-1}\|x-y\| \sup_{x \in K_\varepsilon} \widetilde{f}(x)$. As O is open and $K_\varepsilon \subseteq O$ we may cover K_ε by a finite number of cubes whose union is contained in O. For x in some cube the value of $\widetilde{f}(x)$ does not exceed the maximum of the values of $\widetilde{f}$ at the vertices. Denoting all vertices of the cubes by $\widetilde{b}_1, ..., \widetilde{b}_N$ we get

$$|\widetilde{f}(x) - \widetilde{f}(y)| \leq \frac{1}{\varepsilon}\|x-y\| \sum_{i=1}^{N} \widetilde{f}(\widetilde{b}_i)$$
$$\leq \frac{1}{\varepsilon}\|x-y\| \sum_{i=1}^{N} (|f(\widetilde{b}_i)| + M \sum_{j=1}^{2^d+1} |f(b_j)|).$$

Now use $f = \widetilde{f} + g_f$ and $|g_f(x) - g_f(y)| \leq \|x-y\| \varepsilon^{-1}\left(\sum_{i=1}^{2^d} |f(b_i)| + d|f(c)|\right)$. □

Solution to Problem 7.83: If $C_\varepsilon(\theta) = \{\eta : \rho_\Delta(\eta, \theta) < \varepsilon\}$, then by Lebesgue's theorem $\lim_{\varepsilon \downarrow 0} \mathcal{H}(\theta_0, C_\varepsilon(\theta)) = \mathsf{H}_s(P_{\theta_0}, P_\theta) < 1$. Hence for every $\theta \in B_\varepsilon$ there is some $\varepsilon(\theta)$ with $\mathcal{H}(\theta_0, C_{\varepsilon(\theta)}(\theta)) < 1$. The rest follows from the compactness of B_ε. □

Solution to Problem 7.91: $(A) \Rightarrow (B) \Rightarrow (C)$ is clear. Fix any $\theta \in \Delta^0$ and consider the function $\varphi(s) = \Lambda(\widehat{\theta}(x) + s(\theta - \widehat{\theta}(x)), x)$, which for a sufficiently small $\varepsilon > 0$ is defined on $(-\varepsilon, 1+\varepsilon)$. As

$$\varphi''(s) = -(\theta - \widehat{\theta}(x))^T \nabla\nabla^T K(\widehat{\theta}(x) + s(\theta - \widehat{\theta}(x))(\theta - \widehat{\theta}(x)) < 0,$$

the function φ is strictly concave for $\widehat{\theta}(x) \neq \theta$. If (C) is fulfilled, then $\varphi^{\cdot}(0) = 0$ and the function φ is strictly decreasing for $s > 0$. Hence $\ln L(\widehat{\theta}(x), x) > \ln L(\theta, x)$ which gives (A). The global maximum point is uniquely determined as the function is strictly concave. □

Solution to Problem 7.92: Choose $\theta \in \Delta^0$, and consider the function

$$\kappa^*(\lambda) := K(\lambda\theta + (1-\lambda)\theta_0) - \langle \lambda\theta + (1-\lambda)\theta_0, t\rangle .$$

It holds $\frac{d\kappa^*(\lambda)}{d\lambda} = \langle \nabla K(\lambda\theta + (1-\lambda)\theta_0), \theta - \theta_0\rangle - \langle t, \theta - \theta_0\rangle$. Suppose that $\theta_0 \in \partial\Delta$. Then the steepness condition (7.49) implies $\lim_{\lambda \downarrow 0} \frac{d\kappa^*(\lambda)}{d\lambda} = -\infty$, so that there exists at least one $\lambda_0 \in (0,1)$ with

$$K(\lambda_0\theta + (1-\lambda_0)\theta_0) - \langle \lambda_0\theta + (1-\lambda_0)\theta_0, t\rangle = \kappa^*(\lambda_0) < \kappa^*(0) = K(\theta_0) - \langle \theta_0, t\rangle ,$$

which contradicts $K(\theta_0) - \langle \theta_0, t\rangle \leq K(\theta) - \langle \theta, t\rangle$, $\theta \in \Delta$. □

Solution to Problem 7.99: We use Proposition 2.17. It is sufficient to consider the case of $\mathbb{P}(Z = 0) = 0$. The distribution of Z has a density $f(s)$ that is nonincreasing for $s > 0$. As the distribution of Z is symmetric we may assume that f is symmetric. Hence $\mathbb{P}(|Z - a| < t) = F(a + t) - F(a - t)$, which is the integral over the $\boldsymbol{\lambda}$-a.e. existing derivative $f(a + t) - f(a - t) \leq 0$, $a > 0$. □

Solution to Problem 7.103: $D^{-}\mathsf{v}(\theta_0) \leq 0 \leq D^{+}\mathsf{v}(\theta_0)$ follows from the definition of $D^{-}\mathsf{v}$ and $D^{+}\mathsf{v}$. Conversely, if these inequalities hold, then by (1.58) the function v is nonincreasing for $t \leq \theta_0$ and nondecreasing for $t \geq \theta_0$. The minimizer is unique if and only if this monotonicity is strict. By (1.58) this is equivalent to $D^{-}\mathsf{v}(\theta_0 - \varepsilon) < 0 < D^{+}\mathsf{v}(\theta_0 + \varepsilon)$, $\epsilon > 0$. The strict convexity of v in a neighborhood of θ_0 implies that $D^{+}\mathsf{v}$ is strictly increasing. □

Solution to Problem 7.104: It holds

$$\frac{1}{h}[\mathsf{v}_{\varrho}(a+h) - \mathsf{v}_{\varrho}(a)] = \int \frac{1}{h}[\varrho(t + (a+h)) - \varrho(t+a)]P(dt).$$

To carry out the limit under the integral we remark that by (1.54) for $h \in (0, 1)$,

$$\begin{aligned}
&\varrho(t+a) - \varrho(t+a-1) \leq \frac{1}{h}[\varrho(t+a) - \varrho(t+a-h)] \leq D^{-}\varrho(t+a) \\
&\leq D^{+}\varrho(t+a) \leq \frac{1}{h}[\varrho(t+a+h) - \varrho(t+a)] \leq \varrho(t+a+1) - \varrho(t+a).
\end{aligned}$$

This inequality implies $\mathbb{E}\left|D^{\pm}\varrho(Z+a)\right| < \infty$, and the first statement in (7.58) is obtained by an application of Lebesgue's theorem. To determine $\gamma_{\mathsf{v}_{\varrho}}$ we calculate $\gamma_{\mathsf{v}_{\varrho}}((a, b])$. It holds

$$\begin{aligned}
\gamma_{\mathsf{v}_{\varrho}}((a, b]) &= D^{+}\mathsf{v}_{\varrho}(b) - D^{+}\mathsf{v}_{\varrho}(a) = \mathbb{E}D^{+}\varrho(Z+b) - \mathbb{E}D^{+}\varrho(Z+a) \\
&= \int \gamma_{\varrho}((a, b] + t)P(dt). \quad \square
\end{aligned}$$

Solution to Problem 7.119: In view of (7.81) it suffices to consider the case $n = 1$. If $(\mathsf{P}\Pi)(A) = \int P_{\theta}(A)\Pi(d\theta) = 0$ for some $A \in \mathfrak{A}$, then there exists at least one $\theta \in \{\theta : \mathsf{K}(P_{\theta_0}, P_{\theta}) < \infty\}$ with $P_{\theta}(A) = 0$. As $\mathsf{K}(P_{\theta_0}, P_{\theta}) < \infty$ we get $P_{\theta_0} \ll P_{\theta}$ and therefore $P_{\theta_0}(A) = 0$. □

Solution to Problem 7.129: It is enough to consider the case of $d = 1$. Let $\rho_N : \mathbb{R} \to \mathbb{R}$ be the piecewise linear function which is defined by the following conditions. $0 \leq \rho_N \leq 1$, $\rho_N(t) = 1$ for $|t| \geq N$, and $\rho_N(t) = 0$ for $|t| < N - 1$. Then $\mathcal{L}(T_n|P_n) \Rightarrow \mathcal{L}(T|P)$ implies

$$\lim_{N\to\infty} \limsup_{n\to\infty} P_n(|T_n| > N) \leq \lim_{N\to\infty} \limsup_{n\to\infty} \int \rho_N(T_n)dP_n = \lim_{N\to\infty} \int \rho_N(T)dP = 0,$$

where the last equality follows from Lebesgue's theorem. If $X_n = O_{P_n}(1)$ and $Y_n = o_{P_n}(1)$, then

$$\limsup_{n\to\infty} P_n(|X_nY_n| > \varepsilon) \leq \limsup_{n\to\infty} P_n(|Y_n| > \varepsilon/N) + \limsup_{n\to\infty} P_n(|X_n| > N).$$

The first term on the right-hand side is zero, whereas the second tends to zero as $N \to \infty$. The last statement is clear from the definition of $\mathbf{o}_{P_n}(0)$. □

Solution to Problem 7.130: The mapping $A \mapsto \det(A)$ is continuous. Hence $P_n(A_n) \to 1$, where $A_n = \{\det(\Sigma + o_{P_n}(1)) \neq 0\}$. Hence $V_n I_{A_n} = (\Sigma + o_{P_n}(1))^{-1} U_n I_{A_n} + (\Sigma + o_{P_n}(1))^{-1} o_{P_n}(1) = (\Sigma + o_{P_n}(1))^{-1} U_n I_{A_n} + o_{P_n}(1)$. As U_n is stochastically bounded we have $(\Sigma + o_{P_n}(1))^{-1} U_n = \Sigma^{-1} U_n + o_{P_n}(1)$. Hence $V_n = \Sigma^{-1} U_n + o_{P_n}(1)$. □

Solution to Problem 7.137: It holds $\gamma_\varrho(B) = \int_B \varrho''(x)\boldsymbol{\lambda}(dx)$; see (1.60). Hence by Remark 7.105 $f(x) = \int \varrho''(x+t)P(dt) = \mathbb{E}\varrho''(\varepsilon + x)$ is a Lebesgue density of $\gamma_{\mathsf{v}_\varrho}$ which is strictly positive. The continuity of f follows from the continuity of ϱ'', as by (7.97) we may pull the limit $t \to t_0$ under the integral sign. The Taylor formula in Theorem A.2 gives $\mathbb{E}(\varrho(\varepsilon_1 + a) - \varrho(\varepsilon_1) - a\varrho'(\varepsilon_1))^2 = a^4 \mathbb{E}(\int_0^1 (1-s)\varrho''(\varepsilon + sa)ds)^2 \leq a^4 \int_0^1 \mathbb{E}(\varrho''(\varepsilon + sa))^2 ds = O(a^4) = o(a^2)$. □

Solution to Problem 7.138: To verify (7.94) use Remark 7.105. It holds $\tau_\alpha(\varepsilon_1 + a) - \tau_\alpha(\varepsilon_1) - aD^+\tau_\alpha(\varepsilon_1) = 0$ if $\operatorname{sgn}(\varepsilon_1 + a) = \operatorname{sgn}(\varepsilon_1)$. If $a > 0$, $\varepsilon_1 + a > 0$, then with $\varepsilon_1 < 0$,

$$\begin{aligned}
&\tau_\alpha(\varepsilon_1 + a) - \tau_\alpha(\varepsilon_1) - aD^+\tau_\alpha(\varepsilon_1) I_{(-a,0)}(\varepsilon_1) \\
&= [\alpha(\varepsilon_1 + a) - (1-\alpha)\varepsilon_1 + a(1-\alpha)] I_{(-a,0)}(\varepsilon_1) = [(2\alpha - 1)\varepsilon_1 + a] I_{(-a,0)}(\varepsilon_1), \\
&\mathbb{E}[\tau_\alpha(\varepsilon_1 + a) - \tau_\alpha(\varepsilon_1) - aD^+\tau_\alpha(\varepsilon_1)]^2 I_{(-a,0)}(\varepsilon_1) \\
&\leq 4a^2 \mathbb{E} I_{(-a,0)}(\varepsilon_1) = 4a^2 \int_{-a}^0 f(t)dt = o(a^2). \quad \text{The other cases are similar.} \quad \Box
\end{aligned}$$

Solution to Problem 7.140: It holds

$$\begin{aligned}
\mathsf{E}_{\theta_0}(\psi_{\theta+h} - \psi_\theta) - \mathsf{E}_{\theta_0}(\dot{\psi}_\theta^T h) &= \mathsf{E}_{\theta_0} \int_0^1 (\dot{\psi}_{\theta+sh}^T h - \mathsf{E}_{\theta_0}(\dot{\psi}_\theta^T h)) ds, \\
\| \mathsf{E}_{\theta_0}(\psi_{\theta+h} - \psi_\theta) - \mathsf{E}_{\theta_0} \dot{\psi}_\theta^T h \| &\leq \int_0^1 \mathsf{E}_{\theta_0} \| \dot{\psi}_{\theta+sh}^T h - \dot{\psi}_\theta^T h \| ds = o(\| h \|),
\end{aligned}$$

where the last equality follows from $\mathsf{E}_{\theta_0} \sup_{\theta \in U(\theta_0)} \| \dot{\psi}_\theta \| < \infty$, Lebesgue's theorem, and the continuity of $\dot{\psi}_\theta$. □

Solution to Problem 7.174: If I is invertible, then for every $z = (x^T, y^T)^T \neq 0$

$$0 < z^T \mathsf{I} z = x^T \mathsf{I}_{1,1} x + 2x^T \mathsf{I}_{1,2} y + y^T \mathsf{I}_{1,1} y.$$

$x = 0$ gives that $I_{2,2}$ is invertible. Set $y = -I_{2,2}^{-1} I_{2,1} x$. Then $0 < z^T I z = x^T (I_{1,1} - I_{1,2} I_{2,2}^{-1} I_{2,1}) x$ shows that G is invertible. The representation of I^{-1} follows by the multiplication of the two matrices. For example,

$$\begin{aligned}
&(\mathsf{J}_{1,1}, -\mathsf{J}_{1,1}\mathsf{I}_{1,2}\mathsf{I}_{2,2}^{-1})(\mathsf{I}_{1,1}, \mathsf{I}_{2,1})^T \\
&= (\mathsf{I}_{1,1} - \mathsf{I}_{1,2}\mathsf{I}_{2,2}^{-1}\mathsf{I}_{2,1})^{-1}\mathsf{I}_{1,1} - (\mathsf{I}_{1,1} - \mathsf{I}_{1,2}\mathsf{I}_{2,2}^{-1}\mathsf{I}_{2,1})^{-1}\mathsf{I}_{1,2}\mathsf{I}_{2,2}^{-1}\mathsf{I}_{2,1} = \mathbf{I}. \quad \Box
\end{aligned}$$

Solution to Problem 7.175: $\mathsf{I}_{2,2}$ and $\mathsf{I}_{2,2}^{-1}$ are symmetric and positive definite. Furthermore, $x^T \mathsf{I}_{1,1} x - x^T (\mathsf{I}_{1,1} - \mathsf{I}_{1,2}\mathsf{I}_{2,2}^{-1}\mathsf{I}_{2,1}) x = (\mathsf{I}_{2,1} x)^T \mathsf{I}_{2,2}^{-1} (\mathsf{I}_{2,1} x) \geq 0$. □

8

Testing

The foundation of the framework of testing statistical hypotheses has been laid in Section 2.2. For a statistical model $\mathcal{M} = (\mathcal{X}, \mathfrak{A}, (P_\theta)_{\theta\in\Delta})$ a test $\varphi : \mathcal{X} \to_m [0,1]$ decides, based on an observation from some P_θ with an unknown $\theta \in \Delta$, whether the null hypothesis $\mathsf{H}_0 : \theta \in \Delta_0$ or the alternative $\mathsf{H}_A : \theta \in \Delta_A$ is true, where Δ_0 and Δ_A is a fixed given partition of Δ. A comprehensive presentation of the theory of optimal tests is provided by Lehmann (1959, 1986) and Lehmann and Romano (2005), where also notes on historical developments can be found. Other books in this area are by Ferguson (1967), Hájek and Šidák (1967), Strasser (1985), Witting (1985), and Schervish (1995), where the majority of the subsequent results can be found.

8.1 Best Tests for Exponential Families

8.1.1 Tests for One–Parameter Exponential Families

In this section tests for one-parameter exponential families $(P_\theta)_{\theta\in\Delta}$ are considered. According to Definition 1.1 the $\boldsymbol{\mu}$-density of P_θ is

$$f_\theta(x) = \exp\{\theta T(x) - K(\theta)\}, \quad x \in \mathcal{X},\ \theta \in \Delta. \tag{8.1}$$

Here $\Delta \subseteq \mathbb{R}$ is a convex set and thus an interval that may be open or closed, and finite or infinite, on either side. For (8.1) the likelihood ratio

$$L_{\theta_0,\theta_1}(x) = \exp\{(\theta_1 - \theta_0)T(x) - K(\theta_1) + K(\theta_0)\}, \quad x \in \mathcal{X},$$

of P_{θ_1} with respect to P_{θ_0} is an increasing function of $T(x)$ for every $x \in \mathcal{X}$ and $\theta_0, \theta_1 \in \Delta$ with $\theta_0 < \theta_1$. If we reparametrize the family and introduce a new parameter η by setting $\theta = \kappa(\eta)$, then the MLR property is preserved if κ is nondecreasing. Such reparametrizations have been systematically considered in Chapter 1. We also recall the assumptions (A1) and (A2) for exponential families that have been made at the beginning of Chapter 1. (A1) guarantees

F. Liese, K.-J. Miescke, *Statistical Decision Theory*,
DOI: 10.1007/978-0-387-73194-0_8, © Springer Science+Business Media, LLC 2008

that the generating statistic T is not a constant, P_θ-a.s.; that is, $\mathsf{V}_\theta(T) = K''(\theta) > 0$, $\theta \in \Delta$. (A2) guarantees that $\Delta^0 \neq \emptyset$.

Uniformly best level α tests for one-sided hypotheses under nondecreasing likelihood ratio in T have been established already in Theorem 2.49. In this section we also consider systematically the hypotheses summarized below.

Testing Problem	$\mathsf{H}_0 : \theta \in \Delta_0$	$\mathsf{H}_A : \theta \in \Delta_A$
(I)	$\Delta_0 = (-\infty, \theta_0] \cap \Delta$	$\Delta_A = (\theta_0, \infty) \cap \Delta$
(II)	$\Delta_0 = \{\theta_0\}$	$\Delta_A = \Delta \backslash \{\theta_0\}$
(III)	$\Delta_0 = \Delta \backslash (\theta_1, \theta_2)$	$\Delta_A = (\theta_1, \theta_2) \cap \Delta$
(IV)	$\Delta_0 = [\theta_1, \theta_2] \cap \Delta$	$\Delta_A = \Delta \backslash [\theta_1, \theta_2]$

(8.2)

For testing problem (I) a uniformly best (UMP) level α test has been given in (2.19). Testing problem (IV) is in some way a reflection of testing problem (III). Somewhat surprising, there exists a uniformly best level α test for testing problem (III), but not for the testing problems (II) and (IV). We also remark that (II) is a limiting case of (IV). Tests for (II) and (IV) are used to test whether the true parameter deviates from a norm values or is outside of a norm interval, respectively. In contrast to this the tests for

$$(V) \quad \mathsf{H}_0 : \theta \in \Delta \backslash \{\theta_0\} \quad \text{versus} \quad \mathsf{H}_A : \theta = \theta_0 \in \Delta$$

and (III) are used to show that the data are from a model for which the true parameter is a given standard value or belongs to a standard interval, respectively. Tests for such hypotheses are called *equivalence tests.* If the power function is continuous, which holds, for example, in exponential models, then every level α test for testing problem (V) has at most power α on the alternative. One way to remedy this is to stretch the alternative by turning to testing problem (III). For more details on equivalence tests we refer to Wellek (2003) and Lehmann and Romano (2005).

To deal with the above testing problems we need some general concepts that are introduced and discussed subsequently. Let $(\mathcal{X}, \mathfrak{A}, (P_\theta)_{\theta \in \Delta})$ be a model, and Δ_0 and Δ_A be a fixed given partition of Δ.

Definition 8.1. *A level α test φ for*

$$\mathsf{H}_0 : \theta \in \Delta_0 \quad \textit{versus} \quad \mathsf{H}_A : \theta \in \Delta_A \tag{8.3}$$

is called unbiased if $\inf_{\theta \in \Delta_A} \mathsf{E}_\theta \varphi \geq \alpha$. *A test ψ is called a uniformly best unbiased level α test for (8.3) if it is an unbiased level α test for (8.3) and $\mathsf{E}_\theta \psi \geq \mathsf{E}_\theta \phi$, $\theta \in \Delta_A$, holds for every unbiased level α test ϕ for (8.3).*

Uniformly best unbiased is also called uniformly most powerful unbiased (UMPU). UMPU level α tests are admissible in a special way.

Problem 8.2.* Let $\alpha \in (0,1)$ be fixed. Every uniformly best unbiased level α test ψ for (8.3) is admissible in the following sense. There is no level α test ϕ for (8.3) with $\mathsf{E}_\theta\psi \le \mathsf{E}_\theta\phi$, $\theta \in \Delta_A$, and $\mathsf{E}_{\theta_1}\psi < \mathsf{E}_{\theta_1}\phi$ at some $\theta_1 \in \Delta_A$.

For $\Delta \subseteq \mathbb{R}^d$ the *boundary* of Δ_0 and Δ_A, denoted by J, is the set of all $\theta \in \Delta$ for which there are points from both, Δ_0 and Δ_A, in every open ball $B \subseteq \mathbb{R}^d$ with center θ. Note that this boundary may be empty. This happens, e.g., if the two hypotheses consist of two disjoint closed intervals. A test φ is called *α-similar on the boundary* of Δ_0 and Δ_A if $\mathsf{E}_\theta\varphi = \alpha$ for every $\theta \in J$.

Problem 8.3. If $\Delta \subseteq \mathbb{R}^d$ and $J \neq \emptyset$, then every unbiased level α test for (8.3) for which $\theta \mapsto \mathsf{E}_\theta\varphi$ is continuous is α-similar on the boundary.

Thus in particular every unbiased level α test φ for (8.3) for which $\theta \mapsto \mathsf{E}_\theta\varphi$ is continuous attains the level α, in the sense of Definition 2.30, provided that $J \neq \emptyset$. This holds especially for an exponential family that satisfies (A1) and (A2), as in this case $\theta \mapsto \mathsf{E}_\theta\varphi$ is continuous for every test φ; see Lemma 1.16.

Problem 8.4.* Suppose that $\theta \mapsto \mathsf{E}_\theta\psi$ is continuous for every test ψ. Let φ be a level α test for (8.3) that is α-similar on the boundary J, where $J \neq \emptyset$. If $\mathsf{E}_\theta\phi \le \mathsf{E}_\theta\varphi$, $\theta \in \Delta_A$, for every test ϕ with $\mathsf{E}_\theta\phi = \alpha$, $\theta \in J$, then φ is a uniformly best unbiased level α test for (8.3).

Two-sided testing problems in (8.2) for $\Delta \subseteq \mathbb{R}$ and $\alpha \in (0,1)$ are considered next. First it is shown that there are no uniformly best level α tests for (II) and (IV).

Proposition 8.5. *Let $(P_\theta)_{\theta\in\Delta}$, $\Delta \subseteq \mathbb{R}$, be a one-parameter exponential family in natural form that satisfies (A1) and (A2). Let $\theta_1, \theta_2 \in \Delta^0$ with $\theta_1 < \theta_2$ and $\alpha \in (0,1)$. Then there does not exist a uniformly best level α test for the testing problems (II) and (IV) in (8.2).*

Proof. Take $a, b \in \Delta$ with $[\theta_1, \theta_2] \subseteq (a,b)$ and consider the two testing problems $\mathsf{H}_0 : \theta \le \theta_1$ versus $\mathsf{H}_A : \theta > \theta_1$ and $\mathsf{H}_0 : \theta \le \theta_2$ versus $\mathsf{H}_A : \theta > \theta_2$.

For the first testing problem, let $\mathsf{p}_1(\theta)$ denote the power function of the uniformly best $(1-\alpha)$ level test that appears in Proposition 2.51. Similarly, let $\mathsf{p}_2(\theta)$ denote the power function of the uniformly best level α test for the second testing problem. If ψ is a uniformly best level α test for problem (IV) with power function $\mathsf{p}(\theta) = \mathsf{E}_\theta\psi$, then $\mathsf{p}_2(\theta) = \mathsf{p}(\theta)$, $\theta > \theta_2$, and therefore, in view of Theorem A.3, $\mathsf{p}_2(\theta) = \mathsf{p}(\theta)$ for every $\theta \in (a,b)$ as both functions are analytic; see Lemma 1.16. Similarly it follows that $1 - \mathsf{p}_1(\theta) = \mathsf{p}(\theta)$ for every $\theta \in (a,b)$. This, however, is impossible as p_1 and p_2 are increasing functions; see Proposition 2.51. The proof for problem (II) is analogous. ■

Tests for problem (I) in (8.2) have been studied already in Chapter 2. It has been shown (see Theorem 2.49) that for families with MLR in T every test of the form $I_{(c,\infty)}(T) + \gamma I_{\{c\}}(T)$ is a uniformly best level α test if the constants $c \in \mathbb{R}$ and $\gamma \in [0,1]$ are determined by $\alpha = \mathsf{E}_{\theta_0}(I_{(c,\infty)}(T) + \gamma I_{\{c\}}(T))$. We show in the subsequent theorems that for the two-sided testing problems (II),

(III), and (IV) in (8.2) the optimal tests, for (II) and (IV) restricted to the class of unbiased tests, are also piecewise constant. The display below gives the structure of these optimal tests and the conditions for their constants.

Test	Structure of Test	Constraints
$\varphi_I(T)$	$\varphi_I = I_{(c,\infty)} + \gamma I_{\{c\}}$	$E_{\theta_0}\varphi_I(T) = \alpha$
$\varphi_{II}(T)$	$\varphi_{II} = I_{(-\infty,b_1)} + \varsigma_1 I_{\{b_1\}} + \varsigma_2 I_{\{b_2\}} + I_{(b_2,\infty)}$	$E_{\theta_0}\varphi_{II}(T) = \alpha$ $E_{\theta_0}\varphi_{II}(T)T = \alpha E_{\theta_0}T$
$\varphi_{III}(T)$	$\varphi_{III} = I_{(c_1,c_2)} + \varsigma_1 I_{\{c_1\}} + \varsigma_2 I_{\{c_2\}}$	$E_{\theta_1}\varphi_{III}(T) = \alpha$ $E_{\theta_2}\varphi_{III}(T) = \alpha$
$\varphi_{IV}(T)$	$\varphi_{IV} = I_{(-\infty,b_1)} + \varsigma_1 I_{\{b_1\}} + \varsigma_2 I_{\{b_2\}} + I_{(b_2,\infty)}$	$E_{\theta_1}\varphi_{IV}(T) = \alpha$ $E_{\theta_2}\varphi_{IV}(T) = \alpha$

(8.4)

The following theorem establishes the optimal tests for problem (IV).

Theorem 8.6. *Let $(P_\theta)_{\theta\in\Delta}$ be a one-parameter exponential family in natural form that satisfies (A1) and (A2), and let $\theta_1, \theta_2 \in \Delta^0$ with $\theta_1 < \theta_2$ be fixed.*

(A) For every $\alpha \in (0,1)$ there exist $b_1, b_2 \in \mathbb{R}$ and $\varsigma_1, \varsigma_2 \in [0,1]$ such that

$$\mathsf{E}_{\theta_1}\varphi_{IV}(T) = \mathsf{E}_{\theta_2}\varphi_{IV}(T) = \alpha. \tag{8.5}$$

(B) Every test $\varphi_{IV}(T)$ in (8.4) that satisfies (8.5) is a uniformly best unbiased level α test for the testing problem (IV) in (8.4).
(C) For every test $\varphi_{IV}(T)$ from (8.4) that satisfies (8.5) the power $\mathsf{E}_\theta\varphi_{IV}(T)$ has a minimum at some uniquely determined $\theta_* \in (\theta_1, \theta_2)$, and it increases as θ tends away from θ_* in either direction.

Proof. For a proof of (A) we refer to Lehmann (1959), Ferguson (1967), and Witting (1985). Now we prove statement (B). To this end we remark that it suffices to consider the case where $[0,1] \subseteq \Delta^0$, $\theta_1 = 0$, and $\theta_2 = 1$, as the general case can be reduced to this case by a linear transformation of the parameter set. Using the density f_θ in (8.1) we consider the function

$$\begin{aligned}\rho(t,\rho_1,\rho_2) &= \rho_1\frac{f_0(t)}{f_\theta(t)} + \rho_2\frac{f_1(t)}{f_\theta(t)}\\ &= \rho_1\exp\{-\theta t - K(0) + K(\theta)\} + \rho_2\exp\{(1-\theta)t - K(1) + K(\theta)\},\end{aligned}$$

along with the equations

$$\rho(b_1,\rho_1,\rho_2) = \rho(b_2,\rho_1,\rho_2) = 1. \tag{8.6}$$

The solution ρ_1^*, ρ_2^* is given by

$$\begin{aligned}
\rho_1^* &= \frac{1}{D}\left[\exp\{(1-\theta)b_2\} - \exp\{(1-\theta)b_1\}\right], \\
\rho_2^* &= \frac{1}{D}\left[\exp\{-\theta b_1\} - \exp\{-\theta b_2\}\right], \\
D &= \exp\{(1-\theta)b_2 - \theta b_1\} - \exp\{(1-\theta)b_1 - \theta b_2\} > 0.
\end{aligned} \tag{8.7}$$

The inequality $(1-\theta)b_2 - \theta b_1 - (1-\theta)b_1 + \theta b_2 = b_2 - b_1 > 0$ implies

$$\begin{aligned}
&\rho_1^* > 0 \quad \text{and} \quad \rho_2^* < 0 \quad \text{for} \quad \theta < 0, \\
&\rho_1^* < 0 \quad \text{and} \quad \rho_2^* > 0 \quad \text{for} \quad 1 < \theta.
\end{aligned} \tag{8.8}$$

The derivative $\rho'(t, \rho_1^*, \rho_2^*) = -\rho_1^*\theta \exp\{-\theta t\} + \rho_2^*(1-\theta)\exp\{(1-\theta)t\}$ of the function $\rho(t, \rho_1^*, \rho_2^*)$ has the unique zero

$$t_0 = \ln \frac{\rho_1^*\theta}{\rho_2^*(1-\theta)},$$

which by (8.6) belongs to $[b_1, b_2]$. From (8.8) and $\lim_{t\to\infty} \rho(t, \rho_1^*, \rho_2^*) = -\infty$ we obtain

$$\begin{aligned}
&\rho(t, \rho_1^*, \rho_2^*) \geq 1, \quad \text{for } b_1 \leq t \leq b_2, \\
&\rho(t, \rho_1^*, \rho_2^*) \leq 1, \quad \text{for } t \leq b_1 \text{ and } t \geq b_2.
\end{aligned} \tag{8.9}$$

As $\mathsf{E}_0\varphi_{IV}(T) = \mathsf{E}_1\varphi_{IV}(T) = \alpha$, it holds for every test ψ that satisfies (8.5),

$$\begin{aligned}
\mathsf{E}_\theta(\varphi_{IV}(T) - \psi) &= \mathsf{E}_\theta(\varphi_{IV}(T) - \psi) - \rho_1^*\mathsf{E}_0(\varphi_{IV}(T) - \psi) - \rho_2^*\mathsf{E}_1(\varphi_{IV}(T) - \psi) \\
&= \mathsf{E}_\theta(\varphi_{IV}(T) - \psi)[1 - \rho(T, \rho_1^*, \rho_2^*)].
\end{aligned}$$

By the definition of $\varphi_{IV}(T)$ and (8.9) both brackets are simultaneously positive or negative so that the product is nonnegative in any case. This gives

$$\mathsf{E}_\theta(\varphi_{IV}(T) - \psi) \geq 0, \quad \theta \notin [0, 1].$$

Especially, $\psi \equiv \alpha$ is an unbiased level α test, and thus the inequality $\mathsf{E}_\theta\varphi_{IV}(T) \geq \alpha$, $\theta \notin [0, 1]$, is established. As θ was arbitrary we see that $\varphi_{IV}(T)$ is a uniformly best unbiased test, provided that

$$\mathsf{E}_\theta\varphi_{IV}(T) \leq \alpha, \quad 0 \leq \theta \leq 1.$$

As we have noted at the beginning of the proof the consideration of $\theta_1 = 0$ and $\theta_2 = 1$ is no restriction of generality. Hence we obtain the following remarkable property of the power function $\mathsf{p}(\theta) := \mathsf{E}_\theta\varphi_{IV}(T)$. For any $\eta_1 < \eta_2$ and $\eta \notin [\eta_1, \eta_2]$,

$$0 < \mathsf{E}_{\eta_1}\varphi_{IV}(T) = \mathsf{E}_{\eta_2}\varphi_{IV}(T) < 1 \quad \text{implies} \quad \mathsf{E}_\eta\varphi_{IV}(T) \geq \mathsf{E}_{\eta_1}\varphi_{IV}(T).$$

This property and the continuity of the function $\mathsf{p}(\eta) = \mathsf{E}_\eta\varphi_{IV}(T)$ (see Lemma 1.16) implies that $\mathsf{p}(\eta)$ does not cross any level more than twice in $0 \leq \eta \leq 1$.

Because $\mathsf{p}(\eta)$ is an analytic function that is uniquely determined by its values in any open subset of Δ, and $\varphi_{IV}(T)$ is not a constant (i.e., equal to α, P_θ-a.s.) there is a uniquely determined θ_* at which $\mathsf{p}(\eta)$ attains the minimum, and $\mathsf{p}(\eta)$ increases if η tends away from θ_*. This completes the proof of (B), and also the proof of (C). ∎

Problem 8.7. Suppose that P_θ has the Lebesgue density $f_\theta(x) = \theta x^{\theta-1} I_{[0,1]}(x)$, $x \in \mathbb{R}$, $\theta \in \Delta = (0,\infty)$. For $0 < \theta_1 < \theta_2$ fixed, let $\mathsf{H}_0 : \theta \in [\theta_1, \theta_2]$ and $\mathsf{H}_A : \theta \in (0,\theta_1) \cup (\theta_2, \infty)$. For a fixed $\alpha \in (0,1)$ find a uniformly best unbiased level α test for H_0 versus H_A.

Now we deal with testing problem (III) in (8.2) for $\alpha \in (0,1)$. For $c_1, c_2 \in \mathbb{R}$ with $c_1 < c_2$, and $\gamma_1, \gamma_2 \in [0,1]$, let $\varphi_{III}(T)$ be defined in (8.4). In contrast to the previous theorem this test, subject to

$$\mathsf{E}_{\theta_1} \varphi_{III}(T) = \mathsf{E}_{\theta_2} \varphi_{III}(T) = \alpha, \tag{8.10}$$

turns out to be a uniformly best level α test for testing problem (III), that is, not only in the restricted class of unbiased level α tests, but in the class of all level α tests.

Theorem 8.8. *Let $(P_\theta)_{\theta\in\Delta}$ be a one-parameter exponential family in natural form that satisfies (A1) and (A2), and let $\theta_1, \theta_2 \in \Delta^0$ with $\theta_1 < \theta_2$ be fixed.*

(A) For every $\alpha \in (0,1)$ there exist $c_1, c_2 \in \mathbb{R}$ and $\gamma_1, \gamma_2 \in [0,1]$ such that (8.10) holds.
(B) Every test $\varphi_{III}(T)$ from (8.4) that satisfies (8.10) is a uniformly best level α test for the testing problem (III) in (8.2).
(C) The power $\mathsf{E}_\theta \varphi_{III}(T)$ of every test $\varphi_{III}(T)$ from (8.4) that satisfies (8.10) has a maximum at some $\theta_* \in (\theta_1, \theta_2)$. Moreover, it decreases as θ tends away from θ_* in either direction.

Proof. The proof is similar to that of the previous theorem. For a proof of (A) we refer again to Lehmann (1959), Ferguson (1967), and Witting (1985). Now we prove statement (B) and assume again without loss of generality that $\theta_1 = 0$ and $\theta_2 = 1$. As before we start with the system of equations (8.6) which has the solution (8.7). But now $0 < \theta < 1$, which implies that ρ_1^*, ρ_2^* are positive, and thus $\rho(t, \rho_1^*, \rho_2^*)$ is a strictly convex function. In view of $\mathsf{E}_0 \psi \le \alpha$, $\mathsf{E}_1 \psi \le \alpha$, and $\rho_i^* \ge 0$, $i = 1, 2$, we have

$$\begin{aligned}
&\mathsf{E}_\theta(\varphi_{III}(T) - \psi) \\
&\ge \mathsf{E}_\theta(\varphi_{III}(T) - \psi) - \rho_1^* \mathsf{E}_0(\varphi_{III}(T) - \psi) - \rho_2^* \mathsf{E}_1(\varphi_{III}(T) - \psi) \\
&= \mathsf{E}_\theta(\varphi_{III}(T) - \psi)[1 - \rho(T, \rho_1^*, \rho_2^*)] \ge 0
\end{aligned}$$

by the definition of $\varphi_{III}(T)$ and the property (8.9) of $\rho(t, \rho_1^*, \rho_2^*)$. Statement (C) follows from the analogous statement (C) in Theorem 8.6 by using $\varphi_{III}(T) := 1 - \varphi_{IV}(T)$. ∎

Problem 8.9. As before in Problem 8.7, let P_θ have the Lebesgue density $f_\theta(x) = \theta x^{\theta-1} I_{[0,1]}(x)$, $x \in \mathbb{R}$, $\theta \in \Delta = (0, \infty)$. For $0 < \theta_1 < \theta_2$ fixed, let $\mathsf{H}_0 : \theta \in (0, \theta_1] \cup [\theta_2, \infty)$ and $\mathsf{H}_A : \theta \in (\theta_1, \theta_2)$. For a fixed $\alpha \in (0, 1)$ find a uniformly best level α test for H_0 versus H_A.

Now we discuss the duality between Theorems 8.6 and 8.8. If ψ is any level α test for (IV) in (8.2) that is α-similar on the boundary $\{\theta_1, \theta_2\}$ (i.e., $\mathsf{E}_{\theta_1}\psi = \mathsf{E}_{\theta_2}\psi = \alpha$), then $\overline{\psi} = 1 - \psi$ is a level $1 - \alpha$ test for (III) in (8.2) that is $(1 - \alpha)$-similar on the boundary $\{\theta_1, \theta_2\}$. Especially we have for the tests from (8.4)

$$1 - \varphi_{IV}(T) = I_{[b_1,b_2]}(T) + (1 - \varsigma_1) I_{\{b_1\}}(T) + (1 - \varsigma_2) I_{\{b_2\}}(T) = \varphi_{III}(T) \quad (8.11)$$

by setting $b_i = c_i$ and $\gamma_i = (1 - \varsigma_i)$, $i = 1, 2$. However, if ψ is a uniformly most powerful level α test for testing problem (III), then $1 - \psi$ is only a uniformly most powerful in the class of all unbiased level $1 - \alpha$ test for testing problem (IV). Despite this fact, the relationship between $\varphi_{III}(T)$ and $\varphi_{IV}(T)$ leads to a statement of the power function of $\varphi_{IV}(T)$.

Proposition 8.10. *Suppose that the assumptions of Theorems 8.6 and 8.8 are fulfilled. Then for every unbiased level α test ψ for the testing problem (IV) in (8.2) with $\mathsf{E}_{\theta_1}\psi = \mathsf{E}_{\theta_2}\psi$ it holds that $\mathsf{E}_\theta \varphi_{IV}(T) \le \mathsf{E}_\theta \psi$ for all $\theta \in [\theta_1, \theta_2]$, and $\mathsf{E}_\theta \varphi_{IV}(T) \ge \mathsf{E}_\theta \psi$ for all $\theta \notin [\theta_1, \theta_2]$.*

Proof. The second statement is clear from Theorem 8.6. To prove the first statement we remark that $1 - \varphi_{IV}(T)$ is by assumption a level $1 - \alpha$ test for the testing problem (III) in (8.2). Hence this test is, according to Theorem 8.8 and (8.11), a uniformly best level $1 - \alpha$ test for the testing problem (III) in (8.2). This proves the first statement. ■

Consider the testing problem (II) in (8.2) at the level $\alpha \in (0, 1)$. In view of Lemma 1.16 the power function $\mathsf{p}(\theta) := \mathsf{E}_\theta \psi$ of a test ψ is infinitely often differentiable, and the differentiation can be carried out under the expectation. If ψ is an unbiased level α test, then the power function $\mathsf{p}(\theta) := \mathsf{E}_\theta \psi$ has a local minimum at $\theta = \theta_0$ so that by Lemma 1.16 and $K'(\theta) = \mathsf{E}_\theta T$ from (1.23) it holds

$$\mathsf{p}'(\theta) = \mathsf{E}_\theta(T\psi) - \mathsf{E}_\theta(\psi K'(\theta)) = 0.$$

Hence we see that every unbiased level α test that attains the level α satisfies the conditions

$$\mathsf{E}_{\theta_0}\psi = \alpha, \quad (8.12)$$

$$\mathsf{E}_{\theta_0}(T\psi) = \alpha \mathsf{E}_{\theta_0} T. \quad (8.13)$$

Now we show that tests with the structure $\varphi_{II}(T)$ in (8.4) are uniformly best unbiased level α tests for the testing problem (II) in (8.2).

Theorem 8.11. *Let $(P_\theta)_{\theta \in \Delta}$ be a one-parameter exponential family in natural form that satisfies (A1) and (A2), and let $\theta_0 \in \Delta^0$ be fixed.*

(A) For every $\alpha \in (0,1)$ there exist $b_1, b_2 \in \mathbb{R}$ and $\varsigma_1, \varsigma_2 \in [0,1]$ such that a test $\varphi_{II}(T)$ in (8.4) satisfies (8.12) and (8.13).

(B) Every test $\varphi_{II}(T)$ in (8.4) that satisfies (8.12) and (8.13) is a uniformly best unbiased level α test for testing problem (II) in (8.2).

(C) The power $\mathsf{E}_\theta\varphi_{II}(T)$ of every test $\varphi_{II}(T)$ from (8.4) that satisfies (8.12) and (8.13) has a minimum at θ_0 and increases as θ tends away from θ_0 in either direction.

Proof. The proof is similar to that of Theorem 8.6. For (A) we refer again to Lehmann (1959), Ferguson (1967), and Witting (1985). Now we prove (B) and similarly as before assume without loss of generality that $\theta_0 = 0$. For $\theta \neq 0$ we study the function $\rho(t, \rho_1, \rho_2) = \exp\{\theta t - K(\theta) + K(0)\} + \rho_1 + \rho_2 t$. As the first expression on the right-hand side is a strictly convex function of t we find ρ_1^* and ρ_2^* such that line $\rho_1^* + \rho_2^* t$ crosses this function at the points b_1 and b_2. Hence, $\rho(t, \rho_1, \rho_2) \geq 0$ for $b_1 \leq t \leq b_2$ and $\rho(t, \rho_1, \rho_2) \leq 0$ otherwise. Let ψ be any test that satisfies the conditions (8.12) and (8.13). Then

$$\begin{aligned}\mathsf{E}_\theta(\varphi_{II}(T) - \psi) &= \mathsf{E}_\theta(\varphi_{II}(T) - \psi) + \mathsf{E}_0(\rho_1^* + \rho_2^* T)(\varphi_{II}(T) - \psi) \\ &= \mathsf{E}_0(\varphi_{II}(T) - \psi)\rho(T, \rho_1^*, \rho_2^*).\end{aligned}$$

By the definition of $\varphi_{II}(T)$ and the construction of $\rho(T, \rho_1^*, \rho_2^*)$ the two terms $\varphi_{II}(T) - \psi$ and $\rho(T, \rho_1^*, \rho_2^*)$ have the same sign. Hence $\mathsf{E}_\theta(\varphi_{II}(T) - \psi) \geq 0$ so that $\varphi_{II}(T)$ is uniformly best among all tests that satisfy the conditions (8.12) and (8.13), which are also satisfied by the constant test $\psi \equiv \alpha$. As $\varphi_{II}(T)$ is a level α test we arrive at

$$\mathsf{E}_0\varphi_{II}(T) \leq \alpha \leq \mathsf{E}_\theta\varphi_{II}(T), \quad \theta \in \Delta_A.$$

Hence $\varphi_{II}(T)$ is an unbiased level α test. As the class of all unbiased test level α tests is a subclass of tests that satisfy (8.12) and (8.13), and $\varphi_{II}(T)$ is uniformly best in this class, we see that $\varphi_{II}(T)$ is a uniformly best unbiased level α test. The proof of (C) is similar to that in Theorem 8.6. ■

Problem 8.12. As before in Problem 8.7, let P_θ have the Lebesgue density $f_\theta(x) = \theta x^{\theta-1} I_{[0,1]}(x)$, $x \in \mathbb{R}$, $\theta \in \Delta = (0, \infty)$. For some fixed $\theta_0 > 0$ let $\mathsf{H}_0 : \theta = \theta_0$ and $\mathsf{H}_A : \theta \neq \theta_0$. For a fixed $\alpha \in (0,1)$ find a uniformly best unbiased level α test for H_0 versus H_A.

Problem 8.13. Let $X_1, ..., X_n$ be an i.i.d. sample from a Poisson distribution $\mathsf{Po}(\lambda)$ with $\lambda > 0$. For some fixed $\lambda_0 > 0$ let $\mathsf{H}_0 : \lambda = \lambda_0$ and $\mathsf{H}_A : \lambda \neq \lambda_0$. For a fixed $\alpha \in (0,1)$ find a uniformly best unbiased level α test for H_0 versus H_A.

Fortunately, an occasional symmetry in the model allows for a simpler determination of the constants $b_1, b_2, \varsigma_1, \varsigma_2$ in (8.4) that specify the optimal tests in the Theorems 8.8 and 8.11. Suppose that for every $\theta \in \Delta$ there exists an $m(\theta)$ such that $T - m(\theta)$ and $m(\theta) - T$ have the same distribution under P_θ, i.e., that

$$P_\theta(T - m(\theta) \le t) = P_\theta(m(\theta) - T \le t), \quad t \in \mathbb{R},\ \theta \in \Delta. \tag{8.14}$$

In this case, of course, we also have

$$m(\theta) = \mathsf{E}_\theta T, \quad \theta \in \Delta. \tag{8.15}$$

By utilizing the symmetry that is given by (8.14) we can represent the test $\varphi_{2,T}$ in a symmetric version. Let $F_{\theta_0}(t) := P_{\theta_0}(T - m(\theta_0) \le t)$, $t \in \mathbb{R}$, and put for $\alpha \in (0,1)$

$$\begin{aligned}\psi_{sy}(T) &= I_{(-\infty,-c_{1-\alpha/2})}(T - m(\theta_0)) + \gamma I_{\{-c_{1-\alpha/2}\}}(T - m(\theta_0)) \\ &\quad + \gamma I_{\{c_{1-\alpha/2}\}}(T - m(\theta_0)) + I_{(c_{1-\alpha/2},\infty)}(T - m(\theta_0)), \\ \gamma &= (F_{\theta_0}(c_{1-\alpha/2}) - (1-\alpha/2)) \oslash P_{\theta_0}(T - m(\theta_0) = c_{1-\alpha/2}),\end{aligned} \tag{8.16}$$

where $c_{1-\alpha/2} = F_{\theta_0}^{-1}(1-\alpha/2)$. By this construction the condition (8.12) is fulfilled. Moreover, because the test $\psi_{sy}(T)$ is symmetric in $T - m(\theta_0)$, we also have $\mathsf{E}_{\theta_0}(T - m(\theta_0))\psi_{sy}(T) = 0$, which, together with (8.15), gives the second side condition (8.13).

Proposition 8.14. *Under the conditions of Theorem 8.11 and the symmetry assumption (8.14), the test $\psi_{sy}(T)$ given by (8.16) is identical with test $\varphi_{II}(T)$ in Theorem 8.11, and thus it has the optimality properties stated there.*

In the above theorems we have constructed optimal tests, but the problem of uniqueness has been left open. In general one cannot expect uniqueness in the class of all level α tests. However, if we consider only tests that depend on T, then there is uniqueness, as specified in the following remark.

Remark 8.15. Let $(P_\theta)_{\theta\in\Delta}$ be a one-parameter exponential family in natural form that satisfies (A1) and (A2). If for two tests $\psi_1(T)$ and $\psi_2(T)$, and some interval $(a,b) \subseteq \Delta^0$, it holds $\mathsf{E}_\theta\psi_1(T) = \mathsf{E}_\theta\psi_2(T)$ for every $\theta \in (a,b)$, then $\psi_1(T) = \psi_2(T)$, P_θ-a.s., $\theta \in \Delta$, follows from the completeness of the generating statistic T in an exponential family; see Theorem 4.73.

If the exponential family is represented in reparametrized form $(P_{pe,\eta})_{\eta\in\Lambda}$, as specified in (1.11), then for a strictly monotone mapping $\kappa : \Lambda \to \Delta$ the above hypotheses are transformed to hypotheses that are again intervals.

Remark 8.16. Let $(P_\theta)_{\theta\in\Delta}$ be a one-parameter exponential family with generating statistic T and natural parameter θ. If $\eta = \kappa(\theta)$ and $S = g(T)$ is another test statistic, where κ and g are increasing, then the optimal tests $\varphi_I(T), ..., \varphi_{IV}(T)$ can be found in the following way. Formulate the testing problems in terms of η; find constants so that the tests $\widetilde{\varphi}_I(S), ..., \widetilde{\varphi}_{IV}(S)$ are piecewise constant and satisfy the constraints in (8.4). Then $\widetilde{\varphi}_I(g(T)), ..., \widetilde{\varphi}_{IV}(g(T))$ are optimal level α tests for the corresponding problems. This simple fact shows that we don't have to go back to the original parameter θ and the generating statistic T. If we have increasing functions of these terms, then it is enough to mimic the construction of the optimal tests $\varphi_I(T), ..., \varphi_{IV}(T)$.

In the first of the following three examples symmetry is utilized to simplify two-sided tests.

Example 8.17. We consider the family $(\mathsf{N}^{\otimes n}(\mu, \sigma_0^2))_{\mu \in \mathbb{R}}$, where $\mu \in \mathbb{R}$ is unknown, but $\sigma_0^2 > 0$ is known. By (1.11) $\mathsf{N}^{\otimes n}(\mu, \sigma_0^2)$ is a one parameter exponential family with natural parameter $\theta = \mu/\sigma_0^2$ and generating statistic $T(x) = \sum_{i=1}^n x_i$, $x = (x_1, ..., x_n) \in \mathbb{R}^n$. As θ and μ differ only by a constant positive factor the null hypotheses that appear in (8.2) can be written as follows.

Testing Problem	H_0	H_A
(I)	$\mu \le \mu_0$	$\mu > \mu_0$
(II)	$\mu = \mu_0$	$\mu \ne \mu_0$
(III)	$\lvert\mu - \mu_0\rvert \ge d\sigma_0/\sqrt{n}$	$\lvert\mu - \mu_0\rvert < d\sigma_0/\sqrt{n}$
(IV)	$\lvert\mu - \mu_0\rvert \le d\sigma_0/\sqrt{n}$	$\lvert\mu - \mu_0\rvert > d\sigma_0/\sqrt{n}$

(8.17)

where $d > 0$ is a constant. As the family of normal distributions is generated by location-scale transformations we may obtain the constants that appear in the optimal tests from $\mathsf{N}(0,1)$. For $0 < \gamma < 1$ let u_γ be the γ-quantile of $\mathsf{N}(0,1)$, i.e., $\Phi(u_\gamma) = \gamma$. We note that for every $d > 0$ and $0 < \alpha < 1$ the equation $\Phi(z - d) - \Phi(-z - d) = \alpha$ has a unique positive solution $z_{d,\alpha}$, say. We set

Standard Gauss Tests

$$\begin{aligned} \psi_I(s) &= I_{(u_{1-\alpha},\infty)}(s), & \psi_{II}(s) &= 1 - I_{(-u_{1-\alpha/2},u_{1-\alpha/2})}(s), \\ \psi_{III}(s) &= I_{(-z_{d,\alpha},z_{d,\alpha})}(s), & \psi_{IV}(s) &= 1 - I_{(-z_{d,1-\alpha},z_{d,1-\alpha})}(s), \end{aligned} \tag{8.18}$$

and call the tests $\psi_I, ..., \psi_{IV}$ Gauss tests. We consider the testing problems (8.17) for the model $\mathcal{M}$ and the test statistic S given by

$$(\mathbb{R}^n, \mathfrak{B}_n, (\mathsf{N}^{\otimes n}(\mu, \sigma_0^2))_{\mu \in \mathbb{R}}) \quad \text{and} \quad S = \sqrt{n}(\overline{X}_n - \mu_0)/\sigma_0, \tag{8.19}$$

respectively, where $X_1, ..., X_n : \mathbb{R}^n \to \mathbb{R}$ are the i.i.d. projections and $\overline{X}_n = (1/n)\sum_{i=1}^n X_i$. Then

$$\begin{aligned} \mathsf{E}_{\mu_0}\psi_I(S) &= \mathsf{E}_{\mu_0}\psi_{II}(S) = \alpha, \\ \mathsf{E}_{\mu_0 \pm d\sigma_0/\sqrt{n}}\psi_{III}(S) &= \mathsf{E}_{\mu_0 \pm d\sigma_0/\sqrt{n}}\psi_{IV}(S) = \alpha. \end{aligned}$$

We already know from Example 2.52 that $\psi_I(S)$ is a uniformly best level α test for testing problem (I) in (8.17). By Remark 8.16, Proposition 8.14, Theorems 8.8 and 8.6, the tests $\psi_{II}(S)$, $\psi_{III}(S)$, and $\psi_{IV}(S)$, with S from (8.19) have the following optimality properties.

$$\begin{aligned} &\psi_I(S) \ \text{UMP for } (I), && \psi_{II}(S) \ \text{UMPU for } (II), \\ &\psi_{III}(S) \ \text{UMP for } (III), && \psi_{IV}(S) \ \text{UMPU for } (IV). \end{aligned} \tag{8.20}$$

Problem 8.18. Evaluate the power of the tests $\psi_I(S), ..., \psi_{IV}(S)$ in Example 8.17.

Example 8.19. We consider the family $(\mathsf{N}^{\otimes n}(\mu_0, \sigma^2))_{\sigma^2>0}$, where $\mu_0 \in \mathbb{R}$ is known, but $\sigma^2 > 0$ is unknown. By (1.11) we know that $\mathsf{N}^{\otimes n}(\mu_0, \sigma^2)$ is a one-parameter exponential family with natural parameter $\theta = \kappa(\sigma^2) = -(2\sigma^2)^{-1}$ and generating statistic $T(x) = \sum_{i=1}^n (x_i - \mu_0)^2$, $x = (x_1, ..., x_n) \in \mathbb{R}^n$. Let $\sigma_0^2 > 0$ and $\alpha \in (0,1)$ be fixed. We follow again the advice in Remark 8.16 and use the test statistic $S = \sigma_0^{-2}T$, which under $\mathsf{N}^{\otimes n}(\mu_0, \sigma_0^2)$ has the χ^2-distribution with n degrees of freedom $\mathsf{H}(n)$. Symmetry cannot be utilized to simplify the construction of the optimal test for the testing problem $\mathsf{H}_0 : \sigma^2 = \sigma_0^2$ versus $\mathsf{H}_A : \sigma^2 \neq \sigma_0^2$, which is problem (II) in (8.2). Denote by $\chi^2_{1-\alpha,n}$ the $1-\alpha$ quantile of $\mathsf{H}(n)$. It holds $\mathsf{E}_{\mu_0,\sigma_0^2} T = n\sigma_0^2$. According to the constraints in (8.12) and (8.13) we have to find $\alpha_1, \alpha_2 > 0$ such that for $c_1 = \chi^2_{\alpha_1,n}$ and $c_2 = \chi^2_{1-\alpha_2,n}$,

$$\varphi_{II}(T) = I_{(0,c_1]}(T/\sigma_0^2) + I_{[c_2,\infty)}(T/\sigma_0^2)$$

satisfies $\mathsf{E}_{\mu_0,\sigma_0^2}\varphi_{II}(T) = \alpha_1 + \alpha_2 = \alpha$ and $\mathsf{E}_{\mu_0}\varphi_{II}(T)T = \alpha n \sigma_0^2$. The latter can be written as

$$\int_{\chi^2_{\alpha_1,n}}^{\chi^2_{1-\alpha_2,n}} s\mathsf{h}_n(s)ds = (1-\alpha)n,$$

where h_n is the Lebesgue density of $\mathsf{H}(n)$, given by

$$\mathsf{h}_n(s) = \frac{1}{2^{n/2}\Gamma(\frac{n}{2})} s^{(n/2)-1} \exp\{-\frac{s}{2}\} I_{(0,\infty)}(s), \quad s \in \mathbb{R}.$$

Hence $s\mathsf{h}_n(s) = n\mathsf{h}_{n+2}(s)$, and with $\mathsf{H}_n(t) = \int_0^t \mathsf{h}_n(s)ds$ the side conditions now read

$$\mathsf{H}_{n+2}(\chi^2_{1-\alpha_2,n}) - \mathsf{H}_{n+2}(\chi^2_{\alpha_1,n}) = 1-\alpha,$$
$$\alpha_1 + \alpha_2 = \alpha.$$

To find α_1 and α_2 one has to solve this system of nonlinear equations numerically. The quite often-made choice of $\alpha_1 = \alpha_2 = \alpha/2$ is an approximation for large n as then $\mathsf{H}_{n+2} \approx \mathsf{H}_n$. Finally we remark that the other testing problems studied in Theorems 8.8 and 8.6 can be treated similarly.

The next example studies tests for the scale parameter in a family of gamma distributions. It is a generalization of the result in the previous example as every χ^2-distribution is a special gamma distribution.

Example 8.20. Let $X_1, ..., X_n$ be an i.i.d. sample from a gamma distribution $\mathsf{Ga}(\lambda, \beta)$ with Lebesgue density

$$\mathsf{ga}_{\lambda,\beta}(x) = \frac{\beta^\lambda}{\Gamma(\lambda)} x^{\lambda-1} \exp\{-\beta x\} I_{(0,\infty)}(x), \ x \in \mathbb{R}, \ \beta > 0,$$

where $\lambda > 0$ is known. $\mathsf{Ga}^{\otimes n}(\lambda, \beta)$ is an exponential family with natural parameter $\theta = \kappa(\beta) = -\beta$ and generating statistic $T(x_1, ..., x_n) = \sum_{i=1}^n x_i$. Testing $\mathsf{H}_0 : \beta = \beta_0$ versus $\mathsf{H}_A : \beta \neq \beta_0$ for a given $\beta_0 \in (0, \infty)$ is covered by Theorem 8.11. A uniformly best unbiased level α test is given by

$$\varphi_{2,T} = I_{(0,b_1]}(T) + I_{[b_2,\infty)}(T),$$

where b_1 and b_2 are determined according to (8.12) and (8.13). To calculate the constants b_1, b_2 we remark that T/β_0 has under H_0 the density $\mathsf{ga}_{\lambda,1}^{\otimes n}$, so that

$$\int_{\beta_0 b_1}^{\beta_0 b_2} \mathsf{ga}_{n\lambda,1}(t)\, dt = 1 - \alpha \quad \text{and} \quad \int_{\beta_0 b_1}^{\beta_0 b_2} t\mathsf{ga}_{n\lambda,1}(t)\, dt = (1-\alpha)n\lambda.$$

Also here, in practice quite often the $\alpha/2$-approach is chosen; see Example 8.19.

Example 8.21. Let X follow a binomial distribution $\mathsf{B}(n,p)$, where $p \in (0,1)$ is unknown. We know from Example 1.7 that $\mathsf{B}(n,p)$ is an exponential family with natural parameter $\theta = \kappa(p) = \ln(p/(1-p))$ and generating statistic $T(x) = x$. As κ is increasing we may use the advice in Remark 8.16. By Theorem 8.6 a uniformly best unbiased level α test for $\mathsf{H}_0 : p \in [p_1, p_2]$ versus $\mathsf{H}_A : p \in (0, p_1) \cup (p_2, 1)$ is

$$\varphi_{IV}(T(x)) = I_{\{0,\ldots,b_1-1\}}(x) + \varsigma_1 I_{\{b_1\}}(x) + \varsigma_2 I_{\{b_2\}}(x) + I_{\{b_2+1,\ldots,n\}}(x),$$

$x \in \{0, 1, ..., n\}$, where b_1, b_2 and ς_1, ς_2 are determined by

$$\sum\nolimits_{k=0}^{b_1-1} \mathsf{b}_{n,p_j}(k) + \varsigma_1 \mathsf{b}_{n,p_j}(b_1) + \varsigma_2 \mathsf{b}_{n,p_j}(b_2) + \sum\nolimits_{x=b_2+1}^{n} \mathsf{b}_{n,p_j}(k) = \alpha, \ j = 1, 2.$$

Calculating the values of b_1, b_2 and ς_1, ς_2 in concrete situations usually requires the use of a computer program. When testing $\mathsf{H}_0 : p = p_0$ versus $\mathsf{H}_A : p \neq p_0$ for a given $p_0 \in (0,1)$, in practice quite often the $\alpha/2$-approach is chosen and that, conservatively, without randomization; that is, H_0 is rejected at x whenever $\mathsf{B}(n,p_0)(\{0,1,...,x\}) \leq \alpha/2$ or $\mathsf{B}(n,p_0)(\{x, x+1, ..., n\}) \leq \alpha/2$.

8.1.2 Tests in Multivariate Normal Distributions

We consider the Gaussian model

$$\mathcal{G}_0 = (\mathbb{R}^d, \mathfrak{B}_d, (\mathsf{N}(\theta, \Sigma_0))_{\theta \in \mathbb{R}^d}),$$

where Σ_0 is known. Here we want to test linear hypotheses in terms of $c^T\theta$, where $c \in \mathbb{R}^d$ is a fixed given vector. The four testing problems considered are those given in (8.17), where in the present setting $\mu = c^T\theta$. If X has the distribution $\mathsf{N}(\theta, \Sigma_0)$, then it seems intuitively reasonable to use $c^T X$ as the test statistic and switch to the reduced model $(\mathbb{R}, \mathfrak{B}, (\mathsf{N}(c^T\theta, c^T\Sigma_0 c))_{\theta \in \mathbb{R}^d})$ for which the optimal tests are readily available from Example 8.17. It may be somewhat surprising that this simple idea in fact provides the optimal tests. However, closer inspection reveals that this is the right approach. Indeed, $\tau = c^T\theta$ is the parameter of interest, and the projection of θ on the linear subspace, say $\mathbb{L}_c^\perp$, that is orthogonal to c is a nuisance parameter that may vary freely in the $(d-1)$-dimensional subspace $\mathbb{L}_c^\perp$. Intuitively it is clear that a level α test should not depend on that projection $\Pi_{\mathbb{L}_c^\perp} X$ but only on $c^T X$. First, it should be pointed out that by a linear transformation the testing problem can be reduced to $c = (1, 0, ..., 0)^T$, i.e., where we test the first component of θ. Second, when testing $\tau \leq a_0$, $\xi \in \mathbb{R}^{d-1}$, versus $\tau > a_0$, $\xi \in \mathbb{R}^{d-1}$, it is enough to consider the case $a_0 = 0$.

Denote by U the projection on the first k, and by V the projection on the remaining $d-k$ components in the model $\mathcal{G}_0$. Let $\Sigma_{i,j}$, $1 \le i,j \le 2$, be the block matrices of Σ_0, i.e., $\Sigma_{1,1} = \mathsf{C}_\theta(U,U)$, $\Sigma_{1,2} = \mathsf{C}_\theta(U,V)$, $\Sigma_{2,1} = \mathsf{C}_\theta(V,U)$, and $\Sigma_{2,2} = \mathsf{C}_\theta(V,V)$. Suppose that $\Sigma_{1,1}$ is nonsingular, introduce the linear mapping and its inverse by

$$B(u,v) = \begin{pmatrix} u \\ v - \Sigma_{2,1}\Sigma_{1,1}^{-1}u \end{pmatrix} \quad \text{and} \quad \widetilde{B}(y,z) = \begin{pmatrix} y \\ z + \Sigma_{2,1}\Sigma_{1,1}^{-1}y \end{pmatrix}, \tag{8.21}$$

respectively, and set

$$\begin{pmatrix} Y \\ Z \end{pmatrix} = B(U,V) = \begin{pmatrix} U \\ V - \Sigma_{2,1}\Sigma_{1,1}^{-1}U \end{pmatrix}.$$

Then it holds

$$\mathsf{C}_\theta(Y,Z) = \mathsf{C}_\theta(V - \Sigma_{2,1}\Sigma_{1,1}^{-1}U, U) = \mathsf{C}_\theta(V,U) - \mathsf{C}_\theta(\Sigma_{2,1}U,U)\Sigma_{1,1}^{-1} = 0.$$

As B is a linear mapping and $(Y^T, Z^T)^T$ has a normal distribution we get that Y and Z are independent. The definition of $(Y^T, Z^T)^T$ implies

$$\begin{aligned} \mathsf{N}(\theta, \Sigma_0) \circ B^{-1} &= \mathcal{L}((Y^T, Z^T)^T | \mathsf{N}(\theta, \Sigma_0)) \\ &= \mathsf{N}(\tau, \Sigma_{1,1}) \otimes \mathsf{N}(\xi - \Sigma_{2,1}\Sigma_{1,1}^{-1}\tau, \Gamma) \end{aligned} \tag{8.22}$$

with $\Gamma = \Sigma_{2,2} - \Sigma_{2,1}\Sigma_{1,1}^{-1}\Sigma_{1,2}$. This relation between the family $(\mathsf{N}(\theta, \Sigma_0))_{\theta \in \mathbb{R}^d}$ and the family $(\mathsf{N}(\theta, \Sigma_0) \circ B^{-1})_{\theta \in \mathbb{R}^d}$ is the crucial point in the proof of the next theorem. Parts of this theorem, proved with different arguments, can also be found in van der Vaart (1998) and Lehmann and Romano (2005). We consider the following testing problems.

Testing Problem	H_0	H_A
(I)	$c^T\theta \le 0$	$c^T\theta > 0$
(II)	$c^T\theta = 0$	$c^T\theta \neq 0$
(III)	$\lvert c^T\theta\rvert \ge d\sigma_0$	$\lvert c^T\theta\rvert < d\sigma_0$
(IV)	$\lvert c^T\theta\rvert \le d\sigma_0$	$\lvert c^T\theta\rvert > d\sigma_0$

(8.23)

Theorem 8.22. *For the model $\mathcal{G}_0 = (\mathbb{R}^d, \mathfrak{B}_d, (\mathsf{N}(\theta, \Sigma_0))_{\theta \in \mathbb{R}^d})$ with known covariance matrix Σ_0 and $\sigma_0^2 = c^T\Sigma_0 c > 0$, the hypotheses in (8.23), the Gauss tests $\psi_I, ..., \psi_{IV}$ in (8.18), and $T(x) = c^T x/\sigma_0$, $x \in \mathbb{R}^d$, the subsequent tests are level α tests and optimal as stated below.*

$$\psi_I(T) \quad \textit{UMP for (I)}, \qquad \psi_{II}(T) \quad \textit{UMPU for (II)},$$

$$\psi_{III}(T) \quad \textit{UMP for (III)}, \qquad \psi_{IV}(T) \quad \textit{UMPU for (IV)}.$$

Proof. We have seen already that it is enough to consider the special case of $c = (1, 0, ..., 0)^T$. The assumption $c^T \Sigma_0 c > 0$ gives $\sigma_0^2 = \Sigma_{1,1} > 0$. We set $k = 1$ and note that the mapping B in (8.21) is one-to-one. Therefore we have only to consider tests that depend on (Y, Z), i.e., we consider the model (8.22). In the first part of the proof we consider the testing problem (I) in (8.23). Let φ be any level α test; that is,

$$\int [\int \varphi(y,z) \mathsf{N}(\tau, \Sigma_{1,1})(dy)] \mathsf{N}(\xi - \Sigma_{2,1}\Sigma_{1,1}^{-1}\tau, \Gamma)(dz) \le \alpha, \quad \tau \le 0, \ \xi \in \mathbb{R}^{d-1}.$$

As $\xi \in \mathbb{R}^{d-1}$ is arbitrary we get with $\sigma_0^2 = \Sigma_{1,1}$

$$\int [\int \varphi(y,z) \mathsf{N}(\tau, \sigma_0^2)(dy)] \mathsf{N}(\xi, \Gamma)(dz) \le \alpha, \quad \tau \le 0, \ \xi \in \mathbb{R}^{d-1}.$$

For fixed $\xi \in \mathbb{R}^{d-1}$ the test $\psi_\xi(y) := \int \varphi(y,z) \mathsf{N}(\xi, \Gamma)(dz)$ is a level α test for $\mathsf{H}_0 : \mathsf{N}(\tau, \sigma_0^2)$, $\tau \le 0$, versus $\mathsf{H}_A : \mathsf{N}(\tau, \sigma_0^2)$, $\tau > 0$. As $\psi_I(t/\sigma_0) = I_{(u_{1-\alpha},\infty)}(t/\sigma_0)$ is a uniformly best level α test (see Example 2.52) we get

$$\begin{aligned} &\int [\int \varphi(y,z) \mathsf{N}(\tau, \sigma_0^2)(dy)] \mathsf{N}(\xi, \Gamma)(dz) \\ &\le \int [\int \psi_I(y/\sigma_0) \mathsf{N}(\tau, \sigma_0^2)(dy)] \mathsf{N}(\xi, \Gamma)(dz), \end{aligned}$$

so that $\psi_I(y/\sigma_0)$ has a power not smaller than that of $\varphi(y,z)$ at every $\tau > 0$ and $\xi \in \mathbb{R}^{d-1}$, and therefore at every $\tau > 0$ and $\xi - \Sigma_{2,1}\Sigma_{1,1}^{-1}\tau$. The proofs for the cases (II), (III), and (IV) in (8.17) with $\mu - \mu_0 = c^T\theta$ are similar if we use Example 8.17, where in the cases (II) and (IV) we have to take into account that $\psi_\xi(y)$ is unbiased if φ is. ■

Problem 8.23. The power functions of the tests $\psi_I(T), ..., \psi_{IV}(T)$ are given by

$$\begin{aligned} \mathsf{p}_I(\theta) &= 1 - \Phi(u_{1-\alpha} - (c^T\theta)/\sigma_0), \\ \mathsf{p}_{II}(\theta) &= 1 - \Phi(u_{1-\alpha/2} - (c^T\theta)/\sigma_0) + \Phi(-u_{1-\alpha/2} - (c^T\theta)/\sigma_0), \\ \mathsf{p}_{III}(\theta) &= \Phi(z_{d,\alpha} - (c^T\theta)/\sigma_0) - \Phi(-z_{d,\alpha} - (c^T\theta)/\sigma_0), \\ \mathsf{p}_{IV}(\theta) &= 1 - \Phi(z_{d,1-\alpha} - (c^T\theta)/\sigma_0) + \Phi(-z_{d,1-\alpha} - (c^T\theta)/\sigma_0). \end{aligned}$$

Now we allow the parameter of interest to be k-dimensional and consider maximin tests. We fix $\delta > 0$, assume that the $k \times k$ matrix $\Sigma_{1,1}$ is nonsingular, and consider the testing problem

$$\mathsf{H}_0 : \tau = 0, \ \xi \in \mathbb{R}^{d-k} \quad \text{versus} \quad \mathsf{H}_A : \tau^T \Sigma_{1,1}^{-1} \tau \ge \delta^2, \ \xi \in \mathbb{R}^{d-k}. \tag{8.24}$$

The next statement is a well-known property of the χ^2-test. See, for example, Strasser (1985), Chapter 6. Denote by $\chi^2_{1-\alpha,k}$ the $1 - \alpha$ quantile of the χ^2-distribution with k degrees of freedom, and by H_{k,δ^2} the c.d.f. of the χ^2-distribution with k degrees of freedom and parameter of noncentrality δ^2. As before, let U be the projection on the first k coordinates.

Theorem 8.24. *For the model* $\mathcal{G}_0 = (\mathbb{R}^d, \mathfrak{B}_d, (\mathsf{N}(\theta, \Sigma_0))_{\theta \in \mathbb{R}^d})$, *with known covariance matrix* Σ_0 *and nonsingular* $\Sigma_{1,1}$, *the test*

$$\varphi_{\chi^2, \mathcal{G}_0} = I_{(\chi^2_{1-\alpha,k}, \infty)}(U^T \Sigma_{1,1}^{-1} U) \tag{8.25}$$

is a maximin level α *test for the testing problem (8.24), where the maximin value of the power is given by* $\inf_{\tau^T \Sigma_{1,1}^{-1} \tau \geq \delta^2} \mathsf{E}_\theta \varphi_{\chi^2, \mathcal{G}_0} = 1 - \mathsf{H}_{k,\delta^2}(\chi^2_{1-\alpha,k})$.

Proof. We use the one-to-one mapping B in (8.21) to turn to the family $\mathsf{N}(\tau, \Sigma_{1,1}) \otimes \mathsf{N}(\xi - \Sigma_{2,1}\Sigma_{1,1}^{-1}\tau, \Gamma)$. The minimum power of any test φ satisfies

$$\begin{aligned} &\inf_{\tau^T \Sigma_{1,1} \tau \geq \delta^2, \xi \in \mathbb{R}^{d-k}} \int [\int \varphi(u,v) \mathsf{N}(\tau, \Sigma_{1,1})(du)] \mathsf{N}(\xi - \Sigma_{2,1}\Sigma_{1,1}^{-1}\tau, \Gamma)(dv) \\ \leq &\inf_{\tau^T \Sigma_{1,1} \tau \geq \delta^2} \int [\int \varphi(u,v) \mathsf{N}(0, \Gamma)(dv)] \mathsf{N}(\tau, \Sigma_{1,1})(du). \end{aligned}$$

Hence it remains to search for maximin level α tests that depend only on u. Denote by $\Sigma_{1,1}^{-1/2}$ a positive definite symmetric matrix with $\Sigma_{1,1}^{-1/2}\Sigma_{1,1}^{-1/2} = \Sigma_{1,1}^{-1}$ and set $T = \Sigma_{1,1}^{-1/2} U$. Then $\mathcal{L}(T|\mathsf{N}(\tau, \Sigma_{1,1})) = \mathsf{N}(\Sigma_{1,1}^{-1/2}\tau, \mathbf{I})$. Theorem 5.43 yields that $\varphi_{\chi^2,\alpha}$ is a maximin level α test for the family $\mathsf{N}(\mu, \mathbf{I})$ for testing $\mu = 0$ versus $\|\mu\|^2 \geq \delta^2$. Hence $\varphi_{\chi^2,\alpha}(\|T\|^2)$ is a maximin level α test for the family $\mathsf{N}(\tau, \Sigma_{1,1})$ for testing $\tau = 0$ versus $\| \Sigma_{1,1}^{-1/2}\tau \|^2 = \tau^T \Sigma_{1,1}^{-1} \tau \geq \delta^2$. ■

Example 8.25. Suppose we want to compare the means of two normal distributions with known variances. By a reduction by sufficiency we may assume that the sample size is one for each population, so that we have the model

$$(\mathbb{R}^2, \mathfrak{B}_2, (\mathsf{N}(\mu_1, \sigma_1^2) \otimes \mathsf{N}(\mu_2, \sigma_2^2))_{(\mu_1,\mu_2) \in \mathbb{R}^2}),$$

where $\sigma_1^2, \sigma_2^2 > 0$ are known. Set $\sigma_0^2 = \sigma_1^2 + \sigma_2^2$, $\theta = (\mu_1, \mu_2)^T$, $c = (1, -1)^T$ and $T(x,y) = x - y$. Then the following tests are level α tests, and they are optimal level α test as stated below.

$$\begin{aligned} &\psi_I(T/\sigma_0) && \text{UMP for} && \mu_1 \leq \mu_2 && \text{versus} && \mu_1 > \mu_2, \\ &\psi_{II}(T/\sigma_0) && \text{UMPU for} && \mu_1 = \mu_2 && \text{versus} && \mu_1 \neq \mu_2, \\ &\psi_{III}(T/\sigma_0) && \text{UMP for} && |\mu_1 - \mu_2| \geq d\sigma_0 && \text{versus} && |\mu_1 - \mu_2| < d\sigma_0, \\ &\psi_{IV}(T/\sigma_0) && \text{UMPU for} && |\mu_1 - \mu_2| \leq d\sigma_0 && \text{versus} && |\mu_1 - \mu_2| > d\sigma_0. \end{aligned}$$

8.1.3 Tests for d-Parameter Exponential Families

Now we deal with d-parameter exponential families where the parameter θ is split again into a parameter of interest τ and a nuisance parameter ξ. To construct optimal tests for hypotheses in terms of τ in the model $\mathcal{M} = (\mathcal{X}, \mathfrak{A}, (P_\theta)_{\theta \in \Delta})$ we reduce it with a statistic $T = (U, V)$, where U and

V are measurable mappings from $(\mathcal{X}, \mathfrak{A})$ into the spaces $(\mathcal{U}, \mathfrak{U})$ and $(\mathcal{V}, \mathfrak{V})$, respectively. Then with $Q_\theta = P_\theta \circ T^{-1}$ the reduced model can be written as $\mathcal{N} = (\mathcal{U} \times \mathcal{V}, \mathfrak{U} \otimes \mathfrak{V}, (Q_\theta)_{\theta \in \Delta})$. Occasionally, in favorable situations, the conditional distribution of U, given V, is independent of the nuisance parameter. Then, under the assumption that $(\mathcal{U}, \mathfrak{U})$ is a Borel space, we can find a family of stochastic kernels $\mathsf{K}_\tau : \mathfrak{U} \otimes \mathcal{V} \to_k [0,1]$ such that for $\theta = (\tau, \xi)$ the distribution Q_θ can be written as $Q_\theta = \mathsf{K}_\tau \otimes P_{\theta,V}$, where $P_{\theta,V} = P_\theta \circ V^{-1}$ is the marginal distribution of Q_θ induced by V; see Lemma A.41. This representation of Q_θ is equivalent to the requirement that for every $h : \mathcal{U} \times \mathcal{V} \to_m [0, \infty)$,

$$\mathsf{E}_\theta h(U,V) = \int [\int h(u,v) \mathsf{K}_\tau(du|v)] P_{\theta,V}(dv). \tag{8.26}$$

To construct a level α test for $\mathsf{H}_0 : (\tau, \xi) \in \Delta_0$ for some $\Delta_0 \subseteq \Delta$ we could, in a first step, choose any test $\varphi : \mathcal{U} \times \mathcal{V} \to_m [0,1]$ that satisfies

$$\int \varphi(u,v) \mathsf{K}_\tau(du|v) \le \alpha, \quad P_{(\tau,\xi),V}\text{-a.s.}, \ (\tau,\xi) \in \Delta_0.$$

Such a test $\varphi(\cdot, v)$ is called a *conditional level α test*, given $V = v$. Then (8.26) yields

$$\mathsf{E}_\theta \varphi(U,V) = \int [\int \varphi(u,v) \mathsf{K}_\tau(du|v)] P_{\theta,V}(dv) \le \alpha, \quad \theta = (\tau, \xi) \in \Delta_0,$$

so that $\varphi(U,V)$ is a level α test for H_0. If we turn to a reduced model we may have a loss of information and a best test within a specific class of tests may not be a function of T. This issue disappears if T is sufficient, as it is the case when $T = (U,V)$ is the generating statistic of an exponential family $(P_\theta)_{\theta \in \Delta}$, $\Delta \subseteq \mathbb{R}^d$, in natural form where the parameter vector is split into $\theta = (\tau, \xi)$. A first essential step is to investigate whether the nuisance parameter can be eliminated under the null hypothesis by conditioning on V. Let $(P_\theta)_{\theta \in \Delta}$, $\Delta \subseteq \mathbb{R}^d$, be an exponential family in natural form, with natural parameter θ and generating statistic T, as given by (1.5). Denote by $Q_\theta = P_\theta \circ T^{-1}$ the distribution of T under P_θ. Starting with $\theta = (\theta_1, ..., \theta_d)$ and $T = (T_1, ..., T_d)$, we set

$$\begin{aligned} U &= T_1 \quad \text{and} \quad V = (T_2, ..., T_d), \\ \tau &= \theta_1 \quad \text{and} \quad \xi = (\theta_2, ..., \theta_d). \end{aligned} \tag{8.27}$$

To avoid overly convoluted formulations we impose a weak condition on the parameter set that is satisfied in all subsequent special cases. We assume that $\Delta = \Upsilon \times \Xi$, where Υ is an open interval of $\mathbb{R}$ and Ξ is an open subset of $\mathbb{R}^{d-1}$. We consider the following four testing problems, where $\tau_i \in \Upsilon$, $i = 0, 1, 2$, are fixed.

Testing Problem	$\mathsf{H}_0 : \theta = (\tau, \xi) \in \Delta_0$	$\mathsf{H}_A : \theta = (\tau, \xi) \in \Delta_A$
(I)	$\tau \leq \tau_0,\ \xi \in \Xi,$	$\tau > \tau_0,\ \xi \in \Xi,$
(II)	$\tau = \tau_0,\ \xi \in \Xi,$	$\tau \neq \tau_0,\ \xi \in \Xi,$
(III)	$\tau \notin (\tau_1, \tau_2),\ \xi \in \Xi,$	$\tau \in (\tau_1, \tau_2),\ \xi \in \Xi,$
(IV)	$\tau \in [\tau_1, \tau_2],\ \xi \in \Xi,$	$\tau \notin [\tau_1, \tau_2],\ \xi \in \Xi.$

(8.28)

The boundary J of Δ_0 and Δ_A depends on the testing problem and is

$$\begin{aligned} J &= \{\tau_0\} \times \Xi && \text{for } (I) \text{ and } (II), \\ J &= \{\tau_1\} \times \Xi \cup \{\tau_2\} \times \Xi && \text{for } (III) \text{ and } (IV). \end{aligned} \tag{8.29}$$

Let now $\theta_* = (\tau_*, \xi_*) \in \Delta$ be fixed. Since U takes on values in $\mathbb{R}$ there exists a regular conditional distribution of U, given $V = v$, i.e., a stochastic kernel $\mathsf{K}_{\theta_\star} : \mathfrak{B}_1 \times \mathbb{R}^{d-1} \to_k [0, 1]$ with $Q_{\theta_*} = \mathsf{K}_{\theta_*} \otimes P_{\theta_*, V}$. At any $\theta = (\tau, \xi) \in \Delta$ it holds

$$\begin{aligned} \mathsf{E}_\theta h(U, V) &= \int h(u, v) Q_\theta(du, dv) \\ &= \int h(u, v) \exp\{(\tau - \tau_\star)u + \langle \xi - \xi_\star, v\rangle - K(\theta) + K(\theta_\star)\} Q_{\theta_\star}(du, dv) \\ &= \int [\int h(u, v) \exp\{(\tau - \tau_\star)u\} \mathsf{K}_{\theta_\star}(du|v)] \\ &\qquad \times \exp\{\langle \xi - \xi_\star, v\rangle - K(\theta) + K(\theta_\star)\} P_{\theta_\star, V}(dv). \end{aligned}$$

Hence,

$$\begin{aligned} \frac{dP_{\theta,V}}{dP_{\theta_\star,V}}(v) &= \exp\{\langle \xi - \xi_\star, v\rangle + K_v(\tau) - K(\theta) + K(\theta_\star)\}, \\ K_v(\tau) &= \ln\{\int \exp\{\langle \tau - \tau_\star, u\rangle\} \mathsf{K}_{\theta_\star}(du|v)\}. \end{aligned} \tag{8.30}$$

We note that $K_v(\tau) < \infty$, $P_{\theta_\star,V}$-a.s., take some generic distribution Q on $(\mathbb{R}, \mathfrak{B})$, and introduce a stochastic kernel K_τ by

$$\mathsf{K}_\tau(B|v) = \begin{cases} \int_B \exp\{\langle \tau - \tau_\star, u\rangle - K_v(\tau)\} \mathsf{K}_{\theta_\star}(du|v) & \text{if } K_v(\tau) < \infty \\ Q(B) & \text{otherwise.} \end{cases} \tag{8.31}$$

Then the above representation of $\mathsf{E}_\theta h(U, V)$ and (8.30) yield $Q_\theta = \mathsf{K}_\tau \otimes P_{\theta,V}$ and we see that there exists a regular conditional distribution K_τ of U, given $V = v$, that is independent of the nuisance parameter ξ.

Proposition 8.26. *Let* $(P_\theta)_{\theta\in\Delta}$ *be an exponential family in natural form, with natural parameter* $\theta = (\tau, \xi)$ *and generating statistic* $T = (U, V)$*, where* $\tau = \theta_1$ *and* $U = T_1$*. Then the conditional distribution of* U*, given* $V = v$*, as specified in (8.31) is independent of the nuisance parameter* ξ*.*

For dimension one the constants that appear in the tests (8.4) have been chosen in Theorem 8.6 in such a way that the test attains the level α on the boundary. Thus for constructing conditional tests one could try to construct tests that satisfy

$$\int \varphi(u, v)\mathsf{K}_\tau(du|v) = \alpha, \quad P_{\theta,V}\text{-a.s.}, \ \theta \in J, \tag{8.32}$$

where in J from (8.29) $\tau = \tau_0$, or $\tau = \tau_1$ and $\tau = \tau_2$, depending on which of the four problems of (8.28) is being considered. Tests that satisfy the condition (8.32) are said to have *Neyman structure*. The lemma below shows that this condition is not very restrictive.

Lemma 8.27. *Let* $(P_\theta)_{\theta\in\Delta}$ *be an exponential family that satisfies (A1) and (A2), where* $\Delta = \Upsilon \times \Xi$ *is open, along with the splitting from (8.27), and* $\alpha \in (0, 1)$ *be fixed. Then every unbiased level* α *test* φ *for each of the four testing problems in (8.28) has Neyman structure (8.32) at* τ_0*,* τ_0*,* τ_1 *and* τ_2*, and* τ_1 *and* τ_2*, respectively.*

Proof. For an exponential family the power function $\mathsf{E}_\theta\psi$ of every test ψ is continuous on Δ; see Lemma 1.16. Hence by Problem 8.3 every unbiased level α test is α-similar on J. Thus

$$\int [\int \varphi(u, v)\mathsf{K}_\tau(du|v) - \alpha] P_{(\tau,\xi),V}(dv) = 0, \quad (\tau, \xi) \in J,$$

where in J we have $\tau = \tau_i$, $i \in \{0, 1, 2\}$, depending on the testing problem. As Ξ is open we get from Theorem 4.73 that the exponential family $P_{(\tau_i,\xi),V}$, $\xi \in \Xi$, is complete. Hence $\int \varphi(u, v)\mathsf{K}_\tau(du|v) - \alpha = 0$, $P_{(\tau,\xi),V}$-a.s., for $\tau = \tau_i$. As all distributions $P_{\theta,V}$, $\theta \in \Delta$, are equivalent the statement follows. ∎

To construct, for a fixed given $\alpha \in (0, 1)$, an optimal level α test in the class of all unbiased tests we search for an optimal test in the class of all (all unbiased) level α tests $\varphi(\cdot, v)$ that satisfy (8.32). As $\mathsf{K}_\tau(\cdot|v)$ is an exponential family we employ the results on optimal tests for one-parametric exponential families. The only difference is that the constants that appear in (8.4) now depend on the condition v. The technical arguments are similar. For the given α and any fixed $v \in \mathbb{R}^{d-1}$ we now construct a test $\varphi_{I,U} : \mathbb{R}^d \to_m [0, 1]$ by setting

$$\varphi_{I,U}(u, v) = I_{(c_{1-\alpha}(v),\infty)}(u) + \gamma(v) I_{\{c_{1-\alpha}(v)\}}(u), \quad u \in \mathbb{R}, \tag{8.33}$$

where $c_{1-\alpha}(v) \in \mathbb{R}$ and $0 \le \gamma(v) \le 1$ are chosen in such a way that

$$\int \varphi_{I,U}(u,v)\mathsf{K}_{\tau_0}(du|v) = \alpha.$$

Suitable constants $c_{1-\alpha}(v)$ and $\gamma(v)$ can be found by using

$$F_{\tau_0}(t|v) := \mathsf{K}_{\tau_0}((-\infty, t]|v), \quad t \in \mathbb{R},$$

the c.d.f. of the conditional distribution of U, given $V = v$, taking

$$\begin{aligned} c_{1-\alpha}(v) &= F_{\tau_0}^{-1}((1-\alpha)|v) \quad \text{and} \\ \gamma(v) &= [F_{\tau_0}(c_{1-\alpha}(v)) - (1-\alpha)] \oslash (\mathsf{K}_{\tau_0}(\{c_{1-\alpha}(v)\})|v), \quad v \in \mathbb{R}^{d-1}. \end{aligned} \tag{8.34}$$

Obviously, the mapping $v \to (c_{1-\alpha}(v), \gamma(v))$ is measurable, which implies that $\varphi_{I,U} : \mathbb{R}^d \to_m [0,1]$.

Theorem 8.28. *Let $(P_\theta)_{\theta\in\Delta}$ be an exponential family in natural form that satisfies (A1) and (A2), $\Delta = \Upsilon \times \Xi$ be open, along with the splitting from (8.27), and $\alpha \in (0,1)$ be fixed. For $c_{1-\alpha}(v)$ and $\gamma(v)$ given by (8.34), the test $\varphi_{I,U}(u,v)$ is a uniformly best unbiased level α test for testing problem (I) in (8.28).*

Proof. First we note that by (8.33) and (8.34) the test $\varphi_{I,U}$ is an unbiased level α test for H_0 versus H_A. Let now ψ be any other unbiased level α test for H_0 versus H_A. By Lemma 8.27,

$$\int \psi(u,v)\mathsf{K}_{\tau_0}(du|v) = \alpha, \quad P_{\theta,V}\text{-a.s.}, \theta \in \Delta.$$

In view of Theorem 2.49 we have for every $\tau > \tau_0$,

$$\int \psi(u,v)\mathsf{K}_\tau(du|v) \le \int \varphi_{I,U}(u,v)\mathsf{K}_\tau(du|v), \quad P_{\theta,V}\text{-a.s.}, \theta \in \Delta.$$

Integration of both sides with respect to $P_{\theta,V}$, $\theta = (\tau,\xi) \in \Delta$ with $\tau > \tau_0$, gives by (8.26) the inequality $\mathsf{E}_\theta\psi(U,V) \le \mathsf{E}_\theta\varphi_{I,U}(U,V)$, and the proof is completed. ■

The previous theorem shows the line of how to deal with the other testing problems in (8.28). Subsequently we focus only on the testing problem (II) which is of special importance. Let us discuss the constraints that follow for tests that are supposed to be unbiased. If the test ψ is unbiased, then its power function $\tau \mapsto \mathsf{E}_{(\tau,\xi)}\psi$, $(\tau,\xi) \in \Delta$, has a local minimum at τ_0 for every fixed $\xi \in \Xi$. Both K_τ and $P_{(\tau,\xi),V}$ are exponential families, and thus the power function of ψ has derivatives of all orders in the open set Δ, where we may exchange differentiation with respect to τ with the integrations; see Lemma 1.16. By utilizing this and (8.31) we arrive at

$$\int [\int (u\psi(u,v) - K'_v(\tau_0)\psi(u,v))\mathsf{K}_{\tau_0}(du|v)]P_{(\tau_0,\xi),V}(dv) = 0.$$

The completeness of the family $P_{(\tau_0,\xi),V}$ for $\xi \in \Xi$ implies that $P_{(\tau_0,\xi),V}$-a.s., and consequently $P_{\theta,V}$-a.s., for every $\theta \in \Delta$,

$$\int u\psi(u,v)\mathsf{K}_{\tau_0}(du|v) = K_v'(\tau_0)\int \psi(u,v)\mathsf{K}_{\tau_0}(du|v) = \int u\mathsf{K}_{\tau_0}(du|v)\alpha,$$

where the second equation follows from Corollary 1.19 and (8.32). Thus every unbiased level α test φ satisfies

$$\int \varphi(u,v)\mathsf{K}_{\tau_0}(du|v) = \alpha, \quad P_{\theta,V}\text{-a.s., } \theta \in \Delta, \tag{8.35}$$
$$\int u\varphi(u,v)\mathsf{K}_{\tau_0}(du|v) = \alpha\int u\mathsf{K}_{\tau_0}(du|v), \quad P_{\theta,V}\text{-a.s., } \theta \in \Delta.$$

Analogously to (8.4) we set

$$\varphi_{II,U}(u,v) = I_{(-\infty,b_1(v))}(u) + \varsigma_1(v)I_{\{b_1(v)\}}(u) + \varsigma_2(v)I_{\{b_2(v)\}}(u) + I_{(b_2(v),\infty)}(u)$$

and state the following.

Theorem 8.29. *Let $(P_\theta)_{\theta\in\Delta}$ be an exponential family in natural form that satisfies (A1) and (A2), where $\Delta = \Upsilon \times \Xi$ is open, along with the splitting from (8.27), and $\alpha \in (0,1)$ be fixed.*

(A) There exist measurable functions $b_i(v)$ and $\varsigma_i(v)$, $i = 1, 2$, such that $\varphi_{II,U}$ defined above satisfies $\varphi_{II,U} : \mathbb{R}^d \to_m [0,1]$ and the conditions in (8.35).
(B) The test $\varphi_{II,U}(U,V)$ is a uniformly best unbiased level α test for the testing problem (II) in (8.28).

Proof. For the proof of (A) we refer to Lehmann (1959), Ferguson (1967), and Witting (1985). To prove statement (B), we note that by Theorem 8.11, for every $\tau \neq \tau_0$ and every $v \in \mathbb{R}^{d-1}$,

$$\alpha = \int \varphi_{II,U}(u,v)\mathsf{K}_{\tau_0}(du|v) \le \int \varphi_{II,U}(u,v)\mathsf{K}_{\tau}(du|v), \quad P_{\theta,V}\text{-a.s., } \theta \in \Delta.$$

Thus, integration with respect to $P_{(\tau,\xi),V}$ gives the unbiasedness of $\varphi_{II,U}$. Above, when motivating the relations in (8.35), we have shown that every unbiased level α test φ satisfies (8.35). As $\varphi_{II,U}$ also satisfies these conditions we get that, $P_{\theta,V}$-a.s., the conditional power $\int \varphi(u,v)\mathsf{K}_\tau(du|v)$ of φ does not exceed the conditional power of $\varphi_{II,U}$. Integration with respect to $P_{(\tau,\xi),V}$ completes the proof. ■

In the examples to follow, one-sided tests in the setting of Theorem 8.28 and two-sided tests in the setting of Theorem 8.29 are considered. First, for a normal distribution with unknown mean and variance, the one-sided *Student's t-test for the mean* and the one-sided *chi-square test for the variance* are derived. According to the above considerations we have to deal with the conditional distribution K_{τ_0} and its quantiles, which may be cumbersome.

Helpful hereby is an idea, following below, that can be utilized in this and in similar situations. Let $\alpha \in (0,1)$ be fixed. If Z has the distribution Q on $(\mathbb{R}, \mathfrak{B})$ with c.d.f. G we set $c_{1-\alpha} = G^{-1}(1-\alpha)$ and

$$\begin{aligned} \varphi_{I,G}(z) &= I_{(c_{1-\alpha},\infty}(z) + \gamma_1 I_{\{c_{1-\alpha}\}}(z), \quad \text{where} \\ \gamma_1 &= [G(c_{1-\alpha}) - (1-\alpha)] \oslash Q(\{c_{1-\alpha}\}). \end{aligned}$$

If the distribution Q of Z is symmetric about 0, then we also set

$$\begin{aligned} \varphi_{II,G}(z) &= I_{(c_{1-\alpha/2},\infty)}(z) + \gamma_2 I_{\{c_{1-\alpha/2}\}}(z) \\ &\quad + \gamma_2 I_{\{c_{1-\alpha/2}\}}(-z) + I_{(c_{1-\alpha/2},\infty)}(-z), \quad \text{where} \\ \gamma_2 &= [G(c_{1-\alpha/2}) - (1-\alpha/2)] \oslash Q(\{c_{1-\alpha/2}\}). \end{aligned}$$

Proposition 8.30. *Let $(P_\theta)_\Delta$ be an exponential family with natural parameter θ, where $\Delta = \Upsilon \times \Xi$ is open, along with the splitting from (8.27), and $\tau_0 \in \Upsilon$ be fixed. Suppose that there is a family of mappings $u \mapsto h_v(u)$ from $\mathbb{R}$ into $\mathbb{R}$ that are strictly increasing and continuous in u for $P_{(\tau_0,\xi),V}$-almost all v, and measurable in (u,v). If the distribution Q of $Z := h_V(U)$ under $P_{(\tau_0,\xi)}$ does not depend on ξ, then Z and V are independent under $P_{(\tau_0,\xi)}$. In this case*

$$\psi_I(u,v) = \varphi_{I,G}(h_v(u)) \tag{8.36}$$

is a uniformly best unbiased level α test for $\mathsf{H}_0 : \tau \le \tau_0,\ \xi \in \Xi$, versus $\mathsf{H}_A : \tau > \tau_0,\ \xi \in \Xi$. If in addition the distribution of Z under $P_{(\tau_0,\xi)}$ is symmetric, then

$$\psi_{II}(u,v) = \varphi_{II,G}(h_v(u)) \tag{8.37}$$

is a uniformly best unbiased level α test for $\mathsf{H}_0 : \tau = \tau_0,\ \xi \in \Xi$, versus $\mathsf{H}_A : \tau \ne \tau_0,\ \xi \in \Xi$.

Proof. The statistic Z is ancillary for the family $P_{(\tau_0,\xi)}$, $\xi \in \Xi$. As V is sufficient for the same family and $P_{(\tau_0,\xi),V}$, $\xi \in \Xi$, is complete we get the independence of Z and V under $P_{(\tau_0,\xi)}$ from Basu's theorem; see Theorem 4.82. This means that there is a version of the conditional distribution $\mathsf{K}_{\tau_0}(\cdot|v)$ that does not depend on v, and it holds $Q = \mathsf{K}_{\tau_0}(\cdot|v) \circ h_v^{-1}$. This, and the strict monotonicity and continuity of $u \mapsto h_v(u)$, yield

$$\psi_I(U,V) = \varphi_{I,U}(U,V), \quad P_{(\tau_0,\xi)}\text{-a.s.}, \ \xi \in \Xi,$$

and thus P_θ-a.s. for every $\theta \in \Delta$, where $\varphi_{I,U}(U,V)$ is the test in (8.33). This, in view of Theorem 8.28, proves the first statement. The proof of the second statement is similar and uses Theorem 8.29. ■

Example 8.31. We consider the model $(\mathbb{R}^n, \mathfrak{B}_n, (\mathsf{N}^{\otimes n}(\mu,\sigma^2))_{(\mu,\sigma^2)\in\mathbb{R}\times(0,\infty)})$ for which the projections on the coordinates $X_1, ..., X_n$ are i.i.d. with common distribution $\mathsf{N}(\mu,\sigma^2)$, where $(\mu,\sigma^2) \in \mathbb{R}\times(0,\infty)$ is unknown. We recall the statistics

$$\overline{X}_n = \frac{1}{n}\sum\nolimits_{i=1}^n X_i \quad \text{and} \quad S_n^2 = \frac{1}{n-1}\sum\nolimits_{i=1}^n (X_i - \overline{X}_n)^2.$$

By Example 1.11 $\mathsf{N}^{\otimes n}(\mu, \sigma^2)$ can be represented as an exponential family in natural form with natural parameter $\theta = (\tau, \xi) = (\mu/\sigma^2, -1/(2\sigma^2))$, parameter set $\Delta = \mathbb{R} \times (-\infty, 0)$, and generating statistic $T = (U, V) = (\sum_{i=1}^n X_i, \sum_{i=1}^n X_i^2)$. Suppose we want to test, for a fixed given $\mu_0 \in \mathbb{R}$, $\mathsf{H}_0 : \mu \leq \mu_0$, $\sigma^2 > 0$, versus $\mathsf{H}_A : \mu > \mu_0$, $\sigma^2 > 0$. We assume without loss of generality that $\mu_0 = 0$, as otherwise we may switch from $X_1, ..., X_n$ to $X_1 - \mu_0, ..., X_n - \mu_0$. Expressed in terms of the natural parameters the hypotheses read

$$\mathsf{H}_0 : \tau \leq 0,\ \xi < 0 \quad \text{versus} \quad \mathsf{H}_A : \tau > 0,\ \xi < 0. \tag{8.38}$$

$\Delta = \mathbb{R} \times (-\infty, 0) = \Upsilon \times \Xi$ is open and thus we may apply Theorem 8.28. Instead of constructing the conditional test we apply Proposition 8.30 with

$$h_v(u) = \sqrt{n} \frac{\frac{1}{n} u}{\sqrt{\frac{1}{n-1}(v - \frac{1}{n} u^2)}},$$

where $u = n\overline{x}_n$ and $v = \sum_{i=1}^n x_i^2$ are the generating statistics. It is easy to see that $(\partial/\partial u) h_v(u) > 0$ holds for every fixed $v > 0$, so that the assumptions on $h_v(u)$ in Proposition 8.30 are satisfied. Set $x = (x_1, ..., x_n)$. As the distribution of

$$h_V(U) = \sqrt{n}\, \overline{X}_n / \sqrt{S_n^2}$$

under $P_{(\tau_0, \xi)} = \mathsf{N}^{\otimes n}(0, \sigma^2)$ is $\mathsf{T}(n-1)$ (i.e., Student's t-distribution with $n-1$ degrees of freedom) we see that this distribution is independent of ξ. Hence for the testing problem (8.38) a uniformly best unbiased level α test from (8.36) is given by

$$\varphi_{I,U}(U, V) = I_{[t_{1-\alpha, n-1}, \infty)}(\sqrt{n}\, \overline{X}_n / \sqrt{S_n^2}), \tag{8.39}$$

where $t_{1-\alpha, n-1}$ is the $1 - \alpha$ quantile of $\mathsf{T}(n-1)$. This is the one-sided version of Student's t-test from (2.24).

The testing problem $\mathsf{H}_0 : \mu = \mu_0$, $\sigma^2 > 0$, versus $\mathsf{H}_A : \mu \neq \mu_0$, $\sigma^2 > 0$, can be treated similarly. For $\mu_0 = 0$ the resulting uniformly best unbiased level α test from (8.37) is the two-sided version of Student's t-test, given by

$$\varphi_{II,U}(U, V) = I_{[t_{1-\alpha/2, n-1}, \infty)}(|\ \sqrt{n}\, \overline{X}_n / \sqrt{S_n^2}\ |). \tag{8.40}$$

Suppose now that we want to test, for a fixed given $\sigma_0^2 > 0$, $\mathsf{H}_0 : \sigma^2 \leq \sigma_0^2$, $\mu \in \mathbb{R}$, versus $\mathsf{H}_A : \sigma^2 > \sigma_0^2$, $\mu \in \mathbb{R}$. With respect to the previous testing problem, the roles of U and V and those of τ and ξ are now exchanged. We set $(\widetilde{\tau}, \widetilde{\xi}) := (-1/(2\sigma^2), \mu/\sigma^2)$ and $(\widetilde{U}, \widetilde{V}) = (\sum_{i=1}^n X_i^2, \sum_{i=1}^n X_i)$, and consider

$$h_{\widetilde{v}}(\widetilde{u}) = \frac{1}{\sigma_0^2}(\widetilde{u} - \frac{1}{n}\widetilde{v}^2),$$

which is strictly increasing and continuous in $\widetilde{u}$ for every fixed $\widetilde{v} \in \mathbb{R}$. As $h_{\widetilde{V}}(\widetilde{U})$, defined by

$$h_{\widetilde{V}(x)}(\widetilde{U}) = \frac{1}{\sigma_0^2} \sum\nolimits_{i=1}^n (X_i - \overline{X}_n)^2,$$

has under $P_{\widetilde{\tau}_0, \widetilde{\xi}} = \mathsf{N}^{\otimes n}(\mu, \sigma_0^2)$ a chi-square distribution $\mathsf{H}(n-1)$ with $n-1$ degrees of freedom, which is independent of μ, we may apply the same arguments as above.

This shows that the test (8.36) turns out to be the one-sided χ^2-test, which is given by

$$\widetilde{\varphi}_{\widetilde{U}}(\widetilde{U},\widetilde{V}) = I_{[\chi^2_{1-\alpha,n-1},\infty)}\left(\frac{1}{\sigma_0^2}\sum\nolimits_{i=1}^{n}(X_i-\overline{X}_n)^2\right),$$

where $\chi^2_{1-\alpha,n-1}$ is the $1-\alpha$ quantile of $\mathsf{H}(n-1)$.

In the next example, *Fisher's exact test* is derived.

Example 8.32. Let X and Y be two independent random variables, where $X \sim \mathsf{B}(n,p)$ with $n \in \mathbb{N}$ and $p \in (0,1)$, and $Y \sim \mathsf{B}(m,q)$ with $m \in \mathbb{N}$ and $q \in (0,1)$. The distribution of (X,Y) belongs to the family $(\mathsf{B}(n,p)\otimes\mathsf{B}(m,q))_{p,q\in(0,1)}$ with probability mass function

$$\mathsf{b}_{n,p}(x)\mathsf{b}_{m,q}(y) = \binom{n}{x}p^x(1-p)^{n-x}\binom{m}{y}q^y(1-q)^{m-y},$$

$(x,y) \in \mathcal{X} := \{0,1,...,n\}\times\{0,1,...,m\}$, $p,q \in (0,1)$. Suppose we want to test $\mathsf{H}_0 : p \le q$ versus $\mathsf{H}_A : p > q$. Intuitively, one would represent the family as a two-parameter exponential family in natural form, with natural parameter $\widetilde{\theta}(p,q) = (\ln(p/(1-p)), \ln(q/(1-q)))$ and generating statistic $\widetilde{T}(x,y) = (x,y)$. However, in this form Theorem 8.28 cannot be used to find a uniformly best unbiased level α test for H_0 versus H_A. A more suitable representation has natural parameter

$$\theta(p,q) = \left(\ln\frac{p(1-q)}{(1-p)q}, \ln\frac{q}{1-q}\right)$$

and generating statistic $T(x,y) = (U(x),V(x)) = (x, x+y)$. With respect to the dominating measure $\boldsymbol{\mu}$ with point mass $\binom{n}{x}\binom{m}{y}$ at $(x,y)\in\mathcal{X}$, the density of $P_{\theta(p,q)} := \mathsf{B}(n,p)\otimes\mathsf{B}(m,q)$ at $(x,y)\in\mathcal{X}$ is

$$f_{p,q}(x,y) = \exp\{\theta_1(p,q)x + \theta_2(q)(x+y) + n\ln(1-p) + m\ln(1-q)\},$$

where $(\theta_1,\theta_2) \in \Delta = \mathbb{R}^2$. Testing H_0 versus H_A is equivalent to testing $\mathsf{H}_0 : \theta_1 \le 0$ versus $\mathsf{H}_A : \theta_1 > 0$. By Theorem 8.28 we get a uniformly best unbiased level α test φ_U. To calculate the conditional distribution of U, given V, at $\theta_1 = 0$ (i.e., at $p = q$) we note that in this case $X+Y$ has a binomial distribution with parameters $m+n$ and p. Put $P_p = \mathsf{B}(n,p)\otimes\mathsf{B}(m,p)$, $p\in(0,1)$. Then for $0\vee(v-m)\le u\le n\wedge v$ and $v\in\{0,1,...,n+m\}$,

$$\begin{aligned}\mathsf{K}_0(\{u\}|v) &= P_p(U=u|V=v) = \frac{P_p(X=u)P_p(Y=v-u)}{P_p(V=v)}\\ &= \frac{\mathsf{b}_{n,p}(u)\mathsf{b}_{m,p}(v-u)}{\mathsf{b}_{n+m,p}(v)} = \frac{\binom{n}{u}\binom{m}{v-u}}{\binom{n+m}{v}},\end{aligned}$$

which is a hypergeometric distribution. Given $v\in\{0,1,...,n+m\}$, we set

$$\varphi_{I,U}(u,v) = I_{[b(v)+1,\ldots,n\wedge v]}(u) + \varsigma(v)I_{\{b(v)\}}(u),\ \ 0\vee(v-m)\le u\le n\wedge v,$$

where $b(v)\in[0\vee(v-m), n\wedge v]\cap\mathbb{N}$ and $\varsigma(v)\in[0,1]$ are determined by

$$\sum\nolimits_{u=0\vee(v-m)}^{n\wedge v}\varphi_{I,U}(u,v)\mathsf{K}_0(\{u\}|v) = \alpha.$$

The numerical evaluation of $b(v)$ and $\varsigma(v)$ has to be done on a computer. Further results on this test and its two-sided version can be found in Finner and Strassburger (2002b).

The data in Example 8.32 are typically presented in form of a 2×2 *contingency table*, i.e.,

$x_{1,1}$	$x_{1,2}$	$x_{1,\cdot} = x_{1,1} + x_{1,2}$
$x_{2,1}$	$x_{2,2}$	$x_{2,\cdot} = x_{2,1} + x_{2,2}$
$x_{\cdot,1} = n$	$x_{\cdot,2} = m$	$N = m + n$

(8.41)

where the marginal totals are indicated by dots in the subscripts. Such tables may also arise in other settings, as we show in the next example. In the above Example 8.32, $x_{\cdot,1} = n$ and $x_{\cdot,2} = m$ are fixed given, $N = n + m$, whereas $x_{1,\cdot}$ and $x_{2,\cdot}$ are the outcomes of the random variables V and $n + m - V$, respectively. The optimal test is a conditional test based on $x_{1,1}$, the outcome of X, given the total $X + Y = x_{1,\cdot}$.

Another quite common situation where data in a 2×2 contingency table have to be analyzed is considered in the next example. At this point it should also be mentioned that the results of Examples 8.32 and 8.33 can be extended to similar analyses of $a \times b$ contingency tables by using similar approaches, see Agresti (1992, 2002) for further details.

Example 8.33. Let $W_1, ..., W_N$ be an i.i.d. sample from a distribution on the points $(1,1), (1,2), (2,1), (2,2)$, with probabilities $p_{i,j} > 0$, $i, j = 1, 2$. Assume that with respect to two certain characteristics A and B, say, $(1,1)$ means that both, A and B, are present, $(1,2)$ means that A is present but not B, $(2,1)$ means that A is not present but B is, and $(2,2)$ means that both, A and B, are not present. An interesting question is whether the occurrences of A and B are stochastically independent or not. In the latter case, A and B are said to be *associated.* A test for this purpose is now derived. Let

$$p_{i,\cdot} = p_{i,1} + p_{i,2} \quad \text{and} \quad p_{\cdot,j} = p_{1,j} + p_{2,j}, \quad i, j = 1, 2.$$

The null hypothesis ("independence") is that $p_{i,j} = p_{i,\cdot} p_{\cdot,j}$ holds for all $i, j = 1, 2$, whereas the alternative ("dependence") is its complement. In the present 2×2 layout, the null hypothesis is equivalent to $p_{1,1} = p_{1,\cdot} p_{\cdot,1}$. However, for larger $a \times b$ layouts more than just one equation of the type $p_{i,j} = p_{i,\cdot} p_{\cdot,j}$ would have to be stated in the null hypothesis. Let

$$X_{i,j} := |\{k : W_k = (i,j),\ k = 1, ..., N\}|, \quad i, j = 1, 2.$$

These outcomes can be summarized in a 2×2 contingency table of the form

$x_{1,1}$	$x_{1,2}$	$x_{1,\cdot} = x_{1,1} + x_{1,2}$
$x_{2,1}$	$x_{2,2}$	$x_{2,\cdot} = x_{2,1} + x_{2,2}$
$x_{\cdot,1} = x_{1,1} + x_{2,1}$	$x_{\cdot,2} = x_{1,2} + x_{2,2}$	N

In contrast to the contingency table (8.41) $x_{\cdot,1}$ and $x_{\cdot,2}$ are now realizations of random variables, so that all four marginal totals $x_{1,\cdot}, x_{2,\cdot}, x_{\cdot,1}$, and $x_{\cdot,2}$ in the 2×2 contingency table are the outcomes of random variables. Apparently, $X = (X_{1,1}, X_{1,2}, X_{2,1}, X_{2,2})$ follows a multinomial distribution $\mathsf{M}(N, p)$ with parameters N and $p = (p_{1,1}, p_{1,2}, p_{2,1}, p_{2,2})$ and p.m.f.

$$\frac{N!}{x_{1,1}!x_{1,2}!x_{2,1}!x_{2,2}!} p_{1,1}^{x_{1,1}} p_{1,2}^{x_{1,2}} p_{2,1}^{x_{2,1}} p_{2,2}^{x_{2,2}},$$

where $x_{i,j} \in \{0, ..., N\}$ and $x_{1,1} + x_{1,2} + x_{2,1} + x_{2,2} = N$. We have seen in Example 1.5 that with respect to the dominating measure $\boldsymbol{\mu}$ with point masses $N!/(x_{1,1}!x_{1,2}!x_{2,1}!x_{2,2}!)$ the family $\mathsf{M}(N, p)$ can be represented as an exponential family in natural form with natural parameter

$$\widetilde{\theta}(p_{1,1}, p_{1,2}, p_{2,1}) = (\ln(\frac{p_{1,1}}{p_{2,2}}), \ln(\frac{p_{1,2}}{p_{2,2}}), \ln(\frac{p_{2,1}}{p_{2,2}}))$$

and generating statistic $\widetilde{T}(x_{11}, x_{12}, x_{21}) = (x_{1,1}, x_{1,2}, x_{2,1})$. As before in Example 8.32, however, we need another representation that is more suitable for the testing problem at hand. This can be achieved by using as a natural parameter $\theta = (\tau, \xi)$ with

$$\tau(p_{1,1}, p_{1,2}, p_{2,1}) = \ln(\frac{p_{1,1}p_{2,2}}{p_{1,2}p_{2,1}}) \quad \text{and} \quad \xi(p_{1,1}, p_{1,2}, p_{2,1}) = (\ln(\frac{p_{1,2}}{p_{2,2}}), \ln(\frac{p_{2,1}}{p_{2,2}})),$$

and as generating statistics $T = (U, V)$ with

$$U(x_{1,1}, x_{1,2}, x_{2,1}) = x_{1,1} \quad \text{and} \quad V(x_{1,1}, x_{1,2}, x_{2,1}) = (x_{1,\cdot}, x_{\cdot,1}).$$

Because $p_{1,\cdot}p_{\cdot,1} = p_{1,1} - (p_{1,1}p_{2,2} - p_{1,2}p_{2,1})$, the null hypothesis is equivalent to $\mathsf{H}_0 : \tau = 0$, and the alternative is equivalent to $\mathsf{H}_A : \tau \neq 0$. A uniformly best unbiased level α test $\varphi_{II,U}$ is provided by Theorem 8.29 which is based on the conditional distribution of U given $V = (x_{1,\cdot}, x_{\cdot,1})$ at $\tau = 0$. The distribution is given in the problem below.

Problem 8.34. Show that in Example 8.33,

$$P_{0,\xi_1,\xi_2}(U = u | V_1 = x_{1,\cdot}, V_2 = x_{\cdot,1}) = \frac{\binom{x_{\cdot,1}}{u}\binom{N - x_{\cdot,1}}{x_{1,\cdot} - u}}{\binom{N}{x_{1,\cdot}}},$$

where $0 \vee (x_{1,\cdot} + x_{\cdot,1} - N) \leq u \leq x_{1,\cdot} \wedge x_{\cdot,1}$ and $x_{1,\cdot} \in \{0, 1, ..., N\}$. Compare this with the conditional distribution of U, given V, at $\theta_1 = 0$, in Example 8.32.

It is interesting to note that in Example 8.32, the uniformly best unbiased level α test for testing $\mathsf{H}_0 : p = q$ versus $\mathsf{H}_A : p \neq q$ provided by Theorem 8.29, which has not been derived here for brevity, turns out to be technically identical with the test $\varphi_{II,U}$ in Example 8.33. Likewise, in Example 8.33, the uniformly best unbiased level α test for testing $\mathsf{H}_0 : p_{1,1}/p_{\cdot,1} \leq p_{1,2}/p_{,2}$ versus $\mathsf{H}_A : p_{1,1}/p_{\cdot,1} > p_{1,2}/p_{\cdot,2}$ provided by Theorem 8.28, which has not been derived here, turns out to be technically identical with the test $\varphi_{I,U}$ in Example 8.32. Despite these technical identities, however, Examples 8.32 and 8.33 are based on different statistical models. Therefore the respective tests for one-sided hypotheses provide different decisions, and the same holds for two-sided hypotheses.

Problem 8.35.* Let $X_1, ..., X_n$ be an i.i.d. sample from a Poisson distribution $\mathsf{Po}(\lambda_1)$ with $\lambda_1 > 0$, and $Y_1, ..., Y_n$ be an i.i.d. sample from a Poisson distribution $\mathsf{Po}(\lambda_2)$ with $\lambda_2 > 0$, where the samples are independent. Let $\alpha \in (0,1)$ be fixed given. As an application of Theorem 8.28, find a uniformly best unbiased level α test for $\mathsf{H}_0 : \lambda_1 \le \lambda_2$ versus $\mathsf{H}_A : \lambda_1 > \lambda_2$. As an application of Theorem 8.29, find a uniformly best unbiased level α test for $\mathsf{H}_0 : \lambda_1 = \lambda_2$ versus $\mathsf{H}_A : \lambda_1 \ne \lambda_2$.

8.2 Confidence Regions and Confidence Bounds

For a statistical model $(\mathcal{X}, \mathfrak{A}, (P_\theta)_{\theta\in\Delta})$ with $\Delta \subseteq \mathbb{R}^d$ we have considered the problem of estimating θ under various aspects in Chapter 7. An estimator, by definition, is a mapping $S : \mathcal{X} \to_m \mathbb{R}^d$. After observing $x \in \mathcal{X}$ the estimate for θ is $S(x)$, but it does not provide any information on how close it is to θ. The concept of *confidence regions* is tailored for this need. In this approach the idea of using a point estimator $S : \mathcal{X} \to_m \mathbb{R}^d$ is abandoned in favor of a *region estimator* $\mathcal{C} : \mathcal{X} \to \mathfrak{P}(\Delta)$, where $\mathfrak{P}(\Delta)$ is the set of all subsets of Δ. Here one has to assume that $\{x : \theta \in \mathcal{C}(x)\} \in \mathfrak{A}$ for every $\theta \in \Delta$. After observing $x \in \mathcal{X}$ the set $\mathcal{C}(x) \subseteq \mathbb{R}^d$ is interpreted as the region where θ is estimated to be located. The attractive feature of a confidence region is that it comes along with a lower bound on the probability of including the true θ. For $d = 1$ (i.e., $\Delta \subseteq \mathbb{R}$) and for most of the concrete models studied so far, confidence regions are either finite or one-sided intervals.

Definition 8.36. *Let $(\mathcal{X}, \mathfrak{A}, (P_\theta)_{\theta\in\Delta})$ be a given model with $\Delta \subseteq \mathbb{R}^d$, and let $\alpha \in (0,1)$ be fixed. A mapping $\mathcal{C} : \mathcal{X} \to \mathfrak{P}(\Delta)$ is called a confidence region at the level $1-\alpha$ if $\{x : \theta \in \mathcal{C}(x)\} \in \mathfrak{A}$ for every $\theta \in \Delta$, and*

$$P_\theta(\theta \in \mathcal{C}) \ge 1 - \alpha, \quad \theta \in \Delta.$$

For $\Delta \subseteq \mathbb{R}$ some special forms are distinguished.

$$\begin{array}{ll} \mathcal{C}(x) = [B_{\mathrm{lcb}}(x), \infty) & \text{lower confidence bound } B_{\mathrm{lcb}}(x), \\ \mathcal{C}(x) = [B_{\mathrm{lci}}(x), B_{\mathrm{uci}}(x)] & \text{confidence interval } [B_{\mathrm{lci}}(x), B_{\mathrm{uci}}(x)]. \end{array} \tag{8.42}$$

The upper confidence bounds can be introduced and treated in a similar way as the lower confidence bounds.

The standard technique to construct confidence regions is based on a duality between confidence regions and the acceptance regions of tests. Let $\alpha \in (0,1)$ be fixed given. We consider a family of testing problems with simple null hypotheses for which there is a family of nonrandomized level α tests $\psi = (\psi_\theta)_{\theta\in\Delta}$; that is,

$$\begin{aligned} &\mathsf{H}_0(\theta_0) : \theta = \theta_0 \quad \text{versus} \quad \mathsf{H}_A(\theta_0) : \theta \in \Delta_A(\theta_0), \text{ where} \\ &\Delta_A(\theta_0) \subseteq \Delta \backslash \{\theta_0\} \quad \text{is predetermined, and} \\ &\psi_{\theta_0} : \mathcal{X} \to_m \{0,1\}, \ \ \mathsf{E}_{\theta_0}\psi_{\theta_0} = \alpha, \ \ \theta_0 \in \Delta. \end{aligned}$$

It is important to note that $\Delta_A(\theta_0)$ does not have to be equal to $\Delta\backslash\{\theta_0\}$. The interpretation of $\Delta_A(\theta_0)$ is that it contains the *unacceptable* parameter values when θ_0 is true. For example, in a location parameter problem we may have $\Delta_A(\theta_0) = (\theta_0, \infty) \subset \Delta = \mathbb{R}$. We restrict ourselves to the most common special cases where $\Delta_A(\theta_0) = (-\infty, \theta_0) \cap \Delta$ for all $\theta_0 \in \Delta$, $\Delta_A(\theta_0) = \Delta\backslash\{\theta_0\}$ for all $\theta_0 \in \Delta$, or $\Delta_A(\theta_0) = (\theta_0, \infty) \cap \Delta$ for all $\theta_0 \in \Delta$. This is good enough to establish optimal confidence bounds and intervals for models with one-parameter families of distributions that have MLR, or are even exponential families. For a broader coverage of this topic we refer to Lehmann (1959, 1986). A more general approach to confidence regions that includes composite null hypotheses can be found in Witting (1985). The restriction to nonrandomized tests is made here to avoid technical complications; see Remark 8.37. Let

$$\begin{aligned} \mathcal{A}_\psi(\theta) &= \{x : \psi_\theta(x) = 0\}, \quad \theta \in \Delta, \quad \text{and} \\ \mathcal{C}_\psi(x) &= \{\theta : x \in \mathcal{A}_\psi(\theta)\}, \quad x \in \mathcal{X}. \end{aligned} \tag{8.43}$$

Then $\mathcal{A}_\psi(\theta)$ is the acceptance region of the test ψ_θ, $\theta \in \Delta$, and

$$x \in \mathcal{A}_\psi(\theta) \quad \text{if and only if} \quad \theta \in \mathcal{C}(x), \quad x \in \mathcal{X},\ \theta \in \Delta, \tag{8.44}$$

shows that $\{x : \theta \in \mathcal{C}(x)\} \in \mathfrak{A}$, $\theta \in \Delta$. If for each $\theta \in \Delta$ the test ψ_θ is a level α test for $\mathsf{H}_0(\theta)$, we get

$$P_\theta(\theta \in \mathcal{C}_\psi) = 1 - \mathsf{E}_\theta \psi_\theta \geq 1 - \alpha, \quad \theta \in \Delta. \tag{8.45}$$

Thus we have constructed a confidence region $\mathcal{C}_\psi$ at the level $1 - \alpha$ from the family of tests $\psi = (\psi_\theta)_{\theta \in \Delta}$, where ψ_{θ_0} is a level α test for $\mathsf{H}_0(\theta_0) : \theta = \theta_0$ for every $\theta_0 \in \Delta$. We call $\mathcal{C}_\psi$ the confidence region associated with the family of tests $\psi = (\psi_\theta)_{\theta \in \Delta}$. Apparently, the form of $\Delta_A(\theta_0)$, $\theta_0 \in \Delta$, has not been used hereby.

Conversely, suppose we are given a confidence region $\mathcal{C}$ at the level $1 - \alpha$. Then we set

$$\psi_{\mathcal{C},\theta_0}(x) = 1 - I_{\mathcal{C}(x)}(\theta_0), \quad x \in \mathcal{X},\ \theta_0 \in \Delta. \tag{8.46}$$

If $\mathcal{C}$ is at the level $1 - \alpha$ we get

$$\mathsf{E}_{\theta_0} \psi_{\mathcal{C},\theta_0} = 1 - P_{\theta_0}(\theta_0 \in \mathcal{C}) \leq \alpha, \quad \theta_0 \in \Delta. \tag{8.47}$$

Thus, for every $\theta_0 \in \Delta$ the test $\psi_{\mathcal{C},\theta_0}$ is a level α test for $\mathsf{H}_0(\theta_0) : \theta = \theta_0$ versus $\mathsf{H}_A(\theta_0) : \theta \in \Delta_A(\theta_0)$, where $\Delta_A(\theta_0) \subseteq \Delta\backslash\{\theta_0\}$ may be chosen in any way. We call $\psi_{\mathcal{C},\theta_0}$, $\theta_0 \in \Delta$, the family of tests associated with the confidence region $\mathcal{C}$.

Remark 8.37. If the confidence region in (8.43) were based on a family $\psi = (\psi_\theta)_{\theta \in \Delta}$ of randomized tests, then at any $x \in \mathcal{X}$ and $\theta \in \Delta$ with $\psi_\theta(x) \in (0, 1)$ the parameter value θ would have to be included in $\mathcal{C}(x)$ with probability $1 - \psi_\theta(x)$. To make such an approach rigorous additional concepts would be needed that go beyond the short introduction into confidence regions given in this section.

For a family $\psi = (\psi_\theta)_{\theta\in\Delta}$ of nonrandomized level α tests ψ_{θ_0} for $\mathsf{H}_0(\theta_0)$: $\theta = \theta_0$, $\theta_0 \in \Delta$, let us examine how their power on the respective alternatives $\mathsf{H}_A(\theta_0) : \theta \in \Delta_A(\theta_0) \subseteq \Delta\backslash\{\theta_0\}$, $\theta_0 \in \Delta$, affects the performance of the associated confidence region $\mathcal{C}_\psi$. From (8.44) we get

$$\mathsf{E}_\theta\psi_{\theta_0} = 1 - P_\theta(\theta_0 \in \mathcal{C}_\psi), \quad \theta \neq \theta_0, \ \ \theta_0 \in \Delta. \tag{8.48}$$

This can be utilized to derive optimal confidence regions from optimal tests, and vice versa. The optimality concept for confidence regions is now established.

Definition 8.38. *Let $(\mathcal{X}, \mathfrak{A}, (P_\theta)_{\theta\in\Delta})$ be a model with $\Delta \subseteq \mathbb{R}^d$, and let $\Delta_A(\theta) \subseteq \Delta\backslash\{\theta\}$ be given for very $\theta \in \Delta$. For $\alpha \in (0,1)$ a confidence region $\mathcal{C}^*$ at the level $1-\alpha$ is called a uniformly most accurate confidence region with respect to $(\Delta_A(\theta))_{\theta\in\Delta}$ at the level $1-\alpha$ if for every confidence region $\mathcal{C}$ at the level $1-\alpha$ it holds*

$$P_\theta(\theta_0 \in \mathcal{C}^*) \leq P_\theta(\theta_0 \in \mathcal{C}), \quad \theta \in \Delta_A(\theta_0), \ \ \theta_0 \in \Delta.$$

Especially for $\Delta \subseteq \mathbb{R}$, a lower confidence bound $B^*_{\mathrm{lcb}}(x)$ at the level $1-\alpha$ is called a *uniformly most accurate lower confidence bound* at the level $1-\alpha$ if for every lower confidence bound $B_{\mathrm{lcb}}(x)$ at the level $1-\alpha$ it holds

$$P_\theta(B^*_{\mathrm{lcb}} \leq \theta_0) \leq P_\theta(B_{\mathrm{lcb}} \leq \theta_0), \quad \theta \in (\theta_0, \infty) \cap \Delta, \ \ \theta_0 \in \Delta.$$

The duality of lower confidence bounds for θ at the level $1-\alpha$ and level α tests for $\mathsf{H}_0(\theta_0) : \theta = \theta_0$ versus $\mathsf{H}_A(\theta_0) : \theta \in (\theta_0, \infty) \cap \Delta$, $\theta_0 \in \Delta$, is extended below to include optimality.

Let now $(\mathcal{X}, \mathfrak{A}, (P_\theta)_{\theta\in\Delta})$ be a statistical model with $\Delta = (a,b) \subseteq \mathbb{R}$. Furthermore, let $T : \mathcal{X} \to_m \mathbb{R}$ be a statistic. To avoid too many technical difficulties we assume here for $F_\theta(t) = P_\theta(T \leq t)$, $t \in \mathbb{R}$, that

$$\begin{aligned} &F_\theta(t) \quad \text{is continuous in } t \in \mathbb{R} \text{ for } \theta \in (a,b), \\ &F_\theta(t) \quad \text{is decreasing in } \theta \in (a,b) \text{ for } t \in \mathbb{R} \text{ with } 0 < F_\theta(t) < 1. \end{aligned} \tag{8.49}$$

The next statement is a special case of Theorem 4 on p. 90 in Lehmann (1986).

Theorem 8.39. *Suppose that $(P_\theta)_{\theta\in(a,b)}$ with $(a,b) \subseteq \mathbb{R}$ has MLR in T, and that $F_\theta(t)$ satisfies (8.49). Let $\alpha \in (0,1)$ be fixed. If there exists a $B^*_{lcb} : \mathcal{X} \to_m (a,b)$ with*

$$F_{B^*_{lcb}(x)}(T(x)) = 1-\alpha, \quad P_\theta\text{-a.s.}, \ \theta \in (a,b), \tag{8.50}$$

then $B^*_{lcb}(x)$ is a uniformly most accurate lower confidence bound at the level $1-\alpha$.

Proof. For every lower confidence bound $B_{\mathrm{lcb}}(x)$ at the level $1-\alpha$ and fixed $\theta_0 \in (a,b)$ we introduce the test $\psi_{\theta_0}(x) = I_{(\theta_0,b)}(B_{\mathrm{lcb}}(x))$. By (8.45) $\mathsf{E}_{\theta_0}\psi_{\theta_0} = P_{\theta_0}(\theta_0 < B_{\mathrm{lcb}}) \leq \alpha$, and thus ψ_{θ_0} is a level α test for $\mathsf{H}_0(\theta_0) : \theta = \theta_0$

versus $\mathsf{H}_A(\theta_0) : \theta \in (\theta_0, b)$. For $\theta \in (\theta_0, b)$ we get from (8.48) $\mathsf{E}_\theta \psi_{\theta_0} = 1 - P_\theta(B_{\text{lcb}} \le \theta_0)$. Due to the continuity of F_{θ_0} it holds

$$P_{\theta_0}(F_{\theta_0}^{-1}(1-\alpha) \le T) = P_{\theta_0}(F_{\theta_0}^{-1}(1-\alpha) < T) = \alpha,$$

so that by Theorem 2.49 the test $\psi := I_{(1-\alpha,1]}(F_{\theta_0}(T))$ is a uniformly best level α test for $\mathsf{H}_0(\theta_0) : \theta = \theta_0$ versus $\mathsf{H}_A(\theta_0) : \theta \in (\theta_0, b)$. Because $F_\theta(t)$ is decreasing in $\theta \in (a, b)$ for every $t \in \mathbb{R}$ with $0 < F_\theta(t) < 1$ it holds in view of (8.50),

$$\theta_0 < B_{\text{lcb}}^*(x) \quad \Leftrightarrow \quad F_{\theta_0}(T(x)) > F_{B_{\text{lcb}}^*(x)}(T(x)) = 1 - \alpha, \quad P_\theta\text{-a.s., } \theta \in (a, b).$$

Thus the test $\psi_{\theta_0}^*(x) = I_{(\theta_0,b)}(B_{\text{lcb}}^*(x))$ satisfies $\psi_{\theta_0}^*(x) = \psi(x)$, P_θ-a.s., $\theta \in (a, b)$, and is a uniformly best level α test for $\mathsf{H}_0(\theta_0) : \theta = \theta_0$ versus $\mathsf{H}_A(\theta_0) : \theta \in (\theta_0, b)$. This shows that $1 - \alpha \le 1 - \mathsf{E}_{\theta_0}\psi_{\theta_0}^* = P_{\theta_0}(B_{\text{lcb}}^* \le \theta_0)$ and

$$P_\theta(B_{\text{lcb}}^* \le \theta_0) = 1 - \mathsf{E}_\theta \psi_{\theta_0}^* \le 1 - \mathsf{E}_\theta \psi_{\theta_0} = P_\theta(B_{\text{lcb}} \le \theta_0), \ \theta \in (\theta_0, b), \ \theta_0 \in \Delta,$$

which completes the proof. ■

Example 8.40. We consider the family $(P_\mu)_{\mu\in\mathbb{R}} = (\mathsf{N}^{\otimes n}(\mu, \sigma_0^2))_{\mu\in\mathbb{R}}$, where σ_0^2 is known, and the level α Gauss test $\psi_I(\sqrt{n}(\overline{X}_n - \mu_0)/\sigma_0)$ for testing $\mathsf{H}_0(\mu_0) : \mu \le \mu_0$ versus $\mathsf{H}_A(\mu_0) : \mu > \mu_0$ from (8.18) and (8.19). For

$$\psi = \psi_I(\sqrt{n}(\overline{X}_n - \mu_0)/\sigma_0), \quad \mu_0 \in \mathbb{R},$$

we have $\mathcal{A}_\psi(\mu_0) = \{x : \overline{x}_n < \mu_0 + u_{1-\alpha}\sigma_0/\sqrt{n}, \ x \in \mathbb{R}^n\}$, $\mu_0 \in \mathbb{R}$, and thus

$$\begin{aligned} \mathcal{C}_\psi(x) &= \{\mu_0 : x \in \mathcal{A}_\psi(\mu_0)\} = [B_{\text{lcb}}^*(x), \infty), \quad x \in \mathbb{R}^n, \quad \text{where} \\ B_{\text{lcb}}^*(x) &= \overline{x}_n - u_{1-\alpha}\sigma_0/\sqrt{n}, \end{aligned}$$

and the conditions (8.49) and (8.50) are satisfied. If now $B_{\text{lcb}}(x)$ is another lower confidence bound at the level $1 - \alpha$, then $\widetilde{\psi}_{\mu_0}(x) = I_{(\mu_0,\infty)}(B_{\text{lcb}}(x))$ is a level α test for testing $\mathsf{H}_0 : \mu_0$ versus $\mathsf{H}_A : \mu > \mu_0$ for every $\mu_0 \in \mathbb{R}$. As the Gauss test is uniformly most powerful at the level α we get $\mathsf{E}_\mu \widetilde{\psi}_{\mu_0} \le \mathsf{E}_\mu \psi_I$ for $\mu > \mu_0$, which implies

$$P_\mu(B_{\text{lcb}}^* \le \mu_0) \le P_\mu(B_{\text{lcb}} \le \mu_0), \quad \mu > \mu_0, \ \mu_0 \in \mathbb{R}.$$

Example 8.41. We consider binomial distributions $\mathsf{B}(n, p)$, $p \in (0, 1)$, as in Example 2.53, but this time with the goal to construct a lower confidence bound for the parameter p. The difficulties pointed out in Remark 8.37 are avoided here by utilizing the interpolation technique of Proposition 2.58. Let Q_p be the distribution of $Y = X + U$, where X and U are independent, X follows a binomial distribution $\mathsf{B}(n, p)$, and U is uniformly distributed in $(0, 1)$, $p \in (0, 1)$. By Proposition 2.58 $(Q_p)_{p\in(0,1)}$ has MLR in the identity. According to (8.50) we have to find for $X = x$ and $U = u$ (i.e., $Y = x + u = y$) a solution $p \in (0, 1)$ for the equation

$$\sum\nolimits_{i=0}^{[y]-1} \mathsf{b}_{n,p}(i) + (y - [y])\mathsf{b}_{n,p}([y]) = 1 - \alpha, \tag{8.51}$$

where $[y]$ denotes the integer part of y. As the family of binomial distributions has strict MLR in the identity the expression on the left-hand side is a continuous and decreasing function of $p \in (0,1)$ so that there exists a unique solution $B^*_{\mathrm{lcb}}(y)$ for p in (8.51), which by Theorem 8.39 is a uniformly most accurate lower confidence bound at the level $1-\alpha$.

To simplify matters in practice often the solution $\widetilde{B}_0(y)$ of $\sum_{i=0}^{[y]-1} \mathsf{b}_{n,p}(i) = 1-\alpha$ is used instead. Then, because of $[y] = x$, the random variable U can be ignored altogether. Here we have $\widetilde{B}_0(y) \le B_0(y)$ and thus another lower confidence bound at the level $1-\alpha$, but it is conservative. This is the classical Clopper–Pearson lower confidence bound; see Witting (1985).

Problem 8.42. The Clopper–Pearson lower confidence bounds for the binomial distributions $\mathsf{B}(n,p)$, $p \in (0,1)$, are solutions of the equation

$$\frac{n!}{([y]-1)!(n-[y])!} \int_p^1 t^{[y]-1}(1-t)^{n-[y]} dt = 1-\alpha.$$

Now we apply the results on optimal two-sided tests to the construction of confidence intervals $\mathcal{C}(x) = [B_{\mathrm{lci}}(x), B_{\mathrm{uci}}(x)]$; see (8.42). Suppose that $(P_\theta)_{\theta\in\Delta}$ is a one-parameter family of distributions on $(\mathcal{X}, \mathfrak{A})$, where $\Delta = (a,b) \subseteq \mathbb{R}$ is an open interval. Let $T : \mathcal{X} \to \mathbb{R}$ be a statistic, where $F_\theta(t) = P_\theta(T \le t)$ is continuous in $t \in \mathbb{R}$ for $\theta \in (a,b)$. Let $\alpha \in (0,1)$ be fixed. Suppose that $c_i(\theta_0, \alpha)$, $i = 1,2$, are functions on $\Delta \times (0,1)$ such that for every $\theta_0 \in \Delta$,

$$\psi_{\theta_0}(x) = 1 - I_{[c_1(\theta_0,\alpha), c_2(\theta_0,\alpha)]}(T(x)), \quad x \in \mathcal{X}, \tag{8.52}$$

is a level α test for $\mathsf{H}_0 : \theta = \theta_0$ versus $\mathsf{H}_A : \theta \ne \theta_0$. If the functions $c_i(\theta_0, \alpha)$ are nondecreasing and continuous in $\theta_0 \in \Delta$, then

$$\{\theta_0 : c_1(\theta_0,\alpha) \le T(x) \le c_2(\theta_0,\alpha), \quad \theta_0 \in \Delta\}$$

is an interval with endpoints $B_{\psi,\mathrm{lci}}(x)$ and $B_{\psi,\mathrm{uci}}(x)$. To get a closed interval, i.e., $[B_{\psi,\mathrm{lci}}(x), B_{\psi,\mathrm{uci}}(x)]$, we assume for simplicity that $\lim_{\theta_0\downarrow a} c_1(\theta_0,\alpha) = -\infty$ and $\lim_{\theta_0\uparrow b} c_2(\theta_0,\alpha) = \infty$. It is called the confidence interval associated with the family of tests $\psi = (\psi_\theta)_{\theta\in\Delta}$, and by (8.45) its confidence level is $1-\alpha$; that is, it holds

$$P_{\theta_0}(B_{\psi,\mathrm{lci}}(x) \le \theta_0 \le B_{\psi,\mathrm{uci}}(x)) \ge 1-\alpha, \quad \theta_0 \in \Delta. \tag{8.53}$$

Conversely, if $[B_{\mathrm{lci}}(x), B_{\mathrm{uci}}(x)]$, $x \in \mathcal{X}$, is a confidence interval at the level $1-\alpha$, then for every $\theta_0 \in \Delta$

$$\psi_{B,\theta_0}(x) = 1 - I_{[B_{\mathrm{lci}}(x), B_{\mathrm{uci}}(x)]}(\theta_0), \quad x \in \mathcal{X}, \tag{8.54}$$

by (8.47) is a level α test for $\mathsf{H}_0 : \theta = \theta_0$ versus $\mathsf{H}_A : \theta \ne \theta_0$, called the test associated with the confidence interval $[B_{\mathrm{lci}}, B_{\mathrm{uci}}]$. Similar as for two-sided tests, a restriction to unbiased confidence intervals is needed to get optimality. A confidence interval $[B_{\mathrm{lci}}, B_{\mathrm{uci}}]$ is called *unbiased* at the level $1-\alpha$ if it satisfies (8.53) and

$$P_{\theta_1}(B_{\mathrm{lci}}(x) \le \theta_0 \le B_{\mathrm{uci}}(x)) \le 1-\alpha, \quad \theta_0 \ne \theta_1, \quad \theta_0, \theta_1 \in \Delta.$$

Similarly to confidence bounds (see Theorem 8.39) optimality properties of the tests ψ_{θ_0}, $\theta_0 \in \Delta$, lead to optimality properties of the confidence interval associated with the tests ψ_{θ_0}, $\theta_0 \in \Delta$.

Proposition 8.43. *Suppose that $(P_\theta)_{\theta\in\Delta}$, $\Delta = (a,b) \subseteq \mathbb{R}$, is an exponential family, where $F_\theta(t)$ is continuous in $t \in \mathbb{R}$ for $\theta \in (a,b)$. Let $\alpha \in (0,1)$ be fixed. Assume that for every $\theta_0 \in \Delta$ the test ψ_{θ_0} is an unbiased level α test for $\mathsf{H}_0 : \theta = \theta_0$ versus $\mathsf{H}_A : \theta \neq \theta_0$, where the functions $c_i(\theta_0, \alpha)$ are nondecreasing and continuous in $\theta_0 \in \Delta$ with $\lim_{\theta_0 \downarrow a} c_1(\theta_0, \alpha) = -\infty$ and $\lim_{\theta_0 \uparrow b} c_2(\theta_0, \alpha) = \infty$. Then $[B_{\psi,lci}, B_{\psi,uci}]$ is an unbiased confidence interval at the level $1-\alpha$. Moreover, if for every $\theta_0 \in \Delta$ the test ψ_{θ_0} is a uniformly best unbiased level α test for $\mathsf{H}_0 : \theta = \theta_0$ versus $\mathsf{H}_A : \theta \neq \theta_0$, then $[B_{\psi,lci}, B_{\psi,uci}]$ is uniformly most accurate unbiased in the sense that for every $\theta_0, \theta_1 \in \Delta$ with $\theta_1 \neq \theta_0$ it holds*

$$P_{\theta_1}(B_{\psi,lci}(x) \leq \theta_0 \leq B_{\psi,uci}(x)) \leq P_{\theta_1}(B_{lci}(x) \leq \theta_0 \leq B_{uci}(x))$$

for every unbiased confidence interval $[B_{lci}, B_{uci}]$ at the level $1 - \alpha$.

Proof. The definition of ψ_{θ_0} in (8.52) yields $B_{\psi,\text{lci}}(x) \leq \theta_0 \leq B_{\psi,\text{uci}}(x)$ if and only if $\psi_{\theta_0}(x) = 0$. $\mathsf{E}_{\theta_0}\psi_{\theta_0} \leq \alpha$ implies $P_{\theta_0}(B_{\psi,\text{lci}} \leq \theta_0 \leq B_{\psi,\text{uci}}) \geq 1 - \alpha$, and $\mathsf{E}_{\theta_1}\psi_{\theta_0} \geq \alpha$ implies $P_{\theta_1}(B_{\psi,\text{lci}} \leq \theta_0 \leq B_{\psi,\text{uci}}) \leq 1 - \alpha$.

Now we prove the second statement. Let $[B_{\text{lci}}, B_{\text{uci}}]$ be any confidence interval that is unbiased at the level $1 - \alpha$. Then for every $\theta_0 \in \Delta$ the test ψ_{B,θ_0} in (8.54) satisfies

$$\mathsf{E}_{\theta_1}\psi_{B,\theta_0} = 1 - P_{\theta_1}(B_{\text{lci}}(x) \leq \theta_0 \leq B_{\text{uci}}(x)), \quad \theta_0, \theta_1 \in \Delta,$$

and is thus an unbiased level α test for $\mathsf{H}_0 : \theta = \theta_0$ versus $\mathsf{H}_A : \theta \neq \theta_0$. As ψ_{θ_0} is a uniformly best unbiased level α test for $\mathsf{H}_0 : \theta = \theta_0$ versus $\mathsf{H}_A : \theta \neq \theta_0$ for every $\theta_0 \in \Delta$, the proof is completed. ■

Example 8.44. Consider the family $(\mathsf{N}^{\otimes n}(\mu, \sigma_0^2))_{\mu\in\mathbb{R}}$, where $\sigma_0^2 > 0$ is known. To construct a uniformly most accurate unbiased confidence interval for $\mu \in \mathbb{R}$ at the level $1 - \alpha$ we use the Gauss test $\psi_{II}(\sqrt{n}(\overline{X}_n - \mu_0)/\sigma_0) = I_{(u_{1-\alpha/2},\infty)}(\sqrt{n}|\overline{X}_n - \mu_0|/\sigma_0)$ from Example 8.17. The confidence interval associated with this test is

$$[\overline{X}_n - u_{1-\alpha/2}\sigma_0/\sqrt{n}, \overline{X}_n + u_{1-\alpha/2}\sigma_0/\sqrt{n}].$$

From Proposition 8.43 we get that this is a most accurate unbiased confidence interval for $\mu \in \mathbb{R}$ at the level $1 - \alpha$.

We conclude this section with a brief comment on optimal confidence intervals in other situations where optimal tests are available. One is where a uniformly best unbiased level α test for testing a one-dimensional parameter of interest in a d-parameter exponential family exists. The considerations on one-parameter exponential families can be extended in a straightforward manner to exponential families with nuisance parameters. Here we give only an example. For further results we refer to Section 5.7 in Lehmann (1986). Analogously to Problem 8.44 we get the following.

Example 8.45. For the family $(\mathsf{N}^{\otimes n}(\mu,\sigma^2))_{\mu\in\mathbb{R},\sigma^2>0}$, a uniformly most accurate unbiased confidence interval for $\mu \in \mathbb{R}$ at the level $1-\alpha$ can be derived from the two-sided version of the Student's t-test in Example 8.31. It turns out to be

$$[\overline{X}_n - (1/\sqrt{n})\sqrt{S_n^2}t_{1-\alpha/2,n-1}, \overline{X}_n + (1/\sqrt{n})\sqrt{S_n^2}t_{1-\alpha/2,n-1}].$$

Finally we remark that asymptotic level α tests and locally asymptotically best level α tests can be used to establish asymptotic confidence intervals that have the analogous optimality properties as the tests on which they are based. For details we refer to Chapter 8 in Pfanzagl (1994).

8.3 Bayes Tests

In the previous section we have studied testing problems for a one-dimensional parameter, the parameter of interest. Later, a nuisance parameter was admitted to make the model more flexible for a better fit to the data. However, this approach does not cover all interesting situations, for example, composite hypotheses which typically appear in problems that arise with k independent samples. In this section we start out with a general statistical model and later focus on exponential families. Before constructing Bayes tests for concrete testing problems we recall some notations and useful facts from Example 3.47. Let $(\mathcal{X}, \mathfrak{A}, (P_\theta)_{\theta\in\Delta})$ be a statistical model that satisfies the assumption (A5). We consider the testing problem and the loss function, respectively,

$$\begin{aligned} \mathsf{H}_0 &: \theta \in \Delta_0 \quad \text{versus} \quad \mathsf{H}_A : \theta \in \Delta_A, \\ L(\theta, a) &= a l_0(\theta) I_{\Delta_0}(\theta) + (1-a) l_1(\theta) I_{\Delta_A}(\theta), \end{aligned}$$

where $l_0, l_1 : \Delta \to_m \mathbb{R}_+$. The risk function and the Bayes risk of a test $\varphi : \mathcal{X} \to_m [0,1]$ are

$$\begin{aligned} \mathsf{R}(\theta,\varphi) &= l_0(\theta) I_{\Delta_0}(\theta) \int \varphi(x) f_\theta(x) \boldsymbol{\mu}(dx) \\ &\quad + l_1(\theta) I_{\Delta_A}(\theta) \int [1-\varphi(x)] f_\theta(x) \boldsymbol{\mu}(dx), \quad \theta \in \Delta, \\ \mathsf{r}(\Pi,\varphi) &= \int \mathsf{R}(\theta,\varphi)\Pi(d\theta) = \int \varphi(x) g_0(x) \boldsymbol{\mu}(dx) + \int [1-\varphi(x)] g_1(x) \boldsymbol{\mu}(dx), \end{aligned}$$

where

$$\begin{aligned} g_0(x) &= \int I_{\Delta_0}(\theta) f_\theta(x) l_0(\theta) \Pi(d\theta), \qquad (8.55) \\ g_1(x) &= \int I_{\Delta_A}(\theta) f_\theta(x) l_1(\theta) \Pi(d\theta). \end{aligned}$$

The "sufficient part" of the following proposition has been established already in (3.37) and formulated in terms of the posterior risks.

Proposition 8.46. *Suppose $(\mathcal{X}, \mathfrak{A}, (P_\theta)_{\theta\in\Delta})$ is a statistical model that satisfies the assumption (A5), and Π is a prior with $0 < \Pi(\Delta_0) < 1$. A test φ_Π is a Bayes test for $\mathsf{H}_0 : \theta \in \Delta_0$ versus $\mathsf{H}_A : \theta \in \Delta_A$ if and only if*

$$\varphi_\Pi(x) = \begin{cases} 1 & \text{if } g_1(x) > g_0(x), \\ \gamma(x) & \text{if } g_1(x) = g_0(x), \\ 0 & \text{if } g_1(x) < g_0(x), \end{cases} \tag{8.56}$$

holds $\boldsymbol{\mu}$-a.e., where $\gamma : \mathcal{X} \to_m [0,1]$ may be any measurable function.

Proof. The case $\inf_\varphi \mathsf{r}(\Pi, \varphi) = \infty$ is trivial. Let now φ be a test with $\mathsf{r}(\Pi, \varphi) < \infty$ and assume that φ_Π satisfies (8.56) $\mathsf{P}\Pi$-a.e. Then

$$[\varphi(x) - \varphi_\Pi(x)][g_0(x) - g_1(x)] \geq 0, \quad \boldsymbol{\mu}\text{-a.e.}$$

By integrating both sides with respect to $\boldsymbol{\mu}$, we get $\mathsf{r}(\Pi, \varphi) - \mathsf{r}(\Pi, \varphi_\Pi) \geq 0$. Since $\mathsf{r}(\Pi, \varphi) < \infty$, equality holds if and only if the condition (8.56) holds. ■

Example 8.47. Let $(P_\theta)_{\theta\in\Delta}$, $\Delta = (a, b)$, $a < b$, be a one-parameter exponential family, where $f_\theta(x) = \exp\{\theta T(x) - K(\theta)\}$ is the $\boldsymbol{\mu}$-density of P_θ, $\theta \in \Delta$. Consider the testing problem $\mathsf{H}_0 : \theta \in (a, \theta_0]$ versus $\mathsf{H}_A : \theta \in (\theta_0, b)$, where $\theta_0 \in (a, b)$ is fixed. Let Π be a prior on (a, b) with $0 < \Pi((a, \theta_0]) < 1$. Here we use the piecewise linear loss function $l_0(\theta) = (\theta - \theta_0)I_{(\theta_0,b)}(\theta)$ and $l_1(\theta) = (\theta_0 - \theta)I_{(a,\theta_0]}(\theta)$.

Assume that $\int |\theta| \Pi(d\theta) < \infty$, and thus

$$\int [\int |\theta| \exp\{\theta T(x) - K(\theta)\} \Pi(d\theta)] \boldsymbol{\mu}(dx) < \infty. \tag{8.57}$$

The functions g_0 and g_1 from (8.55) are given by

$$g_0(x) = \int (\theta - \theta_0) I_{(\theta_0,b)}(\theta) \exp\{\theta T(x) - K(\theta)\} \Pi(d\theta), \quad x \in \mathcal{X},$$
$$g_1(x) = \int (\theta_0 - \theta) I_{(a,\theta_0]}(\theta) \exp\{\theta T(x) - K(\theta)\} \Pi(d\theta), \quad x \in \mathcal{X},$$

where the integrals are $\boldsymbol{\mu}$-a.e. finite in view of (8.57). Hence $g_1(x) < g_0(x)$ holds if and only if

$$\int (\theta - \theta_0) \exp\{\theta T(x) - K(\theta)\} \Pi(d\theta) > 0.$$

To analyze this condition in more detail, we set $\nu(d\theta) = \exp\{-K(\theta)\}\Pi(d\theta)$ and consider the function $\phi(t) = \int \exp\{\theta t\} \nu(d\theta), t \in \mathbb{R}$. If ν is not concentrated at one point, which we assume to hold here, then Hölder's inequality yields

$$\phi(\alpha t_1 + (1 - \alpha)t_2) \leq [\phi(t_1)]^\alpha [\phi(t_2)]^{1-\alpha}, \quad t_1, t_2 \in \mathbb{R}, \ \alpha \in (0, 1),$$

where equality holds if and only if $t_1 = t_2$. Hence, $\ln \phi(t)$ is strictly convex, and $\psi(t) = \phi'(t)/\phi(t)$ is a strictly increasing function on (u, v), say, the interior of the interval in which ϕ is finite. If now

$$\lim_{t\downarrow u} \psi(t) = -\infty \quad \text{and} \quad \lim_{t\uparrow v} \psi(t) = \infty,$$

then there exists a uniquely determined c such that $\psi(c) = \theta_0$. In this case c is also the unique solution of the equation

$$\int (\theta - \theta_0) \exp\{\theta t - K(\theta)\} \Pi(d\theta) = 0.$$

This means that $g_1(x) < g_0(x)$ if and only if $T(x) > c$, and $g_1(x) > g_0(x)$ if and only if $T(x) < c$. That $g_1(x) > g_0(x)$ is equivalent to $T(x) < c$ can be shown analogously. To conclude, a test φ_Π is a Bayes test if and only if

$$\varphi_\Pi(x) = \begin{cases} 1 & \text{if } T(x) > c, \\ \gamma(x) & \text{if } T(x) = c, \\ 0 & \text{if } T(x) < c, \end{cases}$$

holds $\boldsymbol{\mu}$-a.e., where $\gamma : \mathcal{X} \to_m [0, 1]$ may be any measurable function.

Problem 8.48. In the setting of Example 8.47, let (X, Θ) have the distribution $\mathsf{P} \otimes \Pi$, where $\mathsf{P}(\cdot|\theta) = P_\theta$, $\theta \in \Delta$. Let $\psi(x) = 1$ if $\mathbb{E}(\Theta|X = x) > \theta_0$, $\psi(x) = 0$ if $\mathbb{E}(\Theta|X = x) < \theta_0$, and $\psi(x) = \gamma(x)$ otherwise, $x \in \mathcal{X}$. Then $\psi \equiv \varphi_\Pi$.

In the remainder of this section we only make use of the zero–one loss (i.e., $l_0 = l_1 = 1$) so that

$$L_{0,1}(\theta, a) = aI_{\Delta_0}(\theta) + (1 - a)I_{\Delta_A}(\theta), \quad \theta \in \Delta, \ a \in \{0, 1\}.$$

Then the functions g_0 and g_1 are closely related to the marginal densities m_0 and m_A that correspond to the priors Π_0 and Π_A, respectively. It follows from (3.36) that

$$\begin{aligned} g_0(x) &= v\mathsf{m}_0(x) = v \int f_\theta(x) \Pi_0(d\theta), \\ g_1(x) &= (1 - v)\mathsf{m}_A(x) = (1 - v) \int f_\theta(x) \Pi_A(d\theta). \end{aligned} \tag{8.58}$$

From Proposition 8.46 it follows that φ_B is a Bayes test for $\mathsf{H}_0 : \theta \in \Delta_0$ versus $\mathsf{H}_A : \theta \in \Delta_A$ under the prior Π with $0 < v = \Pi(\Delta_0) < 1$ if and only if $\boldsymbol{\mu}$-a.e.,

$$\varphi_B(x) = \begin{cases} 1 & \text{if } (1 - v)\mathsf{m}_A(x) > v\mathsf{m}_0(x), \\ \gamma(x) & \text{if } (1 - v)\mathsf{m}_A(x) = v\mathsf{m}_0(x), \\ 0 & \text{if } (1 - v)\mathsf{m}_A(x) < v\mathsf{m}_0(x), \end{cases} \tag{8.59}$$

where $\gamma : \mathcal{X} \to_m [0, 1]$ is arbitrary and m_0 and m_A are defined in (3.36). It should be noted that the test φ_B is a likelihood ratio test for testing $\mathsf{P}\Pi_0$ versus $\mathsf{P}\Pi_A$.

Problem 8.49. Let (X, Θ) have the distribution $\mathsf{P} \otimes \Pi$, where $\mathsf{P}(\cdot|\theta) = P_\theta$, $\theta \in \Delta$. Let ϕ be a test with $\phi(x) = 1$ if $\mathbb{P}(\Theta \in \Delta_1|X = x) > 1/2$ and $\phi(x) = 0$ if $\mathbb{P}(\Theta \in \Delta_1|X = x) < 1/2$, $x \in \mathcal{X}$. Then ϕ is a version, up to γ, of the Bayes test φ_B in (8.59).

Let us now consider testing problems where the null hypothesis is simple:

$$\mathsf{H}_0 : \theta = \theta_0 \quad \text{versus} \quad \mathsf{H}_A : \theta \in \Delta_A = \Delta \backslash \{\theta_0\}, \tag{8.60}$$

i.e., where $\Delta_0 = \{\theta_0\}$ for some fixed $\theta_0 \in \Delta$. Here we choose some $\upsilon \in (0,1)$ and $\Pi = \upsilon \boldsymbol{\delta}_{\theta_0} + (1-\upsilon)\Pi_A$, where Π_A is any distribution on Δ that satisfies

$$\Pi_A(\{\theta_0\}) = 0. \tag{8.61}$$

For a simple null hypothesis we may set $\mathsf{m}_0(x) = f_{\theta_0}(x)$ in (8.59). For any $\alpha \in (0,1)$, by suitable choices of υ and $\gamma(x)$, we can make φ_B be a level α test. Indeed, let F_{θ_0} be the c.d.f. of $\mathsf{m}_A(X)/\mathsf{m}_0(X)$ under H_0, and let $c_{1-\alpha} = F_{\theta_0}^{-1}(1-\alpha)$ be its $1-\alpha$ quantile. Then the test

$$\begin{aligned} \varphi_{B,\alpha} &= I_{(c_{1-\alpha}\mathsf{m}_0(x),\infty)}(\mathsf{m}_A(x)) + \gamma I_{\{c_{1-\alpha}\mathsf{m}_0(x)\}}(\mathsf{m}_A(x)), \quad \text{where} \\ \gamma &= [F_{\theta_0}(c_{1-\alpha}) - (1-\alpha)] \oslash [P_{\theta_0}(\mathsf{m}_A(x) = c_{1-\alpha}\mathsf{m}_0(x))], \end{aligned}$$

satisfies, by construction, $\int \varphi_{B,\alpha} dP_{\theta_0} = \alpha$. Hence if $\upsilon = c_{1-\alpha}/(1+c_{1-\alpha})$, then the test $\varphi_{B,\alpha}$ is a Bayes test, and at the same time a level α test for the simple null hypothesis $\mathsf{H}_0 : \theta = \theta_0$. In the following examples Bayes tests for simple null hypotheses are studied.

Example 8.50. Let $\mathsf{N}(\theta, \Sigma_0)$, $\theta \in \Delta = \mathbb{R}^d$, be the family of all normal distributions where Σ_0 is a known nonsingular covariance matrix. We consider the testing problem $\mathsf{H}_0 : \theta = 0$ versus $\mathsf{H}_A : \theta \neq 0$. Set $\Pi_A = \mathsf{N}(0, \Sigma_1)$, where Σ_1 is nonsingular. Then the condition (8.61) is fulfilled. By construction m_A is the density of $X + \Theta$, where $X \sim \mathsf{N}(\theta, \Sigma_0)$ and $\Theta \sim \mathsf{N}(0, \Sigma_1)$ are independent. This yields

$$\mathsf{m}_A(x) = \varphi_{0,\Sigma_0+\Sigma_1}(x) = (2\pi)^{-d/2}(\det(\Sigma_0 + \Sigma_1))^{-1/2} \exp\{-\frac{1}{2}\langle x, (\Sigma_0 + \Sigma_1)^{-1}x\rangle\}.$$

As $\mathsf{m}_0(x) = \varphi_{0,\Sigma_0}(x)$, the Bayes test φ_B from (8.59) is given by

$$\varphi_B(x) = \begin{cases} 1 & \text{if } \langle x, (\Sigma_0^{-1} - (\Sigma_0 + \Sigma_1)^{-1})x\rangle \geq c, \\ 0 & \text{if } \langle x, (\Sigma_0^{-1} - (\Sigma_0 + \Sigma_1)^{-1})x\rangle < c, \end{cases}$$

where $c = 2\ln(\upsilon/(1-\upsilon)) - \ln(\det(\Sigma_0 + \Sigma_1)) + \ln(\det(\Sigma_0))$. As $(\Sigma_0 + \Sigma_1)^{-1} \preceq \Sigma_0^{-1}$ in the Löwner semiorder, the null hypothesis is rejected for large values of the nonnegative (quadratic form) statistic

$$S(x) = \langle x, (\Sigma_0^{-1} - (\Sigma_0 + \Sigma_1)^{-1})x\rangle.$$

If Σ_0 and Σ_1 are diagonal matrices with diagonal elements $\sigma_{0,i}^2$ and $\sigma_{1,i}^2$, respectively, then $\Sigma_0^{-1} - (\Sigma_0 + \Sigma_1)^{-1}$ is a diagonal matrix, with $\sigma_{1,i}^2/(\sigma_{0,i}^2(\sigma_{0,i}^2 + \sigma_{1,i}^2))$, $i = 1, ..., d$, in the diagonal, so that H_0 is rejected for large values of

$$\sum\nolimits_{i=1}^{d} \frac{\sigma_{1,i}^2}{\sigma_{0,i}^2(\sigma_{0,i}^2 + \sigma_{1,i}^2)} x_i^2.$$

This test may be viewed as a weighted χ^2-test. It reduces to the standard χ^2-test under $\sigma_{0,1}^2 = \cdots = \sigma_{0,d}^2$ and $\sigma_{1,1}^2 = \cdots = \sigma_{1,d}^2$.

A variety of Bayes tests can be obtained for exponential families by using conjugate priors. The use of computers may be necessary for explicit evaluations. If the measure $\boldsymbol{\tau}$ that is associated with the conjugate priors (see Definition 1.34) is atomless (i.e., $\boldsymbol{\tau}(\{\theta\}) = 0$ for every $\theta \in \Delta$), then the conjugate priors have the same property, so that $\Pi_{a,b}(\{\theta\}) = 0$, $\theta \in \Delta$, $(a,b) \in \Upsilon$. For $\mathsf{H}_0 : \theta = \theta_0$, the fixed $\upsilon \in (0,1)$, and some $(a,b) \in \Upsilon$ we introduce the prior by

$$\Pi = \upsilon \boldsymbol{\delta}_{\theta_0} + (1-\upsilon)\Pi_{a,b}. \tag{8.62}$$

We set $S(x) = L(a+1, b+T(x)) - L(a,b) - \langle \theta_0, T(x) \rangle + K(\theta_0)$, where $L(a,b)$ is defined in (1.38).

Proposition 8.51. *Let $(P_\theta)_{\theta\in\Delta}$ be an exponential family as specified by (1.7). If the dominating measure $\boldsymbol{\tau}$ for the conjugate priors is atomless and $(a,b) \in \Upsilon$, then under the zero–one loss every test of the form*

$$\varphi_B(x) = \begin{cases} 1 & \text{if } S(x) > \ln \frac{\upsilon}{1-\upsilon}, \\ \gamma(x) & \text{if } S(x) = \ln \frac{\upsilon}{1-\upsilon}, \\ 0 & \text{if } S(x) < \ln \frac{\upsilon}{1-\upsilon}, \end{cases}$$

where $\gamma : \mathcal{X} \to_m [0,1]$ is arbitrary, is a Bayes test for the testing problem (8.60) under the prior Π from (8.62).

Proof. We apply the test (8.59) with $\mathsf{m}_A(x) = \mathsf{m}_{a,b}(x)$ from (1.40) and with $\mathsf{m}_0(x) = \exp\{\langle \theta_0, T(x)\rangle - K(\theta_0)\}$. ■

Example 8.52. Let $X_1, ..., X_n$ be an i.i.d. sample from an exponential distribution, that is, with the Lebesgue density

$$f_\theta(x) = \theta \exp\{-\theta x\} I_{(0,\infty)}(x), \quad \theta \in \Delta = (0,\infty).$$

We use the gamma distribution $\mathsf{Ga}(\lambda,\beta)$ for Π_A. Then

$$\mathsf{m}_A(x) = I_{(0,\infty)}(x) \int_0^\infty \theta \exp\{-\theta x\} \frac{\beta^\lambda}{\Gamma(\lambda)} \theta^{\lambda-1} \exp\{-\beta\theta\} d\theta = \frac{\lambda \beta^\lambda}{(x+\beta)^{\lambda+1}}.$$

Hence for the prior $\Pi = \upsilon \boldsymbol{\delta}_{\theta_0} + (1-\upsilon)\mathsf{Ga}(\lambda,\beta)$ it holds

$$\frac{\mathsf{m}_A(x)}{\mathsf{m}_0(x)} = \frac{\lambda \beta^\lambda \exp\{\theta_0 x\}}{\theta_0 (x+\beta)^{\lambda+1}} I_{(0,\infty)}(x).$$

From here we get the test (8.59).

Next we deal with testing problems that have composite null hypotheses. We start with a one-sided testing problem in a specific one-parameter exponential family.

Example 8.53. For the family of binomial distributions $\mathsf{B}(n,p)$, $p \in (0,1)$, consider the testing problem $\mathsf{H}_0 : p \leq 1/2$ versus $\mathsf{H}_A : p > 1/2$. Using for Π_0 and Π_A the uniform distributions on $[0,1/2]$ and $(1/2,1]$, respectively, we get

$$\mathsf{m}_0(x) = \binom{n}{x} 2\int_0^{1/2} p^x(1-p)^{n-x}dp \quad \text{and} \quad \mathsf{m}_A(x) = \binom{n}{x} 2\int_{1/2}^1 p^x(1-p)^{n-x}dp\,.$$

Hence,

$$\frac{\mathsf{m}_A(x)}{\mathsf{m}_0(x)} = \frac{\int_0^{1/2} p^{n-x}(1-p)^x dp}{\int_0^{1/2} p^x(1-p)^{n-x}dp}.$$

If we choose $\upsilon = 1/2$ in (8.59), then φ_B is given by

$$\varphi_B(x) = \begin{cases} 1 & \text{if } x > n/2, \\ \gamma & \text{if } x = n/2 \text{ and } n \text{ is even}, \\ 0 & \text{if } x < n/2. \end{cases}$$

Problem 8.54. In the previous example verify the specific form of φ_B for $\upsilon = 1/2$.

Now we consider composite hypotheses in the multivariate setting. Suppose that k independent populations are associated with the distributions $P_{\theta_1}, ..., P_{\theta_k}$ that all belong to an exponential family $(P_\vartheta)_{\vartheta\in\Delta}$ in natural form. We set $\theta = (\theta_1, ..., \theta_k)$. The statistical model is then given by

$$(\mathcal{X}^k, \mathfrak{A}^{\otimes k}, (\bigotimes_{i=1}^k P_{\theta_i})_{\theta\in\Delta^k}).$$

Let the testing problem be $\mathsf{H}_0 : \theta \in \Delta_0$ versus $\mathsf{H}_A : \theta \in \Delta_A$, where

$$\begin{aligned} \Delta_0 &= \{\theta : \theta_1 = \cdots = \theta_k \in \Delta\} \quad \text{and} \\ \Delta_A &= \{\theta : \theta_i \neq \theta_j \text{ for some } i \text{ and } j, \ \ \theta \in \Delta^k\}. \end{aligned}$$

To create $\Pi_0(\cdot) = \Pi(\cdot|\Delta_0)$, let Ξ be a random variable with values in Δ_0 and Π_0 be the distribution of the vector $(\Xi, ..., \Xi)$ which consists of identical components. To define $\Pi_A(\cdot) = \Pi(\cdot|\Delta_A)$ we randomly spread the values of the vector $(\Xi, ..., \Xi)$ in Δ_0 into a vector which takes values in Δ_A. To this end let $\Xi, \Xi_1, ..., \Xi_k$ be independent random variables with values in $\Delta = \mathbb{R}$, say, where the distributions of $\Xi_1, ..., \Xi_k$ are atomless so that $(\Xi_1, ..., \Xi_k)$ belongs to Δ_A with probability one. Then $(\Xi + \Xi_1, ..., \Xi + \Xi_k)$ belongs to Δ_A with probability one. Π_0 is the distribution of $(\Xi, ..., \Xi)$, whereas Π_A is the distribution of $(\Xi + \Xi_1, ..., \Xi + \Xi_k)$.

Example 8.55. Consider the model $(\mathbb{R}^k, \mathfrak{B}_k, (\bigotimes_{i=1}^k \mathsf{N}(\theta_i, \sigma^2))_{\theta\in\mathbb{R}^k})$, where $\sigma^2 > 0$ is known, and where we want to test

$$\mathsf{H}_0 : \theta_1 = \cdots = \theta_k \quad \text{versus} \quad \mathsf{H}_A : \theta_i \neq \theta_j \ \text{ for some } i \text{ and } j.$$

Assume that $\Xi, \Xi_1, ..., \Xi_k$ are independent, where Ξ has the distribution $\mathsf{N}(0, \sigma_0^2)$ and $\Xi_1, ..., \Xi_k$ have the common distribution $\mathsf{N}(0, \sigma_A^2)$. According to (3.36),

$$\begin{aligned} \mathsf{m}_0(x_1, ..., x_k) &= \int \prod\nolimits_{i=1}^k \varphi_{\xi,\sigma^2}(x_i)\varphi_{0,\sigma_0^2}(\xi)d\xi \\ &= (2\pi\sigma^2)^{-k/2}(\tau_0^2/\sigma_0^2)^{1/2}\exp\{-\frac{1}{2\sigma^2}\sum\nolimits_{i=1}^k x_i^2 + \frac{\tau_0^2}{2\sigma^4}(\sum\nolimits_{i=1}^k x_i)^2\}, \\ \mathsf{m}_A(x_1, ..., x_k) &= \int [\prod\nolimits_{i=1}^k \int \varphi_{\xi+\xi_i,\sigma^2}(x_i)\varphi_{0,\sigma_A^2}(\xi_i)d\xi_i]\varphi_{0,\sigma_0^2}(\xi)d\xi \\ &= \int \prod\nolimits_{i=1}^k \varphi_{\xi,\sigma^2+\sigma_A^2}(x_i)\varphi_{0,\sigma_0^2}(\xi)d\xi = (2\pi(\sigma^2+\sigma_A^2))^{-k/2}(\tau_A^2/\sigma_0^2)^{1/2} \\ &\quad \times \exp\{-\frac{1}{2(\sigma^2+\sigma_A^2)}\sum\nolimits_{i=1}^k x_i^2 + \frac{\tau_A^2}{2(\sigma^2+\sigma_A^2)^2}(\sum\nolimits_{i=1}^k x_i)^2\}, \end{aligned}$$

where $\tau_0^2 = \sigma_0^2\sigma^2/(\sigma^2 + k\sigma_0^2)$ and $\tau_A^2 = \sigma_0^2(\sigma^2 + \sigma_A^2)/(\sigma^2 + \sigma_A^2 + k\sigma_0^2)$. With m_0 and m_A as calculated above, the Bayes test is given by (8.59).

The question of which prior should be used is one of the basic questions in Bayesian analysis. For detailed discussions of and solutions to the problem of choosing hierarchical priors we refer to Schervish (1995) and Berger (1985).

8.4 Uniformly Best Invariant Tests

The framework of uniformly best invariant tests has been introduced in Section 5.2, but the topic has been touched on only briefly. In this section we study it in more detail. For a given statistical model $(\mathcal{X}, \mathfrak{A}, (P_\theta)_{\theta\in\Delta})$ we consider the problem of testing $\mathsf{H}_0 : \theta \in \Delta_0$ versus $\mathsf{H}_A : \theta \in \Delta_A$, where Δ_0 and Δ_A is a decomposition of Δ, under the zero–one loss at a fixed level $\alpha \in (0, 1)$. Suppose that the testing problem is invariant, in the sense of Definition 5.25 with (5.21), under a group of measurable transformations $\mathcal{U} = (u_\gamma)_{\gamma\in\mathcal{G}}$. We restrict ourselves to tests φ that are invariant, that is, tests φ that satisfy $\varphi(u_\gamma(x)) = \varphi(x)$, $x \in \mathcal{X}$, $\gamma \in \mathcal{G}$. If there exists a maximal invariant statistic T, then we may write φ as $\varphi = h(T)$, where h may be chosen to be measurable if the conditions of Problem 5.21, or those of Problem 8.56 below, are satisfied. In such a case it suffices to construct an optimal test for the reduced model $(\mathcal{T}, \mathfrak{T}, (P_\theta \circ T^{-1})_{\theta\in\Delta})$. This reduction step is called reduction by invariance. Typically the reduction by invariance simplifies the model and facilitates the construction of uniformly best tests for the reduced model, and consequently uniformly best invariant tests for the original model $(\mathcal{X}, \mathfrak{A}, (P_\theta)_{\theta\in\Delta})$. The following technical result is sometimes useful in the reduction process.

Problem 8.56.* Suppose T is a maximal invariant statistic, $T(\mathcal{X}) = \mathcal{T}$, and there is a mapping $U : \mathcal{T} \to_m \mathcal{X}$ such that $T(U(t)) = t$ for every $t \in \mathcal{T}$. Then for every invariant statistic $S : \mathcal{X} \to_m \mathcal{S}$ it holds $S = h(T)$, where $h = S(U) : \mathcal{T} \to_m \mathcal{S}$.

To illustrate the main ideas of a reduction by invariance we begin with a pivotal model and testing problem, respectively,

$$(\mathbb{R}^n_{\neq 0}, \mathfrak{B}_{n,\neq 0}, (\mathsf{N}(\mu, \mathbf{I}))_{\mu\in\mathbb{R}^n}),$$
$$\mathsf{H}_0 : \mu = 0 \quad \text{versus} \quad \mathsf{H}_A : \mu \neq 0,$$

that has been considered already in Chapter 5. This decision problem is invariant under the group of rotations $\mathcal{U}_{rot}$ for which $\chi^2_n : \mathcal{X} = \mathbb{R}^n_{\neq 0} \to \mathcal{T} = (0, \infty)$, that is, $\chi^2_n(x) = \|x\|^2$, is maximal invariant. According to Problem 8.56 every rotation invariant test is a measurable function of χ^2_n. Hence we may reduce our model by invariance and get the model

$$(\mathcal{T}, \mathfrak{T}, (P_\theta \circ T^{-1})_{\theta\in\Delta}) = ((0, \infty), \mathfrak{B}_{(0,\infty)}, (\mathsf{H}(n, \delta^2))_{\delta^2\geq 0}). \tag{8.63}$$

In Theorem 5.33 it has been shown that the test $\varphi_{\chi_n^2} = I_{[\chi^2_{1-\alpha,n},\infty)}(\chi_n^2)$ is a uniformly best level α test in the class of all tests that are invariant under $\mathcal{U}_{rot}$. The main ingredient of the proof was the fact that the new family $\mathsf{N}(\mu, \mathbf{I}) \circ (\chi_n^2)^{-1} = \mathsf{H}(n, \delta^2)$ of noncentral χ^2-distributions with n degrees of freedom and noncentrality parameter $\delta^2 = \|\mu\|^2$ has MLR in the identity; see Theorem 2.27. Hence for $\mathsf{H}_0 : \delta^2 = 0$ versus $\mathsf{H}_A : \delta^2 > 0$ the test $I_{[\chi^2_{1-\alpha,n},\infty)}$ is a uniformly best level α test for the model (8.63), and $\varphi_{\chi_n^2}$ is a uniformly best rotation invariant level α test. A similar reduction technique can also be applied in other models.

Problem 8.57.* In Example 8.31, the test $\varphi_{I,U}$ in (8.39) is a uniformly best scale-invariant level α test for $\mathsf{H}_0 : \mu \leq 0,\ \sigma^2 > 0$, versus $\mathsf{H}_A : \mu > 0,\ \sigma^2 > 0$, and $\varphi_{II,U}$ in (8.40) is a uniformly best scale-invariant level α test for $\mathsf{H}_0 : \mu = 0,\ \sigma^2 > 0$, versus $\mathsf{H}_A : \mu \neq 0,\ \sigma^2 > 0$.

To construct the analysis of variance (ANOVA) model we choose as the parameter space a k-dimensional linear subspace $\mathbb{L}_k$, say, of $\mathbb{R}^n$. Let $\mathbb{A}_h$ be a linear subspace of $\mathbb{L}_k$ that represents the null hypothesis. Let $\mathbb{B}_{k-h}$ be the $(k-h)$-dimensional linear subspace of $\mathbb{L}_k$ that is orthogonal to $\mathbb{A}_h$. Likewise, let $\mathbb{C}_{n-k}$ be the $(n-k)$-dimensional linear subspace of $\mathbb{R}^n$ that is orthogonal to $\mathbb{L}_k$. This may be expressed by writing $\mathbb{A}_h \oplus \mathbb{B}_{k-h} = \mathbb{L}_k$ and $\mathbb{L}_k \oplus \mathbb{C}_{n-k} = \mathbb{R}^n$. Let $\Pi_{\mathbb{A}_h} x$, $\Pi_{\mathbb{B}_{k-h}} x$, $\Pi_{\mathbb{L}_k} x$, and $\Pi_{\mathbb{C}_{n-k}} x$ denote the (orthogonal) projections of $x \in \mathbb{R}^n$ on $\mathbb{A}_h$, $\mathbb{B}_{k-h}$, $\mathbb{L}_k$, and $\mathbb{C}_{n-k}$, respectively. Thus the model and the hypotheses are related in the following way.

Sample space	Model space	H_0	H_A
$\mathbb{R}^n = \mathbb{L}_k \oplus \mathbb{C}_{n-k}$	$\mathbb{L}_k = \mathbb{A}_h \oplus \mathbb{B}_{k-h}$	$\mathbb{A}_h \subset \mathbb{L}_k$	$\mathbb{L}_k \backslash \mathbb{A}_h$

Let $\mathcal{O}_{n\times n}$ be the group of orthogonal $n \times n$ matrices that leave the spaces $\mathbb{A}_h$, $\mathbb{B}_{k-h}$, and $\mathbb{C}_{n-k}$ invariant. We consider the linear subspace $\mathbb{A}_h$ as an additive group of measurable transformations of $(\mathbb{R}^n, \mathfrak{B}_n)$ and introduce the rule of combination on $\mathcal{G}_{AN} := (0,\infty) \times \mathcal{O}_{n\times n} \times \mathbb{A}_h$ by

$$\gamma_1 \odot \gamma_2 = (\alpha_1, A_1, b_1) \odot (\alpha_2, A_2, b_2) = (\alpha_1\alpha_2, A_1A_2, \alpha_1 A_1 b_2 + b_1)\,.$$

Then $\mathcal{U}_{AN} = \{u_\gamma : \gamma \in \mathcal{G}_{AN}\}$ with $u_\gamma(x) = \alpha O x + b$ is a group of measurable transformations of $(\mathbb{R}^n, \mathfrak{B}_n)$. This group leaves the set

$$\mathcal{X} = \{x : \Pi_{\mathbb{L}_k} x - \Pi_{\mathbb{A}_h} x \neq 0,\ x - \Pi_{\mathbb{L}_k} x \neq 0\}$$

invariant. Putting $\mathfrak{A} = \mathfrak{B}_h \otimes \mathfrak{B}_{k-h,\neq 0} \otimes \mathfrak{B}_{n-k,\neq 0}$, we introduce the ANOVA model and testing problem, respectively, by

$$\mathcal{M}_{AN} = (\mathcal{X}, \mathfrak{A}, (\mathsf{N}(\mu, \sigma^2\mathbf{I}))_{\mu\in\mathbb{L}_k, \sigma^2>0}), \tag{8.64}$$
$$\mathsf{H}_0 : \mu \in \mathbb{A}_h \quad \text{versus} \quad \mathsf{H}_A : \mu \in \mathbb{L}_k \backslash \mathbb{A}_h.$$

By construction the model (8.64) is $\mathcal{U}_{AN}$-invariant, and for $\mathcal{V} = \mathcal{U}_{AN}$ the hypothesis testing problem is invariant. Consider the statistic

$$F(x) = \frac{\frac{1}{k-h} \left\| \Pi_{\mathbb{L}_k} x - \Pi_{\mathbb{A}_h} x \right\|^2}{\frac{1}{n-k} \left\| x - \Pi_{\mathbb{L}_k} x \right\|^2}, \quad x \in \mathcal{X}. \tag{8.65}$$

It holds $\Pi_{\mathbb{L}_k} x - \Pi_{\mathbb{A}_h} x = \Pi_{\mathbb{B}_{k-h}} x$ and $x - \Pi_{\mathbb{L}_k} x = \Pi_{\mathbb{C}_{n-k}} x$, $x \in \mathcal{X}$. By construction $F : \mathcal{X} \to_m (0, \infty)$ is an invariant mapping. Moreover, F is maximal invariant. Indeed, if $F(x) = F(y)$, then there is some $\alpha > 0$ such that

$$\left\| \Pi_{\mathbb{B}_{k-h}} \alpha x \right\|^2 = \left\| \Pi_{\mathbb{B}_{k-h}} y \right\|^2 \quad \text{and} \quad \| \Pi_{\mathbb{C}_{n-k}} \alpha x \|^2 = \| \Pi_{\mathbb{C}_{n-k}} y \|^2 .$$

As the rotations from $\mathcal{O}_{n \times n}$ leave the spaces $\mathbb{A}_h$, $\mathbb{B}_{k-h}$, and $\mathbb{C}_{n-k}$ invariant we find a rotation $O \in \mathcal{O}_{n \times n}$ with $\alpha O \Pi_{\mathbb{B}_{k-h}} x = \Pi_{\mathbb{B}_{k-h}} y$ and $\alpha O \Pi_{\mathbb{C}_{n-k}} x = \Pi_{\mathbb{C}_{n-k}} y$. Hence,

$$\begin{aligned} y &= \Pi_{\mathbb{A}_h} y + \Pi_{\mathbb{B}_{k-h}} y + \Pi_{\mathbb{C}_{n-k}} y = \Pi_{\mathbb{A}_h} y + \alpha O \Pi_{\mathbb{B}_{k-h}} x + \alpha O \Pi_{\mathbb{C}_{n-k}} x \\ &= \alpha O \Pi_{\mathbb{A}_h} x + \alpha O \Pi_{\mathbb{B}_{k-h}} x + \alpha O \Pi_{\mathbb{C}_{n-k}} x + \Pi_{\mathbb{A}_h} y - \alpha O \Pi_{\mathbb{A}_h} x = \alpha O x + b, \end{aligned}$$

where $b = \Pi_{\mathbb{A}_h} y - \alpha O \Pi_{\mathbb{A}_h} x \in \mathbb{A}_h$.

Finally, we construct the mapping $U : (0, \infty) \to_m \mathcal{X}$ in Problem 8.56 by setting $U(r) = r e_0$, where $e_0 \in \mathbb{R}^h \times \mathbb{R}^{k-h}_{\neq 0} \times \mathbb{R}^{n-k}_{\neq 0}$ is any fixed unit vector. Then we see from Problem 8.56 that every $\mathcal{U}_{AN}$-invariant statistic, and especially every invariant test, is a measurable function of F.

As we have seen above for the model (8.64), under the transformation group $\mathcal{G}_{AN}$ the statistic F in (8.65) is maximal invariant, and every invariant test is a measurable function of F. To get the induced distribution we need the following fact. If the random vector X has the distribution $\mathsf{N}(\mu, \sigma^2 \mathbf{I})$, then

$$((X - \Pi_{\mathbb{L}_k} X)^T, (\Pi_{\mathbb{L}_k} X - \Pi_{\mathbb{A}_h} X)^T),$$

being a linear image of X, is again normally distributed. As $(\Pi_{\mathbb{L}_k} X - \Pi_{\mathbb{A}_h} X)^T$ is a vector that belongs to $\mathbb{L}_k$ it is orthogonal to $(X - \Pi_{\mathbb{L}_k} X)^T$ which belongs to $\mathbb{L}_k^{\perp}$. Hence the covariance matrix between these two vectors is zero, and due to the joint normal distribution they are independent. Thus we turn to the reduced model

$$(\mathcal{T}, \mathfrak{T}, (P_\theta \circ T^{-1})_{\theta \in \Delta}) = ((0, \infty), \mathfrak{B}_{(0,\infty)}, (\mathsf{F}(k-h, n-h, \delta^2))_{\delta^2 \geq 0}), \tag{8.66}$$

where we have used the fact that $\mathsf{N}(\mu, \sigma^2 \mathbf{I}) \circ F^{-1} = \mathsf{F}(k-h, n-h, \delta^2)$, the F-distribution with $k-h$ and $n-k$ degrees of freedom and noncentrality parameter $\delta^2 = \left\| \Pi_{\mathbb{L}_k} \mu - \Pi_{\mathbb{A}_h} \mu \right\| / \sigma^2$; see also Example 2.41. In the reduced model the original testing problem $\mathsf{H}_0 : \mu \in \mathbb{L}_k$ versus $\mathsf{H}_A : \mu \in \mathbb{L}_k \backslash \mathbb{A}_h$ has now the form $\mathsf{H}_0 : \delta^2 = 0$ versus $\mathsf{H}_A : \delta^2 > 0$, where $\delta^2 = \left\| \Pi_{\mathbb{L}_k} \mu - \Pi_{\mathbb{A}_h} \mu \right\|^2 / \sigma^2$.

We recall from Theorem 2.27 that $\mathsf{F}(k-h, n-h, \delta^2)$ has MLR in the identity. Thus by Theorem 2.49 we get a uniformly best level α test for the reduced model (8.66), and it rejects H_0 for large values of the identity. Let $F_{1-\alpha,k-h,n-k}$ be the $1-\alpha$ quantile of the F-distribution $\mathsf{F}(k-h,n-k)$. As the set $\mathbb{R}^n\backslash\mathcal{X} = \mathbb{R}^n\backslash(\mathbb{R}^h \times \mathbb{R}^{k-h}_{\neq 0} \times \mathbb{R}^{n-k}_{\neq 0})$ has probability zero under every $\mathsf{N}(\mu,\mathbf{I})$, $\mu \in \mathbb{R}^n$, we may switch back to the original sample space $\mathbb{R}^n$ and obtain the following result.

Theorem 8.58. *Under the model* $(\mathbb{R}^n, \mathfrak{B}_n, (\mathsf{N}(\mu,\sigma^2\mathbf{I}))_{\mu\in\mathbb{R}^n,\sigma^2>0})$, *for testing* $\mathsf{H}_0 : \mu \in \mathbb{A}_h$, $\sigma^2 > 0$, *versus* $\mathsf{H}_A : \mu \in \mathbb{L}_k\backslash\mathbb{A}_h$, $\sigma^2 > 0$, *the F-test defined by*

$$\varphi_F(x) = \begin{cases} 1 \text{ if } & \dfrac{\frac{1}{k-h}\left\|\Pi_{\mathbb{L}_k}x - \Pi_{\mathbb{A}_h}x\right\|^2}{\frac{1}{n-k}\left\|x - \Pi_{\mathbb{L}_k}x\right\|^2} \geq F_{1-\alpha,k-h,n-k} \\ 0 \text{ otherwise,} & \end{cases}$$

is a uniformly best level α test in the class of the $\mathcal{U}_{AN}$-invariant tests.

Example 8.59. Let us consider the one-way layout ANOVA model. It is given by $X_{i,j} = \mu_i + \varepsilon_{i,j}$, $\mu_i \in \mathbb{R}$, where $\varepsilon_{i,j}$, $j = 1,...,n_i$, $i = 1,...,k$, are i.i.d. from $\mathsf{N}(0,\sigma^2)$. Suppose we want to test $\mathsf{H}_0 : \mu_1 = \cdots = \mu_k$, $\sigma^2 > 0$, versus $\mathsf{H}_A : \mu_i \neq \mu_j$ for some $i \neq j$, $\sigma^2 > 0$. Set $n = \sum_{i=1}^k n_i$. The linear subspace $\mathbb{L}_k$ is the linear hull of the orthogonal vectors e_i, $i = 1,...,k$, where the components of e_i at the places $n_1 + \cdots + n_{i-1} + 1, ..., n_1 + \cdots + n_i$ are equal to 1 and 0 elsewhere, $i = 1,...,k$. For example, $e_1 = (1,...,1,0,...,0)^T$ where 1 occurs n_1 times. Furthermore, $h = 1$ and $\mathbb{A}_1$ is the one-dimensional subspace generated by $\mathbf{1} = (1,...,1)^T$. We introduce the standard ANOVA notation

$$\begin{aligned} \overline{x}_{i,\cdot} &= \frac{1}{n_i}\sum\nolimits_{j=1}^{n_i} x_{i,j}, \quad i = 1,...,k \\ \overline{x}_{\cdot,\cdot} &= \sum\nolimits_{i=1}^{k} \frac{n_i}{n}\overline{x}_{i,\cdot} = \frac{1}{n}\sum\nolimits_{i=1}^{k}\sum\nolimits_{j=1}^{n_i} x_{i,j}. \end{aligned}$$

The projections of the vector $x = (x_{1,1},...,x_{1,n_1}, x_{2,1},...,x_{2,n_2},...,x_{k,1},...,x_{k,n_k})$ on $\mathbb{L}_k$ and $\mathbb{A}_1$ are, respectively, given by

$$\begin{aligned} \Pi_{\mathbb{L}_k}x &= \sum\nolimits_{i=1}^{k}\langle x, \frac{1}{\|e_i\|}e_i\rangle\frac{1}{\|e_i\|}e_i = \sum\nolimits_{i=1}^{k}\frac{1}{n_i}\sum\nolimits_{j=1}^{n_i} x_{i,j}e_i = \sum\nolimits_{i=1}^{k}\overline{x}_{i,\cdot}e_i, \\ \Pi_{\mathbb{A}_1}x &= \langle x, \frac{1}{\|\mathbf{1}\|}\mathbf{1}\rangle\frac{1}{\|\mathbf{1}\|}\mathbf{1} = \overline{x}_{\cdot,\cdot}\mathbf{1}. \end{aligned}$$

Hence, by $\mathbf{1} = \sum_{i=1}^k e_i$ and the orthogonality of the e_i,

$$\begin{aligned} F &= \frac{\frac{1}{k-1}\|\Pi_{\mathbb{L}_k}x - \Pi_{\mathbb{A}_1}x\|^2}{\frac{1}{n-k}\|x - \Pi_{\mathbb{L}_k}x\|^2} = \frac{\frac{1}{k-1}\|\sum_{i=1}^k \overline{x}_{i\cdot}e_i - \overline{x}_{\cdot,\cdot}\mathbf{1}\|^2}{\frac{1}{n-k}\|x - \sum_{i=1}^k \overline{x}_{i,\cdot}e_i\|^2} \\ &= \frac{\frac{1}{k-1}\sum_{i=1}^k n_i(\overline{x}_{i,\cdot} - \overline{x}_{\cdot,\cdot})^2}{\frac{1}{n-k}\sum_{i=1}^k\sum_{j=1}^{n_i}(x_{i,j} - \overline{x}_{i,\cdot})^2}. \end{aligned}$$

This means that the test φ_F rejects H_0 if $F \geq F_{1-\alpha,k-1,n-k}$.

That the distribution of the maximal invariant statistic could be evaluated explicitly, and that the family of distributions in the reduced model had MLR, have been the crucial points in the above construction of best invariant tests. In the situations considered above the conditions were favorable as to guarantee that the maximal invariant statistic T generates the σ-algebra $\mathfrak{I}$ of invariant sets, and that every invariant statistic S with values in $\mathbb{R}^q$, or more generally in a Borel space, is a measurable function of T. As long as this is guaranteed we may, equivalently, deal with either the model $(\mathcal{T}, \mathfrak{T}, (P_\theta \circ T^{-1})_{\theta\in\Delta})$ or with $(\mathcal{X}, \mathfrak{I}, (P_\theta^{\mathfrak{I}})_{\theta\in\Delta})$, where $P_\theta^{\mathfrak{I}}$ is the restriction of P_θ to the sub-σ-algebra $\mathfrak{I}$. The following simple fact proves useful when evaluating the likelihood ratio in the family $(P_\theta^{\mathfrak{I}})_{\theta\in\Delta}$.

Problem 8.60.* If $T : \mathcal{X} \to_m \mathcal{T}$ is maximal invariant and generates $\mathfrak{I}$, and $M_{\theta_0,\theta}$ is a version of the likelihood ratio of $P_\theta \circ T^{-1}$ with respect to $P_{\theta_0} \circ T^{-1}$, then $L_{\theta_0,\theta}^{\mathfrak{I}} := M_{\theta_0,\theta}(T)$ is a version of the likelihood ratio of $P_\theta^{\mathfrak{I}}$ with respect to $P_{\theta_0}^{\mathfrak{I}}$.

Next we study some other models where the distribution of the likelihood ratio $L_{\theta_0,\theta}^{\mathfrak{I}}$ can be evaluated explicitly. We start with the binary model $(\mathbb{R}^n, \mathfrak{B}_n, \{P_0, P_1\})$, use the group $\mathcal{U}_l$ in (5.4) that generates location models, and denote by $\mathfrak{I}_l$ the σ-algebra of measurable and $\mathcal{U}_l$-invariant sets.

Problem 8.61.* Suppose the distributions P_0 and P_1 have the Lebesgue densities f_0 and f_1, respectively, and set

$$\begin{aligned}\overline{f}_i(x_1, ..., x_n) &= \int_{-\infty}^{+\infty} f_i(x_1 + s, ..., x_n + s)ds, \quad i = 0, 1,\\ L_{0,1}^{\mathfrak{I}_l} &= (\overline{f}_1/\overline{f}_0)I_{\{\overline{f}_0>0\}} + \infty I_{\{\overline{f}_0=0,\overline{f}_1>0\}}.\end{aligned}$$

Then $L_{0,1}^{\mathfrak{I}_l}$ is a version of the likelihood ratio of $P_1^{\mathfrak{I}_l}$ with respect to $P_0^{\mathfrak{I}_l}$.

We consider again the model $(\mathbb{R}^n, \mathfrak{B}_n, \{P_0, P_1\})$, but now we use the group $\mathcal{U}_{ls}$ in (5.5) that generates location-scale models. Let $\mathfrak{I}_{ls}$ be the σ-algebra of measurable and $\mathcal{U}_{ls}$-invariant sets.

Problem 8.62.* Suppose that the distributions P_0 and P_1 have Lebesgue densities f_0 and f_1, respectively, and set

$$\begin{aligned}\overline{f}_i(x_1, ..., x_n) &= \int_0^\infty w^{n-2}[\int_{-\infty}^\infty f_i(wx_1 + v, ..., wx_n + v)dv]dw, \quad i = 0, 1,\\ L_{0,1}^{\mathfrak{I}_{ls}} &= (\overline{f}_1/\overline{f}_0)I_{\{\overline{f}_0>0\}} + \infty I_{\{\overline{f}_0=0,\overline{f}_1>0\}}. \qquad (8.67)\end{aligned}$$

Then $L_{0,1}^{\mathfrak{I}_{ls}}$ is a version of the likelihood ratio of $P_1^{\mathfrak{I}_{ls}}$ with respect to $P_0^{\mathfrak{I}_{ls}}$.

Set $\mathcal{P}_i = (P_i \circ u_{\alpha,\beta}^{-1})_{\alpha\in\mathbb{R},\beta>0}$, $i = 0, 1$, where $u_{\alpha,\beta}(x_1, ..., x_n) = (\beta x_1 + \alpha, ..., \beta x_n + \alpha)$. In the next theorem we study for the statistical model $(\mathbb{R}^n, \mathfrak{B}_n, \mathcal{P}_0 \cup \mathcal{P}_1)$ the testing problem $\mathsf{H}_0 : P \in \mathcal{P}_0$ versus $\mathsf{H}_A : P \in \mathcal{P}_1$. It should be noted that the distributions induced by the maximal invariant statistic T_{ls} from (5.13) do not dependent on α and β. Indeed, by utilizing the invariance, $T_{ls}(u_{\alpha,\beta}(x)) = T_{ls}(u_{0,1}(x)) = T_{ls}(x)$ implies

$$(P_i \circ u_{\alpha,\beta}^{-1})(T_{ls} \in B) = P_i(T_{ls}(u_{\alpha,\beta}) \in B) = P_i \circ T_{ls}^{-1}(B), \quad i = 0, 1. \qquad (8.68)$$

Theorem 8.63. *If the distributions P_0 and P_1 have Lebesgue densities, then for $L_{0,1}^{\mathfrak{I}_{ls}}$ in (8.67) the test*

$$\varphi_{\mathfrak{I}_{ls}} = I_{(c_{1-\alpha},\infty)}(L_{0,1}^{\mathfrak{I}_{ls}}) + \gamma I_{\{c_{1-\alpha}\}}(L_{0,1}^{\mathfrak{I}_{ls}}),$$

with $\gamma \in [0,1]$ and $\mathsf{E}_0\varphi_{\mathfrak{I}_{ls}} = \alpha$, is a uniformly best $\mathcal{U}_{ls}$-invariant level α test for $\mathsf{H}_0 : P \in \mathcal{P}_0$ versus $\mathsf{H}_A : P \in \mathcal{P}_1$.

Proof. As the P_i have Lebesgue densities it holds $(P_i \circ u_{\alpha,\beta}^{-1})(\mathbb{R}_{\neq}^n) = 1$, $i = 0,1$, so that we may switch to the reduced sample space $\mathbb{R}_{\neq}^n$ and consider the distributions $Q_i = (P_i \circ u_{\alpha_i,\beta_i}^{-1}) \circ T_{ls}^{-1}$, $i = 0,1$. The latter are independent of α_i, β_i in view of (8.68). If $M_{0,1}$ is a version of the likelihood ratio of Q_1 with respect to Q_0, then in view of Problem 8.60 $M_{0,1}(T_{ls})$ is a version of the likelihood ratio of $P_1^{\mathfrak{I}_{ls}} \circ u_{\alpha_1,\beta_1}^{-1}$ with respect to $P_0^{\mathfrak{I}_{ls}} \circ u_{\alpha_0,\beta_0}^{-1}$, which is independent of the α_i, β_i and is, $P_i^{\mathfrak{I}_{ls}}$-a.s., identical with $L_{0,1}^{\mathfrak{I}_{ls}}$, $i = 0,1$. To complete the proof we have only to apply Neyman–Pearson's lemma (see Theorem 2.45) and to use the representation of $L_{0,1}^{\mathfrak{I}_{ls}}$ in Problem 8.62. ■

Problem 8.64.* If $f_0(x) = \prod_{i=1}^n \varphi_{0,1}(x_i)$ is the density of n independent standard normal random variables, then

$$\overline{f}_0(x_1,...,x_n) = c_n[s_n(x_1,...,x_n)]^{-n+1}, \quad \text{where}$$
$$c_n = \frac{1}{2}n^{-1/2}[(n-1)\pi]^{-(n-1)/2}\Gamma((n-1)/2), \quad \text{and}$$
$$s_n(x_1,...,x_n) = (\frac{1}{n-1}\sum\nolimits_{i=1}^n (x_i - \overline{x}_n)^2)^{1/2}.$$

If $f_1(x) = \prod_{i=1}^n \exp\{-x_i\}I_{(0,\infty)}(x_i)$ is the density of n independent random variables from an exponential distribution $\mathsf{Ex}(1)$, then with $x_{[1]} = \min\{x_1,...,x_n\}$,

$$\overline{f}_1(x_1,...,x_n) = \frac{1}{n}\Gamma(n-1)[n(\overline{x}_n - x_{[1]})]^{-n+1}.$$

Next we take $\mathsf{Ex}(\alpha,\beta)$, the shifted exponential distribution with Lebesgue density $\mathsf{ex}_{\alpha,\beta}(t) = \beta\exp\{-\beta(t-\alpha)\}I_{[\alpha,\infty)}(t)$, as the alternative hypothesis to normality.

Example 8.65. To find the uniformly best $\mathcal{U}_{ls}$-invariant level α test for testing $\mathsf{H}_0 : \mathsf{N}^{\otimes n}(\mu,\sigma^2)$, $\mu \in \mathbb{R}$, $\sigma^2 > 0$, versus $\mathsf{H}_A : \mathsf{Ex}^{\otimes n}(\alpha,\beta)$, $\alpha \in \mathbb{R}$, $\beta > 0$, we consider the ratio of the densities in the previous problem. With some constant $d(n)$ that depends only on n we get

$$L_{0,1}^{\mathfrak{I}_{ls}}(x_1,...,x_n) = d(n)\left(\frac{s_n(x_1,...,x_n)}{\overline{x}_n - x_{[1]}}\right)^{n-1}.$$

Note that $T_n(x_1,...,x_n) = s_n(x_1,...,x_n)/(\overline{x}_n - x_{[1]})$ has a distribution with a continuous c.d.f. According to Neyman–Pearson's lemma the uniformly best level α test rejects H_0 for large values of $\overline{f}_1/\overline{f}_0$, which is an increasing function of T_n. Thus the uniformly best level α test is given by $\varphi = I_{(c_{1-\alpha},\infty)}(T_n)$, where $c_{1-\alpha}$ is the $1-\alpha$ quantile of the distribution of T_n under $\mathsf{N}^{\otimes n}(0,1)$.

Problem 8.66.* If $f_0(x) = \prod_{i=1}^{2} \prod_{j=1}^{n_i} \varphi_{0,1}(x_{i,j})$, and

$$f_1(x) = \prod_{a=1}^{n_1} \varphi_{\mu,1}(x_{1,a}) \prod_{b=1}^{n_2} \varphi_{0,1}(x_{2,b}),$$

for $x = (x_{1,1}, ..., x_{2,n_2}) \in \mathbb{R}^n$ and $n = n_1 + n_2$, then

$$L_{0,1}^{\mathfrak{I}_{ls}}(x) = \exp\{-\frac{n_1 n_2 \mu^2}{2n}\} 2^{-(n-3)/2} (\Gamma(\frac{n-1}{2}))^{-1} \times \int_0^\infty w^{n-2} \exp\{-\frac{w^2}{2} + w\mu V(x)\} dw, \tag{8.69}$$

where

$$V(x) = \frac{n_1 n_2}{n} (\overline{x}_{1,\cdot} - \overline{x}_{2,\cdot}) (\sum_{i=1}^{2} \sum_{j=1}^{n_i} (x_{i,j} - \overline{x}_{\cdot,\cdot})^2)^{-1/2}, \tag{8.70}$$

and where $\overline{x}_{1,\cdot}$, $\overline{x}_{2,\cdot}$, and $\overline{x}_{\cdot,\cdot}$ are as in Example 8.59 in the special case of $k = 2$.

It is clear that in a similar manner best invariant tests can be established as long as only location or scale models are considered. The above examples are special cases in the theory of separate families of distributions which was initiated by Cox (1961, 1962). A review is provided in Pereira (1977).

Example 8.67. Consider the following two-sample testing problem for normal populations.

$$\begin{aligned} &\mathsf{H}_0 : \mathsf{N}^{\otimes n_1}(\mu, \sigma^2) \otimes \mathsf{N}^{\otimes n_2}(\mu, \sigma^2), \quad \mu \in \mathbb{R},\ \sigma^2 > 0, \quad \text{versus} \\ &\mathsf{H}_A : \mathsf{N}^{\otimes n_1}(\mu_1, \sigma^2) \otimes \mathsf{N}^{\otimes n_2}(\mu_2, \sigma^2), \quad \mu_1 > \mu_2,\ \sigma^2 > 0. \end{aligned}$$

As we restrict ourselves to $\mathcal{U}_{ls}$-invariant tests we may assume without loss of generality that $\mu_1 = \mu$, $\mu_2 = 0$, and $\sigma^2 = 1$. Thus we consider the testing problem H_0 : $\mu = 0$ versus $\mathsf{H}_A : \mu > 0$. As $\mu > 0$ in the alternative, the likelihood ratio $\overline{f}_1/\overline{f}_0$ is an increasing function of V from (8.70). Thus the uniformly best invariant level α test rejects H_0 for large values of V. This test is equivalent to the one-sided version of the two-sample t-test which rejects H_0 for large values of

$$T_{n-2}(x) = (n_1 n_2 (n-2)/n)^{1/2} (\overline{x}_{1\cdot} - \overline{x}_{2\cdot}) (\sum_{i=1}^{2} \sum_{j=1}^{n_i} (x_{i,j} - \overline{x}_{i\cdot})^2)^{-1/2}.$$

As the distribution of T_{n-2} at $\mu_1 = \mu_2$ is $\mathsf{T}(n-2)$, H_0 is rejected if $T_{n-2} \geq t_{1-\alpha,n-2}$. This test is also a uniformly best unbiased level α test, which can be shown within the framework of Section 8.1.3 with a three-parameter exponential family. A proof of this fact is given in Section 5.3 of Lehmann (1986).

The two-sided version of the two-sample t-test for H_0 versus $\mathsf{H}_A : \mu_1 \neq \mu_2$, $\sigma^2 > 0$, can be obtained as a uniformly best invariant level α test if we extend the group $\mathcal{G}_{ls}$ by including reflections $(x_1, ..., x_n) \to (-x_1, ..., -x_n)$. For details and extensions to the more general settings of linear models we refer to Giri (1996) and Witting (1985).

Problem 8.68. In the settings of the previous example, show that $V(x)$ from (8.70) is an increasing function of $T_{n-2}(x)$. Then establish the one-sided version of the two-sample t-test at a given level $\alpha \in (0, 1)$.

Problem 8.69. In the settings of the previous example, let H_A be replaced by $\overline{\mathsf{H}}_A : \mu_1 \neq \mu_2,\ \sigma^2 > 0$. Then the two-sided two-sample t-test, which rejects H_0 for large values of $|V|$, is equivalent to the test φ_F in Example 8.59 in the special case of $k = 2$.

Remark 8.70. The crucial point in the above examples has been the calculation of the likelihood ratio. A general approach to this for a locally compact group $\mathcal{G}$ with the left invariant Haar measure λ is due to Stein (1956), who established

$$\frac{\overline{f}_1(x)}{\overline{f}_0(x)} = \frac{\int f_1(u_\gamma(x))\lambda(d\gamma)}{\int f_0(u_\gamma(x))\lambda(d\gamma)} \tag{8.71}$$

without stating conditions under which this formula holds. Later on this problem was studied in Schwartz (1967), Anderson (1982), and Wijsman (1969, 1985). For details and further references we refer to Giri (1996) and Wijsman (1990). If $\mathcal{G}$ is the group of all translations, which generates the location model, then the Haar measure is the Lebesgue measure, and $\overline{f}_1/\overline{f}_0$ in (8.71) is $L_{0,1}^{\mathfrak{J}}$ in Problem 8.61. Similarly, for the group in the location-scale model the Haar measure is $w^{-1}dvdw$ and then $\overline{f}_1/\overline{f}_0$ in (8.71) is $L_{0,1}^{\mathfrak{J}}$ in Problem 8.62.

8.5 Exponential Rates of Error Probabilities

Our starting point is a sequence of models $\mathcal{M}_n = (\mathcal{X}_n, \mathfrak{A}_n, (P_{n,\theta})_{\theta\in\Delta})$ with a common parameter set Δ. Suppose we want to test $\mathsf{H}_0 : \theta \in \Delta_0$ versus $\mathsf{H}_A : \theta \in \Delta_A$, where Δ_0 and Δ_A is a decomposition of Δ. By an *asymptotic test* $\{\varphi_n\}$ we mean a sequence of tests $\varphi_n : \mathcal{X}_n \to_m [0,1]$, $n = 1, 2, ...$ For simplicity we often just write φ_n instead of $\{\varphi_n\}$. The first question is whether there are asymptotic tests that separate the null hypothesis and the alternative completely. Such sequences of tests are called completely consistent. Consequently, an asymptotic test φ_n is *completely consistent* if

$$\lim_{n\to\infty} \mathsf{E}_{n,\theta_1}(1-\varphi_n) = 0, \quad \theta_1 \in \Delta_A \quad \text{and} \quad \lim_{n\to\infty} \mathsf{E}_{n,\theta_0}\varphi_n = 0, \quad \theta_0 \in \Delta_0.$$

For i.i.d. observations we have already established conditions in Proposition 7.122 under which both, the error probabilities of the first and of the second kind, tend to zero at an exponential rate. The aim of this section is to establish the exact exponential rate for a simple null hypothesis and a simple alternative.

Recall that $\mathsf{b}_\pi(P, Q)$ is the minimal Bayes risk for testing $\mathsf{H}_0 : P$ versus $\mathsf{H}_A : Q$; see Lemma 1.66. The next problem gives a simple bound for the Bayes risk.

Problem 8.71.* For any distributions P_0, P_1 on $(\mathcal{X}, \mathfrak{A})$ it holds $\mathsf{b}_\pi(P^{\otimes n}, Q^{\otimes n}) \le \pi^s(1-\pi)^{1-s}(\mathsf{H}_s(P,Q))^n$, $0 < s < 1$.

The inequality in Problem 8.71 gives an upper bound for the exponential rate of convergence of $\mathsf{b}_\pi(P^{\otimes n}, Q^{\otimes n})$. It holds

$$\limsup_{n\to\infty} \frac{1}{n} \ln \mathsf{b}_\pi(P^{\otimes n}, Q^{\otimes n}) \le \inf_{0<s<1} \ln \mathsf{H}_s(P,Q). \tag{8.72}$$

The next theorem states that in (8.72) in fact equality holds. This statement is due to Chernoff (1952).

Theorem 8.72. ***(Chernoff's Theorem)*** *Under the prior* $(\pi,1-\pi)$ *the minimum Bayes risk* $\mathsf{b}_\pi(P^{\otimes n}, Q^{\otimes n})$ *for testing* $\mathsf{H}_0 : P^{\otimes n}$ *versus* $\mathsf{H}_A : Q^{\otimes n}$ *satisfies*

$$\lim_{n\to\infty} \frac{1}{n} \ln \mathsf{b}_\pi(P^{\otimes n}, Q^{\otimes n}) = \inf_{0<s<1} \ln \mathsf{H}_s(P,Q), \quad \pi \in (0,1).$$

Corollary 8.73. *The minimax risk* $\mathsf{m}_\rho(P^{\otimes n}, Q^{\otimes n})$ *satisfies*

$$\lim_{n\to\infty} \frac{1}{n} \ln \mathsf{m}_\rho(P^{\otimes n}, Q^{\otimes n}) = \inf_{0<s<1} \ln \mathsf{H}_s(P,Q), \quad \rho \in (0,1).$$

The upper bound for the convergence rate is already established by (8.72). For a proof that it is also a lower bound we refer to Chernoff (1952) and Krafft and Plachky (1970). The statement of the corollary follows from

$$\mathsf{m}_\rho(P^{\otimes n}, Q^{\otimes n}) \le \mathsf{b}_\rho(P^{\otimes n}, Q^{\otimes n}) \le 2\mathsf{m}_\rho(P^{\otimes n}, Q^{\otimes n}).$$

The quantity

$$\mathsf{C}(P,Q) = -\inf_{0<s<1} \ln \mathsf{H}_s(P,Q) \tag{8.73}$$

is called the *Chernoff index* of P and Q. It provides the exponential rate at which the minimum Bayes risk $\mathsf{b}_\pi(P^{\otimes n}, Q^{\otimes n})$ tends to zero. With the Chernoff index the statements of Theorem 8.72 and Corollary 8.73 may be written as

$$\lim_{n\to\infty} \frac{1}{n} \ln \mathsf{b}_\pi(P^{\otimes n}, Q^{\otimes n}) = \lim_{n\to\infty} \frac{1}{n} \ln \mathsf{m}_\rho(P^{\otimes n}, Q^{\otimes n}) = -\mathsf{C}(P,Q). \tag{8.74}$$

Example 8.74. To illustrate the statement above suppose that P_θ, $\theta \in \Delta \subseteq \mathbb{R}^d$, is an exponential family with natural parameter θ and generating statistic $T : \mathcal{X} \to_m \mathbb{R}^d$. Then by Example 1.88,

$$\mathsf{C}(P_{\theta_1}, P_{\theta_2}) = \inf_{0<s<1} \{sK(\theta_1) + (1-s)K(\theta_2) - K(s\theta_1 + (1-s)\theta_2)\}.$$

If $P_\theta = \mathsf{N}(\theta, \sigma^2)$, then by (1.79)

$$\begin{aligned} \ln \mathsf{H}_s(\mathsf{N}(\theta_0,\sigma^2), \mathsf{N}(\theta_1,\sigma^2)) &= -\frac{1}{2\sigma^2} s(1-s)(\theta_0-\theta_1)^2, \quad \text{and} \\ \mathsf{C}(\mathsf{N}(\theta_0,\sigma^2), \mathsf{N}(\theta_1,\sigma^2)) &= (\theta_0-\theta_1)^2/(8\sigma^2). \end{aligned} \tag{8.75}$$

Now we consider for the model $(\mathcal{X}^n, \mathfrak{A}^{\otimes n}, \{P^{\otimes n}, Q^{\otimes n}\})$ the problem of testing $\mathsf{H}_0 : P^{\otimes n}$ versus $\mathsf{H}_A : Q^{\otimes n}$ at a fixed given level $\alpha \in (0,1)$. The exponential rate of the probability of an error of the second kind of the best level α test, as n tends to infinity, can be characterized as follows.

Theorem 8.75. ***(Stein's Theorem)*** *If $\mathsf{K}(P,Q)$, the Kullback–Leibler divergence, is finite, then the error probability of the second kind $\mathsf{g}_\alpha(P^{\otimes n},Q^{\otimes n})$ of the best level α test φ_n for testing $\mathsf{H}_0 : P^{\otimes n}$ versus $\mathsf{H}_A : Q^{\otimes n}$ satisfies*

$$-\lim_{n\to\infty}\frac{1}{n}\ln \mathsf{g}_\alpha(P^{\otimes n},Q^{\otimes n}) = \mathsf{K}(P,Q). \tag{8.76}$$

Proof. $\mathsf{K}(P,Q)<\infty$ and (1.81) imply that $P \ll Q$ and $\mathsf{K}(P,Q)=\mathsf{E}_P \ln f$, where $f = dP/dQ$. Let $L=(1/f)I_{(0,\infty)}(f)+\infty I_{\{0\}}(f)$ be the likelihhood ratio of Q with respect to P and $X_1,...,X_n : \mathcal{X}^n \to \mathcal{X}$ be the projections. If L_n denotes the likelihhood ratio of $Q^{\otimes n}$ with respect to $P^{\otimes n}$, then Definition 1.57 and Proposition A.29 yield $L_n = \exp\{\sum_{i=1}^n \ln L(X_i)\}$ and

$$\int h dQ^{\otimes n} = \int I_{(0,\infty)}(L_n) h L_n dP^{\otimes n} + \int I_{\{\infty\}}(L_n) h dQ^{\otimes n},$$

for every $h : \mathcal{X}^n \to_m \mathbb{R}_+$. For $T_n = n^{-1}\ln L_n$ and $\alpha \in (0,1)$ we find by Theorem 2.45 constants c_n and $\gamma_n \in [0,1]$ such that $\varphi_n = I_{(c_n,\infty)}(T_n) + \gamma_n I_{\{c_n\}}(T_n)$ is a best level α test. On $A_n = \{|T_n + \mathsf{K}(P,Q)| < \varepsilon\}$, $\varepsilon>0$, it holds $L_n \ge \exp\{n(-\varepsilon - \mathsf{K}(P,Q))\}$. Hence with $h = 1-\varphi_n$,

$$\int (1-\varphi_n) dQ^{\otimes n} \ge \exp\{n(-\varepsilon - \mathsf{K}(P,Q))\} \int (1-\varphi_n) I_{A_n} dP^{\otimes n}.$$

The fact that $\mathsf{K}(P,Q) = \mathsf{E}_P \ln f = -\mathsf{E}_P \ln L$ and the law of large numbers yield $\lim_{n\to\infty} P^{\otimes n}(A_n) = 1$ and $\int (1-\varphi_n) I_{A_n} dP^{\otimes n} \to 1-\alpha$. Taking first $n\to\infty$ and then $\varepsilon \to 0$ one arrives at

$$\lim_{n\to\infty}\frac{1}{n}\ln\Big(\int (1-\varphi_n) dQ^{\otimes n}\Big) \ge -\mathsf{K}(P,Q).$$

To prove the opposite inequality, we note that $1-\varphi_n > 0$ implies $T_n \le c_n$. Hence,

$$\int (1-\varphi_n) dQ^{\otimes n} = \int (1-\varphi_n) L_n dP^{\otimes n} \le \exp\{nc_n\} \int (1-\varphi_n) dP^{\otimes n}.$$

As $P^{\otimes n}(A_n) \to 1$ implies $c_n \to -\mathsf{K}(P,Q)$ we get, together with $\int (1-\varphi_n) dP^{\otimes n} = 1-\alpha$, the opposite inequality. ■

In the proof of the previous theorem we have shown that the critical values c_n, that appear in the Neyman–Pearson level α test for testing $P^{\otimes n}$ versus $Q^{\otimes n}$, satisfy $c_n \to -\mathsf{K}(P,Q)$. Thus the statement of Stein's theorem may also be written as

$$\lim_{n\to\infty}\frac{1}{n}\ln \mathsf{g}_\alpha(P^{\otimes n},Q^{\otimes n}) = \lim_{n\to\infty} c_n = -\mathsf{K}(P,Q). \tag{8.77}$$

Theorem 8.75 is called Stein's theorem in the literature. Chernoff (1956) and Kullback (1959) refer to an unpublished paper by Stein for the statement

(8.76). In the theorem above we have assumed that the Kullback–Leibler divergence $\mathsf{K}(P,Q)$ is finite. The case of $\mathsf{K}(P,Q) = \infty$ has been studied in Janssen (1986), where conditions are established that guarantee that the first equality in (8.77) is valid even in the case of $\mathsf{K}(P,Q) = \infty$.

Example 8.76. We illustrate the above theorem by considering two distributions $P = P_{\theta_0}$ and $Q = P_{\theta_1}$ that come from the same exponential family. Then

$$\begin{aligned}\mathsf{K}(P,Q) &= \int (\ln \frac{dP_{\theta_0}}{dP_{\theta_1}}) dP_{\theta_0} = \int \langle T, \theta_0 - \theta_1 \rangle \, dP_{\theta_0} + K(\theta_1) - K(\theta_0) \\ &= \langle \nabla K(\theta_0), \theta_0 - \theta_1 \rangle + K(\theta_1) - K(\theta_0),\end{aligned}$$

where we have used (1.23). If $P_\theta = \mathsf{N}(\theta, \sigma_0^2)$, then

$$\mathsf{K}(\mathsf{N}_{\theta_0,\sigma_0^2}, \mathsf{N}_{\theta_1,\sigma_0^2}) = \int (\ln \frac{\varphi_{\theta_0,\sigma_0^2}(x)}{\varphi_{\theta_1,\sigma_0^2}(x)}) \varphi_{\theta_0,\sigma_0^2}(x) dx = \frac{(\theta_0 - \theta_1)^2}{2\sigma_0^2}.$$

This means that

$$-\lim_{n\to\infty} \frac{1}{n} \ln \mathsf{g}_\alpha(\mathsf{N}^{\otimes n}_{\theta_0,\sigma_0^2}, \mathsf{N}^{\otimes n}_{\theta_1,\sigma_0^2}) = \frac{(\theta_0 - \theta_1)^2}{2\sigma_0^2}. \tag{8.78}$$

We compare this rate with the exact value of $\mathsf{g}_\alpha(\mathsf{N}^{\otimes n}_{\theta_0,\sigma_0^2} \mathsf{N}^{\otimes n}_{\theta_1,\sigma_0^2})$, which is nothing else than the probability of an error of the second kind of the Gauss test. We get from Example 2.37 that for $\theta_0 < \theta_1$ the minimal error probability of a level α test is given by $\mathsf{g}_\alpha(\mathsf{N}^{\otimes n}_{\theta_0,\sigma_0^2}, \mathsf{N}^{\otimes n}_{\theta_1,\sigma_0^2}) = \Phi_{0,1}(u_{1-\alpha} - \sqrt{n}(\theta_1 - \theta_0)/\sigma_0)$. In order to evaluate the right-hand side for large n we use Mill's ratio; see Mitrinovic and Vaic (1970).

$$|x|/(1+x^2) \le \Phi_{0,1}(x)/\varphi_{0,1}(x) \le 1/|x|, \qquad x < 0.$$

Hence with $a_n = u_{1-\alpha} - \sqrt{n}(\theta_1 - \theta_0)/\sigma_0$ and all sufficiently large n,

$$\frac{|a_n|}{(1+a_n^2)\sqrt{2\pi}\sigma_0} \exp\{-\frac{a_n^2}{2\sigma_0^2}\} \le \mathsf{g}_\alpha(\mathsf{N}^{\otimes n}_{\theta_0,\sigma_0^2}, \mathsf{N}^{\otimes n}_{\theta_1,\sigma_0^2}) \le \frac{1}{|a_n|\sqrt{2\pi}\sigma_0} \exp\{-\frac{a_n^2}{2\sigma_0^2}\}.$$

As $\lim_{n\to\infty} n^{-1} \ln |a_n| = 0$ we obtain the result of (8.78) once again. But from this example we see that the exponential rate alone by itself provides an asymptotic expression that ignores the factor $(|a_n|\sqrt{2\pi}\sigma_0)^{-1} \sim n^{-1/2}$. This means that in our example the actual error probability of the second kind tends by the factor $n^{-1/2}$ faster to zero than is indicated by the exponential rate.

Remark 8.77. There are a large number of papers that deal with the exponential rate of the convergence of error probabilities for increasing sample sizes. Without an attempt to give a complete list we refer to Bahadur (1971), Steinebach (1980), Ellis (1985), and other standard books on large deviations. Using the results of Chernoff and Stein one may characterize the efficiency of statistical tests by comparing their exponential rates for the error probabilities with the exponential rates provided by likelihood ratio tests which appear in both the Bayes approach and the approach via level α tests. But the evaluation of the exponential rates requires techniques from the area of large deviation. We do not consider that topic in this book. Instead we refer to Bahadur (1971) and Kester (1987).

Because both types of error probabilities are involved when the efficiency of tests is investigated, one has to establish side conditions to get one numerical value that characterizes the asymptotic quality of a test. In the Chernoff approach the error probabilities of the first kind α_n and of the second kind β_n tend to zero and the common distribution of the i.i.d. sample is fixed. Several other approaches are possible. One that requires that $\alpha_n \to 0$ and $\beta_n \to \beta > 0$ is due to Bahadur. For other approaches we refer to the overview given in Serfling (1980). As pointed out in Lehmann and Romano (2005), p. 539, there is one exceptional approach which is the Pitman efficiency, or the concept of local alternatives. In this case $\alpha_n \to \alpha > 0$ and $\beta_n \to \beta > 0$, where the common distributions depend on n and originate from a localization procedure. In the concept of local alternatives the limiting models that appear there are not degenerate. They are typically Gaussian models, and the optimal solutions of the decision problems provide the asymptotically optimal decisions by substituting the central sequence for the central variable. These facts presumably are the reason for the superiority of the Pitman approach over other efficiency concepts. In the subsequent part of this chapter we systematically use the sketched way to construct sequences of tests that are optimal in sequences of localized models. Finally, we refer to Serfling (1980) and Nikitin (1995) for detailed discussions of different concepts of efficiency in statistics.

8.6 U-Statistics and Rank Statistics

We have seen in Chapter 7 that stochastic Taylor expansions in combination with the third lemma of LeCam are the fundamental tools to study the asymptotic behavior of estimators under local alternatives. For example, using this approach we have established there the local asymptotic optimality of the MLE. In this section we use a similar idea for tests. In a first step we approximate a given test statistic T_n under the null hypothesis by a linear test statistic

$$T_n = \frac{1}{\sqrt{n}} \sum\nolimits_{i=1}^{n} \Psi(X_i) + o_{P_{\theta_0}^{\otimes n}}(1), \quad \Psi \in \mathbb{L}_{2,m}^{0}(P_{\theta_0}). \tag{8.79}$$

Statistics of this type have appeared already at several places where nonlinear statistics of the observations have been approximated by statistics of a linear type. Approximating the log-likelihood in the second lemma of LeCam (see Theorem 6.70), a linearization has been obtained that contains the so-called central sequence which is given by

$$Z_n = \frac{1}{\sqrt{n}} \sum\nolimits_{i=1}^{n} \dot{L}_{\theta_0}(X_i).$$

In Theorem 7.148 we have established a stochastic Taylor expansion for the MLE,

$$\sqrt{n}(\widehat{\theta}_n - \theta_0) = \frac{1}{\sqrt{n}} \sum\nolimits_{i=1}^{n} \mathsf{I}^{-1}(\theta_0)\dot{L}_{\theta_0}(X_i) + o_{P_{\theta_0}^{\otimes n}}(1).$$

We have also established linear approximations for M-estimators; see Theorem 7.142. An application of Corollary 6.74 shows that the limit distribution of T_n under local alternatives $P_{\theta_0+h/\sqrt{n}}^{\otimes n}$ differs from the limit distribution under $P_{\theta_0}^{\otimes n}$ by a shift that depends on the correlation of $\dot{L}_{\theta_0}$ and the influence function Ψ. The test statistic of an asymptotically optimal test produces the maximum shift, and this occurs when Ψ and $\dot{L}_{\theta_0}$ are proportional. Roughly speaking, these steps are the content of the test theory in localized models.

In situations where there is no parametric model one has to analyze statistical tests that are based on nonparametric estimators of special functionals of distributions, e.g., the median. The question arises as to what can be said about the power of such a nonparametric test. To answer this question one may fix a distribution P and choose a one-parameter $\mathbb{L}_2$-differentiable family of distributions $(P_\theta)_{a<\theta<b}$ such that $P_{\theta_0} = P$ belongs to the null hypothesis and P_θ for $\theta \neq \theta_0$ belongs to the alternative. Then this curve characterizes the deviations from the null hypotheses in a specific direction that is given by the $\mathbb{L}_2$-derivative $\dot{L}_{\theta_0}$. The $\mathbb{L}_2$-derivative $\dot{L}_{\theta_0}$ is the influence function of the most efficient test statistics as in the case where the correlation between Ψ and $\dot{L}_{\theta_0}$ becomes maximum. Otherwise, for any test based on the influence function Ψ the correlation between Ψ and $\dot{L}_{\theta_0}$ characterizes the asymptotic efficiency of φ_n in the direction given by $\dot{L}_{\theta_0}$. This approach allows us to find the direction in which the nonparametric test has a good performance, and also directions where its power is poor.

Necessary for the realization of the program sketched above is both a suitable linearization of the log-likelihood and a linearization of the sequence of the relevant nonlinear test statistics. The linearization of the log-likelihood has been established already within the framework of the LAN theory in Chapter 6. It remains to create an asymptotic linearization technique for nonlinear test statistics. By an asymptotic linearization we mean the following. Let $X_1, ..., X_n$ be a sample where the X_i takes on values in $(\mathcal{X}, \mathfrak{A})$ and let $T_n : \mathcal{X}^n \to_m \mathbb{R}^d$ be a sequence of statistics. If there exists $\Psi : \mathcal{X} \to_m \mathbb{R}^d$ with $\mathbb{E}\Psi(X_i) = 0$ and $\mathbb{E}\,\|\Psi(X_i)\|^2 < \infty$ such that (8.79) holds, then we call $n^{-1/2}\sum_{i=1}^{n}\Psi(X_i)$ an *asymptotic linearization* and (8.79) a *stochastic Taylor expansion.* Typically the $X_1, ..., X_n$ are i.i.d. Then we get from Slutsky's lemma and the multivariate central limit theorem that T_n is asymptotically normal.

We start with a classical Taylor expansion technique known as the δ-*method.* Suppose W_n is a sequence of random vectors in $\mathbb{R}^d$ such that

$$\mathcal{L}(\sqrt{n}(W_n - a)) \Rightarrow \mathsf{N}(0, \Sigma). \tag{8.80}$$

Let $U(a) \subseteq \mathbb{R}^d$ be an open neighborhood of a and $g = (g_1, ..., g_k)^T : U(a) \to \mathbb{R}^k$ be a function for which all components are differentiable at a. Let

$$J_g = (\partial g_i/\partial t_j)_{1\le i\le k, 1\le j\le d}$$

denote the Jacobian of g. Then we call $\dot{g}(a) = J_g^T(a)$ the derivative of g. If $k = 1$, then $\dot{g}$ is the gradient. We set $g(W_n) = 0$ if $W_n \notin U(a)$. As $\mathbb{P}(W_n \notin U(a)) \to 0$, it is irrelevant for further asymptotic considerations how $g(W_n)$ is defined for $W_n \notin U(a)$.

Proposition 8.78. ***(δ-Method)*** *Suppose that $g : U(a) \to \mathbb{R}^k$ is differentiable at a. Then (8.80) implies*

$$\mathcal{L}(\sqrt{n}(g(W_n) - g(a))) \Rightarrow \mathsf{N}(0, \dot{g}^T(a)\Sigma\dot{g}(a)).$$

Moreover,

$$\sqrt{n}(W_n - a) = n^{-1/2}\sum\nolimits_{i=1}^{n} \Psi(X_i) + o_{\mathbb{P}}(1)$$

implies

$$\sqrt{n}(g(W_n) - g(a)) = n^{-1/2}\sum\nolimits_{i=1}^{n} \dot{g}^T(a)\Psi(X_i) + o_{\mathbb{P}}(1).$$

Proof. Put $R(x) = g(x) - g(a) - \dot{g}^T(a)(x - a)$. Then $\|R(x)\| = o(\|x - a\|)$ as $x \to a$, by the differentiability of g at a. The condition $\mathcal{L}(\sqrt{n}(W_n - a)) \Rightarrow \mathsf{N}(0, \Sigma)$ implies that $\sqrt{n}\,\|W_n - a\|$ is stochastically bounded and W_n tends stochastically to a. Hence

$$\sqrt{n}(g(W_n) - g(a)) - \dot{g}^T(a)\sqrt{n}(W_n - a) = \sqrt{n}\,\|W_n - a\|\,\frac{R(W_n)}{\|W_n - a\|} = o_{\mathbb{P}}(1)$$

implies the first statement. Plugging in the linear representation of $\sqrt{n}(W_n - a)$ we get the second statement. ■

Problem 8.79. Suppose $X_1, ..., X_n$ are i.i.d. with common distribution $\mathsf{Ex}(\beta)$. Then $W_n = 1/\overline{X}_n$ is a consistent estimator for β. Establish an asymptotic linearization for $\sqrt{n}(W_n - \beta)$ and the limit distribution.

Suppose $X_1, ..., X_n$ are i.i.d. with $\mathbb{E}X_1^{2m} < \infty$, and set $a_i = \mathbb{E}X_1^i$. Let Σ be the matrix with entries $\sigma_{i,j} = \mathbb{E}(X_1^i - a_i)(X_1^j - a_j) = a_{i+j} - a_i a_j$, $1 \le i, j \le m$. Set $M_{n,i} = (1/n)\sum_{l=1}^{n} X_l^i$, $i = 1, ..., m$. Then

$$\mathcal{L}(n^{1/2}((M_{n,1}, ..., M_{n,m}) - (a_1, ..., a_m))^T) \Rightarrow \mathsf{N}(0, \Sigma)$$

by the multivariate central limit theorem; see Theorem A.53. If $g : \mathbb{R}^m \to \mathbb{R}$ is any function that is differentiable at $a = (a_1, ..., a_m)^T$, then the δ-method provides the limit distribution of $\sqrt{n}(g(M_{n,1}, ..., M_{n,m}) - g(a_1, ..., a_m))^T$. This is a way to get the limit distribution of functions of empirical moments. For example,

$$\begin{aligned} g_1(a_1, a_2, a_3) &= \frac{\mathbb{E}(X_1 - \mathbb{E}X_1)^3}{(\mathsf{V}(X_1))^{3/2}} = \frac{a_3 - 3a_2a_1 + 2a_1^3}{(a_2 - a_1^2)^{3/2}} \quad \text{and} \\ g_2(a_1, ..., a_4) &= \frac{\mathbb{E}(X_1 - \mathbb{E}X_1)^4}{(\mathsf{V}(X_1))^2} = \frac{a_4 - 4a_3a_1 + 6a_2a_1^2 - 3a_1^4}{(a_2 - a_1^2)^2} \end{aligned}$$

are called the skewness and kurtosis of the distribution of X_1, respectively. Let $\gamma_1 = g_1(a_1, a_2, a_3)$ and $\gamma_2 = g_1(a_1, ..., a_4)$. Then the limit distribution of

$$\sqrt{n}\begin{pmatrix} G_{1,n} - \gamma_1 \\ G_{2,n} - \gamma_2 \end{pmatrix} := \sqrt{n}\begin{pmatrix} g_1(M_{n,1}, M_{n,2}, M_{n,3}) - \gamma_1 \\ g_2(M_{n,1}, M_{n,2}, M_{n,3}, M_{n,4}) - \gamma_2 \end{pmatrix}$$

is provided by the δ-method. From here one may construct an asymptotic test for normality based on the statistic

$$n(G_{1,n} - \gamma_1)^2 + n(G_{2,n} - \gamma_2)^2.$$

Although this way is not difficult it requires somewhat lengthy calculations for $\dot{g}^T(a)\Sigma\dot{g}(a)$. But we note that with similar considerations one may obtain a large variety of classical asymptotic tests, for example, the test for dependence based on the correlation coefficients; see Serfling (1980). Another way is to simplify the statistic by a direct linearization. An example is given below.

Example 8.80. Suppose again $X_1, ..., X_n$ are i.i.d. with $\mathbb{E}X_1^6 < \infty$. Set $\mu = \mathbb{E}X_1$ and $\sigma^2 = \mathbb{E}(X_1 - \mu)^2$. It holds

$$\begin{aligned} \frac{1}{n}\sum\nolimits_{i=1}^n (X_i - \overline{X}_n)^3 &= \frac{1}{n}\sum\nolimits_{i=1}^n [(X_i - \mu) - (\overline{X}_n - \mu)]^3 \\ &= \frac{1}{n}\sum\nolimits_{i=1}^n (X_i - \mu)^3 - 3(\overline{X}_n - \mu)\frac{1}{n}\sum\nolimits_{i=1}^n (X_i - \mu)^2 + 2(\overline{X}_n - \mu)^3. \end{aligned}$$

The central limit theorem gives $(\overline{X}_n - \mu) = O_{\mathbb{P}}(1/\sqrt{n})$. Hence $(\overline{X}_n - \mu)^3 = O_{\mathbb{P}}(n^{-3/2})$ and by the law of large numbers,

$$\begin{aligned} (\overline{X}_n - \mu)\frac{1}{n}\sum\nolimits_{i=1}^n (X_i - \mu)^2 &= \sigma^2(\overline{X}_n - \mu) + (\overline{X}_n - \mu)(\frac{1}{n}\sum\nolimits_{i=1}^n (X_i - \mu)^2 - \sigma^2) \\ &= \sigma^2(\overline{X}_n - \mu) + o_{\mathbb{P}}(1/\sqrt{n}) \quad \text{and} \\ \frac{1}{\sqrt{n}}\sum\nolimits_{i=1}^n (X_i - \overline{X}_n)^3 &= \frac{1}{\sqrt{n}}\sum\nolimits_{i=1}^n ((X_i - \mu)^3 - 3\sigma^2(X_i - \mu)) + o_{\mathbb{P}}(1). \end{aligned}$$

Set $S_n = n^{-1}\sum_{i=1}^n (X_i - \mu)^2$. If we want to test $\mathsf{H}_0 : \mathbb{E}(X_1 - \mu)^3 = 0$, then under H_0 by Slutsky's lemma we obtain

$$\mathcal{L}(S_n^{-3/2} n^{-1/2} \sum\nolimits_{i=1}^n (X_i - \overline{X}_n)^3) \Rightarrow \mathsf{N}(0, \tau^2), \quad \tau^2 = \frac{1}{\sigma^6}\mathbb{E}((X_1 - \mu)^3 - 3\sigma^2(X_1 - \mu))^2.$$

Especially, if the X_i are normally distributed, then by the well-known formula for central moments for $X_1 \sim \mathsf{N}(\mu, \sigma^2)$, i.e., $\mathbb{E}(X_1 - \mu)^{2k} = 1 \cdot 3 \cdots (2k-1)\sigma^{2k}$, it follows $\tau^2 = 6$.

Now we present a general technique that allows the approximation of any statistic by a linear statistic in the sense of (8.79). This is the famous *projection lemma.* Our representation follows Hájek, Šidák, and Sen (1999) and Witting and Müller-Funk (1995). Recall that for a given probability space $(\Omega, \mathfrak{F}, \mathbb{P})$ the space $\mathbb{L}_2(\mathbb{P})$ is the space of all real-valued random variables X with $\mathbb{E}X^2 < \infty$, where $\mathbb{P}$-a.s. identical random variables are identified. Then $\mathbb{L}_2(\mathbb{P})$ is a Hilbert space that is equipped with the scalar product $\langle X, Y \rangle = \mathbb{E}(XY)$ and thus with the norm $\|X\| = (\mathbb{E}X^2)^{1/2}$. We write $X \perp Y$ if $\langle X, Y\rangle = 0$. If $\mathbb{A} \subseteq \mathbb{L}_2(\mathbb{P})$ is any subset, then $span(\mathbb{A})$ denotes the linear hull of $\mathbb{A}$ and $[\mathbb{A}]$ the closure of $span(\mathbb{A})$. A linear subspace $\mathcal{L} \subseteq \mathbb{L}_2(\mathbb{P})$ is called finite-dimensional if there are $X_1, ..., X_n$ such that $\mathcal{L} = span(\{X_1, ..., X_n\})$. It is a simple but important fact that finite-dimensional linear subspaces are closed; see Problem 8.94.

We recall the concept of an (orthogonal) projection. If $\mathcal{L} \subseteq \mathbb{L}_2(\mathbb{P})$ is a closed linear subspace of $\mathbb{L}_2(\mathbb{P})$ and $Y \in \mathbb{L}_2(\mathbb{P})$, then there exists a uniquely determined element in $\mathcal{L}$, denoted by $\Pi_{\mathcal{L}}Y$, called the *projection of* Y *on* $\mathcal{L}$, such that $\|Y - \Pi_{\mathcal{L}}Y\| = \inf_{X\in\mathcal{L}} \|Y - X\|$; see Problem 8.95. The following characterization of the projection (see Problem 8.96) is often used.

$$Z = \Pi_{\mathcal{L}}Y \iff Z \in \mathcal{L} \quad \text{and} \quad Y - Z \perp X,\ X \in \mathcal{L}. \tag{8.81}$$

If $Y = (Y_1, ..., Y_k)^T$ is a k-dimensional random vector, then we set

$$\Pi_{\mathcal{L}}Y := (\Pi_{\mathcal{L}}Y_1, ..., \Pi_{\mathcal{L}}Y_k)^T \tag{8.82}$$

and call $\Pi_{\mathcal{L}}Y$ the projection of Y on $\mathcal{L}$. Let P be a distribution on $(\mathcal{X}, \mathfrak{A})$ and introduce the subspace $\mathbb{L}_2^0(P)$ of $\mathbb{L}_2(P)$ by

$$\mathbb{L}_2^0(P) = \{a : a \in \mathbb{L}_2(P), \int a dP = 0\}.$$

We use the notation $P_n = P^{\otimes n}$ and consider the Hilbert space $\mathbb{L}_2(P_n)$. Denote by $X_1, ..., X_n$ the projections of $\mathcal{X}^n$ on $\mathcal{X}$. Set

$$\mathcal{L} = \{S : S = a_0 + \sum\nolimits_{i=1}^n a_i(X_i), \quad a_0 \in \mathbb{R}, \quad a_1, ..., a_n \in \mathbb{L}_2^0(P)\}. \tag{8.83}$$

Then $\mathcal{L}$ is a closed linear subspace of $\mathbb{L}_2(P_n)$ and the representation of S by the a_i is unique; see Problems 8.101 and 8.100.

For $T_n \in \mathbb{L}_2(P_n)$ we find a best approximation by elements from $\mathcal{L}$ by taking the projection of T_n on $\mathcal{L}$, which is given by

$$\Pi_{\mathcal{L}}T_n = b_0 + \sum\nolimits_{i=1}^n b_i(X_i) \tag{8.84}$$

for some $b_0 \in \mathbb{R}$ and $b_i \in \mathbb{L}_2^0(P)$. To find the b_i we note that $\mathcal{L}$ is spanned by 1 and $a(X_j)$, $j = 1, ..., n$, $a \in \mathbb{L}_2^0(P)$. Hence by taking the scalar product of $T_n - b_0 - \sum_{i=1}^n b_i(X_i)$ with 1 and the $a(X_j)$ we get from (8.81)

$$\mathsf{E}_{P_n}T_n = b_0, \quad \text{and} \quad \mathsf{E}_{P_n}T_n a(X_j) = \mathsf{E}_{P_n}b_j(X_j)a(X_j), \quad j = 1, ..., n,$$

for every $a \in \mathbb{L}_2^0(P)$. Set $\mathbb{H}_j^0 = \{a(X_j) : a \in \mathbb{L}_2^0(P)\} \subseteq \mathbb{L}_2(P_n)$. Then (8.81) implies that $b_j(X_j)$ is the projection of T_n on $\mathbb{H}_j^0$. Hence,

$$b_j(X_j) = \mathsf{E}_{P_n}(T_n|\sigma(X_j)) - \mathsf{E}_{P_n}T_n \tag{8.85}$$

by Problem 8.99. By Problem 8.98 we have

$$b_j(x_j) = \int T_n(x) P^{\otimes \neq j}(dx_{\neq j}) - \mathsf{E}_{P_n} T_n, \tag{8.86}$$

where $P^{\otimes \neq j}(dx_{\neq j}) = P(dx_1) \cdots P(dx_{j-1}) \cdot P(dx_{j+1}) \cdots P(dx_n)$. The subsequent lemma is due to Hájek (1968); see also Hájek, Šidák, and Sen (1999).

Lemma 8.81. ***(Projection Lemma)*** *If $T_n \in \mathbb{L}_2(P_n)$, where $P_n = P^{\otimes n}$, then for $\mathcal{L}$ in (8.83) it holds with b_j from (8.86),*

$$\Pi_{\mathcal{L}} T_n = \widehat{T}_n := \mathsf{E}_{P_n} T_n + \sum\nolimits_{j=1}^{n} b_j(X_j),$$

$$\mathsf{V}_{P_n}(T_n) = \mathsf{V}_{P_n}(\widehat{T}_n) + \mathsf{V}_{P_n}(T_n - \widehat{T}_n), \tag{8.87}$$

$$\mathsf{E}_{P_n}(T_n - \widehat{T}_n)^2 = \mathsf{V}_{P_n}(T_n) - \mathsf{V}_{P_n}(\widehat{T}_n). \tag{8.88}$$

Proof. The first statement follows from (8.84), (8.85) and (8.86). The relations (8.87) and (8.88) follow from $T_n - \widehat{T}_n \perp \widehat{T}_n$ and $T_n - \widehat{T}_n \perp \mathsf{E}_{P_n} T_n$, and therefore $T_n - \widehat{T}_n \perp \widehat{T}_n - \mathsf{E}_{P_n} T_n$. ■

The b_j in (8.86) depend on j in general. But this dependence disappears if T_n is permutation invariant or, in other words, symmetric. Subsequently we study an important class of such statistics.

Let $\Psi \in \mathbb{L}_2(P^{\otimes m})$ be symmetric; i.e., $\Psi(x_1, ..., x_m) = \Psi(x_{\gamma(1)}, ..., x_{\gamma(m)})$ for every permutation γ. For $m < n$ the statistic

$$\mathcal{U}_n(x_1, ..., x_n) = \frac{1}{\binom{n}{m}} \sum_{1 \le i_1 < \cdots < i_m \le n} \Psi(x_{i_1}, ..., x_{i_m}),$$

is called a *U-statistic of order m with kernel Ψ*. The concept of U-statistics was introduced by Hoeffding (1948) who proved the asymptotic normality. Since then numerous papers on U-statistics have been published; see Koroljuk and Borovskich (1994) for a comprehensive presentation and references.

Set $\mathcal{A}_m = \{A : A \subseteq \{1, ..., n\}, |A| = m\}$. For $A = \{i_1, ..., i_m\} \in \mathcal{A}_m$ let $\Psi(x_A) = \Psi(x_{i_1}, ..., x_{i_m})$, where the order of the x_{i_j} in $\Psi(x_{i_1}, ..., x_{i_m})$ is irrelevant. Hence we may write

$$\mathcal{U}_n(x_1, ..., x_n) = \frac{1}{\binom{n}{m}} \sum\nolimits_{A \in \mathcal{A}_m} \Psi(x_A).$$

We set for $k = 1, ..., m-1$,

$$\begin{aligned} \Psi_k(x_1, ..., x_k) &= \textstyle\int \Psi(x_1, ..., x_m) P^{\otimes(m-k)}(dx_{k+1}, ..., dx_m), \\ \gamma = \mathsf{E}_{P^{\otimes m}} \Psi &= \textstyle\int \Psi(x_1, ..., x_m) P^{\otimes m}(dx_1, ..., dx_m). \end{aligned} \tag{8.89}$$

Proposition 8.82. *If $\mathcal{U}_n$ is a U-statistic, then with $\mathcal{L}$ in (8.83),*

$$\widehat{\mathcal{U}}_n = \Pi_{\mathcal{L}} \mathcal{U}_n = \gamma + \frac{m}{n} \sum\nolimits_{j=1}^{n} (\Psi_1(X_j) - \gamma).$$

Proof. The projection lemma (see Lemma 8.81) and (8.85) give

$$b_j(x_j) = \frac{1}{\binom{n}{m}} \sum\nolimits_{A \in \mathcal{A}_m} \int \Psi(x_A) P^{\otimes \neq j}(dx_{\neq j}) - \gamma.$$

It holds,

$$\int \Psi(x_A) P^{\otimes \neq j}(dx_{\neq j}) = \begin{cases} \gamma & \text{if } j \notin A, \\ \Psi_1(x_j) & \text{if } j \in A. \end{cases}$$

As

$$|\{A : A \in \mathcal{A}_m,\ j \notin A\}| = \tbinom{n-1}{m} \quad \text{and} \quad |\{A : A \in \mathcal{A}_m,\ j \in A\}| = \tbinom{n-1}{m-1},$$

it follows that

$$b_j(x_j) = \frac{1}{\binom{n}{m}} [\tbinom{n-1}{m}\gamma + \tbinom{n-1}{m-1}\Psi_1(x_j)] - \gamma = \frac{m}{n}(\Psi_1(x_j) - \gamma).$$

■

Theorem 8.83. *If* $\Psi \in \mathbb{L}_2(P^{\otimes m})$ *is symmetric, and* Ψ_1 *and* γ *are defined by (8.89), then* $\mathsf{E}_{P^{\otimes n}}(\mathcal{U}_n - \widehat{\mathcal{U}}_n)^2 = o(1/n)$*, and the* U*-statistic* $\mathcal{U}_n$ *admits the representation*

$$\sqrt{n}(\mathcal{U}_n - \gamma) = \frac{m}{\sqrt{n}} \sum\nolimits_{i=1}^{n} (\Psi_1(X_i) - \gamma) + o_{P^{\otimes n}}(1).$$

Proof. Use $\mathsf{E}_{P_n}(\mathcal{U}_n - \widehat{\mathcal{U}}_n)^2 = o(1/n)$, which follows from Problem 8.102 and Proposition 8.82. ■

Subsequently we present a few select examples that demonstrate that many of the frequently used test statistics are U-statistics, or can be transformed easily to U-statistics.

Example 8.84. Let $X_1, ..., X_n$ be i.i.d. with finite fourth moments, and denote by μ, σ^2, and μ_4 the expectation, variance, and fourth central moment of X_1, respectively. It holds

$$S_n^2 = \frac{1}{n-1} \sum\nolimits_{i=1}^{n} (X_i - \overline{X}_n)^2 = \frac{1}{n(n-1)} \sum\nolimits_{1 \le i < j \le n} (X_i - X_j)^2.$$

Hence S_n^2 is a U-statistic of order 2 with kernel $\Psi(x_1, x_2) = \frac{1}{2}(x_1 - x_2)^2$, and it holds

$$\gamma = \int \frac{1}{2}(x_1 - x_2)^2 P^{\otimes 2}(dx_1, dx_2) = \sigma^2,$$

$$\Psi_1(x_1) = \int \frac{1}{2}(x_1 - x_2)^2 P(dx_2) = \frac{1}{2}\left[(x_1 - \mu)^2 + \sigma^2\right],$$

$$\widehat{\mathcal{U}}_n = \sigma^2 + \frac{1}{n} \sum\nolimits_{i=1}^{n} ((X_i - \mu)^2 - \sigma^2), \quad \text{and} \quad \mathsf{V}_{P_n}(\widehat{\mathcal{U}}_n) = \frac{1}{n}(\mu_4 - \sigma^4).$$

Hence $\mathsf{V}_{P_n}(\mathcal{U}_n) - \mathsf{V}_{P_n}(\widehat{\mathcal{U}}_n) = 2\sigma^4/(n(n-1))$ by Problem 8.103, and by Theorem 8.83,

$$\sqrt{n}(S_n^2 - \sigma^2) = \sqrt{n}(\widehat{\mathcal{U}}_n - \sigma^2) + o_{P_n}(1) = n^{-1/2} \sum\nolimits_{i=1}^{n} ((X_i - \mu)^2 - \sigma^2) + o_{P_n}(1).$$

From Slutsky's lemma we get $\mathcal{L}(\sqrt{n}(S_n^2 - \sigma^2)) \Rightarrow \mathsf{N}(0, \mu_4 - \sigma^4)$.

One generalization of the U-statistics considered above concerns the k-sample problem, where a kernel $\Psi(x_{1,1},...,x_{1,m_1},...,x_{k,1},...,x_{k,m_k})$ is given which, for every fixed $i \in \{1,...,k\}$, is symmetric in the variables $x_{i,1},...,x_{i,m_i}$. We consider here only the special case of a two-sample U statistic of order $m_1 = 1$ and $m_2 = 1$. We set

$$\begin{aligned}
\mathcal{U}_{n_1,n_2} &= \frac{1}{n_1 n_2}\sum\nolimits_{i=1}^{n_1}\sum\nolimits_{j=1}^{n_2} \Psi(X_{1,i}, X_{2,j}),\\
\Psi_1(x_1) &= \int \Psi(x_1,x_2)P_2(dx_2), \quad \Psi_2(x_2) = \int \Psi(x_1,x_2)P_1(dx_1),\\
\mathcal{U}^1_{n_1} &= \frac{1}{n_1}\sum\nolimits_{i=1}^{n_1} \Psi_1(X_{1,i}), \quad \mathcal{U}^2_{n_2} = \frac{1}{n_2}\sum\nolimits_{j=1}^{n_2} \Psi_2(X_{2,j}),
\end{aligned}$$

where $X_{i,1},...,X_{i,n_i}$ take on values in $(\mathcal{X}_i, \mathfrak{A}_i)$ and have the common distribution P_i, $i = 1,2$. Put $P_{n_1,n_2} = P_1^{\otimes n_1} \otimes P_2^{\otimes n_2}$. In the special case under consideration the linearization of $\mathcal{U}_{n_1,n_2}$ can be directly obtained. It holds

$$\mathsf{E}_{P_{n_1,n_2}}[(\mathcal{U}_{n_1,n_2} - \gamma) - (\mathcal{U}^1_{n_1} - \gamma) - (\mathcal{U}^2_{n_2} - \gamma)]^2 \le \frac{1}{n_1}\frac{1}{n_2}\sigma^2 \tag{8.90}$$

for $\mathsf{E}_{P_1\otimes P_2}\Psi^2 < \infty$, $\gamma = \mathsf{E}_{P_1\otimes P_2}\Psi$, and $\sigma^2 = \mathsf{V}_{P_1\otimes P_2}(\Psi)$; see Problem 8.104.

Now we let n_1 and n_2 tend to infinity at the same rate; that is, we assume that for $n = n_1 + n_2$ it holds

$$\kappa = \lim_{n\to\infty} \frac{n_1}{n} \quad \text{and} \quad 0 < \kappa < 1. \tag{8.91}$$

The subsequent proposition is the well-known asymptotic Hoeffding decomposition of two-sample U-statistics in a special case.

Proposition 8.85. *Suppose $\Psi : \mathcal{X} \times \mathcal{X} \to_m \mathbb{R}$ satisfies $\mathsf{E}_{P_{n_1,n_2}}\Psi^2(X_{1,1}, X_{2,1}) < \infty$. If $\gamma = \mathsf{E}_{P_{n_1,n_2}}\Psi(X_{1,1}, X_{2,1})$ and the condition (8.91) is satisfied, then*

$$\begin{aligned}
&\sqrt{\frac{n_1 n_2}{n}}(\mathcal{U}_{n_1,n_2} - \gamma) = \sqrt{\frac{n_1 n_2}{n}}[(\mathcal{U}^1_{n_1} - \gamma) + (\mathcal{U}^2_{n_2} - \gamma)] + o_{P_{n_1,n_2}}(1),\\
&\mathcal{L}(\sqrt{\frac{n_1 n_2}{n}}(\mathcal{U}_{n_1,n_2} - \gamma)|P_{n_1,n_2}) \Rightarrow \mathsf{N}(0, (1-\kappa)\sigma_1^2 + \kappa\sigma_2^2).
\end{aligned}$$

Proof. $\mathcal{U}^1_{n_1} - \gamma$ and $\mathcal{U}^2_{n_2} - \gamma$ are independent, and it holds

$$\mathcal{L}(\sqrt{n_i}(\mathcal{U}^i_{n_i} - \gamma)|P_{n_i}) \Rightarrow \mathsf{N}(0, \sigma_i^2), \quad i = 1,2.$$

Hence,

$$\mathcal{L}(\sqrt{\frac{n_1 n_2}{n}}[(\mathcal{U}^1_{n_1} - \gamma) + (\mathcal{U}^2_{n_2} - \gamma)]|P_{n_1,n_2}) \Rightarrow \mathsf{N}(0, (1-\kappa)\sigma_1^2 + \kappa\sigma_2^2).$$

To complete the proof we use (8.90). ∎

Example 8.86. Suppose the observations $X_{i,j}$, $1 \leq j \leq n_i$, $i = 1, 2$, are real-valued random variables. We consider the Wilcoxon statistic in the Mann–Whitney form; that is,

$$W_{n_1,n_2} = \frac{1}{n_1 n_2} \sum\nolimits_{i=1}^{n_1} \sum\nolimits_{j=1}^{n_2} I_{(X_{2,j},\infty)}(X_{1,i}),$$

which is, up to the constant $1/(n_1 n_2)$, the number of inversions of the pairs $(X_{1,i}, X_{2,j})$. W_{n_1,n_2} is a U-statistic with $m_1 = m_2 = 1$ and the kernel $\Psi(s,t) = I_{(t,\infty)}(s)$. Suppose $\mathcal{L}(X_{i,k}) = P_i$ with a continuous c.d.f. F_i, $i = 1, 2$. Then

$$\gamma = \mathsf{E}_{P_1 \otimes P_2} \Psi(X_{1,1}, X_{2,1}) = \int [\int I_{(t,\infty)}(s) P_1(ds)] P_2(dt) = \int [1 - F_1(t)] P_2(dt).$$

If $P_1 = P_2 =: P$, then $F(X_{2,1})$ has a uniform distribution on $(0,1)$, so that $\gamma = \frac{1}{2}$. Moreover,

$$\Psi_1(s) = \int I_{(t,\infty)}(s) P(dt) = F(s), \quad \Psi_2(t) = \int I_{(t,\infty)}(s) P(ds) = 1 - F(t),$$

$$\sigma_1^2 = \sigma_2^2 = \int (1 - F(s) - \frac{1}{2})^2 P(ds) = \frac{1}{12}.$$

From Proposition 8.85 we get under the condition (8.91)

$$\begin{aligned}\sqrt{\frac{n_1 n_2}{n}} (W_{n_1,n_2} - \frac{1}{2}) &= \sqrt{\frac{n_1 n_2}{n}} [(\mathcal{U}^1_{n_1} - \frac{1}{2}) + (\mathcal{U}^2_{n_2} - \frac{1}{2})] + o_{P_{n_1,n_2}}(1) \\ &= \sqrt{\frac{n_1 n_2}{n}} [\frac{1}{n_1} \sum\nolimits_{i=1}^{n_1} (F(X_{1,i}) - \frac{1}{n_2} \sum\nolimits_{j=1}^{n_2} (F(X_{2,j})] + o_{P_{n_1,n_2}}(1),\end{aligned}$$

and

$$\mathcal{L}(\sqrt{\frac{n_1 n_2}{n}} (W_{n_1,n_2} - \frac{1}{2}) | P^{\otimes n_1} \otimes P^{\otimes n_2}) \Rightarrow \mathsf{N}(0, \frac{1}{12}),$$

which is the classical result on the asymptotic distribution of the Mann–Whitney statistic.

Remark 8.87. Theorem 8.83 and Proposition 8.85 deal with special U-statistics and thus provide only first results. A deeper insight into the structure of U-statistics gives the Hoeffding decomposition of $\mathcal{U}_n$ into mutually uncorrelated statistics. For this and other results we refer to Koroljuk and Borovskich (1994).

Often one is interested in a comparison of two or more populations with the goal of finding that one with the stochastically largest distribution. If the data are not explained by a parametric model, then rank methods come into consideration. To study the behavior of tests that are based on rank statistics under both the null hypothesis and local alternatives, we need several technical results that are taken from Hájek, Šidák, and Sen (1999).

We recall the rank statistic $r(x)$ and the order statistic $s(x)$ that have been introduced in (2.5). If $x \in \mathbb{R}^n_{\neq}$, then all components of the vector $r(x)$ are different so that $r(x)$ is a permutation of $(1, ..., n)$. Then the inverse permutation of $r(x)$ is called the vector of *antiranks* and is denoted by $d(x) = (d_1(x), ..., d_n(x))$. It holds

$$x_i = s_{r_i(x)}(x) \quad \text{and} \quad s_i(x) = x_{d_i(x)}, \quad x \in \mathbb{R}^n_{\neq}, \quad i = 1, ..., n. \tag{8.92}$$

Let Π_n be the group of permutations of $(1, ..., n)$. For every $\gamma \in \Pi_n$ we set $u_\gamma(x_1, ..., x_n) = (x_{\gamma(1)}, ..., x_{\gamma(n)})$. We recall that the distribution P on the Borel sets of the real line is called atomless if $P(\{t\}) = 0$ for every $t \in \mathbb{R}$, which is equivalent to the continuity of the c.d.f. $F(t) = P((-\infty, t])$, $t \in \mathbb{R}$. In that case $P^{\otimes n}(\mathbb{R}^n_{\neq}) = 1$ (see, e.g., Problem 8.105). Let $X_1, ..., X_n$ be the coordinate mappings of $\mathbb{R}^n$ onto $\mathbb{R}$ which are i.i.d. random variables on $(\mathbb{R}^n, \mathfrak{B}_n, P^{\otimes n})$. For $X = (X_1, ..., X_n)$ we introduce the antirank statistic D_n by

$$D_n = (D_{n,1}, ..., D_{n,n}) = (d_1(X), ..., d_n(X)). \tag{8.93}$$

We also recall the standard notation, that is, $X_{[\cdot]} = (X_{n,[1]}, ..., X_{n,[n]}) = s(X)$ for the order statistic and $R_n = (R_{n,1}, ..., R_{n,n}) = r(X)$ for the rank statistic; see (2.5).

Proposition 8.88. *If P is atomless, then the rank statistic R_n and the order statistic $X_{[\cdot]}$ are independent. $R_n = (R_{n,1}, ..., R_{n,n})$ has a uniform distribution on Π_n. $X_{[\cdot]} = (X_{n,[1]}, ..., X_{n,[n]})$ has the distribution $n!P^{\otimes n}(\cdot \cap \mathbb{R}^n_{<})$.*

Proof. The product measure $P_n = P^{\otimes n}$ is concentrated on $\mathbb{R}^n_{\neq}$ and is permutation invariant. Hence we get for any $\gamma \in \Pi_n$ and $B \in \mathfrak{B}_{n,<}$ from (8.92) that

$$\begin{aligned} P_n(R_n = \gamma,\ X_{[\cdot]} \in B) &= P_n((X_{\gamma^{-1}(1)}, ..., X_{\gamma^{-1}(n)}) \in B) \\ &= \int I_B(x_1, ..., x_n) P_n(dx_{\gamma(1)}, ..., dx_{\gamma(n)}) = P_n(B). \end{aligned}$$

Taking the sum over all permutations we get $P_n(X_{[\cdot]} \in B) = n!P_n(B)$. On the other hand, for $B = \mathbb{R}_{n,<}$ the permutation invariance of $P^{\otimes n}$ yields

$$P_n(R_n = \gamma) = P_n(\mathbb{R}_{n,<}) = \frac{1}{n!}.$$

Hence,

$$P_n(R_n = \gamma,\ X_{[\cdot]} \in B) = P_n(R_n = \gamma) P_n(X_{[\cdot]} \in B).$$

■

Next we consider the model $(\mathbb{R}^n, \mathfrak{B}_n, P_n)$, where $P_n = P^{\otimes n}$, and P is the uniform distribution on $(0,1)$. To indicate this we denote now the coordinate mappings by $U_1, ..., U_n$. Set

$$\begin{aligned} \mathbb{L}_2((0,1)) &= \{\varphi : \int_0^1 \varphi^2(t)dt < \infty,\ \varphi : (0,1) \to_m \mathbb{R}\}, \\ a_n^\varphi(k) &= \mathsf{E}_{P_n}(\varphi(U_1)|R_{n,1} = k), \quad \varphi \in \mathbb{L}_2((0,1)),\ k = 1, ..., n. \end{aligned}$$

We note that $x_i = s_{r_i(x)}(x)$ in (8.92) implies

$$U_i = U_{n,[R_{n,i}]}, \quad i = 1, ..., n. \tag{8.94}$$

Hence by the independence of $(U_{n,[1]}, ..., U_{n,[n]})$ and $R_n = (R_{n,1}, ..., R_{n,n})$, and Problem 8.98,

$$a_n^\varphi(k) = \mathsf{E}_{P_n}\varphi(U_{n,[k]}), \quad k = 1, ..., n.$$

The next proposition, as well as the other subsequent results on rank statistics, are taken from Hájek, Šidák, and Sen (1999).

Proposition 8.89. *It holds for every $\varphi \in \mathbb{L}_2((0,1))$,*

$$\lim_{n\to\infty} \mathsf{E}_{P_n}(a_n^\varphi(R_{n,1}) - \varphi(U_1))^2 = 0.$$

Proof. Consider all random variables that appear below to be defined on $(\Omega, \mathfrak{F}, \mathbb{P})$. Problem 8.109 implies that $(n+1)^{-1}R_{n,1}$ tends stochastically to U_1. Hence by Proposition A.12 there is a subsequence n_k such that $(n_k+1)^{-1}R_{n_k,1}$ tends to U_1, $\mathbb{P}$-a.s. Hence $U_1 = \widetilde{U}$, $\mathbb{P}$-*a.s.*, for some random variable $\widetilde{U}$ that is measurable with respect to the sub-σ-algebra generated by all $R_{n,k}$, $1 \le k \le n$, $n = 1, 2, ...$ Denote by $\mathfrak{F}_n$ the sub-σ-algebra generated by $R_{m,k}$, $1 \le k \le m \le n$. Then $\mathfrak{F}_1 \subseteq \mathfrak{F}_2 \subseteq \cdots$, and $\mathfrak{F}_\infty$ is generated by $\mathfrak{F}_i$, $i = 1, 2, ...$ We note that by the definition of a_n^φ it holds $\mathbb{E}(\varphi(U_1)|\mathfrak{F}_n) = a_n^\varphi(R_{n,1})$. As $\varphi(\widetilde{U})$ is $\mathfrak{F}_\infty$-measurable and $U_1 = \widetilde{U}$, $\mathbb{P}$-a.s., the martingale convergence theorem (see Theorem A.34) gives the statement. ■

Let $[x]$ denote the the integer part of x. Then $a_n^\varphi(1+[un])$ is a piecewise constant function of u on $[0,1)$ that takes on the value $a_n^\varphi(k)$ on $[(k-1)/n, k/n)$, $k = 1, ..., n$. Put for completeness $a_n^\varphi(n+1) := a_n^\varphi(n)$.

Lemma 8.90. *It holds* $\lim_{n\to\infty} \int_0^1 (a_n^\varphi(1+[un]) - \varphi(u))^2 du = 0$ *for every* $\varphi \in \mathbb{L}_2((0,1))$.

Proof. Set $b_n(u) = [un]$. It holds for $\varphi, \psi \in \mathbb{L}_2((0,1))$,

$$\begin{aligned}
&\int_0^1 (a_n^\varphi(1+b_n(u)) - a_n^\psi(1+b_n(u)))^2 du \\
&= \mathsf{E}_{P_n} \int_0^1 (\mathsf{E}_{P_n}(\varphi(U_1) - \psi(U_1)|R_{n,1} = 1+b_n(u)))^2 du \\
&= \int_0^1 \mathsf{E}_{P_n}(\mathsf{E}_{P_n}(\varphi(U_1) - \psi(U_1)|R_{n,1} = 1+b_n(u)))^2 du \qquad (8.95)\\
&\le \mathsf{E}_{P_n}(\varphi(U_1) - \psi(U_1))^2.
\end{aligned}$$

As the set of all bounded and continuous functions φ is dense in $\mathbb{L}_2((0,1))$ it suffices to show the statements for such a φ. Note that by Problem 8.108 $U_{n,[k]}$ has the distribution $\mathsf{Be}(k, n-k+1)$. By (8.94), the independence of $(U_{n,[1]}, ..., U_{n,[n]})$ and $R_n = (R_{n,1}, ..., R_{n,n})$, and Problem 8.98, we get that the conditional distribution of U_1, given $R_{n,1} = 1 + b_n(u)$, is the distribution of $U_{n,[1+b_n(u)]}$, which is $\mathsf{Be}(1+b_n(u), n - b_n(u))$. As

$$\mathsf{V}_{P_n}(U_{n,[1+b_n(u)]}) = \frac{(1+b_n(u))(n-b_n(u))}{(n+1)^2(n+2)} \to 0, \quad \text{and}$$
$$\mathsf{E}_{P_n} U_{n,[1+b_n(u)]} = (1+b_n(u))/(n+1) \to u,$$

we get that the distributions $\mathsf{Be}(1+b_n(u), n-b_n(u))$ tend weakly to the delta distribution at u. Hence

$$a_n^{\varphi}(1+b_n(u)) = \int \varphi(t)\mathsf{Be}(1+b_n(u), n-b_n(u))(dt) \to \varphi(u), \quad 0<u<1,$$

for every φ that is bounded and continuous on $(0,1)$. The inequality (8.95) gives for $\psi = 0$

$$\int_0^1 (a_n^{\varphi}(1+b_n(u)))^2 du = \mathbb{E}(\varphi(U_1))^2 \le \int_0^1 (\varphi(u))^2 du.$$

Applying Vitali's theorem (see Theorem A.21) completes the proof. ∎

Let $c_{i,n}$, $i=1,...,n$, be constants in $\mathbb{R}$ that satisfy

$$\sum\nolimits_{i=1}^n c_{i,n} = 0 \quad \text{and} \quad \sum\nolimits_{i=1}^n c_{i,n}^2 = 1. \tag{8.96}$$

Lemma 8.91. *If P is the uniform distribution on $(0,1)$, $P_n = P^{\otimes n}$, and (8.96) is satisfied, then for any $a_n(k)$, $k=1,...,n$,*

$$\mathsf{E}_{P_n}(\sum\nolimits_{i=1}^n c_{i,n}a_n(R_{n,i}) - \sum\nolimits_{i=1}^n c_{i,n}\varphi(U_i))^2 \le \frac{n}{n-1}\mathsf{E}_{P_n}(a_n(R_{n,1}) - \varphi(U_1))^2.$$

Proof. As $U_{[\cdot]} = (U_{n,[1]},...,U_{n,[n]})$ and $R_n = (R_{n,1},...,R_{n,n})$ are independent, and $U_i = U_{n,[R_{n,i}]}$, $i=1,...,n$,

$$\begin{aligned}&\mathsf{E}_{P_n}((\sum\nolimits_{i=1}^n c_{i,n}(a_n(R_{n,i}) - \varphi(U_i)))^2 | U_{[\cdot]} = (u_1,...,u_n))\\ &= \mathsf{E}_{P_n}(\sum\nolimits_{i=1}^n c_{i,n}(a_n(R_{n,i}) - \varphi(u_{R_{n,i}})))^2.\end{aligned}$$

As $R_{n,i}$ has a uniform distribution on $\{1,...,n\}$ it holds,

$$\mathsf{E}_{P_n}\sum\nolimits_{i=1}^n c_{i,n}(a_n(R_{n,i}) - \varphi(u_{R_{n,i}})) = \sum\nolimits_{i=1}^n c_{i,n}\frac{1}{n}\sum\nolimits_{k=1}^n (a_n(k) - \varphi(u_k)) = 0.$$

Hence by Problem 8.106 and $\sum_{i=1}^n c_{i,n}^2 = 1$,

$$\mathsf{E}_{P_n}(\sum\nolimits_{i=1}^n c_{i,n}(a_n(R_{n,i}) - \varphi(u_{R_{n,i}})))^2 = \sigma_a^2 \le \frac{1}{n-1}\sum\nolimits_{k=1}^n a^2(k),$$

where $a(k) = a_n(k) - \varphi(u_k)$. This yields

$$\begin{aligned}&\mathsf{E}_{P_n}(\sum\nolimits_{i=1}^n c_{i,n}(a_n(R_{n,i}) - \varphi(U_i)))^2 \le \frac{1}{n-1}\sum\nolimits_{k=1}^n \mathsf{E}_{P_n}(a_n(k) - \varphi(U_{n,[k]}))^2\\ &= \frac{1}{n-1}\sum_{k=1}^n \mathsf{E}_{P_n}(a_n(R_{n,k}) - \varphi(U_{n:R_{n,k}}))^2 = \frac{n}{n-1}\mathsf{E}_{P_n}(a_n(R_{n,1}) - \varphi(U_1))^2,\end{aligned}$$

where we have used that $U_{n,[R_{n,k}]} = U_k$ and $\mathcal{L}(R_{n,k}, U_k) = \mathcal{L}(R_{n,1}, U_1)$. ■

Now we are ready to establish the basic result on the approximation of linear rank statistics. The statement below is Theorem 1 on p. 194 in Hájek, Šidák, and Sen (1999).

Theorem 8.92. *Suppose $U_1, ..., U_n$ are i.i.d. with a common uniform distribution on $(0,1)$, $\varphi \in \mathbb{L}_2((0,1))$, and (8.96) is satisfied. If*

$$\lim_{n\to\infty} \int_0^1 (a_n(1+[un]) - \varphi(u))^2 du = 0, \tag{8.97}$$

then

$$\sum\nolimits_{i=1}^n c_{i,n} a_n(R_{n,i}) = \sum\nolimits_{i=1}^n c_{i,n} \varphi(U_i) + o_{P_n}(1).$$

Proof. In view of Lemma 8.91, with $\varphi = 0$ and a_n replaced by $a_n - a_n^\varphi$, it holds

$$\begin{aligned}
&\mathsf{E}_{P_n}(\sum\nolimits_{i=1}^n c_{i,n} a_n(R_{n,i}) - \sum\nolimits_{i=1}^n c_{i,n} a_n^\varphi(R_{n,i}))^2 \\
&\le \frac{n}{n-1} \mathsf{E}_{P_n}(a_n(R_{n,1}) - a_n^\varphi(R_{n,1}))^2 \\
&\le \frac{n}{n-1} \int_0^1 (a_n(1+[un]) - a_n^\varphi(1+[un]))^2 du \\
&\le \frac{2n}{n-1} [\int_0^1 (a_n(1+[un]) - \varphi(u))^2 du + \int_0^1 (a_n^\varphi(1+[un]) - \varphi(u))^2 du].
\end{aligned}$$

Now we use condition (8.97) and Lemma 8.90 to get

$$\sum\nolimits_{i=1}^n c_{i,n} a_n(R_{n,i}) = \sum\nolimits_{i=1}^n c_{i,n} a_n^\varphi(R_{n,i}) + o_{\mathbb{P}}(1) = \sum\nolimits_{i=1}^n c_{i,n} \varphi(U_i) + o_{\mathbb{P}}(1),$$

where the second equality follows from Lemma 8.91. ■

Next we present some ways of constructing *scores* $a_n(k)$ that satisfy (8.97).

Lemma 8.93. *The condition (8.97) is fulfilled if the $a_n(k)$ satisfy at least one of the following conditions.*

Exact scores	*Averaged scores*	*Approximate scores*	
$a_n(k) = \mathsf{E}_{P_n}\varphi(U_{n,[k]})$	$a_n(k) = n\int_{(k-1)/n}^{k/n} \varphi(t)dt$	$a_n(k) = \varphi(\frac{k}{n+1})$	(8.98)
$\varphi \in \mathbb{L}_2((0,1))$	$\varphi \in \mathbb{L}_2((0,1))$	*φ is a finite sum of monotone functions*	

Proof. The statement for the exact scores has been established already in Lemma 8.90. The statement for the averaged scores is clear for continuous functions. As these functions are dense in $\mathbb{L}_2((0,1))$ there is a sequence of continuous functions φ_m with $\int_0^1 (\varphi_m(t) - \varphi(t))^2 dt \to 0$. If $a_n^{\varphi_m}(k)$ and $a_n^{\varphi}(k)$ are defined by φ_m and φ, respectively, then by (8.95)

$$\int_0^1 (a_n^{\varphi_m}(1 + [un]) - a_n^{\varphi}(1 + [un]))^2 du \le \int_0^1 (\varphi_m(t) - \varphi(t))^2 dt,$$

which completes the proof of the second statement. As to the third statement, we note that φ, and therefore $|\varphi|$, is of bounded variation in every closed subinterval of $(0,1)$, so that we may assume that $|\varphi|$ is nondecreasing. Set $\varphi_n(u) := \varphi((1 + [un])/(n+1))$. Then

$$\begin{aligned}\frac{n}{n+1}\int_0^1 |\varphi_n(u)|^2 du &= \frac{1}{n+1}\sum\nolimits_{j=1}^{n} |\varphi(j/(n+1))|^2 \\ &\le \sum\nolimits_{j=1}^{n} \int I_{(\frac{j}{n+1},\frac{j+1}{n+1}]}(u)|\varphi(u)|^2 du \le \int_0^1 |\varphi(u)|^2 du.\end{aligned}$$

Hence,

$$\limsup_{n\to\infty} \int_0^1 |\varphi_n(u)|^2 du \le \int_0^1 |\varphi(u)|^2 du.$$

The set of points of discontinuity is at most countable, so that $\varphi_n(u) \to \varphi(u)$, a.e. with respect to the Lebesgue measure. An application of Vitali's theorem (see Theorem A.21) completes the proof. ■

Theorem 8.92 gives an approximation of a linear rank statistic by a linear statistic. This allows us to study the asymptotic distribution of the linear rank statistics under both the null hypothesis and local alternatives. Now, given a linear statistic $\sum_{i=1}^n c_{i,n}\Psi(X_i)$, where $X_1, ..., X_n$ are i.i.d. with a continuous c.d.f. F, we use $X_i = F^{-1}(U_i)$, $\mathbb{P}$-a.s., for $U_i = F(X_i)$ to get

$$\sum\nolimits_{i=1}^{n} c_{i,n}\Psi(X_i) = \sum\nolimits_{i=1}^{n} c_{i,n}\varphi(U_i), \quad \text{where}$$
$$\varphi(u) = \Psi(F^{-1}(u)), \quad 0 < u < 1.$$

To rewrite the linear statistic as a linear rank statistic we assume that $\mathbb{E}\Psi^2(X_1) < \infty$, which implies that $\varphi \in \mathbb{L}_2((0,1))$. If the scores a_n are chosen to satisfy condition (8.98), then by Theorem 8.92

$$\sum\nolimits_{i=1}^{n} c_{i,n}\Psi(X_i) = \sum\nolimits_{i=1}^{n} c_{i,n}a_n(R_{n,i}) + o_{\mathbb{P}}(1). \tag{8.99}$$

Statistics of the type $\sum_{i=1}^n c_{i,n}\Psi(X_i)$, with *regression coefficients* $c_{i,n}$ satisfying (8.96), typically appear in two-sample problems. Indeed, if we consider two samples $X_{1,1}, ..., X_{1,n_1}$ and $X_{2,1}, ..., X_{2,n_2}$, where under the null hypothesis the distributions in both populations are the same, then statistics

$$T_n = \sqrt{\frac{n_1 n_2}{n}}[\frac{1}{n_1}\sum\nolimits_{j=1}^{n_1} \Psi(X_{1,j}) - \frac{1}{n_2}\sum\nolimits_{j=1}^{n_2} \Psi(X_{2,j})],$$

where $n = n_1 + n_2$, are often used as test statistics. If we set $X_i = X_{1,i}$, $i = 1, ..., n_1$, and $X_{i+n_1} = X_{2,i}$, $i = 1, ..., n_2$, and

$$\begin{aligned} c_{i,n} &= \sqrt{\frac{n_1 n_2}{n}} \frac{1}{n_1}, \quad 1 \leq i \leq n_1, \\ c_{i,n} &= -\sqrt{\frac{n_1 n_2}{n}} \frac{1}{n_2}, \quad i = n_1 + 1, ..., n_1 + n_2, \end{aligned} \tag{8.100}$$

then (8.96) is satisfied. We get from (8.99)

$$T_n = \sqrt{\frac{n_1 n_2}{n}} [\frac{1}{n_1} \sum\nolimits_{i=1}^{n_1} a_n(R_{n,i}) - \frac{1}{n_2} \sum\nolimits_{i=n_1+1}^{n} a_n(R_{n,i})] + o_{\mathbb{P}}(1), \tag{8.101}$$

F is the c.d.f. of $X_{i,j}$, a_n and $\varphi = \Psi(F^{-1})$ are related by (8.97).

The influence function Ψ often comes into consideration as the $\mathbb{L}_2$-derivative of a one-parameter model. The next display presents linear rank statistics that originate from the location model $f(x - \theta)$ with $\Psi = \dot{L}_0 = -f'/f$.

	Fisher-Yates	Median	Wilcoxon
f	$\frac{1}{\sqrt{2\pi}} \exp\{-\frac{x^2}{2}\}$	$\frac{1}{2} \exp\{-\|x\|\}$	$\frac{\exp\{-x\}}{(1+\exp\{-x\})^2}$
$\varphi = -\frac{f'}{f}(F^{-1})$	$\Phi^{-1}(u)$	$\mathrm{sgn}(2x - 1)$	$2x - 1$
$a_n(k)$	$\mathsf{E}_{P_n} \Phi^{-1}(U_{n,[k]})$	$\mathrm{sgn}(\frac{k}{n+1} - 1)$	$\frac{2k}{n+1} - 1$

(8.102)

Let us consider once more the statistic $T_n = \sum_{i=1}^{n} c_{i,n} \Psi(X_i)$. If T_n is used as the test statistic, then under the null hypothesis the X_i are i.i.d. If the test rejects the null hypothesis for large values of T_n one needs the $1 - \alpha$ quantile of the distribution of T_n which can be hardly calculated in closed form unless the $\Psi(X_i)$ satisfy additional conditions which guarantee that the convolutions involved can be carried out. One way out of this dilemma is large sample approximations, which have the shortcoming that the level α may be exceeded. Another way is the concept of *permutation tests*. Suppose that the X_i are *i.i.d.* with a continuous distribution. Using the ranks $R_{n,i}$ in (8.92) we get

$$T_{\Psi,n} = \sum\nolimits_{i=1}^{n} c_{i,n} \Psi(X_i) = \sum\nolimits_{i=1}^{n} c_{i,n} \Psi(X_{n,[R_{n,i}]}).$$

Now we use the fact that the order statistic $X_{[\cdot]}$ and the rank vector R_n are independent. The conditional distribution of T_n, given $X_{[\cdot]} = (x_{[1]}, ..., x_{[n]})$, is the distribution of $\sum_{i=1}^{n} c_{i,n} \Psi(x_{[R_{n,i}]})$, where the vector R_n is uniformly distributed on the set of all permutations. Then a conditional test, called permutation test, can be established according the concept of a conditional test that we have dicussed in Section 8.1.3. For further details on permutation tests we refer to Lehmann and Romano (2005).

Subsequently, problems are listed whose results have been used in the preceding proofs.

Problem 8.94.* If $X_i \in \mathbb{L}_2(\mathbb{P})$, then $\mathcal{L} = span(\{X_1, ..., X_n\})$ is closed. If $X_1, X_2, ...$are i.i.d. with $X_i \sim \mathsf{N}(\mu, 1)$ and $\mu \neq 0$, then $\mathcal{L} = span(\{X_1, X_2, ...\})$ is not closed.

Problem 8.95.* Prove the existence and uniqueness of the projection $\Pi_{\mathcal{L}} Y$.

Problem 8.96.* For $Z \in \mathbb{L}_2(\mathbb{P})$ it holds $Z = \Pi_{\mathcal{L}} Y$ if and only if $Z \in \mathcal{L}$, and $Y - Z \perp X$, $X \in \mathcal{L}$.

Problem 8.97.* Set $X = (X_1, ..., X_n)^T$, $\mathcal{L} = span(\{X_1, ..., X_n\})$ and suppose that $X_1, ..., X_n \in \mathbb{L}_2(\mathbb{P})$ are linearly independent. Then $\mathbb{E}(XX^T)$ is nonsingular and $\Pi_{\mathcal{L}} Y = (\mathbb{E}(YX^T))(\mathbb{E}(XX^T))^{-1} X$ for $Y \in \mathbb{L}_2(\mathbb{P})$.

Problem 8.98.* If X and Y are independent random variables with values in $(\mathcal{X}, \mathfrak{A})$ and $(\mathcal{Y}, \mathfrak{B})$, respectively, and $h : \mathcal{X} \times \mathcal{Y} \to_m \mathbb{R}$ satisfies $\mathbb{E}|h(X,Y)| < \infty$, then $\mathbb{E}(h(X,Y)|\sigma(X)) = g(X)$, $\mathbb{P}$-a.s., where $g(x) = \mathbb{E}h(x,Y) = \int h(x,y) P_Y(dy)$.

Problem 8.99.* Let $\mathfrak{G}$ be a sub-σ-algebra of $\mathfrak{F}$. Let $\mathbb{H}$ be the subspace of $\mathbb{L}_2(\mathbb{P})$ of all $Z \in \mathbb{L}_2(\mathbb{P})$ which are $\mathfrak{G}$-measurable, and $\mathbb{H}^0 \subseteq \mathbb{H}$ the subspace for which in addition $\mathbb{E}Z = 0$ holds. Then for $\mathbb{E}X^2 < \infty$ it holds $\mathbb{E}(X|\mathfrak{G}) = \Pi_{\mathbb{H}} X$ and $\mathbb{E}(X|\mathfrak{G}) - \mathbb{E}X = \Pi_{\mathbb{H}^0} X$.

Problem 8.100.* For $S \in \mathcal{L}$ it holds $\mathsf{E}_{P_n} S^2 = a_0^2 + \sum_{i=1}^n \mathsf{E}_{P_n} a_i^2(X_i)$, and the representation $S = a_0 + \sum_{i=1}^n a_i(X_i)$ is unique.

Problem 8.101.* The space $\mathcal{L}$ in (8.83) is a closed linear subspace of $\mathbb{L}_2(P^{\otimes n})$.

Problem 8.102.* $\mathsf{V}_{P_n}(\mathcal{U}_n) = \binom{n}{m}^{-1} \sum_{k=1}^m \binom{m}{k}\binom{n-m}{m-k} \sigma_k^2$, where σ_k^2 is given by $\sigma_k^2 = \mathsf{V}_{P_n}(\Psi_k(X_1, ..., X_k))$. $\mathsf{V}_{P_n}(\Psi_k(X_1, ..., X_k)) \leq \mathsf{V}_{P_n}(\Psi_m(X_1, ..., X_m))$, $k \leq m$. $\mathsf{E}_{P_n}(\mathcal{U}_n - \widehat{\mathcal{U}}_n)^2 = \mathsf{V}_{P_n}(\mathcal{U}_n) - \mathsf{V}_{P_n}(\widehat{\mathcal{U}}_n) = o(1/n)$.

Problem 8.103. Let $X_1, ..., X_n$ be i.i.d. with finite fourth moments, and denote by σ^2 the variance and by μ_4 the fourth central moment. Then the variance of $S_n^2 = (n-1)^{-1} \sum_{i=1}^n (X_i - \overline{X}_n)^2$ is $\mathsf{V}(S_n^2) = n^{-1}(\mu_4 - \sigma^4) + 2[n(n-1)]^{-1}\sigma^4$.

Problem 8.104.* Show (8.90) for $\Psi : \mathcal{X}_1 \times \mathcal{X}_2 \to_m \mathbb{R}$, $\sigma^2 = \mathsf{V}_{P_1 \otimes P_2} \Psi^2 < \infty$.

Problem 8.105. For independent random variables X and Y it holds $\mathbb{P}(X = Y) = \int \mathbb{P}(X = t) P_Y(dt) = \sum_{t \in A} \mathbb{P}(X = t)\mathbb{P}(Y = t)$, where the set A of all t with $\mathbb{P}(X = t)\mathbb{P}(Y = t) \neq 0$ is at most countable.

Problem 8.106.* For $S_n = \sum_{i=1}^n c_{i,n} a_n(R_{n,i})$ it holds $\mathsf{E}_{P_n} S_n = \overline{a}_n \sum_{i=1}^n c_{i,n}$ and $\mathsf{V}_{P_n}(S_n) = \sigma_{a_n}^2 \sum_{i=1}^n (c_{i,n} - \overline{c}_n)^2$, where $\overline{a}_n = n^{-1} \sum_{i=1}^n a_n(i)$, $\overline{c}_n = n^{-1} \sum_{i=1}^n c_{i,n}$, and $\sigma_{a_n}^2 = (n-1)^{-1} \sum_{i=1}^n (a(i) - \overline{a}_n)^2$.

Problem 8.107. If $\mathsf{Be}_{a,\beta}$ is the c.d.f. of a beta distribution and $\mathsf{b}_{n,p}$ is the p.m.f. of a binomial distribution, then $\sum_{l=k}^n \mathsf{b}_{n,p}(l) = \mathsf{Be}_{\alpha,\beta}(p)$ for $\alpha = k$ and $\beta = n - k + 1$.

Problem 8.108.* If $U_1, ..., U_n$ are i.i.d. with a uniform distribution on $(0,1)$, then the distribution of $U_{n,[k]}$ is $\mathsf{Be}(k, n-k+1)$, and it holds $\mathsf{E}_{P_n} U_{n,[k]} = k/(n+1)$ and $\mathsf{V}_{P_n}(U_{n,[k]}) = k(n-k+1)/[(n+1)^2(n+2)]$.

Problem 8.109.* If P is the uniform distribution on $(0,1)$, then it holds that $\mathsf{E}_{P_n}(U_1 - (n+1)^{-1} R_{n,1})^2 = n^{-1} \sum_{k=1}^n k(n-k+1)[(n+1)^2(n+2)]^{-1} < 1/n$.

8.7 Statistics with Estimated Parameters

Quite often in testing problems the parameter $\theta \in \Delta \subseteq \mathbb{R}^d$ consists of two components, the parameter of interest τ and a nuisance parameter ξ. For d-parameter exponential families we were able to eliminate the nuisance parameter by conditioning. This technique, however, is restricted to exponential families. For asymptotic tests another idea appears to be intuitively feasible. Suppose we want to test $\mathsf{H}_0 : \tau = \tau_0$ versus $\mathsf{H}_A : \tau \neq \tau_0$. In a first step, we could fix $\xi = \xi_0$, construct an asymptotically optimal test for the simple null hypothesis, and then, in a second step, plug in a consistent estimator for the unknown ξ_0. This technique seems to be the "natural way". However, the problem arises that this plug-in procedure may change the asymptotic size of the test. In other words, the new test may not be an asymptotic level α test, unless additional suitable conditions are satisfied. The aim of this section is to find such conditions, and to present a technique that allows a systematic construction of test statistics that are insensitive to the estimation of nuisance parameters. Here again, the concept of projection is of great importance.

We consider estimators that admit a stochastic Taylor expansion, where the influence function depends on some nuisance parameter ξ. For the model $(\mathcal{X}, \mathfrak{A}, (P_\xi)_{\xi \in \Xi})$ denote by $X_1, ..., X_n$ the projections of $\mathcal{X}^n \to \mathcal{X}$ onto the coordinates, and set $P_{n,\xi_0} = P_{\xi_0}^{\otimes n}$. Let $\Xi \subseteq \mathbb{R}^m$ be open and $\widehat{\xi}_n : \mathcal{X}^n \to_m \Xi$ be a sequence of estimators of ξ_0 that admits the stochastic Taylor expansion

$$\sqrt{n}(\widehat{\xi}_n - \xi_0) = \frac{1}{\sqrt{n}} \sum\nolimits_{i=1}^{n} \Psi(X_i) + o_{P_{n,\xi_0}}(1), \quad \Psi \in \mathbb{L}_{2,m}^0(P_{\xi_0}). \tag{8.103}$$

The next result is a transformation rule for the influence function of statistics with estimated parameters.

Proposition 8.110. ***(Transformation of Influence Function)*** *Assume the function $S : \Xi \times \mathcal{X} \to \mathbb{R}^k$ satisfies $S \in \mathsf{C}_m^{(1)}(U(\xi_0), \mathcal{X})$ and condition (B) in (A9) is satisfied for the model $(\mathcal{X}, \mathfrak{A}, (P_\xi)_{\xi \in \Xi})$. If $\widehat{\xi}_n : \mathcal{X}^n \to_m \Xi$ is a $\sqrt{n}$-consistent estimator at ξ_0, then*

$$\frac{1}{\sqrt{n}} \sum\nolimits_{i=1}^{n} S_{\widehat{\xi}_n}(X_i) = \frac{1}{\sqrt{n}} \sum\nolimits_{i=1}^{n} S_{\xi_0}(X_i) + (\mathsf{E}_{\xi_0} \dot{S}_{\xi_0}^T(X_1)) \sqrt{n}(\widehat{\xi}_n - \xi_0) + o_{P_{n,\xi_0}}(1).$$

If in addition (8.103) holds, then

$$\frac{1}{\sqrt{n}} \sum\nolimits_{i=1}^{n} S_{\widehat{\xi}_n}(X_i) = \frac{1}{\sqrt{n}} \sum\nolimits_{i=1}^{n} \widetilde{S}_{\xi_0}(X_i) + o_{P_{n,\xi_0}}(1), \quad \text{where}$$

$$\widetilde{S}_{\xi_0} = S_{\xi_0} + (\mathsf{E}_{\xi_0} \dot{S}_{\xi_0}^T) \Psi. \tag{8.104}$$

Proof. The Taylor expansion in Theorem A.2 and Lemma 7.141 give

$$\begin{aligned}
&\frac{1}{\sqrt{n}}\sum\nolimits_{i=1}^{n}(S_{\widehat{\xi}_n}(X_i)-S_{\xi_0}(X_i))\\
&=(\frac{1}{n}\sum\nolimits_{i=1}^{n}\int_0^1 \dot{S}^T_{\xi_0+s(\widehat{\xi}_n-\xi_0)}(X_i)ds)\sqrt{n}(\widehat{\xi}_n-\xi_0)+o_{P_{n,\xi_0}}(1)\\
&=(\mathsf{E}_{\xi_0}\dot{S}^T_{\xi_0}(X_1)+o_{P_n}(1))\sqrt{n}(\widehat{\xi}_n-\xi_0)+o_{P_{n,\xi_0}}(1)\\
&=(\mathsf{E}_{\xi_0}\dot{S}^T_{\xi_0}(X_1))\sqrt{n}(\widehat{\xi}_n-\xi_0)+o_{P_{n,\xi_0}}(1),
\end{aligned}$$

where the last equality follows from the fact that $\sqrt{n}(\widehat{\xi}_n-\xi_0)=O_{P_{n,\xi_0}}(1)$. ∎

The above transformation rule admits a simple interpretation. The plug-in procedure produces an additional term that depends on $\mathsf{E}_{\xi_0}\dot{S}_{\xi_0}(X_1)$ and the influence function Ψ that expresses the fluctuation of the estimator. If $\mathsf{E}_{\xi_0}\dot{S}_{\xi_0}(X_1)=0$, then the difference

$$\frac{1}{\sqrt{n}}\sum\nolimits_{i=1}^{n}S_{\widehat{\xi}_n}(X_i)-\frac{1}{\sqrt{n}}\sum\nolimits_{i=1}^{n}S_{\xi_0}(X_i)$$

is asymptotically negligible. Such a situation is typical for the two-sample case. Consider the sequence of models

$$(\mathcal{X}^{n_1}\times\mathcal{X}^{n_2},\mathfrak{A}^{\otimes n_1}\otimes\mathfrak{A}^{\otimes n_2},(P_\xi^{\otimes n_1}\otimes P_\xi^{\otimes n_2})_{\xi\in\Xi}),$$

for which the projections $X_{1,1},...,X_{1,n_1}$ and $X_{2,1},...,X_{2,n_2}$ are i.i.d. with $\mathcal{L}(X_{i,j})=P_\xi$, $j=1,...,n_i$, $i=1,2$. Set $P_{n_1,n_2,\xi}=P_\xi^{\otimes n_1}\otimes P_\xi^{\otimes n_2}$ and $n=n_1+n_2$.

Lemma 8.111. *Assume that the function $S:\Xi\times\mathcal{X}\to\mathbb{R}^k$ satisfies $S\in\mathsf{C}_m^{(1)}(U(\xi_0),\mathcal{X})$ and (B) in (A9) is fulfilled. If there are positive constants c_1 and c_2 such that $c_1\le n_i/n\le c_2$, then for every sequence of $\sqrt{n}$-consistent estimators $\widehat{\xi}_n:\mathcal{X}^n\to_m\Xi$ of ξ_0 it holds*

$$\begin{aligned}
&\frac{1}{n_1}\sum\nolimits_{j=1}^{n_1}S_{\widehat{\xi}_n}(X_{1,j})-\frac{1}{n_2}\sum\nolimits_{j=1}^{n_2}S_{\widehat{\xi}_n}(X_{2,j})\\
&=\frac{1}{n_1}\sum\nolimits_{j=1}^{n_1}S_{\xi_0}(X_{1,j})-\frac{1}{n_2}\sum\nolimits_{j=1}^{n_2}S_{\xi_0}(X_{2,j})+o_{P_{n_1,n_2,\xi_0}}(1/\sqrt{n}).
\end{aligned}$$

Proof. As $\mathsf{E}_{\xi_0}\dot{S}_{\xi_0}(X_{1,1})=\mathsf{E}_{\xi_0}\dot{S}_{\xi_0}(X_{2,1})$ it is, in view of $c_1\le n_i/n\le c_2$, enough to show that for $i=1,2$,

$$\frac{1}{n_i}\sum\nolimits_{j=1}^{n_i}(S_{\widehat{\xi}_n}(X_{i,j})-S_{\xi_0}(X_{i,j})-(\mathsf{E}_{\xi_0}\dot{S}^T_{\xi_0}(X_{1,1}))(\widehat{\xi}_n-\xi_0))=o_{P_{n_1\,n_2,\xi_0}}(1/\sqrt{n}).$$

But this follows from Proposition 8.110. ∎

We have seen in Proposition 8.110 that a linearized statistic is insensitive to plugging in an estimated parameter if $\mathsf{E}_{\xi_0}\dot{S}_{\xi_0}=0$ holds. To analyze this condition in more detail we suppose that $k=m=1$, $S_\xi(x)$ satisfies $S\in$

$\mathsf{C}_m^{(1)}(U(\xi_0), \mathcal{X})$, and (A9) is fulfilled. Suppose the family $(P_\xi)_{\xi \in U(\xi_0)}$ is $\mathbb{L}_2$-differentiable with derivative $\dot{L}_{\xi_0}$. Assume in addition that

$$\mathsf{E}_\xi S_\xi = 0, \quad \xi \in U(\xi_0), \quad \lim_{\xi \to \xi_0} \mathsf{E}_{\xi_0}(S_\xi - S_{\xi_0})^2 = 0.$$

Then by Lemma 7.143 it holds $\mathsf{E}_{\xi_0}\dot{S}_{\xi_0} = -\mathsf{E}_{\xi_0}S_{\xi_0}\dot{L}_{\xi_0}$. This means that $\mathsf{E}_{\xi_0}\dot{S}_{\xi_0} = 0$ if and only if the influence function S_{ξ_0} is orthogonal to $\dot{L}_{\xi_0}$ in the sense of $\mathbb{L}_2(P_{\xi_0})$. Regardles of whether the last condition holds we set

$$W_\xi = S_\xi + (\mathsf{E}_\xi \dot{S}_\xi)\mathsf{I}^{-1}(\xi)\dot{L}_\xi.$$

Then W_{ξ_0} satisfies this orthogonality condition, i.e., it holds $\mathsf{E}_{\xi_0}W_{\xi_0}\dot{L}_{\xi_0} = 0$. Hence we may expect that the new influence function W_ξ leads to a sequence of statistics that is insensitive against plugging in estimators for ξ. This is the content of the next proposition, which we formulate for vector-valued functions S for later purposes.

Proposition 8.112. *Suppose the family $(P_\xi)_{\xi \in U(\xi_0)}$, $U(\xi_0) \subseteq \mathbb{R}^m$, satisfies condition (A10), and the Fisher information matrix $\mathsf{I}(\xi)$ is nonsingular for $\xi \in U(\xi_0)$. Assume that $S : U(\xi_0) \times \mathcal{X} \to \mathbb{R}^k$ satisfies $S \in \mathsf{C}_m^{(1)}(U(\xi_0), \mathcal{X})$ and condition (A9) is fulfilled. If $\xi \mapsto (\mathsf{E}_\xi \dot{S}_\xi)^T \mathsf{I}^{-1}(\xi)$ is continuous and*

$$W_\xi := S_\xi + (\mathsf{E}_\xi \dot{S}_\xi^T)\mathsf{I}^{-1}(\xi)\dot{L}_\xi, \tag{8.105}$$

then

$$\frac{1}{\sqrt{n}}\sum\nolimits_{i=1}^n W_{\widehat{\xi}_n}(X_i) = \frac{1}{\sqrt{n}}\sum\nolimits_{i=1}^n W_{\xi_0}(X_i) + o_{P_{\xi_0}^{\otimes n}}(1) \tag{8.106}$$

holds for every $\sqrt{n}$-consistent sequence of estimators $\widehat{\xi}_n$.

Proof. The distribution of $n^{-1/2}\sum_{i=1}^n \dot{L}_{\xi_0}(X_i)$, by the central limit theorem, tends to a normal distribution. Hence $n^{-1/2}\sum_{i=1}^n \dot{L}_{\xi_0}(X_i)$ is stochastically bounded (see, e.g., Problem 7.129). Thus we get by the continuity of $A(\xi) := (\mathsf{E}_\xi \dot{S}_\xi^T)\mathsf{I}^{-1}(\xi)$,

$$[A(\widehat{\xi}_n) - A(\xi_0)]n^{-1/2}\sum\nolimits_{i=1}^n \dot{L}_{\xi_0}(X_i) = o_{P_{n,\xi_0}}(1).$$

Proposition 8.110 yields

$$\begin{aligned}
&\frac{1}{\sqrt{n}}\sum\nolimits_{i=1}^n (W_{\widehat{\xi}_n}(X_i) - W_{\xi_0}(X_i)) \\
&= \frac{1}{\sqrt{n}}\sum\nolimits_{i=1}^n [(S_{\widehat{\xi}_n}(X_i) - S_{\xi_0}(X_i)) + A(\widehat{\xi}_n)(\dot{L}_{\widehat{\xi}_n}(X_i) - \dot{L}_{\xi_0}(X_i))] + o_{P_{n,\xi_0}}(1) \\
&= [\mathsf{E}_{\xi_0}\dot{S}_{\xi_0}^T + A(\widehat{\xi}_n)\frac{1}{n}\sum\nolimits_{i=1}^n \int_0^1 \ddot{L}_{\xi_0 + s(\widehat{\xi}_n - \xi_0)}(X_i)ds]\sqrt{n}(\widehat{\xi}_n - \xi_0) + o_{P_{n,\xi_0}}(1) \\
&= [\mathsf{E}_{\xi_0}\dot{S}_{\xi_0}^T + A(\xi_0)\mathsf{E}_{\xi_0}\ddot{L}_{\xi_0}(X_1)]\sqrt{n}(\widehat{\xi}_n - \xi_0) + o_{P_{n,\xi_0}}(1),
\end{aligned}$$

where we have used Lemma 7.141 and the continuity of A to get the last equality. Using $\mathsf{E}_{\xi_0}\ddot{L}_{\xi_0}(X_1) = -\mathsf{I}(\xi_0)$ from (7.111), and thus $A(\xi_0)\mathsf{E}_{\xi_0}\ddot{L}_{\xi_0}(X_1) = -\mathsf{E}_{\xi_0}\dot{S}_{\xi_0}^T$, we get the statement. ■

Now we split the parameter $\theta \in \Delta \subseteq \mathbb{R}^d$ into two parts by setting $\theta = (\tau^T, \xi^T)^T$. Part one is τ, which has dimension k, and is the parameter of interest. Part two is ξ, which has dimension $m = d - k$, and is the nuisance parameter. We suppose that $(P_\theta)_{\theta \in U(\theta_0)}$ satisfies (A10). Let

$$\dot{L}_\theta = \begin{pmatrix} U_\theta \\ V_\theta \end{pmatrix}, \quad \theta = \begin{pmatrix} \tau \\ \xi \end{pmatrix} \in \Delta^0, \tag{8.107}$$

be the partition of the $\mathbb{L}_2$-derivative $\dot{L}_\theta$, where U_θ and V_θ are the vectors that contain the derivatives of the log-likelihood with respect to the components of τ and ξ, respectively. The covariance matrix $\mathsf{C}_{\theta_0}(\dot{L}_{\theta_0})$ is the Fisher information matrix, and we partition it analogously by

$$\mathsf{C}_{\theta_0}(\dot{L}_{\theta_0}) = \mathsf{I}(\theta_0) = \begin{pmatrix} \mathsf{I}_{1,1}(\theta_0) & \mathsf{I}_{1,2}(\theta_0) \\ \mathsf{I}_{2,1}(\theta_0) & \mathsf{I}_{2,2}(\theta_0) \end{pmatrix}.$$

Set $S_\xi = U_{\tau_0,\xi}$. Then $\mathsf{I}(\xi) = \mathsf{I}_{2,2}(\tau_0, \xi)$ and (7.111) implies $\mathsf{E}_\xi \dot{S}_\xi^T = -\mathsf{I}_{1,2}(\tau_0, \xi)$. We use the transformation (8.105) that transforms S_ξ into the statistic W_ξ. Then

$$W_{\tau_0,\xi} = U_{\tau_0,\xi} - \mathsf{I}_{1,2}(\tau_0, \xi)\mathsf{I}_{2,2}^{-1}(\tau_0, \xi)V_{\tau_0,\xi}, \quad (\tau_0, \xi) \in U(\theta_0). \tag{8.108}$$

Proposition 8.113. *Suppose $(P_\theta)_{\theta \in U(\theta_0)}$ satisfies (A10) and $\mathsf{I}_{2,2}(\theta_0)$ is nonsingular. If $\widetilde{\xi}_n$ is any $\sqrt{n}$-consistent sequence of estimators for the submodel $(P_{(\tau_0,\xi)})_{(\tau_0^T,\xi^T)^T \in U(\theta_0)}$ for which τ_0 is known, then*

$$\frac{1}{\sqrt{n}} \sum\nolimits_{i=1}^{n} W_{\tau_0,\widetilde{\xi}_n}(X_i) = \frac{1}{\sqrt{n}} \sum\nolimits_{i=1}^{n} W_{\tau_0,\xi_0}(X_i) + o_{P_{\theta_0}^{\otimes n}}(1).$$

Proof. To prove the statement we set $P_\xi = P_{(\tau_0,\xi)}$ in Proposition 8.112. The assumption (A10) implies that $S_\xi = U_{\tau_0,\xi}$ satisfies the conditions in (A9). Furthermore, $\mathsf{I}(\xi) = \mathsf{I}_{2,2}(\tau_0, \xi)$ and $\mathsf{I}(\xi_0)$ is nonsingular. Moreover, $\xi \mapsto (\mathsf{E}_\xi \dot{S}_\xi^T)\mathsf{I}^{-1}(\xi) = -\mathsf{I}_{1,2}(\tau_0, \xi)\mathsf{I}_{2,2}^{-1}(\tau_0, \xi)$ is continuous by assumption (A10) and Lemma 7.147, so that the statement follows from Proposition 8.112. ■

8.8 Asymptotic Null Distribution

In this section we study the asymptotic behavior of the distributions of test statistics under the null hypothesis. These results are used to construct asymptotic tests that have asymptotically under the null hypothesis an error probability that does not exceed a given $\alpha \in (0, 1)$.

Let $\mathcal{M}_n = (\mathcal{X}_n, \mathfrak{A}_n, (P_{n,\theta})_{\theta\in\Delta})$ be a sequence of models. The asymptotic test $\{\varphi_n\}$ is called an *asymptotic level α test* if $\limsup_{n\to\infty} \mathsf{E}_{n,\theta}\varphi_n \le \alpha$, $\theta \in \Delta_0$. An asymptotic level α test $\{\varphi_n\}$ is called an *asymptotically unbiased level α test* if

$$\liminf_{n\to\infty} \mathsf{E}_{n,\theta}\varphi_n \ge \alpha, \quad \theta \in \Delta_A. \tag{8.109}$$

For a simple null hypothesis asymptotic tests are often based on asymptotically normal statistics. We consider the simple null hypothesis $\mathsf{H}_0 : \{\theta_0\}$ and the alternative $\mathsf{H}_A : \Delta\backslash\{\theta_0\}$ in the sequence of models $\mathcal{M}_n = (\mathcal{X}_n, \mathfrak{A}_n, (P_{n,\theta})_{\theta\in\Delta})$. Recall that $\chi^2_{1-\alpha,d}$ is the $1-\alpha$ quantile of the χ^2-distribution with d degrees of freedom.

Proposition 8.114. *Suppose $\Delta \subseteq \mathbb{R}^d$ is open and $\widehat{\theta}_n : \mathcal{X}_n \to_m \Delta$ satisfies $\mathcal{L}(\sqrt{n}(\widehat{\theta}_n - \theta_0)|P_{n,\theta_0}) \Rightarrow \mathsf{N}(0, \Sigma(\theta_0))$, where Σ_0 is nonsingular. Then*

$$\varphi_n = I_{(\chi^2_{1-\alpha,d},\infty)}(n(\widehat{\theta}_n - \theta_0)^T \Sigma^{-1}(\theta_0)(\widehat{\theta}_n - \theta_0)) \tag{8.110}$$

is an asymptotic level α test for $\mathsf{H}_0 : \{\theta_0\}$ versus $\mathsf{H}_A : \Delta\backslash\{\theta_0\}$ for $\alpha \in (0,1)$.

Proof. If the random vector Z in $\mathbb{R}^d$ has the distribution $\mathsf{N}(0, \Sigma(\theta_0))$, where $\Sigma(\theta_0)$ is nonsingular, then it follows from a transformation to principal axes that $Z^T\Sigma^{-1}(\theta_0)Z$ has a χ^2-distribution with d degrees of freedom. The statement follows from $\mathcal{L}(\sqrt{n}(\widehat{\theta}_n - \theta_0)|P_{n,\theta_0}) \Rightarrow \mathcal{L}(Z)$. ■

If $\widehat{\theta}_n$ is the MLE, then under the conditions of Theorem 7.148 it holds

$$\mathcal{L}(\sqrt{n}(\widehat{\theta}_n - \theta_0)|P_{\theta_0}^{\otimes n}) \Rightarrow \mathsf{N}(0, \mathsf{I}^{-1}(\theta_0)),$$

so that the test φ_n in (8.110) turns into the test

$$\varphi_{n,W} = I_{(\chi^2_{1-\alpha,d},\infty)}(n(\widehat{\theta}_n - \theta_0)^T \mathsf{I}(\theta_0)(\widehat{\theta}_n - \theta_0)).$$

$\varphi_{n,W}$ was introduced by Wald who based asymptotic level α tests on the MLE. There is a close connection between the MLE and the central sequence of the localized models. Indeed, if the conditions of Theorem 7.148 are satisfied, then the model is $\mathbb{L}_2$-differentiable, fulfils the ULAN$(Z_n, \mathsf{I}(\theta_0))$ condition with central sequence $Z_n = n^{-1/2}\sum_{i=1}^n \dot{L}_{\theta_0}(X_i)$, and by (7.113)

$$\sqrt{n}(\widehat{\theta}_n - \theta_0) = \mathsf{I}^{-1}(\theta_0)Z_n + o_{P_{\theta_0}^{\otimes n}}(1).$$

Hence we may also operate directly with the statistic Z_n that is already defined if the model is $\mathbb{L}_2$-differentiable, which is a weaker condition than the conditions in Theorem 7.148. As $\mathcal{L}(Z_n|P_{\theta_0}^{\otimes n}) \Rightarrow \mathsf{N}(0, \mathsf{I}(\theta_0))$ we see that the sequence of tests

$$\psi_{n,NR} = \begin{cases} 1 & \text{if} \quad (\sum_{i=1}^n \dot{L}_{\theta_0}(X_i))^T \mathsf{I}^{-1}(\theta_0)(\sum_{i=1}^n \dot{L}_{\theta_0}(X_i)) > n\chi^2_{1-\alpha,d} \\ 0 & \text{else} \end{cases} \tag{8.111}$$

is an asymptotic level α test, which is called the *Neyman–Rao test.*

Often there are nuisance parameters present in the model. In that case we consider the partition (8.107), where U_{θ_0} corresponds to the k derivatives of the log-likelihood with respect to the components of τ and V_{θ_0} corresponds to the $m = d - k$ derivatives with respect to the components of ξ. To make the statistic $U_{\theta_0} = U_{\tau_0,\xi_0}$ insensitive to plugging in an estimator for ξ_0 we have to project U_{τ_0,ξ_0} on the orthogonal complement of the linear subspace $[V_{\theta_0}]$. This projection is just the transformation rule (8.108), which provides the new influence function and test statistic given by, respectively,

$$\begin{aligned} W_{\theta_0} &= U_{\theta_0} - \mathsf{I}_{1,2}(\theta_0)\mathsf{I}_{2,2}^{-1}(\theta_0)V_{\theta_0}, \\ S_{n,NR}(\theta_0) &= \frac{1}{\sqrt{n}}\sum\nolimits_{i=1}^{n} W_{\theta_0}(X_i). \end{aligned} \tag{8.112}$$

We recall the formula for the inverse of a matrix given in block form from Problem 7.174. If $\mathsf{I}^{-1}(\theta_0)$ exists, then by omitting θ_0 for brevity,

$$\begin{pmatrix} \mathsf{I}_{1,1} & \mathsf{I}_{1,2} \\ \mathsf{I}_{2,1} & \mathsf{I}_{2,2} \end{pmatrix}^{-1} = \begin{pmatrix} \mathsf{G}^{-1} & -\mathsf{G}^{-1}\mathsf{I}_{1,2}\mathsf{I}_{2,2}^{-1} \\ -(\mathsf{I}_{1,2}\mathsf{I}_{2,2}^{-1})^T\mathsf{G}^{-1} & \mathsf{I}_{2,2}^{-1}+(\mathsf{I}_{1,2}\mathsf{I}_{2,2}^{-1})^T\mathsf{G}^{-1}(\mathsf{I}_{1,2}\mathsf{I}_{2,2}^{-1}) \end{pmatrix}, \tag{8.113}$$

where $\mathsf{G} = \mathsf{I}_{1,1} - \mathsf{I}_{1,2}\mathsf{I}_{2,2}^{-1}\mathsf{I}_{2,1}$. We get from here that the statistic $S_{n,NR}(\theta_0)$ in (8.112) and the central sequence

$$Z_n = n^{-1/2}\sum\nolimits_{i=1}^{n} \dot{L}_{\theta_0}(X_i) = n^{-1/2}\sum\nolimits_{i=1}^{n} \begin{pmatrix} U_{\theta_0}(X_i) \\ V_{\theta_0}(X_i) \end{pmatrix}$$

are related by

$$S_{n,NR}(\theta_0) = \mathsf{G}(\theta_0)\Pi_k\left(\mathsf{I}^{-1}(\theta_0)Z_n\right), \tag{8.114}$$

where Π_k is the projection on the first k coordinates. We calculate the covariance matrix of W_{θ_0}. It holds

$$\begin{aligned} \mathsf{C}_{\theta_0}(W_{\theta_0}) &= \mathsf{E}_{\theta_0} W_{\theta_0} W_{\theta_0}^T \\ &= \mathsf{I}_{1,1}(\theta_0) - \mathsf{I}_{1,2}(\theta_0)\mathsf{I}_{2,2}^{-1}(\theta_0)\mathsf{I}_{2,1}(\theta_0) = \mathsf{G}(\theta_0). \end{aligned} \tag{8.115}$$

Now we consider the MLE $\widehat{\theta}_n$ in the full model, where we use the partition $\widehat{\theta}_n = (\widehat{\tau}_n^T, \widehat{\xi}_n^T)^T$. Under the assumptions of Theorem 7.148 we have the expansion

$$\sqrt{n}\begin{pmatrix} \widehat{\tau}_n - \tau_0 \\ \widehat{\xi}_n - \xi_0 \end{pmatrix} = \frac{1}{\sqrt{n}}\sum\nolimits_{i=1}^{n} \begin{pmatrix} \mathsf{I}_{1,1}(\theta_0) & \mathsf{I}_{1,2}(\theta_0) \\ \mathsf{I}_{2,1}(\theta_0) & \mathsf{I}_{2,2}(\theta_0) \end{pmatrix}^{-1} \begin{pmatrix} U_{\theta_0}(X_i) \\ V_{\theta_0}(X_i) \end{pmatrix} + o_{P_{\theta_0}^{\otimes n}}(1).$$

The representation of $\mathsf{I}^{-1}(\theta_0)$ in (8.113) gives

$$\sqrt{n}(\widehat{\tau}_n - \tau_0) = \mathsf{G}^{-1}(\theta_0)S_{n,NR}(\theta_0) + o_{P_{\theta_0}^{\otimes n}}(1). \tag{8.116}$$

This means that, up to a normalization by a suitable matrix that depends on $\theta_0 = (\tau_0, \xi_0)$, the statistics $\sqrt{n}(\widehat{\tau}_n - \tau_0)$ and $S_{n,NR}$ are asymptotically equivalent. We introduce a quadratic statistic to construct asymptotic tests by

$$Q_{n,NR}(\tau_0,\xi) := S_{n,NR}(\tau_0,\xi)\mathsf{G}^{-1}(\tau_0,\xi)S_{n,NR}(\tau_0,\xi), \tag{8.117}$$

$$Q_{n,W}(\tau_0,\xi) := n\left(\widehat{\tau}_n - \tau_0\right)^T \mathsf{G}(\tau_0,\xi)\left(\widehat{\tau}_n - \tau_0\right). \tag{8.118}$$

To eliminate the dependence on the nuisance parameter ξ we replace ξ by estimators. Let $\widetilde{\xi}_n$ be any $\sqrt{n}$-consistent estimator for the submodel $(P^{\otimes n}_{(\tau_0,\xi)})_{\xi\in\Xi(\tau_0)}$, and $\widehat{\xi}_n$ be from the partition $\widehat{\theta}_n = (\widehat{\tau}_n^T, \widehat{\xi}_n^T)^T$. The tests

$$\psi_{n,NR} = \begin{cases} 1 & \text{if} \quad Q_{n,NR}(\tau_0,\widetilde{\xi}_n) > \chi^2_{1-\alpha,k} \\ 0 & \text{else} \end{cases} \tag{8.119}$$

$$\psi_{n,W} = \begin{cases} 1 & \text{if} \quad Q_{n,W}(\tau_0,\widehat{\xi}_n) > \chi^2_{1-\alpha,k} \\ 0 & \text{else} \end{cases} \tag{8.120}$$

are called the *Neyman–Rao test* and *Wald test*, respectively, for testing $\tau = \tau_0$ in the presence of nuisance parameters. We consider the hypotheses

$$\mathsf{H}_0 : (\tau,\xi) = (\tau_0,\xi),\ (\tau_0,\xi) \in \Delta, \quad \mathsf{H}_A : (\tau,\xi) \neq (\tau_0,\xi),\ (\tau,\xi) \in \Delta. \tag{8.121}$$

The two test statistics differ in the construction. The idea in the Neyman–Rao statistic is to use any $\sqrt{n}$-consistent estimator $\widetilde{\xi}_n$ under the null hypothesis, i.e., for the submodel $(P^{\otimes n}_{(\tau_0,\xi)})_{\xi\in\Xi(\tau_0)}$, where $\Xi(\tau_0) = \{\xi : (\tau_0^T, \xi^T)^T \in \Delta\}$, to construct the test statistic $S_{n,NR}(\theta_0)$ and to plug in τ_0 and $\widetilde{\xi}_n$ in the normalizing matrix G^{-1} and in $S_{n,NR}(\theta_0)$. The Wald test requires the MLE $\widehat{\theta}_n = (\widehat{\tau}_n^T, \widehat{\xi}_n^T)^T$ for the whole model. The first component $\widehat{\tau}_n^T$ is used to construct the quadratic form, whereas the second component is plugged into the normalizing matrix G.

Proposition 8.115. ***(Neyman–Rao and Wald Tests)*** *Suppose $\Delta \subseteq \mathbb{R}^d$ is open and the model $(\mathcal{X}, \mathfrak{A}, (P_\theta)_{\theta\in\Delta})$ satisfies the assumption (A10). If $\widetilde{\xi}_n : \mathcal{X}^n \to_m \{\xi : (\tau_0^T, \xi^T)^T \in \Delta\}$ is a sequence of $\sqrt{n}$-consistent estimators for the submodels $(\mathcal{X}^n, \mathfrak{A}^{\otimes n}, (P^{\otimes n}_{(\tau_0,\xi)})_{\xi\in\Xi(\tau_0)})$, and $\mathsf{I}(\theta_0)$ is nonsingular, then $\psi_{n,NR}$ is an asymptotic level α test for testing problem (8.121). If in addition the assumptions of Theorem 7.148 are satisfied, then*

$$\begin{aligned} Q_{n,NR}(\tau_0,\widetilde{\xi}_n) &= Q_{n,NR}(\tau_0,\xi_0) + o_{P^{\otimes n}_{\theta_0}}(1) \\ &= Q_{n,W}(\tau_0,\widehat{\xi}_n) + o_{P^{\otimes n}_{\theta_0}}(1) = Q_{n,W}(\tau_0,\xi_0) + o_{P^{\otimes n}_{\theta_0}}(1), \end{aligned} \tag{8.122}$$

so that the Wald test is also an asymptotic level α test for testing problem (8.121).

Proof. The central limit theorem entails, in view of (8.115), the relation $\mathcal{L}(S_{n,NR}|P_{\theta_0}^{\otimes n}) \Rightarrow \mathsf{N}(0, \mathsf{G}(\theta_0))$. Especially the $S_{n,NR}$ are stochastically bounded. By assumption (A10) the matrices $\mathsf{I}_{i,j}(\theta)$ depend continuously on $\theta \in U(\theta_0)$. Hence by Proposition 8.113

$$Q_{n,NR}(\tau_0, \widetilde{\xi}_n) = Q_{n,NR}(\tau_0, \xi_0) + o_{P_{\theta_0}^{\otimes n}}(1).$$

The first statement follows from here as in the proof of Proposition 8.114. The representation (8.116) yields that $\sqrt{n}(\widehat{\tau}_n - \tau_0)$ is stochastically bounded. Hence $Q_{n,W}(\tau_0, \widehat{\xi}_n) = Q_{n,W}(\tau_0, \xi_0) + o_{P_{\theta_0}^{\otimes n}}(1)$ and (8.122) holds. This relation implies that the Wald test is an asymptotic level α test. To complete the proof we remark that $Q_{n,W}(\tau_0, \xi_0) = Q_{n,NR}(\tau_0, \xi_0) + o_{P_{\theta_0}^{\otimes n}}(1)$ by (8.116). ∎

The above tests are based on the central sequence and the MLE, respectively. Now we operate directly with the log-likelihood function $\Lambda_{n,\theta}$ based on n observations. We consider the maximum value of the log-likelihood obtained on Δ (i.e., $\max_{\theta\in\Delta} \Lambda_{n,\theta}$) and compare it with $\max_{\theta\in\Delta_0} \Lambda_{n,\theta}$. If the true value θ_0 belongs to Δ_0, then after the same normalization the two sequences should be of the same order. If θ_0 belongs to Δ_A, then we expect $\max_{\theta\in\Delta_0} \Lambda_{n,\theta}$ to be essentially smaller than $\max_{\theta\in\Delta} \Lambda_{n,\theta}$. This is the motivation for likelihood ratio tests.

We start with a simple hypothesis $\mathsf{H}_0 : \{\theta_0\}$ and evaluate $\max_{\theta\in\Delta} \Lambda_{n,\theta} - \Lambda_{n,\theta_0}$. We consider the sequence of models $(\mathcal{X}^n, \mathfrak{A}^{\otimes n}, (P_\theta^{\otimes n})_{\theta\in\Delta})$, where Δ is an open subset of $\mathbb{R}^d$. Then the coordinate projections $X_1, ..., X_n$ are i.i.d. with distribution P_θ. We assume that the condition (A10) is satisfied and that $\Lambda_n(\theta) = \sum_{i=1}^n \ln f_\theta(X_i)$ is the log-likelihood. The following theorem presents the likelihood ratio test for a simple null hypothesis and is closely related to Theorem 7.148.

Theorem 8.116. ***(Likelihood Ratio Test I)*** *If $\widehat{\theta}_n$ is a consistent asymptotic solution of the likelihood equation (7.109), and the conditions of Theorem 7.148 are satisfied, then*

$$\begin{aligned}\Lambda_n(\widehat{\theta}_n) - \Lambda_n(\theta_0) &= \frac{1}{2n}\left(\sum_{i=1}^n \dot{L}_{\theta_0}(X_i)\right)^T \mathsf{I}^{-1}(\theta_0)\left(\sum_{i=1}^n \dot{L}_{\theta_0}(X_i)\right) + o_{P_{\theta_0}^{\otimes n}}(1) \\ &= \frac{1}{2} n(\widehat{\theta}_n - \theta_0)^T \mathsf{I}(\theta_0)(\widehat{\theta}_n - \theta_0) + o_{P_{\theta_0}^{\otimes n}}(1), \quad (8.123)\\ \mathcal{L}(2[\Lambda_n(\widehat{\theta}_n) - \Lambda_n(\theta_0)]|P_{\theta_0}^{\otimes n}) &\Rightarrow \mathsf{H}(d), \quad (8.124)\end{aligned}$$

where $\mathsf{H}(d)$ is the χ^2-distribution with d degrees of freedom. The sequence of tests

$$\varphi_n = \begin{cases} 1 & \text{if} \quad 2[\Lambda_n(\widehat{\theta}_n) - \Lambda_n(\theta_0)] > \chi^2_{1-\alpha,d} \\ 0 & \text{else} \end{cases}$$

is an asymptotic level α test for $\mathsf{H}_0 : \theta = \theta_0$ versus $\mathsf{H}_A : \theta \neq \theta_0$.

Proof. Set $\Lambda_\theta = \ln f_\theta$ and apply a second-order Taylor expansion with respect to θ, where $\dot{\Lambda}_\theta$ is the gradient and $\ddot{\Lambda}_\theta$ the Hessian matrix. We have $\dot{\Lambda}_n(\widehat{\theta}_n) = \mathbf{o}_{P_{\theta_0}^{\otimes n}}(0)$ in view of (7.109), where $\mathbf{o}_{P_{\theta_0}^{\otimes n}}(0)$ denotes a sequence of random vectors that are nonzero only on events whose probabilities tend to zero. Hence by the Taylor expansion in Theorem A.2 at $\widehat{\theta}_n$,

$$\Lambda_n(\widehat{\theta}_n) - \Lambda_n(\theta_0) = \sqrt{n}(\widehat{\theta}_n - \theta_0)^T(\frac{1}{2}\mathsf{I}(\theta_0) + R_n(\widehat{\theta}_n, \theta_0))\sqrt{n}(\widehat{\theta}_n - \theta_0) + o_{P_{\theta_0}^{\otimes n}}(1),$$

$$R_n(\theta, \theta_0) = \frac{1}{n}\sum\nolimits_{i=1}^{n} \int_0^1 [-(1-s)\ddot{\Lambda}_{\theta_0 + s(\theta - \theta_0)}(X_i) - (1-s)\mathsf{I}(\theta_0)]ds.$$

To show that $R_n(\widehat{\theta}_n, \theta_0) = o_{P_{\theta_0}^{\otimes n}}(1)$ we note that $\mathsf{E}_{\theta_0}\ddot{\Lambda}_{\theta_0} = -\mathsf{I}(\theta_0)$ by Lemma 7.147. Condition (A10) was assumed in Theorem 7.148 and implies that $\dot{\Lambda}_\theta$ satisfies the conditions in (A9) and we may apply Lemma 7.141 to

$$\varphi_\theta(x) = \int_0^1 [-(1-s)\ddot{\Lambda}_{\theta_0 + s(\theta - \theta_0)}(x) - (1-s)\mathsf{I}(\theta_0)]ds.$$

To get (8.123) we note that $\sqrt{n}(\widehat{\theta}_n - \theta_0)$ is by (7.113) stochastically bounded. The statement (8.124) follows as in the proof of Proposition 8.114. ■

Example 8.117. Let X be an observation with d possible outcomes that have probabilities $p_1, ..., p_d$. More specifically, let

$$\begin{aligned} &\mathcal{X} = \{1, ..., d\}, \quad \text{and} \quad P_\theta = \textstyle\sum_{i=1}^d p_i(\theta)\boldsymbol{\delta}_i, \\ &p(\theta) = (p_1(\theta), ..., p_d(\theta)), \quad p_i(\theta) = \theta_i, \quad i = 1, ..., d-1, \\ &p_d(\theta) = 1 - \textstyle\sum_{i=1}^{d-1} \theta_i, \quad \text{and} \quad \theta = (\theta_1, ..., \theta_{d-1}) \in \mathbb{S}_{d-1}. \end{aligned} \tag{8.125}$$

Condition (A10) is obviously fulfilled, the information matrix is given by the elements

$$\sum\nolimits_{i=1}^d \frac{\partial \ln p_i(\theta)}{\partial \theta_k}\frac{\partial \ln p_i(\theta)}{\partial \theta_l} p_i(\theta) = \frac{1}{\theta_k}\delta_{k,l} + (1 - \sum\nolimits_{i=1}^{d-1} \theta_i)^{-1},$$

and it has rank $d-1$; see Problem 1.42. If we have a sample of n observations, then the density with respect to the counting measure is

$$f_{n,\theta}(x_1, ..., x_n) = \prod\nolimits_{i=1}^d (p_i(\theta))^{Y_{i,n}(\mathbf{x}_n)},$$

where $\mathbf{x}_n = (x_1, ..., x_n)$ and $Y_{i,n}(\mathbf{x}_n) = |\{j : 1 \le j \le n,\ x_j = i\}|$. The likelihood equations are given by

$$\frac{\partial}{\partial \theta_k}\sum\nolimits_{i=1}^d Y_{i,n}(\mathbf{x}_n)\ln p_i(\theta) = 0, \quad \text{i.e.,} \quad \frac{Y_{k,n}}{p_k(\theta)} = \frac{Y_{d,n}}{p_d(\theta)}, \quad k = 1, ..., d-1.$$

Denoting the right-hand side by C we get $Cp_k(\theta) = Y_{k,n}(\mathbf{x}_n)$, and by the sum over k and $\sum_{i=1}^d Y_{i,n} = n$ we arrive at

$$p_k(\widehat{\theta}_n) = \frac{Y_{k,n}}{n} \quad k = 1, ..., d. \tag{8.126}$$

Thus we estimate θ by $\widehat{\theta}_n$, so that $p_k(\widehat{\theta}_n)$ are the relative frequencies which are consistent. Hence the $\widehat{\theta}_n$ are also consistent. As in this case the likelihood equation (7.109) is given by (8.126) we see that the conditions in Theorem 7.148 are satisfied for $\theta_0 \in \mathbb{S}_{d-1}$, and we get for $p_{0,k} = p(\theta_0)$,

$$\begin{aligned} 2(\Lambda_n(\widehat{\theta}_n) - \Lambda_n(\theta_0)) &= \sum\nolimits_{k=1}^{d} Y_{k,n}(\ln p_k(\widehat{\theta}_n) - \ln p_{0,k}) \\ &= 2n \sum\nolimits_{k=1}^{d} (Y_{k,n}/n) \ln \frac{(Y_{k,n}/n)}{p_{0,k}}. \end{aligned} \tag{8.127}$$

Let

$$H_{k,n} = \frac{Y_{k,n}}{n}, \quad \widehat{P}_n = \sum\nolimits_{k=1}^{d} H_{k,n}\boldsymbol{\delta}_k, \quad \text{and} \quad P_{\theta_0} = \sum\nolimits_{k=1}^{d} p_{0,k}\boldsymbol{\delta}_k, \tag{8.128}$$

where $\theta_0 = (p_{0,1}, ..., p_{0,d-1}) \in \mathbb{S}_{d-1}$, denote the relative frequencies, empirical distribution, and distribution of the observations, respectively. Then

$$\sum\nolimits_{k=1}^{d} (Y_{k,n}/n) \ln \frac{(Y_{k,n}/n)}{p_{0,k}} = \sum\nolimits_{k=1}^{d} (\frac{H_{k,n}}{p_{0,k}} \ln \frac{H_{k,n}}{p_{0,k}}) p_{0,k} = \mathsf{K}(\widehat{P}_n, P_{\theta_0}), \tag{8.129}$$

which is the Kullback–Leibler distance between $\widehat{P}_n$ and P_{θ_0} that has been introduced in (1.74). This distance is based on the convex function $x \ln x$. This leads to the idea of taking any convex function v and then to use $\mathsf{I}_\mathsf{v}(\widehat{P}_n, P_{\theta_0})$ as test statistics. For example, if we use $\mathsf{v}(x) = (x-1)^2$, then we arrive at the famous χ^2-statistic

$$\chi^2(\widehat{P}_n, P_{\theta_0}) = \sum\nolimits_{k=1}^{d} \frac{(H_{k,n} - p_{0,k})^2}{p_{0,k}}.$$

With other convex functions one arrives at the Hellinger distance and other distances between the empirical and the hypothetical distribution. The use of such statistics is studied systematically in Pardo (2006). As we show subsequently all of these statistics are asymptotically equivalent and lead to the same limiting distribution.

The next lemma shows that all test statistics $\mathsf{I}_\mathsf{v}(\widehat{P}_n, P_{\theta_0})$ are asymptotically equivalent under the null hypothesis provided that v is smooth. Later we see that the same statement continues to hold under local alternatives.

Lemma 8.118. *If $\widehat{P}_n$ is defined as in (8.128), and v_1 and v_2 are convex functions that are twice continuously differentiable in a neighborhood of $x_0 = 1$ and satisfy $\mathsf{v}_i''(1) > 0$, $i = 1, 2$, then*

$$\frac{2n}{\mathsf{v}_1''(1)}[\mathsf{I}_{\mathsf{v}_1}(\widehat{P}_n, P_{\theta_0}) - \mathsf{v}_1(1)] = \frac{2n}{\mathsf{v}_2''(1)}[\mathsf{I}_{\mathsf{v}_2}(\widehat{P}_n, P_{\theta_0}) - \mathsf{v}_2(1)] + o_{P_{\theta_0}^{\otimes n}}(1).$$

Proof. If we set $\mathsf{w}_i(x) = \mathsf{v}_i(x) - \mathsf{v}_i(1) - \mathsf{v}_i'(1)(x-1)$, $i = 1, 2$, then in view of $\sum_{k=1}^{d} (H_{k,n}/p_{0,k})p_{0,k} = 1$ it holds

$$\mathsf{I}_{\mathsf{v}_i}(\widehat{P}_n, P_{\theta_0}) - \mathsf{v}_i(1) = \mathsf{I}_{\mathsf{w}_i}(\widehat{P}_n, P_{\theta_0}), \quad i = 1, 2.$$

For fixed k the random variable $Y_{k,n}$ has a binomial distribution with success probability $p_{0,k}$. Hence $\mathcal{L}(\sqrt{n}(H_{k,n} - p_{0,k})) \Rightarrow \mathsf{N}(0, p_{0,k}(1 - p_{0,k}))$ which implies that the sequence $n(H_{k,n} - p_{0,k})^2$ is stochastically bounded. As $\lim_{x\to 1} \mathsf{w}_2(x)/(x-1)^2 = \mathsf{v}_2''(1)/2$ we see that the sequences $n\mathsf{w}_2(H_{k,n}/p_{0,k})$, $k = 1, ..., d$, are also stochastically bounded. Then by $H_{k,n} \to^{P_{\theta_0}^{\otimes n}} p_{0,k}$,

$$\begin{aligned} n\mathsf{I}_{\mathsf{w}_1}(\widehat{P}_n, P_{\theta_0}) &= \sum\nolimits_{k=1}^{d} (\mathsf{w}_1(H_{k,n}/p_{0,k}))\,(\mathsf{w}_2(H_{k,n}/p_{0,k}))^{-1}\, n\mathsf{w}_2(H_{k,n}/p_{0,k})p_{0,k} \\ &= (\mathsf{v}_2''(1))^{-1}\mathsf{v}_1''(1) n\mathsf{I}_{\mathsf{w}_2}(\widehat{P}_n, P_{\theta_0}) + o_{P_{\theta_0}^{\otimes n}}(1). \end{aligned}$$

■

Theorem 8.119. ***(Divergence Tests)*** *Let $Y_n = (Y_{1,n}, ..., Y_{d,n})$ have a multinomial distribution with parameters $p_{0,1}, ..., p_{0,d} \in (0,1)$. If v is a convex function that is twice continuously differentiable in a neighborhood of $x_0 = 1$ and satisfies $\mathsf{v}''(1) > 0$, then*

$$\mathcal{L}\Big(\frac{2n}{\mathsf{v}''(1)}(\mathsf{I}_{\mathsf{v}}(\widehat{P}_n, P_{\theta_0}) - \mathsf{v}(1))|P_{\theta_0}^{\otimes n}\Big) \Rightarrow \mathsf{H}(d-1).$$

The sequence of tests

$$\varphi_n = \begin{cases} 1 & \text{if} \quad \mathsf{I}_{\mathsf{v}}(\widehat{P}_n, P_{\theta_0}) - \mathsf{v}(1) > \frac{\mathsf{v}''(1)}{2n}\chi^2_{1-\alpha,d-1} \\ 0 & \text{else} \end{cases}$$

is an asymptotic level α test for $\mathsf{H}_0 : (p_1, ..., p_d) = (p_{0,1}, ..., p_{0,d})$ versus $\mathsf{H}_A : (p_1, ..., p_d) \neq (p_{0,1}, ..., p_{0,d})$.

Proof. The model $(\mathcal{X}, \mathfrak{A}, (P_\theta)_{\theta\in\Delta})$ specified by (8.125) satisfies the conditions of Theorem 7.148; see Example 8.117. We get from (8.127) and (8.129) that $2(\Lambda_n(\widehat{\theta}_n) - \Lambda_n(\theta_0)) = 2n\mathsf{K}(\widehat{P}_n, P_{\theta_0})$. Hence $\mathcal{L}(2n\mathsf{K}(\widehat{P}_n, P_{\theta_0})|P_{\theta_0}^{\otimes n}) \Rightarrow \mathsf{H}(d-1)$ by Theorem 8.116. The rest follows from Lemma 8.118. ■

Now we study the case where the null hypothesis $\mathsf{H}_0 : \Delta_0$ is a lower-dimensional subset of Δ. We assume that $\Upsilon \subseteq \mathbb{R}^k$, $k < d$, and $\Delta_0 \subseteq \mathbb{R}^d$ are open sets, and that $\kappa : \Upsilon \to \Delta_0$ is a twice continuously differentiable mapping of Υ on Δ_0. Under H_0 we have now the family $(P_{\kappa(\eta)})_{\eta\in\Upsilon}$. If $\theta_0 \in \Delta_0$ and $(P_\theta)_{\theta\in\Delta}$ satisfies (A10), then it is easy to see that $(P_{\kappa(\eta)})_{\eta\in\Upsilon}$ also satisfies (A10) for the new parameter set Υ. If $g_\eta := f_{\kappa(\eta)}$ and the log-likelihood for $P_{\kappa(\eta)}$ is denoted by $\Gamma(\eta) := \Lambda_{\kappa(\eta)}$, then the $\mathbb{L}_2$-derivative is

$$\dot{M}_\eta = \dot{\kappa}(\eta)\dot{L}_{\kappa(\eta)}, \tag{8.130}$$

where $\dot{\kappa}^T = J_\kappa = (\partial\kappa_i/\partial\eta_j)_{1\le i\le d, 1\le j\le k}$ is the Jacobian; see Proposition 1.112. The information matrix for the family $(P_{k(\eta)})_{\eta\in\Upsilon}$ is given by

$$\widetilde{\mathsf{I}}(\eta_0) = \mathsf{E}_{\eta_0}\dot{M}_{\eta_0}\dot{M}_{\eta_0}^T = \dot{\kappa}(\eta_0)\mathsf{I}(\theta_0)\dot{\kappa}^T(\eta_0),$$

see also Proposition 1.112. Suppose now that $\widehat{\eta}_n$ maximizes $\Gamma_n(\eta) = \Lambda_n(\kappa(\eta))$ on Υ. Then $\widetilde{\theta}_n = \kappa(\widehat{\eta}_n)$ maximizes $\Lambda_n(\theta)$ on Δ_0. Let $\widehat{\theta}_n$ maximize the log-likelihood $\Lambda_n(\theta)$ of the whole model on Δ. Set $f_{n,\theta}(\mathbf{x}_n) = \Pi_{i=1}^n f_\theta(x_i)$. The next theorem gives the asymptotic distribution of the likelihood ratio statistic

$$2\ln\frac{\sup_{\theta\in\Delta} f_{n,\theta}(\mathbf{x}_n)}{\sup_{\theta\in\Delta_0} f_{n,\theta}(\mathbf{x}_n)} = 2[\Lambda_n(\widehat{\theta}_n(\mathbf{x}_n),\mathbf{x}_n) - \Lambda_n(\widetilde{\theta}_n(\mathbf{x}_n),\mathbf{x}_n)].$$

Theorem 8.120. ***(Likelihood Ratio Test II)*** *Suppose $\Delta \subseteq \mathbb{R}^d$ and $\Upsilon \subseteq \mathbb{R}^k$, $k < d$, are open. Let $\kappa : \Upsilon \to \Delta$ be a twice continuously differentiable mapping where $\dot{\kappa}(\eta)$ has rank k for every $\eta \in \Upsilon$. Assume that $(P_\theta)_{\theta\in\Delta}$ satisfies (A10) for every $\theta_0 \in \Delta$, and that the information matrix $\mathsf{I}(\theta_0)$ is nonsingular. Assume that $\widehat{\theta}_n : \mathcal{X}^n \to_m \Delta$ and $\widehat{\eta}_n : \mathcal{X}^n \to_m \Upsilon$ are consistent and asymptotic solutions of the likelihood equations for the families $(P_\theta)_{\theta\in\Delta}$ and $(P_{\kappa(\eta)})_{\eta\in\Upsilon}$, respectively, at every parameter $\theta_0 \in \Delta$ and $\eta_0 \in \Upsilon$. If $\eta_0 \in \Upsilon$ and $\theta_0 = \kappa(\eta_0)$ is the true parameter, then the likelihood ratio statistic $Q_{n,LR} = 2[\Lambda_n(\widehat{\theta}_n) - \Lambda_n(\kappa(\widehat{\eta}_n)]$ satisfies*

$$\mathcal{L}(Q_{n,LR}|P_{\theta_0}^{\otimes n}) \Rightarrow \mathsf{H}(d-k). \tag{8.131}$$

Corollary 8.121. *The likelihood ratio test*

$$\psi_{n,Q_{n,LR}} = \begin{cases} 1 & \textit{if} \quad Q_{n,LR} > \chi^2_{1-\alpha,d-k} \\ 0 & \textit{else} \end{cases}$$

is an asymptotic level α test for $\mathsf{H}_0 : \theta \in \kappa(\Upsilon)$ versus $\mathsf{H}_A : \theta \in \Delta\backslash\kappa(\Upsilon)$.

Proof. Let $\mathsf{I}^{1/2}(\theta_0)$ and $\mathsf{I}^{-1/2}(\theta_0)$ be symmetric matrices that are satisfying $\mathsf{I}^{1/2}(\theta_0)\mathsf{I}^{1/2}(\theta_0) = \mathsf{I}(\theta_0)$ and $\mathsf{I}^{-1/2}(\theta_0)\mathsf{I}^{-1/2}(\theta_0) = \mathsf{I}^{-1}(\theta_0)$. The existence of $\mathsf{I}^{1/2}(\theta_0)$ follows from a principal axes transformation. Put $A_0 = \dot{\kappa}(\eta_0)\mathsf{I}^{1/2}(\kappa(\eta_0))$. Then $\widetilde{\mathsf{I}}(\eta_0) = A_0 A_0^T$. Set

$$S_n = n^{-1/2}\sum\nolimits_{i=1}^n \mathsf{I}^{-1/2}(\theta_0)\dot{L}_{\theta_0}(X_i).$$

Theorems 7.148 and 8.116 yield

$$\begin{aligned}\sqrt{n}(\widehat{\theta}_n - \theta_0) &= \mathsf{I}^{-1/2}(\theta_0)S_n + o_{P_{\theta_0}^{\otimes n}}(1),\\ 2(\Lambda_n(\widehat{\theta}_n) - \Lambda_n(\theta_0)) &= S_n^T S_n + o_{P_{\theta_0}^{\otimes n}}(1).\end{aligned} \tag{8.132}$$

As $\mathsf{I}(\theta_0)$ is nonsingular and $\dot{\kappa}$ has rank k the matrix A_0 has rank k. By Theorems 7.148 and 8.116, applied to $\widehat{\eta}_n$ and (8.130), we get

$$\begin{aligned}&\sqrt{n}(\widehat{\eta}_n - \eta_0) = \frac{1}{\sqrt{n}}\sum\nolimits_{i=1}^n \widetilde{\mathsf{I}}^{-1}(\eta_0)\dot{M}_{\eta_0}(X_i) + o_{P_{\theta_0}^{\otimes n}}(1)\\ &= \widetilde{\mathsf{I}}^{-1}(\eta_0)\dot{\kappa}(\eta_0)\mathsf{I}^{1/2}(\theta_0)S_n + o_{P_{\theta_0}^{\otimes n}}(1) = (A_0A_0^T)^{-1}A_0S_n + o_{P_{\theta_0}^{\otimes n}}(1),\\ &2(\Lambda_n(\widetilde{\theta}_n) - \Lambda_n(\theta_0)) = 2(\Gamma_n(\widehat{\eta}_n) - \Gamma_n(\eta_0)) \qquad (8.133)\\ &= n(\widehat{\eta}_n - \eta_0)^T A_0A_0^T(\widehat{\eta}_n - \eta_0) + o_{P_{\theta_0}^{\otimes n}}(1) = S_n^T A_0^T(A_0A_0^T)^{-1}A_0S_n + o_{P_{\theta_0}^{\otimes n}}(1),\end{aligned}$$

where $\widetilde{\theta}_n = \kappa(\widehat{\eta}_n)$. $B_0 = A_0^T(A_0A_0^T)^{-1}A_0$ is the projection matrix that projects a vector on the subspace spanned by the column vectors of A_0, see Problem 7.26. As $B_0^TB_0 = B_0$ and $B_0^T = B_0$ we get

$$\begin{aligned} 2(\Lambda_n(\widehat{\theta}_n) - \Lambda_n(\widetilde{\theta}_n)) &= S_n^TS_n - S_n^TB_0S_n + o_{P_{\theta_0}^{\otimes n}}(1) \\ &= ((\mathbf{I} - B_0)S_n)^T((\mathbf{I} - B_0)S_n) + o_{P_{\theta_0}^{\otimes n}}(1). \end{aligned}$$

As B_0 has rank k the projection matrix $\mathbf{I}-B_0$ has rank $d-k$. By the definition of S_n and the central limit theorem it follows that $\mathcal{L}(S_n|P_{\theta_0}^{\otimes n}) \Rightarrow \mathsf{N}(0,\mathbf{I})$. To complete the proof we use Slutsky's lemma and the fact that $((\mathbf{I}-B_0)Z)^T((\mathbf{I}-B_0)Z)$ has a χ^2-distribution with $d-k$ degrees of freedom for $Z \sim \mathsf{N}(0,\mathbf{I})$, see Problem 2.38. The corollary is a direct consequence of (8.131). ■

Subsequently we need a special decomposition of the inverse of the Fisher information matrix that follows from the block representation in Problem 7.175.

Problem 8.122.* Suppose we have a block matrix $\mathsf{I} = (\mathsf{I}_{i,j})_{1\le i,j\le 2}$. If I is invertible, then with $\mathsf{G} = \mathsf{I}_{1,1} - \mathsf{I}_{1,2}\mathsf{I}_{2,2}^{-1}\mathsf{I}_{2,1}$,

$$\mathsf{I}^{-1} = \begin{pmatrix} \mathbf{I} \\ -\mathsf{I}_{2,2}^{-1}\mathsf{I}_{1,2}^T \end{pmatrix} \mathsf{G}^{-1} \left(\mathbf{I},\ -\mathsf{I}_{1,2}\mathsf{I}_{2,2}^{-1}\right) + \begin{pmatrix} 0 & 0 \\ 0 & \mathsf{I}_{2,2}^{-1} \end{pmatrix}.$$

Now we compare the likelihood ratio test with the Neyman–Rao test and the Wald test in the special case where the parameter vector is decomposed into the k-dimensional component τ, the parameter of interest, and the $m = (d-k)$-dimensional component ξ, the nuisance parameter. We consider the partition (8.107) and set

$$\begin{pmatrix} U_n(\theta_0) \\ V_n(\theta_0) \end{pmatrix} = \frac{1}{\sqrt{n}}\sum_{i=1}^n \begin{pmatrix} U_{\theta_0}(X_i) \\ V_{\theta_0}(X_i) \end{pmatrix}, \quad \theta_0 = \begin{pmatrix} \tau_0 \\ \xi_0 \end{pmatrix} \in \Delta. \tag{8.134}$$

Then by (8.123),

$$2(\Lambda_n(\widehat{\theta}_n) - \Lambda_n(\theta_0)) = \begin{pmatrix} U_n(\theta_0) \\ V_n(\theta_0) \end{pmatrix}^T \mathsf{I}^{-1}(\theta_0) \begin{pmatrix} U_n(\theta_0) \\ V_n(\theta_0) \end{pmatrix} + o_{P_{\theta_0}^{\otimes n}}(1). \tag{8.135}$$

Similarly, if $\widetilde{\xi}_n$ is a consistent MLE for the submodel $(P_{(\tau_0,\xi)}^{\otimes n})_{\xi\in\Xi(\tau_0)}$, then for $\widetilde{\theta}_n = (\tau_0^T, \widetilde{\xi}_n^T)^T$,

$$2(\Lambda_n(\widetilde{\theta}_n) - \Lambda_n(\theta_0)) = V_n(\theta_0)\mathsf{I}_{2,2}^{-1}(\theta_0)V_n(\theta_0) + o_{P_{\theta_0}^{\otimes n}}(1).$$

Taking the difference, we get from Problem 8.122 and the representation of $\sqrt{n}(\widehat{\tau}_n - \tau_0)$ with the help of (8.112) and (8.116) that

$$\begin{aligned} Q_{n,LR} = 2[\Lambda_n(\widehat{\theta}_n) - \Lambda_n(\widetilde{\theta}_n)] &= n(\widehat{\tau}_n - \tau_0)^T\mathsf{G}(\theta_0)(\widehat{\tau}_n - \tau_0) + o_{P_{\theta_0}^{\otimes n}}(1) \\ &= Q_{n,NR}(\tau_0,\widetilde{\xi}_n) + o_{P_{\theta_0}^{\otimes n}}(1) = Q_{n,W}(\tau_0,\widehat{\theta}_n) + o_{P_{\theta_0}^{\otimes n}}(1). \end{aligned}$$

If $\widetilde{\xi}_n$ is $\sqrt{n}$-consistent, then by (8.112), (8.117), and Proposition 8.113,

$$Q_{n,LR} = Q_{n,NR}(\tau_0, \xi_0) + o_{P_{\theta_0}^{\otimes n}}(1).$$

Similarly, by the representation of $\sqrt{n}(\widehat{\tau}_n - \tau_0)$ in (8.116) and (8.118),

$$Q_{n,LR} = Q_{n,W}(\tau_0, \xi_0) + o_{P_{\theta_0}^{\otimes n}}(1).$$

To summarize, we have obtained the following result for the Neyman–Rao, Wald, and likelihood ratio tests from Proposition 8.115.

Proposition 8.123. *Under the conditions of Proposition 8.115 the Neyman–Rao statistic $Q_{n,NR}(\tau_0, \widetilde{\xi}_n)$ and the Wald statistic $Q_{n,W}(\tau_0, \widehat{\xi}_n)$ differ only by terms $o_{P_{\theta_0}^{\otimes n}}(1)$ as $n \to \infty$. If in addition the assumptions of Theorem 7.148 hold for the subfamily $(P_{(\tau_0,\xi)})_{\xi \in \Xi(\tau_0)}$, then also the likelihood ratio statistic $Q_{n,LR}$ differs only by terms $o_{P_{\theta_0}^{\otimes n}}(1)$ from $Q_{n,NR}(\tau_0, \widetilde{\xi}_n)$ and $Q_{n,W}(\tau_0, \widehat{\xi}_n)$ as $n \to \infty$.*

Now we consider the likelihood ratio tests for special models.

Example 8.124. We consider the situation in Theorem 8.58. Let $\mathbb{A}_h \subseteq \mathbb{L}_k \subseteq \mathbb{R}^n$ be linear subspaces. Suppose we want to test in the model $(\mathbb{R}^n, \mathfrak{B}_n, (\mathsf{N}(\mu, \sigma^2 \mathbf{I}))_{\mu \in \mathbb{L}_k})$ $\mathsf{H}_0 : \mu \in \mathbb{A}_h, \ \sigma^2 > 0$, versus $\mathsf{H}_A : \mu \in \mathbb{L}_k \backslash \mathbb{A}_h, \ \sigma^2 > 0$. The log-likelihood is given by

$$\Lambda_n(\mu, \sigma^2) = \frac{n}{2}[-\ln(2\pi\sigma^2) - \frac{\|x - \mu\|^2}{n\sigma^2}].$$

As for any linear subspace $\mathbb{L} \subset \mathbb{R}^n$ it holds $\mathsf{N}(\mu, \sigma^2 \mathbf{I})(\mathbb{L}) = 0$ we get, $\mathsf{N}(\mu, \sigma^2 \mathbf{I})$-a.s.,

$$\begin{aligned}
\sup_{\mu \in \mathbb{L}, \sigma^2 > 0} \Lambda_n(\mu, \sigma^2) &= \sup_{\sigma^2 > 0} \Lambda_n(\Pi_{\mathbb{L}} x, \sigma^2) \\
= \Lambda_n(\Pi_{\mathbb{L}} x, \frac{1}{n} \|x - \Pi_{\mathbb{L}} x\|^2) &= \frac{n}{2}[-\ln(\frac{2\pi}{n} \|x - \Pi_{\mathbb{L}} x\|^2) - 1], \\
\sup_{\mu \in \mathbb{L}_k, \sigma^2 > 0} \Lambda_n(\mu, \sigma^2) - \sup_{\mu \in \mathbb{A}_h, \sigma^2 > 0} \Lambda_n(\mu, \sigma^2) &= \frac{n}{2} \ln \frac{\|x - \Pi_{\mathbb{A}_h} x\|^2}{\|x - \Pi_{\mathbb{L}_k} x\|^2},
\end{aligned}$$

and H_0 is rejected for large values of $\|x - \Pi_{\mathbb{A}_h} x\|^2 / \|x - \Pi_{\mathbb{L}_k} x\|^2$. As $\|x - \Pi_{\mathbb{A}_h} x\|^2 = \|x - \Pi_{\mathbb{L}_k} x\|^2 + \|\Pi_{\mathbb{L}_k} x - \Pi_{\mathbb{A}_h} x\|^2$ a test statistic that leads to the same test is

$$F(x) = \frac{(n-k)\, \|\Pi_{\mathbb{L}_k} x - \Pi_{\mathbb{A}_h} x\|^2}{(k-h)\, \|x - \Pi_{\mathbb{L}_k} x\|^2}, \quad x \in \mathbb{R}^n.$$

The test that rejects H_0 for large values of F is the F-test in Theorem 8.58.

Example 8.125. Here we construct tests for testing if the random variables A and B are independent, where A and B take on the values $1, ..., a$ and $1, ..., b$, respectively. We set $X = (A, B)$ and use the notation from Example 8.117. For n independent observations $(A_1, B_1), ..., (A_n, B_n)$ we set $\mathbf{x}_n = (a_1, b_1, ..., a_n, b_n)$ and

$$Y_{i,j,n}(\mathbf{x}_n) = |\{r : (a_r, b_r) = (i, j), \ 1 \le r \le n\}|, \ 1 \le i \le a, \ 1 \le j \le b.$$

Then the probability mass function is

$$f_{n,\theta}(a_1,b_1,...,a_n,b_n) = \prod\nolimits_{i=1,j=1}^{a,b} p_{i,j}^{Y_{i,j,n}(\mathbf{x}_n)}(\theta),$$

$$p_{i,j}(\theta) = \theta_{i,j}, \quad (i,j) \neq (a,b), \quad p_{a,b}(\theta) = 1 - \sum\nolimits_{(i,j)\neq(a,b)} \theta_{i,j},$$

$$\theta = (\theta_{1,1},...,\theta_{a,b-1}) \in \mathbb{S}_{ab-1}, \quad \theta_{a,b} = 1 - \sum\nolimits_{(i,j)\neq(a,b)} \theta_{i,j},$$

and the log-likelihood has the form

$$\Lambda_n(\theta) = \sum\nolimits_{i=1,j=1}^{a,b} Y_{i,j,n} \ln p_{i,j}(\theta).$$

The likelihood equations are given by

$$\frac{\partial}{\partial\theta_{k,l}} \sum\nolimits_{i,j=1}^{a,b} Y_{i,j,n}(\mathbf{x}_n) \ln p_{i,j}(\theta) = 0, \quad (k,l) \neq (a,b),$$

so that

$$\frac{Y_{k,l,n}}{p_{k,l}(\theta)} = \frac{Y_{a,b,n}}{p_{a,b}(\theta)}, \quad (k,l) \neq (a,b).$$

Denoting the right-hand side by C we get $Cp_{k,l}(\theta) = Y_{k,l,n}(\mathbf{x}_n)$, and by the sum over k,l and $\sum_{k,l=1}^{a,b} Y_{k,l,n}(\mathbf{x}_n) = n$ we arrive at

$$p_{k,l}(\widehat{\theta}_n) = \frac{1}{n} Y_{k,l,n}, \quad k = 1,...,a, \; l = 1,...,b.$$

Thus we estimate θ by $\widehat{\theta}_n$, so that the $p_{k,l}(\widehat{\theta}_n)$ are the relative frequencies. We want to test if the components A and B are independent. The null hypothesis is given by $\Delta_0 = \{(p_i q_j, 1 \leq i \leq a, 1 \leq j \leq b) : (p_1,...,p_{a-1}) \in \mathbb{S}_{a-1}, \, (q_1,...,q_{b-1}) \in \mathbb{S}_{b-1}\}$. Under the null hypothesis the likelihood equations read $p_i q_j = Y_{i,j,n}/n$. Taking the sum over $j = 1,...,b$ we get $p_i(\widetilde{\theta}_n) = Y_{i,\cdot,n}/n$, and analogously $q_j(\widetilde{\theta}_n) = Y_{\cdot,j,n}/n$. Thus the likelihood ratio statistic is

$$\begin{aligned} Q_{n,LR} = \Lambda_n(\widehat{\theta}_n) - \Lambda_n(\widetilde{\theta}_n) &= \sum\nolimits_{i=1,j=1}^{a,b} Y_{i,j,n}[\ln p_{i,j}(\widehat{\theta}_n) - \ln(p_i(\widetilde{\theta}_n) q_j(\widetilde{\theta}_n))] \\ &= \sum\nolimits_{i,j=1}^{a,b} Y_{i,j,n} \ln \frac{nY_{i,j,n}}{Y_{i,\cdot,n} Y_{\cdot,j,n}}. \end{aligned}$$

We rewrite this expression. Put

$$\widehat{P}_n = \sum\nolimits_{i,j=1}^{a,b} (\frac{1}{n} Y_{i,j,n}) \boldsymbol{\delta}_{i,j}, \; \widehat{P}_{1,n} = \sum\nolimits_{i=1}^{a} (\frac{1}{n} Y_{i,\cdot,n}) \boldsymbol{\delta}_i, \; \widehat{P}_{2,n} = \sum\nolimits_{j=1}^{b} (\frac{1}{n} Y_{\cdot,j,n}) \boldsymbol{\delta}_j. \tag{8.136}$$

Then, similarly as in Example 8.117, it holds $Q_{n,LR} = n\mathsf{K}(\widehat{P}_n, \widehat{P}_{1,n} \otimes \widehat{P}_{2,n})$.

Proposition 8.126. ***(Test of Independence)*** *Let $Y_n = (Y_{1,1,n},...,Y_{a,b,n})$ have a multinomial distribution with parameter $(p_{1,1},...,p_{a,b})$, where $p_{i,j} = p_{i,\cdot} p_{\cdot,j} > 0$. If v is a convex function that is twice continuously differentiable in a neighborhood of $x_0 = 1$ and satisfies $\mathsf{v}''(1) > 0$, and $\widehat{P}_n, \widehat{P}_{1,n}$, and $\widehat{P}_{2,n}$ are defined as in (8.136), then*

$$\mathcal{L}((2n/\mathsf{v}''(1))[\mathsf{I}_{\mathsf{v}}(\widehat{P}_n, \widehat{P}_{1,n} \otimes \widehat{P}_{2,n}) - \mathsf{v}(1)]) \Rightarrow \mathsf{H}((a-1)(b-1)).$$

The sequence of tests

$$\varphi_n = \begin{cases} 1 & \text{if} \quad \frac{2n}{\mathsf{v}''(1)}[\mathsf{I}_\mathsf{v}(\widehat{P}_n, \widehat{P}_{1,n} \otimes \widehat{P}_{2,n}) - \mathsf{v}(1)] > \chi^2_{1-\alpha,(a-1)(b-1)} \\ 0 & \text{else} \end{cases}$$

is an asymptotic level α test for testing

$$\begin{aligned} &\mathsf{H}_0 : p_{i,j} = p_{i,\cdot}p_{\cdot,j} > 0, \quad 1 \le i \le a, \ 1 \le j \le b, \quad \textit{versus} \\ &\mathsf{H}_A : p_{i_0,j_0} \neq p_{i_0,\cdot}p_{\cdot,j_0}, \quad \textit{for at least one } (i_0, j_0). \end{aligned}$$

Proof. If $\mathsf{v}(x) = x \ln x$, then by Example 8.125 and Theorem 8.120 we get $\mathcal{L}(2n\mathsf{K}(\widehat{P}_n, \widehat{P}_{1,n} \otimes \widehat{P}_{2,n})) \Rightarrow \mathsf{H}(d-k)$, where $d - k = (ab-1) - (a+b-2) = (a-1)(b-1)$. Hence the statement holds for the convex function $\mathsf{v}(x) = x \ln x$. If $\mathsf{v}_1, \mathsf{v}_2$ are two twice continuously differentiable functions with $\mathsf{v}_i''(1) > 0$, then the statement

$$\frac{2n}{\mathsf{v}_1''(1)}[\mathsf{I}_{\mathsf{v}_1}(\widehat{P}_n, \widehat{P}_{1,n} \otimes \widehat{P}_{2,n}) - \mathsf{v}_1(1)] = \frac{2n}{\mathsf{v}_2''(1)}[\mathsf{I}_{\mathsf{v}_2}(\widehat{P}_n, \widehat{P}_{1,n} \otimes \widehat{P}_{2,n}) - \mathsf{v}_2(1)] + o_{\mathbb{P}}(1)$$

can be established as in the proof of Lemma 8.118. ■

By specializing the convex function v that appears in the above proposition one obtains classical tests for independence. Putting $\mathsf{v}(x) = (x-1)^2$ we get the χ^2-test of independence. $\mathsf{v}(x) = x \ln x$ gives the likelihood ratio test. The choice of $\mathsf{v}(x) = (\sqrt{x} - 1)^2$ leads to a test statistic that compares the joint empirical distribution $\widehat{P}_n$ with the product of marginals $\widehat{P}_{1,n} \otimes \widehat{P}_{2,n}$ by means of the Hellinger distance.

8.9 Locally Asymptotically Optimal Tests

8.9.1 Testing of Univariate Parameters

Testing a Linear Contrast

Let $(P_\theta)_{\theta \in \Delta}$ be a parametrized family of distributions on $(\mathcal{X}, \mathfrak{A})$ with $\Delta \subseteq \mathbb{R}^d$ where we want to test the hypotheses $\mathsf{H}_0 : \theta \in \Delta_0$ versus $\mathsf{H}_A : \theta \in \Delta_A$ for increasing sample sizes. We assume that the hypotheses are specified by a linear function of the parameter, say $c^T\theta$. If θ has dimension 1, then we set $c = 1$ so that the tests refer directly to the parameter $\theta \in \Delta$. If θ has dimension $d > 1$, then for $c = (1, 0, ..., 0)$ we test the first component of θ, and the remaining components are nuisance parameters. For $c = (1, -1, 0, ..., 0)$ we test the difference between the first and second component of θ, and the remaining components are nuisance parameters. Especially, this case covers the comparison of two one-dimensional parameters in a two-sample problem.

We have seen already in Theorem 8.75 that for a fixed point in Δ_A the power of an asymptotic test typically tends to one, which is due to the fact that for $\theta_0 \neq \theta_1$ the sequences $P_{\theta_0}^{\otimes n}$ and $P_{\theta_1}^{\otimes n}$ are entirely separated. This means that the distributions $P_{\theta_0}^{\otimes n}$ and $P_{\theta_1}^{\otimes n}$ are, for large n, approximately concentrated on disjoint sets. A localization of the model, which has been introduced in Chapter 7 in the study of the efficiency of estimators, is also used here to turn to sequences of models for which the limiting models are not degenerate. To study the above testing problem we use a point θ_0 from the boundary of the null hypothesis as the localization point and consider the localized models with $(P_{\theta_0+h/\sqrt{n}}^{\otimes n})_{h\in\Delta_n}$ and $\Delta_n = \{h : \theta_0+h/\sqrt{n} \in \Delta\}$. If θ_0 is an interior point of Δ, then $\Delta_n \uparrow \mathbb{R}^d$, so that every $h \in \mathbb{R}^d$ belongs to Δ_n for all sufficiently large n. Because of this we may use $\mathbb{R}^d$ as the parameter set for asymptotic testing problems. We imagine that the localization point is known in this section. This is in fact true if we test a simple null hypothesis and we localize the model at the null hypothesis. For other testing problems the power of tests that include a known localization point is used as a benchmark for tests lacking this knowledge.

Regardless of whether the sequence of models originates from a localization procedure, we consider the following asymptotic testing problems.

$$\begin{aligned} &\mathcal{M}_n = (\mathcal{X}_n, \mathfrak{A}_n, (P_{n,h})_{h\in\Delta_n}), \quad \Delta_n \uparrow \mathbb{R}^d, \\ &\mathsf{H}_0 : h \in \Delta_0, \quad \mathsf{H}_A : h \in \Delta_A, \quad \Delta_0 \cap \Delta_A = \emptyset, \quad \Delta_0 \cup \Delta_A = \mathbb{R}^d. \end{aligned} \tag{8.137}$$

We have already introduced the concepts of asymptotic level α tests and asymptotic unbiased level α tests at (8.109).

Definition 8.127. *Given a sequence of models $\mathcal{M}_n$ and the testing problem in (8.137) we call an asymptotic level α test $\varphi_n : \mathcal{X}_n \to_m [0,1]$ a locally asymptotically uniformly best (LAUMP) level α test if for every asymptotic level α test ψ_n and every $h \in \Delta_A$ it holds*

$$\liminf_{n\to\infty}(\mathsf{E}_{n,h}\varphi_n - \mathsf{E}_{n,h}\psi_n) \geq 0. \tag{8.138}$$

An asymptotic unbiased level α test φ_n is called a locally asymptotically uniformly best unbiased (LAUMPU) level α test if for every asymptotic unbiased level α test ψ_n it holds (8.138) for every $h \in \Delta_A$.

Remark 8.128. The abbreviation LAUMP means locally asymptotically uniformly most powerful, where "uniformly" refers to the fact that (8.138) holds for every $h \in \Delta_A$. Similarly as before in the fixed sample size case, we use "best" instead of "most powerful" in the text, but maintain the traditional abbreviations LAUMP and LAUMPU. In all subsequent cases the limiting power $\lim_{n\to\infty} \mathsf{E}_{n,h}\psi_n$ of the best asymptotic test exists. If the ULAN condition holds, then (8.138) implies for every compact subset $C \subseteq \Delta_A$

$$\liminf_{n\to\infty} \inf_{h\in C}(\mathsf{E}_{n,h}\varphi_n - \mathsf{E}_{n,h}\psi_n) \geq 0, \tag{8.139}$$

so that another uniformity comes into consideration. Sequences of tests that satisfy (8.139) are called locally asymptotically uniformly most powerful in Lehmann and Romano (2005).

The construction of locally asymptotically best level α tests consists of two steps. To derive an upper bound for the power of asymptotic tests we treat the testing problem as a decision problem with decision space $\mathcal{D} = \{0,1\}$ under the zero–one loss. Then minimizing the risk on the alternative means just maximizing the power. Hence asymptotic lower bounds for the risk turn into asymptotic upper bounds for the power function. We derive these bounds from the general Hájek–LeCam bound in Proposition 6.89. In a second step we search for tests that achieve these upper bounds. Such tests are typically constructed with the help of the limit model. Under the LAN condition the limit model is a Gaussian model and the central variable is a sufficient statistic, so that every test here is equivalent to a test that depends on the central variable. The construction principle for asymptotically best tests is to replace the central variable in the Gaussian model with the central sequence of the sequence of models.

Suppose that the sequence of models in (8.137) satisfies the LAN(Z_n, I_0) condition (see Definition 6.63) with positive definite Fisher information matrix I_0. Then by Corollary 6.66 we have the weak convergence of models

$$\mathcal{M}_n = (\mathcal{X}_n, \mathfrak{A}_n, (P_{n,h})_{h\in\Delta_n}) \Rightarrow \mathcal{G}_0 = (\mathbb{R}^d, \mathfrak{B}_d, (\mathsf{N}(h, \mathsf{I}_0^{-1})_{h\in\mathbb{R}^d}). \quad (8.140)$$

If Z_n is the central sequence, then the third lemma of LeCam (see Theorem 6.72) implies that

$$\mathcal{L}(\mathsf{I}_0^{-1} Z_n | P_{n,h}) \Rightarrow \mathsf{N}(h, \mathsf{I}_0^{-1}), \quad h \in \mathbb{R}^d, \quad (8.141)$$

where in view of Corollary 6.73 the convergence is locally uniform if instead of the LAN(Z_n, I_0) condition the stronger ULAN(Z_n, I_0) condition is satisfied.

For a fixed vector c, a constant $d > 0$, and $\sigma_0^2 = c^T \mathsf{I}_0^{-1} c$, we consider the following testing problems for the local parameter h.

Testing Problem	H_0	H_A
(I)	$c^T h \le 0$	$c^T h > 0$
(II)	$c^T h = 0$	$c^T h \neq 0$
(III)	$\lvert c^T h \rvert \ge d\sigma_0$	$\lvert c^T h \rvert < d\sigma_0$
(IV)	$\lvert c^T h \rvert \le d\sigma_0$	$\lvert c^T h \rvert > d\sigma_0$

(8.142)

We recall the standard Gauss tests $\psi_I, \psi_{II}, \psi_{III}, \psi_{IV}$ from (8.18), and Problem 8.23 which provides the power of the best tests in Theorem 8.22 for the model $\mathcal{G}_0$ in (8.140). We set $T_n = c^T \mathsf{I}_0^{-1} Z_n / \sigma_0$ and

$$\begin{aligned}
\mathsf{p}_I(h) &= 1 - \Phi(u_{1-\alpha} - (c^T h)/\sigma_0) \\
\mathsf{p}_{II}(h) &= 1 - \Phi(u_{1-\alpha/2} - (c^T h)/\sigma_0) + \Phi(-u_{1-\alpha/2} - (c^T h)/\sigma_0) \\
\mathsf{p}_{III}(h) &= \Phi(z_{d,\alpha} - (c^T h)/\sigma_0) - \Phi(-z_{d,\alpha} - (c^T h)/\sigma_0) \\
\mathsf{p}_{IV}(h) &= 1 - \Phi(z_{d,1-\alpha} - (c^T h)/\sigma_0) + \Phi(-z_{d,1-\alpha} - (c^T h)/\sigma_0).
\end{aligned} \tag{8.143}$$

Problem 8.129.* If the LAN(Z_n, I_0) condition is satisfied, then it holds $\mathsf{p}_i(h) = \lim_{n\to\infty} \mathsf{E}_{n,h}\psi_i(T_n)$ for $i = I, II, III, IV$, where the convergence is locally uniform if the ULAN(Z_n, I_0) condition is satisfied.

Theorem 8.130. *Assume the sequence of models $\mathcal{M}_n = (\mathcal{X}_n, \mathfrak{A}_n, (P_{n,h})_{h\in\Delta_n})$ with $\Delta_n \uparrow \mathbb{R}^d$ satisfies the LAN(Z_n, I_0) condition with a nonsingular Fisher information matrix I_0. Let $\varphi_{I,n}, ..., \varphi_{IV,n}$ be asymptotic level α tests for the testing problems (I)–(IV) in (8.142), respectively. Assume in addition that $\varphi_{II,n}$ and $\varphi_{IV,n}$ are asymptotically unbiased. Then for $i \in \{I, II, III, IV\}$ it holds under H_A,*

$$\limsup_{n\to\infty} \mathsf{E}_{n,h}\varphi_{i,n} \le \mathsf{p}_i(h). \tag{8.144}$$

The tests $\psi_i(T_n)$, with ψ_i in (8.18), $i = I, II, III, IV$, are asymptotic level α tests and attain the respective upper bounds, where the convergence is locally uniform if the ULAN(Z_n, I_0) condition is satisfied.

Corollary 8.131. *The tests $\psi_i(T_n)$, $i = I, II, III, IV$, have the following optimality properties for the testing problems in (8.142).*

$\psi_I(T_n)$ *LAUMP for (I),*	$\psi_{II}(T_n)$ *LAUMPU for (II),*
$\psi_{III}(T_n)$ *LAUMP for (III),*	$\psi_{IV}(T_n)$ *LAUMPU for (IV).*

Proof. Let $i \in \{I, III, IV, II\}$. We use the decision space $\mathcal{D} = \{0, 1\}$ and the zero–one loss function $L(h, a) = (1 - a)I_{\Delta_{i,A}}(h) + aI_{\Delta_{i,0}}(h)$. Let $\mathbb{D}_{i,\mathcal{G}_0}$ denote the set of all decisions $\varphi\boldsymbol{\delta}_1 + (1 - \varphi)\boldsymbol{\delta}_0$, where φ is a level α test for the model $\mathcal{G}_0$. For $i \in \{II, IV\}$ we require in addition that the tests in $\mathbb{D}_{i,\mathcal{G}_0}$ are unbiased. As the weak convergence of such decisions defined by tests is just the convergence of power functions (see Definition 6.83) we get that $\mathbb{D}_{i,\mathcal{G}_0}$ is closed. From the convergence (8.140) and Proposition 6.89 we get for any fixed h from the alternative,

$$\limsup_{n\to\infty} \mathsf{E}_{n,h}\varphi_{i,n} \le \sup_{\varphi\in\mathbb{D}_{i,\mathcal{G}_0}} \int \varphi(x)\mathsf{N}(h, \mathsf{I}_0^{-1})(dx).$$

The convergence $\lim_{n\to\infty} \mathsf{E}_{n,h}\psi_i(T_n) = \mathsf{p}_i(h)$, including the local uniform convergence, was established already in Problem 8.129. To conclude the proof it remains to remark that according to Theorem 8.22 it holds

$$\sup_{\varphi\in\mathbb{D}_{i,\mathcal{G}_0}} \int \varphi(x)\mathsf{N}(h, \mathsf{I}_0^{-1})(dx) = \mathsf{p}_i(h).$$

■

Tests in One-Parameter Families

In this section we consider one-parameter families $(P_\theta)_{\theta\in(a,b)}$, where we want to test if the true parameter is in a certain range, that is related to a given norm value. Depending on the concrete situation we have four different types of testing problems. For a fixed $\theta_0 \in (a,b)$ we consider the localized models

$$\mathcal{M}_n = (\mathcal{X}^n, \mathfrak{A}^{\otimes n}, P^{\otimes n}_{\theta_0+h/\sqrt{n}})_{h\in\Delta_n}, \quad \Delta_n = \{h : a < \theta_0 + h/\sqrt{n} < b\},$$

and for a fixed constant $d > 0$ the following testing problems.

Testing Problem	H_0	H_A
(I)	$h \le 0$	$h > 0$
(II)	$h = 0$	$h \neq 0$
(III)	$\lvert h\rvert \ge d\,\mathsf{I}^{-1/2}(\theta_0)$	$\lvert h\rvert < d\,\mathsf{I}^{-1/2}(\theta_0)$
(IV)	$\lvert h\rvert \le d\,\mathsf{I}^{-1/2}(\theta_0)$	$\lvert h\rvert > d\,\mathsf{I}^{-1/2}(\theta_0)$

(8.145)

The meaning of (I) and (II) is clear. In (III) we want to detect if there is only a small deviation from the norm value θ_0. Tests for (III) are called *equivalence tests*. In (IV) we want to detect if there is more than a small deviation from the norm value θ_0.

We suppose that $(P_\theta)_{\theta\in(a,b)}$ is $\mathbb{L}_2$-differentiable with derivative $\dot{L}_{\theta_0}$ and Fisher information $\mathsf{I}(\theta_0)$, and we set

$$Z_n(\theta_0) = n^{-1/2}\sum\nolimits_{i=1}^{n} \dot{L}_{\theta_0}(X_i) \quad \text{and} \quad T_n(\theta_0) = \mathsf{I}^{-1/2}(\theta_0)Z_n(\theta_0). \qquad (8.146)$$

Proposition 8.132. *Suppose the family $(P_\theta)_{\theta\in(a,b)}$ is $\mathbb{L}_2$-differentiable at θ_0 with positive Fisher information $\mathsf{I}(\theta_0)$. Let $\varphi_{I,n}, ..., \varphi_{IV,n}$ be asymptotic level α tests for the testing problems (I)–(IV) in (8.145), respectively, where $\varphi_{II,n}$ and $\varphi_{IV,n}$ are in addition asymptotically unbiased. Then for $i = I, ..., IV$, under the alternative H_A, it holds*

$$\limsup_{n\to\infty} \mathsf{E}_{n,h}\varphi_{i,n} \le \mathsf{p}_i(h),$$

where $\mathsf{p}_i(h)$ is given by (8.143) with $c = 1$. The asymptotic upper bounds for the power are locally uniformly attained by the respective tests $\psi_i(T_n(\theta_0))$.

Corollary 8.133. *The tests $\psi_i(T_n(\theta_0))$, $i = I, II, III, IV$, have the following optimality properties for the testing problems in (8.145).*

$\psi_I(T_n(\theta_0))$ *LAUMP for (I)*, $\qquad \psi_{II}(T_n(\theta_0))$ *LAUMPU for (II)*,

$\psi_{III}(T_n(\theta_0))$ *LAUMP for (III)*, $\qquad \psi_{IV}(T_n(\theta_0))$ *LAUMPU for (IV)*.

Proof. The $\mathbb{L}_2$-differentiability implies the ULAN$(Z_n, \mathsf{I}(\theta_0))$ condition. The statement follows from Theorem 8.130 and its corollary if we observe that $c = 1$, $\sigma_0^2 = \mathsf{I}_0^{-1}$, and $T_n(\theta_0) = c^T \mathsf{I}_0^{-1} Z_n(\theta_0)/\sigma_0 = \mathsf{I}_0^{-1/2} Z_n$. ■

The test $\psi_{III}(\mathsf{I}^{-1/2}(\theta_0) Z_n(\theta_0))$ is an equivalence test. The asymptotic power of such tests is studied in Lehmann and Romano (2005) and Romano (2005). A nonparametric approach to equivalence tests is given in Janssen (2000). The tests $\psi_I(T_n(\theta_0))$ and $\psi_{II}(T_n(\theta_0))$, known as *Rao's score tests,* were studied in Rao (1947). Wald (1939, 1941a,b, 1943) used the MLE to construct asymptotic level α tests and studied the power.

To discuss the relations between the above tests and the Wald-type tests based on the MLE we suppose that the assumptions of Theorem 7.148 are satisfied. Then the MLE satisfies

$$\sqrt{n}\, \mathsf{I}(\theta_0)(\widehat{\theta}_n - \theta_0) = n^{-1/2} \sum\nolimits_{i=1}^{n} \dot{L}_{\theta_0}(X_i) + o_{P_{\theta_0}^{\otimes n}}(1).$$

We replace the central sequence with $\sqrt{n}\, \mathsf{I}(\theta_0)(\widehat{\theta}_n - \theta_0)$ to get tests that are again asymptotic level α tests and have under local alternatives the same asymptotic behavior as the tests in Proposition 8.132. Although the tests based on Z_n and $\sqrt{n}\, \mathsf{I}(\theta_0)(\widehat{\theta}_n - \theta_0)$ are equivalent under the null hypothesis, and by the contiguity in Corollary 6.67 also under local alternatives, the situation changes if we turn to fixed alternatives. Consider, for example, the case where the MLE is consistent at every θ_0 and the testing problem is $\mathsf{H}_0 : \theta = \theta_0$ versus $\mathsf{H}_A : \theta \neq \theta_0$. It is easy to see that for $\theta \neq \theta_0$ it holds $\mathsf{E}_\theta \psi_{II}(\sqrt{n}\, \mathsf{I}(\theta_0)(\widehat{\theta}_n - \theta_0)) \to 1$ so that the error probabilities of the second kind tend to zero and the sequence of tests is consistent in this sense. On the other hand, Proposition 8.132 needs only the $\mathbb{L}_2$-differentiability, but the consistency of the asymptotic test $\psi_{II}(T_n(\theta_0))$ remains open. Moreover, Rao's score tests can be applied in nonparametric models by considering a one-parameter subfamily. In this case there is no MLE available to construct a Wald-type test.

The following example deals with a one-parameter exponential family.

Example 8.134. Let $(P_\theta)_{\theta \in (a,b)}$ be a one parameter exponential family with generating statistic T. The family $(P_\theta)_{\theta \in (a,b)}$ is $\mathbb{L}_2$-differentiable at every $\theta_0 \in (a, b)$ with derivative $\dot{L}_{\theta_0} = (T - K'(\theta_0))$ and Fisher information $\mathsf{I}(\theta_0) = K''(\theta_0)$, see Example 1.120. Hence the central sequence Z_n is given by

$$Z_n(\theta_0) = \frac{1}{\sqrt{n}} \sum\nolimits_{i=1}^{n} (T(X_i) - K'(\theta_0)).$$

If we use $T_{\oplus n} = \sum_{i=1}^{n} T(X_i)$ instead of T in the Theorems 8.6, 8.8, and 8.11 we see that the tests $\psi_i(T_n(\theta_0))$ are just asymptotic versions of the corresponding tests in these theorems in the sense that the tests are nonrandomized tests and the critical values, at which the null hypothesis is rejected, are determined with the help of the normal approximation $\mathcal{L}(Z_n | P_{\theta_0}^{\otimes n}) \Rightarrow \mathsf{N}(0, \mathsf{I}(\theta_0))$.

The next example deals with the important class of location models.

Example 8.135. We consider location models. Let f be a Lebesgue density on $\mathbb{R}$, and let P_θ be the distribution with Lebesgue density $f(x-\theta)$. If $f(x) > 0$ for every $x \in \mathbb{R}$, f is absolutely continuous, and $\mathsf{I} := \int ((f')^2/f) d\boldsymbol{\lambda} < \infty$, then we know from Lemma 1.121 that the family $(P_\theta)_{\theta\in\mathbb{R}}$ is $\mathbb{L}_2$-differentiable at every $\theta_0 \in \mathbb{R}$ with derivative $\dot{L}_{\theta_0}(x) = -f'(x-\theta_0)/f(x-\theta_0)$ and Fisher information I. We assume without loss of generality that $\theta_0 = 0$. A localization at $\theta_0 = 0$ gives the central sequence and the distribution under local alternatives $P^{\otimes n}_{h/\sqrt{n}}$, respectively,

$$Z_n = -\frac{1}{\sqrt{n}} \sum\nolimits_{i=1}^{n} f'(X_i)/f(X_i) \quad \text{and} \quad \mathcal{L}(Z_n|P^{\otimes n}_{h/\sqrt{n}}) \Rightarrow \mathsf{N}(\mathsf{I}h, \mathsf{I}).$$

Then the tests $\psi_i(\mathsf{I}^{-1/2} Z_n)$ for the local alternatives in (8.145) have under local alternatives the asymptotic power $\mathsf{p}_i(h)$, where $\mathsf{p}_i(h)$ is given by (8.143) with $c = 1$. Moreover they have the optimality properties stated in Corollary 8.133. Some special densities and their score functions are shown below, where we assume that $\beta > 1/2$.

f	$c(\beta)\exp\{-\lvert x\rvert^\beta\}$	$\frac{1}{2}\exp\{-\lvert x\rvert\}$	$\frac{1}{\sqrt{2\pi}}\exp\{-x^2/2\}$	$\frac{1}{\pi(1+x^2)}$
$-f'/f$	$\beta\,\mathrm{sgn}(x)\lvert x\rvert^{\beta-1}$	$\mathrm{sgn}(x)$	x	$\frac{2x}{1+x^2}$

The question arises as to what happens if we use the score function from a model which is different from the model from which the data originate, and why we should do that. For example, the score function for the Laplace distribution is $\mathrm{sgn}(x)$, which is a bounded function. Hence for data from a normal distribution gross error outliers have less influence on the inference than for the exact and unbounded score function $-f'(x)/f(x) = x$. Hence the "false" score function protects us against such outliers. Now to the efficiency. If $X_1, ..., X_n$ is the observed sample we set $\varepsilon_i = \mathrm{sgn}(X_i)$ and $T_n = \sum_{i=1}^n \varepsilon_i$. We consider the one-sided testing problem $\mathsf{H}_0 : \theta \le 0$ versus $\mathsf{H}_A : \theta > 0$ and the test $\psi_n = I_{(u_{1-\alpha},\infty)}(T_n/\sqrt{n})$. By construction $\mathsf{E}_{n,0}\psi_n \to \alpha$. To calculate the power we note that by Corollary 6.74,

$$\mathcal{L}(T_n/\sqrt{n}|\mathsf{N}^{\otimes n}(h/\sqrt{n}, 1)) \Rightarrow \mathsf{N}(\sigma_{1,2}h, 1), \quad \text{where}$$
$$\sigma_{1,2} = \int [\mathrm{sgn}(x)\varphi'_{0,1}(x)/\varphi_{0,1}(x)]\varphi_{0,1}(x)dx = \int \mathrm{sgn}(x) x \varphi_{0,1}(x)dx = \sqrt{2/\pi}.$$

On the other hand, the finite and asymptotically optimal test is the Gauss test $\varphi_n = I_{(u_{1-\alpha},\infty)}(n^{1/2}\overline{X}_n)$. Obviously $\mathcal{L}(\sqrt{n}\,\overline{X}_n|\mathsf{N}^{\otimes n}(h/\sqrt{n}, 1)) = \mathsf{N}(h, 1)$. Hence we obtain the following asymptotic power under local alternatives.

$$\lim_{n\to\infty} \mathsf{E}_{n,h}\psi_n = 1 - \Phi(u_{1-\alpha} - h\sqrt{2/\pi}), \quad \lim_{n\to\infty} \mathsf{E}_{n,h}\varphi_n = 1 - \Phi(u_{1-\alpha} - h), \quad (8.147)$$

so that φ_n has larger asymptotic power than ψ_n, but not by much.

In the previous example we have studied the efficiency of various asymptotic tests for the location model. Now we study the efficiency of tests systematically. We use asymptotically linear statistics to construct tests. Such statistics have appeared already at several places. For asymptotic tests that are based on such statistics the investigation of asymptotic power can be reduced to a comparison of the influence function and the score function $\dot{L}_{\theta_0}$

that appears in the central sequence. More precisely, suppose that the model under consideration is a one-parameter model for which we study the localized models $(\mathcal{X}^n, \mathfrak{A}^{\otimes n}, (P^{\otimes n}_{\theta_0+h/\sqrt{n}})_{h\in\Delta_n})$ and the hypotheses in (8.145). For simplicity of the formulation we confine ourselves to the first testing problem. We fix an influence function $\Psi \in \mathbb{L}_2^0(P_{\theta_0})$, set $\sigma_{1,1} = \mathsf{E}_{\theta_0}\Psi^2$, and construct the sequence of tests

$$\varphi_{\Psi,n} := I_{(u_{1-\alpha},\infty)}(\sigma_{1,1}^{-1/2} n^{-1/2} \sum\nolimits_{i=1}^{n} \Psi(X_i)). \tag{8.148}$$

To calculate the asymptotic power we note that by Corollary 6.74,

$$\begin{aligned} &\mathcal{L}(\sigma_{1,1}^{-1/2} n^{-1/2} \sum\nolimits_{i=1}^{n} \Psi(X_i)|P^{\otimes n}_{\theta_0+h/\sqrt{n}}) \Rightarrow \mathsf{N}(h\sigma_{1,1}^{-1/2}\sigma_{1,2}, 1), \\ &\lim_{n\to\infty} \mathsf{E}_{n,h}\varphi_{\Psi,n} = 1 - \Phi(u_{1-\alpha} - h\sigma_{1,1}^{-1/2}\sigma_{1,2}), \quad \sigma_{1,2} = \mathsf{E}_{\theta_0}(\Psi \dot{L}_{\theta_0}). \end{aligned} \tag{8.149}$$

These expressions for the asymptotic power lead to the concept of asymptotic relative efficiency. Assume that φ_n and $\widetilde{\varphi}_n$ are any asymptotic level α tests for which locally uniform for $h > 0$ and for some nonnegative γ and $\widetilde{\gamma}$,

$$\lim_{n\to\infty} \mathsf{E}_{n,h}\varphi_n = 1 - \Phi(u_{1-\alpha} - h\gamma), \quad \lim_{n\to\infty} \mathsf{E}_{n,h}\widetilde{\varphi}_n = 1 - \Phi(u_{1-\alpha} - h\widetilde{\gamma}). \tag{8.150}$$

Then

$$ARE(\widetilde{\varphi}_n : \varphi_n|P_{n,h}) := \widetilde{\gamma}^2/\gamma^2 \tag{8.151}$$

is called the *asymptotic relative efficiency* (ARE) of the asymptotic level α test $\widetilde{\varphi}_n$ with respect to the asymptotic level α test φ_n. It admits a simple interpretation in localized models $P^{\otimes n}_{n,\theta_0+h/\sqrt{n}}$. Suppose $\beta := \widetilde{\gamma}^2/\gamma^2 < 1$. For every n, we take, instead of the test φ_n based on the sample size n, the test $\varphi_{[\beta n]}$ which uses only the observations $X_1, ..., X_{[\beta n]}$ and neglects $X_{[\beta n]+1}, ..., X_n$. Then we get from (8.150)

$$\lim_{n\to\infty} \mathsf{E}_{n,h}\varphi_{[\beta n]} = 1 - \Phi(u_{1-\alpha} - \sqrt{\beta}h\gamma) = 1 - \Phi(u_{1-\alpha} - h\widetilde{\gamma}).$$

Thus we see that the sequence of tests $\varphi_{[\beta n]}$ based on the sample size $[\beta n]$ has asymptotically the same power as the tests $\widetilde{\varphi}_n$ based on the sample size n. This means that φ_n is more efficient than $\widetilde{\varphi}_n$, as already a sample of size $[\beta n]$ is sufficient to produce the same power as $\widetilde{\varphi}_n$ for $n \to \infty$. For example, we get from (8.147) that for a normal population, and the problem of testing μ, the median test ψ_n has with respect to the Gauss test φ_n the asymptotic relative efficiency $ARE(\psi_n : \varphi_n|P_{n,h}) = 2/\pi$.

Within the class of tests that are based on linearized statistics the ARE is just the squared ratio of the correlations of the influence functions and the derivative of the model. For any two random variables X, Y we denote by $\rho_{\theta_0}(X, Y)$ their correlation under P_{θ_0}.

Proposition 8.136. *Suppose* $(P_\theta)_{\theta\in\Delta}$ *with* $\Delta \subseteq \mathbb{R}$ *is* $\mathbb{L}_2$*-differentiable at* $\theta_0 \in \Delta^0$ *with derivative* $\dot{L}_{\theta_0}$ *and positive Fisher information* $\mathsf{I}(\theta_0)$. *Let* $\Psi, \widetilde{\Psi} \in \mathbb{L}_2^0(P_{\theta_0})$. *If* $\varphi_{\Psi,n}$ *and* $\varphi_{\widetilde{\Psi},n}$ *are the tests which are defined by the influence functions* Ψ *and* $\widetilde{\Psi}$, *see (8.148), then it holds*

$$\lim_{n\to\infty} \mathsf{E}_{n,h}\varphi_{\widetilde{\Psi},n} = 1 - \Phi(u_{1-\alpha} - \rho_{\theta_0}(\widetilde{\Psi}, \dot{L}_{\theta_0})h\ \mathsf{I}^{1/2}(\theta_0)),$$
$$\lim_{n\to\infty} \mathsf{E}_{n,h}\varphi_{\Psi,n} = 1 - \Phi(u_{1-\alpha} - \rho_{\theta_0}(\Psi, \dot{L}_{\theta_0})h\ \mathsf{I}^{1/2}(\theta_0)),$$

and $ARE(\varphi_{\widetilde{\Psi},n} : \varphi_{\Psi,n}|P_{n,h}) = \rho^2_{\theta_0}(\widetilde{\Psi}, \dot{L}_{\theta_0})/\rho^2_{\theta_0}(\Psi, \dot{L}_{\theta_0})$.

Proof. The statement follows from (8.149). ■

Remark 8.137. To illustrate the concept of ARE we have studied only the testing problem I in (8.145). However, from that display one can see that the ratio $\widetilde{\gamma}^2/\gamma^2$ on the right-hand side of (8.151) admits the same interpretation via the sample sizes for the other testing problems II, III, IV.

The ARE was introduced by Pitman (1949). Lehmann and Romano (2005) refer to an unpublished set of lecture notes and point out that Pitman developed the ARE concept and applied it to several examples, including the Wilcoxon test. Noether (1955) generalized Pitman's results.

In the remainder of this section we present the proof of Theorem 7.162. Its proof had been postponed as it requires results on locally optimal tests. The proof below is taken from Witting and Müller-Funk (1995).

Proof. (LeCam–Bahadur's Theorem 7.162) Fix $u \in \mathbb{R}^d$. The assumption on T_n implies $\mathcal{L}(\sqrt{n}u^T(T_n - \theta)|P_\theta^{\otimes n}) \Rightarrow \mathsf{N}(0, u^T\Sigma(\theta)u)$, $\theta \in \Delta$. This means that

$$f_n(\theta, u) := |\frac{1}{2} - P_\theta^{\otimes n}(\sqrt{n}u^T(T_n - \theta) > 0)| \to 0, \quad \theta \in \Delta,\ u \in \mathbb{R}^d.$$

We extend $f_n(\theta, u)$ by setting $f_n(\theta, u) = 0$ if $\theta \in \mathbb{R}^d\backslash\Delta$, $u \in \mathbb{R}^d$. Then

$$\int f_n(\theta + v/\sqrt{n}, u)\mathsf{N}(0, \mathbf{I})(d\theta)$$
$$= \int (2\pi)^{-n/2} f_n(t, u)\exp\{-\frac{1}{2}(t - v/\sqrt{n})^T(t - v/\sqrt{n})\}\boldsymbol{\lambda}_d(dt).$$

As $(t - v/\sqrt{n})^T(t - v/\sqrt{n}) \geq t^Tt - 2t^Tv/\sqrt{n} \geq \frac{1}{2}t^Tt$ for $\|v/\sqrt{n}\| \leq \|t\|/4$ we may apply Lebesgue's theorem to the above integral and get for every $v \in \mathbb{R}^d$

$$f_n(\theta + v/\sqrt{n}) \to^{\mathsf{N}(0,\mathbf{I})} 0.$$

Let $\mathbb{D}$ be a countable dense subset of $\mathbb{R}^d$. As every stochastic convergent sequence contains an a.s. convergent subsequence (see Proposition A.12) and

$\mathbb{D}$ is countable, we may use the diagonal technique to find a universal subsequence n_k and a Borel set N with $\mathsf{N}(0, \mathbf{I})(N) = 0$ such that for every $\theta \in \Delta \backslash N$ and $u, v \in \mathbb{D}$,

$$P_{\theta+v/\sqrt{n_k}}^{\otimes n}(\sqrt{n_k} u^T(T_{n_k} - \theta) > u^T v) \to \frac{1}{2}.$$

We fix $\theta_0 \in \Delta \backslash N$. As Δ is open it holds $\theta_0 + \eta v/\sqrt{n} \in \Delta$ for all sufficiently large n. The family $Q_\eta = P_{\theta_0+\eta v}$ is $\mathbb{L}_2$-differentiable at $\eta_0 = 0$ with derivative $v^T \dot{L}_\theta$ and Fisher information $v^T \mathsf{I}(\theta_0) v$; see Proposition 1.112. We consider the testing problem $\mathsf{H}_0 : \eta \le 0$ versus $\mathsf{H}_A : \eta > 0$. Then by (8.144), every asymptotic level α test φ_n satisfies for $\eta = 1$

$$\limsup_{n\to\infty} \mathsf{E}_{n,1}\varphi_n \le 1 - \Phi(u_{1-\alpha} - (v^T \mathsf{I}(\theta_0) v)^{1/2}). \tag{8.152}$$

The assumption $\mathcal{L}(\sqrt{n}(T_n - \theta)|P_\theta^{\otimes n}) \Rightarrow \mathsf{N}(0, \Sigma(\theta))$ implies that for $u \in \mathbb{D}$, $u \ne 0$ the tests $\psi_n = I_{(u^T v,\infty)}(\sqrt{n} u^T(T_n - \theta_0))$, satisfy

$$\begin{aligned}\lim_{n\to\infty} \mathsf{E}_{n,0}\psi_n &= \lim_{n\to\infty} P_{\theta_0}^{\otimes n}(\sqrt{n} u^T(T_n - \theta_0) > u^T v) = 1 - \Phi_{0,u^T\Sigma(\theta_0)u}(u^T v) \\ &= 1 - \Phi((u^T \Sigma(\theta_0) u)^{-1/2}(u^T v)).\end{aligned}$$

Put $\alpha_0 = 1 - \Phi((u^T \Sigma(\theta_0) u)^{-1/2}(u^T v))$ so that $u_{1-\alpha_0} = (u^T \Sigma(\theta_0) u)^{-1/2}(u^T v)$. Then ψ_n becomes an asymptotic level α_0 test. As $\lim_{k\to\infty} \mathsf{E}_{n_k,1}\psi_{n_k} = 1/2$ by the construction of the subsequence n_k we get from (8.152),

$$\frac{1}{2} \le 1 - \Phi((u^T \Sigma(\theta_0) u)^{-1/2}(u^T v) - (v^T \mathsf{I}(\theta_0) v)^{1/2}),$$

which implies $(u^T v)^2 \le (u^T \Sigma(\theta_0) u)(v^T \mathsf{I}(\theta_0) v)$, for every $u, v \in \mathbb{D}$, $u \ne 0$. As the left- and right-hand terms are continuous functions of u and v, we obtain

$$(u^T v)^2 \le (u^T \Sigma(\theta_0) u)(v^T \mathsf{I}(\theta_0) v), \quad u, v \in \mathbb{R}^d.$$

Put $v = \mathsf{I}^{-1}(\theta_0) u$ to conclude $u^T \mathsf{I}^{-1}(\theta_0) u \le u^T \Sigma(\theta_0) u$ for every $u \in \mathbb{R}^d$, and the proof is completed. ■

Testing Univariate Parameters in Multivariate Models

In this section we consider the case where the parameter vector $\theta = (\tau, \xi^T)^T$ consists of a one-dimensional parameter τ of interest and a $d-1$ dimensional nuisance parameter ξ. We assume that the model is $\mathbb{L}_2$-differentiable at $\theta_0 \in \Delta^0$. According to the partition of the parameter θ we have the following partitions of the $\mathbb{L}_2$-derivative of the model and the central sequence.

$$\dot{L}_{\theta_0} = \begin{pmatrix} U_{\theta_0} \\ V_{\theta_0} \end{pmatrix}, \; Z_n(\theta_0) = \frac{1}{\sqrt{n}} \sum_{i=1}^n \begin{pmatrix} U_{\theta_0}(X_i) \\ V_{\theta_0}(X_i) \end{pmatrix}, \; \theta_0 = \begin{pmatrix} \tau_0 \\ \xi_0 \end{pmatrix} \in \Delta^0. \tag{8.153}$$

The Neyman–Rao statistic

$$S_{n,NR}(\theta_0) = n^{-1/2} \sum\nolimits_{i=1}^{n} W_{\theta_0}(X_i), \quad \text{where}$$
$$W_{\theta_0} = U_{\theta_0} - \mathsf{I}_{1,2}(\theta_0)\mathsf{I}_{2,2}^{-1}(\theta_0)V_{\theta_0}, \tag{8.154}$$

has been introduced in (8.112). We consider the localized models

$$\begin{aligned}\mathcal{M}_n &= (\mathcal{X}^n, \mathfrak{A}^{\otimes n}, (P^{\otimes n}_{\tau_0+h/\sqrt{n},\xi_0+g/\sqrt{n}})_{(h,g)\in\Delta_n}),\\ \Delta_n &= \{(h,g) : (\tau_0 + h/\sqrt{n}, (\xi_0 + g/\sqrt{n})^T)^T \in \Delta\}.\end{aligned} \tag{8.155}$$

It holds

$$\begin{aligned}\mathsf{E}_{\theta_0}(U_{\theta_0} W_{\theta_0}) &= \mathsf{E}_{\theta_0} W_{\theta_0}^2\\ &= \mathsf{I}_{1,1}(\theta_0) - \mathsf{I}_{1,2}(\theta_0)\mathsf{I}_{2,2}^{-1}(\theta_0)\mathsf{I}_{2,1}(\theta_0) = \mathsf{G}(\theta_0).\end{aligned}$$

As $\mathsf{E}_{\theta_0}(W_{\theta_0} V_{\theta_0}^T) = 0$ we get from Corollary 6.73 for $(h, g^T)^T \in \mathbb{R}^d$,

$$\mathcal{L}(S_{n,NR}(\theta_0)|P^{\otimes n}_{\tau_0+h/\sqrt{n},\xi_0+g/\sqrt{n}}) \Rightarrow \mathsf{N}(\mathsf{G}(\theta_0)h, \mathsf{G}(\theta_0)). \tag{8.156}$$

For the sequence of models (8.155) we consider the following hypotheses.

Testing Problem	H_0	H_A
(I)	$h \leq 0,\ g \in \mathbb{R}^{d-1}$	$h > 0,\ g \in \mathbb{R}^{d-1}$
(II)	$h = 0,\ g \in \mathbb{R}^{d-1}$	$h \neq 0,\ g \in \mathbb{R}^{d-1}$

(8.157)

We normalize the Neyman–Rao statistic in (8.154) and set

$$T_n(\theta_0) = \mathsf{G}^{-1/2}(\theta_0)S_{n,NR}(\theta_0) = \frac{1}{\sqrt{n}} \sum\nolimits_{i=1}^{n} \mathsf{G}^{-1/2}(\theta_0)W_{\theta_0}(X_i).$$

The following result goes back to Neyman (1959), who constructed an asymptotically optimal test in the presence of nuisance parameters.

Theorem 8.138. ***(Neyman's Test)*** *Suppose the family $(P_\theta)_{\theta\in\Delta}$, $\Delta \subseteq \mathbb{R}^d$, satisfies condition (A10) and $\mathsf{I}(\theta_0)$ is nonsingular. Let $\theta = (\tau, \xi^T)^T$, and suppose $\widetilde{\xi}_n$ is a $\sqrt{n}$-consistent estimator of ξ for the sequence of submodels $(P^{\otimes n}_{(\tau_0,\xi)})_{\xi\in\Xi(\tau_0)}$. Let $\alpha \in (0,1)$ be fixed. Then*

$$\psi_I(T_n(\tau_0, \widetilde{\xi}_n)), \quad \textit{where } \psi_I(t) = I_{(u_{1-\alpha},\infty)}(t),\ t \in \mathbb{R},$$

is a LAUMP level α test for testing problem (I) in (8.157), and

$$\psi_{II}(T_n(\tau_0, \widetilde{\xi}_n)), \quad \textit{where } \psi_{II}(t) = I_{(-\infty,-u_{1-\alpha/2})}(t) + I_{(u_{1-\alpha/2},\infty)}(t),\ t \in \mathbb{R},$$

is a LAUMPU level α test for testing problem (II) in (8.157).

Proof. By Theorem 8.130 the test $\psi_I(T_n(\tau_0,\xi_0))$ is a LAUMP level α test and $\psi_{II}(T_n(\tau_0,\xi_0))$ is a LAUMPU level α test. It follows from Proposition 8.113 that

$$T_n(\tau_0,\widetilde{\xi}_n) = T_n(\tau_0,\xi_0) + o_{P^{\otimes n}_{(\tau_0,\xi_0)}}(1).$$

The contiguity in Corollary 6.67 and Slutsky's lemma show that $T_n(\tau_0,\widetilde{\xi}_n)$ and $T_n(\tau_0,\xi_0)$ have the same limit distributions under $P^{\otimes n}_{\tau_0+h/\sqrt{n},\xi_0+g/\sqrt{n}}$ and provide tests with the same asymptotic power. ■

We consider a location-scale model $\mathcal{M} = (\mathbb{R},\mathfrak{B},(P_{\mu,\sigma})_{\mu\in\mathbb{R},\sigma>0})$, generated by a distribution with Lebesgue density f that is positive and continuously differentiable, and apply Theorem 8.138. The distribution $P_{\mu,\sigma}$ has the Lebesgue density $f_{\mu,\sigma}(t) = \sigma^{-1}f((t-\mu)/\sigma)$. We set

$$U(x) = -\frac{f'(x)}{f(x)}, \quad V(x) = -1 - x\frac{f'(x)}{f(x)},$$

$$\begin{pmatrix} \mathsf{I} & \mathsf{K} \\ \mathsf{K} & \mathsf{J} \end{pmatrix} = \begin{pmatrix} \int \frac{(f'(x))^2}{f(x)}dx & \int x\frac{(f'(x))^2}{f(x)}dx \\ \int x\frac{(f'(x))^2}{f(x)}dx & \int x^2\frac{(f'(x))^2}{f(x)}dx - 1 \end{pmatrix}.$$

If $\mathsf{I} < \infty$, $\mathsf{J} < \infty$, and $\mathsf{G} = \mathsf{I} - \mathsf{K}^2/\mathsf{J}$, then by Example 1.119 the family $(P_{\mu,\sigma})_{\mu\in\mathbb{R},\sigma>0}$ is $\mathbb{L}_2$-differentiable at $\theta_0 = (\mu_0,\sigma_0)$ with $\mathbb{L}_2$-derivative and information matrix, respectively, given by

$$\dot{L}_{\theta_0}(x) = \begin{pmatrix} U_{\theta_0}(x) \\ V_{\theta_0}(x) \end{pmatrix} = \frac{1}{\sigma_0}\begin{pmatrix} U(\frac{x-\mu_0}{\sigma_0}) \\ V(\frac{x-\mu_0}{\sigma_0}) \end{pmatrix},$$

$$\mathsf{I}(\theta_0) = \begin{pmatrix} \mathsf{I}_{1,1}(\theta_0) & \mathsf{I}_{1,2}(\theta_0) \\ \mathsf{I}_{2,1}(\theta_0) & \mathsf{I}_{2,2}(\theta_0) \end{pmatrix} = \frac{1}{\sigma_0^2}\begin{pmatrix} \mathsf{I} & \mathsf{K} \\ \mathsf{K} & \mathsf{J} \end{pmatrix},$$

$$\mathsf{G}(\theta_0) = \mathsf{I}_{1,1}(\theta_0) - \mathsf{I}_{1,2}(\theta_0)\mathsf{I}_{2,2}^{-1}(\theta_0)\mathsf{I}_{2,1}(\theta_0) = \frac{1}{\sigma_0^2}\mathsf{G}.$$

The statistic W_{θ_0} in (8.154) is given by

$$W_{\theta_0}(x) = U_{\theta_0}(x) - \mathsf{I}_{1,2}(\theta_0)\mathsf{I}_{2,2}^{-1}(\theta_0)V_{\theta_0}(x) = \frac{1}{\sigma_0}(U(\frac{x-\mu_0}{\sigma_0}) - \frac{\mathsf{K}}{\mathsf{J}}V(\frac{x-\mu_0}{\sigma_0})).$$

The next example considers one-sample tests for location-scale models.

Example 8.139. Suppose the family of densities $f_\theta(x) = \sigma^{-1}f((t-\mu)/\sigma)$, $\theta = (\mu,\sigma) \in \mathbb{R}\times(0,\infty)$, satisfies the condition (A10) and consider the two testing problems

$$(I)\ \mathsf{H}_0 : \mu \le \mu_0,\ \sigma^2 > 0 \quad \text{versus} \quad \mathsf{H}_A : \mu > \mu_0,\ \sigma^2 > 0,$$
$$(II)\ \mathsf{H}_0 : \mu = \mu_0,\ \sigma^2 > 0 \quad \text{versus} \quad \mathsf{H}_A : \mu \ne \mu_0,\ \sigma^2 > 0.$$

To this end we introduce the local parameters h and g by setting $\mu = \mu_0 + h/\sqrt{n}$ and $\sigma = \sigma_0 + g/\sqrt{n}$. If $\mathsf{I} < \infty$ and $\mathsf{J} < \infty$, then by Example 1.119 the family $(P_{\mu,\sigma})_{\mu\in\mathbb{R},\sigma>0}$ is $\mathbb{L}_2$-differentiable at $\theta_0 = (\mu_0,\sigma_0)$, and the sequence of models $(\mathbb{R}^n,\mathfrak{B}_n,P^{\otimes n}_{\mu_0+h/\sqrt{n},\sigma_0+g/\sqrt{n}})$ satisfies the ULAN condition with central sequence

$$Z_n = \frac{1}{\sqrt{n}} \sum\nolimits_{i=1}^{n} (\frac{1}{\sigma_0} U(\frac{X_i - \mu_0}{\sigma_0}), \frac{1}{\sigma_0} V(\frac{X_i - \mu_0}{\sigma_0}))^T.$$

The test statistic $S_{n,NR}(\mu_0, \sigma_0)$ in (8.112) is

$$S_{n,NR}(\mu_0, \sigma_0) = \frac{1}{\sqrt{n}} \sum\nolimits_{i=1}^{n} \frac{1}{\sigma_0} [U(\frac{X_i - \mu_0}{\sigma_0}) - \frac{\mathsf{K}}{\mathsf{J}} V(\frac{X_i - \mu_0}{\sigma_0})].$$

For testing problem (I) σ_0 is a nuisance parameter and has to be estimated. Set $\widehat{\sigma}_n^2 = n^{-1} \sum_{i=1}^{n} (X_i - \mu_0)^2$. If X_1 has a finite fourth moment μ_4, then by the central limit theorem $\mathcal{L}(\sqrt{n}(\widehat{\sigma}_n^2 - \sigma_0^2)) \Rightarrow \mathsf{N}(0, \mu_4)$. With the δ-method (see Proposition 8.78) we get $\mathcal{L}(\sqrt{n}(\widehat{\sigma}_n - \sigma_0)) \Rightarrow \mathsf{N}(0, \mu_4/(4\sigma_0^2))$, and thus $\widehat{\sigma}_n$ is $\sqrt{n}$-consistent. For $\tau_0 = \mu_0$ and $\widetilde{\xi}_n = \widehat{\sigma}_n$ the test statistic $T_n(\mu_0, \widehat{\sigma}_n)$ in Theorem 8.138 is

$$\begin{aligned} T_n(\mu_0, \widehat{\sigma}_n) &= \frac{1}{\sqrt{n}} \widehat{\sigma}_n (\mathsf{I} - \mathsf{K}^2/\mathsf{J})^{-1/2} \sum\nolimits_{i=1}^{n} \frac{1}{\widehat{\sigma}_n} [U(\frac{X_i - \mu_0}{\widehat{\sigma}_n}) - \frac{\mathsf{K}}{\mathsf{J}} V(\frac{X_i - \mu_0}{\widehat{\sigma}_n})] \\ &= \frac{1}{\sqrt{n}} (\mathsf{I} - \mathsf{K}^2/\mathsf{J})^{-1/2} \sum\nolimits_{i=1}^{n} [U(\frac{X_i - \mu_0}{\widehat{\sigma}_n}) - \frac{\mathsf{K}}{\mathsf{J}} V(\frac{X_i - \mu_0}{\widehat{\sigma}_n})]. \end{aligned}$$

Theorem 8.138 implies that the test $\psi_I(T_n(\mu_0, \widehat{\sigma}_n)) = I_{(u_{1-\alpha}, \infty)}(T_n(\mu_0, \widehat{\sigma}_n))$ is a LAUMP level α test for testing problem (I), and that $\psi_{II}(T_n(\mu_0, \widehat{\sigma}_n)) = 1 - I_{(-u_{1-\alpha/2}, u_{1-\alpha/2})}(T_n(\mu_0, \widehat{\sigma}_n))$ is a LAUMPU level α test for testing problem (II). If the fourth moments are not finite (e.g., for the Cauchy distribution), then one can find a $\sqrt{n}$-consistent estimator by estimating the interquartile distance, i.e., the difference of the (3/4)th and the (1/4)th quantile with the help of quantiles of the empirical distribution; see Example 7.139 for the limit theorem for quantiles. Finally, for symmetric densities it holds $\mathsf{K} = 0$ and the test statistic $T_n(\mu_0, \widehat{\sigma}_n)$ reduces to

$$T_n(\mu_0, \widehat{\sigma}_n) = \frac{1}{\sqrt{n}} \mathsf{I}^{-1/2} \sum\nolimits_{i=1}^{n} U(\frac{X_i - \mu_0}{\widehat{\sigma}_n}).$$

If f is the standard normal density, then $\mathsf{I} = 1$ and $U(x) = x$. Hence

$$T_n(\mu_0, \widehat{\sigma}_n) = \frac{1}{\widehat{\sigma}_n \sqrt{n}} \sum\nolimits_{i=1}^{n} (X_i - \mu_0) = T_n + o_{\mathsf{N}^{\otimes n}(\mu_0, \sigma_0^2)}(1),$$

where

$$T_n = \frac{1}{\sqrt{n}} [\frac{1}{n-1} \sum\nolimits_{i=1}^{n} (X_i - \overline{X}_n)^2]^{-1/2} \sum\nolimits_{i=1}^{n} (X_i - \mu_0)$$

is the test statistic of the t-test; see (8.40). Hence the sequence of tests $\psi_I(T_n(\mu_0, \widehat{\sigma}_n))$ and the sequence of t-tests for testing problem (I) are both asymptotic level α tests. The mutual contiguity of $\mathsf{N}^{\otimes n}(\mu_0, \sigma_0^2)$ and $\mathsf{N}^{\otimes n}(\mu_0 + h/\sqrt{n}, (\sigma_0 + g/\sqrt{n})^2)$ implies that the tests $\psi_I(T_n(\mu_0, \widehat{\sigma}_n))$ and the sequence of t-tests have the same asymptotic power under the local alternatives $\mathsf{N}^{\otimes n}(\mu_0 + h/\sqrt{n}, (\sigma_0 + g/\sqrt{n})^2)$. As the sequence $\psi_I(T_n(\mu_0, \widehat{\sigma}_n))$ is LAUMP the sequence of t-tests has the same property, which is not surprising as the t-test is already for every fixed n a UMP test. A similar equivalence holds for $\psi_{II}(T_n(\mu_0, \widehat{\sigma}_n))$ and the two-sided t-tests where the optimality is now in terms of LAUMPU. Finally we remark that tests for the scale parameter can be constructed in a similar way by estimating the location nuisance parameter with the sample mean if the second moment is finite, and by the median otherwise.

Testing in Two-Sample Models

For testing the hypothesis $\mathsf{H}_0 : \theta_1 = \theta_2$ in a two-sample model we suppose that from each marginal model there is a sample of size n_i, $i = 1, 2$. If the sample sizes are different, then the families $P_{\theta_1}^{\otimes n_1}$ and $P_{\theta_2}^{\otimes n_2}$, $\theta_i \in \Delta \subseteq \mathbb{R}^d$, contain different information on the respective parameters. We compensate this by including the possibly different sample sizes in the definition of the local parameter. More precisely, we localize θ_1 and θ_2 by setting

$$\theta_1 = \theta_0 + \frac{g}{\sqrt{n}} + d_{1,n}h \quad \text{and} \quad \theta_2 = \theta_0 + \frac{g}{\sqrt{n}} + d_{2,n}h,$$

$$d_{1,n} = \frac{1}{n_1}\sqrt{\frac{n_1 n_2}{n}} \quad \text{and} \quad d_{2,n} = -\frac{1}{n_2}\sqrt{\frac{n_1 n_2}{n}}, \qquad n = n_1 + n_2.$$

The idea of this parametrization is that the parameters contain a joint value $\theta_0 + g/\sqrt{n}$ from which the parameters in the populations deviate in opposite directions with a rate that depends on the sample sizes. We study the models

$$\begin{aligned}
&\mathcal{M}_n = (\mathcal{X}^{n_1} \times \mathcal{X}^{n_2}, \mathfrak{A}^{\otimes n_1} \otimes \mathfrak{A}^{\otimes n_2}, (P_{1,n,g,h} \otimes P_{2,n,g,h})_{(g,h)\in\Delta_n}),\\
&P_{1,n,g,h} = P^{\otimes n_1}_{\theta_0+g/\sqrt{n}+d_{1,n}h} \quad \text{and} \quad P_{2,n,g,h} = P^{\otimes n_2}_{\theta_0+g/\sqrt{n}+d_{2,n}h}, \\
&\Delta_n = \{(g,h) : \theta_0 + g/\sqrt{n} + d_{i,n}h \in \Delta,\ i = 1, 2\}.
\end{aligned} \tag{8.158}$$

Suppose that $n_i = n_i(n)$, $i = 1, 2$, are sequences with $n_1 + n_2 = n$ and

$$\lim_{n\to\infty} \frac{n_1}{n} = \kappa, \quad 0 < \kappa < 1. \tag{8.159}$$

Denote by $X_{1,1}, ..., X_{1,n_1}, X_{2,1}, ... X_{2,n_2}$ the projections of $\mathcal{X}^{n_1} \times \mathcal{X}^{n_2}$ onto $\mathcal{X}$. Suppose $(P_\theta)_{\theta\in\Delta}$ is $\mathbb{L}_2$-differentiable at $\theta_0 \in \Delta^0$. Then

$$\ln(dP^{\otimes n_i}_{\theta_0+g_i/\sqrt{n_i}}/dP^{\otimes n_i}_{\theta_0}) = g_i^T \frac{1}{\sqrt{n_i}} \sum\nolimits_{j=1}^{n_i} \dot{L}_0(X_{i,j}) - \frac{1}{2} g_i^T \mathsf{I}(\theta_0) g_i + o_{P^{\otimes n_i}_{\theta_0}}(1).$$

Replacing $g_i/\sqrt{n_i}$ with $g/\sqrt{n} + d_{i,n}h$ we get

$$\begin{aligned}
&\ln\left(\frac{d(P^{\otimes n_1}_{\theta_0+g/\sqrt{n}+d_{1,n}h} \otimes P^{\otimes n_2}_{\theta_0+g/\sqrt{n}+d_{2,n}h})}{d(P^{\otimes n_1}_{\theta_0} \otimes P^{\otimes n_2}_{\theta_0})}\right)\\
&= h^T U_n + g^T V_n - \frac{1}{2} g^T \mathsf{I}(\theta_0) g - \frac{1}{2} h^T \mathsf{I}(\theta_0) h + o_{P^{\otimes n_1}_{\theta_0} \otimes P^{\otimes n_2}_{\theta_0}}(1), \quad \text{where}\\
&U_n(\theta_0) = \sqrt{\frac{n_1 n_2}{n}}[\frac{1}{n_1}\sum\nolimits_{j=1}^{n_1} \dot{L}_{\theta_0}(X_{1,j}) - \frac{1}{n_2}\sum\nolimits_{j=1}^{n_2} \dot{L}_{\theta_0}(X_{2,j})],\\
&V_n(\theta_0) = \frac{1}{\sqrt{n}} \sum\nolimits_{i=1}^{2} \sum\nolimits_{j=1}^{n_i} \dot{L}_{\theta_0}(X_{i,j}).
\end{aligned} \tag{8.160}$$

This expansion, together with the central limit theorem and condition (8.159), entails that the sequence of models (8.158) satisfies the LAN($Z_n, \widetilde{\mathsf{I}}(\theta_0)$) condition, where the central sequence Z_n and the information matrix $\widetilde{\mathsf{I}}(\theta_0)$ are

$$Z_n(\theta_0) = \begin{pmatrix} U_n(\theta_0) \\ V_n(\theta_0) \end{pmatrix}, \quad \widetilde{\mathsf{I}}(\theta_0) = \begin{pmatrix} \mathsf{I}(\theta_0) & 0 \\ 0 & \mathsf{I}(\theta_0) \end{pmatrix}. \tag{8.161}$$

From now on we consider only univariate parameters, i.e., let $\Delta = (a,b) \subseteq \mathbb{R}$. We consider the following hypotheses for the local parameters $g, h \in \mathbb{R}$.

Testing Problem	H_0	H_A
(I)	$h \le 0,\ g \in \mathbb{R},$	$h > 0,\ g \in \mathbb{R},$
(II)	$h = 0,\ g \in \mathbb{R},$	$h \ne 0,\ g \in \mathbb{R}.$

(8.162)

These hypotheses correspond to the hypotheses (I) and (II) in (8.142) for $c = (1,0)^T$ and $(h,g)^T$ instead of h. Then

$$\sigma_0^2 := c^T(\widetilde{\mathsf{I}}(\theta_0))^{-1}c = \frac{1}{\mathsf{I}(\theta_0)}, \tag{8.163}$$

and the test statistic $T_n(\theta_0)$ in Theorem 8.130 is given by

$$\begin{aligned} T_n(\theta_0) &= c^T(\widetilde{\mathsf{I}}(\theta_0))^{-1}Z_n/\sigma_0 \\ &= \mathsf{I}^{-1/2}(\theta_0)\sqrt{\frac{n_1 n_2}{n}}\left[\frac{1}{n_1}\sum\nolimits_{j=1}^{n_1} \dot{L}_{\theta_0}(X_{1,j}) - \frac{1}{n_2}\sum\nolimits_{j=1}^{n_2} \dot{L}_{\theta_0}(X_{2,j})\right]. \end{aligned}$$

Theorem 8.140. ***(Two-Sample Tests)*** *Suppose that the family $(P_\theta)_{\theta\in(a,b)}$ satisfies condition (A10), condition (8.159) is fulfilled, and $\mathsf{I}(\theta_0)$ is positive. Let $\widehat{\theta}_n : \mathcal{X}^{n_1}\times\mathcal{X}^{n_2} \to_m (a,b)$ be a sequence of estimators that is $\sqrt{n}$-consistent. Then for ψ_I and ψ_{II} in (8.18), and the sequence of models in (8.158), the test $\psi_I(T_n(\widehat{\theta}_n))$ is a LAUMP level α test for testing problem (I) in (8.162), and $\psi_{II}(T_n(\widehat{\theta}_n))$ is a LAUMPU level α test for testing problem (II) in (8.162).*

Proof. The $\mathbb{L}_2$-differentiability gives the LAN($Z_n, \widetilde{\mathsf{I}}(\theta_0)$) condition with Z_n and $\widetilde{\mathsf{I}}(\theta_0)$ in (8.161). Hence Theorem 8.130 implies that the tests $\psi_i(T_n(\theta_0))$ have the stated optimality properties. It remains to show that $\psi_i(T_n(\widehat{\theta}_n))$ has the same power as $\psi_i(T_n(\theta_0))$ under local alternatives. To this end we note that (A10) implies the continuity of the Fisher information $\mathsf{I}(\theta)$. Hence we get from Lemma 8.111 that $T_n(\widehat{\theta}_n) = T_n(\theta_0) + R_n$, where $R_n = o_{P_{\theta_0}^{\otimes n_1}\otimes P_{\theta_0}^{\otimes n_2}}(1)$. The contiguity of the sequences

$$P^{\otimes n_1}_{\theta_0+g/\sqrt{n}+d_{1,n}h} \otimes P^{\otimes n_2}_{\theta_0+g/\sqrt{n}+d_{2,n}h} \quad \text{and} \quad P^{\otimes n_1}_{\theta_0} \otimes P^{\otimes n_2}_{\theta_0}$$

follows from the LAN($Z_n, \widetilde{\mathsf{I}}(\theta_0)$) condition (see Lemma 6.67) and implies

$$R_n = o_{P^{\otimes n_1}_{\theta_0+g/\sqrt{n}+d_{1,n}h} \otimes P^{\otimes n_2}_{\theta_0+g/\sqrt{n}+d_{2,n}h}}(1).$$

Hence by the Slutsky's lemma,

$$\mathcal{L}(T_n(\widehat{\theta}_n)|P^{\otimes n_1}_{\theta_0+g/\sqrt{n}+d_{1,n}h} \otimes P^{\otimes n_2}_{\theta_0+g/\sqrt{n}+d_{2,n}h}) \quad \text{and}$$
$$\mathcal{L}(T_n(\theta_0)|P^{\otimes n_1}_{\theta_0+g/\sqrt{n}+d_{1,n}h} \otimes P^{\otimes n_2}_{\theta_0+g/\sqrt{n}+d_{2,n}h})$$

tend to the same normal distribution. Thus,

$$\lim_{n\to\infty} \int \psi_i(T_n(\theta_0))d(P^{\otimes n_1}_{\theta_0+d_{1,n}h_1} \otimes P^{\otimes n_2}_{\theta_0+d_{2,n}h_2})$$
$$= \lim_{n\to\infty} \int \psi_i(T_n(\widehat{\theta}_n))d(P^{\otimes n_1}_{\theta_0+d_{1,n}h_1} \otimes P^{\otimes n_2}_{\theta_0+d_{2,n}h_2}).$$

■

In the two-sample case, instead of $\dot{L}_{\theta_0}$, one may also use any function $\Psi : \mathcal{X} \to_m \mathbb{R}$ to construct the test statistics

$$S_n = \sqrt{\frac{n_1 n_2}{n}} \left[\frac{1}{n_1} \sum\nolimits_{j=1}^{n_1} \Psi(X_{1,j}) - \frac{1}{n_2} \sum\nolimits_{j=1}^{n_2} \Psi(X_{2,j}) \right]. \tag{8.164}$$

This approach is often used in nonparametric statistics to test whether the distributions in the two populations are the same. More precisely, let $\mathcal{P}$ be any family of distributions on $(\mathcal{X}, \mathfrak{A})$ and consider the sequence of models and the hypotheses

$$\mathcal{M}_{n_1,n_2} = (\mathcal{X}^{n_1} \times \mathcal{X}^{n_2}, \mathfrak{A}^{\otimes n_1} \otimes \mathfrak{A}^{\otimes n_2}, P_1^{\otimes n_1} \otimes P_2^{\otimes n_2}),$$
$$\mathsf{H}_0 : P_1 = P_2 \quad \text{versus} \quad \mathsf{H}_0 : P_1 \neq P_2, \quad P_1, P_2 \in \mathcal{P}.$$

To study the power of a test based on the test statistic S_n, we consider a one-parameter differentiable curve through the null hypothesis, i.e., we suppose that $(P_\theta)_{\theta\in(a,b)}$ is $\mathbb{L}_2$-differentiable and for some θ_0 it holds $P_1 = P_2 = P_{\theta_0}$. To study S_n under the local models 8.158 we set $\mu = \mathsf{E}_{\theta_0}\Psi(X_{1,1})$ and

$$S_{i,n} = \sqrt{\frac{n_1 n_2}{n}} \frac{1}{n_i} \sum\nolimits_{j=1}^{n_i} (\Psi(X_{i,j}) - \mu).$$

Put $\sigma_{1,1} = \mathsf{V}_{\theta_0}(\Psi)$, $\sigma_{2,2} = \mathsf{V}_{\theta_0}(\dot{L}_{\theta_0})$, and $\sigma_{1,2} = \mathsf{C}_{\theta_0}(\Psi, \dot{L}_{\theta_0})$. Then by Corollary 6.74 for every convergent sequence $g_{i,n} \to g_i$ it holds

$$\mathcal{L}(S_{i,n}|P^{\otimes n_i}_{\theta_0+g_{i,n}/\sqrt{n_i}}) \Rightarrow \mathsf{N}(\sigma_{1,2}\kappa_i g_i, \kappa_i^2 \sigma_{1,1}), \quad \text{where}$$
$$\kappa_1 = \lim_{n\to\infty} \sqrt{n_2/n} = (1-\kappa)^{1/2} \quad \text{and} \quad \kappa_2 = \lim_{n\to\infty} \sqrt{n_1/n} = \kappa^{1/2}.$$

If $g_{i,n}/\sqrt{n_i} = g/\sqrt{n_1+n_2} + d_{i,n}h$, $i = 1, 2$, then

$$\lim_{n\to\infty} g_{1,n} = g\kappa^{1/2} + h(1-\kappa)^{1/2} \quad \text{and} \quad \lim_{n\to\infty} g_{2,n} = g(1-\kappa)^{1/2} - h\kappa^{1/2}.$$

Hence by the independence of the populations

$$\begin{aligned}
&\mathcal{L}(S_n|P_{\theta_0+g/\sqrt{n}+d_{1,n}h}^{\otimes n_1} \otimes P_{\theta_0+g/\sqrt{n}+d_{2,n}h}^{\otimes n_2}) \\
&\Rightarrow \mathsf{N}(\sigma_{1,2}((1-\kappa)\kappa)^{1/2}g + (1-\kappa)h), (1-\kappa)\sigma_{1,1}) \\
&\quad * \mathsf{N}(-\sigma_{1,2}((1-\kappa)\kappa)^{1/2}g + \kappa h), \kappa\sigma_{1,1}) = \mathsf{N}(\sigma_{1,2}h, \sigma_{1,1}).
\end{aligned}$$

From here one obtains the asymptotic power $\mathsf{p}_i(h)$ of the tests $\psi_i(S_n/\sqrt{\sigma_{1,1}})$ for the hypotheses in (8.162). For example, if we test $\mathsf{H}_0 : h \leq 0$ versus $\mathsf{H}_A : h > 0$, then $\psi_I(S_n/\sqrt{\sigma_{1,1}}) = I_{(u_{1-\alpha},\infty)}(S_n/\sqrt{\sigma_{1,1}})$, and

$$\lim_{n\to\infty} \mathsf{E}_{n,h,g}\psi_I(S_n/\sqrt{\sigma_{1,1}}) = 1 - \Phi(u_{1-\alpha} - h\sigma_{1,2}/\sqrt{\sigma_{1,1}}).$$

If we use $\dot{L}_{\theta_0}$ instead of Ψ, then S_n turns into U_n in 8.160 and the asymptotic power of the test

$$\psi_I(\mathsf{I}^{-1/2}(\theta_0)U_n(\theta_0)) = I_{(u_{1-\alpha},\infty)}(\mathsf{I}^{-1/2}(\theta_0)U_n(\theta_0))$$

turns out to be

$$\lim_{n\to\infty} \mathsf{E}_{n,h}\psi_I(\mathsf{I}^{-1/2}(\theta_0)U_n(\theta_0)) = 1 - \Phi(u_{1-\alpha} - \mathsf{I}^{-1/2}(\theta_0)h).$$

It follows from Schwarz' inequality that $\sigma_{1,2}/\sqrt{\sigma_{1,1}} \leq \sqrt{\sigma_{2,2}}$, where equality holds if and only if $\Psi = c\dot{L}_{\theta_0}$ for some constant c. Thus we see again that the correlation $\sigma_{1,2}/\sqrt{\sigma_{1,1}\sigma_{2,2}}$ between the influence function Ψ and the score function $\dot{L}_{\theta_0}$ describes the relative efficiency of a two-sample test that is based on Ψ.

Now we study the efficiency of two-sample rank tests. To this end we suppose $\mathcal{X} = \mathbb{R}$ in the sequence of models (8.158), so that we observe real-valued and independent $X_{1,1}, ..., X_{1,n_1}$ with distribution P_1 and $X_{2,1}, ..., X_{2,n_2}$ with distribution P_2. We introduce the pooled sample by setting $X_i = X_{1,i}$, $i = 1, ..., n_1$, and $X_i = X_{2,i}$, $i = n_1 + 1, ..., n$, where $n = n_1 + n_2$. Let $(R_{n,1}, ..., R_{n,n_1}, R_{n,n_1+1}, ..., R_{n,n})$ denote the vector of ranks of the pooled sample. We introduce the regression coefficients $c_{i,n}$, $i = 1, ..., n_1 + n_2$, by

$$c_{i,n} = \frac{1}{n_1}\sqrt{\frac{n_1 n_2}{n}}, \ i \leq n_1, \quad c_{i,n} = -\frac{1}{n_2}\sqrt{\frac{n_1 n_2}{n}}, \ i \geq n_1 + 1. \tag{8.165}$$

Using a sequence of scores a_n as in (8.98) we employ the rank statistic

$$S_n = \sum\nolimits_{i=1}^{n} c_{i,n} a_n(R_{n,i})$$

to test $\mathsf{H}_0 : P_1 = P_2$ versus $\mathsf{H}_A : P_1 \neq P_2$. It is clear that such tests will have a good power only for special deviations from the null hypothesis.

We assume that under H_0 the distribution $P := P_1 = P_2$ is atomless, which is equivalent to P having a continuous c.d.f. F. To construct a rank test for a special direction when deviating from the null hypothesis, we take a smooth curve P_θ through $P = P_1 = P_2$ with $P_{\theta_0} = P$, say. The next theorem shows that a rank test is locally asymptotically best in the direction of the chosen curve if we construct the scores with the help of the tangent (i.e., $\mathbb{L}_2$-derivative) of the given curve. More precisely, we set $\varphi(t) = \dot{L}_{\theta_0}(F^{-1}(t))$,

$$\sigma_{\theta_0}^2 = \int_0^1 \varphi^2(t)dt = \int \dot{L}_{\theta_0}^2 dP_{\theta_0} = \mathsf{I}(\theta_0),$$

and define the scores $a_n(k)$ as in (8.98). For the rank vector $(R_{n,1}, ..., R_{n,n})$ of the pooled sample we introduce the two-sample rank statistic by

$$S_n = \sqrt{\frac{n_1 n_2}{n}} \left[\frac{1}{n_1} \sum\nolimits_{i=1}^{n_1} a_n(R_{n,i}) - \frac{1}{n_2} \sum\nolimits_{i=n_1+1}^{n} a_n(R_{n,i}) \right]. \tag{8.166}$$

Theorem 8.141. ***(Two-Sample Rank Tests)*** *Let P be an atomless distribution on $(\mathbb{R}, \mathfrak{B})$. Suppose $(P_\theta)_{\theta \in (a,b)}$ is an $\mathbb{L}_2$-differentiable family of distributions with $P_{\theta_0} = P$, Fisher information $\mathsf{I}(\theta_0) > 0$, and $\mathbb{L}_2$-derivative $\dot{L}_{\theta_0}$. For $\varphi(t) = \dot{L}_{\theta_0}(F^{-1}(t))$ let $a_n(k)$ be defined by one of the versions in (8.98). Let S_n be defined by (8.166). Then for the model (8.158) the sequence of tests $\psi_I(S_n/\sigma_{\theta_0})$, is a LAUMP level α test for testing problem (I) in (8.162), and $\psi_{II}(S_n/\sigma_{\theta_0})$, is a LAUMPU level α tests for testing problem (II) in (8.162).*

Proof. The continuity of F implies that $U_1 = F(X_1), ..., U_n = F(X_n)$ are i.i.d. under H_0 with a common distribution that is the uniform distribution on $[0,1]$. Moreover it holds $X_i = F^{-1}(U_i)$, P_{θ_0}-a.s. Set $\varphi(t) = \dot{L}_{\theta_0}(F^{-1}(t))$. Then with $c_{i,n}$ in (8.165),

$$\begin{aligned}
T_n &= \sqrt{\frac{n_1 n_2}{n}} \left[\frac{1}{n_1} \sum\nolimits_{j=1}^{n_1} \dot{L}_{\theta_0}(X_{1,j}) - \frac{1}{n_2} \sum\nolimits_{j=1}^{n_2} \dot{L}_{\theta_0}(X_{2,j}) \right] \\
&= \sum\nolimits_{i=1}^{n} c_{i,n} \varphi(U_i) = \sum\nolimits_{i=1}^{n} c_{i,n} a_{n,i}(R_{n,i}) + o_{P_{\theta_0}^{\otimes n_1} \otimes P_{\theta_0}^{\otimes n_2}}(1) \\
&= S_n + o_{P_{\theta_0}^{\otimes n_1} \otimes P_{\theta_0}^{\otimes n_2}}(1),
\end{aligned}$$

where the third equality follows from Theorem 8.92. Similar as in the proof of Theorem 8.140 one can see that the test statistics T_n and S_n differ under local alternatives only by terms that tend stochastically to zero, so that the associated asymptotic level α tests have identical power functions. In the proof of Theorem 8.140 it has been pointed out that the tests that are based on $T_n(\theta_0)$ are asymptotic level α tests with the stated optimality properties.Therefore the tests that are based on T_n have the same properties. This completes the proof. ■

A typical application of rank tests concerns location models for which the Fisher information $\mathsf{I}(\theta_0)$ is a constant value, i.e., does not depend on the point of localization. Two-sample Wilcoxon tests are considered below.

Example 8.142. Let $(P_\theta)_{\theta\in\mathbb{R}}$ be the location family that is generated by the logistic distribution. Suppose we have n_i observations from population i that has the distribution P_{θ_i}, $i = 1, 2$, and set $n = n_1 + n_2$. Here we want to test if $\theta_1 = \theta_2$, or $\theta_1 \le \theta_2$. To this end we turn to the sequence of localized models $\mathcal{M}_n$ in (8.158). As the ranks are invariant under a common translation of all data we may assume that $\theta_0 = 0$, and thus $P_{\theta_0} = P_0$. Then by (8.102) $\varphi = \dot{L}_{\theta_0}(F^{-1}(t)) = 2t - 1$. Using the approximate scores in (8.98) it holds $a_n(k) = 2k/(n+1) - 1$ and

$$S_n = \sqrt{\frac{n_1 n_2}{n}}\left[\frac{1}{n_1}\sum\nolimits_{i=1}^{n_1}(\frac{2R_{n,i}}{n} - 1) - \frac{1}{n_2}\sum\nolimits_{i=n_1+1}^{n}(\frac{2R_{n,i}}{n} - 1)\right].$$

As $R_{1,i} - i$ is the number of j with $X_{2,j} < X_{1,i}$ we get for the Mann–Whitney statistic W_{n_1,n_2} in Example 8.86,

$$\begin{aligned} W_{n_1,n_2} &= \frac{1}{n_1 n_2}\sum\nolimits_{i=1}^{n_1}\sum\nolimits_{j=1}^{n_2} I_{(X_{2,j},\infty)}(X_{1,i}) = \frac{1}{n_1 n_2}\sum\nolimits_{i=1}^{n_1}(R_{n,i} - i) \\ &= \frac{1}{n_1 n_2}\sum\nolimits_{i=1}^{n_1} R_{n,i} - \frac{(n_1+1)}{2n_2}. \end{aligned}$$

As $\sum_{i=1}^{n} R_{n,i} = n(n+1)/2$ it follows that

$$S_n = 2\sqrt{\frac{n_1 n_2}{n}}(W_{n_1,n_2} - \frac{1}{2}).$$

We have proved already that

$$\mathcal{L}(\sqrt{\frac{n_1 n_2}{n}}(W_{n_1,n_2} - \frac{1}{2})|P^{\otimes n_1} \otimes P^{\otimes n_2}) \Rightarrow \mathsf{N}(0, \frac{1}{12})$$

in Example 8.86. Hence $\psi_I(S_n/\sqrt{3})$ is a LAUMP level α test, and $\psi_{II}(S_n/\sqrt{3})$ is a LAUMPU level α test, for the corresponding testing problem in (8.157) for the models (8.158), provided the parent distribution P is the logistic distribution.

8.9.2 Testing of Multivariate Parameters

We have studied testing problems with a one-dimensional parameter of interest in the previous section, where we allowed that the models contain nuisance parameters. Now we use the minimax concept to get, in that sense, asymptotically optimal tests. As we use the zero–one loss function exclusively we deal with the error probabilities, so that the minimax concept for decisions translates into the maximin concept for tests that has been used already in Theorem 8.24.

Suppose that for the model $(\mathcal{X}, \mathfrak{A}, (P_\theta)_{\theta\in\Delta})$, $\Delta \subseteq \mathbb{R}^d$, and $\theta_0 \in \Delta^0$, we want to test the simple null hypothesis $\mathsf{H}_0 : \theta = \theta_0$ against the alternative $\mathsf{H}_A : \theta \ne \theta_0$. Again we turn, by localization, to a sequence of models that satisfies the LAN(Z_n, I_0) condition. For $\delta^2 \le c$ we set

$$\Delta_{c,\delta} = \{h : \delta^2 \le h^T \mathsf{I}_0 h \le c\}.$$

Recall that H_{d,δ^2} is the c.d.f. of the χ^2-distribution with d degrees of freedom and noncentrality parameter δ^2, and that $\chi^2_{1-\alpha,d}$ is the $1-\alpha$ quantile of $\mathsf{H}_d = \mathsf{H}_{d,0}$.

Theorem 8.143. *If the sequence of models* $\mathcal{M}_n = (\mathcal{X}_n, \mathfrak{A}_n, (P_{n,h})_{\theta\in\Delta_n})$, $\Delta_n \uparrow \mathbb{R}^d$, *satisfies the LAN*$(Z_n, \mathsf{I}_0)$ *condition with invertible Fisher information matrix* I_0, *then for every* $0 \le \delta^2 \le c$,

$$\limsup_{n\to\infty} \left(\inf\nolimits_{h\in\Delta_{c,\delta}} \mathsf{E}_{n,h}\varphi_n\right) \le \mathsf{H}_{d,\delta^2}(\chi^2_{1-\alpha,d}), \tag{8.167}$$

for every sequence of asymptotic level α *tests* $\varphi_n : \mathcal{X}_n \to_m [0,1]$ *for* $\mathsf{H}_0 : h = 0$ *versus* $\mathsf{H}_A : h \neq 0$. *If the ULAN*(Z_n, I_0) *condition is satisfied, then the sequence of tests*

$$\varphi_{n,\chi^2} = I_{(\chi^2_{1-\alpha,d},\infty)}(Z_n^T \mathsf{I}_0^{-1} Z_n) \tag{8.168}$$

satisfies

$$\begin{aligned} \lim_{n\to\infty} \mathsf{E}_{n,0}\varphi_{n,\chi^2} &= \alpha, \\ \lim_{n\to\infty} (\inf\nolimits_{h\in\Delta_{c,\delta}} \mathsf{E}_{n,h}\varphi_{n,\chi^2}) &= \mathsf{H}_{d,\delta^2}(\chi^2_{1-\alpha,d}), \end{aligned} \tag{8.169}$$

so that φ_{n,χ^2} *is an asymptotic level* α *test that is locally asymptotically maximin (LAMM) for* $\mathsf{H}_0 : h = 0$ *versus* $\mathsf{H}_A : h \in \Delta_{c,\delta}$ *in the sense that for every further asymptotic level* α *test* ψ_n *it holds*

$$\limsup_{n\to\infty}(\inf\nolimits_{h\in\Delta_{c,\delta}} \mathsf{E}_{n,h}\psi_n) \le \lim_{n\to\infty} (\inf\nolimits_{h\in\Delta_{c,\delta}} \mathsf{E}_{n,h}\varphi_{n,\chi^2}). \tag{8.170}$$

Proof. By the same argument as in the proof of Theorem 8.130 we see that the set $\mathbb{D}_0$ of all level α tests for the model $\mathcal{G}_0 = (\mathbb{R}^d, \mathfrak{B}_d, (\mathsf{N}(h, \mathsf{I}_0^{-1}))_{h\in\mathbb{R}^d})$ is closed. An application of the Hájek–LeCam bound in Proposition 6.89 to the testing problem $\mathsf{H}_0 : h = 0$ versus $\mathsf{H}_A : h \neq 0$ under the zero–one loss function with $C = \Delta_{c,\delta}$ gives

$$\limsup_{n\to\infty} \left(\inf\nolimits_{h\in\Delta_{c,\delta}} \mathsf{E}_{n,h}\varphi_n\right) \le \sup_{\varphi\in\mathbb{D}_0} \inf_{h\in\Delta_{c,\delta}} \int \varphi(x)\mathsf{N}(h, \mathsf{I}_0^{-1})(dx).$$

To evaluate the right-hand term we note that by Theorem 5.43 the test $I_{(\chi^2_{1-\alpha,d},\infty)}(\|x\|^2)$ is a maximin level α test for the family $\mathsf{N}(\mu, \mathbf{I})$ for testing $\mu = 0$ versus $\delta^2 \le \|\mu\|^2$, where the maximin value is attained for $\|\mu\|^2 = \delta^2$. Hence $I_{(\chi^2_{1-\alpha,d},\infty)}(x^T \mathsf{I}_0 x)$ is a maximin level α test for the family $(\mathsf{N}(h, \mathsf{I}_0^{-1}))_{h\in\mathbb{R}^d}$ for testing $h = 0$ versus $\delta^2 \le h^T \mathsf{I}_0 h \le c$, and the maximin value is $\mathsf{H}_{d,\delta^2}(\chi^2_{1-\alpha,d})$. This proves (8.167). The relation (6.90) in the third lemma of LeCam, Theorem 6.72, and Corollary 6.73 yield that for every $\boldsymbol{\lambda}_d$-a.e. continuous and bounded function φ,

$$\mathsf{E}_{n,h}\varphi(Z_n^T \mathsf{I}_0^{-1} Z_n) \to \int \varphi(y^T y)\mathsf{N}(\mathsf{I}_0^{1/2} h, \mathbf{I})(dy),$$

where the convergence is locally uniform. This implies

$$\lim_{n\to\infty}\left(\inf\nolimits_{h\in\Delta_{c,\delta}} \mathsf{E}_{n,h}\varphi_n\right) = \inf_{h\in\Delta_{c,\delta}} \int I_{(\chi^2_{1-\alpha,d},\infty)}(y^T y)\mathsf{N}(\mathsf{I}_0^{1/2}h,\mathbf{I})(dy)$$
$$= \mathsf{H}_{d,\delta^2}(\chi^2_{1-\alpha,d}).$$

■

We apply the above theorem to $\mathbb{L}_2$-differentiable models and consider the localized models

$$\mathcal{M}_n = (\mathcal{X}^n, \mathfrak{A}^{\otimes n}, (P^{\otimes n}_{\theta_0+h/\sqrt{n}})_{h\in\Delta_n}), \quad \Delta_n = \{h : \theta_0 + h/\sqrt{n} \in \Delta\}.$$

Proposition 8.144. ***(Rao's Score Test)*** *If $(P_\theta)_{\theta\in\Delta}$ is $\mathbb{L}_2$-differentiable at $\theta_0 \in \Delta^0$ with derivative $\dot{L}_{\theta_0}$ and nonsingular Fisher information matrix $\mathsf{I}(\theta_0)$, then the Neyman–Rao Test $\psi_{n,NR}$ in (8.111) satisfies*

$$\lim_{n\to\infty} \mathsf{E}_{n,0}\psi_{n,NR} = \alpha,$$
$$\lim_{n\to\infty}\left(\inf\nolimits_{h\in\Delta_{c,\delta}} \mathsf{E}_{n,h}\psi_{n,NR}\right) = \mathsf{H}_{d,\delta^2}(\chi^2_{1-\alpha,d}),$$

so that $\psi_{n,NR}$ is an asymptotic level α test for $\mathsf{H}_0 : h = 0$ versus $\mathsf{H}_A : h \neq 0$, and is LAMM for testing $\mathsf{H}_0 : h = 0$ versus $\mathsf{H}_A : h \in \Delta_{c,\delta}$ in the sense that

$$\limsup_{n\to\infty}(\inf\nolimits_{h\in\Delta_{c,\delta}} \mathsf{E}_{n,h}\varphi_n) \leq \lim_{n\to\infty}(\inf\nolimits_{h\in\Delta_{c,\delta}} \mathsf{E}_{n,h}\psi_{n,NR})$$

for every further asymptotic level α test φ_n.

Proof. The $\mathbb{L}_2$-differentiability implies in view of the second lemma of LeCam (see Theorem 6.70) that the ULAN$(Z_n, \mathsf{I}(\theta_0))$ condition is satisfied with $Z_n = n^{-1/2}\sum_{i=1}^n \dot{L}_{\theta_0}(X_i)$. The statement follows from (8.169). ■

Tests of the above type were introduced and studied by Rao (1947). The statistic $\mathsf{I}_0^{-1/2}Z_n$ is commonly called the *score statistic* and therefore the test in (8.168) is called *Rao's score test.*

As we have mentioned already, the score statistic $\mathsf{I}_0^{-1/2}Z_n$ is closely related to the MLE $\widehat{\theta}_n$, provided the MLE is consistent and additional regularity conditions are satisfied which guarantee that $\widehat{\theta}_n$ admits a stochastic expansion. Indeed, if the assumptions of Theorem 7.148 are satisfied, then the MLE $\widehat{\theta}_n$ satisfies

$$n \parallel \widehat{\theta}_n - \theta_0 \parallel^2 = \parallel \mathsf{I}^{-1/2}(\theta_0)Z_n \parallel^2 + o_{P^{\otimes n}_{\theta_0}}(1). \tag{8.171}$$

However, the assumptions for Rao's score test are weaker, as only the $\mathbb{L}_2$-differentiability is needed. Moreover, it is not necessary to calculate the MLE. On the other hand, in some special models such as in exponential families the MLE can be evaluated explicitly. Then it is more convenient to use the left-hand term in (8.171) as the test statistic. A positive side-effect is that this sequence of tests is consistent in the sense that the error probabilities of the second kind tend to zero whenever the MLE is consistent.

Rao's score test can be used to construct *goodness-of-fit tests* for a simple null hypothesis. The classical way of doing this is to embed the distribution P of the null hypothesis into an exponential family. This idea goes back to Neyman (1937), who constructed a special class of tests that is discussed subsequently.

Let P be a distribution on $(\mathcal{X}, \mathfrak{A})$. We fix d and functions $T_i \in \mathbb{L}_2^0(P)$, $i = 1, ..., d$, such that the set

$$\Delta = \{(\theta_1, ..., \theta_d) : \quad K(\theta) := \ln(\int \exp\{\sum\nolimits_{i=1}^{d} \theta_i T_i\} dP) < \infty\}$$

is nonempty and contains 0 as an inner point. Moreover, we assume that the covariance matrix of the vector $T = (T_1, ..., T_d)^T$ is nonsingular. Then the exponential family $(P_\theta)_{\theta \in \Delta}$ defined by

$$dP_\theta = \exp\{\sum\nolimits_{i=1}^{d} \theta_i T_i - K(\theta)\} dP \tag{8.172}$$

satisfies (A1) and (A2). We consider the sequence of models

$$\mathcal{M}_n = (\mathcal{X}^n, \mathfrak{A}^{\otimes n}, (P_{h/\sqrt{n}}^{\otimes n})_{h \in \Delta_n}), \quad \Delta_n = \{h : h/\sqrt{n} \in \Delta\}. \tag{8.173}$$

Denote by $X_j : \mathcal{X}^n \to \mathcal{X}$, $j = 1, ..., n$, the projections. As the family $(P_\theta)_{\theta \in \Delta}$ is $\mathbb{L}_2$-differentiable at $\theta_0 = 0$ with derivative $\dot{L}_{\theta_0} = (T_1, ..., T_d)^T$ and Fisher information matrix $\mathsf{I}(0) = \nabla \nabla^T K(0) = \mathsf{C}_0(T)$ (see Corollary 1.19) we get that the sequence $\mathcal{M}_n$ satisfies the ULAN$(Z_n, \mathsf{I}(0))$ condition with central sequence

$$Z_n = \frac{1}{\sqrt{n}} \sum\nolimits_{j=1}^{n} (T_1(X_j), ..., T_d(X_j))^T,$$

where we used the assumption $\mathsf{E}_0 T_i = 0$. To simplify the calculation of $\mathsf{I}^{-1}(0)$ one often assumes that the statistics T_i are orthonormal in the sense that

$$\int T_i T_j dP = \delta_{i,j}, \quad i, j = 1, ..., d. \tag{8.174}$$

If this condition is not satisfied, then one can apply the Gram–Schmidt orthogonalization procedure to turn to an orthonormal system. If (8.174) is satisfied, then $\mathsf{C}_0(T) = \mathbf{I}$ and the Rao test, which rejects the null hypothesis for large values of the squared norm of the score statistic, is now called *Neyman's smooth test*, which is given by

$$\psi_{n,N} = I_{(\chi^2_{1-\alpha,d}, \infty)}(\frac{1}{n} \sum\nolimits_{i=1}^{d} [\sum\nolimits_{j=1}^{n} T_i(X_j)]^2).$$

Let us consider the local alternatives $P_{n,h} = P_{h/\sqrt{n}}^{\otimes n}$, $h = (h_1, ..., h_d)$. As by construction $\mathsf{I}(0) = \mathbf{I}$ we get from the third lemma of LeCam (see Theorem 6.72) that $\mathcal{L}(Z_n | P_{n,h}) \Rightarrow \mathsf{N}(h, \mathbf{I})$. Hence

$$\lim_{n\to\infty} \mathsf{E}_{n,h}\psi_{n,N} = \mathsf{H}_{d,\delta^2}(\chi^2_{1-\alpha,d}),$$

where $\delta^2 = \sum_{i=1}^d h_i^2$. This means that Neyman's smooth test distributes the power uniformly to all directions in the space of distributions that are defined by the functions $T_1, ..., T_d$. The following examples taken from Lehmann and Romano (2005) consider tests for uniformity and normality.

Example 8.145. Consider the models $\mathcal{M}_n$ in (8.173), where $\mathcal{X} = [0,1]$ and P is the uniform distribution on $[0,1]$. We consider the functions T_j that are obtained by constructing an orthonormal system starting with power functions x^j. This leads to the Legendre polynomials. It follows from the Gram–Schmidt procedure that $T_0(x) = 1$, $T_1(x) = \sqrt{3}(2x-1)$, $T_2(x) = \sqrt{5}(6x^2 - 6x + 1)$, and $T_3(x) = \sqrt{7}(20x^3 - 30x^2 + 12x - 1)$. Then $T_i \in \mathbb{L}_2^0(P)$, and Neyman's smooth test is given by

$$\psi_{n,N} = I_{(\chi^2_{1-\alpha,3},\infty)}\Big(\frac{1}{n}\sum\nolimits_{i=1}^{3}\Big[\sum\nolimits_{j=1}^{n} T_i(X_j)\Big]^2\Big).$$

This test is LAMM for the sequence of models obtained from (8.172) by a localization at $\theta = 0$.

Example 8.146. Consider the models $\mathcal{M}_n$ in (8.173) where $\mathcal{X} = \mathbb{R}$ and P is the standard normal distribution. We again consider the functions T_j that are obtained by constructing an orthonormal system starting with power functions x^j. This leads to the Hermite polynomials. The first of them are $H_0(x) = 1$, $H_1(x) = x$, $H_2(x) = 2^{-1/2}(x^2 - 1)$, $H_3(x) = 6^{-1/2}(x^3 - 3x)$, and $H_4(x) = 24^{-1/2}(x^4 - 6x^2 + 3)$. It holds $H_i \in \mathbb{L}_2^0(P)$, and Neyman's smooth test is

$$\psi_{n,N} = I_{(\chi^2_{1-\alpha,4},\infty)}\Big(\frac{1}{n}\sum\nolimits_{i=1}^{4}\Big[\sum\nolimits_{j=1}^{n} H_i(X_j)\Big]^2\Big).$$

The four squared terms admit a simple statistical interpretation. $[\sum_{j=1}^n H_1(X_j)]^2$ becomes large if there are deviations from the expectation zero. If the expectation is zero, then the second term indicates deviations from the variance. If the first two moments are 0 and 1, then the third term hints at a possible skewness, whereas the fourth term becomes large if the data from the alternative have a nonzero kurtosis.

In some cases it is more convenient to turn from the score statistic to the likelihood ratio statistic, which differs from the score statistic only by terms $o_{P_{\theta_0}^{\otimes n}}(1)$; see Theorem 8.116.

The following example considers χ^2 goodness-of-fit tests.

Example 8.147. This a continuation of Example 8.117. Suppose that for the multinomial distribution we want to test $\mathsf{H}_0 : p = (p_1, ..., p_d) = p_0 = (p_{0,1}, ..., p_{0,d})$ versus $\mathsf{H}_A : p \neq p_0$. We use the parametrization in Example 8.117. Then we obtain from Theorem 8.116 and (8.127) that

$$\| \mathsf{I}^{-1/2}(\theta_0) Z_n \|^2 = 2n \sum\nolimits_{k=1}^{d} (Y_{k,n}/n) \ln \frac{(Y_{k,n}/n)}{p_{0,k}} + o_{P_{\theta_0}^{\otimes n}}(1).$$

We know that the score test, which rejects the null hypothesis for large values of $\| \mathsf{I}^{-1/2}(\theta_0) Z_n \|^2$, is LAMM for local alternatives that satisfy $\delta^2 \leq h^T \mathsf{I}(\theta_0) h \leq c$. To analyze this condition we note that in view of Example 8.117 it holds

$$\mathsf{I}(\theta_0) = \left(\frac{1}{\theta_{0,k}}\delta_{k,l} + (1 - \sum\nolimits_{i=1}^{d-1}\theta_{0,i})^{-1}\right)_{1\le k,l\le d-1}.$$

Hence with $p_{0,l} = \theta_{0,l}$, $1 \le l \le d-1$, and $p_{0,d} = 1 - \sum_{i=1}^{d-1}\theta_{0,i}$,

$$h^T\mathsf{I}(\theta_0)h = \sum\nolimits_{l=1}^{d-1}\frac{1}{p_{0,l}}h_l^2 + \frac{1}{p_{0,d}}(\sum\nolimits_{l=1}^{d-1}h_l)^2.$$

If we set $h_d = -\sum_{l=1}^{d-1}h_l$, then the local alternatives can be written as

$$p_{n,k} = p_{0,k} + h_k/\sqrt{n}, \quad \sum\nolimits_{l=1}^{d}h_l = 0, \quad \delta^2 \le \sum\nolimits_{l=1}^{d}\frac{1}{p_{0,l}}h_l^2 \le c. \tag{8.175}$$

We get from Theorem 8.143 that the sequence of Rao score tests is an asymptotic level α test and has the LAMM property on the set of local alternatives of the type (8.175). Under the null hypothesis the statistic $\| \mathsf{I}^{-1/2}(\theta_0)Z_n \|^2$ differs from the likelihood ratio, and in view of Lemma 8.118 also from the statistics $(2n/\mathsf{v}''(1))[\mathsf{I}_\mathsf{v}(\widehat{P}_n, P_{\theta_0}) - \mathsf{v}(1)]$, only by terms $o_{P_{\theta_0}^{\otimes n}}(1)$. As the ULAN condition is satisfied the sequence of models is asymptotically equicontinuous in the sense of (6.80). This implies that the power of the three tests tends locally uniformly to the same function. As the sequence of score tests has been verified already to be LAMM we get the same property for the likelihood ratio test and the test that is based on $(2n/\mathsf{v}''(1))[\mathsf{I}_\mathsf{v}(\widehat{P}_n, P_{\theta_0}) - \mathsf{v}(1)]$.

Similarly as in (8.155), we consider again the sequence of localized model

$$\begin{aligned}\mathcal{M}_n &= (\mathcal{X}^n, \mathfrak{A}^{\otimes n}, (P^{\otimes n}_{\tau_0+h/\sqrt{n}, \xi_0+g/\sqrt{n}})_{(h,g)\in\Delta_n}), \\ \Delta_n &= \{(h,g) : ((\tau_0 + h/\sqrt{n})^T, (\xi_0 + g/\sqrt{n})^T)^T \in \Delta\},\end{aligned} \tag{8.176}$$

where the parameter of interest, however, is now of dimension k. To construct a test statistic we use $S_{n,NR}(\theta_0)$ in (8.112) and set

$$Q_{n,NR}(\theta_0) = S^T_{n,NR}(\theta_0)\mathsf{G}^{-1}(\theta_0)S_{n,NR}(\theta_0). \tag{8.177}$$

The subsequent theorem corresponds to Theorem 8.138, but now it concerns the case where the parameter of interest is multivariate and the criterion for optimality is the maximin concept.

Theorem 8.148. *Suppose the family $(P_\theta)_{\theta\in\Delta}$, $\Delta \subseteq \mathbb{R}^d$, satisfies condition (A10) and $\mathsf{I}(\theta_0)$ is nonsingular. Assume that $\widetilde{\xi}_n$ is a $\sqrt{n}$-consistent estimator of ξ for the sequence of submodels $(P^{\otimes n}_{(\tau_0,\xi)})_{\xi\in\Xi(\tau_0)}$. Then the sequence of Neyman–Rao tests*

$$\psi_{n,NR} = I_{(\chi^2_{1-\alpha,k},\infty)}(Q_{n,NR}(\tau_0, \widetilde{\xi}_n))$$

is an asymptotic level α test for the testing problem $\mathsf{H}_0 : h = 0$, $g \in \mathbb{R}^{d-k}$, versus $\mathsf{H}_A : h \neq 0$, $g \in \mathbb{R}^{d-k}$ in the sequence of models $\mathcal{M}_n$ in (8.176), and it holds

$$\lim_{n\to\infty} \inf_{(g,h)\in\Delta_{c,\delta}} \mathsf{E}_{n,g,h}\psi_{n,NR} = \mathsf{H}_{d,\delta^2}(\chi^2_{1-\alpha,k}), \tag{8.178}$$

where $\Delta_{c,\delta} = \{(g,h) : \delta^2 \leq h^T\mathsf{G}(\theta_0)h \leq c, g^Tg \leq c\}$, $0 \leq \delta^2 \leq c$. *The sequence* $\psi_{n,NR}$ *is LAMM for testing* $\mathsf{H}_0 : h = 0$ *versus* $\mathsf{H}_A : h \in \Delta_{c,\delta}$ *in the sense that*

$$\limsup_{n\to\infty}(\inf_{h\in\Delta_{c,\delta}} \mathsf{E}_{n,h}\varphi_n) \leq \lim_{n\to\infty}(\inf_{h\in\Delta_{c,\delta}} \mathsf{E}_{n,h}\psi_{n,NR}).$$

for every further asymptotic level α *test* φ_n.

Proof. Condition (A10) allows the application of Proposition 8.113, which yields, with $\theta_0 = (\tau_0^T, \xi_0^T)^T$,

$$\frac{1}{\sqrt{n}}\sum\nolimits_{i=1}^n W_{\tau_0,\widetilde{\xi}_n}(X_i) = \frac{1}{\sqrt{n}}\sum\nolimits_{i=1}^n W_{\tau_0,\xi_0}(X_i) + o_{P_{\theta_0}^{\otimes n}}(1).$$

We note that

$$\mathsf{C}_{\theta_0}(W_{\theta_0}, V_{\theta_0}) = 0 \quad \text{and} \quad \mathsf{C}_{\theta_0}(W_{\theta_0}, W_{\theta_0}) = \mathsf{G}(\theta_0),$$

by the definition of W_{θ_0}in (8.112) and (8.115), respectively. Hence

$$\mathsf{C}_{\theta_0}(\mathsf{G}^{-1/2}(\theta_0)W_{\theta_0}, \dot{L}_{\theta_0}) = \mathsf{C}_{\theta_0}\left(\mathsf{G}^{-1/2}(\theta_0)W_{\theta_0}, \begin{pmatrix} U_{\theta_0} \\ V_{\theta_0}\end{pmatrix}\right) = (\mathsf{G}^{1/2}(\theta_0), 0).$$

Corollary 6.74 provides the locally uniform convergence

$$\mathcal{L}(\mathsf{G}^{-1/2}(\theta_0)S_{n,NR}(\theta_0)|P^{\otimes n}_{\tau_0+h/\sqrt{n},\xi_0+g/\sqrt{n}}) = \mathsf{N}(\mathsf{G}^{1/2}(\theta_0)h, \mathbf{I}),$$

$$\lim_{n\to\infty}\int \psi_{n,NR}dP^{\otimes n}_{\tau_0+h/\sqrt{n},\xi_0+g/\sqrt{n}} = \mathsf{H}_{k,h^T\mathsf{G}(\theta_0)h}(\chi^2_{1-\alpha,k}),$$

which shows that $\psi_{n,NR}$ is an asymptotic level α test and (8.178) holds. Using Theorem 8.143 it remains to prove that $\mathsf{H}_{d,\delta^2}(\chi^2_{1-\alpha,k})$ is the maximin value for the limit model $\mathcal{G}_0 = (\mathbb{R}^d, \mathfrak{B}_d, (\mathsf{N}(\theta, \mathsf{I}^{-1}(\theta_0)))_{\theta\in\mathbb{R}^d})$. But this follows from Theorem 8.24, where we have to use $\Sigma_0 = \mathsf{I}^{-1}(\theta_0)$ so that $\Sigma_{1,1} = \mathsf{G}^{-1}(\theta_0)$ in view of (8.113) and $\Sigma_{1,1}^{-1} = \mathsf{G}(\theta_0)$. ■

Now we investigate how the Neyman–Rao test $\varphi_{n,\chi^2}(Q_{n,NR}(\tau_0, \widetilde{\xi}_n))$ is related to the Wald test and the likelihood ratio test. We assume that the conditions in Proposition 8.115 hold, so that especially $\widehat{\theta}_n = (\widehat{\tau}_n, \widehat{\xi}_n)$ is the MLE for the total model, and $\widetilde{\xi}_n$ is a $\sqrt{n}$-consistent estimator for ξ in the submodel with $(P_{(\tau_0,\xi)})_{\xi\in\Xi(\tau_0)}$. Then under the assumptions of Proposition 8.123 the test statistics $Q_{n,NR}(\tau_0, \widetilde{\xi}_n)$ in (8.119), $Q_{n,W}(\tau_0, \widehat{\xi}_n)$ in (8.120), and the likelihood ratio statistic $Q_{N,LR} = 2(\Lambda_n(\widehat{\theta}_n) - \Lambda_n(\widetilde{\theta}_n))$ differ only by terms $o_{P_{\theta_0}^{\otimes n}}(1)$, so that that the pointwise limits of the power functions of the associated tests are identical. Due to (6.80) the convergence is locally uniform. This gives the following statement.

Proposition 8.149. *If the conditions of Proposition 8.123 hold, then the Neyman–Rao test, the Wald test, and the likelihood ratio test are asymptotic level* α *tests that have the LAMM property for the testing Problem* $\mathsf{H}_0 : h = 0$, $g \in \mathbb{R}^{d-k}$, *versus* $\mathsf{H}_A : \delta^2 \leq h^T\mathsf{G}(\theta_0)h \leq c$, $g^Tg \leq c$.

8.10 Solutions to Selected Problems

Solution to Problem 8.2: $\mathsf{E}_\theta\psi \le \mathsf{E}_\theta\phi$, $\theta \in \Delta_A$, implies that the test ϕ is unbiased. $\mathsf{E}_{\theta_1}\psi < \mathsf{E}_{\theta_1}\phi$ at some $\theta_1 \in \Delta_A$ would contradict that ψ is a uniformly best unbiased level α test. □

Solution to Problem 8.4: The special choice of $\phi \equiv \alpha$ shows that the test φ is unbiased. By Problem 8.3 every unbiased level α test ψ for which $\theta \to \mathsf{E}_\theta\psi$ is continuous satisfies $\mathsf{E}_\theta\psi = \alpha$, $\theta \in J$. □

Solution to Problem 8.35: $(X, Y) = (\sum_{i=1}^n X_i, \sum_{i=1}^n Y_i)$ is sufficient for $(\lambda_1, \lambda_2) \in (0,\infty)^2$, where X and Y are independent and follow a Poisson distribution $\mathsf{Po}(n\lambda_1)$ and $\mathsf{Po}(n\lambda_2)$, respectively. Let $\eta_j = n\lambda_j$, $j = 1, 2$. The p.m.f. of (X, Y) is

$$\begin{aligned}\mathsf{po}_{\eta_1}(x)\mathsf{po}_{\eta_2}(y) &= \frac{(\eta_1)^x}{x!}\exp\{-\eta_1\}\frac{(\eta_2)^y}{y!}\exp\{-\eta_2\}\\ &= \exp\{x\ln(\frac{\eta_1}{\eta_2}) + (x+y)\ln(\eta_2) - \eta_1 - \eta_2\}\frac{1}{x!y!}\\ &= \exp\{U(x,y)\tau + V(x,y)\xi - K(\tau,\xi)\}\frac{1}{x!y!},\end{aligned}$$

where $U(x, y) = x$ and $V(x, y) = x + y$, $(x, y) \in \mathbb{N}^2$. This complies with (8.27). As $\tau = \ln(\eta_1/\eta_2) = \ln(\lambda_1/\lambda_2)$, $\mathsf{H}_0 : \lambda_1 \le \lambda_2$ is equivalent to $\mathsf{H}_0 : \tau \le 0$, and $\mathsf{H}_A : \lambda_1 > \lambda_2$ is equivalent to $\mathsf{H}_A : \tau > 0$. Now Theorem 8.28 can be applied which provides the uniformly best unbiased level α test $\varphi_{1,U}(u, v)$ for H_0 versus H_A. At $\tau = 0$ the conditional distribution of U, given $V = v$, is the binomial distribution $\mathsf{B}(v, \lambda_1/(\lambda_1 + \lambda_2))$. Testing $\mathsf{H}_0 : \lambda_1 = \lambda_2$ versus $\mathsf{H}_A : \lambda_1 \ne \lambda_2$ goes analogously. □

Solution to Problem 8.56: As T is maximal invariant the sample space $\mathcal{X}$ can be written as a union of the disjoint orbits $\{x : T(x) = t\}$, $t \in \mathcal{T}$, on which the statistic S takes on the constant value $S(x) = S(U(T(y)))$ for every $y \in \{x : T(x) = t\}$. □

Solution to Problem 8.57: As $(\overline{X}_n, S_n^2)$ is sufficient for $(\mu, \sigma^2) \in \mathbb{R} \times (0,\infty)$ we consider only tests that are based on $(\overline{X}_n, S_n^2)$. The hypotheses remain invariant under scale transformations. Scale invariance leads to the maximal invariant statistic $T = \sqrt{n}\ \overline{X}_n/\sqrt{S_n^2}$, which follows a noncentral t-distribution $\mathsf{T}(n-1, \sqrt{n}\mu/\sigma)$; see Example 2.37. By Theorem 2.27 it has MLR in the identity. The rest follows from Theorem 2.49. □

Solution to Problem 8.60: As T generates $\mathfrak{I}$ every $A \in \mathfrak{I}$ may be written as $A = T^{-1}(B)$, $B \in \mathfrak{T}$. Hence by the definition of the likelihood ratio $M_{\theta_0,\theta}$,

$$\begin{aligned}P_\theta(A) = (P_\theta \circ T^{-1})(B) &= \int_B M_{\theta_0,\theta} d(P_{\theta_0} \circ T^{-1}) + (P_\theta \circ T^{-1})(\{M_{\theta_0,\theta} = \infty\} \cap B)\\ &= \int_A M_{\theta_0,\theta}(T) dP_{\theta_0} + P_\theta(\{M_{\theta_0,\theta}(T) = \infty\} \cap B). \quad \square\end{aligned}$$

Solution to Problem 8.61: If the distribution of $(X_1, ..., X_n)$ has the Lebesgue density f, then the distribution of $(X_1, X_2 - X_1 ..., X_n - X_1)$ has the Lebesgue density $f(t_1, t_2 + t_1, ..., t_n + t_1)$. Integration over t_1 gives the marginal density of $(X_2 - X_1, ..., X_n - X_1)$. □

Solution to Problem 8.62: In a first step we calculate the density g of

$$(X_1, X_2 - X_1, \frac{X_3 - X_1}{X_2 - X_1}, ..., \frac{X_n - X_1}{X_2 - X_1}),$$

where $(X_1, ..., X_n)$ has the Lebesgue density f. The mapping

$$(t_1, ..., t_n) \to (t_1, t_2 - t_1, \frac{t_3 - t_1}{t_2 - t_1}, ..., \frac{t_n - t_1}{t_2 - t_1})$$

is a diffeomorphism $D(t_1, ..., t_n) : \mathbb{R}^n_{\neq 0} \to \mathbb{R} \times \mathbb{R}_{\neq 0} \times \mathbb{R}^{n-2}$ with the inverse mapping $D^{-1}(s_1, ..., s_n) = (s_1, s_1 + s_2, s_1 + s_3 s_2, ..., s_1 + s_n s_2)$, which has the Jacobian

$$J_{D^{-1}}(s_1, ..., s_n) = \begin{pmatrix} 1 & 0 & \cdot & \cdot & \cdot & 0 \\ 1 & 1 & 0 & \cdot & \cdot & 0 \\ 1 & s_3 & s_2 & 0 & \cdot & 0 \\ \cdot & \cdot & 0 & s_2 & & \cdot \\ \cdot & \cdot & \cdot & \cdot & \cdot & \cdot \\ \cdot & \cdot & \cdot & & \cdot & \cdot \\ 1 & s_n & 0 & \cdot & \cdot & s_2 \end{pmatrix}.$$

The absolute value of the determinant of $J_{D^{-1}}$ is $|s_2|^{n-2}$ so that the density g is, according to the transformation theorem for Lebesgue densities, given by

$$g(s) = |J_{D^{-1}}(s)| f(D^{-1}(s))| = |s_2|^{n-2} f(s_1, s_1 + s_2, s_1 + s_3 s_2, ..., s_1 + s_n s_2),$$

where $s = (s_1, ..., s_n)$. Hence the density h of

$$\left(X_2 - X_1, \frac{X_3 - X_1}{X_2 - X_1}, ..., \frac{X_n - X_1}{X_2 - X_1}\right)$$

is the marginal density

$$h(s_2, ..., s_n) = \int |s_2|^{n-2} f(s_1, s_1 + s_2, s_1 + s_3 s_2, ..., s_1 + s_n s_2) ds_1.$$

Set $\varepsilon = \operatorname{sgn}(X_2 - X_1)$, $Y_i = (X_i - X_1)/(X_2 - X_1)$, $i = 3, ..., n$, and fix a Borel set $B \in \mathfrak{B}_{n-2}$. Then with $u = (s_3, ..., s_n)$,

$$\mathbb{P}(\varepsilon = \pm 1, (Y_3, ..., Y_n) \in B) = \int I_{(0,\infty)}(\pm s_2)[\int_B h(s_2, u) \boldsymbol{\lambda}_{n-2}(du)] \boldsymbol{\lambda}(ds_2).$$

Let ϖ denote the counting measure on $\{-1, 1\}$ and $h_{\pm}$ the integrand. Then the distribution of $T_{ls} = (\varepsilon, Y_3, ..., Y_n)$ has, with respect to $\varpi \otimes \boldsymbol{\lambda}_{n-2}$, the density

$$\begin{aligned} k(\delta, s_3, ..., s_n) &= I_{\{-1\}}(\delta) h_-(s_3, ..., s_n) + I_{\{1\}}(\delta) h_+(s_3, ..., s_n) \\ &= I_{\{-1\}}(\delta) \int_{-\infty}^{0} [\int_{-\infty}^{\infty} |s_2|^{n-2} f(s_1, s_1 + s_2, s_1 + s_3 s_2, ..., s_1 + s_n s_2) ds_1] ds_2 \\ &\quad + I_{\{1\}}(\delta) \int_{0}^{\infty} [\int_{-\infty}^{\infty} |s_2|^{n-2} f(s_1, s_1 + s_2, s_1 + s_3 s_2, ..., s_1 + s_n s_2) ds_1] ds_2 \\ &= \int_0^{\infty} [\int_{-\infty}^{\infty} s_2^{n-2} f(s_1, s_1 + \delta s_2, s_1 + s_3 \delta s_2, ..., s_1 + s_n \delta s_2) ds_1] ds_2. \end{aligned}$$

Set $\delta(x_1, x_2) = 1$ if $x_1 < x_2$, and $\delta(x_1, x_2) = -1$ if $x_1 > x_2$. Then we have $\delta(x_1, x_2)(x_2 - x_1) = |x_2 - x_1|$, and $k(T_{ls}(x_1, ..., x_n))$ is given by

$$\begin{aligned}
&k(T_{ls}(x_1, ..., x_n)) \\
&= \int_0^\infty [\int_{-\infty}^\infty s_2^{n-2} f(s_1, s_1 + s_2 \frac{x_2 - x_1}{|x_2 - x_1|}, ..., s_1 + \frac{x_n - x_1}{|x_2 - x_1|} s_2) ds_1] ds_2 \\
&= |x_2 - x_1|^{n-1} \int_0^\infty [\int_{-\infty}^\infty w^{n-2} f(s_1, s_1 + w(x_2 - x_1), ..., \\
&\qquad\qquad s_1 + w(x_n - x_1)) ds_1] dw \\
&= |x_2 - x_1|^{n-1} \int_0^\infty [\int_{-\infty}^\infty w^{n-2} f(v + wx_1, v + wx_2, ..., v + wx_n) dv] dw.
\end{aligned}$$

To complete the proof we have only to apply Problem 8.60. □

Solution to Problem 8.64: If $f_0(x_1, ..., x_n) = (2\pi)^{-n/2} \exp\{-\frac{1}{2} \sum_{i=1}^n x_i^2\}$, then $\sum_{i=1}^n (wx_i + v)^2 = (n-1)w^2 s_n^2 + n(w\overline{x}_n + v)^2$, and with $x = (x_1, ..., x_n)$,

$$\begin{aligned}
\overline{f}_0(x) &= (2\pi)^{-n/2} \int_0^\infty w^{n-2} [\int_{-\infty}^\infty \exp\{-\frac{1}{2} \sum\nolimits_{i=1}^n (wx_i + v)^2\}) dv] dw \\
&= (2\pi)^{-n/2} \int_0^\infty w^{n-2} \exp\{-\frac{1}{2}(n-1)w^2 s_n^2\} \\
&\qquad \times [\int_{-\infty}^\infty \exp\{-\frac{1}{2} n(w\overline{x}_n + v)^2\} dv] dw \\
&= (2\pi)^{-n/2} (2\pi)^{1/2} n^{-1/2} \int_0^\infty w^{n-2} \exp\{-\frac{1}{2}(n-1)w^2 s_n^2\} dw \\
&= (2\pi)^{-(n-1)/2} n^{-1/2} [(n-1)s_n^2]^{-(n-1)/2} \int_0^\infty u^{n-2} \exp\{-\frac{1}{2} u^2\} du \\
&= \frac{1}{2} n^{-1/2} [(n-1)\pi]^{-(n-1)/2} \Gamma(\frac{n-1}{2}) s_n^{-n+1}.
\end{aligned}$$

If $f_1(x_1, ..., x_n) = \exp\{-\sum_{i=1}^n x_i\} I_{(0,\infty)}(x_{[1]})$, then with $x = (x_1, ..., x_n)$,

$$\begin{aligned}
\overline{f}_1(x) &= \int_0^\infty w^{n-2} [\int_{-\infty}^\infty I_{(0,\infty)}(wx_{[1]} + v) \exp\{-nw\overline{x}_n - nv\} dv] dw \\
&= \int_0^\infty w^{n-2} \exp\{-nw\overline{x}_n\} [\int_{-wx_{[1]}}^\infty \exp\{-nv\} dv] dw \\
&= n^{-1} \Gamma(n-1) [n(\overline{x}_n - x_{[1]})]^{-n+1}. \quad \square
\end{aligned}$$

Solution to Problem 8.66: If $f_1(x) = \prod_{j=1}^{n_1} \varphi_{\mu,1}(x_{1,j}) \prod_{j=1}^{n_2} \varphi_{0,1}(x_{2,j})$, $x = (x_{1,1}, ..., x_{2,n_2}) \in \mathbb{R}^n$, where $n = n_1 + n_2$, then

$$\begin{aligned}
\overline{f}_1(x) = (2\pi)^{-n/2} \int_0^\infty w^{n-2} [\int_{-\infty}^\infty \exp\{-\frac{1}{2} \sum\nolimits_{j=1}^{n_1} (wx_{1,j} + v - \mu)^2 \\
-\frac{1}{2} \sum\nolimits_{j=2}^{n_2} (wx_{2,j} + v)^2\} dv] dw.
\end{aligned}$$

The inner integral can be solved by rearranging the quadratic terms. This leads to

$$\overline{f}_1(x) = n^{-1/2}(2\pi)^{-(n-1)/2}[\sum_{i=1}^{2}\sum_{j=1}^{n_i}(x_{i,j} - \overline{x}_{i\cdot})^2]^{-(n-1)/2}$$
$$\times \exp\{-\frac{n_1 n_2 \mu^2}{2n}\} \int_0^\infty w^{n-2} \exp\{-\frac{1}{2}w^2 + w\mu V(x)\}dw.$$

For $\mu = 0$ we get $\overline{f}_0(x)$. The ratio $\overline{f}_1(x)/\overline{f}_0(x)$ turns out to be (8.69). □

Solution to Problem 8.71: Using the elementary inequality $a \wedge b \leq a^s b^{1-s}$ we get the inequality from (1.82). □

Solution to Problem 8.94: Use the Gram–Schmidt orthogonalization procedure to find an orthonormal basis $Z_1, ..., Z_n$. The mapping $X \to (\langle X, Z_1\rangle, ..., \langle X, Z_n\rangle)$ is an isometry between $\mathcal{L}$ and $\mathbb{R}^n$, i.e., it is linear, preserves the scalar product, and is one-to-one. Hence for a convergent sequence $X_m \to X$ the vectors $(\langle X_m, Z_1\rangle, ..., \langle X_m, Z_n\rangle)$ converge to the vector $(\langle X, Z_1\rangle, ..., \langle X, Z_n\rangle)$ so that we have $X = \sum_{i=1}^n \langle X, Z_i\rangle Z_i \in \mathcal{L}$. As to the second statement, we remark that every $X \in span(\{X_1, X_2, ...\})$ is either zero or has a positive variance. $\mu = \lim_{n\to\infty} \overline{X}_n$ does not have this property but belongs to the closure of $\mathcal{L}$. □

Solution to Problem 8.95: Use the same arguments as in Problem 3.79. □

Solution to Problem 8.96: Suppose $Z = \Pi_{\mathcal{L}} Y$ and consider for $X \in \mathcal{L}$ the function $f(t) = \|Y - \Pi_{\mathcal{L}} Y + tX\|_2^2 = t^2 \|X\|^2 + 2t\langle Y - \Pi_{\mathcal{L}} Y, X\rangle + \|Y - \Pi_{\mathcal{L}} Y\|^2$ which has the minimum at $t = 0$. Then $2\langle Y - \Pi_{\mathcal{L}} Y, X\rangle = f'(0) = 0$. Conversely, if (8.81) holds, then for every $X \in \mathcal{L}$, and thus for $Y - \Pi_{\mathcal{L}} Y \perp \Pi_{\mathcal{L}} Y - X$, it holds $\|Y - X\|^2 = \|Y - \Pi_{\mathcal{L}} Y\|^2 + \|\Pi_{\mathcal{L}} Y - X\|^2 \geq \|Y - \Pi_{\mathcal{L}} Y\|^2$. □

Solution to Problem 8.97: Put $a^T = (\mathbb{E}(YX^T))(\mathbb{E}(XX^T))^{-1}$ and $Z = a^T X$. Then $Z \in \mathcal{L}$ and for every $b \in \mathbb{R}^n$ it holds

$$\mathbb{E}(Y - a^T X)b^T X = \mathbb{E}(YX^T)b - a^T \mathbb{E}(XX^T)b = 0. \qquad \square$$

Solution to Problem 8.98: If $A \in \sigma(X)$, then $A = X^{-1}(B)$ for some $B \in \mathfrak{B}$ and $\mathbb{E}I_A g(X) = \mathbb{E}I_B(X)g(X) = \int[\int I_B(x)h(x,y)P_Y(dy)]P_X(dx) = \mathbb{E}I_B(X)h(X,Y) = \mathbb{E}I_A h(X,Y)$. □

Solution to Problem 8.99: For $Z \in \mathbb{H}$ the relations (a) and (b) in Proposition A.31 imply $\mathbb{E}(X - \mathbb{E}(X|\mathfrak{G}))Z = 0$ so that $X - \mathbb{E}(X|\mathfrak{G}) \perp \mathbb{H}$ and $\mathbb{E}(X|\mathfrak{G}) \in \mathbb{H}$. The first statement follows from Problem 8.96. The second follows from $\mathbb{E}(X|\mathfrak{G}) - \mathbb{E}X \in \mathbb{H}^0$ and $\mathbb{E}Z(X - \mathbb{E}(X|\mathfrak{G}) + \mathbb{E}X) = 0$, $Z \in \mathbb{H}^0$, and Problem 8.96 again. □

Solution to Problem 8.100: The independence of X_i and X_j for $i \neq j$ implies $\mathsf{E}_{P_n}(a_i(X_i)a_j(X_j)) = \mathsf{E}_{P_n} a_i(X_i)\mathsf{E}_{P_n} a_j(X_j) = 0$ which gives the statement on $\mathsf{E}_{P_n} S^2$. If $T = b_0 + \sum_{i=1}^n b_i(X_i)$, then $\mathsf{E}_{P_n}(S - T)^2 = (a_0 - b_0)^2 + \sum_{i=1}^n \mathsf{E}_{P_n}(a_i(X_i) - b_i(X_i))^2$ which gives the uniqueness. □

Solution to Problem 8.101: Consider a sequence $S_m = a_{m,0} + \sum_{i=1}^n a_{m,i}(X_i)$ which converges to some $S \in \mathbb{L}_2(P^{\otimes n})$. Then S_m is a Cauchy sequence and it follows from Problem 8.100 that the $a_{m,0} \in \mathbb{R}$ and the $a_{m,i} \in \mathbb{L}_2^0(P)$ form a Cauchy sequence and thus converge to some $a_0 \in \mathbb{R}$ and $a_i \in \mathbb{L}_2^0(P)$. Then by the first statement in Problem 8.100 S_m converges in the sense of $\mathbb{L}_2^0(P)$ to $a_0 + \sum_{i=1}^n a_i(X_i)$ which is S. $\square$

Solution to Problem 8.102: Without loss of generality assume $\gamma = 0$. Then $\mathsf{V}_{P_n}(\mathcal{U}_n) = (\binom{n}{m}\binom{n}{m})^{-1} \sum_{A,B \in \mathcal{A}_m} \mathsf{E}_{P_n}\Psi(X_A)\Psi(X_B)$. It holds $\mathsf{E}_{P_n}\Psi(X_A)\Psi(X_B) = 0$ if $A \cap B = \emptyset$ and $\mathsf{E}_{P_n}\Psi(X_A)\Psi(X_B) = \sigma_k^2$ if $|A \cap B| = k$. The number of such pairs of A and B are obtained as follows. There are $\binom{n}{k}$ selections of the k joint elements from $1, ..., n$. Select from the set of size $n-k$ the $m-k$ elements for A and from the $n-m$ elements the remaining $m-k$ for B. Hence we get

$$\binom{n}{k}\binom{n-k}{m-k}\binom{n-m}{m-k} = \binom{n}{m}\binom{m}{k}\binom{n-m}{m-k}$$

possibilities. As $\gamma = 0$ we have with $X = (X_1, ..., X_k)$,

$$\begin{aligned}
\mathsf{V}_{P_n}(\Psi_k(X)) &= \int [\int \Psi(x_1, ..., x_m) P^{\otimes(m-k)}(dx_{k+1}, ..., dx_m)]^2 P^{\otimes k}(dx_1, ..., dx_k) \\
&\le \int \Psi(x_1, ..., x_m)^2 P^{\otimes m}(dx_1, ..., dx_m) = \sigma_m^2.
\end{aligned}$$

$\mathsf{V}_{P_n}(\widehat{\mathcal{U}}_n) = (m^2/n)\mathsf{V}_{P_n}(\Psi_1(X_1)) = (m^2/n)\sigma_1^2$ yields

$$\begin{aligned}
&0 \le \mathsf{V}_{P_n}(\mathcal{U}_n) - \mathsf{V}_{P_n}(\widehat{\mathcal{U}}_n) = \frac{1}{\binom{n}{m}} \sum\nolimits_{k=1}^m \binom{m}{k}\binom{n-m}{m-k} \mathsf{V}_{P_n}(\Psi_k(X_1, ..., X_k)) - \frac{m^2}{n}\sigma_1^2 \\
&= \frac{1}{\binom{n}{m}}\binom{m}{1}\binom{n-m}{m-1}\sigma_1^2 - \frac{m^2}{n}\sigma_1^2 + \frac{1}{\binom{n}{m}} \sum\nolimits_{k=2}^m \binom{m}{k}\binom{n-m}{m-k} \mathsf{V}_{P_n}(\Psi_k(X_1, ..., X_k)) \\
&\le \left| \frac{m^2}{n} - \frac{m^2(n-m)!(n-m)!}{n!(n-2m+1)!} \right| \sigma_m^2 + \sigma_m^2 \frac{1}{\binom{n}{m}} \sum\nolimits_{k=2}^m \binom{m}{k}\binom{n-m}{m-k}.
\end{aligned}$$

It holds

$$\begin{aligned}
&\frac{m^2}{n} - \frac{m^2(n-m)!(n-m)!}{n!(n-2m+1)!} = \frac{m^2}{n}\left(1 - \frac{(n-m)\cdots(n-2m+2)}{(n-1)(n-2)\cdots(n-m+1)}\right) = o(1/n), \\
&\frac{1}{\binom{n}{m}} \sum\nolimits_{k=2}^m \binom{m}{k}\binom{n-m}{m-k} = o(1/n).
\end{aligned}$$

As $\widehat{\mathcal{U}}_n$ is the projection of $\mathcal{U}_n$ on $\mathcal{L}$ it holds $\mathcal{U}_n - \widehat{\mathcal{U}}_n \perp \widehat{\mathcal{U}}_n$ and $\mathsf{E}_{P_n}(\mathcal{U}_n - \widehat{\mathcal{U}}_n)^2 = \mathsf{E}_{P_n}\mathcal{U}_n^2 - \mathsf{E}_{P_n}\widehat{\mathcal{U}}_n^2$. $\square$

Solution to Problem 8.104: We may assume $\gamma = 0$. Put $V(X_{1,i}, X_{2,j}) = \Psi(X_{1,i}, X_{2,j}) - \Psi_1(X_{1,i}) - \Psi_2(X_{2,j})$. By the independence of $X_{1,i}, X_{2,j}, X_{1,k}, X_{2,l}$, $i \ne k$, $j \ne l$, and the independence of $X_{1,i}, X_{2,j}, X_{2,l}$, $i, j \ne l$,

$$\begin{aligned}
&\mathsf{E}_{P_{n_1,n_2}} V(X_{1,i}, X_{2,j}) V(X_{1,k}, X_{2,l}) = 0, \quad i \ne k, \ j \ne l, \\
&\mathsf{E}_{P_{n_1,n_2}} V(X_{1,i}, X_{2,j}) V(X_{1,i}, X_{2,l}) = \mathsf{E}_{P_{n_1,n_2}} \Psi(X_{1,i}, X_{2,j}) \Psi(X_{1,i}, X_{2,l}) \\
&- \mathsf{E}_{P_{n_1,n_2}} \Psi(X_{1,i}, X_{2,j}) \Psi_1(X_{1,i}) - \mathsf{E}_{P_{n_1,n_2}} \Psi_1(X_{1,i}) \Psi(X_{1,i}, X_{2,l}) + \mathsf{E}_{P_{n_1,n_2}} \Psi_1^2(X_{1,i}) \\
&= \mathsf{E}_{P_{n_1,n_2}} \Psi_1^2(X_{1,i}) - \mathsf{E}_{P_{n_1,n_2}} \Psi_1^2(X_{1,i}) - \mathsf{E}_{P_{n_1,n_2}} \Psi_1^2(X_{1,i}) + \mathsf{E}_{P_{n_1,n_2}} \Psi_1^2(X_{1,i}) = 0,
\end{aligned}$$

if $j \neq l$. Hence,

$$\begin{aligned}
&\mathsf{E}_{P_{n_1,n_2}}(\frac{1}{n_1}\frac{1}{n_2}\sum\nolimits_{i=1}^{n_1}\sum\nolimits_{j=1}^{n_2} V(X_{1,i},X_{2,j}))^2 \\
&= \frac{1}{n_1^2}\frac{1}{n_2^2}\sum\nolimits_{i=1}^{n_1}\sum\nolimits_{j=1}^{n_2} \mathsf{E}_{P_{n_1,n_2}} V(X_{1,i},X_{2,j})V(X_{1,i},X_{2,j}) \\
&= \frac{1}{n_1}\frac{1}{n_2}\mathsf{E}_{P_{n_1,n_2}} V^2(X_{1,1},X_{2,1}), \\
&\mathsf{E}_{P_{n_1,n_2}} V^2(X_{1,1},X_{2,1}) = \mathsf{E}_{P_{n_1,n_2}}(\Psi^2(X_{1,1},X_{2,1}) + \Psi_1^2(X_{1,1}) + \Psi_2^2(X_{2,1})) \\
&- 2\mathsf{E}_{P_{n_1,n_2}}(\Psi(X_{1,1},X_{2,1})\Psi_1(X_{1,1}) + \Psi(X_{1,1},X_{2,1}))\Psi_2(X_{2,1}) + \Psi_1(X_{1,1})\Psi_2(X_{2,1})) \\
&= \mathsf{E}_{P_{n_1,n_2}}(\Psi^2(X_{1,1},X_{2,1}) - \Psi_1^2(X_{1,1}) - \Psi_2^2(X_{2,1})) \leq \sigma^2. \quad \square
\end{aligned}$$

Solution to Problem 8.106: It holds

$$\mathsf{E}_{P_n} S_n = \sum\nolimits_{i=1}^{n} c_{i,n}\mathsf{E}_{P_n} a_n(R_{n,i}) = (\sum\nolimits_{i=1}^{n} c_{i,n})(\frac{1}{n}\sum\nolimits_{j=1}^{n} a_n(j)) = \overline{a}_n\sum\nolimits_{i=1}^{n} c_{i,n}.$$

As $S_n - n\overline{a}_n\overline{c}_n = \sum_{i=1}^n c_{i,n}(a_n(R_{n,i}) - \overline{a}_n) = \sum_{i=1}^n (c_{i,n} - \overline{c}_n)(a_n(R_{n,i}) - \overline{a}_n)$ it suffices to consider the case of $\overline{a}_n = \overline{c}_n = 0$ to calculate the variance. Using $\sum_{i:i\neq j} c_{i,n} = -c_{n,j}$ and $\sum_{k:k\neq l} a_n(k) = -a_n(l)$ we get

$$\begin{aligned}
\mathsf{E}_{P_n} S_n^2 &= \sum\nolimits_{i,j=1}^{n} \mathsf{E}_{P_n} c_{i,n}a_n(R_{n,i})c_{n,j}a_n(R_{n,j}) \\
&= \sum\nolimits_{i\neq j} c_{i,n}c_{j,n}\mathsf{E}_{P_n} a_n(R_{n,i})a_n(R_{n,j}) + \sum\nolimits_{i=1}^{n} c_{i,n}^2\mathsf{E}_{P_n} a_n^2(R_{n,i}) \\
&= \frac{1}{n(n-1)}\sum\nolimits_{i\neq j} c_{i,n}c_{j,n}\sum\nolimits_{k\neq l} a_n(k)a_n(l) + \frac{1}{n}\sum\nolimits_{i=1}^{n} c_{i,n}^2\sum\nolimits_{k=1}^{n} a_n^2(k) \\
&= \frac{1}{n-1}\sum\nolimits_{i=1}^{n} c_{i,n}^2\sum\nolimits_{k=1}^{n} a_n^2(k). \quad \square
\end{aligned}$$

Solution to Problem 8.108: Set $\varepsilon_i = I_{(0,t]}(U_i)$, $0 < t < 1$. $\sum_{i=1}^n \varepsilon_i$ has a binomial distribution $\mathsf{B}(n,t)$. Then $P_n(U_{n,[k]} \leq t) = P_n(\sum_{i=1}^n \varepsilon_i \geq k) = \mathsf{Be}_{a,\beta}(t)$ in view of Problem 8.107. The mean and variance of $\mathsf{Be}(\alpha,\beta)$ are $\alpha/[\alpha+\beta]$ and $\alpha\beta/[(\alpha+\beta)^2(\alpha+\beta+1)]$, which imply the expressions for $\mathsf{E}_{P_n} U_{n,[k]}$ and $\mathsf{V}_{P_n}(U_{n,[k]})$. $\square$

Solution to Problem 8.109: If $(U_{n,[1]},...,U_{n,[n]})$ is the order statistic, then $U_1 = U_{n,[R_{n,1}]}$. Hence by the independence of $(U_{n,[1]},...,U_{n,[n]})$ and $R_n = (R_{n,1},...,R_{n,n})$, and Problem 8.98,

$$\begin{aligned}
&\mathsf{E}_{P_n}(U_1 - \frac{1}{n+1}R_{n,1})^2 = \mathsf{E}_{P_n}(\mathsf{E}_{P_n}((U_{n,[R_{n,1}]} - \frac{1}{n+1}R_{n,1})^2|R_{n,1})) \\
&= \frac{1}{n}\sum_{k=1}^{n}\mathsf{E}_{P_n}(U_{n,[k]} - \frac{k}{n+1})^2 = \frac{1}{n}\sum_{k=1}^{n}\frac{k(n-k+1)}{(n+1)^2(n+2)} \leq \frac{n^2}{(n+1)^2(n+2)}. \quad \square
\end{aligned}$$

Solution to Problem 8.122: Use the representation of I^{-1} in (8.113). $\square$

Solution to Problem 8.129: Use (6.90) and Corollary 6.73. $\square$

9

Selection

9.1 The Selection Models

Competition, determination of the winner, and subsequent consequences are parts of normal life. The variety of ways selection problems have been formulated for k competing populations is great. For example, the goal could be to find a best population, r best populations, r populations that include a best population, a subset of random size of populations that includes a best population, or a complete ranking of all k populations. Because of the latter, this research area was originally given the name *ranking and selection*. Since ranking plays only a minor role in the literature "ranking" is rarely mentioned anymore. Another earlier name, *multiple decision procedures*, has lost its popularity as well because its meaning has become too broad. It covers also multiple comparisons, and even inferential methods based on a sample from one population where three decisions (e.g., a small, medium, or large mean) or more are made.

Among the early contributions to the literature of *selection procedures*, called *selection rules* here, are the papers by Paulson (1949, 1952), Bahadur (1950), Bahadur and Robbins (1950), Bahadur and Goodman (1952), Bechhofer (1954), Bechhofer, Dunnett, and Sobel (1954), Dunnett (1955, 1960), Gupta (1956, 1965), Sobel (1956), Lehmann (1957a,b, 1961, 1963, 1966), Hall (1959), and Eaton (1967a,b). The first research monograph was written by Bechhofer, Kiefer, and Sobel (1968) with the focus on a sequential approach for exponential (Koopman–Darmois) families. The dramatic developments that would follow in the field inspired Gupta and Panchapakesan (1979) to write their classical monograph that provides an up to the time complete overview of the entire related literature. Soon after, an extension of this overview followed with Gupta and Huang (1981). A categorized guide to selection and ranking procedures was provided by Dudewicz and Koo (1982). Collections of research papers on selection rules are included in Gupta and Yackel (1971), Gupta and Moore (1977), Gupta (1977), Dudewicz (1982), Santner and Tamhane (1984), Gupta and Berger (1982, 1988), Hoppe (1993), Miescke and Herrendörfer

F. Liese, K.-J. Miescke, *Statistical Decision Theory*,
DOI: 10.1007/978-0-387-73194-0_9, © Springer Science+Business Media, LLC 2008

(1993, 1994), Miescke and Rasch (1996a,b), Panchapakesan and Balakrishnan (1997), and Balakrishnan and Miescke (2006). Several books that emphasize the methodology of selection rules are by Dudewicz (1976), Gibbons, Olkin, and Sobel (1977), Büringer, Martin, and Schriever (1980), Mukhopadhyay and Solanky (1994), Bechhofer, Santner, and Goldsman (1995), Rasch (1995), and Horn and Volland (1995).

Selection problems in their various settings are not only statistically highly relevant, but also theoretically challenging, with technical parts that are quite different from those of estimation and testing problems. Although often intuitively assumed to be appropriate, it simply is not good enough to compare populations just in terms of optimal estimators, or pairwise with optimal two-sample tests, of the relevant parameters to end up with an optimal selection decision.

Inasmuch as this book is restricted to decision theory, many of the research publications on selection rules are not mentioned here due to a missing link to decision theory. Readers who are interested in such publications are referred to the references given above. The purpose of this chapter is to provide an outline of such a theory. However, we focus on the main branches of this field, leaving out others that are nevertheless interesting and worth being studied further. Even in the main branches that are covered here many theoretical questions remain open, as the readers will notice. It is hoped that this will stimulate the readers' interest in pursuing their own research work in this regard.

Let $X = (X_1, ..., X_k)$ be the vector of observations from the k populations that take on values in $(\mathcal{X}_i, \mathfrak{A}_i)$ and have the distribution P_{i,θ_i}, $i = 1, ..., k$, where the parameters $\theta_1, ..., \theta_k$ belong to the same parameter set Δ. Let $\kappa : \Delta \to \mathbb{R}$ be a given functional. Each population is addressed either by P_{i,θ_i} or for simplicity just by its index i, $i = 1, ..., k$. The typical goal in the area of selection theory is to find a *best population*. This is a population i_0, say, for which $i_0 \in \mathsf{M}_\kappa(\theta)$, where $\theta = (\theta_1, ..., \theta_k) \in \Delta^k$ and

$$\mathsf{M}_\kappa(\theta) = \underset{i \in \{1,...,k\}}{\arg\max}\ \kappa(\theta_i) = \{i : \kappa(\theta_i) = \max_{1 \le l \le k} \kappa(\theta_l)\}. \tag{9.1}$$

In many selection problems $\Delta \subseteq \mathbb{R}$ is one-dimensional, and in this case κ is usually taken as the identical mapping. On the other hand, if $\Delta \subseteq \mathbb{R}^d$ for some $d > 1$, then to be able to establish the concept of a "best population" on a one-dimensional scale, a suitable functional κ has to be chosen. For example, κ could be the projection onto one specific coordinate of the parameter vector in Δ, where the remaining $d - 1$ coordinates are treated as nuisance parameters. To give another example, in nonparametric models, where Δ may be a family of distributions, κ could be the median or some other nonparametric functional.

Whether we select one population and declare it to be a best population, or we select a subset of populations and declare that it contains a best population, depends on the given decision-theoretic framework. This is discussed in more detail later on.

We do not require that the populations be independent. This means that P_{i,θ_i}, $i = 1, ..., k$, are the marginal distributions of the distribution of $(X_1, ..., X_k)$ which is denoted by P_θ, $\theta = (\theta_1, ..., \theta_k) \in \Delta^k$. The *general selection model* is

$$\mathcal{M}_s = (\mathsf{X}_{i=1}^k \mathcal{X}_i, \bigotimes_{i=1}^k \mathfrak{A}_i, (P_\theta)_{\theta \in \Delta^k}). \tag{9.2}$$

Throughout this chapter it is assumed that $\mathsf{P} = (P_\theta)_{\theta \in \Delta^k}$ is a stochastic kernel. This allows us to utilize Bayes techniques for finding optimal selection rules.

Often the populations are independent so that the joint distribution P_θ is the product distribution of the marginal distributions P_{i,θ_i}. Then we have the *independent selection model*

$$\mathcal{M}_{is} = (\mathsf{X}_{i=1}^k \mathcal{X}_i, \bigotimes_{i=1}^k \mathfrak{A}_i, (\bigotimes_{i=1}^k P_{i,\theta_i})_{\theta \in \Delta^k}), \quad \theta = (\theta_1, ..., \theta_k), \tag{9.3}$$

which has been considered previously in (3.10). We call the independent selection model *balanced* if the sample spaces $(\mathcal{X}_i, \mathfrak{A}_i)$ as well as the families $(P_{i,\theta_i})_{\theta_i \in \Delta}$ are identical. The *balanced selection model* is given by

$$\mathcal{M}_{bs} = (\mathcal{X}^k, \mathfrak{A}^{\otimes k}, (\bigotimes_{i=1}^k P_{\theta_i})_{\theta \in \Delta^k}), \quad \theta = (\theta_1, ..., \theta_k). \tag{9.4}$$

A typical situation where we have to deal with an *unbalanced selection model* occurs when the sampling design is unbalanced. More precisely, let $X_{i,1}, ..., X_{i,n_i}$ be observed from population P_{θ_i}, $i = 1, ..., k$, where the observations are altogether independent. Then we have the selection model

$$\mathcal{M}_{us} = (\mathsf{X}_{i=1}^k \mathcal{X}^{n_i}, \bigotimes_{i=1}^k \mathfrak{A}^{\otimes n_i}, (\bigotimes_{i=1}^k P_{\theta_i}^{\otimes n_i})_{\theta \in \Delta^k}). \tag{9.5}$$

It is of the form (9.3) if we identify $\mathcal{X}^{n_i}$ with $\mathcal{X}_i$, $\mathfrak{A}^{\otimes n_i}$ with $\mathfrak{A}_i$, and $P_{\theta_i}^{\otimes n_i}$ with P_{i,θ_i}. It is clear that $\mathcal{M}_{us}$ is balanced if and only if $n_1 = \cdots = n_k$.

Often we reduce the model $\mathcal{M}_s$ in a first step by means of a statistic $V : \mathsf{X}_{i=1}^k \mathcal{X}_i \to_m \mathbb{R}^k$ and obtain the reduced model

$$\mathcal{M}_{ss} = (\mathbb{R}^k, \mathfrak{B}_k, (Q_\theta)_{\theta \in \Delta^k}), \tag{9.6}$$

where $Q_\theta = P_\theta \circ V^{-1}$. Usually such a statistic V is sufficient, and then the models $\mathcal{M}_{ss}$ and $\mathcal{M}_s$ are equivalent; see Remark 4.68. Regardless of whether $(Q_\theta)_{\theta \in \Delta^k}$ in $\mathcal{M}_{ss}$ has been obtained via $Q_\theta = P_\theta \circ V^{-1}$ or has been introduced in any other way we call $\mathcal{M}_{ss}$ the *standard selection model.*

To set up a selection model that is based on a one-parameter exponential family let $\mathcal{M}_{ne} = (\mathcal{X}, \mathfrak{A}, (P_\vartheta)_{\vartheta \in \Delta})$, $\Delta \subseteq \mathbb{R}$, be from (1.7) with $d = 1$ and satisfy (A1) and (A2). Denote by $X_{i,j} : \mathsf{X}_{i=1}^k \mathcal{X}^{n_i} \to \mathcal{X}$, $j = 1, ..., n_i$, $i = 1, ..., k$, the projections on the coordinates. Similarly as in Proposition 1.4 one obtains from the structure of the density in (1.6) that $(\bigotimes_{i=1}^k P_{\theta_i}^{\otimes n_i})_{\theta \in \Delta^k}$

is again an exponential family with natural parameter $\theta = (\theta_1, ..., \theta_k)$ and generating statistic

$$V = (T_{1,\oplus n_1}, ..., T_{k,\oplus n_k}), \quad T_{i,\oplus n_i} = \sum\nolimits_{j=1}^{n_i} T(X_{i,j}),\ i = 1, ..., k, \tag{9.7}$$

which by Theorem 4.50 is sufficient for the family $(\bigotimes_{i=1}^k P_{\theta_i}^{\otimes n_i})_{\theta \in \Delta^k}$. This means that for the exponential family $(P_\vartheta)_{\vartheta \in \Delta}$ the model (9.5) can be reduced by sufficiency and is, with $\theta = (\theta_1, ..., \theta_k)$, equivalent to the model

$$\begin{aligned} \mathcal{M}_{sel} &= (\mathbb{R}^k, \mathfrak{B}_k, (\textstyle\bigotimes_{i=1}^k Q_{n_i,\theta_i})_{\theta \in \Delta^k}), \quad \text{where} \\ Q_{n_i,\theta_i} &= \mathcal{L}(T_{i,\oplus n_i} | \textstyle\bigotimes_{i=1}^k P_{\theta_i}^{\otimes n_i}), \quad i = 1, ..., k. \end{aligned} \tag{9.8}$$

As $T_{i,\oplus n_i}$ is the sum of the random variables $T(X_{i,j})$, $j = 1, ..., n_i$, which are i.i.d. under $\bigotimes_{i=1}^k P_{\theta_i}^{\otimes n_i}$ with distribution $Q_{\theta_i} := \mathcal{L}(T(X_{i,j})|P_{\theta_i})$, it holds

$$Q_{n_i,\theta_i} = \mathcal{L}(T_{i,\oplus n_i} | \textstyle\bigotimes_{i=1}^k P_{\theta_i}^{\otimes n_i}) = Q_{\theta_i}^{*n_i}, \quad i = 1, ..., k,$$

where $*$ denotes the convolution of distributions. Hence with $\theta = (\theta_1, ..., \theta_k)$,

$$\mathcal{M}_{sel} = (\mathbb{R}^k, \mathfrak{B}_k, (\textstyle\bigotimes_{i=1}^k Q_{\theta_i}^{*n_i})_{\theta \in \Delta^k}). \tag{9.9}$$

With $\boldsymbol{\nu}_i := \boldsymbol{\nu}^{\otimes n_i} \circ S_i^{-1}$, $i = 1, ..., k$, where $S_i := T_{i,\oplus n_i}$, $i = 1, ..., k$, the density of $S = (S_1, ..., S_k)$ with respect to $\bigotimes_{i=1}^k \boldsymbol{\nu}_i$ is given by

$$h_\theta(s) = \exp\{\langle \theta, s \rangle - \textstyle\sum_{i=1}^k n_i K(\theta_i)\}, \quad s \in \mathbb{R}^k,\ \theta \in \Delta^k. \tag{9.10}$$

Whenever the sample sizes $n_1, ..., n_k$ are not equal the distribution of S lacks symmetry, which is reflected in $\sum_{i=1}^k n_i K(\theta_i)$ and $\bigotimes_{i=1}^k \boldsymbol{\nu}_i = \bigotimes_{i=1}^k \boldsymbol{\nu}^{\otimes n_i} \circ S_i^{-1}$.

Example 9.1. We consider the problem of selecting from k normal populations $\mathsf{N}(\mu_i, \sigma_i^2)$, $i = 1, ..., k$, one that has the largest mean $\mu_{[k]} = \max\{\mu_1, ..., \mu_k\}$, where the σ_i^2 are known. Hence the model of the type $\mathcal{M}_{us}$ in (9.5) is given by

$$\mathcal{M}_{Normal1} = \textstyle\bigotimes_{i=1}^k (\mathbb{R}^{n_i}, \mathfrak{B}_{n_i}, (\mathsf{N}^{\otimes n_i}(\mu_i, \sigma_i^2))_{\mu_i \in \mathbb{R}}). \tag{9.11}$$

We recall that in view of $\varphi_{\mu,\sigma^2}(x) \propto \exp\{-\mu x/\sigma^2\}$, $\mathsf{N}(\mu, \sigma^2)$ is an exponential family with generating statistic $T(x) = x/\sigma^2$ and natural parameter $\theta = \mu$. The sufficient statistic V in (9.7) and the induced model in (9.9) are given, respectively, by

$$\begin{aligned} V &= (\sigma_1^{-2} \sum\nolimits_{j=1}^{n_1} X_{1,j}, ..., \sigma_k^{-2} \sum\nolimits_{j=1}^{n_k} X_{k,j}), \\ \mathcal{M}_{Normal2} &= (\mathbb{R}^k, \mathfrak{B}_k, (\textstyle\bigotimes_{i=1}^k \mathsf{N}(n_i \mu_i \sigma_i^{-2}, n_i \sigma_i^{-2}))_{\mu \in \mathbb{R}^k}), \end{aligned}$$

which is obviously equivalent to

$$\mathcal{M}_{Normal3} = (\mathbb{R}^k, \mathfrak{B}_k, (\textstyle\bigotimes_{i=1}^k \mathsf{N}(\mu_i, \sigma_i^2/n_i))_{\mu \in \mathbb{R}^k}). \tag{9.12}$$

The reduced model based on $\overline{T} = (\overline{T}_1, ..., \overline{T}_k) = (T_{1,\oplus n_1}/n_1, ..., T_{k,\oplus n_k}/n_k)$, is also of the type $\mathcal{M}_{ss}$.

Example 9.2. Suppose that we want to find one of the k Bernoulli populations $(1-p_i)\boldsymbol{\delta}_0 + p_i\boldsymbol{\delta}_1$ that has the largest parameter $p_{[k]} = \max\{p_1, ..., p_k\}$, where n_i observations are taken from the population with index i, $i = 1, ..., k$. Hence the model of the type $\mathcal{M}_{us}$ in (9.5), with $n = n_1 + \cdots + n_k$, is given by

$$\mathcal{M}_{us} = (\mathbb{R}^n, \mathfrak{B}_n, (\bigotimes_{i=1}^k ((1-p_i)\boldsymbol{\delta}_0 + p_i\boldsymbol{\delta}_1)^{\otimes n_i})_{(p_1,...,p_k)\in(0,1)^k}).$$

According to Problem 1.7, the family $((1-p)\boldsymbol{\delta}_0 + p\boldsymbol{\delta}_1)_{p\in(0,1)}$ is a one-parameter exponential family with generating statistic $T(x) = x$ and $\theta = \ln(p/(1-p))$. As $((1-p_i)\boldsymbol{\delta}_0 + p_i\boldsymbol{\delta}_1)^{*n_i} = \mathsf{B}(n_i, p_i)$ we see that that model in (9.9) turns into

$$\mathcal{M}_{Binomial} = (\mathbb{R}^k, \mathfrak{B}_k, (\bigotimes_{i=1}^k \mathsf{B}(n_i, p_i))_{(p_1,...,p_k)\in(0,1)^k}).$$

9.2 Optimal Point Selections

9.2.1 Point Selections, Loss, and Risk

Point selection rules are decisions on which of the k populations is best. To specify what is meant by "best", in the general selection model $\mathcal{M}_s$ from (9.2) we assume that, according to the goal of the experimenter, a functional $\kappa : \Delta \to \mathbb{R}$ has been chosen where a population i_0 is considered to be best if $i_0 \in \mathsf{M}_\kappa(\theta)$ with $\mathsf{M}_\kappa(\theta)$ from (9.1). Although there may be more than one best population, a point selection rule (see Definition 3.8) selects exactly one population, and thus the decision space is $\mathcal{D}_{pt} = \{1, ..., k\}$. Given the model $\mathcal{M}_s$ from (9.2) a point selection rule D is a stochastic kernel $\mathsf{D}(A|x)$, $A \in \mathfrak{P}(\{1, ..., k\})$, $x = (x_1, ..., x_k) \in \mathsf{X}_{i=1}^k \mathcal{X}_i$. Setting

$$\begin{aligned} \varphi_i(x) &= \mathsf{D}(\{i\}|x), \quad x \in \mathsf{X}_{i=1}^k \mathcal{X}_i,\ i = 1, ..., k, \\ \mathsf{D}(A|x) &= \sum\nolimits_{i=1}^k \varphi_i(x)\boldsymbol{\delta}_i(A), \quad A \subseteq \{1, ..., k\},\ x \in \mathsf{X}_{i=1}^k \mathcal{X}_i, \end{aligned}$$

we may identify the stochastic kernel D with $\varphi = (\varphi_1, ..., \varphi_k)$, where

$$\varphi_i : \mathsf{X}_{i=1}^k \mathcal{X}_i \to_m [0,1], \quad \sum\nolimits_{i=1}^k \varphi_i(x) = 1. \tag{9.13}$$

An equivalent condition is $\varphi : \mathcal{X} \to_m \mathbf{S}_k^c$ where $\mathbf{S}_k^c$ is the closed unit simplex from (1.13) whose elements are the vectors $(p_1, ..., p_k)$ with $p_i \geq 0$ and $\sum_{i=1}^k p_i = 1$. For brevity φ is also called a selection rule or a selection.

Let $L : \Delta^k \times \mathcal{D}_{pt} \to \mathbb{R}$ be any loss function. As the integration with respect to $\mathsf{D}(\cdot|x)$ is a weighted sum the risk of a selection rule φ under L is given by

$$\mathsf{R}(\theta, \varphi) = \sum\nolimits_{i=1}^k L(\theta, i) \int \varphi_i(x) P_\theta(dx) = \sum\nolimits_{i=1}^k L(\theta, i)\mathsf{E}_\theta\varphi_i, \ \theta \in \Delta^k. \tag{9.14}$$

In view of the general goal, which is based on (9.1), the loss functions used for selections depend on $\theta = (\theta_1, ..., \theta_k) \in \Delta^k$ only through $(\kappa(\theta_1), ..., \kappa(\theta_k))$. As the decision space is finite our standard assumption that $L(\theta, \cdot)$ be bounded from below for every $\theta \in \Delta^k$ is automatically fulfilled here.

In the special case of $k = 2$ we see that $\varphi = \varphi_1$ is a test and $\varphi_2 = 1 - \varphi$. In this case the selection problem reduces to the problem of testing the hypotheses $\mathsf{H}_0 : \kappa(\theta_2) \geq \kappa(\theta_1)$ versus $\mathsf{H}_A : \kappa(\theta_2) < \kappa(\theta_1)$, where φ is the probability of rejecting H_0.

If there exists a sufficient statistic V, then we may restrict our search for optimal selections to selections that depend on the data only through V. The following proposition is a direct consequence of Theorem 4.66; see also Remark 4.68.

Proposition 9.3. *If the model $\mathcal{M}_s$ in (9.2) is dominated and $V : \mathcal{X} \to_m \mathbb{R}^k$ is a sufficient statistic, then for every selection rule $\varphi : \mathcal{X} \to_m \mathbf{S}_k^c$ for the model $\mathcal{M}_s$ there exists a selection rule $\psi : \mathbb{R}^k \to_m \mathbf{S}_k^c$ for the model (9.6) with*

$$\mathsf{R}(\theta, \varphi) = \sum\nolimits_{i=1}^{k} L(\theta, i)\mathsf{E}_\theta \varphi_i = \sum\nolimits_{i=1}^{k} L(\theta, i)\mathsf{E}_\theta \psi_i(V) = \mathsf{R}(\theta, \psi(V)).$$

Example 9.4. Let $\mathcal{M}_{ne} = (\mathcal{X}, \mathfrak{A}, (P_\vartheta)_{\vartheta \in \Delta})$, $\Delta \subseteq \mathbb{R}$, be an exponential family from (1.7) with generating statistic T. Then $V = (T_{1,\oplus n_1}, ..., T_{k,\oplus n_k})$ in (9.7) is sufficient, so that all selection rules can be based on V or, equivalently, on $\overline{T} = (\overline{T}_1, ..., \overline{T}_k) = (T_{1,\oplus n_1}/n_1, ..., T_{k,\oplus n_k}/n_k)$.

In the literature, since the beginning in the 1950s, the favorite loss function for selections has been the *zero–one loss*,

$$L_{0,1}(\theta, i) = 1 - I_{\mathsf{M}_\kappa(\theta)}(i), \quad \theta = (\theta_1, ..., \theta_k) \in \Delta^k, \; i = 1, ..., k, \tag{9.15}$$

with $\mathsf{M}_\kappa(\theta)$ from (9.1). For this loss the risk of a selection rule φ is

$$\mathsf{R}(\theta, \varphi) = 1 - \sum\nolimits_{i=1}^{k} I_{\mathsf{M}_\kappa(\theta)}(i)\mathsf{E}_\theta \varphi_i, \quad \theta \in \Delta^k. \tag{9.16}$$

Denote by $\kappa_{[1]}(\theta) \leq \cdots \leq \kappa_{[k]}(\theta)$ the ordered values of $\kappa(\theta_1), ..., \kappa(\theta_k)$.

As in (3.1) we suppose that A and $X = (X_1, ..., X_k)$ are random variables on $(\Omega, \mathfrak{F}, \mathbb{P}_\theta)$ such that $\mathcal{L}(A, X) = \mathsf{D} \otimes P_\theta$, where $\mathsf{D}(B|x) = \sum_{i=1}^{k} \varphi_i(x)\boldsymbol{\delta}_i(B)$ is the conditional distribution of A, given $X = x$. Then

$$\mathbb{P}_\theta(A \in \mathsf{M}_\kappa(\theta)) = \sum\nolimits_{i=1}^{k} I_{\mathsf{M}_\kappa(\theta)}(i) \int \varphi_i(x) P_\theta(dx), \quad \theta \in \Delta^k.$$

Therefore, as in (3.13),

$$P_{cs}(\theta, \varphi) := \sum\nolimits_{i=1}^{k} I_{\mathsf{M}_\kappa(\theta)}(i)\mathsf{E}_\theta \varphi_i, \quad \theta \in \Delta^k, \tag{9.17}$$

is called the *probability of a correct selection* (PCS).

Because $P_{cs}(\theta, \varphi) = 1 - \mathsf{R}(\theta, \varphi)$, maximizing the PCS is equivalent to minimizing the risk. Let

$$\Delta^* = \{\theta : \kappa_{[k-1]}(\theta) < \kappa_{[k]}(\theta), \quad \theta \in \Delta^k\}$$

be the set of parameters where the best population is unique. For $\theta \in \Delta^*$ the set $\mathsf{M}_\kappa(\theta)$ is a singleton, say $\mathsf{M}_\kappa(\theta) = \{i^*(\theta)\}$, and it holds

$$P_{cs}(\theta, \varphi) = \mathsf{E}_\theta \varphi_{i^*(\theta)}, \quad \theta \in \Delta^*.$$

In the *indifference zone approach* of Bechhofer (1954) the PCS of a selection rule is required to be at least P^* on the so-called *preference zone*

$$\Delta_\delta = \{\theta : \kappa_{[k-1]}(\theta) \leq \kappa_{[k]}(\theta) - \delta, \quad \theta \in \Delta^k\}, \tag{9.18}$$

where $\delta > 0$ and $P^* \in (1/k, 1)$ are fixed given. On the set $\Delta^k \backslash \Delta_\delta$, which is called the *indifference zone*, the performance of a selection rule is considered to be of no importance.

One way to construct selection rules is to start with a statistic $S = (S_1, ..., S_k) : \mathcal{X} \to_m \mathbb{R}^k$ for which $\kappa(\theta_i) > \kappa(\theta_j)$ implies that $S_i > S_j$ is in some way more likely to be expected than $S_i < S_j$. Typically S_i is an estimator of $\kappa(\theta_i)$, $i = 1, ..., k$. Set

$$\mathsf{M}(s) = \underset{i \in \{1,...,k\}}{\arg\max}\, s_i, \quad s = (s_1, ..., s_k) \in \mathbb{R}^k, \tag{9.19}$$

and let $|\mathsf{M}(s)|$ denote the number of elements in $\mathsf{M}(s)$.

Definition 9.5. *For the model (9.2) the selection rule*

$$\varphi_S^{nat}(x) = \frac{1}{|\mathsf{M}(S(x))|}(I_{\mathsf{M}(S(x))}(1), ..., I_{\mathsf{M}(S(x))}(k)), \quad x \in \mathsf{X}_{i=1}^k \mathcal{X}_i, \tag{9.20}$$

is called the natural selection rule based on the statistic S. For $\mathsf{X}_{i=1}^k \mathcal{X}_i = \mathbb{R}^k$ and $S(t) \equiv t$ we call

$$\varphi^{nat}(t) = \frac{1}{|\mathsf{M}(t)|}(I_{\mathsf{M}(t)}(1), ..., I_{\mathsf{M}(t)}(k)), \quad t \in \mathbb{R}^k,$$

the natural selection rule.

The natural selection rule φ_S^{nat} has been considered previously, within limited settings, in Example 3.11 regarding its optimality, and in Example 5.32 regarding its permutation invariance. In the following example the risk of the natural selection rule is studied for a normal location model with a common sample size n for the k populations.

Problem 9.6.* If $Z_1, ..., Z_r$ are i.i.d. with continuous c.d.f. F and distribution P, then for every $h : \mathbb{R} \to_m \mathbb{R}_+$ it holds

$$\mathbb{E}h(\max_{1 \leq j \leq r} Z_j) = r \int h(t) F^{r-1}(t) P(dt).$$

For any subset $\Delta \subseteq \mathbb{R}$ of the real line we denote by

$$\Delta_r^k = \{\theta : \theta \in \Delta^k,\ \theta_{[k]} = \theta_{[k-1]} = \cdots = \theta_{[k-r+1]} > \theta_{[k-r]} \geq \cdots \geq \theta_{[1]}\} \quad (9.21)$$

the set of all k-dimensional vectors for which r components are tied for the largest value. Especially Δ_1^k is the set of all vectors for which there exists exactly one largest value, and

$$\Delta_k^k = \{\theta : \theta \in \Delta^k,\ \theta_1 = \cdots = \theta_k\}$$

is the diagonal. We study the zero–one risk or the PCS of a natural selection rule that is based on the statistic S.

Proposition 9.7. *Consider the model (9.4) and $S = (S_1, ..., S_k)$ with $S_i(x) = T(x_i)$, where $T : \mathcal{X} \to_m \mathbb{R}$ and $x = (x_1, ..., x_k) \in \mathcal{X}^k$. Set $Q_\theta = P_\theta \circ T^{-1}$ and suppose that the c.d.f. $F_\theta(t) = P_\theta(T \leq t)$ is continuous for every $\theta \in \Delta$. Then*

$$P_{cs}(\theta, \varphi_S^{nat}) = r \int (\prod\nolimits_{j \neq [k]} F_{\theta_{[j]}}(t)) Q_{\theta_{[k]}}(dt), \quad \theta \in \Delta_r^k. \quad (9.22)$$

Proof. As the natural selection rule and the zero–one loss are permutation invariant we may assume that $\theta_1 \leq \cdots \leq \theta_k$. If $\theta \in \Delta_r^k$, then $\mathsf{M}(\theta) = \{k - r + 1, ..., k\}$.

By the independence of the populations and the continuity of the c.d.f. the natural selection rule

$$\varphi_{i,S}^{nat} = \frac{1}{|\mathsf{M}(S_1, ..., S_k)|} I_{\mathsf{M}(S_1,...,S_k)}(i)$$

takes on only the values 0 or 1, $\bigotimes_{i=1}^k P_{\theta_i}$-a.s., and it holds

$$\begin{aligned} P_{cs}(\theta, \varphi_S^{nat}) &= \sum\nolimits_{i=1}^k I_{\mathsf{M}(\theta)}(i) \mathsf{E}_\theta \varphi_{i,S}^{nat} \\ &= \sum\nolimits_{i=k-r+1}^k (\bigotimes\nolimits_{i=1}^k P_{\theta_i})(S_i > \max\nolimits_{j \neq i} S_j). \end{aligned}$$

Using $\{S_{i_1} > \max_{j \neq i_1} S_j\} \cap \{S_{i_2} > \max_{j \neq i_2} S_j\} = \emptyset$ for $i_1 \neq i_2$ we get

$$P_{cs}(\theta, \varphi_S^{nat}) = (\bigotimes\nolimits_{i=1}^k P_{\theta_i})(\max\nolimits_{1 \leq i \leq k-r} S_i < \max\nolimits_{k-r+1 \leq j \leq k} S_j).$$

Let Q be the distribution of $\max_{k-r+1 \leq j \leq k} S_j$. Then by the Problems 3.9 and 9.6

$$\begin{aligned} &(\bigotimes\nolimits_{i=1}^k P_{\theta_i})(\max\nolimits_{1 \leq i \leq k-r} S_i < \max\nolimits_{k-r+1 \leq j \leq k} S_j) = \int (\prod\nolimits_{j=1}^{k-r} F_{\theta_j}(t)) Q(dt) \\ &= r \int (\prod\nolimits_{j=1}^{k-r} F_{\theta_j}(t)) F_{\theta_k}^{r-1}(t)(P_{\theta_k} \circ T^{-1})(dt) \\ &= r \int (\prod\nolimits_{j=1}^{k-1} F_{\theta_{[j]}}(T(x))) P_{\theta_{[k]}}(dx). \end{aligned}$$

■

Example 9.8. We consider the selection model (9.12) in the balanced case, that is, where $n_1 = \cdots = n_k = n$ and $\sigma_1^2 = \cdots = \sigma_k^2 = \sigma^2$. A reduction by sufficiency shows that it suffices to consider the model

$$\mathcal{M} = (\mathbb{R}^k, \mathfrak{B}_k, \bigotimes_{i=1}^k (\mathsf{N}(\mu_i, \sigma^2))_{\mu \in \mathbb{R}^k}), \quad \mu = (\mu_1, ..., \mu_k).$$

To apply the previous proposition we denote by $X_i : \mathbb{R}^k \to \mathbb{R}$ the projections and set $\mathcal{X} = \mathbb{R}$, $T(x_i) = x_i$, and $S = (X_1, ..., X_k)$. Note that X_i has the distribution $\mathsf{N}(\mu_i, \sigma^2)$ and

$$\Delta_r^k = \mathbb{R}_r^k = \{\mu : \mu = (\mu_1, ..., \mu_k),\ \mu_{[k]} = \cdots = \mu_{[k-r+1]} > \mu_{[k-r]}\}.$$

The natural selection rule φ^{nat} in Definition 9.5 selects the population with the largest X_i. We get from (9.22) that the probability of a correct selection is given by

$$\begin{aligned} P_{\mathsf{N},cs}(\mu, \sigma^2, \varphi^{nat}) &= r \int \prod\nolimits_{i=1}^{k-1} \Phi_{\mu_{[i]},\sigma^2}(t) \varphi_{\mu_{[k]},\sigma^2}(t) dt \\ &= r \int \prod\nolimits_{i=1}^{k-1} \Phi((\mu_{[k]} - \mu_{[i]})/\sigma + s) \varphi(s) ds, \quad \mu \in \mathbb{R}_r^k. \end{aligned} \tag{9.23}$$

This yields

$$\inf\nolimits_{\mu \in \mathbb{R}_r^k} P_{\mathsf{N},cs}(\mu, \sigma^2, \varphi^{nat}) = r \int \Phi^{k-1}(s) \varphi(s) ds = \frac{r}{k}, \tag{9.24}$$

$$\inf\nolimits_{\mu \in \mathbb{R}^k} P_{\mathsf{N},cs}(\mu, \sigma^2, \varphi^{nat}) = \min_{1 \le r \le k} \inf\nolimits_{\mu \in \mathbb{R}_r^k} P_{\mathsf{N},cs}(\mu, \sigma^2, \varphi^{nat}) = \frac{1}{k}, \tag{9.25}$$

where the value of the integral follows from Problem 9.6 with $h = 1$. If we have n observations in each population, then by a reduction by sufficiency we have only to switch to σ^2/n. Let now $\delta > 0$ and $P^* \in (1/k, 1)$ be fixed given. Then according to (9.23) the infimum PCS of φ^{nat} on $\mathbb{R}_{1,\delta}^k = \{\mu : \mu_{[k-1]} \le \mu_{[k]} - \delta,\ \mu \in \mathbb{R}^k\}$ is attained at the least favorable parameter configuration (LFC) $\mu_{[1]} = \cdots = \mu_{[k-1]} = \mu_k - \delta$, and the P^*-condition turns out to be

$$\int \Phi(\sqrt{n}\delta/\sigma + z)^{k-1} \varphi(z) dz \ge P^*.$$

The left-hand side is increasing in n and tends to 1 as $n \to \infty$. This can be utilized to determine the smallest common sample size n for which the P^*-condition is met.

The no-data rule which selects each population with probability $1/k$ has a PCS that is always equal to r/k whenever r populations are tied for the best, $r = 1, ..., k$. This may serve as a justification for requiring that $P^* \in (1/k, 1)$ and for adopting the indifference zone approach.

Another loss function for selections that is often used is the *linear loss*,

$$L_{lin}(\theta, i) = \kappa_{[k]}(\theta) - \kappa(\theta_i), \quad \theta \in \Delta^k,\ i = 1, ..., k. \tag{9.26}$$

The risk of a selection rule φ under this loss is

$$\mathsf{R}(\theta, \varphi) = \kappa_{[k]}(\theta) - \sum\nolimits_{i=1}^k \kappa(\theta_i) \mathsf{E}_\theta \varphi_i, \quad \theta \in \Delta^k,$$

which is the difference between the value of κ of the parameter of a best population and its expected value of the parameter of the selected population. In this case an optimal selection rule would maximize $\sum_{i=1}^{k} \kappa(\theta_i)\mathsf{E}_\theta\varphi_i$, $\theta \in \Delta^k$.

Next we consider minimum average risk and especially Bayes selection rules ψ, utilizing the general framework of Section 3.4. According to Remark 3.30 we may restrict ourselves to Bayes selection rules.

If $(\Delta, \mathfrak{B}_\Delta)$ is a Borel space, then $(\Delta^k, \mathfrak{B}_\Delta^{\otimes k})$ is also a Borel space (see Lemma B.41 in Schervish (1995)) and we can find a stochastic kernel $\boldsymbol{\Pi}$: $\mathfrak{B}_\Delta^{\otimes k} \times (\mathsf{X}_{i=1}^k \mathcal{X}_i) \to_k [0,1]$ such that

$$\mathsf{P}(dx|\theta)\Pi(d\theta) = \boldsymbol{\Pi}(d\theta|x)(\mathsf{P}\Pi)(dx).$$

For a statistic $V : \mathsf{X}_{i=1}^k \mathcal{X}_i \to_m \mathbb{R}^k$ and the reduced model $(\mathbb{R}^k, \mathfrak{B}_k, (Q_\theta)_{\theta\in\Delta^k})$ from (9.6) we introduce for the stochastic kernel $\mathsf{Q} = (Q_\theta)_{\theta\in\Delta^k}$ the posterior $\boldsymbol{\Lambda}_\mathsf{Q}$ by

$$\mathsf{Q}(dt|\theta)\Pi(d\theta) = \boldsymbol{\Lambda}_\mathsf{Q}(d\theta|t)(\mathsf{Q}\Pi)(dt). \tag{9.27}$$

For any sufficient statistic V the posterior for the original model can be factorized by V. This is Bayes sufficiency which has been introduced by Definition 4.60.

If the populations are independent so that $P_\theta = \bigotimes_{i=1}^k P_{i,\theta_i}$, the kernels P and P_i are defined by P_θ and P_{i,θ_i}, respectively, and the prior Π is a product distribution $\Pi = \bigotimes_{i=1}^k \Pi_i$, then we introduce the posteriors $\boldsymbol{\Pi}_i : \mathfrak{B}_\Delta \times \mathcal{X}_i \to_k [0,1]$ of the populations by

$$\mathsf{P}_i(dx_i|\theta_i)\Pi_i(d\theta_i) = \boldsymbol{\Pi}_i(d\theta_i|x_i)(\mathsf{P}_i\Pi_i)(dx_i), \quad i = 1, ..., k. \tag{9.28}$$

Problem 9.9. For the independent model (9.3), that is, where $P_\theta = \bigotimes_{i=1}^k P_{i,\theta_i}$ and $\Pi = \bigotimes_{i=1}^k \Pi_i$, it holds $\mathsf{P} \otimes \Pi = \bigotimes_{i=1}^k (\mathsf{P}_i \otimes \Pi_i) = \bigotimes_{i=1}^k (\boldsymbol{\Pi}_i \otimes (\mathsf{P}_i\Pi_i))$, and $\boldsymbol{\Pi} : \mathfrak{B}_\Delta^{\otimes k} \times (\mathsf{X}_{i=1}^k \mathcal{X}_i) \to_k [0,1]$ satisfies, with $x = (x_1, ..., x_k)$,

$$\boldsymbol{\Pi}(d\theta|x) = \bigotimes_{i=1}^k \boldsymbol{\Pi}_i(d\theta_i|x_i), \quad \mathsf{P}\Pi\text{-a.s.} \tag{9.29}$$

We call (see Definition 3.33)

$$r(\Pi, i|x) = \int L(\theta, i)\boldsymbol{\Pi}(d\theta|x) \tag{9.30}$$

the posterior risk of selecting population $i \in \{1, ..., k\}$ at $x \in \mathsf{X}_{i=1}^k \mathcal{X}_i$. According to (3.27) the posterior risk at x for a selection rule ψ is given by

$$\mathsf{r}(\Pi, \psi|x) = \sum_{i=1}^k r(\Pi, i|x)\psi_i(x). \tag{9.31}$$

For $r(\Pi, i|x)$ from (9.30) we set

$$M_\Pi^{pt}(x) = \underset{i\in\{1,...,k\}}{\arg\min}\, r(\Pi, i|x).$$

The next proposition is a direct consequence of Theorem 3.37 and the representation of the posterior risk in (9.31).

Proposition 9.10. *Let $(\Delta, \mathfrak{B}_\Delta)$ be a Borel space and $L(\cdot, i) : \Delta \to_m \mathbb{R}_+$, $i = 1, ..., k$. A selection rule $\psi : \mathcal{X} \to_m \mathbf{S}_k^c$ is a Bayes rule for the selection model (9.2) if and only if*

$$\sum\nolimits_{i \in M_\Pi^{pt}(x)} \psi_i(x) = 1, \quad \mathsf{P}\Pi\text{-a.s.}$$

Remark 9.11. Under the assumptions of the above proposition there are of course nonrandomized Bayes selection rules. For example, $\psi^{nr} = (\psi_1^{nr}, ..., \psi_k^{nr})$, defined by $\psi_j^{nr}(x) = 1$ if $j = \min\{i : i \in M_\Pi^{pt}(x)\}$ and $\psi_j^{nr}(x) = 0$ otherwise, is a nonrandomized Bayes selection rule.

Next we evaluate the set $M_\Pi^{pt}(x)$ for special loss functions.

Corollary 9.12. *Under the assumptions of Proposition 9.10, for any prior $\Pi \in \mathcal{P}(\mathfrak{B}_\Delta^{\otimes k})$ and any $\kappa : \Delta \to_m \mathbb{R}$, it holds,*

$$M_\Pi^{pt}(x) = \underset{i \in \{1,...,k\}}{\arg\max}\, \boldsymbol{\Pi}(\{\theta : \kappa(\theta_i) = \kappa_{[k]}(\theta)\}|x), \quad \text{if } L = L_{0,1}, \tag{9.32}$$

$$M_\Pi^{pt}(x) = \underset{i \in \{1,...,k\}}{\arg\max} \int \kappa(\theta_i) \boldsymbol{\Pi}(d\theta|x), \quad \text{if } L = L_{lin}, \tag{9.33}$$

where $\theta = (\theta_1, ..., \theta_k)$, and where in the second case $\int |\kappa(\theta_i)| \Pi(d\theta) < \infty$, $i = 1, ..., k$, is assumed to hold.

Proof. If $L = L_{0,1}$, then $L(\theta, i) = 1 - I_{\mathsf{M}_\kappa(\theta)}(i)$, and

$$r(\Pi, i|x) = 1 - \boldsymbol{\Pi}(\{\theta : i \in \mathsf{M}_\kappa(\theta)\}|x) = 1 - \boldsymbol{\Pi}(\{\theta : \kappa(\theta_i) = \kappa_{[k]}(\theta)\}|x),$$

which proves the first statement. The proof of the second statement is similar. ■

To deal with (9.32) in concrete situations the following technical tool for stochastically ordered distributions proves useful.

Proposition 9.13. *Let $\Theta_1, ..., \Theta_k$ be independent random variables with values in $\mathbb{R}$ and $\mathcal{L}(\Theta_1) \preceq \cdots \preceq \mathcal{L}(\Theta_k)$. Then*

$$\mathbb{P}(\Theta_1 = \Theta_{[k]}) \le \cdots \le \mathbb{P}(\Theta_k = \Theta_{[k]}).$$

Proof. Let Q_i be the distribution of Θ_i, $i = 1, ..., k$. For $a < b$ it holds

$$\begin{aligned}
\mathbb{P}(\Theta_a = \Theta_{[k]}) &= \int \prod\nolimits_{i=1, i\neq a}^k Q_i((-\infty, t]) Q_a(dt) \\
&\le \int \prod\nolimits_{i=1, i\neq a,b}^k Q_i((-\infty, t]) Q_a((-\infty, t]) Q_a(dt) \\
&\le \int \prod\nolimits_{i=1, i\neq b}^k Q_i((-\infty, t]) Q_b(dt) = \mathbb{P}(\Theta_b = \Theta_{[k]}),
\end{aligned}$$

where the first inequality follows from $\mathcal{L}(\Theta_a) \preceq \mathcal{L}(\Theta_b)$, and the second from Proposition 2.7. ■

Applications to some specific parametric families are considered next. The first example is taken from Gupta and Miescke (1988).

Example 9.14. We consider the problem of selecting the population with the largest mean in the selection model (9.11). For fixed $\nu_i \in \mathbb{R}$ and $\delta_i^2 > 0$, let the prior for Θ, the random version of $\mu = (\mu_1, ..., \mu_k) \in \mathbb{R}^k$, be $\Pi = \bigotimes_{i=1}^k \mathsf{N}(\nu_i, \delta_i^2)$, that is, the product of the conjugate prior from Lemma 1.37. The posterior distribution $\boldsymbol{\Lambda}$, in view of Lemma 1.37 and (9.29), is

$$\boldsymbol{\Pi} = \bigotimes_{i=1}^k \mathsf{N}\left(\alpha_i + \beta_i T_{i,\oplus n_i}, \tau_i^2\right), \quad T_{i,\oplus n_i}(x) = \sum_{j=1}^{n_i} x_{i,j},$$

$$\alpha_i = \frac{\sigma_i^2 \nu_i}{\sigma_i^2 + n_i \delta_i^2}, \quad \beta_i = \frac{\delta_i^2}{\sigma_i^2 + n_i \delta_i^2}, \quad \tau_i^2 = \frac{\sigma_i^2 \delta_i^2}{\sigma_i^2 + n_i \delta_i^2}, \quad i = 1, ..., k.$$

Then under the loss $L_{0,1}$, the set (9.32) is given by

$$M_{\Pi}^{pt}(x) = \underset{1 \le i \le k}{\arg\max} \mathbb{P}(\widetilde{\Theta}_i = \widetilde{\Theta}_{[k]}), \quad \text{where}$$

$$(\widetilde{\Theta}_1, ..., \widetilde{\Theta}_k) \sim \bigotimes_{i=1}^k \mathsf{N}\left(\alpha_i + \beta_i T_{i,\oplus n_i}, \tau_i^2\right).$$

If for some $\tau^2 < \sigma_i^2/n_i$, $i = 1, ..., k$, we choose δ_i^2 so as to satisfy

$$\frac{1}{\delta_i^2} + \frac{n_i}{\sigma_i^2} = \frac{1}{\tau^2}, \quad i = 1, ..., k, \tag{9.34}$$

i.e., that the sum of the precisions provided by the prior and the sample mean are equal for the k populations, then $\tau_1^2 = \cdots = \tau_k^2 = \tau^2$, and by Proposition 9.13,

$$\underset{1 \le i \le k}{\arg\max} \mathbb{P}(\widetilde{\Theta}_i = \widetilde{\Theta}_{[k]}) = \underset{1 \le i \le k}{\arg\max} (\alpha_i + \beta_i T_{i,\oplus n_i}).$$

In this case the natural selection rule φ_S^{nat} based on the statistic

$$S = (\alpha_1 + \beta_1 \sum_{j=1}^{n_1} x_{1,j}, ..., \alpha_k + \beta_k \sum_{j=1}^{n_k} x_{k,j})$$

is a Bayes selection rule. Specializing further, if $n_i = n$, $\sigma_i^2 = \sigma^2$, $\nu_i = \nu$, and $\delta_i^2 = \delta^2$, $i = 1, ..., k$, then φ_S^{nat} selects the population with the largest mean $n^{-1} \sum_{j=1}^n x_{i,j}$.

Other related work for normal distributions has been done by Berger and Deely (1988), Bansal and Gupta (1997), Bansal and Misra (1999), and Bansal and Miescke (2002).

Problem 9.15.* Consider the problem of selecting the binomial population with the largest success probability in the selection model $(\mathcal{X}^k, \mathfrak{P}(\mathcal{X}^k), \bigotimes_{i=1}^k \mathsf{B}(n, p_i))$, where the observations are $x_1, ..., x_k \in \mathcal{X} = \{0, 1, ..., n\}$. For fixed $\alpha, \beta > 0$, let the prior of Ξ, the random version of $p = (p_1, ..., p_k) \in (0,1)^k$, be $\mathsf{Be}(\alpha, \beta)^{\otimes k}$, that is, the k-fold product of the conjugate prior from Example 1.45. Show that under either of the loss functions $L_{0,1}$ or L_{lin} every Bayes selection rule selects in terms of the largest value of $x_1, ..., x_k$.

Results for the unbalanced model $(\mathsf{X}_{i=1}^k \mathcal{X}_i, \mathfrak{P}(\mathsf{X}_{i=1}^k \mathcal{X}_i), \bigotimes_{i=1}^k \mathsf{B}(n_i, p_i))$, with $\mathcal{X}_i = \{0, 1, ..., n_i\}$, $i = 1, ..., k$, by Abughalous and Miescke (1989) are presented in Example 9.41.

9.2.2 Point Selections in Balanced Models

When the sizes of the independent samples from the k populations are equal, and the underlying family of distributions has monotone likelihood ratio, or is even an exponential family, then the natural selection rule is the best permutation invariant selection rule under any loss function L that satisfies (9.42) and (9.43). This is essentially the content of the *Bahadur–Goodman–Lehmann–Eaton theorem.* It was established, starting with Bahadur (1950) and Bahadur and Goodman (1952), by Lehmann (1966) and Eaton (1967a).

We recall that Π_k denotes the group of permutations γ of $(1, ..., k)$. For any measurable space $(\mathcal{T}, \mathfrak{T})$ we set $u_\gamma(t) = (t_{\gamma(1)}, ..., t_{\gamma(k)})$, $t = (t_1, ..., t_k) \in \mathcal{T}^k$, $\gamma \in \Pi_k$. We call a measure μ defined on $(\mathcal{T}^k, \mathfrak{T}^{\otimes k})$ *permutation invariant* if $\mu \circ u_\gamma^{-1} = \mu$, $\gamma \in \Pi_k$. Let $(\mathcal{S}, \mathfrak{S})$ be another measurable space, where we set $u_\gamma(B) = \{u_\gamma(s) : s \in B\}$, $B \in \mathfrak{S}^{\otimes k}$, $\gamma \in \Pi_k$. Assume $\boldsymbol{\Gamma} : \mathfrak{S}^{\otimes k} \times \mathcal{T}^k \to_k [0,1]$ is a stochastic kernel and that $\mu \in \mathcal{M}^\sigma(\mathfrak{T}^{\otimes k})$ is permutation invariant.

Definition 9.16. *We say that the stochastic kernel $\boldsymbol{\Gamma}$ is μ-a.e. permutation invariant if for every fixed $\gamma \in \Pi_k$, $B \in \mathfrak{S}^{\otimes k}$,*

$$\boldsymbol{\Gamma}(u_\gamma(B)|u_\gamma(t)) = \boldsymbol{\Gamma}(B|t) \tag{9.35}$$

holds μ-a.e. with respect to t. We say that $\boldsymbol{\Gamma}$ is permutation invariant if (9.35) holds for every $t \in \mathcal{T}^k$, $B \in \mathfrak{S}^{\otimes k}$.

It follows from (9.35) and the standard extension technique that for every $h : \mathbb{R}^k \to_m \mathbb{R}_m$,

$$\int h(u_\gamma(s))\boldsymbol{\Gamma}(ds|t) = \int h(s)\boldsymbol{\Gamma}(ds|u_\gamma(t)), \quad \gamma \in \Pi_k. \tag{9.36}$$

Kernels with the above invariance properties are of importance in this section.

We consider now situations where larger observations are in some way more likely to occur at larger parameters. To make this idea precise, let $\mathcal{S}$ and $\mathcal{T}$ be Borel subsets of $\mathbb{R}$, and let $t^{(i,j)} \in \mathbb{R}^k$ be the vector that is obtained from $t \in \mathbb{R}^k$ by exchanging the coordinates t_i and t_j.

Definition 9.17. *Let μ be a permutation invariant probability measure on $(\mathcal{T}^k, \mathfrak{T}^{\otimes k})$, and $\boldsymbol{\Gamma} : \mathfrak{B}_{\mathcal{S}}^{\otimes k} \times \mathcal{T}^k \to_k [0,1]$ be a stochastic kernel that is permutation invariant. We say that $\boldsymbol{\Gamma}$ has the DT property if for every $g : \mathbb{R}^k \to_m \mathbb{R}_+$, every $i, j \in \{1, ..., k\}$, and every $t \in \mathcal{T}^k$ it holds,*

$$\begin{aligned} &\int g(s)I_{[t_i,\infty)}(t_j)I_{[s_i,\infty)}(s_j)\boldsymbol{\Gamma}(ds|t^{(i,j)}) \\ &\leq \int g(s)I_{[t_i,\infty)}(t_j)I_{[s_i,\infty)}(s_j)\boldsymbol{\Gamma}(ds|t). \end{aligned} \tag{9.37}$$

If the latter condition only holds μ-a.s. for every g, then we say that $\boldsymbol{\Gamma}$ has the DT property μ-a.s.

The next lemma is of central importance. It shows that for a random vector (S,T) with $\mathcal{L}(S,T) = \boldsymbol{\Gamma} \otimes \Pi$, and the posterior $\boldsymbol{\Pi}$ defined by $\mathcal{L}(T,S) = \boldsymbol{\Pi} \otimes \boldsymbol{\Gamma}\Pi$, the kernel $\boldsymbol{\Pi}$ has the same permutation invariance and DT properties as $\boldsymbol{\Gamma}$ in $\mathcal{L}(S,T)$. Typically in applications, S represents the data or a function of them, whereas T represents the random parameter in the Bayes model. By establishing this type of symmetry between S and T the task of minimizing the posterior risk of selection rules is facilitated.

Lemma 9.18. *Let $\mathcal{S}$ and $\mathcal{T}$ be Borel subsets of $\mathbb{R}$, and let $\Pi \in \mathcal{P}(\mathfrak{B}_{\mathcal{T}}^{\otimes k})$ be permutation invariant. If $\boldsymbol{\Gamma} : \mathfrak{B}_{\mathcal{S}}^{\otimes k} \times \mathcal{T}^k \to_k [0,1]$ is permutation invariant and has the DT property, then there exists a kernel $\boldsymbol{\Pi}$ that is permutation invariant, has the DT property $\boldsymbol{\Gamma}\Pi$-a.s., and satisfies*

$$\int [\int h(s,t)\boldsymbol{\Gamma}(ds|t)]\Pi(dt) = \int [\int h(s,t)\boldsymbol{\Pi}(dt|s)](\boldsymbol{\Gamma}\Pi)(ds) \tag{9.38}$$

for every $h : \mathcal{S} \times \mathcal{T} \to_m \mathbb{R}_+$.

Proof. Let $\boldsymbol{\Pi}_0$ be any kernel that satisfies (9.38). It follows from the permutation invariance of $\boldsymbol{\Gamma}$ and Π that the permutation invariant kernel

$$\boldsymbol{\Pi}(B|t) = \frac{1}{k!} \sum\nolimits_{\gamma \in \Pi_k} \boldsymbol{\Pi}_0(u_\gamma(B)|u_\gamma(t))$$

also satisfies (9.38). With γ as the permutation that exchanges i and j, and leaves the other $k-2$ arguments unchanged, we see that condition (9.37) implies

$$\begin{aligned}
&\int [\int h(s,t) I_{[t_j,\infty)}(t_i) I_{[s_i,\infty)}(s_j) \boldsymbol{\Gamma}(ds|t)]\Pi(dt) \\
&\leq \int [\int h(s,t) I_{[t_i,\infty)}(t_j) I_{[s_i,\infty)}(s_j) \boldsymbol{\Gamma}(ds|t)]\Pi(dt)
\end{aligned}$$

for every $h : \mathcal{S}^k \times \mathcal{T}^k \to_m \mathbb{R}_+$. Hence

$$\begin{aligned}
&\int [\int h(s,t) I_{[t_j,\infty)}(t_i) I_{[s_i,\infty)}(s_j) \boldsymbol{\Pi}(dt|s)(\boldsymbol{\Gamma}\Pi)(ds) \\
&\leq \int [\int h(s,t) I_{[t_i,\infty)}(t_j) I_{[s_i,\infty)}(s_j) \boldsymbol{\Pi}(dt|s)(\boldsymbol{\Gamma}\Pi)(ds).
\end{aligned}$$

As h is arbitrary we get that for every $g : \mathcal{T}^k \to_m \mathbb{R}_+$ there is a $\boldsymbol{\Gamma}\Pi$-null set, say N_g, such that

$$\int g(t) I_{[t_j,\infty)}(t_i) I_{[s_i,\infty)}(s_j) \boldsymbol{\Pi}(dt|s) \leq \int g(t) I_{[t_i,\infty)}(t_j) I_{[s_i,\infty)}(s_j) \boldsymbol{\Pi}(dt|s)$$

for all $s \notin N_g$. ∎

The kernel $\boldsymbol{\Gamma}$ is often defined by densities with respect to a σ-finite permutation invariant measure. In this case the conditions (9.35) and (9.37) are consequences of properties of the corresponding densities.

Problem 9.19.* Let $\mathcal{S}$ and $\mathcal{T}$ be Borel subsets of $\mathbb{R}$. Let μ be a σ-finite permutation invariant measure on $(\mathcal{S}^k, \mathfrak{B}_{\mathcal{S}}^{\otimes k})$ and $(g_t)_{t\in\mathcal{T}^k}$ a family of probability densities such that $g_t(s)$ is measurable in (s,t) and satisfies

$$g_{u_\gamma(t)}(u_\gamma(s)) = g_t(s), \quad s \in \mathcal{S}^k, t \in \mathcal{T}^k, \gamma \in \Pi_k. \tag{9.39}$$

Then the kernel $\boldsymbol{\Gamma}(A|t) := \int I_A(s) g_t(s)\mu(ds)$, $A \in \mathfrak{B}_{\mathcal{S}}^{\otimes k}$, is permutation invariant. Moreover, if

$$g_{t(i,j)}(s) \le g_t(s), \quad t_i \le t_j,\ s_i \le s_j,\ s \in \mathcal{S}^k,\ t \in \mathcal{T}^k,\ i,j \in \{1,...,k\}, \tag{9.40}$$

then the kernel $\boldsymbol{\Gamma}$ has the DT property.

The typical situation where (9.39) and (9.40) occurs is the following.

Problem 9.20. Let $\mathcal{S}$ and $\mathcal{T}$ be Borel subsets of $\mathbb{R}$. Let $g_t(s) = \prod_{i=1}^k f_{t_i}(s_i)$, $s \in \mathcal{S}^k$, $t \in \mathcal{T}^k$, where $f_b(a)$, $a \in \mathcal{S}$, $b \in \mathcal{T}$, is positive and has the following property. $f_{b_2}(a)/f_{b_1}(a)$ is nondecreasing in $a \in \mathcal{S}$ for every $b_1, b_2 \in \mathcal{T}$ with $b_1 < b_2$. Then $g_t(s)$, $s \in \mathcal{S}^k$, $t \in \mathcal{T}^k$, satisfies (9.39) and (9.40). As a one-parameter exponential family has MLR (see Example 2.13) the density $h_\theta(s)$ in (9.10) satisfies (9.39) and (9.40) if and only if $n_1 = \cdots = n_k$.

Remark 9.21. The combination of (9.39) and (9.40) is called *property M* in Eaton (1967a), and *decreasing in transposition property* (DT) in Hollander, Proschan, and Sethuraman (1977).

Some special cases are considered below. We start with the case of independent populations. For that we need a special property of families with the MLR property. The following is a generalized version of Problem 2.22.

Problem 9.22.* If $(P_\theta)_{\theta\in\Delta}$ is a family of distributions on $(\mathbb{R}, \mathfrak{B})$ that has the MLR property in the identity, then for every $\theta_1, \theta_2 \in \Delta$ with $\theta_1 \le \theta_2$ and every function $h : \mathbb{R}^2 \to_m \mathbb{R}_+$ it holds for $x = (x_1, x_2)$,

$$\int h(x) I_{(x_1,\infty)}(x_2)(P_{\theta_2} \otimes P_{\theta_1})(dx) \le \int h(x) I_{(x_1,\infty)}(x_2)(P_{\theta_1} \otimes P_{\theta_2})(dx). \tag{9.41}$$

Problem 9.23.* Let $(P_\vartheta)_{\vartheta\in\Delta}$ be a one-parameter family of distributions on $(\mathbb{R}, \mathfrak{B})$ that is a stochastic kernel, and let $P_\theta = \bigotimes_{i=1}^k P_{\theta_i}$, $\theta = (\theta_1, ..., \theta_k) \in \Delta^k$. Then the stochastic kernel $\mathsf{P} = (P_\theta)_{\theta\in\Delta^k}$ satisfies (9.35). Moreover, if the stochastic kernel $(P_\vartheta)_{\vartheta\in\Delta}$ has the MLR property in the identity, then the stochastic kernel $\mathsf{P} = (P_\theta)_{\theta\in\Delta^k}$ has the DT property.

Now we consider cases where the populations are not independent.

Example 9.24. Suppose that $S_i : \mathbb{R} \to_m \mathbb{R}$, $i = 1, ..., k$, and $\kappa : \Delta \to \mathbb{R}$, with $\Delta \subseteq \mathbb{R}$, are nondecreasing, and that $\boldsymbol{\nu}$ is a σ-finite symmetric measure on $(\mathbb{R}^k, \mathfrak{B}_k)$. If there exists a nonnegative nondecreasing function f and a permutation symmetric function g such that for $x = (x_1, ..., x_k) \in \mathbb{R}^k$ and $\theta = (\theta_1, ..., \theta_k) \in \Delta^k$,

$$\frac{dQ_\theta}{d\boldsymbol{\nu}}(x) = g(\theta) f(\textstyle\sum_{i=1}^k S_i(x_i)\kappa(\theta_i)),$$

then $(dQ_\theta/d\boldsymbol{\nu})(x)$ satisfies the conditions (9.39) and (9.40).

Problem 9.25.* The kernel $\mathsf{K}(B|\theta) = \mathsf{N}(\theta, \Sigma)(B)$ defined by the normal distribution $\mathsf{N}(\theta, \Sigma)$, $\theta \in \mathbb{R}^k$, with a known and nonsingular Σ, is permutation invariant and has the DT property if and only if

$$\Sigma^{-1} = \alpha I + \beta \mathbf{1}\mathbf{1}^T, \quad \alpha > 0, \ \alpha + k\beta > 0.$$

For further results we refer to Eaton (1967a).

Problem 9.26.* The multinomial distributions $\mathsf{M}(n, p)$ with $p \in \mathbf{S}_k^c$ satisfy the conditions (9.39) and (9.40).

Regarding the loss we consider the large class of loss functions $L : \Delta^k \times \mathcal{D}_{pt} \to_m \mathbb{R}_+$ that are permutation invariant, that is,

$$L(\theta, \gamma(i)) = L(u_\gamma(\theta), i), \quad i = 1, ..., k, \ \theta \in \Delta^k, \ \gamma \in \Pi_k, \tag{9.42}$$

and that favor selections of larger values of κ; that is, for $\theta = (\theta_1, ..., \theta_k)$,

$$L(\theta, i) \geq L(\theta, j), \quad \kappa(\theta_i) \leq \kappa(\theta_j), \ i, j \in \{1, ..., k\}, \ \theta \in \Delta^k. \tag{9.43}$$

The key lemma of this subsection is as follows.

Lemma 9.27. *Let $\mathcal{S}$ and $\mathcal{T}$ be Borel subsets of $\mathbb{R}$, and let the loss function L satisfy (9.42) and (9.43), where $\Delta = \mathcal{T}$ and κ is the identical mapping. Assume that $\mathsf{K} : \mathfrak{B}_{\mathcal{T}}^{\otimes k} \times \mathcal{S}^k \to_k [0, 1]$ and $\mu \in \mathcal{P}(\mathfrak{B}_{\mathcal{S}}^{\otimes k})$. If K is μ-a.s. permutation invariant and μ-a.s. DT, then*

$$\mathsf{M}(s) \subseteq \underset{i \in \{1, ..., k\}}{\arg\min} \int L(t, i) \mathsf{K}(dt|s), \quad \mu\text{-a.s.} \tag{9.44}$$

Proof. Let $i, j \in \{1, ..., k\}$ be fixed. It follows from $\kappa(\vartheta) = \vartheta$ and condition (9.43) that $L(t, i) - L(t, j) = 0$ for $t = (t_1, ..., t_k)$ and $t_i = t_j$. Hence

$$\begin{aligned}\int [L(t, i) - L(t, j)] \mathsf{K}(dt|s) &= \int [L(t, i) - L(t, j)] I_{(t_i, \infty)}(t_j) \mathsf{K}(dt|s) \\ &\quad + \int [L(t, i) - L(t, j)] I_{(t_j, \infty)}(t_i) \mathsf{K}(dt|s).\end{aligned}$$

Let $t^{(i,j)} \in \mathbb{R}^k$ be the vector that is obtained from $t \in \mathbb{R}^k$ by exchanging the coordinates t_i and t_j. On $A = \{s : s = (s_1, ..., s_k), \ s_i \leq s_j\}$ it holds μ-a.s.

$$\begin{aligned}&\int [L(t, i) - L(t, j)] I_{(t_j, \infty)}(t_i) \mathsf{K}(dt|s) \\ &= \int [L(t^{(i,j)}, j) - L(t^{(i,j)}, i)] I_{(t_j, \infty)}(t_i) \mathsf{K}(dt|s) \\ &= \int [L(t, j) - L(t, i)] I_{(t_i, \infty)}(t_j) \mathsf{K}(dt|s^{(i,j)}) \\ &\geq \int [L(t, j) - L(t, i)] I_{(t_i, \infty)}(t_j) \mathsf{K}(dt|s),\end{aligned}$$

where we have used in the first equation (9.42), in the second equation (9.35), and in the inequality (9.43) and (9.37). Consequently, $\int [L(t,i) - L(t,j)]\mathsf{K}(dt|s) \geq 0$, and the proof is completed. ■

As we have explained already in Example 3.11, one cannot find uniformly best selection rules in the class of all selection rules. Consequently we restrict ourselves to selection rules that are permutation invariant, which have been discussed already in a special case in Example 3.11.

Definition 9.28. *For the model in (9.2) a selection rule* $\psi : \mathsf{X}_{i=1}^{k}\mathcal{X}_i \to_m \mathbf{S}_k^c$ *is called permutation invariant if*

$$\psi_i(x_{\gamma(1)},...,x_{\gamma(k)}) = \psi_{\gamma(i)}(x_1,...,x_k), \quad i = 1,...,k,$$

holds for every permutation $\gamma \in \Pi_k$ *and every* $(x_1,...,x_k) \in \mathsf{X}_{i=1}^{k}\mathcal{X}_i$.

The following example is given for clarification.

Example 9.29. For a permutation invariant selection rule ψ the expressions $\psi_{\gamma(i)}(x_{\gamma(1)},...,x_{\gamma(k)})$ and $\psi_i(x_1,...,x_k)$ are in general not equal. This can be seen for $k = 3$. Let $\gamma(1) = 3$, $\gamma(2) = 1$, and $\gamma(3) = 2$. Then $\psi_{\gamma(1)}(x_{\gamma(1)}, x_{\gamma(2)}, x_{\gamma(3)}) = \psi_3(x_3, x_1, x_2)$, which by the permutation invariance of ψ is equal to $\psi_2(x_1, x_2, x_3)$.

Problem 9.30. Verify that the natural selection rule φ_S^{nat} based on a statistic S (see Definition 9.5) is permutation invariant.

The fundamental Bahadur–Goodman–Lehmann–Eaton theorem can now be stated.

Theorem 9.31. ***(Bahadur–Goodman–Lehmann–Eaton)*** *Let* $\Delta \subseteq \mathbb{R}$ *be a Borel set and* κ *the identical mapping. Suppose that for the selection model* $\mathcal{M}_s$ *in (9.2) there exists a sufficient statistic* $V : \mathcal{X} \to_m \mathbb{R}^k$ *such that for* $Q_\theta = P_\theta \circ V^{-1}$ *the family* $\mathsf{Q} = (Q_\theta)_{\theta\in\Delta^k}$ *is a stochastic kernel that is permutation invariant and has the DT property. If the loss function satisfies (9.42) and (9.43), then the natural selection rule* φ_V^{nat} *based on* V *has the following optimality properties.*

(A) Bayes:

φ_V^{nat} *is a Bayes selection rule under every discrete permutation invariant prior.*

(B) Uniformly best invariant:

φ_V^{nat} *is a uniformly best permutation invariant selection rule.*

(C) Minimax:

$\sup_{\theta\in\mathbb{A}} \mathsf{R}(\theta, \varphi_V^{nat}) \leq \sup_{\theta\in\mathbb{A}} \mathsf{R}(\theta, \varphi)$ *for every selection rule* φ *and every permutation invariant* $\mathbb{A} \subseteq \Delta^k$.

(D) Admissible:

φ_V^{nat} *is admissible in the class of all selection rules* φ.

Corollary 9.32. *Suppose that for the balanced selection model $\mathcal{M}_{bs}$ in (9.4) the family of distributions $(P_\vartheta)_{\vartheta\in\Delta}$ on $(\mathcal{X},\mathfrak{A})$ with $\Delta \subseteq \mathbb{R}$ is dominated and has MLR in T, and that the loss function L satisfies (9.42) and (9.43). Then for $\mathcal{M}_{bs}$ the natural selection rule based on $S = (S_1, ..., S_k)$, where $S_i(x) = T(x_i)$, $i = 1, ...k$, $x = (x_1, ..., x_k) \in \mathcal{X}^k$, has the optimality properties stated in the theorem.*

Proof. Due to the reduction by sufficiency in Proposition 9.3 we have only to deal with the model $\mathcal{M}_{ss}$. To show (A) we fix any permutation invariant prior Π. Then in view of Lemma 9.18 the posterior $\boldsymbol{\Pi}$ is $\mathsf{Q}\Pi$-a.s. permutation invariant and $\mathsf{Q}\Pi$-a.s. DT. This means that the kernel $\mathsf{K} = \boldsymbol{\Pi}$ satisfies the conditions in Lemma 9.27 with $\mu = \mathsf{Q}\Pi$ so that (A) follows from (9.44) and Proposition 9.10. (B) follows from (A) and Proposition 5.39.

Next we prove (C). In the class of all permutation invariant selection rules the natural selection rule φ_V^{nat} based on the identity is uniformly best, and thus also minimax, in this class of selection rules for the model $(\mathbb{R}^k, \mathfrak{B}_k, (Q_\theta)_{\theta\in\mathbb{A}})$. (C) follows now from Proposition 5.41 if we take $\mathbb{D}_0$ as the set of all selection rules and Q as the uniform distribution on the set of all permutations.

As to (D), suppose that φ_V^{nat} is not admissible. Then there exists a selection rule ϕ with $\mathsf{R}(\theta,\phi) \le \mathsf{R}(\theta,\varphi_V^{nat})$, $\theta\in\Delta$, and $\mathsf{R}(\widetilde{\theta},\phi) < \mathsf{R}(\widetilde{\theta},\varphi_V^{nat})$ for at least one $\widetilde{\theta}\in\Delta$. We fix one such $\widetilde{\theta}$ and denote by μ the uniform distribution on the orbit $\Delta_{\widetilde{\theta}} = \{u_\gamma(\widetilde{\theta}) : \gamma \in \Pi_k\}$. Apparently, μ is a permutation invariant prior on $\Delta_{\widetilde{\theta}}$ and $\mathsf{r}(\mu,\phi) < \mathsf{r}(\mu,\varphi_V^{nat})$, which is a contradiction to (A).

To prove the corollary, we first remark that in view of Problem 4.54 and the independence of the populations the statistic $S(x) = (T(x_1), ..., T(x_k))$ is sufficient. The permutation invariance and the DT property of Q follow from Problem 9.23. ■

Example 9.33. Suppose that the k populations are independent, and that we have n_i observations from population i, $i = 1, ..., k$. The model is then given in (9.5). Assume that $(P_\vartheta)_{\vartheta\in\Delta}$ is a one-parameter exponential family with generating statistic T. Hence $dP_\vartheta/d\boldsymbol{\mu} = \exp\{\vartheta T - K(\vartheta)\}$. The reduced model

$$(\mathbb{R}^k, \mathfrak{B}_k, \bigotimes_{i=1}^k Q_{i,\theta_i}), \quad Q_{i,\theta_i} = P_{\theta_i}^{\otimes n_i} \circ T_{i,\oplus n_i}^{-1},$$

in (9.8) is equivalent to the original model by Proposition 9.3. We see from (9.9) and (9.10), respectively, that the stochastic kernel $\mathsf{Q} = (Q_\theta)_{\theta\in\Delta^k}$ with $Q_\theta = \bigotimes_{i=1}^k Q_{i,\theta_i}$ is permutation invariant in the sense of Definition 9.16 if and only if the model is balanced (i.e., $n_1 = \cdots = n_k$). If the sample sizes are all equal to n, say, then by Problem 9.23 the kernel Q is, according to Definition 9.17, DT as well since the exponential family $(P_\vartheta^{\otimes n})_{\vartheta\in\Delta}$ has MLR in $T_{\oplus n}$. In this case the natural selection rule φ_S^{nat} based on the statistic $S = (T_{1,\oplus n}, ..., T_{k,\oplus n})$ has the optimality properties (A)–(D). Usually the parameter of interest is not the natural parameter itself. Let Λ be an interval and $\kappa : \Lambda \leftrightarrow_m \Delta$. If κ is nondecreasing, then $P_{\kappa(\eta)}^{\otimes n}$ has again MLR, so that the natural selection rule based on S in turn is optimal in the sense of (A)–(D) for selecting the population with the largest value of $\eta_1, ..., \eta_k$.

Example 9.34. Let us reconsider Examples 9.14 and 9.8, i.e., the problem of selecting the population with the largest mean in the model $(\mathbb{R}^k, \mathfrak{B}_k, \bigotimes_{i=1}^k \mathsf{N}(\mu_i, \sigma^2))$, but where now $\sigma^2 > 0$ is unknown. Because for every fixed σ^2 the assumptions of Corollary 9.32 are satisfied, whereas φ_S^{nat} does not depend on σ^2, it follows that φ_S^{nat} has the optimality properties (A)–(D), even if σ^2 is unknown.

Another aspect of the optimality of the natural selection rule is that, under the assumptions of Theorem 9.31, it is *most economical* in the sense that no other permutation invariant selection rule can have the same PCS with a smaller common sample size. Details in this regard can be found in Hall (1959) and Miescke (1979a).

Now we examine the minimax statement (C) in Theorem 9.31 in more detail. For a fixed permutation invariant set $\mathbb{A} \subseteq \Delta^k \subseteq \mathbb{R}^k$ we call $\theta_0 \in \mathbb{A}$ a *least favorable configuration* (LFC) on $\mathbb{A}$ for the selection rule φ if $\sup_{\theta \in \mathbb{A}} \mathsf{R}(\theta, \varphi) = \mathsf{R}(\theta_0, \varphi)$.

Proposition 9.35. *Assume the assumptions of Corollary 9.32 are satisfied, Δ is an interval, and the zero–one loss $L_{0,1}$ is used for the natural selection rule φ_S^{nat} based on S. Suppose that $F_\vartheta(t) := P_\vartheta(T \le t)$, $t \in \mathbb{R}$, is continuous for every $\vartheta \in \Delta$. For every $\delta > 0$, $\vartheta^* \in \Delta$, and $\vartheta^* - \delta \in \Delta$ we set*

$$\Delta_{r,\delta}^k = \{\theta : \theta \in \Delta^k, \theta_{[k]} = \cdots = \theta_{[k-r+1]} \ge \theta_{[k-r]} + \delta\}.$$

Then

$$\begin{aligned} &\sup_{\theta \in \Delta_{r,\delta}^k} \mathsf{R}(\theta, \varphi_S^{nat}) \qquad (9.45)\\ &= 1 - r[\inf_{\vartheta^*, \vartheta^* - \delta \in \Delta} \int F_{\vartheta^* - \delta}^{k-r}(T(x)) F_{\vartheta^*}^{r-1}(T(x)) P_{\vartheta^*}(dx)]. \end{aligned}$$

Proof. Apply Proposition 9.7 and use the fact that F_ϑ is stochastically nondecreasing in ϑ; see Theorem 2.10. ■

In location models the above minimization can be carried out explicitly. As an example we consider the normal distribution.

Example 9.36. We reconsider the selection problem of Example 9.8 and establish the minimaxity of the natural selection rule in the indifference zone approach. The model is $(\mathbb{R}^k, \mathfrak{B}_k, \bigotimes_{i=1}^k \mathsf{N}(\mu_i, \sigma^2))$ and the loss is the $L_{0,1}$. Then

$$\Delta_{r,\delta}^k = \mathbb{R}_{r,\delta}^k = \{\mu : \mu \in \mathbb{R}^k,\ \mu_{[k]} = \cdots = \mu_{[k-r+1]} \ge \mu_{[k-r]} + \delta\},$$

where $\delta > 0$ is fixed given. As $\Delta_{r,\delta}^k$ is permutation invariant, by Theorem 9.31 (C) the natural selection rule φ^{nat} is minimax on $\Delta_{r,\delta}^k$. As we use the zero–one loss we may operate with $P_{cs}(\mu, \varphi^{nat})$. It follows from (9.23),

$$\begin{aligned} \inf_{\mu \in \mathbb{R}_{r,\delta}^k} P_{\mathsf{N},cs}(\mu, \sigma^2, \varphi^{nat}) &= r \inf_{\mu \in \mathbb{R}_{r,\delta}^k} \int \prod\nolimits_{i=1}^{k-1} \Phi((\mu_{[k]} - \mu_{[i]})/\sigma + s)\varphi(s)ds,\\ &= r \int \Phi^{k-r}(\frac{\delta}{\sigma} + s)\Phi^{r-1}(s)\varphi(s)ds, \quad \mu \in \mathbb{R}_{r,\delta}^k. \qquad (9.46) \end{aligned}$$

This means that every vector for which for any real number ν, $k-r$ components have the value $\nu-\delta$ and the remaining r have the value ν, is a least favorable configuration on $\mathbb{R}^k_{r,\delta}$. We may also consider the probability of a correct selection for $\mu \in \mathbb{R}^k$. Then φ^{nat} is again minimax and by (9.23) the minimax value for the risk or the maximin value for $P_{\mathsf{N},cs}$ is given by

$$\inf_{\mu\in\mathbb{R}^k} P_{\mathsf{N},cs}(\mu,\sigma^2,\varphi^{nat}) = \inf_{\mu\in\mathbb{R}^k} \int \prod_{i=1}^{k-1} \Phi((\mu_{[k]}-\mu_{[i]})/\sigma + s)\varphi(s)ds$$
$$= \int \Phi^{k-1}(s)\varphi(s)ds = \frac{1}{k}.$$

The above proposition can be utilized to find the LFC of φ_S^{nat} on $\Delta_\delta = \{\theta : \theta_{[k]} \geq \theta_{[k-1]} + \delta,\ \ \theta \in \Delta^k\}$. This may still be a difficult task as one has to go through all $\vartheta^* \in \Delta$. Once the LFC has been found then the minimum common sample size n can be determined for which the natural selection rule meets the P^*-condition on Δ_δ, similarly as it has been done in Example 9.8. The LFC of the natural selection rule for stochastically ordered families has been established in Bofinger (1976). In an alternative approach, a lower confidence bound on the PCS of the natural selection rule for location parameter families with MLR has been provided by Kim (1986).

We conclude this section with some historical remarks. In the earlier times of the development, a selection problem was usually formulated for a model with a specific parametric family of distributions such as normal or binomial. A selection rule was then proposed, implemented under some basic requirement such as the P^*-condition in Example 9.8, studied in detail, and perhaps also compared to other existing selection rules. Such an ad hoc rule typically selects in terms of the largest of k estimates, based on an optimal single-sample estimator for the parameter of interest, and is then simply called *the natural selection rule*. This works well and indeed often leads to optimal selection rules. According to Corollary 9.32 this is the case when the populations are independent, the underlying family of distributions has monotone likelihood ratio (see Definition 2.11), and the k samples sizes are equal. However, in other settings this concept is questionable. Let us briefly look at two scenarios in support of the decision theoretic approach.

In the first scenario, suppose that the sample sizes are equal but there is no MLR. This occurs, for example, if the underlying family of distributions is a location parameter family $(Q_\theta)_{\theta\in\mathbb{R}}$ with Lebesgue densities $q_\vartheta(t) = q(t-\vartheta)$, $t,\vartheta \in \mathbb{R}$, where $q(t) > 0$, $t \in \mathbb{R}$, but q is not log-concave; see Proposition 2.20. In this case the statement of Corollary 9.32 is no longer valid. On the other hand, as the family $(Q_\theta)_{\theta\in\mathbb{R}}$ is stochastically nondecreasing (see Example 2.8), Proposition 9.35 can be easily extended to the present case. This means that the "natural selection rule" based on S from Corollary 9.32 can be implemented under the P^*-condition in the indifference zone approach. Again, however, there is no guarantee of optimality. Selection rules that may be reasonable could be based on other sufficient statistics or functions of them, especially estimators of θ that are optimal in some way. References to nonparametric selection rules for location parameter families can be found on p. 68 in Bechhofer, Santner, and Goldsman (1995), and at the end of this chapter.

In the second scenario, suppose that the sample sizes are not all equal, but MLR is present. Here we restrict ourselves to the normal means case with a common variance. The first who cast doubt on using the "natural selection rule" based on the sample means under unequal sample sizes were Lam and Chiu (1976). They showed that if the means are close together its PCS decreases if more observations are taken from the best population. There is a simple explanation for this effect. Using reduction by sufficiency (see (9.12)), the case of unequal sample sizes is equivalent to that of finding the population with largest mean among k normal distributions with different variances. However, such distributions are not comparable in terms of the stochastic semiorder of distributions. A discussion of the behavior of the natural selection rule of not being most economical, and further references, can be found in Tong and Wetzell (1979), Bofinger (1985), and Gupta and Miescke (1988). In the latter it has been shown that the natural selection rule is not minimax under the zero–one loss if the sample sizes are not equal, whereas the no-data rule that selects each population with probability $1/k$ is minimax. This is explained in detail in the next section.

Miescke and Park (1999) have studied the performance of the natural selection rule under normality, i.e., in the setting of Example 9.1. It turns out that under the linear loss (9.26) and a specific prior, utilizing (9.33), it is the unique, up to Lebesgue null sets, Bayes rule and thus admissible. On the other hand, under the zero–one loss it could not be shown whether it is admissible. The method of Blyth (1951), in the form given in Berger (1985), has been utilized. A minimum average risk (generalized Bayes) selection rule has also been considered where the weight measure is the Lebesgue measure.

9.2.3 Point Selections in Unbalanced Models

In this section we consider selection rules for the unbalanced selection model (9.5). Theorem 9.31 is no longer applicable here as the permutation invariance in (9.35) is not given. Now the samples from the k populations contain different types of information on the respective parameters. If we turn to the Bayes approach, then there is an opportunity to counterbalance this by switching to a suitable prior which adjusts the posterior risk accordingly. Such a prior, however, would no longer be permutation invariant. The alternative approach of switching to a suitable loss function (see Remark 3.30), is not pursued here.

To sketch the construction of such a suitable prior we recall that in the proof of part (A) in Theorem 9.31 we had $\mathcal{L}(V, \Theta) = \mathsf{Q} \otimes \Pi$, where Π was a discrete permutation invariant prior and the stochastic kernel Q was permutation invariant and DT. The Bayes risk in this setting is

$$
\begin{aligned}
\mathsf{r}(\Pi,\psi) &= \sum\nolimits_{i=1}^{k} \mathbb{E}\psi_i(V)L(\Theta|i) \\
&= \sum\nolimits_{i=1}^{k} \int [\int \psi_i(t)L(\theta|i)\boldsymbol{\Pi}(d\theta|t)](\mathsf{Q}\Pi)(dt).
\end{aligned}
$$

Under the above conditions on Q and Π, by Lemma 9.18, the posterior $\boldsymbol{\Pi}$ turned out to be permutation invariant and DT. The crucial point in the proof of (A) in Theorem 9.31 was the minimization of $i \mapsto \int L(\theta|i)\boldsymbol{\Pi}(d\theta|t)$ which was taken care of by Lemma 9.27. If now Q fails to be permutation invariant and DT, then we may try to find a suitable prior Π, no longer permutation invariant, such that the posterior $\boldsymbol{\Pi}$ is permutation invariant and DT. If such a prior can be found, then Lemma 9.27 can be utilized in the search for a Bayes selection rule.

Theorem 9.37. *Let $\Delta \subseteq \mathbb{R}$ be a Borel set. Suppose that for the model $\mathcal{M}_s$ in (9.2) there exists a sufficient statistic $V : \bigotimes_{i=1}^{k} \mathcal{X}_i \to_m \mathbb{R}^k$ and that for $Q_\theta = P_\theta \circ V^{-1}$ the family $\mathsf{Q} = (Q_\theta)_{\theta\in\Delta^k}$ is a stochastic kernel. Suppose that there is a prior $\Pi \in \mathcal{P}(\mathfrak{B}_{\Delta^k})$ such that the posterior $\boldsymbol{\Lambda}_{\mathsf{Q}}$ for the reduced model in (9.27) admits the representation*

$$
\boldsymbol{\Lambda}_{\mathsf{Q}}(B|t) = \mathsf{K}(B|S(t)), \quad \mathsf{Q}\Pi\text{-a.s.},\ B \in \mathfrak{B}_{\Delta^k}, \tag{9.47}
$$

with a stochastic kernel $\mathsf{K} : \mathfrak{B}_k \times \Delta^k \to_k [0,1]$ and a statistic $S : \mathbb{R}^k \to \Delta^k$. If K is permutation invariant and DT, and the loss function satisfies (9.42) and (9.43), then the natural selection based on $S(V)$ is a Bayes selection rule.

Proof. The statement follows from Proposition 9.10 and Lemma 9.27. ■

We consider now the problem of finding priors that satisfy the condition in Theorem 9.37 for one-parameter exponential families. Let $(P_\theta)_{\theta\in\Delta}$ be an exponential family on $(\mathcal{X},\mathfrak{A})$ with natural parameter θ and generating statistic T, where $\Delta \subseteq \mathbb{R}$. The set Δ is convex and thus an interval. We assume that the variance of T is positive and that Δ has an inner point, which are the standard assumptions (A1) and (A2), respectively. We recall the family of conjugate priors introduced in Section 1.2. For a σ-finite measure $\boldsymbol{\tau}$ on Δ we set

$$
\begin{aligned}
\Upsilon &= \{(a,b) : a,b \in \mathbb{R},\ L(a,b) = \ln(\int \exp\{\theta b - aK(\theta)\}\boldsymbol{\tau}(d\theta)) < \infty\}, \\
\pi_{a,b}(\theta) &= \exp\{b\theta - aK(\theta) - L(a,b)\}, \quad \theta \in \Delta \subseteq \mathbb{R},\ (a,b) \in \Upsilon, \\
\Pi_{a,b}(B) &= \int_B \exp\{b\theta - aK(\theta) - L(a,b)\}\boldsymbol{\tau}(d\theta), \quad B \in \mathfrak{B}_\Delta,\ (a,b) \in \Upsilon.
\end{aligned}
$$

From Proposition 1.4 we know that $P_\theta^{\otimes n}$ is again an exponential family, now with generating statistic $T_{\oplus n}$. The conjugate priors are again $\Pi_{a,b}$, $(a,b) \in \Upsilon$; see Lemma 1.35. According to (1.43) the posterior distributions under these priors are given by

$$\boldsymbol{\Pi}_{n,a,b}(B|x) = \int_B \pi_{a+n,b+T_{\oplus n}(x)}(\theta)\boldsymbol{\tau}(d\theta) \tag{9.48}$$
$$= \Pi_{a+n,b+T_{\oplus n}(x)}(B), \quad B \in \mathfrak{B}_\Delta,\ x \in \mathcal{X}^n,\ (a,b) \in \Upsilon,$$

where by Lemma 1.35 for every $(a,b) \in \Upsilon$ it holds

$$(a+n, b+T_{\oplus n}(x)) \in \Upsilon,\ \boldsymbol{\mu}^{\otimes n}\text{-a.e.} \tag{9.49}$$

For k populations we set $\theta = (\theta_1, ..., \theta_k)$, $a = (a_1, ..., a_k)$, $b = (b_1, ..., b_k)$, and $n = (n_1, ..., n_k)$, and consider the selection model and the prior, respectively,

$$\mathcal{M}_{se} = (\mathsf{X}_{i=1}^k \mathcal{X}^{n_i}, \bigotimes_{i=1}^k \mathfrak{A}^{\otimes n_i}, (\bigotimes_{i=1}^k P_{\theta_i}^{\otimes n_i})_{\theta \in \Delta^k}), \tag{9.50}$$
$$\Pi_{a,b} = \bigotimes_{i=1}^k \Pi_{a_i,b_i}, \quad (a,b) \in \Upsilon^k.$$

As the populations, as well as the parameters under the prior, are independent the posterior distribution, given $X = x$, for any $x \in \mathsf{X}_{i=1}^k \mathcal{X}^{n_i}$, turns out to be

$$\Pi_{a+n,b+T_\oplus(x)} = \bigotimes_{i=1}^k \Pi_{a_i+n_i,b_i+T_{i,\oplus n_i}(x_i)}, \quad (a,b) \in \Upsilon^k, \tag{9.51}$$

where $T_\oplus(x) = (T_{1,\oplus n_1}(x_1), ..., T_{k,\oplus n_k}(x_k))$ with $T_{i,\oplus n_i}(x_i) = \sum_{j=1}^{n_i} T(x_{i,j})$, $i = 1, ..., k$.

In the next proposition we study Bayes selection rules for the above exponential selection model and its conjugate priors. Let P be the stochastic kernel $(\bigotimes_{i=1}^k P_{\theta_i}^{\otimes n_i})_{\theta \in \Delta^k}$.

Proposition 9.38. *For the selection model (9.50) a selection rule ψ is Bayes with respect to the prior $\Pi_{a,b} = \bigotimes_{i=1}^k \Pi_{a_i,b_i}$, $(a,b) \in \Upsilon^k$, if and only if*

$$\mathsf{P}\Pi_{a,b}(\{x : \sum_{i \in M_{\Pi_{a,b}}^{pt}(x)} \psi_i(x) = 1,\ x \in \mathsf{X}_{i=1}^k \mathcal{X}^{n_i}\}) = 1,$$

where $M_{\Pi_{a,b}}^{pt}(x)$ denotes the set

$$\underset{i \in \{1,...,k\}}{\arg\min} \int L(\theta, i) \exp\{\langle b + T_\oplus(x), \theta\rangle - \sum_{j=1}^k (a_j + n_j)K(\theta_j)\}\boldsymbol{\tau}^{\otimes k}(d\theta).$$

Proof. The statement follows from Proposition 9.10, (9.51), and (9.48), as the factor

$$\exp\{-\sum_{j=1}^k L(a_j + n_j, b_j + T_{j,\oplus n_j}(x_j))\}$$

is independent of θ and therefore irrelevant for the minimization. ■

The next problem deals with the question of under which circumstances the posterior $\bigotimes_{i=1}^k \Pi_{a_i+n_i,b_i+t_i}$ satisfies the conditions in Theorem 9.37.

Problem 9.39.* Suppose that

$$a_1 + n_1 = \cdots = a_k + n_k =: a_0, \tag{9.52}$$

and set $\Upsilon_{a_0} = \{b : (a_0, b) \in \Upsilon\}$. Then the kernel

$$\mathsf{K}(\cdot|t) = (\bigotimes_{i=1}^k \Pi_{a_0,t_i})(\cdot), \quad t = (t_1, ..., t_k) \in \Upsilon_{a_0}^k, \tag{9.53}$$

is permutation invariant and DT.

Now we are ready to deal with the selection model for unequal sample sizes.

Theorem 9.40. *For the selection model (9.50), suppose that (9.52) holds for some a_0 and $b_1, ..., b_k \in \Upsilon_{a_0}$ are fixed. If the loss function satisfies (9.42) and (9.43), then the natural selection rule based on*

$$W := (b_1 + T_{1,\oplus n_1}, ..., b_k + T_{k,\oplus n_k})$$

is a Bayes selection rule for the selection model (9.50) under the prior $\bigotimes_{i=1}^k \Pi_{a_i,b_i}$. In particular, if $b_1 = \cdots = b_k$, then selecting in terms of the largest $T_{i,\oplus n_i}$, $i = 1, ..., k$, is a Bayes selection rule.

Proof. For $\Upsilon_{a_0} = \{b : (a_0, b) \in \Upsilon\}$ it holds $b_i + T_{i,\oplus n_i} \in \Upsilon_{a_0}$, $\boldsymbol{\mu}^{\otimes n_i}$-a.e., by (9.49). Then $S \in \Upsilon_{a_0}^k$ holds $\boldsymbol{\mu}^{\otimes(n_1+\cdots+n_k)}$-a.e., so that S and K in (9.53) satisfy (9.47). By Problem 9.39 the kernel K is permutation invariant and DT. An application of Theorem 9.37 with $S(V) = W$ completes the proof. ∎

When dealing with exponential families that are not given in the natural form a direct application of Theorem 9.37 could also be used instead of first transforming to natural parameters and then using Theorem 9.40.

Example 9.41. Let $X_i \sim \mathsf{B}(n_i, p_i)$, $p_i \in (0, 1)$, $i = 1, ..., k$, be independent binomial random variables, and set $X = (X_1, ..., X_k)$. Suppose we want to select from the populations $\mathsf{B}(n_i, p_i)$, $p_i \in (0, 1)$, $i = 1, ..., k$, one with the largest parameter $p_{[k]} = \max\{p_1, ..., p_k\}$. Hence our selection model is

$$(\mathbb{R}^k, \mathfrak{B}_k, (\bigotimes_{i=1}^k \mathsf{B}(n_i, p_i))_{p\in(0,1)^k}).$$

Let the prior Π of the random version of the parameter $p = (p_1, ..., p_k)$ be the product of conjugate priors; that is, $\Pi = \bigotimes_{i=1}^k \mathsf{Be}(\alpha_i, \beta_i)$ with $\alpha_i, \beta_i > 0$, $i = 1, ..., k$. Using this fact and Example 1.45 we get that under this prior the posterior, given $X_1 = x_1, ..., X_k = x_k$, is

$$\boldsymbol{\Pi}(\cdot|x) = \bigotimes_{i=1}^k \mathsf{Be}(\alpha_i + x_i, \beta_i + n_i - x_i) = \bigotimes_{i=1}^k \mathsf{Be}(s_i, \alpha_i + \beta_i + n_i - s_i),$$

where $x = (x_1, ..., x_k)$ and $s_i = \alpha_i + x_i$, $i = 1, ..., k$. For any fixed $\delta > n_{[k]}$ let the parameters a_i, β_i be chosen as to satisfy

$$\alpha_i + \beta_i + n_i = \delta, \quad i = 1, ..., k. \tag{9.54}$$

Then

$$\boldsymbol{\Pi}(\cdot|x) = \bigotimes\nolimits_{i=1}^{k} \mathsf{Be}(\alpha_i + x_i, \delta - (\alpha_i + x_i)). \tag{9.55}$$

As the Lebesgue density $\mathsf{be}_{s,\delta-s}$ of $\mathsf{Be}(s, \delta - s)$satisfies

$$\mathsf{be}_{s,\delta-s}(p) \propto I_{(0,1)}(p)p^{s-1}(1-p)^{\delta-s-1}$$

we get that the Lebesgue density of $Q_\theta = \bigotimes_{i=1}^{k} \mathsf{Be}(\alpha_i + x_i, \delta - (\alpha_i + x_i))$ with $\theta_i = \alpha_i + x_i$ satisfies the condition in Example 9.24 for $S_i(x_i) = \alpha_i + x_i$. Problem 9.19 implies that $\boldsymbol{\Pi}(\cdot|x)$ in (9.55) has the DT property. Theorem 9.37 yields that the natural selection rule based on $S(x_1, .., x_k) = (\alpha_1 + x_1, ..., \alpha_k + x_k)$ is a Bayes selection rule for the prior $\Pi = \bigotimes_{i=1}^{k} \mathsf{Be}(\alpha_i, \beta_i)$ if the parameters α_i, β_i satisfy (9.54). In particular, if $\alpha_1 = \cdots = \alpha_k$, then selecting in terms of the largest x_i, $i = 1, ..., k$, is a Bayes selection rule. Apparently, for unequal sample sizes, these selection rules differ dramatically from the natural selection rule which selects in terms of the largest value of x_i/n_i, $i = 1, ..., k$. These and further results for selection rules in the binomial case can be found in Abughalous and Miescke (1989).

Problem 9.42. Compare the normal means selection problem in Example 9.14 with the binomial selection problem in Example 9.41.

We now explain, following Gupta and Miescke (1988), the loss of minimaxity under the zero–one loss of the natural selection rule for k normal populations with a common variance $\sigma^2 > 0$ when the sample sizes $n_1, ..., n_k$ are unequal. Technically, these parameters appear only in the form σ^2/n_i, $i = 1, ..., k$, and thus we consider the more general selection model

$$\mathcal{M}_{Normal4} = (\mathbb{R}^k, \mathfrak{B}_k, (\bigotimes\nolimits_{i=1}^{k} \mathsf{N}(\mu_i, \sigma_i^2))_{\mu \in \mathbb{R}^k}), \tag{9.56}$$

where $\sigma_i^2 > 0$, $i = 1, ..., k$. In practical applications, however, one usually deals with the special case of $\sigma_i^2 = \sigma^2/n_i$, $i = 1, ..., k$.

The following auxiliary result is due to Tong and Wetzell (1979), but formulated and proved here differently.

Lemma 9.43. *The function*

$$H(\gamma_1, ..., \gamma_{k-1}) = \int \prod\nolimits_{i=1}^{k-1} \Phi(\gamma_i z)\varphi(z)dz$$

is strictly increasing in $\gamma_i > 0$, $i = 1, ..., k-1$.

Proof.

$$\frac{\partial}{\partial \gamma_1} H(\gamma_1, ..., \gamma_{k-1}) = \int \prod\nolimits_{i=2}^{k-1} \Phi(\gamma_i z) z \varphi(\gamma_1 z)\varphi(z)dz.$$

Combining the two φ-functions, and then integrating by parts, leads to

$$\frac{\partial}{\partial \gamma_1} H(\gamma_1, ..., \gamma_{k-1}) = \frac{1}{\sqrt{2\pi}} \frac{1}{1+\gamma_1^2} \int M(y)\varphi(y)dy,$$

where

$$M(y) = \frac{\partial}{\partial y} \prod\nolimits_{i=2}^{k-1} \Phi(\frac{\gamma_i}{\sqrt{1+\gamma_1^2}} y), \quad y \in \mathbb{R}.$$

Obviously, $M(y) > 0$, $y \in \mathbb{R}$. ■

Now we state the following result.

Theorem 9.44. *For the model in (9.56) consider the problem of selecting a population with the largest mean under the zero–one loss. The natural selection rule* φ^{nat} *is minimax if and only if* $\sigma_1^2 = \sigma_2^2 = \cdots = \sigma_k^2$. *The minimax value of the problem is* $1 - 1/k$.

Proof. Let $\sigma_1^2 = \sigma_2^2 = \cdots = \sigma_k^2$. We know already from Example 9.36 that φ^{nat} is minimax, and that the supremum of the risk of φ^{nat} is $1 - 1/k$.

Suppose now that $\min_{1 \le i \le k} \sigma_i^2 < \max_{1 \le i \le k} \sigma_i^2$, and assume without loss of generality that $\sigma_k^2 = \max_{1 \le i \le k} \sigma_i^2$. Then for $\mu_n = (0, ..., 0, 1/n)$ by Proposition 9.7,

$$\begin{aligned}\lim_{n \to \infty} \mathsf{R}(\mu_n, \varphi^{nat}) &= 1 - \lim_{n \to \infty} \int \prod\nolimits_{i=1}^{k-1} \Phi\Big(\frac{z + 1/n}{\sigma_i}\Big) \frac{1}{\sigma_k} \varphi\Big(\frac{z}{\sigma_k}\Big) dz \\ &= 1 - \int \prod\nolimits_{i=1}^{k-1} \Phi\Big(\frac{z}{\sigma_i}\Big) \frac{1}{\sigma_k} \varphi\Big(\frac{z}{\sigma_k}\Big) dz > 1 - 1/k,\end{aligned}$$

where the inequality follows from Lemma 9.43. On the other hand, the no-data rule $\phi = (1/k, ..., 1/k)$ satisfies $\sup_\mu \mathsf{R}(\mu, \phi) = 1 - 1/k$, and thus the natural selection rule φ^{nat} is not minimax.

Now we show that the minimax value of the problem remains $1 - 1/k$ under $\min_{1 \le i \le k} \sigma_i^2 < \max_{1 \le i \le k} \sigma_i^2 = \sigma_k^2$. Assume without loss of generality that $\sigma_1 = \min_{1 \le i \le k} \sigma_i^2$, and consider the two models

$$\begin{aligned}\mathcal{M} &= (\mathbb{R}^k, \mathfrak{B}_k, (\bigotimes\nolimits_{i=1}^k \mathsf{N}(\mu_i, \sigma_1^2))_{\mu \in \mathbb{R}^k}) \\ \widetilde{\mathcal{M}} &= (\mathbb{R}^k, \mathfrak{B}_k, (\bigotimes\nolimits_{i=1}^k \mathsf{N}(\mu_i, \sigma_i^2))_{\mu \in \mathbb{R}^k}).\end{aligned}$$

Then $\widetilde{\mathcal{M}}$ is a randomization of $\mathcal{M}$, see Problem 4.7. Hence the corresponding minimax values satisfy

$$\inf_\varphi \sup_\mu \mathsf{R}_{\mathcal{M}}(\mu, \varphi) \le \inf_\varphi \sup_\mu \mathsf{R}_{\widetilde{\mathcal{M}}}(\mu, \varphi).$$

φ^{nat} is minimax in the balanced model $\mathcal{M}$ and $\sup_\mu \mathsf{R}_{\mathcal{M}}(\mu, \varphi) = 1 - 1/k$. On the other hand, the no-data rule $\phi = (1/k, ..., 1/k)$ satisfies $\mathsf{R}_{\widetilde{\mathcal{M}}}(\mu, \phi) = 1 - 1/k$, and thus the proof is completed. ∎

For normal populations with unequal variances, sample size allocations have been studied in Dudewicz and Dalal (1975) and Bechhofer, Hayter, and Tamhane (1991). Nonminimaxity of natural decision rules under heteroscedasticity has been shown in Dhariyal and Misra (1994). Selection of the best of k exponential families has been studied by Abughalous and Bansal (1995). Although restricted to families with a quadratic variance function, many of their results can be extended to general one-parameter exponential families. Selecting the best treatment in a generalized linear model has been studied in Bansal and Miescke (2006).

9.2.4 Point Selections with Estimation

After a population has been selected some natural follow-up questions may arise. The first is how large the parameter of the selected population is. A second is how far the parameter of the selected population is away from the largest parameter of the k populations. A third is how large the largest parameter of the k populations is. In this section we mainly deal with the first question, and briefly with the second. References for work on the third question can be found in Gupta and Panchapakesan (1979). Answers to the first question have been provided for specific statistical models by Cohen and Sackrowitz (1988), Gupta and Miescke (1990, 1993), Bansal and Miescke (2002, 2005), and Misra, van der Meulen, and Branden (2006). All of these works have been done in the Bayes approach which is also utilized here. We deal slightly more generally with a functional of the parameters rather than with the parameters.

We start with the general selection model (9.2) and assume that a functional $\kappa : \Delta \to \mathbb{R}$ is given. The new decision space is $\mathcal{D}_{pt,es} = \{1,...,k\} \times \mathbb{R}$ which we equip with the σ-algebra $\mathfrak{D} = \mathfrak{P}(\{1,...,k\}) \otimes \mathfrak{B}$. For any decision $\mathsf{D} : \mathfrak{D} \times (\mathsf{X}_{i=1}^k \mathcal{X}_i) \to_k [0,1]$ and $x = (x_1,...,x_n) \in \mathsf{X}_{i=1}^k \mathcal{X}_i$ we set $\varphi_i(x) = \mathsf{D}(\{i\} \times \mathbb{R}|x)$, and for any fixed kernel $\mathsf{K} : \mathfrak{B} \times (\mathsf{X}_{i=1}^k \mathcal{X}_i) \to_k [0,1]$ we set

$$\mathsf{K}_i(E|x) = \begin{cases} \frac{1}{\varphi_i(x)} \mathsf{D}(\{i\} \times E|x) & \text{if} \quad \varphi_i(x) > 0 \\ \mathsf{K}(E|x) & \text{if} \quad \varphi_i(x) = 0 \end{cases}, \qquad E \in \mathfrak{B},$$

$i = 1,...,k$. Obviously, the $\mathsf{K}_i : \mathfrak{B} \times (\mathsf{X}_{i=1}^k \mathcal{X}_i) \to_k [0,1]$ are stochastic kernels, $\varphi = (\varphi_1,...,\varphi_k)$ is a point selection rule, and it holds

$$\mathsf{D}(C \times E|x) = \sum\nolimits_{i=1}^k \varphi_i(x) \boldsymbol{\delta}_i(C) \mathsf{K}_i(E|x), \quad C \subseteq \{1,...,k\},\ E \in \mathfrak{B}. \tag{9.57}$$

It is clear that conversely, every point selection rule φ, along with a sequence of kernels $\mathsf{K}_1,...,\mathsf{K}_k$, leads to a decision D that is defined by (9.57).

Let $A : \Delta^k \times \{1,...,k\} \to \mathbb{R}_+$ and $B : \mathbb{R} \times \mathbb{R} \to \mathbb{R}_+$, and introduce the loss function

$$L(\theta,(i,t)) = A(\theta,i) + B(\kappa(\theta_i),t_i), \quad \theta \in \Delta^k,\ i = 1,...,k,\ t \in \mathbb{R}^k, \tag{9.58}$$

where $\theta = (\theta_1,...,\theta_k)$ and $t = (t_1,...,t_k)$.

Example 9.45. For the loss function in (9.58) one could take for part A the zero–one loss from (9.15) or the linear loss from (9.26). For part B one could take $B(\theta_i,t_i) = |\kappa(\theta_i) - t_i|^q$, $i = 1,...,k$, for some fixed $q > 0$.

From (9.57) we see that the risk of a decision D is

$$\begin{aligned} \mathsf{R}(\theta,\mathsf{D}) = {} & \sum\nolimits_{i=1}^k A(\theta,i) \int \varphi_i(x) P_\theta(dx) \\ & + \sum\nolimits_{i=1}^k \int \varphi_i(x) [\int B(\kappa(\theta_i),t_i) \mathsf{K}_i(dt_i|x)] P_\theta(dx), \quad \theta \in \Delta^k. \end{aligned}$$

The special structure of this risk allows us to minimize the risk, as a function of D, in two consecutive steps. At any fixed $x \in \mathsf{X}_{i=1}^{k}\mathcal{X}_i$, in the first step best estimators $\mathsf{K}_i^0(\cdot|x)$ for $i = 1, ..., k$ are determined, provided there are any. Set

$$C_i(\theta, x) = \int (B(\kappa(\theta_i), t_i)\mathsf{K}_i^0(dt_i|x), \quad i = 1, ..., k,$$

and consider

$$\arg\min_{i \in \{1,\ldots,k\}} [A(\theta, i) + C_i(\theta, x)], \quad \theta \in \Delta^k.$$

As this set depends on θ we cannot construct an optimal selection rule φ by concentrating the discrete distribution $\varphi(x)$ on this set. To overcome the dependence on θ we use the Bayes approach. Suppose that $\Delta \subseteq \mathbb{R}$ is a Borel set, κ is the identical mapping, and the nonnegative functions A and B in (9.58) are measurable. We assume that the populations are independent so that $P_\theta = \bigotimes_{i=1}^{k} P_{i,\theta_i}$. Suppose that $\mathsf{P}_i = (P_{i,\theta_i})_{\theta_i \in \Delta}$ is a stochastic kernel, $i = 1, ..., k$. Let the prior Π on $(\Delta^k, \mathfrak{B}_{\Delta}^{\otimes k})$ be given by $\Pi = \bigotimes_{i=1}^{k} \Pi_i$, so that the posteriors of the populations are determined by (9.28). Due to the independence of the populations and priors we may restrict ourselves to estimators K_i that depend on x only through x_i, $i = 1, ..., k$. The Bayes risk is given by

$$\begin{aligned}
\mathsf{r}(\Pi, \mathsf{D}) &= \sum\nolimits_{i=1}^{k} A(\theta, i) \int [\int \varphi_i(x) P_\theta(dx)] \Pi(d\theta) \\
&+ \sum\nolimits_{i=1}^{k} \int [\int \varphi_i(x) (\int B(\theta_i, t_i) \mathsf{K}(dt_i|x)) P_\theta(dx)] \Pi(d\theta) \\
&= \sum\nolimits_{i=1}^{k} A(\theta, i) \int [\int \varphi_i(x) \bigotimes\nolimits_{i=1}^{k} \boldsymbol{\Pi}_i(d\theta_i|x_i)] \bigotimes\nolimits_{j=1}^{k} (\mathsf{P}_j \Pi_j)(dx_j) \\
&+ \sum\nolimits_{i=1}^{k} \int [\int \varphi_i(x) (\int B(\theta_i, t_i) \mathsf{K}(dt_i|x)) (\bigotimes\nolimits_{i=1}^{k} P_{i,\theta_i})(dx)] \Pi_i(d\theta_i).
\end{aligned}$$

As we have seen in (3.5) there is usually no need for considering randomized estimators. In this regard for Bayes estimators we refer to Corollary 3.40. Let, for every $i = 1, ..., k$, $S_i^0 : \mathcal{X}_i \to_m \mathbb{R}$ be a nonrandomized estimator that minimizes the posterior risk, so that $\mathsf{P}_i \Pi_i$-a.s. for every K_i it holds

$$\begin{aligned}
\int B(\theta_i, S_i^0(x_i)) \boldsymbol{\Pi}_i(d\theta_i|x_i) &\leq \int [\int B(\theta_i, t_i) \mathsf{K}_i(dt_i|x)] \boldsymbol{\Pi}_i(d\theta_i|x_i) \qquad (9.59) \\
&= \int [\int B(\theta_i, t_i) \boldsymbol{\Pi}_i(d\theta_i|x_i)] \mathsf{K}_i(dt_i|x).
\end{aligned}$$

Then for every decision D we have

$$\begin{aligned}
\mathsf{r}(\Pi, \mathsf{D}) \geq \sum\nolimits_{i=1}^{k} \int \varphi_i(x) [\int (A(\theta, i) \bigotimes\nolimits_{j=1}^{k} \boldsymbol{\Pi}_j(d\theta_j|x_j) \qquad (9.60) \\
+ \int B(\theta_i, S_i^0(x_i)) \boldsymbol{\Pi}_i(d\theta_i|x_i)] \bigotimes\nolimits_{j=1}^{k} (\mathsf{P}_j \Pi_j)(dx_j).
\end{aligned}$$

Introduce, for $i = 1, ..., k$,

$$V_i(x) = \int A(\theta, i) \bigotimes_{j=1}^{k} \boldsymbol{\Pi}_j(d\theta_j|x_j) + \int B(\theta_i, S_i^0(x_i))\boldsymbol{\Pi}_i(d\theta_i|x_i), \tag{9.61}$$

and let

$$M_{\Pi}^{pt,es}(x) = \underset{i\in\{1,...,k\}}{\arg\min} V_i(x), \quad x \in \mathsf{X}_{i=1}^{k}\mathcal{X}_i. \tag{9.62}$$

Fix any mapping $\mathsf{p}^0 : \mathsf{X}_{i=1}^{k}\mathcal{X}_i \to_m \{1, ..., k\}$ with $\mathsf{p}^0(x) \in M_{\Pi}^{pt,es}(x)$, $x \in \mathsf{X}_{i=1}^{k}\mathcal{X}_i$. Then

$$V_{\mathsf{p}^0(x)}(x) = \min_{i\in\{1,...,k\}} V_i(x_i), \quad x \in \mathsf{X}_{i=1}^{k}\mathcal{X}_i. \tag{9.63}$$

Finally, let the nonrandomized decision $\mathsf{d}^0 : \mathsf{X}_{i=1}^{k}\mathcal{X}_i \to \{1, ..., k\} \times \mathbb{R}^k$ be given by

$$\mathsf{d}^0(x) = (\mathsf{p}^0(x), S_1^0(x_1), ..., S_k^0(x_k)), \quad x \in \mathsf{X}_{i=1}^{k}\mathcal{X}_i. \tag{9.64}$$

Then $S_{\mathsf{p}^0(x)}^0(x)$ is the estimator of the parameter of the selected population. For the class of all point selections with estimation,

$$\mathbb{D}_0 = \{\mathsf{D} : \mathsf{D}(C \times E|x) = \sum\nolimits_{i=1}^{k} \varphi_i(x)\boldsymbol{\delta}_i(C)\mathsf{K}_i(E|x), \ C \subseteq \{1, ..., k\},$$
$$E \in \mathfrak{B}, \ \varphi : \mathsf{X}_{i=1}^{k}\mathcal{X}_i \to_m \mathbf{S}_k^c, \ \mathsf{K}_i : \mathfrak{B} \times \mathcal{X}_i \to_k [0,1], \ i = 1, ..., k\},$$

the subsequent statement follows directly from (9.60).

Theorem 9.46. *Assume that $\Delta \subseteq \mathbb{R}$ is a Borel set, the family $P_i = (P_{i,\theta_i})_{\theta_i \in \mathbb{R}}$ is a stochastic kernel, $i = 1, ..., k$, and A and B from (9.58) are measurable. If there are estimators $S_1^0, ..., S_k^0$ that satisfy (9.59), then for every point selection rule φ^0 that satisfies*

$$\sum\nolimits_{i=1}^{k} I_{M_{\Pi}^{pt,es}(x)}(i)\varphi_i^0(x) = 1, \quad \mathsf{P}\Pi\text{-a.s.}, \tag{9.65}$$

the decision

$$\mathsf{D}^0(C \times E|x) = \sum\nolimits_{i=1}^{k} \varphi_i^0(x)\boldsymbol{\delta}_i(C)\boldsymbol{\delta}_{S_i^0(x_i)}(E),$$

$C \subseteq \{1, ..., k\}$, $E \in \mathfrak{B}$, $x \in \mathsf{X}_{i=1}^{k}\mathcal{X}_i$, satisfies $\mathsf{r}(\Pi, \mathsf{D}^0) \leq \mathsf{r}(\Pi, \mathsf{D})$ for every decision $\mathsf{D} \in \mathbb{D}_0$. Especially the nonrandomized decision d^0 in (9.64) satisfies $\mathsf{r}(\Pi, \mathsf{d}^0) \leq \mathsf{r}(\Pi, \mathsf{D})$.

As we have seen above, each of the decisions D^0 and d^0 has to be evaluated at $X = x$ in two consecutive steps. First one has to find the Bayes estimators $S_1^0(x), ..., S_k^0(x)$. Then $V_1(x), ..., V_k(x)$ are determined from which an optimal point selection $\varphi^0(x)$ can be derived. Hence we see that rather than estimating after selection, the Bayes approach leads to selecting after estimation. The statement of the theorem is taken from Gupta and Miescke (1990), which is a straightforward extension of the work by Cohen and Sackrowitz (1988).

The following example is taken from Gupta and Miescke (1990), where the case of k independent normal populations has been studied in detail.

Example 9.47. In the setting of Examples 9.1 and 9.14, let us consider Bayes simultaneous selection and estimation rules under the same prior $\Pi = \bigotimes_{i=1}^{k} \mathsf{N}(\nu_i, \delta_i^2)$ with $\nu_i \in \mathbb{R}$ and $\delta_i^2 > 0$, $i = 1, ..., k$. Let the loss function be given by

$$L_1(\mu, (i, t_i)) = a(\mu_{[k]} - \mu_i) + \rho(\mu_i - t_i), \quad \mu \in \mathbb{R}^k,\ t_i \in \mathbb{R},\ i = 1, ..., k,$$

where $a > 0$ is fixed given and ρ is a nonnegative function with $\rho(-x) = \rho(x)$, $\rho(0) = 0$, and ρ is nondecreasing for $x > 0$. We know from Anderson's lemma (see Proposition 3.62, or (3.45)) that for every $Y \sim \mathsf{N}(\mu, \sigma^2)$ the function $a \to \mathbb{E}\rho(Y - a)$ attains its minimum at μ; that is, that

$$\mu \in \arg\min_{a \in \mathbb{R}} \mathbb{E}\rho(Y - a). \tag{9.66}$$

We know from Example 9.14 that at any $x = (x_1, ..., x_k)$ with $x_i = (x_{i,1}, ..., x_{i,n_i}) \in \mathbb{R}^{n_i}$, $i = 1, ..., k$,

$$\begin{aligned}
&\boldsymbol{\Pi}(\cdot|x) = \bigotimes_{i=1}^{k} \boldsymbol{\Pi}_i(\cdot|x_i) = \bigotimes_{i=1}^{k} \mathsf{N}\left(S_i^0(x_i), \tau_i^2\right), \quad \text{where} \\
&\tau_i^2 = \frac{\sigma_i^2 \delta_i^2}{\sigma_i^2 + n_i \delta_i^2}, \quad S_i^0(x_i) = \frac{\sigma_i^2}{\sigma_i^2 + n_i \delta_i^2}\nu_i + \frac{n_i \delta_i^2}{\sigma_i^2 + n_i \delta_i^2} \frac{1}{n_i} \sum\nolimits_{j=1}^{n_i} x_{i,j},
\end{aligned}$$

$i = 1, ..., k$. By (9.66) the S_i^0 are the Bayes estimators in (9.60) and

$$\begin{aligned}
\int B(s_i, S_i^0(x_i)) \boldsymbol{\Pi}_i(ds_i|x_i) &= \int \rho(s_i - S_i^0(x_i)) \varphi_{S_i^0(x_i), \tau_i^2}(s_i) ds_i \\
&= \int \rho(\frac{s_i}{\tau_i}) \varphi_{0,1}(s_i) ds_i.
\end{aligned}$$

Hence the statistics V_i in (9.61) are

$$\begin{aligned}
V_i(x) = &\int a[\max_{j=1,\dots,k} \frac{s_j - S_j^0(x_j)}{\tau_j} - \frac{s_i - S_i^0(x_i)}{\tau_i}] \mathsf{N}^{\otimes k}(0,1)(ds_1, ..., ds_k) \\
&+ \int \rho(\frac{s_i}{\tau_i}) \varphi_{0,1}(s_i) ds_i, \quad i = 1, ..., k.
\end{aligned}$$

The optimal nonrandomized decision d^0 is then determined by (9.63) and (9.64).

Some interesting special cases of the above example are considered in the following problems.

Problem 9.48.* In Example 9.47, let $\sigma_1^2 = \cdots = \sigma_1^2 = \sigma^2$, say, and in the loss function L_1 let $\rho(t) = |t|$, $t \in \mathbb{R}$. Consider the following three special cases. (a) The limiting case of the prior where $\delta_i^2 \to \infty$, $i = 1, ..., k$. This leads to Bayes decisions under a noninformative prior. (b) The case where the prior variances are proportional to the respective variances of the sample means, i.e., where $\delta_i^2 = b\sigma^2/n_i$, $i = 1, ..., k$, for some fixed $b > 0$. (c) The case where the posterior is permutation invariant and DT, that is, where (9.34) holds.

Problem 9.49. In Example 9.47, consider the same three special cases of Problem 9.48 for $\sigma_1^2 = \cdots = \sigma_1^2 = \sigma^2$, but now under the loss function

$$L_2(\mu, (i, t_i)) = a(\mu_{[k]} - \mu_i)^2 + (\mu_i - t_i)^2, \quad \mu \in \mathbb{R}^k,\ t_i \in \mathbb{R},\ i = 1, ..., k.$$

The following example is taken from Gupta and Miescke (1993), where the case of k binomial populations has been studied in detail.

Example 9.50. In the setting of Example 9.41, let us consider Bayes simultaneous selection and estimation rules under the same prior $\Pi = \bigotimes_{i=1}^{k} \mathsf{Be}(\alpha_i, \beta_i)$ with $\alpha_i, \beta_i > 0$, $i = 1, ..., k$. Let the loss function be given by

$$L(p, (i, t_i)) = p_{[k]} - p_i + b(p_i - t_i)^2, \quad p \in (0,1)^k,\ t_i \in \mathbb{R},\ i = 1, ..., k,$$

where $b > 0$ is fixed given. It follows from Example 1.45 that

$$\boldsymbol{\Pi}(\cdot|x) = \bigotimes_{i=1}^{k} \boldsymbol{\Pi}_i(\cdot|x_i) = \bigotimes_{i=1}^{k} \mathsf{Be}(\alpha_i + x_i, \beta_i + n_i - x_i).$$

Recall that the expectation and variance of $\mathsf{Be}(\alpha, \beta)$ are

$$\frac{\alpha}{\alpha+\beta} \quad \text{and} \quad \frac{\alpha\beta}{(\alpha+\beta)^2(\alpha+\beta+1)},$$

respectively. For $i \in \{1, ..., k\}$, under the given loss, the Bayes estimator S_i^0 and its associated posterior risk $\int B(p_i, S_i^0(x_i))\boldsymbol{\Pi}_i(dp_i|x_i)$ are the conditional expectation and, up to the factor b, the conditional variance.

$$S_i^0(x_i) = \frac{\alpha_i + x_i}{\alpha_i + \beta_i + n_i},$$

$$\int B(p_i, S_i^0(x_i))\boldsymbol{\Pi}_i(dp_i|x_i) = b\frac{(\alpha_i + x_i)(\beta_i + n_i - x_i)}{(\alpha_i + \beta_i + n_i)^2(\alpha_i + \beta_i + n_i + 1)}.$$

Hence the statistic V_i in (9.61) is given by

$$V_i(x) = \int p_{[k]} \left(\bigotimes_{j=1}^{k} \mathsf{Be}(\alpha_j + x_j, \beta_j + n_j - x_j)\right)(dp_1, ..., dp_k)$$
$$- \frac{\alpha_i + x_i}{\alpha_i + \beta_i + n_i} + b\frac{(\alpha_i + x_i)(\beta_i + n_i - x_i)}{(\alpha_i + \beta_i + n_i)^2(\alpha_i + \beta_i + n_i + 1)}.$$

As the first term is independent of i the point selection rule $\varphi^0(x)$ from (9.65) is concentrated on

$$\underset{i \in \{1,...,k\}}{\arg\max} \frac{(\alpha_i + x_i)[(\alpha_i + \beta_i + n_i)(\alpha_i + \beta_i + n_i + 1) - b(\beta_i + n_i - x_i)]}{(\alpha_i + \beta_i + n_i)^2(\alpha_i + \beta_i + n_i + 1)}.$$

Together with $S_1^0, ..., S_k^0$ the optimal decision rule $\mathsf{d}^0(x)$ in (9.64) is completely determined.

Some interesting special cases and modifications of the above example are considered in the following problems.

Problem 9.51. In Example 9.50, suppose that $n_1 = \cdots = n_k$, $\alpha_1 = \cdots = \alpha_k$, and $\beta_1 = \cdots = \beta_k$ hold altogether. Then the point selection φ^0 turns out to be the natural selection rule based on the identity.

Problem 9.52. In Example 9.50, examine the special case where $b > \alpha_i + \beta_i + n_i + 1$, $i = 1, ..., k$. If all k estimates are close to zero, then $\varphi^0(x)$ would favor smaller estimates because of a smaller posterior risk due to estimation.

Problem 9.53.* In Example 9.50, replace the loss function by

$$L^*(p,(i,t_i)) = p_{[k]} - p_i + b\frac{(p_i - t_i)^2}{p_i(1-p_i)}, \quad p \in (0,1)^k,\ t_i \in \mathbb{R},\ i = 1,...,k,$$

and derive the Bayes simultaneous selection and estimation rules for the same prior $\Pi = \bigotimes_{i=1}^k \mathsf{Be}(\alpha_i, \beta_i)$ with $\alpha_i, \beta_i > 0$, $i = 1, ..., k$.

For general linear models, Bansal and Miescke (2002) have derived Bayes simultaneous selection and estimation rules for proper and improper (noninformative) priors. The problem of finding Bayes designs (i.e., designs that have minimum Bayes risk) within a given class of designs is also discussed there. Similar work for comparisons with a control has been done in Bansal and Miescke (2005).

9.3 Optimal Subset Selections

9.3.1 Subset Selections, Loss, and Risk

Subset selection rules are decisions on subsets of the set of k populations, and a selected subset should contain good or best populations in some specified way. If the subsets are restricted to a fixed size t, then usually it is desired that it contain t best populations, that is, t populations that have the t largest values of $\kappa(\theta_1), ..., \kappa(\theta_k)$ for a given functional $\kappa : \Delta \to \mathbb{R}$. This type of problem can be treated within a moderate extension of the framework of Section 9.2.2, as shown later on. Slightly more involved is the problem of partitioning the k populations into q groups, of fixed sizes $t_1, ..., t_q$, of increasingly better populations, which has been treated under the assumption of DT by Eaton (1967a). Minimax results for problems of this type can be found in Bansal, Misra, and van der Meulen (1997).

In this section we consider the selection of a subset of populations that may also be of random size. Here several goals are reasonable. The classical approach, due to Gupta (1956, 1965), is to select a nonempty subset, of preferably small size, that contains a best population with a probability of at least P^*, where $P^* \in (1/k, 1)$ is fixed given.

In subset selection problems where the subset size may be random the decision space is given by

$$\mathcal{D}_{su} = \{A : A \subseteq \{1, ..., k\},\ A \neq \emptyset\},$$

where a decision for $A \in \mathcal{D}_{su}$ means that exactly the populations with $i \in A$ are selected. The decision space $\mathcal{D}_{su}$ consists of $2^k - 1$ elements.

Given the model $\mathcal{M}_s$ in (9.2) we call every stochastic kernel $\mathsf{K} : \mathfrak{P}(\mathcal{D}_{su}) \times \mathsf{X}_{i=1}^k \mathcal{X}_i \to_k [0,1]$ a subset selection rule. Putting $\varphi_A(x) = \mathsf{K}(\{A\}|x)$ every subset selection rule can be represented, analogously to (9.13), by

$$\varphi_A : \mathsf{X}_{i=1}^k \mathcal{X}_i \to_m [0,1], \quad A \in \mathcal{D}_{su}, \ \sum\nolimits_{A \in \mathcal{D}_{su}} \varphi_A(x) = 1, \ x \in \mathsf{X}_{i=1}^k \mathcal{X}_i,$$

or equivalently by $\varphi = (\varphi_A)_{A \in \mathcal{D}_{su}} : \mathsf{X}_{i=1}^k \mathcal{X}_i \to_m \mathbf{S}^c_{2^k-1}$.

It is clear that there is a trade-off between the probability of a best population being included in a selected subset A and the size $|A|$ of A. Several loss functions are presented below that are suitable for handling this. For $A \in \mathcal{D}_{su}$, $\theta \in \Delta^k$, $\kappa_{[k]}(\theta) = \max\{\kappa(\theta_1), ..., \kappa(\theta_k)\}$, and a fixed $c > 0$, let

$$\begin{aligned}
L_{1a}(\theta, A) &= \sum\nolimits_{i \in A} [\kappa_{[k]}(\theta) - \kappa(\theta_i)], \quad L_{1b}(\theta, A) = \frac{1}{|A|} L_{1a}(\theta, A), \quad (9.67)\\
L_2(\theta, A) &= \kappa_{[k]}(\theta) - \max_{i \in A} \kappa(\theta_i) + c\,|A|\,,\\
L_3(\theta, A) &= \sum\nolimits_{i \in A} [\kappa_{[k]}(\theta) - \kappa(\theta_i) - \varepsilon],\\
L_{4a}(\theta, A) &= c\,|A| - I_{\{\kappa_{[k]}(\theta)\}}(\max_{i \in A} \kappa(\theta_i)),\\
L_{4b}(\theta, A) &= c\,|A| - \sum\nolimits_{i \in A} I_{\{\kappa_{[k]}(\theta)\}}(\kappa(\theta_i)),\\
L_5(\theta, A) &= \sum\nolimits_{i=1}^k [l_1(\theta, i) I_{\overline{A}}(i) + l_2(\theta, i) I_A(i)], \quad l_1, l_2 \geq 0.
\end{aligned}$$

From the structure of each of these loss functions it can be seen easily in which particular way they are managing the trade-off mentioned above. In the literature usually $\Delta \subseteq \mathbb{R}$ is a Borel set and κ is the identical mapping. $L_{1,a}$ and $L_{1,b}$ have been introduced by Deely and Gupta (1968). Goel and Rubin (1977) have utilized L_2. Miescke (1979b) has made use of L_3, and L_{4a} and L_{4b} appear in Gupta and Miescke (2002). The loss L_{4a} combines the zero–one loss for including a best population with the loss of including any more populations. Note that the inclusion of any more than one best population, if such a population exists, is penalized under L_{4a}, but not under L_{4b}. The loss function L_5 goes back to Lehmann (1957a,b, 1961). A discussion and further references can be found in Bjørnstad (1981) and Gupta and Miescke (2002). An alternative approach can be found in Liu (1995).

Let the loss function now be given by some $L : \Delta^k \times \mathcal{D}_{su} \to_m \mathbb{R}_+$. That L is a measurable function of (θ, A) is equivalent to $L(\cdot, A) : \Delta^k \to_m \mathbb{R}_+$ for every $A \in \mathcal{D}_{su}$. For any subset selection rule $\varphi = (\varphi_A)_{A \in \mathcal{D}_{su}}$ the associated risk $\mathsf{R}(\theta, \varphi)$ under the selection model (9.2), with the stochastic kernel $\mathsf{P} = (P_\theta)_{\theta \in \Delta^k}$, is

$$\mathsf{R}(\theta, \varphi) = \sum\nolimits_{A \in \mathcal{D}_{su}} L(\theta, A) \int \varphi_A(x) P_\theta(dx), \quad \theta \in \Delta^k.$$

If $(\Delta, \mathfrak{B}_\Delta)$ a Borel space, Π a prior, and $\boldsymbol{\Pi}$ the posterior, then the Bayes risk of a subset selection rule is

$$\begin{aligned}\mathsf{r}(\Pi,\varphi) &= \int \mathsf{R}(\theta,\varphi)\Pi(d\theta) \\ &= \int \sum\nolimits_{A\in\mathcal{D}_{su}} L(\theta,A)[\int \varphi_A(x)P_\theta(dx)]\Pi(d\theta) \\ &= \int \sum\nolimits_{A\in\mathcal{D}_{su}} \varphi_A(x)[\int L(\theta,A)\boldsymbol{\Pi}(d\theta|x)](\mathsf{P}\Pi)(dx).\end{aligned}$$

We call φ a *Bayes subset selection rule* if it minimizes $\mathsf{r}(\Pi,\varphi)$. Set

$$M_\Pi^{su}(x) = \underset{A\in\mathcal{D}_{su}}{\arg\min} \int L(\theta,A)\boldsymbol{\Pi}(d\theta|x), \quad x \in \mathsf{X}_{i=1}^k \mathcal{X}_i. \tag{9.68}$$

Similarly as in the case of point selections, which has been treated in the previous section, the following characterization of Bayes subset selections is a direct consequence of Theorem 3.37.

Proposition 9.54. *Under any loss function $L : \Delta^k \times \mathcal{D}_{su} \to_m \mathbb{R}_+$ a subset selection rule φ is a Bayes subset selection rule if and only if*

$$\sum\nolimits_{A\in M_\Pi^{su}(x)} \varphi_A(x) = 1, \quad \mathsf{P}\Pi\text{-}a.s.$$

It should be pointed out that under the assumptions of the proposition above, every $A \in M_\Pi^{su}(x)$ can be chosen as a nonrandomized Bayes subset selection rule at x by simply taking $\varphi_A(x) = 1$. The extension of Proposition 9.54 to minimum average risk subset selection rules is straightforward.

Example 9.55. Let $\kappa : \Delta \to_m \mathbb{R}$ be a given functional and consider the loss function L_{4a} in (9.67) for a fixed $c > 0$. Then by (9.68),

$$M_\Pi^{su}(x) = \underset{A\in\mathcal{D}_{su}}{\arg\min}[c\,|A| + \boldsymbol{\Pi}(\{\theta : \max_{i\in A}\kappa(\theta_i) < \kappa_{[k]}(\theta)\}|x)].$$

This means that under a prior Π a Bayes subset selection rule selects in terms of the smallest weighted average of the posterior probability of not including a best population and the subset size.

The corresponding set $M_\Pi^{su}(x)$ under the loss function L_{4b} is given by

$$M_\Pi^{su}(x) = \underset{A\in\mathcal{D}_{su}}{\arg\min}[(c-1)\,|A| + \sum\nolimits_{i\in A} \boldsymbol{\Pi}(\{\theta : \kappa(\theta_i) < \kappa_{[k]}(\theta)\}|x)].$$

Example 9.56. Let the loss function be given by L_3, where $\varepsilon > 0$ is fixed. Then the set $M_\Pi^{su}(x)$ in (9.68) turns out to be

$$M_\Pi^{su}(x) = \underset{A\in\mathcal{D}_{su}}{\arg\min} \sum\nolimits_{i\in A} \int [\kappa_{[k]}(\theta) - \kappa(\theta_i) - \varepsilon]\boldsymbol{\Pi}(d\theta|x).$$

This time, in contrast to (9.33), it also depends on the term $\kappa_{[k]}(\theta)$ that appears in the loss function.

Problem 9.57.* Assume that $\int |\kappa_{[k]}(\theta)|\Pi(d\theta) < \infty$. In the previous example, that is, under the loss function L_3, the following holds. The set $M_{\Pi}^{su}(x)$ consists of all nonempty sets $B \subseteq \{1, ..., k\}$ with

$$\{i : \int \kappa(\theta_i)\boldsymbol{\Pi}(d\theta|x) > \int \kappa_{[k]}(\theta)\boldsymbol{\Pi}(d\theta|x) - \varepsilon\}$$
$$\subseteq B \subseteq \{i : \int \kappa(\theta_i)\boldsymbol{\Pi}(d\theta|x) \geq \int \kappa_{[k]}(\theta)\boldsymbol{\Pi}(d\theta|x) - \varepsilon\},$$

as long as the set on the right-hand side is not empty. Otherwise, the set $M_{\Pi}^{su}(x)$ consists of all singletons $\{i_0\} \subseteq \{1, ..., k\}$ with

$$\int \kappa(\theta_{i_0})\boldsymbol{\Pi}(d\theta|x) = \max_{j\in\{1,\ldots,k\}} \int \kappa(\theta_j)\boldsymbol{\Pi}(d\theta|x).$$

Similarly as done for point selections with (9.42) and (9.43) let us choose now a large class of loss functions that reflect our goals of subset selection. Again, there are two natural assumptions that are also made here. The first is that the loss function is permutation invariant.

$$L(\theta, u_\gamma(A)) = L(u_\gamma(\theta), A), \quad A \in \mathcal{D}_{su}, \ \theta \in \Delta^k, \ \gamma \in \Pi_k, \tag{9.69}$$

where $u_\gamma(\theta) = (\theta_{\gamma(1)}, ..., \theta_{\gamma(k)})$ and $u_\gamma(A) = \{\gamma(i) : i \in A\}$. The second is that populations with larger parameter values are preferred for selections. More precisely, we assume that for any $i, j \in \{1, ..., k\}$ and $C \subseteq \{1, ..., k\}$ with $\{i, j\} \cap C = \emptyset$,

$$L(\theta, C \cup \{i\}) \geq L(\theta, C \cup \{j\}), \quad \kappa(\theta_i) \leq \kappa(\theta_j), \ \theta \in \Delta^k. \tag{9.70}$$

Problem 9.58. Verify that the loss functions in (9.67) have the properties (9.69) and (9.70).

For the above class of loss functions and permutation invariant models with DT the following natural property of Bayes subset selections can be established.

Proposition 9.59. *Let $\Delta \subseteq \mathbb{R}$ be a Borel set, $\kappa(\theta) = \theta$, and $\Pi \in \mathcal{P}(\mathfrak{B}_{\Delta^k})$ be a permutation invariant prior. Suppose L satisfies (9.69) and (9.70). Assume that for the selection model $\mathcal{M}_s$ in (9.2) $\mathsf{P} = (P_\theta)_{\theta\in\Delta^k}$ is a stochastic kernel, $V : \mathsf{X}_{i=1}^k \mathcal{X}_i \to_m \mathbb{R}^k$ a sufficient statistic, and $Q_\theta = P_\theta \circ V^{-1}$. Assume that $\mathsf{Q} = (Q_\theta)_{\theta\in\Delta^k}$, is a permutation invariant stochastic kernel that is DT at every $\theta \in \Delta^k$ and $M_{\Pi}^{su}(x)$ is from (9.68). Let $\{i, j\}, C \subseteq \{1, ..., k\}$ with $\{i, j\} \cap C = \emptyset$. Then $\mathsf{P}\Pi$-a.s.*

$$(C \cup \{i\}) \in M_{\Pi}^{su}(x) \quad \textit{and} \quad V_i(x) \leq V_j(x) \quad \textit{implies} \quad (C \cup \{j\}) \in M_{\Pi}^{su}(x).$$

Proof. The proof follows along the lines of the proof of Lemma 9.27. For $\{i, j\} \cap C = \emptyset$ and $V_i(x) \leq V_j(x)$ let $C_i = C \cup \{i\}$, $C_j = C \cup \{j\}$, and $V(x) = v$. From (9.70) it follows that $L(\theta, C_i) - L(\theta, C_j) = 0$ if $\theta_i = \theta_j$. With $Q_\theta(dv)\Pi(d\theta) = \boldsymbol{\Lambda}_{\mathsf{Q}}(d\theta|v)(\mathsf{Q}\Pi)(dv)$ from (9.27) we have

$$\begin{aligned}
&\int [L(\theta, C_i) - L(\theta, C_j)] \boldsymbol{\Lambda}_{\mathsf{Q}}(d\theta|v) \\
&= \int [L(\theta, C_i) - L(\theta, C_j)] I_{[\theta_i,\infty)}(\theta_j) \boldsymbol{\Lambda}_{\mathsf{Q}}(d\theta|v) \\
&\quad + \int [L(\theta, C_i) - L(\theta, C_j)] I_{[\theta_j,\infty)}(\theta_i) \boldsymbol{\Lambda}_{\mathsf{Q}}(d\theta|v).
\end{aligned}$$

The last integral satisfies, in view of (9.69),

$$\begin{aligned}
&\int [L(\theta, C_i) - L(\theta, C_j)] I_{[\theta_j,\infty)}(\theta_i) \boldsymbol{\Lambda}_{\mathsf{Q}}(d\theta|v) \\
&= \int [L(\theta^{(i,j)}, C_j) - L(\theta^{(i,j)}, C_i)] I_{[\theta_j,\infty)}(\theta_i) \boldsymbol{\Lambda}_{\mathsf{Q}}(d\theta|v) \\
&= \int [L(\theta, C_j) - L(\theta, C_i)] I_{[\theta_i,\infty)}(\theta_j) \boldsymbol{\Lambda}_{\mathsf{Q}}(d\theta|v^{(i,j)}) \\
&\geq \int [L(\theta, C_j) - L(\theta, C_i)] I_{[\theta_i,\infty)}(\theta_j) \boldsymbol{\Lambda}_{\mathsf{Q}}(d\theta|v),
\end{aligned}$$

where the inequality follows from (9.70) and (9.37). Therefore

$$\int [L(\theta, C_i) - L(\theta, C_j)] \boldsymbol{\Lambda}_{\mathsf{Q}}(d\theta|v) \geq 0,$$

and by $\boldsymbol{\Lambda}_{\mathsf{Q}}(d\theta|v) = \boldsymbol{\Pi}(d\theta|x)$ the proof is completed. ■

For $\gamma \in \Pi_k$, $x \in \mathbb{R}^k$, and $A \subseteq \{1, ..., k\}$, we set $\gamma(x) = (x_{\gamma(1)}, ..., x_{\gamma(k)})$ and $\gamma(A) = \{\gamma(i) : i \in A\}$.

Definition 9.60. *We call a subset selection $\varphi = (\varphi_A)_{A \in \mathcal{D}_{su}}$ permutation invariant if $\varphi_{\gamma(A)}(x) = \varphi_A(\gamma(x))$ for every $\gamma \in \Pi_k$, $x \in \mathbb{R}^k$, and $A \subseteq \{1, ..., k\}$.*

Corollary 9.61. *Suppose that the subset to be selected is restricted to be of size t, say, where t is fixed given. Then under the assumption of Proposition 9.59 the natural subset selection rule $\varphi_{V,t}^{nat}$ which selects in term of the t largest values of $V_1(x), ..., V_k(x)$, breaking ties with equal probabilities, is a uniformly best permutation invariant subset selection rule.*

Proof. Let $t < k$ be fixed. Let $D \subset \{1, ..., k\}$ with $|D| = t$ satisfy $V_i(x) \leq V_j(x)$ for every $i \notin D$ and $j \in D$. Then by the same arguments as in the proof of the above proposition it follows that $\mathsf{P}\Pi$-a.s.

$$D \in \underset{A \in \mathcal{D}_{su}, |A|=t}{\arg\min} \int L(\theta, A) \boldsymbol{\Pi}(d\theta|x), \quad x \in \times_{i=1}^k \mathcal{X}_i.$$

Thus, $\varphi_{V,t}^{nat}$ is a Bayes rule for every permutation invariant prior $\Pi \in \mathcal{P}(\mathfrak{B}_{\Delta^k})$. The rest follows as in the proof of (B) in Theorem 9.31. ■

Remark 9.62. A more general problem is to partition the k populations into t_1 with the t_1 smallest, t_2 with the t_2 next smallest, ..., t_q with the t_q largest parameters. This can be solved analogously; see Eaton (1967a).

According to the previous proposition Bayes selections under a permutation invariant prior are made in terms of the largest values of $V_1(x), ..., V_k(x)$. This result is, of course, not completely satisfactory because the question regarding the optimal size of the selected subset remains open. Obviously, that depends on the particular loss chosen and indeed may be a difficult problem.

To get more results on the structure of Bayes selection rules we consider loss functions that are additive, such as $L_5(\theta, A)$ in (9.67). The nonnegative functions l_1 and l_2 represent losses due to the exclusion and inclusion, respectively, of an individual population. Denote by

$$\psi_i(x) = \sum\nolimits_{A \in \mathcal{D}_{su}} I_A(i)\varphi_A(x), \quad x \in \mathsf{X}_{i=1}^k \mathcal{X}_i, \tag{9.71}$$

the probability that population i is included in the selected subset of the subset selection rule φ at x, $i = 1, ..., k$. Note that the vector of inclusion probabilities $(\psi_1(x), ..., \psi_k(x))$ of φ is not a probability distribution. In contrast to (9.13) we have now the following.

Problem 9.63.* The inclusion probabilities in (9.71) of a subset selection rule φ satisfy $\sum_{i=1}^k \psi_i(x) \geq 1$. There may be more than one subset selection rule that has the same inclusion probabilities, unless $\psi_i(x) \in \{0, 1\}$, $i = 1, ..., k$.

Remark 9.64. To construct subset selection rules we use the following simple fact. If $\psi_i : \mathsf{X}_{i=1}^k \mathcal{X}_i \to_m \{0, 1\}, i = 1, ..., k$, and $\sum_{i=1}^k \psi_i(x) > 0$ for $x \in \mathsf{X}_{i=1}^k \mathcal{X}_i$, then

$$\mathsf{d}(x) := \{i : \psi_i(x) = 1\}. \tag{9.72}$$

is a nonrandomized subset selection rule.

Under the additive loss function L_5 the risk of a subset selection rule φ depends only on its inclusion probabilities.

Lemma 9.65. *Under the additive loss function L_5 in (9.67) the risk of a subset selection rule φ with inclusion probabilities ψ_i, $i = 1, ..., k$, is given by*

$$\begin{aligned}\mathsf{R}(\theta, \varphi) &= \sum\nolimits_{i=1}^k [l_2(\theta, i) - l_1(\theta, i)]\mathsf{E}_\theta \psi_i + \sum\nolimits_{i=1}^k l_1(\theta, i) \\ &= \sum\nolimits_{i=1}^k [l_2(\theta, i)\mathsf{E}_\theta \psi_i + l_1(\theta, i)(1 - \mathsf{E}_\theta \psi_i)], \quad \theta \in \Delta^k.\end{aligned}$$

Proof. It holds,

$$\begin{aligned}\mathsf{R}(\theta, \varphi) &= \sum\nolimits_{A \in \mathcal{D}_{su}} [\sum\nolimits_{i \in A} l_2(\theta, i) + \sum\nolimits_{j \notin A} l_1(\theta, j)] \int \varphi_A(x) P_\theta(dx) \\ &= \sum\nolimits_{i=1}^k [l_2(\theta, i) - l_1(\theta, i)] \int \sum\nolimits_{A \in \mathcal{D}_{su}} I_A(i)\varphi_A(x) P_\theta(dx) + \sum\nolimits_{j=1}^k l_1(\theta, j) \\ &= \sum\nolimits_{i=1}^k [l_2(\theta, i) - l_1(\theta, i)] \int \psi_i(x) P_\theta(dx) + \sum\nolimits_{j=1}^k l_1(\theta, j).\end{aligned}$$

■

The problem of minimizing the risk $\mathsf{R}(\theta,\varphi)$ has been reduced to solving k separate testing problems, and then to comply with the requirement that the selected subset must not be empty. Because there is no test that minimizes both types of error probabilities, except in trivial cases (see Proposition 2.29), we take recourse to the Bayes approach.

Suppose that $(\Delta, \mathfrak{B}_\Delta)$ is a standard Borel space, $l_1(\theta, i)$ and $l_2(\theta, i)$ are measurable in θ, and the prior Π satisfies $\int l_m(\theta, i)\Pi(d\theta) < \infty$, $m = 1, 2$, $i = 1, ..., k$. Then by the last equation in the proof above,

$$\begin{aligned}\mathsf{r}(\Pi,\varphi) &= \int \mathsf{R}(\theta,\varphi)\Pi(d\theta) \\ &= \int (\sum\nolimits_{i=1}^{k} \psi_i(x)[\int [l_2(\theta,i) - l_1(\theta,i)]\boldsymbol{\Pi}(d\theta|x)])(\mathsf{P}\Pi)(dx) \\ &\quad + \sum\nolimits_{j=1}^{k} \int l_1(\theta,j)\Pi(d\theta).\end{aligned}$$

Due to the additivity of the posterior risk all Bayes subset selections can be found by first optimizing the inclusion probabilities separately at every fixed $x \in \times_{i=1}^k \mathcal{X}_i$. If that leads to an empty set, then an adjustment has to be made that is described below; see also Problem 9.57. If there exists a Bayes solution at $x \in \times_{i=1}^k \mathcal{X}_i$, then there is always one with inclusion probabilities $\psi_i(x) \in \{0, 1\}$, $i = 1, ..., k$, which in turn determine uniquely the selected subset at x. Let

$$\begin{aligned}S_<(x) &= \{i : \int [l_2(\theta,i) - l_1(\theta,i)]\boldsymbol{\Pi}(d\theta|x) < 0\}, \\ S_=(x) &= \{i : \int [l_2(\theta,i) - l_1(\theta,i)]\boldsymbol{\Pi}(d\theta|x) = 0\}, \\ S_>(x) &= \{i : \int [l_2(\theta,i) - l_1(\theta,i)]\boldsymbol{\Pi}(d\theta|x) > 0\}.\end{aligned}$$

Three cases have to be considered. The first is $S_<(x) \neq \emptyset$. Here the Bayes subset selections are the sets B with $S_<(x) \subseteq B \subseteq S_<(x) \cup S_=(x)$. The second case is $S_<(x) = \emptyset$ and $S_=(x) \neq \emptyset$. Here the Bayes subset selections are all nonempty sets B with $B \subseteq S_=(x)$. The third case is $S_<(x) = \emptyset$ and $S_=(x) = \emptyset$. Here the Bayes subset selections are all singletons $\{i_0\} \subseteq \{1, ..., k\}$ with

$$\int [l_2(\theta,i_0) - l_1(\theta,i_0)]\,\boldsymbol{\Pi}(d\theta|x) = \min_{j\in\{1,\ldots,k\}} \int [l_2(\theta,j) - l_1(\theta,j)]\,\boldsymbol{\Pi}(d\theta|x).$$

The above findings can be summarized by stating that the set $M_\Pi^{su}(x)$ from (9.68) is

$$M_\Pi^{su}(x) = \underset{A\in\mathcal{D}_{su}}{\arg\min} \sum\nolimits_{i\in A} \int [l_2(\theta,i) - l_1(\theta,i)]\boldsymbol{\Pi}(d\theta|x).$$

Let now $l_1(\theta, i)$ and $l_2(\theta, i)$ be permutation invariant in the sense of (9.42). Then (9.69) holds. Moreover, let $-l_1(\theta, i)$ and $l_2(\theta, i)$ satisfy (9.43). Then (9.70) holds as well. Therefore, according to Proposition 9.59, under the assumptions stated there, every Bayes subset selection at $x \in \mathsf{X}_{i=1}^k \mathcal{X}_i$ consists of populations that are associated with the largest values of $V_1(x), ..., V_k(x)$.

Example 9.66. Let $l_1(\theta, i) = L_{1,i} I_{(\theta_{0,i},\infty)}(\theta_i)$ and $l_2(\theta, i) = L_{2,i} I_{(-\infty,\theta_{0,i}]}(\theta_i)$, $\theta = (\theta_1, ..., \theta_k) \in \Delta^k$, where $L_{1,i}, L_{2,i} \geq 0$ and $\theta_{0,i} \in \Delta \subseteq \mathbb{R}$ are fixed given, $i = 1, ..., k$. Let $B_i = \{\theta : \theta_i \leq \theta_{0,i},\ \theta \in \Delta\}$, $i = 1, ..., k$. Then

$$\begin{aligned} M_{\Pi}^{su}(x) &= \underset{A \in \mathcal{D}_{su}}{\arg\min} \sum\nolimits_{i \in A} [L_{2,i} \boldsymbol{\Pi}(B_i|x) - L_{1,i}(1 - \boldsymbol{\Pi}(B_i|x))] \\ &= \underset{A \in \mathcal{D}_{su}}{\arg\min} \sum\nolimits_{i \in A} [(L_{1,i} + L_{2,i}) \boldsymbol{\Pi}(B_i|x) - L_{1,i}]. \end{aligned}$$

The Bayes subset selections can be constructed as follows. Each population i with $\boldsymbol{\Pi}(B_i|x) < L_{1,i}/(L_{1,i} + L_{2,i})$ is included. The inclusion of any subset of populations i with $\boldsymbol{\Pi}(B_i|x) = L_{1,i}/(L_{1,i} + L_{2,i})$ is optional. If that does not lead to a nonempty subset, then exactly one population i_0 is selected with

$$i_0 \in \underset{i \in \{1,...,k\}}{\arg\min} [(L_{1,i} + L_{2,i}) \boldsymbol{\Pi}(B_i|x) - L_{1,i}].$$

Problem 9.67. In the previous example the Bayes subset selections at $x \in \mathsf{X}_{i=1}^k \mathcal{X}_i$ are based on Bayes tests for $\mathsf{H}_0^{(i)} : \theta_i \leq \theta_{0,i}$ versus $\mathsf{H}_A^{(i)} : \theta_i > \theta_{0,i}$, $i = 1, ..., k$. Such tests have been studied in Example 3.47. Consider also the special case of $L_{1,i} = L_{2,i}$, $i = 1, ..., k$.

In the remainder of this section we discuss Gupta's (1956, 1965) approach to subset selection. It was introduced in the setting of k normal populations (i.e., in the setting of Example 9.1), but with equal sample sizes and equal variances. After a reduction by sufficiency we only have to deal with the k sample means, say, $X_i \sim \mathsf{N}(\mu_i, \sigma^2/n)$, $i = 1, ..., k$. We consider the class of subset selection rules φ for which the minimum probability of a correct selection (PCS), which means here the probability of including a best population, is at least P^*, where $P^* \in (1/k, 1)$ is fixed given. The PCS in the balanced selection model

$$\mathcal{M} = (\mathbb{R}^k, \mathfrak{B}_k, (\bigotimes\nolimits_{i=1}^k \mathsf{N}(\mu_i, \sigma^2/n))_{\mu \in \mathbb{R}^k}),$$

with $\mu = (\mu_1, ..., \mu_k) \in \mathbb{R}^k$, is defined here by

$$P_{cs}(\mu, \varphi) = \sum\nolimits_{A: A \cap \mathsf{M}_\kappa(\mu) \neq \emptyset} \int \varphi_A(x) (\bigotimes\nolimits_{i=1}^k \mathsf{N}(\mu_i, \sigma^2/n))(dx), \qquad (9.73)$$

where $\mathsf{M}_\kappa(\mu)$ is from (9.1) with κ as the identical mapping, $\theta = \mu$, and $x = (x_1, ..., x_k) \in \mathbb{R}^k$. The basic idea of Gupta's approach is that the selected subset should include a best population with the P^*-guarantee while being as small as possible. One way toward this goal would be to search for a subset selection rule that meets the P^*-requirement and has a minimum expected subset size, uniformly in $\mu \in \mathbb{R}^k$. Unfortunately, however, no such rule exists; see Deely and Johnson (1997).

Remark 9.68. The case of $P^* \le 1/k$ is trivial. The no-data selection rule which selects each $\{i\} \subseteq \{1, ..., k\}$ with probability $1/k$, $i = 1, ..., k$, would meet the P^*-requirement and would produce a subset of minimum size 1.

Gupta's subset selection rule φ^{gup} is defined as follows.

$$\varphi_A^{gup}(x) = 1, \quad A = \{i : x_i \ge x_{[k]} - d\sigma/\sqrt{n}\}, \;\; x \in \mathbb{R}^k,$$

where the constant d is determined by

$$\int \Phi^{k-1}(t+d)\varphi(t)dt = P^*.$$

Proposition 9.69. *The infimum of the PCS of Gupta's subset selection rule on $\mathbb{R}^k$ is P^*.*

Proof. First we assume that exactly one population has the largest mean. As Gupta's subset selection rule is permutation invariant we may assume without loss of generality that $\mu_1, ..., \mu_{k-1} < \mu_k$. Let $Z_1, ..., Z_k$ be generic i.i.d. standard normal random variables. Then by

$$(\sigma n^{-1/2} Z_1 + \mu_1, ..., \sigma n^{-1/2} Z_k + \mu_k) \sim \bigotimes\nolimits_{i=1}^k \mathsf{N}(\mu_i, \sigma^2/n),$$

together with (9.73) and $\{A : A \cap \mathsf{M}_\kappa(\mu) \neq \emptyset\} = \{A : k \in A\}$, we get that the PCS in (9.73) of Gupta's rule is

$$\begin{aligned}
&\mathbb{P}(Z_i < Z_k + (\sqrt{n}/\sigma)[\mu_k - \mu_i] + d, \;\; i < k) \\
&= \int \textstyle\prod_{i=1}^{k-1} \Phi(t + (\sqrt{n}/\sigma)[\mu_k - \mu_i] + d)\varphi(t)dt \ge \int \Phi^{k-1}(t+d)\varphi(t)dt = P^*,
\end{aligned}$$

where the first term tends to P^* as $\mu_k - \mu_i \to 0$ for $i = 1, ..., k-1$. To complete the proof we note that on $\mathbb{R}^k \backslash \mathbb{R}_1^k$, where $\mathbb{R}_1^k = \{\mu : \mu_{[k-1]} < \mu_{[k]}, \; \mu \in \mathbb{R}^k\}$, apparently the infimum PCS is bigger than P^*. ■

Establishing an ad hoc subset selection rule at the P^*-condition in various specific distribution models has been the topic of many earlier papers on subset selection problems; see Gupta and Panchapakesan (1979).

Questions regarding the quality of the performance of Gupta's subset selection rule, also for its extensions to the case of unequal sample sizes, have been considered by various authors; see Gupta and Panchapakesan (1979). The minimax aspect has been studied by Berger and Gupta (1980). In Hsu (1981) an alternative approach has been introduced for location parameter families that provides simultaneous confidence intervals for all distances from the best population. See also Hsu (1982), Hsu and Edwards (1983), and Hsu (1996). Another approach by Finner and Giani (1996), based on the duality of multiple testing and selection, has been shown to be in favor of Gupta's subset selection rule. Subset selections for certain two-factor normal models can be found in Santner and Pan (1997).

An interesting class of subset selection rules has been considered by Seal (1955, 1957). A rule in this class selects population $i \in \{1, ..., k\}$ if $x_i \geq \alpha_1 y_{i:1} + \cdots + \alpha_{k-1} y_{i:k-1} + c$, where $y_{i:1} \leq \cdots \leq y_{i:k-1}$ are the ordered values of $x_1, ..., x_{i-1}, x_{i+1}, ..., x_k$. Each choice of nonnegative weights $\alpha_1, ..., \alpha_{k-1}$ with $\alpha_1 + \cdots + \alpha_{k-1} = 1$ is hereby allowed, and the constant c is determined by setting the infimum of the PCS equal to P^*. At least it could be shown that for $k = 3$ and $P^* \in [2/3, 1)$ there exist constants $a, b \geq 0$ such that Gupta's subset selection rule has, within Seal's class, a minimum expected subset size on

$$M = \{\mu : \mu_{[2]} - \mu_{[1]} > a, \ \ \mu_{[3]} - \mu_{[2]} > b, \quad \mu \in \mathbb{R}^3\}.$$

The proof and further details in this regard can be found in Gupta and Miescke (1981). In Bjørnstad (1984) it has been shown that Gupta's rule is the only rule in Seal's class that can be asymptotically consistent. An extension of this work can be found in Bjørnstad (1986).

Under a class of additive loss functions, which includes L_3 from (9.67), φ^{gup} has been shown in Miescke (1979b) to be the pointwise limit of Bayes rules as $n \to \infty$. Under the loss functions L_3 and L_{4b} from (9.67) Gupta and Miescke (2002) have made an attempt to determine whether φ^{gup} is admissible. In the course of that work two respective generalized Bayes rules have been derived with the Lebesgue measure acting as noninformative prior. In a simulation study by Miescke and Ryan (2006) φ^{gup} has then been compared with these two generalized Bayes rules under their respective loss functions and under a common P^*-condition. The results have been favorable for φ^{gup}.

9.3.2 Γ-Minimax Subset Selections

Subset selection problems with a standard or control have not been considered so far. A *standard* is a given value $\theta_{0,i} \in \Delta \subseteq \mathbb{R}$ that separates Δ into "good" and "bad" parameter values for population $i \in \{1, ..., k\}$. If $\theta_{0,i}$ is not known and $P_{\theta_{0,i}}$ has to be sampled to gain information on $\theta_{0,i}$, then it is called a *control*. This is an important area, and some remarks and references are given at the end of this section. For brevity only one type of such problems is presented here in the Γ-minimax approach, which has been considered previously in Section 3.6. The results in this subsection are taken from Randles and Hollander (1971) and Miescke (1981), which are based on Lehmann (1957a, 1961) and Blum and Rosenblatt (1967).

We begin with a general subset selection problem, under an additive loss function, that includes problems with a standard or control. Let the decision space be $\mathcal{D}_{sc} = \{A | A \subseteq \{1, ..., k\}\}$, where the empty set may also be selected. The subset selection rules in this section are represented by $\varphi = (\varphi_A)_{A \in \mathcal{D}_{sc}}$ with

$$\varphi_A : \mathsf{X}_{i=1}^k \mathcal{X}_i \to_m [0, 1], \ A \in \mathcal{D}_{sc}, \ \text{ and } \sum\nolimits_{A \in \mathcal{D}_{sc}} \varphi_A(x) = 1, \ \ x \in \mathsf{X}_{i=1}^k \mathcal{X}_i.$$

The loss function is assumed to be additive, of the form as L_5 in (9.67), but now also being defined on the empty set.

$$L(\theta, A) = \sum\nolimits_{j \notin A} l_1(\theta, j) + \sum\nolimits_{i \in A} l_2(\theta, i), \quad A \in \mathcal{D}_{sc}, \quad \theta \in \Delta^k. \tag{9.74}$$

where $l_1(\cdot, i) : \Delta^k \to_m \mathbb{R}_+$ represents the loss of not selecting population i when it is good, and $l_2(\cdot, i) : \Delta^k \to_m \mathbb{R}_+$ represents the loss of selecting population i when it is bad, $i = 1, ..., k$. The sum over an empty set is assumed to be zero.

The risk and the Bayes risk of a subset selection rule $\varphi = (\varphi_A)_{A \in \mathcal{D}_{sc}}$ under the loss function (9.74) can be seen to depend only on the k inclusion probabilities

$$\psi_i(x) = \sum\nolimits_{A \in \mathcal{D}_{sc}} I_A(i) \varphi_A(x), \quad x \in \mathsf{X}_{i=1}^k \mathcal{X}_i, \; i = 1, ..., k. \tag{9.75}$$

The inclusion probabilities in (9.71) and (9.75) are actually identical since $\mathcal{D}_{su} = \mathcal{D}_{sc} \backslash \emptyset$. Lemma 9.65 and its proof can be extended in a straightforward manner to subset selection rules that may also select the empty set. Thus,

$$\mathsf{R}(\theta, \varphi) = \sum\nolimits_{i=1}^k [l_2(\theta, i) \mathsf{E}_\theta \psi_i + l_1(\theta, i)(1 - \mathsf{E}_\theta \psi_i)], \quad \theta \in \Delta^k.$$

Let now $\Gamma \subseteq \mathcal{P}(\mathfrak{B}_{\Delta^k})$ be a given set of priors. The Bayes risk of φ under a prior $\Pi \in \Gamma$ is

$$\mathsf{r}(\Pi, \varphi) = \sum\nolimits_{i=1}^k \mathsf{r}^{(i)}(\Pi, \psi_i), \quad \text{where} \tag{9.76}$$
$$\mathsf{r}^{(i)}(\Pi, \psi_i) = \int [l_2(\theta, i) \mathsf{E}_\theta \psi_i + l_1(\theta, i)(1 - \mathsf{E}_\theta \psi_i)] \Pi(d\theta) \quad i = 1, ..., k.$$

As before in the previous subsection, if there exists a Bayes solution at $x \in \mathsf{X}_{i=1}^k \mathcal{X}_i$, then there exists a nonrandomized Bayes subset selection rule d which has inclusion probabilities $\psi_i(x) \in \{0, 1\}$, $i = 1, ..., k$. Moreover, by (9.72) the inclusion probabilities uniquely determine d.

The simplest way of constructing a subset selection rule φ^Γ that has the inclusion probabilities ψ^Γ is to make the k inclusion decisions, given $x \in \mathsf{X}_{i=1}^k \mathcal{X}_i$, mutually independent; that is,

$$\varphi_A^\Gamma(x) = \prod\nolimits_{i \in A} \psi_i^\Gamma(x), \quad x \in \mathsf{X}_{i=1}^k \mathcal{X}_i, \quad A \in \mathcal{D}_{sc}.$$

Of course, if the k inclusion probabilities at some $x \in \mathsf{X}_{i=1}^k \mathcal{X}_i$ are all either 0 or 1, then the selected subset is uniquely determined at that x. Fixing this construction we may identify a selection rule φ with a vector $\psi = (\psi_1, ..., \psi_k)$: $\mathsf{X}_{i=1}^k \mathcal{X}_i \to_m [0, 1]^k$, $i = 1, ..., k$. Let $\mathcal{P}_{sc}$ be the class of all such vectors.

We consider now a special case of the above problem where the selection model is based on a one-parameter exponential family and a standard is given for every population. The model considered here is $\mathcal{M}_{sel}$ from (9.8). Suppose that for each population $i \in \{1, ..., k\}$ a standard $\theta_{0,i} \in \Delta^0$ is given, along with some distance value $d_i > 0$ with $\theta_{0,i} + d_i \in \Delta^0$. The loss is assumed to be a special case of (9.74).

$$L(\theta, A) = \sum\nolimits_{j \notin A} L_{1,j} I_{[\theta_{0,j}+d_j,\infty)}(\theta_i) + \sum\nolimits_{i \in A} L_{2,i} I_{(-\infty,\theta_{0,i}]}(\theta_i), \tag{9.77}$$

$A \in \mathcal{D}_{sc}$, $\theta \in \Delta^k$, where $L_{1,i}, L_{2,i} \geq 0$, $i = 1, ..., k$, are fixed given constants. Let Δ be a Borel subset of $\mathbb{R}$ and denote by $\Pi_1, ..., \Pi_k$ the marginal distributions of a prior $\Pi \in \mathcal{P}(\mathfrak{B}_{\Delta^k})$. Let $\Gamma \subseteq \mathcal{P}(\mathfrak{B}_{\Delta^k})$ be the class of priors $\Pi \in \Gamma$ that satisfy

$$\Pi_i([\theta_{0,i} + d_i, \infty) \cap \Delta) \leq \pi_i \quad \text{and} \quad \Pi_i((-\infty, \theta_{0,i}] \cap \Delta) \leq \pi_i', \tag{9.78}$$

where $\pi_i, \pi_i' \geq 0$ with $\pi_i + \pi_i' \leq 1$, $i = 1, ..., k$, are fixed given. These constraints prevent too much mass of the prior from being concentrated on either side. Then the following holds.

Theorem 9.70. *Suppose the observations $X_{i,j}, j = 1, ..., n_i$, $i = 1, ..., k$, are from the model (9.5), where $(P_\theta)_{\theta \in \Delta}$ is a one-parameter exponential family with generating statistic T. Let the loss be given by (9.77), and let $\psi^\Gamma = (\psi_1^\Gamma, ..., \psi_k^\Gamma) \in \mathcal{P}_{sc}$, where $\psi_i^\Gamma = I_{[c_i,\infty)}(T_{\oplus n_i})$ with*

$$c_i = \frac{1}{d_i}[\ln \frac{L_{2,i}\pi_i'}{L_{1,i}\pi_i} + n_i K(\theta_{0,i} + d_i) - n_i K(\theta_{0,i})], \quad i = 1, ..., k.$$

Then, ψ^Γ is Γ-minimax in the class $\mathcal{P}_{sc}$. Every subset selection rule φ^Γ that has the inclusion probabilities ψ^Γ is Γ-minimax in the class of all subset selection rules that are allowed to select also the empty set.

Proof. By the definition of the Bayes risk in (9.76) we see that for $y_i = (x_{i,1}, ..., x_{i,n_i})$, $i = 1, ..., k$,

$$\begin{aligned} \mathsf{r}(\Pi, \varphi) &= \sum\nolimits_{i=1}^k \mathsf{r}^{(i)}(\Pi_i, \psi_i) \text{ where} \\ \mathsf{r}^{(i)}(\Pi_i, \psi_i) &= \int [L_{1,i} I_{[\theta_{0,j}+d_j,\infty)}(\theta_i)[1 - \int \psi_i(y_i) P_{\theta_i}^{\otimes n_i}(dy_i)] \\ &\quad + L_{2,i} I_{(-\infty,\theta_{0,i}]}(\theta_i) \int \psi_i(y_i) P_{\theta_i}^{\otimes n_i}(dy_i)] \Pi_i(d\theta_i). \end{aligned}$$

Denote by Γ_i the set of priors that satisfy (9.78). Then

$$\inf_{\psi^\Gamma \in \mathcal{P}_{sc}} \sup_{\Pi \in \Gamma} \mathsf{r}(\Pi, \varphi) = \sum\nolimits_{i=1}^k \inf_{\psi_i} \sup_{\Pi_i \in \Gamma_i} \mathsf{r}^{(i)}(\Pi_i, \psi_i),$$

because $\mathsf{r}^{(i)}(\Pi_i, \psi_i)$ is completely restricted to the ith population's Γ-minimax problem, $i = 1, ..., k$. To complete the proof we have only to apply Example 3.69. ■

A special case is the following.

Problem 9.71. Let $X_{i,j} \sim \mathsf{N}(\mu_i, \sigma_i^2)$, $j = 1, ..., n_i$, $i = 1, ..., k$, be independent, where $\sigma_i^2 > 0$, $i = 1, ..., k$, are known. Let $\bar{x}_i = (1/n_i)\sum_{j=1}^{n_i} x_{i,j}$, $i = 1, ..., k$, be the sample means. Then under the loss function from (9.77) the subset selection rule φ^Γ that has the inclusion probabilities

$$\psi^\Gamma(\bar{x}_1, ..., \bar{x}_k) = (\psi_1^\Gamma(\bar{x}_1), ..., \psi_k^\Gamma(\bar{x}_k)),$$

where $\psi_i^\Gamma(t) = I_{[b_i,\infty)}(t)$, $t \in \mathbb{R}$, and

$$b_i = \theta_{0,i} + \frac{d_i}{2} + \frac{1}{n_i d_i}\sigma_i^2 \ln \frac{L_{2,i}\pi_i'}{L_{1,i}\pi_i}, \quad i = 1, ..., k,$$

is Γ-minimax in the class of all subset selection rules that are allowed to select also the empty set.

Now we consider situations where there are unknown control parameters $\theta_{0,1}, ..., \theta_{0,k}$. The distance values $d_i > 0$, $i = 1, ..., k$, remain, however, fixed given and are used in the same way as before.

If the control consists of k independent populations, one for each treatment population, where the $2k$ populations are independent, then the selection model becomes

$$\mathcal{M}_{ctrl1} = (\mathbb{R}^{2k}, \mathfrak{B}_{2k}, (\textstyle\bigotimes_{i=1}^k (Q_{n_i,\theta_i} \otimes Q_{m_i,\theta_{0,i}}))_{\theta,\eta \in \Delta^k}),$$

where $\theta = (\theta_1, ..., \theta_k)$ and $\eta = (\theta_{0,1}, ..., \theta_{0,k})$. Under this model, and the loss from (9.77), a Γ-minimax subset selection rule can still be found by solving a Γ-minimax problem individually for each of the k pairings of a treatment population and its control population. The arguments are very similar to those that have been used above and thus they are omitted here for brevity.

Quite a different situation arises when the control parameters are known to be equal, i.e., $\theta_{0,1} = \cdots = \theta_{0,k} = \theta_0$, say, and one single control population has to be shared by the k treatment populations. In this case the model is

$$\mathcal{M}_{ctrl2} = (\mathbb{R}^{k+1}, \mathfrak{B}_{k+1}, (\textstyle\bigotimes_{i=0}^k Q_{n_i,\theta_i})_{\theta \in \Delta^{k+1}}),$$

which is the model $\mathcal{M}_{sel}$ from (9.8), augmented by the control population with the parameter θ_0. Now we have $\theta = (\theta_0, \theta_1, ..., \theta_k)$ and $S = (S_0, S_1, ..., S_k)$, where $\theta_0 \in \Delta$ is the unknown control parameter, common to the k treatment populations, and S_0 is the statistic of the observations from the control population. Under this model, and the loss function from (9.77) with $\theta_{0,1} = \cdots = \theta_{0,k} = \theta_0$, that is,

$$L_0(\theta, A) = \sum\nolimits_{j \notin A} L_{1,j} I_{[\theta_0 + d_j,\infty)}(\theta_j) + \sum\nolimits_{i \in A} L_{2,i} I_{(-\infty,\theta_0]}(\theta_i), \tag{9.79}$$

$A \in \mathcal{D}_{sc}$, $\theta \in \Delta^k$, a Γ-minimax subset selection rule is harder to find. Results are only known for a restricted class of subset selection rules. Hereby the $k + 1$ populations may be of different types, but each must be a location parameter family with MLR; see Example 2.20. Let now Γ be the set of all priors $\Pi \in \mathcal{P}(\mathfrak{B}_{\Delta^{k+1}})$ that satisfy

$$\Pi(\{\theta : \theta_0 + d_i \leq \theta_i\} \cap \Delta) \leq \pi_i \quad \text{and} \quad \Pi(\{\theta : \theta_i \leq \theta_0\} \cap \Delta) \leq \pi_i',$$

where $\pi_i, \pi_i' \geq 0$ with $\pi_i + \pi_i' \leq 1$, $i = 1, ..., k$, are fixed given.

Theorem 9.72. *Let $Z_0, Z_1, ..., Z_k$ be independent random variables and $W_i = Z_i - Z_0$. Suppose that Z_i has the Lebesgue density*

$$f_{i,\theta_i}(z_i) = f_i(z_i - \theta_i), \quad z_i,\ \theta_i \in \mathbb{R},$$

that has MLR in the identity, $i = 0, 1, ..., k$. Let $g_i(w_i - [\theta_i - \theta_0])$, $w_i \in \mathbb{R}$, be the Lebesgue density of $W_i = Z_i - Z_0$. Let the loss function be from (9.79) and $\tilde{\psi}^\Gamma = (I_{[a_1,\infty)}(z_1 - z_0), ..., I_{[a_k,\infty)}(z_k - z_0))$, $z = (z_0, z_1, ..., z_k) \in \mathbb{R}^{k+1}$, where a_i is chosen such that

$$[L_{2,i}\pi_i' g_i(a) - L_{1,i}\pi_i g_i(a - d_i)](a_i - a) \geq 0, \quad a \in \mathbb{R},\ i = 1, ..., k.$$

Then $\tilde{\psi}^\Gamma$ is Γ-minimax in the class of all $\tilde{\psi}(z) = (\tilde{\psi}_1(z_0, z_1), ..., \tilde{\psi}_k(z_0, z_k))$, $z \in \mathbb{R}^{k+1}$, with $\tilde{\psi}_i : \mathbb{R}^2 \to_m [0, 1]$, $i = 1, ..., k$. Every subset selection rule φ^Γ that has the inclusion probabilities $\tilde{\psi}^\Gamma$ is Γ-minimax in the class of all subset selection rules φ that are allowed to select also the empty set and have inclusion probabilities of the form $\tilde{\psi}(z) = (\tilde{\psi}_1(z_0, z_1), ..., \tilde{\psi}_k(z_0, z_k))$, $z \in \mathbb{R}^{k+1}$, with $\tilde{\psi}_i : \mathbb{R}^2 \to_m [0, 1]$, $i = 1, ..., k$.

The above theorem was first stated in Randles and Hollander (1971). An essential argument is that for every $i \in \{1, ..., k\}$,

$$g_i(w_i - [\theta_i - \theta_0]) = \int f_i(w_i + t - \theta_i) f_0(t - \theta_0) dt, \quad w_i \in \mathbb{R},$$

has MLR if $\theta_i - \theta_0$ is treated as a location parameter of W_i. This result is due to Schoenberg (1951). A proof of the above theorem, which is based on Proposition 3.71, can be found in Miescke (1981). The sequence of priors utilized there is as follows. For every $n = 1, 2, ...$, $\Theta_0 \sim \mathsf{U}(-n, n)$, and given $\Theta_0 = \theta_0$, $\Theta_1, ..., \Theta_k$ are independent, where Θ_i assumes the values $\theta_0 + d_i$, θ_0, and $\theta_0 + d_i/2$ with probabilities π_i, π_i', and $1 - \pi_i - \pi_i'$, respectively, $i = 1, ..., k$.

A special case of the above theorem is the following.

Problem 9.73. Let $X_{i,,j} \sim \mathsf{N}(\mu_i, \sigma^2)$, $j = 1, ..., n_i$, $i = 0, 1, ..., k$, be independent, where the common variance $\sigma^2 > 0$ is known. Let $\bar{x}_i = (1/n_i)\sum_{j=1}^{n_i} x_{i,j}$, $i = 0, 1, ..., k$, be the sample means. Then under the loss function from (9.79) a Γ-minimax subset selection rule φ^Γ in the class of subset selection rules specified in Theorem 9.72 is given by the inclusion probabilities

$$\tilde{\psi}^\Gamma(\bar{x}_0, \bar{x}_1, ..., \bar{x}_k) = (\tilde{\psi}_1^\Gamma(\bar{x}_1 - \bar{x}_0), ..., \tilde{\psi}_k^\Gamma(\bar{x}_k - \bar{x}_0)),$$

where $\tilde{\psi}_i^\Gamma(t) = I_{[e_i,\infty)}(t)$, $t \in \mathbb{R}$, and

$$e_i = \frac{d_i}{2} + \frac{\sigma^2}{d_i}\left(\frac{1}{n_i} + \frac{1}{n_0}\right) \ln \frac{L_{2,i}\pi_i'}{L_{1,i}\pi_i}, \quad i = 1, ..., k.$$

Finally, it should be mentioned that in the setting of Problem 9.73, but with $\sigma^2 > 0$ unknown, minimax subset selection rules under both a common standard and a common control have been derived in Gupta and Miescke (1985) for the loss function

$$L_0^*(\theta, A) = \sum\nolimits_{j \notin A} L_{1,j} I_{[\theta_0,\infty)}(\theta_j) + \sum\nolimits_{i \in A} L_{2,i} I_{(-\infty,\theta_0)}(\theta_i),$$

$A \in \mathcal{D}_{sc}$, $\theta \in \Delta^k$. In this approach Proposition 3.71 has been used with $\Gamma = \mathcal{P}(\mathfrak{B}_{\Delta^k})$. References to other papers on comparing k normal populations with a standard or a control can also be found there.

There are other ways of separating good and bad populations with respect to a standard or control. In the frequentist approach recent work has been done by Huang, Panchapakesan, and Tseng (1984), Finner and Giani, (1994), Giani and Strassburger (1994, 1997, 2000), and Finner, Giani, and Strassburger (2006). The latter is based on a partition principle by Finner and Strassburger (2002a). In the Bayes or empirical Bayes approach recent work has been done by Liang (1997), Gupta and Liang (1999a,b), Gupta and Liese (2000), Gupta and Li (2005), Liang (2006), and Huang and Chang (2006). Further references can be found in these papers.

9.4 Optimal Multistage Selections

Let the selection model $\mathcal{M}_{us}$ from (9.5) be extended in such a way that independent sequences of i.i.d. observations $X_{i,1}, X_{i,2}, ...$ from population P_{θ_i}, $i = 1, ..., k$, become available in the search for a best population. In this section selection models are considered that are based on a one-parameter exponential family that satisfies (A1) and (A2). Let the reduced model $\mathcal{M}_{sel}$ from (9.8) be extended correspondingly. The goal is to find a best population, i.e., a population that has the largest parameter $\theta_{[k]} = \max\{\theta_1, ..., \theta_k\}$. A great variety of sequential and multistage selection rules has been considered in the literature, mainly under the P^*-condition for the PCS, but some also in the decision-theoretic and especially in the Bayes approach. The motivation for developing such procedures is to reduce the expected total number of observations required to make a terminal point or subset selection under a given performance or optimality requirement. Books that include results on multistage selection rules are Bechhofer, Kiefer, and Sobel (1968), Gibbons, Olkin, and Sobel (1977), Gupta and Panchapakesan (1979), Mukhopadhyay and Solanky (1994), and Bechhofer, Santner, and Goldsman (1995).

9.4.1 Common Sample Size per Stage and Hard Elimination

In this section we consider multistage selection rules. These are truncated sequential selection rules where the number of stages is limited to some predetermined number q, say. However, the classical sequential selection rule by

Bechhofer, Kiefer, and Sobel (1968) for selecting t best populations should be at least mentioned here, and subsequently some related rules. Below it is presented for the case of $t = 1$. The generalization to $t \geq 1$ is straightforward.

Example 9.74. Let $\delta^* > 0$ and $P^* \in (1/k, 1)$ be fixed given. The following sequential selection rule for the extended version of model $\mathcal{M}_{us}$ from (9.5), where $(P_\vartheta)_{\vartheta \in \Delta}$ is a one-parameter exponential family, consists of the following three parts.

SAMPLING RULE: At Stage j observe $(X_{1,j}, ..., X_{k,j})$, $j = 1, 2, ...$

STOPPING RULE: Stop at the first Stage m with

$$\sum\nolimits_{l=1}^{k-1} \exp\{-\delta^*(T_{[k]}^{(m)} - T_{[l]}^{(m)})\} \leq (1 - P^*)/P^*,$$

where $T_{[1]}^{(m)} \leq \cdots \leq T_{[k]}^{(m)}$ are the ordered values of

$$T_i^{(m)} = \sum\nolimits_{r=1}^{m} T(X_{i,r}), \quad i = 1, ..., k.$$

TERMINAL DECISION RULE: Select the population i with the largest value of $T_i^{(m)}$, $i = 1, ..., k$. If r populations are tied for the largest value, then select each of them with probability $1/r$.

For this sequential selection rule the probability of correctly selecting the population with the largest parameter $\theta_{[k]}$ is at least P^* whenever $\theta_{[k]} - \theta_{[k-1]} \geq \delta^*$. It stops almost surely in finitely many stages; see Bechhofer, Kiefer, and Sobel (1968) p. 258.

For k normal populations with a common known variance a truncated version (i.e., where the number of stages is restricted by a given number q) has been provided by Bechhofer and Goldsman (1989) which has a smaller expected number of observations. References to previous related work by these authors are given there. Optimality considerations of the above rules should involve, besides the probability of a correct selection (PCS), the expected total number of observations and/or the expected number of stages. Comparisons of several rules in this regard that are based on simulations can be found in Bechhofer, Santner, and Goldsman (1995).

Related sequential selection rules for finding t of k coins with the largest success probabilities can be found in Levin and Robbins (1981), Leu and Levin (1999a,b, 2007), and Levin and Leu (2007).

The earliest multistage selection rule is due to Paulson (1964).

Example 9.75. Let $X_{i,1}, X_{i,2}, ...$ be an i.i.d. sequence from $\mathsf{N}(\mu_i, \sigma^2)$, $i \in S_1 = \{1, ..., k\}$, where $\sigma^2 > 0$ is known. The sequences are assumed to be independent. Let P^*, δ^*, and λ be given constants with $P^* \in (1/k, 1)$ and $0 < \lambda < \delta^*$. For the following multistage selection rule the probability of correctly selecting the population with the largest mean is at least P^* whenever $\mu_{[k]} - \mu_{[k-1]} \geq \delta^*$. Let

$$a_\lambda = \frac{\sigma^2}{\delta^* - \lambda} \ln \frac{k-1}{1 - P^*}$$

and w_λ be the largest integer less than a_λ/λ.

STAGE 1: Observe $X_{i,1}$ with $i \in S_1$. Eliminate all populations $i \in S_1$ with

$$X_{i,1} < \max_{r \in S_1} X_{r,1} - a_\lambda + \lambda.$$

Let the remaining populations be $S_2 \subseteq S_1$. If S_2 contains only one population, then stop and select that population. Otherwise proceed to Stage 2.

The subsequent Stages $m \in \{2, ..., w_\lambda\}$ are set up as follows.

STAGE m: Observe $X_{i,m}$ with $i \in S_m$. Eliminate all populations $i \in S_m$ with

$$\sum_{j=1}^{m} X_{i,j} < \max_{r \in S_m} \sum_{j=1}^{m} X_{r,j} - a_\lambda + m\lambda.$$

Let the remaining populations be $S_{m+1} \subseteq S_m$. If S_{m+1} contains only one population, then stop and select that population. Otherwise proceed to Stage $m+1$.

STAGE $w_\lambda + 1$: Observe $X_{i,w_\lambda+1}$ with $i \in S_{w_\lambda+1}$. Select the population with the largest value of $\sum_{j=1}^{w_\lambda+1} X_{i,j}$, $i \in S_{w_\lambda+1}$. Ties occur only with probability zero.

For this rule, apparently, we have $q = w_\lambda + 1$. Some questions regarding optimality, starting with that of an optimum choice of λ, still remain open. References can be found in Gupta and Panchapakesan (1979). An improved version, due to Paulson, can be found in Bechhofer, Santner, and Goldsman (1995); see also Paulson (1994).

A multistage selection rule for finding a best population consists of four types of decisions that have to be made throughout the q stages. At the beginning it has to be decided which observations to draw from the k populations (*sampling rule*). At every stage, after the observations have been drawn, based on all observations drawn up to that point, it has to be decided whether to stop (*stopping rule*). In the case of stopping a point (or subset) selection has to be made (*terminal decision rule*). In the case of not stopping populations may be eliminated from further sampling (*elimination rule*) and a decision has to be made which observations to draw from the not eliminated populations at the next stage (*sampling rule*).

The simplest stopping rule is never to stop until all q stages have been completed with sampling. Many other more sophisticated stopping rules, such as those in the previous two examples, are possible. The sampling rule may, for example, assign a common fixed sample size to each stage which, however, may differ from stage to stage. In the previous two examples the common sample size for each stage is 1. More generally, the sampling rule for a next stage may depend on the observations drawn thus far. We call this an *adaptive* sampling rule. Such a rule could, for example, decide from which population to draw the next single observation. There are two types of *elimination rules*. A *soft elimination* occurs if a population is eliminated from further sampling but still remains in the pool of populations that can be selected eventually. A *hard elimination* occurs if a population is eliminated from further sampling and selection. The elimination rule could be, for example, to keep the t_m best performing populations from Stage m, $m = 1, ..., q$, where $t_1 \geq t_2 \geq \cdots \geq t_q = 1$ are fixed given. It could also be at every Stage m with $m < q$ some subset selection rule, such as a Gupta's type rule. The optimal terminal decision rule turns out to be the natural selection rule whenever the entire selection problem is permutation invariant and the loss favors selections of populations with larger parameters. With adaptive sampling, however, other terminal decision rules may be optimal, such as those considered in Section 9.2.3. An overview of the literature in this area is provided in Miescke (1984a).

Multistage selection rules can be based on the selection model $\mathcal{M}_{us}$ from (9.5), and especially on $\mathcal{M}_{sel}$ from (9.8). Those which employ terminal point selections can be viewed as competitors to the point selection rules that have been considered in Section 9.2 under such models. Similarly, those with terminal subset selections may compete with the subset selection rules considered in Section 9.3. The basic idea is that for some point or subset selection rule there may exist a multistage selection rule that can complete the same task equally well, but with less observations on average. The majority of papers in this area deal with two-stage selection rules, and some also include decision-theoretic results. Results on two-stage selection rules for Weibull populations with type-II censored data can be found in Gupta and Miescke (1987). A two-stage selection rule for normal populations (see Example 9.76), which utilizes a Gupta-type subset selection rule for screening at the first stage, has been considered by Cohen (1959), Alam (1970), Tamhane and Bechhofer (1977, 1979), Miescke and Sehr (1980), Gupta and Miescke (1982a), Sehr (1988), Bhandari and Chaudhuri (1990), Santner and Hayter (1992), and Hayter (1994). A variant can be found in Santner and Behaxeteguy (1992).

Example 9.76. Let $X_{i,1},...,X_{i,n_1+n_2}$ be an i.i.d. sample from $\mathsf{N}(\mu_i,\sigma^2)$, $i \in S_1 = \{1,...,k\}$, where $\sigma^2 > 0$ is known. The k samples are assumed to be independent. Let $c > 0$ be a fixed given constant.

STAGE 1: Observe $(X_{i,1},...,X_{i,n_1})$, $i = 1,...,k$, and set

$$S_2 = \{r : \sum\nolimits_{j=1}^{n_1} X_{r,j} \geq \max_{i\in S_1} \sum\nolimits_{j=1}^{n_1} X_{i,j} - c, \quad r \in \{1,...,k\}\}.$$

If S_2 contains only one population, then stop and select that population. Otherwise proceed to the next stage.

STAGE 2: Observe $(X_{i,n_1+1},...,X_{i,n_1+n_2})$ with $i \in S_2$ and select the population with the largest value of $\sum_{j=1}^{n_1+n_2} X_{i,j}$, $i \in S_2$.

To guarantee that the probability of correctly selecting the population with the largest mean is at least P^* whenever $\mu_{[k]} - \mu_{[k-1]} \geq \delta^*$, where $P^* \in (1/k, 1)$ and $\delta^* > 0$ are given constants, the least favorable parameter configuration (LFC) for $\mu = (\mu_1,...,\mu_k) \in \mathbb{R}^k$ with $\mu_{[k]} - \mu_{[k-1]} \geq \delta^*$ has to be found to determine the right value of c. A long standing conjecture that the LFC is $(t,...,t,t+\delta^*)$, $t \in \mathbb{R}$, has been proved by Miescke and Sehr (1980) to be correct for $k = 3$. Further results in this respect can be found in the references given above.

A variant, taken from Gupta and Miescke (1982a), is the following.

Example 9.77. In the setting of Example 9.76 other subset selection rules could be used for screening. Instead of the S_2 there we take

$$S_2 = \{r : \sum\nolimits_{j=1}^{n_1} X_{r,j} \text{ is one of the } t \text{ largest values of } \sum\nolimits_{j=1}^{n_1} X_{i,j}, \ i \in S_1\},$$

where $t \in \{2, ..., k-1\}$ is fixed given. S_2 is almost surely uniquely determined. To guarantee that the PCS is at least P^* for $\mu_{[k]} - \mu_{[k-1]} \geq \delta^*$, where $P^* \in (1/k, 1)$ and $\delta^* > 0$ are given constants, the LFC for $\mu = (\mu_1, ..., \mu_k) \in \mathbb{R}^k$ with $\mu_{[k]} - \mu_{[k-1]} \geq \delta^*$ has to be found to determine appropriate sample sizes n_1 and n_2. Also here the LFC is the slippage configuration $(t, ..., t, t + \delta^*)$, $t \in \mathbb{R}$.

Another variant, also from Gupta and Miescke (1982a), this time with a standard, is as follows.

Example 9.78. In the setting of Example 9.76 let $\mu_0 \in \mathbb{R}$ be a given standard. Instead of the S_2 there we take

$$S_2 = \{r : \sum\nolimits_{j=1}^{n_1} X_{r,j} \geq b_r, \ \ r \in S_1\},$$

where $b_1, ..., b_k \in \mathbb{R}$ are fixed given constants. The screening device is now a multiple testing procedure, simultaneously testing each population if it is better than the standard. The set S_2 may turn out to be empty, in which case it is decided that none of the populations is better than the control.

We set up the requirement that the probability of S_2 being empty should be at least β^* whenever $\mu_1, ..., \mu_k \leq \mu_0$, where $\beta^* \in (0, 1)$ is fixed given. Moreover, we require that the PCS on all $\mu = (\mu_1, ..., \mu_k) \in \mathbb{R}^k$ with $\mu_{[k]} - \mu_{[k-1]} \geq \delta^*$ and $\mu_{[k]} > \mu_0$ be at least P^*, where $P^* \in (1/k, 1)$ and $\delta^* > 0$ are given constants. Here the LFC is $(t, ..., t, t + \delta^*)$ with $t + \delta^* > \mu_0$, $t \in \mathbb{R}$.

The problem of finding a best population with respect to a standard or control in two stages has also been treated in Gupta and Miescke (1982b) and Miescke (1984b). For models with the DT property, if the selection problem is permutation invariant and the loss favors selections of larger parameters, two-stage selection rules that select in terms of the largest sufficient statistics at both stages are optimal. Such a generalization of the Bahadur–Goodman–Lehmann–Eaton theorem 9.31 is intuitively to be expected, and it has in fact been proved in Gupta and Miescke (1982b). More work in this direction, for models with and without standard or control, has been done by Gupta and Miescke (1983, 1984b). The most general results in relation to Theorem 9.31 have been established in Gupta and Miescke (1984a).

In Gupta and Miescke (1984a) sequential selection rules for exponential families have been studied in the decision-theoretic approach. Let $\mathcal{M}_{ne} = (\mathcal{X}, \mathfrak{A}, (P_\vartheta)_{\vartheta \in \Delta})$, $\Delta \subseteq \mathbb{R}$, from (1.7) with $d = 1$ satisfy (A1) and (A2) and be chosen as the basic underlying natural exponential model. We restrict here our considerations to the results on multistage selection rules with q stages. In the presence of a standard, terminally selected subsets may be allowed to be empty. Suppose that at Stage m there are n_m observations $X_{i,j}^{(m)}$, $j = 1, ..., n_m$, available from population $i \in \{1, ..., k\}$, $m = 1, ..., q$, where $n_1, ..., n_q$ are fixed given. All $N_q = k(n_1 + \cdots + n_q)$ observations are assumed to be independent. The model is a version of $\mathcal{M}_{bs}$ from (9.4).

$$\mathcal{M}_{mss1} = (\mathcal{X}^{N_q}, \mathfrak{A}^{\otimes N_q}, (\bigotimes\nolimits_{i=1}^{k} \bigotimes\nolimits_{m=1}^{q} P_{\theta_i}^{\otimes n_m})_{\theta \in \Delta^k}), \tag{9.80}$$
$$\frac{dP_{\theta_i}}{d\boldsymbol{\mu}}(x) = \exp\{\theta_i T(x) - K(\theta_i)\}, \quad x \in \mathcal{X}.$$

Denote by $X_{i,j}^{(m)}$ the projections of $\mathcal{X}^{N_q}$ on the coordinates. A reduction by sufficiency at each stage, restricted to the observations drawn at that stage leads, as in Section 9.1, to the statistics

$$U_{i,m} = \sum\nolimits_{j=1}^{n_m} T(X_{i,j}^{(m)}), \quad m = 1, ..., q, \quad i = 1, ..., k. \tag{9.81}$$

If $Q_\theta = P_\theta \circ T^{-1}$, then $Q_{\theta_i}^{*n_m} = \mathcal{L}(U_{i,m}|P_{\theta_i}^{\otimes n_m})$, $m = 1, ..., q$, $i = 1, ..., k$, and the resulting model, analogously to $\mathcal{M}_{sel}$ in (9.8), is

$$\begin{aligned} \mathcal{M}_{mss} &= (\mathbb{R}^{kq}, \mathfrak{B}_{kq}, (\bigotimes\nolimits_{i=1}^{k} \bigotimes\nolimits_{m=1}^{q} Q_{n_m,\theta_i})_{\theta \in \Delta^k}), \\ Q_{n_m,\theta_i} &= Q_{\theta_i}^{*n_m}. \end{aligned} \tag{9.82}$$

Problem 9.79.* Set $\boldsymbol{\nu}_{n_m} = \boldsymbol{\mu}^{\otimes n_m} \circ T_{\oplus n_m}^{-1}$ and

$$f_{N_q,\theta}(\mathbf{t}) = \exp\{\mathbf{t}^T\theta - \sum\nolimits_{i=1}^{k} (n_1 + \cdots + n_q)K(\theta_i)\}. \tag{9.83}$$

Then $f_{N_q,\theta}(\sum_{m=1}^{q} \mathbf{u}_m)$ is the density of $\bigotimes_{i=1}^{k} \bigotimes_{m=1}^{q} Q_{n_m,\theta_i}$ with respect to $\bigotimes_{m=1}^{q} \boldsymbol{\nu}_{n_m}^{\otimes k}$, where $\mathbf{u}_m = (u_{1,m}, ..., u_{k,m})^T$, $\theta = (\theta_1, ..., \theta_k)^T$.

For $m = 1, ..., q$ let

$$\mathbf{U}_m = (U_{1,m}, ..., U_{k,m}), \quad \mathbf{V}_m = (\mathbf{U}_1, ..., \mathbf{U}_m), \quad \mathbf{W}_m = \mathbf{U}_1 + \cdots + \mathbf{U}_m. \tag{9.84}$$

The multistage selection rules considered here are first described briefly before a formal definition is presented. Such a rule decides at every stage, after sampling at that stage has been completed, either to stop (ξ), how many populations to retain (φ), and which populations to select as a terminal decision (ψ), or not to stop ($1 - \xi$), how many populations to retain ($\tilde{\varphi}$), and which populations to select for further examination at the next stage ($\tilde{\psi}$). At Stage q only stopping is allowed. These rules are assumed to use hard elimination. Once a population has been eliminated it can never be sampled or selected at any subsequent stage. Hard elimination guarantees that the various decisions that are made in the course of a multistage selection rule are based on an equal number of observations from each of the populations that are still in the competition. This allows us to utilize optimization techniques similar to those used to prove Theorem 9.31 and Proposition 9.59.

Because of the complexity of the setup of q-stage selection rules we consider first the special case of $q = 2$ in full detail, and then report the general results without proofs. The optimization steps and technical tools can be developed and explained already with two-stage selection rules. Stage 1 starts with all populations $S_1 = \{1, ..., k\}$ and ends with populations $S_2 \subseteq S_1$. Stage 2, if entered, starts with populations S_2 and ends with populations $S_3 \subseteq S_2$. At Stage 2 observations are made only from populations S_2. To deal with that restriction on the observations we utilize the projection $p_{S_2} : \mathbb{R}^k \to \mathbb{R}^{|S_2|}$ onto the coordinates i with $i \in S_2$.

The decision space $\mathcal{D}_{2st}$ is chosen to consist of two components, one for decisions that are made without ever entering Stage 2, and all others. We set $\mathcal{D}_{2st} = \mathcal{D}_1 \cup \mathcal{D}_2$, where

$$\begin{aligned}\mathcal{D}_1 &= \{(1, r_2, S_2) : S_2 \subseteq S_1,\ |S_2| = r_2,\ 0 \le r_2 \le k\} \quad \text{and} \\ \mathcal{D}_2 &= \{(2, r_2, S_2, r_3, S_3) : S_3 \subseteq S_2 \subseteq S_1,\ |S_2| = r_2,\ |S_3| = r_3, \\ &\qquad 0 \le r_3 \le r_2,\ 1 \le r_2 \le k\}.\end{aligned}$$

The two-stage selection rules are now introduced as distributions on $\mathcal{D}_{2st}$ in dependence of the observations $\mathbf{U}_1 = \mathbf{u}_1$ and $\mathbf{U}_2 = \mathbf{u}_2$. They are constructed by means of several components. Let $\xi : \mathbb{R}^k \to_m [0,1]$, which acts as the stopping rule. We consider functions

$$\varphi_{r_2}, \psi_{S_2|r_2} : \mathbb{R}^k \to_m [0,1]$$

for $(1, r_2, S_2) \in \mathcal{D}_1$, and functions

$$\widetilde{\varphi}_{r_2}, \widetilde{\psi}_{S_2|r_2} : \mathbb{R}^k \to_m [0,1], \quad \varphi_{r_3|S_2}, \psi_{S_3|r_3,S_2} : \mathbb{R}^k \times \mathbb{R}^{|S_2|} \to_m [0,1]$$

for $(2, r_2, S_2, r_3, S_3) \in \mathcal{D}_2$. We assume that

$$\sum\nolimits_{r_2=0}^{k} \varphi_{r_2} = \sum\nolimits_{r_2=1}^{k} \widetilde{\varphi}_{r_2} = 1, \quad \sum\nolimits_{r_3=0}^{|S_2|} \varphi_{r_3|S_2} = 1,\ \emptyset \ne S_2 \subseteq S_1, \tag{9.85}$$

and

$$\begin{aligned}&\sum\nolimits_{S_2 \subseteq S_1,\ |S_2| = r_2} \psi_{S_2|r_2} = 1, \quad 0 \le r_2 \le k, \\ &\sum\nolimits_{S_2 \subseteq S_1,\ |S_2| = r_2} \widetilde{\psi}_{S_2|r_2} = 1, \quad 1 \le r_2 \le k, \\ &\sum\nolimits_{S_3 \subseteq S_2,\ |S_3| = r_3} \psi_{S_3|r_3,S_2} = 1, \quad 0 \le r_3 \le |S_2|,\ \emptyset \ne S_2 \subseteq S_1.\end{aligned} \tag{9.86}$$

Definition 9.80. *A two-stage selection rule* $(\xi, \varphi, \widetilde{\varphi}, \psi, \widetilde{\psi})$ *is, for every fixed* $(\mathbf{u}_1, \mathbf{u}_2) \in \mathbb{R}^{2k}$*, a distribution* $\mathsf{D}(\cdot|\mathbf{u}_1, \mathbf{u}_2)$ *on* $\mathcal{D}_{2st}$ *that satisfies*

$$\mathsf{D}(\{(1, r_2, S_2)\}|\mathbf{u}_1, \mathbf{u}_2) = \xi(\mathbf{u}_1)\varphi_{r_2}(\mathbf{u}_1)\psi_{S_2|r_2}(\mathbf{u}_1),$$

for $(1, r_2, S_2) \in \mathcal{D}_1$*, and*

$$\begin{aligned}&\mathsf{D}(\{(2, r_2, S_2, r_3, S_3)\}|\mathbf{u}_1, \mathbf{u}_2) \\ &= (1 - \xi(\mathbf{u}_1))\widetilde{\varphi}_{r_2}(\mathbf{u}_1)\widetilde{\psi}_{S_2|r_2}(\mathbf{u}_1)\varphi_{r_3|S_2}(\mathbf{u}_1, p_{S_2}(\mathbf{u}_2))\psi_{S_3|r_3,S_2}(\mathbf{u}_1, p_{S_2}(\mathbf{u}_2)),\end{aligned}$$

for $(2, r_2, S_2, r_3, S_3) \in \mathcal{D}_2$*, where the components of* $(\xi, \varphi, \widetilde{\varphi}, \psi, \widetilde{\psi})$ *are as specified above and satisfy (9.85) and (9.86).*

The interpretation of a two-stage selection rule $(\xi, \varphi, \widetilde{\varphi}, \psi, \widetilde{\psi})$ is as follows.

STAGE 1: At the beginning of this stage $\mathbf{U}_1 = \mathbf{u}_1 = (u_{1,1}, ..., u_{k,1})$ is observed. The rule decides with probability $\xi(\mathbf{u}_1)$ to stop, and with probability $1-\xi(\mathbf{u}_1)$ not to stop.

If it stops, then it decides with probability $\varphi_{r_2}(\mathbf{u}_1)$ that $r_2 \in \{0, 1, ..., k\}$ populations should be selected from S_1. After that it selects with probability $\psi_{S_2|r_2}(\mathbf{u}_1)$ any subset $S_2 \subseteq S_1$ of size $|S_2| = r_2$, which is the terminal decision.

If it doesn't stop, then it decides with probability $\widetilde{\varphi}_{r_2}(\mathbf{u}_1)$ that $r_2 \in \{1, ..., k\}$ populations should be selected from S_1. After that it selects with probability $\widetilde{\psi}_{S_2|r_2}(\mathbf{u}_1)$ any subset $S_2 \subseteq S_1$ of size $|S_2| = r_2$. Now Stage 2 is entered with S_2.

STAGE 2: At the beginning of this stage $p_{S_2}(\mathbf{U}_2) = p_{S_2}(\mathbf{u}_2) = (u_{i_1,2}, ..., u_{i_{r_2},2})$ is observed, where $1 \le i_1 \le \cdots \le i_{r_2} \le k$ and $\{i_1, ..., i_{r_2}\} = S_2$.

It decides with probability $\varphi_{r_3|S_2}(\mathbf{u}_1, p_{S_2}(\mathbf{u}_2))$ that $r_3 \in \{0, 1, ..., r_2\}$ populations should be selected from S_2. After that it selects with probability $\psi_{S_3|r_3,S_2}(\mathbf{u}_1, p_{S_2}(\mathbf{u}_2))$ any subset $S_3 \subseteq S_2$ of size $|S_3| = r_3$, which is the terminal decision.

Remark 9.81. The above definition includes one-stage selection rules as a special case. They are obtained by setting $\xi \equiv 1$.

Remark 9.82. The above definition includes two-stage selection rules for problems with standards. In such cases it makes sense to allow entering Stage 2 with a subset S_2 of size 1. This would simply mean that one population has not been decided yet to be good enough and further sampling on it is needed. On the other hand, without a standard every subset S_2 for Stage 2 should contain at least two populations. This can be enforced by imposing restrictions on $\widetilde{\varphi}_{r_2}$, or by including cost of sampling in the loss.

Problem 9.83. Verify that the selection rules in Examples 9.76, 9.77, and 9.78 are two-stage selection rules in the sense of Definition 9.80.

We focus now on the goal of optimizing the two components ψ and $\widetilde{\psi}$, thereby generalizing Theorem 9.31 and Proposition 9.59. Hereby we have to restrict ourselves to permutation invariant two-stage selection rules, and to loss functions that are permutation invariant and favor selections of populations with larger parameters.

For every $\gamma \in \Pi_k$ and $S_m \subseteq S_1$ we set $\gamma(S_m) = \{\gamma(i) : i \in S_m\}$, $m = 1, 2, 3$, and $\gamma(\mathbf{u}_m) = (u_{\gamma(1),m}, ..., u_{\gamma(k),m})$, $m = 1, 2$.

Definition 9.84. *A two-stage selection rule $(\xi, \varphi, \tilde{\varphi}, \psi, \tilde{\psi})$ is called permutation invariant if the five types of decision functions are permutation invariant in the following sense. For any $\gamma \in \Pi_k$ it holds $\mathbb{P}_{(\mathbf{U}_1,\mathbf{U}_2)}$-a.s.*

$$\begin{aligned}
\xi(\mathbf{u}_1) &= \xi(\gamma(\mathbf{u}_1)), \\
\varphi_{r_2}(\mathbf{u}_1) &= \varphi_{r_2}(\gamma(\mathbf{u}_1)), \\
\varphi_{r_3|\gamma(S_2)}(\mathbf{u}_1, p_{\gamma(S_2)}(\mathbf{u}_2)) &= \varphi_{r_3|S_2}(\gamma(\mathbf{u}_1), p_{S_2}(\gamma(\mathbf{u}_2))) \\
\psi_{\gamma(S_2)|r_2}(\mathbf{u}_1) &= \psi_{S_2|r_2}(\gamma(\mathbf{u}_1)) \\
\psi_{\gamma(S_3)|r_3,\gamma(S_2)}(\mathbf{u}_1, p_{\gamma(S_2)}(\mathbf{u}_2)) &= \psi_{S_3|r_3,S_2}(\gamma(\mathbf{u}_1), p_{S_2}(\gamma(\mathbf{u}_2))),
\end{aligned}$$

and $\tilde{\varphi}$ and $\tilde{\psi}$ have the same properties as φ and ψ, respectively.

Problem 9.85. Verify that the selection rules in Examples 9.76, 9.77, and 9.78 are permutation invariant two-stage selection rules in the sense of Definition 9.84.

Finally, we adopt a class of loss functions that are permutation invariant and favor selections of populations with larger parameters. Let $L_1(\theta, S_2)$ be the loss that occurs at $\theta \in \Delta^k$ if at Stage 1 the procedure stops and selects S_2, where $L_1(\cdot, S_2) : \Delta^k \to_m \mathbb{R}_+$. Let $L_2(\theta, S_2, S_3)$ be the loss that occurs at $\theta \in \Delta^k$ if at Stage 1 subset $S_2 \subseteq S_1$, and at Stage 2 subset $S_3 \subseteq S_2$, is selected, where $L_2(\cdot, S_2, S_3) : \Delta^k \to_m \mathbb{R}_+$. Analogously to (9.69) and (9.70) we assume that for all $S_3 \subseteq S_2 \subseteq S_1$,

$$\begin{aligned} L_1(\theta, \gamma(S_2)) &= L_1(\gamma(\theta), S_2), \\ L_2(\theta, \gamma(S_2), \gamma(S_3)) &= L_2(\gamma(\theta), S_2, S_3), \quad \gamma \in \Pi_k,\ \theta \in \Delta^k, \end{aligned} \tag{9.87}$$

and, moreover,

$$\begin{aligned} L_1(\theta, \widetilde{S}_2) &\le L_1(\theta, S_2), \\ L_2(\theta, \widetilde{S}_2, \widetilde{S}_3) &\le L_2(\theta, S_2, S_3), \end{aligned} \tag{9.88}$$

if at $\theta \in \Delta^k$ for some $i, j \in \{1, ..., k\}$ with $\theta_i \le \theta_j$ the following holds. For every $c \in \{2, 3\}$ with $i \in S_c$ and $j \notin S_c$, $\widetilde{S}_c = (S_c \backslash \{i\}) \cup \{j\}$, and $\widetilde{S}_c = S_c$, otherwise. This means that a worse population should be eliminated at an earlier stage than a better population.

Problem 9.86. Determine the loss functions used in Examples 9.76, 9.77, and 9.78, and verify that they satisfy (9.87) and (9.88).

In this permutation invariant selection problem a two-stage selection rule $(\xi, \varphi, \tilde{\varphi}, \psi, \tilde{\psi})$ can often be improved by replacing ψ by ψ^{nat} and $\tilde{\psi}$ by $\tilde{\psi}^{nat}$, where ψ^{nat} and $\tilde{\psi}^{nat}$ are the natural selections defined below.

Definition 9.87. *For every fixed $\mathbf{u}_1$ let $\psi^{nat}_{S_2|r_2}(\mathbf{u}_1)$ be the uniform distribution on the sets $S_2 \subseteq S_1$ with $|S_2| = r_2$ and $\max\{u_{i,1} : i \in S_1 \backslash S_2\} \le \min\{u_{i,1} : i \in S_2\}$. Let $\tilde{\psi}^{nat}_{S_2|r_2}(\mathbf{u}_1) = \psi^{nat}_{S_2|r_2}(\mathbf{u}_1)$. For every fixed $(\mathbf{u}_1, \mathbf{u}_2)$ let $\psi^{nat}_{S_3|r_3,S_2}(\mathbf{u}_1, p_{S_2}(\mathbf{u}_2))$ be the uniform distribution on the sets $S_3 \subseteq S_2$ with $|S_3| = r_3$ and $\max\{u_{i,1} + u_{i,2} : i \in S_2 \backslash S_3\} \le \min\{u_{i,1} + u_{i,2} : i \in S_3\}$.*

In a first step we consider the switch from $(\xi, \varphi, \tilde{\varphi}, \psi, \tilde{\psi})$ to $(\xi, \varphi, \tilde{\varphi}, \psi^{nat}, \tilde{\psi})$ which affects only the terminal decisions. Under a loss with the properties (9.87) and (9.88) a permutation invariant two-stage selection rule should be modified by making the terminal decisions with the natural selections ψ^{nat}. This is the content of the next theorem.

The properties (9.87) and (9.88) of the loss carry over to the posterior risks under favorable circumstances. To show and to utilize this we introduce a class of auxiliary functions. For this purpose we represent (S_1, S_2) with $S_2 \subseteq S_1$ by (T_1, T_2), where $T_1 = S_1 \backslash S_2$ and $T_2 = S_2$. Let $\mathfrak{T}_2$ denote the set of all decompositions (T_1, T_2) of S_1. Likewise, let us represent (S_1, S_2, S_3) with $S_3 \subseteq S_2 \subseteq S_1$ by (T_1, T_2, T_3), where $T_1 = S_1 \backslash S_2$, $T_2 = S_2 \backslash S_3$, and $T_3 = S_3$. Let $\mathfrak{T}_3$ denote the set of all decompositions (T_1, T_2, T_3) of S_1. Now, for a fixed Borel set $A \subseteq \mathbb{R}$ let $\mathfrak{L}_1(\cdot, T_1, T_2) : A^k \to_m \mathbb{R}_+$, $(T_1, T_2) \in \mathfrak{T}_2$, and $\mathfrak{L}_2(\cdot, T_1, T_2, T_3) : A^k \to_m \mathbb{R}_+$, $(T_1, T_2, T_3) \in \mathfrak{T}_3$.

Definition 9.88. *The function $\mathfrak{L}_1$ is said to have the property $\mathfrak{D}(1, A)$ if for every $a \in A^k$, $\gamma \in \Pi_k$, and $(T_1, T_2) \in \mathfrak{T}_2$ the following holds.*

$$\begin{aligned} \mathfrak{L}_1(a, \gamma(T_1), \gamma(T_2)) &= \mathfrak{L}_1(\gamma(a), T_1, T_2) \quad \textit{and} \\ \mathfrak{L}_1(a, \widetilde{T}_1, \widetilde{T}_2) &\leq \mathfrak{L}_1(a, T_1, T_2), \end{aligned}$$

if for some $i, j \in \{1, ..., k\}$ with $a_i \leq a_j$, $i \in T_2$, $j \in T_1$, $\widetilde{T}_1 = (T_1 \backslash \{j\}) \cup \{i\}$, and $\widetilde{T}_2 = (T_2 \backslash \{i\}) \cup \{j\}$.

The function $\mathfrak{L}_2$ is said to have the property $\mathfrak{D}(2, A)$ if for every $a \in A^k$, $\gamma \in \Pi_k$, and $(T_1, T_2, T_3) \in \mathfrak{T}_3$ the following holds.

$$\begin{aligned} \mathfrak{L}_2(a, \gamma(T_1), \gamma(T_2), \gamma(T_3)) &= \mathfrak{L}_2(\gamma(a), T_1, T_2, T_3) \quad \textit{and} \\ \mathfrak{L}_2(a, \widetilde{T}_1, \widetilde{T}_2, \widetilde{T}_3) &\leq \mathfrak{L}_2(a, T_1, T_2, T_3), \end{aligned}$$

if for some $i, j \in \{1, ..., k\}$ with $a_i \leq a_j$, and $1 \leq \alpha < \beta \leq 3$, $j \in T_\alpha$, $i \in T_\beta$, $\widetilde{T}_\alpha = (T_\alpha \backslash \{j\}) \cup \{i\}$, and $\widetilde{T}_\beta = (T_\beta \backslash \{i\}) \cup \{j\}$, and $\widetilde{T}_\rho = T_\rho$ for $\rho \notin \{\alpha, \beta\}$.

Problem 9.89. Suppose that for $\theta \in \Delta^k$, $\mathfrak{L}_1(\theta, T_1, T_2) = L_1(\theta, T_2)$, $(T_1, T_2) \in \mathfrak{T}_2$, and $\mathfrak{L}_2(\theta, T_1, T_2, T_3) = L_2(\theta, T_2 \cup T_3, T_3)$, $(T_1, T_2, T_3) \in \mathfrak{T}_3$. Then for $m = 1, 2$ the following holds. L_m satisfies the respective conditions (9.87) and (9.88) if and only if $\mathfrak{L}_m$ has the property $\mathfrak{D}(m, \Delta)$.

The key lemma of this section is as follows.

Lemma 9.90. *Suppose $\mathsf{K} : \mathfrak{B}_{A^k} \times B^k \to_k [0, 1]$ is permutation invariant and has the DT property; see Definition 9.17. If $\mathfrak{L}_1$ has the property $\mathfrak{D}(1, A)$ and*

$$\widetilde{\mathfrak{L}}_1(b, T_1, T_2) = \int \mathfrak{L}_1(a, T_1, T_2) \mathsf{K}(da|b), \quad b \in B^k, \ (T_1, T_2) \in \mathfrak{T}_2, \tag{9.89}$$

then $\widetilde{\mathfrak{L}}_1$ has the property $\mathfrak{D}(1, B)$. If $\mathfrak{L}_2$ has the property $\mathfrak{D}(2, A)$ and

$$\widetilde{\mathfrak{L}}_2(b, T_1, T_2, T_3) = \int \mathfrak{L}_2(a, T_1, T_2, T_3) \mathsf{K}(da|b), \ b \in B^k, \ (T_1, T_2, T_3) \in \mathfrak{T}_3, \tag{9.90}$$

then $\widetilde{\mathfrak{L}}_2$ has the property $\mathfrak{D}(2, B)$.

Proof. The proof can be reduced to the proof of Lemma 9.27. For fixed $i, j \in \{1, ..., k\}$ and $a = (a_1, ..., a_k) \in A^k$ we denote by $a^{(i,j)}$ the vector for which the components a_i and a_j are interchanged whereas the other components remain unchanged. Let M, L be two nonnegative measurable functions on A^k that have the following properties

$$L(a^{(i,j)}) = M(a) \quad \text{and} \quad L(a)I_{[a_i,\infty)}(a_j) \geq M(a)I_{[a_i,\infty)}(a_j). \tag{9.91}$$

We have shown in the proof of Lemma 9.27 that for $b_i \leq b_j$ it holds

$$\int L(a)\mathsf{K}(da|b) \geq \int M(a)\mathsf{K}(da|b).$$

Let T_1, T_2 be a partition of S_2. Suppose that $a_i \leq a_j$ for some $i \in S_1$ and $j \neq S_1$. Set $L(a) = \mathfrak{L}_1(a, T_1, T_2)$. Then $L(a^{(i,j)}) = \mathfrak{L}_1(a, \widetilde{T}_1, \widetilde{T}_2) =: M(a)$.The property $\mathfrak{D}(1, A)$ for $\mathfrak{L}_1$ implies that (9.91) is satisfied and the first statement follows. The proof of the second statement is similar. ■

The risk of a two-stage selection rule $(\xi, \varphi, \tilde{\varphi}, \psi, \tilde{\psi})$ at $\theta \in \Delta^k$ is defined by

$$\begin{aligned}
&\mathsf{R}(\theta, (\xi, \varphi, \tilde{\varphi}, \psi, \tilde{\psi})) = \sum\nolimits_{S_2 \subseteq S_1} L_1(\theta, S_2)\mathsf{E}_\theta[\xi(\mathbf{U}_1)\varphi_{|S_2|}(\mathbf{U}_1)\psi_{S_2||S_2|}(\mathbf{U}_1)] \\
&+ \sum\nolimits_{\emptyset \neq S_2 \subseteq S_1} \sum\nolimits_{S_3 \subseteq S_2} L_2(\theta, S_2, S_3)\mathsf{E}_\theta[(1 - \xi(\mathbf{U}_1))\widetilde{\varphi}_{|S_2|}(\mathbf{U}_1)\widetilde{\psi}_{S_2||S_2|}(\mathbf{U}_1) \\
&\qquad \times \varphi_{|S_3||S_2}(\mathbf{U}_1, p_{S_2}(\mathbf{U}_2))\psi_{S_3||S_3|,S_2}(\mathbf{U}_1, p_{S_2}(\mathbf{U}_2))],
\end{aligned}$$

which, apparently, is finite. The Bayes risk under a permutation invariant prior Π on the Borel sets of Δ^k is used to establish ψ^{nat} and $\tilde{\psi}^{nat}$ as improvements over ψ and $\tilde{\psi}$, respectively, of a permutation invariant two-stage selection rule, similarly as has been done in Proposition 9.59. Again, also here, the question regarding the optimization of φ and $\tilde{\varphi}$, which decide on the sizes of the subsets to be selected, remains open as this depends on the form of the loss functions chosen for the individual stages and on the prior; see Gupta and Miescke (1984b). The Bayes risk is defined by

$$\mathsf{r}(\Pi, (\xi, \varphi, \tilde{\varphi}, \psi, \tilde{\psi})) = \int \mathsf{R}(\theta, (\xi, \varphi, \tilde{\varphi}, \psi, \tilde{\psi}))\Pi(d\theta).$$

To evaluate this risk we turn to the Bayes model

$$\begin{aligned}
(\Omega, \mathfrak{F}, \mathbb{P}) &= (\mathcal{X}^{N_q} \times \Delta^k, \mathfrak{A}^{\otimes N_q} \otimes \mathfrak{B}_{\Delta^k}, \mathsf{P}_{N_q} \otimes \Pi), \\
\mathsf{P}_{N_q}(\cdot|\theta) &= \bigotimes\nolimits_{i=1}^k \bigotimes\nolimits_{m=1}^q P_{\theta_i}^{\otimes n_m},
\end{aligned}$$

and denote by Θ the projection of $\mathcal{X}^{N_q} \times \Delta^k$ on Δ^k. Then by Problem 9.79

$$\frac{d((\bigotimes_{i=1}^k Q_{\theta_i}^{*n_1}) \otimes (\bigotimes_{i=1}^k Q_{\theta_i}^{*n_2}))}{d(\boldsymbol{\nu}_{n_1}^{\otimes k} \otimes \boldsymbol{\nu}_{n_2}^{\otimes k})}(\mathbf{u}_1, \mathbf{u}_2) = f_{N_2,\theta}(\mathbf{u}_1 + \mathbf{u}_2).$$

Introduce the marginal density, conditional density, and conditional distribution by

$$\mathsf{m}_{n_1,n_2}(\mathbf{w}) = \int f_{N_2,\theta}(\mathbf{w})\Pi(d\theta) \quad \text{and} \quad f(\theta|\mathbf{w}) = \frac{f_{N_2,\theta}(\mathbf{w})}{\mathsf{m}_{n_1,n_2}(\mathbf{w})}, \quad \mathbf{w} \in \mathbb{R}^k,$$

$$\mathsf{K}(B|\mathbf{w}) = \int I_B(\theta) f(\theta|\mathbf{w})\Pi(d\theta), \quad \mathbf{w} \in \mathbb{R}^k. \tag{9.92}$$

The expression for $f_{N_2,\theta}(\mathbf{w})$ in (9.83) gives

$$f(\theta|\mathbf{w}) = (\int \exp\{\mathbf{w}^T\theta - N_2 \sum\nolimits_{i=1}^k K(\theta_i)\}\Pi(d\theta))^{-1} \tag{9.93}$$
$$\times \exp\{\mathbf{w}^T\theta - N_2 \sum\nolimits_{i=1}^k K(\theta_i)\}.$$

Furthermore,

$$\mathbb{E}h(\Theta, \mathbf{U}_1, \mathbf{U}_2) = \mathbb{E} \int h(\theta, \mathbf{U}_1, \mathbf{U}_2)\mathsf{K}(d\theta|\mathbf{U}_1 + \mathbf{U}_2), \tag{9.94}$$

for every $h : \Delta^k \times \mathbb{R}^k \times \mathbb{R}^k \to_m \mathbb{R}_+$. This implies that $\mathbf{U}_1 + \mathbf{U}_2$ is Bayes sufficient and the kernel K in (9.92) is a regular conditional distribution of Θ given $\mathbf{U}_1, \mathbf{U}_2$. Then the Bayes risk of a two-stage selection rule $(\xi, \varphi, \tilde{\varphi}, \psi, \tilde{\psi})$ under a prior $\Pi \in \mathcal{P}(\mathfrak{B}_{\Delta^k})$ can be represented as follows.

$$\begin{aligned} &\mathsf{r}(\Pi, (\xi, \varphi, \tilde{\varphi}, \psi, \tilde{\psi})) \qquad (9.95)\\ &= \mathbb{E}[\xi(\mathbf{U}_1) \sum\nolimits_{r_2=0}^k \varphi_{r_2}(\mathbf{U}_1) \sum\nolimits_{S_2 \subseteq S_1, |S_2|=r_2} \psi_{S_2|r_2}(\mathbf{U}_1)\mathbb{E}(L_1(\Theta, S_2)|\mathbf{U}_1)\\ &+ (1 - \xi(\mathbf{U}_1)) \sum\nolimits_{r_2=1}^k \tilde{\varphi}_{r_2}(\mathbf{U}_1) \sum\nolimits_{S_2 \subseteq S_1, |S_2|=r_2} \tilde{\psi}_{S_2|r_2}(\mathbf{U}_1)\\ &\times \mathbb{E}\{\sum\nolimits_{r_3=0}^{r_2} \varphi_{r_3|S_2}(\mathbf{U}_1, p_{S_2}(\mathbf{U}_2))\\ &\times \sum\nolimits_{S_3 \subseteq S_2, |S_3|=r_3} \psi_{S_3|r_3,S_2}(\mathbf{U}_1, p_{S_2}(\mathbf{U}_2))\mathbb{E}(L_2(\Theta, S_2, S_3)|\mathbf{U}_1, \mathbf{U}_2)|\mathbf{U}_1\}]. \end{aligned}$$

Now we can state the following.

Theorem 9.91. *Given the model $\mathcal{M}_{mss1}$ in (9.80) and the decision space $\mathcal{D}_{2st}$ let the loss L_1 and L_2 satisfy (9.87) and (9.88). Then for every permutation invariant two-stage selection rule $(\xi, \varphi, \tilde{\varphi}, \psi, \tilde{\psi})$ it holds*

$$\mathsf{R}(\theta, (\xi, \varphi, \tilde{\varphi}, \psi^{nat}, \tilde{\psi})) \le \mathsf{R}(\theta, (\xi, \varphi, \tilde{\varphi}, \psi, \tilde{\psi})), \quad \theta \in \Delta^k.$$

Proof. For a fixed permutation invariant prior $\Pi \in \mathcal{P}(\mathfrak{B}_{\Delta^k})$ it follows from (9.94) that for $\mathbf{W}_2 = \mathbf{U}_1 + \mathbf{U}_2$ and every $S_3 \subseteq S_2 \subseteq S_1$,

$$\mathbb{E}(L_2(\Theta, S_2, S_3)|\mathbf{U}_1, \mathbf{U}_2) = \int L_2(\theta, S_2, S_3)\mathsf{K}(d\theta|\mathbf{W}_2), \quad \mathbb{P}\text{-a.s.}$$

It follows from (9.48) and (9.92) that the kernel K is permutation invariant. Taking in addition into account Problem 9.19 we see from (9.48) that K has the DT property. Consider now $\mathfrak{L}_2$ that has been introduced in Problem 9.89. According to the statement there, because L_2 has the properties (9.87) and (9.88), $\mathfrak{L}_2$ has the property $\mathfrak{D}(2, \Delta)$. Let

$$\tilde{\mathfrak{L}}_2(\mathbf{w}, T_1, T_2, T_3) = \int_{\Delta^k} \mathfrak{L}_2(\theta, T_1, T_2, T_3)\mathsf{K}(d\theta|\mathbf{w}). \tag{9.96}$$

Then by Lemma 9.90 $\tilde{\mathfrak{L}}_2$ has the property $\mathfrak{D}(2, \mathbb{R})$. Let $S_2 \subseteq S_1$ and $r_3 \leq |S_2|$ be fixed, and set $(T_1, T_2, T_3) = (S_1 \backslash S_2, S_2 \backslash S_3, S_3)$. $\tilde{\mathfrak{L}}_2(w, T_1, T_2, T_3)$ is minimized, subject to $S_3 \subseteq S_2$ and $|S_3| = r_3$, for those S_3 that are associated with r_3 of the largest w_i, $i \in S_2$. It should be noted here that the w_j with $j \in S_1 \backslash S_2$ do not play any role at this point. Since at every $(\mathbf{u}_1, \mathbf{u}_2)$ with $\mathbf{u}_1 + \mathbf{u}_2 = \mathbf{w}$ we have $\psi^{nat}_{\cdot|r_3,S_2}(\mathbf{u}_1, p_{S_2}(\mathbf{u}_2))$ as the uniform distribution on these sets (see Definition 9.87), the minimum posterior risk is achieved if every $\psi_{S_3|r_3,S_2}$ in (9.95) is replaced with the corresponding $\psi^{nat}_{S_3|r_3,S_2}$.

At Stage 1, the same arguments used above hold analogously, just in a simpler setting with $\mathbf{W}_1 = \mathbf{U}_1$, and the minimum posterior risk is achieved if every $\psi_{S_2|r_2}$ in (9.95) is replaced with the corresponding $\psi^{nat}_{S_2|r_2}$. Thus, altogether, we get

$$\mathsf{r}(\Pi, (\xi, \varphi, \tilde{\varphi}, \psi^{nat}, \tilde{\psi})) \leq \mathsf{r}(\Pi, (\xi, \varphi, \tilde{\varphi}, \psi, \tilde{\psi})).$$

Finally, let $\theta_0 \in \Delta^k$ be fixed and let Λ_{θ_0} be the prior that contributes mass $1/k!$ to every $\gamma(\theta_0)$, $\gamma \in \Pi_k$. Note that if some coordinates in θ_0 are equal, then $\gamma_1(\theta_0) = \gamma_2(\theta_0)$ may occur for some $\gamma_1, \gamma_2 \in \Pi_k$. In this case $\Lambda_{\theta_0}(\gamma_1(\theta_0))$ is the appropriate multiple of $1/k!$. Because for every permutation invariant two-stage rule $(\xi, \varphi, \tilde{\varphi}, \psi, \tilde{\psi})$, by (5.20) in Lemma 5.27, we have

$$\mathsf{r}(\Lambda_\theta, (\xi, \varphi, \tilde{\varphi}, \psi, \tilde{\psi})) = \mathsf{R}(\theta, (\xi, \varphi, \tilde{\varphi}, \psi, \tilde{\psi})), \quad \theta \in \Delta^k, \tag{9.97}$$

and Λ_θ is permutation invariant for every $\theta \in \Delta^k$, the proof is completed. ■

Remark 9.92. Suppose we replace in (9.95) the decisions $\psi_{S_3|r_3,S_2}(\mathbf{U}_1, p_{S_2}(\mathbf{U}_2))$ by decisions of the form $\psi_{S_3|r_3,S_2}(\mathbf{U}_1, \mathbf{U}_2)$. Then, according to the above proof, the natural decision functions $\psi^{nat}_{S_3|r_3,S_2}(\mathbf{U}_1, p_{S_2}(\mathbf{U}_2))$ would still be optimal. That for $S_2 \subseteq S_1$ and $r_3 \leq r_2$ only the observations $p_{S_2}(\mathbf{U}_1)$ and $p_{S_2}(\mathbf{U}_2)$ are relevant for the optimization of S_3 at Stage 2, and in fact only $p_{S_2}(\mathbf{U}_1) + p_{S_2}(\mathbf{U}_2)$ is used, is due to the special structure of the two-stage selection rules and the loss.

Problem 9.93. Show that Proposition 9.59 is a special case of Theorem 9.91.

The two-stage selection rules in Examples 9.76, 9.77, and 9.78 employ ψ^{nat}, which is under any loss that satisfies (9.87) and (9.88), according to the last theorem, the optimal choice.

Next, we consider several applications and extensions of Theorem 9.91, mainly for situations where in addition to ψ also $\tilde{\psi}$ can be improved. This turns out to be a more challenging task. To optimize a Bayes procedure, or some part of it, at Stage 1 we have to know the Bayes posterior risk to be expected at Stage 2. This means that the optimization process has to go backwards, starting with Stage 2. Once Stage 2 has been optimized, then we have to move back to Stage 1 and optimize that as well. This process is called *backward optimization.*

To prepare for the next results we need some results on log-concave exponential families. We assume that the dominating measure $\boldsymbol{\mu}$ for the one-parameter exponential family P_θ is absolutely continuous with respect to the Lebesgue measure $\boldsymbol{\lambda}$. It should be pointed out that discrete exponential families can also be treated in this setting by utilizing the interpolation technique from Proposition 2.58. In the case of Lebesgue densities that is considered here we suppose that

$$\frac{dQ_\theta}{d\boldsymbol{\lambda}}(t) = \exp\{\theta t - K(\theta)\}d(t), \quad d = \frac{d\boldsymbol{\mu}}{d\boldsymbol{\lambda}}.$$

Without loss of generality we may assume that d is a probability density and $0 \in \Delta$. Otherwise we turn to $\exp\{\theta_0 T(x) - K(\theta_0)\}d(t)$ for some fixed θ_0. The exponential family is called log-concave if $d(t)$ is positive on some interval and $\ln d(t)$ is concave. Obviously $\ln d(t)$ is concave if and only if $\ln(dQ_\theta/d\boldsymbol{\lambda})$ is concave.

Problem 9.94.* If the density $d(t)$ is log-concave, then for every m the distribution Q_θ^{*m} has the $\boldsymbol{\lambda}$-density

$$f_\theta^{(m)}(u) = \exp\{\theta u - rK(\theta)\}d_r(u),$$

which is again log-concave. Here $d_1 = d$, and for $r = 2, 3, ..., d_r(u)$ denotes the r-fold convolution of d which is defined by

$$d_r(u) = \int d_{r-1}(u-s)d(s)ds.$$

We assume now that the distribution of $T(X_{i,j}^{(m)})$ has the Lebesgue density

$$q_{\theta_i}(t) = \exp\{\theta_i t - K(\theta_i)\}d(t), \quad t \in \mathbb{R}, \ \theta_i \in \Delta,$$

$j = 1, ..., n_i$, $i = 1, ..., k$, $m = 1, 2$. Then the Lebesgue density of $\mathbf{U}_m$ is

$$\begin{aligned} &f_\theta^{(m)}(\mathbf{u}) = \exp\{\theta^T \mathbf{u} - K_{n_m}(\theta)\}\mathbf{d}_{n_m}(\mathbf{u}), \quad \mathbf{u} \in \mathbb{R}^k, \theta \in \Delta^k, \quad \text{where} \\ &K_{n_m}(\theta) = n_m \sum\nolimits_{i=1}^{n_m} K(\theta_i), \quad \mathbf{d}_{n_m}(\mathbf{u}) = \Pi_{i=1}^k d_{n_m}(u_i), \quad m = 1, 2. \end{aligned}$$

Lemma 9.95. *Let Π be a permutation invariant prior on $\mathfrak{B}_{\Delta^k}$. Based on the joint distribution of $(\Theta, \mathbf{W}_1, \mathbf{W}_2)$, where $\mathbf{W}_1 = \mathbf{U}_1$ and $\mathbf{W}_2 = \mathbf{U}_1 + \mathbf{U}_2$, let P_w^Π be the conditional distribution of $\mathbf{W}_2$, given $\mathbf{W}_1 = w$. If the function $d : B \to_m \mathbb{R}_+$ is positive and log-concave, then there is a regular version of the conditional distribution P_w^Π so that the stochastic kernel $\mathsf{P}^\Pi = (P_w^\Pi)_{w \in \mathbb{R}^k}$ is permutation invariant and DT.*

Proof. The translation invariance of $\boldsymbol{\lambda}_k$ yields for every $h : \mathbb{R}^{2k} \to \mathbb{R}_+$

$$\begin{aligned}
&\mathbb{E}h(\mathbf{U}_1, \mathbf{U}_1 + \mathbf{U}_2) \\
&= \int (\int [\int h(\mathbf{u}_1, \mathbf{u}_1 + \mathbf{u}_2) f_\theta^{(1)}(\mathbf{u}_1) f_\theta^{(2)}(\mathbf{u}_2) \Pi(d\theta)] \boldsymbol{\lambda}_k(d\mathbf{u}_1)) \boldsymbol{\lambda}_k(d\mathbf{u}_2) \\
&= \int (\int [\int h(\mathbf{s}_1, \mathbf{s}_2) f_\theta^{(1)}(\mathbf{s}_1) f_\theta^{(2)}(\mathbf{s}_2 - \mathbf{s}_1) \Pi(d\theta)] \boldsymbol{\lambda}_k(d\mathbf{s}_1)) \boldsymbol{\lambda}_k(d\mathbf{s}_2).
\end{aligned}$$

This shows that the conditional density of $\mathbf{U}_1 + \mathbf{U}_2$, given $\mathbf{U}_1$, is given by

$$\xi^{(2)}(\mathbf{s}_2|\mathbf{s}_1) = \frac{\int f_\theta^{(1)}(\mathbf{s}_1) f_\theta^{(2)}(\mathbf{s}_2 - \mathbf{s}_1) \Pi(d\theta)}{\int [\int f_\theta^{(1)}(\mathbf{s}_1) f_\theta^{(2)}(\mathbf{s}_2 - \mathbf{s}_1) \Pi(d\theta)] \boldsymbol{\lambda}_k(d\mathbf{s}_2)} =: \frac{\varphi(\mathbf{s}_1, \mathbf{s}_2)}{\psi(\mathbf{s}_1)}.$$

It holds

$$\begin{aligned}
\varphi(\mathbf{s}_1, \mathbf{s}_2) &= \int \exp\{\theta^T \mathbf{s}_1 - K_{n_1}(\theta)\} \mathbf{d}_{n_1}(\mathbf{s}_1) \\
&\qquad \times \exp\{\theta^T (\mathbf{s}_2 - \mathbf{s}_1) - K_{n_2}(\theta)\} \mathbf{d}_{n_2}(\mathbf{s}_2 - \mathbf{s}_1) \Pi(d\theta) \\
&= \int \exp\{\theta^T \mathbf{s}_2 - K_{n_1}(\theta) - K_{n_2}(\theta)\} \Pi(d\theta) \mathbf{d}_{n_2}(\mathbf{s}_2 - \mathbf{s}_1) \mathbf{d}_{n_1}(\mathbf{s}_1), \\
\psi(\mathbf{s}_1) &= \int \exp\{\theta^T \mathbf{s}_1 - K_{n_1}(\theta)\} \\
&\quad \times [\int \exp\{\theta^T (\mathbf{s}_2 - \mathbf{s}_1) - K_{n_2}(\theta)\} \mathbf{d}_{n_2}(\mathbf{s}_2 - \mathbf{s}_1) \boldsymbol{\lambda}_k(d\mathbf{s}_2)] \Pi(d\theta) \mathbf{d}_{n_1}(\mathbf{s}_1).
\end{aligned}$$

As $\boldsymbol{\lambda}_k$ is translation invariant and $\exp\{\theta^T \mathbf{w} - K_{n_2}(\theta)\} \mathbf{d}_{n_2}(\mathbf{w})$ is a density we get

$$\psi(\mathbf{s}_1) = \mathbf{d}_{n_1}(\mathbf{s}_1) \int \exp\{\theta^T \mathbf{s}_1 - K_{n_1}(\theta)\} \Pi(d\theta),$$

and

$$\xi^{(2)}(\mathbf{s}_2|\mathbf{s}_1) = g(\mathbf{s}_1, \mathbf{s}_2) \mathbf{d}_{n_1}(\mathbf{s}_2 - \mathbf{s}_1),$$

where

$$g(\mathbf{s}_1, \mathbf{s}_2) = \frac{\int \exp\{\theta^T \mathbf{s}_2 - K_{N_2}(\theta)\} \Pi(d\theta)}{\int \exp\{\theta^T \mathbf{s}_1 - K_{N_1}(\theta)\} \Pi(d\theta)},$$

$N_1 = n_1$, $N_2 = n_1 + n_2$, and $g(\mathbf{s}_1, \mathbf{s}_2)$ is permutation invariant, i.e., $g(\mathbf{s}_1, \mathbf{s}_2) = g(\gamma(\mathbf{s}_1), \gamma(\mathbf{s}_2))$. By considering $s_{1,i}$ as a parameter we get from Problem 9.94 and Proposition 2.20 that $d_{N_1}(s_{2,i} - s_{1,i})$ as a function of $s_{2,i}$ has MLR in $s_{2,i}$. Hence $\mathbf{d}_{N_1}(\mathbf{s}_2 - \mathbf{s}_1)$ satisfies the conditions (9.39) and (9.40). The permutation invariance of g shows that $\xi^{(2)}(\mathbf{s}_2|\mathbf{s}_1)$ also satisfies (9.39) and (9.40). An application of Problem 9.19 completes the proof. ■

For the backward optimization we also need the following.

Lemma 9.96. *Let $A \subseteq \mathbb{R}$ be a fixed Borel set. Suppose that the function $\mathfrak{L}_2$ has the property $\mathfrak{D}(2, A)$. For every $a \in A^k$ and $(T_1, T_2) \in \mathfrak{T}_2$ let*

$$\mathfrak{L}_1(a, T_1, T_2) = \min\{\mathfrak{L}_2(a, (T_1, \widehat{T}_2, \widehat{T}_3) : \widehat{T}_2 \cup \widehat{T}_3 = T_2,\ \widehat{T}_2 \cap \widehat{T}_3 = \emptyset\}. \quad (9.98)$$

Then $\mathfrak{L}_1$ has the property $\mathfrak{D}(1, A)$.

Proof. Let $a \in A^k$, $(T_1, T_2) \in \mathfrak{T}_2$, and $\gamma \in \Pi_k$ be fixed. As $\mathfrak{L}_2$ is permutation invariant in the sense of Definition 9.88,

$$\begin{aligned} &\mathfrak{L}_1(\gamma(a), T_1, T_2) \\ &= \min\{\mathfrak{L}_2(a, \gamma(T_1), T_2^*, T_3^*) : T_2^* \cup T_3^* = \gamma(T_2),\ T_2^* \cap T_3^* = \emptyset\} \\ &= \mathfrak{L}_1(a, \gamma(T_1), \gamma(T_2)), \end{aligned}$$

which means that $\mathfrak{L}_1$ in permutation invariant.

Suppose that for some fixed $i, j \in \{1, ..., k\}$ with $a_i \leq a_j$ we have $j \in T_1$ and $i \in T_2$. Let $\widetilde{T}_1 = (T_1 \backslash \{j\}) \cup \{i\}$ and $\widetilde{T}_2 = (T_2 \backslash \{i\}) \cup \{j\}$. Moreover, we distinguish between two types of decompositions $\widehat{T}_2, \widehat{T}_3$ of T_2. The first is where $i \in \widehat{T}_2$. Here we set $\overline{T}_2 = (\widehat{T}_2 \backslash \{i\}) \cup \{j\}$ and $\overline{T}_3 = \widehat{T}_3$. The second is where $i \in \widehat{T}_3$. Here we set $\overline{T}_2 = \widehat{T}_2$ and $\overline{T}_3 = (\widehat{T}_3 \backslash \{i\}) \cup \{j\}$. For either type $\overline{T}_2, \overline{T}_3$ is a decomposition of $\widetilde{T}_2$, and

$$\mathfrak{L}_2(a, \widetilde{T}_1, \overline{T}_2, \overline{T}_3) \leq \mathfrak{L}_2(a, (T_1, \widehat{T}_2, \widehat{T}_3),$$

which implies that $\mathfrak{L}_1(a, \widetilde{T}_1, \widetilde{T}_2) \leq \mathfrak{L}_1(a, T_1, T_2)$. ■

Remark 9.97. Sometimes restrictions on the sizes of the subsets to be selected are imposed on a two-stage selection rule which affects φ and $\tilde{\varphi}$. Customized versions of Lemma 9.96 have then to be created and proved in a similar manner. The minimum in (9.98) has to be restricted and the proof adjusted accordingly.

The next result is derived by backward optimization.

Theorem 9.98. *Suppose that the underlying exponential family is strongly unimodal with an everywhere positive density. Let the loss functions L_1 and L_2 satisfy (9.87) and (9.88). Let $(\xi, \varphi, \tilde{\varphi}, \psi, \tilde{\psi})$ be a permutation invariant two-stage selection rule that does not stop at Stage 1 and has a fixed predetermined subset size $R_3 \geq 1$ for selections at Stage 2. Then it holds*

$$\mathsf{R}(\theta, (\xi, \varphi, \tilde{\varphi}, \psi^{nat}, \tilde{\psi}^{nat})) \leq \mathsf{R}(\theta, (\xi, \varphi, \tilde{\varphi}, \psi, \tilde{\psi})), \quad \theta \in \Delta^k.$$

Proof. Let Π be any permutation invariant prior on $\mathfrak{B}_{\Delta^k}$ that has a finite support, and thus the Bayes risk of $(\xi, \varphi, \tilde{\varphi}, \psi, \tilde{\psi})$ is finite. In (9.95), by assumption, we have $\xi = 0$ and $\varphi_{R_3|S_2} = 1$.

At Stage 2, by Theorem 9.91, the component of ψ^{nat} associated with Stage 2 is optimal. Let $r_2 \in \{1, ..., k\}$ and $S_2 \subseteq S_1$ with $|S_2| = r_2$ be fixed. Set $T_1 = S_1 \backslash S_2$ and $T_2 = S_2$. Within $\mathbb{E}\{\cdot|\mathbf{U}_1\}$ of (9.95) we insert $\varphi_{R_3|S_2} = 1$ and $\psi_{S_3|R_3,S_2} = \psi^{nat}_{S_3|R_3,S_2}$. Then the conditional expectation $\mathbb{E}\{\cdot|\mathbf{U}_1\}$ in (9.95) turns out to be

$$\tilde{\mathfrak{H}}(\mathbf{U}_1, T_1, T_2) = \mathbb{E}(\mathfrak{H}(\mathbf{W}_2, T_1, T_2)|\mathbf{U}_1), \quad \text{where}$$

$$\mathfrak{H}(w,T_1,T_2)=\min\{\tilde{\mathfrak{L}}_2(w,T_1,\widehat{T}_2,\widehat{T}_3):\widehat{T}_2\cup\widehat{T}_3=T_2,\widehat{T}_2\cap\widehat{T}_3=\emptyset,\ |\widehat{T}_3|=R_3\}$$

for $w\in\mathcal{R}(\mathbf{W}_2)$, and where $\tilde{\mathfrak{L}}_2$ is defined in (9.96). The crucial point is that the component of ψ^{nat} at Stage 2 remains optimal even if the component of ψ at Stage 2 were allowed to make use of $(\mathbf{U}_1,\mathbf{U}_2)$, and thus in particular of $\mathbf{W}_2=\mathbf{U}_1+\mathbf{U}_2$.

As stated in the proof of Theorem 9.91, $\tilde{\mathfrak{L}}_2$ has, by Lemma 9.90, the property $\mathfrak{D}(2,\mathcal{R}(W_{1,2}))$. By Lemma 9.96 and Remark 9.97 it follows that $\mathfrak{H}$ has the property $\mathfrak{D}(1,\mathcal{R}(W_{1,2}))$. Lemma 9.95 states that the conditional distribution of $\mathbf{W}_2$, given $\mathbf{W}_1$, is permutation invariant and DT. Therefore another application of Lemma 9.90 implies that $\tilde{\mathfrak{H}}$ has the property $\mathfrak{D}(1,\mathcal{R}(W_{1,1}))$. As the Bayes risk has been reduced to

$$\mathbb{E}[\sum\nolimits_{r_2=1}^{k}\widetilde{\varphi}_{r_2}(\mathbf{U}_1)\sum\nolimits_{S_2\subseteq S_1,|S_2|=r_2}\widetilde{\psi}_{S_2|r_2}(\mathbf{U}_1)\tilde{\mathfrak{H}}(\mathbf{U}_1,S_1\backslash S_2,S_2)],$$

apparently $\tilde{\psi}^{nat}_{S_2|r_2}$ is optimal. In view of (9.97) the proof is completed. ■

The above results on two-stage selection rules have been presented in slightly different form in Gupta and Miescke (1983). They are special cases of results in Gupta and Miescke (1984a) on q-stage selection rules. The proofs for the latter are similar to those given above, but more involved. These more general results are reported below without proofs. We recall that the model is assumed to be from (9.82) and the observations are explained by (9.81) and (9.84). For the $m+1$ subsets of populations that are selected up to the end of Stage m we introduce the notation

$$\mathbf{S}_{m+1}=(S_1,...,S_{m+1}),\quad S_1\supseteq S_2\supseteq\cdots\supseteq S_{m+1},\ \ m=1,...,q. \tag{9.99}$$

$\mathbf{S}_1=S_1=\{1,...,k\}$ are the populations available at the beginning of Stage 1. $\mathbf{S}_2=(S_1,S_2)$ means that at Stage 1 we start with populations S_1, sample from populations S_1 and select populations $S_2\subseteq S_1$, either as the terminal decision or to be sampled at Stage 2, and so on. $\mathbf{S}_{q+1}=(S_1,...,S_{q+1})$ means that at the end of Stage q the terminal decision is the selection of S_{q+1}.

The decision space $\mathcal{D}_{qst}$ is chosen to consist of q components, one for decisions that make their terminal decision at the end of Stage m, $m=1,...,q$. We set $\mathcal{D}_{qst}=\bigcup_{m=1}^{q}\mathcal{D}_m$, where for $m=1,...,q$,

$$\begin{aligned}\mathcal{D}_m=\{(m,r_2,S_2,...,r_{m+1},S_{m+1}):S_1\supseteq S_2\supseteq\cdots\supseteq S_{m+1},\ |S_i|=r_i,\\ i=2,...,m+1,\ k\geq r_2\geq r_3\geq\cdots\geq r_m\geq 1,\ 0\leq r_{m+1}\leq r_m\}.\end{aligned}$$

The q-stage selection rules are now introduced as distributions on $\mathcal{D}_{qst}$ in dependence of the observations $\mathbf{V}_q=(\mathbf{U}_1,...,\mathbf{U}_q)$ from (9.84). We consider functions $\widehat{\xi}_{\mathbf{S}_m},\widehat{\varphi}_{r_{m+1}|\mathbf{S}_m},\widehat{\psi}_{S_{m+1}|r_{m+1},\mathbf{S}_m}:\mathsf{X}_{j=1}^{m}\mathbb{R}^{|S_j|}\to_m[0,1]$ and set

$$\begin{aligned}\xi_{\mathbf{S}_m}(\mathbf{v}_m)&=\widehat{\xi}_{\mathbf{S}_m}(\mathbf{u}_1,p_{S_2}(\mathbf{u}_2),...,p_{S_m}(\mathbf{u}_m)),\\ \varphi_{r_{m+1}|\mathbf{S}_m}(\mathbf{v}_m)&=\widehat{\varphi}_{r_{m+1}|\mathbf{S}_m}(\mathbf{u}_1,p_{S_2}(\mathbf{u}_2),...,p_{S_m}(\mathbf{u}_m)),\\ \psi_{S_{m+1}|r_{m+1},\mathbf{S}_m}(\mathbf{v}_m)&=\widehat{\psi}_{S_{m+1}|r_{m+1},\mathbf{S}_m}(\mathbf{u}_1,p_{S_2}(\mathbf{u}_2),...,p_{S_m}(\mathbf{u}_m)),\end{aligned}$$

for $(m, r_2, S_2, ..., r_{m+1}, S_{m+1}) \in \mathcal{D}_m$, $\mathbf{v}_m = (\mathbf{u}_1, ..., \mathbf{u}_m) \in \mathbb{R}^{km}$ and $m = 1, ..., q$. $\tilde{\varphi}_{r_{m+1}|\mathbf{S}_m}$ and $\tilde{\psi}_{S_{m+1}|r_{m+1},\mathbf{S}_m}$ are set up analogously to $\varphi_{r_{m+1}|\mathbf{S}_m}$ and $\psi_{S_{m+1}|r_{m+1},\mathbf{S}_m}$, respectively, for $m = 1, ..., q-1$. We assume, analogously to (9.85) and (9.86), that for every fixed m and $\mathbf{v}_m = (\mathbf{u}_1, ..., \mathbf{u}_m) \in \mathbb{R}^{km}$ each function introduced above sums up to 1 over the range of objects on which it is defined.

Definition 9.99. *A q-stage selection rule $(\xi, \varphi, \tilde{\varphi}, \psi, \tilde{\psi})$ is, for every fixed $\mathbf{v}_q = (\mathbf{u}_1, ..., \mathbf{u}_q) \in \mathbb{R}^{kq}$, a distribution $\mathsf{D}(\cdot|\mathbf{v}_q)$ on $\mathcal{D}_{qst}$ that satisfies*

$$\begin{aligned}
&\mathsf{D}(\{(m, r_2, S_2, ..., r_{m+1}, S_{m+1})\}|\mathbf{v}_q) \\
&= \prod\nolimits_{j=1}^{m-1} (1 - \xi_{\mathbf{S}_j}(\mathbf{v}_j))\tilde{\varphi}_{r_{j+1}|\mathbf{S}_j}(\mathbf{v}_j)\tilde{\psi}_{S_{j+1}|r_{j+1},\mathbf{S}_j}(\mathbf{v}_j) \\
&\qquad \times \xi_{\mathbf{S}_m}(\mathbf{v}_m)\varphi_{r_{m+1}|\mathbf{S}_m}(\mathbf{v}_m)\psi_{S_{m+1}|r_{m+1},\mathbf{S}_m}(\mathbf{v}_m)
\end{aligned}$$

for $(m, r_2, S_2, ..., r_{m+1}, S_{m+1}) \in \mathcal{D}_m$, $m = 1, ..., q$, and also $\xi_{\mathbf{S}_q}(\mathbf{v}_q) = 1$.

The interpretation of a q-stage selection rule $(\xi, \varphi, \tilde{\varphi}, \psi, \tilde{\psi})$ is analogous to that given after Definition 9.80 and omitted here for brevity. The approach above can be extended to include sequential selection rules by letting $q \to \infty$. The Bechhofer–Kiefer–Sobel sequential selection rule of Example 9.74 is an example for such a rule.

Problem 9.100. Verify that the selection rules in Examples 9.75, 9.76, 9.77, and 9.78 are q-stage selection rules in the sense of Definition 9.99.

We focus again on the goal of optimizing the two components ψ and $\tilde{\psi}$ of q-stage selection rules. Also here we have to restrict ourselves to permutation invariant q-stage selection rules, and to loss functions that are permutation invariant and favor selections of populations with larger parameters.

Let $m \in \{1, ..., q\}$, $\mathbf{S}_m = (S_1, ..., S_m)$, $\mathbf{v}_m = (\mathbf{u}_1, ..., \mathbf{u}_m) \in \mathbb{R}^{km}$, and $\gamma \in \Pi_k$, be fixed. We set $\gamma(\mathbf{S}_m) = (\gamma(S_1), ..., \gamma(S_m))$, where $\gamma(S_r) = \{\gamma(i) : i \in S_r\}$, $r = 1, ..., m$. Moreover, we set $\gamma(\mathbf{v}_m) = (\gamma(\mathbf{u}_1), ..., \gamma(\mathbf{u}_m))$, where $\gamma(\mathbf{u}_r) = (u_{\gamma(1),r}, ..., u_{\gamma(k),r})$, $r = 1, ..., m$.

Definition 9.101. *A q-stage selection rule $(\xi, \varphi, \tilde{\varphi}, \psi, \tilde{\psi})$ is called permutation invariant if the five types of decision functions are permutation invariant in the following sense. For any $\gamma \in \Pi_k$ and $m \in \{1, ..., q\}$ it holds $\mathbb{P}_{\mathbf{V}_m}$-a.s.*

$$\begin{aligned}
\xi_{\gamma(\mathbf{S}_m)}(\mathbf{v}_m) &= \xi_{\mathbf{S}_m}(\gamma(\mathbf{v}_m)), \\
\varphi_{r_{m+1}|\gamma(\mathbf{S}_m)}(\mathbf{v}_m) &= \varphi_{r_{m+1}|\mathbf{S}_m}(\gamma(\mathbf{v}_m)), \\
\psi_{\gamma(S_{m+1})|r_{m+1},\gamma(\mathbf{S}_m)}(\mathbf{v}_m) &= \psi_{S_{m+1}|r_{m+1},\mathbf{S}_m}(\gamma(\mathbf{v}_m)),
\end{aligned}$$

and for any $\gamma \in \Pi_k$ and $m \in \{1, ..., q-1\}$ $\tilde{\varphi}$ and $\tilde{\psi}$ have the same properties as φ and ψ, respectively.

Problem 9.102. Verify that the selection rules in Examples 9.75, 9.76, 9.77, and 9.78 are permutation invariant q-stage selection rules in the sense of Definition 9.101.

Finally, we adopt a class of loss functions that are permutation invariant and favor selections of populations with larger parameters. For $m = 1, ..., q$, let $L_m(\theta, \mathbf{S}_{m+1})$ be the loss that occurs at $\theta \in \Delta^k$ if at Stage m the procedure stops and produces the subset configuration $\mathbf{S}_{m+1}$, where $L_m(\cdot, \mathbf{S}_{m+1}) : \Delta^k \to_m \mathbb{R}_+$. Analogously to (9.87) and (9.88) we assume that

$$L_m(\theta, \gamma(\mathbf{S}_{m+1})) = L_m(\gamma(\theta), \mathbf{S}_{m+1}), \quad \gamma \in \Pi_k, \, \theta \in \Delta^k, \tag{9.100}$$

and, moreover,

$$L_m(\theta, \widetilde{\mathbf{S}}_{m+1}) \le L_m(\theta, \mathbf{S}_{m+1}) \tag{9.101}$$

if for some $i, j \in \{1, ..., k\}$ with $\theta_i \le \theta_j$ the following holds. For every $c \in \{1, ..., m+1\}$ with $i \in S_c$ and $j \notin S_c$, $\widetilde{S}_c = (S_c \backslash \{i\}) \cup \{j\}$, and $\widetilde{S}_c = S_c$, otherwise. Thus, according to (9.101), a worse population should be eliminated at an earlier stage than a better population.

In permutation invariant selection problems, under a loss that satisfies (9.100) and (9.101), a q-stage selection rule $(\xi, \varphi, \tilde{\varphi}, \psi, \tilde{\psi})$ can be improved by replacing ψ by ψ^{nat} and $\tilde{\psi}$ by $\tilde{\psi}^{nat}$, where ψ^{nat} and $\tilde{\psi}^{nat}$ are defined below.

Definition 9.103. *For every fixed $m \in \{1, ..., q\}$, $\mathbf{S}_m$, $r_{m+1} \le |S_m|$, $\mathbf{v}_m = (\mathbf{u}_1, ..., \mathbf{u}_m) \in \mathbb{R}^{km}$, and $\mathbf{w}_m = (w_{1,m}, ..., w_{k,m}) = \mathbf{u}_1 + \cdots + \mathbf{u}_m$, let $\psi^{nat}_{S_{m+1}|r_{m+1},\mathbf{S}_m}(\mathbf{v}_m)$ be equal to a positive constant value for all $S_{m+1} \subseteq S_m$ with $|S_{m+1}| = r_{m+1}$ that satisfy*

$$\max\{w_{i,m} : i \in S_m \backslash S_{m+1}\} \le \min\{w_{j,m} : j \in S_{m+1}\},$$

and be zero otherwise. For $m \in \{1, ..., q-1\}$ let $\tilde{\psi}^{nat}$ be defined in the same way; that is, $\tilde{\psi}^{nat} = \psi^{nat}$.

Switching from a q-stage selection rule $(\xi, \varphi, \tilde{\varphi}, \psi, \tilde{\psi})$ to $(\xi, \varphi, \tilde{\varphi}, \psi^{nat}, \tilde{\psi}^{nat})$ enforces that in the transition from any S_m to $S_{m+1} \subseteq S_m$ only populations with the largest values of $w_{i,m}$, $i \in S_m$, are included in S_{m+1}.

In a first step we consider the switch from $(\xi, \varphi, \tilde{\varphi}, \psi, \tilde{\psi})$ to $(\xi, \varphi, \tilde{\varphi}, \psi^{nat}, \tilde{\psi})$ which affects only the terminal decisions. Under a loss with the properties (9.100) and (9.101) a permutation invariant q-stage selection rule should be modified by making the terminal decisions with natural selection rules. This is the content of the next theorem, which has been proved in Gupta and Miescke (1984a). The technical tools used in the proof are straightforward generalizations of Definition 9.88, Problem 9.89, and Lemma 9.90 to the q-stage setting.

The risk of a q-stage selection rule $(\xi, \varphi, \tilde{\varphi}, \psi, \tilde{\psi})$ at $\theta \in \Delta^k$ is given by

$$\begin{aligned} \mathsf{R}(\theta, (\xi, \varphi, \tilde{\varphi}, \psi, \tilde{\psi})) &= \sum\nolimits_{m=1}^{q} \sum\nolimits_{\mathbf{S}_{m+1}} L_m(\theta, \mathbf{S}_{m+1}) \\ &\times \mathsf{E}_\theta[\prod\nolimits_{j=1}^{m-1} (1 - \xi_{\mathbf{S}_j}(\mathbf{V}_j)) \tilde{\varphi}_{|S_{j+1}||\mathbf{S}_j}(\mathbf{V}_j) \tilde{\psi}_{S_{j+1}||S_{j+1}|,\mathbf{S}_j}(\mathbf{V}_j) \\ &\times \xi_{\mathbf{S}_m}(\mathbf{V}_m) \varphi_{|S_{m+1}||\mathbf{S}_m}(\mathbf{V}_m) \psi_{S_{m+1}||S_{m+1}|,\mathbf{S}_m}(\mathbf{V}_m)], \end{aligned} \tag{9.102}$$

where the second sum is with respect to $\mathbf{S}_{m+1} = (S_1, ..., S_{m+1})$ with $S_{m+1} \subseteq S_m \subseteq \cdots \subseteq S_1 = \{1, ..., k\}$ and $S_m \neq \emptyset$, and where $\xi_{\mathbf{S}_q} = 1$.

Remark 9.104. It should be pointed out that (9.102) holds also for untruncated (open sequential) procedures as long as they stop almost surely in finitely many steps and the risk is finite. One has to drop the assumption that $\xi_{\mathbf{S}_q} = 1$ and then take $q = \infty$ in (9.102).

The Bayes risk under a permutation invariant prior Π on the Borel sets of Δ^k can be used to establish ψ^{nat} and $\tilde{\psi}^{nat}$ (see Definition 9.103) as improvements over ψ and $\tilde{\psi}$, respectively, in a permutation invariant q-stage selection rule, similarly as has been done in Proposition 9.59. Again, also here, the question regarding the optimization of φ and $\tilde{\varphi}$, which decide on the sizes of the subsets to be selected, remains open as this depends on the form of the loss functions chosen for the individual stages and on the prior. The generalization of Theorem 9.91 to q-stage selection rules is as follows.

Theorem 9.105. *Let the loss $L_m(\theta, S_{m+1})$ for all S_{m+1} from (9.99), $m = 1, ..., q$, and $\theta \in \Delta^k$, satisfy (9.100) and (9.101). Then for every permutation invariant q-stage selection rule $(\xi, \varphi, \tilde{\varphi}, \psi, \tilde{\psi})$ it holds*

$$\mathsf{R}(\theta, (\xi, \varphi, \tilde{\varphi}, \psi^{nat}, \tilde{\psi})) \leq \mathsf{R}(\theta, (\xi, \varphi, \tilde{\varphi}, \psi, \tilde{\psi})), \quad \theta \in \Delta^k.$$

The q-stage selection rules in Examples 9.75, 9.76, 9.77, and 9.78 employ ψ^{nat}, which is the optimal choice, according to Theorem 9.105, under any loss that satisfies (9.100) and (9.101).

Problem 9.106. Suppose that a permutation invariant q-stage selection rule is not allowed to eliminate any population at Stages 1 through $q-1$. This means that $\tilde{\varphi}_{r_{m+1}|\mathbf{S}_m} = 1$ for $r_{m+1} = k$ and every $\mathbf{S}_m$ from (9.99), $m = 1, ..., q-1$. In this case, no matter which stopping rule is used, the natural terminal decision rule ψ^{nat} is optimal. This can be extended to the case of $q \to \infty$, and thus in particular the sequential selection rule in Example 9.74 employs the optimal terminal decision.

To conclude this section, several applications and extensions of Theorem 9.105 are considered in situations where in addition to ψ also $\tilde{\psi}$ can be improved. This turns out to be a more challenging task. To optimize, in the Bayes approach, a q-stage selection rule, or some part of it, at Stage $m, m+1, ..., q-1$ we have to know the Bayes posterior risk to be expected at Stage $m+1, m+2, ..., q$, respectively. This means that the optimization process has to go backwards, starting with Stage q. Once Stage q has been optimized, we have to move back to Stage $q-1$ and optimize that, and so on. This process is called *backward optimization.* If a specific optimization technique is used repeatedly through the stages, then we can utilize *backward induction.* From this point on, as before with two-stage selection rules, we assume that the underlying exponential family is strongly unimodal. The technical tools used to prove the results below are straightforward generalizations of Lemma 9.95, and Lemma 9.96 to the q-stage setting.

The following generalization of Theorem 9.98 to q-stage selection rules has been proved in Gupta and Miescke (1984a) by backward optimization and backward induction.

Theorem 9.107. *Suppose that the underlying exponential family is strongly unimodal. Let the loss $L_m(\theta, S_{m+1})$ for all S_{m+1} from (9.99), $m = 1, ..., q$, and $\theta \in \Delta^k$, satisfy (9.100) and (9.101). Let $(\xi, \varphi, \tilde{\varphi}, \psi, \tilde{\psi})$ be a permutation invariant q-stage selection rule that has fixed predetermined subset sizes $r_2 \geq \cdots \geq r_{q+1}$ for selections at Stages 1 through q, respectively, and does not stop at Stages 1 through $q-1$. Then it holds*

$$\mathsf{R}(\theta, (\xi, \varphi, \tilde{\varphi}, \psi^{nat}, \tilde{\psi}^{nat})) \leq \mathsf{R}(\theta, (\xi, \varphi, \tilde{\varphi}, \psi, \tilde{\psi})), \quad \theta \in \Delta^k.$$

If the assumptions of Theorem 9.107 are relaxed by allowing the subset sizes $R_2 \geq \cdots \geq R_q$ for selections at Stages 1 through $q-1$, respectively, to be random, then the situation becomes more difficult. What can be shown, however, is the following.

Corollary 9.108. *If the assumptions of Theorem 9.107 are relaxed by allowing $r_2 \geq \cdots \geq r_q$ to be random, then*

$$\mathsf{R}(\theta, (\xi, \varphi, \tilde{\varphi}, \psi^{nat}, \tilde{\psi}^{\bullet})) \leq \mathsf{R}(\theta, (\xi, \varphi, \tilde{\varphi}, \psi, \tilde{\psi})), \quad \theta \in \Delta^k,$$

where the components of $\tilde{\psi}^{\bullet}$ are the same as those of $\tilde{\psi}$, except for Stage $q-1$ where the component of $\tilde{\psi}^{\bullet}$ is that of $\tilde{\psi}^{nat}$.

The statement of the corollary can be stretched somewhat further. More precisely, the main tool for optimizing a permutation invariant q-stage selection rule in terms of $\tilde{\psi}$ appears to work only s stages backwards from the last stage, provided that $r_{q-s+2} \geq \cdots \geq r_{q+1}$ are fixed predetermined. Further optimization of $\tilde{\psi}$ under permutation invariant priors seems to be infeasible under a random r_{q-s+1}, $s = 1, ..., q-1$. In view of this shortcoming Bayes rules under i.i.d. priors have been studied in Gupta and Miescke (1984a). This leads to the following result, which has also been proved by backward optimization and backward induction.

Theorem 9.109. *Suppose that the underlying exponential family is strongly unimodal. Let the loss $L_m(\theta, \mathbf{S}_{m+1})$ for all $\mathbf{S}_{m+1}$ from (9.99), and $\theta \in \Delta^k$, satisfy (9.100) and (9.101), and in addition let it be a function of only those θ_i with $i \in S_{m+1}$, $m = 1, ..., q$. Let $\Pi \in \mathcal{P}(\mathfrak{B}_\Delta)$. If under the prior $\Upsilon = \Pi^{\otimes k}$ for $\Theta = (\Theta_1, ..., \Theta_k)$ there exists a Bayes permutation invariant q-stage selection rule, then there also exists, under the same prior Υ, a Bayes permutation invariant q-stage selection rule of the form $(\xi, \varphi, \tilde{\varphi}, \psi^{nat}, \tilde{\psi}^{nat})$.*

Problem 9.110. Determine which of the selection rules in Examples 9.75, 9.76, 9.77, and 9.78 are optimized in the sense of Theorems 9.105, 9.107, and 9.109.

9.4.2 Bayes Sampling Designs for Adaptive Sampling

In this section we consider multistage selection rules that are quite different from the q-stage selection rules in the previous subsection. This time a predetermined total number N of observations is taken in a sequential fashion from the k populations and then a terminal decision is made in form of a point selection; see Section 9.2. At certain stages of the sampling process decisions are made regarding the populations to be sampled next, utilizing all of the the information gathered so far, which includes the information provided by the prior. In other words, *adaptive sampling* is employed. As shown, ideally one should take one observation at a time and then allocate the next observation to a population in an optimal way. In practical applications, however, this may not always be feasible, and then one has to have recourse to approximately optimal allocations.

The Bayes multistage selection rules are developed heuristically, starting with one-stage rules, moving on to two-stage rules, and eventually establishing the general case. The basic idea of the *Bayes look-ahead* method that is used hereby has been taken from Berger (1985) and utilized for multistage selection problems in Miescke (1990, 1993), Gupta and Miescke (1994, 1996a,b), Miescke and Shi (1995), Miescke and Park (1997), and Miescke (1999).

Let the statistical model be given by

$$\mathcal{M}_{msa} = (\mathcal{X}^{kN}, \mathfrak{A}^{\otimes kN}, (\bigotimes_{i=1}^{k} P_{\theta_i}^{\otimes N})_{\theta \in \Delta^k}),$$

where $\theta = (\theta_1, ..., \theta_k)$, which allows us to take N independent observations $X_{i,j}$, $j = 1, ..., n_i$, $i = 1, ..., k$, with $n_1 + \cdots + n_k = N$ in any fashion from the k populations. The parameter set Δ is assumed to be a Borel subset of $\mathbb{R}$. As we are utilizing the Bayes approach we assume that the standard condition (A3) is satisfied so that the family $(P_\theta)_{\theta\in\Delta}$ is a stochastic kernel, say P. Our goal is to find a population with the largest parameter value.

We begin with the study of one-stage selection rules. Let the N observations be allocated in such a way that n_i are drawn from population i, $i = 1, ..., k$, where $n_1 + \cdots + n_k = N$. Let $\Pi \in \mathcal{P}(\mathfrak{B}_{\Delta}^{\otimes k})$ be a given prior for $\Theta = (\Theta_1, ..., \Theta_k)$. Altogether, our Bayes decision problem is modeled by the probability space

$$(\Omega, \mathfrak{F}, \mathbb{P}) =((\mathsf{X}_{i=1}^{k} \mathcal{X}^{n_i}) \times \Delta^k, (\bigotimes_{i=1}^{k} \mathfrak{A}^{\otimes n_i}) \otimes \mathfrak{B}_{\Delta}^{\otimes k}, (\bigotimes_{i=1}^{k} \mathsf{P}^{\otimes n_i}) \otimes \Pi),$$

where $\otimes$ denotes the product of the stochastic kernels; see Proposition A.39. The means here that

$$\begin{aligned}&((\bigotimes_{i=1}^{k} \mathsf{P}^{\otimes n_i}) \otimes \Pi)(dx_{1,1}, ..., dx_{1,n_1}, ..., dx_{k,1}, ..., dx_{k,n_k}, d\theta_1, ..., d\theta_k)\\ &= (\bigotimes_{i=1}^{k} \bigotimes_{j=1}^{n_i} P_{\theta_i}(dx_{i,j})) \otimes \Pi(d\theta_1, ..., d\theta_k).\end{aligned}$$

To simplify notation let $X_{1,1}, ..., X_{k,n_k}, \Theta_1, ..., \Theta_k$ denote the projections on the coordinates. Moreover, we set $\mathbf{X}_{i,n_i} = (X_{i,1}, ..., X_{i,n_i})$, $\Theta = (\Theta_1, ..., \Theta_k)$, $\mathbf{x}_{i,n_i} = (x_{i,1}, ..., x_{i,n_i})$, and $\theta = (\theta_1, ..., \theta_k)$.

As to the loss, cost of sampling is of no concern because N observations are always drawn altogether. Elimination of populations is not allowed here and thus we assume that the loss function is of the form $L : \Delta^k \times \mathcal{D}_{pt} \to_m \mathbb{R}_+$ and satisfies (9.42) and (9.43) with κ as the identity. For any nonrandomized selection rule $\mathsf{d}_{\mathbf{n}} : \times_{i=1}^k \mathcal{X}^{n_i} \to_m \{1, ..., k\}$, where $\mathbf{n} = (n_1, ..., n_k)$, the Bayes risk is

$$\begin{aligned} &\mathsf{r}(\Pi, \mathsf{d}_{\mathbf{n}}) = \mathbb{E}L(\Theta, \mathsf{d}_{\mathbf{n}}(\mathbf{X}_{1,n_1}, ..., \mathbf{X}_{k,n_k})) \\ &= \int [\int L(\theta, \mathsf{d}_{\mathbf{n}}(\mathbf{x}_{1,n_1}, ..., \mathbf{x}_{k,n_k}))(\textstyle\bigotimes_{i=1}^k \mathsf{P}^{\otimes n_i})(d\mathbf{x}_{1,n_1}, ..., d\mathbf{x}_{k,n_k}|\theta)]\Pi(d\theta). \end{aligned}$$

We note that in view of Remark 9.11 it is sufficient to deal with nonrandomized selection rules. Turning to the posterior risk at $i \in \{1, ..., k\}$, that is,

$$\mathsf{r}(\Pi, i|\mathbf{x}_{1,n_1}, ..., \mathbf{x}_{k,n_k}) = \mathbb{E}(L(\Theta, i)|\mathbf{X}_{1,n_1} = \mathbf{x}_{1,n_1}, ..., \mathbf{X}_{k,n_k} = \mathbf{x}_{k,n_k}),$$

we get

$$\mathsf{r}(\Pi, \mathsf{d}_{\mathbf{n}}) = \mathbb{E}\sum\nolimits_{i=1}^k \mathsf{r}(\Pi, i|\mathbf{X}_{1,n_1}, ..., \mathbf{X}_{k,n_k}) I_{\{i\}}(\mathsf{d}_{\mathbf{n}}(\mathbf{X}_{1,n_1}, ..., \mathbf{X}_{k,n_k})).$$

Hence the minimum Bayes risk is attained by any nonrandomized selection rule $\mathsf{d}_{\mathbf{n}}^B$ that satisfies

$$\mathsf{d}_{\mathbf{n}}^B(\mathbf{x}_{1,n_1}, ..., \mathbf{x}_{k,n_k}) \in \arg\min_{1 \le i \le k} \{\mathsf{r}(\Pi, i|\mathbf{x}_{1,n_1}, ..., \mathbf{x}_{k,n_k})\},$$

and it holds

$$\begin{aligned} \mathsf{r}(\Pi, \mathsf{d}_{\mathbf{n}}^B) &= \mathbb{E}(\min_{1 \le i \le k} \mathsf{r}(\Pi, i|\mathbf{X}_{1,n_1}, ..., \mathbf{X}_{k,n_k})) \\ &= \mathbb{E}(\min_{1 \le i \le k} \mathbb{E}(L(\Theta, i)|\mathbf{X}_{1,n_1}, ..., \mathbf{X}_{k,n_k})). \end{aligned}$$

Up to this point the sample sizes $n_1, ..., n_k$ for the samples taken from the k populations have been assumed to be fixed. Now we allow an additional optimization step by requiring only that the total number of observations $|\mathbf{n}| := n_1 + \cdots + n_k = N$ is fixed. A sampling allocation $\mathbf{n}^*$ with $|\mathbf{n}^*| = N$ that provides the smallest Bayes risk satisfies

$$\mathsf{r}(\Pi, \mathsf{d}_{\mathbf{n}^*}^B) = \min_{|\mathbf{n}|=N} \mathbb{E}(\min_{1 \le i \le k} \mathbb{E}(L(\Theta, i)|\mathbf{X}_{1,n_1}, ..., \mathbf{X}_{k,n_k})), \tag{9.103}$$

and we call $\mathbf{n}^*$ a *Bayes design* for the one-stage selection problem.

Example 9.111. According to Corollary 9.12, with κ as the identity, an optimal sampling allocation $\mathbf{n}^*$ under the zero–one loss from (9.15) is determined by finding

$$\max_{|\mathbf{n}|=N} \mathbb{E}(\max_{1 \le i \le k} \mathbb{P}(\Theta_i = \Theta_{[k]}|\mathbf{X}_{1,n_1}, ..., \mathbf{X}_{k,n_k})),$$

and under the linear loss from (9.26) by finding

$$\max_{|\mathbf{n}|=N} \mathbb{E}(\max_{1 \le i \le k} \mathbb{E}(\Theta_i|\mathbf{X}_{1,n_1}, ..., \mathbf{X}_{k,n_k})).$$

Problem 9.112. Evaluate (9.103) in the settings of Examples 9.14 and 9.41 under the zero–one loss and under the linear loss.

Moving on to two-stage selection rules, let $\mathbf{n}_l = (n_{l,1}, ..., n_{l,k})$, $l = 1, 2$, be the sample sizes at Stages 1 and 2, respectively. Let N_1 and N_2 with $N_1 + N_2 = N$ be fixed. We consider now sampling designs $(\mathbf{n}_1, \mathbf{n}_2)$ with $|\mathbf{n}_1| = N_1$ and $|\mathbf{n}_2| = N_2$. To utilize again the Bayes approach we set

$$\mathcal{Y}_{\mathbf{n}_l} = \mathsf{X}_{i=1}^k \mathcal{X}^{n_{l,i}}, \ \mathfrak{B}_{\mathbf{n}_l} = \bigotimes\nolimits_{i=1}^k \mathfrak{A}^{\otimes n_{l,i}}, \ \mathsf{P}_{\mathbf{n}_l} = \bigotimes\nolimits_{i=1}^k \mathsf{P}^{\otimes n_{l,i}}, \quad l = 1, 2,$$

and consider the probability space

$$(\Omega, \mathfrak{F}, \mathbb{P}) = (\mathcal{Y}_{\mathbf{n}_1} \times \mathcal{Y}_{\mathbf{n}_2} \times \Delta^k, \mathfrak{B}_{\mathbf{n}_1} \otimes \mathfrak{B}_{\mathbf{n}_2} \otimes \mathfrak{B}_\Delta^{\otimes k}, (\mathsf{P}_{\mathbf{n}_1} \otimes \mathsf{P}_{\mathbf{n}_2}) \otimes \Pi). \quad (9.104)$$

Denote by

$$\mathbf{Y}_{l,\mathbf{n}_l} = (\mathbf{X}_{l,1,n_1}, ..., \mathbf{X}_{l,k,n_k}) = ((X_{l,1,1}, ..., X_{l,1,n_1}), ..., (X_{l,k,1}, ..., X_{l,k,n_k})),$$

$l = 1, 2$, and $\Theta = (\Theta_1, ..., \Theta_k)$ the corresponding projections. The crucial point is now that

$$\begin{aligned}(\mathsf{P}_{\mathbf{n}_1} \otimes \mathsf{P}_{\mathbf{n}_2}) \otimes \Pi)(d\mathbf{y}_{1,\mathbf{n}_1}, d\mathbf{y}_{2,\mathbf{n}_2}, d\theta) \\ = (\bigotimes\nolimits_{i=1}^k P_{\theta_i}^{\otimes n_{1,i}})(d\mathbf{y}_{1,\mathbf{n}_1})(\bigotimes\nolimits_{i=1}^k P_{\theta_i}^{\otimes n_{2,i}})(d\mathbf{y}_{2,\mathbf{n}_2})\Pi(d\theta).\end{aligned} \quad (9.105)$$

Let us consider the joint distribution of $\mathbf{Y}_{1,\mathbf{n}_1}$ and Θ, which is given by

$$(\bigotimes\nolimits_{i=1}^k P_{\theta_i}^{\otimes n_{1,i}})(d\mathbf{y}_{1,\mathbf{n}_1})\Pi(d\theta),$$

and denote by $P_{\mathbf{Y}_{1,\mathbf{n}_1}}$ the distribution of $\mathbf{Y}_{1,\mathbf{n}_1}$ under $\mathbb{P}$ in (9.104). As Δ is assumed to be a Borel subset of $\mathbb{R}$ we may choose the conditional distribution of Θ, given $\mathbf{Y}_{1,\mathbf{n}_1} = \mathbf{y}_{1,\mathbf{n}_1}$, as a stochastic kernel, say $\boldsymbol{\Pi}_1$, so that

$$(\bigotimes\nolimits_{i=1}^k P_{\theta_i}^{\otimes n_{1,i}})(d\mathbf{y}_{1,\mathbf{n}_1})\Pi(d\theta) = \boldsymbol{\Pi}_1(d\theta|\mathbf{y}_{1,\mathbf{n}_1})P_{\mathbf{Y}_{1,\mathbf{n}_1}}(d\mathbf{y}_{1,\mathbf{n}_1}).$$

Then it holds

$$\begin{aligned}((\mathsf{P}_{\mathbf{n}_1} \otimes \mathsf{P}_{\mathbf{n}_2}) \otimes \Pi)(d\mathbf{y}_{1,\mathbf{n}_1}, d\mathbf{y}_{2,\mathbf{n}_2}, d\theta) \\ = (\bigotimes\nolimits_{i=1}^k P_{\theta_i}^{\otimes n_{2,i}})(d\mathbf{y}_{2,\mathbf{n}_2})\boldsymbol{\Pi}_1(d\theta|\mathbf{y}_{1,\mathbf{n}_1})P_{\mathbf{Y}_{1,\mathbf{n}_1}}(d\mathbf{y}_{1,\mathbf{n}_1}).\end{aligned}$$

That the conditional distribution of $\mathbf{Y}_{2,\mathbf{n}_2}$, given $\Theta = \theta$ and $\mathbf{Y}_{1,\mathbf{n}_1} = \mathbf{y}_{1,\mathbf{n}_1}$, depends only on θ is due to the fact that according to (9.105) $\mathbf{Y}_{1,\mathbf{n}_1}$ and $\mathbf{Y}_{2,\mathbf{n}_2}$ are conditionally independent, given $\Theta = \theta$, so that according to Problem 1.99 $\mathbf{Y}_{1,\mathbf{n}_1}, \Theta, \mathbf{Y}_{2,\mathbf{n}_2}$ is a Markov chain, and the conditional distribution of $\mathbf{Y}_{2,\mathbf{n}_2}$, given $\Theta = \theta$ and $\mathbf{Y}_{1,\mathbf{n}_1} = \mathbf{y}_{1,\mathbf{n}_1}$, is independent of $\mathbf{y}_{1,\mathbf{n}_1}$. This Markov property admits the following statistical interpretation. After the observations at the first stage have been made the value of $\mathbf{Y}_{1,\mathbf{n}_1} = \mathbf{y}_{1,\mathbf{n}_1}$ is fixed. Then the further sampling and inference depend only on the conditional distribution of $\mathbf{Y}_{2,\mathbf{n}_2}$, given Θ, which is the original conditional distribution $(\bigotimes_{i=1}^{k} P_{\theta_i}^{\otimes n_{2,i}})(d\mathbf{y}_{2,\mathbf{n}_2})$, and on a new prior $\boldsymbol{\Pi}_1(d\theta|\mathbf{y}_{1,\mathbf{n}_1})$. The latter is called the *updated prior* based on the observation $\mathbf{Y}_{1,\mathbf{n}_1} = \mathbf{y}_{1,\mathbf{n}_1}$ which, of course, is just the posterior for the observations in the first step.

For any nonrandomized two-stage selection rule $\mathsf{d}_{\mathbf{n}_1,\mathbf{n}_2} : \mathsf{X}_{i=1}^{k}\mathcal{X}^{n_{1,i}} \times \mathsf{X}_{i=1}^{k}\mathcal{X}^{n_{2,i}} \to_m \{1,...k\}$, where $\mathbf{n}_1 = (n_{1,1},...,n_{1,k})$ and $\mathbf{n}_2 = (n_{2,1},...,n_{2,k})$ are fixed, the Bayes risk can be written as follows.

$$\begin{aligned}
&\mathsf{r}(\Pi,\mathsf{d}_{\mathbf{n}_1,\mathbf{n}_2}) = \mathbb{E}L(\Theta,\mathsf{d}_{\mathbf{n}_1,\mathbf{n}_2}(\mathbf{Y}_{1,\mathbf{n}_1},\mathbf{Y}_{2,n_2}))\\
&= \mathbb{E}(\mathbb{E}(L(\Theta,\mathsf{d}_{\mathbf{n}_1,\mathbf{n}_2}(\mathbf{Y}_{1,\mathbf{n}_1},\mathbf{Y}_{2,\mathbf{n}_2})|\Theta,\mathbf{Y}_{1,\mathbf{n}_1}))\\
&= \int \mathbb{E}(L(\Theta,\mathsf{d}_{\mathbf{n}_1,\mathbf{n}_2}(\mathbf{y}_{1,\mathbf{n}_1},\mathbf{Y}_{2,\mathbf{n}_2})|\Theta)P_{\mathbf{Y}_{1,\mathbf{n}_1}}(d\mathbf{y}_{1,\mathbf{n}_1})\\
&= \int[\int[\int L(\theta,\mathsf{d}_{\mathbf{n}_1,\mathbf{n}_2}(\mathbf{y}_{1,\mathbf{n}_1},\mathbf{y}_{2,\mathbf{n}_2}))\mathsf{P}_{\mathbf{n}_2}(d\mathbf{y}_{2,\mathbf{n}_2}|\theta)]\boldsymbol{\Pi}_1(d\theta|\mathbf{y}_{1,\mathbf{n}_1})]\\
&\qquad \times P_{\mathbf{Y}_{1,\mathbf{n}_1}}(d\mathbf{y}_{1,\mathbf{n}_1}).
\end{aligned}$$

Let $\boldsymbol{\Pi}_2(d\theta|\mathbf{y}_{1,\mathbf{n}_1},\mathbf{y}_{2,\mathbf{n}_2})$ be the second-stage posterior, which is a regular conditional distribution of Θ, given $\mathbf{Y}_{1,\mathbf{n}_1} = \mathbf{y}_{1,\mathbf{n}_1}$ and $\mathbf{Y}_{2,\mathbf{n}_2} = \mathbf{y}_{2,\mathbf{n}_2}$. Furthermore let the stochastic kernel Q represent the conditional distribution of $\mathbf{Y}_{2,\mathbf{n}_2}$, given $\mathbf{Y}_{1,\mathbf{n}_1}$. Then

$$\begin{aligned}
\mathsf{r}(\Pi,\mathsf{d}_{\mathbf{n}_1,\mathbf{n}_2}) = &\int[\int[\int L(\theta,\mathsf{d}_{\mathbf{n}_1,\mathbf{n}_2}(\mathbf{y}_{1,\mathbf{n}_1},\mathbf{y}_{2,\mathbf{n}_2}))\boldsymbol{\Pi}_2(d\theta|\mathbf{y}_{1,\mathbf{n}_1},\mathbf{y}_{2,\mathbf{n}_2})]\\
&\times \mathsf{Q}(d\mathbf{y}_{2,\mathbf{n}_2}|\mathbf{y}_{1,\mathbf{n}_1})]P_{\mathbf{Y}_{1,\mathbf{n}_1}}(d\mathbf{y}_{1,\mathbf{n}_1}). \qquad (9.106)
\end{aligned}$$

Denote by $\mathsf{d}_{\mathbf{n}_1,\mathbf{n}_2}^B$ the Bayes selection rule with fixed sample size design $(\mathbf{n}_1,\mathbf{n}_2)$; that is, $\mathsf{d}_{\mathbf{n}_1,\mathbf{n}_2}^B$ is defined by

$$\mathsf{r}(\Pi,\mathsf{d}_{\mathbf{n}_1,\mathbf{n}_2}^B) = \min_{\mathsf{d}_{\mathbf{n}_1,\mathbf{n}_2}} \mathsf{r}(\Pi,\mathsf{d}_{\mathbf{n}_1,\mathbf{n}_2}).$$

If we minimize $\mathsf{r}(\Pi,\mathsf{d}_{\mathbf{n}_1,\mathbf{n}_2})$, not only over the selection rules $\mathsf{d}_{\mathbf{n}_1,\mathbf{n}_2}$, but also over all sampling designs given by $\mathbf{n}_1 = (n_{1,1},...,n_{1,k})$ and $\mathbf{n}_2 = (n_{2,1},...,n_{2,k})$ subject to $|\mathbf{n}_1| = N_1$ and $|\mathbf{n}_2| = N_2$, where N_1 and N_2 are fixed with $N_1+N_2 = N$, then we get

$$\min_{|\mathbf{n}_1|=N_1,|\mathbf{n}_2|=N_2} \mathsf{r}(\Pi,\mathsf{d}_{\mathbf{n}_1,\mathbf{n}_2}^B) = \min_{|\mathbf{n}_1+\mathbf{n}_2|=N} \mathsf{r}(\Pi,\mathsf{d}_{\mathbf{n}_1+\mathbf{n}_2}^B) = \mathsf{r}(\Pi,\mathsf{d}_{\mathbf{n}^*}^B), \qquad (9.107)$$

where $\mathsf{r}(\Pi,\mathsf{d}_{\mathbf{n}^*}^B)$ is from (9.103), and apparently nothing has been gained. However, the Bayes risk can be improved further by an optimization of $\mathbf{n}_2$ in dependence of $\mathbf{n}_1$ and $\mathbf{y}_{1,\mathbf{n}_1}$, followed by an optimization of $\mathbf{n}_1$. We call the resulting design $(\mathbf{n}_1^*,\mathbf{n}_2^*(\mathbf{n}_1^*,\mathbf{Y}_{1,\mathbf{n}_1}))$, in short $(\mathbf{n}_1^*,\mathbf{n}_2^*)$, an *adaptive Bayes design*. It is derived by the *look-ahead* method, or in other words by backward optimization. Let $\mathsf{d}_{\mathbf{n}_1^*,\mathbf{n}_2^*}^B$ denote the associated Bayes selection rule. The next theorem shows that the minimization can be carried out stepwise under the integrals, starting with θ, going back to $\mathbf{y}_{2,\mathbf{n}_2}$, and then going back to $\mathbf{y}_{1,\mathbf{n}_1}$.

Theorem 9.113. *For the Bayes model (9.104), under the constraints* $|\mathbf{n}_1| = N_1$, $|\mathbf{n}_2| = N_2$, *and* $N_1 + N_2 = N$, *the Bayes risk of the adaptive Bayes design* $(\mathbf{n}_1^*, \mathbf{n}_2^*)$ *and the associated Bayes selection rule* $\mathsf{d}^B_{\mathbf{n}_1^*,\mathbf{n}_2^*}$ *satisfy*

$$\mathsf{r}(\Pi, \mathsf{d}^B_{\mathbf{n}_1^*,\mathbf{n}_2^*}) \le \mathsf{r}(\Pi, \mathsf{d}^B_{\mathbf{n}^*}),$$

where $\mathsf{r}(\Pi, \mathsf{d}^B_{\mathbf{n}^*})$ *is from (9.103).*

Proof. The statement follows from

$$\begin{aligned}
&\mathsf{r}(\Pi, \mathsf{d}^B_{\mathbf{n}_1^*,\mathbf{n}_2^*})\\
&= \min_{|\mathbf{n}_1|=N_1} \int \min_{|\mathbf{n}_2|=N_2} \Big[\int \min_{1\le i\le k} \Big[\int L(\theta, i)\boldsymbol{\Pi}_2(d\theta|\mathbf{y}_{1,\mathbf{n}_1}, \mathbf{y}_{2,\mathbf{n}_2})\Big]\mathsf{Q}(d\mathbf{y}_{2,\mathbf{n}_2}|\mathbf{y}_{1,\mathbf{n}_1})\Big]\\
&\qquad \times P_{\mathbf{Y}_{1,\mathbf{n}_1}}(d\mathbf{y}_{1,\mathbf{n}_1})\\
&\le \min_{|\mathbf{n}_1|=N_1} \min_{|\mathbf{n}_2|=N_2} \int \Big[\int \min_{1\le i\le k} \Big[\int L(\theta, i)\boldsymbol{\Pi}_2(d\theta|\mathbf{y}_{1,\mathbf{n}_1}, \mathbf{y}_{2,\mathbf{n}_2})\Big]\mathsf{Q}(d\mathbf{y}_{2,\mathbf{n}_2}|\mathbf{y}_{1,\mathbf{n}_1})\Big]\\
&\qquad \times P_{\mathbf{Y}_{1,\mathbf{n}_1}}(d\mathbf{y}_{1,\mathbf{n}_1})\\
&= \mathsf{r}(\Pi, \mathsf{d}^B_{\mathbf{n}^*}),
\end{aligned}$$

where the last equation follows from (9.107). ■

Remark 9.114. The process of breaking up one stage with N observations into Stage 1 with N_1 and Stage 2 with N_2 observations, subject to $N_1 + N_2 = N$, can be iterated with either of the two stages, and each breakup leads to a new Bayes risk that is less than or equal to the previous one. The overall smallest Bayes risk is achieved by taking the N observations one at a time and using a Bayes N-stage selection rule.

The practical use of the above approach is rather limited if $\mathbf{Y}_{1,\mathbf{n}_1}$ is not a discrete random variable. Although in $\mathsf{r}(\Pi, \mathsf{d}^B_{\mathbf{n}_1^*,\mathbf{n}_2^*})$ the minimization in terms of $\mathbf{n}_2$ may be carried out for every $\mathbf{y}_{1,\mathbf{n}_1}$ and $\mathbf{n}_1$ with $|\mathbf{n}_1| = N_1$, the evaluation of the outer integral with respect to $P_{\mathbf{Y}_{1,\mathbf{n}_1}}$, and thus its minimization, may not be feasible. For such cases approximations to the adaptive Bayes design can be found in the literature, with references given at the beginning of this subsection.

Example 9.115. Let $(X_{i,1}, ..., X_{,i,n_i}, \Theta_i)$, $i = 1, ..., k$, $n_1 + \cdots + n_k = N$, be independent random vectors, where $X_{i,1}, ..., X_{i,n_i}$ are conditionally independent Bernoulli variables with success probabilities θ_i, given $\Theta_i = \theta_i$, $i = 1, ..., k$. Let the prior be given by $\bigotimes_{i=1}^k \mathsf{Be}(\alpha_i, \beta_i)$ with $\alpha_i, \beta_i > 0$, $i = 1, ..., k$. Due to the independence of the vectors $(X_{i,1}, ..., X_{i,n_i}, \Theta_i)$, $i = 1, ..., k$, we may calculate the posterior distribution for each population i and then take the product to get the posterior. According to Example 7.114 the conditional distribution of Θ_i, given $X_{i,1} = x_{i,1}, ..., X_{i,n_i} = x_{i,n_i}$, is $\mathsf{Be}(\alpha_i + \sum_{j=1}^{n_i} x_{i,j}, \beta_i + n_i - \sum_{j=1}^{n_i} x_{i,j})$. Thus with $x_i := \sum_{j=1}^{n_i} x_{i,j}$, $i = 1, ..., k$, utilizing Bayes sufficiency, (9.103) turns out to be

$$\mathsf{r}(\Pi, \mathsf{d}^B_{\mathbf{n}^*}) = \min_{|\mathbf{n}|=N} \sum\nolimits_{x_1=0}^{n_1} \cdots \sum\nolimits_{x_k=0}^{n_k} [\min_{i=1,\ldots,k} \int L(\theta, i) \bigotimes\nolimits_{i=1}^{k} \mathsf{Be}_{\alpha_i+x_i,\beta_i+n_i-x_i}(d\theta_i)] \\ \times \bigotimes\nolimits_{i=1}^{k} \mathsf{pe}_{n_i,\alpha_i,\beta_i}(x_i),$$

where $\mathsf{pe}_{n_i,\alpha_i,\beta_i}(x_i)$ is the probability mass function of the Pólya–Eggenberger distribution; see Gupta and Miescke (1993).

Moving on to two-stage selection rules, let $\mathbf{n}_l = (n_{l,1}, ..., n_{l,k})$, $l = 1, 2$, be the sample sizes at Stages 1 and 2, respectively. Let N_1 and N_2 with $N_1 + N_2 = N$ be fixed. We consider now sampling designs $(\mathbf{n}_1, \mathbf{n}_2)$ with $|\mathbf{n}_1| = N_1$ and $\mathbf{n}_2| = N_2$. Let $X_{l,i,j}$, $j = 1, ..., n_{l,i}$, $i = 1, ..., k$, be the observations at Stage l, $l = 1, 2$. If we use the posterior of Stage 1 as the prior for Stage 2, then the posterior of Stage 2 is

$$\bigotimes\nolimits_{i=1}^{k} \mathsf{Be}(\alpha_i + \sum\nolimits_{l=1}^{2} \sum\nolimits_{j=1}^{n_{l,i}} x_{l,i,j}, \beta_i + n_{1,i} + n_{2,i} - \sum\nolimits_{l=1}^{2} \sum\nolimits_{j=1}^{n_{l,i}} x_{l,i,j}).$$

Then with $x_i := \sum_{j=1}^{n_i} x_{1,i,j}$ and $y_i := \sum_{l=1}^{2} \sum_{j=1}^{n_{l,i}} x_{l,i,j}$, $i = 1, ..., k$, utilizing Bayes sufficiency, we get

$$\begin{aligned} &\mathsf{r}(\Pi, \mathsf{d}^B_{\mathbf{n}_1^*,\mathbf{n}_2^*}) \\ &= \min_{|\mathbf{n}_1|=N_1} \sum\nolimits_{x_1=1}^{n_{1,1}} \cdots \sum\nolimits_{x_k=1}^{n_{1,k}} [\min_{|\mathbf{n}_2|=N_2} \sum\nolimits_{y_1=x_1}^{x_1+n_{2,1}} \cdots \sum\nolimits_{y_k=x_k}^{x_k+n_{2,k}} [\min_{1\le i\le k} \int L(\theta, i) \\ &\times \bigotimes\nolimits_{i=1}^{k} \mathsf{Be}_{\alpha_i+y_i,\beta_i+n_{1,i}+n_{2,i}-y_i}(d\theta_i)] \bigotimes\nolimits_{i=1}^{k} \mathsf{H}_{n_{2,i}|n_{1,i}}(y_i|x_i)] \bigotimes\nolimits_{i=1}^{k} \mathsf{pe}_{n_{1,i},\alpha_i,\beta_i}(x_i), \end{aligned}$$

where $\mathsf{H}_{n_{2,i}|n_{1,i}}(y_i|x_i)$ is the p.m.f. of the distribution of $\sum_{l=1}^{2} \sum_{j=1}^{n_{l,i}} X_{l,i,j}$, given $\sum_{j=1}^{n_{1,i}} X_{1,i,j} = x_i$. Dealing with it can be avoided by using the method of updating the prior at the end of Stage 1.

The minimizations that appear in the previous example have been carried out numerically for $k = 3$ in Miescke and Park (1997), where the optimal allocation of observations, taken one at a time, has been determined by backward optimization. Numerical comparisons with other multistage selection rules, including the look-ahead one stage selection rule, can also be found there.

The problem of finding a best binomial population in several stages, or sequentially, with adaptive sampling has been studied by many authors in various settings in the past. The best-known allocation method, due to Robbins (1956), is *play-the-winner*, which has been utilized in Sobel and Weiss (1972a,b). For an overview of the work done in this area we refer to Gupta and Panchapakesan (1979).

9.5 Asymptotically Optimal Point Selections

9.5.1 Exponential Rate of Error Probabilities

In Chapter 8 we have studied, for binary models with independent observations, the exponential rate at which the error probabilities of the second kind of best level α tests tend to zero. Similar results have been established for the Bayes error probabilities and for the maximum of the error probabilities.

In this section we study the exponential rate of error probabilities in *classification problems*, slightly different from those considered previously, which is due to Krafft and Puri (1974). The results are then utilized to investigate the rate at which the error probabilities of point selection rules tend to zero.

Let $\mathcal{M} = (\mathcal{X}, \mathfrak{A}, (P_\theta)_{\theta\in\Delta})$ be a statistical model with m distinct distributions, i.e., $\Delta = \{1, ..., m\}$. Suppose we want to estimate the parameter $\theta \in \Delta$, i.e., find the true distribution. Then the decision space is $\mathcal{D} = \{1, ..., m\}$ and $\mathfrak{D} = \mathfrak{P}(\{1, ..., m\})$. For any decision $\mathsf{D} : \mathfrak{D} \times \mathcal{X} \to_k [0, 1]$ and $x \in \mathcal{X}$ we set $\varphi_i(x) = \mathsf{D}(\{i\}|x)$, $i = 1, ..., m$, and call $\varphi(x) = (\varphi_1(x), ..., \varphi_m(x))$ *a classification rule*. By construction, $\varphi_i(x)$ is the probability of deciding in favor of $i \in \{1, ..., m\}$ after x has been observed, and it holds $\sum_{i=1}^m \varphi_i(x) = 1$.

We use the zero–one loss function $L(\theta, i) = 1 - I_{\{\theta\}}(i)$, $i, \theta \in \mathcal{D}$. The risk is then given by

$$\mathsf{R}(\theta, \varphi) = \int \sum\nolimits_{i=1}^m L(\theta, i)\varphi_i(x)P_\theta(dx) = \int (1 - \varphi_\theta(x))P_\theta(dx),$$

which is the probability of a false classification. For $\theta_1 \neq \theta_2$ it holds $1 - \varphi_{\theta_2}(x) \geq \varphi_{\theta_1}(x)$ and thus

$$\begin{aligned} \max\{\mathsf{R}(\theta_1, \varphi), \mathsf{R}(\theta_2, \varphi)\} &= \max\{\int (1 - \varphi_{\theta_1})dP_{\theta_1}, \int (1 - \varphi_{\theta_2})dP_{\theta_2}\} \quad (9.108)\\ &\geq \max\{\int (1 - \varphi_{\theta_1})dP_{\theta_1}, \int \varphi_{\theta_1} dP_{\theta_2}\} \geq 2\mathsf{m}_{1/2}(P_{\theta_1}, P_{\theta_2}), \end{aligned}$$

where $\mathsf{m}_{1/2}(P_{\theta_1}, P_{\theta_2})$ is the minimax value from (3.62).

Suppose now that an i.i.d. sample of size n is available, so that we deal with the sequence of statistical models $(\mathcal{X}^n, \mathfrak{A}^{\otimes n}, (P_\theta^{\otimes n})_{\theta\in\Delta})$. If $\varphi^{(n)}$ is any sequence of classification rules, then by (9.108)

$$\max_{1\leq\theta\leq m} \mathsf{R}(\theta, \varphi^{(n)}) \geq 2 \max_{1\leq\theta_1\neq\theta_2\leq m} \mathsf{m}_{1/2}(P_{\theta_1}^{\otimes n}, P_{\theta_2}^{\otimes n}).$$

We apply Theorem 8.72 to the right-hand side and get

$$\begin{aligned} &\liminf_{n\to\infty} \frac{1}{n} \ln(\max_{1\leq\theta\leq m} \mathsf{R}(\theta, \varphi^{(n)}))\\ &\geq \liminf_{n\to\infty} \frac{1}{n} \ln(\max_{1\leq\theta_1\neq\theta_2\leq m} \mathsf{m}_{1/2}(P_{\theta_1}^{\otimes n}, P_{\theta_2}^{\otimes n}))\\ &\geq \max_{1\leq\theta_1\neq\theta_2\leq m} (-\mathsf{C}(P_{\theta_1}, P_{\theta_2})) = -\min_{1\leq\theta_1\neq\theta_2\leq m} \mathsf{C}(P_{\theta_1}, P_{\theta_2}), \quad (9.109) \end{aligned}$$

where $\mathsf{C}(P_{\theta_1}, P_{\theta_2})$ is the Chernoff index of P_{θ_1} and P_{θ_2} from (8.73).

The natural question arises as to whether we can find a classification rule that attains asymptotically the lower bound for the risk in (9.109). In the sequel it is shown that the maximum likelihood classification rule below has that property. Fix a $\mu \in \mathcal{M}^\sigma(\mathfrak{A})$ that dominates all P_θ, let $f_{n,\theta} = dP_\theta^{\otimes n}/d\mu^{\otimes n}$, and

$$B_n(x) = \{\eta : \eta \in \{1, ..., m\},\ f_{n,\eta}(x) = \max_{1 \le \theta \le m} f_{n,\theta}(x)\},\ x \in \mathcal{X}^n.$$

The *maximum likelihood classification rule* is defined to be

$$\varphi_{ml}^{(n)}(x) = \frac{1}{|B_n(x)|}(I_{B_n(x)}(1), ..., I_{B_n(x)}(m)), \quad x \in \mathcal{X}^n. \tag{9.110}$$

For any $x \in \mathcal{X}^n$, if $B_n(x) \subseteq \{1, ..., m\}$ is a singleton, then $\varphi_{ml}^{(n)}$ decides for this value. Otherwise, $\varphi_{ml}^{(n)}$ selects a point from $B_n(x)$ at random by using the uniform distribution on $B_n(x)$.

To derive an upper bound for the risk of this rule we note that $I_{B_n(x)}(i) > 0$ implies that $f_{n,i}(x) = \max_{\eta \in \Delta} f_{n,\eta}(x) \ge f_{n,\theta}(x)$, $\theta \in \{1, ..., m\}$. Hence,

$$\begin{aligned} \mathsf{R}(\theta, \varphi_{ml}^{(n)}) &= \sum\nolimits_{i \ne \theta} \int \frac{1}{|B_n(x)|} I_{B_n(x)}(i) P_\theta^{\otimes n}(dx) \le \sum\nolimits_{i \ne \theta} P_\theta^{\otimes n}(f_{n,\theta} \le f_{n,i}) \\ &\le \sum\nolimits_{i \ne \theta} \int (f_{n,\theta} \wedge f_{n,i}) d\mu^{\otimes n} \le 2 \sum\nolimits_{i,j: i \ne j} \mathsf{b}_{1/2}(P_j^{\otimes n}, P_i^{\otimes n}), \end{aligned}$$

where $\mathsf{b}_{1/2}(P_0, P_1)$ has been introduced in Lemma 1.66.

Problem 9.116.* For any sequences $a_n, b_n \ge 0$ it holds

$$\limsup_{n\to\infty} \frac{1}{n} \ln(a_n + b_n) = \max\{\limsup_{n\to\infty} \frac{1}{n} \ln a_n, \limsup_{n\to\infty} \frac{1}{n} \ln b_n\}. \tag{9.111}$$

An application of Theorem 8.72 and (9.111) to the above inequality yields

$$\limsup_{n\to\infty} \frac{1}{n} \ln(\max_{1 \le \theta \le m} \mathsf{R}(\theta, \varphi_{ml}^{(n)})) \le - \min_{1 \le \theta_1 \ne \theta_2 \le m} \mathsf{C}(P_{\theta_1}, P_{\theta_2}).$$

By combining this with (9.109) we get the following result which has been established in Krafft and Puri (1974).

Theorem 9.117. *If $\varphi^{(n)}$ is a sequence of classification rules for the sequence of models $(\mathcal{X}^n, \mathfrak{A}^{\otimes n}, (P_\theta^{\otimes n})_{\theta \in \Delta})$, where $\Delta = \{1, ..., m\}$, then*

$$\liminf_{n\to\infty} \frac{1}{n} \ln(\max_{1 \le \theta \le m} \mathsf{R}(\theta, \varphi^{(n)})) \ge - \min_{1 \le \theta_1 \ne \theta_2 \le m} \mathsf{C}(P_{\theta_1}, P_{\theta_2}). \tag{9.112}$$

The maximum likelihood classification rule $\varphi_{ml}^{(n)}$ in (9.110) attains the lower bound of the probability of an incorrect classification; that is,

$$\lim_{n\to\infty} \frac{1}{n} \ln(\max_{1 \le \theta \le m} \mathsf{R}(\theta, \varphi_{ml}^{(n)})) = - \min_{1 \le \theta_1 \ne \theta_2 \le m} \mathsf{C}(P_{\theta_1}, P_{\theta_2}).$$

Example 9.118. Let $(P_\theta)_{\theta \in \Delta}$ be an exponential family with natural parameter θ, and let $\{\theta_1, ..., \theta_m\}$ be a finite subset of Δ. From Example 8.74 we get the exponential rate of the maximum error probabilities of the asymptotically optimal classification rule

$$\inf_{1 \leq \theta_i \neq \theta_j \leq m} \mathsf{C}(P_{\theta_i}, P_{\theta_j})$$
$$= \inf_{1 \leq \theta_i \neq \theta_j \leq m, 0 < s < 1} \{sK(\theta_i) + (1-s)K(\theta_j) - K(s\theta_i + (1-s)\theta_j)\}.$$

Especially, if $P_{\theta_j} = \mathsf{N}(\theta_j, \sigma^2)$, then by (8.75),

$$\inf_{1 \leq \theta_i \neq \theta_j \leq m} \mathsf{C}(\mathsf{N}(\theta_i, \sigma^2), \mathsf{N}(\theta_j, \sigma^2)) = \frac{1}{8\sigma^2} \min_{i \neq j} (\theta_i - \theta_j)^2.$$

Now we turn to the error probabilities in another type of problem with a finite decision space. Let $Q_1, ..., Q_k$ be given distinct distributions on the sample space $(\mathcal{X}, \mathfrak{A})$, where one of them, say Q_k, is singled out as the *best population*. We take a sample from each of the k distributions, say $X_1, ..., X_k$, and assume that they are independent. The model assumption is that we know that $\{\mathcal{L}(X_1), ..., \mathcal{L}(X_k)\} = \{Q_1, ..., Q_k\}$, but not the actual pairing. Such a problem is called an *identification problem*, see, for example, Bechhofer, Kiefer, and Sobel (1968) and Miescke (1979a). This is the simplest form of a selection problem. We take $\Delta = \Pi_k$, i.e., the set of all $k!$ permutations θ of $(1, ..., k)$, and $P_\theta = \bigotimes_{i=1}^k Q_{\theta(i)}$, where $Q_{\theta(i)} = \mathcal{L}(X_i)$, $i = 1, ..., k$, at $\theta \in \Delta$. Our goal is to identify that particular $i \in \{1, ..., k\}$ for which $\mathcal{L}(X_i) = Q_k$, and thus the decision space is $\mathcal{D} = \{1, ..., k\}$. Each decision $\mathsf{D}(A|x)$, $A \subseteq \mathcal{D}$, $x \in \mathcal{X}^k$, can be reduced to elementary probabilities by switching to $\varphi_i(x) := \mathsf{D}(\{i\}|x)$, the probability of selecting the ith population, $i = 1, ..., k$, after $x \in \mathcal{X}^k$ has been observed. Because $\theta^{-1}(k)$ is the position of the best population if θ is the true parameter, the zero–one loss function, which we adopt here, has the form $L(\theta, i) = 1 - I_{\{\theta^{-1}(k)\}}(i)$, $\theta \in \Delta$, $i = 1, ..., k$. The risk is $\mathsf{R}(\theta, \varphi) = 1 - P_{cs}(\theta, \varphi)$, where, analogously to (9.17),

$$P_{cs}(\theta, \varphi) = \int \varphi_{\theta^{-1}(k)}(x) P_\theta(dx), \quad \theta \in \Delta,$$

is the probability of a correct selection (PCS). To deal with $\mathsf{R}(\theta, \varphi)$ we use similar ideas as in (9.108). For $\theta_1, \theta_2 \in \Delta$ with $\theta_1^{-1}(k) \neq \theta_2^{-1}(k)$ it holds $1 - \varphi_{\theta_2^{-1}(k)} \geq \varphi_{\theta_1^{-1}(k)}$, and thus as in (9.108),

$$\max(\mathsf{R}(\theta_1, \varphi), \mathsf{R}(\theta_2, \varphi)) \geq 2\mathsf{m}_{1/2}(P_{\theta_1}, P_{\theta_2}). \tag{9.113}$$

To study the rate of error probabilities for increasing sample size n we replace $P_\theta = \bigotimes_{i=1}^k Q_{\theta(i)}$ by $P_\theta^{\otimes n} = \bigotimes_{i=1}^k Q_{\theta(i)}^{\otimes n}$ and obtain from (9.112) that

$$\liminf_{n \to \infty} \frac{1}{n} \ln(\max_{\theta \in \Delta} \mathsf{R}(\theta, \varphi^{(n)})) \geq - \min_{\theta_1, \theta_2 : \theta_1^{-1}(k) \neq \theta_2^{-1}(k)} \mathsf{C}(P_{\theta_1}, P_{\theta_2}).$$

Denote by $x_{n,i} \in \mathcal{X}^n$ the vector of the n observations from the ith population, and by $\mathbf{x}_n = (x_{n,1}, ..., x_{n,k}) \in \mathcal{X}^{nk}$ the collection of all observations. We dominate $Q_1, ..., Q_m$ by a σ-finite measure μ, say, put $f_i = dQ_i/d\mu$, $f_i^{(n)} = dQ_i^{\otimes n}/d\mu^{\otimes n}$, $i = 1, ..., k$, and

$$f_{n,\theta}(\mathbf{x}_n) = \prod\nolimits_{i=1}^{k} f_{\theta(i)}^{(n)}(x_{n,i}).$$

To construct a maximum likelihood selection rule we recall that Δ is the set Π_k of all permutations of $(1, ..., k)$ and denote by

$$B_n(\mathbf{x}_n) = \{\gamma : f_{n,\gamma}(\mathbf{x}_n) = \max_{\theta \in \Delta} f_{n,\theta}(\mathbf{x}_n),\ \gamma \in \Delta\}$$

the set of all permutations γ that maximize the likelihood function at $\mathbf{x}_n \in \mathcal{X}^{nk}$. Set

$$C_n(\mathbf{x}_n) = \{j : \text{there exists a permutation } \gamma \in B_n(\mathbf{x}_n) \text{ with } j = \gamma^{-1}(k)\}.$$

The *maximum likelihood selection rule* $\varphi_{ml}^{(n)}$ is defined to be the uniform distribution on $C_n(\mathbf{x}_n)$; that is,

$$\varphi_{ml}^{(n)}(\mathbf{x}_n) = \frac{1}{|C_n(\mathbf{x}_n)|}(I_{C_n(\mathbf{x}_n)}(1), ..., I_{C_n(\mathbf{x}_n)}(k)), \quad \mathbf{x}_n \in \mathcal{X}^{nk}.$$

If the true permutation θ satisfies $\theta^{-1}(k) \notin C_n(\mathbf{x}_n)$, then there is a permutation γ with $\theta^{-1}(k) \neq \gamma^{-1}(k)$ such that $f_{n,\gamma}(\mathbf{x}_n) \geq f_{n,\theta}(\mathbf{x}_n)$. Hence,

$$\begin{aligned} \mathsf{R}(\theta, \varphi_{ml}^{(n)}) &\leq \sum_{i \neq \theta^{-1}(k)} \int \frac{1}{|C_n(\mathbf{x}_n)|} I_{C_n(\mathbf{x}_n)}(i) P_\theta^{\otimes n}(d\mathbf{x}_n) \qquad (9.114) \\ &\leq \sum_{\gamma:\gamma^{-1}(k) \neq \theta^{-1}(k)} P_\theta^{\otimes n}(f_{n,\theta} \leq f_{n,\gamma}) \leq 2 \sum_{\gamma:\gamma^{-1}(k) \neq \theta^{-1}(k)} \mathsf{b}_{1/2}(P_\theta^{\otimes n}, P_\gamma^{\otimes n}). \end{aligned}$$

The following result has been established in Liese and Miescke (1999a).

Theorem 9.119. *Let $\varphi^{(n)}$ be a sequence of respective selection rules for the sequence of models $(\mathcal{X}^n, \mathfrak{A}^{\otimes n}, (P_\theta^{\otimes n})_{\theta \in \Delta})$, where $P_\theta^{\otimes n} = \bigotimes_{i=1}^k Q_{\theta(i)}^{\otimes n}$. Then*

$$\liminf_{n \to \infty} \frac{1}{n} \ln(\max_{\theta \in \Delta} \mathsf{R}(\theta, \varphi^{(n)})) \geq 2 \max_{j \neq k}(\ln \mathsf{H}_{1/2}(Q_j, Q_k)).$$

The maximum likelihood selection rule in (9.114) attains the lower bound of the probability of an incorrect selection; that is,

$$\lim_{n \to \infty} \frac{1}{n} \ln(\max_{\theta \in \Delta} \mathsf{R}(\theta, \varphi_{ml}^{(n)})) = 2 \max_{j \neq k}(\ln \mathsf{H}_{1/2}(Q_j, Q_k)).$$

Proof. Similar as in the proof of the previous theorem an application of Theorem 8.72 to the inequalities (9.113) and (9.114) yields, respectively,

$$\liminf_{n \to \infty} \frac{1}{n} \ln(\max_{\theta \in \Delta} \mathsf{R}(\theta, \varphi^{(n)})) \geq - \inf_{\theta_1, \theta_2 : \theta_1^{-1}(k) \neq \theta_2^{-1}(k)} \mathsf{C}(P_{\theta_1}, P_{\theta_2}),$$

$$\limsup_{n \to \infty} \frac{1}{n} \ln(\max_{\theta \in \Delta} \mathsf{R}(\theta, \varphi_{ml}^{(n)})) \leq - \inf_{\theta_1, \theta_2 : \theta_1^{-1}(k) \neq \theta_2^{-1}(k)} \mathsf{C}(P_{\theta_1}, P_{\theta_2}).$$

To complete the proof we have to evaluate the right-hand side. It holds

$$\inf_{\theta_1,\theta_2:\theta_1^{-1}(k)\neq\theta_2^{-1}(k)} \mathsf{C}(P_{\theta_1},P_{\theta_2}) = \inf_{\theta:\theta^{-1}(k)\neq k} \mathsf{C}(\bigotimes\nolimits_{i=1}^k Q_{\theta(i)}, \bigotimes\nolimits_{i=1}^k Q_i)$$

$$= \inf_{\theta:\theta^{-1}(k)\neq k} [\sup_{0<s<1} \sum\nolimits_{i=1}^k (-\ln(\mathsf{H}_s(Q_{\theta(i)},Q_i)))].$$

As $0 \leq \mathsf{H}_s(Q_{\theta(i)},Q_i) \leq 1$, we have to take the infimum only over permutations that exchange k with another j, but do not change any other of the $i \in \{1,...,k\}$. Similar as in (1.4) one can use Hölder's inequality to show that the function $s \longmapsto \ln \mathsf{H}_s(Q_j,Q_k)$ is convex. As $\mathsf{H}_{1-s}(Q_j,Q_k) = \mathsf{H}_s(Q_k,Q_j)$ we get that $s \longmapsto (\ln \mathsf{H}_s(Q_j,Q_k) + \ln \mathsf{H}_s(Q_k,Q_j))$ is a convex function that is symmetric about 1/2. Hence

$$\inf_{\theta:\theta^{-1}(k)\neq k} \mathsf{C}(\bigotimes\nolimits_{i=1}^k Q_{\theta(i)}, \bigotimes\nolimits_{i=1}^k Q_i)$$

$$= \inf_{j\neq k}[\sup_{0<s<1} (-\ln(\mathsf{H}_s(Q_j,Q_k) - \ln(\mathsf{H}_s(Q_k,Q_j))] = -2\max_{j\neq k}(\ln \mathsf{H}_{1/2}(Q_j,Q_k)),$$

which completes the proof. ∎

Example 9.120. Assume that Q_θ, $\theta \in (a,b)$, is a one-parameter exponential family with natural parameter θ and generating statistic T. Suppose that $\theta_1 < \cdots < \theta_k$ are given values from (a,b), and that we want to select the population with the largest parameter value. From Theorem 9.119 we know that the maximum likelihood selection rule has the maximum exponential rate, which is given by

$$\liminf_{n\to\infty} \frac{1}{n} \ln(\max_{\theta\in\Delta} \mathsf{R}(\theta,\varphi_{ml}^{(n)})) = 2\max_{j\neq k}(\ln \mathsf{H}_{1/2}(Q_{\theta_j},Q_{\theta_k})).$$

It follows from (1.122) that

$$\ln(\mathsf{H}_{1/2}(Q_{\theta_j},Q_{\theta_k})) = K(\frac{1}{2}(\theta_j+\theta_k)) - \frac{1}{2}(K(\theta_j)+K(\theta_k)).$$

The convexity of K yields that the right-hand term as function of θ_j is increasing, so that

$$2\max_{j\neq k}(\ln \mathsf{H}_{1/2}(Q_{\theta_j},Q_{\theta_k})) = 2\ln \mathsf{H}_{1/2}(Q_{\theta_{k-1}},Q_{\theta_k}).$$

This means that the population $Q_{\theta_{k-1}}$ that is closest to Q_{θ_k} determines the exponential rate.

9.5.2 Locally Asymptotically Optimal Point Selections

After considering the exponential rate of error probabilities of selection procedures for k given distributions, where one is tagged as the best, we turn now to selection models with k families of distributions. Here we utilize the approach based on model localization to compare asymptotically the quality of point selection rules, and to find locally optimal point selection rules. In a first step we deal with the limiting model which again is a Gaussian model. We consider the model

$$(\mathbb{R}^k, \mathfrak{B}_k, (\bigotimes_{i=1}^k \mathsf{N}(\mu_i, \sigma_0^2))_{\mu \in \mathbb{R}^k}).$$

Suppose we want to find at the true but unknown $\mu = (\mu_1, ..., \mu_k) \in \mathbb{R}^k$ a population that is associated with $\mu_{[k]}$, where $\mu_{[1]} \leq \mu_{[2]} \leq \cdots \leq \mu_{[k]}$ denote the ordered values of $\mu_1, ..., \mu_k$. The decision space is $\mathcal{D} = \{1, ..., k\}$. Let $L : \mathbb{R}^k \times \mathcal{D} \to \mathbb{R}_+$ be a given loss function. As in (9.14) the risk of a selection rule ψ under L is

$$\mathsf{R}(\mu, \psi) = \sum\nolimits_{i=1}^k L(\mu, i) \int \psi_i(x) \mathsf{N}(\mu, \sigma_0^2 \mathbf{I})(dx).$$

Set $\mathsf{M}(\mu) = \{i : \mu_i = \mu_{[k]}\}$. Especially if $L_{0,1}(\mu, i) = 1 - I_{\mathsf{M}(\mu)}(i)$, then

$$\mathsf{R}(\mu, \psi) = 1 - P_{cs}(\mu, \psi), \tag{9.115}$$

where the probability of a correct selection $P_{cs}(\mu, \psi)$ (see (9.17)) is given by

$$P_{\mathsf{N},cs}(\mu, \sigma_0^2, \psi) := \sum\nolimits_{i=1}^k I_{\mathsf{M}(\mu)}(i) \int \psi_i(x) \mathsf{N}(\mu, \sigma_0^2 \mathbf{I})(dx).$$

We recall the natural selection rule

$$\begin{aligned} \varphi^{nat}(x_1, ..., x_k) &= (\varphi_1^{nat}(x_1, ..., x_k), ..., \varphi_k^{nat}(x_1, ..., x_k)) \qquad (9.116) \\ &= \frac{1}{|\mathsf{M}(x_1, ..., x_k)|} (I_{\mathsf{M}(x_1,...,x_k)}(1), ..., I_{\mathsf{M}(x_1,...,x_k)}(k)), \end{aligned}$$

and note that by the definition of $\mathsf{M}(x_1, ..., x_k) = \{i : x_i = x_{[k]}\}$ it holds

$$\varphi_i^{nat}(x_1, ..., x_k) = \varphi_i^{nat}(0, x_2 - x_1, ..., x_k - x_1). \tag{9.117}$$

It holds $|\mathsf{M}(x_1, ..., x_k)| = 1$, $\mathsf{N}(\mu, \sigma_0^2 \mathbf{I})$-a.s. Hence $\mathsf{E}_\mu \varphi_i^{nat}$ is the probability that the ith component of a vector with distribution $\mathsf{N}(\nu, \mathbf{I})$, $\nu \in \mathbb{R}^k$, is the largest component of this vector. According to Problem 3.9 this probability is

$$\begin{aligned} &\mathsf{E}_\mu \varphi_i^{nat} = \int I_{\mathsf{M}(x)}(i) \mathsf{N}(\mu, \sigma_0^2 \mathbf{I})(dx) = \\ &\int \prod\nolimits_{j:j \neq i} \Phi(\frac{\mu_i - \mu_j}{\sigma_0} + t) \varphi(t) dt =: \gamma_{i,\mathsf{N}}(\sigma_0^{-1} \mu), \quad \text{where} \qquad (9.118) \\ &\gamma_{i,\mathsf{N}}(\nu) = \int \prod\nolimits_{j=1, j \neq i}^k \Phi(\nu_i - \nu_j + t) \varphi(t) dt, \quad \nu = (\nu_1, ..., \nu_k) \in \mathbb{R}^k. \end{aligned}$$

We recall that for a permutation γ of $(1, ..., k)$ the mapping u_γ is defined by $u_\gamma(x) = (x_{\gamma(1)}, ..., x_{\gamma(k)})$, and that by Definition 9.28 a selection rule ψ is permutation invariant if $\psi(u_\gamma(x)) = u_\gamma(\psi(x))$. Suppose that the loss function satisfies the conditions in (9.42) and (9.43); that is,

$$L(u_\gamma(\mu), i) = L(\mu, \gamma(i)) \quad \text{and} \quad L(\mu, i) \geq L(\mu, j) \text{ for } \mu_i \leq \mu_j, \tag{9.119}$$

where $\mu = (\mu_1, ..., \mu_k) \in \mathbb{R}^k$ and $\gamma \in \Pi_k$. By Theorem 9.31 the selection rule φ^{nat} in (9.116) is a uniformly best permutation invariant selection rule. This means that for every permutation invariant selection rule ψ it holds

$$\mathsf{R}(\mu, \psi) \geq \mathsf{R}(\mu, \varphi^{nat}) = \sum\nolimits_{i=1}^{k} L(\mu, i)\gamma_{i,\mathsf{N}}(\sigma_0^{-1}\mu). \tag{9.120}$$

For the zero–one loss function the risk can be expressed by the probability of a correct selection. Then by (9.115) and (9.23)

$$\begin{aligned} P_{\mathsf{N},cs}(\mu, \sigma^2, \psi) &\leq P_{\mathsf{N},cs}(\mu, \sigma^2, \varphi^{nat}) \\ &= r \int \prod\nolimits_{i=1}^{k-1} \Phi((\mu_{[k]} - \mu_{[i]})/\sigma + s)\varphi(s)ds, \quad \mu \in \mathbb{R}_r^k. \end{aligned} \tag{9.121}$$

Moreover, it follows from Theorem 9.31 that for every permutation invariant subset C of $\mathbb{R}^k$, for every not necessarily permutation invariant ψ, and for any loss function L that satisfies (9.119) it holds

$$\begin{aligned} \sup\nolimits_{\mu \in C} \mathsf{R}(\mu, \psi) &\geq \sup\nolimits_{\mu \in C} \mathsf{R}(\mu, \varphi^{nat}) \\ \inf\nolimits_{\mu \in C} P_{\mathsf{N},cs}(\mu, \sigma^2, \psi) &\leq \inf\nolimits_{\mu \in C} P_{\mathsf{N},cs}(\mu, \sigma^2, \varphi^{nat}). \end{aligned} \tag{9.122}$$

Now we deal with sequences of models for which we want to construct locally optimal selection rules. To this end we use the LAN concept introduced in Definition 6.63. Assume that $(\mathcal{X}_n, \mathfrak{A}_n, (P_{n,h})_{h \in \Delta_n})$ with $\Delta_n \uparrow \mathbb{R}$ satisfies the LAN(Z_n, I_0) condition. As the selection problem is a k sample problem we turn to the product space model

$$\mathcal{M}_n = (\mathcal{X}_n^k, \mathfrak{A}_n^{\otimes k}, (P_{n,h})_{h \in \Delta_n^k}), \quad \text{with } P_{n,h} = \bigotimes\nolimits_{i=1}^{k} P_{n,h_i}, \tag{9.123}$$

$h = (h_1, ..., h_k) \in \Delta_n^k$, and denote by $X_{n,i} : \mathcal{X}_n^k \to \mathcal{X}_n$ the projection onto the ith coordinate, $i = 1, ..., k$. Subsequently we use the following simple fact.

Remark 9.121. Suppose the sequence of models $(\mathcal{X}_n, \mathfrak{A}_n, (P_{n,g})_{g \in \Delta_n})$ with $\Delta_n \uparrow \mathbb{R}$ satisfies the LAN(Z_n, I_0)-condition with central sequence $Z_n : \mathcal{X}_n \to_m \mathbb{R}$ and Fisher information I_0. Now we replace g with h_i and turn to the k sample model $\mathcal{M}_n$ in (9.123). Then it follows directly from Definition 6.63 that $\mathcal{M}_n$ satisfies the LAN$(Z_{n,\otimes k}, \mathcal{I}_0)$-condition, where the central sequence and the Fisher information matrix are given by

$$\begin{aligned} &Z_{n,\otimes k} = (Z_n(X_{n,1}), ..., Z_n(X_{n,k}))^T, \quad \text{central sequence}, \\ &\mathcal{I}_0 := \mathsf{I}_0 \mathbf{I}, \quad \text{Fisher information matrix}, \end{aligned} \tag{9.124}$$

and $\mathbf{I}$ is the $k \times k$ unit matrix. Similarly, if the sequence $(\mathcal{X}_n, \mathfrak{A}_n, (P_{n,g})_{g \in \Delta_n})$ satisfies the ULAN(Z_n, I_0)-condition, then $\mathcal{M}_n$ satisfies the ULAN$(Z_{n,\otimes k}, \mathcal{I}_0)$-condition, where the central sequence and the Fisher information matrix are given by (9.124).

Our goal is to select a population that is associated with $h_{[k]}$. Let $\varphi_n = (\varphi_{n,1}, ..., \varphi_{n,k})$ be a sequence of selection rules for $\mathcal{M}_n$, and let $\varphi = (\varphi_1, ..., \varphi_k)$ be a selection rule for the Gaussian model

$$\mathcal{G} = (\mathbb{R}^k, \mathfrak{B}_k, (\mathsf{N}(\mathcal{I}_0 h, \mathcal{I}_0)_{h\in\mathbb{R}^k}), \quad \mathcal{I}_0 := \mathsf{I}_0 \mathbf{I}, \, \mathsf{I}_0 > 0. \tag{9.125}$$

If the LAN$(Z_{n,\otimes k}, \mathcal{I}_0)$-condition is satisfied, then we have the weak convergence of models $\mathcal{M}_n \Rightarrow \mathcal{G}$, see Theorem 6.65. As the decision space $\mathcal{D} = \{1, ..., k\}$ is finite we get from Definition 6.83 that the weak convergence of the decisions φ_n to φ is equivalent to

$$\lim_{n\to\infty} \int \varphi_{n,i}(x_n) P_{n,h}(dx_n) = \int \varphi_i(x) \mathsf{N}(\mathcal{I}_0 h, \mathcal{I}_0)(dx), \quad h \in \mathbb{R}^k, \tag{9.126}$$

$i = 1, ..., k$. If $\mathsf{I}_0 > 0$, then by $\mathcal{I}_0 = \mathsf{I}_0 \mathbf{I}$ the family $(\mathsf{N}(\mathcal{I}_0 h, \mathcal{I}_0)_{h\in\mathbb{R}^k})$ is complete, and the distributions $\mathsf{N}(\mathcal{I}_0 h, \mathcal{I}_0)$ are equivalent to $\mathsf{N}(0, \mathbf{I})$. We see that every limit $\varphi = (\varphi_1, ..., \varphi_k)$ is $\mathsf{N}(0, \mathbf{I})$-a.s. uniquely determined.

Problem 9.122.* If the sequence of models $\mathcal{M}_n$ satisfies the LAN$(Z_{n,\otimes k}, \mathcal{I}_0)$-condition, it holds $\mathsf{I}_0 > 0$, and φ_n is a sequence of permutation invariant selection rules that converges weakly to φ, then there exists a permutation invariant selection rule ψ such that $\varphi = \psi$, $\mathsf{N}(0, \mathbf{I})$-a.s.

Problem 9.123.* Let $\mathbb{D}_0$ be the set of all selection rules φ for the model $\mathcal{G}$ in (9.125), where $\mathsf{I}_0 > 0$, for which for every $\varphi \in \mathbb{D}_0$ there exists a permutation invariant selection rule ψ with $\varphi = \psi$, $\mathsf{N}(0, \mathbf{I})$-a.s. Then the set $\mathbb{D}_0$ is closed with respect to the weak convergence of decisions in the sense of Definition 3.19.

Next we introduce the concept of locally asymptotically uniformly best selection rules and focus on permutation invariant selection rules. As these concepts originate from the LAN theory we use the term "local" despite the fact that the sequence of models is arbitrary in the first step. The "local" character of the parameter becomes clear later on when we deal with the localization of differentiable models.

Definition 9.124. *Given a loss function $L : \mathbb{R}^k \times \{1, ..., k\} \to \mathbb{R}_+$ that satisfies (9.119), a sequence of selection rules $\varphi_n : \mathcal{X}_n^k \to_m \mathbf{S}_k^c$ for the models $\mathcal{M}_n$ in (9.123) is called locally asymptotically best permutation invariant (LABP) if φ_n is permutation invariant for every n, and for every further sequence of permutation invariant selection rules $\psi_n : \mathcal{X}_n^k \to_m \mathbf{S}_k^c$ it holds*

$$\limsup_{n\to\infty} [\mathsf{R}(h, \varphi_n) - \mathsf{R}(h, \psi_n)] \le 0, \quad h = (h_1, ..., h_k) \in \mathbb{R}^k.$$

The sequence φ_n is called locally asymptotically minimax (LAMM) if for every permutation invariant compact set $C \subseteq \mathbb{R}^k$ it holds

$$\limsup_{n\to\infty} [\sup_{h\in C} \mathsf{R}(h, \varphi_n) - \sup_{h\in C} \mathsf{R}(h, \psi_n)] \le 0,$$

for every other sequence of selection rules $\psi_n : \mathcal{X}_n^k \to_m \mathbf{S}_k^c$.

If the LAN(Z_n, I_0)-condition holds, then the LAN$(Z_{n,\otimes k}, \mathcal{I}_0)$-condition is also fulfilled; see Remark 9.121. For φ^{nat} in (9.116) we denote by

$$\varphi^{nat}_{Z_{n,\otimes k}} = \varphi^{nat}(Z_{n,\otimes k}) \tag{9.127}$$

the natural selection rule based on the central sequence $Z_{n,\otimes k}$; see (9.124). Beside the loss function the essential ingredient of the risk is the probability that the ith population is selected by the rule under consideration. For the natural selection rule $\varphi^{nat}_{Z_{n,\otimes k}} = (\varphi^{nat}_{1,Z_{n,\otimes k}}, ..., \varphi^{nat}_{k,Z_{n,\otimes k}})$ this probability is $\int \varphi^{nat}_{i,Z_{n,\otimes k}} dP_{n,h}$ and the risk given by

$$\mathsf{R}(h, \varphi^{nat}_{Z_{n,\otimes k}}) = \sum\nolimits_{i=1}^{k} L(h,i) \int \varphi^{nat}_{i,Z_{n,\otimes k}} dP_{n,h}. \tag{9.128}$$

As the functions $\varphi^{nat}_i = I_{\mathsf{M}(x)}(i)$ in (9.116), $i = 1, ..., k$, are bounded and $\boldsymbol{\lambda}_k$-a.e. continuous it follows from (6.90) that

$$\lim_{n\to\infty} \int \varphi^{nat}_{i,Z_{n,\otimes k}} dP_{n,h} = \int \varphi^{nat}_i(x) \mathsf{N}(\mathcal{I}_0 h, \mathcal{I}_0)(dx) = \gamma_{i,\mathsf{N}}(\mathsf{I}_0^{1/2} h), \tag{9.129}$$

where according to (9.118)

$$\gamma_{i,\mathsf{N}}(\mathsf{I}_0^{1/2} h) = \int \prod\nolimits_{j=1, j\neq i}^{k} \Phi(t + \mathsf{I}_0^{1/2}(h_i - h_j))\varphi(t)dt. \tag{9.130}$$

We recall that for any sequence of selection rules $\varphi_n = (\varphi_{n,1}, ..., \varphi_{n,k})$ for the models $\mathcal{M}_n$ in (9.123) the risk and the probability of a correct selection, respectively, are given by

$$\mathsf{R}(h, \varphi_n) = \sum\nolimits_{i=1}^{k} L(h,i) \int \varphi_{n,i} dP_{n,h}, \quad \text{and}$$
$$P_{cs}(h, \varphi_n) = \sum\nolimits_{i=1}^{k} \mathsf{I}_{\mathsf{M}(h)}(i) \int \varphi_{n,i} dP_{n,h},$$

where $\mathsf{M}(h) = \{i : h_i = h_{[k]}\}$. Now we establish the pointwise lower Hájek–LeCam bound for selection rules.

Theorem 9.125. *Suppose the sequence of models $(\mathcal{X}_n, \mathfrak{A}_n, (P_{n,g})_{g\in\Delta_n})$, with $\Delta_n \uparrow \mathbb{R}$, satisfies the LAN$(Z_n, \mathsf{I}_0)$ condition, and it holds $\mathsf{I}_0 > 0$. Suppose the loss function $L : \mathbb{R}^k \times \{1, ..., k\} \to \mathbb{R}_+$ satisfies the conditions*

$$L(u_\gamma(h), i) = L(h, \gamma(i)) \quad \textit{and} \quad L(h,i) \geq L(h,j) \textit{ for } h_i \leq h_j, \tag{9.131}$$

$h \in \mathbb{R}^k$, $i, j = 1, ..., k$. Then for the models $\mathcal{M}_n$ in (9.123) every sequence of permutation invariant selection rules φ_n satisfies, with $\gamma_{i,\mathsf{N}}(\mathsf{I}_0^{1/2} h)$ in (9.130),

$$\liminf_{n\to\infty} \mathsf{R}(h, \varphi_n) \geq \mathsf{R}(h, \varphi^{nat}) = \sum\nolimits_{i=1}^{k} L(h,i)\gamma_{i,\mathsf{N}}(\mathsf{I}_0^{1/2} h), \quad h \in \mathbb{R}^k.$$

The sequence of natural selection rules $\varphi^{nat}_{Z_{n,\otimes k}}$ *based on the central sequence* $Z_{n,\otimes k}$ *in (9.124) for the models* $\mathcal{M}_n$ *in (9.123) is LABP, and it holds*

$$\lim_{n\to\infty} \mathsf{R}(h, \varphi^{nat}_{Z_{n,\otimes k}}) = \mathsf{R}(h, \varphi^{nat}), \quad h \in \mathbb{R}^k. \tag{9.132}$$

If the ULAN(Z_n, I_0)*-condition is satisfied and*

$$D_C := \sup\nolimits_{h\in C,\, i=1,\dots,k} L(h,i) < \infty, \tag{9.133}$$

for the compact set C*, then the convergence is uniform on* C*.*

Corollary 9.126. *Under the assumptions of Theorem 9.125 it holds*

$$\begin{aligned}
&\limsup_{n\to\infty} P_{cs}(h, \varphi_n) \le P_{\mathsf{N},cs}(\mathsf{I}_0 h, \mathsf{I}_0, \varphi^{nat}),\\
&\lim_{n\to\infty} P_{cs}(h, \varphi^{nat}_{Z_{n,\otimes k}}) = P_{\mathsf{N},cs}(\mathsf{I}_0 h, \mathsf{I}_0, \varphi^{nat}), \quad h \in \mathbb{R}^k,
\end{aligned}$$

where for $h \in \mathbb{R}^k_r = \{h : h \in \mathbb{R}^k,\ h_{[1]} \le \cdots \le h_{[k-r]} < h_{[k-r+1]} = \cdots = h_{[k]}\}$, $r = 1, ..., k$, *according to (9.23) it holds*

$$P_{\mathsf{N},cs}(\mathsf{I}_0 h, \mathsf{I}_0, \varphi^{nat}) = r \int \prod\nolimits_{i=1}^{k-1} \Phi(\mathsf{I}_0^{1/2}(h_{[k]} - h_{[i]}) + s)\varphi(s)ds. \tag{9.134}$$

The convergence is uniform in h *on compact sets if the ULAN*(Z_n, I_0)*-condition is satisfied.*

Proof. In Remark 9.121 we have pointed out already that the sequence of models $\mathcal{M}_n$ in (9.123) fulfils the LAN$(Z_{n,\otimes k}, \mathcal{I}_0)$ condition, so that the sequence $\mathcal{M}_n$ converges weakly to the model $\mathcal{G}$ in (9.125). Denote by $\mathbb{D}_\mathcal{G}$ the set of all selection rules ϕ for which there exists a permutation invariant selection ψ such that $\phi = \psi$, $\mathsf{N}(h, \mathbf{I})$-a.s. Then by Problems 9.122 and 9.123 $\mathbb{D}_\mathcal{G}$ is closed and contains all accumulation points of the sequence φ_n. As for every $\phi \in \mathbb{D}_\mathcal{G}$ there is a permutation invariant rule ψ with the same risk function, we get from (9.120) and (9.118) that

$$\inf\nolimits_{\phi\in\mathbb{D}_\mathcal{G}} \mathsf{R}(h,\phi) = \mathsf{R}(h, \varphi^{nat}) = \sum\nolimits_{i=1}^k L(h,i)\gamma_{i,\mathsf{N}}(\mathsf{I}_0^{1/2} h).$$

Hence by (6.109),

$$\liminf_{n\to\infty} \mathsf{R}(h, \varphi_n) \ge \sum\nolimits_{i=1}^k L(h,i)\gamma_{i,\mathsf{N}}(\mathsf{I}_0^{1/2} h).$$

The statement (9.132) follows from (9.128) and (9.129). As $\varphi^{nat}_{Z_{n,\otimes k}}$ is permutation invariant by definition the property LABP follows. It holds

$$\begin{aligned}
&\sup\nolimits_{h\in C} \mid \mathsf{R}(h, \varphi^{nat}_{Z_{n,\otimes k}}) - \mathsf{R}(h, \varphi^{nat}) \mid\\
&= \sup\nolimits_{h\in C} \mid \sum\nolimits_{i=1}^k L(h,i)(\int \varphi^{nat}_{i,Z_{n,\otimes k}} dP_{n,h} - \gamma_{i,\mathsf{N}}(\mathsf{I}_0^{1/2} h)) \mid \qquad (9.135)\\
&\le D_C \sum\nolimits_{i=1}^k \sup_{h\in C} \mid \int \varphi^{nat}_{i,Z_{n,\otimes k}} dP_{n,h} - \gamma_{i,\mathsf{N}}(\mathsf{I}_0^{1/2} h) \mid,
\end{aligned}$$

with D_C in (9.133). As the functions $\varphi_i^{nat} = I_{\mathsf{M}(x)}(i)$ in (9.116) are bounded and $\boldsymbol{\lambda}_k$-a.e. continuous it follows from (6.90) and Corollary 6.73 that under the ULAN(Z_n, I_0)-condition the convergence in (9.129) is uniform on compact subsets. The corollary follows from the theorem and (9.121). To prove the uniform convergence we use the inequality

$$\begin{aligned}
&\mid P_{cs}(h, \varphi_{Z_{n,\otimes k}}^{nat}) - P_{\mathsf{N},cs}(\mathsf{I}_0 h, \mathsf{I}_0, \varphi^{nat}) \mid \\
&= \mid \sum\nolimits_{i=1}^{k} I_{\mathsf{M}(h)}(i)(\int \varphi_{i,Z_{n,\otimes k}}^{nat} dP_{n,h} - \gamma_{i,\mathsf{N}}(\mathsf{I}_0^{1/2} h)) \mid \\
&\le \sum\nolimits_{i=1}^{k} \mid \int \varphi_{i,Z_{n,\otimes k}}^{nat} dP_{n,h} - \gamma_{i,\mathsf{N}}(\mathsf{I}_0^{1/2} h)) \mid,
\end{aligned}$$

and the fact the convergence in (9.129) is uniform on compact subsets under the ULAN(Z_n, I_0)-condition. ■

The Hájek–LeCam bound in Theorem 9.125 is a pointwise inequality that holds for every sequence of permutation invariant selection rules. Now we turn to arbitrary selection rules, but consider the maximum risk instead of the pointwise risk.

Theorem 9.127. *Suppose the conditions of Theorem 9.125 are satisfied. Then for every permutation invariant subset $C \subseteq \mathbb{R}^k$ and every sequence of selection rules φ_n it holds*

$$\liminf_{n\to\infty} \sup_{h\in C} \mathsf{R}(h, \varphi_n) \ge \sup_{h\in C} \sum\nolimits_{i=1}^{k} L(h,i)\gamma_{i,\mathsf{N}}(\mathsf{I}_0^{1/2} h). \tag{9.136}$$

If in addition the ULAN(Z_n, I_0)-condition is satisfied, C is compact and permutation invariant, and condition (9.133) is fulfilled, then the sequence of natural selection rules $\varphi_{Z_{n,\otimes k}}^{nat}$ in (9.127), which is based on the central sequence $Z_{n,\otimes k}$ in (9.124) for the models $\mathcal{M}_n$ in (9.123), satisfies

$$\lim_{n\to\infty} \sup_{h\in C} \mathsf{R}(h, \varphi_{Z_{n,\otimes k}}^{nat}) = \sup_{h\in C} \mathsf{R}(h, \varphi^{nat}) = \sup_{h\in C} \sum\nolimits_{i=1}^{k} L(h,i)\gamma_{i,\mathsf{N}}(\mathsf{I}_0^{1/2} h)$$

and has therefore the LAMM property.

Corollary 9.128. *If the conditions of Theorem 9.127 are satisfied and C is permutation invariant, then*

$$\limsup_{n\to\infty} \inf_{h\in C} P_{cs}(h, \varphi_n) \le \inf_{h\in C} P_{\mathsf{N},cs}(\mathsf{I}_0 h, \mathsf{I}_0, \varphi^{nat}).$$

If in addition C is compact and the ULAN(Z_n, I_0)-condition is fulfilled, then with $P_{\mathsf{N},cs}(\mathsf{I}_0 h, \mathsf{I}_0, \varphi^{nat})$ in (9.134),

$$\lim_{n\to\infty} \inf_{h\in C} P_{cs}(h, \varphi_{Z_{n,\otimes k}}^{nat}) = \inf_{h\in C} P_{\mathsf{N},cs}(\mathsf{I}_0 h, \mathsf{I}_0, \varphi^{nat}).$$

Corollary 9.129. *For* $K_{c,\delta} = \{h : h \in \mathbb{R}^k,\ \|h\| \le c,\ h_{[1]} \le \cdots \le h_{[k-1]} < h_{[k]} - \delta\}$ *and* $0 < \delta < c < \infty$ *it holds*

$$\limsup_{n\to\infty} \inf_{h\in K_{c,\delta}} P_{cs}(h, \varphi_n) \le \int \Phi^{k-1}(t + \mathsf{I}_0^{1/2}\delta)\varphi(t)dt$$

$$\lim_{n\to\infty} \inf_{h\in K_{c,\delta}} P_{cs}(h, \varphi^{nat}_{Z_n,\otimes k}) = \int \Phi^{k-1}(t + \mathsf{I}_0^{1/2}\delta)\varphi(t)dt.$$

Proof. The proof is almost identical with the proof of the previous theorem. In contrast to the proof there we denote now by $\mathbb{D}_{\mathcal{G}}$ the set of all selection rules. The stated inequality follows from Proposition 6.89 and (9.122). Finally, the fact that $\varphi^{nat}_{Z_n,\otimes k}$ attains the asymptotic lower minimax bound follows from the locally uniform convergence of the risk if the ULAN(Z_n, I_0)-condition holds; see Theorem 9.125. The first corollary follows from the uniform convergence of $P_{cs}(h, \varphi^{nat}_{Z_n,\otimes k})$ established in Corollary 9.126. The second corollary follows from the first one, the relation (9.134) for $r = 1$, and

$$\inf_{h\in K_{c,\delta}} \int \prod\nolimits_{i=1}^{k-1} \Phi(t + \mathsf{I}_0^{1/2}(h_{[k]} - h_{[i]}))\varphi(t)dt = \int \Phi^{k-1}(t + \mathsf{I}_0^{1/2}\delta)\varphi(t)dt.$$

■

Let $(\mathcal{X}, \mathfrak{A}, (P_\theta)_{\theta\in(a,b)})$ be a model for which the family $(P_\theta)_{\theta\in(a,b)}$ is $\mathbb{L}_2$-differentiable at $\theta_0 \in (a,b)$ with derivative $\dot{L}_{\theta_0}$ and Fisher information $\mathsf{I}(\theta_0)$. We consider the sequence of models

$$\mathcal{M}_n = (\mathcal{X}^{kn}, \mathfrak{A}^{\otimes kn}, (\bigotimes\nolimits_{i=1}^{k} P^{\otimes n}_{\theta_0+h_i/\sqrt{n}})_{h\in\Delta_n^k}), \tag{9.137}$$

where $h = (h_1, ..., h_k)$ and $\Delta_n = (\sqrt{n}(a - \theta_0), \sqrt{n}(b - \theta_0))$. We denote by $X_{i,j}$ the projections of $\mathcal{X}^{kn}$ onto the respective coordinates, omitting the dependence of $X_{i,j}$ on n for simplicity. It holds for $j = 1, ..., n$ and $i = 1, ..., k$,

$$\mathcal{L}(X_{i,j}|P_{n,h}) = P_{\theta_0+h_i/\sqrt{n}}, \quad \text{where} \quad P_{n,h} = \bigotimes\nolimits_{i=1}^{k} P^{\otimes n}_{\theta_0+h_i/\sqrt{n}}. \tag{9.138}$$

We have seen in Theorems 9.125 and 9.127 that the natural selection rule based on the central sequence is the asymptotically best selection rule, in a sense that is specified in these theorems. There is, however, one shortcoming. The central sequence $n^{-1/2}\sum_{j=1}^{n} \dot{L}_{\theta_0}(X_{i,j})$ (see Corollary 6.71) depends on θ_0 and provides an optimal selection rule for populations with parameters close to some fictive common central point θ_0. The latter is, of course, unknown so that $\varphi^{nat}_{Z_n,\otimes k}$ is not a selection rule in the strict sense. It only provides a benchmark with which other selection rules can be compared. The following two examples deal with point selections that are based on given influence functions. We study the efficiency of such selection rules.

Example 9.130. We construct a point selection rule by using a linear statistic. Fix a function $\Psi \in \mathbb{L}_2^0(P_{\theta_0})$ and set $V_n = (V_{n,1}, ..., V_{n,k})$ with

$$V_n = (n^{-1/2} \sum\nolimits_{j=1}^{n} \Psi(X_{1,j}), ..., n^{-1/2} \sum\nolimits_{j=1}^{n} \Psi(X_{k,j})). \tag{9.139}$$

We study the efficiency of the natural selection rule based on V_n under the assumption that the family $(P_\theta)_{\theta \in (a,b)}$ is $\mathbb{L}_2$-differentiable at $\theta_0 \in (a,b)$. By the second lemma of LeCam (see Corollary 6.71) and Remark 9.121 the sequence of models $\mathcal{M}_n$ in (9.123) satisfies the ULAN$(Z_{n,\otimes k}, \mathcal{I}_0)$-condition with central sequence

$$Z_{n,\otimes k} = (n^{-1/2} \sum\nolimits_{j=1}^{n} \dot{L}_{\theta_0}(X_{1,j}), ..., n^{-1/2} \sum\nolimits_{j=1}^{n} \dot{L}_{\theta_0}(X_{k,j})) \tag{9.140}$$

and Fisher information matrix $\mathcal{I}_0 = \mathsf{I}(\theta_0)\mathbf{I}$, where $\mathsf{I}(\theta_0) = \mathsf{E}_{\theta_0} \dot{L}_{\theta_0}^2$ is the Fisher information in the marginal model $(P_\theta)_{\theta \in (a,b)}$. Let

$$\Sigma = \mathsf{C}_{\theta_0}((\Psi, \dot{L}_{\theta_0})^T) = \begin{pmatrix} \sigma_{1,1} & \sigma_{1,2} \\ \sigma_{2,1} & \sigma_{2,2} \end{pmatrix} \tag{9.141}$$

be the covariance matrix of the two-dimensional vector $(\Psi, \dot{L}_{\theta_0})$. Then especially $\sigma_{2,2} = \mathsf{I}(\theta_0)$. We assume that $\sigma_{1,1} > 0$ and $\sigma_{2,2} > 0$. Denote by Γ the $(2k) \times (2k)$ matrix

$$\Gamma = \begin{pmatrix} \sigma_{1,1}\mathbf{I} & \sigma_{1,2}\mathbf{I} \\ \sigma_{2,1}\mathbf{I} & \sigma_{2,2}\mathbf{I} \end{pmatrix},$$

where $\mathbf{I}$ is the $k \times k$ unit matrix. The statistics V_n can be represented in the form (6.93) if we replace $\Psi(X_i)$ with $(\Psi(X_{1,i}), ..., \Psi(X_{k,i}))$. As

$$\mathsf{C}_{\theta_0}((\Psi(X_{1,1}), ..., \Psi(X_{k,1}))^T, \dot{L}_{\theta_0}) = \sigma_{1,2}\mathbf{I}$$

we get from Corollary 6.74, locally uniformly in $h = (h_1, ..., h_k) \in \mathbb{R}^k$,

$$\lim_{n\to\infty} \int \varphi(V_n) dP_{n,h} = \int \varphi(x) \mathsf{N}(\sigma_{1,2}h, \sigma_{1,1}\mathbf{I})(dx), \tag{9.142}$$

for every bounded and $\boldsymbol{\lambda}_k$-a.e. continuous function $\varphi : \mathbb{R}^k \to \mathbb{R}$. Using $\varphi = \varphi_i^{nat}$ in (9.116) we get for the natural selection rule $\varphi_{V_n}^{nat}$ based on V_n,

$$\lim_{n\to\infty} \mathsf{R}(h, \varphi_{V_n}^{nat}) = \sum\nolimits_{i=1}^{k} L(h,i)\gamma_{i,\mathsf{N}}(\sigma_{1,1}^{-1/2}\sigma_{1,2}h), \quad h \in \mathbb{R}^k,$$

with $\gamma_{i,\mathsf{N}}$ in (9.118). Especially for the zero–one loss function,

$$\lim_{n\to\infty} P_{cs}(h, \varphi_{V_n}^{nat}) = P_{\mathsf{N},cs}(\sigma_{1,2}h, \sigma_{1,1}, \varphi^{nat}), \tag{9.143}$$

where according to (9.23)

$$P_{\mathsf{N},cs}(\sigma_{1,2}h, \sigma_{1,1}, \varphi^{nat}) = \tag{9.144}$$
$$r\int \prod\nolimits_{i=1}^{k-1} \Phi(t + \frac{\sigma_{1,2}}{\sqrt{\sigma_{1,1}}}(h_{[k]} - h_{[i]}))\varphi(t)dt, \quad h \in \mathbb{R}_r^k$$

and the convergence is locally uniform in both cases. From here we see that the asymptotic efficiency of the selection rule $\varphi_{V_n}^{nat}$, measured by $P_{cs}(h, \varphi_{V_n}^{nat})$, depends only on the correlation between the influence function Ψ and the score function $\dot{L}_{\theta_0}$. It follows from the Schwarz inequality that

$$\frac{\sigma_{1,2}}{\sqrt{\sigma_{1,1}}} \le \sqrt{\sigma_{2,2}} = \mathsf{I}^{1/2}(\theta_0), \tag{9.145}$$

where equality holds if $\Psi = c\dot{L}_{\theta_0}$, P_{θ_0}-a.s., for some constant c. To summarize, we can say that the loss of efficiency, when using Ψ instead of the optimal influence function $\dot{L}_{\theta_0}$, is measured by the correlation $\sigma_{1,2}/\sqrt{\sigma_{1,1}\sigma_{2,2}}$ of Ψ and $\dot{L}_{\theta_0}$.

Example 9.131. We consider the location family $(P_\theta)_{\theta\in\mathbb{R}}$ that is generated by the positive and absolutely continuous Lebesgue density f with finite Fisher information I. According to Lemma 1.121 the model is $\mathbb{L}_2$-differentiable at every θ_0 with derivative $\dot{L}_{\theta_0}(x) = -f'(x-\theta_0)/f(x-\theta_0)$. Suppose that $\int xf(x)dx = 0$ and $\sigma^2 = \int x^2 f(x)dx < \infty$. Then the location parameter θ_0 is the expectation. Set

$$V_n = \sqrt{n}(\overline{X}_{n,1}-\theta_0,...,\overline{X}_{n,k}-\theta_0), \quad \text{where } \overline{X}_{n,i} = \frac{1}{n}\sum\nolimits_{j=1}^{n} X_{i,j},\ i=1,...,k.$$

Then the natural selection rule based on V_n selects the population with the largest value of the $\overline{X}_{n,i}$. Moreover, $\Psi(x) = x-\theta_0$ is the influence function in (9.139). We have seen in Example 9.130 that the efficiency of the selection rule is characterized by the ratio $\sigma_{1,2}/\sqrt{\sigma_{1,1}\sigma_{2,2}}$. For the natural selection rule $\varphi_{V_n}^{nat}$ this ratio is given by

$$\frac{\sigma_{1,2}}{\sqrt{\sigma_{1,1}\sigma_{2,2}}} = -\frac{1}{\sigma\mathsf{I}^{1/2}}\int (x-\theta_0)\frac{f'(x-\theta_0)}{f(x-\theta_0)}f(x-\theta_0)dx = -\frac{1}{\sigma\mathsf{I}^{1/2}}\int xf'(x)dx \le 1,$$

where the inequality is from (9.145). If $f = \varphi_{0,\sigma^2}$ is the density of $\mathsf{N}(0,\sigma^2)$, then $f'(x)/f(x) = -x/\sigma^2$, $-\int xf'(x)dx = 1$, and $\mathsf{I}_0 = \sigma^{-2}$, so that we have equality in the last inequality. But this is not surprising, as for normal distributions the natural selection rule based on the arithmetic mean is, according to Theorem 9.31, the uniformly best permutation invariant selection rule.

As mentioned already, the locally asymptotically optimal selection rule still depends on the localization point θ_0 which is an unknown nuisance parameter. To get rid of it in the decision rule we make use of the MLEs $\widehat{\theta}_{n,1},...,\widehat{\theta}_{n,k}$ in the k populations. If the assumptions of Theorem 7.148 are satisfied, then it follows from (7.113) that for $i=1,...,k$,

$$\sqrt{n}(\widehat{\theta}_{n,i}-\theta_0) = \frac{1}{\sqrt{n}}\sum\nolimits_{j=1}^{n} \mathsf{I}^{-1}(\theta_0)\dot{L}_{\theta_0}(X_{i,j}) + o_{P_{n,0}}(1). \tag{9.146}$$

From here and (9.140) we see that, up to terms $o_{P_{n,0}}(1)$, the selection rule $\varphi_{Z_{n,\otimes k}}^{nat}$ is identical with the natural selection rule $\varphi_{\widehat{\theta}_n}^{nat}$ based on the vector $\widehat{\theta}_n = (\widehat{\theta}_{n,1},...,\widehat{\theta}_{n,k})$ of the MLEs from the individual populations.

Theorem 9.132. *If the assumptions of Theorem 7.148 are met for the model* $(\mathcal{X},\mathfrak{A},(P_\theta)_{\theta\in(a,b)})$ *at* $\theta_0\in(a,b)$, *then the natural selection rule* $\varphi_{\widehat{\theta}_n}^{nat} = (\varphi_{1,\widehat{\theta}_n}^{nat},...,\varphi_{k,\widehat{\theta}_n}^{nat})$ *based on* $\widehat{\theta}_n$ *is consistent in the sense that in the sequence of models*

$$\mathcal{M}_n = (\mathcal{X}^{kn},\mathfrak{A}^{\otimes kn},(\textstyle\bigotimes_{i=1}^{k} P_{\theta_i}^{\otimes n})_{\theta\in(a,b)^k}),$$

where $\theta = (\theta_1,...,\theta_k)$, *it holds*

$$\lim_{n\to\infty} P_{cs}(\theta,\varphi_{\widehat{\theta}_n}^{nat}) = 1,$$

for every $\theta = (\theta_1, ..., \theta_k) \in (a,b)^k$ *with* $\theta_{[k]} > \theta_{[k-1]}$. *Moreover, if (9.133) is satisfied, then in the localized model with* $P_{n,h}$ *from (9.138) it holds, locally uniformly in* h, *with* $\gamma_{i,\mathsf{N}}(\mathsf{I}_0^{1/2}h)$ *in (9.130),*

$$\lim_{n\to\infty} \mathsf{R}(h, \varphi^{nat}_{\widehat{\theta}_n}) = \sum\nolimits_{i=1}^{k} L(h,i)\gamma_{i,\mathsf{N}}(\mathsf{I}_0^{1/2}h),$$

and $\varphi^{nat}_{\widehat{\theta}_n}$ *is LABP and LAMM, provided that the loss function satisfies (9.131) and (9.133).*

Proof. Without loss of generality we may assume that $\theta_1 \le \cdots \le \theta_{k-1} < \theta_k$. Then

$$\sum_{i\neq k} \varphi^{nat}_{i,\widehat{\theta}_n} = \sum_{i\neq k} \frac{1}{|\mathsf{M}(\widehat{\theta}_n)|} I_{\mathsf{M}(\widehat{\theta}_n)}(i) \le I_{A_N}, \quad \text{where}$$

$$A_N = \{\max_{1\le j\le k-1} \widehat{\theta}_{n,j} \ge \widehat{\theta}_{n,k}\} = \bigcup\nolimits_{j=1}^{k-1} \{\widehat{\theta}_{n,j} \ge \widehat{\theta}_{n,k}\}.$$

Choose an $\varepsilon > 0$ such that $\varepsilon \le \theta_k - \theta_{k-1}$. Then for every $j \neq k$,

$$\begin{aligned}(\textstyle\bigotimes_{i=1}^{k} P_{\theta_i}^{\otimes n})(\{\widehat{\theta}_{n,j} \ge \widehat{\theta}_{n,k}\}) &\le (\textstyle\bigotimes_{i=1}^{k} P_{\theta_i}^{\otimes n})(\{\widehat{\theta}_{n,j} - \theta_j \ge \widehat{\theta}_{n,k} - \theta_k + \varepsilon\}) \\ &\le (\textstyle\bigotimes_{i=1}^{k} P_{\theta_i}^{\otimes n})(\widehat{\theta}_{n,j} - \theta_j \ge \varepsilon/2) + (\textstyle\bigotimes_{i=1}^{k} P_{\theta_i}^{\otimes n})(\widehat{\theta}_{n,k} - \theta_k \le -\varepsilon/2) \to 0,\end{aligned}$$

by the consistency of $\widehat{\theta}_n$. The condition (A10) that is assumed in Theorem 7.148 implies the $\mathbb{L}_2$-differentiability and thus, as in Example 9.130, the ULAN$(Z_{n,\otimes k}, \mathcal{I}_0)$-condition, where $\mathcal{I}_0 = \mathsf{I}(\theta_0)\mathbf{I}$ and $Z_{n,\otimes k}$ is given in (9.140). We see from the definition of $\varphi^{nat}(\widehat{\theta}_n)$, the representation (9.146) and Corollary 6.74 that both

$$\mathsf{R}(h, \varphi^{nat}_{\widehat{\theta}_n}) = \sum\nolimits_{i=1}^{k} L(h,i) \int \varphi^{nat}_{i,\widehat{\theta}_n}(x_n) P_{n,h}(dx_n)$$

and

$$\mathsf{R}(h, \varphi^{nat}_{Z_{n,\otimes k}}) = \sum\nolimits_{i=1}^{k} L(h,i) \int \varphi^{nat}_{i,Z_{n,\otimes k}}(x_n) P_{n,h}(dx_n),$$

with $Z_{n,\otimes k}$ in (9.140), converge locally uniform to $\sum_{i=1}^{k} L(h,i)\gamma_{i,\mathsf{N}}(\mathsf{I}_0^{1/2}h)$. Hence the LABP and LAMM property of $\varphi^{nat}_{\widehat{\theta}_n}$ follow from the corresponding property of $\varphi^{nat}_{Z_{n,\otimes k}}$ which was established in Theorems 9.125 and 9.127. ■

Now we return to selection rules that are based on a given influence function, which may be, for example, the score function $\dot{L}_{\theta_0}$. As we have pointed out already, the dependence of the selection rule on the localization point is undesirable and thus we want to get rid of it by plugging in an estimator. As in concrete applications (e.g., in location-scale families) the score function may depend on additional nuisance parameters we now study selection rules that utilize estimated parameters in a general setting. Similarly as in Section

8.7 we operate in a first step with two types of parameters. The first, $\theta \in \Delta$, specifies the model whereas the second, $\eta \in \Lambda$, specifies the influence function. We suppose that Λ is an open subset of $\mathbb{R}^m$ and recall the class of functions introduced in (7.98). We assume that

$$\begin{array}{lll} \Psi : \Lambda \times \mathcal{X} \to_m \mathbb{R} & \text{and} & \Psi_{\cdot}(x) \in \mathsf{C}_m^{(1)}(U(\eta_0), \mathcal{X}), \\ \Psi_{\eta_0}(\cdot) \in \mathbb{L}_2^0(P_{\theta_0}) & \text{and} & \mathsf{E}_{P_{\theta_0}} \sup_{\eta \in U(\eta_0)} \parallel \dot{\Psi}_\eta \parallel < \infty. \end{array} \tag{9.147}$$

The family of statistics $V_{n,\eta} = (V_{n,1,\eta}, ..., V_{n,k,\eta})$ with

$$V_{n,i,\eta} = n^{-1/2} \sum\nolimits_{j=1}^{n} \Psi_\eta(X_{i,j}), \quad i = 1, ..., k,$$

provides the family of natural selection rules $\varphi_{V_{n,\eta}}^{nat}$. It is an important fact that under the mild regularity conditions (9.147) we may plug in a $\sqrt{n}$-consistent estimator into $\varphi_{V_{n,\eta}}^{nat}$ without changing the local asymptotic risk.

Proposition 9.133. *Suppose that the family $(P_\theta)_{\theta \in (a,b)}$ is $\mathbb{L}_2$-differentiable at $\theta_0 \in (a,b)$. If the condition (9.147) is satisfied and $\widehat{\eta}_n : \mathcal{X}^{kn} \to_m \Lambda$ is under $P_{\theta_0}^{\otimes kn}$ a $\sqrt{n}$-consistent estimator at η_0, then the following convergence is locally uniform. With $\gamma_{i,\mathsf{N}}$ in (9.118),*

$$\begin{aligned} &\lim_{n\to\infty} \mathsf{R}(h, \varphi_{V_{n,\widehat{\eta}_n}}^{nat}) = \lim_{n\to\infty} \sum\nolimits_{i=1}^{k} L(h,i) \int \varphi_{i,V_{n,\widehat{\eta}_n}}^{nat}(x_n) P_{n,h}(dx_n) \\ &= \lim_{n\to\infty} \mathsf{R}(h, \varphi_{V_{n,\eta_0}}^{nat}) = \sum\nolimits_{i=1}^{k} L(h,i) \gamma_{i,\mathsf{N}}(\sigma_{1,1}^{-1/2} \sigma_{1,2} h), \quad h \in \mathbb{R}^k. \\ &\lim_{n\to\infty} P_{cs}(h, \varphi_{V_{n,\widehat{\eta}_n}}^{nat}) = \lim_{n\to\infty} P_{cs}(h, \varphi_{V_{n,\eta_0}}^{nat}) = P_{\mathsf{N},cs}(\sigma_{1,2}h, \sigma_{1,1}, \varphi^{nat}) \\ &= r \int \prod\nolimits_{i=1}^{k-1} \Phi(t + \frac{\sigma_{1,2}}{\sqrt{\sigma_{1,1}}}(h_{[k]} - h_{[i]})) \varphi(t) dt, \quad h \in \mathbb{R}_r^k, \end{aligned}$$

where the $\sigma_{i,j}$ are the elements of the matrix (9.141) with $\Psi = \Psi_{\eta_0}$.

Proof. First of all we remark that Lemma 8.111 yields

$$V_{n,i,\widehat{\eta}_n} - V_{n,1,\widehat{\eta}_n} = V_{n,i,\eta_0} - V_{n,1,\eta_0} + o_{P_{\theta_0}^{\otimes kn}}(1). \tag{9.148}$$

Hence,

$$\begin{pmatrix} V_{n,2,\widehat{\eta}_n} - V_{n,1,\widehat{\eta}_n} \\ \vdots \\ V_{n,k,\widehat{\eta}_n} - V_{n,1,\widehat{\eta}_n} \end{pmatrix} = \frac{1}{\sqrt{n}} \sum_{j=1}^{n} \begin{pmatrix} \Psi_{\eta_0}(X_{2,j}) - \Psi_{\eta_0}(X_{1,j}) \\ \vdots \\ \Psi_{\eta_0}(X_{k,j}) - \Psi_{\eta_0}(X_{1,j}) \end{pmatrix} + o_{P_{\theta_0}^{\otimes kn}}(1).$$

Thus by Corollary 6.74, and for every bounded and $\boldsymbol{\lambda}_{k-1}$-a.e. continuous function φ, it holds locally uniformly in h,

$$\begin{aligned} &\lim_{n\to\infty} \int \varphi(V_{n,2,\widehat{\eta}_n} - V_{n,1,\widehat{\eta}_n}, ..., V_{n,k,\widehat{\eta}_n} - V_{n,1,\widehat{\eta}_n}) dP_{n,h} \\ &= \lim_{n\to\infty} \int \varphi(V_{n,2,\eta_0} - V_{n,1,\eta_0}, ..., V_{n,k,\eta_0} - V_{n,1,\eta_0}) dP_{n,h} = \mathsf{N}(\mu(h), \Sigma), \end{aligned}$$

where in view of (9.142) $\mathsf{N}(\mu(h), \Sigma)$ is the distribution of $(V_2 - V_1, ..., V_k - V_1)$, and $(V_1, ..., V_k)$ has the distribution $\mathsf{N}(\sigma_{1,2}h, \sigma_{1,1}\mathbf{I})$. Hence the relation $\varphi_i^{nat}(t_1, t_2, ..., t_k) = \varphi_i^{nat}(0, t_2 - t_1, ..., t_k - t_1)$ yields locally uniformly in h,

$$\lim_{n\to\infty} \int \varphi_i^{nat}(0, V_{n,2,\widehat{\eta}_n} - V_{n,1,\widehat{\eta}_n}, ..., V_{n,k,\widehat{\eta}_n} - V_{n,1,\widehat{\eta}_n}) dP_{n,h} = \gamma_{i,\mathsf{N}}(\sigma_{1,1}^{-1/2}\sigma_{1,2}h),$$

with $\gamma_{i,\mathsf{N}}$ in (9.118), and the first statement is proved. The statement concerning the probability of a correct selection follows from (9.115) and (9.134). ■

Now we assume that the parameter $\theta \in \Delta \subseteq \mathbb{R}^d$ consists of two parts. One is τ, which has dimension one and is the parameter of interest, and the other one is ξ, which has dimension $m = d - 1$ and is a nuisance parameter. Thus we have the partitions $\theta = (\tau, \xi^T)^T$ and $\dot{L}_\theta = (U_\theta, V_\theta^T)^T$ as in (8.107). We use the component U_θ as influence function Ψ to construct a selection rule, and any $\sqrt{n}$-consistent estimator to estimate θ. Recall that $X_{i,j} : \mathcal{X}^{kn} \to \mathcal{X}$, $i = 1, ..., k$, $j = 1, ..., n$, are the projections, and that $P_{n,h} = \bigotimes_{i=1}^k P_{\tau_0 + h_i/\sqrt{n}, \xi_0}^{\otimes n}$. We consider the sequence of localized models for k samples with a common nuisance parameter ξ_0; that is,

$$\mathcal{M}_n = (\mathcal{X}^{kn}, \mathfrak{A}^{\otimes kn}, (\bigotimes_{i=1}^k P_{\tau_0 + h_i/\sqrt{n}, \xi_0}^{\otimes n})_{h \in \mathbb{R}^k}). \tag{9.149}$$

If we use any estimator $\widehat{\theta}_n : \mathcal{X}^{kn} \to_m \Delta$ and plug it into the statistic

$$S_n(\theta) = (n^{-1/2} \sum\nolimits_{j=1}^n U_\theta(X_{1,j}), ..., n^{-1/2} \sum\nolimits_{j=1}^n U_\theta(X_{k,j})), \tag{9.150}$$

then the natural selection rule $\varphi_{S_n(\widehat{\theta}_n)}^{nat}$ based on the statistic $S_n(\widehat{\theta}_n)$ is not necessarily permutation invariant, unless the estimator $\widehat{\theta}_n$ is invariant under permutations of the populations; that is,

$$\widehat{\theta}_n(x_{1,1}, ..., x_{i,j}, ..., x_{k,n}) = \widehat{\theta}_n(x_{\gamma(1),1}, ..., x_{\gamma(i),j}, ..., x_{\gamma(k),n}), \tag{9.151}$$

for every $(x_{1,1}, ..., x_{i,j}, ..., x_{k,n}) \in \mathcal{X}^{kn}$ and every permutation γ of $(1, ..., k)$. This condition is not very restrictive. One could start, for example, with a sequence $\widetilde{\theta}_n : \mathcal{X}^n \to_m \Delta$ and put

$$\widehat{\theta}_n(x_{1,1}, ..., x_{i,j}, ..., x_{k,n}) = \frac{1}{k} \sum\nolimits_{r=1}^k \widetilde{\theta}_n(x_{r,1}, ..., x_{r,n}).$$

Then $\widehat{\theta}_n$ is obviously permutation invariant.

Theorem 9.134. *Suppose that the family of distributions $(P_\theta)_{\theta \in \Delta}$ satisfies condition (A10), where $\dot{L}_\theta = (U_\theta, V_\theta^T)^T$ is the $\mathbb{L}_2$-derivative and $\mathsf{I}(\theta_0)$ is the invertible information matrix with block matrices $\mathsf{I}_{i,j}(\theta_0)$. Suppose that $\widehat{\theta}_n : \mathcal{X}^{kn} \to_m \Delta$ is a sequence of $\sqrt{n}$-consistent estimators for θ_0 and the loss* *function satisfies (9.131) and (9.133). Then for the risk of the natural selection rule* $\varphi^{nat}_{S_n(\widehat{\theta}_n)}$ *based on* $S_n(\widehat{\theta}_n)$ *with* $S_n(\theta)$ *in (9.150) it holds*

$$\lim_{n\to\infty}\sum\nolimits_{i=1}^{k} L(h,i)\int \varphi^{nat}_{i,S_n(\widehat{\theta}_n)}dP_{n,h} = \sum\nolimits_{i=1}^{k} L(h,i)\gamma_{i,\mathsf{N}}(\mathsf{I}_{1,1}^{1/2}(\theta_0)h), \quad h\in\mathbb{R}^k,$$

locally uniformly in h*, and* $\varphi^{nat}_{S_n(\widehat{\theta}_n)}$ *has the LAMM property. If in addition the condition (9.151) is satisfied, then* $\varphi^{nat}_{S_n(\widehat{\theta}_n)}$ *is LABP.*

Proof. The condition (A10) implies the $\mathbb{L}_2$-differentiability at $\theta_0\in\Delta$ which yields the ULAN$(Z_n,\mathsf{I}(\theta_0))$-condition by the second lemma of LeCam; see Corollary 6.71. Hence the ULAN$(Z_{n,\otimes k},\mathsf{I}_{1,1}(\theta_0)\mathbf{I})$-condition for the sequence of models (9.149) is satisfied, where $S_n(\theta_0)$ in (9.150) is the central sequence. We set $\eta=(\tau,\xi)$ and $\Psi_\eta=U_{\tau,\xi}$. The condition (A10) implies that Ψ_η satisfies (9.147). To use Proposition 9.133 we set $V_n=Z_{n,\otimes k}$. Then $\sigma_{1,1}=\sigma_{1,2}=\mathsf{I}_{1,1}(\theta_0)$, and it holds locally uniformly in h,

$$\lim_{n\to\infty}\mathsf{R}(h,\varphi^{nat}_{S_n(\theta_0)}) = \lim_{n\to\infty}\mathsf{R}(h,\varphi^{nat}_{S_n(\widehat{\theta}_n)}) = \sum\nolimits_{i=1}^{k} L(h,i)\gamma_{i,\mathsf{N}}(\mathsf{I}_{1,1}^{1/2}(\theta_0)h),$$

with $\gamma_{i,\mathsf{N}}$ in (9.118). This means that the maximum risk over permutation invariant and compact sets attains the asymptotic lower bound in (9.136) and $\varphi^{nat}_{S_n(\widehat{\theta}_n)}$ is therefore LAMM. If in addition (9.151) is satisfied, then $\varphi^{nat}_{S_n(\widehat{\theta}_n)}$ is a sequence of permutation invariant selection rules which attains the lower bound for the risk of sequences of permutation invariant selection rules in Theorem 9.125 . Hence $\varphi^{nat}_{S_n(\widehat{\theta}_n)}$ is LABP. ■

In the previous theorem we have shown that a nuisance parameter can be replaced with a $\sqrt{n}$-consistent estimator without reducing the efficiency. Similar results were obtained in Liese and Miescke (1999b) for the semiparametric random censorship model and by Liese (1996) for the semiparametric location model.

The next example deals with point selections in location-scale models with nuisance scale parameters.

Example 9.135. Let f be a positive twice continuously differentiable Lebesgue density so that the family of densities $f_\theta(x)=\sigma^{-1}f((x-\mu)/\sigma)$, $\theta=(\mu,\sigma)\in\mathbb{R}\times(0,\infty)$, satisfies (A10). Then the $\mathbb{L}_2$-derivative with respect to the parameter μ is

$$U_\theta(x) = -\sigma^{-2}\mathsf{U}((x-\mu)/\sigma), \quad \text{where} \quad \mathsf{U}(x) = -f'(x)/f(x).$$

To construct the statistic $S_n(\widehat{\theta}_n)$ from (9.150) we need $\sqrt{n}$-consistent estimators for μ and σ. If the second moment of f is finite, then we may assume without loss of generality that $\int xf(x)dx=0$ and $\int x^2f(x)dx=1$. Then the parameter μ in the family of densities $\sigma^{-1}f((x-\mu)/\sigma)$ is the expectation and σ^2 the variance. We estimate $\theta=(\mu,\sigma)$ by

$$\widehat{\mu}_n = \frac{1}{kn}\sum\nolimits_{i,j} X_{i,j}, \quad \text{and} \quad \widehat{\sigma}_n = (\frac{1}{kn}\sum\nolimits_{i,j}(X_{i,j}-\widehat{\mu}_n)^2)^{1/2}.$$

Then both $\widehat{\mu}_n$ and $\widehat{\sigma}_n^2$ are permutation invariant. Moreover, if $\int x^4 f(x)dx < \infty$, then both estimators are $\sqrt{n}$-consistent. The statement for $\widehat{\mu}_n$ is clear. The statement for $\widehat{\sigma}_n$ follows from Example 8.84 and the δ-method; see Proposition 8.78. The statistic S_n in (9.150) leads to

$$S_n(\widehat{\theta}_n) = \frac{1}{\sqrt{n}\widehat{\sigma}_n^2}(\sum\nolimits_{j=1}^{n} \mathsf{U}((X_{1,j} - \widehat{\mu}_n)/\widehat{\sigma}_n), ..., \sum\nolimits_{j=1}^{n} \mathsf{U}((X_{k,j} - \widehat{\mu}_n)/\widehat{\sigma}_n)). \quad (9.152)$$

As the factor $1/(\sqrt{n}\widehat{\sigma}_n^2)$ is irrelevant for the natural selection rule based on $S_n(\widehat{\theta}_n)$ we see that $\varphi^{nat}_{S_n(\widehat{\theta}_n)}$ is the uniform distribution on

$$\mathsf{M}_n = \{l : \sum\nolimits_{j=1}^{n} \mathsf{U}((X_{l,j} - \widehat{\mu}_n)/\widehat{\sigma}_n) = \max_{1 \le i \le k} \sum\nolimits_{j=1}^{n} \mathsf{U}((X_{i,j} - \widehat{\mu}_n)/\widehat{\sigma}_n)\}.$$

We know from Theorems 9.125 and 9.127 that $\varphi^{nat}_{S_n(\widehat{\theta}_n)}$ has the properties LABP and LAMM. If f is the density of $\mathsf{N}(0,1)$, then $\mathsf{U}(x) = x$ and we see that

$$\begin{aligned}\mathsf{M}_n &= \{l : \sum\nolimits_{j=1}^{n} (X_{l,j} - \widehat{\mu}_n)/\widehat{\sigma}_n = \max_{1 \le i \le k} \sum\nolimits_{j=1}^{n} (X_{i,j} - \widehat{\mu}_n)/\widehat{\sigma}_n\} \\ &= \{l : \overline{X}_{n,l} = \max_{1 \le i \le k} \overline{X}_{n,i}\}.\end{aligned}$$

This means that $\varphi^{nat}_{S_n(\widehat{\theta}_n)}$ selects the population with the largest sample mean. We know from Theorem 9.31 that this selection rule is already for every fixed sample size n a uniformly best permutation invariant selection rule.

If the above moment conditions are not fulfilled one has to turn to other estimators of μ and σ. This is necessary, for example, if we consider a location-scale model generated by the Cauchy distribution. We use sample quantiles to estimate μ and σ. To this end let F be the c.d.f. with the density f. Put $F_{\mu,\sigma}(t) = F((t-\mu)/\sigma)$ and note that F is strictly increasing as f has been assumed to be positive everywhere. We fix $0 < \alpha < \beta < 1$ and set $z_\alpha = F^{-1}(\alpha)$ and $z_\beta = F^{-1}(\beta)$. Then $\sigma z_\alpha + \mu = F^{-1}_{\mu,\sigma}(\alpha)$ and $\sigma z_\beta + \mu = F^{-1}_{\mu,\sigma}(\beta)$. Hence

$$\sigma = \frac{F^{-1}_{\mu,\sigma}(\beta) - F^{-1}_{\mu,\sigma}(\alpha)}{z_\beta - z_\alpha} \quad \text{and} \quad \mu = F^{-1}_{\mu,\sigma}(\alpha) - z_\alpha \frac{F^{-1}_{\mu,\sigma}(\beta) - F^{-1}_{\mu,\sigma}(\alpha)}{z_\beta - z_\alpha}.$$

Let $(X_1, ..., X_N) = (X_{1,1}, ..., X_{k,n})$ be the pooled sample and $\widehat{F}_N$ the empirical c.d.f. Then it follows from Example 7.139 that

$$\mathcal{L}(\sqrt{n}(\widehat{F}_N^{-1}(\alpha) - F^{-1}_{\mu,\sigma}(\alpha))) \Rightarrow \mathsf{N}(0, \alpha(1-\alpha)/f^2_{\mu,\sigma}(z_\alpha)).$$

Hence $\widehat{F}_N^{-1}(\alpha)$ is a $\sqrt{n}$-consistent and permutation invariant estimator of z_α. The same holds analogously for $\widehat{F}_N^{-1}(\beta)$. We set

$$\widehat{\sigma}_n = \frac{\widehat{F}_N^{-1}(\beta) - \widehat{F}_N^{-1}(\alpha)}{z_\beta - z_\alpha} \quad \text{and} \quad \widehat{\mu}_n = \widehat{F}_N^{-1}(\alpha) - z_\alpha \widehat{\sigma}_n.$$

This representation shows that $\widehat{\mu}_n$ and $\widehat{\sigma}_n$ are $\sqrt{n}$-consistent and permutation invariant estimators that can be used to construct the statistic $S_n(\widehat{\theta}_n)$ in (9.152).

In the location-scale model studied above the location parameter was the parameter of interest and the scale parameter the nuisance parameter. It is clear that more complex models, that include, for example, a shape parameter for the density, may be treated in a similar manner. One may even include the density f itself as an infinite-dimensional parameter in the location model. This would mean that $f(x-\theta)$ constitutes a semiparametric location model, where f is common for all populations, but the parameter θ takes on different values. Asymptotically optimal selection rules for this semiparametric location model have been constructed in Liese (1996).

9.5.3 Rank Selection Rules

If the k populations are stochastically ordered, and we want to select the stochastically largest population, then the use of ranks seems to be an adequate approach. Suppose that $X_{i,j}$, $j=1,...,n$, is an i.i.d. sample from population i, $i=1,...,k$, where the k samples are independent. Set $N=kn$, denote by $X_1,...,X_N$ the pooled sample of all observations, and let $R_{N,1},...,R_{N,N}$ be the ranks of the pooled sample. Selection rules based on such ranks have been considered in Lehmann (1963), Puri and Puri (1968, 1969), Bhapkar and Gore (1971), and Büringer, Martin, and Schriever (1980). Further references can be found in Gupta and Panchapakesan (1979).

To set up a suitable statistic for selection, let us first assume that all N observations have a common distribution P with a continuous c.d.f. F. We set $U_r=F(X_r)$, $r=1,...,N$. For a sequence of score functions a_N we use the same score function a_N in each population and introduce the k-dimensional rank statistic by

$$S_n=(\sum\nolimits_{i=1}^{n} a_N(R_{N,i}),...,\sum\nolimits_{i=(k-1)n+1}^{kn} a_N(R_{N,i})). \tag{9.153}$$

If the sequence a_N satisfies the condition (8.97) for some $\varphi:(0,1)\to_m\mathbb{R}$ with $\varphi\in\mathbb{L}_2(\boldsymbol{\lambda})$, and the regression coefficients $c_{i,N}$, $i=1,...,N$, satisfy the condition (8.96), then by Theorem 8.92,

$$\sum\nolimits_{i=1}^{N} c_{i,N}a_N(R_{N,i})=\sum\nolimits_{i=1}^{N} c_{i,N}\varphi(U_i)+o_{P^{\otimes N}}(1).$$

This implies that for $i=2,...,k$,

$$\begin{aligned} & n^{-1/2}[\sum\nolimits_{j=(i-1)n+1}^{in} a_N(R_{N,j})-\sum\nolimits_{j=1}^{n} a_N(R_{N,j})] \\ &= n^{-1/2}[\sum\nolimits_{j=(i-1)n+1}^{in} \varphi(F(X_{i,j}))-\sum\nolimits_{i=1}^{n} \varphi(F(X_{1,j}))]+o_{P^{\otimes N}}(1), \end{aligned} \tag{9.154}$$

where F is the c.d.f. of $X_{i,j}$ under P, and F is assumed to be continuous.

The next theorem studies the asymptotic behavior of selection rules generated by linear rank statistics under local alternatives. We set $P=P_{\theta_0}$ and

$$P_{n,h} = \bigotimes_{i=1}^{k} P_{\theta_0+h_i/\sqrt{n}}^{\otimes n},$$

$$\sigma_{1,1} = \int \varphi^2(F^{-1}(x))P_{\theta_0}(dx), \quad \sigma_{1,2} = \sigma_{2,1} = \int \varphi(F^{-1}(x))\dot{L}_{\theta_0}(x)P_{\theta_0}(dx),$$

$$\sigma_{2,2} = \mathsf{I}(\theta_0) = \int \dot{L}_{\theta_0}^2(x)P_{\theta_0}(dx).$$

Theorem 9.136. *Suppose the sequence a_N satisfies the condition (8.97) for some $\varphi : (0,1) \to_m \mathbb{R}$ with $\varphi \in \mathbb{L}_2^0(\boldsymbol{\lambda})$. Assume that $(P_\theta)_{\theta\in\mathbb{R}}$ is a one-parameter family of distributions on $(\mathbb{R}, \mathfrak{B})$ that is $\mathbb{L}_2$-differentiable at θ_0 with derivative $\dot{L}_{\theta_0}$ and Fisher information $\mathsf{I}(\theta_0) > 0$, where P_{θ_0} has a continuous c.d.f. Then the natural selection rule $\varphi_{S_n}^{nat}$ based on S_n in (9.153) satisfies locally uniform in $h \in \mathbb{R}^k$,*

$$\begin{aligned} \lim_{n\to\infty} \mathsf{R}(\varphi_{S_n}^{nat}, h) &= \lim_{n\to\infty} \sum\nolimits_{i=1}^{k} L(h,i) \int \varphi_{i,S_n}^{nat} dP_{n,h} \\ &= \sum\nolimits_{i=1}^{k} L(h,i)\gamma_{i,\mathsf{N}}(\sigma_{1,1}^{-1/2}\sigma_{1,2}h), \quad h \in \mathbb{R}^k, \end{aligned} \tag{9.155}$$

with $\gamma_{i,\mathsf{N}}$ in (9.118).

Corollary 9.137. *It holds locally uniform in $h \in \mathbb{R}^k$,*

$$\begin{aligned} &\lim_{n\to\infty} P_{cs}(\varphi_{S_n}^{nat}, h) = P_{\mathsf{N},cs}(\sigma_{1,2}h, \sigma_{1,1}, \varphi^{nat}) \\ &= r \int \prod\nolimits_{i=1}^{k-1} \Phi(t + \frac{\sigma_{1,2}}{\sqrt{\sigma_{1,1}}}(h_{[k]} - h_{[i]}))\varphi(t)dt, \quad h \in \mathbb{R}_r^k. \end{aligned}$$

Proof. The proof is almost identical with the proof of Proposition 9.133. One has only to use (9.154) instead of (9.148) and to replace $V_{n,i,\widehat{\eta}_n} - V_{n,1,\widehat{\eta}_n}$ with

$$n^{-1/2} \sum\nolimits_{j=(i-1)n+1}^{in} a_N(R_{N,j}) - n^{-1/2} \sum\nolimits_{j=1}^{n} a_N(R_{N,j}),$$

and $V_{n,i,\eta_0} - V_{n,1,\eta_0}$ with

$$n^{-1/2} \sum\nolimits_{j=(i-1)n+1}^{in} \varphi(F(X_{i,j})) - n^{-1/2} \sum\nolimits_{j=1}^{n} \varphi(F(X_{1,j})).$$

■

Problem 9.138.* If the loss function L satisfies (9.131), then $\sum_{i=1}^k L(h,i)\gamma_{i,\mathsf{N}}(\gamma h)$ is decreasing in $\gamma \geq 0$, for every $h \in \mathbb{R}^k$.

If the loss function L satisfies the condition (9.131) in Theorem 9.125, then by the above problem we see that the right-hand term in (9.155) becomes minimal if $\sigma_{1,2} = \sqrt{\sigma_{1,1}\sigma_{2,2}}$. By similar arguments as in Example 9.130 we get that equality holds if and only if

$$\varphi(F(x)) = c\dot{L}_{\theta_0}(x), \quad P_{\theta_0}\text{-a.s.},$$

for some nonnegative constant c. This means that whenever we have a distribution P with a continuous c.d.f. and we want to construct a rank selection rule that is optimal in the sense that it has the properties LABP and LAMM in Definition 9.124 in a special direction that is defined by an influence function $\Psi \in \mathbb{L}_2^0(P)$, we set $\varphi(t) = \Psi(F^{-1}(t))$ and define the $a_N(k)$ as in Lemma 8.93. Then the condition in (8.97) is satisfied. Furthermore, for any $\mathbb{L}_2$-differentiable curve $(P_\theta)_{\theta\in\mathbb{R}}$ with $P_{\theta_0} = P$ and $\mathbb{L}_2$-derivative $\dot{L}_{\theta_0} = \Psi$ Theorem 9.136, in combination with Theorems 9.125 and 9.127, shows that the rank selection rule $\varphi_{S_n}^{nat}$, with S_n from (9.153), has the properties LABP and LAMM.

A typical field of application of rank selection procedures concerns the location models. The following example deals with point selections based on Wilcoxon rank sum statistics.

Example 9.139. Let

$$F_\theta(x) = \frac{\exp\{x-\theta\}}{1+\exp\{x-\theta\}}, \quad x, \theta \in \mathbb{R},$$

be the c.d.f. of the logistic distribution with location parameter θ. Then

$$\begin{aligned} f_\theta(x) &= \frac{\exp\{x-\theta\}}{(1+\exp\{x-\theta\})^2} \\ \dot{L}_{\theta_0}(x) &= -1 + 2\frac{\exp\{x-\theta\}}{1+\exp\{x-\theta\}} = -1 + 2F_\theta(x), \\ \varphi(t) &= \dot{L}_{\theta_0}(F_{\theta_0}^{-1}(t)) = -1 + 2t. \end{aligned}$$

Set $N = kn$. The approximate scores are given by $a_N(k) = -1 + 2k/(N+1)$. Then the statistic S_n in (9.153) is given by

$$S_n = (-n + 2\sum\nolimits_{i=1}^{n} R_{N,i}, ..., -n + 2\sum\nolimits_{i=(k-1)n+1}^{kn} R_{N,i}).$$

It is clear that $\varphi_{S_n}^{nat}$ is identical with the selection rule that selects the population with the largest rank sum, and this selection rule has the properties LABP and LAMM if the data are from the family of logistic distributions.

We conclude this section with the remark that the construction of asymptotically optimal selection rules via the convergence of models also works if the limit model is not necessarily Gaussian. In Liese (2006) the selection problem for thinned point processes under the sparse conditions has been studied. The limit model is then a Poisson point process.

9.6 Solutions to Selected Problems

Solution to Problem 9.6: Let $U_1, ..., U_r$ be i.i.d. with common uniform distribution on $(0,1)$. Then $F^{-1}(U_1), ..., F^{-1}(U_r)$ are i.i.d. with common distribution P. As $\max_{1\le j\le r} U_j$ has the Lebesgue density $rs^{r-1}I_{(0,1)}(s)$ it holds

$$\mathbb{E}h(\max_{1\le j\le r} F^{-1}(U_j)) = \mathbb{E}h(F^{-1}(\max_{1\le j\le r} U_j)) = \int h(F^{-1}(s))rs^{r-1}ds.$$

Let U be uniformly distributed in $(0,1)$. Then $F^{-1}(U)$ has the distribution P and

$$\int h(F^{-1}(s))rs^{r-1}ds = \mathbb{E}h(F^{-1}(U))r(F(F^{-1}(U)))^{r-1} = r\int h(t)F^{r-1}(t)P(dt). \quad \square$$

Solution to Problem 9.15: The posterior distribution of $\varXi$, the random version of $p = (p_1, ..., p_k)$, given $x_1, ..., x_k$, is $\bigotimes_{i=1}^k \mathsf{Be}(\alpha + x_i, \beta + n - x_i)$.

For $y \in \{0, 1, ..., n\}$ the Lebesgue density of $\mathsf{Be}(\alpha + y, \beta + n - y)$ is

$$\mathsf{be}_{\alpha+y,\beta+n-y}(t) = \frac{\Gamma(\alpha+\beta+n)}{\Gamma(\alpha+y)\Gamma(\beta+n-y)} t^{\alpha+y}(1-t)^{\beta+n-y} I_{(0,1)}(t), \quad t \in \mathbb{R}.$$

If $y \in \{0, 1, ..., n\}$ is treated as a parameter, and α, β, and n are held fixed, then $\mathsf{be}_{\alpha+y,\beta+n-y}(t)$, $t \in \mathbb{R}$, has a nondecreasing MLR in the identity. Thus, by Problem 2.22, $\mathsf{Be}(\alpha + y, \beta + n - y)$ is stochastically nondecreasing in $y \in \{0, 1, ..., n\}$. The statement for the loss $L_{0,1}$ follows now from (9.32) and Proposition 9.13.

The expectation of $\mathsf{Be}(\alpha + x_i, \beta + n - x_i)$ is $(\alpha + x_i)/(\alpha + \beta + n)$, $i = 1, ..., k$. Thus under the loss L_{lin}, according to (9.33), every Bayes selection rule again selects in terms of the largest value of $x_1, ..., x_k$. $\square$

Solution to Problem 9.19: Let $h \ge 0$ be measurable. It holds

$$\begin{aligned}\int h(u_\gamma(s))\boldsymbol{\Gamma}(ds|t) &= \int h(u_\gamma(s))g_t(s)\mu(ds) = \int h(u_\gamma(s))g_{u_\gamma(t)}(u_\gamma(s))\mu(ds)\\ &= \int h(s)g_{u_\gamma(t)}(s)\mu(ds) = \int h(s)\boldsymbol{\Gamma}(ds|u_\gamma(t)),\end{aligned}$$

and thus (9.35). To establish (9.37) we note that for $t_i < t_j$

$$\begin{aligned}&\int h(s)I_{[s_i,\infty)}(s_j)\boldsymbol{\Gamma}(ds|t^{(i,j)}) = \int h(s)I_{[s_i,\infty)}(s_j)g_{t^{(i,j)}}(s)\mu(ds)\\ &\le \int h(s)I_{[s_i,\infty)}(s_j)g_t(s)\mu(ds) = \int h(s)I_{[s_i,\infty)}(s_j)\boldsymbol{\Gamma}(ds|t). \quad \square\end{aligned}$$

Solution to Problem 9.22: Let $\theta_1 \le \theta_2$. An application on the monotone convergence theorem shows that it suffices to deal with bounded nonnegative functions h that vanish for $x_2 \le x_1$. First we assume that $P_{\theta_2} \ll P_{\theta_1}$. The likelihood ratio L_{θ_1,θ_2} is then the P_{θ_1}-density $dP_{\theta_2}/dP_{\theta_1}$ and we have

$$\begin{aligned}&\int\int h(x_1,x_2)[P_{\theta_1}(dx_1)P_{\theta_2}(dx_2) - P_{\theta_2}(dx_1)P_{\theta_1}(dx_2)] =\\ &\int[\int h(x_1,x_2)[L_{\theta_1,\theta_2}(x_2) - L_{\theta_1,\theta_2}(x_1)]P_{\theta_1}(dx_1)]P_{\theta_1}(dx_2) \ge 0,\end{aligned}$$

as $h(x_1, x_2) = 0$ for $x_2 \le x_1$, and $L_{\theta_1,\theta_2}(x_2) - L_{\theta_1,\theta_2}(x_1) \ge 0$ and $h(x_1, x_2) \ge 0$ for $x_1 < x_2$. If $P_{\theta_2} \ll P_{\theta_1}$ is not satisfied we set $\overline{P}_{\theta_1} = (P_{\theta_1} + P_{\theta_2})/2$. It holds $P_{\theta_2} \ll \overline{P}_{\theta_1}$ and

$$\frac{dP_{\theta_2}}{d\overline{P}_{\theta_1}} = \begin{cases} \frac{2L_{\theta_1,\theta_2}}{L_{\theta_1,\theta_2}+1} & \text{if } L_{\theta_1,\theta_2} < \infty, \\ 2 & \text{if } L_{\theta_1,\theta_2} = \infty. \end{cases}$$

Thus the density $dP_{\theta_2}/d\overline{P}_{\theta_1}$ is a nondecreasing function of the identity. Hence by $\overline{P}_{\theta_1} = (P_{\theta_1} + P_{\theta_2})/2$ and the first step of the proof,

$$\begin{aligned} &\frac{1}{2}\int [\int h(x_1,x_2)P_{\theta_2}(dx_1)]P_{\theta_1}(dx_2) + \frac{1}{2}\int [\int h(x_1,x_2)P_{\theta_2}(dx_1)]P_{\theta_2}(dx_2) \\ \le &\frac{1}{2}\int [\int h(x_1,x_2)P_{\theta_1}(dx_1)]P_{\theta_2}(dx_2) + \frac{1}{2}\int [\int h(x_1,x_2)P_{\theta_2}(dx_1)]P_{\theta_2}(dx_2). \quad \square \end{aligned}$$

Solution to Problem 9.23: The first statement is clear. In view of the permutation invariance we may assume $i = 1$ and $j = 2$. Suppose $\theta_1 \le \theta_2$ and let g be nonnegative and measurable. An application of (9.41) to

$$h(x_1, x_2) = I_{(x_1,\infty)}(x_2) \int g(x_1, x_2, x_3, ..., x_k)(\bigotimes\nolimits_{i=3}^k P_{\theta_i})(dx_3, ..., dx_k).$$

completes the proof of the second statement. $\square$

Solution to Problem 9.25: Let $\varphi_{\theta,\Sigma}(x)$, $x \in \mathbb{R}^k$, be the Lebesgue density of a multivariate normal distribution. The Lebesgue measure $\mu = \boldsymbol{\lambda}_k$ is permutation invariant. The density $\varphi_{\theta,\Sigma}(x)$ is permutation invariant in the sense of (9.39) if and only if $u_\gamma(z)^T\Sigma^{-1}u_\gamma(z) = z^T\Sigma^{-1}z$, $z \in \mathbb{R}^k$, $\gamma \in \Pi_k$. We show now that this holds if and only if Σ^{-1} is of the form $\Sigma^{-1} = \alpha I + \beta\mathbf{1}\mathbf{1}^T$ with $\alpha > 0$ and $\alpha + k\beta > 0$, where $\mathbf{1}^T = (1, ..., 1) \in \mathbb{R}^k$. If Σ^{-1} is of the form $\Sigma^{-1} = \alpha I + \beta\mathbf{1}\mathbf{1}^T$, then clearly $u_\gamma(z)^T\Sigma^{-1}u_\gamma(z) = z^T\Sigma^{-1}z$, $z \in \mathbb{R}^k$, $\gamma \in \Pi_k$. Conversely, suppose that $u_\gamma(z)^T\Sigma^{-1}u_\gamma(z) = z^T\Sigma^{-1}z$, $z \in \mathbb{R}^k$, $\gamma \in \Pi_k$. Let $\Sigma^{-1} = (c_{i,j})_{i,j=1,...,k}$. Let $e_{(i)} \in \mathbb{R}^k$ be the vector with 1 in position i, and 0 in all other positions, $i = 1, ..., k$. It holds $e_{(i)}^T\Sigma^{-1}e_{(i)} = c_{i,i}$, $i = 1, ..., k$, and thus $c_{1,1} = c_{2,2} = \cdots = c_{k,k}$. Let $e_{(i,j)}$ be the vector with 1 in the two positions i and j, and 0 in all other positions, $1 \le i < j \le k$. It holds $e_{(i,j)}^T\Sigma^{-1}e_{(i,j)} = c_{i,i} + c_{i,j} + c_{j,i} + c_{j,j} = 2c_{1,1} + 2c_{i,j}$, and thus $c_{i,j} = c_{1,2}$, $1 \le i < j \le k$. This means that Σ^{-1} is of the form $\Sigma^{-1} = \alpha I + \beta\mathbf{1}\mathbf{1}^T$. Now we determine the range of α and β for which $\alpha I + \beta\mathbf{1}\mathbf{1}^T$ is positive definite. The case of $\beta = 0$ is trivial. As to the case of $\beta \ne 0$, let $y \in \mathbb{R}^k\backslash\{0\}$ and set $\overline{y} = (1/k)\sum_{i=1}^k y_i$. It holds $(\alpha I + \beta\mathbf{1}\mathbf{1}^T)y = \lambda y$ if and only if $\beta k\overline{y}\mathbf{1} = (\lambda - \alpha)y$. The solutions are $\lambda = \alpha$ along with $\overline{y} = 0$, and $\lambda = \alpha + k\beta$ along with $y = \overline{y}\mathbf{1}$. Thus, $\alpha I + \beta\mathbf{1}\mathbf{1}^T$ is positive definite if and only if $\alpha > 0$ and $\alpha + k\beta > 0$. If $\Sigma^{-1} = \alpha I + \beta\mathbf{1}\mathbf{1}^T$, $\alpha > 0$, and $\alpha + k\beta > 0$, then $\varphi_{\theta,\Sigma}(x)$ satisfies (9.40). This can be seen as follows.

$$(x-\theta)^T\Sigma^{-1}(x-\theta) = \alpha\sum\nolimits_{i=1}^k (x_i - \theta_i)^2 + \beta[\sum\nolimits_{i=1}^k (x_i - \theta_i)]^2, \quad x, \theta \in \mathbb{R}^k.$$

For any $i, j \in \{1, ..., k\}$ let $x, \theta \in \mathbb{R}^k$ with $x_i \le x_j$ and $\theta_i \le \theta_j$. Let $\theta^{(i,j)}$ be the vector that is obtained from θ by exchanging the coordinates θ_i and θ_j. Then

$$\begin{aligned} &(x - \theta^{(i,j)})^T\Sigma^{-1}(x - \theta^{(i,j)}) - (x-\theta)^T\Sigma^{-1}(x-\theta) \\ = &\alpha[(x_i - \theta_j)^2 + (x_j - \theta_i)^2 - (x_i - \theta_i)^2 - (x_j - \theta_j)^2] = 2\alpha(\theta_j - \theta_i)(x_j - x_i) \ge 0, \end{aligned}$$

which implies $\varphi_{\theta^{(i,j)},\Sigma}(x) \le \varphi_{\theta,\Sigma}(x)$. Moreover, $\Sigma = \widetilde{\alpha}I + \widetilde{\beta}\mathbf{1}\mathbf{1}^T$, where $\widetilde{\alpha} = \alpha^{-1}$ and $\widetilde{\alpha} + k\widetilde{\beta} = (\alpha + k\beta)^{-1}$. $\square$

Solution to Problem 9.26: Obviously, $W = \{x : x \in \{0,1,...,n\}^k, \sum_{i=1}^k x_i = n\}$ and $C = \{p : p_i \in (0,1), i = 1,...,k, \sum_{i=1}^k p_i = 1\}$ are symmetric, and the counting measure $\boldsymbol{\kappa}_d$ on $\mathfrak{B}_W = \mathfrak{P}(W)$ is permutation invariant. The $\boldsymbol{\kappa}_d$-density of $\mathsf{M}(n,p)$ is

$$\mathsf{m}_{n,p}(x) = \frac{n!}{x_1! \cdots x_k!} \prod\nolimits_{i=1}^k p_i^{x_i}, \quad x \in W, \ p \in C.$$

It obviously satisfies (9.39). To verify (9.40) let $x \in W$, where for some $i,j \in \{1,...,k\}$ with $p_i - p_j \le 0$ we have $x_i - x_j \ge 0$. Then

$$\frac{\mathsf{m}_{n,p}(x)}{\mathsf{m}_{n,p}(x^{(i,j)})} = (p_i/p_j)^{x_i - x_j} \le 1. \qquad \square$$

Solution to Problem 9.39: Use Problem 9.19 and Example 9.24. □

Solution to Problem 9.48: Let $y_i = \sum_{j=1}^{n_i} x_{i,j}$, $i = 1,...,k$. The set $M_{\Pi}^{pt,es}(x)$ from (9.62) turns out to be here

$$M_{\Pi}^{pt,es}(x) = \underset{i \in \{1,...,k\}}{\arg\max}\, [S_i^0(x_i) - \frac{1}{a}(\frac{2}{\pi}\frac{\sigma^2\delta_i^2}{\sigma^2 + n_i\delta_i^2})^{1/2}], \quad x \in \bigotimes\nolimits_{i=1}^k \mathbb{R}^{n_i}.$$

(a): If $\delta_i^2 \to \infty$, then $S_i^0(x_i) \to y_i/n_i$, $i = 1,...,k$, and

$$M_{\Pi}^{pt,es}(x) \to \underset{i \in \{1,...,k\}}{\arg\max}\, [\frac{y_i}{n_i} - \frac{1}{a}(\frac{2}{\pi}\frac{\sigma^2}{n_i})^{1/2}].$$

It is interesting to note here that the selection is not made strictly in terms of the largest of the sample means y_i/n_i, $i = 1,...,k$, but with an adjustment that depends on the sample sizes $n_1,...,n_k$.
(b): Let $\delta_i^2 = b\sigma^2/n_i$, $i = 1,...,k$, for some fixed $b > 0$. Then

$$S_i^0(x_i) = \frac{b(y_i/n_i) + \nu_i}{b+1}, \quad i = 1,...,k, \quad \text{and}$$

$$M_{\Pi}^{pt,es}(x) = \underset{i \in \{1,...,k\}}{\arg\max}\, \{\frac{b(y_i/n_i) + \nu_i}{b+1} - \frac{1}{a}(\frac{2}{\pi}\frac{b}{b+1}\frac{\sigma^2}{n_i})^{1/2}\}.$$

Especially if $\nu_1 = \cdots = \nu_k$, then

$$M_{\Pi}^{pt,es}(x) = \underset{i \in \{1,...,k\}}{\arg\max}\, \{\frac{y_i}{n_i} - \frac{1}{a}(\frac{2}{\pi}\frac{b+1}{b}\frac{\sigma^2}{n_i})^{1/2}\}.$$

(c): Suppose that $(1/\delta_i^2) + (n_i/\sigma^2) = 1/c^2$, $i = 1,...,k$, holds for some constant $c^2 > 0$. Then

$$S_i^0(x_i) = c^2(\frac{y_i}{\sigma^2} + \frac{\nu_i}{\delta_i^2}), \quad i = 1,...,k, \quad \text{and } M_{\Pi}^{pt,es}(x) = \underset{i \in \{1,...,k\}}{\arg\max}\, \{\frac{y_i}{\sigma^2} + \frac{\nu_i}{\delta_i^2}\}.$$

Especially if $\nu_1 = \cdots = \nu_k = \nu$, say, then $M_{\Pi}^{pt,es}(x) = \underset{i \in \{1,...,k\}}{\arg\max}\, \{n_i((y_i/n_i) - \nu)\}$. Thus if a population has a sample mean larger (smaller) than the common prior mean ν, then a larger sample size is an advantage (a handicap) for a population to be selected. □

Solution to Problem 9.53: Let $x \in \bigotimes_{i=1}^{k}\{0,1,...,n_i\}$ be fixed. First let us assume that $\alpha_i, \beta_i > 1$, $i = 1,...,k$. Then the Bayes estimate for population $i \in \{1,...,k\}$ is

$$\begin{aligned} S_i^0(x_i) &= \underset{w_i \in [0,1]}{\arg\min} \int_{[0,1]} \frac{(t-w_i)^2}{t(1-t)} \mathsf{be}_{\alpha_i+x_i,\beta_i+n_i-x_i}(t)dt \\ &= c(\alpha_i,\beta_i,x_i) \underset{w_i \in [0,1]}{\arg\min} \int_{[0,1]} (t-w_i)^2 \mathsf{be}_{\alpha_i+x_i-1,\beta_i+n_i-x_i-1}(t)dt, \end{aligned}$$

where $c(\alpha_i,\beta_i,x_i)$ compensates for the switch of the normalizing factors of the two beta densities. It is

$$\begin{aligned} c(\alpha_i,\beta_i,x_i) &= \frac{\Gamma(\alpha_i+\beta_i+n_i)}{\Gamma(\alpha_i+x_i)\Gamma(\beta_i+n_i-x_i)} \frac{\Gamma(\alpha_i+x_i-1)\Gamma(\beta_i+n_i-x_i-1)}{\Gamma(\alpha_i+\beta_i+n_i-2)} \\ &= \frac{(\alpha_i+\beta_i+n_i-1)(\alpha_i+\beta_i+n_i-2)}{(\alpha_i+x_i-1)(\beta_i+n_i-x_i-1)}. \end{aligned}$$

Similarly to Example 9.50 we see that $S_i^0(x_i) = (\alpha_i+x_i-1)/(\alpha_i+\beta_i+n_i-2)$. Next we have to determine the posterior expected loss due to estimation when population i is selected. It is

$$\begin{aligned} &\int_{[0,1]} \frac{(t-w_i)^2}{t(1-t)} \mathsf{be}_{\alpha_i+x_i,\beta_i+n_i-x_i}(t)dt \\ &= c(\alpha_i,\beta_i,x_i) \int_{[0,1]} (t-w_i)^2 \mathsf{be}_{\alpha_i+x_i-1,\beta_i+n_i-x_i-1}(t)dt \\ &= c(\alpha_i,\beta_i,x_i) \frac{(\alpha_i+x_i-1)(\beta_i+n_i-x_i-1)}{(\alpha_i+\beta_i+n_i-2)^2(\alpha_i+\beta_i+n_i-1)} = \frac{1}{(\alpha_i+\beta_i+n_i-2)}. \end{aligned}$$

Thus, every Bayes rule $\varphi^B(x)$ satisfies $\varphi_i^B(x) = 0$, $i \notin M_\Pi^{pt,es}(x)$, where

$$M_\Pi^{pt,es}(x) = \underset{i \in \{1,\dots,k\}}{\arg\min} \left[\frac{\alpha_i+x_i}{\alpha_i+\beta_i+n_i} + \frac{\rho}{(\alpha_i+\beta_i+n_i-2)}\right]. \tag{9.156}$$

Adjustments to priors with $\alpha_i, \beta_i > 0$, $i = 1,...,k$, are straightforward. If $\alpha_i \le 1$ and $x_i = 0$, then $S_i^0(x_i) = 0$, and the last summand in (9.156) for that particular i changes to $\rho\alpha_i/(\beta_i+n_i-1)$. If $\beta_i \le 1$ and $x_i = n_i$, then $S_i^0(x_i) = 1$, and the last summand in (9.156) for that particular i changes to $\rho\beta_i/(\alpha_i+n_i-1)$. □

Solution to Problem 9.57: By Example 9.56 we have for every $x \in \mathcal{X}$,

$$\begin{aligned} M_\Pi^{su}(x) &= \underset{A \in \mathcal{D}_{su}}{\arg\min} \sum\nolimits_{i \in A} \int [\kappa_{[k]}(\theta) - \kappa(\theta_i) - \varepsilon] \boldsymbol{\Pi}(d\theta|x) \\ &= \underset{A \in \mathcal{D}_{su}}{\arg\min} \sum\nolimits_{i \in A} \left[\int \kappa_{[k]}(\theta)\boldsymbol{\Pi}(d\theta|x) - \varepsilon - \int \kappa(\theta_i)\boldsymbol{\Pi}(d\theta|x)\right]. \end{aligned}$$

From here the statements follow easily. □

Solution to Problem 9.63: At any $x \in \mathcal{X}$, the posterior expected size of the selected subset $\mathbf{A}$, say, under a selection rule φ satisfies

$$1 \le \mathbb{E}(|\mathbf{A}| \; |x) = \sum\nolimits_{A \in \mathcal{D}_{su}} |A| \, \varphi_A(x) = \sum\nolimits_{A \in \mathcal{D}_{su}} [\sum\nolimits_{i=1}^{k} I_A(i)] \varphi_A(x)$$
$$= \sum\nolimits_{i=1}^{k} [\sum\nolimits_{A \in \mathcal{D}_{su}} I_A(i) \varphi_A(x)] = \sum\nolimits_{i=1}^{k} \psi_i(x).$$

The inequality follows from the fact that the empty set cannot be selected, and the last equation follows from (9.71).

Let $k = 3$ and $s \in \mathbb{R}^3$ be fixed. Set $\varphi_{\{i\}}(x) = 1/4$, $i = 1, 2, 3$, and $\varphi_{\{1,2,3\}}(x) = 1/4$. Set $\tilde{\varphi}_{\{i\}}(x) = 1/6$, $i = 1, 2, 3$, and $\tilde{\varphi}_{\{1,2\}}(x) = \tilde{\varphi}_{\{1,3\}}(x) = \tilde{\varphi}_{\{2,3\}}(x) = 1/6$. For both, φ and $\tilde{\varphi}$, the inclusion probabilities are $\psi_i(x) = 1/2$, $i = 1, 2, 3$. □

Solution to Problem 9.79: Apply Proposition A.29 and the substitution rule Lemma A.15 to the exponential structure of $dP_\theta / d\boldsymbol{\mu}$. □

Solution to Problem 9.94: It is enough to consider the case $\theta = 0$. The statement then follows from the fact that, according to Barndorff-Nielsen (1978), the m-fold convolution of d is again a log-concave function. This fact follows also from Proposition 2.19 which states that a density is log-concave or strictly unimodal if the convolution with an unimodal density is again unimodal. □

Solution to Problem 9.116: It holds,

$$\frac{1}{n} \ln a_n, \frac{1}{n} \ln b_n \le \frac{1}{n} \ln(a_n + b_n) \le \frac{1}{n} \ln(2(a_n \vee b_n)),$$
$$\limsup_{n \to \infty} \frac{1}{n} \ln(2(a_n \vee b_n)) = (\limsup_{n \to \infty} \frac{1}{n} \ln a_n) \vee (\limsup_{n \to \infty} \frac{1}{n} \ln b_n). \quad \square$$

Solution to Problem 9.122: The conditions (9.126), $\varphi_n(u_\gamma(x)) = u_\gamma(\varphi_n(x))$, and the completeness of the family $(\mathsf{N}(\mathcal{I}_0 h, \mathcal{I}_0)_{h \in \mathbb{R}^k})$ imply the existence of N_γ with $\varphi(u_\gamma(x)) = u_\gamma(\varphi(x))$, $x \notin N_\gamma$, and $\mathsf{N}(0, \mathbf{I})(N_\gamma) = 0$. Set $N = \cup_\gamma N_\gamma$ and define $\psi(x) = \varphi(x)$ for $x \notin N$, and as the uniform distribution on $\{1, ..., k\}$ for $x \in N$. The equivalence of the distributions $\mathsf{N}(\mathcal{I}_0 h, \mathcal{I}_0)$, $h \in \mathbb{R}^k$, and $\mathsf{N}(0, \mathbf{I})$ completes the proof. □

Solution to Problem 9.123: The statement follows from Problem 9.122 if we set $\mathcal{M}_n = \mathcal{G}$. □

Solution to Problem 9.138: Without loss of generality let $h_1 \le \cdots \le h_k$. Then $L(h, 1) \ge \cdots \ge L(h, k)$. Let $Z_1, ..., Z_k$ be generic i.i.d. standard normal random variables. Set $a_i = \mathbb{P}(Z_j + \gamma h_j \le Z_i + \gamma h_i, \, j \ne i)$, $\gamma \ge 0$, $i = 1, ..., k$. Then

$$\sum\nolimits_{i=1}^{k} L(h, i) \int I_{\mathsf{M}(x)}(i) \mathsf{N}(\gamma h, \mathbf{I})(dx) = \sum\nolimits_{i=1}^{k} L(h, i) a_i$$
$$= \sum\nolimits_{q=1}^{k-1} [L(h, q) - L(h, q+1)] \sum\nolimits_{i=1}^{q} a_i + L(h, k) \sum\nolimits_{i=1}^{k} a_i.$$

Now for $q = 1, ..., k$,

$$\sum\nolimits_{i=1}^{q} a_i = \mathbb{P}(\max_{a \ge q+1} (Z_a + \gamma h_a) \le \max_{b \le q} (Z_b + \gamma h_b))$$
$$= \mathbb{P}(\max_{a \ge q+1} (Z_a + \gamma (h_a - h_q)) \le \max_{b \le q} (Z_b + \gamma (h_b - h_q))).$$

Because $h_a \ge h_q$ for $a \ge q + 1$ and $h_b \le h_q$ for $b \le q$, the proof is completed. □

A

Appendix: Topics from Analysis, Measure Theory, and Probability Theory

In the following three sections we collect results from analysis, measure theory, and probability theory that are used in the book. We emphasize that these results are not necessarily established in their most possible generality. Sometimes we present only special cases of general results that are sufficient for our purposes. This allows us to avoid unnecessarily convoluted formulations, extended notations, and additional concepts. We do not give proofs unless a suitable version of a result that is needed is not directly available.

A.1 Topics from Analysis

Let $\mathbb{R} = (-\infty, +\infty)$ be the real line and denote by $\overline{\mathbb{R}} = [-\infty, +\infty]$ the extended real line. The sum, product, and ratio of elements of $\overline{\mathbb{R}}$ are defined by the standard conventions that can be found, for example, in Hewitt and Stromberg (1965), pp. 54–55. Set $\psi(x) = (1 + |x|)^{-1}x$, $x \in \mathbb{R}$, $\psi(-\infty) = -1$, and $\psi(\infty) = 1$. Put $\overline{\rho}(x, y) = |\psi(x) - \psi(y)|$, $x, y \in \overline{\mathbb{R}}$. For the product space $\overline{\mathbb{R}}^d = [-\infty, +\infty]^d$, $x = (x_1, ..., x_d) \in \overline{\mathbb{R}}^d$, and $y = (y_1, ..., y_d) \in \overline{\mathbb{R}}^d$ we set $\overline{\rho}_d(x, y) = \sum_{i=1}^d \overline{\rho}(x_i, y_i)$.

Remark A.1. $(\overline{\mathbb{R}}, \overline{\rho})$ and $(\overline{\mathbb{R}}^d, \overline{\rho}_d)$ are compact metric spaces.

Let $A \subseteq \mathbb{R}^d$ be an open set. The function $f : A \to \mathbb{R}$ is called *differentiable* at $x_0 \in A$ if there is a vector, called a *gradient* and denoted by column vector $\nabla f(x_0)$, such that

$$f(x_0 + u) - f(x_0) = \langle u, \nabla f(x_0) \rangle + o(\|u\|),$$

where $\|u\|$ is the Euclidean norm and $o(\|u\|)/\|u\| \to 0$ for $u \to 0$. We also use the notations $(\frac{\partial f}{\partial t_1}(x_0), ..., \frac{\partial f}{\partial t_d}(x_0))^T$ and $\dot{f}(x_0)$ for the gradient $\nabla f(x_0)$. If the partial derivatives $\frac{\partial f}{\partial t_i}$, $i = 1, ..., d$, of f exist, and are continuous in a neighborhood of x_0, then f is differentiable at x_0 with gradient $(\frac{\partial f}{\partial t_1}(x_0), ..., \frac{\partial f}{\partial t_d}(x_0))^T$.

If $A \subseteq \mathbb{R}^d$ is a open set and $f : A \to \mathbb{R}^k$, then we call $f = (f_1, ..., f_k)^T$ differentiable at x_0 if every f_i is differentiable at x_0. In this case the $k \times d$ matrix

$$J_f(x_0) := (\nabla f_1(x_0), ..., \nabla f_k(x_0))^T = \left(\frac{\partial f_i}{\partial t_j}(x_0) \right)_{1 \le i \le k, 1 \le j \le d}$$

is called the *Jacobian* of f, and the first-order Taylor expansion can be written as $f(x_0 + u) - f(x_0) = J_f(x_0)u + o(\|u\|)$. We call f *continuously differentiable* in A if f is differentiable at every $x_0 \in A$ and $J_f(x)$ is continuous on A. The corresponding multiple differentiability of vector-valued functions is defined componentwise. If $A, B \subseteq \mathbb{R}^d$ are open sets, $f : A \to B \subseteq \mathbb{R}^k$ is one-to-one, and both f and its inverse mapping g are continuously differentiable, then we call f a *diffeomorphism*.

The subsequent Taylor expansion with remainder term follows from Theorem 8.14.3 in Dieudonné (1960) by an application to the vector-valued function f.

Theorem A.2. *If $A \subseteq \mathbb{R}^d$ is an open set and $f : A \to \mathbb{R}^k$ is continuously differentiable in a neighborhood $U(x)$ of $x \in A$, then for $x + sh \in U(x)$, $0 \le s \le 1$,*

$$f(x+h) - f(x) = \int_0^1 J_f(x + sh) h ds.$$

If $f : A \to \mathbb{R}$ is twice continuously differentiable in a neighborhood $U(x)$ of $x \in A$, $\dot{f}$ is the gradient, and $\ddot{f}$ the matrix of the second derivatives, then for $x + sh \in U(x)$, $0 \le s \le 1$,

$$f(x+h) - f(x) = \dot{f}^T(x)h + \int_0^1 (1-s)h^T \ddot{f}(x + sh) h ds.$$

Let $\mathbb{C} = \{z : z = u + iv, \quad u, v \in \mathbb{R}\}$ denote the space of the complex numbers and $A \subseteq \mathbb{C}^d$ be open. A function $f : A \to \mathbb{C}$ is called *analytic* if for every $(z_{1,0}, ..., z_{d,0}) \in A$ it can be expanded in a power series

$$f(z) = \sum_{k=0}^{\infty} \sum_{m_1 + \cdots + m_d = k} \frac{1}{m_1! \cdots m_d!} a_{m_1, \ldots, m_d} \prod\nolimits_{j=1}^{d} (z_j - z_{j,0})^{m_j}, \quad z \in A,$$

which is absolutely convergent in some neighborhood of $(z_{1,0}, ..., z_{d,0})$. The uniqueness theorem below is a special case of 9.4.2 in Dieudonné (1960).

Theorem A.3. *Suppose $A \subseteq \mathbb{C}^d$ is an open connected set that contains an open real rectangle $\mathsf{X}_{j=1}^d (a_i, b_i)$. If $f, g : A \to \mathbb{C}$ are analytic on A and $f(u_1, ..., u_d) = g(u_1, ..., u_d)$ for every $(u_1, ..., u_d) \in \mathsf{X}_{j=1}^d (a_i, b_i)$, then $f(z) = g(z)$ for every $z \in A$.*

Let $(\mathcal{X}, \rho_{\mathcal{X}})$ be a metric space. We denote by $\mathsf{C}(\mathcal{X})$ the space of all real-valued continuous functions on $\mathcal{X}$, by $\mathsf{C}_b(\mathcal{X})$ the subspace of all bounded functions in $\mathsf{C}(\mathcal{X})$, and by $\mathsf{C}_{u,b}(\mathcal{X})$ the subspace of all uniformly continuous functions in $\mathsf{C}_b(\mathcal{X})$. It is clear that for a compact metric space $\mathsf{C}(\mathcal{X}) = \mathsf{C}_b(\mathcal{X}) = \mathsf{C}_{u,b}(\mathcal{X})$. We set $\|f\|_u = \sup_{x\in\mathcal{X}} |f(x)|$ for every $f : \mathcal{X} \to \mathbb{R}$. It is easy to see that $\rho_{\mathsf{C}}(f, g) := \|f - g\|_u$ is a metric on $\mathsf{C}_b(\mathcal{X})$.

A special class of uniformly continuous functions is the Lipschitz functions. We set for $f : \mathcal{X} \to \mathbb{R}$

$$\|f\|_{\mathsf{Lip}} = \|f\|_u + L_f, \quad \text{where } L_f = \sup_{s\neq t} \frac{1}{\rho_{\mathcal{X}}(s,t)} |f(s) - f(t)|, \tag{A.1}$$

and denote by $\mathsf{Lip}(\mathcal{X}) = \{f : \|f\|_{\mathsf{Lip}} < \infty\}$ the space of all bounded Lipschitz functions. It is clear that $\mathsf{Lip}(\mathcal{X}) \subseteq \mathsf{C}_{u,b}(\mathcal{X})$. The next statement is Theorem 11.2.4. and Corollary 11.2.5 in Dudley (2002) to which we refer for a proof.

Proposition A.4. *If $(\mathcal{X}, \rho_{\mathcal{X}})$ is a compact metric space, then $\mathsf{C}(\mathcal{X})$ and $\mathsf{Lip}(\mathcal{S})$ are separable metric spaces and the space $\mathsf{Lip}(\mathcal{X})$ is dense in $\mathsf{C}(\mathcal{X})$ for $\|\cdot\|_u$.*

The next lemma concerns the pointwise approximation of lower semicontinuous functions.

Lemma A.5. *If $(\mathcal{X}, \rho_{\mathcal{X}})$ is a metric space and $f : \mathcal{X} \to [0, \infty]$ is a lower semicontinuous function, then there is a nondecreasing sequence of Lipschitz functions $f_n : \mathcal{X} \to [0, \infty)$ such that $f(x) = \lim_{n\to\infty} f_n(x)$ for every $x \in \mathcal{X}$.*

Proof. If $f \equiv \infty$, then set $f_n = n$. Otherwise set $f_n(x) = \inf_{y\in\mathcal{X}}(f(y) + n\rho_{\mathcal{X}}(y, x))$. Then $f_n(x) \le f_{n+1}(x) < \infty$ and $f_n(x) \le f(x)$ for every x and $|f_n(x_1) - f_n(x_2)| \le n\rho_{\mathcal{X}}(x_1, x_2)$. It remains to show that $\lim_{n\to\infty} f_n(x) = f(x)$. If $\lim_{n\to\infty} f_n(x_0) < f(x_0)$ for some x_0, then by the lower semicontinuity of f there is some β with $\lim_{n\to\infty} f_n(x) < \beta < f(x_0)$ and some $\delta > 0$ such that $f(z) > \beta$ for every y with $\rho_{\mathcal{X}}(z, x_0) < \delta$. If $\rho_{\mathcal{X}}(y, x_0) \ge \delta$, then by $f \ge 0$ it holds $f(y) + n\rho_{\mathcal{X}}(y, x_0) \ge \beta$ for all sufficiently large n. If $\rho_{\mathcal{X}}(y, x_0) \le \delta$, then $f(y) + n\rho_{\mathcal{X}}(y, x_0) \ge \beta$. Hence $f_n(x_0) \ge \beta$ for all sufficiently large n and the proof is completed. ∎

A.2 Topics from Measure Theory

For a nonempty set $\mathcal{X}$ let $\mathfrak{P}(\mathcal{X})$ denote the system of all subsets of $\mathcal{X}$. If $\mathfrak{G} \subseteq \mathfrak{P}(\mathcal{X})$, then $\sigma(\mathfrak{G})$ denotes the smallest σ-algebra containing $\mathfrak{G}$. If $\mathfrak{A} \subseteq \mathfrak{P}(\mathcal{X})$ is a σ-algebra, then the pair $(\mathcal{X}, \mathfrak{A})$ is called a *measurable space*. For $A \in \mathfrak{A}$ we call $\mathfrak{A}_A := \{A \cap B : B \in \mathfrak{A}\}$ the *trace σ-algebra* of subsets of A.

For a mapping $f : \mathcal{X} \to \mathcal{Y}$ the set $f^{-1}(B) := \{x : f(x) \in B\}$ is called the *inverse image of B*. Given two measurable spaces $(\mathcal{X}, \mathfrak{A})$ and $(\mathcal{Y}, \mathfrak{B})$ and a mapping $f : \mathcal{X} \to \mathcal{Y}$ the system

$$\sigma(f) := \{f^{-1}(B) : B \in \mathfrak{B}\} \subseteq \mathfrak{P}(\mathcal{X})$$

is a σ-algebra called the *σ-algebra generated by f*. If $\sigma(f) \subseteq \mathfrak{A}$, then the mapping f is called $\mathfrak{A}$-$\mathfrak{B}$ *measurable*. Whenever it is clear which σ-algebras are involved, we just call f *measurable* and express it by $f : \mathcal{X} \to_m \mathcal{Y}$. If f is a bijection and the inverse mapping, denoted by f^{-1}, is $\mathfrak{B}$-$\mathfrak{A}$ measurable, then we say that f is $\mathfrak{A}$-$\mathfrak{B}$ *bimeasurable* and write $f : \mathcal{X} \leftrightarrow_m \mathcal{Y}$. In this case we say that $(\mathcal{X}, \mathfrak{A})$ and $(\mathcal{Y}, \mathfrak{B})$ are *Borel isomorphic.*

If $(\mathcal{X}, \rho)$ is a metric space, and $\mathfrak{O}$ denotes the system of all open subsets, then we call $\mathfrak{B} = \sigma(\mathfrak{O})$ the *σ-algebra of Borel sets.* This σ-algebra is sometimes also denoted by $\mathfrak{B}_{\mathcal{X}}$. For $\mathcal{X} = \mathbb{R}$ and $\mathcal{X} = \overline{\mathbb{R}}$ the σ-algebra of Borel sets $\mathfrak{B}$ and $\overline{\mathfrak{B}}$, respectively, is generated by the open intervals. If $\mathcal{Y} = \mathbb{R}, \overline{\mathbb{R}}, \mathbb{R}_+$, or $\overline{\mathbb{R}}_+$, then $\mathcal{X} \to_m \mathcal{Y}$ means $\mathfrak{A}$-$\mathfrak{B}$, $\mathfrak{A}$-$\overline{\mathfrak{B}}$, $\mathfrak{A}$-$\mathfrak{B}_+$, or $\mathfrak{A}$-$\overline{\mathfrak{B}}_+$ measurability, respectively.

At many places we utilize the so-called *standard extension technique* which establishes a statement step by step from indicator functions to linear combinations of indicator functions and then by monotone convergence to all nonnegative measurable functions. This technique is based on the fact that every nonnegative measurable function is the pointwise limit of a nondecreasing sequence of nonnegative step functions; see, e.g., Kallenberg (1997).

Lemma A.6. ***(Standard Extension Technique)*** *Suppose $(\mathcal{X}, \mathfrak{A})$ is a measurable space and $\mathbb{F}$ is a set of measurable functions $f : \mathcal{X} \to_m \overline{\mathbb{R}}_+$ with the following three properties. (a) $I_A \in \mathbb{F}$ for every $A \in \mathfrak{A}$. (b) $f, g \in \mathbb{F}$, $a, b \geq 0$, implies $af + bg \in \mathbb{F}$. (c) $f_n \in \mathbb{F}$ and $f_n \uparrow f$ implies $f \in \mathbb{F}$. Then $\mathbb{F}$ is the set of all $f : \mathcal{X} \to_m \overline{\mathbb{R}}_+$.*

For several more advanced technical issues in probability theory the concept of a Borel space is useful.

Definition A.7. *A measurable space $(\mathcal{Z}, \mathfrak{C})$ is called a Borel space if there exists a Borel set $B \subseteq [0,1]$ and a bijection $b : \mathcal{Z} \leftrightarrow B$ which is a $\mathfrak{C}$-$\mathfrak{B}_B$ bimeasurable mapping.*

Lemma A.8. *If $\mathcal{Z}$ is a complete separable metric space, and $\mathfrak{C}$ is the σ-algebra of Borel sets, then $(\mathcal{Z}, \mathfrak{C})$ is a Borel space.*

For a proof we refer to Theorem A.1.6 in Kallenberg (1997).

Lemma A.9. ***(Factorization Lemma)*** *Let $(\mathcal{X}, \mathfrak{A}), (\mathcal{Y}, \mathfrak{B})$, and $(\mathcal{Z}, \mathfrak{C})$ be measurable spaces and $g : \mathcal{X} \to_m \mathcal{Y}$. If $(\mathcal{Z}, \mathfrak{C})$ is a Borel space, then $f : \mathcal{X} \to \mathcal{Z}$ is $\sigma(g)$-$\mathfrak{C}$ measurable if and only if there exists a $\mathfrak{B}$-$\mathfrak{C}$ measurable function $h : \mathcal{Y} \to \mathcal{Z}$ such that $f(x) = h(g(x))$, $x \in \mathcal{X}$.*

For a proof we refer to Lemma 1.13 in Kallenberg (1997).

Theorem A.10. ***(Measurable Selection Theorem)*** *Suppose $(\mathcal{X}, \mathfrak{A})$ is a measurable space, $\mathcal{Y}$ is a Polish space, and $f : \mathcal{X} \times \mathcal{Y} \to \mathbb{R}$ fulfils the following* *conditions:* $x \mapsto f(x,y)$ *is measurable for every* $y \in \mathcal{Y}$ *and* $y \mapsto f(x,y)$ *is continuous for every* $x \in \mathcal{X}$. *If for every* x *there exists some* $g(x) \in \mathcal{Y}$ *such that* $f(x, g(x)) = \inf_{y \in \mathcal{Y}} f(x,y)$, *then there exists a measurable mapping* $\widehat{\theta} : \mathcal{X} \to_m \mathcal{Y}$ *such that* $f(x, \widehat{\theta}(x)) = \inf_{y \in \mathcal{Y}} f(x,y)$ *for every* $x \in \mathcal{X}$.

For a proof and references we refer to Theorem 6.7.22 in Pfanzagl (1994).

A measure μ defined on $(\mathcal{X}, \mathfrak{A})$ is called *σ-finite* if there is an increasing sequence $B_i \in \mathfrak{A}$ with $\mu(B_i) < \infty$, $i = 1, 2, ...$ and $\cup_{i=1}^{\infty} B_i = \mathcal{X}$. We denote by $\mathcal{M}(\mathfrak{A})$, $\mathcal{M}^{\sigma}(\mathfrak{A})$, and $\mathcal{P}(\mathfrak{A})$ the set of all measures, σ-finite measures, and probabilities on $(\mathcal{X}, \mathfrak{A})$, respectively. $(\mathcal{X}, \mathfrak{A}, \mu)$ is called a *measure space.*

Definition A.11. *Let* $(\mathcal{X}, \mathfrak{A}, \mu)$ *be a measure space and* $f, f_n : \mathcal{X} \to_m \mathbb{R}$. *We say the following.*

f_n *converges almost everywhere to* f, *in short* $f_n \to f$, μ-*a.e., if* $\mu(\{x : f_n(x) \nrightarrow f(x)\}) = 0$.

f_n *converges in measure to* f, *in short* $f_n \to^{\mu} f$, *if* $\lim_{n\to\infty} \mu(\{x : |f_n(x) - f(x)| > \varepsilon\}) = 0$ *for every* $\varepsilon > 0$.

f_n *converges locally in measure to* f, *in short* $f_n \to^{\mu}_{loc} f$, *if* $I_A f_n \to^{\mu} I_A f$ *for every* $A \in \mathfrak{A}$ *with* $\mu(A) < \infty$.

For a finite measure μ the local convergence in measure is identical with the convergence in measure. Especially for a probability measure μ we call the convergence in measure *stochastic convergence*, and the almost everywhere convergence *almost sure (*a.s.*) convergence.*

Let $\mathcal{X}$ be a separable metric space with metric ρ, and let $\mathfrak{A}$ be the σ-algebra of Borel sets. If $X, Y : \Omega \to_m \mathcal{X}$ are random variables defined on $(\Omega, \mathfrak{F}, \mathbb{P})$, then $\omega \mapsto \rho(X(\omega), Y(\omega))$ is $\mathfrak{F}$-$\mathfrak{B}_{[0,\infty)}$ measurable. For $X, X_n : \Omega \to_m \mathcal{X}$ we write $X_n \to^{\mathbb{P}} X$ if $\rho(X, X_n) \to^{\mathbb{P}} 0$, and $X_n \to X$, $\mathbb{P}$-a.s., if $\rho(X, X_n) \to 0$, $\mathbb{P}$-a.s.

Proposition A.12. *It holds* $\rho(X, X_n) \to^{\mathbb{P}} 0$ *if and only if for every subsequence* X_{n_k} *there exits a subsequence* $X_{n_{k_l}}$ *with* $\rho(X, X_{n_{k_l}}) \to 0$, $\mathbb{P}$-*a.s.. The* $\mathbb{P}$-*a.s. convergence of random variables with values in a separable metric space implies the stochastic convergence with respect to* $\mathbb{P}$.

For a proof we refer to Kallenberg (1997).

Given a measurable space $(\mathcal{X}, \mathfrak{A})$ and $\mu \in \mathcal{M}(\mathfrak{A})$ we set $\int g(x)\mu(dx) = \sum_{i=1}^{n} a_i \mu(A_i)$ for a nonnegative step function $g(x) = \sum_{i=1}^{n} a_i I_{A_i}(x)$, $a_i \in \overline{\mathbb{R}}_+$, $A_i \in \mathfrak{A}$, $i = 1, ..., n$. Note that here and in the sequel we tacitly use the standard convention for calculations on the extended real line. If $f : \mathcal{X} \to_m \overline{\mathbb{R}}_+ = [0, \infty]$, then we set

$$\int f d\mu = \int f(x)\mu(dx) = \sup \int g d\mu,$$

where the supremum is taken over all nonnegative step functions g that satisfy $g(x) \leq f(x)$ for every $x \in \mathcal{X}$. For $f : \mathcal{X} \to_m \mathbb{R}$, we introduce $f^+(x) =$

$\max\{f(x), 0\}$, $f^-(x) = -\min\{f(x), 0\}$. It holds $\int |f| d\mu < \infty$ if and only if $\int f^+ d\mu < \infty$ and $\int f^- d\mu < \infty$. If $\min(\int f^+ d\mu, \int f^- d\mu) < \infty$ we set $\int f d\mu = \int f^+ d\mu - \int f^- d\mu$. The spaces $\mathbb{L}_p(\mu)$ are defined by

$$\mathbb{L}_p(\mu) = \{f : f : \mathcal{X} \to_m \overline{\mathbb{R}}, \quad \int |f|^p d\mu < \infty\}, \quad p > 0.$$

For $p \geq 1$ the pseudonorm norm $\|\cdot\|_p$ is given by

$$\|f\|_p = (\int |f|^p d\mu)^{1/p}, \quad f \in \mathbb{L}_p(\mu), \ p \geq 1.$$

The functions in $\mathbb{L}_1(\mu)$ are called *μ-integrable*, or just *integrable*, if it is clear which measure is used. If $(\Omega, \mathfrak{F}, \mathbb{P})$ is a probability space and $X : \Omega \to_m \overline{\mathbb{R}}_+$ is a random variable, then $\mathbb{E}X := \int X d\mathbb{P}$ is called the *expectation* or *mean* of X. If $X : \Omega \to_m \overline{\mathbb{R}}$ and $\min(\mathbb{E}X^+, \mathbb{E}X^-) < \infty$ we set $\mathbb{E}X = \mathbb{E}X^+ - \mathbb{E}X^-$ and call $\mathbb{E}X$ the *expectation* of X.

For proofs of the following classical inequalities we refer to Dudley (2002), Theorems 5.1.2 and 5.1.5.

Lemma A.13. *For $f_i, f, g : \mathcal{X} \to_m \mathbb{R}$, and $p_i > 1$, $i = 1, ..., k$, the following inequalities hold.*

$$\int |fg| d\mu \leq \|f\|_p \|g\|_q, \quad \frac{1}{p} + \frac{1}{q} = 1, \ p > 1, \quad \textit{Hölder}$$

$$\int |\prod\nolimits_{i=1}^k f_i| d\mu \leq \prod\nolimits_{i=1}^k \|f_i\|_{p_i}, \quad \sum\nolimits_{i=1}^k \frac{1}{p_i} = 1, \quad \textit{generalized Hölder}$$

$$\int |fg| d\mu \leq \|f\|_2 \|g\|_2, \quad \textit{Schwarz}$$

$$\|f + g\|_p \leq \|f\|_p + \|g\|_p, \ p \geq 1, \quad \textit{Minkowski.}$$

Minkowski's inequality shows that the spaces $\mathbb{L}_p(\mu)$, $p \geq 1$, are linear spaces and that $\|f\|_p$ is a pseudonorm. If we identify functions that differ only on μ-nullsets, then the pseudonorm $\|f\|_p$ becomes a norm. The spaces $\mathbb{L}_p(\mu)$, $p \geq 1$, are complete and thus Banach spaces. For a proof of the completeness statement below we refer to Dudley (2002), Theorem 5.2.1.

Theorem A.14. *If $f_n \in \mathbb{L}_p(\mu)$, $p \geq 1$, satisfy $\lim_{m,n\to\infty} \int |f_m - f_n|^p d\mu = 0$, then there is an $f \in \mathbb{L}_p(\mu)$ with $\lim_{n\to\infty} \int |f_n - f|^p d\mu = 0$.*

Let $(\mathcal{X}, \mathfrak{A})$ and $(\mathcal{Y}, \mathfrak{B})$ be measurable spaces and $T : \mathcal{X} \to \mathcal{Y}$ be $\mathfrak{A}$-$\mathfrak{B}$ measurable. If $\mu \in \mathcal{M}(\mathfrak{A})$, then $\mu \circ T^{-1}$, defined by $(\mu \circ T^{-1})(B) = \mu(T^{-1}(B)) = \mu(\{x : T(x) \in B\})$, $B \in \mathfrak{B}$, is easily seen to be a measure on $\mathfrak{B}$ which is called the *induced measure.* The following *substitution rule* is a simple consequence of the standard extension technique in Lemma A.6.

Lemma A.15. ***(Substitution Rule)*** *If $T : \mathcal{X} \to \mathcal{Y}$ is $\mathfrak{A}$-$\mathfrak{B}$ measurable, then it holds that $\int h(y)(\mu \circ T^{-1})(dy) = \int h(T(x))\mu(dx)$ for every $h : \mathcal{Y} \to_m \overline{\mathbb{R}}_+$.*

Next we present the basic limit theorems of measure theory.

Theorem A.16. ***(Monotone Convergence Theorem)*** *Let* $f_n, g_n : \mathcal{X} \to_m \overline{\mathbb{R}}_+$ *with* $0 \leq f_1(x) \leq f_2(x) \leq \cdots$ *and* $g_1(x) \geq g_2(x) \geq \cdots$ *for every* $x \in \mathcal{X}$. *If* $\int g_{n_0} d\mu < \infty$ *for some* n_0, *then*

$$\lim_{n\to\infty} \int f_n d\mu = \int \lim_{n\to\infty} f_n d\mu \quad and \quad \lim_{n\to\infty} \int g_n d\mu = \int \lim_{n\to\infty} g_n d\mu.$$

Lemma A.17. ***(Lemma of Fatou)*** *If* $f_n : \mathcal{X} \to_m \overline{\mathbb{R}}_+$, $n = 1, 2, ...,$ *and* $\mu \in \mathcal{M}(\mathfrak{A})$, *then*

$$\liminf_{n\to\infty} \int f_n d\mu \geq \int (\liminf_{n\to\infty} f_n) d\mu.$$

Theorem A.18. ***(Theorem of Lebesgue)*** *Let* $f_n, f : \mathcal{X} \to_m \overline{\mathbb{R}}$, $n = 1, 2, ...,$ $\mu \in \mathcal{M}(\mathfrak{A})$, *and* $g \in \mathbb{L}_1(\mu)$ *with* $|f_n| \leq g$. *If at least one of the conditions, (a)* $f_n \to f$, *μ-a.e., or (b)* $f_n \to^{\mu}_{loc} f$ *and μ is σ-finite, is fulfilled, then* $f \in \mathbb{L}_1(\mu)$,

$$\lim_{n\to\infty} \int |f_n - f| d\mu = 0, \quad and \quad \lim_{n\to\infty} \int f_n d\mu = \int f d\mu.$$

For a proof of the monotone convergence theorem, Fatou's lemma, and Lebesgue's theorem under the condition (a) we refer to Section 4.3 in Dudley (2002). If (b) is fulfilled, then we use Proposition A.12 and a subsequent argument.

The next statement is Lemma 21.6 in Bauer (2001).

Lemma A.19. ***(Lemma of Scheffé)*** *Let* $f_n, f : \mathcal{X} \to_m \overline{\mathbb{R}}_+$, $n = 1, 2, ...,$ *and* $\mu \in \mathcal{M}^{\sigma}(\mathfrak{A})$. *If* $\lim_{n\to\infty} \int f_n d\mu = \int f d\mu < \infty$ *and* $f_n \to^{\mu}_{loc} f$, *then* $\lim_{n\to\infty} \int |f_n - f| d\mu = 0$.

In some situations the assumption of $|f_n| \leq g$ in Lebesgue's theorem is too strong and not flexible enough.

Definition A.20. $\mathbb{M} \subseteq \mathbb{L}_1(\mu)$ *is called uniformly integrable if for every* $\varepsilon > 0$ *there is an* $f_\varepsilon \in \mathbb{L}_1(\mu)$ *with* $f_\varepsilon \geq 0$ *so that* $\sup_{f\in\mathbb{M}} \int_{\{|f|>f_\varepsilon\}} |f| d\mu < \varepsilon$.

Theorem A.21. ***(Theorem of Vitali)*** *If* $\mu \in \mathcal{M}^{\sigma}(\mathfrak{A})$, f_n, $f \in \mathbb{L}_p(\mu)$, *and* $0 < p < \infty$, *then the following conditions are equivalent.*

(a) $f_n \to^{\mu}_{loc} f$ *and* $\{|f_n|^p\}$ *is uniformly integrable.*

(b) $f_n \to^{\mu}_{loc} f$ *and* $\lim_{n\to\infty} \int |f_n|^p d\mu = \int |f|^p d\mu$.

(c) $f_n \to^{\mu}_{loc} f$ *and* $\limsup_{n\to\infty} \int |f_n|^p d\mu \leq \int |f|^p d\mu$.

(d) $\lim_{n\to\infty} \int |f_n - f|^p d\mu = 0$.

Proof. (a) $\leftrightarrow$ (d) is Theorem 21.4 in Bauer (2001). (b) $\to$ (d): Set $g_n = |f_n|^p$ and $g = |f|^p$. Scheffé's lemma yields $\lim_{n\to\infty} \int |g_n - g| d\mu = 0$, so that by the equivalence of (a) and (d) for g_n and g the sequence $|f_n|^p$ is uniformly integrable. Then the sequence $|f_n - f|^p$ is also uniformly integrable and (d) follows from (a) applied to $|f_n - f|^p$. (d) $\to$ (b) is clear. (b) $\leftrightarrow$ (c) follows from Fatou's lemma. ■

The next proposition is a version of Lebesgue's theorem where the condition of the sequence of functions to be bounded in absolute value by an integrable function is replaced by the condition that the sequence of functions is uniformly integrable. This is a direct consequence of Vitali's theorem.

Proposition A.22. *Let $\mu \in \mathcal{M}^\sigma(\mathfrak{A})$ and $f_n, f \in \mathbb{L}_1(\mu)$. If $f_n \to^{\mu}_{loc} f$ and the sequence f_n, $n = 1, 2, ...$, is uniformly integrable, then $\lim_{n\to\infty} \int f_n d\mu = \int f d\mu$.*

The following is a standard transformation technique for the Lebesgue measure $\boldsymbol{\lambda}_d$; see Bauer (2001).

Theorem A.23. *If $A, B \subseteq \mathbb{R}^d$ are open sets and $f : A \to B$ is a diffeomorphism with Jacobian J_f, then for every $h : B \to_m \overline{\mathbb{R}}_+$*

$$\int_B h(y)\boldsymbol{\lambda}_d(dy) = \int_A h(f(x))|J_f(x)|\boldsymbol{\lambda}_d(dx).$$

If $[a, b]$ is a finite interval a function $f : [a, b] \to \mathbb{R}$ is called *absolutely continuous* if for every ε there exists a $\delta > 0$ such that $\sum_{k=1}^n |f(d_k) - f(c_k)| < \varepsilon$ for every pairwise disjoint open subintervals (c_k, d_k) with $\sum_{k=1}^n (d_k - c_k) < \delta$. $f : \mathbb{R} \to \mathbb{R}$ is called absolutely continuous if $f : [a, b] \to \mathbb{R}$ is absolutely continuous on $[a, b]$ for every $a < b$. For a proof of the following theorem we refer to Hewitt and Stromberg (1965).

Theorem A.24. *The function $f : [a, b] \to \mathbb{R}$ is absolutely continuous if and only if there exists a $g : [a, b] \to_m \mathbb{R}$ with $\int I_{[a,b]}(t)|g(t)|\boldsymbol{\lambda}(dt) < \infty$ such that $f(x) - f(a) = \int I_{[a,x]}(t)g(t)\boldsymbol{\lambda}(dt)$ for every $a < x \le b$. The function f is in this case $\boldsymbol{\lambda}$-a.e. differentiable with derivative f' that satisfies $f' = g$, $\boldsymbol{\lambda}$-a.e.*

Now we consider measures on product spaces.

Definition A.25. *Let $(\mathcal{X}, \mathfrak{A})$ and $(\mathcal{Y}, \mathfrak{B})$ be measurable spaces. Set $\mathfrak{A} \otimes \mathfrak{B} = \sigma\{A \times B : A \in \mathfrak{A}, B \in \mathfrak{B}\}$. We call $\mathfrak{A} \otimes \mathfrak{B}$ the product σ-algebra, and $(\mathcal{X} \times \mathcal{Y}, \mathfrak{A} \otimes \mathfrak{B})$ the product space.*

Let $\mu \in \mathcal{M}(\mathfrak{A})$ and $\nu \in \mathcal{M}(\mathfrak{B})$. For every $C \in \mathfrak{A} \otimes \mathfrak{B}$ the function $x \mapsto \int (I_C(x, y)\nu(dy)$ is measurable and

$$(\mu \otimes \nu)(C) := \int [\int (I_C(x, y)\nu(dy)]\mu(dx), \quad C \in \mathfrak{A} \otimes \mathfrak{B},$$

is a measure on $\mathfrak{A} \otimes \mathfrak{B}$ that satisfies $(\mu \otimes \nu)(A \times B) = \mu(A)\nu(B)$, $A \in \mathfrak{A}$, $B \in \mathfrak{B}$. The measure $\mu \otimes \nu$ and is called the *product measure of μ and ν.*

Theorem A.26. ***(Theorem of Fubini)*** *If $\mu \in \mathcal{M}^\sigma(\mathfrak{A})$ and $\nu \in \mathcal{M}^\sigma(\mathfrak{B})$, then $\mu\otimes\nu$ is the uniquely determined measure on $\mathfrak{A}\otimes\mathfrak{B}$ with $(\mu\otimes\nu)(A\times B) = \mu(A)\nu(B)$, $A \in \mathfrak{A}$, $B \in \mathfrak{B}$. Moreover, for every $h : \mathcal{X} \times \mathcal{Y} \to_m \overline{\mathbb{R}}_+$ it holds*

$$\int hd(\mu \otimes \nu) = \int [\int h(x,y)\nu(dy)]\mu(dx) = \int [\int (h(x,y)\mu(dx)]\nu(dy). \quad \text{(A.2)}$$

For a proof of the uniqueness we refer to Theorem B on p. 144 in Halmos (1974). The statement A.2 follows from the standard extension technique; see Lemma A.6.

Now we formulate the Radon–Nikodym theorem. Let $\mu, \nu \in \mathcal{M}(\mathfrak{A})$. If $\nu(A) = 0$ implies $\mu(A) = 0$, $A \in \mathfrak{A}$, then we say that μ is *absolutely continuous* with respect to ν, or ν *dominates* μ, and write $\mu \ll \nu$. If $\mu \ll \nu$ and $\nu \ll \mu$, then we call μ and ν *equivalent* and write $\mu \ll\gg \nu$.

Theorem A.27. ***(Theorem of Radon–Nikodym)*** *If $\mu, \nu \in \mathcal{M}^\sigma(\mathfrak{A})$ and $\mu \ll \nu$, then there is a ν-a.e. uniquely determined $f : \mathcal{X} \to_m \mathbb{R}_+$, called the density of μ with respect to ν, or the ν-density of μ, such that for every $A \in \mathfrak{A}$ and every $h : \mathcal{X} \to_m \overline{\mathbb{R}}_+$ it holds*

$$\mu(A) = \int_A f(x)\nu(dx) \quad \text{and} \quad \int hd\mu = \int hfd\nu.$$

The function f is also denoted by $d\mu/d\nu$ and called the *Radon–Nikodym derivative* of μ with respect to ν. For a proof of the existence and uniqueness of $d\mu/d\nu$ we refer to Theorem B, §31, in Halmos (1974). The second statement follows from the standard extension technique; see Lemma A.6.

Proposition A.28. ***(Chain Rule)*** *If $\mu, \nu, \rho \in \mathcal{M}^\sigma(\mathfrak{A})$ and $\mu \ll \nu \ll \rho$, then*

$$\frac{d\mu}{d\rho} = \frac{d\mu}{d\nu} \times \frac{d\nu}{d\rho}, \quad \rho\text{-a.e.}$$

Proposition A.29. *If $(\mathcal{X}_i, \mathfrak{A}_i)$ are measurable spaces and $\mu_i, \nu_i \in \mathcal{M}^\sigma(\mathfrak{A}_i)$ with $\mu_i \ll \nu_i$, $i = 1, 2$, then*

$$\frac{d(\mu_1 \otimes \mu_2)}{d(\nu_1 \otimes \nu_2)}(x_1, x_2) = \frac{d\mu_1}{d\nu_1}(x_1)\frac{d\mu_2}{d\nu_2}(x_2), \quad \nu_1 \otimes \nu_2\text{-a.e. } (x_1, x_2) \in \mathcal{X}_1 \times \mathcal{X}_2.$$

A.3 Topics from Probability Theory

Let $(\Omega, \mathfrak{F}, \mathbb{P})$ be a probability space and X a nonnegative random variable on it with $\mathbb{E}X < \infty$. Then $\mu(A) := \mathbb{E}I_A X$, $A \in \mathfrak{F}$, is a finite measure on $(\Omega, \mathfrak{F})$ that is dominated by $\mathbb{P}$, and $X = d\mu/d\mathbb{P}$. Let $\mathfrak{G} \subseteq \mathfrak{F}$ be a sub-σ-algebra and denote by $\mu^{\mathfrak{G}}$ and $\mathbb{P}^{\mathfrak{G}}$ the restrictions of μ and $\mathbb{P}$ on $\mathfrak{G}$, respectively. Then $\mu^{\mathfrak{G}} \ll \mathbb{P}^{\mathfrak{G}}$, so that by the Radon–Nikodym theorem there exists a $\mathbb{P}$-a.s. uniquely determined random variable Z that is $\mathfrak{G}$-measurable and satisfies $\int_B X d\mathbb{P} = \int_B Z d\mathbb{P}$, $B \in \mathfrak{G}$. We set $\mathbb{E}(X|\mathfrak{G}) := Z$.

Definition A.30. *Given a sub-σ-algebra $\mathfrak{G}$ of $\mathfrak{F}$ and a nonnegative random variable X with $\mathbb{E}X < \infty$, the conditional expectation of X, given $\mathfrak{G}$, is the $\mathbb{P}$-a.s. uniquely determined $\mathfrak{G}$-measurable random variable $\mathbb{E}(X|\mathfrak{G})$ that satisfies $\int_B X d\mathbb{P} = \int_B \mathbb{E}(X|\mathfrak{G}) d\mathbb{P}$ for every $B \in \mathfrak{G}$. If $X \geq 0$ and $\mathbb{E}X = \infty$, then $\mathbb{E}(X|\mathfrak{G}) := \lim_{N\to\infty} \mathbb{E}(X \wedge N|\mathfrak{G})$. If $Y : \Omega \to_m \mathbb{R}$ and $(\mathbb{E}Y^+) \wedge (\mathbb{E}Y^-) < \infty$, then $\mathbb{E}(Y|\mathfrak{G}) := \mathbb{E}(Y^+|\mathfrak{G}) - \mathbb{E}(Y^-|\mathfrak{G})$.*

Let $(\mathcal{X}, \mathfrak{A})$ be a measurable space and X a random variable with values in $\mathcal{X}$. Let $\sigma(X)$ be the σ-algebra generated by X. If Y is a random variable with values in $\mathbb{R}$ and $\mathbb{E}|Y| < \infty$, and $\mathfrak{B}$ is the σ-algebra of Borel sets of $\mathbb{R}$, then by Lemma A.9 there exists an $\mathfrak{A}$-$\mathfrak{B}$ measurable function $g : \mathcal{X} \to_m \mathbb{R}$ such that $\mathbb{E}(Y|\sigma(X)) = g(X)$. This $\mathbb{P}\circ X^{-1}$-a.s. unique function g is called the *regression function* of Y with respect to X. It is common to denote $g(x)$ by $\mathbb{E}(Y|X = x)$ and called it the *conditional expectation of Y, given $X = x$.*

Below we present some standard properties of the conditional expectation which follow directly from its definition. For details we refer to Theorem 5.1 in Kallenberg (1997).

Proposition A.31. *Suppose that $\mathbb{E}|X| < \infty$ and $\mathbb{E}|X_i| < \infty$, $i = 1, 2$. Let $\mathfrak{G}$ be a sub-σ-algebra of $\mathfrak{F}$. Then the following hold $\mathbb{P}$-a.s.*

(a) $\mathbb{E}(\mathbb{E}(X|\mathfrak{G})) = \mathbb{E}X$. (b) $\mathbb{E}(X|\mathfrak{G}) \geq 0$ if $X \geq 0$. (c) $\mathbb{E}(a_1X_1 + a_2X_2|\mathfrak{G}) = a_1\mathbb{E}(X_1|\mathfrak{G}) + a_2\mathbb{E}(X_2|\mathfrak{G})$. (d) If X_2 is $\mathfrak{G}$-measurable and $\mathbb{E}|X_1X_2| < \infty$, then $\mathbb{E}(X_1X_2|\mathfrak{G}) = X_2\mathbb{E}(X_1|\mathfrak{G})$.

Lemma A.32. *If the sequence of random variables X_n, $n = 1, 2, ...$, is uniformly integrable, then $Z_n = \mathbb{E}(X_n|\mathfrak{G})$ is also uniformly integrable.*

Proof. It follows from Definition A.20 that a sequence of random variables Y_n is uniformly integrable if and only if $\lim_{n\to\infty} \mathbb{E}|Y_n|I_{B_n} = 0$ for every sequence $B_n \in \mathfrak{F}$ with $\lim_{n\to\infty} P(B_n) = 0$. As Z_n is $\mathfrak{G}$-measurable we may assume that $B_n \in \mathfrak{G}$. Then $\mathbb{E}|Z_n|I_{B_n} = \mathbb{E}(|\mathbb{E}(X_n|\mathfrak{G})|)I_{B_n} = \mathbb{E}(|\mathbb{E}(I_{B_n}X_n|\mathfrak{G})|) \leq \mathbb{E}(\mathbb{E}(I_{B_n}|X_n||\mathfrak{G})) = \mathbb{E}I_{B_n}|X_n| \to 0$, as $n \to \infty$. ■

For a random vector X with $\mathbb{E}\|X\| < \infty$ the conditional expectation $\mathbb{E}(X|\mathfrak{G})$ is defined componentwise. For a proof of the next lemma we refer to Dudley (2002), pp. 348–349.

Lemma A.33. (Jensen's Inequality) *Let C be an open and convex subset of $\mathbb{R}^m$ and $L : C \to \mathbb{R}$ be a convex function. If X is a random vector with $\mathbb{P}(X \in C) = 1$ and $\mathbb{E}\|X\| < \infty$, then $\mathbb{E}X \in C$ and $L(\mathbb{E}X) \leq \mathbb{E}L(X)$. If $\mathfrak{G}$ is a sub-σ-algebra of $\mathfrak{F}$, then $\mathbb{P}$-a.s. $\mathbb{E}(X|\mathfrak{G}) \in C$ and $L(\mathbb{E}(X|\mathfrak{G})) \leq \mathbb{E}(L(X)|\mathfrak{G}))$.*

Now we formulate Levy's convergence theorem for martingales in a version as given by Corollary 6.22 and Theorem 6.23 in Kallenberg (1997).

Theorem A.34. (Convergence Theorem of Levy) *Let X be a random variable defined on $(\Omega, \mathfrak{F}, \mathbb{P})$ that satisfies $\mathbb{E}|X|^p < \infty$ for some $p \geq 1$. Assume* *that* $\mathfrak{F}_0 \subseteq \mathfrak{F}_1 \subseteq \cdots$ *is a nondecreasing sequence of sub-*σ*-algebras of* $\mathfrak{F}$. *Let* $\mathfrak{F}_\infty = \sigma(\cup_{i=0}^{\infty}\mathfrak{F}_i)$. *Then* $\mathbb{P}$*-a.s., as* $n \to \infty$,

$$\mathbb{E}(X|\mathfrak{F}_n) \to \mathbb{E}(X|\mathfrak{F}_\infty) \quad \text{and} \quad \mathbb{E}|\mathbb{E}(X|\mathfrak{F}_n) - \mathbb{E}(X|\mathfrak{F}_\infty)|^p \to 0.$$

$\mathbb{P}(A|\mathfrak{G}) := \mathbb{E}(I_A|\mathfrak{G})$ is called the *conditional probability of* A *given* $\mathfrak{G}$. Let X and Y be random variables on the probability space $(\Omega, \mathfrak{F}, \mathbb{P})$ with values in $(\mathcal{X}, \mathfrak{A})$ and $(\mathcal{Y}, \mathfrak{B})$, respectively. Denote by $\mathbb{P}_{(X,Y)} := \mathbb{P} \circ (X,Y)^{-1} \in \mathcal{P}(\mathfrak{A} \otimes \mathfrak{B})$ the joint distribution of X and Y, and by $\mathbb{P}_X := \mathbb{P} \circ X^{-1} \in \mathcal{P}(\mathfrak{A})$ and $\mathbb{P}_Y := \mathbb{P} \circ Y^{-1} \in \mathcal{P}(\mathfrak{B})$ the distributions of X and Y, respectively. $\mathbb{P}_X$ and $\mathbb{P}_Y$ are called the *marginal distributions* of $\mathbb{P}_{(X,Y)}$. The set function $B \mapsto \mathbb{P}(Y \in B|X = x) := \mathbb{E}(I_B(Y)|X = x)$, beeing the conditional probability for fixed x, is in general not a measure for every x. The concept of *regular conditional distributions* eliminates this difficulty. It is a regular version of the conditional probability $\mathbb{P}(Y \in B|X = x)$. The basic tool is the *stochastic kernel.*

Definition A.35. *For two measurable spaces* $(\mathcal{X}, \mathfrak{A})$ *and* $(\mathcal{Y}, \mathfrak{B})$ *a mapping* $\mathsf{K} : \mathfrak{B} \times \mathcal{X} \to [0,1]$ *is called a stochastic kernel if for every* $B \in \mathfrak{B}$ *the function* $x \mapsto \mathsf{K}(B|x)$ *is* $\mathfrak{A}$-$\mathfrak{B}_{[0,1]}$ *measurable, and for every* $x \in \mathcal{X}$, *it holds* $\mathsf{K}(\cdot|x) \in \mathcal{P}(\mathfrak{B})$. *In short we write* $\mathsf{K} : \mathfrak{B} \times \mathcal{X} \to_k [0,1]$.

Definition A.36. *Let* X *and* Y *be random variables on the probability space* $(\Omega, \mathfrak{F}, \mathbb{P})$ *with values in* $(\mathcal{X}, \mathfrak{A})$ *and* $(\mathcal{Y}, \mathfrak{B})$,*respectively. The kernel* $\mathsf{K} : \mathfrak{B} \times \mathcal{X} \to_k [0,1]$ *is called a regular conditional distribution of* Y *given* X *if*

$$\mathbb{P}_{X,Y}(A \times B) = \int_A \mathsf{K}(B|x)\mathbb{P}_X(dx), \quad A \in \mathfrak{A},\ B \in \mathfrak{B},$$

where $\mathbb{P}_{X,Y} = \mathbb{P}\circ(X,Y)^{-1}$ *is the joint distribution of* X *and* Y. *We write in short* $\mathbb{P}_{X,Y} = \mathsf{K} \otimes \mathbb{P}_X$, *and we denote the marginal distribution of* Y *by* $\mathsf{K}\mathbb{P}_X$.

For the existence of regular conditional distributions additional assumptions on the measurable space $(\mathcal{Y}, \mathfrak{B})$ have to be made. For a proof of the next result we refer to Theorem 5.3 in Kallenberg (1997).

Theorem A.37. ***(Existence of a Regular Conditional Distribution)*** *If* X *and* Y *are random variables with values in* $(\mathcal{X}, \mathfrak{A})$ *and* $(\mathcal{Y}, \mathfrak{B})$, *respectively, and* $(\mathcal{Y}, \mathfrak{B})$ *is a Borel space in the sense of Definition A.7, then there exists a regular conditional distribution* $\mathsf{K} : \mathfrak{B} \times \mathcal{X} \to_k [0,1]$ *in the sense of Definition A.36.*

The existence of conditional distributions can be established easily, without additional assumptions on the sample spaces, if there exist *conditional densities* in the following sense. Suppose that X and Y are random variables with values in $(\mathcal{X}, \mathfrak{A})$ and $(\mathcal{Y}, \mathfrak{B})$, respectively. Assume that there are $\mu \in \mathcal{M}^\sigma(\mathfrak{A})$ and $\nu \in \mathcal{M}^\sigma(\mathfrak{B})$ so that $\mathbb{P}_{X,Y} \ll \mu \otimes \nu$. Set $f_{X,Y} = \frac{d\mathbb{P}_{X,Y}}{d\mu\otimes\nu}$. It is easy to see that $\mathbb{P}_{X,Y} \ll \mu \otimes \nu$ implies $\mathbb{P}_X \ll \mu$ and $\mathbb{P}_Y \ll \nu$, and the corresponding densities are given by $f_X(x) := \frac{dP_X}{d\mu}(x) = \int f(x,y)\nu(dy)$, $x \in \mathcal{X}$, and $f_Y(y) := \frac{dP_Y}{d\mu}(y) = \int f(x,y)\mu(dx)$, $y \in \mathcal{Y}$, which are called the *marginal densities*.

Definition A.38. *The function*

$$f_{Y|X}(y|x) = \begin{cases} \frac{f_{X,Y}(x,y)}{f_X(x)} & \text{if } f_X(x) > 0, \\ f_Y(y) & \text{if } f_X(x) = 0, \end{cases}$$

is called the conditional density of Y, given $X = x$. The stochastic kernel $\mathsf{K}(B|x) = \int_B f_{Y|X}(y|x)\nu(dy)$ is called the regular conditional distribution of Y, given $X = x$, based on the conditional density.

The fact that K is indeed a stochastic kernel follows from the additivity of the integral and Theorem A.16.

Let $(\mathcal{X}, \mathfrak{A})$, $(\mathcal{Y}, \mathfrak{B})$, and $(\mathcal{Z}, \mathfrak{C})$ be measurable spaces, and $\mathsf{K} : \mathfrak{B} \times \mathcal{X} \to_k [0,1]$ and $\mathsf{L} : \mathfrak{C} \times (\mathcal{X} \times \mathcal{Y}) \to_k [0,1]$ be stochastic kernels. Set

$$(\mathsf{L} \otimes \mathsf{K})(D|x) = \int [\int I_D(y,z)\mathsf{L}(dz|x,y)]\mathsf{K}(dy|x), \quad D \in \mathfrak{B} \otimes \mathfrak{C}.$$

Proposition A.39. *Let $\mathsf{K} : \mathfrak{B} \times \mathcal{X} \to_k [0,1]$ and $\mathsf{L} : \mathfrak{C} \times (\mathcal{X} \times \mathcal{Y}) \to_k [0,1]$ be stochastic kernels, $f : \mathcal{X} \times \mathcal{Y} \to_m \overline{\mathbb{R}}_+$, and $g : \mathcal{X} \times \mathcal{Y} \to_m \mathcal{Z}$. Then the following hold. (a) $x \mapsto \int f(x,y)\mathsf{K}(dy|x)$ is measurable. (b) $\mathsf{M} = \mathsf{K}(g^{-1}(x,\cdot)|x) : \mathfrak{C} \times \mathcal{X} \to [0,1]$ is a stochastic kernel. (c) $\mathsf{L} \otimes \mathsf{K} : (\mathfrak{B} \otimes \mathfrak{C}) \times \mathcal{X} \to [0,1]$ is a stochastic kernel.*

For a proof we refer to Lemma 1.38 in Kallenberg (1997).

The next theorem may be considered as a generalization of Fubini's theorem to stochastic kernels.

Proposition A.40. ***(Theorem of Fubini for Kernels)*** *Given a stochastic kernel $\mathsf{K} : \mathfrak{B} \times \mathcal{X} \to_k [0,1]$ and some $\mu \in \mathcal{M}(\mathfrak{A})$, the set function*

$$(\mathsf{K} \otimes \mu)(C) := \int [\int I_C(x,y)\mathsf{K}(dy|x)]\mu(dx), \quad C \in \mathfrak{A} \otimes \mathfrak{B}, \tag{A.3}$$

is a measure on $\mathfrak{A} \otimes \mathfrak{B}$ with $\mu(A) = (\mathsf{K} \otimes \mu)(A \times \mathcal{Y})$, $A \in \mathfrak{A}$, and it holds for every $h : \mathcal{X} \times \mathcal{Y} \to_m \overline{\mathbb{R}}_+$,

$$\int h d(\mathsf{K} \otimes \mu) = \int [\int h(x,y)\mathsf{K}(dy|x)]\mu(dx). \tag{A.4}$$

Proof. The fact that $\mathsf{K} \otimes \mu$ is a measure follows from a twofold application of the monotone convergence theorem; see Theorem A.16. The statement (A.4) follows from the standard extension technique; see Lemma A.6. ■

Sometimes it is more convenient to formulate a statement in terms of random variables.

Lemma A.41. ***(Disintegration Lemma)*** *If X and Y are random variables and there exists a regular conditional distribution $\mathsf{K} : \mathfrak{B} \times \mathcal{X} \to_k [0,1]$ of Y given X, then for every $C \in \mathfrak{A} \otimes \mathfrak{B}$ and $f : \mathcal{X} \times \mathcal{Y} \to_m \overline{\mathbb{R}}_+$ it holds*

$$\mathbb{P}_{(X,Y)}(C) = \int [\int I_C(x,y)\mathsf{K}(dy|x)]\mathbb{P}_X(dx),$$

$$\mathbb{E}f(X,Y) = \int [\int f(x,y)\mathsf{K}(dy|x)]\mathbb{P}_X(dx),$$

$$\mathbb{E}(f(X,Y)|X=x) = \int f(x,y)\mathsf{K}(dy|x), \quad \mathbb{P}_X\text{-a.s. } x \in \mathcal{X}.$$

Let $(\mathcal{X}, \rho)$ be a metric space with metric ρ, and denote by $\mathfrak{B}$ the σ-algebra of Borel sets in $\mathcal{X}$. Recall C_b, the space of all bounded continuous $f : \mathcal{X} \to \mathbb{R}$. We also need the space $\mathsf{C}_0 \subseteq \mathsf{C}_b$ of all continuous functions f with compact support, which means that the closure of the set $\{x : f(x) \neq 0\}$ is compact.

Definition A.42. *Let $(\mathcal{X}, \rho)$ be a metric space and $P, P_n \in \mathcal{P}(\mathfrak{B})$. The sequence P_n, $n = 1, 2, ...,$ is said to converge weakly to P if $\int f(x)P_n(dx) \to \int f(x)P(dx)$ as $n \to \infty$ for every $f \in \mathsf{C}_b$. In such a case we write $P_n \Rightarrow P$.*

If X and X_n are random variables defined on $(\Omega, \mathfrak{F}, \mathbb{P})$ and $(\Omega_n, \mathfrak{F}_n, \mathbb{P}_n)$, respectively, $n = 1, 2, ...,$ with values in $\mathcal{X}$, then the distributions $\mathcal{L}(X) = \mathbb{P} \circ X^{-1}$ and $\mathcal{L}(X_n) = \mathbb{P}_n \circ X_n^{-1}$ satisfy $\mathcal{L}(X_n) \Rightarrow \mathcal{L}(X)$ if and only if $\lim_{n\to\infty} \mathbb{E}f(X_n) = \mathbb{E}f(X)$ for every $f \in \mathsf{C}_b$. Lebesgue's theorem provides that the a.s. convergence of random variables implies the weak convergence of distributions. A subsequence argument and Proposition A.12 give the following statement.

Lemma A.43. *Let $(\mathcal{X}, \rho)$ be a separable metric space and $Z, Z_1, Z_2, ...$ be random variables defined on the probability space $(\Omega, \mathfrak{F}, \mathbb{P})$ with values in $\mathcal{X}$. Then $Z_n \to^{\mathbb{P}} Z$ implies $\mathcal{L}(Z_n) \Rightarrow \mathcal{L}(Z)$, and especially $Z_n \to Z$, $\mathbb{P}$-a.s., implies $\mathcal{L}(Z_n) \Rightarrow \mathcal{L}(Z)$.*

To derive the convergence $\lim_{n\to\infty} \mathbb{E}f(X_n) = \mathbb{E}f(X)$ from $\mathcal{L}(X_n) \Rightarrow \mathcal{L}(X)$ for unbounded continuous functions f one needs the uniform integrability of f under the distributions of X_n.

Proposition A.44. *Let $(\mathcal{X}, \rho)$ be a separable metric space and $P, P_n \in \mathcal{P}(\mathfrak{B})$. If $P_n \Rightarrow P$, and for a continuous function $f : \mathcal{X} \to_m \mathbb{R}$ it holds $\int |f(x)|P(dx) < \infty$ and*

$$\limsup_{N\to\infty} \limsup_{n\to\infty} \int |f(x)| I_{[N,\infty)}(|f(x)|) P_n(dx) = 0, \tag{A.5}$$

then $\lim_{n\to\infty} \int f(x)P_n(dx) = \int f(x)P(dx)$.

Proof. Set $g_N(x) = -N$ if $f(x) < -N$, $g_N(x) = f(x)$ if $-N \le f(x) \le N$, and $g_N(x) = N$ if $N < f(x)$. Then

$$\begin{aligned} |\int f(x)P_n(dx) - \int f(x)P(dx)| \ &\le \ |\int g_N(x)P_n(dx) - \int g_N(x)P(dx)| \\ &+ \int |f(x)|I_{[N,\infty)}(|f(x)|)P(dx) + \int |f(x)|I_{[N,\infty)}(|f(x)|)P_n(dx). \end{aligned}$$

For $n \to \infty$ the first term vanishes. Then take $N \to \infty$ to get the statement from A.5, $\int |f(x)|P(dx) < \infty$, and Lebesgue's theorem; see Theorem A.18. ■

The next proposition concerns the uniformity of weak convergence on classes of continuous functions. It is Conclusion 2.7 in Bhattacharya and Rao (1976).

Proposition A.45. *Let $(\mathcal{X}, \rho)$ be a separable metric space. Let $\mathbb{F}$ be a family of continuous functions $f : \mathcal{X} \to \mathbb{R}$ with $\sup_{f \in \mathbb{F}} \sup_{x \in \mathcal{X}} |f(x)| < \infty$ and $\lim_{\delta \downarrow 0} \sup_{f \in \mathbb{F}} \sup_{x:\rho(x,x_0) \le \delta} |f(x) - f(x_0)| = 0$, $x_0 \in \mathcal{X}$. If $P, P_n \in \mathcal{P}(\mathfrak{B})$ and $P_n \Rightarrow P$, then*

$$\lim_{n \to \infty} \sup_{f \in \mathbb{F}} |\int f(x)P_n(dx) - \int f(x)P(dx) | = 0.$$

The next statement is Theorem 4.1 in Billingsley (1968).

Lemma A.46. ***(Lemma of Slutsky)*** *Let $(\mathcal{X}, \rho)$ be a separable metric space and X, X_n, Y_n, $n = 1, 2, ...$, be random variables on $(\Omega, \mathfrak{F}, \mathbb{P})$ with values in $\mathcal{X}$. If $\mathcal{L}(X_n) \Rightarrow \mathcal{L}(X)$ and $\rho(X_n, Y_n) \to^{\mathbb{P}} 0$, then $\mathcal{L}(Y_n) \Rightarrow \mathcal{L}(X)$.*

It is clear from the definition of weak convergence that for any continuous mapping h the relation $\mathcal{L}(X_n) \Rightarrow \mathcal{L}(X)$ implies $\mathcal{L}(h(X_n)) \Rightarrow \mathcal{L}(h(X))$. The next theorem provides conditions, weaker than the continuity of h, under which $\mathcal{L}(h_n(X_n)) \Rightarrow \mathcal{L}(h(X))$ follows. This is Theorem 3.27 in Kallenberg (1997).

Theorem A.47. ***(Continuous Mapping Theorem)*** *Let $\mathcal{X}$ and $\mathcal{Y}$ be metric spaces, $C \subseteq \mathcal{X}$ be a Borel set, and $h, h_n : \mathcal{X} \to_m \mathcal{Y}$, $n = 1, 2, ...$, so that $h_n(x_n) \to h(x)$ as $x_n \to x$, $x \in C$. If X, X_n, $n = 1, 2, ...$, are random variables on $(\Omega, \mathfrak{F}, \mathbb{P})$ with values in $\mathcal{X}$, and $X \in C$, $\mathbb{P}$-a.s., then $\mathcal{L}(X_n) \Rightarrow \mathcal{L}(X)$ implies $\mathcal{L}(h_n(X_n)) \Rightarrow \mathcal{L}(h(X))$.*

The next theorem characterizes the sequential compactness of sequences of distributions.

Theorem A.48. ***(Theorem of Prohorov)*** *Let $\mathcal{X}$ be a complete and separable metric space and $P_n \in \mathcal{P}(\mathfrak{B})$, $n = 1, 2, ...$ For every subsequence P_{n_k} there exists a subsequence that converges weakly to some $P \in \mathcal{P}(\mathfrak{B})$ if and only if the sequence P_n is tight; that is, for every $\varepsilon > 0$ there is a compact set $K_\varepsilon \subseteq \mathcal{X}$ such that $P_n(K_\varepsilon) \ge 1 - \varepsilon$ for every n.*

For a proof we refer to Chapter 1, Section 6, in Billingsley (1968).

Let ∂A denote the boundary of a set A. A Borel set A is called a *P-continuity set* if $P(\partial A) = 0$. Let $\mathsf{C}_{b,P}$ be the space of all bounded measurable functions for which the set of all discontinuity points has P-measure zero. In the following theorem the equivalence of conditions (A) through (E) is known in the literature as the *Portmanteau theorem*; see Theorem 2.1 in Billingsley (1968). The equivalence of (F) follows from Theorem 5.2 in Billingsley (1968), and the equivalence of (G) from Lemma A.5 and the fact that I_O is lower semicontinuous for open sets O.

Theorem A.49. ***(Portmanteau Theorem)*** *Let $\mathcal{X}$ be a metric space, $\mathfrak{B}$ the σ-algebra of Borel sets, and $P, P_1, P_2, ...$ distributions on $(\mathcal{X}, \mathfrak{B})$. Then the following statements are equivalent.*

(A) $P_n \Rightarrow P$	
(B) $\lim_{n\to\infty} \int f dP_n = \int f dP$	*for every bounded and uniformly continuous f.*
(C) $\limsup_{n\to\infty} P_n(A) \le P(A)$	*for every closed set A.*
(D) $\liminf_{n\to\infty} P_n(O) \ge P(O)$	*for every open set O.*
(E) $\lim_{n\to\infty} P_n(B) = P(B)$	*for every P-continuity set B.*
(F) $\lim_{n\to\infty} \int f dP_n = \int f dP$	*for every $f \in \mathsf{C}_{b,P}$.*
(G) $\liminf_{n\to\infty} \int f dP_n \ge \int f dP$	*for every lower semicontinuous $f : \mathcal{X} \to [0, \infty]$.*

For several purposes it is convenient to metricize the weak convergence and to consider the space of all distributions as a new metric space. There are several ways to introduce such a metric. As it turned out the metric introduced by and named after Dudley is one of the best tractable metrics. The reason is that it is a dual norm of a function space. More precisely, for any metric space $\mathcal{X}$ let

$$\|P - Q\|_{\mathsf{D}} = \sup | \int f dP - \int f dQ|, \tag{A.6}$$

where the supremum is taken over all Lipschitz functions f with $L_f \le 1$ and $\|f\|_u \le 1$, and L_f is from (A.1). $\|P - Q\|_{\mathsf{D}}$ is called the *Dudley metric*. The following theorem is established in Dudley (2002), Proposition 11.3.2 and Theorem 11.3.3.

Theorem A.50. *If $\mathcal{X}$ is a complete separable metric space and $\mathfrak{B}$ the σ-algebra of Borel sets, then $\|P - Q\|_{\mathsf{D}}$ is a metric on $\mathcal{P}(\mathfrak{B})$, and it holds $P_n \Rightarrow P$ if and only if $\|P_n - P\|_{\mathsf{D}} \to 0$.*

We recall that for any random vector X with values in $\mathbb{R}^d$, $\varphi_X(t) = \mathbb{E}\exp\{i\langle t, X\rangle\}$, $t \in \mathbb{R}^d$, is the *characteristic function* of X.

Theorem A.51. ***(Uniqueness and Continuity Theorem for Characteristic Functions)*** *Let U, V, X, and X_n, $n = 1, 2, ...$, be random vectors on $(\Omega, \mathfrak{F}, \mathbb{P})$ with values in $\mathbb{R}^d$. Then $\mathcal{L}(U) = \mathcal{L}(V)$ if and only if $\varphi_U(t) = \varphi_V(t)$ for every $t \in \mathbb{R}^d$, and $\mathcal{L}(X_n) \Rightarrow \mathcal{L}(X)$ if and only if $\lim_{n\to\infty} \varphi_{X_n}(t) = \varphi_X(t)$ for every $t \in \mathbb{R}^d$.*

For a proof we refer to Theorem 7.6 in Billingsley (1968).

The following *Cramér–Wold device* states that the weak convergence of the distributions of random vectors in $\mathbb{R}^d$ can be reduced to the weak convergence of distributions of random variables with values in $\mathbb{R}$.

Criterion A.52 ***(Cramér–Wold Device)*** *Let $X, X_1, X_2, ...$ be random vectors on $(\Omega, \mathfrak{F}, \mathbb{P})$ with values in $\mathbb{R}^d$. Then $\mathcal{L}(X_n) \Rightarrow \mathcal{L}(X)$ if and only if $\mathcal{L}(\langle X_n, t\rangle) \Rightarrow \mathcal{L}(\langle X_n, t\rangle)$ for every $t \in \mathbb{R}^d$.*

The proof follows from the continuity theorem for characteristic functions; see Theorem A.51.

In conclusion we formulate suitable versions of the *central limit theorem.*

Theorem A.53. ***(Central Limit Theorem for i.i.d. Random Vectors)*** *Let $X_1, X_2, ...$ be i.i.d. random vectors on $(\Omega, \mathfrak{F}, \mathbb{P})$ with values in $\mathbb{R}^d$ and $\mathbb{E}\,\|X_1\|^2 < \infty$. Set $\mu = \mathbb{E}X_1$ and $\Sigma = \mathbb{E}(X_1 - \mu)(X_1 - \mu)^T$. Then*

$$\mathcal{L}(n^{-1/2} \sum\nolimits_{i=1}^{n} (X_i - \mu)) \Rightarrow \mathsf{N}(0, \Sigma).$$

The case of $d = 1$ follows from the subsequent more general theorem. The case $d > 1$ can be reduced to the case $d = 1$ with the help of the Cramér–Wold device; see Criterion A.52.

Theorem A.54. ***(Central Limit Theorem for Double Arrays)*** *Let $X_{n,i}$, $i = 1, ..., m_n$, $n = 1, 2, ...$, be a double array of random variables on $(\Omega, \mathfrak{F}, \mathbb{P})$ with values in $\mathbb{R}$ so that for every n the random variables $X_{n,1}, ..., X_{n,m_n}$ are independent with $\mathbb{E}X_{n,i} = 0$, $i = 1, ..., m_n$, and $\sum_{i=1}^{m_n} \mathbb{E}X_{n,i}^2 = 1$. If the Lindeberg condition, that is,*

$$\lim_{n\to\infty} \sum\nolimits_{i=1}^{m_n} \mathbb{E}X_{n,i}^2 I_{(\varepsilon,\infty)}(|X_{n,i}|) = 0, \quad \varepsilon > 0,$$

is satisfied, then $\mathcal{L}(\sum_{i=1}^{m_n} X_{n,i}) \Rightarrow \mathsf{N}(0, 1)$.

For a proof we refer to Theorem 4.12 in Kallenberg (1997).

B

Appendix: Common Notation and Distributions

B.1 Common Notation

Euclidean Space

$\mathbb{N} = \{0, 1, 2, ...\}$
$\mathbb{Q}$ rational numbers
$\mathbb{R}$ real numbers
$\mathbb{C}$ complex numbers
$\mathbb{R}_+ = [0, \infty)$
$\mathbb{R}_{\neq 0} = \mathbb{R}\backslash\{0\}$
$\overline{\mathbb{R}} = [-\infty, \infty]$
$\overline{\mathbb{R}}_+ = [0, \infty]$
$\mathbb{R}^n$ Euclidean space
$\mathbb{R}^n_< = \{x : x_1 < \cdots < x_n, x \in \mathbb{R}^n\}$
$\mathbb{R}^n_\leq = \{x : x_1 \leq \cdots \leq x_n, x \in \mathbb{R}^n\}$
$\mathbb{R}^n_{\neq} = \{x : x_i \neq x_j, i < j, x \in \mathbb{R}^n\}$
$\mathbb{R}^n_{\neq 0} = \{x : x \neq 0, x \in \mathbb{R}^n\}$
$\mathbb{R}^k_* = \{x : x_{[k-1]} < x_{[k]}, x \in \mathbb{R}^k\}$
$\mathbb{R}^k_r$ p.524
$\mathbb{R}^k_{r,\delta}$ p.534
$(x_1, ..., x_n)$ vector in $\mathbb{R}^n$
$(x_1, ..., x_n)^T$ column vector

$x_{[1]} \leq \cdots \leq x_{[n]}$ order statistics
$X_{i,\cdot}, X_{\cdot,\cdot}, \theta.$ p.88
$\mathbf{1} = (1, ..., 1)^T$
$a \leq b$ in $\mathbb{R}^d$ componentwise
$\mathcal{X} \times \mathcal{Y}$ product of sets
$\times_{i=1}^d \mathcal{X}_i$ product of d sets
$\mathbb{S}_{d-1}$, $\mathbf{S}^0_d$, $\mathbf{S}^c_d$ simplex p.6
$\mathcal{F}$ finite set
$|\mathcal{F}|$ size of $\mathcal{F}$
$\mathcal{S}_m$ simplex p.168
$\mathbb{R}_\oplus$ additive group on $\mathbb{R}$
$\mathbb{R}_{\oplus n}$ additive group on $\mathbb{R}^n$
$\mathbb{R}^+_\bullet$ multiplicative group on $\mathbb{R}_+$
$\mathbb{L}_d$ p.199
$\mathcal{M}^r_{n\times n}$ p.199
$\mathcal{O}_{n\times n}$ p.199
$A \succeq B : A - B$ pos. semidefinite
$\Pi_{\mathbb{L}_k}$ projection p.444

Measurable Spaces

$\mathfrak{P}(\mathcal{X})$ all subsets of $\mathcal{X}$
$\sigma(\mathfrak{G})$ σ-algebra generated by $\mathfrak{G}$
$\mathfrak{B}_n$: Borel sets in $\mathbb{R}^n$
$\mathfrak{B} = \mathfrak{B}_1$
$\mathfrak{B}_\mathcal{X}$ Borel sets of metric space $\mathcal{X}$

$\mathfrak{B}_+$ Borel sets in $\mathbb{R}_+$
$\overline{\mathfrak{B}}$ Borel sets in $\overline{\mathbb{R}}$
$\overline{\mathfrak{B}}_+$ Borel sets in $\overline{\mathbb{R}}_+$
$\mathfrak{B}_{n,<}$ Borel sets in $\mathbb{R}^n_<$
$\mathfrak{B}_{n,\neq}$ Borel sets in $\mathbb{R}^n_{\neq}$

$\mathfrak{B}_{n,\neq 0}$ Borel sets in $\mathbb{R}^n_{\neq 0}$
$\mathfrak{B}_{k,*}$ Borel sets in $\mathbb{R}^k_*$
$\mathfrak{F}(\Delta)$ p.146
$\mathfrak{I}$ p.202
$(\mathcal{X}, \mathfrak{A})$ measurable space
$\mathfrak{A} \otimes \mathfrak{C}$ product-σ-algebra
$\bigotimes_{i=1}^d \mathfrak{A}_i$ product-σ-algebra
$\mathfrak{A}^{\otimes d}$ d-fold product-σ-algebra
$\bigotimes_{i=1}^d (\mathcal{X}_i, \mathfrak{A}_i)$ product space
$S : \mathcal{X} \to_m \mathcal{Y}$ measurable mapping
$S : \mathcal{X} \leftrightarrow_m \mathcal{Y}$ bimeasurable map.
$S^{-1}(C) = \{x : S(x) \in C\}$
$\sigma(S)$ σ-algebra generated by S
$f \propto g : f = cg$ for some $c > 0$
I_A indicator function

Measures and Kernels

$\boldsymbol{\lambda}_d$ Lebesgue measure on $\mathfrak{B}_d$
$\boldsymbol{\lambda} = \boldsymbol{\lambda}_1$
$\boldsymbol{\kappa}_d$ counting measure on $\mathfrak{P}(\mathbb{N}^d)$
$\boldsymbol{\kappa} = \boldsymbol{\kappa}_1$
$\boldsymbol{\delta}_x$ delta measure on x
$\boldsymbol{\mu}$ p.2
$\boldsymbol{\nu}$ p.4
γ_{v} p.32
ρ_{v} p.39
$\mu_F, \mu_{n,F}$ p.243
$\mathbb{P}$, $\mathbb{P}_\theta$, P_θ probability
$P_0^{\mathfrak{G}}, P_1^{\mathfrak{G}}$ p.42
$\overline{P}$ p.44
$Q_\theta = P_\theta \circ T^{-1}$ distribution of T
$Q_1 \preceq Q_2$ stochastic semiorder
F, F^{-1} p.75
$\mathcal{L}(X)$ distribution of X
$X \sim P : \mathcal{L}(X) = P$
$\mathcal{L}(T|P)$ distribution of T under P
$P_{pe,\eta}$ p.5
Π prior distribution
$\mathbf{\Pi}_{a,b}$ conjugate prior
Υ p.20
Υ_n p.22
$\mathbf{\Pi}(B|x)$ posterior distribution
$\pi(\theta|x)$ posterior density
$\{P_0, P_1\}$-a.s. p.34
$P_1 * P_2$ convolution of P_1 andP_2
$\mathcal{M}(\mathfrak{A})$ measures on $\mathfrak{A}$
$\mathcal{M}^\sigma(\mathfrak{A})$ σ-finite measures on $\mathfrak{A}$
$\mathcal{P}(\mathfrak{A})$ probabilities on $\mathfrak{A}$
$\mathcal{P}_c(\mathfrak{A})$ atomless probabilities on $\mathfrak{A}$
$\mu \ll \nu : \nu$ dominates μ
$\mu \ll\gg \nu : \mu \ll \nu$ and $\nu \ll \mu$
$\mu \perp \nu$ mutually singular p.34
$\mu \otimes \nu$ product measure
$\bigotimes_{i=1}^d \mu_i$ product measure
$\mu^{\otimes d}$ d-fold product measure
$P_{\theta_0}^{\otimes\infty}$ infinite product measure
$\mu_f = \mu \circ f^{-1}$ induced measure
$P * Q$ convolution
$\mathsf{P} : \mathfrak{A} \times \Delta \to_k [0,1]$ stoch. kernel
$\mathsf{D}, \mathsf{K}, \mathsf{L}$ stochastic kernel
$\mathsf{P} \otimes \Pi$ p.17
$\mathsf{D} \otimes \mathsf{P} \otimes \Pi$ p.130
$\mathsf{P}\Pi$ p.17
$\mathsf{M} = \mathsf{P}\Pi$

Numerical Characteristics

$a \vee b = \max(a, b)$
$a \wedge b = \min(a, b)$
$b \oslash a$ p.85
$[x]$ the integer part of x
$\arg\max$ p.110
$\arg\min$ p.126
$\langle x, y \rangle = x^T y$
$\|x\|^2 = \langle x, x \rangle$
$\chi^2(x) = \|x\|^2$
ρ_Δ metric p.120
$\rho_{\mathcal{T}}$ metric p.280
v, v^* p.33
$\mathsf{v}, \mathsf{I}_{\mathsf{v}}(P_0, P_1)$ p.36
$\mathsf{m}_s, \chi^s(P_0, P_1)$ p.36

$\mathsf{v}_s, \mathsf{H}_s(P_0, P_1)$ p.36
$\mathsf{w}_s, \mathsf{K}_s(P_0, P_1), \mathsf{I}(P_0, P_1)$ p.36
$\mathsf{k}_\pi, \mathsf{B}_\pi(P_0, P_1)$ p.36
$\mathsf{I}_\mathsf{v}(X\|Y)$ p.52
$\mathsf{I}(X\|Y) = \mathsf{I}_{x \ln x}(X\|Y)$
$\mathsf{I}_\mathsf{v}(P_0, P_1)$ v-divergence p.35
$\mathsf{K}(P_0, P_1)$ Kullback-Leibler dist.
$\mathsf{C}(P_{\theta_1}, P_{\theta_2})$ Chernoff index
$\mathsf{H}_s(P_0, P_1)$ Hellinger integral
$\mathsf{H}_s(P_1, ..., P_m)$ Hellinger transform
$\mathsf{D}(P_0, P_1)$ Hellinger distance
$\|P_0 - P_1\|$ variational distance
$\|P_0 - P_1\|_\mathsf{D}$ Dudley metric
$\mathsf{I}(\theta_0)$ Fisher information p.59
$\mathsf{S}_\nu(P)$ Shannon entropy
$C_\mathsf{v}(\mathsf{P}, \mathcal{P})$ p.54
τ_α p.312
$\mathbb{E}$, $\mathbb{E}_\theta$, E_θ, E_P expectation
$\mathsf{E}_{\overline{P}}$ p.44
$\mathsf{V}_\theta(T_1)$ variance
$\mathsf{cov}(T_1, T_2)$ covariance
$\mathsf{C}_\theta(S, T)$ covariance matrix
$\mathsf{C}_\theta(T) = \mathsf{C}_\theta(T, T)$
$\mathbb{E}(Y|X)$ conditional expectation
$\mathbb{E}(Y|\mathfrak{A})$ conditional expectation
$\mathsf{E}_P(f|\mathfrak{A})$ conditional expectation

Spaces of Functions

$\mathbb{H}$ Hilbert space
$\mathbb{L}_r(P_{\theta_0})$ p.58
$\mathbb{L}_{2,d}(\mu)$ p.59
$\mathbb{L}^0_{2,d}(P)$ p.66
$\mathfrak{L}_d$ p.138
$\mathsf{C}(\mathcal{X})$ continuous functions
$\mathsf{C}_b(\mathcal{X})$ bounded functions in $\mathsf{C}(\mathcal{X})$
$\mathsf{C}^{(k)}(O)$ 367
$\mathsf{C}_m(\Delta, \mathcal{X})$ 344
$\mathsf{C}^{(k)}_m(O, \mathcal{X})$ 367
$\| f \|_u = \sup_{x \in \mathcal{X}} |f(x)|$
$\| f \|_p = (\int |f|^p d\mu)^{1/p}$

Derivatives

$\frac{d}{dx}$ derivative
$\frac{\partial}{\partial x_i}$ partial derivative
∇, $\nabla\nabla^T$ p.12
D^-, D^+ p.31
D^α p.10
J_κ Jacobian p.62
$\dot{g}(\theta) = J^T_g(\theta)$ p.296
$\dot{f}_\theta = \nabla f_\theta$
$\ddot{f}_\theta = \nabla\nabla^T f_\theta$
$L_{0,1}$ likelihood ratio p.34
$L_{\theta_0,\theta}$ likelihood ratio
$\dot{L}_{\theta_0}$ $\mathbb{L}_2$-derivative at θ_0
$\Lambda_\theta = \ln f_\theta$

Convergence

$P_n \Rightarrow P$ weak for distributions
$\mathsf{D}_n \Rightarrow \mathsf{D}$ weak for kernels p.117
$\mathcal{M}_n \Rightarrow \mathcal{M}$ weak for models p.242
$f_n \to f$, μ-a.e., almost everywhere
$f_n \to f$, P-a.s., almost surely
$f_n \to^\mu f$ in measure
$f_n \to^\mu_{loc} f$ locally in measure
$L_{n,h}, Z_n, r_{n,h}$ p.270
$\{P_n\} \lhd \{Q_n\}$ p.250
$\{P_n\} \lhd\rhd \{Q_n\}$ p.250
$\{P_n\} \vartriangle \{Q_n\}$ p.257
O, o Landau symbols
$O_{P_n}, o_{P_n}, \mathbf{o}_{P_n}$ p.360

Statistical Models

$(\Omega, \mathfrak{F}, \mathbb{P})$ generic probability space
$(\mathcal{X}, \mathfrak{A}, P)$ probability space
$\mathcal{P}$ family of distributions
$(\mathcal{X}, \mathfrak{A}, \mathcal{P})$ statistical model
$(\mathcal{X}, \mathfrak{A}, (P_\theta)_{\theta \in \Delta})$ statistical model
$\mathcal{G}$ Gaussian model p.247
$\mathcal{G}_0$ Gaussian model p.268
$\mathcal{M}_n = (\mathcal{X}_n, \mathfrak{A}_n, (P_{n,\theta})_{\theta \in \Delta_n})$ p.246
$\mathcal{M}_{1,F}$ p.235
$\mathcal{M} \succeq^{\varepsilon} \mathcal{N}$ p.160
$\mathcal{M} \succeq \mathcal{N}, \mathcal{M} \sim \mathcal{N}$ p.160
$\mathsf{D}_{\mathcal{N}}$ p.159
$\delta(\mathcal{M}_1, \mathcal{M}_2)$ p.235
$d(\mathcal{M}_1, \mathcal{M}_2)$ p.235
$\Delta(\mathcal{M}_{1,F}, \mathcal{M}_{2,F})$ p.236
$D(\mathcal{M}_1, \mathcal{M}_2)$ p.236
$\delta(\mathcal{M}_{1,F}, \mathcal{M}_{2,F})$ p.235

Special Statistics

$X_{[\cdot]}, X_{[i]}$ order statistics
$X_{n,[\cdot]}, X_{n,[i]}$ order statistics
R, R_i rank statistics
$R_n, R_{n,i}$ rank statistics
$R_1(x), ..., R_n(x)$ rank statistics
Λ_n log-likelihood
χ^2 chi-square statistic
$T_{\oplus n}$ p.5
$\mathcal{U}_n(x_1, ..., x_n)$ U-statistic
$\sum_{i=1}^n c_{i,n} a_n(R_{n,i})$ rank statistic
$S_{n,NR}(\theta_0)$ p.475
$Q_{n,NR}(\tau_0, \widetilde{\xi}_n)$ p.476
$Q_{n,W}(\tau_0, \widehat{\theta}_n)$ p.476

Decisions, Loss, and Risk

$(\mathcal{X}, \mathfrak{A})$ sample space
$(\mathcal{D}, \mathfrak{D})$ decision space
$\mathsf{D} : \mathfrak{D} \times \mathcal{X} \to_k [0, 1]$ decision
$\mathbb{D}$ class of all decisions
Δ parameter space
$\mathfrak{B}_\Delta$ Borel sets in Δ
Δ_r^k p.524
$\Delta_{r,\delta}^k$ p.534
γ_{m}, κ_{m} p.13
$\odot$ group operation p.200
$u_\gamma, v_\gamma, w_\gamma$ p.204
$A\gamma$ p.199
$L : \Delta \times \mathcal{D} \to \mathbb{R}_+$ loss function
l loss function p.107
$B(f, g)$ p.114
$\mathsf{R}(\theta, \mathsf{D})$ risk function
$\mathsf{r}(\rho, \mathsf{D})$ average risk
$\mathsf{r}(\Pi, \mathsf{D})$ Bayes risk
$\mathsf{r}(\Pi, \mathsf{D}|x)$ posterior risk
$r(\Pi, a|x)$ posterior risk at $a \in \mathcal{D}$
$\mathsf{b}_\pi(P_0, P_1)$ p.97
$\mathsf{g}_\alpha(P_0, P_1)$ p.99
$\mathsf{m}_\rho(P_0, P_1)$ p.152
$\mathcal{R}_{0,1}$ p.90
$\hat{\theta}_n$ estimator
$\mathfrak{E}$ the equivariant estimators
$\mathcal{E}$ equivariant estimator
$\mathcal{P}(x)$ Pitman estimator p.224
$T_{nat} : T_{nat}(x) = x$
$\varphi_{T,\alpha}$ level α test
$\varphi_I(T), ..., \varphi_{IV}(T)$ p.409
$\psi_I, ..., \psi_{IV}$ standard Gauss tests
$\psi_{\mathcal{C},\theta_0}$ p.432
H_0, H_A hypotheses
Δ_0, Δ_A decomposition of Δ
$\mathcal{C}(x)$ p.431
$B_{lcb}(x), B_{lci}(x), B_{uci}(x)$ p.431
ψ_i selection probability
φ^{nat} natural selection
$P_{\mathsf{N},cs}(\mu, \sigma^2, \varphi^{nat})$ p.524
$P_{cs}(\theta, \psi)$ p.111
$\gamma_{i,\mathsf{N}}(v)$ p.593

B.2 Common Distributions

Discrete Type

$\mathsf{B}(n,p)$: **binomial** with $n \in \{1,2,...\}$ and $p \in (0,1)$.
$\mathsf{b}_{n,p}(x)$: the probability mass function of $\mathsf{B}(n,p)$,

$$\mathsf{b}_{n,p}(x) = \binom{n}{x} p^x (1-p)^{n-x} I_{\{0,1,\ldots,n\}}(x), \quad x \in \mathbb{N}.$$

$\mathsf{Ge}(p)$: **geometric** with $p \in (0,1]$.
$\mathsf{ge}_p(x)$: the probability mass function of $\mathsf{Ge}(p)$,

$$\mathsf{ge}_p(x) = p(1-p)^x, \quad x \in \mathbb{N}.$$

$\mathsf{Hg}(n,m,v)$: **hypergeometric** with $n,m,v \in \{1,2,...\}$ and $v \leq n+m$.
$\mathsf{hg}_{n,m,v}(x)$: the probability mass function of $\mathsf{Hg}(n,m,v)$,

$$\mathsf{hg}_{n,m,v}(x) = \frac{\binom{n}{x}\binom{m}{v-x}}{\binom{n+m}{v}} I_{[0\vee(v-m),\ n\wedge v]}(x), \quad x \in \mathbb{N}.$$

$\mathsf{M}(n,p)$: **multinomial** with $n \in \{1,2,...\}$ and $p \in \mathbf{S}_d^0$.
$\mathsf{m}_{n,p}(x_1,...,x_d)$: the probability mass function of $\mathsf{M}(n,p)$,

$$\mathsf{m}_{n,p}(x_1,...,x_d) = \frac{n!}{\prod_{i=1}^d x_i!} \prod\nolimits_{i=1}^d p_i^{x_i} I_{\mathbf{S}_{n,d}}(x), \quad x \in \mathbb{N}^d, \quad \text{where}$$

$$\mathbf{S}_{n,d} = \{(x_1,...,x_n) : x_1,...,x_n \in \{0,1,...,n\},\ \sum\nolimits_{i=1}^d x_i = n\}.$$

- $\mathbf{S}_d^0 = \{(p_1,...,p_d) : p_1,...,p_d > 0,\ \sum_{j=1}^d p_j = 1\}$.
- $\mathcal{L}(X_1,...,X_d) = \mathsf{M}(n,p) \Rightarrow \mathcal{L}(X_1) = \mathsf{B}(n,p_1)$.

$\mathsf{Nb}(k,p)$: **negative binomial** with $k \in \{1,2,...\}$ and $p \in (0,1]$.
$\mathsf{nb}_{k,p}(x)$: the probability mass function $\mathsf{Nb}(k,p)$,

$$\mathsf{nb}_{k,p}(x) = \binom{x+k-1}{k-1} p^k (1-p)^x, \quad x \in \mathbb{N}.$$

- $\mathsf{Nb}(1,p) = \mathsf{Ge}(p)$.

$\mathsf{Po}(\lambda)$: **Poisson** with $\lambda > 0$.
$\mathsf{po}_\lambda(x)$: the probability mass function of $\mathsf{Po}(\lambda)$,

$$\mathsf{po}_\lambda(x) = \frac{\lambda^x}{x!} \exp\{-\lambda\}, \quad x \in \mathbb{N}.$$

Continuous Type Standard

$\mathsf{Be}(\alpha,\beta)$: **beta** with $\alpha,\beta>0$.
$\mathsf{Be}_{\alpha,\beta}$: the c.d.f. of $\mathsf{Be}(\alpha,\beta)$.
$\mathsf{be}_{\alpha,\beta}$: the $\boldsymbol{\lambda}$-density of $\mathsf{Be}(\alpha,\beta)$,

$$\mathsf{be}_{\alpha,\beta}(t)=\frac{\Gamma(\alpha+\beta)}{\Gamma(\alpha)\Gamma(\beta)}t^{\alpha-1}(1-t)^{\beta-1}I_{(0,1)}(t),\quad t\in\mathbb{R}.$$

$\mathsf{Ca}(\alpha,\beta)$: **Cauchy** with $\alpha\in\mathbb{R}$ and $\beta>0$.
$\mathsf{Ca}_{\alpha,\beta}$: the c.d.f. of $\mathsf{Ca}(\alpha,\beta)$.
$\mathsf{ca}_{\alpha,\beta}$: the $\boldsymbol{\lambda}$-density of $\mathsf{Ca}(\alpha,\beta)$,

$$\mathsf{ca}_{\alpha,\beta}(t)=\frac{1}{\pi}\frac{\beta}{\beta^2+(x-\alpha)^2},\quad t\in\mathbb{R}.$$

$\mathsf{Di}(\alpha_1,...,\alpha_d)$: **Dirichlet** with $\alpha_i>0$, $i=1,...,d$.
$\mathsf{Di}_{\alpha_1,...,\alpha_d}$: the c.d.f. of $\mathsf{Di}(\alpha_1,...,\alpha_d)$.
$\mathsf{di}_{\alpha_1,...,\alpha_d}$: the $\boldsymbol{\lambda}_{d-1}$-density of $\mathsf{Di}(\alpha_1,...,\alpha_d)$,

$$\mathsf{di}_{\alpha_1,...,\alpha_d}(t_1,...,t_{d-1})=\frac{\Gamma(\sum_{i=1}^d\alpha_i)}{\prod_{i=1}^d\Gamma(\alpha_i)}\prod\nolimits_{i=1}^d t_i^{\alpha_i-1}I_{\mathbb{S}_{d-1}}(t_1,...,t_{d-1}),$$
$$\text{where } t_d=1-\sum\nolimits_{i=1}^{d-1}t_i,\quad (t_1,...,t_{d-1})\in\mathbb{S}_{d-1}.$$

- $\mathbb{S}_{d-1}=\{(t_1,...,t_{d-1}):t_1,...,t_{d-1}>0,\ \sum_{i=1}^{d-1}t_i<1\}$.

$\mathsf{Ex}(\beta)$: **exponential** with $\beta>0$.
Ex_{β} : the c.d.f. of $\mathsf{Ex}(\beta)$.
ex_{β} : the $\boldsymbol{\lambda}$-density of $\mathsf{Ex}(\beta)$,

$$\mathsf{ex}_{\beta}(t)=\beta\exp\{-\beta t\}I_{(0,\infty)}(t),\quad t\in\mathbb{R}.$$

$\mathsf{Ex}(\theta,\beta)$: **shifted exponential** with $\theta\in\mathbb{R}$ and $\beta>0$.
$\mathsf{Ex}_{\theta,\beta}$: the c.d.f. of $\mathsf{Ex}(\theta,\beta)$.
$\mathsf{ex}_{\theta,\beta}$: the $\boldsymbol{\lambda}$-density of $\mathsf{Ex}(\theta,\beta)$,

$$\mathsf{ex}_{\theta,\beta}(t)=\beta\exp\{-\beta(t-\theta)\}I_{(\theta,\infty)}(t),\quad t\in\mathbb{R}.$$

- $\mathsf{Ex}(0,\beta)=\mathsf{Ex}(\beta)$.

$\mathsf{F}(n_1, n_2)$: **F** with $n_1, n_2 \in \{1, 2, ...\}$ degrees of freedom.

F_{n_1,n_2} : the c.d.f. of $\mathsf{F}(n_1, n_2)$.

f_{n_1,n_2} : the $\boldsymbol{\lambda}$-density of $\mathsf{F}(n_1, n_2)$,

$$\mathsf{f}_{n_1,n_2}(t) = \frac{\Gamma(\frac{n_1+n_2}{2})}{\Gamma(\frac{n_1}{2})\Gamma(\frac{n_2}{2})}(\frac{n_1}{n_2})^{n_1/2}\frac{t^{(n_1/2)-1}}{(1+\frac{n_1}{n_2}t)^{(n_1+n_2)/2}}I_{[0,\infty)}(t), \quad t \in \mathbb{R}.$$

- $\mathcal{L}(W) = \mathsf{T}(n) \Rightarrow \mathcal{L}(W^2) = \mathsf{F}(1, n)$.
- Y_1, Y_2 independent, $\mathcal{L}(Y_i) = \mathsf{H}(n_i) \Rightarrow \mathcal{L}(\frac{n_2 Y_1}{n_1 Y_2}) = \mathsf{F}(n_1, n_2)$.

$\mathsf{Ga}(\lambda, \beta)$: **gamma** with $\lambda, \beta > 0$.

$\mathsf{Ga}_{\lambda,\beta}$: the c.d.f. of $\mathsf{Ga}(\lambda, \beta)$.

$\mathsf{ga}_{\lambda,\beta}$: the $\boldsymbol{\lambda}$-density of $\mathsf{Ga}(\lambda, \beta)$,

$$\mathsf{ga}_{\lambda,\beta}(t) = \frac{\beta^\lambda}{\Gamma(\lambda)}t^{\lambda-1}\exp\{-\beta t\}I_{(0,\infty)}(t), \quad t \in \mathbb{R}.$$

- $\mathsf{Ga}(n/2, 1/2) = \mathsf{H}(n)$ and $\mathsf{Ga}(1, \beta) = \mathsf{Ex}(\beta)$.

$\mathsf{Gi}(\lambda, m)$: **inverse Gaussian** with $\lambda > 0$ and $m \in (0, \infty]$.

$\mathsf{Gi}_{\lambda,m}$: the c.d.f. of $\mathsf{Gi}(\lambda, m)$.

$\mathsf{gi}_{\lambda,m}$: the $\boldsymbol{\lambda}$-density of $\mathsf{Gi}(\lambda, m)$,

$$\mathsf{gi}_{\lambda,m}(t) = \sqrt{\frac{\lambda}{2\pi t^3}}\exp\{-\frac{\lambda}{2m^2}\frac{(t-m)^2}{t}\}I_{(0,\infty)}(t), \quad t \in \mathbb{R}.$$

$\mathsf{H}(n)$: **chi-square**, or χ^2, with $n \in \{1, 2, ...\}$ degrees of freedom.

H_n : the c.d.f. of $\mathsf{H}(n)$.

h_n : the $\boldsymbol{\lambda}$-density of $\mathsf{H}(n)$,

$$\mathsf{h}_n(t) = \frac{1}{2^{n/2}\Gamma(\frac{n}{2})}t^{(n/2)-1}\exp\{-\frac{t}{2}\}I_{(0,\infty)}(t), \quad t \in \mathbb{R}.$$

- $X_1, ..., X_n$ i.i.d., $\mathcal{L}(X_i) = \mathsf{N}(0, 1) \Rightarrow \mathcal{L}(\sum_{i=1}^n X_i^2) = \mathsf{H}(n)$.

$\mathsf{Ig}(\lambda, \beta)$: **inverse gamma** with $\lambda, \beta > 0$.

$\mathsf{Ig}_{\lambda,\beta}$: the c.d.f. of $\mathsf{Ig}(\lambda, \beta)$.

$\mathsf{ig}_{\lambda,\beta}$: the $\boldsymbol{\lambda}$-density of $\mathsf{Ig}(\lambda, \beta)$,

$$\mathsf{ig}_{\lambda,\beta}(t) = \frac{\beta^\lambda}{\Gamma(\lambda)}t^{-\lambda-1}\exp\{-\frac{\beta}{t}\}I_{(0,\infty)}(t), \quad t \in \mathbb{R}.$$

- $\mathcal{L}(V) = \mathsf{Ga}(\lambda, \beta) \Rightarrow \mathcal{L}(1/V) = \mathsf{Ig}(\lambda, \beta)$.

$\mathsf{Lp}(\theta)$: **Laplace**, or **double exponential**, with $\theta \in \mathbb{R}$.

Lp_θ : the c.d.f. of $\mathsf{Lp}(\beta)$.

lp_θ : the $\boldsymbol{\lambda}$-density of $\mathsf{Lp}(\theta)$,

$$\mathsf{lp}_\theta(t) = \frac{1}{2}\exp\{-|t-\theta|\}, \quad t \in \mathbb{R}.$$

$\mathsf{Lo}(\theta)$: **logistic** with $\theta \in \mathbb{R}$.

Lo_θ : the c.d.f. of $\mathsf{Lo}(\theta)$.

lo_θ : the $\boldsymbol{\lambda}$-density of $\mathsf{Lo}(\theta)$,

$$\mathsf{lo}_\theta(t) = \frac{\exp\{t-\theta\}}{(1+\exp\{t-\theta\})^2}, \quad t \in \mathbb{R}.$$

$\mathsf{N}(\mu,\sigma^2)$: **normal** with $\mu \in \mathbb{R}$ and $\sigma^2 > 0$.

Φ_{μ,σ^2} : the c.d.f. of $\mathsf{N}(\mu,\sigma^2)$, and especially $\Phi_{0,1} = \Phi$.

φ_{μ,σ^2} : the $\boldsymbol{\lambda}$-density of $\mathsf{N}(\mu,\sigma^2)$, and especially $\varphi_{0,1} = \varphi$,

$$\varphi_{\mu,\sigma^2}(t) = \frac{1}{\sqrt{2\pi\sigma^2}}\exp\{-\frac{1}{2}(\frac{t-\mu}{\sigma})^2\}, \quad t \in \mathbb{R}.$$

$\mathsf{N}(\mu,\Sigma)$: **multivariate normal** on $\mathfrak{B}_n$ with $\mu \in \mathbb{R}^n$ and symmetric positiv definite $n \times n$ matrix Σ.

$\varphi_{\mu,\Sigma}$: the $\boldsymbol{\lambda}_n$-density of $\mathsf{N}(\mu,\Sigma)$ in $\mathbb{R}^n$,

$$\varphi_{\mu,\Sigma}(t) = \frac{1}{(2\pi)^{n/2}\,|\Sigma|^{1/2}}\exp\{-\frac{1}{2}(t-\mu)^T\Sigma^{-1}(t-\mu)\}, \quad t \in \mathbb{R}^n.$$

$\mathsf{T}(n)$: **t** with $n \in \{1,2,...\}$ degrees of freedom.

T_n : the c.d.f. of $\mathsf{T}(n)$.

t_n : the $\boldsymbol{\lambda}$-density of $\mathsf{T}(n)$,

$$\mathsf{t}_n(t) = \frac{\Gamma(\frac{n+1}{2})}{\sqrt{n\pi}\,\Gamma(\frac{n}{2})}(1+\frac{t^2}{n})^{-(n+1)/2}, \quad t \in \mathbb{R}.$$

- X, Y indep., $\mathcal{L}(X) = \mathsf{N}(0,1)$, $\mathcal{L}(Y) = \mathsf{H}(n) \Rightarrow \mathcal{L}(\frac{X}{\sqrt{Y/n}}) = \mathsf{T}(n)$.

$\mathsf{U}(a,b)$: **uniform** with $a < b$.

$\mathsf{U}_{a,b}$: the c.d.f. of $\mathsf{U}(a,b)$.

$\mathsf{u}_{a,b}$: the $\boldsymbol{\lambda}$-density of $\mathsf{U}(a,b)$,

$$\mathsf{u}_{a,b}(t) = \frac{1}{b-a}I_{(a,b)}(t), \quad t \in \mathbb{R}.$$

Continuous Type Noncentral

$\mathsf{F}(n_1, n_2, \delta^2)$: **noncentral F** with $n_1, n_2 \in \{1, 2, ...\}$ degrees of freedom and noncentrality $\delta^2 > 0$.

$\mathsf{F}_{n_1,n_2,\delta^2}$: the c.d.f. of $\mathsf{F}(n_1, n_2, \delta^2)$.

$\mathsf{f}_{n_1,n_2,\delta^2}$: the $\boldsymbol{\lambda}$-density of $\mathsf{F}(n_1, n_2, \delta^2)$,

$$\mathsf{f}_{n_1,n_2,\delta^2}(t) = \sum\nolimits_{k=0}^{\infty} \mathsf{po}_{\delta^2/2}(k)\mathsf{f}_{n_1+2k,n_2}(t), \quad t \in \mathbb{R}.$$

- $\mathsf{F}(n_1, n_2, 0) = \mathsf{F}(n_1, n_2)$.
- $\mathcal{L}(W) = \mathsf{T}(n, \mu) \Rightarrow \mathcal{L}(W^2) = \mathsf{F}(1, n, \mu^2)$.
- Y_1, Y_2 indep., $\mathcal{L}(Y_1) = \mathsf{H}(n_1, \delta^2)$, $\mathcal{L}(Y_2) = \mathsf{H}(n_2, 0) \Rightarrow \mathcal{L}(\frac{n_2 Y_1}{n_1 Y_2}) = \mathsf{F}(n_1, n_2, \delta^2)$.

$\mathsf{H}(n, \delta^2)$: **noncentral chi-square**, or χ^2, with $n \in \{1, 2, ...\}$ degrees of freedom and noncentrality $\delta^2 > 0$.

H_{n,δ^2} : the c.d.f. of $\mathsf{H}(n, \delta^2)$.

h_{n,δ^2} : the $\boldsymbol{\lambda}$-density of $\mathsf{H}(n, \delta^2)$,

$$\mathsf{h}_{n,\delta^2}(t) = \sum\nolimits_{k=0}^{\infty} \mathsf{po}_{\delta^2/2}(k)\mathsf{h}_{n+2k}(t), \quad t \in \mathbb{R}.$$

- $\mathsf{H}(n, 0) = \mathsf{H}(n)$.
- $X_1, ..., X_n$ i.i.d., $\mathcal{L}(X_i) = \mathsf{N}(\mu_i, 1) \Rightarrow \mathcal{L}(\sum_{i=1}^n X_i^2) = \mathsf{H}(n, \sum_{i=1}^n \mu_i^2)$.

$\mathsf{T}(n, \mu)$: **noncentral t** with $n \in \{1, 2, ...\}$ degrees of freedom and noncentrality $\mu \in \mathbb{R}$.

$\mathsf{T}_{n,\mu}$: the c.d.f. of $\mathsf{T}(n, \mu)$.

$\mathsf{t}_{n,\mu}$: the $\boldsymbol{\lambda}$-density of $\mathsf{T}(n, \mu)$,

$$\begin{aligned} \mathsf{t}_{n,\mu}(t) &= \frac{1}{2^{(n+1)/2}\sqrt{n\pi}\Gamma(\frac{n}{2})} \\ &\times \int_0^\infty y^{(n-1)/2} \exp\{-\frac{y}{2}\} \exp\{-\frac{1}{2}(t\sqrt{\frac{y}{n}} - \mu)^2\} dy, \quad t \in \mathbb{R}. \end{aligned}$$

- $\mathsf{T}(n, 0) = \mathsf{T}(n)$.
- X, Y indep., $\mathcal{L}(X) = \mathsf{N}(\mu, 1)$, $\mathcal{L}(Y) = \mathsf{H}(n, 0) \Rightarrow \mathcal{L}(\frac{X}{\sqrt{Y/n}}) = \mathsf{T}(n, \mu)$.

References

Abughalous, M.M. and Bansal, N.K. (1995). On selecting the best natural exponential family with quadratic variance function. *Statist. Probab. Lett.* **25**, 341–349.

Abughalous, M.M. and Miescke, K.-J. (1989). On selecting the largest success probability under unequal sample sizes. *J. Statist. Plann. Inference* **21**, 53–68.

Agresti, A. (1992). A survey of exact inference for contingency tables (with discussion). *Statistical Science* **7**, 131–177.

Agresti, A. (2002). *Categorical Data Analysis.* 2nd ed. Wiley, New York.

Akahira, M. and Takeuchi, K. (1981). *Asymptotic Efficiency of Statistical Estimators.* Springer, New York.

Alam, K. (1970). A two-sample procedure for selecting the population with the largest mean from k normal populations. *Ann. Inst. Statist. Math.* **22**, 127–136.

Ali, M.S. and Silvey, D. (1966). A general class of coefficients of divergence of one distribution from another. *J. Roy. Statist. Soc.* **B-28**, 131–140.

Amari, S. (1985). *Differential Geometrical Methods in Statistics. Lecture Notes in Statistics* **28**, Springer, New York.

Andersen, P.K. and Gill, R.D. (1982). Cox's regression model for counting processes: a large sample study. *Ann. Statist.* **10**, 1100–1120.

Anderson, S.A. (1982). Distribution of maximal invariants using quotient measures. *Ann. Statist.* **10**, 955–961.

Anderson, T.W. (1984). *An Introduction to Multivariate Statistical Analysis.* 2nd edition. Wiley, New York.

Andrews, D.F., Bickel, P.J., Hampel, F.R., Huber, P.J., Rogers, W.H., and Tukey, J.W. (1972). *Robust Estimation of Location. Survey and Advances.* Princeton Univ. Press, Princeton.

Bahadur, R.R. (1950). On a problem in the theory of k populations. *Ann. Math. Statist.* **21**, 362–375.

Bahadur, R.R. (1958). Examples of inconsistency of maximum likelihood estimates. *Sankhyā* **20**, 207–210.

Bahadur, R.R. (1960). On the asymptotic efficiency of tests and estimators. *Sankhyā* **22**, 229–256.

Bahadur. R.R. (1964). On Fisher's bound for asymptotic variances. *Ann. Math. Statist.* **38**, 303–324.

Bahadur. R.R. (1971). *Some Limit Theorems in Statistics. CBS-NSF Regional Conference Series in Applied Math.* **4**, SIAM.

Bahadur, R.R. and Goodman, L. (1952). Impartial decision rules and sufficient statistics. *Ann. Math. Statist.* **23**, 553–562.

Bahadur, R.R. and Robbins, H. (1950). The problem of the greater mean. *Ann. Math. Statist.* **21**, 469–487. Correction (1951) **22**, 310.

Balakrishnan, N. and Miescke, K.-J., eds. (2006). *In Memory of Dr. Shanti Swarup Gupta.* Special Issue. *J. Statist. Plann. Inference* **136**.

Bansal, N.K. and Gupta, S. (1997). On the natural selection rule in general linear models. *Metrika* **46**, 59–69.

Bansal, N.K. and Gupta, S. (2000). On the role of the natural decision rule in ranking and selecting the best component in multivariate populations. *Proc. Ntl. Sem. on Bayesian Statistics and Appl.*, S.K. Upadhyay and U. Singh, eds., Banaras Hindu Univ., Varanasi, 49–57.

Bansal, N.K. and Miescke, K.-J. (2002). Simultaneous selection and estimation in general linear models. *J. Statist. Plann. Inference* **104**, 377–390.

Bansal, N.K. and Miescke, K.-J. (2005). Simultaneous selection and estimation of the best treatment with respect to a control in general linear models. *J. Statist. Plann. Inference* **129**, 387–404.

Bansal, N.K. and Miescke, K.-J. (2006). On selecting the best treatment in a generalized linear model. *J. Statist. Plann. Inference* **136**, 2070–2086.

Bansal, N.K. and Misra, N. (1999). On ranking treatment effects in a general linear model. *J. Statist. Plann. Inference* **78**, 1–11.

Bansal, N.K., Misra, N., and van der Meulen, E.C. (1997). On the minimax decision rules in ranking problems. *Stat. and Prob. Letters* **34**, 179–186.

Baranchik, A.J. (1970). A family of minimax estimators of the mean of a multivariate normal distribution. *Ann Math. Statist.* **41**, 642–645.

Barankin, E.W. (1949). Locally best unbiased estimates. *Ann. Math. Statist.* **20**, 477–501.

Barankin, E.W. (1950). Extension of a theorem of Blackwell. *Ann. Math. Statist.* **21**, 280–284.

Barndorff-Nielsen, O. (1978). *Information and Exponential Families.* Wiley, New York.

Basawa, I.V. and Prakasa Rao, B.L.S. (1980). *Statistical Inference for Stochastic Processes.* Academic Press, New York.

Basawa, I.V. and Scott, D.J. (1983). *Asymptotic Optimal Inference for Non-Ergodic Models.* Springer, New York.

Basu, D. (1955). On statistics independent of a complete sufficient statistic. *Sankhyā* **15**, 377–380.

Basu, D. (1958). On statistics independent of a sufficient statistic. *Sankhyā* **20**, 223–226.

Bauer, H. (2001). *Measure and Integration Theory.* W. de Gruyter, Berlin.

Bechhofer, R.E. (1954). A single-sample multiple decision procedure for ranking means of normal populations with known variances. *Ann. Math. Statist.* **25**, 16–39.

Bechhofer, R.E. and Goldsman, D.M. (1989). Truncation of the Bechhofer-Kiefer-Sobel sequential procedure for selecting the normal population which has the largest mean (III): Supplementary truncation numbers and resulting performance characteristics. *Commun. Stat. - Simul. Comput.* **B-18**, 63–81.

Bechhofer, R.E., Dunnett, C.W., and Sobel, M. (1954). A two-sample multiple decision procedure for ranking means of normal populations with a common unknown variance. *Biometrika* **41**, 170–176.

Bechhofer, R.E., Hayter, A.J., and Tamhane, A.C. (1991). Designing experiments for selecting the largest normal mean when the variances are known and unequal: Optimal sample size allocation. *J. Statist. Plann. Inference* **28**, 271–289.

Bechhofer, R.E., Kiefer, J., and Sobel, M. (1968). *Sequential Identification and Ranking Procedures.* Univ. of Chicago Press, Chicago.

Bechhofer, R.E., Santner, T.J., and Goldsman, D.M. (1995). *Design and Analysis of Experiments for Statistical Selection, Screening, and Multiple Comparisons.* Wiley, New York.

Behnen, K. and Neuhaus, G. (1989). *Rank Tests With Estimated Scores and Their Application.* Teubner, Stuttgart.

Berger, J.O. (1985). *Statistical Decision Theory and Bayesian Analysis.* 2nd edition. Springer, New York.

Berger, J.O. and Deely, J.J. (1988). A Bayesian approach to ranking and selection of related means with alternatives to AOV methodology. *J. Amer. Statist. Assoc.* **83**, 364–373.

Berger, R.L. and Gupta, S.S. (1980). Minimax subset selection rules with applications to unequal variance (unequal sample size) problems. *Scand. J. Stat.* **7**, 21–26.

Berk, R.H. (1972). A note on invariance and sufficiency. *Ann. Math. Stat.* **43**, 647–650.

Berk, R.H. and Bickel, P.J. (1968). On invariance and almost invariance. *Ann. Math. Stat.* **39**, 1573–1576.

Berlinet, A., Liese, F., and Vajda, I. (2000). Necessary and sufficient conditions for consistency of M-estimates in regression models with general errors. *J. Statist. Plann. Inference* **89**, 243–267.

Bernardo, J.M. and Smith, A.F.M. (1994). *Bayesian Theory.* Wiley, New York.

Bernstein, S. (1917). *Theory of Probability.* (In Russion.) 4th edition. Gostekhizdat, Moscow–Leningrad.

Bhandari, S.K. and Chaudhuri, A.R. (1990). On two conjectures about two-stage selection procedures. *Sankhyā* **A-52**, 131–141.

Bhapkar, V.P. and Gore, A.P. (1971). Some selection procedures based on U-statistics for the location and scale problems. *Ann. Inst. Statist. Math.* **23**, 375–386.

Bhattacharya, R.N. and Rao, R.R. (1976). *Normal Approximation and Asymptotic Expansions.* Wiley, New York.

Bhattacharyya, A. (1946). On some analogues to the amount of information and their uses in statistical estimation I. *Sankhyā* **8**, 1–14.

Bhattacharyya, A. (1947a). On some analogues to the amount of information and their uses in statistical estimation II. *Sankhyā* **8**, 201–218.

Bhattacharyya, A. (1947b). On some analogues to the amount of information and their uses in statistical estimation III. *Sankhyā* **8**, 315–328.

Bickel, P.J. and Doksum, K.A. (1977). *Mathematical Statistics.* Prentice Hall, Englewood Cliffs, NJ.

Bickel, P.J. and Yahav, J.A. (1969). Some contributions to the asymptotic theory of Bayes solutions. *Zeitschr. Wahrschth. verw. Geb.* **11**, 257–276.

Bickel, P.J., Klaassen, C.A.J., Ritov, Y., and Wellner, J.A. (1993). *Efficient and Adaptive Estimation for Semiparametric Models.* Johns Hopkins Univ. Press, Baltimore.

Billingsley, P. (1968). *Convergence of Probability Measures.* Wiley, New York.

Birnbaum, A. (1964). Median-unbiased estimators. *Bull. Math. Statist.* **11**, 25–34.

Bischoff, W., Fieger, W., and Wulfert, S. (1995). Minimax- and G-minimax estimation of the bounded normal mean under LINEX-loss. *Statistics and Decisions* **13**, 287–298.

Bjørnstad, J.F. (1981). A decision theoretic approach to subset selection. *Commun. Statist. - Theor. Meth.* **A-10**, 2411–2433.

Bjørnstad, J.F. (1984). A general theory of asymptotic consistency for subset selection with applications. *Ann. Statist.* **12**, 1058–1070.

Bjørnstad, J.F. (1986). Asymptotic consistency for subset selection procedures satisfying the P^*-condition. *J. Statist. Plann. Inference* **13**, 319–335.

Blackwell, D. (1947). Conditional expectation and unbiased sequential estimation. *Ann. Math. Statist.* **18**, 105–110.

Blackwell, D. (1951). On the translation parameter problem for discrete variables. *Ann. Math. Statist.* **22**, 393–399.

Blackwell, D. (1953). Equivalent comparisons of experiments. *Ann. Math. Statist.* **24**, 265–272.

Blackwell, D. and Girshick, M.A. (1954). *Theory of Games and Statistical Decisions.* Wiley, New York.

Blackwell, D. and Ramamoorthi, R.V. (1982). A Bayes but not classically sufficient statistic. *Ann. Statist.* **10**, 1025–1026.

Blum, J.R. and Rosenblatt, J. (1967). On partial a priori information in statistical inference. *Ann. Math. Statist.* **38**, 1671–1678.

Blyth, C.R. (1951). On minimax statistical decision procedures and their admissibility. *Ann. Math. Statist.* **22**, 22–42.

Boekee, B.E. (1978). The D_f information of order s. *Trans. 8th Prague Conf. on Inf. Theory,* Vol. C, 55–65.

Bofinger, E. (1976). Least favorable configuration when ties are broken. *J. Amer. Statist. Assoc.* **71**, 423–424.

Bofinger, E. (1985). Monotonicity of the probability of correct selection or are bigger samples better? *J. Roy. Statist. Soc.* **B-47**, 84–89.

Boll, C. (1955). *Comparison of Experiments in the Infinite Case and the Use of Invariance in Establishing Sufficiency.* Ph.D. thesis, Stanford Univ., Stanford.

Bondar, J.V. and Milnes, P. (1981). Amenability: A survey for statistical applications of Hunt-Stein and related conditions on groups. *Zeitschr. Wahrschth. verw. Geb.* **57**, 103–128.

Borges, R. and Pfanzagl, J. (1963). A characterization of the one parameter exponential family of distributions by monotonicity of likelihood ratios. *Zeitschr. Wahrschth. verw. Geb.* **2**, 11–117.

Brown, G.W. (1947). On small-sample estimation. *Ann. Math. Statist.* **18**, 582–585.

Brown, L.D. (1964). Sufficient statistics in the case of independent random variables. *Ann. Math. Statist.* **35**, 1456–1474.

Brown, L.D. (1971). Admissible estimators, recurrent diffusions, and insoluble boundary value problems. *Ann. Math. Statist.* **42**, 855–904.

Brown, L.D. (1986). *Fundamentals of Statistical Exponential Families. IMS Lecture Notes - Monograph Series* **9**.

Brown, L.D. and Hwang, J.T. (1982). A unified admissibility proof. *Statistical Decision Theory and Related Topics III*, S.S. Gupta and J.O. Berger, eds., Academic Press, New York, Vol. **1**, 205–230.

Büringer, H., Martin, H., and Schriever, K.H. (1980). *Nonparametric Sequential Selection Procedures.* Birkhäuser, Boston.

Čenvoc, N.N. (1982). Statistical Decision Rules and Optimal Inference. *Trans. Math. Monographs* **53**. Amer. Math. Soc., Providence.

Chapman, D.G. and Robbins, H. (1951). Minimum variance estimation without regularity assumptions. *Ann. Math. Statist.* **22**, 581–586.

Chaudhuri, A. and Mukerjee, R. (1988). *Randomized Response.* Dekker, New York.

Chen, M.-H., Shao, Q.-M., and Ibrahim, J.G. (2000). *Monte Carlo Methods in Bayesian Computation.* Springer, New York.

Chernoff, H. (1952). A measure of asymptotic efficiency for tests of a hypothesis based on the sum of observations. *Ann. Math. Statist.* **23**, 493–507.

Chernoff, H. (1956). Large sample theory: "Parametric case". *Ann. Math. Statist.* **27**, 1–22.

Chikara, R.S. and Folks, J.L. (1989). *The Inverse Gaussian Distribution.* Dekker, New York.

Christensen, R. (1987). *Plane Answers to Complex Questions: The Theory of Linear Models.* 2nd edition. Springer, New York.

Cohen, A. and Sackrowitz, H.B. (1988). A decision theory formulation for population selection followed by estimating the mean of the selected population. *Statistical Decision Theory and Related Topics IV.* J.O. Berger and S.S. Gupta, eds., Springer, New York, Vol. **2**, 33–36.

Cohen, D.S. (1959). *A two-sample decision procedure for ranking means of normal populations with a common known variance.* Unpublished M.S. Thesis, Dept. of Operations Research, Cornell Univ., Ithaca, NY.

Cover, T.M. and Thomas, J.A. (1991). *Elements of Information Theory.* Wiley, New York.

Cox, D.R. (1961). Tests of separate families of hypotheses. *Proc. Fourth Berkeley Symp. Math. Statist. Probab.*, J. Neyman, ed., Univ. of California Press, Berkeley, Vol. **1**, 105–123.

Cox, D.R. (1962). Further results on tests of separate families of hypotheses. *J. Roy. Statist. Soc.* **B-24**, 406–423.

Cramér, H. (1946). A contribution to the theory of statistical estimation. *Skand. Aktuar. Tidskrift* **29**, 85–94.

Csiszár, I. (1963). Eine informationstheoretische Ungleichung und ihre Anwendung auf den Beweis der Ergodizität von Markoffschen Ketten. *Publ. Math. Inst. Hungar. Acad. Sci.* **A-8**, 85–108.

Darmois, G. (1945). Sur lois limites de la dispersion de certains estimations. *Rev. Inst. Int. Statist* **13**, 9–15.

Davies, R. (1985). Asymptotic inference when the amount of information is random. *Proc. Neyman Kiefer Conference*, Vol. **II**, 841–864.

Deely, J.J. and Gupta, S.S. (1968). On the properties of subset selection procedures. *Sankhyā* **A-30**, 37–50.

Deely, J.J. and Johnson, W. (1997). Normal means revisited. *Advances in Statistical Decision Theory and Related Topics.* Festschrift in honor of S.S. Gupta. S. Panchapakesan and N. Balakrishnan, eds., Birkhäuser, Boston, 19–30.

DeGroot, M.H. (1970). *Optimal Statistical Decisions.* McGraw-Hill, New York.

Denny, J.L. (1964). On continuous sufficient statistics. *Ann. Math. Statist.* **35**, 1229–1233.

Denny, J.L. (1970). Cauchy's equation and sufficient statistics on arcwise connected spaces. *Ann. Math. Statist.* **41**, 401–411.

Dhariyal, I.D. and Misra, N. (1994). Non-minimaxity of natural decision rules under heteroscedasticity. *Statistics and Decisions* **12**, 79–98.

Diaconis, P. and Freedman, D. (1986a). On the consistency of Bayes estimates. (With a discussion and a rejoinder by the authors.) *Ann. Statist.* **14**, 1–67.

Diaconis, P. and Freedman, D. (1986b). On inconsistent Bayes estimates of location. *Ann. Statist.* **14**, 68–87.

Diaconis, P. and Stein, C. (1983). *Lectures on Statistical Decision Theory.* Unpublished Lecture Notes, Stanford Univ., Stanford.

Diaconis, P. and Ylvisaker, D. (1979). Conjugate priors for exponential families. *Ann. Statist.* **7**, 269–281.

Dieudonné, J. (1960). *Foundations of Modern Analysis.* Academic Press, New York.

Dieudonné, J. (1974). *Elements d'Analyse*, Tome II, Chapitres XII à XV, 2e edition, revue et augmentée, Gauthiers-Villars, Paris.

Doob, J.L. (1949). Applications of the theory of martingales. *Le Calcule des Probabilités et ses Applications* , 23–27. Colloques Internationaux du Centre National de la Recherche Scientifique, Paris.

Droste, W. and Wefelmeyer, W. (1984). On Hájek's convolution theorem. *Statistics and Decisions* **2**, 131–144.

Dudewicz, E.J. (1976). *Introduction to Statistics and Probability.* Holt, Rinehart, and Winston, New York.

Dudewicz, E.J., ed. (1982). *The Frontiers of Modern Statistical Inference Procedures.* Proceedings and Discussions of The IPASRAS Conference 1982. American Sciences Press, Columbus, OH.

Dudewicz, E.J. and Dalal, S.R. (1975). Allocation of observations in ranking and selection with unequal variances. *Sankhyā* **B-37**, 28–78.

Dudewicz, E.J. and Koo, J.O. (1982). *The Complete Categorized Guide to Statistical Selection and Ranking Procedures.* American Sciences Press, Columbus, OH.

Dudley, R.M. (2002). *Real Analysis and Probability.* Cambridge Univ. Press, Cambridge, U.K.

Duistermaat, J.J. and Kolk, J.A.C. (2004). *Multidimensional Real Analysis I: Differentiation.* Cambridge Univ. Press, Cambridge, U.K.

Dunnett, C.W. (1955). A multiple comparison procedure for comparing several treatments with a control. *J. Amer. Statist. Assoc.* **50**, 1096–1121.

Dunnett, C.W. (1960). On selecting the largest of k normal population means (with Discussion). *J. Roy. Statist. Soc.* **B-22**, 1–40.

Dynkin, E.B. (1951). Necessary and sufficient statistics for a family of probability distributions. (In Russian). English Translation in: *Select. Trans. Math. Statist. Prob.* **1**, (1961), 23–41.

Eaton, M.L (1967a). Some optimum properties of ranking procedures. *Ann. Math. Statist.* **38**, 124–137.

Eaton, M.L. (1967b). The generalized variance: testing and ranking problem. *Ann. Math. Statist.* **38**, 941–943.

Ehrman, C.M. and Miescke, K.-J. (1989). Structured decision rules for ranking and selecting mailing lists and creative packages for direct marketing. *J. Direct Marketing* **3**, 47–59.

Ehrman, C.M., Krieger, A., and Miescke, K.-J. (1987). Subset selection toward optimizing the best performance at a second stage. *J. Business and Economic Statist.* **5**, 295–303.

Ellis, R.S. (1985). *Entropy, Large Deviations, and Statistical Mechanics.* Springer, New York.

Fabian, V. and Hannan, J. (1977). On the Cramér-Rao inequality. *Ann. Statist.* **5**, 197–205.

Falk, M. and Liese, F. (1998). LAN of thinned empirical processes with an application to fuzzy set density estimation. *Extremes* **1**, 323–349.

Farrell, R.H. (1968). Towards a theory of generalized Bayes tests. *Ann. Math. Statist.* **38**, 1–22.

Feller, W. (1971). *An Introduction to Probability Theory and Its Applications.* Volume II. 2nd edition. Wiley, New York.

Ferguson, T.S. (1967). *Mathematical Statistics: A Decision Theoretic Approach.* Academic Press, New York.

Ferguson, T.S. (1986). *A Course in Large Sample Theory.* Chapman & Hall, London.

Finner, H. and Giani, G. (1994). Closed subset selection procedures for selecting good populations. *J. Statist. Plann. Inference* **38**, 179–200.

Finner, H. and Giani, G. (1996). Duality between multiple testing and selection. *J. Statist. Plann. Inference* **54**, 201–227.

Finner, H. and Strassburger, K. (2002a). The partition principle: A powerful tool in multiple decision theory. *Ann. Statist.* **30**, 1194–1213.

Finner, H. and Strassburger, K. (2002b). Structural properties of UMPU-tests for 2×2 tables and some applications. *J. Statist. Plann. Inference* **104**, 103–120.

Finner, H., Giani, G., and Strassburger, K. (2006). Partitioning principle and selection of good treatments. *J. Statist. Plann. Inference* **136**, 2053–2069.

Fisher, R.A. (1920). A mathematical examination of the methods of determining the accuracy of an observation by the mean error, and by the mean square error. *Monthly Not. Roy. Astr. Soc.* **80**, 758–770.

Fisher, R.A. (1922). On the mathematical foundation of theoretical statistics. *Philos. Trans. Roy. Soc. London* **A-222**, 309–368.

Fisher, R.A. (1934). Two new properties of mathematical likelihood. *Proc. Royal Soc.* **A-144**, 285–307.

Fréchet, M. (1943). Sur l'extension de certaines evaluations statistique de petit-sechantillons. *Rev. Int. Statist.* **11**, 182–205.

Ghosh, J.K. and Ramamoorthi, R.V. (2003). *Bayesian Nonparametrics.* Springer, New York.

Ghosh, J.K, Sinha, B.K., and Joshi, S.N. (1982). Expansions for posterior probability and integrated Bayes risk. *Statistical Decision Theory and Related Topics III*, S.S. Gupta and J.O. Berger, eds., Academic Press, New York, Vol. **1**, 403–456.

Giani G. and Strassburger, K. (1994). Testing and selecting for equivalence with respect to a control. *J. Amer. Statist. Assoc.* **89**, 320–329.

Giani G. and Strassburger, K. (1997). Optimum partition procedures for separating good and bad treatments. *J. Amer. Statist. Assoc.* **92**, 291–298.

Giani G. and Strassburger, K. (2000). Multiple comparison procedures for optimally discriminating between good, equivalent, and bad treatments with respect to a control. *J. Statist. Plann. Inference* **83**, 413–440.

Gibbons, J.D., Olkin, I., and Sobel, M. (1977). *Selecting and Ordering Populations: A New Statistical Methodology.* Wiley, New York. Second (corrected) printing 1999 by SIAM: *Classics in Appl. Math.* **26**.

Giri, N.C. (1977). *Multivariate Statistical Inference.* Academic Press, New York.

Giri, N.C. (1996). *Group Invariance in Statistical Inference.* World Scientific Press, Singapore.

Giri, N.C. and Das, M.N. (1986). *Design and Analysis of Experiments.* Wiley, New York.

Girshick, M.A. and Savage, L.J. (1951). Bayes and minimax estimates for quadratic loss functions. *Proc. Second Berkeley Symp. Math. Statist. Probab.*, J. Neyman, ed., Univ. of California Press, Berkeley, Vol. **1**, 53–73.

Godambe, V.P. (1991). *Estimating Functions.* Clarendon Press, UK.

Goel, P.K. and DeGroot, M.H. (1981). Information about hyperparameters in hierarchical models. *J. Amer. Statist. Assoc.* **76**, 140–147.

Goel, P.K. and Rubin, H. (1977). On selecting a subset containing the best population - A Bayesian approach. *Ann. Statist.* **5**, 969–983.

Greenwood, P. and Shiryaev, A.N. (1985). *Contiguity and the Statistical Invariance Principle.* Gordon and Breach, New York.

Gupta, S.S. (1956). On a decision rule for a problem in ranking means. Ph.D. Thesis, *Mimeo. Ser. 150,* Institute of Statistics, Univ. of North Carolina, Chapel Hill.

Gupta, S.S. (1965). On some multiple decision (selection and ranking) rules. *Technometrics* **7**, 225–245.

Gupta, S.S., ed. (1977). Special Issue on Selection and Ranking Problems. *Commun. Statist. - Theor. Meth.* **A-6**, No. 1.

Gupta, S.S. and Berger, J.O., eds. (1982). *Statistical Decision Theory and Related Topics III*, Vol. **1** and **2**. Proceedings of the Third Purdue Symposium 1981. Academic Press, New York.

Gupta, S.S. and Berger, J.O., eds. (1988). *Statistical Decision Theory and Related Topics IV*, Vol. **1** and **2**. Proceedings of the Fourth Purdue Symposium 1986. Springer, New York.

Gupta, S.S. and Huang, D.-Y. (1981). *Multiple Statistical Decision Theory: Recent Developments. Lecture Notes in Statistics* **6**, Springer, New York.

Gupta, S.S. and Li, J. (2005). On empirical Bayes procedures for selecting good populations in a positive exponential family. *J. Statist. Plann. Inference* **129**, 3–18.

Gupta, S.S. and Liang, T. (1999a). On empirical Bayes simultaneous selection procedures for comparing normal populations with a standard. *J. Statist. Plann. Inference* **77**, 73–88.

Gupta, S.S. and Liang, T. (1999b). Selecting good exponential populations compared with a control: a nonparametric Bayes approach. *Sankhyā* **B-61**, 289–304.

Gupta, S.S. and Liese, F. (2000). Asymptotic distribution of the conditional regret risk for selecting good exponential populations. *Kybernetika* **36**, 571–588.

Gupta, S.S. and Miescke, K.-J. (1981). Optimality of subset selection procedures for ranking means of three normal populations. *Sankhyā* **B-43**, 1–17.

Gupta, S.S. and Miescke, K.-J. (1982a). On the least favorable configurations in certain two-stage selection procedures. *Essays in Probability and Statistics.* In honor of C.R. Rao. G. Kallianpur, P.R. Krishnaiah, and J.K. Ghosh, eds., North-Holland, Amsterdam, 295–305.

Gupta, S.S. and Miescke, K.-J. (1982b). On the problem of finding the best population with respect to a control in two stages. *Statistical Decision Theory and Related Topics III.* J.O. Berger and S.S. Gupta, eds., Academic Press, New York, Vol. **1**, 473–496.

Gupta, S.S. and Miescke, K.-J. (1983). An essentially complete class of two-stage selection procedures with screening at the first stage. *Statistics and Decisions* **1**, 427–439.

Gupta, S.S. and Miescke, K.-J. (1984a). Sequential selection procedures: A decision theoretic approach. *Ann. Statist.* **12**, 336–350.

Gupta, S.S. and Miescke, K.-J. (1984b). On two-stage Bayes selection procedures. *Sankhyā* **B-46**, 123–134.

Gupta, S.S. and Miescke, K.-J. (1985). Minimax multiple t-tests for comparing k normal populations with a control. *J. Statist. Plann. Inference* **12**, 161–169.

Gupta, S.S. and Miescke, K.-J. (1987). Optimum two-stage selection procedures for Weibull populations. *J. Statist. Plann. Inference* **15**, 147–156.

Gupta, S.S. and Miescke, K.-J. (1988). On the problem of finding the largest normal mean under heteroscedasticity. *Statistical Decision Theory and Related Topics IV.* J.O. Berger and S.S. Gupta, eds., Springer, New York, Vol. **2**, 37–49.

Gupta, S.S. and Miescke, K.-J. (1989). On selecting the best of k lognormal populations. *Metrika* **36**, 233–247.

Gupta, S.S. and Miescke, K.-J. (1990). On finding the largest normal mean and estimating the selected mean. *Sankhyā* **B-52**, 144–157.

Gupta, S.S. and Miescke, K.-J. (1993). On combining selection and estimation in the search for the largest binomial parameter. *J. Statist. Plann. Inference* **36**, 129–140.

Gupta, S.S. and Miescke, K.-J. (1994). Bayesian look-ahead one-stage sampling allocations for selecting the largest normal mean. *Statistical Papers* **35**, 169–177.

Gupta, S.S. and Miescke, K.-J. (1996a). Bayesian look-ahead one-stage sampling allocations for selection of the best population. *J. Statist. Plann. Inference* **54**, 229–244.

Gupta, S.S. and Miescke, K.-J. (1996b). Bayesian look-ahead sampling allocations for selecting the best Bernoulli population. *Research Developments in Probability and Statistics.* Festschrift in honor of M.L. Puri. E. Brunner and M. Denker, eds., VSP International Publishers, Zeist, The Netherlands, 353–369.

Gupta, S.S. and Miescke, K.-J. (2002). On the performance of subset selection procedures under normality. *J. Statist. Plann. Inference* **103**, 101–115.

Gupta, S.S. and Moore, D.S., eds. (1977). *Statistical Decision Theory and Related Topics II.* Proceedings of the Second Purdue Symposium 1976. Academic Press, New York.

Gupta, S.S. and Panchapakesan, S. (1979). *Multiple Decision Procedures: Theory and Methodology of Selecting and Ranking Populations.* Wiley, New York.

Gupta, S.S. and Yackel, J., eds. (1971). *Statistical Decision Theory and Related Topics.* Proceedings of the First Purdue Symposium 1971. Academic Press, New York.

Gutiérrez-Peña, E. and Smith, A.F.M. (1997). Exponential and Bayesian conjugate families: Review and extensions (with discussion). *Test* **6**, 1–90.

Haberman, S.J. (1989). Concavity and estimation. *Ann. Statist.* **17**, 1631–1661.

Hájek, J. (1962). Asymptotically most powerful rank order tests. *Ann. Math. Statist.* **33**, 1124–1147.

Hájek, J. (1968). Asymptotic normality of simple linear rank statistics under alternatives. *Ann. Math. Statist.* **39**, 325–346.

Hájek, J. (1970). A characterization of limiting distributions of regular estimates. *Zeitschr. Wahrschth. verw. Geb.* **14**, 323–330.

Hájek, J. (1972). Local asymptotic minimax and admissibility in estimation. *Proc. Sixth Berkeley Symp. Math. Statist. Probab.*, L. LeCam, J. Neyman, and E.L. Scott, eds., Univ. of California Press, Berkeley, Vol. **1**, 175–194.

Hájek, J. and Šidák, Z. (1967). *Theory of Rank Tests.* Academia, Publishing House of the Czechoslovak Academy of Sciences, Prague.

Hájek, J., Šidák, Z., and Sen, P.K. (1999). *Theory of Rank Tests.* Academic Press, New York.

Hall, W.J. (1959). The most economical character of Bechhofer and Sobel decision rules. *Ann. Math. Statist.* **30**, 964–969.

Halmos, P.R. (1946). The theory of unbiased estimation. *Ann. Math. Stat.* **17**, 34–43.

Halmos, P.R. (1974). *Measure Theory.* Springer, New York.

Halmos, P.R. and Savage, L.J. (1949). Application of the Radon-Nikodym theorem to the theory of sufficient statistics. *Ann. Math. Statist.* **20**, 225–241.

Hampel, F.R., Ronchetti, E.M., Rousseeuw, P.J., and Stahel, W.A. (1986). *Robust Statistics: The Approach Based on Influence Functions.* Wiley, New York.

Hayter, A.J. (1994). On the selection probabilities of two-stage decision procedures. *J. Statist. Plann. Inference* **38**, 223–236.

Herrendörfer, G. and Miescke, K.-J., eds. (1993). *Selection Procedures I.* Proceedings of the Third Schwerin Conference on Mathematical Statistics, Bad Doberan 1993. *FBN Schriftenreihe* **1**, Forschungsinstitut für die Biologie Landwirtschaftlicher Nutztiere, Dummerstorf.

Herrendörfer, G. and Miescke, K.-J., eds. (1994). *Selection Procedures II.* Proceedings of the Third Schwerin Conference on Mathematical Statistics, Bad Doberan 1993. *FBN Schriftenreihe* **2**, Forschungsinstitut für die Biologie Landwirtschaftlicher Nutztiere, Dummerstorf.

Hewitt, H. and Stromberg, K. (1965). *Real and Abstract Analysis.* Springer, New York.

Heyer, H. (1982). *Theory of Statistical Experiments.* Springer, New York.

Hipp, C. (1974). Sufficient statistics and exponential families. *Ann. Statist.* **2**, 1283–1292.

Hjort, N.L. and Pollard, D. (1993). Asymptotic for minimisers of convex processes, *Preprint*, Dept. of Statistics, Yale Univ.

Hodges, J.L. and Lehmann, E.L. (1950). Some problems in minimax point estimation. *Ann. Math. Statist.* **21**, 182–197.

Hoeffding, W. (1948). A class of statistics with asymptotically normal distribution. *Ann. Math. Statist.* **19**, 293–325.

Hoeffding, W. (1963). Probability inequalities for sums of boundend random variables. *J. Amer. Statist. Assoc.* **58**, 13–30.

Hoffmann, K. (1992). *Improved Estimation of Distribution Parameters: Stein-Type Estimators.* Teubner, Leipzig.

Hoffmann-Jørgensen, J. (1994). *Probability with a View Toward Statistics, I and II.* Chapman and Hall, New York.

Hollander, M., Proschan, F., and Sethuraman, J. (1977). Functions decreasing in transposition and their applications in ranking problems. *Ann. Stat.* **5**, 722–733.

Höpfner, R., Jacod, J., and Ladell, L. (1990). Local asymptotic normality and mixed normality for Markov statistical models. *Prob. Theory Rel. Fields* **86**, 105–129.

Hoppe, F.M., ed. (1993). *Multiple Comparisons, Selection, and Applications in Biometry.* Festschrift in Honor of Charles W. Dunnett. Dekker, New York.

Horn, M. and Volland, R. (1995). *Multiple Tests und Auswahlverfahren.* Biometrie, Gustav Fischer, Stuttgart.

Hsu, J.C. (1981). Simultaneous confidence intervals for all distances from the "best". *Ann. Statist.* **9**, 1026–1034.

Hsu, J.C. (1982). Simultaneous inference with respect to the best treatment in block designs. *J. Amer. Statist. Assoc.* **77**, 461–467.

Hsu, J.C. (1996). *Multiple Comparisons: Theory and Methods.* Chapman and Hall, New York.

Hsu, J.C. and Edwards, D.G. (1983). Sequential multiple comparisons with the best. *J. Amer. Statist. Assoc.* **78**, 958–964.

Huang, D.-Y., Panchapakesan, S., and Tseng, S.-T. (1984). Some locally optimal subset selection rules for comparison with a control. *J. Statist. Plann. Inference* **9**, 63–72.

Huang, W.-T. and Chang, Y.-P. (2006). Some empirical Bayes rules for selecting the best population with multiple criteria. *J. Statist. Plann. Inference* **136**, 2129–2143.

Huber, P.J. (1967). The behavior of the maximum likelihood estimator under nonstandard conditions. *Proc. Fifth Berkeley Symp. Math. Statist. Probab.*, J. Neyman and E.L. Scott, eds., Univ. of California Press, Berkeley, Vol. **1**, 221–233.

Huber, P.J. (1981). *Robust Statistics.* Wiley, New York.

Hunt, G. and Stein, C. (1946). Most stringent tests of statistical hypotheses. Unpublished.

Ibragimov, I.A. (1956). On the composition of unimodal distributions. *Theory Prob. Appl.* **1**, 255–260.

Ibragimov, I.A. (1956). On the composition of unimodal distributions (In Russian). *Teoriya Veroyatnostey* **1**, 283–288.

Ibragimov, I.A. and Has'minskii, R.Z. (1972). Asymptotic behavior of statistical estimators. II. Limit theorems for the a posteriori density and Bayes' estimators. *Theory Prob. Applic.* **18**, 76–91.

Ibragimov, I.A. and Has'minskii, R.Z. (1981). *Statistical Estimation: Asymptotic Theory.* Springer, New York.

Inagaki, N. (1970). On the limiting distribution of a sequence of estimators with uniformity property. *Ann. Inst. Statist. Math.* **22**, 1–13.

Jacod, J. and Shiryaev, A.N. (1987). *Limit Theorems for Stochastic Processes.* Springer, New York.

Jacod, J. and Shiryaev, A.N. (2002). *Limit Theorems for Stochastic Processes.* 2nd edition. Springer, New York.

James, W. and Stein, C. (1960). Estimation with quadratic loss. *Proc. Fourth Berkeley Symp. Math. Statist. Probab.*, J. Neyman, ed., Univ. of California Press, Berkeley, Vol. **1**, 361–380.

Janssen, A. (1986). Asymptotic properties of Neyman-Pearson tests for infinite Kullback-Leibler information. *Ann. Math. Stat.* **14**, 1068–1079.

Janssen, A. (1998). *Zur Asymptotik nichtparametrischer Tests.* Lecture Notes. *Skripten zur Stochastik* **29**. Gesellschaft zur Förderung der Mathematischen Statistik, Münster.

Janssen, A. (1999a). Nonparametric symmetry tests for statistical functionals. *Math. Meth. Stat.* **8**, 320–343.

Janssen, A. (1999b). Testing nonparametric statistical functionals with application to rank tests. *J. Statist. Plann. Inference* **81**, 71–93. Erratum (2001) **92**, 297.

Janssen, A. (2000). Nonparametric bioequivalence for tests for statistical functionals and their efficient power functions. *Statistics and Decisions* **18**, 49–78.

Janssen, A. (2004). Asymptotic relative efficiency of tests at the boundary of regular statistical models. *J. Statist. Plann. Inference* **126**, 461–477.

Janssen, A. and Mason, D.M. (1990). *Non-Standard Rank Tests. Lectures Notes in Statistics* **65**, Springer, New York.

Janssen, A., Milbrodt, H., and Strasser, H. (1985). *Infinitely Divisible Statistical Experiments. Lecture Notes in Statistics* **27**, Springer, New York.

Jeganathan, P. (1980a). *Asymptotic theory of estimation when the limit of the log-likelihood ratios is mixed normal.* Ph.D. Thesis, Indian Statistical Institute.

Jeganathan, P. (1980b). An extension of a result of LeCam concerning asymptotic normality. *Sankhyā* **A-42**, 146–160.

Johansen, S. (1979). *Introduction to the Theory of Regular Exponential Families. Lecture Notes* **3**, Institute of Mathematical Statistics, Univ. of Copenhagen.

Johnson, N.L. and Kotz, S. (1969). *Distributions in Statistics: Discrete Distributions.* Houghton Mifflin, Boston.

Johnson, N.L. and Kotz, S. (1970). *Distributions in Statistics: Continuous Univariate Distributions - 1 and 2.* Houghton Mifflin, Boston.

Jongbloed, G. (2000). Minimax lower bounds and moduli of continuity. *Stat. and Prob. Letters* **50**, 279–284.

Joshi, V.M. (1976). On the attainment of the Cramér-Rao lower bound. *Ann. Statist.* **4**, 998–1002.

Jurečková, J. (1977). Asymptotic relations of M-estimates and R-estimates in linear regression models. *Ann. Statist.* **5**, 464–472.

Jurečková, J. (1992). Estimation in a linear model based on regression rank scores. *Nonparametric Statistics* **1**, 197–203.

Jurečková, J. and Sen, P.K. (1996). *Robust Statistical Procedures: Asymptotics and Interrelations.* Wiley, New York.

Kailath, T. (1967). The divergence and Bhattacharyya distance in signal selection. *Trans. IEEE* COM-**15**, 52–60.

Kakutani, S. (1948). On equivalence of infinite product measures. *Ann. Math.* **49**, 214–224.

Kallenberg, O. (1997). *Foundations of Modern Probability.* Springer, New York.

Kallenberg, O. (2002). *Foundations of Modern Probability.* 2nd edition, Springer, New York.

Karatzas, I. and Shreve, S.S. (1988). *Brownian Motion and Stochastic Calculus.* Springer, New York.

Karlin, S. (1957). Pólya type distributions, II. *Ann. Math. Stat.* **28**, 281–308.

Karlin, S. and Rubin, H. (1956). The theory of decision procedures for distributions with monotone likelihood ratio. *Ann. Math. Stat.* **27**, 272–299.

Kerstan, J. and Matthes, K. (1968). Gleichverteilungseigenschaften von Faltungen von Verteilungsgesetzen auf lokal-kompakten Abelschen Gruppen I. *Math. Nachr.* **37**, 267–312.

Kerstan, J. and Matthes, K. (1969). Gleichverteilungseigenschaften von Faltungen von Verteilungsgesetzen auf lokal-kompakten Abelschen Gruppen II. *Math. Nachr.* **41**, 121–132.

Kester, A.D.M. (1987). Some large deviation results in statistics. *J. Amer. Statist. Assoc.* **82**, 343–344.

Kester, A.D.M. and Kallenberg, W.C.M. (1986). Large deviations of estimators. *Ann. Statist.* **14**, 648–664.

Kim, W.-C. (1986). A lower confidence bound on the probability of a correct selection. *J. Amer. Statist. Assoc.* **81**, 1012–1017.

Koenker, R. and Bassat, G. (1978). Regression quantiles. *Econometrika* **46**, 33–50.

Kolmogorov, A.N. (1950). *Foundation of the Theory of Probability.* Chelsea Press, New York.

Kolomiets, E.I. (1987). On asymptotical behavior of probabilities of the second type error for Neyman-Pearson test. *Theory Prob. Appl.* **32**, 503–522.

Koopman, L.H. (1936). On distributions admitting a sufficient statistic. *Trans. Am. Math. Soc.* **39**, 399–409.

Koroljuk, V.S. and Borovskich, Yu.V. (1994). *Theory of U-Statistics.* Kluwer, Dordrecht.

Kozek, A. (1977a). Efficiency and Cramér-Rao type inequalities for convex loss functions. *J. Multiv. Anal.* **7**, 89–106.

Kozek, A. (1977b). On the theory of estimation with convex loss functions. *Proc. Hon. J. Neyman*, R. Bartoszynski, E. Fidelis, and W. Klonecki, eds., PWN-Polish Scientific Publishers, Warszawa, 177–202.

Krafft, O. and Plachky, D. (1970). Bounds for the power of likelihood ratio test and their asymptotic properties. *Ann. Math. Statist.* **41**, 1646–1654.

Krafft, O. and Puri, M.L. (1974). The asymptotic behaviour of the minimax risk for multiple decision problems. *Sankhyā* **36**, 1–12.

Kraft, C. (1955). Some conditions for consistency and uniform consistency of statistical procedures. *Univ. Calif. Publ.* **1**, 125–142.

Küchler, U. and Sørensen, M. (1997). *Exponential Families of Stochastic Processes.* Springer, New York.

Kudo, A. (1959). The classificatory problem viewed as a two-decision problem. *Mem. Fac. Sci. Kyushu Univ.* **A-13**, 96–125.

Kühn, T. and Liese, F. (1978). A short proof of the Hájek-Feldmann theorem. *Teoriya Veroyatnostey* **23**, 449–450.

Kullback, S. (1959). *Information Theory and Statistics.* Wiley, New York.

Kullback, S. and Leibler, R. (1951). On information and sufficiency. *Ann. Math. Stat.* **22**, 79–86.

Kutoyanc, J.A. (2004). *Statistical Inference for Ergodic Diffusion Processes.* Springer, New York.

Lam, K. and Chiu, W.K. (1976). On the probability of correctly selecting the best of several normal populations. *Biometrika* **63**, 410–411.

Landers, D. and Rogge, L. (1973). On sufficiency and invariance. *Ann. Statist.* **1**, 543–544.

Lang, R. (1986). A note on measurability of convex sets. *Arch. Math.* **47**, 90–92.

Laplace, P.S. (1774). Mémoire sur les probabilités des causes par les événements. *Mémoires de l'Académie royale des Sciences de Paris* (Savants étrangers) **6**, 621–656. (*Œuvres Complètes*, Vol. **8** (1891), 27–65, Paris).

Laplace, P.S. de (1820). *Théorie analytique des probabilités.* 3rd edition. Courcier, Paris.

LeCam, L. (1953) On some asymptotic properties of maximum likelihood estimates and related Bayes estimates. *Univ. Calif. Publ. in Statist.* **1**, 277–330.

LeCam, L. (1955). An extension of Wald's theory of statistical decision functions. *Ann. Math. Stat.* **26**, 69–81.

LeCam, L. (1956). On the asymptotic theory of estimation and testing hypotheses. *Proc. Third Berkeley Symp. Math. Statist. Probab.*, J. Neyman, ed., Univ. of California Press, Berkeley, Vol. **1**. 129–156.

LeCam, L. (1958). Les propriétés asymptotiques des solutions de Bayes. *Publ. Inst. Statist. l'Univ. Paris VII*, Fasc. 3–4, 17–35.

LeCam, L. (1960). Locally asymptotically normal families of distributions. *Univ. Calif. Publ. Statist.* **3**, 27–98.

LeCam, L. (1964). Sufficiency and approximate sufficiency. *Ann. Math. Stat.* **35**, 1419–1455.

LeCam, L. (1969). *Theorie Asymptotique de la Decision Statistique.* Univ. of Montral Press, Montreal.

LeCam, L. (1970). On the assumption used to prove asymptotic normality of maximum likelihood estimators. *Ann. Math. Statist.* **41**, 802–828.

LeCam, L. (1972). Limit of experiments. *Proc. Fourth Berkeley Symp. Math. Statist. Probab.*, J. Neyman, ed., Univ. of California Press, Berkeley, Vol. **1**, 245–261.

LeCam, L. (1974). On the information contained in additional observations. *Ann. Statist.* **2**, 630–649.

LeCam, L. (1979). On a theorem of J. Hájek. *Contributions to Statistics - Hájek Memorial Volume*, J. Jurečková, ed., Reidel, Dordrecht, 119–135.

LeCam, L. (1986). *Asymptotic Methods in Statistical Decision Theory.* Springer, New York.

LeCam, L. (1990). Maximum likelihood: An introduction. *Int. Statist. Rev.* **58**, 153–171.

LeCam, L. and Yang, G.L. (1990). *Asymptotics in Statistics: Some Basic Concepts.* Springer, New York.

LeCam, L. and Yang, G.L. (2000). *Asymptotics in Statistics: Some Basic Concepts.* 2nd edition. Springer, New York.

Lehmann, E.L. (1951). A general concept of unbiasedness. *Ann. Math. Statist.* **22**, 587–592.

Lehmann, L.E. (1957a). A theory of some multiple decision problems, I. *Ann. Math. Statist.* **28**, 1–25.

Lehmann, L.E. (1957b). A theory of some multiple decision problems, II. *Ann. Math. Statist.* **28**, 547–572.

Lehmann, E.L. (1959). *Testing Statistical Hypotheses.* Wiley, New York.

Lehmann, E.L. (1961). Some model I problems of selection. *Ann. Math. Statist.* **32**, 990–1012.

Lehmann, E.L. (1963). A class of selection procedures based on ranks. *Math. Ann.* **150**, 268–275.

Lehmann, E.L. (1966). On a theorem of Bahadur and Goodman. *Ann. Math. Statist.* **37**, 1–6.

Lehmann, E.L. (1983). *Theory of Point Estimation.* Wiley, New York.

Lehmann, E.L. (1986). *Testing Statistical Hypotheses.* 2nd edition. Wiley, New York.

Lehmann, E.L. (1988). Comparing location experiments. *Ann. Statist.* **16**, 521–533.

Lehmann, E.L. (1998). *Elements of Large-Sample Theory.* Springer, New York.

Lehmann, E.L. and Casella, G. (1998). *Theory of Point Estimation.* 2nd edition. Springer, New York.

Lehmann, E.L. and Romano, J.P. (2005). *Testing Statistical Hypotheses.* 3rd edition. Springer, New York.

Lehmann, E.L. and Scheffé, H. (1947). On the problem of similar regions. *Proc. Nat. Acad. Sci. USA* **33**, 382–386.

Lehmann, E.L. and Scheffé, H. (1950). Completeness, similar regions, and unbiased estimation: Part I. *Sankhyā* **10**, 305–340.

Lehmann, E.L. and Scheffé, H. (1955). Completeness, similar regions, and unbiased estimation: Part II. *Sankhyā* **15**, 219–236. Correction (1956) **17**, 250.

Leu, C.-S. and Levin, B. (1999a). On the probability of correct selection in the Levin-Robbins sequential elimination procedure. *Statistica Sinica* **9**, 879–891.

Leu, C.-S. and Levin, B. (1999b). Proof of a lower bound formula for the expected reward in the Levin-Robbins sequential elimination procedure. *Sequential Analysis* **18**, 81–105.

Leu, C.-S. and Levin, B. (2007). A generalization of the Levin-Robbins procedure for binomial subset selection and recruitment problems. *Statistica Sinica* (in press).

Levin, B. and Leu, C.-S. (2007). A comparison of two procedures to select the best binomial population with sequential elimination of inferior populations. *J. Statist. Plann. Inference* **137**, 245–263.

Levin, B. and Robbins, H. (1981). Selecting the highest probability in binomial or multinomial trials. *Proc. Natl. Acad. Sci. USA* **78**, 4663–4666.

Li, J., Gupta, S.S., and Liese, F. (2005). Convergence rates of empirical Bayes estimation in exponential family. *J. Statist. Plann. Inference* **131**, 101–115.

Liang, T. (1997). Simultaneously selecting normal populations close to a control. *J. Statist. Plann. Inference* **61**, 297–316.

Liang, T. (2006). Simultaneous inspection of variable equivalence for finite populations. *J. Statist. Plann. Inference* **136**, 2112–2128.

Liese, F. (1975). On the existence of f-projections. *Colloq. Math. Soc. J. Bolyai* **16**, 431–446.

Liese, F. (1982). Hellinger integrals of Gaussian processes with independent increments. *Stochastics* **6**, 81–96.

Liese, F. (1986). Hellinger integrals of diffusion processes. *Statistics* **17**, 63–78.

Liese, F. (1987a). Hellinger integrals, contiguity, and entire separation. *Kybernetika* **23**, 104–123.

Liese, F. (1987b). Estimates of Hellinger integrals of infinitely divisible distributions. *Kybernetika* **23**, 227–238.

Liese, F. (1988). A Rao-Cramér type inequality for a convex loss function. *Trans. Tenth Prague Conf. on Inf. Theory, Prague 1986*, 121–128.

Liese, F. (1996). Adaptive selection of the best population. *J. Statist. Plann. Inference* **54**, 245–269.

Liese, F. (2006). Selection procedures for sparse data. *J. Statist. Plann. Inference* **136**, 2035–2052.

Liese, F. and Lorz, U. (1999). Contiguity and LAN-property of sequences of Poisson processes. *Kybernetika* **35**, 281–308.

Liese, F. and Miescke, K.-J. (1999a). Exponential rates for the error probabilities in selection procedures. *Kybernetika* **35**, 309–332.

Liese, F. and Miescke, K.-J. (1999b). Selection of the best population in models with nuisance parameters. *Math. Meth. of Stat.* **8**, 371–396.

Liese, F. and Vajda, I. (1987). *Convex Statistical Distances.* Teubner, Leipzig.

Liese, F. and Vajda, I. (1994). Consistency of M-estimates in general regression models. *J. Multiv. Anal.* **50**, 93–114.

Liese, F. and Vajda, I. (1995). Necessary and sufficient conditions for consistency of generalized M-estimates. *Metrika* **42**, 291–324.

Liese, F. and Vajda, I. (1999). M-Estimators of structural parameters in pseudolinear models. *Applicat. of Math.* **44**, 245–270.

Liese, F. and Vajda, I. (2003a). On $\sqrt{n}$-consistency and asymptotic normality of consistent estimators in models with independent observations. *Rostocker Math. Kolloqu.* **57**, 3–51.

Liese, F. and Vajda, I. (2003b). A general asymptotic theory of M-estimators. Part I. *Mathem. Meth. of Stat.* **12**, 454–477.

Liese, F. and Vajda, I. (2004). A general asymptotic theory of M-estimators. Part II. *Mathem. Meth. of Stat.* **13**, 82–95.

Liese, F. and Vajda, I. (2006). On divergences and informations in statistics and information theory. *IEEE Trans. Inf. Th.* **52**, 4394–4412.

von der Linden, W., Dose, V., Fischer, R., and Preuss, R., eds. (1999). *Maximum Entropy and Bayesian Methods.* Proceedings of the 18th International Workshop. Kluwer, Dordrecht.

Linkov, Yu. N. (2001). *Asymptotic Statistical Methods for Stochastic Processes.* AMS Vol. **196**, Providence.

Liu, W. (1995). A random-size subset approach to the selection problem. *J. Statist. Plann. Inference* **48**, 153–164.

Luschgy, H. (1992a). Local asymptotic mixed normality for semimartingale experiments. *Prob. Theory Rel. Fields* **92**, 151–176.

Luschgy, H. (1992b). Comparison of location models for stochastic processes. *Prob. Theory Rel. Fields* **93**, 39–66.

Mattenklott, A., Miescke, K.-J., and Sehr, J. (1982). A stochastic model for paired comparisons of social stimuli. *J. Math. Psychology* **26**, 149–168.

Mattner, L. (1993). Some incomplete but boundedly complete location families. *Ann. Statist.* **21**, 2158–2162.

Matusita, K. (1955). Decision rules, based on the distance, for problems of fit, two samples, and estimation. *Ann. Math. Stat.* **26**, 631–640.

Miescke, K.-J. (1979a). Identification and selection procedures based on tests. *Ann. Statist.* **7**, 207–219.

Miescke, K.-J. (1979b). Bayesian subset selection for additive and linear loss functions. *Commun. Statist. - Theor. Meth.* **A-8**, 1205–1226.

Miescke, K.-J. (1981). Gamma-minimax selection procedures in simultaneous testing problems. *Ann. Statist.* **9**, 215–220.

Miescke, K.-J. (1984a). Recent results on multi-stage selection procedures. *Proc. of the Seventh Conference on Probability Theory, Brasov, Romania, 1982.* Editura Academiei Republicii Socialiste Romania, 259–268.

Miescke, K.-J. (1984b). Two-stage selection procedures based on tests. *Design of Experiments: Ranking and Selection.* Essays in honor of R.E. Bechhofer. T.J. Santner and A.C. Tamhane, eds., Dekker, New York, 165–178.

Miescke, K.-J. (1990). Optimum replacement policies for using the most reliable components. *J. Statist. Plann. Inference* **26**, 267–276.

Miescke, K.-J. (1993). Bayesian look-ahead sampling allocations for selection procedures. *Proc. Third Schwerin Conf. Mathematical Statistics - Selection Procedures, 1993.* G. Herrendörfer and K.-J. Miescke, eds., *FBN Schriftenreihe* **1**, 124–133.

Miescke, K.-J. (1999). Bayes sampling designs for selection procedures. Chapter in: *Multivariate Analysis, Design of Experiments, and Survey Sampling.* A Tribute to Jagdish N. Srivastava. S. Ghosh, ed., Dekker, New York, 93–117.

Miescke, K.-J. (2003). Selections with unequal sample sizes. *Proc. 54th Session of the ISI, Berlin 2003.* 2 pages.

Miescke, K.-J. and Park, H. (1997). Bayesian m-truncated sampling allocations for selecting the best Bernoulli population. *Advances in Statistical Decision Theory and Related Topics.* Festschrift in honor of S.S. Gupta. S. Panchapakesan and N. Balakrishnan, eds., Birkhäuser, Boston, 31–47.

Miescke, K.-J. and Park, H. (1999). On the natural selection rule under normality. *Statistics and Decisions* Supplemental Issue **4**, 165–178.

Miescke, K.-J. and Pöppel, E. (1982). A nonparametric procedure to detect periods in time series. *Stoch. Processes and Appl.* **13**, 319–325.

Miescke, K.-J. and Rasch, D., eds. (1996a). 40 Years of Statistical Selection Theory, Part I. Special Issue. *J. Statist. Plann. Inference* **54**, No. 2.

Miescke, K.-J. and Rasch, D., eds. (1996b). 40 Years of Statistical Selection Theory, Part II. Special Issue. *J. Statist. Plann. Inference* **54**, No. 3.

Miescke, K.-J. and Ryan, K.J. (2006). On the performance of Gupta's subset selection procedure. *J. Statist. Plann. Inference* **136**, 2004–2019.

Miescke, K.-J. and Sehr, J. (1980). On a conjecture concerning least favorable configurations in certain two-stage selection procedures. *Commun. Statist. - Theor. Meth.* **A-9**, 1609–1617.

Miescke, K.-J. and Shi, D. (1995). Optimum Markov replacement policies for using the most reliable components. *J. Statist. Plann. Inference* **45**, 331–346.

Millar, P.W. (1983). The minimax principle in asymptotic statistical theory. *Lecture Notes in Math.* **976**, 76–267. Springer, New York.

von Mises, R. (1931). *Wahrscheinlichkeitsrechnung und ihre Anwendung in der Statistik und theoretischen Physik.* Deuticke, Leipzig.

Misra, N., van der Meulen, E.C., and Branden, K.V. (2006). On some inadmissibility results for the scale parameters of selected gamma populations. *J. Statist. Plann. Inference* **136**, 2340–2351.

Mitrinovic, D.S. and Vaic, P.M. (1970). *Analytic Inequalities.* Springer, New York.

Mukhopadhyay, N. and Solanky, T.K.S. (1994). *Multistage Selection and Ranking Procedures: Second-Order Asymptotics.* Dekker, New York.

Müller-Funk, U., Pukelsheim, F., and Witting, H. (1989). On the attainment of the Cramér-Rao bound in $\mathbb{L}_r$-differentiable families of distributions. *Ann. Statist.* **17**, 1742–1748.

Nachbin, L. (1965). *The Haar Integral.* Van Nostrand-Reinhold, Princeton, NJ.

Nachbin, L. (1976), *The Haar Integral.* Krieger, Huntington.

Nemetz, T. (1967). Information theory and testing hypotheses. *Proc. Coll. Inform. Theory, Debrecen*, 283–293.

Nemetz, T. (1974). Equivalence-orthogonality dichotomies of probability measures. *Colloq. Math. Soc. J. Bolyai* **11**, 183–191.

Neyman, J. (1935). Sur un teorema concernente le cosidette statistiche sufficienti. *Giorn. Ist. Ital. Att.* **6**, 320–334.

Neyman, J. (1937). Smooth test for goodness of fit. *Skand. Aktuar. Tidskrift* **20**, 150–199.

Neyman, J. (1959). Optimal asymptotic tests of composite statistical hypotheses. *Probability and Statistics (The Harald Cramér Volume)*, Grenander, U., ed., Wiley, New York, 213–234.

Niemiro, W. (1992). Asymptotics for M-estimators defined by convex minimization. *Ann. Statist.* **20**, 1514–1533.

Nikitin, Y. (1995). *Asymptotic Efficiency of Nonparametric Tests.* Cambridge Univ. Press, Cambridge, U.K.

Noether, G. (1955). On a theorem of Pitman. *Ann. Math. Statist.* **26**, 64–68.

Nölle, G. and Plachky, D. (1967). Zur schwachen Folgenkompaktheit von Testfunktionen. *Zeitschr. Wahrschth. verw. Geb.* **8**, 182–184.

Oosterhoff, J. and van Zwet, W.R. (1979). A note on contiguity and Hellinger distance. *Contributions to Statistics: Jaroslav Hájek Memorial Volume*, J. Jurečková, ed., Reidel, Dordrecht, 157–166.

Österreicher, F. (1978). On the dimensioning of tests for composite hypothesis and not necessarily independent observations. *Probl. of Control and Inf. Th.* **7**, 333–343.

Österreicher, F. and Feldman, D. (1981). Divergenzen von Wahrscheinlichkeitsverteilungen - integralgeometrisch betrachtet. *Acta Math. Sci. Hungar.* **37**, 329–337.

Panchapakesan, S. and Balakrishnan, N., eds. (1997). *Advances in Statistical Decision Theory and Applications.* In Honor of Shanti S. Gupta. Birkhäuser, Boston.

Pardo, L. (2006). *Statistical Inference Based on Divergences Measures.* Chapman and Hall, New York.

Paulson, E. (1949). A multiple decision procedure for certain problems in the analysis of variance. *Ann. Math. Statist.* **20**, 95–98.

Paulson, E. (1952). On the comparison of several experimental categories with a control. *Ann. Math. Statist.* **23**, 239–246.

Paulson, E. (1964). A sequential procedure for selecting the population with the largest mean from k normal populations. *Ann. Math. Statist.* **35**, 174–180.

Paulson, E. (1994). Sequential procedures for selecting the best one of k Koopman-Darmois populations. *Sequential Analysis* **13**, 207–220.

Pereira, B.d.B. (1977). Discriminating among separate models: A bibliography. *Internat. Stat. Rev.* **45**, 163–172.

Perlman, M. (1972). On the strong consistency of approximate maximum likelihood estimators. *Proc. Sixth Berkeley Symp. Math. Statist. Probab.*, L. LeCam, J. Neyman, and E.L. Scott, eds., Univ. of California Press, Berkeley, Vol. **1**, 263–281.

Perng, S.K. (1970). Inadmissibility of various 'good' statistical procedures which are translation invariant. *Ann. Math. Statist.* **41**, 1311–1321.

Pfaff, T. (1982). Quick consistency of quasi maximum likelihood estimators. *Ann. Statist.* **10**, 990–1005.

Pfanzagl, J. (1968). A characterization of the one parameter exponential family by existence of uniformly most powerful tests. *Sankhyā* **A-30**, 147–156.

Pfanzagl. J. (1969). On the measurability and consistency of minimum contrast estimates. *Metrika* **14**, 249–272.

Pfanzagl, J. (1970). Median unbiased estimates for M.L.R. families. *Metrika* **15**, 30–39.

Pfanzagl, J. (1971). On median unbiased estimates. *Metrika* **18**, 154–173.

Pfanzagl, J. (1972). Transformation groups and sufficient statistics. *Ann. Math. Stat.* **43**, 553–568.

Pfanzagl, J. (1974). A characterization of sufficiency by power functions. *Metrika* **21**, 197–199.

Pfanzagl, J. (1994). *Parametric Statistical Theory.* W. de Gruyter, Berlin.

Pfanzagl, J. and Wefelmeyer, W. (1982). *Contribution to a General Asymptotic Statistical Theory. Lecture Notes in Statistics* **13**, Springer, New York.

Pfanzagl, J. and Wefelmeyer, W. (1985). *Asymptotic Expansions for General Statistical Models. Lecture Notes in Statistics* **31**, Springer, New York.

Pitman, E.J.G. (1936). Sufficient statistics and intrinsic accuracy. *Proc. Camb. Phil. Soc.* **32**, 567–579.

Pitman, E.J.G. (1949). Lecture notes on nonparametric statistical inference. Unpublished.

Pollard, D. (1990). *Empirical Processes: Theory and Applications. NSF-CBMS Regional Conf. Series Probability and Statistics* **2**, IMS, Hayward.

Pollard, D. (1991). Asymptotics for least absolute deviation regression estimators. *Econometric Theory* **7**, 186–199.

Pratt, J.W. (1976). F.Y. Edgeworth and R.A. Fisher on the efficiency of maximum likelihood estimation. *Ann. Statist.* **4**, 501–514.

Puri, M.L. and Puri, P.S. (1969). Multiple decision procedures based on ranks for certain problems in analysis of variance. *Ann. Math. Statist.* **40**, 619–632.

Puri, P.S. and Puri, M.L. (1968). Selection procedures based on ranks: scale parameter case. *Sankhyā* **A-30**, 291–302.

Randles, H.R. and Hollander, M. (1971). Γ-minimax selection procedures in treatment versus control problems. *Ann. Math. Statist.* **42**, 330–341.

Rao, C.R. (1945). Information and accuracy attainable in the estimation of statistical parameters. *Bull. Cal. Math. Soc.* **37**, 81–91.

Rao, C.R. (1947). Large sample tests of statistical hypotheses concerning several parameters with applications to problems of estimation. *Proc. Camb. Phil. Soc.* **44**, 50–57.

Rasch, D. (1995). *Mathematische Statistik.* J.A. Barth, Heidelberg.

Read, T.R.C. and Cressie, N.A.C. (1988). *Goodness of Fit Statistics for Discrete Multivariate Data.* Springer, New York.

Reiss, R.-D. (1989). *Approximate Distributions of Order Statistics.* Springer, New York.

Reiss, R.-D. (1993). *A Course on Point Processes.* Springer, New York.

Rényi, A. (1960). On measures of entropy and information. *Proc. Fourth Berkeley Symp. Math. Statist. Probab.*, J. Neyman, ed., Univ. of California Press, Berkeley, Vol. **1**, 547–561.

Rieder, H. (1994). *Robust Asymptotic Statistics.* Springer, New York.

Robbins, H. (1955). An empirical Bayes approach to statistics. *Proc. Third Berkeley Symp. Math. Statist. Probab.*, J. Neyman, ed., Univ. of California Press, Berkeley, Vol. **1**, 157–164.

Robbins, H. (1956). A sequential decision problem with a finite memory. *Proc. Nat. Acad. Sci. U.S.A.* **42**, 920–923.

Robert, C.P. (2001). *The Bayesian Choice: From Decision-Theoretic Foundations to Computational Implementation.* 2nd edition. Springer, New York.

Roberts, A.W. and Varberg, D.E. (1973). *Convex Functions.* Academic Press, New York.

Rockafellar, R.T. (1970). *Convex Analysis.* Princeton Univ. Press, Princeton.

Romano (2005). Optimal testing of equivalence hypotheses. *Ann. Statist.* **33**, 1036–1047.

Roussas, G.G. (1972). *Contiguity of Probability Measures: Some Applications in Statistics.* Cambridge Univ. Press, Cambridge, U.K.

Rukhin, A.L. (1986). Admissibility and minimaxity results in estimation of exponential quantiles. *Ann. Statist.* **14**, 220–231.

Rüschendorf, L. (1984). On the minimum discrimination information theorem. *Recent Results in Estimation Theory,* E.J. Dudewicz, D. Plachky, and P.K. Sen, eds., *Statistics and Decisions* Supplemental Issue **1**, 263–283.

Rüschendorf, L. (1988). *Asymptotische Statistik.* Teubner, Stuttgart.

Santner, T.J. and Behaxeteguy, M. (1992). A two-stage procedure for selecting the largest normal mean whose first stage selects a bounded random number of populations. *J. Statist. Plann. Inference* **31**, 147–168.

Santner, T.J. and Hayter, A.J. (1992). The least favorable configuration of a two-stage procedure for selecting the largest normal mean. *Multiple Comparisons in Biostatistics: Current Research in the topics of C.W. Dunnett.* F.M. Hoppe, ed., Dekker, New York.

Santner, T.J. and Pan, G. (1997). Subset selection in two-factor experiments using randomization restricted designs. *J. Statist. Plann. Inference* **62**, 339–363.

Santner, T.J. and Tamhane, A.C., eds. (1984). *Design of Experiments: Ranking and Selection.* Essays in honor of Robert E. Bechhofer. Dekker, New York.

Savage, L.J. (1954). *The Foundations of Statistics.* Wiley, New York. Revision 1972 by Dover Publications.

Scheffé, H. (1943). Statistical inference in the non-parametric case. *Ann. Math. Statist.* **14**, 305–332.

Schervish, M.J. (1995). *Theory of Statistics.* Springer, New York.

Schervish, M.J. and Seidenfels, T. (1990). An approach to consensus and certainty with increasing evidence. *J. Statist. Plann. Inference* **25**, 401–414.

Schoenberg, T.S. (1951). On Pólya frequency functions. *J. Analyse Math.* **1**, 331–374.

Schwartz, L. (1965). On Bayes procedures. *Zeitschr. Wahrschth. verw. Geb.* **4**, 10–26.

Schwartz, R. (1967). Locally minimax tests. *Ann. Math. Stat.* **38**, 340–359.

Seal, K.C. (1955). On a class of decision procedures for ranking means of normal populations. *Ann. Math. Stat.* **26**, 387–398.

Seal, K.C. (1957). An optimum decision rule for ranking means of normal populations. *Calcutta Statist. Assoc. Bull.* **7**, 131–150.

Searle, S.R., Casella,G., and McCulloch, C.E. (1992). *Variance Components.* Wiley, New York.

Sehr, J. (1988). On a conjecture concerning the least favorable configuration of a two-stage selection procedure. *Commun. Statist. - Theor. Meth.* **A-17**, 3221–3233.

Serfling, R.J. (1980). *Approximation Theorems of Mathematical Statistics.* Wiley, New York.

Seshadri, V. (1993). *The Inverse Gaussian Distribution.* Clarendon Press, Oxford.

Shiryaev, A.N. and Spokoiny, V.G. (2000). *Statistical Experiments and Decisions. Asymptotic Theory.* World Scientific, Singapore.

Shorack, G.R. and Wellner J.A. (1986). *Empirical Processes with Applications to Statistics.* Wiley, New York.

Skorohod, A.V. (1956). Limit theorems for stochastic processes. *Th. Prob. Applic.* **1**, 261–290.

Sobel, M. (1956). Sequential procedures for selecting the best exponential population. *Proc. Third Berkeley Symp. Math. Statist. Probab.*, J. Neyman, ed., Univ. of California Press, Berkeley, Vol. **5**, 99–110.

Sobel, M. and Weiss, G.H. (1972a). Play-the-winner rule and inverse sampling for selecting the best of $k \geq 3$ binomial populations. *Ann. Math. Stat.* **43**, 1808–1826.

Sobel, M. and Weiss, G.H. (1972b). Recent results on using the play-the-winner sampling rule with binomial selection problems. *Proc. Sixth Berkeley Symp. Math. Statist. Probab.*, L. LeCam, J. Neyman, and E.L. Scott, eds., Univ. of California Press, Berkeley, Vol. **1**, 717–736.

Srivastava, S.M. (1998). *A Course on Borel Sets.* Springer, New York.

Stein, C. (1945). A two-sample test for a linear hypothesis whose power is independent of the variance. *Ann. Math. Statist.* **16**, 243–258.

Stein, C. (1955a). Inadmissibility of the usual estimator for the mean of a multivariate normal distribution. *Proc. Third Berkeley Symp. Math. Statist. Probab.*, J. Neyman, ed., Univ. of California Press, Berkeley, Vol. **1**, 197–206.

Stein, C. (1955b). A necessary and sufficient condition for admissibility. *Ann. Math. Statist.* **26**, 518–522.

Stein, C. (1956). Multivariate Analysis I. *Technical Report* **42**, Dept. of Stat., Stanford Univ., Stanford.

Stein, C. (1959). The admissibility of Pitman's estimator for a single location parameter. *Ann. Math. Statist.* **30**, 970–979.

Steinebach, J. (1980). Large deviations and some related topics. *Carleton Math. Lect. Notes* **28**, Carleton Univ., Ottawa.

Strasser, H. (1975). The asymptotic equivalence of Bayes and maximum likelihood estimation. *J. Multiv. Anal.* **5**, 206–226.

Strasser, H. (1976). Asymptotic properties of posterior distributions. *Zeitschr. Wahrschth. verw. Geb.* **35**, 269–282.

Strasser, H. (1977a). Improved bounds for the equivalence of Bayes and maximum likelihood estimation. *Theory Probab. Appl.* **22**, 349–361.

Strasser, H. (1977b). Asymptotic expansions for Bayes procedures. *Recent Dev. Stat., Proc. Eur. Meet. Stat. Grenoble 1976*, 9–35.

Strasser, H. (1981a). Consistency of maximum likelihood and Bayes estimation. *Ann. Statist.* **9**, 1107–1113.

Strasser, H. (1981b). Convergence of estimates: Part 1 and Part 2. *J. Multiv. Anal.* **11**, 127–151 and 152–172.

Strasser, H. (1982). Local asymptotic minimax properties of Pitman estimates. *Zeitschr. Wahrschth. verw. Geb.* **60**, 223–247.

Strasser, H. (1985). *Mathematical Theory of Statistics.* W. de Gruyter, Berlin.

Strasser, H. (1988). Differentiability of statistical experiments. *Statistics and Decisions* **6**, 113–130.

Strasser, H. (1989). Tangent vectors for models with independent but non-identically distributed observations. *Statistics and Decisions* **7**, 127–152.

Strasser, H. (1997). Asymptotic admissibility and uniqueness of efficient estimates in semiparametric models. *Research Papers in Probability and Statistics.* Festschrift for Lucien LeCam. D. Pollard and G.L. Yang, eds., Springer, New York, 369–376.

Strasser, H. and Becker, C. (1986). Local asymptotic admissibility of Pitman estimates. *Statistics and Decisions* **4**, 61–74.

Sverdrup, E. (1953). Similarity, unbiasedness, minimaxibility, and admissibility of statistical test procedures. *Skand. Aktuar. Tidskrift* **36**, 64–86.

Sverdrup, E. (1966). The present state of the decision theory and the Neyman-Pearson theory. *Rev. Int. Statist. Inst.* **34**, 309–333.

Tamhane, A.C. and Bechhofer, R.E. (1977). A two-stage minimax procedure with screening for selecting the largest normal mean. *Commun. Statist. - Theor. Meth.* **A-6**, 1003–1033.

Tamhane, A.C. and Bechhofer, R.E. (1979). A two-stage minimax procedure with screening for selecting the largest normal mean (II): An improved PCS lower bound and associated tables. *Commun. Statist. - Theor. Meth.* **A-8**, 337–358.

Tong, Y.L. and Wetzell, D.E. (1979). On the behaviour of the probability function for selecting the best normal population. *Biometrika* **66**, 174–176.

Torgersen, E. (1970). Comparision of experiments when the parameter space is finite. *Zeitschr. Wahrschth. verw. Geb.* **16**, 219–249.

Torgersen, E. (1991). *Comparison of Statistical Experiments.* Cambridge Univ. Press, Cambridge, U.K.

Tweedie, M.C.K. (1957). Statistical properties of inverse Gaussian distributions. *Ann. Math. Statist.* **28**, 362–377.

van der Vaart, A.W. (1998). *Asymptotic Statistics.* Cambridge Univ. Press, Cambridge, U.K.

van der Vaart, A.W. and Wellner, J.A. (1996). *Weak Convergence and Empirical Processes. With Applications to Statistics.* 2nd printing 2000. Springer, New York.

van der Vaart, H.R. (1961). Some extensions of the idea of bias. *Ann. Math. Statist.* **32**, 436–447.

Vajda, I. (1971). Limit theorems for total variation of Cartesian product measures. *Period. Math. Hungar.* **2**, 223.234.

Vajda, I. (1973). χ^α-divergences and generalized Fisher's information. *Trans. 6th Prague Conf. on Inf. Th.*, Academia, Praha, 873–886.

Vajda, I. (1989). *Theory of Statistical Inference and Information.* Kluwer, Boston.

Vajda, I. and Österreicher, F. (2003). A new class of metric divergences on probability spaces and its applicability in statistics. *Ann. Inst. Statist. Math.* **55**, 639–653.

Wald, A. (1939). Contributions to the theory of statistical estimation and testing hypotheses. *Ann. Math. Statist.* **10**, 299–326.

Wald, A. (1941a). Asymptotically most powerful tests of statistical hypotheses. *Ann. Math. Statist.* **12**, 1–19.

Wald, A. (1941b). Some examples of asymptotically most powerful tests. *Ann. Math. Statist.* **12**, 396–408.

Wald, A. (1943). Tests of statistical hypotheses concerning several parameters when the number of observations is large. *Trans. Amer. Math. Soc.* **54**, 426–482.

Wald, A. (1947). *Sequential Analysis.* Wiley, New York.

Wald, A. (1949). Note on the consistency of the maximum likelihood estimate. *Ann. Math. Statist.* **20**, 595–601.

Wald, A. (1950). *Statistical Decision Functions.* Wiley, New York.

Wellek, S. (2003). *Testing Statistical Hypotheses of Equivalence.* Chapman and Hall/CRC, Boca Raton.

Wets, R.J.-B. (1989). Stochastic programming. *Handbooks in Operation Research and Management Science, Optimization.* G.L. Nemhauser, A.H.G. Rinnooy Kan, and M.J. Todd, eds., North Holland, Vol. **1**, 573–629.

Wijsman, R.A. (1969). General proof of termination with probability one of invariant sequential probability ratio test based an multivariate observation. *Ann. Math. Stat.* **38**, 8–24.

Wijsman, R.A. (1973). On the attainment of the Cramér-Rao lower bound. *Ann. Statist.* **1**, 538–542.

Wijsman, R.A. (1985). Proper action in steps, with application to density ratios of maximal invariants. *Ann. Statist.* **13**, 395–402.

Wijsman, R.A. (1990). *Invariant Measures on Groups and Their Use in Statistics. IMS Lect. Notes Ser.* **14**, IMS, Hayward.

Witting, H. (1985). *Mathematische Statistik I, Parametrische Verfahren bei festem Stichprobenumfang.* Teubner, Stuttgart.

Witting, H. and Müller-Funk U. (1995). *Mathematische Statistik II, Asymptotische Statistik: Parametrische Modelle und nichtparametrische Funktionale.* Teubner, Stuttgart.

Zacks, S. (1971). *The Theory of Statistical Inference.* Wiley, New York.

Zinzius, E. (1981). Minimaxschätzer für den Mittelwert ϑ einer normalverteilten Zufallsgröße mit bekannter Varianz bei vorgegebener oberer und unterer Schranke für ϑ. *Math. Operationsforsch. Statist. Ser. Statistics.* **12**, 551–557.

Author Index

Subject Index

Springer Series in Statistics

(continued from page ii)

Ibrahim/Chen/Sinha: Bayesian Survival Analysis
Jiang: Linear and Generalized Linear Mixed Models and Their Applications
Jolliffe: Principal Component Analysis, 2nd edition
Konishi/Kitagawa: Information Criteria and Statistical Modeling
Knottnerus: Sample Survey Theory: Some Pythagorean Perspectives
Kosorok: Introduction to Empirical Processes and Semiparametric Inference
Küchler/Sørensen: Exponential Families of Stochastic Processes
Kutoyants: Statistical Inference for Ergodic Diffusion Processes
Lahiri: Resampling Methods for Dependent Data
Lavallée: Indirect Sampling
Le Cam: Asymptotic Methods in Statistical Decision Theory
Le Cam/Yang: Asymptotics in Statistics: Some Basic Concepts, 2nd edition
Le/Zidek: Statistical Analysis of Environmental Space-Time Processes
Liese/Miescke: Statistical Decision Theory: Estimation, Testing, and Selection
Liu: Monte Carlo Strategies in Scientific Computing
Manski: Partial Identification of Probability Distributions
Mielke/Berry: Permutation Methods: A Distance Function Approach, 2nd edition
Molenberghs/Verbeke: Models for Discrete Longitudinal Data
Mukerjee/Wu: A Modern Theory of Factorial Designs
Nelsen: An Introduction to Copulas, 2nd edition
Pan/Fang: Growth Curve Models and Statistical Diagnostics
Politis/Romano/Wolf: Subsampling
Ramsay/Silverman: Applied Functional Data Analysis: Methods and Case Studies
Ramsay/Silverman: Functional Data Analysis, 2nd edition
Reinsel: Elements of Multivariate Time Series Analysis, 2nd edition
Rosenbaum: Observational Studies, 2nd edition
Rosenblatt: Gaussian and Non-Gaussian Linear Time Series and Random Fields
Särndal/Swensson/Wretman: Model Assisted Survey Sampling
Santner/Williams/Notz: The Design and Analysis of Computer Experiments
Schervish: Theory of Statistics
Shaked/Shanthikumar: Stochastic Orders
Shao/Tu: The Jackknife and Bootstrap
Simonoff: Smoothing Methods in Statistics
Song: Correlated Data Analysis: Modeling, Analytics, and Applications
Sprott: Statistical Inference in Science
Stein: Interpolation of Spatial Data: Some Theory for Kriging
Taniguchi/Kakizawa: Asymptotic Theory for Statistical Inference for Time Series
Tanner: Tools for Statistical Inference: Methods for the Exploration of Posterior Distributions and Likelihood Functions, 3rd edition
Tillé: Sampling Algorithms
Tsaitis: Semiparametric Theory and Missing Data
van der Laan/Robins: Unified Methods for Censored Longitudinal Data and Causality
van der Vaart/Wellner: Weak Convergence and Empirical Processes: With Applications to Statistics
Verbeke/Molenberghs: Linear Mixed Models for Longitudinal Data
Weerahandi: Exact Statistical Methods for Data Analysis

Printed in the United States of America